# Practical Handbook of
# Materials
# Science

*Edited by*

## Charles T. Lynch, Ph.D.
Senior Engineer
Vitro Corporation
Washington, District of Columbia

CRC Press, Inc.
Boca Raton, Florida

*Cover photograph:* very high purity high density
refractory oxide (mold pressed and sintered) surface
(yttria-stabilized zirconia). Characterization
Facility ASWAL, Materials Laboratory W-PAFB, Ohio.
Based on materials prepared by K. S. Mazdiyasni,
J. S. Smith, II, and C. T. Lynch.

**Library of Congress Cataloging-in-Publication Data**
Practical handbook of materials science / editor, Charles T. Lynch.
     p.  cm.
   Bibliography: p.
   Includes index.
   ISBN 0-8493-3702-X
  1. Materials—Handbooks, manuals, etc.  I. Lynch, Charles T.
TA403.4.P7  1989
620.1'1—dc19                                         890912
                                                     CIP

# PREFACE

The original *CRC Handbook of Materials Science* was a compilation in four volumes of materials properties, reasonably comprehensive in scope, and critical in selection of coverage, attempting to include both materials of commerce and experimental materials where the properties had been reasonably established through comparative data from reliable sources. Materials of questionable composition, structure, and properties were not generally covered. The Handbook Series contained one volume devoted to elemental and general properties of the elements and selected compounds; one volume on metals, composites, and refractory materials; one volume on nonmetallic materials and applications, including polymeric, electronic, graphitic, and nuclear materials; and one volume on wood. The books were published over a period of several years.

It has become apparent in recent years that there is need, also, in addition to a comprehensive Handbook Series, for a smaller publication that could easily be carried from desk to desk, room to room, etc., which would be comprehensive enough to be useful to most workers in materials science and related fields, and yet be small enough for more general use. It has been a goal also to make such a book more affordable to the average advanced undergraduate and graduate student who needs this type of information either for reference in study or research.

Much of the information contained in this *Practical Handbook of Materials Science* is identical or similar to that contained in the original four-volume series. It has been necessary, however, to update some of the materials sections, notably those dealing with composite materials, to stay abreast of the rapid developments in this field in the years since the original series was published. Important advances in reinforcements and composite materials have resulted in their increasing use in commerce, particularly in automobiles, aerospace structures, and sporting goods. In adding new data, the same restrictions have been applied, that the materials properties be reasonably established through comparative data from reliable sources.

The most notable new advance not included involves the new high-temperature superconductors which need more comprehensive research and development to accurately determine their properties, and considerably more time to become materials of commerce, despite their obviously tremendous potential for advancing technology. To keep from expanding this book beyond a satisfactory size, it has also been necessary to make other difficult decisions to eliminate tables and sections.

A complete revision of the materials information section has been included to provide the reader with current information on the Federal laboratory and information services in the field of mateials science and engineering. The author believes that users of this *Practical Handbook of Materials Science* will find the addition of a comprehensive list of Federal Materials Research Centers and Federal Materials Information Centers extremely useful for any general information search, whether for a quick overview or for a comprehensive evaluation of some aspect of this important field.

Materials science and engineering is a pacing science for modern technology. Many of the problems associated with modern technology are materials limited, depending on the development of either new materials or new structures and microstructures of existing materials for their solution. It is an interdisciplinary field which requires that the modern researcher know the general properties, and sometimes the very specific physical and chemical properties, of many diverse materials. It is not enough to be a conversant polymer chemist to understand resin selection for a new composite structure, for example. One must also understand the selection of reinforcement materials, both graphitic and nongraphitic, and also the nature of some of the materials that the composites will be used with, joined to, and/or replace, such as structural metals, or fiberglass composites. It is to provide a useful, convenient, economical, and portable information source on materials that this *Practical Handbook of Materials Science* is intended.

# THE EDITOR

**Charles T. Lynch, Ph.D.,** is a senior engineer with Vitro Corporation in Washington, D.C., conducting systems studies related to the NASA Space Station Program. Dr. Lynch received an Associate in Arts degree in 1953 and a B.S. in Chemistry in 1955 from the George Washington University, Washington, D.C. He received the M.S. degree in 1957 and in 1960 received the Ph.D. degree in Analytical Chemistry from the University of Illinois, Urbana.

Following service in the U.S. Air Force (1960 to 1962) as a First Lieutenant and Project Officer at Wright-Patterson Air Force Base, Ohio, he joined the Air Force Materials Laboratory as a civilian scientist and administrator, serving in a number of positions over a period of 20 years. These included serving as Group Leader for Ceramics and Graphite R & D; Chief, Advanced Metallurigical Studies Branch; and Senior Scientist for Environmental Effects. He was also a Lecturer in Chemistry at Wright-State University for several years. In 1981 he became the Head, Materials Division, of the Office of Naval Research in Arlington, Virginia. He was responsible for administration of the Navy basic materials research program on metals, alloys, composites, dielectrics, missile domes, welding, processing and rapid solidification technology, corrosion, etc., including all phases of synthesis, properties, failure reliability, and quality control.

Dr. Lynch is a graduate of the Senior Executive Education Program of the Federal Executive Institute in Charlottesville, Virginia. He is the author and co-author of over 100 journal and book publications, 14 patents, over 100 presentations at major scientific meetings, and is the Editor of the *CRC Handbook of Materials Science* four-volume series. He is past secretary and chairman of the Metallurigical Society of the AIME Corrosion and Environmental Effects Committee, Liason Member to several panels of the National Materials Advisory Board, and a U.S. representative of the Structures and Materials Panel, AGARD (NATO). He has received a number of professional achievement awards, including a USAF Commendation Medal, USAF Nominee for the Arthur S. Flemming Award, the USAF Laboratory Achievement Award, and patent awards.

He is the co-developer of a new refractory ceramic *Zyttrite,* a high-density, fine-grained, high-purity, translucent yttria-stabilized zirconia made by the decomposition of mixed alkoxides to mixed oxides by thermal or hydrolytic decomposition and sintering or hot pressing the resultant oxides. He pioneered the general approach of decomposition of organometallic compounds as precursors to making a variety of refractory inorganic nonmetallic materials of high purity, controlled particle size, and specified composition. He provided technical leadership to the USAF for new R & D efforts in the fields of corrosion prediction, prevention, and control, which led to scheduling depot maintenance to optimize corrosion repairs based on need rather than fixed intervals for major aircraft systems. Under his direction a new corrosion severity index to determine/estimate corrosion severity based on atmospheric conditions was established. This was used to determine the need for washing, painting, and estimation of corrosion damage for Depot and Base-level Maintenance for the Air Force Logistics Command and operational commands. Research on corrosion inhibitors led to the discovery and development of a new class of multifunctional inhibitors effective against both localized and general corrosion in aerospace vehicles. The new inhibitor systems were incorporated in a test program in the Automated Rinse Facility at MacDill AFB, FL, for use on fighter aircraft.

# ACKNOWLEDGEMENTS

The Author wishes to acknowledge his many colleagues who have contributed to the earlier Handbook Series, which provided the basis for this latest effort. Many people have provided suggestions for what should be included in a practical handbook of manageable size. I thank them all, and particularly thank Professor Robert Summitt of Michigan State University for his suggestions and advice. I also greatly appreciate the efforts of Mrs. Laura Lynch in working on portions of the manuscript. My thanks also to Mrs. Amy G. Skallerup of CRC Press, Inc., for her editorial advice and patience throughout this endeavor, which is always more involved than one anticipates. Finally, my thanks to my wife, Betty Ann, and son, Thomas E., for their help, support, and willingness to do many things so that this Handbook could be completed.

## Sources

Much of the material in this volume is based upon a group of basic CRC Press references. For the sake of simplicity, they will be cited in this section as follows:

1. **Lynch, C. T.,** Ed., *CRC Handbook of Materials Science,* Vol. 1, CRC Press, Inc., Boca Raton, Fla., 1974.
2. **Lynch, C. T.,** Ed., *CRC Handbook of Materials Science,* Vol. 2, CRC Press, Boca Raton, Fla., 1975.
3. **Lynch, C. T.,** Ed., *CRC Handbook of Materials Science,* Vol. 3, CRC Press, Inc. Boca Raton, Fla., 1975.
4. **Bolz, R. E. and Tuve, G. L.,** Eds., *CRC Handbook of Tables for Applied Engineering Science,* 2nd ed., CRC Press, Inc., Boca Raton, Fla., 1973.
5. **Weast, R. C.,** Ed., CRC Handbook of Chemistry and Physics, 55th ed., CRC Press, Inc., Cleveland, 1974.
6. **Weast, R.C. et al.,** Eds., *CRC Handbook of Chemistry and Physics,* 69th ed., CRC Press, Inc., Boca Raton, Fla., 1988.

New material in the present work is identified as such. Credit lines are given where required. Sources of individual sections are listed:

**Section 1** — modified from ''The Elements'' by C. R. Hammond in Reference 6.

**Section 2** — except for Tables 2—1 through 2—3, this section is based upon material in Reference 1. In some cases, an updated table from Reference 6 replaces the original. (A credit line indicates this substitution.)

**Section 3** — based upon Reference 1.

**Section 4** — based upon Reference 1.

**Section 5** — based upon Reference 1.

**Section 6** — consists of selections from Subsections ''Glasses and Glass-Ceramics'' and ''Alumina and Other Refractory Materials'' from Reference 2.

**Section 7** — Subsection 7.1 is derived from a variety of commerical sources, current literature and handbook values, and Reference 2; Subsection 7.2 was compiled from the literature and current handbook values; Subsection 7.3 is taken from ''Metal Matrix Composites'' (by F. S. Galasso) in Reference 2; and Subsection 7.4 is compiled from current literature, commercial sources, handbooks, etc.

**Section 8** — Tables 8—1 and 8—2 appeared in Reference 1; the section on superconductivity is excerpted from Reference 6. The balance of the material is drawn from Reference 3.

**Section 9** — excerpted from Reference 3.

**Section 10** — subsections were derived from similarly named subsections of Reference 2. Sources of the subsections are identified in the text.

**Section 11** — derived from "Nuclear Materials" in Reference 3.

**Section 12** — composed for this volume.

**Section 13** — Subsection 13.1 revises and updates material in Reference 3; while Subsections 13.2 and 13.3 are based upon the Directory of Federal Laboratory and Technology Resources, 1988—1989, PB 88-100011, U.S. Department of Commerce, NTIS, Springfield, Va., February 1988.

# TABLE OF CONTENTS

# Section 1

## The Elements

# Section 1

# THE ELEMENTS*

One of the most striking facts about the elements is their unequal distribution and occurrence in nature. Present knowledge of the chemical composition of the universe, obtained from the study of the spectra of stars and nebulae, indicates that hydrogen is by far the most abundant element and may account for more than 90% of the atoms or about 75% of the mass of the universe. Helium atoms make up most of the remainder. All of the other elements together contribute only slightly to the total mass.

The chemical composition of the universe is undergoing continuous change. Hydrogen is being converted into helium, and helium is being changed into heavier elements. As time goes on, the ratio of heavier elements increases relative to hydrogen. Presumably, the process is not reversible.

Burbidge, Burbidge, Fowler, and Hoyle, and more recently, Peebles, Penzias, and others have studied the synthesis of elements in stars. To explain all of the features of the nuclear abundance curve — obtained by studies of the composition of the earth, meteorites, stars, etc. — it is necessary to postulate that the elements were originally formed by at least eight different processes: (1) hydrogen burning, (2) helium burning, (3) $\chi$ process, (4) e process, (5) s process, (6) r process, (7) p process, and (8) the X process. The X process is thought to account for the existence of light nuclei such as D, Li, Be, and B. Common metals such as Fe, Cr, Ni, Cu, Ti, Zn, etc. were likely produced early in the history of our galaxy. Probably, most of the heavy elements on earth and elsewhere in the universe were originally formed in supernovae or in the hot interior of the stars.

Studies of the solar spectrum have led to the identification of 67 elements in the sun's atmosphere; however, all elements cannot be identified with the same degree of certainty. Other elements may be present in the sun, although not yet detected spectroscopically. Helium was discovered on the sun before it was found on earth. Some elements such as scandium are relatively more plentiful in the sun and stars than on earth.

Minerals in lunar rocks brought back from the moon on the Apollo missions consist predominantly of plagioclase [(Ca,Na) (Al,Si)$O_4O_8$] and pyroxene [(Ca,Mg,Fe)$_2$Si$_2$O$_6$] — two minerals common in terrestrial volcanic rock. No new elements have been found on the moon; however, three minerals, armalcolite [(Fe,Mg)Ti$_2$O$_5$], pyroxferroite [CaFe$_6$(SiO$_3$)$_7$], and tranquillityite [Fe$_8$(Zr,Y)Ti$_3$Si$_3$O$_2$], are new. The oldest known terrestrial rocks are about 3.75 billion years old. One rock, known as the "Genesis Rock", brought back from the Apollo 15 Mission, is about 4.15 billion years old. This is only one half billion years younger than the supposed age of the moon and solar system. Lunar rocks appear to be relatively enriched in refractory elements such as chromium, titanium, zirconium, and the rare earths, and impoverished in volatile elements such as the alkali metals, in chlorine, and in noble metals such as nickel, platinum, and gold.

Older than the "Genesis Rock" are carbonaceous chondrites, a type of meteorite that has fallen to earth and has been studied. These are some of the most primitive objects of the solar system yet found. The grains making up these objects probably condensed directly out of the gaseous nebula from which the sun and planets were born. Most of the condensation of the grains probably was completed within 50,000 years of the time the disk of the nebula was first formed — about 4.6 billion years ago. This type of meteorite may contain a small percentage of presolar dust grains. The relative abundances of the elements in these meteorites are about the same as the abundances found in the solar chromosphere.

The X-ray fluorescent spectrometer sent with the Viking I spacecraft to Mars shows that the Martian soil contains about 12 to 16% iron, 14 to 15% silicon, 3 to 8% calcium, 2 to 7% aluminum, and 0.5 to 2% titanium. The gas chromatograph-mass spectrometer on Viking II found no trace of the organic compounds that should be present if life ever existed there.

F. W. Clarke and others have carefully studied

* Modified from Hammond, C. R., in *Handbook of Chemistry and Physics,* 69th edition, Weast, R. C., Ed., CRC Press, Boca Raton, Fla., 1988, B-5.

the composition of rocks making up the crust of the earth. Oxygen accounts for about 47% of the crust, by weight, while silicon comprises about 28%, and aluminum about 8%. These elements, plus iron, calcium, sodium, potassium, and magnesium, account for about 99% of the composition of the crust.

Many essential elements such as tin, copper, zinc, lead, mercury, silver, platinum, antimony, arsenic, and gold, are among some of the rarest elements in the earth's crust. These are made available to us only by the processes of concentration in ore bodies. The rare earth elements have been found to be much more plentiful than originally thought and are about as abundant as uranium, mercury, lead, or bismuth. The least abundant rare earth or lanthanide element, thulium, is now believed to be more plentiful on earth than silver, cadmium, gold, or iodine. Rubidium, the 16th most abundant element, is more plentiful than chlorine, while its compounds are little known in chemistry and commerce.

It is now thought that at least 24 elements are essential to living matter. The four most abundant elements in the human body are hydrogen, oxygen, carbon, and nitrogen. The seven next most common, in order of abundance, are calcium, phosphorus, chlorine, potassium, sulfur, sodium, and magnesium. Iron, copper, zinc, silicon, iodine, cobalt, manganese, molybdenum, fluorine, tin, chromium, selenium, and vanadium are needed to play a role in living matter. Boron is thought essential for some plants, and it is possible that aluminum, nickel, and germanium may turn out to be necessary.

Ninety-one elements occur naturally on earth. Minute traces of plutonium-244 have been discovered in rocks mined in Southern California. This discovery supports the theory that heavy elements were produced during creation of the solar system. While technetium and promethium have not been found naturally on earth, they have been found to be present in stars. Technetium has been identified in the spectra of certain "late" type stars, and promethium lines have been identified in the spectra of a faintly visible star HR465 in Andromeda. Promethium must have been made very recently near the surface of the stars, for no known isotope of this element has a half-life longer than 17.7 years. It has been suggested that californium is present in certain stellar explosions known as supernovae; however, this has not been proved. At present, no elements are found elsewhere in the universe that cannot be accounted for here on earth.

All atomic mass numbers from 1 to 238 are found naturally on earth, except for masses 5 and 8. About 280 stable and 67 naturally radioactive isotopes occur on earth, totaling 347. In addition, the neutron, technetium, promethium, and the transuranic elements (those lying beyond uranium) up to element 107 have been produced artifically. West German scientists in late 1982 confirmed the existence of element 107 made by Soviet scientists and announced synthesis of element 109. Laboratory processes have now extended the radioactive mass numbers beyond 238 to 266. Each element from atomic number 1 to 107 is known to have at least one radioactive isotope. About 1700 different nuclides (the number given to different kinds of nuclei, whether they are of the same or different elements) are now recognized. Many unstable and radioactive isotopes are now produced and distributed by the Oak Ridge National Laboratory, Oak Ridge, Tenn., U.S., to customers licensed by the U.S. Department of Energy.

The nucleus of an atom is characterized by the number of protons it contains, usually denoted by Z, and by the number of neutrons, N. Isotopes of an element have the same value of Z, but different values of N. The mass number A is the sum of Z and N. For example, uranium-238 has a mass number of 238 and would contain 92 protons and 146 neutrons.

There is evidence that the definition of chemical elements must be broadened to include the electon. Several compounds known as electrides have recently been made of alkaline metal elements and electrons. A relatively stable combination of a positron and electron, known as positronium, has also been studied.

In addition to the proton, neutron, and electron, there are considerably more than 100 other fundamental particles which have been discovered or hypothesized to exist. The majority of these fall into one of two classes, leptons or hadrons. The leptons comprise just four known particles, the electron, the muon ($\mu$ meson), and two kinds of neutrinos. The muon is essentially similar to the electron and has a charge of $-1$, but it is 200 times heavier. The neutrino is either of two stable particles of small (probably zero) rest mass, carrying no charge. Also, there are four antileptons, identical to the corresponding leptons in some respects, such as mass, but they have properties exactly opposite those of the leptons. The positron, for example, is an antilepton, with a charge of $+1$. Leptons cannot be broken into smaller units and are considered to be elementary. On the other hand, hadrons are complex and thought to have internal structure. Protons and neutrons, which make up atomic nuclei, are hadrons.

Elementary particle physics is not yet clearly understood, but groupings and arrangements of these particles have been made resembling the periodic table of chemical elements. This has led to the speculation that hadrons are composed of three (or possibly more) simpler components called quarks. Quarks are presumed to be elementary particles. There is at present no evidence that quarks exist in isolation. Many physicists now hold that all the matter and energy in the universe is controlled by four fundamental natural forces: the electromagnetic force, gravity, a weak nuclear force, and a strong nuclear force. Each of these natural forces is passed back and forth among the basic particles of matter by unique force-carrying particles. The electromagnetic force is carried by the photon, the weak nuclear force by the intermediate vector boson, and gravity by the gravitron. There is now evidence of the existence of a new particle, known as the gluon, that binds quarks together by carrying the strong nuclear force.

Available evidence leads to the conclusion that elements 89 (actinium) through 103 (lawrencium) are chemically similar to the rare earth or lanthanide elements (elements 57 to 71, inclusive). These elements therefore have been named actinides after the first member of this series. Those elements beyond uranium that have been produced artificially have the following names and symbols: neptunium, 93 (Np); plutonium, 94 (Pu); americium, 95 (Am); curium, 96 (Cm); berkelium, 97 (Bk); californium, 98 (Cf); einsteinium, 99 (Es); fermium, 100 (Fm); mendelevium, 101 (Md); nobelium, 102 (No); and lawrencium 103 (Lr). It is now claimed that elements 104, 105, 106, 107, and 109 have been produced and identified. Names and chemical symbols have been suggested for elements 104 and 105, but have not been officially adopted. Names for elements 106, 107, and 109 have not yet been suggested.

Element 104 is expected to have chemical properties similar to those of hafnium and would not be a member of the actinide series. Element 105 probably would have chemical properties similar to those of tantalum, element 106 is similar to tungsten, element 107 similar to rhenium, and element 109 similar to iridium.

There is still thought to be a possibility of producing elements beyond element 109, either by bombardment of heavy isotopic targets with heavy ions, or by the irradiation of uranium or transuranic elements with the instantaneous high flux of neutrons produced by underground nuclear explosions. The limit will be set by the yields of the nuclear reactions and by the half-lives of radioactive decay. It has been suggested that elements 102 and 103 have abnormally short lives only because they are in a pocket of instability and that this region of instability might begin to "heal" around element 105. If so, it may be possible to produce heavier isotopes with longer half-lives. It has also been suggested that element 114, with a mass number of 298, and element 126, with a mass number of 310, may be sufficiently stable to make discovery and identification possible. Calculations indicate that element 110, a homolog of platinum, may have a half-life of as long as 100 million years. Searches have already been made in a number of laboratories for element 110 and its neighboring elements in naturally occurring platinum. Recent studies of the xenon component ($^{131-136}$Xe) of certain carbonaceous chronditic meteorites suggest that elements 113, 114, or 115 may have been its progenitor.

There are many claims in the literature of the existence of various allotropic modifications of the elements, some of which are based on doubtful or incomplete evidence. Also, the physical properties of an element may change drastically by the presence of small amounts of impurities. With new methods of purification, which are now able to produce elements with 99.9999% purity, it has been necessary to restudy the properties of the elements. For example, the melting point of thorium changes by several hundred degrees by the presence of a small percentage of $ThO_2$ as an impurity. Ordinary commercial tungsten is brittle and can be worked only with difficulty. Pure tungsten, however, can be cut with a hacksaw, forged, spun, drawn, or extruded. In general, the value of physical properties given here applies to the pure element, when it is known.

Many of the chemical elements and their compounds are toxic and should be handled with due respect and care. In recent years there has been a greatly increased knowledge and awareness of the health hazards associated with chemicals, radioactive materials, and other agents. Anyone working with the elements and certain of their compounds should become thoroughly familiar with the proper safeguards to be taken. Reference should be made to publications such as the following:

1. *Code of Federal Regulations,* Title 29, Labor, Chapter XVII, Section 1910.93 of Subpart G, redesignated as 1910.1000 at 40 FR (Federal Register) 23072, May 28, 1975; amended at 41 FR 11505, March 19, 1976; 41 FR 35184, August 20, 1976; FR 46784, October 22, 1976; 42 FR 3304, January 18, 1977 (corrections and additional amendments and corrections as issued, U.S. Government Printing Office, Superintendent of Documents, Washington, D.C.)
2. *Code of Federal Regulations,* Title 10, Energy, Chaper 1, Nuclear Regulatory Commission, Section 20.103—20.108; 20.201—207; 20.301—305; 20.401—409; 20.501—2; 20.601; appendices, corrections, and amendments.
3. *Occupational Safety and Health Reporter* (latest edition with amendments and corrections), Bureau of National Affairs, Washington, D.C.
4. *Maximum Permissible Body Burdens and Maximum Permissible Concentrations of Radionuclides in Air and in Water for Occupational Exposure,* with addenda, N.B.S. Handbook No. 69 (NCRP Report No. 22), latest edition, National Council on Radiation Protection and Measurements (NCRP), U.S. Department of Commerce, Bethesda, MD.; also refer to Permissible Quarterly Intakes of Radionuclides, *Handbook of Chemistry and Physics,* 69th Edition, CRC Press, Boca Raton, FL, 1988, B-448.
5. *TLVs® Threshold Limit Values for Chemical Substances and Physical Agents in Workroom Environment with Intended Changes,* latest edition, American Conference of Governmental Industrial Hygienists, Cincinnati, Ohio.

**Actinium** — (Gr. *aktis, aktinos,* beam or ray), Ac; at wt (227.0482); at no 89; mp 1050°C; bp 3200 ± 300°C (est.); sp gr 10.07 (calc). Discovered by Andre Debierne in 1899 and independently by F. Giesel in 1892. Occurs naturally in association with uranium minerals. Actinium-227, a decay product of uranium-235, is a beta emitter with a 21.6-year half-life). Its principal decay products are thorium-227 (18.5-day half-life), radium-223 (11.4-day half-life), and a number of short-lived

products including radon, bismuth, polonium, and lead isotopes. In equilibrium with its decay products, it is a powerful source of alpha rays. Actinium metal has been prepared by the reduction of actinium fluoride with lithium vapor at about 1100 to 1300°C. The chemical behavior of actinium is similar to that of the rare earths, particularly lanthanum. Purified actinium comes into equilibrium with its decay products at the end of 185 days, and then decays according to its 21.6-year half-life. It is about 150 times as active as radium, making it of value in the production of neutrons.

**Aluminum** — (L. *alumen, alum*), Al; at wt 26.98154; at no 13; mp 660.37°C; bp 2467°C; sp gr 2.6989 (20°C); valence 3. The ancient Greeks and Romans used alum in medicine as an astringent and as a mordant in dyeing. In 1761 de Morveau proposed the name *alumine* for the base in alum, and Lavoisier, in 1787, thought this to be the oxide of a still undiscovered metal. Wohler is generally credited with having isolated the metal in 1827, although an impure form was prepared by Oersted 2 years earlier. In 1807, Davy proposed the name *alumium* for the undiscovered metal and later agreed to change it to *aluminum.* Shortly thereafter, the name *aluminium* was adopted to conform with the ''ium'' ending of most elements, and this spelling is now in use elsewhere in the world. In 1925, the American Chemical Society officially decided to use the name ''aluminum'' thereafter in its publications. The method of obtaining aluminum metal by the electrolysis of alumina dissolved in cryolite was discovered in 1886 by Hall in the U.S. and about the same time by Heroult in France. Cryolite, a natural ore, is no longer widely used in commercial production, but has been replaced by an artificial mixture of sodium, aluminum, and calcium fluorides. Bauxite, an impure hydrated oxide ore, is found in large deposits in Jamaica, Australia, Surinam, Guyana, Arkansas, and elsewhere. The Bayer process is most commonly used today to refine bauxite so it can be accommodated in the Hall-Heroult refining process, used to produce most aluminum. Aluminum can now be produced from clay, but the process is not yet economical. Aluminum is the most abundant metal to be found in the crust of the earth (8.1%), but is never found free in nature. It is found in bauxite, feldspars, granite, and in many other common minerals. Pure aluminum, a silvery-white metal, possesses many desirable characteristics. It is light, nontoxic, has a pleasing appearance, is easily formed, machined, or cast, has a high thermal conductivity, and has excellent corrosion resistance. It is nonmagnetic and nonsparking, and stands second among metals in malleability and sixth in ductility. It is extensively used for kitchen utensils, outside building decoration, and in thousands of industrial applications where a strong, light, easily constructed material is needed. Although its electrical conductivity is only about 60% that of copper per area of cross section, it is used in electrical transmission lines because of its light weight. Pure aluminum is soft and lacks strength, but it can be alloyed with small amounts of copper, magnesium, silicon, manganese, and other elements to impart a variety of useful properties. These alloys are of vital importance in the aerospace industry. Aluminum, evaporated in a vacuum, forms a highly reflective coating for both visible light and radiant heat. These coatings soon form a thin layer of the protective oxide and do not deteriorate as do silver coatings. The oxide, alumina, occurs naturally as ruby, sapphire, corundum, and emery. Synthetic ruby and sapphire have found application in the construction of lasers for producing coherent light. In 1852, the price of aluminum was about $5545/lb. and just before Hall's discovery in 1886, about $11.00. The price rapidly dropped to 30¢ and has been as low as 15¢/lb.

**Americium** — (the Americas), Am; at wt 243; at no 95; mp 994 ± 4°C; bp 2607°C; sp gr 13.67 (20°C); valence 2, 3, 4, 5, or 6. Americium was the fourth transuranium element to be discovered; the isotope $^{241}$Am was identified by Seaborg, James, Morgan, and Ghiorso late in 1944 at the wartime Metallurgical Laboratory (now the Argonne National Laboratory) of the University of Chicago as the result of successive neutron capture reactions by plutonium isotopes in a nuclear reactor:

$$^{239}\text{Pu (n}\gamma) \ ^{240}\text{Pu (n}\gamma) \ ^{241}\text{Pu} \xrightarrow{\beta} \ ^{241}\text{Am}$$

Since the isotope $^{241}$Am can be prepared in relatively pure form by extraction as a decay product over a period of years from strongly neutron-bombarded plutonium $^{241}$Pu, this isotope is used for much of the chemical investigation of this element. Better suited is the isotope $^{243}$Am, due to its longer half-life ($8.8 \times 10^3$ years as compared to 470 years for $^{241}$Am). A mixture of the isotopes $^{241}$Am, $^{242}$Am, and $^{243}$Am can be prepared by intense neutron irradiation of $^{241}$Am according to the reactions $^{241}$Am (n, γ) $^{242}$Am (n, γ) $^{243}$Am. Nearly isotopically pure

$^{243}$Am can be prepared by a sequence of neutron bombardments and chemical separations as follows: neutron bombardment of $^{241}$Am yields $^{242}$Pu by the reactions $^{241}$Am (n, γ) $^{242}$Am → $^{242}$Pu. After chemical separation, the $^{242}$Pu can be transformed to $^{243}$Am via the reactions $^{242}$Pu (n,γ) $^{243}$Pu → $^{243}$Am, and the $^{243}$Am can be chemically separated. Fairly pure $^{242}$Pu can be prepared more simply by very intense neutron irradiation of $^{239}$Pu as the result of successive neutron-capture reactions. Americium metal has been prepared by reducing the trifluoride with barium vapor at 1000 to 1200°C or the dioxide by lanthanum metal. The luster of freshly prepared americium metal is whiter and more silvery than plutonium or neptunium prepared in the same manner. It appears to be more malleable than uranium or neptunium and tarnishes slowly in dry air at room temperature. Americium is thought to exist in two forms: an alpha form which has a double hexagonal close-packed structure and a loose-packed cubic beta form. Americium must be handled with great care. As little as 0.03 μg of $^{241}$Am is the maximum permissible total body burden. The alpha activity from $^{241}$Am is about three times that of radium. When gram quantities of $^{241}$Am are handled, the intense gamma activity makes exposure a serious problem. Americium dioxide, $AmO_2$, is the most important oxide. $AmF_3$, $AmF_4$, $AmCl_3$, $AmBr_3$, $AmI_3$, and other compounds have been prepared. The isotope $^{241}$Am has been used as a portable source for gamma radiography and as a source of ionization for smoke detectors. Americum-241 is available from the A.E.C. at a cost of $1600/g and americium-243 at a cost of $100/mg.

**Antimony** — (Gr. *anti* plus *monos,* a metal not found alone), Sb; at wt 121.75 ± 3; at no 51; mp 630.74°C; bp 1950°C; sp gr 6.691 (20°C); valence 0, −3, +3, or +5. Antimony was recognized in compounds by the ancients and was known as a metal at the beginning of the 17th century and possibly much earlier. It is not abundant, but is found in over 100 mineral species. It is sometimes found native, but more frequently as the sulfide, stibnite ($Sb_2S_3$); it is also found as antimonides of the heavy metals and as oxides. It is extracted from the sulfide by roasting to the oxide, which is reduced by salt and scrap iron; from its oxides it is also prepared by reduction with carbon. Two allotropic forms of antimony exist: the normal stable, metallic form, and the amorphous gray form. The so-called explosive antimony is an ill-defined material always containing an appreciable amount of halogen. The yellow form, obtained by oxidation of stibine, $SbH_3$, is probably impure. Metallic antimony is an extremely brittle metal of a flaky, crystalline texture. It is bluish white and has a metallic luster. It is not acted on by air at room temperature, but burns brilliantly when heated, with the formation of white fumes of $Sb_2O_3$. It is a poor conductor of heat and electricity, and has a hardness of 3 to 3.5. Antimony, available commercially with a purity of 99.999 + %, is finding use in semiconductor technology. Commercial-grade antimony is widely used in alloys, with percentages ranging from 1 to 20. It greatly increases the hardness and mechanical strength of lead. Batteries, antifriction alloys, type metal, small arms and tracer bullets, cable sheathing, and minor products use about half the metal produced. Compounds taking up the other half are oxides, sulfides, sodium antimonate, and antimony trichloride. Tartar emetic (hydrated potassium antimonyltartate) has been used in medicine. Antimony and many of its compounds are toxic. Exposure to antimony and its compounds should not exceed 0.5 mg/m$^3$ (8-hr time-weighted average, 40-hr work week).

**Argon** — (Gr. *argos,* inactive), Ar; at wt 39.948; at no 18; freezing pt $-189.2°C$; bp $-185.7°C$; density 1.7837 g/l. Its presence in air was suspected by Cavendish in 1785, discovered by Lord Rayleigh and Sir William Ramsay in 1894. The gas is prepared by fractionation of liquid air, the atmosphere containing 0.94% argon. The atmosphere of Mars contains 1.6% of $^{40}Ar$ and 5 ppm of $^{36}Ar$. Argon is two and one half times as soluble in water as nitrogen, having about the same solubility as oxygen. It is recognized by the characteristic lines in the red end of the spectrum. It is used in electric light bulbs and in fluorescent tubes at a pressure of about 3 mm. Argon is also used as an inert gas shield for arc welding and cutting, as a blanket for the production of reactive elements, and as a protective atmosphere for growing silicon and germanium crystals. Argon is colorless and odorless, both as a gas and liquid. It is available in high-purity form. Commercial argon is available at a cost of about 3¢/ft$^3$. Argon is considered to be a very inert gas and is not known to form true chemical compounds, as do krypton, xenon, and radon. However, it does form a hydrate having a dissociation pressure of 105 atm at 0°C. Ion molecules such as $(ArKr)^+$, $(ArXe)^+$, and $(NeAr)^+$ have been observed spectroscopically. Argon also forms a

clathrate with β hydroquinone. Van der Waals' forces act to hold the argon. Naturally occurring argon is a mixture of three isotopes. Five other radioactive isotopes are now known to exist.

**Arsenic** — (L. *arsenicum,* Gr. *arsenikon,* yellow orpiment, identified with arsenikos, male, from the belief that metals were different sexes; Arab, *az-zernikh,* the orpiment from Persian *zerni-zar,* gold), As; at wt 74.9216; at no 33; valence $-3.0$, $+3$, or $+5$. Elemental arsenic occurs in two solid modifications: yellow, and gray or metallic, with specific gravities of 1.97, and 5.73, respectively. Gray arsenic, the ordinary stable form, has a mp of 817°C (28 atm) and sublimes at 613°C. Several other allotropic forms of arsenic are reported in the literature. It is believed that Albertus Magnus obtained the element in 1250 A.D. In 1649 Schroeder published two methods of preparing the element. It is found native, in the sulfides realgar and orpiment, as arsenides and sulfarsenides of heavy metals, as the oxide, and as arsenates. Mispickel or arsenopyrite (FeSAs) is the most common mineral, from which on heating the arsenic sublimes leaving ferrous sulfide. The element is a steel gray, very brittle, crystalline, semimetallic solid; it tarnishes in air, and when heated is rapidly oxidized to arsenous oxide ($As_2O_3$) with the odor of garlic. Arsenic and its compounds are poisonous. Exposure to arsenic and its compounds (as/As) should not exceed 0.5 mg/m$^3$ (8-hr time-weighted average, 40-hr work week.) These values, however, are being studied, and may be lowered. Arsenic is also used in bronzing, pyrotechny, and for hardening and improving the sphericity of shot. The most important compounds are white arsenic ($As_2O_3$), the sulfide, Paris green $3Cu(AsO_2)_2$ $(Cu(C_2H_3O_2)_2$, calcium arsenate, and lead arsenate; the last three have been used as agricultural insecticides and poisons. Marsh's test makes use of the formation and ready decomposition of arsine ($AsH_3$). Arsenic is available in high-purity form. It is finding increasing uses as a doping agent in solid-state devices. Gallium arsenide is used as a laser material to convert electricity directly into coherent light.

**Astatine** — (Gr. *astatos,* unstable), At; at wt (210); at no 85; mp 302°C; bp 337°C (est.); valence probably 1, 3, 5, or 7. Synthesized in 1940 by D. R. Corson, K. R. MacKenzie, and E. Segre at the University of California by bombarding bismuth with alpha particles. The longest-lived isotope, $^{218}At$, has a half-life of only 8.3 hr. Twenty isotopes are now known. Minute quantities of $^{215}At$,

$^{216}$At, and $^{219}$At exist in equilibrium in nature with naturally occurring uranium and thorium isotopes, and traces of $^{217}$At are in equilibrium with $^{233}$U and $^{239}$Np resulting from interaction of thorium and uranium with naturally produced neutrons. The total amount of astatine present in the earth's crust, however, totals less than 1 oz. Astatine can be produced by bombarding bismuth with energetic alpha particles to obtain the relatively long-lived $^{209-211}$At, which can be distilled from the target by heating it in air. Only about 0.05 μg of astatine have been prepared to date. The "time of flight" mass spectrometer has been used to confirm that this highly radioactive halogen behaves chemically very much like other halogens, particularly iodine. The interhalogen compounds AtI, AtBr, and AtCl are known to form. HAt and $CH_3At$ (methyl astatide) have been detected. Astatine is said to be more metallic than iodine, and, like iodine, it probably accumulates in the thyroid gland. Workers at the Brookhaven National Laboratory have recently used reactive scattering in crossed molecular beams to identify and measure elementary reactions involving astatine.

**Barium** — (Gr. *barys,* heavy) Ba; at wt 137.33, at no 56; mp 725°C; bp 1640°C; sp gr 3.5 (20°C); valence 2. Baryta was distinguished from lime by Scheele in 1774; the element was discovered by Sir Humphrey Davy in 1808. It is found only in combination with other elements, chiefly in barite or heavy spar (sulfate) and witherite (carbonate) and is prepared by electrolysis of the chloride. Barium is a metallic element, soft, and when pure is silvery white like lead; it belongs to the alkaline earth group, resembling calcium chemically. The metal oxidizes very easily and should be kept under petroleum or other suitable oxygen-free liquids to exclude air. It is decomposed by water or alcohol. The metal is used as a "getter" in vacuum tubes. The most important compounds are the peroxide ($BaO_2$), chloride, sulfate, carbonate, nitrate, and chlorate. Lithopone, a pigment containing barium sulfate and zinc sulfide, has good covering power, and does not darken in the presence of sulfides. The sulfate, as permanent white or 'blanc fixe', is also used in paint and in X-ray diagnostic work. The carbonate has been used as a rat poison, while the nitrate and chlorate give colors in pyrotechny. The impure sulfide phosphoresces after exposure to the light. The compounds and the metal are not expensive. Barium metal (99.5 + % pure) costs about $20/lb. All barium compounds that are water

or acid soluble are poisonous. Naturally occurring barium is a mixture of seven stable isotopes. Thirteen other radioactive isotopes are known to exist.

**Berkelium** — (*Berkeley,* home of the University of California), Bk; at wt (247); at no 97; valence 3 or 4; sp gr 14 (est.). Berkelium, the eighth member of the actinide transition series, was discovered in December 1949 by Thompson, Ghiorso, and Seaborg, and was the fifth transuranium element synthesized. It was produced by cyclotron bombardment of milligram amounts of $^{241}$Am with helium ions at Berkeley, California. The first isotope produced had a mass number of 243 and decayed with a half-life of 4.6 hr. Eight isotopes are now known and have been synthesized. The existence of $^{249}$Bk, with a half-life of 314 days, makes it feasible to isolate berkelium in weighable amounts so that its properties can be investigated with macroscopic quantities. One of the first visible amounts of a pure berkelium compound, berkelium chloride, was produced in 1962. It weighed 3 billionth of a gram. Berkelium probably has not been prepared in elemental form, but it is expected to be a silvery metal, easily soluble in dilute mineral acids, and readily oxidized by air oxygen at elevated temperatures to form the oxide. X-ray diffraction methods have been used to identify the following compounds: $BkO_2$, $BkO_3$, $BkF_3$, $BkCl_3$, and BkOCl. As with other actinide elements, berkelium tends to accumulate in the skeletal system. The maximum permissible body burden of $^{249}$Bk in the human skeleton is about 0.0004 μ. Because of its rarity, berkelium has at present no commercial or technological use. Berkelium-249 is available from O.R.N.L. at a cost of $100/mg.

**Beryllium** — (Gr. *beryllos, beryl;* also called Glucinium or Glucinum, Gr. *glykys,* sweet), Be; at wt 9.01218; at no 4; mp 1278 ± 5°C; bp 2970°C; sp gr 1.848 (20°C); valence 2. Discovered as the oxide by Vauquelin in beryl and in emeralds in 1798. The metal was isolated in 1828 by Wohler and by Bussy independently by the action of potassium on beryllium chloride. Beryllium is found in some 30 mineral species, the most important of which are bertrandite, beryl, chrysoberyl, and phenacite. Aquamarine and emerald are precious forms of beryl. Beryl ($3BeO \cdot Al_2O_2 \cdot 6SiO_2$) and bertrandite ($4BeO \cdot 2SiO_2 \cdot H_2O$) are the most important commercial sources of the element and its compounds. Most of the metal is now prepared by reducing beryllium fluoride with magnesium metal. Beryllium metal did not become readily available to industry until 1957. The metal, steel gray in color, has many desirable properties. It is one of the lightest of all metals, and has one of the highest melting points of the light metals. Its modulus of elasticity is about one third greater than that of steel. It resists attack by concentrated nitric acid, has excellent thermal conductivity, and is non-magnetic. It has a high permeability to X-rays, and when bombarded by alpha particles, as from radium or polonium, neutrons are produced in the ratio of about 30 neutrons per million alpha particles. At ordinary temperatures beryllium resists oxidation in air, although its ability to scratch glass is probably due to the formation of a thin layer of the oxide. Beryllium is used as an alloying agent in producing beryllium copper; which is extensively used for springs, electrical contacts, spot-welding electrodes, and nonsparking tools. It has found application as an aerospace structural material.

Beryllium is used in nuclear reactors as a reflector or moderator, for it has a low thermal neutron absorption cross section. It is used in gyroscopes, computer parts, and instruments where lightness, stiffness, and dimensional stability are required. The oxide has a very high melting point and is also used in nuclear work and ceramic applications. Beryllium and its salts are toxic and should be handled with the greatest of care. Beryllium and its compounds should not be tasted to verify the sweetish nature of beryllium (as did early experimenters). Exposure to beryllium dust in air should be limited to 2 $\mu$ g/m$^3$ (8-hr time-weighted average, 40-hr week), with a ceiling concentration of 5$\mu$ g/m$^3$. These values are being reviewed and studied. Beryllium metal in vacuum-cast billet form is priced roughly at $150/lb. Fabricated forms are more expensive.

**Bismuth** — (Ger. *weisse Masse,* white mass; later *Wisuth* and *Bisemutum*), Bi; at wt 208.9804; at no 83; mp 271.3°C; bp 1560 ± 5°C; sp gr 9.747 (20°C); valence 3 or 5. In early times, bismuth was confused with tin and lead. Claude Geoffroy the Younger showed it to be distinct from lead in 1753. It is a white crystalline, brittle metal with a pinkish tinge. It occurs native. The most important ores are bismuthinite or bismuth glance ($Bi_2S_3$) and bismite ($Bi_2O_3$). Peru, Japan, Mexico, Bolivia, and Canada are major bismuth producers. Much of the bismuth produced in the U.S. is obtained as a by-product in refining lead, copper, tin, silver, and gold ores. Bismuth is the most diamagnetic of all metals, and the thermal conductivity is lower than

any metal, except mercury. It has a high electrical resistance, and has the highest Hall effect of any metal (i.e., the greatest increase in electrical resistance when placed in a magnetic field). "Bismanol" is a permanent magnet of high coercive force, made of MnBi, by the U.S. Naval Surface Weapons Center. Bismuth expands 3.32% on solidification. This property makes bismuth alloys particularly suited to the making of sharp casting of objects subject to damage by high temperatures. With other metals such as tin, cadmium, etc., bismuth forms low-melting alloys which are extensively used for safety devices used in fire detection and extinguishing systems. When bismuth is heated in air it burns with a blue flame, forming yellow fumes of the oxide. The metal is also used as a thermocouple material (has highest negativity known). Its soluble salts are characterized by forming insoluble basic salts on the addition of water, a property sometimes used in detection work. Bismuth oxychloride is used extensively in cosmetics. Bismuth subnitrate and subcarbonate are used in medicine. Bismuth metal costs about $8/lb in large quantities.

**Boron** — (Ar. *Buraq,* Pers. *Burah*), B; at wt 10.81; at no 5; mp 2079°C; bp sublimes 2550°C; sp gr of crystals 2.34, of amorphous variety 2.37; valence 3. Boron compounds have been known for thousands of years, but the element was not discovered until 1808 by Sir Humphry Davy and by Gay-Lussac and Thenard. The element is not found free in nature, but occurs as orthoboric acid, usually in certain volcanic spring waters and as borates in borax and colemanite. Ulexite, another boron mineral, is interesting as it is nature's own version of "fiber optics". By far the most important source of boron is the mineral rasorite, also known as kernite, found in the Mojave desert of California. Extensive borax deposits are also found in Turkey. Boron exists naturally as 19.75% $_5^{10}$B isotope and 80.22% $_5^{11}$B isotope. High-purity crystalline boron may be prepared by the vapor phase reduction of boron trichloride or tribromide with hydrogen on electrically heated filaments. The impure, or amorphous boron, a brownish-black powder, can be obtained by heating the trioxide with magnesium powder. Boron of 99.9999% purity has been produced and is available commercially. Elemental boron has an energy band gap of 1.50 to 1.56 eV, which is higher than that of either silicon or germanium. It has interesting optical characteristics, transmitting portions of the infrared, and is a poor conductor of electricity at room temperature, but a good conductor at high temperature. Amorphous boron is used in pyrotechnic flares to provide a distinctive green color, and in rockets as an igniter. The most important compounds of boron are boric, or boracic, acid, widely used as a mild antiseptic, and borax ($Na_2B_4O_9 \cdot 10H_2O$), which serves as a cleaning flux in welding and as a water softener in washing powders. Boron compounds show promise in treating arthritis. The isotope $^{10}B$ is used as a control for nuclear reactors, as a shield for nuclear radiation, and in instruments used for detecting neutrons. Boron nitride has remarkable properties and can be used to make a material as hard as diamond. The nitride also behaves like an electrical isulator but conducts heat like a metal. It also has lubricating properties similar to graphite. High strength boron filaments have been utilized in advanced composite structures. Boron is similar to carbon in the capacity to form stable covalently bonded molecular networks. Carboranes, metalloboranes, and other families comprise thousands of compounds. Crystalline boron (99%) costs about $5/g. Amorphous boron costs about $2/g. Elemental boron is not considered to be a poison, but assimilation of its compounds has a cumulative poisonous effect. Care must be taken in handling these.

**Bromine** — (Gr. *bromos*, stench), Br; at wt 79.904; at no 35; mp $-7.2°C$; bp $58.78°C$; density of gas 7.59 g/l, liquid 3.12 (20°C); valence 1, 3, 5, or 7. Discovered by Balard in 1826, but not prepared in quantity until 1860. A member of the halogen group of elements, it is obtained from natural brines from wells in Michigan and Arkansas. Little bromine is extracted today from seawater, which contains only about 85 ppm. Bromine is the only liquid nonmetallic element. It is a heavy, mobile, reddish-brown liquid, volatizing readily at room temperature to a red vapor with a strong disagreeable odor, resembling chlorine, and having a very irritating effect on the eyes and throat; it is readily soluble in water or carbon disulfide, forming a red solution; it is less active than chlorine, but more so than iodine; it unites readily with many elements and has a bleaching action; when spilled on the skin it produces painful sores. It presents a serious health hazard, and maximum safety precautions should be taken when handling it. Much of the bromine output in the U.S. is used in the production of ethylene dibromide, a lead scavenger used in making gasoline antiknock compounds.

Lead in gasoline, however, is presently being drastically reduced, due to environmental considerations. Bromine is alo used in making fumigants, flameproofing agents, water purification compounds, dyes, medicinals, sanitizers, inorganic bromides for photography, etc. Organic bromides are also important.

**Cadmium** — (L. *cadmia;* Gr. *kadmeia* — ancient name for calamine, zinc carbonate), Cd; at wt 112.41; at no 48; mp 320.9°C; bp 765°C; sp gr 8.65 (20°C); valence 2. Discovered by Stromeyer in 1817 from an impurity in zinc carbonate. Cadmium most often occurs in small quantities associated with zinc ores, such as sphalerite (ZnS). Greenockite (CdS) is the only mineral of any consequence bearing cadmium. Almost all cadmium is obtained as a by-product in the treatment of zinc, copper, and lead ores. It is a soft, bluish-white metal which is easily cut with a knife. It is similar in many respects to zinc. It is a component of some of the lowest melting alloys; it is used in bearing alloys with low coefficients of friction and great resistance to fatigue; it is used extensively in electroplating, which accounts for about 60% of its use. It is also used in many types of solder, for standard EMF cells, for Ni-Cd batteries, and as a barrier to control atomic fission. Cadmium compounds are used in black and white television phosphors and in blue and green phosphors for color TV tubes. It forms a number of salts, of which the sulfate is the most common; the sulfide is used as a yellow pigment. Cadmium and solutions of its compounds are toxic. Serious toxicity problems have been found from long-term exposure and work with cadmium-plating baths. Exposure to cadmium dust should not exceed 0.05 mg/m$^3$ (8-hr time-weighted average, 40-hr week). The ceiling concentration (maximum), for a period of 15 min, should not exceed 0.15 mg/m$^3$. Cadmium oxide fume exposure (8-hr, 40-hr week) should not exceed 0.05 mg/m$^3$. These values are presently being restudied and recommendations have been made to reduce the exposure. In 1927 the International Conference on Weights and Measures redefined the meter in terms of the wavelength of the red cadmium spectral line (i.e., 1 m = 1,553,164.13 wavelengths). This definition has been changed (see Krypton). The current price of cadmium is about $12/lb. It is available in high purity form.

**Calcium** — (L. *calx,* lime), Ca; at wt 40.08; at no 20; mp 839 $\pm$ 2°C; bp 1484°C; sp gr 1.55 (20°C); valence 2. Though lime was prepared by the Romans in the first century under the name calx, the metal was not discovered until 1808. After learning that Berzelius and Pontin prepared calcium amalgam by electrolyzing lime in mercury, Davy was able to isolate the impure metal. Calcium is a metallic element, fifth in abundance in the earth's crust, of which it forms more than 3%. It is an essential constituent of leaves, bones, teeth, and shells. Never found in nature uncombined, it occurs abundantly as limestone ($CaCO_3$), gypsum ($CaSO_4 \cdot 2H_2O$), and fluorite ($CaF_2$); apatite is the fluophosphate or chlorophosphate of calcium. The metal has a silvery color, is rather hard, and is prepared by electrolysis of the fused chloride to which calcium fluoride is added to lower the melting point. Chemically it is one of the alkaline earth elements; it readily forms a white coating of nitride in the air, reacts with water, and burns with a yellow-red flame, forming largely the nitride. The metal is used as a reducing agent in preparing other metals such as throium, uranium, zirconium, etc., and is used as a deoxidizer, desulfurizer, or decarburizer for various ferrous and nonferrous alloys. It is also used as an alloying agent for aluminum, beryllium, copper, lead, and magnesium alloys, and serves as a "getter" for residual gases in vacuum tubes, etc. Quicklime ($CaO$), made by heating limestone and changed into slaked lime by the careful addition of water, is the great cheap base of chemical industry with countless uses. Mixed with sand, it hardens as mortar and plaster by taking up carbon dioxide from the air. Calcium from limestone is an important element in Portland cement. The solubility of the carbonate in water containing carbon dioxide causes the formation of caves with stalactites and stalagmites and hardness in water. Other important compounds are the carbide ($CaC_2$), chloride ($CaCl_2$), cyanamide ($CaCN_2$), hypochlorite [$Ca(OCl)_2$], nitrate [$Ca(NO_3)_2$], and sulfide ($CaS$).

**Californium** — (Named for the State and University of California), Cf; at wt (251); at no 98. Californium, the sixth transuranium element to be discovered, was produced by Thompson, Street, Ghiorso, and Seaborg in 1950 by bombarding microgram quantities of $^{242}Cm$ with 35 MeV helium ions in the Berkeley 60-in. cyclotron. Californium(III) is the only ion stable in aqueous solutions, all attempts to reduce or oxidize californium(III) having failed. The isotope $^{249}Cf$ results from the beta decay of $^{249}Bk$, while the heavier isotopes are produced by intense neutron irradiation by the reactions:

$$^{249}Bk \; (n\gamma) \; ^{250}Bk \xrightarrow{\beta}$$

$$^{250}Cf \quad \text{and} \quad ^{249}Cf \; (n\gamma) \; ^{250}Cf$$

followed by

$$^{250}Cf \; (n\gamma) \; ^{251}Cf \; (n\gamma) \; ^{252}Cf$$

The existence of the isotopes $^{249}Cf$, $^{250}Cf$, $^{251}Cf$, and $^{252}Cf$ makes it feasible to isolate californium in weighable amounts so that its properties can be investigated with macroscopic quantities. Californium-252 is a very strong neutron emitter. One microgram releases 170 million neutrons per minute, which presents biological hazards. Proper safeguards should be used in handling californium. In 1960, a few tenths of a microgram of californium trichloride $Cf_2Cl_3$, californium oxychloride, CfOCl, and californium oxide, $Cf_2O_3$, were first prepared. Reduction of californium to its metallic state has not yet been accomplished. Because californium is a very efficient source of neutrons, many new uses are expected for it. It is being used as a portable neutron source for discovery of metals such as gold or silver by on-the-spot activation analysis. $^{252}Cf$ is now being offered for sale by the O.R.N.L. at a cost of $10/\mu g$. As of May 1975, more than 63 mg had been produced and sold. It has been suggested that californium may be produced in certain stellar explosions, called supernovae, for the radioactive decay of $^{254}Cr$ (55-day half-life) agrees with the characteristics of the light curves of such explosions observed through telescopes.

**Carbon** — (L. *carbo,* charcoal), C; at wt 12 exactly ($^{12}C$); at wt (natural carbon) 12.011; at no 6; mp $\sim$ 3550°C, graphite sublimes at 3367 $\pm$ 25°C; triple point: (graphite-liquid-gas), 3627 $\pm$ 50°C at a pressure of 10.1 MPa and (graphite-diamond-liquid), 3830—3930° at a pressure of 12—13 GPa; sp gr amorphous 1.8 to 2.1, graphite 1.9 to 2.3, diamond 3.15 to 3.53 (depending on variety); gem diamond 3.513 (25°C); valence 2, 3, or 4. Carbon, an element of prehistoric discovery, is very widely distributed in nature. It is found in abundance in the sun, stars, comets, and atmospheres of most planets. Carbon in the form of microscopic diamonds is found in some meteorites. Natural diamonds are found in kimberlite of ancient volcanic "pipes", such as are found in South Africa, Arkansas, and elsewhere. Diamonds are now also being recovered from the ocean floor off the Cape of Good Hope. About 30% of all industrial diamonds used in the U.S. are now made synthetically. The energy of the sun and stars can be attributed at least in part to the well-known carbon-nitrogen cycle. Carbon is found free in nature in three allotropic forms: amorphous, graphite, and diamond. A fourth form, known as "white" carbon, is now thought to exist. Graphite is one of the softest known materials, while diamond is one of the hardest. Graphite exists in two forms: alpha and beta. These have identifcal physical properties, except for their crystal structure. Naturally occurring graphites are reported to contain as much as 30% of the rhombohedral (beta) form, whereas synthetic materials contain only the alpha form. The hexagonal alpha type can be converted to the beta by mechanical treatment, and the beta form reverts to the alpha when it is heated above 1000°C. In 1969 a new allotropic form of carbon was produced during the sublimation of pyrolytic graphite at low pressures. Under free-vaporization conditions above ~2550 K, "white" carbon forms as small transparent crystals on the edges of the basal planes of graphite. The interplanar spacings of "white" carbon are identical to those of a carbon form noted in the graphitic gneiss from the Ries (meteoritic) Crater in Germany. "White" carbon is a transparent birefringent material. Little information is presently available about this allotrope. In combination, carbon is found as $CO_2$ in the atmosphere of the earth and dissolved in all natural waters. It is a component of great rock masses in the form of carbonates of calcium (limestone), magnesium, and iron. Coral, petroleum, and natural gas are chiefly hydrocarbons. Carbon is unique among the elements in the vast number of variety of compounds it can form. With hydrogen, oxygen, and nitrogen, and other elements, it forms a very large number of compounds, carbon atom often being linked to carbon atom. There are upwards of a million or more known carbon compounds, many thousands of which are vital to organic and life processes. Without carbon, the basis for life would be impossible. While it has been thought that silicon might take the place of carbon in forming a host of similar compounds, it is not now possible to form stable compounds with very long chains of silicon atoms. The atmosphere of Mars contains 96.2% $CO_2$. Some of the most important compounds of carbon are carbon dioxide ($CO_2$), carbon monoxide (CO), carbon disulfide ($CS_2$), chloroform ($CHCl_3$), carbon tetrachloride ($CCl_4$), methane ($CH_4$), ethylene ($C_2H_4$), acetylene ($C_2H_2$), benzene ($C_6H_6$), ethyl alcohol ($C_2H_5OH$), acetic acid ($CH_3COOH$), and their derivatives. Carbon has seven isotopes. In 1961 the International Union of Pure and Applied Chemistry adopted the isotope carbon-12 as the basis for atomic weight. Carbon-14, an isotope with a half-life of 5730 years, has been widely used to date such materials as wood, archeological specimens, etc. Carbon-13 is now commercially available at a cost of $700/g.

**Cerium** — (named for the asteroid Ceres, which was discovered in 1801, only 2 years before the element), Ce; at wt 140.12; at no 58; mp 799°C; bp 3426°C; sp gr 6.657 (25°C), valence 3 or 4. Discovered in 1803 by Klaproth and by Berzelius and Hisinger; metal prepared by Hillebrand and Norton in 1875. Cerium is the most abundant of the metals of the so-called rare earths. It is found in a number of minerals including allanite (also known as orthite), monazite, bastnasite, cerite, and samarskite. Monazite and bastnasite are the most important sources of cerium. Large deposits of monazite found on the beaches of Travancore, India, in river sands in Brazil, and deposits of allanite in the western United States, and bastnasite in Southern California will supply cerium, thorium, and the other rare earth metals for many years to come. Metallic cerium is prepared by metallothermic reduction techniques, such as by reducing cerous fluoride with calcium, or by electrolysis of molten cerous chloride or other cerous halides. The metallothermic technique is used to produce high-purity cerium. Cerium is especially interesting because of its variable electronic structure. The energy of the inner 4f level is nearly the same as that of the outer or valence electrons, and only small amounts of energy are required to change the relative occupancy of these electronic levels. This gives rise to dual valency states. For example, a volume change of about 10% occurs when cerium is subjected to high pressures or low temperatures. It appears that the valence changes from about 3 to 4 when it is cooled or compressed. The low temperature behavior of cerium is complex. Four allotropic modifications are thought to exist: cerium at room temperature and at atmospheric pressure is known as γ cerium. Upon cooling to −23°C, γ cerium changes to β cerium. The remaining γ cerium starts to change to α cerium when cooled to −158°C, and the transformation is complete at −196°C. α Cerium has a density of 8.24; δ cerium exists above 726°C. At atmospheric pressure, liquid cerium is more dense than its solid form at the melting point. Cerium is an iron-gray lustrous metal.

It is malleable, and oxidizes very readily at room temperature, especially in moist air. Except for europium, cerium is the most reactive of the "rare earth" metals. It slowly decomposes in cold water, and rapidly in hot water. Alkali solutions and dilute and concentrated acids attack the metal rapidly. The pure metal is likely to ignite if scratched with a knife. Ceric salts are orange red or yellowish; cerous salts are usually white. Cerium is a component of misch metal, which is extensively used in the manufacture of pyrophoric alloys for cigarette lighters, etc. While cerium is not radioactive, the impure commercial grade may contain traces of radioactive thorium. The oxide is an important constituent of incandescent gas mantles and it is emerging as a hydrocarbon catalyst in "self-cleaning" ovens. In this application it can be incorporated into oven walls to prevent the collection of cooking residues. As ceric sulfate it finds extensive use as a volumetric oxidizing agent in quantitative analysis. Cerium, with other rare earths, is used in carbon-arc lighting, especially in the motion picture industry. It is also finding use as an important catalyst in petroleum refining. In small lots, 99.9% cerium costs about $2/g.

**Cesium** — (L. *caesius*, sky blue), Cs; at wt 132.9054; at no 55; mp 28.40 ± 0.01°C; bp 669.3°C; sp gr 1.873 (20°C); valence 1. Cesium was discovered spectroscopically by Bunsen and Kirchhoff in 1860 in mineral water from Dürcheim. Cesium, an alkali metal, occurs in lepidolite, pollucite (a hydrated silicate of aluminum and cesium), and in other sources. One of the richest sources of cesium in the world is located at Bernic Lake, Manitoba. The deposits are estimated to contain 300,000 tons of pollucite, averaging 20% cesium. It can be isolated by electrolysis of the fused cyanide and by a number of other methods. Very pure, gas-free cesium can be prepared by thermal decomposition of cesium azide. The metal is characterized by a spectrum containing two bright lines in the blue along with several others in the red, yellow, and green. It is silvery white, soft, and ductile. It is the most electropositive and most alkaline element. Cesium, gallium, and mercury are the only three metals that are liquid at room temperature. Cesium reacts explosively with cold water, and reacts with ice at temperatures above −116°C. Cesium hydroxide, the strongest base known, attacks glass. Because of its great affinity for oxygen, the metal is used as a "getter" in electron tubes. It is also used in photoelectric cells,

as well as a catalyst in the hydrogenation of certain organic compounds. The metal has recently found application in ion propulsion systems. Although these are not usable in the earth's atmosphere, 1 lb of cesium in outer space theoretically will propel a vehicle 140 times as far as the burning of the same amount of any known liquid or solid. Cesium is used in atomic clocks, which are accurate to 5 sec in 300 years. Its chief compounds are the chloride and the nitrate. The present price of cesium is about $25/g.

**Chlorine** — (Gr. *chloros*, greenish yellow), Cl; at wt 35.453; at no 17; mp −100.98°C; bp −34.6°C; density 3.214 g/l; sp gr 1.56 (−33.6°C); valence 1, 3, 5, or 7. Discovered in 1774 by Scheele, who thought it contained oxygen; named in 1810 by Davy, who insisted it was an element. In nature it is found in the combined state only, chiefly with sodium as common salt (NaCl), carnallite (KMgCl$_3$ · 6H$_2$O), and sylvite (KCl). It is a member of the halogen (salt-forming) group of elements and is obtained from chlorides by the action of oxidizing agents and more often by electrolysis; it is a greenish yellow gas, combining directly with nearly all elements. At 10°C, one volume of water dissolves 3.10 volumes of chlorine, at 30°C, only 1.77 volumes. Chlorine is widely used in making many everyday products. It is used for producing safe drinking water the world over. Even the smallest water supplies are now usually chlorinated. It is also extensively used in the production of paper products, dyestuffs, textiles, petroleum products, medicines, antiseptics, insecticides, foodstuffs, solvents, paints, plastics, and many other consumer products. Most of the chlorine produced is used in the manufacture of chlorinated compounds of sanitation, pulp bleaching, disinfectants, and textile processing. Further use is in the manufacture of chlorates, chloroform, carbon tetrachloride, and in the extraction of bromine. Organic chemistry demands much from chlorine, both as an oxidizing agent and in substitution, since it often brings desired properties in an organic compound when substituted for hydrogen, as in one form of synthetic rubber. Chlorine is a respiratory irritant. The gas irritates the mucous membranes and the liquid burns the skin. As little as 3.5 ppm can be detected as an odor, and 1000 ppm is likely to be fatal after a few deep breaths. It was used as a war gas in 1915. Exposure to chlorine should not exceed 1 ppm (8-hr time-weighted average, 40-hr week).

**Chromium** — (Gr. *chroma*, color), Cr; at wt

51.996; at no 24; mp 1857 ± 20°C; bp 2672°C; sp gr 7.18 to 7.20 (20°C); valence chiefly 2, 3, or 6. Discovered in 1797 by Vauquelin, who prepared the metal the next year. Chromium is a steel gray, lustrous, hard metal that takes a high polish. The principal ore is chromite ($FeCr_2O_4$), which is found in Zimbabwe, the U.S.S.R., the Transvaal, Turkey, Iran, Albania, Finland, the Democratic Republic of Madagascar, and the Phillipines. The metal is usually produced by reducing the oxide with aluminum. Chromium is used to harden steel, to manufacture stainless steel, and to form many useful alloys. Much is used in plating to produce a hard, beautiful surface and to prevent corrosion. Chromium is used to give glass an emerald green color. It finds wide use as a catalyst. All compounds of chromium are colored; the most important are the chromates of sodium and potassium ($K_2CrO_4$), and the dichromates ($K_2Cr_2O_7$) and the potassium and ammonium chrome alums, such as $KCr(SO_4)_2 \cdot 12H_2O$. The dichromates are used as oxidizing agents in quantitative analysis, and also in tanning leather. Other compounds are of industrial value; lead chromate is chrome yellow, a valued pigment. Chromium compounds are used in the textile industry as mordants, and by the aircraft and other industries for anodizing aluminum. Chromium compounds are toxic and should be handled with proper safeguards.

**Cobalt** — (*Kobald,* from the German, goblin or evil spirit; *cobalos,* Greek, mine), Co; at wt 58.9332; at no 27; mp 1495°C; bp 2870°C; sp gr 8.9 (20°C); valence 2 or 3. Discovered by Brandt about 1735. Cobalt occurs in the minerals cobaltite, smaltite, and erythrite, and is often associated with nickel, silver, lead, copper, and iron ores, from which it is most frequently obtained by a by-product. It is also present in meteorites. Important ore deposits are found in Zaire, Morocco, and Canada. The U.S. Geological Survey has announced that the bottom of the north central Pacific Ocean may have cobalt-rich deposits at relatively shallow depths in waters close to the Hawaiian Islands and other U.S. Pacific territories. Cobalt is a brittle, hard metal, closely resembling iron and nickel in appearance. It has a magnetic permeability of about two thirds that of iron. Cobalt tends to exist as a mixture of two allotropes over a wide temperature range; the β-form predominates below 400°C, and the α above that temperature. The transformation is sluggish and accounts in part for the wide variation in data on physical properties of cobalt. It

is alloyed with iron, nickel, and other metals to make Alnico, an alloy of unusual magnetic strength. Stellite® alloys, containing cobalt, chromium, and tungsten, are used for high-speed, heavy-duty, high-temperature cutting tools and dies. Cobalt is also used in other magnet steels and stainless steels, and in alloys used in turbine engines and generators. The metal is used in electroplating because of its appearance, hardness, and resistance to oxidation. The salts have been used for centuries for the production of brilliant and permanent blue colors in porcelain, glass, pottery, tiles, and enamels. It is the principal ingredient in Sevre's and Thenard's blue. Cobalt carefully used in the form of the chloride, sulfate, acetate, or nitrate has been found effective in correcting a certain mineral deficiency disease in animals. Soils should contain 0.13 to 0.30 ppm of cobalt for proper animal nutrition. Cobalt-60, an artificial isotope, is an important gamma-ray source, and is extensively used as a tracer and a radiotherapeutic agent. Single compact sources of Cobalt-60 are readily available and have an equivalent gamma-ray output equal to thousands of grams of radium. The cost of cobalt-60 varies from about $1 to $10/curie, depending on quantity and specific activity. Exposure to cobalt (metal fumes and dust) should be limited to 0.05 mg/m³ (8-hr time-weighted average, 40-hr week).

**Columbium** — see **Niobium.**

**Copper** — (L. *cuprum,* from the island of Cyprus), Cu; at wt 63.546 ± 3; at no 29; mp 1083.4 ± 0.2°C; bp 2567°C; sp gr 8.96 (20°C); valence 1 or 2. The discovery of copper dates from prehistoric times; it is said to have been mined for more than 5000 years. It is one of man's most important metals. Copper is reddish colored, takes on a bright metallic luster, and is malleable, ductile, and a good conductor of heat and electricity (second only to silver in electrical conductivity). It also has excellent corrosion resistance, as do its alloys. The electrical industry is one of the greatest users of copper. Copper occasionally occurs native, and is found in many minerals such as cuprite, malachite, azurite, chalcopyrite, and bornite. Large copper ore deposits are found in the U.S., Chile, Zambia, Zaire, Peru, and Canada. The most important copper ores are the sulfides, oxides, and carbonates. From these copper is obtained by smelting, leaching, and by electrolysis. Its alloys, brass and bronze, long used, are still very important; all American coins are now copper alloys; monel and gun metals also contain copper. The most important compounds are the oxide and the sulfate, blue vitriol; the latter has wide use as an agricultural poison and as an algicide in water purification. Copper compounds such as Fehling's solution are widely used in analytical chemistry in tests for sugar. High-purity copper (99.999 + %) is available commercially.

**Curium** — (Pierre and Marie Curie), Cm; at wt (247); at no 96; mp 1340 ± 40°C; sp gr 13.51 (calc); valence 3 and 4. Although curium follows americium in the periodic system, it was actually known before americium and was the third transuranium element to be discovered. It was identified by Seaborg, James, and Shiorso in 1944 at the wartime Metallurgical Laboratory in Chicago as a result of helium-ion bombardment of $^{239}$Pu in the Berkeley, California, 60-in. cyclotron. Visible amounts (30 μg) of $^{242}$Cm, in the form of the hydroxide, were first isolated by Werner and Perlman of the University of California in 1947. In 1950, Crane, Wallmann, and Cunningham found that the magnetic susceptibility of microgram samples of $CmF_3$ was of the same magnitude as that of $GdF_3$. In 1951, the same workers prepared curium in its elemental form for the first time. Thirteen isotopes of curium are now known. The most stable, $^{247}$Cm, with a half-life of 16 million years, is so short-lived compared to the age of the earth that any primordial curium must have disappeared long ago from the natural scene. Minute amounts of curium probably exist in natural deposits of uranium, as a result of a sequence of neutron captures and β decays sustained by the very low flux of neutrons naturally present in uranium ores. The presence of natural curium, however, has never been detected. $^{242}$Cm and $^{244}$Cm are available in multigram quantities. $^{248}$Cm has been produced only in milligram amounts. Curium is similar in some regards to gadolinium, its rare-earth homolog, but is has a more complex crystal structure. Curium is silver in color, is chemically reactive, and is more electropositive than aluminum. $CmO_2$, $Cm_2O_3$, $CmF_3$, $CmF_4$, $CmCl_3$, $CmBr_3$, and $CmI_3$ have been prepared. Most compounds of trivalent curium are faintly yellow in color. The A.E.C. is attempting to produce several kilograms of $^{244}$Cm, an isotope with a 17.6-year half-life, by neutron irradiation of plutonium in a nuclear reactor. $^{242}$Cm generates about three thermal watts of energy per gram. This compares to one-half thermal watt per gram of $^{238}$Pu. This suggests use for curium as an isotope power source. $^{244}$Cm is now offered for sale by the A.E.C. at $100/mg. Curium absorbed into the body accumulates in the bones, and is therefore very toxic because its radiation destroys the red cell-forming mechanism. The maximum permissible total body burden of $^{244}$Cm (soluble) in a human being is 0.3 μCi.

**Deuterium,** an isotope of hydrogen — see **Hydrogen.**

**Dysprosium** — (Gr. *dysprositos,* hard to get at), Dy; at wt 162.50 ± 3; at no 66; mp 1412°C; bp 2562°C; sp gr 8.550 (25°C); valence 3. Dysprosium was discovered in 1886 by Lecoq de Boisbaudran, but not isolated. Neither the oxide nor the metal was available in relatively pure form until the development of ion-exchange separation and metallographic reduction techniques by Spedding and associates about 1950. Dysprosium occurs along with other so-called rare earth or lanthanide elements in a variety of minerals such as xenotime, fergusonite, gadolinite, euxenite, polycrase, and blomstrandine. The most important sources, however, are from monazite and bastnasite. Dysprosium can be prepared by reduction of the trifluoride with calcium. The element has a metallic, bright silver luster. It is relatively stable in air at room temperature, and is readily attacked and dissolved, with the evolution of hydrogen, by dilute and concentrated mineral acids. The metal is soft enough to be cut with a knife. Small amounts of impurities can greatly affect its physical properties. A dysprosium oxide-nickel cermet has found use in cooling nuclear reactor control rods. In combination with vanadium and other rare earths, dysprosium has been used in making laser materials. The cost of dysprosium metal has dropped in recent years since the development of ion-exchange and solvent extraction techniques and the discovery of large ore bodies. The metal is still expensive, however, and costs about $3/g in purities of 99 + %.

**Einsteinium** — (Albert Einstein), Es; at wt (252); at no 99. Einsteinium, the seventh transuranic element of the actinide series to be discovered, was identified by Ghiorso and co-workers at Berkeley in December 1952 in debris from the first large thermonuclear or "hydrogen" bomb explosion, which took place in the Pacific in November 1952. The isotope produced was the 20-day $^{253}$Es isotope. In 1961, a sufficient amount of einsteinium was produced to permit separation of a macroscopic amount of $^{253}$Es. This sample weighed about 0.01 μg. A special magnetic-type balance was used in making this determination. $^{253}$Es so produced was used to produce mendelevium (element 101). About 3 μg of einsteinium has been produced at Oak Ridge National Laboratories by irradiating kilogram quantities of $^{239}$Pu in a reactor for several years to produce $^{242}$Pu. This was then fabricated into pellets of plutonium oxide and aluminum powder and loaded into target rods for an initial 1-year irradiation at the A.E.C. Savannah River Plant, followed by irradiation in a HFIR (high flux isotopic reactor). After 4 months in the HFIR, the targets were removed for chemical separation of the einsteinium from californium. Eleven isotopes of einsteinium are now recognized. $^{254}$Es has the longest half-life (276 days). Tracer studies using $^{253}$Es show that einsteinium has chemical properties typical of a heavy trivalent, actinide element.

**Element 104** — In 1964, workers of the Joint Nuclear Research Institute at Dubna (U.S.S.R.) bombarded plutonium with accelerated 113 to 115 MeV neon ions. By measuring fission tracks in a special glass with a microscope, they detected an isotope that decays by spontaneous fission. They suggested that this isotope, which had a half-life of 0.3 ± 0.1 sec might be $^{260}$104, produced by the following reaction:

$$^{242}_{94}\text{Pu} + {}^{22}_{10}\text{Ne}^{22} \rightarrow {}^{260}104 + 4n$$

Element 104, the first transactinide element, is expected to have chemical properties similar to those of hafnium. It would, for example, form a relatively volatile compound with chlorine (a tetrachloride). The Soviet scientists have performed experiments aimed at chemical identification, and have attempted to show that the 0.3-sec activity is more volatile than that of the relatively nonvolatile actinide trichlorides. This experiment does not fulfill the test of chemically separating the new element from all others, but it provides important evidence for evaluation. New data, reportedly issued by Soviet scientists, have reduced the half-life of the isotope they worked with from 0.3 to 0.15 sec. The Dubna scientists suggest the name *kurchatovium* and symbol Ku for element 104, in honor of Igor Vasilevich Kurchatov (1903 to 1960), late Head of Soviet Nuclear Research. In 1969, Ghiorso, Nurmia, Harris, K. A. Y Eskola, and P. L. Eskola of the University of California at Berkeley reported they had positively identified two, and possibly three, isotopes of element 104. The group also indicated that after repeated attempts, so far they have been unable to produce isotope

$^{260}$104 reported by the Dubna group in 1964. The discoveries at Berkeley were made by bombarding a target of $^{249}$Cf with $^{12}$C nuclei of 71 MeV, and $^{13}$C nuclei of 69 MeV. The combination of $^{12}$C with $^{249}$Cf, followed by instant emission of four neutrons, produced element $^{257}$104. This isotope has a half-life of 4 to 5 sec, decaying by emitting an alpha particle into No. 253, with a half-life of 105 sec. The same reaction, except with the emission of three neutrons, was thought to have produced $^{258}$104, with a half-life of about 1/100 sec. Element $^{259}$104 is formed by the merging of a $^{13}$C nuclei with $^{249}$Cf, followed by emission of three neutrons. This isotope has a half-life of 3 to 4 sec, and decays by emitting an alpha particle into No. 255, which has a half-life of 185 sec. Thousands of atoms of $^{257}$104 and $^{259}$104 have been detected. The Berkeley group believe their identification of $^{258}$104 is correct, but they do not attach the same degree of confidence to this work as to their work on $^{257}$104 and $^{259}$104. The Berkeley group proposes for the new element the name rutherfordium (symbol Rf), in honor of Ernest R. Rutherford, New Zealand physicist. The claims for discovery and the naming of element 104 are still in question.

**Element 105** — In 1967, G. N. Flerov reported that a Soviet team working at the Joint Institute for Nuclear Research at Dubna may have produced a few atoms of $^{260}$105 and $^{261}$105 by bombarding $^{243}$Am with $^{22}$Ne. Their evidence was based on time-coincidence measurements of alpha energies. More recently, it was reported that early in 1970, Dubna scientists synthesized element 105 and that by the end of April 1970 they "had investigated all the types of decay of the new element and had determined its chemical properties." The Soviet group has not proposed a name for element 105. In late April 1970, it was announced that Ghiorso, Nurmia, Harris, K. A. Y. Eskola, and P. L. Eskola, working at the University of California at Berkeley, had positively identified element 105. The discovery was made by bombarding a target of $^{249}$Cf with a beam of 84 MeV nitrogen nuclei in the heavy ion linear accelerator (HILAC). When a $^{15}$N nucleus is absorbed by a $^{249}$Cf nucleus, four neutrons are emitted and a new atom of $^{260}$105 with a half-life of 1.6 sec is formed. While the first atoms of element 105 are said to have been detected conclusively on March 5, 1970, there is evidence that element 105 had been formed in Berkeley experiments a year earlier by the method described. Ghiorso and his associates have attempted to confirm Soviet findings by more sophisticated methods without success. The Berkeley Group proposes the name hahnium, after the late German scientist Otto Hahn (1879 to 1968), and Ha for the chemical symbol.

More recently, in October 1971, it was announced that two new isotopes of element 105 were synthesized with the heavy ion linear accelerator by A. Ghiorso and co-workers at Berkeley. Element 105 was produced both by bombarding $^{250}$Cf with $^{15}$N and by bombarding $^{249}$Bk with $^{16}$O. The isotope emits 8.93-MeV α particles and decays to $^{257}$Lr with a half-life of about 1.8 sec. Element $^{262}$105 was produced by bombarding $^{249}$Bk with $^{18}$O. It emits 8.45 MeV α particles and decays to $^{258}$Lr with a half-life of about 40 sec.

**Element 106** — In June 1974, members of the Joint Institute for Nuclear Research in Dubna, U.S.S.R., reported their discovery of Element 106, which they claim to have synthesized. In September 1974, workers of the Lawrence Berkeley and Livermore Laboratories also reported creation of element 106 "without any scientific doubt". The LBL and LLL Group used the SuperHILAC to accelerate $^{18}$O ions onto a $^{249}$Cf target. Element 106 was created by the reaction $^{249}$Cf ($^{18}$O 4n) $^{263}_{106}$X, which decayed by α emission to rutherfordium, and then by α emission to nobelium, which in turn further decayed by α emission. An elaborate detection system not only looked for correlations between the new element and its daughter, but also between daughter and granddaughter. The element so identified had α energies of 9.06 and 9.25 MeV, with a half-life of 0.9 ± 0.2 sec. At Dubna, 280-MeV ions of $^{54}$Cr from the 310-cm cyclotron were used to strike targets of $^{206}$Pb, $^{207}$Pb, and $^{208}$Pb, in separate runs. Foils exposed to a rotating target disc were used to detect spontaneous fission activities, the foils being etched and examined microscopically to detect the number of fission tracts and the half-life of the fission activity. Other experiments were made to aid in confirmation of the discovery. Neither the Dubna team nor the Berkeley-Livermore Group has proposed a name as yet for element 106.

**Element 107** — In 1976, Soviet scientists at Dubna announced they had synthesized element 107 by bombarding $^{204}$Bi with heavy nuclei of $^{54}$Cr. It is reported that earlier experiments in 1975 had allowed scientists "to glimpse" the new element for 2/1000 sec. A rapidly rotating cylinder, coated with a thin layer of the bismuth metal, was used

as the target. This was bombarded by a stream of $^{54}$Cr ions fired tangentially. The existence of element 107 was confirmed by a team of West German physicists at the Heavy Ion Research Laboratory at Darmstadt, who created and identified six nuclei of element 107.

**Element 109** — On August 29, 1982, element 109 was made and discovered by physicists of the Heavy Ion Research Laboratory, Darmstadt, West Germany, by bombing a target of $^{209}$Bi with accelerated nuclei of $^{58}$Fe. If the combined energy of two nuclei is sufficiently high, the repulsive forces between the nuclei can be overcome. In this experiment it took a week of target bombardment to produce a single fused nucleus. The team confirmed the existence of element 109 by four independent measurements. The newly formed atom recoiled from the target at a predicted velocity and was separated from smaller, faster nuclei by a newly developed velocity filter. The time of flight to the detector and the striking energy were measured and found to match predicted values. The nucleus of $^{266}_{109}$X started to decay 5 msec after striking the detector. A high-energy α particle was emitted, producing $^{262}_{107}$X. This is turn emitted an α particle, becoming $^{258}_{105}$Ha, which in turn captured an electron and became $^{258}_{104}$Rf. This in turn decayed into other nuclides. This experiment demonstrated the feasibility of using fusion techniques as a method of making new, heavy nuclei. West German scientists, as well as those at Berkeley, will soon attempt to use cold fusion to make element 116 by bombarding curium-248 with calcium-48 nuclei. The fission product is expected to be element 116, which should decay through a whole series of hitherto unknown nuclides. Names have not yet been proposed for elements 106, 107, or 109.

**Erbium** — (Ytterby, a town in Sweden), Er; at wt 167.26 ± 3; at no 68; mp 159°C; bp 2863°C; sp gr 9.066 (25°C); valence 3. Erbium, one of the so-called rare earth elements of the lanthanide series, is found in the minerals mentioned under dysprosium. In 1842, Mosander separated "yttria", found in the mineral gadolinite, into three fractions which he called yttria, erbia, and terbia. The names erbia and terbia became confused in this early period. After 1860, Mosander's terbia was known as erbia, and after 1877, the earlier known erbia became terbia. The erbia of this period was later shown to consist of five oxides, now known as erbia, scandia, holmia, thulia, and ytterbia. By 1905, Urbain and James independently succeeded in isolating fairly pure $Er_2O_3$. Klemm and Bommer first produced reasonably pure erbium metal in 1934 by reducing the anhydrous chloride with potassium vapor. The pure metal is soft and malleable and has a bright, silvery, metallic luster. As with other rare earth metals, its properties depend to a certain extent on the impurities present. The metal is fairly stable in air and does not oxidize as rapidly as some of the other rare earth metals. Naturally occurring erbium is a mixture of six isotopes, all of which are stable. Nine radioactive isotopes of erbium are also recognized. Recent production techniques, using ion-exchange reactions, have resulted in much lower prices of rare earth metals and their compounds. The cost of $99+\%$ erbium metal is about \$3/g in small quantities. Erbium is finding nuclear and metallurgical uses. Added to vanadium, for example, erbium lowers the hardness and improves workability. Erbium oxide gives a pink color and has been used as a colorant in glasses and porcelain enamel glazes.

**Europium** — (Europe), Eu; at wt 151.96; at no 63; mp 822°C; bp 1597°C; sp gr 5.243 (25°C); valence 2 or 3. In 1890, Boisbaudran obtained basic fractions from samarium-gadolinium concentrates which had spark spectral lines not accounted for by samarium or gadolinium. These lines subsequently have been shown to belong to europium. The discovery of europium is generally credited to Demarcay, who separated the rare earth in reasonably pure form in 1901. The pure metal was not isolated until recent years. Europium is now prepared by mixing $Eu_2O_3$ with a 10% excess of lanthanum metal and heating the mixture in a tantalum crucible under high vacuum. The element is collected as a silvery white metallic deposit on the walls of the crucible. As with other rare earth metals, except for lanthanum, europium ignites in air at about 150 to 180°C. Europium is about as hard as lead and is quite ductile. It is the most reactive of the rare earth metals, quickly oxidizing in air. It resembles calcium in its reaction with water. Bastnasite and monazite are the principal ores containing europium. Europium has been identified spectroscopically in the sun and in certain stars. Seventeen isotopes are now recognized. Europium isotopes are good neutron absorbers and are being studied for use in nuclear control applications. Europium oxide is now widely used as a phosphor activator and europium-activated yttrium vanadate is in commercial use as the red phosphor in color TV tubes. With the development of ion-exchange techniques and special processes, the cost of the metal has been greatly reduced, but europium remains one of the rarest and most costly of the rare earth metals. It is priced at about \$50/g.

**Fermium** — (Enrico Fermi), Fm; at wt (257); at no 100. Fermium, the eighth transuranium element of the actinide series to be discovered, was identified by Ghiorso and co-workers in 1952 in the debris from a thermonuclear explosion in the Pacific in work involving the University of California Radiation Laboratory, the Argonne National Laboratory, and the Los Alamos Scientific Laboratory. The isotope produced was the 20-hr $^{255}$Fm. During 1953 and early 1954, while discovery of elements 99 and 100 was withheld from publication for security reasons, a group from the Nobel Institute of Physics in Stockholm bombarded $^{238}$U with $^{16}$O ions, and isolated a 30-min α-emitter, which they ascribed to $^{250}$100, without claiming discovery of the element. This isotope has since been identified positively, and the 30-min half-life confirmed. The chemical properties of fermium have been studied solely with tracer amounts, and in normal aqueous media only the (III) oxidation state appears to exist. The isotope $^{254}$Fm and heavier isotopes can be produced by intense neutron irradiation of lower elements such as plutonium by a process of successive neutron capture interspersed with beta decays until these mass numbers and atomic numbers are reached. Ten isotopes of fermium are known to exist. $^{257}$Fm, with a half-life of about 80 days, is the longest lived. $^{250}$Fm, with a half-life of 30 min, has been shown to be a product of decay of element $^{254}$102. It was by chemical identification of $^{250}$Fm that it was certain that element 102 (nobelium) had been produced.

**Fluorine** — (L. F. *fluere,* flow, or flux), F; at wt 18.998403; at no 9; mp −219.62°C (1 atm); bp −188.14°C (1 atm); density 1.696 g/l (0°C, 1 atm); sp gr of liquid 1.108 at bp; valence 1. In 1529, Georigius Agricola described the use of fluorspar as a flux, and as early as 1670, Schwandhard found that glass was etched when exposed to fluorspar treated with acid. Scheele and many later investigators, including Davy, Gay-Lussac, Lavoisier, and Thenard, experimented with hydrofluoric acid, some experiments ending in tragedy. The element was finally isolated in 1886 by Moisson after nearly 74 years of continuous effort. Fluorine occurs chiefly in fluorspar ($CaF_2$) and cryolite ($Na_3AlF_6$), but is rather widely distributed in other minerals. It is a member of the halogen family of elements and is obtained by electrolyzing a solution of potassium hydrogen fluoride in anhydrous hydrogen fluoride in a vessel of metal or transparent fluorspar. Modern commercial production methods are essentially variations on the procedures first used by Moisson. Fluorine is the most electronegative and reactive of all elements. It is a pale yellow, corrosive gas, which reacts with practically all organic and inorganic substances. Finely divided metals, glass, ceramics, carbon, and even water burn in fluorine with a bright flame. Until World War II, there was no commercial production of elemental fluorine. The atom bomb project and nuclear energy applications, however, made it necessary to produce large quantities. Safe handling techniques have now been developed, and it is possible at present to transport liquid fluorine by the ton. Fluorine and its compounds are used in producing uranium (from the hexafluoride) and more than 100 commercial fluorochemicals, including many well-known high-temperature plastics. Hydrofluoric acid is extensively used for etching the glass of light bulbs, etc. Fluorochloro hydrocarbons are extensively used in air conditioning and refrigeration. It has been suggested that fluorine can be substituted for hydrogen wherever it occurs in organic compounds, which could lead to an astronomical number of new fluorine compounds. The presence of soluble fluorides in drinking water at 2 ppm may cause mottled enamel in teeth, when used by children acquiring permanent teeth; in smaller amounts, however, fluorides are beneficial and used in water supplies to prevent dental cavities. Compounds of fluorine with rare gases have now been confirmed. Fluorides of xenon, radon, and krypton are among those reported. Elemental fluorine and the fluoride ion are highly toxic. The free element has a characteristic pungent odor, detactable in concentrations as low as 20 ppb, which is below the safe working level. The recommended maximum allowable concentration for a daily 8-hr time-weighted exposure is 0.1 ppm.

**Francium** — (France), Fr; at no 87; at wt (223); mp 27°C; bp 677°C; valence 1. Discovered in 1939 by Mlle. Marguerite Perey of the Curie Institute, Paris. Francium, the heaviest known member of the alkali metal series, occurs as a result of an alpha disintegration of actinium. It can also be made artificially by bombarding thorium with protons. While it occurs naturally in uranium minerals, there is probably less than an ounce of francium at any time in the total crust of the earth. It has the highest equivalent weight of any element, and is the most unstable of the first 101 elements of the periodic system. Twenty isotopes of francium are recognized. The longest lived, $^{223}$Fr (AcK), a daughter of $^{227}$Ac, has a half-life of 22 min. This is the only isotope of francium occurring in nature. Because all known isotopes of francium are highly unstable, knowledge of the chemical properties of this element comes from radiochemical techniques. No weighable quantity of the element has been prepared or isolated. The chemical properties of francium most closely resemble cesium.

**Gadolinium** — (*gadolinite*, a mineral named for Gadolin, a Finnish chemist), Gd; at wt 157.25 ± 3; at no 64; mp 1313 ± 1°C; bp 3266°C; sp gr 7.9004 (25°C); valence 3. Gadolinia, the oxide of gadolinium, was separated by Marignac in 1880, and Lecoq de Boisbaudran independently isolated the element from Mosander's "yttria" in 1886. The element was named for the mineral gadolinite from which this rare earth was originally obtained. Gadolinium is found in several other minerals, including monazite and bastnasite, which are of commercial importance. The element has been isolated only in recent years. With the development of ion-exchange and solvent extraction techniques, the availability and price of gadolinium and other rare earth metals have greatly improved. Seventeen isotopes of gadolinium are now recognized; seven occur naturally. The metal can be prepared by the reduction of the anhydrous fluoride with metallic calcium. As with other related rare earth metals, it is silvery white, has a metallic luster, and is malleable and ductile. At room temperature, gadolinium crystallizes in the hexagonal, close-packed α form. Upon heating to 1262°C, α gadolinium transforms into the β form, which has a body-centered cubic structure. The metal is relatively stable in dry air, but in moist air it tarnishes, with the formation of a loosely adhering oxide film that spalls off and exposes more surface to oxidation. The metal reacts slowly with water and is soluble in dilute acid. Gadolinium has the highest thermal neutron capture cross-section of any known element (49,000 barns). Natural gadolinium is a mixture of seven isotopes. Two of these $^{155}$Gd and $^{157}$Gd, have excellent capture characteristics, but they are present naturally in low concentrations. As a result, gadolinium has a very fast burnout rate and has limited use as a nuclear control rod material. It has been used in making gadolinium yttrium garnets, which have microwave applications. Compounds of gadolinium are used in making phosphors for color TV tubes. The metal has unusual superconductive properties. As little as 1% gadolinium has been found to improve the workability and resistance of iron, chromium, and related alloys to high temperatures and oxidation. The metal is ferromagnetic. Gadolinium is unique for its high magnetic moment and for its special Curie temperature (above which ferromagnetism vanishes) lying just at room temperature. The price of the metal is $4/g.

**Gallium** — (L. *Gallia*, France; also from L. *gallus*, a translation of *Lecoq*, a cook); Ga; at wt 69.72; at no 31; mp 29.78°C; bp 2403°C; sp gr 5.904 (29.6°C) solid; sp gr 6.095 (29.6°C) liquid; valence 2 or 3. Predicted and described by Mendeleev as ekaaluminum, and discovered spectroscopically in 1875 by Lecoq de Boisbaudran, who in the same year obtained the free metal by electrolysis a solution of the hydroxide in KOH. Gallium is often found as a trace element in diaspore, sphalerite, germanite, bauxite, and coal. Some flue dusts from burning coal have been shown to contain as much as 1.5% gallium. It is the only metal, except for mercury, cesium, and rubidium, which can be liquid near room temperatures; this makes possible its use in high-temperature thermometers. It has one of the longest liquid ranges of any metal and has a low vapor pressure even at high temperatures. There is a strong tendency for gallium to supercool below its freezing point. Therefore, seeding may be necessary to initiate solidification. Ultrapure gallium has a beautiful, silvery appearance, and the solid metal exhibits a conchoidal fracture similar to glass. The metal expands 3.1% on solidifying; therefore, it should not be stored in glass or metal containers, as they may break as at the metal solidifies. Gallium wets glass or porcelain, and forms a brilliant mirror when it is painted on glass. It has found recent use in doping semiconductors. High-purity gallium is attacked only slowly by mineral acids. Magnesium gallate containing divalent impurities such as $Mn^{2+}$ is finding use in commercial ultraviolet activated powder phosphors. Gallium arsenide is capable of converting electricity directly into coherent light. Gallium readily alloys with most metals, and has been used as a component in low-melting alloys. Its toxicity appears to be of a low order, but it should be handled with care until more data are forthcoming. The metal can be supplied in ultrapure form (99.99999 + %). The cost is about $3/g.

**Germanium** — (L. *Germania,* Germany), Ge; at wt 72.59 ± 3; at no 32; mp 937.4°C; bp 2,830°C; sp gr 5.323 (25°C); valence 2 and 4. Predicted by Mendeleev in 1871 as ekasilicon, and discovered by Winkler in 1886. The metal is found in argyrodite, a sulfide of germanium and silver; in germanite, which contains 8% of the element; in zinc ores; in coal; and in other minerals. The element is frequently obtained commercially from flue dusts of smelters processing zinc ores, and has been recovered from the by-products of combustion of certain coals. Its presence in coal insures a large reserve of the element. Germanium can be separated from other metals by fractional distillation of its volatile tetrachloride. The tetrachloride may then be hydrolyzed to give $GeO_2$; then the dioxide can be reduced with hydrogen to give the metal. Recently developed zone-refining techniques permit the production of germanium of ultrahigh purity. The element is a gray-white metalloid, and in its pure state is crystalline and brittle, retaining its luster in air at room temperature. It is a very important semiconductor material. Zone-refining techniques have led to production of crystalline germanium for semiconductor use with an impurity of only one part in $10^{10}$. Doped with arsenic, gallium, or other elements, it is used as a transistor element in thousands of electronic applications. Its application as a semiconductor element provides the largest use for germanium. Germanium and germanium oxide are transparent to the infrared and are used in extremely sensitive infrared detectors. The high index of refraction and dispersion of germanium oxide has made it useful as a component of glasses used in wide-angle camera lenses and microscopic objectives. The field of organogermanium chemistry is becoming increasingly important. Certain germanium compounds have a low mammalian toxicity, but a marked activity against certain bacteria, which makes them of interest as chemotherapeutic agents. The cost of germanium is about $300/lb.

**Gold** — (Sanskrit *Jval;* Anglo-Saxon *gold*), Au (L. *aurum,* shining dawn) at wt 196.9665; at no 79; mp 1064.43°C; bp 3080°C; sp gr ~ 19.3(20°C); valence 1 or 3. Known and highly valued from earliest times, gold is found in nature as the free metal and in tellurides; it is very widely distributed and is almost always associated with quartz or pyrite. It occurs in veins and alluvial deposits, and is often separated from rocks and other minerals by sluicing or panning operations. About two thirds of the world gold output now comes from South Africa, and about two thirds of the total U.S. production comes from South Dakota and Nevada. The metal is recovered from its ores by cyaniding, amalgamating, and smelting processes. Refining is also frequently done by electrolysis. Gold occurs in sea water to the extent of 0.1 to 2 mg/ton, depending on the location where the sample is taken. As yet, no method has been found for recovering gold from sea water profitably. It is estimated that all the gold in the world, so far refined, could be placed in a single cube 60 ft on a side. Of all the elements, gold in its pure state is undoubtedly the most beautiful. It is metallic, having a yellow color when in a mass, but when finely divided it may be black, ruby, or purple. The Purple of Cassius is a delicate test for auric gold. It is the most malleable and dictile metal; 1 oz of gold can be beaten out to 300 ft$^2$. It is a soft metal and is usually alloyed to give it more strength. It is a good conductor of heat and electricity, and is unaffected by air and most reagents. It is used in coinage and is a standard for monetary systems in many countries. It is also extensively used for jewelry, decoration, dental work, and for plating. It is used for coating certain space satellites, as it is a good reflector of infrared and is inert. Gold, like other precious metals, is measured in troy weight; when gold is alloyed with other metals, the term "carat" is used to express the amount of gold present, 24 carats being pure gold. For many years, the value of gold was set by the U.S. at $20.67/troy ounce (oz t); in 1934 this value was fixed by law at $35.00/oz t ounce, 9/10th fine. On March 17, 1968, because of a gold crisis, a two-tiered pricing system was established whereby gold was still used to settle international accounts at the old $35.00/oz t price while the price of gold on the private market would be allowed to fluctuate. Since this time, the price of gold on the free market has fluctuated widely. On March 19, 1968, President Johnson signed into law a bill removing the last statutory requirement for a gold backing against U.S. currency. On August 15, 1971, President Nixon announced an embargo on U.S. gold to settle international accounts, and on May 18, 1972, U.S. monetary gold was revalued at $38.00/oz t. In February 1973, the U.S. in effect devalued the dollar by another 10%, decreasing the dollar in relation to gold from $38.00 to $42.22/oz t. On September 21, 1973, President Nixon signed a bill ratifying the action taken earlier in February; the bill also restored to private U.S. citizens the right to own gold after December 31, 1973 (private possession, except for gold in the form of jewelry, certain coins, etc., had been prohibited since 1933 when the U.S. went off the gold standard). The final version of the bill gave the President discretion to lift the ban when he determined that private ownership would not impair the monetary position during this period of instability. President Ford signed final legislation on August 14, 1974, lifting the 41-year ban, to be effective after December 31, 1974. The price of gold on the free market reached a price of $620/oz t in January 1980 and $480/oz t in January 1988. The most common gold compounds are auric chloride ($AuCl_9$) and chlorauric acid ($HAuCl_4$), the latter being used in photography for toning the silver image. Gold has 18 isotopes; $^{198}Au$, with a half-life of 2.7 days, is used for treating cancer and other diseases. Disodium aurothiomalate is administered intramuscularly as a treatment for arthritis. A mixture of one part nitric acid with three of hydrochloric acid is called aqua regia (because it dissolved gold, the "King of Metals"). Gold is available commercially with a purity of 99.999 + %. For many years, the temperature assigned to the freezing point of gold has been 1063.0°C; this has served as a calibration point for the International Temperature Scales (ITS-27 and ITS-48) and the International Practical Temperature Scale (IPTS-48). In 1968, a new International Practical Temperature Scale (IPTS-68) was adopted, which demands that the freezing point of gold be changed to 1064.43°C. Many of the scale changes are of minor significance to the routine user. IPTS-68 has defined several other fixed temperature points, among which are the boiling points of hydrogen, neon, oxygen, and sulfur, and the freezing points of zinc, silver, tin, lead antimony, and aluminum. The specific gravity of gold has been found to vary considerably, depending on temperature, how the metal is precipitated, and cold-worked.

**Hafnium** — (*Hafinia*, Latin name for Copenhagen), Hf; at wt 178.49 ± 3; at no 72; mp 2227 ± 20°C; bp 4602°C; sp gr 13.31 (?)20°C; valence 4. Hafnium was thought to be present in various minerals and concentrations many years prior to its discovery, in 1923, credited to D. Coster and G. von Hevesey. On the basis of the Bohr theory, the new element was expected to be associated with zirconium. It was finally identified in zircon from Norway, by means of X-ray spectroscopic analysis. It was named in honor of the city in which the discovery was made. Most zirconium minerals contain 1 to 5% hafnium. It was originally separated from zirconium by repeated recrystallization of the double ammonium or potassium fluorides by von Hevesey and Jantzen. Metallic hafnium was first prepared by van Arkel and deBoer by passing the vapor of the tetraiodide over a heated tungsten filament. Almost all hafnium metal now produced is made by reducing the tetrachloride with magnesium or with sodium (Kroll process). Hafnium is a ductile metal with a brilliant silver luster. Its properties are considerably influenced by the impurities of zirconium present. Of all the elements, zirconium and hafnium are two of the most difficult to separate. Their chemistry is almost identical; however, the density of zirconium is about half that of hafnium. Very pure hafnium has been produced, with zirconium being the major impurity. Because hafnium has a good absorption cross section for thermal neutrons (almost 600 times that of zirconium), has excellent mechanical properties, and is extremely corrosion resistant, it is used for reactor control rods. Such rods are used in nuclear submarines. Hafnium has been successfully alloyed with iron, titanium, niobium, tantalum, and other metals. Hafnium carbide is the most refractory binary composition known, and the nitride is the most refractory of all known metal nitrides (mp 3310°C). Hafnium is used in gas-filled and incandescent lamps and is an efficient "getter" for scavenging oxygen and nitrogen. Exposure to hafnium should not exceed 0.5 mg/m³ (8-hr time-weighted average, 40-hr week). The price of the metal is in the broad range of $100 to $500/lb, depending on purity and quantity. The yearly demand for hafnium in the U.S. is now in excess of 100,000 lb.

**Hahnium** — see **Element 105.**

**Helium** — (Gr. *helios,* the sun) He; at wt 4.00260; at no 2; mp below −272.2°C (26 atm); bp −268.934°C; density 0.1785 g/l (0°C, 1 atm); liquid density 7.62 lb/ft³ at bp; valence usually 0. Evidence of the existence of helium was first obtained by Janssen during the solar eclipse of 1868, when he detected a new line in the solar spectrum; Lockyer and Frankland suggested the name helium for the new element; in 1895, Ramsay discovered helium in the uranium mineral clevite, and it was independently discovered in clevite by the Swedish chemists Cleve and Langlet about the same time. Rutherford and Royds in 1907 demonstrated that α particles are helium nuclei. Except for hydrogen, helium is the most abundant element found throughout the universe. It has been detected spectroscopically in great abundance, especially in the hotter stars, and it is an important component in both the proton-proton reaction and the carbon cycle, which account for the energy of the sun and stars. The fusion of hydrogen into helium provides the energy of the hydrogen bomb. The helium content of the atmosphere is about 1 part in 200,000. While it is present in various radioactive minerals as a decay product, the bulk of the Western World's supply is obtained from wells in Texas, Oklahoma, and Kansas. The only known helium extraction plants in 1984 were in Eastern Europe and the U.S.S.R. The cost of helium fell from $2500/ft³ in 1915 to 1.5¢/ft³ in 1940. The U.S. Bureau of Mines has set the price of Grade A helium at $35/1000 ft³. Helium has the lowest melting point of any element and has found wide use in cryogenic research, since its boiling point is close to absolute zero. Its use in the study of superconductivity is vital. Using liquid helium, Kurti and co-workers, and others, have succeeded in obtaining temperatures of a few microdegrees K by the adiabatic demagnetization of copper nuclei, starting from about 0.01 K. Five isotopes of helium are known. Liquid helium ($^4$He) exists in two forms; $^4$I He, and $^4$II He, with a sharp transition point at 2.174 K (3.83 cm Hg). $^4$I He (above this temperature) is a normal liquid, but $^4$II He (below it) is unlike any other known substance. It expands on cooling; its conductivity for heat is enormous; and neither its heat conduction nor viscosity obeys normal rules. It has other peculiar properties. Helium is the only liquid that cannot be solidified by lowering the temperature. It remains liquid down to absolute zero at ordinary pressures, but it can readily be solidified by increasing the pressure. Solid $^3$He and $^4$He are unusual in that both can readily be changed in volume by more than 30% by application of pressure. The specific heat of helium gas is unusually high. The density of helium vapor at the normal boiling point is also very high, with the vapor expanding greatly when heated to room temperature. Containers filled with helium gas at 5 to 10 K should be treated as though they contained liquid helium due to the large increase in pressure resulting from warming the gas to room temperature. While helium normally has a 0 valence, it has a weak tendency to combine with certain other elements. Means of preparing helium difluoride have been studied, and species such as HeNe and the molecular ions He$^+$ and He$^{2+}$ have been investigated. Helium is widely used as an inert gas shield for arc welding; as a protective gas in growing silicon and germanium crystals, and in titanium and zirconium production; and as a cooling medium for nuclear reactors. A mixture of 80% helium and 20% oxygen is used as an artificial atmosphere for divers and others working under pressure. Helium is extensively used for filling balloons as it is a much safer gas than hydrogen. At sea level, 1000 ft$^3$ of helium lifts 68.5 lb.

**Holmium** — (L. *Holmia*, for Stockholm), Ho; at wt 164.9304; at no 67; mp 1474°C; bp 2695°C; sp gr 8.795 (25°C); valence +3. The spectral absorption bands of holmium were noticed in 1878 by Swiss chemists Delafontaine and Soret, who announced the existence of an "Element X". Cleve, of Sweden, later independently discovered the element while working on erbia earth. The element is named after Cleve's native city. Pure holmia, the yellow oxide, was prepared by Homberg in 1911. Holmium occurs in gadolinite, monazite, and in other rare earth minerals. It is commercially obtained from monazite, occurring in that mineral to the extent of about 0.05%. It has been isolated in pure form only in recent years. It can be separated from other rare earths by ion-exchange and solvent extraction techniques, and isolated by the reduction of its anhydrous chloride or fluoride with calcium metal. Pure holmium has a metallic to bright silver luster. It is relatively soft and malleable, and is stable in dry air at room temperature, but rapidly oxidizes in moist air and at elevated temperatures. The metal has unusual magnetic properties. Few uses have yet been found for the element. The element, as with other rare earths, seems to have a low acute toxic rating. The price of 99 + % holmium metal is about $8/g.

**Hydrogen** — Gr. *hydro*, water, and *genes*, forming), H; at wt (natural) 1.00794 ± 7; at wt ($^1$H) 1.007822; at no 1; mp −254.14°C; bp −252.87°C; density 0.08988 g/l; density (liquid) 70.8 g/l (−253°C); density (solid) 70.6 g/l (−262°C); valence 1. Hydrogen was prepared many years before it was recognized as a distinct substance by Cavendish in 1766. It was named by Lavoisier. Hydrogen is the most abundant of all elements in the universe, and it is thought that the heavier elements were, and still are, being built from hydrogen and helium. It has been estimated that hydrogen makes up more than 90% of all the atoms or three quarters of the mass of the universe. It is found in the sun and most stars, and plays an important part in the proton-proton reaction and carbon-nitrogen cycle, which accounts for the energy of the sun and stars. It is thought that hydrogen is a major component of the planet Jupiter and that at some depth in the interior of the planet the pressure is so great that solid molecular hydrogen is converted into solid metallic hydrogen. In 1973, it was reported that a group of Russian experimenters may have produced metallic hydrogen at a pressure of 2.8 Mbar. At the transition the density changed from 1.08 to 1.3 g/cm$^3$. Earlier, in 1972, a Livermore (California) group also reported on a similar experiment in which they observed a pressure-volume point centered at 2 Mbar. It has been predicted that metallic hydrogen may be metastable; others have predicted it would be a superconductor at room temperature. On earth, hydrogen occurs chiefly in combination with oxygen in water, but it is also present in organic matter such as living plants, petroleum, coal, etc. It is present as the free element in the atmosphere, but only to the extent of less than 1 ppm, by volume. It is the lightest of all gases, and combines with other elements, sometimes explosively, to form compounds. Great quantities of hydrogen are required commercially for the fixation of nitrogen from the air in the Haber ammonia process and for the hydrogenation of fats and oils. It is used in large quantities in methanol production, in hydrodealkylation, hydrocracking, and hydrodesulfurization. It is used as a rocket fuel, for welding, for production of hydrochloric acid, for the reduction of metallic ores, and for filling balloons. The lifting power of 1 ft$^3$ of hydrogen gas is about 0.076 lb at 0°C, 760 mm pressure. Production of hydrogen in the U.S. alone now amounts to about 3 billion ft$^3$/year. It is prepared by the action of steam on heated carbon, by decomposition of certain hydrocarbons with heat, by the electrolysis of water, or by the displacement from acids by certain metals. It is also produced by the action of sodium or potassium hydroxide on aluminum. Liquid hydrogen is important in cryogenics and in the study of superconductivity as its melting point is only a few degrees above absolute zero. The ordinary isotope of hydrogen, $^1_1$H, is known as protium. In 1932, Urey announced the preparation of a stable isotope, deuterium ($^2_1$H or D) with an atomic weight of 2. Two years later an unstable isotope, tritium($^3_1$H), with an atomic weight of 3 was discovered. Tritium has a half-life of about 12.5 years. One atom of deuterium is found mixed in with abut 6000 ordinary hydrogen atoms. Tritium atoms are also present but in much smaller proportion. Tritium is readily produced in nuclear reactors and is used in the production of the hydrogen bomb. It is also used as a radioactive agent in making luminous paints, and as a tracer. The current price of tritium, to authorized personnel, is about $2/Ci; deuterium gas is readily available, without permit, at about $1/1. Heavy water, deuterium oxide ($D_2O$), which is used as a moderator to slow down neutrons, is available without permit at a cost of 6¢ to $1/g, depending on quantity and purity. Quite apart from isotopes, it has been shown that hydrogen gas under ordinary conditions is a mixture of two kinds of molecules, known as *ortho-* and *para*-hydrogen, which differ from one another by the spins of their electrons and nuclei. Normal hydrogen at room temperature contains 25% of the *para* form and 75% of the *ortho* form. The *ortho* form cannot be prepared in the pure state. Since the two forms differ in energy, the physical properties also differ. The melting and boiling points of *para*-hydrogen are about 0.1°C lower than those of normal hydrogen.

**Indium** — (from the brilliant indigo line in its spectrum), In; at wt 114.82; at no 49; mp 156.61°C; bp 2080°C; sp gr 7.31 (20°C); valence 1, 2, or 3. Discovered by Reich and Richter, who later isolated the metal. Indium is most frequently associated with zinc minerals, and it is from these that most commercial indium is now obtained; however, it is also found in iron, lead, and copper ores. Until 1924, a gram or so constituted the world supply of this element in isolated form. It is probably about as abundant as silver. About 4 million troy ounces of indium are now produced annually in the non-Communist countries. Canada is presently producing more than 1,000,000 troy ounces annually. The present cost of indium is about $1 to $5/g, depending on quantity and purity. It is available in ultrapure form. Indium is a very soft, silvery white metal with a brilliant luster. The pure metal gives a high-pitched "cry" when bent. It wets glass, as does gallium. It has found application in making low-melting alloys; an alloy of 24% indium—76% gallium is liquid at room temperature. It is used in making bearing alloys, germanium transistors, rectifiers, thermistors, and photoconductors. It can be plated onto metal and evaporated onto glass, forming a mirror as good as that made with silver but with more resistance to atmospheric corrosion. There is evidence that indium has a low order of toxicity; however, care

should be taken until further information is available.

**Iodine** — (Gr. *iodes*, violet), I; at wt 126.9045; at no 53; mp 113.5°C; bp 184.35°C; density of the gas 11.27 g/l; sp gr solid 4.93 (20°C); valence 1, 3, 5, or 7. Discovered by Courtois in 1811. Iodine, a halogen, occurs sparingly in the form of iodides in sea water, from which it is assimilated by seaweeds; in Chilean saltpeter and nitrate-bearing earth, known as caliche; in brines from old sea deposits; and in brackish waters from oil and salt wells. Ultrapure iodine can be obtained from the reaction of potassium iodide with copper sulfate. Iodine is a bluish black, lustrous solid, volatilizing at ordinary temperatures into a blue violet gas with an irritating odor; it forms compounds with many elements, but is less active than the other halogens, which displace it from iodides. Iodine exhibits some metallic-like properties. It dissolves readily in chloroform, carbon tetrachloride, or carbon disulfide to form beautiful purple solutions. It is only slightly soluble in water. Iodine compounds are important in organic chemistry and very useful in medicine. Twenty-three isotopes are recognized. Only one stable isotope, $^{127}I$, is found in nature. The artificial radioisotope $^{131}I$, with a half-life of 8 days, has been used in treating the thyroid gland. The most common compounds are the iodides of sodium and potassium (KI) and the iodates ($KIO_3$). Lack of iodine is the cause of goiter. The iodide, and thyroxin which contains iodine, are used internally in medicine, and a solution of KI and iodine in alcohol is used for external wounds. Potassium iodide finds use in photography. The deep blue color with starch solution is characteristic of the free element. Care should be taken in handling and using iodine, as contact with the skin can cause lesions; iodine vapor is intensely irritating to the eyes and mucous membranes. The maximum allowable concentration of iodine in air should not exceed 1 mg/m³ (8-hr time-weighted average, 40-hr).

**Iridium** — (L. *iris*, rainbow), Ir; at wt 192.22 ± 3; at no 77; mp 2410°C; bp 4130°C; sp gr 22.42 (17°C); valence 3 or 4. Discovered in 1803 by Tennant in the residue left when crude platinum is dissolved by aqua regia. The name iridium is appropriate, for its salts are highly colored. Iridium, a metal of the platinum family, is white, similar to platinum, but with a slight yellowish cast. It is very hard and brittle, making it very hard to machine, form, or work. It is the most corrosion-resistant metal known, and was used in making the standard meter bar of Paris, which is a 90% platinum-10% iridium alloy. This meter bar was replaced in 1960 as a fundamental unit of length (see Krypton). Iridium is not attacked by any of the acids nor by aqua regia, but is attacked by molten salts, such as NaCl and NaCN. Iridium occurs uncombined in nature with platinum and other metals of this family in alluvial deposits. It is recovered as a by-product from the nickel mining industry. Iridium has found use in making crucibles and apparatus for use at high temperatures. It is also used for electrical contacts. Its principal use is as a hardening agent for platinum. With osmium, it forms an alloy which is used for tipping pens and compass bearings. The specific gravity of iridium is only very slightly lower than that of osmium and from the space lattices gives values of 22.65 and 22.61 g/cm$^3$, respectively. These values may be more reliable than actual physical measurements. Iridium costs about $300/troy ounce.

**Iron** — (Anglo-Saxon, *iron*), Fe; (L. *ferrum*), at wt 55.847 ± 3; at no 26; mp 1535°C; bp 2750°C; sp gr 7.874 (20°C); valence 2, 3, 4, or 6. The use of iron is prehistoric. Genesis mentions that Tubal-Cain, seven generations from Adam, was "an instructer of every artificer in brass and iron." Homer mentions a ball of iron as the prize awarded to Achilles for athletic prowess. The metallurgy of iron began widespread development around 1400 to 1000 B.C., ushering in the Iron Age, with evidence of the ability to use steelworking and hardening to make weapons and agricultural implements. Today, iron is the most useful metal and, the one most widely used in modern civilization. After many years of dominating the world in the mining of iron ore and the production of steel, the U.S. in 1970 fell behind the Soviet Union in the output of iron ore, production of pig iron, and output of steel. In 1985 the world output of crude steel was estimated at 678.9 million metric tons. The leading producer was the Soviet Union at 154.7 million metric tons, followed by Japan at 105.3, the U.S. at 80.1, Mainland China at 46.8, and West Germany at 39.2. A remarkable iron pillar, dating to about A.D. 400, remains standing today in Delhi, India. This solid shaft of wrought iron is about 7 1/4 m high by 40 cm in diameter. Corrosion to the pillar has been minimal, although it has been exposed to the weather since its erection. Iron is a relatively abundant element in the universe. It is found in the sun and many types of stars in considerable quantity. Its nuclei are very stable. Iron is found native as a principal component of a class of meteorites known as siderites, and is a minor constituent of the other two classes. The core of the earth, 2150 miles in radius, is thought to be largely composed of iron, with about 10% occluded hydrogen. The metal is the fourth most abundant element, by weight, making up the crust of the earth. The most common ore is hematite ($Fe_2O_3$), from which the metal is obtained by reduction with carbon. Iron is found in other widely distributed minerals, such as magnetite which is frequently seen as black sands along beaches and banks of streams. In addition to magnetite ($Fe_3O_4$), limonite (hydrate iron oxide), siderite ($FeCO_3$), and a magnetite-bearing ore called taconite are important. Taconite is becoming increasingly important as a commercial ore. Common iron is a mixture of four isotopes, 56, 54, 57, and 58. Six other isotopes are known to exist. Iron is a vital constituent of plant and animal life, and appears in hemoglobin. The pure metal is not often encountered in commerce, but is usually alloyed with carbon or other metals. The pure metal is very reactive chemically, and rapidly corrodes, especially in moist air or at elevated temperatures. It has four allotropic forms, or ferrites, known as α, β, γ, and δ, with transition points at 770°, 928°, and 1530°C. The α form is magnetic but when transformed into the β form, the magnetism disappears although the lattice remains unchanged. The relations of these forms are peculiar. Iron is smelted from the ore from a mixture with limestone and coke at high temperatures up to 1300°C, to produce cast iron or pig iron. Pig iron is an alloy of pure iron or ferrite containing about 3% of a glassy slag which is a complex high iron silicate containing S, Mn, and P, and other elements. It is hard, brittle, fairly fusible, and is used to produce other alloys. Wrought iron is obtained by oxidizing and fluxing the impurities from pig iron. It contains only a few tenths of % carbon, is tough, malleable, less fusible, and usually has a "fibrous" structure. Steel was originally made by long-term heat treatment of wrought iron with charcoal to increase the carbon content to nearly 1%. In the Bessemer process introduced in 1856, an air blast was blown directly through the melted pig iron to produce a high quality, relatively cheap product in large quantities. It revolutionized the industry and was followed by other large scale methods such as the open-hearth process. More recently, the basic-oxygen process which uses 99.5% pure oxygen has been utilized by a growing number of firms. Carbon steel is an alloy of iron with carbon. Carbon steels are divided into dead-soft steel (0.08 to 0.18% carbon), structural grade or mild steel (0.15 to 0.25% carbon), medium grade steel (0.25 to 0.35% carbon), medium-hard steel (0.35 to 0.65% carbon), and hard steel (0.65 to 0.85% carbon), the spring steels (0.85 to 1.05% carbon), and the high carbon tool steels (1.05 to 1.20% carbon). The addition of carbon to $\alpha$-iron decreases the transition temperature to $\gamma$-iron to a minimum of 723°C at 0.8% carbon. Above this temperature, the bcc structure of ferrite transforms to an fcc form called austenite which holds much greater amounts of carbon in solid solution than $\alpha$-iron, which forms ferrite containing a maximum of 0.025% carbon in solid solution at 723°C. When a steel containing more than 0.025% carbon is cooled below 723°C, the excess carbon forms a hard, brittle compound called cementite, $Fe_3C$. At the eutectoid point at 0.8% carbon, 723°C, the austenite decomposes on cooling to ferrite and cementite in a lamellar structure called pearlite. Controlled heat treating of austenite containing varying amounts of carbon, followed by various cooling rates, produces a wide variety of metallurgical structures. Very rapid quenching produces a ferrite supersaturated with carbon, called martensite. Steels produced with a mixture of cementite and martensite give a maximum hardness. Alloy steels are carbon steels with major alloying elements added. These are principally nickel, chromium, manganese, vanadium, molybdenum, boron, tungsten, and cobalt. Stainless steels contain greater than 10% chromium. The most important of these is a steel containing 18% chromium and 8% nickel.

**Krypton** — (Gr. *kryptos*, hidden), Kr; at wt 83.80; at no 36; mp −156.6°C; bp −152.30 ± 0.10°C; density 3.733 g/l (0°C); valence usually 0. Discovered in 1898 by Ramsay and Travers in the residue left after liquid air had nearly boiled away. Krypton is present in the air to the extent of about 1 ppm. The atmosphere of Mars has been found to contain 0.3 ppm of krypton. It is one of the "noble" gases. It is characterized by its brilliant green and orange spectral lines. Naturally occurring krypton contains six stable isotopes. Fifteen other unstable isotopes are now recognized. The spectral lines of krypton are easily produced and some are very sharp. In 1960, it was internationally agreed that the fundamental unit of length, the meter, should be defined in terms of the orange red spectral line of $^{86}$Kr, corresponding to the transition $5p[O_{1/2}]$, as follows: 1 m = 1,650,763.73 wavelengths (in vacuo) of the orange red line of $^{86}$Kr. This replaced the standard meter of Paris, which was defined in terms of a bar made of a platinum-iridium alloy. In October 1983, the meter, which originally was defined as being one ten millionth of a quadrant of the earth's polar circumference, was again redefined by the International Bureau of Weight and Measures as being the length of path traveled by light in a vacuum during a time interval of 1/299,792,458 of a second. Solid krypton is a white crystalline substance with a face-centered cubic structure which is common to all the "rare gases." While krypton is generally thought of as a rare gas that normally does not combine with other elements to form compounds, it now appears that the existence of some krypton compounds is established. Krypton difluoride has been prepared in gram quantities and can be made by several methods. A higher fluoride of krypton and a salt of an oxyacid of krypton also have been reported. Molecule ions of $ArKr^+$ and $KrH^+$ have been identified and investigated, and evidence is provided for the formation of KrXe or $KrXe^+$. Krypton clathrates have been prepared with hydroquinone and phenol. $^{85}$Kr has found recent application in chemical analysis. By imbedding the isotope in various solids, kryptonates are formed. The activity of these kryptonates is sensitive to chemical reactions at the surface. Estimates of the concentration of reactants are therefore made possible. Krypton uses thus far have been limited because of its high cost. Krypton gas presently costs about $30/l.

**Kurchatovium** — see **Element 104.**

**Lanthanum** — (Gr. *lanthanein*, to lie hidden), La; at wt 138.9055 ± 3; at no 57; mp 921°C; bp 3457°C; sp gr 6.145 (25°C); valence 3. Mosander in 1839 extracted a new earth, lanthana, from impure cerium nitrate, and recognized the new element. Lanthanum is found in rare earth minerals such as cerite, monazite, allanite, and bastnasite. Monazite and bastnasite are principal ores in which lanthanum occurs in percentages up to 25 and 38%, respectively. Misch metal, used in making lighter flints, contains about 25% lanthanum. Lanthanum was isolated in relatively pure form in 1923. Ion-exchange and solvent extraction techniques have led to much easier isolation of the so-called "rare earth" elements. The availability of lanthanum and other rare earths has improved greatly in recent years. The metal can be produced by reducing the anhydrous fluoride with calcium. Lanthanum is silvery white, malleable, ductile, and soft enough to be cut with a knife. It is one of the most reactive of the rare earth metals. It oxidizes rapidly when exposed to air. Cold water attacks lanthanum slowly, and hot water attacks it much more rapidly. The metal reacts directly with elemental carbon, nitrogen, boron, selenium, silicon, phosphorus, sulfur, and with halogens. At 310°C, lanthanum changes from a hexagonal to a face-centered cubic structure, and at 865°C it again transforms into a body-centered cubic structure. Natural lanthanum is a mixture of two stable isotopes, [138]La and [139]La. Seventeen other radioactive isotopes are recognized. Rare-earth compounds containing lanthanum are extensively used in carbon lighting applications, especially by the motion picture industry for studio lighting and projection. $La_2O_3$ improves the alkali resistance of glass and is used in making special optical glasses. Lanthanum and its compounds have a low to moderate acute toxicity rating; therefore, care should be taken in handling them. The metal costs about $25/oz.

**Lawrencium** — (for Ernest O. Lawrence, inventor of the cyclotron), Lr; at no 103; at mass no (260); valence +3(?). This member of the 5f transition elements (actinide series) was discovered in March 1961 by A. Ghiorso, T. Sikkeland, A. E. Larsh, and R. M. Latimer. A 3-μg californium target, consisting of a mixture of isotopes of mass number 249, 250, 251, and 252, was bombarded with either [10]B or [11]B. The electrically charged trasmutation nuclei recoiled with an atmosphere of helium and were collected on a thin copper conveyor tape which was then moved to place collected atoms in front of a series of solid-state detectors. The isotope of element 103 produced in this way decayed by emitting an 8.6-MeV alpha particle with a half-life of 8 sec. In 1967, Flerov and associates of the Dubna Laboratory reported their inability to detect an alpha emitter with a half-life of 8 sec which was assigned by the Berkeley group to [257]103. This assignment has been changed to [258]Lr or [259]Lr. In 1965, the Dubna workers found a longer-lived lawrencium isotope, [256]Lr, with a half-life of 35 sec. In 1968, Ghiorso and associates at Berkeley were able to use a few atoms of this isotope to study the oxidation behavior of lawrencium. Using solvent extraction techniques and working very rapidly, they extracted lawrencium ions from a buffered aqueous solution into an organic solvent, completing each extraction in about 30 sec. It was found that lawrencium behaves differently from dipositive nobelium and more like the tripositive elements earlier in the actinide series.

**Lead** — (Anglo-Saxon *lead*), Pb; (L. *plumbum*) at wt 207.2; at no 82; mp 327.502°C; bp 1740°C; sp gr 11.35 (20°C); valence 2 or 4. Long known, mentioned in Exodus. Alchemists believed lead to be the oldest metal and associated it with the planet Saturn. Native lead occurs in nature, but it is rare. Lead is obtained chiefly from galena (PbS) by a roasting process. Anglesite ($PbSO_4$), cerussite ($PbCO_3$), and minim ($Pb_3O_4$) are other common lead minerals. Lead is a bluish white metal of bright luster, is very soft, highly malleable, ductile, and a poor conductor of electricity. It is very resistant to corrosion; lead pipes bearing the insignia of Roman emperors, used as drains from the baths, are still in service. It is used in containers for corrosive liquids (such as in sulfuric acid chambers) and may be toughened by the addition of a small percentage of antimony or other metals. Natural lead is a mixture of four stable isotopes: [204]Pb (1.48%), [206]Pb (23.6%), [207]Pb (22.6%), and [208]Pb (52.3%). Lead isotopes are the end products of each of the three series of naturally occurring radioactive elements: [206]Pb for the uranium series, [207]Pb for the actinium series, and [208]Pb for the thorium series. Seventeen other isotopes of lead, all of which are radioactive, are recognized. Its alloys include solder, type metal, and various antifriction metals. Great quantities of lead, both as the metal and as the dioxide, are used in storage batteries. Much metal also goes into cable covering, plumbing, ammunition, and in the manufacture of lead tetraethyl, used as an antiknock compound in gasoline. The metal is very effective as a sound absorber, is used as a radiation shield around X-ray equipment and nuclear reactors, and is used to absorb vibration. White lead, the basic carbonate, sublimed white lead ($PbSO_4$), chrome yellow ($PbCrO_4$), red lead ($Pb_3O_4$), and other lead compounds are used extensively in paints, although in recent years the use of lead in paints has been drastically curtailed to eliminate or reduce health hazards. Lead oxide is used in producing fine "crystal glass" and "flint glass" of a high index of refraction for achromatic lenses. The nitrate and the acetate are soluble salts. Lead salts such as lead arsenate have been used as insecticides, but their use in recent years has been practically eliminated in favor of less harmful organic compounds. Care must be used in handling lead as it is a cumulative poison. Environmental concern with lead poisoning has resulted in a national program to reduce the concentration of lead in gasoline, paints, and drinking water.

**Lithium** — (Gr. *lithos*, stone), Li; at wt 6.941;

at no 3; mp 180.54°C; bp 1342°C; sp gr 0.534 (20°C); valence 1. Discovered by Arfvedson in 1817. Lithium is the lightest of all metals, with a density only about half that of water. It does not occur free in nature; combined it is found in small amounts in nearly all igneous rocks and in the waters of many mineral springs. Lepidolite, spodumene, petalite, and amblygonite are the more important minerals containing it. Lithium is presently being recovered from brines of Searles Lake, in California, and from those in Nevada. Large deposits of spodumene are found in North Carolina. The metal is produced electrolytically from the fused chloride. Lithium is silvery in appearance, much like Na and K, other members of the alkali metal series. It reacts with water, but not as vigorously as sodium. Lithium imparts a beautiful crimson color to a flame, but when the metal burns strongly, the flame is a dazzling white. Since World War II, production of lithium metal and its compounds has increased greatly. Because the metal has the highest specific heat of any solid element, it has found use in heat transfer applications; however, it is corrosive and requires special handling. The metal has been used as an alloying agent, is of interest in synthesis of organic compounds, and has nuclear applications. It ranks as a leading contender as a battery anode material since it has a high electrochemical potential. Lithium is used in special glasses and ceramics. The glass for the 200-in. telescope at Mt. Palomar contains lithium as a minor ingredient. Lithium chloride is one of the most hygroscopic materials known, and it, as well as lithium bromide, is used in air conditioning and industrial drying systems. Lithium stearate is used as an all-purpose and high-temperature lubricant. Other lithium compounds are used in dry cells and storage batteries. The metal is priced at about $20/lb.

**Lutetium** — (*Lutetia*, ancient name for Paris; sometimes called *cassiopeium* by the Germans), Lu; at wt 174.967; at no 71; mp 1663°C; bp 3395°C; sp gr 9.840 (25°C); valence 3. In 1907, Urbain described a process by which Marignac's ytterbium (1879) could be separated into the two elements ytterbium (neoytterbium) and lutetium. These elements were identical with "aldebaranium" and "cassiopeium", independently discovered by von Welsbach about the same time. Charles James of the University of New Hampshire also independently prepared the very pure oxide, lutecia, at this time. The spelling of the element was changed from lutecium to lutetium in 1949. Lutetium occurs in very small amounts in nearly all minerals containing yttrium and is present in monazite, which is a commercial source, to the extent of about 0.003%. The pure metal has been isolated only in recent years and is one of the most difficult to prepare. It can be prepared by the reduction of anhydrous $LuCl_3$ or $LuF_3$, by an alkali or alkaline earth metal. The metal is silvery white and relatively stable in air. While new techniques, including ion-exchange reactions, have been developed to separate the various rare earth elements, lutetium is still the most costly of all naturally occurring rare earths. It is slightly more abundant than thulium. It is now priced at about $74/g or $30,000/lb. $^{176}Lu$ occurs naturally (2.6%) with $^{175}Lu$ (97.4%). It is radioactive, with a half-life of about $3 \times 10^{10}$ years. Stable lutetium nuclides, which emit pure beta radiation after thermal neutron activation, can be used as catalysts in cracking, alkylation, hydrogenation, and polymerization. Virtually no other commercial uses have been found yet for lutetium. While lutetium, like other rare earth metals, is thought to have a low toxicity rating, it should be handled with care until more information is available.

**Magnesium** — (Magnesia, district in Thessaly) Mg; at wt 24.305; at no 12; mp 648.8 ± 0.5°C; bp 1090°C; sp gr 1.738 (20°C); valence 2. Compounds of magnesium have long been known. Black recognized magnesium as an element in 1755. It was isolated by Davy in 1808, and prepared in coherent form by Bussy in 1831. Magnesium is the eighth most abundant element in the crust of the earth. It does not occur uncombined, but is found in large deposits in the form of magnesite, dolomite, and other minerals. The metal is now principally obtained in the U.S. by electrolysis of fused magnesium chloride derived from brines, wells, and sea water. Magnesium is a light, silvery white, and fairly tough metal. It tarnishes slightly in air, and finely divided magnesium readily ignites upon heating in air and burns with a dazzling white flame. It is used in flashlight photography, flares, and pyrotechnics, including incendiary bombs. It is one third lighter than aluminum, and in alloys is essential for airplane and missile construction. The metal improves the mechanical, fabrication, and welding characteristics of aluminum when used as an alloying agent. Magnesium is used in producing nodular graphite in cast iron and is used as an additive to conventional propellants. It is also used as a reducing agent in the production of pure uranium and other metals from their salts. The hydroxide (milk of magnesia), chloride, sulfate (Epsom salts), and citrate are used in medicine. Deadburned magnesite is employed for refractory purposes such as brick and liners in furnaces and converters. Organic magnesium compounds (Grignard's reaction) are important. Magnesium is an important element in both plant and animal life. Chlorophylls are magnesium-centered porphyrins. The adult daily requirement of magnesium is about 300 mg/day. Great care should be taken in handling magnesium metal, since serious fires can occur. Water should not be used on burning magnesium or on magnesium fires.

**Manganese** — (L. *magnes*, magnet, from magnetic properties of pyrolusite; It. *manganese*, corrupt form of magnesia), Mn; at wt 54.9380; at no 25; mp 1244 $\pm$ 3°C; bp 1962°C; sp gr 7.21 to 7.44, depending on allotropic form; valence 1, 2, 3, 4, 6, or 7. Recognized by Scheele, Bergman, and others as an element and isolated by Gahn in 1774 by reduction of the dioxide with carbon. Manganese minerals are widely distributed; oxides, silicates, and carbonates are the most common. The discovery of large quantities of manganese nodules on the floor of the ocean holds promise as a source of manganese. These nodules contain about 24% manganese. Most manganese today is obtained from ores found in the U.S.S.R., Brazil, Australia, Republic of South Africa, Gabon, and India. Pyrolusite ($MnO_2$) and rhodochrosite ($MnCO_3$) are among the most common manganese minerals. The metal is obtained by reduction of the oxide with sodium, magnesium, aluminum, or by electrolysis. It is gray-white, resembling iron, but is harder and very brittle. The metal is reactive chemically and decomposes cold water slowly. Manganese is used to form many important alloys. In steel, manganese improves the rolling and forging qualities, strength, toughness, stiffness, wear resistance, hardness, and hardenability. With aluminum and antimony, especially with small amounts of copper, it forms highly ferromagnetic alloys. Manganese metal is ferromagnetic only after special treatment. The pure metal exists in four allotropic forms. The alpha form is stable at ordinary temperature; gamma manganese, which changes to alpha at ordinary temperatures, is flexible, soft, easily cut, and capable of being bent. The dioxide (pyrolusite) is used as a depolarizer in dry cells, and is used to "decolorize" glass that is colored green by impurities of iron. Manganese by itself colors glass an amethyst color, and is responsible for the color

of true amethyst. The permanganate is a powerful oxidizing agent and is used in quantitative analysis and in medicine. Manganese is widely distributed throughout the animal kingdom. It is an important trace element and may be essential for utilization of vitamin $B_1$. Exposure to manganese dusts, fume, and compounds (as Mn) should not exceed the ceiling value of 5 mg/m$^3$ for even short periods because of its toxicity.

**Mendelevium** — (Dmitri Mendeleev), Md; at wt 257; at no 101; valence $+2$, $+3$. Mendelevium, the ninth transuranium element of the actinide series to be discovered, was first identified by Ghiorso, Harvey, Choppin, Thompson, and Seaborg early in 1955 as a result of the bombardment of the isotope $^{253}$Es with helium ions in the Berkeley 60-in. cyclotron. The isotope produced was $^{256}$Md, which has a half-life of 77 min. Four isotopes are now recognized. $^{258}$Md has a half-life of 2 months. This isotope has been produced by the bombardment of an isotope of einsteinium with ions of helium. $^{256}$Md has been used to elucidate some of the chemical properties of mendelevium in aqueous solution. Experiments seem to show that the element possesses a moderately stable dipositive (II) oxidation state in addition to the tripositive (III) oxidation state, which is characteristic of actinide elements.

**Mercury** — (from planet *Mercury*), Hg; (L. hydrargyrum, liquid silver) at wt 200.59 $\pm$ 3; at no 80; mp $-38.842$°C; bp 356.58°C; sp gr 13.546 (20°C); valence 1 or 2. Known to ancient Chinese and Hindus; found in Egyptian tombs of 1500 B.C. Mercury is the only common metal liquid at ordinary temperatures. It rarely occurs free in nature. The chief ore is cinnabar (HgS). Spain and Italy produce about 50% of the world supply of the metal. The commercial unit for handling mercury is the "flask", which weighs 76 lb. The metal is obtained by heating cinnabar in a current of air and by condensing the vapor. It is a heavy, silvery white metal; a poor conductor of heat, compared with other metals; and a fair conductor of electricity. It easily forms alloys with many metals, such as gold, silver, and tin, which are called amalgams. Amalgamating with gold is used in the recovery of gold from its ores. The metal is widely used in laboratory work for making thermometers, barometers, diffusion pumps, and many other instruments. It is used in making mercury-vapor lamps, mercury switches and other electrical apparatus, pesticides, mercury cells for caustic soda and chlorine production, dental preparations, and antifouling paint. The most important salts are mercuric chloride ($HgCl_2$) (corrosive sublimate, a violent poison), mercurous chloride $Hg_2Cl_2$ (calomel), mercury fulminate ($Hg(ONC)_2$, a detonator widely used in explosives), and mercuric sulfide (HgS, vermillion, a high-grade paint pigment). Organic mercury compounds are important. An electrical discharge causes mercury vapor to combine with neon, argon, krypton, and xenon forming van der Waals' bonded HgNe, HgAr, HgKr, and HgXe. Mercury is a virulent poison and is readily absorbed through the respiratory tract, the gastrointestinal tract, or through broken skin. It acts as a cumulative poison since only small amounts of the element can be eliminated at a time by humans. Since mercury is a very volatile element, dangerous levels are readily attained in air. Air saturated with mercury vapor at 20°C contains a concentration that exceeds the toxic limit. The danger increases at higher temperatures. Containers of mercury should be securely covered and spillage should be avoided. The National Bureau of Standards has recently redetermined the triple point of mercury at −38.84168°C.

**Molybdenum** — (Gr. *molybdos*, lead), Mo; at wt 95.94; at no 42; mp 2617°C; bp 4612°C; sp gr 10.22 (20°C); valence 2, 3, 4?, 5?, or 6. Before Scheele recognized molybdenite as a distinct ore of a new element in 1778, it was confused with graphite and lead ore. The metal was prepared in an impure form in 1782 by Hjelm. Molybdenum does not occur native, but is obtained principally from molybdenite ($MoS_2$). Wulfenite ($PbMoO_4$), and powellite [$Ca(MoW)O_4$] are also minor commercial ores. Molybdenum is also recovered as a by-product of copper- and tungsten-mining operations. The metal is prepared from the powder made by the hydrogen reduction of purified molybdic trioxide or ammonium molybdate. The metal is silvery white, very hard, but is softer and more ductile than tungsten. It has a high elastic modulus, and only tungsten and tantalum, of the more readily available metals, have higher melting points. It is a valuable alloying agent, as it contributes to the hardenability and toughness of quenched and tempered steels, and improves the strength of steel at high temperatures. It is used in nickel-based alloys, such as the "Hastelloys®" which are heat-resistant and corrosion-resistant. Molybdenum oxidizes at elevated temperatures. The metal is used in nuclear energy applications and for missile and aircraft parts.

Molybdenum is valuable as a catalyst in the refining of petroleum. Molybdenum is an essential trace element in plant nutrition. Some lands are barren for lack of this element in the soil. Molybdenum sulfide is used as a lubricant, especially at high temperatures. Almost all ultrahigh strength steels with minimum yield points up to 300,000 psi (lb/in². ) contain molybdenum in amounts from 0.25 to 8%.

**Neodymium** — (Gr. *neos*, new, and *didymos*, twin), Nd; at wt 144.24 ± 3; at no 60; mp 1021°C; bp 3068°C; sp gr 6.80 and 7.007, depending on allotropic form; valence 3. In 1841, Mosander extracted from cerite a new rose-colored oxide, which he believed contained a new element. He named the element didymium, as it was "an inseparable twin brother of lanthanum." In 1885 von Welsbach separated didymium into two new elemental components, neodymia and praseodymia, by repeated fractionation of ammonium didymium nitrate. While the free metal is in misch metal, long known and used as a pyrophoric alloy for lighter flints, the element was not isolated in relatively pure form until 1925. Neodymium is present in misch metal to the extent of about 18%. It is present in the minerals monazite and bastnasite, which are principal sources of rare earth metals. The element may be obtained by separating neodymium salts from other rare earths by ion exchange or solvent extraction techniques, and by reducing anhydrous halides such as $NdF_3$ with calcium metal. The metal has a bright silvery metallic luster. Neodymium is one of the more reactive rare earth metals and quickly tarnishes in air, forming an oxide that spalls off and exposes metal to further oxidation. Neodymium exists in two allotropic forms, with a transformation from a double hexagonal to a body-centered cubic structure taking place at 860°C. Natural neodymium is a mixture of seven stable isotopes. Seven other radioactive isotopes are recognized. Didymium, of which neodymium is a component, is used for coloring glass to make welder's goggles. Neodymium glass has been used in astronomical work to produce sharp bands by which spectral lines may be calibrated. Glass containing neodymium also can be used as a laser material in place of ruby to produce coherent light. The price of the metal is about $2/g. Neodymium has a low-to-moderated acute toxic rating. As with other rare earths, neodymium should be handled with care.

**Neon** — (Gr. *neos*, new), Ne; at wt 20.179; at

no 10; mp $-248.67°C$; bp $-248.048°C$ (1 atm); density of gas 0.89990 g/l (1 atm, 0°C); density of liquid at bp 1.207 g/cm³; valence 0. Discovered by Ramsay and Travers in 1898. Neon is a rare gaseous element present in the atmosphere to the extent of 1 part in 65,000 of air. It is obtained by liquifaction of air and separated from the other gases by fractional distillation. Natural neon is a mixture of three isotopes. Five other unstable isotopes are known. It is a very inert element; however, it is said to form a compound with fluorine. It is still questionable if true compounds of neon exist, but evidence is mounting in favor of their existence. The following ions are known from optical and mass spectrometric studies: $Ne^+$, $(NeAr)^+$, $(NeH)^+$, and $(HeNe)^+$. Neon also forms an unstable hydrate. In a vacuum discharge tube, neon glows reddish orange. Of all the rare gases, the discharge of neon is the most intense at ordinary voltages and currents. Neon is used in making the common neon advertising signs, which accounts for its largest use. Neon and helium are used in making gas lasers. Liquid neon is now commercially available and is finding important application as an economical cryogenic refrigerant. It has over 40 times more refrigerating capacity per unit volume than liquid helium and more than three times that of liquid hydrogen. It is compact, inert, and is less expensive than helium when it meets refrigeration requirements. Neon costs about $2.00/1.

**Neptunium** — (from planet *Neptune*), Np; at wt 237.0482; at no 93; mp 640 ± 1°C; bp 3902°C (est.); sp gr 20.25 (20°C); valence 3, 4, 5, and 6. Neptunium was the first synthetic transuranium element of the actinide series discovered; the isotope $^{239}Np$ was produced by McMillan and Abelson in 1940 at Berkeley, California, as the result of bombarding uranium with cyclotron-produced neutrons. The isotope $^{237}Np$ (half-life of $2.14 \times 10^6$ years) is currently obtained in gram quantities as a by-product from nuclear reactors in the production of plutonium. Trace quantities of the element are actually found in nature due to transmutation reactions in uranium ores produced by the neutrons which are present. Neptunium is prepared by the reduction of $NpF_3$ with barium or lithium vapor at about 1200°C. Neptunium metal has a silvery appearance, is chemically reactive, and exists in at least three structural modifications: α-neptunium, orthorhombic, density 20.25 g/cm³; β-neptunium (above 280°C), tetragonal, density (313°C) 19.36 g/cm³; γ-neptunium (above 577°C), cubic, density

(600°C) 18.0 g/cm³. Neptunium has four ionic oxidation states in solution: $Np^{3+}$ (pale purple), analogous to the rare earth ion $Pm^{3+}$, $Np^{4+}$ (yellow green); $NpO^+$ (green blue); and $NpO^{2+}$ (pale pink). These latter oxygenated species are in contrast to the rare earths which exhibit only simple ions of the (II), (III), and (IV) oxidation states in aqueous solution. The element forms tri- and tetrahalides such as $NpF_3$, $NpF_4$, $NpCl_4$, $NpBr_3$, $NpI_3$, and oxides of various compositions such as are found in the uranium-oxygen system, icluding $Np_3O_8$ and $NpO_2$. Thirteen isotopes of neptunium are now recognized. The O.R.N.L. has $^{237}Np$ available for sale to its licensees and for export. This isotope can be used as a component in neutron detection instruments. It is offered at a price of $280/g.

**Nickel** — (Ger, *Nickel*, Satan or Old Nick's, and from *Kupfernickel*, Old Nick's copper), Ni; at wt 58.69; at no 28; mp 1453°C; bp 2732°C; sp gr 8.902 (25°C); valence 0, 1, 2, and 3. Discovered by Cronstedt in 1751 in kupfernickel (niccolite). Nickel is found as a constituent in most meteorites and its presence often serves as one of the criteria for distinguishing a meteorite from other minerals. Iron meteorites, or siderites, may contain iron alloyed with from 5 to nearly 20% nickel. Nickel is obtained commercially from pentlandite and pyrrhotite of the Sudbury region of Ontario, a district that produces about 30% of the nickel for the non-Communist World. Other deposits are found in New Caledonia, Australia, Cuba, Indonesia, and elsewhere. Nickel is silvery white and takes on a high polish. It is hard, malleable, ductile, somewhat ferromagnetic, and a fair conductor of heat and electricity. It belongs to the iron-cobalt group of metals and is chiefly valuable for the alloys it forms. It is extensively used for making stainless steel and other corrosion-resistant alloys such as Invar®, Monel®, Inconel®, and the Hastelloys®. Tubing made of a copper-nickel alloy is extensively used in making desalination plants for converting sea water into fresh water. Nickel is also now used extensively in coinage and in making nickel steel for armor plate and burglar-proof vaults, and is a component in Nichrome®, Permalloy®, and constantan. Nickel added to glass gives a green color. Nickel plating is often used to provide a protective coating for other metals, and finely divided nickel is a catalyst for hydrogenating vegetable oils. It is also used in ceramics, in the manufacture of Alnico magnets, and in the Edison® storage battery. Natural nickel is a mixture of five stable isotopes;

seven other unstable isotopes are known. Exposure to nickel metal and soluble compounds (as Ni) should not exceed 1 mg/m³ (8-hr time-weighted average 40-hr week). Nickel carbonyl exposure, however, should not exceed 0.007 mg/m³, and is considered to be a very toxic material. Nickel sulfide fume and dust are recognized as having carcinogenic potential.

**Niobium** — (*Niobe*, daughter of Tantalus), or Columbium (*Columbia*, name for America), Nb; at wt 92.9064; at no 41; mp 2468 ± 10°C; bp 4742°C; sp gr 8.57 (20°C); valence 2, 3, 4?, or 5. Discovered in 1801 by Hatchett in an ore sent to England more than a century before by John Winthrop the Younger, first governor of Connecticut. The metal was first prepared in 1864 by Blomstrand, who reduced the chloride by heating it in a hydrogen atmosphere. The name "niobium" was adopted by the International Union of Pure and Applied Chemistry in 1950 after 100 years of controversy. Many leading chemical societies and government organizations refer to it by this name. Many metallurgists, leading metal societies, and leading U.S. commercial producers, however, still refer to the metal as "columbium". The element is found in niobite (or columbite), niobite-tantalite, pyrochlore, and euxenite. Large deposits of niobium have been found associated with carbonatites (carbon-silicate rocks), as a constituent of pyrochlore. Extensive ore reserves are found in Canada, Brazil, Nigeria, Zaire, and in the U.S.S.R. The metal can be isolated from tantalum, and prepared in several ways. It is a shiny, white, soft, and ductile metal, and takes on a bluish cast when exposed to air at room temperature for a long time. The metal starts to oxidize in air at 200°C. It is used as an alloying agent in carbon and alloy steels and in nonferrous metals. The metal has a low-capture cross-section for thermal neutrons. Niobium has been used in advanced air frame systems such as were used in the Gemini space program. The element has superconductive properties; superconductive magnets have been made with Nb-Zr wire, which retains its superconductivity in strong magnetic fields. This type of application offers hope of direct large-scale generation of electric power. Sixteen isotopes of niobium are known. Niobium metal (99.5% pure) is priced at about $30/lb.

**Nitrogen** — (L. *nitrum*; Gr. *nitron*, native soda, *genes*, forming,) N; at wt 14.0067; at no 7; mp −209.86°C; bp −195.8°C; density 1.2506 g/l; sp gr liquid 0.808 (−195.8°C), solid 1.026 (−252°C); valence 3 or 5. Discovered by Daniel Rutherford in 1772, but Scheele, Cavendish, Priestley, and others at about the same time studied "burnt or dephlogisticated air", as air without oxygen was then called. Nitrogen makes up 78% by volume of the air. The atmosphere of Mars, by comparison, is 2.6% nitrogen. The estimated amount of this element in our atmosphere is more than 4000 trillion tons. From this inexhaustible source it can be obtained by liquefaction and fractional distillation. Nitrogen molecules give the orange red, blue green, blue violet, and deep violet shades to the aurora. The element is so inert that Lavoisier named it "azote", meaning without life, yet its compounds are so active as to be most important in foods, poisons, fertilizers, and explosives. Nitrogen can be also easily prepared by heating a water solution of ammonium nitrite. Nitrogen, as a gas, is colorless, odorless, and a generally inert element. As a liquid it is also colorless and odorless, and is similar in appearance to water. Two allotropic forms of solid nitrogen exist, with the transition from the α to the β form taking place at −237°C. When nitrogen is heated, it combines directly with magnesium, lithium, or calcium; when mixed with oxygen and subjected to electric sparks, it forms first nitric oxide (NO) and then the dioxide ($NO_2$); when heated under pressure with a catalyst with hydrogen, ammonia is formed (Haber process). The ammonia thus formed is used in fertilizers, and it can be oxidized to nitric acid (Ostwald process). The ammonia industry is the largest consumer of nitrogen. Large amounts of the gas are also used by the electronics industry, the steel industry, and the drug industry. Nitrogen is used as a refrigerant both for the immersion freezing of food products and for transportation of foods. Sodium and potassium nitrates are formed by the decomposition of organic matter with compounds of the metals present. In certain dry areas of the world these saltpeters are found in quantity. Nitrogen gas prices vary from 2¢ to $2.75 per 100 ft³, depending on purity, etc. Production of elemental nitrogen in the U.S. is more than 9 million short tons per year.

**Nobelium** — (for Alfred Nobel, discoverer of dynamite), No; at wt (259); at no 102; valence +2 or +3. Nobelium was unambiguously discovered and identified in April 1958 at Berkeley by A. Ghiorso, T. Sikkeland, J. R. Walton, and G. T. Seaborg, who used a new double-recoil technique. A heavy-ion linear accelerator (HILAC) was used to bombard a thin target of curium (95% ²⁴⁴Cm and 4.5% $^{246}$Cm) with $^{12}$C ions to produce $^{254}$102 according to the $^{246}$Cm ($^{12}$C 4n) reaction. Earlier in 1957, workers of the U.S., Britain, and Sweden announced the discovery of an isotope of element 102 with a 10-min half-life at 8.5 MeV, as a result of bombarding $^{244}$Cm with $^{13}$C nuclei. On the basis of this experiment, the name *nobelium* was assigned and accepted by the Commission on Atomic Weights of the International Union of Pure and Applied Chemistry. The acceptance of the name was premature, for both Russian and American efforts now completely rule out the possibility of any isotope of element 102 having a half-life of 10-min in the vicinity of 8.5 MeV. Early work in 1957 on the search for this element (in Russia at the Kurchatov Institute), was marred by the assignment of 8.9 ± 0.4 MeV alpha radiation with a half-life of 2 to 40 sec, which was too indefinite to support claim to discovery. Confirmatory experiments at Berkeley in 1966 have shown the existence of $^{254}$102 with a 55-sec half-life, $^{252}$102 with a 2.3-sec half-life, and $^{257}$102 with a 23-sec half-life. Four other isotopes are now recognized, one of which ($^{255}$102) has a half-life of 3 min. In view of the discoverer's traditional right to name an element, the Berkeley group, in 1967, suggested that the hastily given name "nobelium", along with the symbol No, be retained.

**Osmium** — (Gr. *osme*, a smell), Os; at wt 190.2; at no 76; mp 3045 ± 30°C; bp 5027 ± 100°C; sp gr 22.57; valence 0 to +8. Discovered in 1803 by Tennant in the residue left when crude platinum is dissolved by aqua regia. Osmium occurs in iridosmine and in platinum-bearing river sands of the Urals, North America, and South America. It is also found in the nickel-bearing ores of the Sudbury, Ontario region along with other platinum metals. While the quantity of platinum metals in these ores is very small, the large tonnages of nickel ores processed make commercial recovery possible. The metal is lustrous, bluish white, extremely hard, and brittle even at high temperatures. It has the highest melting point and lowest vapor pressure of the platinum group. The metal is very difficult to fabricate, but the powder can be sintered in a hydrogen atmosphere at a temperature of 2000°C. The solid metal is not affected by air at room temperature, but the powdered or spongy metal slowly gives osmium tetroxide, which is a powerful oxidizing agent and has a strong smell. The tetroxide is highly toxic, and boils at 130°C (760 mm). Concentrations in air as low as $10^{-7}$ g/m$^3$ can cause

lung congestion, skin damage, or eye damage. Exposure to osmium tetroxide should not exceed 0.002 mg/m$^3$ (8-hr time-weighted average, 40-hr work week). The tetroxide has been used to detect fingerprints and to stain fatty tissue for microscope slides. The metal is almost entirely used to produce very hard alloys, with other metals of the platinum group. The price of 99% pure osmium powder — the form usually supplied commercially — is about $20/g or $1000/oz t, depending on quantity and supplier. The measured densities of iridium and osmium indicate that osmium is slightly more dense than iridium, so osmium has generally been credited with being the heaviest known element. Calculations of the density from the space lattice, which may be more reliable for these elements than actual measurements, however, give a density of 22.65 for iridium compared to 22.61 for osmium.

**Oxygen** — (Gr. *oxys*, sharp, acid, and *genes*, forming, acid former), O; at wt (natural) 15.9994 ± 3; at no 8; mp −218.4°C; bp −182.962°C; density 1.429 g/l (0°C); sp gr liquid 1.14 (−182.96°C); valence 2. For many centuries, workers occasionally realized air was composed of more than one component. The behavior of oxygen and nitrogen as components of air led to the advancement of the phlogiston theory of combustion, which captured the minds of chemists for a century. Oxygen was prepared by several workers, including Bayen and Borch, but they did not know how to collect it, did not study its properties, and did not recognize it as an elementary substance. Priestley is generally credited with its discovery, although Scheele also discovered it independently. Oxygen is the third most abundant element found in the sun and it plays a part in the carbon-nitrogen cycle, one process thought to give the sun and stars their energy. Oxygen under excited conditions is responsible for the bright red and yellow green colors of the aurora. Oxygen, as a gaseous element, forms 21% of the atmosphere by volume from which it can be obtained by liquefaction and fractional distillation. The atmosphere of Mars contains about 0.15% oxygen. The element and its compounds make up 49.2%, by weight, of the earths crust. About 2/3 of the human body and 9/10 of water is oxygen. In the laboratory, it can be prepared by the electrolysis of water or by heating potassium chlorate with manganese dioxide as a catalyst. The gas is colorless, odorless, and tasteless. The liquid and solid forms are a pale blue color and are strongly paramagnetic. Ozone ($O_3$), a highly active allotropic form of oxygen, is formed by the action of an electrical discharge or ultraviolet light on oxygen. The presence of ozone in the atmosphere (amounting to the equivalent of a layer 3 mm thick at ordinary pressures and temperatures) is of vital importance in preventing harmful ultraviolet rays of the sun from reaching the surface of the earth. There has been recent concern that aerosols in the atmosphere may have a detrimental effect on this ozone layer. Ozone is toxic and exposure should not exceed $0.2 \text{ mg/m}^3$ (8-hr time-weighted average, 40-hr work week). Undiluted ozone has a bluish color. Liquid ozone is bluish black, and solid ozone is violet black. Oxygen is very reactive and is capable of combining with most elements. It is a component of hundreds of thousands of organic compounds. It is essential for respiration of all plants and animals and for practically all combustion. Its atomic weight was used as a standard of comparison for each of the other elements until 1961 when the International Union of Pure and Applied Chemistry adopted carbon 12 as the new basis. Oxygen has eight isotopes. Natural oxygen is a mixture of three isotopes. $^{18}O$ occurs naturally, is stable, and is available commercially. Water ($H_2O$ with 1.5% $^{18}O$) is also available. Commercial oxygen consumption in the U.S. is estimated to be 20 million short tons per year. Oxygen enrichment in steel blast furnaces accounts for the greatest use. Large quantities are also used in making synthesis gas for ammonia and methanol, ethylene oxide, and for oxyacetylene welding. Air separation plants produce about 99% of the gas, electrolysis plants about 1%. The gas costs $5¢/\text{ft}^3$ in small quantities, and about $15/ton.

**Palladium** — (named after the asteroid Pallas, discovered about the same time; Gr *Pallas*, goddess of wisdom), Pd; at wt 106.42; at no 46; mp 1554°C; bp 2970°C; sp gr 12.02 (20°C); valence 2, 3, or 4. Discovered in 1803 by Wollaston. Palladium is found along with platinum and other metals of the platinum group in placer deposits of the U.S.S.R., South and North America, Ethiopia, and Australia. It is also found associated with the nickel-copper deposits of South Africa and Ontario. Its separation from the platinum metals depends upon the type of ore in which it is found. It is a steel-white metal, does not tarnish in air, and is the least dense and lowest melting of the platinum group of metals. When annealed, it is soft and ductile; cold working greatly increases its strength and hardness. Palladium is attacked by nitric and sulfuric acid. At room temperatures the metal has the unusual property of absorbing up to 900 times its own volume of hydrogen, possibly forming $Pd_2H$. It is not clear if this is a true compound. Hydrogen readily diffuses through heated palladium and this provides a means of purifying the gas. Finely divided palladium is a good catalyst and is used for hydrogenation and dehydrogenation reactions. It is alloyed and used in jewelry trades. White gold is an alloy of gold decolorized by the addition of palladium. Like gold, palladium can be beaten into leaf as thin as 1/250,000 in. The metal is used in dentistry, watchmaking, and in making surgical instruments and electrical contacts. The metal sells for about $130/oz t.

**Phosphorus** — (Gr. *phosphoros*, light bearing; ancient name for the planet Venus when appearing before sunrise), P; at wt 30.97376; at no 15; mp (white) 44.1°C; bp (white) 280°C; sp gr (white) 1.82, (red) 2.20, (black) 2.25 to 2.69; valence 3 or 5. Discovered in 1669 by Brand, who prepared it from urine. Phosphorus exists in four or more allotropic forms: white (or yellow), red, and black (or violet). White phosphorus has two modifications, $\alpha$ and $\beta$, with a transition temperature at $-3.8$°C. Never found free in nature, it is widely distributed in combination with minerals. Phosphate rock, which contains the mineral apatite, an impure tricalcium phosphate, is an important source of the element. Large deposits are found in the U.S.S.R., in Morocco, and in Florida, Tennessee, Utah, Idaho, and elsewhere. Phosphorus is an essential ingredient of all cell protoplasm, nervous tissue, and bones. Ordinary phosphorus is a waxy white solid; when pure it is colorless and transparent. It is insoluble in water, but soluble in carbon disulfide. It takes fire spontaneously in air, burning to the pentoxide. It is very poisonous, 50 mg constituting an approximate fatal dose. Exposure to white phosphorus should not exceed 0.1 $\text{mg/m}^3$ (8-hr time-weighted average, 40-hr work week). White phosphorus should be kept under water, and should be handled with forceps, since contact with the skin may cause severe burns. When exposed to sunlight or when heated in its own vapor to 250°C, it is converted to the red variety, which does not phosphoresce in air as does the white variety. This form does not ignite spontaneously and it is not as dangerous as white phosphorus. It should, however, be handled with care because it emits highly toxic fumes of the oxides of phosphorus when heated. The red modification is fairly stable, sublimes with a vapor pressure of 1 atm at 417°C, and is used in the manufacture of safety matches, pyrotechnics, pesticides, incendiary shells, smoke bombs, tracer bullets, etc. White phosphorus may be made by several methods. By one process, tricalcium phosphate, the essential ingredient of phosphate rock, is heated in the presence of carbon and silica in an electric furnace or fuel-fired blast furnace. Elementary phosphorus is liberated as vapor and collected under water. Phosphorus vapor and carbon monoxide produced by the reaction can be oxidized at once in the presence of moisture or water to produce phosphoric acid, an important compound in making superphosphate fertilizers. World wide demand for fertilizers has caused record phosphate production in recent years. Bone-ash, calcium phosphate, is also used to produce fine chinaware and to produce monocalcium phosphate used in baking powder. Trisodium phosphate is important as a cleaning agent, as a water softener, and for preventing boiler scale and corrosion of pipes and boiler tubes.

**Platinum** — (Sp. *platina*, silver), Pt; at wt 195.08 ± 3; at no 78; mp 1772°C; bp 3827 ± 100°C; sp gr 21.45 (20°C); valence 1?, 2, 3, or 4. Discovered in South America by Ulloa in 1735 and by Wood in 1741. The metal was used by pre-Columbian Indians. Platinum occurs native, accompanied by small quantities of iridium, osmium, palladium, ruthenium, and rhodium, all belonging to the same group of metals. These are found in the alluvial deposits of the Ural mountains, of Columbia, and of certain western American states. Sperrylite ($PtAs_2$), occurring with the nickel-bearing deposits of Sudbury, Ontario, is the source of a considerable amount of the metal. The large production of nickel offsets there being only one part of the platinum metals in 2 million parts of ore. Platinum is a beautiful silvery white metal, when pure, and is malleable and ductile. It has a coefficient of expansion almost equal to that of soda-lime-silica glass, and is therefore used to make sealed electrodes in glass systems. The metal does not oxidize in air at any temperature, but is corroded by halogens, cyanides, sulfur, and caustic alkalis. It is insoluble in hydrochloric and nitric acid, but dissolves when they are mixed as aqua regia, forming chloroplatinic acid ($H_2PtCl_6$), an important compound. The metal is extensively used in jewelry, wire, and vessels for laboratory use, and in many valuable instruments including thermocouple elements, and in dentistry. Platinum-cobalt alloys have magnetic

properties. One such alloy made of 76.7% Pt and 23.3% Co, by weight, is an extremely powerful magnet that offers a B-H (max) almost twice that of Alnico V. Platinum resistance wires are used for constructing high-temperature electric furnaces. The metal, like palladium, absorbs large volumes of hydrogen, retaining it at ordinary temperatures but giving it up at red heat. In the finely divided state, platinum is an excellent catalyst, having long been used in the contact process for producing sulfuric acid and in cracking petroleum products. There is also much current use of platinum as a catalyst in fuel cells and in antipollution devices for automobiles. Fine platinum wire will glow red hot when placed in the vapor of ethyl alcohol. It acts as a catalyst, converting the alcohol to formaldehyde. This phenomenon has been used commercially to produce cigarette lighters and hand warmers. Hydrogen and oxygen explode in the presence of platinum. The price of platinum has varied widely; more than a century ago it was used to adulterate gold. It was nearly eight times as valuable as gold in 1920. The price in December 1985 was about $360/oz t.

**Plutonium** — (planet Pluto), Pu; at wt (244); at no 94; isotopic mass $^{239}Pu$ 239.0522; sp gr (α modification) 19.84 (25°C); mp 641°C; bp 3232°C; valence 3, 4, 5, or 6. Plutonium was the second transuranium element of the actinide series to be discovered. The isotope $^{238}Pu$ was produced in 1940 by Seaborg, McMillan, Kennedy, and Wahl by deuteron bombardment of uranium in the 60-in. cyclotron at Berkeley, California. Plutonium also exists in trace quantities in naturally occurring uranium ores. It is formed in much the same manner as neptunium, by irradiation of natural uranium with the neutrons which are present. By far of the greatest importance is the isotope $^{239}Pu$, with a half-life of 24,400 years, produced in extensive quantities in nuclear reactors from natural uranium.

$$^{238}U \ (n\gamma) \ ^{239}U \xrightarrow{\beta} \ ^{239}Np \xrightarrow{\beta} \ ^{239}Pu$$

Fifteen isotopes of plutonium are known. Plutonium has assumed the position of dominant importance among the transuranium elements because of its successful use as an explosive ingredient in nuclear weapons and the place that it holds as a key material in the development of industrial use of nuclear power. One kilogram is equivalent to about 22 million kilowatt hours of heat energy. The complete detonation of a kilogram of plutonium produces an explosion equal to about 20,000 tons of chemical explosive. Its importance depends on the nuclear property of being readily fissionable with neutrons and its availability in quantity. The world's nuclear power reactors are now producing about 20,000 kg of plutonium per year. By 1982 it was estimated that about 300,000 kg had accumulated. The various nuclear applications of plutonium are well known. $^{238}$Pu was used in the Apollo lunar missions to power equipment on the lunar surface. As with neptunium and uranium, plutonium metal can be prepared by reduction of the trifluoride with alkaline-earth metals. The metal has a silvery appearance and takes on a yellow tarnish when slightly oxidized. It is chemically reactive. A relatively large piece of plutonium is warm to the touch because of the energy given off in alpha decay. Larger pieces will produce enough heat to boil water. The metal readily dissolves in concentrated hydrochloric acid, hydroiodic acid, or perchloric acid with formation of the $Pu^{3+}$ ion. The metal exhibits six allotropic modifications having various crystalline structures. The densities of these vary from 16.00 to 19.86 g/cm$^3$. Plutonium also exhibits four ionic valence states in aqueous solutions; $Pu^{3+}$ (blue lavender), $Pu^{4+}$ (yellow brown), $PuO^+$ (pink?), and $PuO^{2+}$ (pink orange). The ion $PuO^+$ is unstable in aqueous solutions, disproportionating into $Pu^{4+}$ and $PuO^{2+}$. The $Pu^{4+}$ thus formed, however, oxidizes the $PuO^+$ into $PuO^{2+}$, itself being reduced to $Pu^{3+}$, giving finally $Pu^{3+}$ and $PuO^{2+}$. Plutonium forms binary compounds with oxygen: $PuO$, $PuO_2$, and intermediate oxides of variable composition; with the halides: $PuF_3$, $PuF_4$, $PuCl_3$, $PuBr_3$, $PuI_3$; with carbon, nitrogen, and silicon: $PuC$, $PuN$, $PuSi_2$. Oxyhalides are also well known; $PuOCl$, $PuOBr$, $PuOI$. Because of the high rate of emission of alpha particles and the element being specifically absorbed by bone marrow, plutonium, as well as all of the other transuranium elements except neptunium, is a radiological poison and must be handled with special equipment and precautions. Plutonium is a very dangerous radiobiological hazard. Scientists at Berkeley believe that linear catechoylamide carboxylate has potential for removing plutonium from living tissue and bone as well as from nuclear reactor wastes. Precautions must also be taken to prevent the unintentional formation of a critical mass. Plutonium in liquid solution is more likely to become critical than solid plutonium. The shape of the mass must also be considered where criticality is concerned. Plutonium-238 is available

from the A.E.C. at a cost of about \$700/g (80 to 89% enriched).

**Polonium** — (*Poland*, native country of Mme. Curie), Po: at mass (209); at no 84; mp 254°C; bp 962°C; sp gr (alpha modification) 9.32; valence $-2$, 0, $+2$, $+3$(?), $+4$, and $+6$. Polonium was the first element discovered by Marie Curie in 1898, while she was seeking the cause of radioactivity of pitchblende from Joachimsthal, Bohemia. The electroscope showed it separating with bismuth. Polonium is also called ''radium F''. Polonium is a very rare natural element. Uranium ores contain only about 100 μg of the element per ton. Its abundance is only about 0.2% of that of radium. In 1934 it was found that when natural bismuth ($^{209}$Bi) was bombarded by neutrons, $^{210}$Bi, the parent of polonium, was obtained. Milligram amounts of polonium may now be prepared in this way, by using the high neutron fluxes of nuclear reactors. Polonium-210 is a low-melting, fairly volatile metal, 50% of which is vaporized in air in 45 hr at 55°C. It is an alpha emitter with a half-life of 138.39 days. A milligram emits as many alpha particles as 5 g of radium. The energy released by its decay is so large (27.5 cal/Ci/day or 140 W/g) that a capsule containing about half a gram reaches a temperature above 500°C. The capsule also presents a contact gamma-ray dose rate of 1.2 R/hr. A few curies of polonium exhibit a blue glow, caused by excitation of the surrounding gas. Because almost all alpha radiation is stopped within the solid source and its container, giving up its energy, polonium has attracted attention for uses as a lightweight heat source for thermoelectric power in space satellites. Polonium has more isotopes than any other element. Twenty-seven isotopes of polonium are known, with atomic masses ranging from 192 to 218. Polonium-210 is the most readily available. Isotopes of mass 209 (half-life 103 years) and mass 208 (half-life 2.9 years) can be prepared by alpha, proton, or deuteron bombardment of lead or bismuth in a cyclotron, but these are expensive to produce. Metallic polonium has been prepared from polonium hydroxide and some other polonium compounds in the presence of concentrated aqueous or anhydrous liquid ammonia. Two allotropic modifications are known to exist. Polonium is readily dissolved in dilute acids, but is only slightly soluble in alkalis. Polonium can be mixed or alloyed with beryllium to provide a source of neutrons. Polonium-210 is very dangerous to handle in even microgram amounts, and special equipment and strict control is necessary. Damage arises from the complete absorption of the energy of the alpha particle into tissue. The maximum permissible body burden for ingested polonium is only 0.03 μCi, which represents a particle weighing only $6.8 \times 10^{-12}$ g. Weight for weight it is about $2.5 \times 10^{11}$ times as toxic as hydrocyanic acid. The maximum allowable concentration for soluble polonium compounds in air is about $2 \times 10^{-11}$ μCi/cm$^3$. Polonium is available commercially on special order with an A.E.C. permit from the Oak Ridge National Laboratory.

**Potassium** — (English, *potash* — pot ashes; L. *kalium*, Arab. *qali*, alkali), K; at wt 39.0983; at no 19; mp 63.25°C; bp 760°C; sp gr 0.862 (20°C); valence 1. Discovered in 1807 by Davy, who obtained it from caustic potash (KOH); this was the first metal isolated by electrolysis. The metal is the seventh most abundant and makes up about 2.4% by weight of the crust of the earth. Most potassium minerals are insoluble and the metal is obtained from them only with great difficulty. Certain minerals, however, such as sylvite, carnallite, langbeinite, and polyhalite are found in ancient lake and sea beds and form rather extensive deposits from which potassium and its salts can readily be obtained. Potash is mined in Germany, New Mexico, California, Utah, and elsewhere. Large deposits of potash, found at a depth of some 3000 ft in Saskatchewan, promise to be important in coming years. Potassium is also found in the ocean, but is present only in relatively small amounts, compared to sodium. The greatest demand for potash has been in its use for fertilizers. Potassium is an essential constituent for plant growth and it is found in most soils. Potassium is never found free in nature, but is obtained by electrolysis of the hydroxide, much in the same manner as prepared by Davy. Thermal methods also are commonly used to produce potassium (such as by reduction of potassium compounds with CaC$_2$, C, Si, or Na). It is one of the most reactive and electropositive of metals; except for lithium, it is the lightest known metal. It is soft, easily cut with a knife, and is silvery in appearance immediately after a fresh surface is exposed. It rapidly oxidizes in air and must be preserved in a mineral oil such as kerosene. As with other metals of the alkali group, it decomposes in water with the evolution of hydrogen. It catches fire spontaneously on water. Potassium and its salts impart a violet color to flames. Nine isotopes of potassium are known. Ordinary potassium is composed of three isotopes, one of which is $^{40}$K (0.0118%), a radioactive isotope with a half-life of $1.28 \times 10^9$ years. The radioactivity presents no appreciable hazard. An alloy of sodium and potassium (NaK) is used as a heat-transfer medium. Metallic potassium is available commercially for about $15/oz in small quantities.

**Praseodymium** — (Gr. *prasios*, green, and *didymos*, twin), Pr; at wt 140.9077; at no 59; mp 931°C; bp 3512°C; sp gr (α) 6.773 (β) 6.64; valence 3 or 4. In 1841, Mosander extracted the rare earth didymia from lanthana; in 1879, Lecoq de Boisbaudran isolated a new earth, samaria, from didymia obtained from the mineral samarskite. Six years later, in 1885, von Welsbach separated didymia into two other earths, praseodymia and neodymia, which gave salts of different colors. As with other rare earths, compounds of these elements in solution have distinctive sharp spectral absorption bands or lines, some of which are only a few Angstroms wide. The element occurs along with other rare-earth elements in a variety of minerals. Monazite and bastnasite are the two principal commercial sources of the rare earth metals. Ion-exchange and solvent extraction techniques have led to much easier isolation of the rare earths and the cost has dropped greatly in the past few years. Praseodymium can be prepared by several methods, such as by calcium reduction of the anhydrous chloride or fluoride. Misch metal, used in making cigarette lighters, contains about 5% praseodymium metal. Praseodymium is soft, silvery, malleable, and ductile. It was prepared in relatively pure form in 1931. It is somewhat more resistant to corrosion in air than europium, lanthanum, cerium, or neodymium, but it does develop a green oxide coating that spalls off when exposed to air. As with other rare-earth metals, it should be kept under a light mineral oil or sealed in plastic. The rare earth oxides, including Pr$_2$O$_3$, are among the most refractory substances known. Salts of praseodymium are used to color glasses and enamels. Didymium glass, of which praseodymium is a component, is a colorant for welder's goggles. The metal (99 + % pure) is priced at about $1.00/g.

**Promethium** — (*Prometheus* who, according to mythology, stole fire from heaven), Pm; at no 61; at wt (145); mp 1168 ± 6°C; bp 2460°C; sp gr 7.22 ± 0.02 (25°C); valence 3. In 1902 Branner predicted the existence of an element between neodymium and samarium, and this was confirmed by Moseley in 1914. In 1941, workers at Ohio State

University irradiated neodymium and praseodymium with neutrons, deuterons, and alpha particles, respectively, and produced several new radioactivities, which most likely were those of element 61. Wu and Segre, and Bethe, in 1942, confirmed the formation; however, chemical proof of the production of element 61 was lacking because of the difficulty in separating the rare earths from each other at that time. In 1945, Marinsky, Glendenin, and Coryell made the first chemical identification by use of ion-exchange chromatography. Their work was done by fission of uranium and by neutron bombardment of neodymium. Searches for the element on earth have been fruitless, and it now appears that promethium is completely missing from the earth's crust. Promethium, however, has been identified in the spectrum of the star HR[465] in Andromeda. This element is being or has been formed recently near the surface of the star, for no known isotope of promethium has a half-life longer than 17.7 years. Thirteen isotopes of promethium, with atomic masses from 141 to 154, are now known. Promethium-147, with a half-life of 2.5 years, is the most generally useful. Promethium-145 is the longest lived, and has a specific activity of 940 Ci/g. It is a soft beta emitter; although no gamma rays are emitted, X-radiation can be generated when beta particles impinge on elements of a high atomic number, and great care must be taken in handling it. Promethium salts luminesce in the dark with a pale blue or greenish glow, due to their high radioactivity. Ion-exchange methods led to the preparation of about 10 g of promethium from atomic reactor fuel processing wastes in early 1963. Little is yet generally known about the properties of metallic promethium. Two or more allotropic modifications are thought to exist. The element has applications as a beta source for thickness gages, and it can be absorbed by a phosphor to produce light. It can be used as a nuclear-powered battery by capturing light in photocells which convert it into electric current. Such a battery, using [147]Pm, would have a useful life of about 5 years. More than 30 promethium compounds have been prepared, most are colored. Promethium-147 is available to A.E.C. licensees at a cost of about 50¢/Ci.

**Protactinium** — (Gr. *protos*, first), Pa; at wt 231.0359; at no 91; mp < 1600°C; bp — ; sp gr 15.37 (calc); valence 4 or 5. The first isotope of element 91 to be discovered was [234m]Pa, also known as UX$_2$, a short-lived member of the naturally occurring [238]U decay series. It was identified by K.

Fajans and O. H. Gohring in 1913 and they named the new element brevium. When the longer-lived isotope [231]Pa was identified by Hahn and Meitner in 1918, the name protactinium was adopted as being more consistent with the characteristics of the most abundant isotope. Soddy, Cranston, and Fleck were also active in this work. The name protoactinium was shortened to protactinium in 1949. In 1927, Grosse prepared 2 mg of a white powder, which was shown to be Pa$_2$O$_5$. Later, in 1934, from 0.1 g of pure Pa$_2$O$_5$ he isolated the element by two methods, one of which was by converting the oxide to an iodide and "cracking" it in a high vacuum by an electrically heated filament by the reaction

$$2PaI_5 \rightarrow 2Pa + 5I_2$$

Protactinium has a bright metallic luster which it retains for some time in air. The element occurs in pitchblende to the extent of about 1 part [231]Pa to 10 million parts of ore. Ores from Zaire have about 3 ppm. Protactinium has 13 isotopes, the most common of which is [231]Pr with a half-life of 32,500 years. A number of protactinium compounds are known, some of which are colored. The element is superconductive below 1.4 K. An indirect measurement indicates that protactinium has a vapor pressure of 5.1 × 10$^{-5}$ at 1927°C. The element is a dangerous toxic material and requires precautions similar to those used when handling plutonium. In 1959 and 1961, it was announced that the Great Britain Atomic Energy Authority extracted by a 12-stage process 125 g of 99.9% protactinium, the world's only stock of the metal for many years to come. The extraction was made from 60 tons of waste material at a cost of about $500,000. Protactinium is one of the rarest and most expensive naturally occurring elements. O.R.N.L. supplies [231]Pr at a cost of about $280/g. The element is an alpha emitter (5.0 MeV), and is a radiological hazard similar to polonium.

**Radium** — (L. *radius*, ray), Ra; at wt 226.0254; at no 88; mp 700°C; bp 1140°C; sp gr 5?; valence 2. Radium was discovered in 1898 by M. and Mme. Curie in the pitchblende or uraninite of North Bohemia, in which it occurs. There is about 1 g of radium in 7 tons of pitchblende. The element was isolated in 1911 by Mme. Curie and Debierne by the electrolysis of a solution of pure radium chloride, employing a mercury cathode; on distillation in an atmosphere of hydrogen this amalgam yielded the pure metal. Originally, radium was obtained from the rich pitchblende ore found at Joachimsthal, Bohemia. The carnotite sands of Colorado furnish some radium, but richer ores are found in the Republic of Zaire and the Great Bear Lake region in Canada. Radium is present in all uranium minerals, and could be extracted from the extensive wastes of uranium processing. Large uranium deposits are located in Ontario, New Mexico, Utah, Australia, and elsewhere. Radium is obtained commercially as the bromide or chloride; it is doubtful if any appreciable stock of the isolated element now exists. The pure metal is brilliant white when freshly prepared, but blackens on exposure to air, probably due to formation of the nitride. It exhibits luminescence, as do its salts; it decomposes in water and is somewhat more volatile than barium. It is a member of the alkaline-earth group of metals. Radium imparts a carmine red color to a flame. Radium emits alpha, beta, and gamma rays and when mixed with beryllium produces neutrons. One gram of $^{226}$Ra undergoes $3.7 \times 10^{10}$ disintegrations per sec. The curie (Ci) is defined as that amount of radioactivity which has the same disintegration rate as 1 g of $^{226}$Ra. Sixteen isotopes are now known; radium-226, the common isotope, has a half-life of 1620 years. One gram of radium produces about 0.0001 ml(stp) of emanation, or radon gas, per day. This is pumped from the radium and sealed in minute tubes, which are used in the treatment of cancer and other diseases. More recently discovered radioisotopes, such as $^{60}$Co, are now being used in place of radium. Radium loses about 1% of its activity in 25 years, being transformed into elements of lower atomic weight. Lead is a final product of disintegration. The study of radium greatly altered our ideas of the structure of the atom. Radium is a radiological hazard. (Stored radium should be ventilated to prevent build-up of radon.) Inhalation, injection, or body exposure to radium can cause cancer and other body disorders. The maximum permissible burden in the total body for $^{226}$Ra is 0.2 μCi (microcuries).

**Radon** — (from radium; called niton at first, L. *nitens*, shining), Rn; at wt (~222); at no 86; mp −71°C; bp −61.8°C; density of gas 9.73 g/l; sp gr liquid 4.4 at −62°C, solid 4; valence usually 0. The element was discovered in 1900 by Dorn, who called it radium emanation. In 1908 Ramsay and Gray, who named it niton, isolated the element and determined its density, finding it to be the heaviest known gas. It is essentially inert and oc-

cupies the last place in the zero group of gases in the Periodic Table. Since 1923, it has been called radon. Twenty isotopes are known. Radon-222, coming from radium, has a half-life of 3.823 days and is an alpha emitter; radon-220, emanating naturally from thorium and called thoron, has a half-life of 54.5 sec and is also an alpha emitter. Radon-219 emanates from actinium and is called actinon. It has a half-life of 3.92 sec and is also an alpha emitter. It is estimated that every square mile of soil to a depth of 6 in. contains about 1 g radium, which releases radon in tiny amounts to the atmosphere. Radon is present in some spring waters, such as those as Hot Springs, Arkansas. On the average, one part of radon is present to 1 sextillion parts of air. At ordinary temperatures radon is a colorless gas; when cooled below the freezing point, radon exhibits a brilliant phosphorescence which becomes yellow as the temperature is lowered and orange-red at the temperature of liquid air. It has been reported that fluorine reacts with radon, forming radon fluoride. Radon clathrates have also been reported. Radon is still produced for therapeutic use by a few hospitals by pumping it from a radium source and sealing it in minute tubes (seeds). This practice has now been largely discontinued as hospitals can order the seeds directly from suppliers, who make up the seeds with the desired activity for the day of use. Radon is available at a cost of about $4/mCi. Care must be taken in handling radon, as with other radioactive materials. The main hazard is from inhalation of the element and its solid daughters, which are collected on dust in the air. The maximum permissible concentration of $^{222}$Rn in air has been set at $3 \times 10^{-8}$ μCi/cc (lung) for an 8-hr day, 40-hr work week. Good ventilation should be provided where radium, thorium, or actinium is stored to prevent build-up of this element. Radon build-up is also a health consideration in uranium mines. Recently, concern for radon build-up in homes as a health hazard has led to widespread public interest and measurement of radon levels with sealing and ventilation in homes urged where safe concentration limits are exceeded. The current safe limit is 4 pCi/1.

**Rhenium** — (L. *Rhenus*, Rhine), Re; at wt 186.207; at no 75; mp 3180°C; bp 5627°C (est.); sp gr 21.02 (20°C); valence −1, +1, 2, 3, 4, 5, 6, and 7. Discovery of rhenium is generally attributed to Noddack, Tacke, and Berg, who announced in 1925 they had detected the element in platinum ores and columbite. They also found the element in gadolinite and molybdenite. By working up 660 kg of molybdenite, they were able in 1928 to extract 1 g of rhenium. The price in 1928 was $10,000/g. Rhenium does not occur free in nature or as a compound in a distinct mineral species. It is, however, widely spread throughout the earth's crust to the extent of about 0.001 ppm. Commercial rhenium in the U.S. today is obtained from molybdenite roaster-flue dusts obtained from copper-sulfide ores mined in the vicinity of Miami, Arizona, and elsewhere in Arizona and Utah. Some molybdenites contain from 0.002 to 0.2% rhenium. More than 150,000 troy ounces of rhenium are now being produced yearly in the United States. The total estimated non-Communist World reserve of rhenium metal is 3500 tons. Natural rhenium is a mixture of two stable isotopes. Sixteen other unstable isotopes are recognized. Rhenium metal is prepared by reducing ammonium perrhenate with hydrogen at elevated temperatures. The element is silvery white with a metallic luster; its density is exceeded only by that of platinum, iridium, and osmium, and its melting point is exceeded only by that of tungsten and carbon. It has other useful properties. The usual commercial form of the element is a powder, but it can be consolidated by pressing and resistance-sintering in a vacuum or hydrogen atmosphere. This produces a compact shape in excess of 90% of the density of the metal. Annealed rhenium is very ductile, and can be bent, coiled, or rolled. Rhenium is used as an additive to tungsten and molybdenum-based alloys. It is used for filaments for mass spectrographs and ion gages. Rhenium-molybdenum alloys are superconductive at 10 K. Rhenium is also used as an electrical contact material as it has good wear resistance and withstands arc corrosion. Thermocouples made of Re-W are used for measuring temperatures up to 2200°C, and rhenium wire is used in photoflash lamps for photography. Rhenium catalysts are exceptionally resistant to poisoning from nitrogen, sulfur, and phosphorus, and are used for hydrogenation of fine chemicals, hydrocracking, reforming, and the disproportionation of olefins. Rhenium costs about $250/oz t. Little is known of its toxicity; therefore, it should be handled with care until more data are available.

**Rhodium** — (Gr. *rhodon*, rose), Rh; at wt 102.9055; at no 45; mp 1966 ± 3°C; bp 3727 ± 100°C; sp gr 12.41 (20°C); valence 2, 3, 4, 5, and 6. Wollaston discovered rhodium in 1803—1804 in crude platinum ore he presumably obtained from South America. Rhodium occurs native with other platinum metals in river sands of the Urals and in North and South America. It is also found with other platinum metals in the copper-nickel sulfide ores of the Sudbury, Ontario region. Although the quantity occurring here is very small, the large tonnages of nickel processed make the recovery commercially feasible. The annual world production of rhodium is only 7 or 8 tons. The metal is silvery white and at red heat slowly changes in air to the sesquioxide. At higher temperatures, it converts back to the element. Rhodium has a higher melting point and lower density than platinum. Its major use is as an alloying agent to harden platinum and palladium. It is useful as an electrical contact material because it has low electrical resistance, and is highly resistant to corrosion. Plated rhodium, produced by electroplating or evaporation, is exceptionally hard and is used for optical instruments. It has a high reflectance and is hard and durable. Rhodium is also used as a catalyst. Exposure to rhodium (metal fume and dust, as Rh) should not exceed 0.1 mg/m³ (8-hr time-weighted average, 40-hr week). Soluble salts should not exceed 0.001 mg/m³. Rhodium costs about $800/oz t.

**Rubidium** — (L. *rubidus*, deepest red), Rb; at wt 85.4678 ± 3; at no 37; mp 38.89°C; bp 686°C; sp gr (solid) 1.532 (20°C), (liquid) 1.475 (39°C); valence 1, 2, 3, 4. Discovered in 1861 by Bunsen and Kirchoff in the mineral lepidolite by use of the spectroscope. The element is much more abundant than it was thought to be several years ago. It is now considered to be the 16th most abundant element in the earth's crust. Rubidium occurs in pollucite, carnallite, leucite, and zinnwaldite, which contains traces up to 1%, in the form of the oxide. It is found in lepidolite to the extent of about 1.5%, and is recovered commercially from this source. Potassium minerals, such as those found at Searles Lake, California, and potassium chloride recovered from brines in Michigan also contain the element and are commercial sources. It is found along with cesium in the extensive deposits of pollucite at Bernic Lake, Manitoba. Rubidium can be liquid at room temperature. It is a soft, silver white metallic element of the alkali group and is the second most electropositive and alkaline element. It ignites spontaneously in air and reacts violently in water, setting fire to the liberated hydrogen. As with other alkali metals, it forms amalgams with mercury and it alloys with gold, cesium, sodium, and potassium. It colors a flame yellowish violet. Rubidium metal can be prepared by reducing rubidium chloride with calcium. It must be kept under a dry mineral oil or in a vacuum or inert atmosphere. Seventeen isotopes of rubidium are known. Naturally occurring rubidium is made of two isotopes, $^{85}$Rb and $^{87}$Rb. Rubidium-87 is present to the extent of 27.85% in natural rubidium and is a beta emitter with a half-life of $5 \times 10^{11}$ years. Ordinary rubidium is sufficiently radioactive to expose a photographic film in about 30 to 60 days. Rubidium forms four oxides: $Rb_2O$, $Rb_2O_2$, $Rb_2O_3$, $Rb_2O_4$. Because rubidium can be easily ionized, it has been considered for use in "ion engines" for space vehicles; however, cesium is somewhat more efficient for this purpose. Rubidium is used as a getter in vacuum tubes and as a photocell component. It has been used in making special glasses. $RbAg_4I_5$ is important, as it has the highest conductivity of any known ionic crystal. At 20°C its conductivity is about the same as dilute sulfuric acid. The present cost in small quantities is about $20/g.

**Ruthenium** — (L. *Ruthenia*, Russia), Ru; at wt 101.07 ± 3; at no 44; mp 2310°C; bp 3900°C; sp gr 12.41 (20°C); valence 0, 1, 2, 3 4, 5, 6, 7, or 8. Berzelius and Osann in 1827 examined the residues left after dissolving crude platinum from the Ural mountains in aqua regia. While Berzelius found no unusual metals, Osann thought he found three new metals, one of which he named ruthenium. In 1844, Klaus, generally recognized as the discoverer, showed that Osann's ruthenium oxide was very impure and that it contained a new metal. Klaus obtained 6 g of ruthenium from the portion of crude platinum that is insoluble in aqua regia. A member of the platinum group, ruthenium occurs native with other members of the group in ores found in the Ural mountains and in North and South America. It is also found along with other platinum metals in small but commercial quantities in pentlandite of the Sudbury, Ontario, nickel-mining region, and in pyroxinite deposits of South Africa. The metal is isolated commercially by a complex chemical process, the final stage of which is the hydrogen reduction of ammonium ruthenium chloride, which yields a powder. The powder is consolidated by powder metallurgy techniques or by argon-arc welding. Ruthenium is a hard, white metal and has four crystal modifications. It does not tarnish at room temperatures, but oxidizes in air at about 800°C. The metal is not attacked by hot or cold acids or aqua regia, but when potassium chlor-

ate is added to the solution, it oxidizes explosively. It is attacked by halogens, hydroxides, etc. Ruthenium can be plated by electrodeposition or by thermal decomposition methods. The metal is one of the most effective hardeners for platinum and palladium and is alloyed with these metals. A ruthenium-molybdenum alloy is said to be superconductive at 10.6 K. The corrosion resistance of titanium is improved a hundredfold by addition of 0.1% ruthenium. It is a versatile catalyst. Hydrogen sulfide can be split catalytically by light, using an aqueous suspension of CdS particles loaded with ruthenium dioxide. Compounds in at least eight oxidation states have been found, but of these, the +2, +3, and +4 states are the most common. Ruthenium tetroxide, like osmium tetroxide, is highly toxic. In addition, it may explode. Ruthenium compounds show a marked resemblance to those of osmium. The metal is priced at about $4/g or $60/oz t.

**Rutherfordium** — See **Element 104.**

**Samarium** — (*Samarskite*, a mineral), Sm; at wt 150.36 ± 3; at no 62; mp 1077 ± 5°C; bp 1791°C; sp gr (α) 7.520 (β) 7.40; valence 2 or 3. Discovered spectroscopically by its sharp absorption lines in 1879 by Lecoq de Boisbaudran in the mineral samarskite, named in honor of a Russian mine official, Col. Samarski. Samarium is found along with other members of the rare earth elements in many minerals, including monazite and bastnasite, which are commercial sources. It occurs in monazite to the extent of 2.8%. While misch metal, containing about 1% of samarium metal, has long been used, samarium had not been isolated in relatively pure form until recent years. Ion-exchange and solvent extraction techniques have simplified separation of the rare earths from one another. Recently, electrochemical deposition, using an electrolytic solution of lithium citrate and a mercury electrode has been reported to be a simple, fast, and highly specific way to separate the rare earths. Samarium metal can be produced by reducing the oxide with barium or lanthanum. Samarium has a bright silver luster and is reasonably stable in air. Two crystal modifications of the metal exist, with a transformation point at 917°C. The metal ignites in air at about 150°C. Sixteen isotopes of samarium exist. Natural samarium is a mixture of seven isotopes, three of which are unstable with long half-lives. The sulfide has excellent high-temperature stability and good thermoelectric efficiencies up to 1100°C. $SmCo_5$ has been used in making a new permanent magnet material with the highest resistance to demagnetization of any known material. It has an intrinsic coercive force as high as 28,000 oersteds. Samarium oxide has been used in optical glass to absorb the infrared. Samarium is used to dope calcium fluoride crystals for use in optical masers or lasers. Compounds of the metal act as sensitizers for phosphors excited in the infrared. The metal is priced at about $5/g. Little is known of the toxicity of samarium; therefore, it should be handled carefully.

**Scandium** — (L. *Scandia*, Scandinavia), Sc; at wt 44.9559; at no 21; mp 1541°C; bp 2831°C; sp gr 2.989 (25°C); valence 3. On the basis of the Periodic System, Mendeleev predicted the existence of *ekaboron*, which would have an atomic weight between the 40 of calcium and 48 of titanium. The element was discovered by Nilson in 1876 in the minerals euxenite and gadolinite, which had not yet been found anywhere except in Scandinavia. By processing 10 kg of euxenite and other residues of rare earth minerals, Nilson was able to prepare about 2 g of scandium oxide of high purity. Cleve later pointed out that Nilson's scandium was identical with Mendeleev's ekaboron. Scandium is apparently a much more abundant element in the sun and certain stars than here on earth. It is about the 23rd most abundant element in the sun, compared to the 50th most abundant on earth. It is widely distributed on earth, occurring in very minute quantities in over 800 mineral species. The blue color of beryl (aquamarine variety) is said to be due to scandium. It occurs as a principal component in the rare mineral thortveitite, found in Scandinavia and Malagasy. It is also found in the residues remaining after the extraction of tungsten from Zinnwald wolframite, and in wiikite and bazzite. Most scandium is presently being recovered from thortveitite or is extracted as a by-product from uranium mill tailings. Metallic scandium was first prepared in 1937 by Fischer, Brunger, and Grieneisen, who electrolyzed a eutectic melt of potassium, lithium, and scandium chlorides at 700 to 800°C. Pure scandium is now produced by reducing scandium fluoride with calcium metal. The production of the first pound of 99% pure scandium metal was announced in 1960 as having been made under a U.S. Air Force contract. Scandium is a silvery white metal which develops a slightly yellowish or pinkish cast upon exposure to air. It is relatively soft, and resembles yttrium and the rare-earth metals more than it resembles aluminum or titanium. It is a very light metal and has a higher melting point than aluminum. Scandium is not attacked by a 1:1 mixture of conc $HNO_3$ and 48% HF, but reacts rapidly with many other acids. Eleven isotopes of scandium are recognized. The metal is expensive, costing about $50/g with a purity of about 99.9%. Scandium oxide costs about $15/g. About 20 kg of scandium (as $Sc_2O_3$) is now being used yearly in the U.S. to produce high-intensity lights and the radioactive isotope $^{46}Sc$ is used as a tracing agent in refinery crackers. Scandium iodide added to mercury vapor lamps produces a highly efficient light source resembling sunlight, which is important for indoor or nighttime color TV transmission. Little is yet known about the toxicity of scandium; therefore, it should be handled with care.

**Selenium** — (Gr. *Selene*, moon), Se; at wt 78.96 ± 3; at no 34; mp (gray) 217°C; bp (gray) 684.9 ± 1.0°C; sp gr (gray) 4.79, (vitreous) 4.28; valence $-2$, $+4$, or $+6$. Discovered in 1817 by Berzelius, who found it associated with tellurium, named for the earth. Selenium is fund in a few rare minerals, such as crooksite and clausthalite. In years past it has been obtained from flue dusts remaining from processing copper sulfide ores, but the anode muds from electrolytic copper refineries now provide the source of most of the world's selenium. Selenium is recovered by roasting the muds with soda or sulfuric acid, or by smelting them with soda and niter. Selenium exists in several allotropic forms. Three are generally recognized, but as many as six have been claimed. Selenium can be prepared with either an amorphous or a crystalline structure. The color of amorphous selenium is either red, in powder form, or black, in vitreous form. Crystalline monoclinic selenium is a deep red; crystalline hexagonal selenium, the most stable variety, is the metallic gray. Natural selenium contains six stable isotopes. Fourteen other nuclides and isomers have been characterized. The element is a member of the sulfur family and resembles sulfur both in its various forms and in its compounds. Selenium exhibits both photovoltaic action, where light is converted directly into electricity, and photoconductive action, where the electrical resistance decreases with increased illumination. These properties make selenium useful in the production of photocells and exposure meters for photographic use, as well as in solar cells. Selenium is also able to convert AC electricity to DC and is extensively used in rectifiers. Below its melting point, selenium is a p-type semiconductor and has many uses in electronic and solid-state applications. It is used in xerography for reproducing and copying and in the glass industry to decolorize glass and to make ruby-colored glasses and enamels. Elemental selenium has been said to be practically nontoxic and is considered to be an essential trace element; however, hydrogen selenide and other selenium compounds are extremely toxic, and resemble arsenic in their physiological reactions. Hydrogen selenide in a concentration of 1.5 ppm is intolerable to man. Selenium occurs in some soils in amounts sufficient to produce serious effects on animals feeding on plants, such as locoweed. Exposure to selenium compounds (as Se) in air should not exceed 0.2 mg/m$^3$ (8-hr time-weighted average, 40-hr week). Selenium is priced at about $20/lb. It is also available in high-purity form at a somewhat higher cost.

**Silicon** — (L. *silex, silicis,* flint), Si; at wt 28.0855 ± 3; at no 14; mp 1410°C; bp 2355°C; sp gr 2.33 (25°C); valence 4. Davy in 1800 thought silica to be a compound and not an element; later, in 1811, Gay Lussac and Thenard probably prepared impure amorphous silicon by heating potassium with silicon tetrafluoride. Berzelius, generally credited with the discovery, in 1824 succeeded in preparing amorphous silicon by the same general method as used earlier, but he purified the product by removing the fluosilicates by repeated washings. Deville in 1854 first prepared crystalline silicon, the second allotropic form of the element. Silicon is present in the sun and stars and is a principal component of a class of meteorites known as aerolites. It is also a component of tektites, a natural glass of uncertain origin. Silicon makes up 25.7% of the earth's crust, by weight, and is the second most abundant element, being exceeded only by oxygen. Silicon is not found free in nature, but occurs chiefly as the oxide, and as silicates. Sand, quartz, rock crystal, amethyst, agate, flint, jasper, and opal are some of the forms in which the oxide appears. Granite, hornblende, asbestos, feldspar, clay, mica, etc. are but a few of the numerous silicate minerals. Silicon is prepared commercially by heating silica and carbon in an electric furnace, using carbon electrodes. Amorphous silicon can be prepared as a brown powder, which can be easily melted or vaporized. Crystalline silicon has a metallic luster and grayish color. The Czochralski process is commonly used to produce single crystals of silicon used for solid-state or semiconductor devices. Hyperpure silicon can be prepared by the thermal decomposition of ultrapure trichlorosilane in a hydrogen atmosphere, and by a vacuum float zone process. This product can be doped with boron, gallium, phosphorus, or arsenic, etc. to produce silicon for use in transistors, solar cells, rectifiers, and other solid-state devices which are used extensively in the electronics and space-age industries. Silicon is a relatively inert element, but it is attacked by halogens and dilute alkali. Most acids, except hydrofluoric, do not affect it. Silicones are important products of silicon. They may be prepared by hydrolyzing a silicon organic chloride, such as dimethyl silicon chloride. Hydrolysis and condensation of various substituted chlorosilanes can be used to produce a very great number of polymeric products, or silicones, ranging from liquids to hard, glasslike solids with many useful properties. Elemental silicon transmits more than 95% of all wavelengths of infrared, from 1.3 to 6.7 μm. Silicon is one of man's most useful elements. In the form of sand and clay it is used to make concrete and brick; it is a useful refractory material for high-temperature work, and in the form of silicates it is used in making enamels, pottery, etc. Silica, as sand, is a principal ingredient of glass, one of the most inexpensive of materials with excellent mechanical, optical, thermal, and electrical properties. Silicon is important in plant and animal life. Diatoms in both fresh and salt water extract silica from the water to build up their cell walls. Silica is present in ashes of plants and in the human skeleton. Silicon is an important ingredient in steel; silicon carbide is one of the most important abrasives and has been used in lasers to produce coherent light of 4560 Å. Regular grade silicon (97%) costs about 20¢/lb. Silicon 99.7% pure costs about $7/lb; hyperpure silicon may cost as much as $100/lb. Miners, stonecutters, and others engaged in work where siliceous dust is breathed in large quantities may develop a serious lung disease known as silicosis unless using precautions such as air filtration masks.

**Silver** — (Anglo-Saxon, *Seolfor siolfur*; L. *argentum*) Ag; at wt 107.8682 ± 3; at no 47; mp 961.93°C; bp 2212°C; sp gr 10.50 (20°C); valence 1, 2. Silver has been known since ancient times. It is mentioned in Genesis. Slag dumps in Asia Minor and on islands in the Aegean Sea indicate that man learned to separate silver from lead as early as 3000 B.C. Silver occurs native and in ores such as argentite ($Ag_2S$) and horn silver (AgCl); lead, lead-zinc, copper, gold, and copper-nickel ores are principal sources. Mexico, Canada, Peru, and the U.S. are the principal silver producers in the western hemisphere. Silver is also recovered during electrolytic refining of copper. Commercial fine silver contains at least 99.9% silver. Purities of 99.999 + % are available commercially. Pure silver has a brilliant white metallic luster. It is a little harder than gold and is very ductile and malleable, being exceeded only by gold and perhaps palladium. Pure silver has the highest electrical and thermal conductivity of all metals, and possesses the lowest contact resistance. It is stable in pure air and water, but tarnishes when exposed to ozone, hydrogen sulfide, or air containing sulfur. The alloys of silver are important. Sterling silver is used for jewelry, silverware, etc. where appearance is paramount. This alloy contains 92.5% silver, the remainder being copper or some other metal. Silver is of utmost importance in photography, about 30% of the U.S. industrial consumption going into this application. It is used for dental alloys. Silver is used in making solder and brazing alloys, electrical contacts, and high capacity silver-zinc and silver-cadmium batteries. It is used in mirror production and may be deposited on glass or metals by chemical deposition, electrodeposition, or by evaporation. When freshly deposited, it is the best reflector of visible light known, but it rapidly tarnishes and loses much of its reflectance. It is a poor reflector of ultraviolet. Silver fulminate ($Ag_2C_2N_2O_2$), a powerful explosive, is sometimes formed during the silvering process. Silver iodide is used in seeding clouds to produce rain. Silver chloride has interesting optical properties as it can be made transparent. Silver nitrate, or lunar caustic, the most important silver compound, is used extensively in photography. While silver itself is not considered to be toxic, most of its salts are poisonous due to the anions present. Exposure to silver (metal and soluble compounds, as Ag) in air should not exceed 0.01 mg/m$^3$, (8-hr time-weighted average, 40-hr week). Silver compounds can be absorbed in the circulatory system and reduced silver deposited in the various tissues of the body. A condition known as argyria results, with a grayish pigmentation of the skin and mucous membranes. Silver has germicidal effects and kills many lower organisms effectively without harm to higher animals. Silver, for centuries, has been used traditionally for coinage by many countries of the world. In recent times, however, consumption of silver has exceeded the output. In 1939, the price of silver was fixed by the U.S. Treasury at 71¢/oz t, and at 90.5¢/oz t in

1946. In November 1961 the U.S. Treasury suspended sales of nonmonetized silver, and the price stabilized for a time at about $1.29, the melt-down value of silver U.S. coins. The Coinage Act of 1965 authorized a change in the metallic composition of the three U.S. subsidiary denominations to clad or composite type coins. This was the first change in U.S. coinage since the monetary system was established in 1792. Clad dimes and quarters are made of an outer layer of 75% Cu and 25% Ni bonded to a central core of pure Cu. The composition of the one- and five-cent pieces remains unchanged. One-cent coins are 95% Cu and 5% Zn. Five-cent coins are 75% Cu and 25% Ni. Old silver dollars are 90% Ag and 10% Cu. Earlier subsidiary coins of 90% Ag and 10% Cu officially were to circulate alongside the clad coins; however, in practice they have largely disappeared (Gresham's Law), as the value of the silver is now greater than their exchange value. Silver coins of other countries have largely been replaced with coins made of other metals. On June 24, 1968, the U.S. Government ceased to redeem U.S. silver certificates with silver. Since that time, the price of silver has fluctuated widely. The U.S. Government discontinued selling silver to domestic users and foreign buyers on November 10, 1970. As of January 1988, the price of silver was about $6.70/oz t.

**Sodium** — (English, *soda;* Medieval Latin, *sodanum*, headache remedy; L. *natrium*), Na; at wt 22.98977; at no 11; mp 97.81 ± 0.03°C; bp 882.9°C; sp gr 0.971 (20°C); valence 1. Long recognized in compounds, sodium was first isolated by Davy in 1807 by electrolysis of caustic soda. Sodium is present in fair abundance in the sun and stars. The D lines of sodium are among the most prominent in the solar spectrum. Sodium is the sixth most abundant element on earth, comprising about 2.6% of the earth's crust; it is the most abundant of the alkali group of metals of which it is a member. The most common compound is sodium chloride, but it occurs in many other minerals, such as soda niter, cryolite, amphibole, zeolite, sodalite, etc. It is a very reactive element and is never found free in nature. It is now obtained commercially by the electrolysis of absolutely dry fused sodium chloride. This method is much cheaper than that of electrolyzing sodium hydroxide, the method used several years ago. Sodium is a soft, bright, silvery metal which floats on water, decomposing it with the evolution of hydrogen and the formation of the hydroxide. It may or may not ignite spontaneously on water, depending on the amount of oxide and metal exposed to the water. It normally does not ignite in air at temperature below 115°C. Metallic sodium is vital in the manufacture of sodamide and sodium cyanide, sodium peroxide, and sodium hydride. It is used in preparing tetraethyl lead, in the reduction of organic esters, and in the preparation of organic compounds. An alloy of sodium with potassium, NaK, is an important heat transfer agent. Sodium compounds are important to the paper, glass, soap, textile, petroleum, chemical, and metal industries. Soap is generally a sodium salt of fatty acids. The importance of common salt to animal nutrition has been recognized since prehistoric times. Among the many compounds that are of the greatest industrial importance are common salt (NaCl), soda ash ($Na_2CO_3$), baking soda (Na-$HCO_3$), caustic soda (NaOH), Chile saltpeter ($NaNO_3$), di- and trisodium phosphates, sodium thiosulfate (hypo, $Na_2S_2O_3 \cdot 5H_2O$), and borax ($Na_2B_4O_7 \cdot 10H_2O$). Seven isotopes of sodium are recognized. Metallic sodium is priced at about 15 to 20¢/lb in quantity. On a volume basis, it is the cheapest of all metals. Sodium metal should be handled with great care. It should be maintained in an inert atmosphere and contact with water and other substances with which it reacts should be avoided.

**Strontium** — (*Strontian*, town in Scotland), Sr; at wt 87.62; at no 38; mp 769°C; bp 1384°C; sp gr 2.54; valence 2. Isolated by Davy by electrolysis in 1808; however, Adair Crawford in 1790 recognized a new mineral (strontianite) as differing from other barium minerals (baryta). Strontium is found chiefly as celestite ($SrSO_4$) and strontianite ($SrCO_3$). The metal can be prepared by electrolysis of the fused chloride mixed with potassium chloride, or is made by reducing strontium oxide with aluminum in a vacuum at a temperature at which strontium distills off. Three allotropic forms of the metal exist, with transition points at 235 and 540°C. Strontium is softer than calcium and decomposes water more vigorously. It does not absorb nitrogen below 380°C. It should be kept under kerosene to prevent oxidation. Freshly cut strontium has a silvery appearance, but rapidly turns a yellowish color with the formation of the oxide. The finely divided metal ignites spontaneously in air. Volatile strontium salts impart a beautiful crimson color to flames, and these salts are used in pyrotechnics and in the production of flares. Natural strontium is a mixture of four stable isotopes. Twelve other unstable isotopes are known to exist. Of greatest importance is $^{90}S$ with a half-life of 28 years. It is a product of nuclear fallout and presents a health problem. This isotope is one of the best long-lived high-energy beta emitters known, and is used in SNAP (systems for nuclear auxiliary power) devices. These devices hold promise for use in space vehicles, remote weather stations, navigational buoys, etc., where a lightweight, long-lived, nuclear-electric power source is needed. The major use for strontium at present is in producing glass for color television picture tubes. It has also found use in producing ferrite magnets and in refining zinc. Strontium titanate is an interesting optical material as it has an extremely high refractive index and an optical dispersion greater than that of diamond. Strontium metal costs about $6 to $8/lb.

**Sulfur** — (Sanskrit, *sulvere;* L. *sulphurium*), S; at wt 32.06; at no 16; mp (rhombic) 112.8°C, (monoclinic) 119.0°C; bp 444.674°C; sp gr (rhombic) 2.07, (monoclinic) 1.957 (20°C); valence 2, 4, or 6. Known to the ancients; referred to in Genesis as brimstone. Sulfur is found in meteorites. A dark area near the crater Aristarchus on the moon has been studied with ultraviolet light by R. W. Wood. This study suggests strongly that it is a sulfur deposit. Sulfur occurs native in the vicinity of volcanoes and hot springs. It is widely distributed in nature as iron pyrites, galena, sphalerite, cinnabar, stibnite, gypsum, Epsom salts, celestite, barite, etc. Sulfur is commercially recovered from wells sunk into the salt domes along the Gulf Coast of the U.S. It is obtained from these wells by the Frasch process, which forces heated water into the wells to melt the sulfur that is then brought to the surface. Sulfur also occurs in natural gas and petroleum crudes and must be removed from these products. Formerly this was done chemically, which wasted the sulfur. New processes now permit recovery, and these sources promise to be very important. Large amounts of sulfur are being recovered from Alberta gas fields. Sulfur is a pale yellow, odorless, brittle solid, which is insoluble in water but soluble in carbon disulfide. In every state, whether gas, liquid or solid, elemental sulfur occurs in more than one allotropic form or modification; these present a confusing multitude of forms whose relations are not yet fully understood. Amorphous or ''plastic'' sulfur is obtained by fast cooling of the crystalline form. X-ray studies indicate that amorphous sulfur may have a helical structure with eight atoms per spiral. Crystalline sulfur seems to be made of rings, each containing eight sulfur atoms, which fit together to give a normal X-ray pattern. Ten isotopes of sulfur exist. Four occur in natural sulfur, none of which is radioactive. A finely divided form of sulfur, known as flowers of sulfur, is obtained by sublimation. Sulfur readily forms sulfides with many elements. Sulfur is a component of black gunpowder, and is used in the vulcanization of natural rubber and as a fungicide. It is also used extensively in making phosphatic fertilizers. A tremendous tonnage is used to produce sulfuric acid, the most important manufactured chemical. It is used in making sulfite paper and other papers, as a fumigant, and in the beaching of dried fruits. Organic compounds containing sulfur are very important. Sulfur is essential to life. It is a minor constituent of fats, body fluids, and skeletal minerals. Carbon disulfide, hydrogen sulfide, and sulfur dioxide should be handled carefully. Hydrogen sulfide in small concentrations can be metabolized, but in higher concentrations it quickly can cause death by respiratory paralysis. It is insidious in that it quickly deadens the sense of smell. Sulfur dioxide is a dangerous component in atmospheric air pollution. In 1975, University of Pennsylvania scientists reported synthesis of polymeric sulfur nitride, which has the properties of a metal, although it contains no metal atoms. High-purity sulfur is commercially available in purities of 99.999 + %.

**Tantalum** — (from Gr. *Tantalos*, mythological character, father of Niobe), Ta; at wt 180.9479; at no 73; mp 2996°C; bp 5425 ± 100°C; sp gr 16.654; valence 2?, 3, 4?, or 5. Discovered in 1802 by Ekeberg, but many chemists thought niobium and tantalum were identical elements until Rose, in 1844, and Marignac, in 1866, indicated and showed that niobic and tantalic acids were two different acids. The early investigators isolated only the impure metal. The first relatively pure ductile tantalum was produced by von Bolton in 1903. Tantalum occurs principally in the mineral columbite-tantalite (Fe, Mn) (Nb, Ta)$_2$O$_6$. Tantalum ores are found in Australia, Brazil, Mozambique, Thailand, Portugal, Nigeria, Zaire, and Canada. Separation of tantalum from niobium requires several complicated steps. Several methods are used to produce the element commercially, including electrolysis of molten potassium fluotantalate, reduction of potassium fluotantalate with sodium, or reacting tantalum carbide with tantalum oxide. Sixteen isotopes of tantalum are known to exist. Natural tantalum contains two isotopes; one of these,

$^{180}$Ta, is present in very small quantity (0.0123%) and is unstable with a very long half-life of $>10^{13}$ years. Tantalum is a gray, heavy, and very hard metal. When pure, it is ductile and can be drawn into fine wire, which is used as a filament for evaporating metals such as aluminum. Tantalum is almost completely immune to chemical attack at temperatures below 150°C, and is attacked only by hydrofluoric acid, acidic solutions containing the fluoride ion, and free sulfur trioxide. Alkalis attack it only slowly. At high temperatures, tantalum becomes much more reactive. The element has a melting point exceeded only by tungsten and rhenium. Tantalum is used to make a variety of alloys with desirable properties such as high melting points. Scientists at Los Alamos have produced a tantalum carbide graphite composite material, which is said to be one of the hardest materials ever made. The compound has a melting point of 6760°F. Tantalum has good "gettering" ability at high temperatures, and tantalum oxide films are stable. Tantalum is used to make electrolytic capacitors and vacuum furnace parts, which account for about 60% of its use. Tantalum is completely immune to body liquids and is a nonirritating metal. It has, therefore, found use in making surgical appliances. In powder form tantalum costs $35 or more per lb, and is much higher priced in sheet and fabricated form.

**Technetium** — (Gr. *technetos*, artificial), Tc; at wt (98); at no 43; mp 2172°C; bp 4877°C; sp gr 11.50 (calc); valence 0, +2, +4, +5, +6, and +7. Element 43 was predicted on the basis of the periodic table, and was erroneously reported as having been discovered in 1925, at which time it was named masurium. The element was actually discovered by Perrier and Segre in Italy in 1937. It was found in a sample of molybdenum, which was bombarded by deuterons in the Berkeley cyclotron, and which E. Lawrence sent to these investigators. Technetium was the first element to be produced artificially. Since its discovery, searches for the element in terrestrial materials have been made without success. If it does exist, the concentration must be very small. Technetium has been found in the spectrum of S-, M-, and N-type stars, and its presence in stellar matter is leading to new theories of the production of heavy elements in the stars. Sixteen isotopes of technetium, with atomic masses ranging from 92 to 107, are known. $^{97}$Tc has a half-life of 2.6 × 10$^6$ years. $^{98}$Tc has a half-life of 1.5 × 10$^6$ years. The isomeric isotope $^{95}$Tc, with a half-life of 61 days, is useful for tracer work, as it produces energetic gamma rays. Technetium metal has been produced in kilogram quantities. The metal was first prepared by passing hydrogen gas at 1100°C over $Tc_2S_7$. It is now conveniently prepared by the reduction of ammonium pertechnetate with hydrogen. Technetium is a silver gray metal that tarnishes slowly in moist air. Until 1960, technetium was available only in small amounts and the price was as high as $2800/g. It is now offered commercially to holders of O.R.N.L. permits at a price of $60/g. The chemistry of technetium is said to be similar to that of rhenium. Technetium dissolves in nitric acid, aqua regia, and conc sulfuric acid, but is not soluble in hydrochloric acid of any strength. The element is a remarkable corrosion inhibitor for steel. It is reported that mild carbon steels may be effectively protected by as little as 5 ppm of $KTcO_4$ in aerated distilled water at temperatures up to 250°C. This corrosion protection is limited to closed systems, since technetium is radioactive and must be confined. $^{99}Tc$ has a specific activity of $6.2 \times 10^8$ disintegrations per sec/g. Activity of this level must not be allowed to spread. $^{99}Tc$ is a contamination hazard and should be handled in a glove box. The metal is an excellent superconductor at 11 K and below.

**Tellurium** — (L. *tellus*, earth), Te; at wt 127.60 ± 3; at no 52; mp 449.50 ± 0.3°C; bp 989.8 ± 3.8°C; sp gr 6.24 (20°C); valence 2, 4, or 6. Discovered by Muller von Reichenstein in 1782; named by Klaproth, who isolated it in 1798. Tellurium is occasionally found native, but is more often found as the telluride of gold (calaverite), and combined with other metals. It is recovered commercially from the anode muds produced during the electrolytic refining of blister copper. The U.S., Canada, Peru, and Japan are the largest producers of the element among non-Communist nations. Crystalline tellurium has a silvery white appearance, and when pure exhibits a metallic luster. It is brittle and easily pulverized. Amorphous tellurium is formed by precipitating tellurium from a solution of telluric or tellurous acid. Whether this form is truly amorphous, or made of minute crystals, is open to question. Tellurium is a p-type semiconductor, and shows greater conductivity in certain directions, depending on alignment of the atoms. Its conductivity increases slightly with exposure to light. It can be doped with silver, copper, gold, tin, or other elements. In air, tellurium burns with a greenish blue flame, forming the dioxide. Molten tellurium corrodes iron, copper, and stainless steel. Tellurium and its compounds are probably toxic and should be handled with care. Workmen exposed to as little as 0.01 $mg/m^3$ or less, in air, develop "tellurium breath", which has a garlic-like odor. Twenty-one isotopes of tellurium are known, with atomic masses ranging from 115 to 135. Natural tellurium consists of eight isotopes, one of which, $^{127}Te$, is unstable. It is present to the extent of 0.87% and has a half-life of 12 × $10^{13}$ years. Tellurium improves the machinability of copper and stainless steel, and its addition to lead decreases the corrosive action of sulfuric acid to lead and improves its strength and hardness. Tellurium costs about $20/lb, with a purity of about 99.5%.

**Terbium** (*Ytterby*, village in Sweden), Tb; at wt 158.9254; at no 65; mp 1356°C; bp 3123°C; sp gr 8.229; valence 3 or 4. Discovered by Mosander in 1843. Terbium is a member of the lanthanide or "rare earth" group of elements. It is found in cerite, gadolinite, and other minerals, along with other rare earths. It is recovered commercially from monazite, in which it is present to the extent of 0.03%, from xenotime, and from euxenite, a complex oxide containing 1% or more of terbia. Terbium has been isolated only in recent years with the development of ion-exchange techniques for separating the rare earth elements. As with other rare earths, it can be produced by reducing the anhydrous chloride of fluoride with calcium metal in a tantalum crucible. Calcium and tantalum impurities can be removed by vacuum remelting. Terbium is reasonably stable in air. It is a silvery gray metal, and is malleable, ductile, and soft enough to be cut with a knife. Two crystal modifications exist, with a transformation temperature of 1315°C. Nineteen isotopes with atomic masses ranging from 147 to 164 are recognized. The oxide is a chocolate or dark maroon color. Sodium terbium borate is used as a laser material and emits coherent light at 5460 Å. Terbium is used to dope calcium fluoride, calcium tungstate, and strontium molybdate, used in solid-state devices. The element is priced at about $3/g or $900/lb. Little is known of the toxicity of terbium. It should be handled with care as with other lanthanide elements.

**Thallium** — (Gr. *thallos*, a green shoot or twig), Tl; at wt 204.383; at no 81; mp 303.5°C; bp 1457 ± 10°C; sp gr 11.85 (20°C); valence 1 or 3. Thallium was discovered spectroscopically in 1861 Crookes. The element was named after the beautiful green spectral line that identified the element. The metal was isolated both by Crookes and by Lamy in 1862 about the same time. Thallium occurs in crooksite, lorandite, and hutchinsonite. It is also present in pyrites and is recovered from the roasting of this ore in connection with the production of sulfuric acid. It is also obtained from the smelting of lead and zinc ores. Extraction is somewhat complex and depends on the source of the thallium. Manganese nodules, found on the ocean floor, contain thallium. When freshly exposed to air, thallium exhibits a metallic luster, but soon develops a bluish gray tinge, resembling lead in appearance. A heavy oxide builds up on thallium if left in air, and in the presence of water the hydroxide is formed. The metal is very soft and malleable. It can be cut with a knife. Twenty isotopic forms of thallium, with atomic masses ranging from 191 to 210, are recognized. Natural thallium is a mixture of two isotopes. The element and its compounds are toxic and should be handled carefully. Contact of the metal with the skin is dangerous, and when melting the metal adequate ventilation should be provided. Exposure to thallium (soluble compounds) — skin, as Tl, should not exceed 0.1 mg/m$^3$ (8-hr time weighted average, 40-hr work week). Thallium is suspect of having carcinogenic potential for man. Thallium sulfate has been widely employed as a rodenticide and ant killer. It is odorless and tasteless, giving no warning of its presence. Its use, however, has been prohibited in the U.S. since 1975 as a household insecticide and rodenticide. The electrical conductivity of thallium sulfide changes with exposure to infrared light, and this compound is used in photocells. Thallium bromide-iodide crystals have been used as infrared detectors. Thallium has been used, with sulfur or selenium and arsenic, to produce low melting glasses which become fluid between 125 and 150°C. Thallium has been used in treating ringworm and other skin infections; however, its use has been limited because of the narrow margin between toxicity and therapeutic benefits. A mercury-thallium alloy, which forms a eutectic at 8.5% thallium, is reported to freeze at −60°C, some 20°C below the freezing point of mercury. Commercial thallium metal costs about $8/lb.

**Thorium** — (*Thor*, Scandinavian god of war), Th; at wt 232.0381; at no 90; mp 1750°C; bp ~4790°C; sp gr 11.72; valence +2(?), +3(?), or +4. Discovered by Berzelius in 1828. Thorium occurs in thorite ($ThSiO_4$) and in thorianite ($ThO_2$ + $UO_2$). Large deposits of thorium minerals have been reported in New England and elsewhere, but these have not been exploited. Thorium is now thought to be about three times as abundant as uranium and about as abundant as lead or molybdenum. The metal is a potential source of nuclear power. There is probably more energy available for use from thorium in the minerals of the crust of the earth than from both uranium and fossil fuels. Any sizable demand for thorium as a nuclear fuel is still several years in the future. Work has been done in developing thorium cycle converter-reactor systems. Several prototypes, including the HTGR (high-temperature gas-cooled reactor) and MSRE (molten salt coverter reactor experiment), have operated. While the HTGR reactors are efficient, they are not expected to become important commercially soon because of certain operating difficulties. Thorium is recovered commercially from the mineral monazite, which contains from 3 to 9% $ThO_2$ along with most rare earth minerals. Much of the internal heat of the earth has been attributed to thorium and uranium. Several methods are available for producing thorium metal; it can be obtained by reducing thorium oxide with calcium, by electrolysis of anhydrous thorium chloride in a fused mixture of sodium and potassium chlorides, by calcium reduction of thorium tetrachloride mixed with anhydrous zinc chloride, and by reduction of thorium tetrachloride with an alkali metal. Thorium was originally assigned a position in Group IV of the periodic table. Because of its atomic weight, valence, etc., it is now considered to be the second member of the actinide series of elements. When pure, thorium is a silvery white metal which is airstable and retains its luster for several months. When contaminated with the oxide, thorium slowly tarnishes in air, becoming gray and finally black. The physical properties of thorium are greatly influenced by the degree of contamination with the oxide. High-purity thorium is soft and very ductile. Thorium is dimorphic, changing at 1400°C from a cubic to a body-centered cubic structure. Thorium oxide has a melting point of 3300°C, which is the highest of all oxides. Only a few elements, such as tungsten, and a few compounds, such as tantalum carbide, have higher melting points. Thorium is slowly attacked by water, but does not dissolve readily in most common acids, except hydrochloric. Powdered thorium metal is often pyrophoric and should be carefully handled. When heated in air, thorium turnings ignite and burn brilliantly with a white light. The principal use of thorium has been in the preparation of the Welsbach mantle, used for portable gas lights. These mantels, consisting of thorium oxide with about 1% cerium oxide and other ingredients, glow with a dazzling light when heated in a gas flame. Because thorium has a low work-function and high electron emission, it is used to coat tungsten wire used in electronic equipment. The oxide is also used to control the grain size of tungsten used for electric lamps. Thorium oxides has also found use as a catalyst in the conversion of ammonia to nitric acid, in petroleum cracking, and in producing sulfuric acid. Twelve isotopes of thorium are known, with atomic masses ranging from 223 to 234. All are unstable. $^{232}$Th occurs naturally and has a half-life of $1.41 \times 10^{10}$ years. It is an alpha emitter. $^{232}$Th goes through six alpha and four beta decay steps before becoming the stable isotope $^{208}$Pb. $^{232}$Th is sufficiently radioactive to expose a photographic plate in a few hours. Thorium disintegrates with the production of thoron ($^{220}$radon), which is an alpha emitter and presents a radiation hazard. Good ventilation of areas where thorium is stored or handled is therefore essential. Thorium and its compounds are subject to licensing and control by the U.S. Atomic Energy Commission. Thorium metal (99.9%) costs about $10/lb.

**Thulium** — (*Thule*, the earliest name for Scandinavia), Tm; at wt 168.9342; at no 69; mp 1545 ± 15°C; bp 1947°C; sp gr 9.321 (25°C); valence 2 and 3. Discovered in 1879 by Cleve. Thulium occurs in small quantities along with other rare earths in a number of minerals. It is obtained commercially from monazite, which contains about 0.007% of the element. Thulium is the least abundant of the rare earth elements, but with new sources recently discovered, it is now considered to be about as rare as silver, gold, or cadmium. Ion-exchange and solvent extraction techniques have recently permitted much easier separation of the rare earths, with much lower costs. Only a few years ago, thulium metal was not obtainable at any cost; in 1950 the oxide sold for $450/g. Thulium metal now costs from $3 to $20/g, depending on the purity, quantity, and supplier. Thulium can be isolated by reduction of the oxide with lanthanum metal or by calcium reduction of the anhydrous fluoride. The pure metal has a bright, silvery luster. It is reasonably stable in air, but the metal should be protected from moisture in a closed container. The element is silver gray, soft, malleable, and ductile,

and can be cut with a knife. Sixteen isotopes are known, with atomic masses ranging from 161 to 176. Natural thulium, $^{169}$Tm, is stable. Because of the relatively high price of the metal, thulium has not yet found many practical applications. $^{169}$Tm bombarded in a nuclear reactor can be used as a radiation source in portable X-ray equipment. $^{171}$Tm is potentially useful as an energy source. Natural thulium also has possible use in ferrites (ceramic magnetic materials) used in microwave equipment. As with other lanthanides, thulium has a low-to-moderate acute toxic rating. It should be handled with care.

**Tin** — (Anglo-Saxon, *tin*), (L. *stannum*), Sn; at wt 118.69 ± 3; at no 50; mp 231.9681°C; bp 2270°C; sp gr (gray) 5.75, (white) 7.31; valence 2, 4. Known to the ancients. Tin is found chiefly in cassiterite ($SnO_2$). Most of the world supply comes from Malaya, Bolivia, Indonesia, Zaire, Thailand, and Nigeria. The U.S. produces almost none, although deposits have been found in Alaska and California. Tin is obtained by reducing the ore with coal in a reverberatory furnace. Ordinary tin is composed of nine stable isotopes; 13 unstable isotopes are also known. Ordinary tin is a silvery white metal, is malleable, somewhat ductile, and has a highly crystalline structure. Due to the breaking of these crystals, a "tin cry" is heard when a bar is bent. The element has two or perhaps three allotropic forms. On warming, gray, or α tin, with a cubic structure, changes at 13.2°C into white, or β tin, the ordinary form of the metal. White tin has a tetragonal structure. Some authorities believe a γ form exists between 161°C and the melting point; however, other authorities discount its existence. When tin is cooled below 13.2°C, it changes slowly from white to gray. This change is affected by impurities such as aluminum and zinc and can be prevented by small additions of antimony or bismuth. This change from the α to β form is called the tin pest. There are few if any uses for gray tin. Tin takes a high polish and is used to coat other metals to prevent corrosion or other chemical action. Such tin plate over steel is used in the so-called tin can for preserving food. Alloys of tin are very important. Soft solder, type metal, fusible metal, pewter, bronze, bell metal, Babbitt metal, white metal, die casting alloy, and phosphor bronze are some of the important alloys using tin. Tin resists distilled, sea, and soft tap water, but is attacked by strong acids, alkalis, and acid salts. Oxygen in solution accelerates the attack. When heated in air, tin forms $SnO_2$, which is weakly acidic, forming stannate salts with basic oxides. The most important salt is the chloride ($SnCl_2 \cdot H_2O$), which is used as a reducing agent and as a mordant in calico printing. Tin salts sprayed onto glass are used to produce electrically conductive coatings. These have been used for panel lighting and for frost-free windshields. Most window glass is now made by floating molten glass on molten tin (float glass) to produce a flat surface (Pilkington process). Of recent interest is a crystalline tin-niobium alloy that is superconductive at very low temperature. This promises to be important in the construction of superconductive magnets that generate enormous field strengths but use practically no power. Such magnets, made of tin-niobium wire, weigh but a few pounds and produce magnetic fields that, when started with a small battery, are comparable to that of a 100-ton electromagnet operated continuously with a large power supply. The small amount of tin used in canned foods is quite harmless. The agreed limit of tin content in U.S. foods is 300 mg/kg. The trialkyl and triaryl tin compounds are used as biocides and must be handled carefully. Over the past 25 years the price of tin has varied from 50¢/lb to its present price of $6/lb.

**Titanium** — (L. *Titans*, the first of the Earth, myth), Ti; at wt 47.88 ± 3; at no 22; mp 1660 ± 10°C; bp 3287°C; sp gr 4.54; valence 2, 3, or, 4. Discovered by Gregor in 1791; named by Klaproth in 1795. Impure titanium was prepared by Nilson and Pettersson in 1887; however, the pure metal (99.9%) was not made until 1910 by Hunter by heating $TiCl_4$ with sodium in a steel bomb. Titanium is present in meteorites and in the sun. Rock obtained during the Apollo 17 lunar mission showed presence of 12.1% $TiO_2$. Analyses of rocks obtained during earlier Apollo missions show lower percentages. Titanium oxide bands are prominent in the spectra of M type stars. The element is the ninth most abundant in the crust of the earth. Titanium is almost always present in igneous rocks and in the sediments derived from them. It occurs in the minerals rutile, ilmenite, and sphene, and is present in titanates and in many iron ores. Titanium is present in the ash of coal, in plants, and in the human body. The metal was a laboratory curiosity until Kroll, in 1946, showed that titanium could be produced commercially by reducing titanium tetrachloride with magnesium. This method is largely used for producing the metal today. The metal can be purified by decomposing the iodide.

Titanium, when pure, is a lustrous, white metal. It has a low density, good strength, is easily fabricated, and has excellent corrosion resistance. It is ductile only when it is free of oxygen. The metal burns in air and is the only element that burns in nitrogen. Titanium is resistant to dilute sulfuric and hydrochloric acid, most organic acids, moist chlorine gas, and chloride solutions. Natural titanium consists of five isotopes with atomic masses from 46 to 50. All are stable. Four other unstable isotopes are known. Natural titanium is reported to become very radioactive after bombardment with deuterons. The emitted radiations are mostly positrons and hard gamma rays. The metal is dimorphic. The hexagonal α form changes to the cubic β form very slowly at about 880°C. The metal combines with oxygen at red heat, and with chlorine at 550°C. Titanium is important as an alloying agent with aluminum, molybdenum, manganese, iron, and other metals. Alloys of titanium are principally used for aircraft, missiles, and jet engines where light-weight strength, and ability to withstand extremes of temperature are important. Titanium is a strong as steel, but 45% lighter. It is 60% heavier than aluminum, but twice as strong. The alloy Ti-6Al-4V is one of several important materials in advanced aerospace construction. Titanium has potential use in desalination plants for converting sea water into fresh water. The metal has excellent resistance to sea water and is used for propellor shafts, rigging, and other parts of ships exposed to salt water. Titanium metal is considered to be physiologically inert. When pure, titanium dioxide is relatively clear and has an extremely high index of refraction with an optical dispersion higher than diamond. It is produced artificially for use as a gemstone, but is relatively soft. Star sapphires and rubies exhibit their asterism as a result of the presence of $TiO_2$. Titanium dioxide is extensively used for both house paint and artist's paint, and accounts for the largest use of the element. Titanium tetrachloride fumes strongly in air and has been used to produce smoke screens. The price of titanium metal powder (99.7%) is about $25/lb.

**Tungsten** — (Swedish, *tung sten*, heavy stone); also known as wolfram (from *wolframite*, said to be named from *wolf rahm* or *spumi lupi*, because the ore interfered with the smelting of tin and was supposed to devour the tin), W; at wt 183.85 ± 3; at no 74; mp 3410 ± 20°C; bp 5660°C; sp gr 19.3 (20°C); valence 2, 3, 4, 5, or 6. In 1779 Peter

Woulfe examined the mineral now known as wolframite and concluded it must contain a new substance. Scheele, in 1781, found that a new acid could be made from "tung sten" (a name first applied about 1758 to a mineral now known as scheelite). Scheele and Bergman suggested the possibility of obtaining a new metal by reducing this acid. The de Elhuyar brothers found an acid in wolframite in 1783 that was identical to the "acid of tungsten" (tungstic acid) of Scheele, and in that year they succeeded in obtaining the element by reduction of this acid with charcoal. Tungsten occurs in wolframite, $(Fe,Mn)WO_4$; scheelite, $CaWO_4$; huebnerite, $MnWO_4$; and ferberite, $FeWO_4$. Important deposits of tungsten occur in California, Colorado, South Korea, Bolivia, the U.S.S.R., and Portugal. China is reported to have about 75% of the tungsten resources of the world. Natural tungsten contains five stable isotopes. Twelve other unstable isotopes are recognized. The metal is obtained commercially by reducing tungsten oxide with hydrogen or carbon. Pure tungsten is a steel-gray to tin-white metal. Very pure tungsten can be cut with a hacksaw, is ductile, and can be drawn into wire. The impure metal is brittle and can be worked only with difficulty. Tungsten has the highest melting point and lowest vapor pressure of all metals, and at temperatures over 1650°C has the highest tensile strength. The metal oxidizes in air and must be protected at elevated temperatures. It has excellent corrosion resistance in solution and is attacked only slightly by most mineral acids. The thermal expansion is about the same as borosilicate glass, which makes tungsten useful for glass-to-metal seals. Tungsten and its alloys are used extensively for filaments for electric lamps, electron and television tubes, and for metal evaporation work; for electrical contact points for automobile distributors; X-ray targets; windings and heating elements for electrical furnaces; and for numerous space missile and high-temperature applications. High-speed tool steels, Hastelloy® Stellite®, and many other alloys contain tungsten. Tungsten carbide is of great importance to the metalworking, mining, and petroleum industries. Tungsten disulfide is a dry, high-temperature lubricant, stable to 500°C. Tungsten powder costs about $15/lb.

**Uranium** — (planet *Uranus*), U; at wt 238.0289; at no 92; mp 1132 ± 0.8°C; bp 3818°C; sp gr ~ 18.95; valence 2, 3, 4, 5, or 6. Yellow-colored glass, containing more than 1% uranium oxide and

dating back to 79 A.D., has been found near Naples, Italy. Klaproth recognized an unknown element in pitchblende and attempted to isolate the metal in 1789. The metal apparently was first isolated in 1841 by Peligot, who reduced the anhydrous chloride with potassium. Uranium is not as rare as it was once thought. It is now considered to be more plentiful than mercury, antimony, silver, or cadmium, and is about as abundant as molybdenum or arsenic. It occurs in numerous minerals such as pitchblende, uraninite, carnotite, autunite, uranophane, davidite, and tobernite. It is also found in phosphate rock, lignite, and monazite sands, and can be recovered commercially from these sources. The A.E.C. purchases uranium in the form of acceptable $U_3O_8$ concentrates. This incentive program has greatly increased the known uranium reserves. Uranium can be prepared by reducing uranium halides with alkali or alkaline earth metals or by reducing uranium oxides by calcium, aluminum, or carbon at high temperatures. The metal can also be produced by electrolysis of $KUF_5$ or $UF_4$, dissolved in a molten mixture of $CaCl_2NaCl$. High-purity uranium can be prepared by the thermal decomposition or uranium halides on a hot filament. Uranium exhibits three crystallographic modifications; α, β, and γ, with transformation temperatures of 667°C and 772°C. Uranium is a heavy, silvery white metal which is pyrophoric when finely divided. It is a little softer than steel and is attacked by cold water in a finely divided state. It is malleable, ductile, and slightly paramagnetic. In air, the metal becomes coated with a layer of oxide. Acids dissolve the metal, but it is unaffected by alkalis. Uranium has fourteen isotopes, all of which are radioactive. Naturally occurring uranium nominally contains 99.2830% by weight $^{238}U$, 0.7110% $^{235}U$, and 0.0054% $^{234}U$. Studies show that the percentage weight of $^{235}U$ in natural uranium varies by as much as 0.1%, depending on the source. The D.O.E. has adopted the value of 0.711 as being their "official" percentage of $^{235}U$ in natural uranium. Natural uranium is sufficiently radioactive to expose a photographic plate in an hour or so. Much of the internal heat of the earth is thought to be attributable to the presence of uranium and thorium. $^{238}U$ with a half-life of $4.51 \times 10^9$ years, has been used to estimate the age of igneous rocks. The origin of uranium, the highest member of the naturally occurring elements — except perhaps for traces of neptunium or plutonium — is not clearly understood, although it may be presumed that uranium is a decay product of elements of higher atomic weight that may have once been present. These original elements may have been created as a result of a primordial "creation", known as "the big bang", in a supernova, or in some other stellar processes. Uranium is of great importance as a nuclear fuel. $^{238}$U can be converted into fissionable plutonium by the following reactions;

$$^{238}\text{U} \ (\text{n}\gamma) \ ^{239}\text{U} \xrightarrow{\beta} \ ^{239}\text{Np} \xrightarrow{\beta} \ ^{239}\text{Pu}$$

This nuclear conversion can be brought about in "breeder" reactors where it is possible to produce more new fissionable material than the fissionable material used in maintaining the chain reaction. $^{235}$U is of even greater importance, for it is the key to the utilization of uranium. $^{235}$U, while occurring in natural uranium to the extent of only 0.71%, is so fissionable with slow neutrons that a self-sustaining fission chain reaction can occur in a reactor constructed from natural uranium and a suitable moderator, such as heavy water of graphite. $^{235}$U can be concentrated by gaseous diffusion, and used directly as a nuclear fuel, instead of natural uranium, or used as an explosive. Natural uranium, slightly enriched with $^{235}$U by a small percentage, is used to fuel nuclear power reactors for the generation of electricity. Natural thorium can be irradiated with neutrons as follows to produce the important isotope $^{233}$U.

$$^{232}\text{Th} \ (\text{n}\gamma) \ ^{233}\text{Th} \xrightarrow{\beta} \ ^{233}\text{Pa} \xrightarrow{\beta} \ ^{233}\text{U}$$

While thorium is not fissionable, $^{233}$U is, and may be used as a nuclear fuel. One pound of completely fissioned uranium has the fuel value of over 1500 tons of coal. The uses of nuclear fuels to generate electrical power, to make isotopes for peaceful purposes, and to make explosives are well known. The estimated worldwide capacity of the 189 nuclear power reactors in operation in 1979 amounted to about 110 million kilowatts. Uranium in the U.S. is controlled by the Nuclear Regulatory Commission. New uses are being found for "depleted" uranium, i.e., uranium with the percentage of $^{235}$U lowered to about 0.2%, in inertial guidance devices, gyro compasses, counterweights for aircraft control surfaces, as ballast for missile reentry vehicles, and as a shielding material. Uranium salts have been used for producing yellow "vaseline" glass and glazes. Uranium and its compounds are highly toxic, both from a chemical and radiological standpoint. Finely divided uranium metal, being pyrophoric, presents a fire hazard. The maximum recommended allowable concentration of soluble uranium compounds in air (based on chemical toxicity) is 0.05 mg/m$^3$ (8-hr time-weighted average, 40-hr week); for insoluble compounds the concentration is set at 0.25 mg/m$^3$ of air. The maximum permissible total body burden of natural uranium (based on radiotoxicity) is 0.2 µCi for soluble compounds.

**Vanadium** — (Scandinavian goddess, *Vanadis*), V; at wt 50.9415; at no 23; mp 1890 ± 10°C; bp 3380°C; sp gr 6.11 (18.7°C); valence 2, 3, 4, or 5. Vanadium was first discovered by del Rio in 1801. Unfortunately, a French chemist incorrectly declared that del Rio's new element was only impure chromium; del Rio thought himself to be mistaken and accepted the French chemist's statement. The element was rediscovered in 1830 by Sefstrom, who named the element in honor of the Scandinavian goddess Vanadis because of its beautiful multicolored compounds. It was isolated in nearly pure form by Roscoe, in 1867, who reduced the chloride with hydrogen. Vanadium of 99.3 to 99.8% purity was not produced until 1927. Vanadium is found in about 65 different minerals among which are carnotite, roscoelite, vanadinite, and patronite. Vanadium is also found in phosphate rock and certain iron ores, and is present in some crude oils in the form of organic complexes. It is also found in small percentages in meteorites. Commercial production from petroleum ash holds promise as an important source of the element. High-purity ductile vanadium can be obtained by reduction of vanadium trichloride with magnesium or with magnesium-sodium mixtures. Much of the vanadium metal being produced is now made by calcium reduction of $V_2O_5$ in a pressure vessel, an adaption of a process developed by McKechnie and Seybolt. Natural vanadium is a mixture of two isotopes, $^{50}$V (0.24%) and $^{51}$V (99.76%). $^{50}$V is slightly radioactive, having a half-life of 6 × 10$^{15}$ years. Seven other unstable isotopes are recognized. Pure vanadium is a bright white metal, and is soft and ductile. It has good corrosion resistance to alkalis, sulfuric and hydrochloric acids, and salt waters, but the metal oxidizes readily above 660°C. The metal has good structural strength and a low-fission neutron cross section, making it useful in nuclear applications. Vanadium is used in producing rust-resistant, spring, and high-speed tool steels. It is an important carbide stabilizer in making steels. About 80% of the vanadium now produced is used as ferrovanadium or as a steel additive. Vanadium pentoxide is used in ceramics and as a catalyst. Vanadium-gallium tape has been used to produce a superconductive magnet with a field of 175,000 gauss. Vanadium and its compounds are toxic and should be handled with care. Exposure to $V_2O_2$ dust (as V) should not exceed the ceiling value of 0.05 mg/m$^3$, and exposure to $V_2O_2$ fume (as V) should not exceed 0.1 mg/m$^3$ (8-hr time-weighted average, 40-hr week). Commercial vanadium metal, of about 95% purity, costs about $10/lb.

**Wolfram** — see **Tungsten**.

**Xenon** — (Gr. *xenon*, stranger), Xe; at wt 131.29 ± 3; at no 54; mp −111.9°C; bp −107.1 ± 3°C; density (gas) 5.887 ± 0.009 g/l; sp gr (liquid) 3.52 (−109°C); valence usually 0. Discovered by Ramsay and Travers in 1898 in the residue left after evaporating liquid air components. Xenon is a member of the so-called noble or "inert" gases. It is present in the atmosphere to the extent of about one part in twenty million. Xenon is present in the Martian atmosphere to the extent of 0.08 ppm. The element is found in the gases evolved from certain mineral springs, and is commercially obtained by extraction from liquid air. Natural xenon is composed of nine stable isotopes. In addition to these, 22 unstable nuclides and isomers have been characterized. Before 1962, it had generally been assumed that xenon and other noble gases were unable to form compounds. Evidence has been mounting in the past few years that xenon, as well as other members of the zero valence elements, do form compounds. Among the "compounds" of xenon now reported are xenon hydrate, sodium perxenate, xenon deuterate, difluoride, tetrafluoride, hexafluoride, and $XePtF_6$ and $XeRhF_6$. Xenon trioxide, which is highly explosive, has been prepared. More than 80 xenon compounds have been made with xenon chemically bonded to fluorine or oxygen. Some xenon compounds are colored. Metallic xenon has been produced, using several hundred kilobars of pressure. Xenon in a vacuum tube produces a beautiful blue glow when excited by an electrical discharge. The gas is used in making electron tubes, stroboscopic lamps, bactericidal lamps, and lamps used to excite ruby lasers for generating coherent light. Xenon is used in bubble chambers, probes, and other applications where its high molecular weight is of value. The perxenates are used in analytical chemistry as oxidizing agents. $^{133}$Xe and $^{135}$Xe are produced by neutron irradiation in air-cooled nuclear reactors. $^{133}$Xe has useful applications as a radioisotope. The element is available in sealed glass containers for about $20/l of gas at standard pressure. Xenon is not toxic, but its compounds are highly toxic because of their strong oxidizing characteristics.

**Ytterbium** — (*Ytterby*, village in Sweden), Yb; at wt 173.04 ± 3; at no 70; mp 819°C; bp 1194°C; sp gr (α) 6.965 (β) 6.54; valence 2, 3. Marignac in 1878 discovered a new component, which he called ytterbia, in the earth then known as erbia. In 1907, Urbain separated ytterbia into two components, which he called neoytterbia and lutecia. The elements in these earths are now known as ytterbium and lutetium, respectively. These elements are identical with aldebaranium and cassiopeium, discovered independently and at about the same time by von Welsbach. Ytterbium occurs along with other rare earths in a number of rare minerals. It is commercially recovered principally from monazite sand, which contains about 0.03%. Ion-exchange and solvent extraction techniques developed in recent years have greatly simplified the separation of the rare earths. The element was first prepared by Klemm and Bonner in 1937 by reducing ytterbium trichloride with potassium. Their metal was mixed, however, with KCl. In 1953, Daane, Dennison, and Spedding prepared a much purer form from which the chemical and physical properties of the element could be determined. Ytterbium has a bright silvery luster, is soft, malleable, and quite ductile. While the element is fairly stable, it should be kept in closed containers to protect it from air and moisture. Ytterbium is readily attacked and dissolved by dilute and concentrated mineral acids and reacts slowly with water. Ytterbium normally has two allotropic forms with a transformation point at 798°C. The alpha form is a room-temperature, face-centered, cubic modification, while the high-temperature beta form is body-centered cubic. Another body-centered cubic phase has recently been found to be stable at high pressures at room temperatures. The alpha form ordinarily has metallic-type conductivity, but reportedly becomes a semiconductor when the pressure is increased above 16,000 atm. Natural ytterbium is a mixture of seven stable isotopes. Seven other unstable isotopes are known. Ytterbium metal has possible use in improving the grain refinement, strength, and other mechanical properties of stainless steel. Ytterbium metal is commercially available with a purity of about 99 + % for about $1.00/g or $300/lb. Ytterbium has a low acute toxic rating.

**Yttrium** — (from *Ytterby*, a village in Sweden near Vauxholm), Y; at wt 88.9059; at no 39; mp 1522 ± 8°C; bp 3338°C; sp gr 4.469 (25°C); valence 3. Yttria, which is an earth containing yttrium, was discovered by Gadolin in 1794. Ytterby is the site of a quarry which yielded many unusual minerals containing rare earths and other elements. This small town, near Stockholm, bears the honor of giving names to erbium, terbium, and ytterbium as well as yttrium. In 1843 Mosander showed that yttria could be resolved into the earths of three elements. The name yttria was reserved for the most basic one; the others were named erbia and terbia. Yttrium occurs in nearly all of the rare-earth minerals. Analysis of lunar rock samples obtained during the Apollo missions show a relatively high yttrium content. It is recovered commercially from monazite sand, which contains about 3%, and from bastnasite, which contains about 0.2%. Wohler obtained the impure element in 1828 by reduction of the anhydrous chloride with potassium. The metal is now produced commercially by reduction of the fluoride with calcium metal. Yttrium has a silvery metallic luster and is relatively stable in air. Finely divided yttrium is very unstable in air. Turnings of the metal ignite in air if their temperature exceeds 400°C. Yttrium oxide is one of the most important compounds of yttrium and accounts for the largest use. It is widely used in making $YVO_4$-europium, and $Y_2O_3$-europium phosphors to give the red color in color television tubes. Many hundreds of thousands of pounds are now used in this application. Yttrium oxide also is used to produce yttrium-iron garnets, which are very effective microwave filters. Yttrium iron, aluminum, and gadolinium garnets, with formulas such as $Y_3Fe_5O_{12}$ and $Y_3Al_5O_{12}$, have interesting magnetic properties. Yttrium iron garnet is also exceptionally efficient as both a transmitter and transducer of acoustic energy. Yttrium aluminum garnet, with a hardness of 8.5, is also finding use as a gemstone (simulated diamond). Small amounts of yttrium (0.1 to 0.2%) can be used to reduce the grain size in chromium, molybdenum, zirconium, and titanium, and to increase strength of aluminum and magnesium alloys. The metal has a low cross section for nuclear capture. $^{90}Y$, one of the isotopes of yttrium, exists in equilibrium with its parent $^{90}Sr$, a product

of atomic explosions. It has potential use in ceramic and glass formulas as the oxide has a high melting point and imparts shock resistance and low expansion characteristics to glass and as a stabilizing additive for zirconium oxide. Natural yttrium contains but one isotope, $^{89}Y$. Twenty other unstable nuclides and isomers have been characterized. Yttrium metal of 99 + % purity is commercially available at a cost of about 60¢/g or $150/lb.

**Zinc** — (Ger. *Zink*, of obscure origin), Zn; at wt 65.38; at no 30; mp 419.58°C; bp 907°C; sp gr 7.133 (25°C); valence 2. Centuries before zinc was recognized as a distinct element, zinc ores were used for making brass. Tubal-Cain, seven generations from Adam, is mentioned as being an "instructor in every artificer in brass and iron." An alloy containing 87% zinc has been found in prehistoric ruins in Transylvania. Metallic zinc was produced in the 13th century A.D. in India by reducing calamine with organic substances such as wool. The metal was rediscovered in Europe by Marggraf in 1746, who showed that it could be obtained by reducing calamine with charcoal. The principal ores of zinc are sphalerite or blende (sulfide), smithsonite (carbonate), calamine (silicate), and franklinite (zinc, manganese, iron oxide). Zinc can be obtained by roasting its ores to form the oxide and by reduction of the oxide with coal or carbon, with subsequent distillation of the metal. Naturally occurring zinc contains five stable isotopes. Ten other unstable nuclides and isomers are recognized. Zinc is a bluish white, lustrous metal. It is brittle at ordinary temperatures but malleable at 100 to 150°C. It is a fair conductor of electricity, and burns in air at high red heat with evolution of white clouds of the oxide. The metal is employed to form numerous alloys with other metals. Brass, nickel silver, typewriter metal, commercial bronze, spring brass, German silver, soft solder, and aluminum solder are some of the more important alloys. Large quantities of zinc are used to produce die castings, used extensively by the automotive, electrical, and hardware industries. An alloy called Prestal®, consisting of 78% zinc and 22% aluminum, is reported to be almost as strong as steel but as easy to mold as plastic. It exhibits superplasticity. Zinc is also extensively used to galvanize other metals such as iron to prevent corrosion. Neither zinc nor zirconium is ferromagnetic, but $ZrZn_2$ exhibits ferromagnetism at temperatures below 35 K. Zinc oxide is widely used in the manufacture of paints, rubber products, cosmetics, pharmaceuticals, floor coverings, plastics, printing inks, soap, storage batteries, textiles, electrical equipment, and other products. Lithopone, a mixture of zinc sulfide and barium sulfate, is an important pigment. Zinc sulfide is used in making luminous dials, X-ray and TV screens, and fluorescent lights. Zinc is an essential element in the growth of human beings and animals. Tests show that zinc-deficient animals require 50% more food to gain the same weight of an animal supplied with sufficient zinc. Zinc is not considered to be toxic, but when freshly formed ZnO is inhaled a disorder known as the oxide shakes or zinc chills sometimes occurs. It is recommended that where zinc oxide is encountered good ventilation be provided to avoid concentrations exceeding 5 mg/m$^3$, (time-weighted over an 8-hr exposure, 40-hr work week). The price of zinc was roughly 45¢/lb in December 1984.

**Zirconium** — (Persian *zargun*, gold-like), Zr; at wt 91.22; at no 40; mp 1852 ± 2°C; bp 4377°C; sp gr 6.506 (20°C); valence +2, +3, and +4. The name *zircon* probably originated from the Persian word *zargun*, which describes the color of the gemstone now known as zircon, jargon, hyacinth, jacinth, or ligure. This mineral, or its variations, is mentioned in biblical writings. The mineral was not known to contain a new element until Klaproth, in 1789, analyzed a jargon from Ceylon and found a new earth, which Werner named "zircon" (*silex circonius*), and Klaproth called "Zirkonerde (zirconia)". The impure metal was first isolated by Berzelius in 1824 by heating a mixture of potassium and potassium zirconium fluoride in a small iron tube. Pure zirconium was first prepared in 1914. Very pure zirconium was first produced in 1925 by van Arkel and de Boer by an iodide decomposition process they developed. Zirconium is found in abundance in S type stars, and has been identified in the sun and meteorites. Analyses of lunar rock samples obtained during the various Apollo missions to the moon show a surprisingly high zirconium oxide content, compared with terrestial rocks. Naturally occurring zirconium contains five isotopes, one of which, $^{96}$Zr (abundant to the extent of 2.80%) is unstable with a very long

half-life of $> 3.6 \times 10^{17}$ years. Fifteen other unstable nuclides and isomers of zirconium have been characterized. Zircon, $ZrSiO_4$, the principal ore, is found in deposits in Florida, South Carolina, Australia, and Brazil. Baddeleyite, found in Brazil, is an important zirconium mineral. It is principally pure $ZrO_2$ in crystalline form, having a hafnium content of about 1%. Zirconium also occurs in some 30 other recognized mineral species. Zirconium is produced commercially by reduction of the chloride with magnesium (the Kroll process). It is a grayish white lustrous metal. When finely divided, the metal may ignite spontaneously in air, especially at elevated temperatures. The inherent toxicity of zirconium compounds is low. Hafnium is invariably found in zirconium ores, and the separation is difficult. Commercial-grade zirconium contains from 1 to 3% hafnium. Zirconium has a low absorption cross-section for neutrons, and is therefore used for nuclear energy applications, such as for cladding fuel elements. Zirconium has been found to be extremely resistant to the corrosive environment inside atomic reactors, and it allows neutrons to pass through the internal zirconium construction material without appreciable absorption of energy. Commercial nuclear power generation now requires more than 90% of the zirconium metal production. Reactor-grade zirconium is essentially free of hafnium. Zircaloy® is an important alloy developed specifically for nuclear applications. Zirconium is exceptionally resistant to corrosion by many common acids and alkalis, by sea water, and by other agents and is used extensively by the chemical industry. Zirconium is used as a getter in vacuum tubes and as an alloying agent in steel. It is used in poison ivy lotions in the form of the carbonate since it combines with urushiol. With niobium, zirconium is superconductive at low temperatures and is used to make superconductive magnets. Alloyed with zinc, zirconium becomes magnetic at temperatures below 35 K. Zirconium oxide (zircon) has a high index of refraction and is used as a gem material. The impure oxide, zirconia, is used as a refractory material. Commercial grade zirconium metal sponge is priced at about $7/lb. Fabricated zirconium parts are higher in cost.

# Section 2

# Elemental Properties

# ELEMENTAL PROPERTIES

## Table 2—1
### PERIODIC TABLE OF THE ELEMENTS

New notation
Previous IUPAC form
CAS version

**KEY TO CHART**

| | |
|---|---|
| Atomic Number → | 50  +2 |
| Symbol → | Sn  +4 |
| 1987 Atomic Weight → | 118.71 |
| | 18 18 4 |

→ Oxidation States
→ Electron Configuration

| Orbit | 1 IA | 2 IIA | 3 IIIA IIIB | 4 IVA IVB | 5 VA VB | 6 VIA VIB | 7 VIIA VIIB | 8 VIIIA VIII | 9 VIIIA VIII | 10 | 11 IB | 12 IIB | 13 IIIB IIIA | 14 IVB IVA | 15 VB VA | 16 VIB VIA | 17 VIIB VIIA | 18 VIIIA |
|---|---|---|---|---|---|---|---|---|---|---|---|---|---|---|---|---|---|---|
| K | 1 H +1 −1 1.00794 1 | | | | | | | | | | | | | | | | | 2 He 0 4.0020602 2 |
| K-L | 3 Li +1 6.941 2-1 | 4 Be +2 9.012182 2-2 | | | | | | | | | | | 5 B +3 10.811 2-3 | 6 C +2 +4 −4 12.011 2-4 | 7 N +1 +2 +3 +4 +5 −1 −2 −3 14.00674 2-5 | 8 O −2 15.9994 2-6 | 9 F −1 18.9984032 2-7 | 10 Ne 0 20.1797 2-8 |
| K-L-M | 11 Na +1 22.989768 2-8-1 | 12 Mg +2 24.3050 2-8-2 | | | | | | | | | | | 13 Al +3 26.981539 2-8-3 | 14 Si +2 +4 −4 28.0855 2-8-4 | 15 P +3 +5 −3 30.97362 2-8-5 | 16 S +4 +6 −2 32.066 2-8-6 | 17 Cl +1 +5 +7 −1 35.4527 2-8-7 | 18 Ar 0 39.948 2-8-8 |
| -L-M-N | 19 K +1 39.0983 -8-8-1 | 20 Ca +2 40.078 -8-8-2 | 21 Sc +3 44.955910 -8-9-2 | 22 Ti +2 +3 +4 47.88 -8-10-2 | 23 V +2 +3 +4 +5 50.9415 -8-11-2 | 24 Cr +2 +3 +6 51.9961 -8-13-1 | 25 Mn +2 +3 +4 +7 54.93085 -8-13-2 | 26 Fe +2 +3 55.847 -8-14-2 | 27 Co +2 +3 58.93320 -8-15-2 | 28 Ni +2 +3 58.69 -8-16-2 | 29 Cu +1 +2 63.546 -8-18-1 | 30 Zn +2 65.39 -8-18-2 | 31 Ga +3 69.723 -8-18-3 | 32 Ge +2 +4 72.61 -8-18-4 | 33 As +3 +5 −3 74.92159 -8-18-5 | 34 Se +4 +6 −2 78.96 -8-18-6 | 35 Br +1 +5 −1 79.904 -8-18-7 | 36 Kr 0 83.80 -8-18-8 |
| -M-N-O | 37 Rb +1 85.4678 -18-8-1 | 38 Sr +2 87.62 -18-8-2 | 39 Y +3 88.90585 -18-9-2 | 40 Zr +4 91.224 -18-10-2 | 41 Nb +3 +5 92.90638 -18-12-1 | 42 Mo +6 95.94 -18-13-1 | 43 Tc (98) -18-13-2 | 44 Ru +3 101.07 -18-15-1 | 45 Rh +3 102.90550 -18-16-1 | 46 Pd +2 +4 106.42 -18-18-0 | 47 Ag +1 107.8682 -18-18-1 | 48 Cd +2 112.411 -18-18-2 | 49 In +3 114.82 -18-18-3 | 50 Sn +2 +4 118.710 -18-18-4 | 51 Sb +3 +5 −3 121.75 -18-18-5 | 52 Te +4 +6 −2 127.60 -18-18-6 | 53 I +1 +5 +7 −1 126.90447 -18-18-7 | 54 Xe 0 131.29 -18-18-8 |
| -N-O-P | 55 Cs +1 132.90543 -18-8-1 | 56 Ba +2 137.327 -18-8-2 | 57** La +3 138.9055 -18-9-2 | 72 Hf +4 178.49 -32-10-2 | 73 Ta +5 180.9479 -32-11-2 | 74 W +6 183.85 -32-12-2 | 75 Re +4 +6 +7 186.207 -32-13-2 | 76 Os +3 +4 190.2 -32-14-2 | 77 Ir +3 +4 192.22 -32-15-2 | 78 Pt +2 +4 195.08 -32-16-2 | 79 Au +1 +3 196.96654 -32-18-1 | 80 Hg +1 +2 200.59 -32-18-2 | 81 Tl +1 +3 204.3833 -32-18-3 | 82 Pb +2 +4 207.2 -32-18-4 | 83 Bi +3 +5 208.98037 -32-18-5 | 84 Po +2 +4 (209) -32-18-6 | 85 At (210) -32-18-7 | 86 Rn 0 (222) -32-18-8 |
| O P Q | 87 Fr +1 (223) -18-8-1 | 88 Ra +2 226.025 -18-8-2 | 89** Ac +3 227.028 -18-9-2 | 104 Unq +4 (261) -32-10-2 | 105 Unp (262) -32-11-2 | 106 Unh (263) -32-12-2 | 107 Uns (262) -32-13-2 | | | | | | | | | | | |

| Orbit | | | | | | | | | | | | | | | | |
|---|---|---|---|---|---|---|---|---|---|---|---|---|---|---|---|---|
| N O P | *Lanthanides | 58 Ce +3 +4 140.115 -20-8-2 | 59 Pr +3 +4 140.90765 -21-8-2 | 60 Nd +3 144.24 -22-8-2 | 61 Pm +3 (145) -23-8-2 | 62 Sm +2 +3 150.36 -24-8-2 | 63 Eu +2 +3 151.965 -25-8-2 | 64 Gd +3 157.25 -25-9-2 | 65 Tb +3 158.92534 -27-8-2 | 66 Dy +3 162.50 -28-8-2 | 67 Ho +3 164.93032 -29-8-2 | 68 Er +3 167.26 -30-8-2 | 69 Tm +3 168.93421 -31-8-2 | 70 Yb +2 +3 173.04 -32-8-2 | 71 Lu +3 174.967 -32-9-2 |
| O P Q | **Actinides | 90 Th +4 232.0381 -18-10-2 | 91 Pa +5 +4 231.03588 -20-9-2 | 92 U +3 +4 +5 +6 238.0289 -21-9-2 | 93 Np +3 +4 +5 +6 237.048 -22-9-2 | 94 Pu +3 +4 +5 +6 (244) -24-8-2 | 95 Am +3 +4 +5 +6 (243) -25-8-2 | 96 Cm +3 (247) -25-9-2 | 97 Bk +3 +4 (247) -27-8-2 | 98 Cf +3 +4 (251) -28-8-2 | 99 Es (252) -29-8-2 | 100 Fm (257) -30-8-2 | 101 Md +2 +3 (258) -31-8-2 | 102 No +2 +3 (259) -32-8-2 | 103 Lr +3 (260) -32-9-2 |

Numbers in parentheses are mass numbers of most stable isotope of that element

This format numbers the groups 1 to 18. Reprinted with permission from *Chem. Eng. News*, February 4, 1985, 63(5), pp. 26—27. Published 1985 by the American Chemical Society.

## Table 2—2
## ELECTRONIC CONFIGURATION OF THE ELEMENTS

### By Laurence S. Foster

References: F. H. Spedding and A. H. Daane, Editors, *The Rare Earths,* John Wiley and Sons, Inc. Publishers, New York, 1961. R. F. Gould, Editor, *Lanthanide-Actinide Chemistry,* Advances in Chemistry Series, No. 71, American Chemical Society, Washington, D.C., 1967; Paper No. 14, Mark Fred, *Electronic Structure of the Actinide Elements.*

| Atomic No. | Element | K | L | M | N | O | P | Q |
|---|---|---|---|---|---|---|---|---|
| | | 1 s | 2 s p | 3 s p d | 4 s p d f | 5 s p d f | 6 s p d f | 7 s p d f |
| 1 | H | 1 | | | | | | |
| 2 | He | 2 | | | | | | |
| 3 | Li | 2 | 1 | | | | | |
| 4 | Be | 2 | 2 | | | | | |
| 5 | B | 2 | 2 1 | | | | | |
| 6 | C | 2 | 2 2 | | | | | |
| 7 | N | 2 | 2 3 | | | | | |
| 8 | O | 2 | 2 4 | | | | | |
| 9 | F | 2 | 2 5 | | | | | |
| 10 | Ne | 2 | 2 6 | | | | | |
| 11 | Na | 2 | 2 6 | 1 | | | | |
| 12 | Mg | 2 | 2 6 | 2 | | | | |
| 13 | Al | 2 | 2 6 | 2 1 | | | | |
| 14 | Si | 2 | 2 6 | 2 2 | | | | |
| 15 | P | 2 | 2 6 | 2 3 | | | | |
| 16 | S | 2 | 2 6 | 2 4 | | | | |
| 17 | Cl | 2 | 2 6 | 2 5 | | | | |
| 18 | Ar | 2 | 2 6 | 2 6 | | | | |
| 19 | K | 2 | 2 6 | 2 6 .. | 1 | | | |
| 20 | Ca | 2 | 2 6 | 2 6 .. | 2 | | | |
| 21 | Sc | 2 | 2 6 | 2 6 1 | 2 | | | |
| 22 | Ti | 2 | 2 6 | 2 6 2 | 2 | | | |
| 23 | V | 2 | 2 6 | 2 6 3 | 2 | | | |
| 24 | Cr | 2 | 2 6 | 2 6 5* | 1 | | | |
| 25 | Mn | 2 | 2 6 | 2 6 5 | 2 | | | |
| 26 | Fe | 2 | 2 6 | 2 6 6 | 2 | | | |
| 27 | Co | 2 | 2 6 | 2 6 7 | 2 | | | |
| 28 | Ni | 2 | 2 6 | 2 6 8 | 2 | | | |
| 29 | Cu | 2 | 2 6 | 2 6 10* | 1 | | | |
| 30 | Zn | 2 | 2 6 | 2 6 10 | 2 | | | |
| 31 | Ga | 2 | 2 6 | 2 6 10 | 2 1 | | | |
| 32 | Ge | 2 | 2 6 | 2 6 10 | 2 2 | | | |
| 33 | As | 2 | 2 6 | 2 6 10 | 2 3 | | | |
| 34 | Se | 2 | 2 6 | 2 6 10 | 2 4 | | | |
| 35 | Br | 2 | 2 6 | 2 6 10 | 2 5 | | | |
| 36 | Kr | 2 | 2 6 | 2 6 10 | 2 6 | | | |
| 37 | Rb | 2 | 2 6 | 2 6 10 | 2 6 .. | 1 | | |
| 38 | Sr | 2 | 2 6 | 2 6 10 | 2 6 .. | 2 | | |
| 39 | Y | 2 | 2 6 | 2 6 10 | 2 6 1 | 2 | | |
| 40 | Zr | 2 | 2 6 | 2 6 10 | 2 6 2 .. | 2 | | |
| 41 | Nb | 2 | 2 6 | 2 6 10 | 2 6 4*.. | 1 | | |
| 42 | Mo | 2 | 2 6 | 2 6 10 | 2 6 5 .. | 1 | | |
| 43 | Tc | 2 | 2 6 | 2 6 10 | 2 6 5 .. | 2 | | |
| 44 | Ru | 2 | 2 6 | 2 6 10 | 2 6 7 .. | 1 | | |
| 45 | Rh | 2 | 2 6 | 2 6 10 | 2 6 8 .. | 1 | | |
| 46 | Pd | 2 | 2 6 | 2 6 10 | 2 6 10*.. | 0 | | |
| 47 | Ag | 2 | 2 6 | 2 6 10 | 2 6 10 .. | 1 | | |
| 48 | Cd | 2 | 2 6 | 2 6 10 | 2 6 10 .. | 2 | | |
| 49 | In | 2 | 2 6 | 2 6 10 | 2 6 10 .. | 2 1 | | |
| 50 | Sn | 2 | 2 6 | 2 6 10 | 2 6 10 .. | 2 2 | | |
| 51 | Sb | 2 | 2 6 | 2 6 10 | 2 6 10 .. | 2 3 | | |
| 52 | Te | 2 | 2 6 | 2 6 10 | 2 6 10 .. | 2 4 | | |
| 53 | I | 2 | 2 6 | 2 6 10 | 2 6 10 .. | 2 5 | | |
| 54 | Xe | 2 | 2 6 | 2 6 10 | 2 6 10 .. | 2 6 | | |

* Note irregularity.

**Table 2—2 (continued)**
## ELECTRONIC CONFIGURATION OF THE ELEMENTS

| Atomic No | Element | K (1) s | L (2) s p | M (3) s p d | N (4) s p d f | O (5) s p d f | P (6) s p d f | Q (7) s p d f |
|---|---|---|---|---|---|---|---|---|
| 55 | Cs | 2 | 2 6 | 2 6 10 | 2 6 10 .. | 2 6 .. .. | 1 | |
| 56 | Ba | 2 | 2 6 | 2 6 10 | 2 6 10 .. | 2 6 .. .. | 2 | |
| 57 | La | 2 | 2 6 | 2 6 10 | 2 6 10 .. | 2 6 1 .. | 2 | |
| 58 | Ce | 2 | 2 6 | 2 6 10 | 2 6 10 2* | 2 6 .. .. | 2 | |
| 59 | Pr | 2 | 2 6 | 2 6 10 | 2 6 10 3 | 2 6 .. .. | 2 | |
| 60 | Nd | 2 | 2 6 | 2 6 10 | 2 6 10 4 | 2 6 .. .. | 2 | |
| 61 | Pm | 2 | 2 6 | 2 6 10 | 2 6 10 5 | 2 6 .. .. | 2 | |
| 62 | Sm | 2 | 2 6 | 2 6 10 | 2 6 10 6 | 2 6 .. .. | 2 | |
| 63 | Eu | 2 | 2 6 | 2 6 10 | 2 6 10 7 | 2 6 .. .. | 2 | |
| 64 | Gd | 2 | 2 6 | 2 6 10 | 2 6 10 7 | 2 6 1 .. | 2 | |
| 65 | Tb | 2 | 2 6 | 2 6 10 | 2 6 10 9* | 2 6 .. .. | 2 | |
| 66 | Dy | 2 | 2 6 | 2 6 10 | 2 6 10 10 | 2 6 .. .. | 2 | |
| 67 | Ho | 2 | 2 6 | 2 6 10 | 2 6 10 11 | 2 6 .. .. | 2 | |
| 68 | Er | 2 | 2 6 | 2 6 10 | 2 6 10 12 | 2 6 .. .. | 2 | |
| 69 | Tm | 2 | 2 6 | 2 6 10 | 2 6 10 13 | 2 6 .. .. | 2 | |
| 70 | Yb | 2 | 2 6 | 2 6 10 | 2 6 10 14 | 2 6 .. .. | 2 | |
| 71 | Lu | 2 | 2 6 | 2 6 10 | 2 6 10 14 | 2 6 1 .. | 2 | |
| 72 | Hf | 2 | 2 6 | 2 6 10 | 2 6 10 14 | 2 6 2 .. | 2 | |
| 73 | Ta | 2 | 2 6 | 2 6 10 | 2 6 10 14 | 2 6 3 .. | 2 | |
| 74 | W | 2 | 2 6 | 2 6 10 | 2 6 10 14 | 2 6 4 .. | 2 | |
| 75 | Re | 2 | 2 6 | 2 6 10 | 2 6 10 14 | 2 6 5 .. | 2 | |
| 76 | Os | 2 | 2 6 | 2 6 10 | 2 6 10 14 | 2 6 6 .. | 2 | |
| 77 | Ir | 2 | 2 6 | 2 6 10 | 2 6 10 14 | 2 6 7 .. | 2 | |
| 78 | Pt | 2 | 2 6 | 2 6 10 | 2 6 10 14 | 2 6 9 .. | 1 | |
| 79 | Au | 2 | 2 6 | 2 6 10 | 2 6 10 14 | 2 6 10 .. | 1 | |
| 80 | Hg | 2 | 2 6 | 2 6 10 | 2 6 10 14 | 2 6 10 .. | 2 | |
| 81 | Tl | 2 | 2 6 | 2 6 10 | 2 6 10 14 | 2 6 10 .. | 2 1 | |
| 82 | Pb | 2 | 2 6 | 2 6 10 | 2 6 10 14 | 2 6 10 .. | 2 2 | |
| 83 | Bi | 2 | 2 6 | 2 6 10 | 2 6 10 14 | 2 6 10 .. | 2 3 | |
| 84 | Po | 2 | 2 6 | 2 6 10 | 2 6 10 14 | 2 6 10 .. | 2 4 | |
| 85 | At | 2 | 2 6 | 2 6 10 | 2 6 10 14 | 2 6 10 .. | 2 5 | |
| 86 | Rn | 2 | 2 6 | 2 6 10 | 2 6 10 14 | 2 6 10 .. | 2 6 | |
| 87 | Fr | 2 | 2 6 | 2 6 10 | 2 6 10 14 | 2 6 10 .. | 2 6 .. .. | 1 |
| 88 | Ra | 2 | 2 6 | 2 6 10 | 2 6 10 14 | 2 6 10 .. | 2 6 .. .. | 2 |
| 89 | Ac | 2 | 2 6 | 2 6 10 | 2 6 10 14 | 2 6 10 .. | 2 6 1 .. | 2 |
| 90 | Th | 2 | 2 6 | 2 6 10 | 2 6 10 14 | 2 6 10 .. | 2 6 2 .. | 2 |
| 91 | Pa | 2 | 2 6 | 2 6 10 | 2 6 10 14 | 2 6 10 2* | 2 6 1 .. | 2 |
| 92 | U | 2 | 2 6 | 2 6 10 | 2 6 10 14 | 2 6 10 3 | 2 6 1 .. | 2 |
| 93 | Np | 2 | 2 6 | 2 6 10 | 2 6 10 14 | 2 6 10 4 | 2 6 1 .. | 2 |
| 94 | Pu | 2 | 2 6 | 2 6 10 | 2 6 10 14 | 2 6 10 6 | 2 6 .. .. | 2 |
| 95 | Am | 2 | 2 6 | 2 6 10 | 2 6 10 14 | 2 6 10 7 | 2 6 .. .. | 2 |
| 96 | Cm | 2 | 2 6 | 2 6 10 | 2 6 10 14 | 2 6 10 7 | 2 6 1 .. | 2 |
| 97 | Bk | 2 | 2 6 | 2 6 10 | 2 6 10 14 | 2 6 10 9* | 2 6 .. .. | 2 |
| 98 | Cf | 2 | 2 6 | 2 6 10 | 2 6 10 14 | 2 6 10 10 | 2 6 .. .. | 2 |
| 99 | Es | 2 | 2 6 | 2 6 10 | 2 6 10 14 | 2 6 10 11 | 2 6 .. .. | 2 |
| 100 | Fm | 2 | 2 6 | 2 6 10 | 2 6 10 14 | 2 6 10 12 | 2 6 .. .. | 2 |
| 101 | Md | 2 | 2 6 | 2 6 10 | 2 6 10 14 | 2 6 10 13 | 2 6 .. .. | 2 |
| 102 | No | 2 | 2 6 | 2 6 10 | 2 6 10 14 | 2 6 10 14 | 2 6 .. .. | 2 |
| 103 | Lr | 2 | 2 6 | 2 6 10 | 2 6 10 14 | 2 6 10 14 | 2 6 1 .. | 2 |
| 104 | — | 2 | 2 6 | 2 6 10 | 2 6 10 14 | 2 6 10 14 | 2 6 2 .. | 2 |

From Weast, R. C. Ed., *Handbook of Chemistry and Physics,* 69th ed., CRC Press, Boca Raton, Fla., 1988, B-4.

# Table 2—3
## STANDARD ATOMIC WEIGHTS 1981

The changes referred to above are incorporated in the 1981 Table of Standard Atomic Weights. A change in the 1981 Tables which the Commission has been debating for some years concerns a general policy regarding the annotations and wording of footnotes. The basic need for annotations to the Atomic Weights Tables and the Table of Isotopic Compositions arises from the necessity to impart to users additional information that is relevant to one or more elements but that cannot be made readable from numerical data in the columns. Any desire to maximize that additional information conveyed by these Tables is tempered by the need to preserve a compact format and a style that can alert the casual, yet possibly affected reader, who would look up neither the last full element by element review statement nor even the full text of a current Report.

The existing footnotes fail to give some details, such as the magnitudes or signs of differences between normal and abnormal atomic weight values or ranges, geological locations, abundancy of commercially available but abnormal sources, etc. Such additional information to be conveyed to users of these Tables will multiply in future years as materials from nuclear technologies, extra-terrestrial sources, and interest in trace-element compositions of isotopes and products from vacuum, vapor-path, and other fractionations become more widespread.

**Standard Atomic Weights 1981**
**(Scaled to the relative atomic mass, $A_r(^{12}C) = 12$)**

| Names | Symbol | Atomic number | Atomic weight | Footnotes | Names | Symbol | Atomic number | Atomic weight | Footnotes |
|---|---|---|---|---|---|---|---|---|---|
| Actinium | Ac | 89 | 227.0278 | L | Mendelevium | Md | 101 | (258) | |
| Aluminum | Al | 13 | 26.98154 | | Mercury | Hg | 80 | 200.59 ± 3 | |
| Americium | Am | 95 | (243) | | Molybdenum | Mo | 42 | 95.94 | g |
| Antimony (Stibium) | Sb | 51 | 121.75 ± 3 | | Neodymium | Nd | 60 | 144.24 ± 3 | g |
| Argon | Ar | 18 | 39.948 | g, r | Neon | Ne | 10 | 20.179 | g, m |
| Arsenic | As | 33 | 74.9216 | | Neptunium | Np | 93 | 237.0482 | L |
| Astatine | At | 85 | (210) | | Nickel | Ni | 28 | 58.69 | |
| Barium | Ba | 56 | 137.33 | g | Niobium | Nb | 41 | 92.9064 | |
| Berkelium | Bk | 97 | (247) | | Nitrogen | N | 7 | 14.0067 | |
| Beryllium | Be | 4 | 9.01218 | | Nobelium | No | 102 | (259) | |
| Bismuth | Bi | 83 | 208.9804 | | Osmium | Os | 76 | 190.2 | g |
| Boron | B | 5 | 10.81 | m, r | Oxygen | O | 8 | 15.9994 ± 3 | g, r |
| Bromine | Br | 35 | 79.904 | | Palladium | Pd | 46 | 106.42 | g |
| Cadmium | Cd | 48 | 112.41 | g | Phosphorus | P | 15 | 30.97376 | |
| Caesium | Cs | 55 | 132.9054 | | Platinum | Pt | 78 | 195.08 ± 3 | |
| Calcium | Ca | 20 | 40.08 | g | Plutonium | Pu | 94 | (244) | |
| Californium | Cf | 98 | (251) | | Polonium | Po | 84 | (209) | |
| Carbon | C | 6 | 12.011 | r | Potassium (Kalium) | K | 19 | 39.0983 | |
| Cerium | Ce | 58 | 140.12 | g | Praseodymium | Pr | 59 | 140.9077 | |
| Chlorine | Cl | 17 | 35.453 | | Promethium | Pm | 61 | (145) | |
| Chromium | Cr | 24 | 51.996 | | Protactinium | Pa | 91 | 231.0359 | L |
| Cobalt | Co | 27 | 58.9332 | | Radium | Ra | 88 | 226.0254 | g, L |
| Copper | Cu | 29 | 63.546 ± 3 | r | Radon | Rn | 86 | (222) | |
| Curium | Cm | 96 | (247) | | Rhenium | Re | 75 | 186.207 | |
| Dysprosium | Dy | 66 | 162.50 ± 3 | | Rhodium | Rh | 45 | 102.9055 | |
| Einsteinium | Es | 99 | (252) | | Rubidium | Rb | 37 | 85.4678 ± 3 | g |
| Erbium | Er | 68 | 167.26 ± 3 | | Ruthenium | Ru | 44 | 101.07 ± 3 | g |
| Europium | Eu | 63 | 151.96 | g | Samarium | Sm | 62 | 150.36 ± 3 | g |
| Fermium | Fm | 100 | (257) | | Scandium | Sc | 21 | 44.9559 | |
| Fluorine | F | 9 | 18.998403 | | Selenium | Se | 34 | 78.96 ± 3 | |
| Francium | Fr | 87 | (223) | | Silicon | Si | 14 | 28.0855 ± 3 | |
| Gadolinium | Gd | 64 | 157.25 ± 3 | g | Silver | Ag | 47 | 107.8682 ± 3 | g |
| Gallium | Ga | 31 | 69.72 | | Sodium (Natrium) | Na | 11 | 22.98977 | |
| Germanium | Ge | 32 | 72.59 ± 3 | | Strontium | Sr | 38 | 87.62 | g |
| Gold | Au | 79 | 196.9665 | | Sulfur | S | 16 | 32.06 | r |
| Hafnium | Hf | 72 | 178.49 ± 3 | | Tantalum | Ta | 73 | 180.9479 | |
| Helium | He | 2 | 4.00260 | g | Technetium | Tc | 43 | (98) | |
| Holmium | Ho | 67 | 164.9304 | | Tellurium | Te | 52 | 127.60 ± 3 | g |
| Hydrogen | H | 1 | 1.00794 ± 7 | g, m, r | Terbium | Tb | 65 | 158.9254 | |
| Indium | In | 49 | 114.82 | g | Thallium | Tl | 81 | 204.383 | |
| Iodine | I | 53 | 126.9045 | | Thorium | Th | 90 | 232.0381 | g, L |
| Iridium | Ir | 77 | 192.22 ± 3 | | Thulium | Tm | 69 | 168.9342 | |
| Iron | Fe | 26 | 55.847 ± 3 | | Tin | Sn | 50 | 118.69 ± 3 | |
| Krypton | Kr | 36 | 83.80 | g, m | Titanium | Ti | 22 | 47.88 ± 3 | |
| Lanthanum | La | 57 | 138.9055 ± 3 | g | Tungsten (Wolfram) | W | 74 | 183.85 ± 3 | |
| Lawrencium | Lr | 103 | (260) | | (Unnilhexium) | (Unh) | 106 | (263) | |
| Lead | Pb | 82 | 207.2 | g, r | (Unnilpentium) | (Unp) | 105 | (262) | |
| Lithium | Li | 3 | 6.941 ± 3 | g, m, r | (Unnilquadium) | (Unq) | 104 | (261) | |
| Lutetium | Lu | 71 | 174.967 | | Uranium | U | 92 | 238.0289 | g, m |
| Magnesium | Mg | 12 | 24.305 | g | Vanadium | V | 23 | 50.9415 | |
| Manganese | Mn | 25 | 54.9380 | | Xenon | Xe | 54 | 131.29 ± 3 | g, m |

## Table 2—3 (continued)

| Names | Symbol | Atomic number | Atomic weight | Footnotes | Names | Symbol | Atomic number | Atomic weight | Footnotes |
|---|---|---|---|---|---|---|---|---|---|
| Ytterbium | Yb | 70 | 173.04 ± 3 | | Zinc | Zn | 30 | 65.38 | |
| Yttrium | Y | 39 | 88.9059 | | Zirconium | Zr | 40 | 91.22 | g |

| Atomic Number | Names | Symbol | Atomic weight | Footnotes | Atomic Number | Names | Symbol | Atomic weight | Footnotes |
|---|---|---|---|---|---|---|---|---|---|
| 1 | Hydrogen | H | 1.00794 ± 7 | g, m, r | 54 | Xenon | Xe | 131.29 ± 3 | g, m |
| 2 | Helium | He | 4.00260 | g | 55 | Caesium | Cs | 132.9054 | |
| 3 | Lithium | Li | 6.941 ± 3 | g, m, r | 56 | Barium | Ba | 137.33 | g |
| 4 | Beryllium | Be | 9.01218 | | 57 | Lanthanum | La | 138.9055 ± 3 | g |
| 5 | Boron | B | 10.81 | M, r | 58 | Cerium | Ce | 140.12 | g |
| 6 | Carbon | C | 12.011 | r | 59 | Praseodymium | Pr | 140.9077 | |
| 7 | Nitrogen | N | 14.0067 | | 60 | Neodymium | Nd | 144.24 ± 3 | g |
| 8 | Oxygen | O | 15.9994 ± 3 | g, r | 61 | Promethium | Pm | (145) | |
| 9 | Fluorine | F | 18.998403 | | 62 | Samarium | Sm | 150.36 ± 3 | g |
| 10 | Neon | Ne | 20.179 | g, m | 63 | Europium | Eu | 151.96 | g |
| 11 | Sodium (Natrium) | Na | 22.98977 | | 64 | Gadolinium | Gd | 157.25 ± 3 | g |
| 12 | Magnesium | Mg | 24.305 | g | 65 | Terbium | Tb | 158.9254 | |
| 13 | Aluminium | Al | 26.98154 | | 66 | Dysprosium | Dy | 162.50 ± 3 | |
| 14 | Silicon | Si | 28.0855 ± 3 | | 67 | Holmium | Ho | 164.9304 | |
| 15 | Phosphorus | P | 30.97376 | | 68 | Erbium | Er | 167.26 ± 3 | |
| 16 | Sulfur | S | 32.06 | r | 69 | Thulium | Tm | 168.9342 | |
| 17 | Chlorine | Cl | 35.453 | | 70 | Ytterbium | Yb | 173.04 ± 3 | |
| 18 | Argon | Ar | 39.948 | g, r | 71 | Lutetium | Lu | 174.967 | |
| 19 | Potassium (Kalium) | K | 39.0983 | | 72 | Hafnium | Hf | 178.49 ± 3 | |
| 20 | Calcium | Ca | 40.08 | g | 73 | Tantalum | Ta | 180.9479 | |
| 21 | Scandium | Sc | 44.9559 | | 74 | Wolfram (Tungsten) | W | 183.85 ± 3 | |
| 22 | Titanium | Ti | 47.88 ± 3 | | 75 | Rhenium | Re | 186.207 | |
| 23 | Vanadium | V | 50.9415 | | 76 | Osmium | Os | 190.2 | g |
| 24 | Chromium | Cr | 51.996 | | 77 | Iridium | Ir | 192.22 ± 3 | |
| 25 | Manganese | Mn | 54.9380 | | 78 | Platinum | Pt | 195.08 ± 3 | |
| 26 | Iron | Fe | 55.847 ± 3 | | 79 | Gold | Au | 196.9665 | |
| 27 | Cobalt | Co | 58.9332 | | 80 | Mercury | Hg | 200.59 ± 3 | |
| 28 | Nickel | Ni | 58.69 | | 81 | Thallium | Tl | 204.383 | |
| 29 | Copper | Cu | 63.546 ± 3 | r | 82 | Lead | Pb | 207.2 | g, r |
| 30 | Zinc | Zn | 65.38 | | 83 | Bismuth | Bi | 208.9804 | |
| 31 | Gallium | Ga | 69.72 | | 84 | Polonium | Po | (209) | |
| 32 | Germanium | Ge | 72.59 ± 3 | | 85 | Astatine | At | (210) | |
| 33 | Arsenic | As | 74.9216 | | 86 | Radon | Rn | (222) | |
| 34 | Selenium | Se | 78.96 ± 3 | | 87 | Francium | Fr | (223) | |
| 35 | Bromine | Br | 79.904 | | 88 | Radium | Ra | 226.0254 | g, L |
| 36 | Krypton | Kr | 83.80 | g, m | 89 | Actinium | Ac | 227.0278 | L |
| 37 | Rubidium | Rb | 85.4678 ± 3 | g | 90 | Thorium | Th | 232.0381 | g, L |
| 38 | Strontium | Sr | 87.62 | g | 91 | Protactinium | Pa | 231.0359 | L |
| 39 | Yttrium | Y | 88.9059 | | 92 | Uranium | U | 238.0289 | g, m |
| 40 | Zirconium | Zr | 91.22 | g | 93 | Neptunium | Np | 237.0482 | L |
| 41 | Niobium | Nb | 92.9064 | | 94 | Plutonium | Pu | (244) | |
| 42 | Molybdenum | Mo | 95.94 | g | 95 | Americium | Am | (243) | |
| 43 | Technetium | Tc | (98) | | 96 | Curium | Cm | (247) | |
| 44 | Ruthenium | Ru | 101.07 ± 3 | g | 97 | Berkelium | Bk | (247) | |
| 45 | Rhodium | Rh | 102.9055 | | 98 | Californium | Cf | (251) | |
| 46 | Palladium | Pd | 106.42 | g | 99 | Einsteinium | Es | (252) | |
| 47 | Silver | Ag | 107.8682 ± 3 | g | 100 | Fermium | Fm | (257) | |
| 48 | Cadmium | Cd | 112.41 | g | 101 | Mendelevium | Md | (258) | |
| 49 | Indium | In | 114.82 | g | 102 | Nobelium | No | (259) | |
| 50 | Tin | Sn | 118.69 ± 3 | g | 103 | Lawrencium | Lr | (260) | |
| 51 | Antimony (Stibium) | Sb | 121.75 ± 3 | g | 104 | (Unnilquadium) | (Unq) | (261) | |
| 52 | Tellurium | Te | 127.60 ± 3 | g | 105 | (Unnilpentium) | (Unp) | (262) | |
| 53 | Iodine | I | 126.9045 | | 106 | (Unnilhexium) | (Unh) | (263) | |

*Note:* The atomic weights of many elements are not invariant but depend on the origin and treatment of the material. The footnotes to this Table elaborate the types of variation to be expected for individual elements. The values of $A_r(E)$ given here apply to elements as they exist naturally on earth and to certain artificial elements. When used with due regard to the footnotes they are considered reliable to ± 1 in the last digit, unless otherwise noted. Values in parentheses are used for radioactive elements whose atomic weights cannot be quoted precisely without knowledge of the origin of the elements; the value given is the atomic mass number of the isotope of that element of longest known half life.

g    geologically exceptional specimens are known in which the element has an isotopic composition outside the limits for normal material. The difference between the atomic weight of the element in such specimens and that given in the Table may exceed considerably the implied uncertainty.

m    modified isotopic compositions may be found in commercially available material because it has been subjected to an undisclosed or inadvertent isotopic separation. Substantial deviations in atomic weight of the element from that given in the Table can occur.

r    range in isotopic composition of normal terrestrial material prevents a more precise atomic weight being given; the tabulated $A_r(E)$ value should be applicable to any normal material.

L    Longest half-life isotope mass is chosen for the tabulated $A_r(E)$ value.

From Weast, R. C. Ed., *Handbook of Chemistry and Physics,* 69th ed., CRC Press, Boca Raton, Fla., 1988, B-2.

## Table 2—4
## PROPERTIES OF THE CHEMICAL ELEMENTS[a]

Atmospheric Pressure at Room Temperature

| Name | Symbol | Atomic number | International at. wt.[b] | Specific gravity (or density) | Melting point, °C | Boiling point, °C | Specific heat at 25° C | Thermal conductivity, watt/cm °C |
|---|---|---|---|---|---|---|---|---|
| Actinium | Ac | 89 | (227) | (10.02) | 1050. | 3200. | — | — |
| Aluminum | Al | 13 | 26.9815 | 2.70 | 660. | 2441. | 0.215 | 2.37 |
| Americium | Am | 95 | (243) | 11.7 | 994. | 2607. | — | — |
| Antimony (Stibium) | Sb | 51 | 121.75 | 6.69 | 630. | 1750. | 0.050 | 0.185 |
| Argon | Ar | 18 | 39.948 | 1.78 g/l | −189. | −186. | 0.125 | $1.75 \times 10^{-4}$ |
| Arsenic | As | 33 | 74.9216 | 5.73 (gray) | 815[c] | 613. (subl.) | 0.079 | — |
| Astatine | At | 85 | (210) | — | 729. | 2125. | — | — |
| Barium | Ba | 56 | 137.34 | 3.5 | 725. | 1630. | 0.046 | — |
| Berkelium | Bk | 97 | (247) | — | — | — | — | — |
| Beryllium | Be | 4 | 9.0122 | 1.85 | 1285. | 2475. | 0.436 | 2.18 |
| Bismuth | Bi | 83 | 208.980 | 9.75 | 271.4 | 1560. | 0.030 | 0.084 |
| Boron | B | 5 | 10.811 | 2.35 | 2300. | 2550. | 0.245 | — |
| Bromine | Br | 35 | 79.904 | 3.12 (liq.) | −7.2 | 58.8 | 0.11 | $0.45 \times 10^{-4}$ |
| Cadmium | Cd | 48 | 112.40 | 8.65 | 321. | 767. | 0.055 | 0.92 |
| Calcium | Ca | 20 | 40.08 | 1.55 | 840. | 1485. | — | 1.3 |
| Californium | Cf | 98 | (251) | — | — | — | — | — |
| Carbon | C | 6 | 12.01115 | | | | | |
| Diamond | | | | 3.5 | > 3800. | 4827. | 0.124 | 1.5 (0°) |
| Graphite | | | | 2.1 | > 3500. | 4200. | 0.170 | 0.24 |
| Cerium | Ce | 58 | 140.12 | 6.77 | 798. | 3257. | 0.047 | 0.11 |
| Cesium | Cs | 55 | 132.905 | 1.87 | 28.6 | 678. | 0.057 | — |
| Chlorine | Cl | 17 | 35.453 | 3.21 g/l | −101. | −34.6 | 0.114 | $0.86 \times 10^{-4}$ |
| Chromium | Cr | 24 | 51.996 | 7.2 | 1860. | 2670. | 0.110 | 0.91 |
| Cobalt | Co | 27 | 58.9332 | 8.9 | 1495. | 2870. | 0.10 | 0.69 |
| Copper | Cu | 29 | 63.546 | 8.96 | 1084. | 2575. | 0.092 | 3.98 |
| Curium | Cm | 96 | (247) | — | — | — | — | — |
| Dysprosium | Dy | 66 | 162.50 | 8.54 | 1409. | 2335. | 0.0414 | 0.10 |
| Einsteinium | Es | 99 | (254) | — | — | — | — | — |
| Erbium | Er | 68 | 167.26 | 9.05 | 1522. | 2510. | 0.04 | 0.096 |
| Europium | Eu | 63 | 151.96 | 5.25 | 822. | 1597. | 0.042 | — |
| Fermium | Fm | 100 | (257) | — | — | — | — | — |
| Fluorine | F | 9 | 18.9984 | 1.11 (liq.) | −219.6 | −188. | 0.197 | $2.63 \times 10^{-4}$ |
| Francium | Fr | 87 | (223) | — | 27. | 677. | — | — |
| Gadolinium | Gd | 64 | 157.25 | 7.90 | 1311. | 3233. | 0.055 | 0.088 |
| Gallium | Ga | 31 | 69.72 | 5.91 | 29.8 | 2300. | 0.089 | 0.29 − 0.38 |
| Germanium | Ge | 32 | 72.59 | 5.32 | 937. | 2830. | 0.077 | 0.59 |
| Gold (Aurum) | Au | 79 | 196.967 | 19.32 | 1063. | 2857. | 0.031 | 3.15 |
| Hafnium | Hf | 72 | 178.49 | 13.29 | 2220. | 4700. | 0.035 | 0.220 |
| Helium | He | 2 | 4.0026 | 0.177 g/l | — | −269. | 1.24 | $14.8 \times 10^{-4}$ |
| Holmium | Ho | 67 | 164.930 | 8.78 | 1470. | 2720. | 0.039 | — |
| Hydrogen | H | 1 | 1.00797 | 0.0899 g/l | −259. | −253. | 3.41 | $18.4 \times 10^{-4}$ |
| Indium | In | 49 | 114.82 | 7.31 | 156. | 2050. | 0.056 | 0.24 |
| Iodine | I | 53 | 126.9044 | 4.93 | 113.5 | 184.4 | 0.102 | $43.5 \times 10^{-4}$ |
| Iridium | Ir | 77 | 192.2 | 22.42 | 2450. | 4390. | 0.031 | 1.47 |
| Iron (Ferrum) | Fe | 26 | 55.847 | 7.87 | 1536. | 2870. | 0.108 | 0.803 |
| Krypton | Kr | 36 | 83.80 | 3.73 g/l | −157. | −152. | 0.059 | $0.94 \times 10^{-4}$ |
| Lanthanum | La | 57 | 138.91 | 6.17 | 920. | 3454. | 0.047 | 0.14 |
| Lawrencium | Lr | 103 | (257) | — | — | — | — | — |
| Lead (Plumbum) | Pb | 82 | 207.19 | 11.35 | 327.5 | 1750. | 0.031 | 0.352 |
| Lithium | Li | 3 | 6.939 | 0.53 | 180. | 1342. | 0.84 | 0.71 |
| Lutetium | Lu | 71 | 174.97 | 9.84 | 1656. | 3315. | 0.037 | — |
| Magnesium | Mg | 12 | 24.312 | 1.74 | 650. | 1090. | 0.243 | 1.56 |

## Table 2—4 (continued)
## PROPERTIES OF THE CHEMICAL ELEMENTS

| Name | Symbol | Atomic number | International at. wt.[b] | Specific gravity (or density) | Melting point, °C | Boiling point, °C | Specific heat at 25° C | Thermal conductivity, watt/cm °C |
|---|---|---|---|---|---|---|---|---|
| Manganese | Mn | 25 | 54.9380 | 7.21–7.44 | 1244. | 2060. | 0.114 | — |
| Mendelevium | Md | 101 | (256) | — | — | — | — | — |
| Mercury (Hydrargyrum) | Hg | 80 | 200.59 | 13.546 | −38.86 | 356.55 | 0.033 | 0.0839 |
| Molybdenum | Mo | 42 | 95.94 | 10.22 | 2620. | 4651. | 0.060 | 1.38 |
| Neodymium | Nd | 60 | 144.24 | 7.00 | 1010. | 3127. | 0.049 | 0.13 |
| Neon | Ne | 10 | 20.183 | 0.90 g/l | −249. | −246. | 0.246 | $4.77 \times 10^{-4}$ |
| Neptunium | Np | 93 | (237) | 18.0–20.45 | 640. | 3902. | 0.296 | — |
| Nickel | Ni | 28 | 58.71 | 8.90 | 1453. | 2914. | 0.106 | 0.905 |
| Niobium (Columbium) | Nb | 41 | 92.906 | 8.57 | 2467. | 4740. | 0.064 | 0.53 |
| Nitrogen | N | 7 | 14.0067 | 1.251 g/l | −210. | −196. | 0.249 | $2.55 \times 10^{-4}$ |
| Nobelium | No | 102 | (254) | — | — | — | — | — |
| Osmium | Os | 76 | 190.2 | 22.57 | 3025. | 4225. | 0.031 | 0.61 |
| Oxygen | O | 8 | 15.9994 | 1.43 g/l | −218.4 | −183. | 0.220 | $2.61 \times 10^{-4}$ |
| Palladium | Pd | 46 | 106.4 | 12.02 | 1550. | 2927. | 0.058 | 0.71 |
| Phosphorus, white | P | 15 | 30.9738 | 1.82 | 44.1 | 280. | 0.18 | — |
| Platinum | Pt | 78 | 195.09 | 21.45 | 1770. | 3825. | 0.032 | 0.73 |
| Plutonium | Pu | 94 | (244) | 19.84 | 640. | 3230. | 0.032 | 0.08 |
| Polonium | Po | 84 | (209) | 9.32 | 254. | 962. | 0.030 | — |
| Potassium (Kalium) | K | 19 | 39.102 | 0.86 | 63.3 | 760. | 0.180 | 0.99 |
| Praseodymium | Pr | 59 | 140.907 | 6.77 | 931. | 3212. | 0.046 | 0.12 |
| Promethium | Pm | 61 | (145) | — | 1080. | 2460. | 0.044 | — |
| Protactinium | Pa | 91 | (231) | (15.37) | — | — | 0.029 | — |
| Radium | Ra | 88 | (226) | — | 700. | 1700. | 0.029 | — |
| Radon | Rn | 86 | (222) | 9.73 g/l | −71. | −62. | 0.0224 | — |
| Rhenium | Re | 75 | 186.2 | 21.0 | 3180. | 5650. | 0.033 | 0.71 |
| Rhodium | Rh | 45 | 102.905 | 12.41 | 1965. | 3700. | 0.058 | 1.50 |
| Rubidium | Rb | 37 | 85.47 | 1.532 | 39. | 700. | 0.086 | — |
| Ruthenium | Ru | 44 | 101.07 | 12.4 | 2400. | 4100. | 0.057 | — |
| Samarium | Sm | 62 | 150.35 | 7.54 | 1072. | 1778. | 0.047 | — |
| Scandium | Sc | 21 | 44.956 | 2.99 | 1539. | 2832. | 0.135 | — |
| Selenium | Se | 34 | 78.96 | 4.8 | 217. | 700. | 0.077 | 0.005 |
| Silicon | Si | 14 | 28.086 | 2.33 | 1411. | 3280. | 0.17 | 0.835 |
| Silver (Argentum) | Ag | 47 | 107.868 | 10.50 | 961. | 2212. | 0.057 | 4.27 |
| Sodium (Natrium) | Na | 11 | 22.9898 | 0.97 | 97.83 | 884. | 0.293 | 1.34 |
| Strontium | Sr | 38 | 87.62 | 2.55 | 770. | 1375. | 0.072 | — |
| Sulfur | S | 16 | 32.064 | 1.96–2.07 | 113. | 445. | 0.175 | $26.4 \times 10^{-4}$ |
| Tantalum | Ta | 73 | 180.948 | 16.6 | 2980. | 5365. | 0.034 | 0.575 |
| Technetium | Tc | 43 | (97) | (11.50) | 2172. | 4877. | 0.058 | — |
| Tellurium | Te | 52 | 127.60 | 6.24 | 450. | 990. | 0.05 | 0.059 |
| Terbium | Tb | 65 | 158.924 | 8.23 | 1360. | 3041. | 0.0435 | — |
| Thallium | Tl | 81 | 204.37 | 11.85 | 304. | 1480. | 0.031 | 0.39 |
| Thorium | Th | 90 | 232.038 | 11.7 | 1750. | 4800. | 0.03 | 0.41 |
| Thulium | Tm | 69 | 168.934 | 9.31 | 1545. | 1727. | 0.0385 | — |
| Tin (Stannum) | Sn | 50 | 118.69 | 7.31 | 232. | 2600. | 0.054 | 0.67 |
| Titanium | Ti | 22 | 47.90 | 4.54 | 1670. | 3290. | 0.125 | 0.22 |
| Tungsten (Wolfram) | W | 74 | 183.85 | 19.3 | 3400. | 5550. | 0.032 | 1.78 |
| Uranium | U | 92 | 238.03 | 18.8 | 1132. | 4140. | 0.028 | 0.25 |
| Vanadium | V | 23 | 50.942 | 6.1 | 1900. | 3400. | 0.116 | 0.60 |
| Xenon | Xe | 54 | 131.30 | 5.89 g/l | −112. | −107. | 0.038 | $5.2 \times 10^{-4}$ |
| Ytterbium | Yb | 70 | 173.04 | 6.97 | 824. | 1193. | 0.071 | — |
| Yttrium | Y | 39 | 88.905 | 4.46 | 1523. | 3337. | 0.0925 | 0.15 |
| Zinc | Zn | 30 | 65.37 | 7. | 419.5 | 910. | 0.093 | 1.21 |
| Zirconium | Zr | 40 | 91.22 | 6.53 | 1852. | 4400. | 0.067 | 0.227 |

## Table 2—4 (continued)
### PROPERTIES OF THE CHEMICAL ELEMENTS

a    Table 2—5 gives additional properties of the chemical elements. See also Weast, R. C., Ed., *Handbook of Chemistry and Physics*, 69th ed., CRC Press, Boca Raton, Fla., 1988.

b    A value in parentheses is the mass number of the most stable isotope of the element.

c    At 28 atm.

From Bolz, R. E. and Tuve, G. L., Eds., *Handbook of Tables for Applied Engineering Science*, 2nd ed., CRC Press, Cleveland, 1973, 329.

## Table 2—5
### ADDITIONAL PROPERTIES OF THE CHEMICAL ELEMENTS [a]

| Name | Atomic number | Latent heat of fusion, cal/g | Coef of linear thermal expansion $\times 10^6$, K$^{-1}$ | | | Elasticity modulus, psi $\times 10^{-6}$ | First ionization potential, eV | Thermal neutron absorption cross section, barns[b] |
|------|------|------|------|------|------|------|------|------|
| | | | 100 | 300 | 500 | | | |
| Actinium (227) | 89 | 11 | — | — | — | — | 6.9 | 510. |
| Aluminum | 13 | 95 | 12.5 | 24 | 27 | 10.0 | 5.984 | 0.24 |
| Americium (243) | 95 | 10 | — | — | — | — | — | — |
| Antimony | 51 | 38.5 | 9 | 9.5 | 10.5 | 11.3 | 8.639 | 5.7 |
| Argon | 18 | 6.7 | — | — | — | — | 15.755 | 0.66 |
| Arsenic | 33 | 88.5 | — | 4.7 | — | — | 9.81 | 4.3 |
| Astatine | 85 | — | — | — | — | — | 9.5 | — |
| Barium | 56 | 13.4 | — | 16 | 24 | — | 5.21 | 1.2 |
| Berkelium | 97 | — | — | — | — | — | — | — |
| Beryllium | 4 | 324 | — | 12 | 15 | 40—44 | 9.32 | 0.01 |
| Bismuth | 83 | 12.4 | 12 | 13 | 13.5 | 4.6 | 7.287 | 0.034 |
| Boron | 5 | 400 | — | 2 | — | 64 | 8.296 | 755 |
| Bromine | 35 | 16.2 | — | — | — | — | 11.84 | 6.7 |
| Cadmium | 48 | 13.2 | 26 | 30 | 38 | 8 | 8.991 | 2 450. |
| Calcium | 20 | 52 | 17.5 | 23 | 26 | 3.2—3.8 | 6.111 | 0.44 |
| Californium (251) | 98 | — | — | — | — | — | — | — |
| Carbon (Graphite) | 6 | — | — | — | — | 0.7 | 11.256 | 0.004 |
| Cerium | 58 | 9 | — | 8 | — | 4.4 | 5.6 | 0.73 |
| Cesium | 55 | 3.8 | — | 97 | — | — | 3.893 | 30.0 |
| Chlorine | 17 | 2.16 | — | — | — | — | 13.01 | 34. |
| Chromium | 24 | 79 | 3.5 | 6 | 9.5 | 36 | 6.764 | 3.1 |
| Cobalt | 27 | 66 | — | 12 | 13 | 30 | 7.86 | 38. |
| Columbium See Niobium | | | | | | | | |
| Copper | 29 | 49 | 10.5 | 16.5 | 18 | 17 | 7.724 | 3.8 |
| Curium (247) | 96 | — | — | — | — | — | — | — |
| Dysprosium | 66 | 26.4 | — | 9.0 | — | 9.2 | 6.8 | 950. |
| Einsteinium (254) | 99 | — | — | — | — | — | — | — |
| Erbium | 68 | 24.6 | — | 9.0 | — | 10.6 | 6.08 | 170. |
| Europium | 63 | 16.9 | — | 26 | — | 2.1 | 5.67 | 4 300. |
| Fermium | — | — | — | — | — | — | — | — |
| Fluorine | 9 | 10.1 | — | — | — | — | 17.418 | 0.01 |
| Francium | 87 | — | — | — | — | — | 4 | — |
| Gadolinium | 64 | 16.4 | — | 4 | — | 8.1 | 6.16 | 46 000 |
| Gallium | 31 | 19.2 | — | 18 | — | — | 6 | 2.8 |
| Germanium | 32 | 114 | 2.5 | 5.6 | 6.5 | — | 7.88 | 2.45 |
| Gold | 79 | 15 | 11.5 | 14 | 15 | 10.8 | 9.22 | 98.8 |
| Hafnium | 72 | 34 | — | 6 | — | 20 | 7 | 105. |
| Helium | 2 | 1.2 | — | — | — | — | 24.481 | 0.007 |

**Table 2—5 (continued)**
## ADDITIONAL PROPERTIES OF THE CHEMICAL ELEMENTS

| Name | Atomic number | Latent heat of fusion, cal/g | Coef of linear thermal expansion $\times 10^6$, $K^{-1}$ | | | Elasticity modulus, psi $\times 10^{-6}$ | First ionization potential, eV | Thermal neutron absorption cross section, barns[b] |
|------|------|------|------|------|------|------|------|------|
| | | | 100 | 300 | 500 | | | |
| Holmium | 67 | — | — | — | — | 9.7 | — | 65. |
| Hydrogen | 1 | 15.0 | — | — | — | — | 13.595 | 0.33 |
| Indium | 49 | 6.8 | 25 | 33 | — | — | 5.785 | 191. |
| Iodine | 53 | 15 | — | 93 | — | — | 10.454 | 7.0 |
| Iridium | 77 | 33 | 4 | 6.5 | 7.5 | 75 | 9 | 425. |
| Iron | 26 | 65 | 6 | 12 | 14.5 | 28.5 | 7.87 | 2.6 |
| Krypton | 36 | 4.7 | — | — | — | — | 13.996 | 31. |
| Lanthanum | 57 | 10 | — | 5 | 6.5 | 5.5 | 5.61 | 8.9 |
| Lawrencium | 103 | — | — | — | — | — | — | — |
| Lead | 82 | 5.5 | 25 | 29 | 32 | 2.0 | 7.415 | 0.18 |
| Lithium | 3 | 103 | 23 | 50 | — | — | 5.39 | 71. |
| Lutetium | 71 | 26.4 | — | — | — | 12.2 | — | 112. |
| Magnesium | 12 | 88.0 | 15 | 25 | 29 | 6.4 | 7.644 | 0.07 |
| Manganese | 25 | 64 | 11.5 | 23 | 28 | 23 | 7.432 | 13.3 |
| Mendelevium | 101 | — | — | — | — | — | — | — |
| Mercury | 80 | 2.7 | — | — | — | — | 10.43 | 375. |
| Molybdenum | 42 | 69 | 3 | 5 | 5.5 | 40 | 7.10 | 2.7 |
| Neodymium | 60 | 13 | — | 7 | 7.5 | 5.5 | 5.51 | 46. |
| Neon | 10 | 4.0 | — | — | — | — | 21.559 | <2.8 |
| Neptunium (237) | 93 | 9.7 | — | — | — | — | — | (170) |
| Nickel | 28 | 71 | 6.5 | 13 | 15.5 | 31 | 7.633 | 4.6 |
| Niobium | 41 | 68 | 5 | 7 | 7.5 | 15 | 6.88 | 1.15 |
| Nitrogen | 7 | 6.2 | — | — | — | — | 14.53 | 1.9 |
| Nobelium | 102 | — | — | — | — | — | — | — |
| Osmium | 76 | 34 | — | 5 | 5.5 | 80 | 8.5 | 15.3 |
| Oxygen | 8 | 3.3 | — | — | — | — | 13.614 | <0.000 2 |
| Palladium | 46 | 38 | 8.5 | 12 | 13 | 17 | 8.33 | 8. |
| Phosphorus | 15 | 4.8 | — | 125 | — | — | 10.484 | 0.2 |
| Platinum | 78 | 24 | 6.8 | 8.9 | 9.5 | 21.3 | 9.0 | 8.8 |
| Plutonium (244) | 94 | 3 | — | 54 | — | 14 | 5.1 | — |
| Polonium | 84 | 11 | — | — | — | — | 8.43 | — |
| Potassium | 19 | 14.5 | — | 83 | — | — | 4.339 | 2.1 |
| Praseodymium | 59 | 17 | — | 5. | 5.3 | 4.7 | 5.46 | 11.3 |
| Promethium | 61 | — | — | — | — | 6.1 | — | — |
| Protactinium (231) | 91 | 17 | — | — | — | — | — | (200) |
| Radium (226) | 88 | 10 | — | — | — | — | 5.277 | (20) |
| Radon (222) | 86 | 3.1 | — | — | — | — | 10.746 | (0.7) |
| Rhenium | 75 | 42 | — | 7 | — | 66.7 | 7.87 | 85. |
| Rhodium | 45 | 50 | 5.0 | 8.3 | 9.3 | 42 | 7.46 | 150. |
| Rubidium | 37 | 6.3 | — | 90 | — | — | 4.176 | 0.7 |
| Ruthenium | 44 | 60 | — | 9 | — | 60 | 7.364 | 2.6 |
| Samarium | 62 | 24.7 | — | — | — | 4.9 | 5.6 | 5 600. |
| Scandium | 21 | 87 | — | — | — | 11.5 | 6.54 | 24. |
| Selenium | 34 | 16 | — | 35 | — | 8.4 | 9.75 | 12.3 |
| Silicon | 14 | 430 | — | 2.5 | 3.5 | 16 | 8.149 | 0.160 |
| Silver | 47 | 26.5 | 14.3 | 19.0 | 20.6 | 10.5 | 7.574 | 63. |
| Sodium | 11 | 27 | 45.7 | 70.0 | — | — | 5.138 | .53 |
| Strontium | 38 | 25 | — | — | — | — | 5.692 | 1.21 |
| Sulfur | 16 | 9.2 | 42 | 63 | — | — | 10.357 | 0.52 |
| Tantalum | 73 | 41 | 5.2 | 6.6 | 6.9 | 27 | 7.88 | 21. |
| Technetium | 43 | 56.7 | — | — | — | — | 7.28 | 22. |
| Tellurium | 52 | 33 | — | 17 | — | 17 | 9.01 | 4.7 |

## Table 2—5 (continued)
## ADDITIONAL PROPERTIES OF THE CHEMICAL ELEMENTS

| Name | Atomic number | Latent heat of fusion, cal/g | Coef of linear thermal expansion × 10⁶, K⁻¹ | | | Elasticity modulus, psi × 10⁻⁶ | First ionization potential, eV | Thermal neutron absorption cross section, barns[b] |
|---|---|---|---|---|---|---|---|---|
| | | | 100 | 300 | 500 | | | |
| Terbium | 65 | 23.6 | — | 7.0 | — | 8.3 | 5.98 | 46. |
| Thallium | 81 | 5.0 | 24 | 29 | 32 | — | 6.106 | 3.4 |
| Thorium | 90 | 17 | 8.7 | 11.4 | 12.5 | 8.5 | 6.95 | 7.5 |
| Thulium | 69 | 26.0 | — | — | — | 11.0 | 5.81 | 127. |
| Tin | 50 | 14.1 | 15.5 | 21 | 27.5 | 6 | 7.342 | 0.63 |
| Titanium | 22 | 100 | 4.4 | 8.6 | 9.8 | 16 | 6.82 | 5.8 |
| Tungsten | 74 | 46 | 2.7 | 4.4 | 4.6 | 50 | 7.98 | 19. |
| Uranium | 92 | 12 | 10.6 | 13.5 | 17 | 24 | 6.08 | 7.7 |
| Vanadium | 23 | 98 | 4 | 8 | — | 19 | 6.74 | 5. |
| Xenon | 54 | 4.2 | — | — | — | — | 12.127 | 35. |
| Ytterbium | 70 | 12.7 | — | 25 | 26.3 | 2.6 | 6.2 | 37. |
| Yttrium | 39 | 45 | — | — | — | 9.4 | 6.38 | 1.3 |
| Zinc | 30 | 27 | 23 | 30 | 32 | 12 | 9.391 | 1.10 |
| Zirconium | 40 | 54 | 3.9 | 5.5 | 6.2 | 13.7 | 6.84 | 0.18 |

[a]  See also Weast, R. C., Ed., *Handbook of Chemistry and Physics,* 69th ed., CRC Press, Boca Raton, Fla, 1988.

[b]  Values in parentheses apply to that isotope for which the mass number is given following the name of the element. All other values of neutron cross section apply to the naturally occurring mixture of isotopes.

From Bolz, R. E. and Tuve, G. L., Eds., *Handbook of Tables for Applied Engineering Science,* 2nd ed., CRC Press, Cleveland, 1973, 331.

## Table 2—6
## VAPOR PRESSURE OF THE ELEMENTS

This table lists the temperature in degrees Celsius (Centigrade) at which an element has a vapor pressure indicated by the headings of the columns. To convert pressures to SI units, 1 mm Hg (torr) = 133.3 N/m$^2$ and 1 atm = 0.1013 MN/m$^2$.

| Element | | mm Hg 1 | 10 | 100 | 400 | 760 | atm 2 | 5 | 10 | 20 | 40 |
|---|---|---|---|---|---|---|---|---|---|---|---|
| Aluminum | Al | 1540 | 1780 | 2080 | 2320 | 2467 | 2610 | 2850 | 3050 | 3270 | 3530 |
| Antimony | Sb | | 960 | 1280 | 1570 | 1750 | 1960 | 2490 | | | |
| Arsenic | As | 380 | 440 | 510 | 580 | 610 | | | | | |
| Barium | Ba | 860 | 1050 | 1300 | 1520 | 1640 | 1790 | 2030 | 2230 | | |
| Beryllium | Be | 1520 | 1860 | 2300 | 2770 | 2970 | 3240 | 3730 | 4110 | 4720 | 5610 |
| Bismuth | Bi | | 1060 | 1280 | 1450 | 1560 | 1660 | 1850 | 2000 | 2180 | |
| Boron | B | 2660 | 3030 | 3460 | 3810 | 4000 | | | | | |
| Bromine | Br | −60 | −30 | +9 | 39 | 59 | 78 | 110 | | | |
| Cadmium | Cd | 393 | 486 | 610 | 710 | 765 | 830 | 930 | 1030 | 1120 | 1240 |
| Calcium | Ca | 800 | 970 | 1200 | 1390 | 1490 | 1630 | 1850 | 2020 | 2290 | |
| Cesium | Cs | | 373 | 513 | 624 | 690 | | | | | |
| Chlorine | Cl | −123 | −101 | −71 | −46 | −34 | −17 | +9 | 30 | 55 | 97 |
| Chromium | Cr | 1610 | 1840 | 2140 | 2360 | 2480 | 2630 | 2850 | 3010 | 3180 | |
| Cobalt | Co | 1910 | 2170 | 2500 | 2760 | 2870 | 3040 | 3270 | | | |
| Copper | Cu | | 1870 | 2190 | 2440 | 2600 | 2760 | 3010 | 3500 | 3460 | 3740 |
| Fluorine | F | | −203 | −193 | −188 | −180.7 | −169.1 | −159.6 | | | |
| Gallium | Ga | 1350 | 1570 | 1850 | 2060 | 2180 | 2320 | 2560 | 2730 | | |
| Germanium | Ge | | 2080 | 2440 | 2710 | 2830 | 2970 | 3200 | 3430 | | |
| Gold | Au | 1880 | 2160 | 2520 | 2800 | 2940 | 3120 | 3490 | 3630 | 3890 | |
| Indium | In | | | 1960 | 2080 | 2230 | 2440 | 2600 | | | |
| Iodine | I | 40 | 72 | 115 | 160 | 185 | 216 | 265 | | | |
| Iridium | Ir | 2830 | 3170 | 3630 | 3960 | 4130 | 4310 | 4650 | | | |
| Iron | Fe | 1780 | 2040 | 2370 | 2620 | 2750 | 2900 | 3150 | 3360 | 3570 | |
| Lanthanum | La | | | | 3230 | 3420 | 3620 | 3960 | 4270 | | |
| Lead | Pb | 970 | 1160 | 1420 | 1630 | 1740 | 1880 | 2140 | 2320 | 2620 | |
| Lithium | Li | 750 | 890 | 1080 | 1240 | 1310 | 1420 | 1518 | | | |
| Magnesium | Mg | 620 | 740 | 900 | 1040 | 1110 | 1190 | 1330 | 1430 | 1560 | |
| Manganese | Mn | | 1510 | 1810 | 2050 | 2100 | 2360 | 2580 | 2850 | | |
| Mercury | Hg | | | 260 | 330 | 356.9 | 398 | 465 | 517 | 581 | 657 |
| Molybdenum | Mo | 3300 | 3770 | 4200 | 4580 | 4830 | 5050 | 5340 | 5680 | 5980 | |
| Neodymium | Nd | | | | 2870 | 3100 | 3300 | 3680 | 3990 | | |
| Nickel | Ni | 1800 | 2090 | 2370 | 2620 | 2730 | 2880 | 3120 | 3300 | 3310 | |
| Palladium | Pd | 1470 | 2290 | 2670 | 2950 | 3140 | 3270 | 3560 | 3840 | | |
| Phosphorus | P | | 127 | 199 | 253 | 283 | 319 | | | | |
| Platinum | Pt | 2600 | 2940 | 3360 | 3650 | 3830 | 4000 | 4310 | 4570 | 4860 | |
| Polonium | Po | 472 | 587 | 752 | 890 | 960 | 1060 | 1200 | 1340 | | |
| Potassium | K | | | 590 | 710 | 770 | 850 | 950 | 1110 | 1240 | 1420 |
| Rhodium | Rh | 2530 | 2850 | 3260 | 3590 | 3760 | 3930 | 4230 | 4440 | | |
| Rubidium | Rb | | 390 | 527 | 640 | 700 | | | | | |
| Selenium | Se | | 429 | 547 | 640 | 685 | 750 | 850 | 920 | 1010 | 1120 |
| Silver | Ag | 1310 | 1540 | 1850 | 2060 | 2210 | 2360 | 2600 | 2850 | 3050 | 3300 |
| Sodium | Na | 440 | 546 | 700 | 830 | 890 | 980 | 1120 | 1230 | 1370 | |
| Strontium | Sr | 740 | 900 | 1100 | 1280 | 1380 | 1480 | 1670 | 1850 | 2030 | |
| Sulfur | S | | 246 | 333 | 407 | 445 | 493 | 574 | 640 | 720 | |
| Tellurium | Te | 520 | 633 | 792 | 900 | 962 | 1030 | 1160 | 1250 | | |
| Thallium | Tl | | 1000 | 1210 | 1370 | 1470 | 1560 | 1750 | 1900 | 2050 | 2260 |
| Tin | Sn | 1610 | 1890 | 2270 | 2580 | 2750 | 2950 | 3270 | 3540 | 3890 | |
| Titanium | Ti | 2180 | 2480 | 2860 | 3100 | 3260 | 3400 | 3650 | 3800 | | |
| Tungsten | W | 3980 | 4490 | 5160 | 5470 | 5940 | 6260 | 6670 | 7250 | 7670 | |
| Uranium | U | 2450 | 2800 | 3270 | 3620 | 3800 | 4040 | 4420 | | | |
| Vanadium | V | 2290 | 2570 | 2950 | 3220 | 3380 | 3540 | 3800 | | | |
| Zinc | Zn | | 590 | 730 | 840 | 907 | 970 | 1090 | 1180 | 1290 | |

From Weast, R. C. Ed., *Handbook of Chemistry and Physics,* 69th ed., CRC Press, Boca Raton, Fla., 1988, D-215.

## Table 2—7
### CRYSTAL IONIC RADII OF THE ELEMENTS

Numerical values of the radii of the ions may vary, depending on how they were measured. They may have been calculated from wave functions or determined from the lattice spacings or crystal structure of various salts. Different values are obtained, depending on the kind of salt used or on the method of calculation. Data for many of the rare-earth ions were furnished by F. H. Spedding and K. Gschneidner.

| Element | Charge | Atomic number | Radius in A | Element | Charge | Atomic number | Radius in A | Element | Charge | Atomic number | Radius in A |
|---|---|---|---|---|---|---|---|---|---|---|---|
| Ac | +3 | 89 | 1.18 | Dy | +3 | 66 | 0.908 | Nb | +1 | 41 | 1.00 |
| Ag | +1 | 47 | 1.26 | Er | +3 | 68 | 0.881 |  | +4 |  | 0.74 |
|  | +2 |  | 0.89 | Eu | +3 | 63 | 0.950 |  | +5 |  | 0.69 |
| Al | +3 | 13 | 0.51 |  | +2 |  | 1.09 | Nd | +3 | 60 | 0.995 |
| Am | +3 | 95 | 1.07 | F | −1 | 9 | 1.33 | Ne | +1 | 10 | 1.12 |
|  | +4 |  | 0.92 |  | +7 |  | 0.08 | Ni | +2 | 28 | 0.69 |
| Ar | +1 | 18 | 1.54 | Fe | +2 | 26 | 0.74 | Np | +3 | 93 | 1.10 |
| As | −3 | 33 | 2.22 |  | +3 |  | 0.64 |  | +4 |  | 0.95 |
|  | +3 |  | 0.58 | Fr | +1 | 87 | 1.80 |  | +7 |  | 0.71 |
|  | +5 |  | 0.46 | Ga | +1 | 31 | 0.81 | O | −2 | 8 | 1.32 |
| At | +7 | 85 | 0.62 |  | +3 |  | 0.62 |  | −1 |  | 1.76 |
| Au | +1 | 79 | 1.37 | Gd | +3 | 64 | 0.938 |  | +1 |  | 0.22 |
|  | +3 |  | 0.85 | Ge | −4 | 32 | 2.72 |  | +6 |  | 0.09 |
| B | +1 | 5 | 0.35 |  | +2 |  | 0.73 | Os | +4 | 76 | 0.88 |
|  | +3 |  | 0.23 |  | +4 |  | 0.53 |  | +6 |  | 0.69 |
| Ba | +1 | 56 | 1.53 | H | −1 | 1 | 1.54 | P | −3 | 15 | 2.12 |
|  | +2 |  | 1.34 | Hf | +4 | 72 | 0.78 |  | +3 |  | 0.44 |
| Be | +1 | 4 | 0.44 | Hg | +1 | 80 | 1.27 |  | +5 |  | 0.35 |
|  | +2 |  | 0.35 |  | +2 |  | 1.10 | Pa | +3 | 91 | 1.13 |
| Bi | +1 | 83 | 0.98 | Ho | +3 | 67 | 0.894 |  | +4 |  | 0.98 |
|  | +3 |  | 0.96 | I | −1 | 53 | 2.20 |  | +5 |  | 0.89 |
|  | +5 |  | 0.74 |  | +5 |  | 0.62 | Pb | +2 | 82 | 1.20 |
| Br | −1 | 35 | 1.96 |  | +7 |  | 0.50 |  | +4 |  | 0.84 |
|  | +5 |  | 0.47 | In | +3 | 49 | 0.81 | Pd | +2 | 46 | 0.80 |
|  | +7 |  | 0.39 | Ir | +4 | 77 | 0.68 |  | +4 |  | 0.65 |
| C | −4 | 6 | 2.60 | K | +1 | 19 | 1.33 | Pm | +3 | 61 | 0.979 |
|  | +4 |  | 0.16 | La | +1 | 57 | 1.39 | Po | +6 | 84 | 0.67 |
| Ca | +1 | 20 | 1.18 |  | +3 |  | 1.016 | Pr | +3 | 59 | 1.013 |
|  | +2 |  | 0.99 | Li | +1 | 3 | 0.68 |  | +4 |  | 0.90 |
| Cd | +1 | 48 | 1.14 | Lu | +3 | 71 | 0.85 | Pt | +2 | 78 | 0.80 |
|  | +2 |  | 0.97 | Mg | +1 | 12 | 0.82 |  | +4 |  | 0.65 |
| Ce | +1 | 58 | 1.27 |  | +2 |  | 0.66 | Pu | +3 | 94 | 1.08 |
|  | +3 |  | 1.034 | Mn | +2 | 25 | 0.80 |  | +4 |  | 0.93 |
|  | +4 |  | 0.92 |  | +3 |  | 0.66 | Ra | +2 | 88 | 1.43 |
| Cl | −1 | 17 | 1.81 |  | +4 |  | 0.60 | Rb | +1 | 37 | 1.47 |
|  | +5 |  | 0.34 |  | +7 |  | 0.46 | Re | +4 | 75 | 0.72 |
|  | +7 |  | 0.27 | Mo | +1 | 42 | 0.93 |  | +7 |  | 0.56 |
| Co | +2 | 27 | 0.72 |  | +4 |  | 0.70 | Rh | +3 | 45 | 0.68 |
|  | +3 |  | 0.63 |  | +6 |  | 0.62 | Ru | +4 | 44 | 0.67 |
| Cr | +1 | 24 | 0.81 | N | −3 | 7 | 1.71 | S | −2 | 16 | 1.84 |
|  | +2 |  | 0.89 |  | +1 |  | 0.25 |  | +2 |  | 2.19 |
|  | +3 |  | 0.63 |  | +3 |  | 0.16 |  | +4 |  | 0.37 |
|  | +6 |  | 0.52 |  | +5 |  | 0.13 |  | +6 |  | 0.30 |
| Cs | +1 | 55 | 1.67 | NH₄ | +1 |  | 1.43 | Sb | −3 | 51 | 2.45 |
| Cu | +1 | 29 | 0.96 | Na | +1 | 11 | 0.97 |  | +3 |  | 0.76 |
|  | +2 |  | 0.72 |  |  |  |  |  | +5 |  | 0.62 |

## Table 2—7 (continued)
## CRYSTAL IONIC RADII OF THE ELEMENTS

| Element | Charge | Atomic number | Radius in A | Element | Charge | Atomic number | Radius in A | Element | Charge | Atomic number | Radius in A |
|---|---|---|---|---|---|---|---|---|---|---|---|
| Sc | +3 | 21 | 0.732 | Ta | +5 | 73 | 0.68 | Tm | +3 | 69 | 0.87 |
| Se | −2 | 34 | 1.91 | Tb | +3 | 65 | 0.923 | U | +4 | 92 | 0.97 |
|    | −1 |    | 2.32 |    | +4 |    | 0.84 |    | +6 |    | 0.80 |
|    | +1 |    | 0.66 | Tc | +7 | 43 | 0.979 | V | +2 | 23 | 0.88 |
|    | +4 |    | 0.50 | Te | −2 | 52 | 2.11 |    | +3 |    | 0.74 |
|    | +6 |    | 0.42 |    | −1 |    | 2.50 |    | +4 |    | 0.63 |
| Si | −4 | 14 | 2.71 |    | +1 |    | 0.82 |    | +5 |    | 0.59 |
|    | −1 |    | 3.84 |    | +4 |    | 0.70 | W | +4 | 74 | 0.70 |
|    | +1 |    | 0.65 |    | +6 |    | 0.56 |    | +6 |    | 0.62 |
|    | +4 |    | 0.42 | Th | +4 | 90 | 1.02 | Y | +3 | 39 | 0.893 |
| Sm | +3 | 62 | 0.964 | Ti | +1 | 22 | 0.96 | Yb | +2 | 70 | 0.93 |
| Sn | −4 | 50 | 2.94 |    | +2 |    | 0.94 |    | +3 |    | 0.858 |
|    | −1 |    | 3.70 |    | +3 |    | 0.76 | Zn | +1 | 30 | 0.88 |
|    | +2 |    | 0.93 |    | +4 |    | 0.68 |    | +2 |    | 0.74 |
|    | +4 |    | 0.71 | Tl | +1 | 81 | 1.47 | Zr | +1 | 40 | 1.09 |
| Sr | +2 | 38 | 1.12 |    | +3 |    | 0.95 |    | +4 |    | 0.79 |

From Weast, R. C. Ed., *Handbook of Chemistry and Physics,* 69th ed., CRC Press, Boca Raton, Fla., 1988, F-164.

## Table 2—8
## ELECTRICAL RESISTIVITY AND TEMPERATURE COEFFICIENTS OF THE ELEMENTS

The units of resistivity, ρ, are $10^{-8}$ ωm. The data were collected from several sources. Those from the *Journal of Physical and Chemical Reference Data*, 8, 339, 439, 1147 1979 and 13, 1097 and 1131, 1984 are used with permission of the copyright owners, the American Institute of Physics and the American Chemical Society. The names of the elements and alloys for which data were obtained from this journal have the superscript (r). Users of this table are referred to that journal for resistivities values listed for temperatures other than those listed in the table below. Data on the rare earth elements were furnished by the Ames Laboraory at Iowa State University.

The table for Conversion Factors for Units of Electrical Resistivity is from the *Journal of Physical and Chemical Reference Data*, 8, 442, 1979.

### A. Conversion Factors for Units of Electrical Resistivity

| To convert from multiply by appropriate factor to Obtain→ | abΩcm | μΩcm | Ωcm | StatΩcm | Ωm | Ωcir. mil ft$^{-1}$ | Ωin. | Ωft |
|---|---|---|---|---|---|---|---|---|
| abohm-centimeter | 1 | $1 \times 10^{-3}$ | $10^{-9}$ | $1.113 \times 10^{-21}$ | $10^{-11}$ | $6.015 \times 10^{-3}$ | $3.937 \times 10^{-10}$ | $3.281 \times 10^{-11}$ |
| microohm-centimeter | $10^3$ | 1 | $10^{-6}$ | $1.113 \times 10^{-18}$ | $10^{-6}$ | 6.015 | $3.937 \times 10^{-7}$ | $3.281 \times 10^{-6}$ |
| ohm centimeter | $10^8$ | $10^6$ | 1 | $1.113 \times 10^{-12}$ | $1 \times 10^{-2}$ | $6.015 \times 10^6$ | $3.937 \times 10^{-1}$ | $3.281 \times 10^{-2}$ |
| statohm-centimeter (esu) | $8.987 \times 10^{20}$ | $8.987 \times 10^{17}$ | $8.987 \times 10^{11}$ | 1 | $8.987 \times 10^9$ | $5.406 \times 10^{18}$ | $3.538 \times 10^{11}$ | $2.949 \times 10^{10}$ |
| ohm-meter | $10^{11}$ | $10^8$ | $10^2$ | $1.113 \times 10^{-10}$ | 1 | $6.015 \times 10^8$ | $3.937 \times 10^1$ | 3.281 |
| ohm-circular mil per foot | $1.662 \times 10^2$ | $1.662 \times 10^{-1}$ | $1.662 \times 10^{-7}$ | $1.850 \times 10^{-19}$ | $1.662 \times 10^{-9}$ | 1 | $6.54 \times 10^{-6}$ | $5.45 \times 10^{-9}$ |
| ohm-inch | $2.54 \times 10^9$ | $2.54 \times 10^6$ | 2.54 | $2.827 \times 10^{-12}$ | $2.54 \times 10^{-2}$ | $1.528 \times 10^7$ | 1 | $8.3 \times 10^{-2}$ |
| ohm-foot | $3.048 \times 10^{10}$ | $3.048 \times 10^7$ | $3.048 \times 10^{-1}$ | $3.3924 \times 10^{-11}$ | $3.048 \times 10^{-1}$ | $1.833 \times 10^8$ | 12 | 1 |

### B. Electrical Resistivity of the Elements

| Element | Temp (K) | ρ ($10^{-8}$Ωm) | Element | Temp (K) | ρ ($10^{-8}$Ωm) |
|---|---|---|---|---|---|
| Aluminum[r] | 273 | 2.417 | αGadolinium | 298 | 131.0 |
| | 293 | 2.650 | Gallium[w] | 293 | 17.4 |
| | 300 | 2.733 | Germanium | 297 | $46 \times 10^6$ |
| | 350 | 3.305 | Gold[r] | 273 | 2.051 |
| | 400 | 3.875 | | 283 | 2.214 |
| Antimony | 273 | 39.0 | | 300 | 2.271 |
| Arsenic | 293 | 33.3 | | 350 | 2.685 |
| Barium[r] | 273 | 30.2 | | 400 | 3.107 |
| | 293 | 33.2 | Hafnium[r] | 273 | 30.39 |
| | 300 | 34.3 | | 293 | 33.08 |
| | 350 | 42.4 | | 300 | 34.03 |
| | 400 | 51.4 | | 350 | 40.97 |
| Beryllium[r,p] | 273 | 3.02 | | 400 | 48.11 |
| | 293 | 3.56 | Holmium | 298 | 81.4 |
| | 300 | 3.76 | Indium | 293 | 8.37 |
| | 350 | 5.82 | Iodine | 293 | $1.3 \times 10^{15}$ |
| | 400 | 6.76 | Iridium | 293 | 5.3 |
| Bismuth | 273 | 106.8 | Iron[r] | 273 | 8.57 |
| Boron | 273 | $1.8 \times 10^{12}$ | | 293 | 9.61 |
| Cadmium | 273 | 6.83 | | 300 | 9.98 |
| Calcium[r] | 273 | 3.11 | | 350 | 16.1 |
| | 293 | 3.36 | | 400 | 23.7 |
| | 300 | 3.45 | αLanthanum[n] | 298 | 61.5 |
| | 350 | 4.09 | Lead | 293 | 20.65 |
| | 400 | 4.73 | Lithium[r] | 273 | 8.53 |
| Carbon | 273 | 1375. | | 293 | 9.28 |
| αCerium[p] | 298 | 74.4 | | 300 | 9.55 |
| βCerium[p] | 298 | 82.8 | | 360 | 11.45 |
| Cesium[r] | 273 | 18.75 | | 400 | 13.40 |
| | 293 | 20.46 | Lutetium | 298 | 58.2 |
| | 300 | 21.04 | Magnesium[r,p] | 273 | 4.05 |
| | 301.55 | 21.66 | | 293 | 4.38 |
| | 301.55 | 36.93ʸ | | 300 | 4.50 |
| Chromium | 273 | 12.9 | | 350 | 5.35 |
| Cobalt | 293 | 6.24 | | 400 | 6.18 |
| Copper[r] | 273 | 1,534 | Manganese[r] | 273 | 143.0 |
| | 293 | 1,678 | | 293 | 144.0 |
| | 300 | 1,725 | | 300 | 144.2 |
| | 350 | 2,063 | | 350 | 146.1 |
| | 400 | 2,402 | | 400 | 147.7 |
| αDysprosium[p] | 298 | 92.6 | Mercury | 323 | 98.4 |
| Erbium | 298 | 86.0 | Molybdenum | 273 | 4.85 |
| Europium | 298 | 90.0 | | 293 | 5.34 |

**Table 2—8 (continued)**
# ELECTRICAL RESISTIVITY AND TEMPERATURE COEFFICIENTS OF THE ELEMENTS

| Element | Temp (K) | ρ (10⁻⁸Ωm) | Element | Temp (K) | ρ (10⁻⁸Ωm) |
|---|---|---|---|---|---|
| | 300 | 5.52 | | 300 | 4.93 |
| | 350 | 6.76 | | 350 | 6.23 |
| | 400 | 8.02 | | 371 | 6.86 |
| α Neodymium | 298 | 64.3 | | 371 | 9.43$^y$ |
| Nickel$^r$ | 273 | 6.16 | Strontium$^r$ | 273 | 12.3 |
| | 293 | 6.03 | | 293 | 13.2 |
| | 300 | 7.20 | | 300 | 13.5 |
| | 350 | 9.34 | | 350 | 15.7 |
| | 400 | 11.78 | | 400 | 17.8 |
| Niobium | 273 | 12.5 | Sulfur (yellow) | 293 | $2 \times 10^{23}$ |
| Osmium | 293 | 9.5 | Tantalum | 273 | 12.20 |
| Palladium$^r$ | 273 | 9.78 | | 293 | 13.15 |
| | 293 | 10.54 | | 300 | 13.48 |
| | 300 | 10.80 | | 350 | 15.82 |
| | 350 | 12.67 | | 400 | 18.21 |
| | 400 | 14.48 | Tellurium | 298 | $4.36 \times 10^6$ |
| Phosphorus, white | 284 | $1 \times 10^{17}$ | α Terbium | 298 | 115.0 |
| Platinum | 293 | 10.6 | Thallium | 273 | 18.0 |
| Potassium | 273 | 6.49 | Thorium | 273 | 13.0 |
| | 293 | 7.20 | Thulium | 298 | 67.6 |
| | 300 | 7.47 | Tin | 273 | 11.0 |
| | 336.35 | 9.22 | Titanium | 293 | 42.0 |
| | 336.35$^y$ | 13.95$^y$ | Tungsten$^r$ | 273 | 4.82 |
| α Praseodymium | 298 | 70.0 | | 293 | 5.28 |
| α Promethium | 298 | 75 (est) | | 300 | 5.44 |
| Rhenium | 293 | 19.3 | | 400 | 7.83 |
| Rhodium | 293 | 4.51 | Vanadium$^r$ | 273 | 18.14 |
| Rubidium$^r$ | 273 | 11.54 | | 293 | 19.68 |
| | 293 | 12.84 | | 300 | 20.21 |
| | 300 | 13.32 | | 350 | 24.2 |
| | 312.64 | 14.21 | | 400 | 28.0 |
| | 312.64 | 36.93$^y$ | Ytterbium | 298 | 25.0 |
| Ruthenium | 273 | 7.6 | Yttrium | 298 | 59.6 |
| α Samarium | 298 | 94.0 | Zinc$^r$ | 273 | 5.479 |
| α Scandium$^r$ | 295 | 56.3 | | 293 | 5.964 |
| Selenium$^x$ | 273 | $10^6$ | | 300 | 6.13 |
| Silicon | 273 | $3.3 \times 10^{6x}$ | | 350 | 7.37 |
| Silver$^r$ | 273 | 1.467 | | 400 | 8.64 |
| | 293 | 1.587 | Zirconium$^r$ | 273 | 38.8 |
| | 300 | 1.629 | | 293 | 42.1 |
| | 350 | 1.932 | | 300 | 43.3 |
| | 400 | 2.241 | | 350 | 51.9 |
| Sodium$^r$ | 273 | 4.33 | | 400 | 60.3 |
| | 293 | 4.77 | | | |

| | |
|---|---|
| **a** | Uncertainty in resistivity is ±2%. |
| **b** | Uncertainty in resistivity is ±3%. |
| **c** | Uncertainty in resistivity is ±5%. |
| **d** | Uncertainty in resistivity is ±7% below 300 K and ±5% at 300 and 400 K. |
| **e** | Uncertainty in resistivity is ±7%. |
| **f** | Uncertainty in resistivity is ±8%. |
| **g** | Uncertainty in resistivity is ±10%. |
| **h** | Uncertainty in resistivity is ±12%. |
| **i** | Uncertainty in resistivity is ±4%. |
| **j** | Uncertainty in resistivity is 1%. |
| **k** | Uncertainty in resistivity is ±3% up to 300 K and ±4% above 300 K. |
| **m** | Uncertainty in resistivity is ±2% up to 300 K and ±4% above 300 K. |
| **n** | Crystal usually a mixture of α-hcp and fcc lattice. |
| **p** | Polycrystalline. |
| **r** | Data from *J. Phys. Chem. Ref. Data.* |
| **s** | Very sensitive to purity. |
| **w** | Hard wire. |
| **x** | Crystalline. |
| **y** | Liquid. |
| **z** | Zone refined bar. |
| **aa** | In temperature range where no experimental data are available. |

From Weast, R. C. Ed., *Handbook of Chemistry and Physics,* 69th ed., CRC Press, Boca Raton, Fla., 1988, F-125.

## Table 2—9

## BOND LENGTHS BETWEEN CARBON AND OTHER ELEMENTS

The following tables are based on bond distance determinations by experimental methods, mainly X-ray and electron diffraction, and include values published up to January 1, 1956. Values are given in Ångstrom units. In the present tables, for the sake of completeness, individual values of bond distances of lower accuracy are quoted, with limits of error indicated where possible. Values for tungsten and bismuth should be treated with particular caution.

According to the statistical theory of errors, if an average quantity $\bar{\mu}$ and a standard deviation $\sigma$ can be evaluated, there is a 95% probability that the true value lies within the interval $\bar{\mu} \pm 2\sigma$. Too much reliance should, however, not be placed on $\sigma$ values in bond distance determinations since the derivation of these certain sources of error may have been neglected. Values of bond lengths and limits of error are given in Ångstrom units.

| Group | Bond type | Element | | | | |
|---|---|---|---|---|---|---|
| I | All types | H[a] 1.056–1.115 | | | | |
| II | | Be 1.93 | | | | Hg 2.07 ± 0.01 |
| III | | B 1.56 ± 0.01 | Al 2.24 ± 0.04 | | In 2.16 ± 0.04 | |
| IV | All types | C[a] 1.54 –1.20 | Si 1.865 ± 0.008 | Ge 1.98 ± 0.03 | Sn 2.143 ± 0.008 | Pb 2.29 ± 0.05 |
| | Alkyls $(CH_3XH_3)$ | | Si 1.84 ± 0.01 | | Sn 2.18 ± 0.02 | |
| | Aryl $(C_6H_5XH_3)$ | | Si 1.88 ± 0.01 | | | |
| | Negatively substituted $(CH_3XCl_3)$ | | | | | |
| V | All types | N[a] 1.47 –1.1 | P 1.87 ± 0.02 | As 1.98 ± 0.02 | Sb 2.202 ± 0.016 | Bi 2.30[b] |
| | Paraffinic $(CH_3)_3X$ | | | | | |
| VI | | O[a] 1.43 –1.15 | S[a] 1.81 –1.55 | Se 1.98 –1.71; Cr 1.92 ± 0.04 | Te 2.05 ± 0.14; Mo 2.08 ± 0.04 | W 2.06 ± 0.01[b] |

## Table 2—9 (continued)
### BOND LENGTHS BETWEEN CARBON AND OTHER ELEMENTS

| Group | Bond type | Element | | | |
|---|---|---|---|---|---|
| | | F | Cl | Br | I |
| VII | Paraffinic (monosubstituted) (CH$_3$X) | 1.381 ± 0.005 | 1.767 ± 0.002 | 1.937 ± 0.003 | 2.13$_5$ ± 0.01 |
| | Paraffinic (disubstituted) (CH$_2$X$_2$) | 1.334 ± 0.004 | 1.767 ± 0.002 | 1.937 ± 0.003 | 2.13$_5$ ± 0.1 |
| | Olefinic (CH$_2$:CHX) | 1.32$_5$ ± 0.1 | 1.72 ± 0.01 | 1.89 ± 0.01 | 2.092 ± 0.005 |
| | Aromatic (C$_6$H$_5$X) | 1.30 ± 0.01 | 1.70 ± 0.01 | 1.85 ± 0.01 | 2.05 ± 0.01 |
| | Acetylenic (HC:CX) | | 1.635 ± 0.004 | 1.79$_5$ ± 0.01 | 1.99 ± 0.02 |
| | | Fe | Co | Ni | Pd |
| VIII | | 1.84 ± 0.02 | 1.83 ± 0.02 | 1.82 ± 0.03 | 2.27 ± 0.04 |

[a]See following individual tables.
[b]Error uncertain.

Reproduced from International Tables for X-ray Crystallography.

**Table 2—9 (continued)**
## BOND LENGTHS BETWEEN CARBON AND OTHER ELEMENTS

### Carbon—Carbon

**Single Bond**

| | | |
|---|---|---|
| 1. Paraffinic | 1.541 | ± 0.003 |
| 2. In diamond (at 18°C) | 1.54452 | ± 0.00014 |

**Partial Double Bond**

| | | |
|---|---|---|
| 1. Shortening of a single bond in the presence of one carbon−carbon double bond, e.g., $(CH_3)_2.C{:}CH_2$, or of an aromatic ring, e.g., $C_6H_5.CH_3$ | 1.53 | ± 0.01 |
| 2. Shortening in the presence of two carbon−oxygen double bond, e.g., $CH_3CHO$ | 1.516 | ± 0.005 |
| 3. Shortening in th presence of two carbon−oxygen double bonds, e.g., $(CO_2H)_2$ | 1.49 | ± 0.01 |
| 4. Shortening in the presence of one carbon−carbon triple bond, e.g., $CH_3.C{:}CH$ | 1.460 | ± 0.003 |
| 5. In compounds with a tendency to dipole formation, e.g., $C{:}C.C{:}N$ | 1.44 | ± 0.01 |
| 6. In graphite (at 15°C) | 1.4210 | ± 0.0001 |
| 7. In aromatic compounds | 1.395 | ± 0.003 |
| 8. In the presence of two carbon−carbon triple bonds, e.g., $HC{:}C.C{:}CH$ | 1.373 | ± 0.004 |

**Double Bond**

| | | |
|---|---|---|
| 1. Simple | 1.337 | ± 0.006 |
| 2. Partial triple bond, e.g., $CH_2{:}C{:}CH_2$ | 1.309 | ± 0.005 |

**Triple Bond**

| | | |
|---|---|---|
| 1. Simple, e.g., $C_2H_2$ | 1.204 | ± 0.002 |
| 2. Conjugated, e.g., $CH_3.(C{:}C)_2.H$ | 1.206 | ± 0.004 |

### Carbon—Hydrogen

| | | |
|---|---|---|
| 1. Paraffinic | | |
| (a) in methane | 1.091 | |
| (b) in monosubstituted carbon | 1.101 | ± 0.003 |
| (c) in disubstituted carbon | 1.073 | ± 0.004 |
| (d) in trisubstituted carbon | 1.070 | ± 0.007 |
| 2. Olefinic, e.g., $CH_2{:}CH_2$ | 1.07 | ± 0.01 |
| 3. Aromatic in $C_6H_6$ | 1.084 | ± 0.006 |
| 4. Acetylenic, e.g., $CH{:}C.X$ | 1.056 | ± 0.003 |
| 5. Shortening in the presence of a carbon triple bond, e.g., $CH_3CN$ | 1.115 | ± 0.004 |
| 6. In small rings, e.g., $(CH_2)_2S$ | 1.081 | ± 0.007 |

### Carbon—Nitrogen

**Single Bond**

| | | |
|---|---|---|
| 1. Paraffinic | | |
| (a) 4-covalent nitrogen | 1.479 | ± 0.005 |
| (b) 3-covalent nitrogen | 1.472 | ± 0.005 |
| 2. In C−N=, e.g., $CH_3NO_2$ | 1.475 | ± 0.010 |
| 3. Aromatic in $C_6H_5NHCOCH_3$ | 1.426 | ± 0.012 |

**Table 2—9 (continued)**
# BOND LENGTHS BETWEEN CARBON AND OTHER ELEMENTS

## Carbon—Nitrogen (continued)

### Single Bond (continued)

| | | |
|---|---|---|
| 4. Shortened (partial double bond) in heterocyclic systems, e.g., $C_5H_5N$ | 1.352 | ± 0.005 |
| 5. Shortened (partial double bond) in N–C=O, e.g., HCONH$_2$ | 1.322 | ± 0.003 |

### Triple Bond

| | | |
|---|---|---|
| 1. In R.C:N | 1.158 | ± 0.002 |

## Carbon—Oxygen

### Single Bond

| | | |
|---|---|---|
| 1. Paraffinic | 1.43 | ± 0.01 |
| 2. Strained, e.g., epoxides | 1.47 | ± 0.01 |
| 3. Shortened (partial double bond) as in carboxylic acids or through the influence of an aromatic ring, e.g., salicylic acid | 1.36 | ± 0.01 |

### Double Bond

| | | |
|---|---|---|
| 1. In aldehydes, ketones, carboxylic acids, esters | 1.23 | ± 0.01 |
| 2. In zwitterion forms, e.g., DL-serine | 1.26 | ± 0.01 |
| 3. Shortened (partial triple bond) as in conjugated systems | 1.207 | ± 0.006 |
| 4. Partial triple bond as in acyl halides or isocyanates | 1.17 | ± 0.01 |

## Carbon—Sulfur

### Single Bond

| | | |
|---|---|---|
| 1. Paraffinic, e.g., $CH_3SH$ | 1.81(5) | ± 0.01 |
| 2. Lengthened in the presence of fluorine, e.g., $(CF_3)_2S$ | 1.83(5) | ± 0.01 |
| 3. Shortened (partial double bond) as in heterocyclic systems, e.g., | 1.73 | ± 0.01 |

### Double Bond

| | | |
|---|---|---|
| 1. In ethylene thiourea | 1.71 | ± 0.02 |
| 2. Shortened (partial triple bond) in the presence of a second carbon double bond, e.g., COS | 1.558 | ± 0.003 |

From Weast, R. C. Ed., *Handbook of Chemistry and Physics*, 69th ed., CRC Press, Boca Raton, Fla., 1988, F-165.

## Table 2—10
## BOND LENGTHS OF ELEMENTS

| Element | Bond | Bond length (Å) |
|---|---|---|
| Ac | Ac–Ac | 3.756 |
| Ag (25°C) | Ag–Ag | 2.8894 |
| Al (25°C) | Al–Al | 2.863 |
| As | As–As | 2.49 |
| As$_4$ | As–As | 2.44 ± 0.03 |
| Au (25°C) | Au–Au | 2.8841 |
| B$_2$ | B–B | 1.589 |
| Ba (room temperature) | Ba–Ba | 4.347 |
| Be | Be–Be | |
| α-form (20°C) | | 2.2260 |
| Bi (25°C) | Bi–Bi | 3.09 |
| Br$_2$ | Br–Br | 2.290 |
| Ca | Ca–Ca | |
| α-form (18°C) | | 3.947 (f.c.c.) |
| β-form (500°C) | | 3.877 (b.c.c.) |
| Cd (21°C) | Cd–Cd | 2.9788 |
| Cl$_2$ | Cl–Cl | 1.988 |
| Ce | Ce–Ce | 3.650 |
| Co | Co–Co | 2.5061 |
| Cr | Cr–Cr | |
| α-form (20°C) | | 2.4980 |
| β-form (>1,850°C) | | 2.61 |
| Cs (–10°C) | Cs–Cs | 5.309 |
| Cu (20°C) | Cu–Cu | 2.5560 |
| Dy | Dy–Dy | 3.503 |
| Er | Er–Er | 3.468 |
| Eu | Eu–Eu | 3.989 |
| F$_2$ | F–F | 1.417 ± 0.001 |
| Fe | Fe–Fe | |
| α-form (20°C) | | 2.4823 (b.c.c.) |
| γ-form (916°C) | | 2.578 (f.c.c.) |
| δ-form (1,394°C) | | 2.539 (b.c.c.) |
| Ga (20°C) | Ga–Ga | 2.442 |
| Gd (20°C) | Gd–Gd | 3.573 |
| Ge (20°C) | Ge–Ge | 2.4498 |
| H$_2$ | H–H in H$_2$ | 0.74611 |
| | H–D in HD | 0.74136 |
| | D–D in D$_2$ | 0.74164 |
| He | He–He in (He$_2$)$^+$ | 1.08$_0$ |
| Hf | Hf–Hf | |
| α-form (24°C) | | 3.1273 (h.c.p.) |
| Hg (–46°C) | Hg–Hg | 3.005 |
| Ho | Ho–Ho | 3.486 |
| I$_2$ | I–I | 2.662 |
| In (20°C) | In–In | 3.2511 |
| Ir (room temperature) | Ir–Ir | 2.714 |
| K (195°C) | K–K | 4.544 |
| La | La–La | |
| α-form | | 3.739 (h.c.p.) |
| β-form | | 3.745 (f.c.c.) |
| Li (20°C) | Li–Li | 3.0390 |
| Lu | Lu–Lu | 3.435 |
| Mg (25°C) | Mg–Mg | 3.1971 |
| Mn | Mn–Mn | |
| γ-form (1,095°C) | | 2.7311 (f.c.c.) |
| δ-form (1,134°C) | | 2.6679 (b.c.c.) |

## Table 2—10 (continued)
## BOND LENGTHS OF ELEMENTS

| Element | Bond | Bond length (Å) |
|---|---|---|
| Mo (20°C) | Mo—Mo | 2.7251 |
| $N_2$ | N—N | $1.0975_8 \pm 0.0001$ |
| Na (20°C) | Na—Na | 3.7157 |
| Nb (20°C) | Nb—Nb | 2.8584 |
| Nd | Nd—Nd | 3.628 |
| Ni (18°C) | Ni—Ni | 2.4916 |
| Np | Np—Np | |
| α-form (20°C) | | 2.60 (orthorh.) |
| β-form (313°C) | | 2.76 (tetra.) |
| γ-form 600°C | | 3.05 (b.c.c.) |
| $O_2$ | O—O | 1.208 |
| $O_3$ (angle = 116.8 ± 0.5°) | O—O | 1.278 ± 0.003 |
| Os (20°C) | Os—Os | 2.6754 |
| P (black) | P—P | 2.18 |
| $P_4$ | P—P | 2.21 ± 0.02 |
| Pa | Pa—Pa | 3.212 |
| Pb (25°C) | Pb—Pb | 3.5003 |
| Pd (25°C) | Pd—Pd | 2.7511 |
| Po | Po—Po | |
| α-form (10°C) | | 3.345 (cub.) |
| β-form (75°C) | | 3.359 (rhbdr.) |
| Pr | Pr—Pr | |
| α-form | | 3.640 (tetra.) |
| β-form | | 3.649 (f.c.c.) |
| Pt (20°C) | Pt—Pt | 2.746 |
| Pu | Pu—Pu | |
| γ-form (235°C) | | 3.026 (f.c.c.) |
| δ-form (313°C) | | 3.279 (f.c.c.) |
| ε-form | | 3.150 (b.c.c.) |
| Rb (20°C) | Rb—Rb | 4.95 |
| Re (room temperature) | Re—Re | 2.741 |
| Rh (20°C) | Rh—Rh | 2.6901 |
| Ru (25°C) | Ru—Ru | 2.6502 |
| $S_2$ | S—S | 1.887 |
| $S_8$ | S—S | 2.07 ± 0.02 |
| Sb (25°C) | Sb—Sb | 2.90 |
| Sc (room temperature) | Sc—Sc | 3.212 |
| Se (20°C) | Se—Se | 2.321 |
| $Se_2$ | Se—Se | 2.152 ± 0.003 |
| $Se_8$ | Se—Se | 2.32 ± 0.003 |
| Si (20°C) | Si—Si | 2.3517 |
| Sn | Sn—Sn, diamond-type lattice | |
| α-form (20°C) | | 2.8099 |
| β-form (25°C) | | 3.022 (tetra.) |
| Sr | Sr—Sr | |
| α-form (25°C) | | 4.302 (f.c.c.) |
| β-form (248°C) | | 4.32　(h.c.p.) |
| γ-form (614°C) | | 4.20 (b.c.c.) |
| Ta (20°C) | Ta—Ta | 2.86 |
| Tb | Tb—Tb | 3.525 |
| Tc (room temperature) | Tc-Tc | 2.703 |
| Te (25°C) | Te—Te | 2.846 |
| Th | Th—Th | |
| α-form (25°C) | | 3.595 (f.c.c.) |
| β-form (1,450°C) | | 3.56 (b.c.c.) |

Table 2—10 (continued)

## BOND LENGTHS OF ELEMENTS

| Element | Bond | Bond length (Å) |
|---|---|---|
| Ti | Ti–Ti | |
| α-form (25°C) | | 2.8956 (h.c.p.) |
| β-form (900°C) | | 2.8636 (b.c.c. |
| Tl | Tl–Tl | |
| α-form (18°C) | | 3.4076 (h.c.p.) |
| β-form (262°C) | | 3.362 (b.c.c.) |
| Tm | Tm–Tm | 3.447 |
| U | U–U | |
| α-form | | 2.77 |
| β-form (805°C) | | 3.058 (b.c.c.) |
| V (30°C) | V–V | 2.6224 |
| W (25°C) | W–W | 2.7409 |
| Y | Y–Y | 3.551 |
| Yb | Yb–Yb | 3.880 |
| Zn (25°C) | Zn–Zn | 2.6694 |
| Zr | Zr–Zr | 3.179 |

From Kennard, O., in *Handbook of Chemistry and Physics,* 69th ed., Weast, R. C., Ed., CRC Press, Boca Raton, Fla., 1988, F-166.

Table 2—11

## BOND LENGTH AND ANGLE VALUES BETWEEN ELEMENTS

| Elements | In | Bond length (Å) | Bond angle (°) |
|---|---|---|---|
| | | **Boron** | |
| B–B | $B_2H_6$ | 1.770  ± 0.013 | H–B–H 121.5 ± 7.5 |
| B–Br | B Br | $1.88_7$ | – |
| | B Br$_3$ | 1.87  ± 0.02 | Br–B–Br 120 ± 6 |
| B–Cl | BCl | $1.715_7$ | – |
| | BCl$_3$ | 1.72  ± 0.01 | Cl–B–Cl 120 ± 3 |
| B–F | BF | 1.262 | – |
| | BF$_3$ | $1.29_5$  ± 0.01 | F–B–F 120 |
| B–H | Hydrides | 1.21  ± 0.02 | – |
| B–H bridge | Hydrides | 1.39  ± 0.02 | – |
| B–N | (BClNH)$_3$ | 1.42  ± 0.01 | B–N–B 121 |
| B–O | BO | 1.2049 | – |
| | B(OH)$_3$ | 1.362  ± 0.005 (av.) | O–B–O 119.7 |
| | | **Nitrogen** | |
| N–Cl | $NO_2Cl$ | 1.79  ± 0.02 | – |
| N–F | $NF_3$ | 1.36  ± 0.02 | F–N–F 102.5 ± 1.5 |
| N–H | $[NH_4]^+$ | 1.034  ± 0.003 | – |
| | NH | 1.038 | – |
| | ND | 1.041 | – |
| | HNCS | 1.013  ± 0.005 | H–N–C 130.25 ± 0.25 |
| N–N | $N_3H$ | 1.02  ± 0.01 | H–N–N′ 112.65 ± 0.5 |
| | $N_2O$ | 1.126  ± 0.002 | – |
| | $[N_2]^+$ | $1.116_2$ | – |
| N–O | $NO_2Cl$ | 1.24  ± 0.01 | O–N–O 126 ± 2 |
| | $NO_2$ | 1.188  ± 0.005 | O–N–O 134.1 ± 0.25 |

## Table 2—11 (continued)
## BOND LENGTH AND ANGLE VALUES BETWEEN ELEMENTS

| Elements | In | Bond length (Å) | Bond angle (°) |
|---|---|---|---|
| | | **Nitrogen (continued)** | |
| N=O | $N_2O$ | 1.186 ± 0.002 | – |
| | $[NO]^+$ | 1.0619 | – |
| N–Si | SiN | 1.572 | – |
| | | **Oxygen** | |
| O–H | $[OH]^+$ | 1.0289 | – |
| | OD | 0.9699 | – |
| | $H_2O_2$ | 0.960 ± 0.005 | O–O–H 100 ± 2 |
| O–O | $H_2O_2$ | 1.48 ± 0.01 | – |
| | $[O_2]^+$ | 1.227 | – |
| | $[O_2]^-$ | 1.26 ± 0.02 | – |
| | $[O_2]^{--}$ | 1.49 ± 0.02 | – |
| | | **Phosphorus** | |
| P–D | PD | 1.429 | – |
| P–H | $[PH_4]^+$ | 1.42 ± 0.02 | – |
| P–N | PN | 1.4910 | – |
| P–S | $PSBr_3(Cl_3, F_3)$ | 1.86 ± 0.02 | – |
| | | **Sulfur** | |
| S–Br | $SOBr_2$ | 2.27 ± 0.02 | Br–S–Br 96 ± 2 |
| S–F | $SOF_2$ | 1.585 ± 0.005 | F–S–F 92.8 ± 1 |
| S–D | SD | 1.3473 | – |
| | $SD_2$ | 1.345 | – |
| S–O | $SO_2$ | 1.4321 | O–S–O 119.54 |
| | $SOCl_2$ | 1.45 ± 0.02 | |
| S–S | $S_2Cl_2$ | 2.04 ± 0.01 | – |
| | | **Silicon** | |
| Si–Br | $SiBr_4$ | 2.17 ± 1.01 | – |
| Si–Cl | $SiCl_4$ | 2.03 ± 1.01 (av.) | – |
| Si–F | $SiF_4$ | 1.561 ± 0.003 (av.) | – |
| SiH | $SiH_4$ | 1.480 ± 0.005 | – |
| Si–O | $[SiO]^+$ | 1.504 | – |
| Si–Si | $Si_2Cl_2$ | 2.30 ± 0.02 | – |

From Kennard, O., in *Handbook of Chemistry and Physics,* 69th ed., Weast, R. C., Ed., CRC Press, Boca Raton, Fla., 1988, F-167.

## Table 2—12
## THERMODYNAMIC PROPERTIES OF ELEMENTS

Thermodynamic calculations over a wide range of temperatures are generally made with the aid of algebraic equations representing the characteristic properties of the substances being considered. The necessary integrations and differentiations, or other mathematical manipulations, are then most easily effected.

The most convenient starting point in making such calculations for a given substance is the heat capacity at constant pressure. From this quantity and a knowledge of the properties of any phase transitions, the other thermodynamic properties may be computed by the well-known equations given in standard texts on thermodynamics.

Users of the following equations and relevant tables are cautioned that the units for a, b, c, and d are cal/g mole, whereas those for A are kcal/g mole. The necessary adjustment must be made when the data are substituted into the equations.

Empirical heat capacity equations are generally in the form of a power series, with the absolute temperature T as the independent variable:

$$C_p = a' + (b' \times 10^{-3})T + (c' \times 10^{-6})T_2$$

or

$$C_p = a'' + (b'' \times 10^{-3})T + \frac{d \times 10^5}{T^2}.$$

Since both forms are used in the ensuing, let

$$C_p = a + (b \times 10^{-3})T + (c \times 10^{-6})T^2 + \frac{d \times 10^5}{T^2}. \qquad (1)$$

The constants a, b, c, and d are to be determined either experimentally or by some theoretical or semi-empirical approach.

The heat content, or enthalpy (H), is determined from the heat capacity by a simple integration of the range of temperatures for which (1) is applicable. Thus, if 298K is taken as a reference temperature,

$$H_T - H_{298} = \int_{298}^{T} C_p dT \qquad (2)$$

$$= a(T - 298) + \tfrac{1}{2}(b \times 10^{-3})(T_2 - 298^2) + \tfrac{1}{3}(c \times 10^{-6})(T^3 - 298^3) - (d \times 10^5)\left(\frac{1}{T} - \frac{1}{298}\right)$$

$$= aT + \tfrac{1}{2}(b \times 10^{-3})T^2 + \tfrac{1}{3}(c \times 10^{-6})T^3 - \frac{d \times 10^5}{T} - A,$$

where all the constants on the right-hand side of the equation have been incorporated in the term −A.

In general, the enthalpy is given by a sum of terms such as (2) for each phase of the substance involved in the temperature range considered plus terms that represent the heats of transitions:

$$H_T - H_{298} = \Sigma \int_{T_1}^{T_2} C_p dT + \Sigma \Delta H_{tr}.$$

In a similar manner, the entropy S is obtained from (1) by performing the integration

$$S_T - S_{298} = \int_{298}^{T} (C_p/T)dt \qquad (3)$$

$$= a\ln(T/298) + (b \times 10^{-3})(T - 298) + \tfrac{1}{2}(c \times 10^{-6})(T^2 - 298^2) - \tfrac{1}{2}(d \times 10^5)\left(\frac{1}{T^2} - \frac{1}{298^2}\right)$$

$$= a\ln T + (b \times 10^{-3})T + \tfrac{1}{2}(c \times 10^{-6})T^2 - \frac{\tfrac{1}{2}(d \times 10^5)}{T^2} - B'$$

or

$$S_T = 2.303\, a \log T + (b \times 10^{-3})T + \tfrac{1}{2}(c \times 10^{-6})T^2 - \frac{\tfrac{1}{2}(d \times 10^5)}{T^2} - B \qquad (4)$$

**Table 2—12 (continued)**
**THERMODYNAMIC PROPERTIES OF ELEMENTS**

where

$$B = B' - S_{298}. \tag{5}$$

the quantity

$$F_T - H_{298} = (H_T - H_{298}) - TS_T$$

From the definition of free energy (F):

is obtained from (2) and (4):

$$F = H - TS$$

$$F_T - H_{298} = -2.303aT \log T - \tfrac{1}{2}(b \times 10^{-3})T^2 - \frac{1}{6}(c \times 10^{-6})T^3 - \frac{\tfrac{1}{2}(d \times 10^5)}{T} + (B + a)T - A \tag{6}$$

and also the free energy function

$$\frac{F_T - H_{298}}{T} = -2.303a \log T - \tfrac{1}{2}(b \times 10^{-3})T - \frac{1}{6}(c \times 10^{-6})T^2 - \frac{\tfrac{1}{2}(d \times 10^5)}{T^2} + (B + a) - \frac{A}{T} \tag{7}$$

Values of the constants for elements are given in Table 2—24. The first column in each table lists the element. The second column gives the phase to which they are applicable. The third, fourth, and fifth columns specify the thermodynamic properties for the transition to the succeeding phase. In column 6, the value of the entropy at 298.15K, the reference temperature, is given. The remaining columns, except for the last, give the values of the constants a,b, c, d, A, and B required in the thermodynamic equations.

All values that represent estimates are enclosed in parentheses. The heat capacities at temperatures beyond the range of experimental determination were estimated by extrapolation. Where no experimental values were found, analogy with compounds of neighboring elements in the Periodic Table was employed.

From Weast, R. C., Ed., *Handbook of Chemistry and Physics,* 55th ed., CRC Press, Cleveland, 1974, D-55.

Table 2—12

THERMODYNAMIC PROPERTIES OF THE ELEMENTS

| Element | Phase | Temperature of transition (K) | Heat of transition (kcal/g mole) | Entropy of transition (e.u.) | Entropy at 298K (e.u.) | a (cal/g mole) | b (cal/g mole) | c (cal/g mole) | d (cal/g mole) | A (kcal/g mole) | B (e.u.) |
|---|---|---|---|---|---|---|---|---|---|---|---|
| Ac | solid | (1090) | (2.5) | (2.3) | (13) | (5.4) | (3.0) | — | — | (1.743) | (18.7) |
|    | liquid | (2750) | (70) | (25) | — | (8) | — | — | — | (0.295) | (31.3) |
| Ag | solid | 1234 | 2.855 | 2.313 | 10.20 | 5.09 | 1.02 | — | 0.36 | 1.488 | 19.21 |
|    | liquid | 2485 | 60.72 | 24.43 | — | 7.30 | — | — | — | 0.164 | 30.12 |
|    | gas | — | — | — | — | (4.97) | — | — | — | (−66.34) | (−12.52) |
| Al | solid | 931.7 | 2.57 | 2.76 | 6.769 | 4.94 | 2.96 | — | — | 1.604 | 22.26 |
|    | liquid | 2600 | 67.9 | 26 | — | 7.0 | — | — | — | 0.33 | 30.83 |
| Am | solid | (1200) | (2.4) | (2.0) | (13) | (4.9) | (4.4) | — | — | (1.657) | (16.2) |
|    | liquid | 2733 | 51.7 | 18.9 | — | (8.5) | — | — | — | (0.409) | (34.5) |
| As | solid | 883 | 31/4 | 35.1/4 | 8.4 | 5.17 | 2.34 | — | — | 1.646 | 21.8 |
| Au | solid | 1336.16 | 3.03 | 2.27 | 11.32 | 6.14 | −0.175 | 0.92 | — | 1.831 | 23.65 |
|    | liquid | 2933 | 74.21 | 25.30 | — | 7.00 | — | — | — | −0.631 | 26.99 |
| B | solid | 2313 | (3.8) | (1.6) | 1.42 | 1.54 | 4.40 | — | — | 0.655 | 8.67 |
|   | liquid | 2800 | 75 | 27 | — | (6.0) | — | — | — | (−4.599) | (31.4) |
| Ba | solid, α | 648 | 0.14 | 0.22 | 16 | 5.55 | 4.50 | — | — | 1.722 | 16.1 |
|    | solid, β | 977 | 1.83 | 1.87 | — | 5.55 | 1.50 | — | — | 1.582 | 15.9 |
|    | liquid | 1911 | 35.665 | 18.63 | — | (7.4) | — | — | — | (0.843) | (25.3) |
|    | gas | — | — | — | — | (4.97) | — | — | — | (−39.65) | (−11.7) |
| Be | solid | 1556 | 2.919 | 1.501 | 2.28 | 5.07 | 1.21 | — | −1.15 | 1.951 | 27.62 |
|    | liquid | — | — | — | — | 5.27 | — | — | — | −1.611 | 25.68 |
| Bi | solid | 544.2 | 2.63 | 4.83 | 13.6 | 5.38 | 2.60 | — | — | 1.720 | 17.8 |
|    | liquid | 1900 | 41.1 | 21.6 | — | 7.60 | — | — | — | −0.087 | 25.6 |
|    | gas | — | — | — | — | (4.97) | — | — | — | (−46.19) | (−15.9) |
| C | solid | — | — | — | 1.3609 | 4.10 | 1.02 | — | −2.10 | 1.972 | 23.484 |

**Table 2—12 (continued)**
**THERMODYNAMIC PROPERTIES OF THE ELEMENTS**

| Element | Phase | Temperature of transition (K) | Heat of transition (kcal/g mole) | Entropy of transition (e.u.) | Entropy at 298K (e.u.) | a (cal/g mole) | b (cal/g mole) | c (cal/g mole) | d (cal/g mole) | A (kcal/g mole) | B (e.u.) |
|---|---|---|---|---|---|---|---|---|---|---|---|
| Ca | solid, α | 723 | 0.24 | 0.33 | 9.95 | 5.24 | 3.50 | — | — | 1.718 | 20.95 |
| | solid, β | 1123 | 2.2 | 1.96 | — | 6.29 | 1.40 | — | — | 1.689 | 26.01 |
| | liquid | 1755 | 38.6 | 22.0 | — | 7.4 | — | — | — | -0.147 | 30.28 |
| | gas | — | — | — | — | (4.97) | — | — | — | (-43.015;) | (-9.88) |
| Cd | solid | 594.1 | 1.46 | 2.46 | 12.3 | 5.31 | 2.94 | — | — | 1.714 | 18.8 |
| | liquid | 1040 | 23.86 | 22.94 | — | 7.10 | — | — | — | 0.798 | 26.1 |
| | gas | — | — | — | — | (4.97) | — | — | — | (-25.28) | (-11.7) |
| Ce | solid | 1048 | 2.1 | 2.0 | 13.8 | 4.40 | 6.0 | — | — | 1.579 | 13.1 |
| | liquid | 2800 | 73 | 26 | — | (7.9) | — | — | — | (-0.148) | (29.1) |
| $Cl_2$ | gas | — | — | — | 53.286 | 8.76 | 0.27 | — | -0.65 | 2.845 | -2.929 |
| Co | solid, α | 723 | 0.005 | 0.007 | 6.8 | 4.72 | 4.30 | — | — | 1.598 | 21.4 |
| | solid, β | 1398 | 0.095 | 0.068 | — | 3.30 | 5.86 | — | — | 0.974 | 3.1 |
| | solid, γ | 1766 | 3.7 | 2.1 | — | 9.60 | — | — | — | 3.961 | 50.5 |
| | liquid | 3370 | 93 | 28 | — | 8.30 | — | — | — | -2.034 | 38.7 |
| Cr | solid | 2173 | 3.5 | 1.6 | 5.68 | 5.35 | 2.36 | — | — | 1.848 | 25.75 |
| | liquid | 2495 | 72.97 | 29.25 | — | 9.40 | — | — | — | 1.556 | 50.13 |
| | gas | — | — | — | — | (4.97) | — | — | -0.44 | (-82.47) | (-13.8) |
| Cs | solid | 301.9 | 0.50 | 1.7 | 19.8 | 7.42 | — | — | — | 2.212 | 22.5 |
| | liquid | 963 | 16.32 | 17.0 | — | 8.00 | — | — | — | 1.887 | 24.1 |
| | gas | — | — | — | — | (4.97) | — | — | — | (-17.35) | (-13.6) |
| Cu | solid | 1356.2 | 3.11 | 2.29 | 7.97 | 5.41 | 1.50 | — | — | 1.680 | 23.30 |
| | liquid | 2868 | 72.8 | 25.4 | — | 7.50 | — | — | — | 0.024 | 34.05 |
| $F_2$ | gas | — | — | — | 48.58 | 8.29 | 0.44 | — | -0.80 | 2.760 | -0.76 |
| Fe | solid, α | 1033 | 0.410 | 0.397 | 6.491 | 3.37 | 7.10 | — | 0.43 | 1.176 | 14.59 |
| | solid, β | 1180 | 0.217 | 0.184 | — | 10.40 | — | — | — | 4.281 | 55.66 |
| | solid, γ | 1673 | 0.15 | 0.084 | — | 4.85 | 3.00 | — | — | 0.396 | 19.76 |
| | solid, δ | 1808 | 3.86 | 2.14 | — | 10.30 | — | — | — | 4.382 | 55.11 |
| | liquid | 3008 | 84.62 | 28.1 | — | 10.00 | — | — | — | -0.021 | 50.73 |

Table 2—12 (continued)
THERMODYNAMIC PROPERTIES OF THE ELEMENTS

| Element | Phase | Temperature of transition (K) | Heat of transition (kcal/g mole) | Entropy of transition (e.u.) | Entropy at 298K (e.u.) | a (cal/g mole) | b (cal/g mole) | c (cal/g mole) | d (cal/g mole) | A (kcal/g mole) | B (e.u.) |
|---|---|---|---|---|---|---|---|---|---|---|---|
| Ga | solid | 302.94 | 1.335 | 4.407 | 9.82 | 5.237 | 3.33 | — | — | 1.710 | 21.01 |
|    | liquid | 2700 | — | — | — | (6.645) | — | — | — | (0.648) | (23.64) |
| Ge | solid | 1232 | 8.3 | 6.7 | 10.1 | 5.90 | 1.13 | — | — | 1.764 | 23.8 |
|    | liquid | 2980 | 68 | 23 | — | (7.3) | — | — | — | (-5.668) | (25.7) |
| $H_2$ | gas | — | — | — | 31.211 | 6.62 | 0.81 | — | — | 2.010 | 6.75 |
| Hf | solid | (2600) | (6.0) | (2.3) | 13.1 | (6.00) | (0.52) | — | — | (1.812) | (21.2) |
| Hg | liquid | 629.73 | 13.985 | 22.208 | 18.46 | — | — | — | — | 1.971 | 19.20 |
|    | gas | — | — | — | — | 4.969 | — | — | — | -13.048 | -13.54 |
| In | solid | 430 | 0.775 | 1.80 | 13.88 | 5.81 | 2.50 | — | — | 1.844 | 19.97 |
|    | liquid | 2440 | 53.8 | 22.0 | — | 7.50 | — | — | — | 1.564 | 27.34 |
|    | gas | — | — | — | — | (4.97) | — | — | — | (-58.42) | (-14.46) |
| Ir | solid | 2727 | 6.6 | 2.4 | 8.7 | 5.56 | 1.42 | — | — | 1.721 | 23.4 |
| K | solid | 336.4 | 0.5575 | 1.657 | 15.2 | 1.3264 | 19.405 | 2.9369 | — | 1.258 | -1.86 |
|    | liquid | 1052 | 18.88 | 17.95 | — | 8.8825 | -4.565 | — | — | 1.923 | 32.55 |
|    | gas | — | — | — | — | (4.97) | — | — | — | (-19.689) | (-9.46) |
| La | solid | 1153 | (2.3) | (2.0) | 13.7 | 6.17 | 1.60 | — | — | 1.911 | 21.9 |
|    | liquid | 3000 | 80 | 27 | — | (7.3) | — | — | — | (-0.15) | (26.0) |
| Li | solid | 459 | 0.69 | 1.5 | 6.70 | 3.05 | 8.60 | — | — | 1.292 | 12.92 |
|    | liquid | 1640 | 32.48 | 19.81 | — | 7.0 | — | — | — | 1.509 | 32.00 |
|    | gas | — | — | — | — | (4.97) | — | — | — | (-34.30) | (-2.84) |
| Mg | solid | 923 | 2.2 | 2.4 | 7.77 | 5.33 | 2.45 | — | -0.103 | 1.733 | 23.39 |
|    | liquid | 1393 | 31.5 | 22.6 | — | (8.0) | — | — | — | 0.942 | 36.967 |
|    | gas | — | — | — | — | (4.97) | — | — | — | (-34.78) | (-7.60) |

## Table 2—12 (continued)
## THERMODYNAMIC PROPERTIES OF THE ELEMENTS

| Element | Phase | Temperature of transition (K) | Heat of transition (kcal/g mole) | Entropy of transition (e.u.) | Entropy at 298K (e.u.) | a (cal/g mole) | b (cal/g mole) | c (cal/g mole) | d (cal/g mole) | A (kcal/g mole) | B (e.u.) |
|---|---|---|---|---|---|---|---|---|---|---|---|
| Mn | solid, α | 1000 | 0.535 | 0.535 | 7.59 | 6.70 | 3.38 | — | -0.37 | 1.974 | 26.11 |
|  | solid, β | 1374 | 0.545 | 0.397 | — | 8.33 | 0.66 | — | — | 2.672 | 41.02 |
|  | solid, γ | 1410 | 0.430 | 0.305 | — | 10.70 | — | — | — | 4.760 | 56.84 |
|  | solid, δ | 1517 | 3.5 | 2.31 | — | 11.30 | — | — | — | 5.176 | 60.88 |
|  | liquid | 2368 | 53.7 | 22.7 | — | 11.00 | — | — | — | 1.221 | 56.38 |
|  | gas | — | — | — | — | 6.26 | — | — | — | -63.704 | -3.13 |
| Mo | solid | 2883 | (5.8) | (2.0) | 6.83 | 5.48 | 1.30 | — | — | 1.692 | 24.78 |
| $N_2$ | gas | — | — | — | 45.767 | 6.76 | 0.606 | 0.13 | — | 2.044 | -7.064 |
| Na | solid | 371 | 0.63 | 1.7 | 12.31 | 5.657 | 3.252 | 0.5785 | — | 1.836 | 20.92 |
|  | liquid | 1187 | 23.4 | 20.1 | — | 8.954 | -4.577 | 2.540 | — | 1.924 | 36.0 |
|  | gas | — | — | — | — | (4.97) | — | — | — | (-24.40) | (-8.7) |
| Nb | solid | 2760 | (5.8) | (2.1) | 8.3 | 5.66 | 0.96 | — | — | 1.730 | 24.24 |
| Nd | solid | 1297 | (2.55) | (1.97) | 13.9 | 5.61 | 5.34 | — | — | 1.910 | 19.7 |
|  | liquid | (2750) | (61) | (22) | — | (9.1) | — | — | — | (-0.606) | 35.8 |
| Ni | solid, α | 626 | 0.092 | 0.15 | 7.137 | 4.06 | 7.04 | — | — | 1.523 | 18.095 |
|  | solid, β | 1728 | 4.21 | 2.44 | — | 6.00 | 1.80 | — | — | 1.619 | 27.16 |
|  | liquid | 3110 | 90.48 | 29.0 | — | 9.20 | — | — | — | 0.251 | 45.47 |
| Np | solid | 913 | (2.3) | (2.5) | (14) | (5.3) | (3.4) | — | — | (1.731) | (17.9) |
|  | liquid | (2525) | (55) | (22) | — | (9.0) | — | — | — | (1.392) | (37.5) |
| $O_2$ | gas | — | — | — | 49.003 | 8.27 | 0.258 | — | -1.877 | 3.007 | -0.750 |
| Os | solid | 2970 | (6.4) | (2.2) | 7.8 | 5.69 | 0.88 | — | — | 1.736 | 24.9 |
| $P_4$ | solid, white | 317.4 | 0.601 | 1.89 | 42.4 | 13.62 | 28.72 | — | — | 5.338 | 43.8 |
|  | liquid | 553 | 11.9 | 21.5 | — | 19.23 | 0.51 | — | -2.98 | 6.035 | 66.7 |
|  | gas | — | — | — | — | (19.5) | (-0.4) | (1.3) | — | (-6.32) | (46.1) |
| Pa | solid | (18.25) | (4.0) | (2.2) | (13.5) | (5.2) | (4.0) | — | — | (1.728) | (17.3) |
|  | liquid | (4500) | (115) | (26) | — | (8.0) | — | — | — | (-3.823) | (28.8) |

Table 2—12 (continued)
THERMODYNAMIC PROPERTIES OF THE ELEMENTS

| Element | Phase | Temperature of transition (K) | Heat of transition (kcal/g mole) | Entropy of transition (e.u.) | Entropy at 298K (e.u.) | a (cal/g mole) | b (cal/g mole) | c (cal/g mole) | d (cal/g mole) | A (kcal/g mole) | B (e.u.) |
|---|---|---|---|---|---|---|---|---|---|---|---|
| Pb | solid | 600.6 | 1.141 | 1.900 | 15.49 | 5.64 | 2.30 | — | — | 1.784 | 17.33 |
|    | liquid | 2023 | 42.5 | 21.0 | — | 7.75 | -0.73 | — | — | 1.362 | 27.11 |
|    | gas | — | — | — | — | (4.97) | — | — | — | (-45.25) | (-13.6) |
| Pd | solid | 1828 | 4.12 | 2.25 | 8.9 | 5.80 | 1.38 | — | — | 1.791 | 24.6 |
|    | liquid | 3440 | 89 | 26 | — | (9.0) | — | — | — | (1.215) | (43.8) |
| Po | solid | 525 | (2.4) | (4.6) | 13 | (5.2) | (3.2) | — | — | (1.693) | (17.6) |
|    | liquid | (1235) | (24.6) | (19.9) | — | (9.0) | — | — | — | (0.847) | (35.2) |
|    | gas | — | — | — | — | (4.97) | — | — | — | (-28.73) | (-13.5) |
| Pr | solid | 1205 | (2.5) | (2.1) | (13.5) | (5.0) | (4.6) | — | — | (1.705) | (16.4) |
|    | liquid | 3563 | — | — | — | (8.0) | — | — | — | (-0.519) | (30.0) |
| Pt | solid | 2042.5 | 5.2 | 2.5 | 10.0 | 5.74 | 1.34 | — | 0.10 | 1.737 | 23.0 |
|    | liquid | 4100 | 122 | 29.8 | — | (9.0) | — | — | — | (0.406) | (42.6) |
| Pu | solid | 913 | (2.26) | (2.48) | (13.0) | (5.2) | (3.6) | — | — | (1.710) | (17.7) |
|    | liquid | — | — | — | — | (8.0) | — | — | — | (0.506) | (31.0) |
| Ra | solid | 1233 | (2.3) | (1.9) | (17) | (5.8) | (1.2) | — | — | (1.783) | (16.4) |
|    | liquid | (1700) | (35) | (21) | — | (8.0) | — | — | — | (1.284) | (28.6) |
|    | gas | — | — | — | — | (4.97) | — | — | — | (-38.87) | (-14.5) |
| Rb | solid | 312.0 | 0.525 | 1.68 | 16.6 | 3.27 | 13.1 | — | — | 1.557 | 5.9 |
|    | liquid | 952 | 18.11 | 19.0 | — | 7.85 | — | — | — | 1.814 | 26.5 |
|    | gas | — | — | — | — | (4.97) | — | — | — | (-19.04) | (-12.3) |
| Re | solid | 3440 | (7.9) | (2.3) | (8.89) | (5.85) | (0.8) | — | — | (1.780) | (24.7) |
| Rh | solid | 2240 | (5.2) | (2.3) | 7.6 | 5.40 | 2.19 | — | — | 1.707 | 23.8 |
|    | liquid | 4150 | 127 | 30.7 | — | (9.0) | — | — | — | (-0.923) | (44.4) |

Table 2—12 (continued)
## THERMODYNAMIC PROPERTIES OF THE ELEMENTS

| Element | Phase | Temperature of transition (K) | Heat of transition (kcal/g mole) | Entropy of transition (e.u.) | Entropy at 298K (e.u.) | a (cal/g mole) | b (cal/g mole) | c (cal/g mole) | d (cal/g mole) | A (kcal/g mole) | B (e.u.) |
|---|---|---|---|---|---|---|---|---|---|---|---|
| Ru | solid, α | 1308 | 0.034 | 0.026 | 6.9 | 5.25 | 1.50 | — | — | 1.632 | 23.5 |
| | solid, β | 1473 | 0 | — | — | 7.20 | — | — | — | 2.867 | 35.5 |
| | solid, γ | 1773 | 0.23 | 0.13 | — | 7.20 | — | — | — | 2.867 | 35.5 |
| | solid, δ | 2700 | (6.1) | (2.3) | — | 7.50 | — | — | — | 3.169 | 37.6 |
| S | solid, α | 368.6 | 0.088 | 0.24 | 7.62 | 3.58 | 6.24 | — | — | 1.345 | 14.64 |
| | solid, β | 392 | 0.293 | 0.747 | — | 3.56 | 6.95 | — | — | 1.298 | 14.54 |
| | liquid | 717.76 | 2.5 | 3.5 | — | 5.4 | 5.0 | — | — | 1.576 | 24.02 |
| ½S₂ | gas | — | — | — | — | (4.25) | (0.15) | — | (−1.0) | (−2.859) | (9.57) |
| Sb | solid, α, β, γ | 903.7 | 4.8 | 5.3 | 10.5 | 5.51 | 1.74 | — | — | 1.720 | 21.4 |
| | liquid | 1713 | 46.665 | 27.3 | — | 7.50 | — | — | — | −1.992 | 28.1 |
| ½Sb₂ | gas | — | — | — | — | 4.47 | — | — | −0.11 | −53.876 | −21.7 |
| Sc | solid | 1670 | (4.0) | (2.4) | (9.0) | (5.13) | (3.0) | — | — | 1.663 | 21.1 |
| | liquid | 3000 | 80 | 27 | — | (7.50) | — | — | — | (−2.563) | 31.3 |
| Se | solid | 490.6 | 1.25 | 2.55 | 10.144 | 3.30 | 8.80 | — | — | 1.375 | 11.28 |
| | liquid | 1000 | 14.27 | 14.27 | — | 7.0 | — | — | — | 0.881 | 27.34 |
| Si | solid | 1683 | 11.1 | 6.60 | 4.50 | 5.70 | 1.02 | — | −1.06 | 2.100 | 28.88 |
| | liquid | 2750 | 71 | 26 | — | 7.4 | — | — | — | −7.646 | 33.17 |
| Sm | solid | 1623 | 3.7 | 2.3 | (15) | (6.7) | (3.4) | — | — | (2.149) | (24.2) |
| | liquid | (2800) | (70) | (25) | — | (9.0) | — | — | — | (−2.296) | (33.4) |
| Sn | solid, α, β | 505.1 | 1.69 | 3.35 | 12.3 | 4.42 | 6.30 | — | — | 1.598 | 14.8 |
| | liquid | 2473 | (55) | (22) | — | 7.30 | — | — | — | 0.559 | 26.2 |
| | gas | — | — | — | — | (4.97) | — | — | — | (−60.21) | (−14.3) |
| Sr | solid | 1043 | 2.2 | 2.1 | 13.0 | (5.60) | (1.37) | — | — | (1.731) | (19.3) |
| | liquid | 1657 | 33.61 | 20.28 | — | (7.7) | — | — | — | (0.976) | (30.4) |
| | gas | — | — | — | — | (4.97) | — | — | — | (−37.16) | (−10.2) |

## Table 2—12 (continued)
## THERMODYNAMIC PROPERTIES OF THE ELEMENTS

| Element | Phase | Temperature of transition (K) | Heat of transition (kcal/g mole) | Entropy of transition (e.u.) | Entropy at 298K (e.u.) | a (cal/g mole) | b (cal/g mole) | c (cal/g mole) | d (cal/g mole) | A (kcal/g mole) | B (e.u.) |
|---|---|---|---|---|---|---|---|---|---|---|---|
| Ta | solid | 3250 | 7.5 | 2.3 | 9.9 | 5.82 | 0.78 | – | – | 1.770 | 23.4 |
| Tc | solid | (2400) | (5.5) | (2.3) | (8.0) | (5.6) | (2.0) | – | – | (1.759) | (24.5) |
|  | liquid | (3800) | (120) | (32) | – | (11) | – | – | – | (3.459) | (59.4) |
| Te | solid, $\alpha$ | 621 | 0.13 | 0.21 | 11.88 | 4.58 | 5.25 | – | – | 1.599 | 15.78 |
|  | solid, $\beta$ | 723 | 4.28 | 5.92 | – | 4.58 | 5.25 | – | – | 1.469 | 15.57 |
|  | liquid | 1360 | 11.9 | 8.75 | – | 9.0 | – | – | – | –0.988 | 34.96 |
| $\frac{1}{2}Te_2$ | gas | – | – | – | – | 4.47 | – | – | – | –19.048 | –6.47 |
| Th | solid | 2173 | (4.6) | (2.1) | 12.76 | 8.2 | –0.77 | 2.04 | – | 2.591 | 33.64 |
|  | liquid | 4500 | (130) | (29) | – | (8.0) | – | – | – | (–7.602) | (26.84) |
| Ti | solid, $\alpha$ | 1155 | 0.950 | 0.822 | 7.334 | 5.25 | 2.52 | – | – | 1.677 | 23.33 |
|  | solid, $\beta$ | 2000 | (4.6) | (2.3) | – | 7.50 | – | – | – | 1.645 | 35.46 |
|  | liquid | 3550 | (101) | (28) | – | (7.8) | – | – | – | (–2.355) | (35.45) |
| Tl | solid, $\alpha$ | 508.3 | 0.082 | 0.16 | 15.4 | 5.26 | 3.46 | – | – | 1.722 | 15.6 |
|  | solid, $\beta$ | 576.8 | 1.03 | 1.79 | – | 7.30 | – | – | – | 2.230 | 26.4 |
|  | liquid | 1730 | 38.81 | 22.4 | – | 7.50 | – | – | – | 1.315 | 25.9 |
|  | gas | – | – | – | – | (4.97) | – | – | – | (–41.88) | (–15.4) |
| U | solid, $\alpha$ | 938 | 0.665 | 0.709 | 12.03 | 3.25 | 8.15 | – | 0.80 | 1.063 | 8.47 |
|  | solid, $\beta$ | 1049 | 1.165 | 1.111 | – | 10.28 | – | – | – | 3.493 | 48.27 |
|  | solid, $\gamma$ | 1405 | (3.0) | (2.1) | – | 9.12 | – | – | – | 1.110 | 39.09 |
|  | liquid | 3800 | – | – | – | (8.99) | – | – | – | (–2.073) | 36.01 |
| V | solid | 2003 | (4.0) | (2.0) | 7.05 | 5.57 | 0.97 | – | – | 1.704 | 24.97 |
|  | liquid | 3800 | – | – | – | (8.6) | – | – | – | 1.827 | 44.06 |
| W | solid | 3650 | 8.42 | 2.3 | 8.0 | 5.74 | 0.76 | – | – | 1.745 | 24.9 |

**Table 2—12 (continued)**
**THERMODYNAMIC PROPERTIES OF THE ELEMENTS**

| Element | Phase | Temperature of transition (K) | Heat of transition (kcal/g mole) | Entropy of transition (e.u.) | Entropy at 298K (e.u.) | a (cal/g mole) | b (cal/g mole) | c (cal/g mole) | d (cal/g mole) | A (kcal/g mole) | B (e.u.) |
|---|---|---|---|---|---|---|---|---|---|---|---|
| Y | solid | 1750 | (4.0) | (2.3) | (11) | (5.6) | (2.2) | — | — | (1.767) | (21.6) |
|   | liquid | 3500 | (90) | (26) | — | (7.5) | — | — | — | (−2.277) | (29.6) |
| Zn | solid | 692.7 | 1.595 | 2.303 | 9.95 | 5.35 | 2.40 | — | — | 1.702 | 21.25 |
|   | liquid | 1180 | 27.43 | 23.24 | — | 7.50 | — | — | — | 1.020 | 31.35 |
|   | gas | — | — | — | — | (4.97) | — | — | — | (−29.407) | (−9.81) |
| Zr | solid, $\alpha$ | 1135 | 0.920 | 0.811 | 9.29 | 6.83 | 1.12 | — | −0.87 | 2.378 | 30.45 |
|   | solid, $\beta$ | 2125 | (4.9) | (2.3) | — | 7.27 | — | — | — | 1.159 | 31.43 |
|   | liquid | (3900) | (100) | (26) | — | (8.0) | — | — | — | (−2.190) | (34.7) |

From Weast, R. C. Ed., *Handbook of Chemistry and Physics*, 69th ed., CRC Press, Boca Raton, Fla., 1988, D-44.

# Section 3

# **Physical Properties of Compounds**

# Section 3

# PHYSICAL PROPERTIES OF COMPOUNDS

### Table 3–1
### PHYSICAL CONSTANTS OF MINERALS

The following table presents data for many of the more common minerals.

In order to avoid duplication and save space, very few cross references are given in the body of the table. If the name sought is not found in the table, consult the synonym index given below.

Specific gravities are given at normal atmospheric temperatures, a more precise statement being valueless considering the large variations in natural minerals.

Hardness is given in terms of Mohs' scale (see under Hardness).

Indices of refraction for the sodium line: $\lambda = 5893$ Å, unless otherwise indicated. Li: $\lambda = 6708$ Å. Indices will invariably be given in the order $\omega$, $\epsilon$, or $\alpha$, $\beta$, $\gamma$. Uniaxial crystals are considered positive if $\epsilon > \omega$ and negative if $\omega > \epsilon$. Biaxial crystals are considered positive if $\beta$ is nearer $\alpha$ in value than it is $\gamma$, and negative if $\beta$ is nearer $\gamma$ than $\alpha$.

### Abbreviations

| Abbreviation | Meaning of abbreviation | Abbreviation | Meaning of abbreviation | Abbreviation | Meaning of abbreviation |
|---|---|---|---|---|---|
| bl............... | blue | grn............... | green | rhbdr............ | rhombohedral |
| blk.... | black | grnsh............ | greenish | rhomb........ | rhombic |
| blksh............. | blackish | hex.............. | hexagonal | somet............ | sometimes |
| blsh............. | bluish | iridesc........... | iridescent | tarn............. | tarnishes |
| br............... | brown | monocl.......... | monoclinic | tetr............. | tetragonal |
| brnsh............ | brownish | oft.............. | often | tricl............ | triclinic |
| ccl.............. | colorless | pa.............. | pale | vlt............. | violet |
| cub............. | cubic | purp............ | purple | wh............. | white |
| dk.............. | dark | (R)............ | radioactive | yel............. | yellow |
| Fe............... | Fe, ferrous iron | redsh............ | redish | yelsh............ | yellowish |
| Fe+³........... | Fe, ferric iron | | | | |

### Synonym Index

| Compound sought | Listed | Compound sought | Listed |
|---|---|---|---|
| Acmite............... | Aegirine | Fibrolite.............. | Sillimanite |
| Agate................ | Quartz (impure) | Flint................ | Quartz (impure) |
| Aluminum hydroxide... | Boehmite, Diaspore, Gibbsite | Fluorapatite.......... | Apatite |
| Amphibole........... | Actinolite, Anthophyllite, | Fluorspar........... | Fluorite |
| | Cummingtonite, Glaucophane, | Garnet.............. | Almandine, Pyrope, Spessartite, |
| | Hornblende, Riebeckite, Tremolite | | Andradite, Grossularite, |
| Antimony oxide....... | Senarmontite, Valentinite | | Uvarovite, Hydrogrossularite |
| Antimony sulfide....... | Stibnite | Garnierite........... | Serpentine (Ni-bearing) |
| Arsenic oxide.......... | Arsenolite, Claudetite | Glauber salt.......... | Mirabilite |
| Arsenic sulfide ........ | Orpiment, Realgar | Hyacinth.............. | Zircon |
| Barium carbonate..... | Witherite | Iceland spar.......... | Calcite |
| Barium sulfate ........ | Barite | Idocrase............. | Vesuvianite |
| Barytes.............. | Barite | Iron carbonate ....... | Siderite |
| Bauxite.............. | Gibbsite, Boehmite, Diaspore | Iron hydroxide ....... | Goethite, Lepidocrocite |
| Brimstone............ | Sulfur | Iron oxide ........... | Hematite, Magnetite |
| Bronzite.............. | Orthopyroxene | Iron spinel .......... | Hercynite |
| Cadmium sulfide ...... | Greenockite | Iron sulfide .......... | Marcasite, Pyrite, Pyrrhotite |
| Calamine............. | Hemimorphite | Lapis lazuli .......... | Lazurite |
| Calcium carbonate .... | Aragonite, Calcite, Vaterite | Lead carbonate ....... | Cerussite |
| Calcium sulfate ....... | Anhydrite, Gypsum | Lead chloride ........ | Cotunnite |
| Calcium sulfide ....... | Oldhamite | Lead chromate ....... | Crocoite |
| Carborundum........ | Moissanite | Lead oxide .......... | Litharge, Minium |
| Chalcedony.......... | Quartz (impure, fibrous) | Lead sulfate ......... | Anglesite |
| Chinaclay........... | Kaolinite | Lead sulfide ......... | Galena |
| Chloanthite.......... | Skutterodite | Limonite............ | Goethite (impure) |
| Chromespinel........ | Chromite | Lithiophyllite........ | Triphylite |
| Chrysolite........... | Serpentine | Lithium mica......... | Lepidolite |
| Clinoptolite.......... | Heulandite | Lodestone........... | Magnetite |
| Clayminerals......... | Illite, Kaolinite, Montmorillonite | Magnesium carbonate . | Magnesite |
| Clinochlore.......... | Chlorite | Magnesium hydroxide.. | Brucite |
| Cobaltbloom........ | Erythrite | Magnesium oxide ..... | Periclase |
| Copper chloride ...... | Nantokite | Magnesium sulfate .... | Kieserite |
| Copper oxide ........ | Cuprite | Manganese carbonate . | Rhodochrosite |
| Copper sulfide ....... | Chalcocite, Covellite, Digenite | Manganese hydroxide . | Pyrochroite |
| Emerald............. | Beryl | Manganese oxide .... | Hausmannite, Manganosite, |
| Emery.............. | Mixture of Corundum, Magnetite | | Pyrolusite |
| | and other minerals | Manganese sulfide .... | Alabandite |
| Epsom salt .......... | Epsomite | Meerschaum.......... | Serpentine |
| Feldspar............. | Orthoclase, Microcline, | Mica................ | Muscovite, Paragonite, Phlogopite, |
| | Anorthoclase, Albite, Oligoclase, | | Biotite, Lepidolite |
| | Andesine. Anorthite | | |

## Table 3–1 (continued)
## PHYSICAL CONSTANTS OF MINERALS

### Synonym Index

| Compound sought | Listed | Compound sought | Listed |
|---|---|---|---|
| Native copper | Copper | Smalltite | Skutterotite |
| Native gold | Gold | Soapstone | Mixture of Talc and other minerals |
| Nickel oxide | Bunsenite | Sodium chloride | Halite |
| Nickel sulfide | Millerite | Sodium sulfate | Thenardite |
| Orthite | Allanite | Strontium carbonate | Strontianite |
| Penninite | Chlorite | Strontium sulfate | Celestite |
| Peridote | Olivine | Thorium oxide | Thorianite |
| Pistacite | Epidote | Tin oxide | Cassiterite |
| Pitchblende | Uraninite | Titanite | Sphene |
| Plagioclase | Albite, Oligoclase, Andesine, Anorthite | Titanium oxide | Anatase, Brookite, Rutile |
| | | Uranium oxide | Uraninite |
| Potassium chloride | Sylvite | Zeolite | Natrolite, Mesolite, Scolecite, Thomasonite, Harmatome, Eddingtonite, Heulandite, Stilbite, Phillipsite, Chabazite, Gmelinite, Levyn, Laumontite, Mordenite |
| Potassium sulfate | Arcanite | | |
| Pyroxene | Diopsite, Angite, Aegirine, Jadeite, Pigeonite, Eustatite, Orthopyroxene | | |
| Rocksalt | Halite | Zincblende | Sphalerite |
| Ruby | Corundum | Zinc carbonate | Smithsonite |
| Sapphire | Corundum | Zinc oxide | Zincite |
| Silica | Christobalite, Quartz, Tridymite | Zinc spinel | Gahnite |
| Silver chloride | Cerargyrite | Zinc sulfide | Sphalerite, Wurtzite |
| Silver iodide | Jodyrite, Miersite | Zirconium oxide | Baddeleyite |
| Silver sulfide | Acanthite, Argentite | | |

## Table 3–1 (continued)
## PHYSICAL CONSTANTS OF MINERALS

| Name | Formula | Sp. gr. | Hardness | Crystalline form and color | Index of refraction (Na) η; ω ε; α β γ |
|---|---|---|---|---|---|
| Acanthite | $Ag_2S$ | 7.2–7.3 | 2–2.5 | rhomb.(?), iron-blk. | |
| Actinolite | $Ca_2(Mg,Fe)_5Si_8O_{22}(OH,F)_2$ | 3.02–3.44 | 5–6 | monocl., pa. to dk. grn. | 1.590–1.688, 1.612–1.697, 1.622–1.705 |
| Aegirine | $NaFe^{+3}Si_2O_6$ | 3.55–3.60 | 6 | monocl., dk. grn. to grnsh. blk. | 1.750–1.776, 1.780–1.820, 1.800–1.836 |
| Åkermanite | $Ca_2MgSi_2O_7$ | 2.944 | 5–6 | tetr., col., gray-grn., br. | 1.632, 1.640 |
| Alabandite | $MnS$ | 4.050 | 3.5–4 | cub., iron-blk., tarn., br. | |
| Albite | $NaAlSi_3O_8$ | 2.63 | 6–6.5 | tricl., col., wh., somet. yel., pink, grn. | 1.527, 1.531, 1.538 |
| Allanite | $(Ca,Mn,Ce,La,Y,Th)_2(Fe,Fe^{+2},Ti)(Al,Fe^{+3})_2Si_3O_{12}(OH)$ | 3.4–4.2 | 5–6.5 | monocl., pa. br. to blk. | 1.690–1.791, 1.700–1.815, 1.706–1.828 |
| Allemontite | $AsSb$ | 5.8–6.2 | 3–4 | hex., tin-wh. to redsh., gray, tarn. gray-brnsh. blk. | |
| Almandine | $Fe_3Al_2Si_3O_{12}$ | 4.318 | 6–7.5 | cub., red, dk. red, blk. | 1.830 |
| Altaite | $PbTe$ | 8.15 | 3 | cub., tin-wh., yelsh., tarn. bronze-yel. | |
| Aluminite | $Al_2(SO_4)(OH)_4 \cdot 7H_2O$ | 1.66–1.82 | 1–2 | monocl.(?), wh. | 1.459, 1.464, 1.470 |
| Alunite | $(K,Na)Al_3(SO_4)_2(OH)_6$ | 2.6–2.9 | 3.5–4 | rhbdr., wh., gray, yel., redsh., br. | 1.572, 1.592 |
| Alunogen | $Al_2(SO_4)_3 \cdot 18H_2O$ | 1.77 | 1.5–2 | tricl., col., wh., yelsh. wh., redsh. wh. | 1.459–1.475, 1.461–1.478, 1.470–1.485 |
| Amblygonite | $(Li,Na)Al(PO_4)(F,OH)$ | 3.0–3.1 | 5.5–6 | tricl., wh., yelsh. wh., grnsh. wh., blsh. wh., gray | 1.591, 1.604, 1.613 |
| Analcite | $NaAlSi_2O_6 \cdot H_2O$ | 2.24–2.29 | 5.5 | cub., wh., pink, gray | 1.479–1.493 |
| Anatase | $TiO_2$ | 3.90 | 5.5–6 | tetr., br., yelsh. br., redsh. br., bl., blk., grn., gray | 2.5612, 2.4880 |
| Andalusite | $Al_2OSiO_4$ | 3.13–3.16 | 6.5–7.5 | rhomb., pink, wh., red | 1.629–1.640, 1.633–1.644, 1.638–1.650 |
| Andesine | $[(Na,Si)_{0.7-0.5}(Ca,Al)_{0.3-0.5}]AlSi_2O_8$ | 2.65–2.68 | 6–6.5 | tricl., wh., gray, grn. | 1.544–1.555, 1.548–1.558, 1.551–1.563 |
| Andorite | $PbAgSb_3S_6$ | 5.33–5.37 | 3–3.5 | rhomb., dk. steel gray, somet. tarn. yel. or iridesc. | |
| Andradite | $Ca_3Fe_2^{+3}Si_3O_{12}$ | 3.859 | 6.5–7.5 | cub., brnsh. red, blk., somet. yel., grn. | 1.887 |
| Anglesite | $PbSO_4$ | 6.37–6.39 | 2.5–3 | rhomb., col., wh., somet. gray, yelsh., grn. tinge | 1.8771, 1.8826, 1.8937 |
| Anhydrite | $CaSO_4$ | 2.98 | 3.5 | rhomb., col., blsh. wh., vit. | 1.5698, 1.5754, 1.6136 |
| Ankerite | $Ca(Fe,Mg,Mn)(CO_3)_2$ | 2.8–3.1 | 3.5–4 | rhbdr., br., yelsh. br., grnsh. br., pink | 1.690–1.750, 1.510–1.548 |
| Anorthite | $CaAl_2Si_2O_8$ | 2.76 | 6–6.5 | tricl., wh., yel., grn., blk. | 1.577, 1.585, 1.590 |
| Anorthoclase | $(Na,K)AlSi_3O_8$ | 2.56–2.60 | 6 | tricl., col., wh. | 1.523, 1.528, 1.529 |
| Anthophyllite | $(Mg,Fe)_7Si_8O_{22}(OH,F)_2$ | 2.85–3.57 | 5.5–6 | rhomb., wh., gray, grn., br., yelsh. br., dk. br. | 1.596–1.694, 1.605–1.710, 1.615–1.722 |
| Antimony | $Sb$ | 6.61–6.72 | 3–3.5 | hex., tin-wh. | |
| Apatite | $Ca_5(PO_4)_3(OH,F,Cl)$ | 3.1–3.35 | 5 | hex., grn., wh., yel., br. red, bl. | 1.629–1.667, 1.624–1.666 |
| Apophyllite | $KFCa_4Si_8O_{20} \cdot 8H_2O$ | 2.33–2.37 | 4.5–5 | tetr., col., wh., pink, pa. yel., pa. grn. | 1.534–1.535, 1.535–1.537 |
| Aragonite | $CaCO_3$ | 2.94–2.95 | 3.5–4 | rhomb., col., wh. | 1.530–1.531, 1.680–1.681, 1.685–1.686 |
| Arcanite | $K_2SO_4$ | 2.663 | 2–2.5 | rhom., col., wh. | 1.4935, 1.4947, 1.4973 |
| Argentite | $Ag_2S$ | 7.2–7.4 | 2–2.5 | cub., blkish, lead gray | |
| Arsenic | $As$ | 5.63–5.78 | 3.5 | hex., tin-wh., tarn. dk. gray | |
| Arsenolite | $As_2O_3$ | 3.86–3.88 | 1.5 | cub., wh., somet. blsh., yelsh., redsh. tinge | 1.755 |
| Arsenopyrite | $FeAsS$ | 5.9–6.2 | 5.5–6 | monocl., silver-wh., to steel gray | |
| Atacamite | $Cu_2(OH)_3Cl$ | 3.74–3.78 | 3–3.5 | rhomb., grn., dk. grn., blksh. grn. | 1.831, 1.861, 1.880 |
| Augite | $(Ca,Mg,Fe,Fe^{+3},Ti,Al)_2(Si,Al)_2O_6$ | 3.23–3.52 | 5–6 | monocl., pa. br., br., purp. br., grn., blk. | 1.671–1.735, 1.672–1.741, 1.703–1.761 |
| Autunite | $Ca(UO_2)_2(PO_4)_2 \cdot 10\text{-}12H_2O$ | 3.1–3.2 | 2–2.5 | tetr., yel., somet. grnsh. yel. to pa. grn. | 1.577, 1.553 |
| Axinite | $(Ca,Mn,Fe)_3Al_2BO_3Si_4O_{12}(OH)$ | 3.26–3.36 | 6.5–7 | tricl., br., yelsh. | 1.674–1.693, 1.681–1.701, 1.684–1.704 |
| Azurite | $Cu_3(OH)_2(CO_3)_2$ | 3.77 | 3.5–4 | monocl., azure bl., dk. bl., pa. bl. | 1.730, 1.758, 1.838 |
| Baddeleyite | $ZrO_2$ | 5.4–6.02 | 6.5 | monocl., col., gr., redsh. br., br., blk. | 2.13, 2.19, 2.20 |
| Barite | $BaSO_4$ | 4.50 | 3–3.5 | rhomb., col., wh., somet. br., dk. br., gray | 1.6362, 1.6373, 1.6482 |
| Benitoite | $BaTi(Si_3O_9)$ | 3.65 | 6–6.5 | rhbdr., bl., purp., col. | 1.757, 1.804 |
| Bertrandite | $Be_4Si_2O_7(OH)_2$ | 2.6 | 6–7 | rhomb., col. | 1.589, 1.602, 1.613 |
| Beryl | $Be_3Al_2Si_6O_{18}$ | 2.66–2.83 | 7.5–8 | hex., col., wh., blsh. grn., grnsh. yel., yel., bl. | 1.565–1.590, 1.567–1.598 |
| Beryllonite | $NaBe(PO_4)$ | 2.81 | 5.5–6 | monocl., col., wh., pa. yel. | 1.5520, 1.5579, 1.561 |
| Biotite | $K(Mg,Fe)_3AlSi_3O_{10}(OH,F)_2$ | 2.7–3.3 | 2.5–3 | monocl., blk., dk. br., br., redsh. br. | 1.565–1.625, 1.605–1.696, 1.605–1.696 |
| Bismuth | $Bi$ | 9.70–9.83 | 2–2.5 | rhbdr., silver-wh. to tin-wh., tarn. yel. or iridesc. | |
| Bismuthinite | $Bi_2S_3$ | 6.75–6.81 | 2 | rhomb., lead gray to tin-wh., tarn. yel. or iridesc. | |

## Table 3–1 (continued)
## PHYSICAL CONSTANTS OF MINERALS

| Name | Formula | Sp. gr. | Hardness | Crystalline form and color | Index of refraction (Na) $\alpha$, $\beta$, $\gamma$ / $\eta$; $\omega$, $\epsilon$ |
|---|---|---|---|---|---|
| Bixbyite | $(Mn,Fe)_2O_3$ | 4.945 | 6-6.5 | cub., blk. | |
| Bloedite | $Na_2Mg(SO_4)_2.4N_2O$ | 2.22-2.28 | 2.5-3 | monocl., col., somet. blsh-grn. or redsh. | 1.483, 1.486, 1.487 |
| Boehmite | $AlO(OH)$ | 3.01-3.06 | 3.5-4 | rhomb., wh. | 1.64-1.65, 1.65-1.66, 1.65-1.67 |
| Boracite | $Mg_3B_7O_{13}Cl$ | 2.91-2.97 | 7-7.5 | rhomb., col., wh., gray, yel., blsh-grn., grn. | 1.66, 1.66, 1.67 |
| Borax | $Na_2B_4O_7.10H_2O$ | 1.715 | 2-2.5 | monocl., col., wh., gray, blsh. or grnsh-wh. | 1.4466, 1.4687, 1.4717 |
| Bornite | $Cu_5FeS_4$ | 5.06-5.08 | 3 | cub., copper red to pinchbeck br., tarn. purp., iridesc. | |
| Boulangerite | $Pb_5Sb_4S_{11}$ | 6.0-6.2 | 2.5-3 | monocl., blsh. lead gray, oft. with yel. spots | |
| Bournonite | $PbCuSbS_3$ | 5.80-5.86 | 2.5-3 | rhomb., steel gray to blk. | |
| Braggite | $PtS$ | 10.0 | | tetr., steel gray | |
| Braunite | $(Mn,Si)_2O_3$ | 4.72-4.83 | 6-6.5 | tetr., brns. blk. to steel gray | |
| Bravoite | $(Ni,Fe)S_2$ | 4.62 | 5.5-6 | cub., steel gray | |
| Breithauptite | $NiSb$ | 8.23 | 5.5 | hex., pa. copper red to vlt. tarn. | |
| Brochantite | $Cu_4(SO_4)(OH)_6$ | 3.79 | 3.5-4 | monocl., emerald-grn. to blksh. grn., pa. grn. | 1.728, 1.771, 1.800 |
| Bromyrite | $AgBr$ | 6.47 | 2.5 | cub., col., gray, yelsh., grnsh-br. | 2.253 |
| Brookite | $TiO_2$ | 4.08-4.20 | 5.5-6 | rhomb., br., yelsh. br., redsh. br., blk. | 2.5831, 2.5843, 2.7004 |
| Brucite | $Mg(OH)_2$ | 2.38-3.40 | 2.5 | hex., wh., pa. grn., gray, bl., yel., br. | 1.560-1.590, 1.580-1.600 |
| Bunsenite | $NiO$ | 6.898 | 5.5 | cub., dk. pistachio-grn. | (Li) 2.37 |
| Cacoxenite | $Fe_4(PO_4)_3(OH)_3.12H_2O$ | 2.2-2.4 | 3-4 | hex., yel. to brnsh-yel., redsh. yel., somet. grnsh. | 1.575-1.585, 1.635-1.656 |
| Calcite | $CaCO_3$ | 2.715-2.94 | 3 | rhdbr., col., wh., somet. gray, yel., pink, bl. | 1.658-1.740, 1.486-1.550 |
| Caledonite | $Cu_2Pb_5(SO_4)_3(CO_3)(OH)_6$ | 5.75-5.77 | 2.5-3 | rhomb., dk. grn., blsh. grn. | 1.815-1.821, 1.863-1.869, 1.906-1.912 |
| Calomel | $HgCl$ | 7.15 | 1.5 | tetr., col., wh., gray, yelsh. wh., br. | 1.973, 2.656 |
| Cancrinite | $(Na,Ca)_{7-8}Al_6Si_6O_{24}(CO_3,SO_4,Cl)_{1-1.5}.1-5H_2O$ | 2.42-2.51 | 5-6 | hex., col., wh., pa. bl., pa. grn., yel., redsh. | 1.528-1.507, 1.503-1.495 |
| Carnallite | $KMgCl_3.6H_2O$ | 1.602 | 2.5 | rhomb., col., wh., oft. redsh., somet. yel., bl. | 1.466, 1.475, 1.494 |
| Carnotite | $K_2(UO_2)_2(VO_4)_2.3H_2O$ | | 1-2 | rhomb. or monocl., bright yel., yel., grnsh. yel. | 1.75, 1.92, 1.95 |
| Cassiterite | $SnO_2$ | 6.99 | 6-7 | tetr., yelsh. or redsh. br., brnsh-blk. | 2.006, 2.0972 |
| Celestite | $SrSO_4$ | 3.96 | 3-3.5 | rhomb., col., wh., pa. bl., redsh., grnsh., brnsh. | 1.621-1.622, 1.623-1.624, 1.630-1.631 |
| Celsian | $BaAl_2Si_2O_8$ | 3.10-3.39 | 6-6.5 | monocl., col., wh., yel. | 1.579-1.587, 1.583-1.593, 1.588-1.600 |
| Cervantite | $Sb_2O_4(?)$ | 6.64 | 4-5 | rhomb.(?), yel., wh., somet. redsh-wh. | |
| Cerargyrite | $AgCl$ | 5.55 | 2.5 | cub., col., gray, grnsh-br., tarn. purp., yelsh. | 2.071 |
| Cerussite | $PbCO_3$ | 6.53-6.57 | 3-3.5 | rhomb., col., wh., gray, somet. bl., blk., grn. | 1.8036, 2.0765, 2.0786 |
| Chabazite | $(Ca,Na_2)Al_2Si_4O_{12}.6H_2O$ | 2.05-2.10 | 4-5 | rhdbr., redsh-wh., wh., yelsh., grnsh. | 1.470-1.494 |
| Chalcocite | $Cu_2S$ | 5.5-5.8 | 2.5-3 | rhomb., blksh., lead gray | |
| Chalcanthite | $CuSO_4.5H_2O$ | 2.28 | 2.5 | tricl., dk. bl. to sky bl., somet. grnsh. | 1.514, 1.537, 1.543 |
| Chalcopyrite | $CuFeS_2$ | 4.1-4.3 | 3.5-4 | tetr., brass-yel., tarn., iridesc. | |
| Chiolite | $Na_5Al_3F_{14}$ | 3.00 | 2-3 | tetr., wh. to col. | 1.349, 1.342 |
| Chlorite | $(Mg,Al,Fe)_{12}(Si,Al)_8O_{20}(OH)_{16}$ | 2.6-3.3 | 2-2.5 | monocl., tricl., wh., yel., pink, br., red | 1.57-1.66, 1.57-1.67, 1.57-1.67 |
| Chloritoid | $(Fe,Mg,Mn)_2(Al,Fe^{3+})Al_3O_2SiO_4(OH)_4$ | 3.51-3.80 | 6.5 | | 1.713-1.730, 1.719-1.734, 1.723-1.740 |
| Chondrodite | $Mg(OH,F)_2.2MgSiO_4$ | 3.16-3.26 | 6-6.5 | monocl., yel., br., red | 1.592-1.615, 1.602-1.627, 1.621-1.646 |
| Chromite | $FeCr_2O_4$ | 4.5-5.1 | 5.5 | cub., blk. | 2.16 |
| Chrysoberyl | $BeAl_2O_4$ | 3.65-3.85 | 8.5 | rhomb., grn., yel. | 1.746, 1.748, 1.756 |
| Chrysocolla | $CuSiO_3.2H_2O$ | ~2.4 | 2-4 | rhomb.(?), grn., bl., br., blk. | 1.575, 1.597, 1.598 |
| Cinnabar | $HgS$ | 8.090 | 2-2.5 | hex., red, brnsh. red., gray | (Li) 2.814, 3.143 |
| Claudetite | $As_2O_3$ | 4.15 | 2.5 | monocl., col. to wh. | 1.87, 1.92, 2.01 |
| Clinozoisite | $Ca_2Al_3Si_3O_{12}(OH)$ | 3.21-3.38 | 6-6.5 | monocl., col., pa. yel., gray, grn. | 1.670-1.715, 1.674-1.725, 1.690-1.734 |
| Cobaltite | $CoAsS$ | 6.33 | 5.5 | cub., silver wh., redsh. steel gray, blk. | |
| Colemanite | $Ca_2B_6O_{11}.5H_2O$ | 2.42-2.43 | 4.5 | monocl., col., wh., yelsh. wh., gray | 1.586, 1.592, 1.614 |
| Columbite | $(Fe,Mn)(Cb,Ta)_2O_6$ | 5.15-5.25 | 6 | rhomb., iron blk. to br. blk. | |
| Connellite | $Cu_{19}(SO_4)Cl_4(OH)_{32}.3H_2O(?)$ | 3.36 | | hex., azure bl. | 1.724-1.738, 1.746-1.758 |
| Copiapite | $(Fe,Mg)Fe_4^{3+}(SO_4)_6(OH)_2.20H_2O$ | 2.08-2.17 | 2.5-3 | tricl., yel., grnsh. yel. | 1.51-1.53, 1.53-1.55, 1.58-1.60 |
| Copper | $Cu$ | 8.95 | 2.5-3 | cub., red | |
| Coquimbite | $Fe_2(SO_4)_3.9H_2O$ | 2.10-2.12 | 2.5 | hex., pa. vlt. to dk. amethystine, yelsh., grnsh. | 1.53-1.55, 1.55-1.57 |
| Cordierite | $Al_3(Mg,Fe)_2Si_5AlO_{18}$ | 2.53-2.78 | 7 | rhomb., gray-bl., bl., dk. bl. | 1.522-1.558, 1.524-1.574, 1.527-1.578 |
| Corundum | $Al_2O_3$ | 4.022 | 9 | hex., col. bl., yel., purp., grn., pink, red | 1.767-1.772, 1.759-1.763 |
| Cotunnite | $PbCl_2$ | 5.80 | 2.5 | rhomb., col. to wh., somet. yelsh., grnsh. | 2.199, 2.217, 2.260 |
| Covellite | $CuS$ | 4.6-4.76 | 1.5-2 | hex., indigo bl., dk. bl., iridesc. brass yel. to red | |

## Table 3–1 (continued)
## PHYSICAL CONSTANTS OF MINERALS

| Name | Formula | Sp. gr. | Hardness | Crystalline form and color | Index of refraction (Na) η; ω ε / α β γ |
|---|---|---|---|---|---|
| Cristobalite | SiO₂ | 2.33 | 6–7 | tetr.(?), col., wh., yel. | 1.487, 1.484 |
| Crocoite | PbCrO₄ | 5.96–6.02 | 2.5–3 | monocl., red, orange red, orange yel. | 2.29, 2.36, 2.66 |
| Cryolite | Na₃AlF₆ | 2.96–2.98 | 2.5 | monocl., col. to wh., brnsh., redsh., blk. | 1.338, 1.338, 1.339 |
| Cryolithionite | Na₃Li₃Al₂F₁₂ | 2.77 | 2.5 | cub., col. to wh. | 1.3395 |
| Cubanite | CuFe₂S₃ | 4.03–4.18 | 3.5 | rhomb., brass to bronze yel. | |
| Cummingtonite | (Mg,Fe)₇Si₈O₂₂(OH)₂ | 3.2–3.5 | 5–6 | monocl., dk. grn., br. | 1.635–1.665, 1.644–1.675, 1.655–1.698 |
| Cuprite | Cu₂O | 6.14 | 3.5–4 | cub., red, somet. blk. | |
| Danburite | CaSi₂B₂O₈ | 3.0 | 7 | rhomb., pa. yel., col., dk. yel., yelsh. br. | 1.63, 1.63–1.64, 1.63–1.64 |
| Datolite | CaBSiO₄(OH) | 2.96–3.00 | 5–5.5 | monocl., col., wh., yelsh., grnsh., pinksh. | 1.622–1.626, 1.649–1.654, 1.666–1.670 |
| Daubreelite | Cr₂FeS₄ | 3.80–3.82 | 5–5.5 | cub., blk. | |
| Derbylite | Fe₄Ti₃Sb₂O₁₃(?) | 4.53 | 5 | rhomb., pitch blk. | 2.45, 2.45, 2.51 |
| Diamond | C | 3.50–3.53 | 10 | cub., col., pa. yel. to dk. yel., pa. br. to dk. br., wh., blsh. wh. | 2.4175 |
| Diaspore | AlO(OH) | 3–3.5 | 6.5–7 | rhomb., wh., graysh. wh., col. | 1.682–1.706, 1.705–1.725, 1.730–1.752 |
| Digenite | Cu₉S₅ | 5.546 | 2.5–3 | cub., bl. to blk. | |
| Diopside | CaMgSi₂O₆ | 3.22–3.38 | 5.5–6.5 | monocl., wh., pa. grn., dk. grn. | 1.664–1.695, 1.672–1.701, 1.695–1.721 |
| Dioptase | Cu₆Si₆O₁₈·6H₂O | 3.5 | 5 | rhbdr., emerald grn. | 1.64–1.66, 1.70–1.71 |
| Dolomite | CaMg(CO₃)₂ | 2.86 | 3.5–4 | rhbdr., wh., oft. yel. or br. tinge, col. | 1.679, 1.500 |
| Douglasite | K₂FeCl₄·2H₂O(?) | 2.16 | | pa. grn., tarn. brnsh. red | 1.485–1.491, 1.497–1.503 |
| Dyscrasite | Ag₃Sb | 9.67–9.81 | 3.5–4 | rhomb., silver wh., tarn. gray, yelsh. or blksh. | |
| Eddingtonite | BaAl₂Si₃O₁₀·4H₂O | 2.7–2.8 | 4–4.5 | rhomb. or monocl., col., pink, br. wh. | 1.541, 1.553, 1.557 |
| Eglestonite | Hg₂OCl₂ | 8.4 | 2.5 | cub., yel., orange-yel. to dk. brnsh., tarn. bl. | 2.47–2.51 |
| Emplectite | CuBiS₂ | 6.38 | 2 | rhomb., gray to tin wh. | |
| Empressite | AgTe | 7.510 | 3–3.5 | pa. bronze | |
| Enargite | Cu₃AsS₄ | 4–4.5 | 3 | rhomb., gray-blk. to iron-blk. | |
| Enstatite | MgSiO₃ | 3.209 | 5–6 | rhomb., col., gray, gray, yel., brn. | 1.650–1.662, 1.653–1.671, 1.658–1.680 |
| Epidote | Ca₂Fe⁺³Al₂Si₃O₁₂(OH) | 3.38–3.49 | 6 | monocl., grn., gray, yel. | 1.715–1.751, 1.725–1.784, 1.734–1.797 |
| Epsomite | MgSO₄·7H₂O | 1.675–1.679 | 2–2.5 | rhomb., col., wh. pink, grn. | 1.4325, 1.4554, 1.4609 |
| Erythrite | (Co,Ni)₃(AsO₄)₂·8H₂O | 3.06 | 1.5–2.5 | monocl., crimson-red, red, pa. pink | 1.626, 1.661, 1.699 |
| Eucairite | CuAgSe | 7.6–7.8 | 2.5 | silver wh. to lead gray | |
| Euclase | BeAlSiO₄(OH) | 3.0–3.1 | 7.5 | monocl., col., pa. grn., bl. | 1.651, 1.655, 1.671 |
| Eudialyte | (Na,Ca,Fe)₄ZrSi₆O₁₈(OH,Cl)(?) | 2.8–3.1 | 5–6 | hex., pa. pink, red, br. | 1.59–1.61, 1.59–1.61 |
| Eulytite | Bi₄Si₃O₁₂ | 6.6 | 4.5 | cub., br., yel., gray | 2.05 |
| Euxenite | (Y,Ca,Ce,U,Th)(Cb,Ta,Ti)₂O₆ | 5.0–5.9 | 5.5–6.5 | rhomb., blk., grnsh. or brnsh. tint. | ~2.2 |
| Fayalite | Fe₂SiO₄ | 4.392 | 6.5 | rhomb., grnsh., yelsh. | 1.827, 1.869, 1.879 |
| Ferberite | FeWO₄ | 7.51 | 4–4.5 | monocl., br. to blk. | |
| Fergusonite | (Y,Er,Ce,Fe)(Cb,Ta,Ti)O₄ | 5.6–5.8 | 5.5–6.5 | tetr., gray, yel., br., dk. br. | 2.1 |
| Fluorite | CaF₂ | 3.18 | 4 | cub., bl., purp., wh., col., yel., grn. | 1.433–1.435 |
| Forsterite | Mg₂SiO₄ | 3.222 | 7 | rhomb., wh., grnsh., yelsh. | 1.635, 1.651, 1.670 |
| Franklinite | ZnFe₂O₄ | 5.07–5.34 | 5.5–6.5 | cub., blk. | (Li) ~2.36 |
| Gahnite | ZnAl₂O₄ | 4.62 | 7.5–8 | cub., dk. bl-grn., somet. yelsh. or brnsh. | 1.79–1.81 |
| Galena | PbS | 7.57–7.59 | 2.5–3 | cub. lead gray | |
| Galenabismuthite | PbBi₂S₄ | 7.04 | 2.5–3 | rhomb., pa. gray to tin-wh., lead gray, somet. tarn., yel. or irid. | |
| Ganomalite | (Ca,Pb)₁₀(OH,Cl)₂(Si₂O₇)₃ | 5.4–5.7 | 3–4 | hex., col., gray | 1.910, 1.945 |
| Gaylussite | Na₂Ca(CO₃)₂·5H₂O | 1.991 | 2.5–3 | monocl., col. to yelsh. wh., graysh. wh., wh. | 1.4435, 1.5156, 1.5233 |
| Gehlenite | Ca₂Al₂SiO₇ | 3.038 | 5–6 | tetr., col. gray-grn., br. | 1.669, 1.658 |
| Geikielite | MgTiO₃ | 4.05 | 5–6 | rhbdr., brnsh blk., blsh. | 2.31, 1.95 |
| Gibbsite | Al(OH)₃ | 2.38–2.42 | 2.5–3.5 | monocl., wh., graysh., grnsh. or redsh.-wh. | 1.56–1.58, 1.56–1.58, 1.58–1.60 |
| Glauberite | Na₂Ca(SO₄)₂ | 2.75–2.85 | 2.5–3 | monocl., gray, yelsh., somet. col., redsh. | 1.515, 1.535, 1.536 |
| Glauconite | (K,Na,Ca)₁.₂₋₂.₅(Fe⁺³,Al,Fe,Mg)₄Si₇₋₇.₆Al₁₋₀.₄O₂₀(OH)₄·nH₂O | 2.4–2.95 | 2 | monocl., col., ye.lsh. grn., grn., blsh. gray | 1.592–1.610, 1.614–1.641, 1.614–1.641 |
| Glaucophane | Na₂Mg₃Al₂Si₈O₂₂(OH)₂ | 3.08–3.30 | 6 | monocl., gray, lavender bl. | 1.606–1.661, 1.622–1.667, 1.627–1.670 |
| Gmelinite | (Ca,Na₂)Al₂Si₄O₁₂·6H₂O | ~2.1 | 4.5 | rhbdr., wh., redsh.-wh., yelsh., grnsh. | 1.476–1.494, 1.474–1.480 |
| Goethite | FeO(OH) | 3.3–4.3 | 5–5.5 | rhomb., blksh.-br., yelsh. or redsh.-br., yel. | 2.260–2.275, 2.393–2.409, 2.398–2.515 |
| Gold | Au | 19.3 | 2.5–3 | cub., yel. | |
| Goslarite | ZnSO₄·7H₂O | 1.978 | 2–2.5 | rhomb., col., wh., somet. br., grn., bl. | 1.4568, 1.4801, 1.4844 |
| Graphite | C | 2.09–2.23 | 1–2 | hex., iron-blk. to steel gray | |
| Greenockite | CdS | 4.9 | 3–3.5 | hex., yel. to orange | 2.506, 2.529 |

Table 3–1 (continued)

## PHYSICAL CONSTANTS OF MINERALS

| Name | Formula | Sp. gr. | Hardness | Crystalline form and color | Index of refraction (Na) $\eta$; $\omega$ $\beta$ $\gamma$ / $\epsilon$ |
|---|---|---|---|---|---|
| Grossularite | $Ca_3Al_2Si_3O_{12}$ | 3.594 | 6-7.5 | cub., wh., yel., grn., br., red | 1.734 |
| Gummite (R) | $UO_3.H_2O$ | 3.9-6.4 | 2.5-5 | yel., orange, redsh.-yel., red, br. blk. | |
| Gypsum | $CaSO_4.2H_2O$ | 2.30-2.37 | 2 | monoel., wh., col., somet. gray, red, yel., br. | 1.519-1.521, 1.523-1.526, 1.529-1.531 |
| Halite | $NaCl$ | 2.16-2.17 | 2.5 | cub., col., wh., orange, red | 1.544 |
| Hambergite | $Be_2(OH)(BO_3)$ | 2.36 | 7.5 | rhomb., col. to gray, wh., yel. | 1.56, 1.59, 1.63 |
| Hanksite | $Na_{22}K(SO_4)_9(CO_3)_2Cl$ | 2.562 | 3-3.5 | hex., col., somet. pa.-yelsh. or gray | 1.481, 1.461 |
| Harmotome | $BaAl_2Si_6O_{16}.6H_2O$ | 2.41-2.47 | 4.5 | monocl. or rhomb., col., wh., pink, gray, yel. | 1.503-1.508, 1.505-1.509, 1.508-1.514 |
| Hausmannite | $Mn_3O_4$ | 4.83-4.85 | 5.5 | tetr., brnsh.-blk. | (Li) 2.46, 2.15 |
| Hauyne | $(Na,Ca)_{4-8}Al_6Si_6O_{24}(SO_4,S)_{1-2}$ | 2.44-2.50 | 5.5-6 | cub., wh., gray, grn., bl. | 1.496-1.505 |
| Hedenbergite | $CaFeSi_2O_6$ | 3.50-3.56 | 6 | monocl., brnsh.-grn., dk. grn., blk. | 1.716-1.726, 1.723-1.730, 1.741-1.751 |
| Helvite | $Mn_8Be_6Si_6O_{18}S$ | 3.20-3.44 | 6 | cub., yel., br., redsh.-brn. | 1.728-1.749 |
| Hematite | $Fe_2O_3$ | 5.26 | 5-6 | rhbdr., steel gray, dull red to bright red | 3.22, 2.94 |
| Hemimorphite | $Zn_4Si_2O_7(OH)_2.H_2O$ | 3.45 | 4.5-5 | rhomb., col., wh., pa. bl., pa. grn., br. | 1.614, 1.617, 1.636 |
| Hercynite | $FeAl_2O_4$ | 4.40 | 7.5-8 | cub., blk. | 1.835 |
| Herderite | $CaBe(PO_4)(F,OH)$ | 2.95-3.01 | 5-5.5 | monocl., col. to pa. wh. or grnsh.-wh. | 1.592, 1.612, 1.621 |
| Hessite | $Ag_2Te$ | 8.24-8.45 | 2-3 | monocl., (<149.5°), cub. (>149.5°), gray | |
| Heulandite | $(Ca,Na)Al_2Si_7O_{18}.6H_2O$ | 2.1-2.2 | 3-4 | pseudo-monocl., col., wh., yel., pink, red, gray, br. | 1.491-1.505, 1.493-1.503, 1.500-1.512 |
| Hopeite | $Zn_3(PO_4)_2.4H_2O$ | 3.0-3.1 | 3.25 | rhomb., col. to grayish-wh., pa. yel. | 1.57-1.59, 1.58-1.60, 1.58-1.60 |
| Hornblende | $(Ca,Na,K)_{2-3}(Mg,Fe,Fe^{+3}Al)_5Si_6(Si,Al)_2O_{22}(OH,F)_2$ | 3.02-3.45 | 5-6 | monocl., grn., dk. grn., blk. | 1.615-1.705, 1.618-1.714, 1.632-1.730 |
| Huebnerite | $MnWO_4$ | 7.12 | 4-4.5 | monocl., yel.-br. to red br., somet. br., blk. | 2.17, 2.22, 2.32 |
| Humite | $Mg_7(OH,F)_2.3Mg_2SiO_4$ | 3.2-3.32 | 6 | rhomb., yel., orange | 1.607-1.643, 1.619-1.653, 1.639-1.675 |
| Huntite | $Mg_3Ca(CO_3)_4$ | 2.696 | | rhomb.(?), wh. | |
| Hydrogrossularite | $Ca_3Al_2Si_3O_8(SiO_4)_{1-m}(OH)_{4m}$ | 3.594-3.13 | 6-7.5 | cub., wh., buff, pa. grn., gray, pink | 1.734-1.675 |
| Hydromagnesite | $Mg_4(OH)_2(CO_3)_3.3H_2O$ | 2.236 | 3.5 | monocl., col. to wh. | 1.520-1.526, 1.524-1.530, 1.544-1.546 |
| Illite | $K_{1-1.5}Al_4(Si_{6.5-7}Al_{1-1.5}O_{20}(OH)_4$ | 2.6-2.9 | 1-2 | monocl., wh. | 1.54-1.57, 1.57-1.61, 1.57-1.61 |
| Ilmenite | $FeTiO_3$ | 4.68-4.76 | 5-6 | rhbdr., iron-blk. | |
| Iodyrite | $AgI$ | 5.69 | 1.5 | hex., col. on exposure to light, yel., br. | 2.21, 2.22 |
| Jadeite | $NaAlSi_2O_6$ | 3.24-3.43 | 6 | monocl., col., wh., grn., grnsh. bl. | 1.640-1.658, 1.645-1.663, 1.652-1.673 |
| Jamesonite | $Pb_4FeSb_6S_{14}$ | 5.63 | 2.5 | monocl., gray-blk., somet. tarn. iridesc. | |
| Jarosite | $KFe_3(SO_4)_2(OH)_6$ | 2.91-3.26 | 2.5-3.5 | rhbdr., ocherous, amber yel. to dk. br. | 1.820, 1.715 |
| Kainite | $KMg(SO_4)Cl.3H_2O$ | 2.15 | 2.5-3 | monocl., col., gray bl., vlt., yelsh., redsh. | 1.494, 1.505, 1.516 |
| Kaliophyllite | $KAlSiO_4$ | 2.61 | 6 | hex., col. | 1.532, 1.537 |
| Kaolinite | $Al_4Si_4O_{10}(OH)_8$ | 2.61-2.68 | 2-2.5 | triel. or monocl., wh., redsh.-wh., grnsh.-wh. | 1.553-1.565, 1.559-1.569, 1.560-1.570 |
| Kernite | $Na_2B_4O_7.4H_2O$ | 1.908 | 2.5 | monocl., col., wh. | 1.454, 1.472, 1.488 |
| Kieserite | $MgSO_4.H_2O$ | 2.571 | 3.5 | monocl., col., gray, wh., yelsh. | 1.520, 1.533, 1.584 |
| Kyanite | $Al_2OSiO_4$ | 3.53-3.65 | 5-7 | triel., bl., wh., gray, grn., yel., pink | 1.712-1.718, 1.721-1.723, 1.727-1.734 |
| Lanarkite | $Pb_2(SO_4)O$ | 6.92 | 2-2.5 | monocl., gray to grnsh.-wh., pa. yel. | 1.925-1.931, 2.004-2.010, 2.033-2.039 |
| Lanthanite | $(La,Ce)_2(CO_3)_3.8H_2O$ | 2.2-2.3 | 2.5-3 | rhomb., col. to wh., pink, yelsh. | 1.51-1.53, 1.584-1.590, 1.610-1.616 |
| Laumontite | $CaAl_2Si_4O_{12}.4-3.5H_2O$ | 2.2-2.3 | 3-3.5 | monocl., col. to wh., red, yel., brn. | 1.502-1.514, 1.512-1.522, 1.514-1.525 |
| Laurionite | $Pb(OH)Cl$ | 6.24 | 3-3.5 | rhomb., col. to wh. | 2.08, 2.12, 2.16 |
| Lawsonite | $CaAl_2(OH)_2Si_2O_7.H_2O$ | 3.05-3.10 | 6 | rhomb., col., wh. | 1.655, 1.674-1.675, 1.684-1.686 |
| Lazulite | $(Mg,Fe)Al_2(PO_4)_2(OH)_2$ | 3.08-3.38 | 5-5.5 | monocl., bl., blsh. wh., dk. bl., blsh. grn. | 1.604-1.626, 1.626-1.654, 1.637-1.663 |
| Lazurite | $Na_4SSi_3Al_3O_{12}$ | 2.38-2.45 | 5-5.5 | cub., berlin bl., azure bl., grnsh. bl., vlt. | 1.500 |
| Leadhillite | $Pb_2(SO_4)(CO_3)_2(OH)_2$ | 6.55 | 2.5-3 | monocl., col. to wh., gray, pa. grn., pa. bl., yelsh. | 1.87, 2.00, 2.01 |
| Lepidocrocite | $FeO(OH)$ | 4.05-4.31 | 5 | rhomb., ruby-red to red-br. | 1.94, 2.20, 2.51 |
| Lepidolite | $K_2(Li,Al)_{5-6}Si_{6-7}Al_{2-1}O_{20}(OH,F)_4$ | 2.80-2.90 | 2.5-4 | monocl., col., pa. pink, pa. purp. | 1.525-1.548, 1.551-1.585, 1.554-1.587 |
| Leucite | $KAlSi_2O_6$ | 2.47-2.50 | 5.5-6 | tetr. (pseudo-cub.) wh., gray | 1.508-1.511 |
| Levyne | $(Ca,Na_2)Al_2Si_4O_{12}.6H_2O$ | ~2.1 | 4-5 | rhbdr., wh., redsh. wh., yelsh., grnsh. | 1.496-1.505, 1.491-1.500 |
| Litharge | $PbO$ | 9.14 | | tetr., red | (Li) 2.665, 2.535 |
| Loellingite | $FeAs_2$ | 7.39-7.41 | 5-5.5 | rhomb., silver wh. to steel-gray | |
| Magnesite | $MgCO_3$ | 2.98-3.44 | 3.5-4.5 | rhbdr., wh., col., somet. yel., br. | 1.700-1.782, 1.509-1.563 |
| Magnetite | $Fe_3O_4$ | 5.175 | 5.5-6.5 | cub., blk. to br.-blk. | 2.42 |
| Malachite | $Cu_2(OH)_2(CO_3)$ | 4.03-4.07 | 3.5-4 | monocl., bright grn. to dk. grn., blksh. grn. | 1.652-1.658, 1.872-1.878, 1.906-1.912 |
| Manganite | $MnO(OH)$ | 4.32-4.43 | 4 | monocl., dk. steel-gray to iron-blk. | (Li) 2.25, 2.25, 2.53 |
| Manganosite | $MnO$ | 5.364 | 5-6 | cub., emerald grn., tarn. bl. | 2.16 |
| Marcasite | $FeS_2$ | 4.887 | 6-6.5 | rhomb., pa. bronze-yel., tin-wh. | |
| Marialite | $Na_4Si_3Al_3O_{12}Cl$ | 2.50-2.62 | 5-6 | tetr., col., wh., pa. grnsh.-yel., gray, br. | 1.546-1.550, 1.540-1.541 |
| Marshite | $CuI$ | 5.68 | 2.5 | cub., col. to pa. yel., on exposure to light, red | 2.346 |

Table 3–1 (continued)

## PHYSICAL CONSTANTS OF MINERALS

| Name | Formula | Sp. gr. | Hardness | Crystalline form and color | Index of refraction (Na) η; ω ε / α β γ |
|---|---|---|---|---|---|
| Mascagnite | (NH₄)₂SO₄ | 1.768 | 2–2.5 | rhomb., col., gray, yelsh. | 1.5202, 1.5230, 1.5330 |
| Matlockite | PbFCl | 7.12 | 2.5–3 | tetr., col. or yel. to pa. amber, grnsh. | 2.145, 2.006 |
| Meionite | Ca₄Al₆Si₆O₂₄CO₃ | 2.78 | 5–6 | tetr., col., wh., pa. grnsh. yel., gray, br. | 1.590–1.600, 1.556–1.562 |
| Melanterite | FeSO₄·7H₂O | 1.898 | 2 | monocl., grn., grnsh. bl., grnsh. wh. | 1.47–1.48, 1.49 |
| Melilite | (Ca,Na,K)₂(Mg,Fe,Fe⁺³,Al,Si)₇O₂₂ | 2.95–3.05 | 5–6 | tetr., yelsh., br., grn.-br. | 1.624–1.666, 1.616–1.661 |
| Mellite | Al₂C₁₂O₁₂·18H₂O | 1.64 | 2–2.5 | rhomb., col. to wh., gray, oft. yel., red, bl. tinge | 1.5393, 1.5110 |
| Mendipite | Pb₃O₂Cl₂ | 7.24 | 2.5 | | 2.22–2.26, 2.25–2.29, 2.29–2.33 |
| Mesolite | Na₂Ca₂(Al₂Si₃O₁₀)₃·8H₂O | ~2.26 | 5 | monocl., col., wh., gray, yel., pink, red | β = 1.504–1.508 |
| Metacinnabar | HgS | 7.65 | 3 | cub., grayish-blk. | |
| Microcline | KAlSi₃O₈ | 2.56–2.63 | 6–6.5 | tricl., col., wh., pink, red, yel., grn. | 1.514–1.529, 1.518–1.533, 1.521–1.539 |
| Microlite | (Na,Ca)₂Ta₂O₆(O,OH,F) | 5.64–6.4 | 5–5.5 | cub., pa. yel. to br., somet. red, grn. | ~2.0 |
| Miersite | AgI | 5.64–5.68 | 2.5 | cub., canary-yel. | 2.18–2.22 |
| Millerite | NiS | 5.3–5.7 | 3–3.5 | hex., pa. brass-yel. to bronze-yel., gray, tarn. iridesc. | |
| Mimetite | Pb₅(AsO₄)₃Cl | 7.24 | 3.5–4 | hex., pa. yel. to yelsh. br., orange-yel., wh. | 2.147, 2.128 |
| Minium | Pb₃O₄ | 8.9–9.2 | 2.5 | scarlet red, bl. red, somet. yel. tint. | (Li) 2.40–2.44 |
| Mirabilite | Na₂SO₄·10H₂O | 1.490 | 1.5–2 | monocl., col. to wh. | 1.391–1.397, 1.393–1.399, 1.395–1.401 |
| Moissanite | SiC | 3.218 | 9.5 | hex., grn. to blk., somet. blsh., red | 2.647–2.649, 2.689–2.693 |
| Molybdenite | MoS₂ | 4.62–4.73 | 1–1.5 | hex., lead-gray | |
| Monazite | (Ce,La,Th)PO₄ | 5.0–5.3 | 5 | monocl., yel., br., redsh. br. | 1.774–1.800, 1.777–1.801, 1.828–1.851 |
| Monetite | CaH(PO₄) | 2.929 | 3.5 | tricl., wh., pa. yelsh.-wh. | 1.587, ~1.615, 1.640 |
| Monticellite | CaMgSiO₄ | 3.08–3.27 | 5.5 | rhomb., col. | 1.639–1.654, 1.646–1.664, 1.653–1.674 |
| Montmorillonite | (0.5Ca,Na)₀.₇(Al,Mh,Fe)₄(Si,Al)₈O₂₀(OH)₄·nH₂O | 2–3 | 1–2 | monocl., wh., yel., grn. | 1.48–1.61, 1.50–1.64, 1.50–1.64 |
| Montroydite | HgO | 11.23 | 2.5 | rhomb., dk. red to brnish. red, br. | (Li) 2.37, 2.5, 2.65 |
| Mordenite | (Na₂,K₂Ca)Al₂Si₁₀O₂₄·7H₂O | 2.12–2.15 | 3–4 | rhomb., col., wh., red, yel., br. | 1.472–1.483, 1.475–1.485, 1.477–1.487 |
| Muscovite | KAl₂AlSi₃O₁₀(OH,F)₂ | 2.77–2.88 | 2–2.5 | monocl., col., pa. grn., pa. red, pa. br. | 1.552–1.574, 1.582–1.610, 1.587–1.616 |
| Nantokite | CuCl | 4.136 | 2.5 | cub., col. to wh., grayish, grn. | 1.925–1.935 |
| Natrolite | Na₂Al₂Si₃O₁₀·2H₂O | 2.20–2.26 | 5–5.5 | rhomb., col., wh., grayish, grn. | 1.473–1.483, 1.476–1.486, 1.485–1.496 |
| Nepheline | Na₃KAl₄Si₄O₁₆ | 2.56–2.665 | 5.5–6 | hex., col., wh., gray | 1.529–1.546, 1.526–1.542 |
| Newberyite | MgH(PO₄)·3H₂O | 2.10 | 3.0–3.5 | rhomb., col. | 1.511–1.517, 1.514–1.520, 1.530–1.536 |
| Nicolite | NiAs | 7.784 | 5–5.5 | hex., pa. copper-red, tarn. gray to blk. | |
| Nosean | Na₈Al₆Si₆O₂₄SO₄ | 2.30–2.40 | 5.5 | cub., gray, bl., br. | 1.495 |
| Oldhamite | CaS | 2.58 | 4 | cub., pa. chestnut-br. | 2.137 |
| Oligoclase | (NaSi)₀.₇₋₀.₅(CaAl)₀.₁₋₀.₃)AlSi₂O₈ | 2.63–2.65 | 6–6.5 | tricl., col., wh., gray, grnsh., pink | 1.533–1.544, 1.537–1.548, 1.543–1.552 |
| Olivenite | Cu₂(AsO₄)(OH) | 4.3 | 3 | rhomb., olive grn., grnsh.-br., br., gray | 1.75–1.78, 1.79–1.82, 1.83–1.87 |
| Olivine | (Mg,Fe)₂SiO₄ | 3.22–4.39 | 6.5–7 | rhomb., olive grn., grayish grn. to yelsh. br. | 1.63–1.83, 1.65–1.87, 1.67–1.88 |
| Opal | SiO₂·nH₂O | 1.73–2.16 | ~6 | col., wh., yel., br., red, grn., bl., blk., amorp. | 1.41–1.46 |
| Orpiment | As₂S₃ | 3.49 | 1.5–2 | monocl., yel., brnsh., red. | (Li) 2.4, 2.81, 3.02 |
| Orthoclase | KAlSi₃O₈ | 2.55–2.63 | 6–6.5 | monocl., col., wh., pink, red, yel., grn. | 1.518–1.529, 1.522–1.533, 1.522–1.539 |
| Orthopyroxene | (Mg,Fe)SiO₃ | 3.209–3.96 | 5–6 | rhomb., col., gray, grn., yel., dk. brn. | 1.650–1.768, 1.653–1.770, 1.658–1.788 |
| Paragonite | NaAl₂AlSi₃O₁₀(OH)₂ | 2.85 | 2.5 | monocl., col., pa. yel. | 1.564–1.580, 1.594–1.609, 1.600–1.609 |
| Parisite | (Ce,La,Na)FCO₃·CaCO₃ | 4.42 | 4.5 | hex., brnsh., yel. | 1.672, 1.771 |
| Pectolite | Ca₂NaH(SiO₃)₃ | 2.86–2.90 | 4.5–5 | tricl., col., wh. | 1.595–1.610, 1.605–1.615, 1.632–1.645 |
| Penfieldite | Pb₂Cl₃(OH)? | 6.6 | 3.5–4 | hex., wh. | 2.13, 2.21 |
| Pentlandite | (Fe,Ni)₉S₈ | 4.6–5.0 | 3.5–4 | cub., pa. bronze-yel. | 2.04–2.06 |
| Percylite | PbCuCl₂(OH)₂(?) | 6.6 | 2.5 | cub.(?), sky bl. | |
| Periclase | MgO | 3.55–3.68 | 5.5 | cub., col. to gray-wh., yel., brnsh. yel., grn., bl. | 7350 |
| Perovskite | CaTiO₃ | 3.97–4.26 | 5.5 | pseudo cub., blk., gray-blk., brnish. bl., redsh. br., br., yel. | 2.30–2.38 |
| Petalite | LiAlSi₄O₁₀ | 2.412–2.422 | 6.5 | monocl., wh., gray, somet. pink, grn. | 1.504–1.507, 1.510–1.513, 1.516–1.523 |
| Pharmacosiderite | Fe₃(AsO₄)₂(OH)₃·5H₂O | 2.797 | 2.5 | cub., olive-grn. to yel., br., redsh. | 1.676–1.704 |
| Phenakite | Be₂SiO₄ | 2.98 | 7.5 | rhbder., col., rose, yel., br. | 1.654, 1.670 |
| Phillipsite | (0.5Ca,Na,K)₃Al₃Si₅O₁₆·6H₂O | 2.2 | 4–4.5 | monocl. or rhomb., col., wh., pink, gray, yel. | 1.483–1.504, 1.484–1.509, 1.496–1.514 |
| Phlogopite | KMg₃AlSi₃O₁₀(OH,F)₂ | 2.76–2.90 | 2–2.5 | monocl., col., yelsh.-br., grn., redsh.-br., br. | 1.530–1.590, 1.557–1.637, 1.558–1.637 |
| Phosgenite | Pb₂(CO₃)₂Cl₂ | 6.133 | 2–3 | tetr., yelsh. wh. to yelsh. br., br., somet. wh., rose, gray | 2.1181, 2.1446 |

Table 3–1 (continued)

## PHYSICAL CONSTANTS OF MINERALS

| Name | Formula | Sp. gr. | Hardness | Crystalline form and color | Index of refraction (Na) $\eta$; $\omega$ $\epsilon$ / $\alpha$ $\beta$ $\gamma$ |
|---|---|---|---|---|---|
| Piemontite | $Ca_2(Mn,Fe^{+3},A^{l})_3AlSi_3O_{12}(OH)$ | 3.45–3.52 | 6 | monocl., redsh. brn. blk. | 1.732–1.794, 1.750–1.807, 1.762–1.829 |
| Pigeonite | $(Mg,Fe,Ca)(Mg,Fe)Si_2O_6$ | 3.30–3.46 | 6 | monocl., br., grnsh. br., blk. | 1.682–1.722, 1.684–1.722, 1.705–1.751 |
| Platinum | $Pt$ | 14–19 | 4–4.5 | cub., whitish, steel gray to dk. gray | |
| Pollucite | $CsAlSi_2O_6$ | 2.9 | 6.5 | tetr., (pseudo-cub.) col. | 1.507–1.527 |
| Polybasite | $(Ag,Cu)_{16}Sb_2S_{11}$ | 6.0–6.2 | 2–3 | monocl., iron-blk. | |
| Powellite | $Ca(Mo,W)O_4$ | 4.21–4.25 | 3.5–4 | tetr., straw-yel., br., grnsh., somet. gray, bl. blk. | 1.959–1.982, 1.967–1.993 |
| Prehnite | $Ca_2Al_2Si_3O_{10}(OH)_2$ | 2.90–2.95 | 6–6.5 | rhomb., pa. grn., yel., gray, wh. | 1.611–1.632, 1.615–1.642, 1.632–1.665 |
| Proustite | $Ag_3AsS_3$ | 5.57 | 2–2.5 | rhbdr., scarlet-vermillion | (Li) 3.0877, 2.7924 |
| Pseudobrookite | $Fe_2TiO_5$ | 4.33–4.39 | 6 | rhomb., dk. red-br. to brnsh. blk. and blk. | (Li) 2.38, 2.39, 2.42 |
| Psilomelane | $BaMn^{+2}Mn_8^{+4}O_{16}(OH)_4$ | 4.71 | 5–6 | rhomb., iron-blk. to steel-gray | |
| Pumpellyite | $Ca_4(Mg,Fe,Mn)(Al,Fe^{+3},Ti)_6(OH)_7Si_6O_{23} \cdot 2H_2O$ | 3.18–3.23 | 6 | monocl., grn., blsh. grn., br. | 1.674–1.702, 1.675–1.715, 1.688–1.722 |
| Pyrargyrite | $Ag_3SbS_3$ | 5.85 | 2.5 | rhbdr., deep red | (Li) 3.084, 2.881 |
| Pyrite | $FeS_2$ | 5.018 | 6–6.5 | cub., pa. brass-yel., tarn. iridesc. | |
| Pyrochlore | $NaCaCb_2O_6F$ | 4.2–6.4 | 5–5.5 | cub., br. to blk., yelsh., redsh. or blksh. br. | |
| Pyrochroite | $Mn(OH)_2$ | 3.258 | 2.5 | hex., col. to pa. grn. or bl., tarn. br. to blk. | 1.72, 1.68 |
| Pyrolusite | $MnO_2$ | 5.04–5.08 | 6–6.5 | tetr., pa. steel-gray, iron-gray, blk., blsh. | |
| Pyromorphite | $Pb_5(PO_4,AsO_4)_3Cl$ | 7.00–7.08 | 3.5–4 | hex., grn., yel., br., orange, brnsh. red., gray | 2.058, 2.048 |
| Pyrope | $Mg_3Al_2Si_3O_{12}$ | 3.582 | 6–7.5 | cub., red, pink | 1.714 |
| Pyrophyllite | $Al_2Si_4O_{10}(OH)_2$ | 2.65–2.90 | 1–2 | monocl., wh., yel., pa. bl., gray-grn., brnsh. grn. | 1.534–1.556, 1.568–1.589, 1.596–1.601 |
| Pyrrhotite | $Fe_{1-x}S$ | 4.58–4.65 | 3.5–4.5 | hex., bronze-yel. to br., tarn., somet. iridesc. | |
| Quartz | $SiO_2$ | 2.65 | 7 | rhbdr., col., wh., blk., purp., grn., bl., rose | 1.544, 1.553 |
| Rammelsbergite | $NiAs_2$ | 7.0–7.2 | 5.5–6 | rhbdr., col., wh., redsh. tinge | |
| Raspite | $PbWO_4$ | 8.46 | 2.5–3 | monocl., yelsh. br., pa. yel., gray | 1.25–1.29, 1.25–1.29, 1.28–1.32 |
| Realgar | $AsS$ | 3.56 | 1.5–2 | monocl., aurora-red to orange-yel. | 2.538, 2.684, 2.704 |
| Riebeckite | $Na_2Fe_3Fe_2^{+3}Si_8O_{22}(OH,F)_2$ | 3.02–3.42 | 5 | monocl., dk. bl., bl. | 1.654–1.701, 1.662–1.711, 1.688–1.717 |
| Rhodochrosite | $MnCO_3$ | 3.70 | 3.5–4 | rhbdr., pink, red, br., brnsh.-yel. | 1.816, 1.597 |
| Rhodonite | $(Mn,Fe,Ca)SiO_3$ | 3.57–3.76 | 5.5–6.5 | triel., pink to brnsh. red | 1.711–1.738, 1.716–1.741, 1.724–1.751 |
| Rutile | $TiO_2$ | 4.23–5.5 | 6–6.5 | tetr., redsh. brn. to red, somet. yelsh., bish. | 2.605–2.613, 2.899–2.901 |
| Safflorite | $(Co,Fe)As_2$ | 7.0–7.5 | 4.5–5 | rhomb., tin-wh., tarn. dk. gray | ~2.20 |
| Samarskite | $(Y,Er,Ce,U,Ca,Fe,Pb,Th)(Cb,Ta,Ti,Sn)_2O_6$ | 5.69 | 5–6 | rhomb., velvet blk., somet. brnsh. tint | 1.701–1.717, 1.703–1.720, 1.705–1.724 |
| Sapphirine | $(Mg,Fe)_2Al_4O_6SiO_4$ | 3.40–3.58 | 7.5 | monocl., pa. bl., pa. grn. | 1.546–1.600, 1.540–1.562 |
| Scapolite | $(Na,Ca)_4Al_3(Al,Si)_3Si_6O_{24}(Cl,F,OH,CO_3,SO_4)$ | 2.50–2.78 | 5–6 | tetr., col., wh., pa. grnsh. yel., gray, bl. | |
| Scheelite | $CaWO_4$ | 6.08–6.12 | 4.5–5 | tetr., yelsh. wh., pa. yel., brnsh., col., wh., gray | 1.920, 1.936 |
| Scolecite | $CaAl_2Si_3O_{10} \cdot 3H_2O$ | 2.25–2.29 | 5 | monocl., col., wh., gray, yel., pink, red | 1.507–1.513, 1.516–1.520, 1.517–1.521 |
| Scorodite | $Fe^{+3}(AsO_4) \cdot 2H_2O$ | 3.28 | 3.5–4 | rhomb., pa. gray grn., br. somet. col., blsh., yel. | 1.784, 1.795, 1.814 |
| Sellaite | $MgF_2$ | 3.15 | 5 | tetr., col. to wh. | 1.378, 1.390 |
| Senarmontite | $Sb_2O_3$ | 5.50 | 2–2.5 | pseudo-cub., col., gray-wh. | 2.087 |
| Serpentine | $Mg_3Si_2O_5(OH)_4$ | ~2.55 | 2.5–3.5 | monocl., wh., gray, grn., blsh. grn. | 1.53–1.57, 1.56, 1.54–1.57 |
| Siderite | $FeCO_3$ | 3.96 | 4–4.5 | rhbdr., yelsh. br., br., dk. br. | 1.875, 1.635 |
| Sillimanite | $Al_2SiO_5$ | 3.23–3.27 | 6.5–7.5 | rhomb., col., wh., yelsh., br., grnsh. | 1.654–1.661, 1.658–1.662, 1.637–1.683 |
| Silver | $Ag$ | 10.1–11.1 | 2.5–3 | cub., wh., tarn. gray or blk. | |
| Skutterudite | $(Co,Ni)As_3$ | 6.1–6.9 | 5.5–6 | cub., between tin-wh. and silver-gray, tarn. gray or iridesc. | |
| Smithsonite | $ZnCO_3$ | 4.42–4.44 | 4–4.5 | rhbdr., grayish wh. to dk. gray, grnsh., brnsh. wh. | 1.848, 1.621 |
| Sodalite | $Na_4Al_3Si_3O_{12}Cl$ | 2.27–2.33 | 5.5–6 | cub., bl., grn., yel., gray, pink | 1.483–1.487 |
| Sperrylite | $PtAs_2$ | 10.58 | 6–7 | cub., tin-wh. | |
| Spessartite | $Mn_3Al_2Si_3O_{12}$ | 4.190 | 6–7.5 | cub., blk., dk. red, brnsh. red., bl., yelsh. orange | 1.800 |
| Sphalerite | $ZnS$ | 3.9–4.1 | 3.5–4 | cub., br., blk., yel., red, wh. | 2.369 |
| Sphene | $CaTiSiO_4(O,OH,F)$ | 3.45–3.55 | 5 | monocl., col., yel., grn., br., blk. | 1.843–1.950, 1.870–2.034, 1.943–2.110 |
| Spinel | $MgAl_2O_4$ | 3.55 | 7.5–8 | cub., grn., red, bl., br. to col. | 1.719 |

## Table 3–1 (continued)
## PHYSICAL CONSTANTS OF MINERALS

| Name | Formula | Sp. gr. | Hardness | Crystalline form and color | Index of refraction (Na) $\eta$; $\omega$ $\beta$ $\gamma$ |
|---|---|---|---|---|---|
| Spodumene | $LiAlSi_2O_6$ | 3.03–3.22 | 6.5–7 | monocl., col., gray-wh., pa. bl., pa. grn., yelsh. | 1.648–1.663, 1.655–1.669, 1.662–1.679 |
| Stannite | $Cu_2FeSn_4$ | 4–4.5 | 4 | tetr., steel gray to iron blk. | |
| Staurolite | $(Fe,Mg)_2(AlFe^{3+})_9O_6SiO_4(O,OH)_2$ | 3.74–3.83 | 7.5 | monocl., brn., redsh., yelsh. | 1.739–1.747, 1.745–1.753, 1.752–1.761 |
| Stercorite | $Na(NH_4)H(PO_4).4H_2O$ | 1.615 | 2 | tricl., wh., yelsh., brnsh. | 1.439, 1.442, 1.469 |
| Stibiotantalite | $Sb(Ta,Cb)O_4$ | 5.7–7.5 | 5.5 | rhomb., dk. br. to pa. yel.-br., red-br., grnsh.-yel. | 2.38, 2.41, 2.46 |
| Stibnite | $Sb_2S_3$ | 4.61–4.65 | 2 | rhomb., lead-gray to steel-gray | 1.494–1.500, 1.492–1.507, 1.494–1.513 |
| Stibite | $(Ca,Na_2,K_2)Al_2Si_7O_{18}.7H_2O$ | 2.1–2.2 | 3.5–4 | monocl., col., wh., yel., pink, red, gray, br. | 1.543–1.634, 1.576–1.745, 1.576–1.745 |
| Stilpnomelane | $(K,Na,Ca)_{0-1}(Fe^{3+}Fe,Mg,Al)_{5-6}Si_8O_{20}(OH)_4(O,OH,H_2O)_{1-8}$ | 2.59–2.96 | 3–4 | monocl., br., dk. br., redsh. br., blk., dk. grn. | |
| Stolzite | $PbWO_4$ | 7.9–8.4 | 2.5–3 | tetr., redsh. br., yelsh. gray, straw-yel., grnsh. | 2.26–2.28, 2.18–2.20 |
| Strengite | $Fe^{3+}(PO_4).2H_2O$ | 2.90 | 3.5 | rhomb., red, carmine, vlt., near col. | 1.707, 1.719, 1.741 |
| Strontianite | $SrCO_3$ | 3.72 | 3.5 | rhomb., col., wh., yel., grnsh., brnsh. | 1.516–1.520, 1.664–1.667, 1.666–1.669 |
| Struvite | $Mg(NH_4)(PO_4).6H_2O$ | 1.71 | 2 | rhomb., col., somet. yelsh., brnsh. | 1.495, 1.496, 1.504 |
| Sulfur | $S$ | 2.07 | 1.5–2.5 | rhomb., yel., brnsh., grnsh., redsh., gray | 1.9579, 2.0377, 2.2452 |
| Sylvanite | $(Ag,Au)Te_4$ | 8.161 | 1.5–2 | monocl., steel-gray to silver-wh. | |
| Sylvite | $KCl$ | 1.99 | 2 | cub., col., wh., somet. grayish, blsh., yelsh., red | 1.49031 |
| Talc | $Mg_3Si_4O_{10}(OH)_2$ | 2.58–2.83 | 1 | monocl., col., wh., pa. grn., dk. grn., br. | 1.539–1.550, 1.589–1.594, 1.589–1.600 |
| Tantalite | $(Fe,Mn)(Ta,Cb)_2O_6$ | 7.90–8.00 | 6.5 | rhomb., iron-bl. to br.-blk. | 2.26, 2.32, 2.43 |
| Tapiolite | $FeTa_2O_6$ | 7.9 | 6–6.5 | tetr., blk. | (Li) 2.27, 2.42 |
| Tellurobismuthite | $Bi_2Te_3$ | 7.800–7.830 | 1.5–2 | rhbdr., pa. lead-gray | |
| Terlinguaite | $Hg_2OCl$ | 8.725 | 2–3 | monocl., yel. to grnsh.-yel., somet. br. | (Li) 2.33–2.37, 2.62–2.66, 2.64–2.68 |
| Tetrahedrite | $(Cu,Fe)_{12}Sb_4S_{13}$ | 4.6–5.1 | 3–4.5 | cub., flint-gray to iron-blk. to dull-blk. | |
| Thenardite | $Na_2SO_4$ | 2.664 | 2.5–3 | rhomb., col., grayish-wh., yelsh., yelsh. br., redsh. | 1.464–1.471, 1.473–1.477, 1.481–1.485 |
| Thermonatrite | $Na_2CO_3.H_2O$ | 2.255 | 1–1.5 | rhomb., col. to wh., grayish, yelsh. | 1.420, 1.506, 1.524 |
| Thomsenolite | $NaCaAlF_6.H_2O$ | 2.981 | 2 | monocl., col. to wh., somet. brnsh., redsh. | 1.4072, 1.4136, 1.4150 |
| Thomsonite | $NaCa_2(Al,Si)_5O_{10}).6H_2O$ | 2·10–2.39 | 5–5.5 | rhomb., col., wh., pink, br. | 1.497–1.530, 1.513–1.533, 1.518–1.544 |
| Thorianite (R) | $ThO_2$ | 9.7 | 6.5 | cub., dk. gray to brnsh.-blk., blk. | ~2.20 |
| Thorite (R) | $ThSiO_4$ | 5–2.5.4 | 4.5–5 | tetr., orange-yel., brnsh. to blk. | ~1.8 |
| Topaz | $Al_2SiO_4(OH,F)_2$ | 3.49–3.57 | 8 | rhomb., col. to wh., yel., gray, grn., red, bl. | 1.606–1.629, 1.609–1.631, 1.616–1.638 |
| Torbernite (R) | $Cu(UO_2)_2(PO_4)_2.8-12H_2O$ | 3.22 | 2–2.5 | tetr., various shades of grn. | 1.592, 1.582 |
| Tourmaline (R) | $Na(Mg,Fe,Mn,Li,Al)_3Al_6Si_6O_{18}(BO_3)_3(OH,F)_4$ | 3.03–3.25 | 7 | rhbdr., blk., bl., grn., yel., red, col., br. | 1.635–1.675, 1.610–1.650 |
| Tremolite | $Ca_2Mg_5Si_8O_{22}(OH,F)_2$ | 3.0 | 5–6 | monocl., col., gray, wh. | 1.599, 1.612, 1.622 |
| Tridymite | $SiO_2$ | 2.27 | 7 | rhomb., col., wh. | 1.471–1.479, 1.472–1.480, 1.474–1.483 |
| Triphyllite–Lithio-phyllite | $Li(Fe,Mn)PO_4$ | 3.34–3.58 | 4–5 | rhomb., blsh. or grnsh. gray to yelsh. br., br. | 1.66–1.70, 1.67–1.70, 1.68–1.71 |
| Troegerite (R) | $(UO_2)_3(AsO_4)_2.12H_2O$ | 2.645 | 2–3 | tetr., lemon-yel. | 1.58–1.59, 1.625–1.635 |
| Trona | $Na_3H(CO_3)_2.2H_2O$ | 2.14 | 2.5–3 | monocl. gray or grnsh. wh., col. | 1.412, 1.492, 1.540 |
| Turquois | $Cu(Al,Fe^{3+})_6(PO_4)_4(OH)_8.4H_2O$ | 2.6–3.2 | 4.5–5.6 | tricl., bl., grn., grnish-gray | 1.61–1.78, 1.62–1.84, 1.65–1.84 |
| Ullmannite | $NiSbS$ | 6.61–6.69 | 5–5.5 | cub., steel-blk., brnsh.-blk., grayish, grn. | |
| Uraninite (R) | $UO_2$ | 8.0–11 | 5–6 | cub., steel-blk., brnsh.-blk., grayish, grn. | 1.86 |
| Uvarovite | $Ca_3Cr_2Si_3O_{12}$ | 3.90 | 6.5–7 | cub., emerald-grn. | 1.86 |
| Valentinite | $Sb_2O_3$ | 5.76 | 2.5–3 | rhomb., col. to wh., somet. yelsh., redsh., gray, br. | 2.18, 2.35, 2.35 |
| Vanadinite | $Pb_5(VO_4)_3Cl$ | 6.5–7.1 | 2.75–3 | hex., orange-red, red, brnsh.-red, br., brnsh., yel., yel. | 2.416, 2.350 |
| Variscite–Strengite | $(AlFe^{3+})(PO_4).2H_2O$ | 2.57–2.87 | 3.5–4.5 | rhomb., pa. grn., grn., blsh.-grn., red, vlt., col. | 1.563–1.707, 1.588–1.719, 1.594–1.741 |
| Vaterite | $CaCO_3$ | 2.645 | | hex., col. | 1.550, 1.640–1.650 |
| Vermiculite | $(Mg,Ca)_{0.7}(Mg,Fe^{3+},Al)_6(Al,Si)_8O_{20}(OH)_4.8H_2O$ | ~2.3 | ~1.5 | monocl., col., yel., grn., br. | 1.525–1.564, 1.545–1.583, 1.545–1.583 |
| Vesuvianite | $Ca_{10}(Mg,Fe)_2Al_4(Si_2O_7)_2(SiO_4)_5(OH,F)_4$ | 3.33–3.43 | 6–7 | tetr., yel., grn., br. | 1.700–1.746, 1.703–1.752 |
| Villiaumite | $NaF$ | 2.79 | 2–2.5 | cub., carmine, (nat.), col. (artif.) | 1.327 |
| Vivianite | $Fe_3(PO_4)_2.8H_2O$ | 2.67–2.69 | 1.5–2 | monocl., col., tarn. pa. bl., grnsh. bl., dk. bl., blsh. blk. | 1.579–1.616, 1.602–1.656, 1.629–1.675 |

### Table 3–1 (continued)
### PHYSICAL CONSTANTS OF MINERALS

| Name | Formula | Sp. gr. | Hardness | Crystalline form and color | Index of refraction (Na) $\eta$; $\omega$ $\epsilon$ / $\alpha$ $\beta$ $\gamma$ |
|---|---|---|---|---|---|
| Wagnerite | $Mg_2(PO_4)F$ | 3 15 | 5–5.5 | monocl., yel., gray, somet. red, grn. | 1.568, 1.572, 1.582 |
| Wavellite | $Al_3(OH)_3(PO_4)_2 \cdot 5H_2O$ | 2 36 | 3.25–4 | rhomb., grnsh. wh., grn. to yel., somet. br., bl., wh. | 1.520–1.535, 1.526–1.543, 1.545–1.561 |
| Whewellite | $Ca(C_2O_4) \cdot H_2O$ | 2 23 | 2.5–3 | monocl., ccl., somet. yelsh., brnsh. | 1.491, 1.554, 1.650 |
| Willemite | $Zn_2SiO_4$ | 3 9–4.1 | 5.5 | rhbdr., wh., yel., grn., red, gray, br. | 1.691, 1.719 |
| Witherite | $BaCO_3$ | 4.29–4.30 | 3.5 | rhomb., col., wh., gray, yelsh. br. | 1.529, 1.676, 1.677 |
| Wolframite | $(Fe,Mn)WO_4$ | 7.12–7.51 | 4–4.5 | monocl., dk. gray, brnsh. blk. to iron blk. | (Li) ~2.26, 2.32, 2.42 |
| Wollastonite | $CaSiO_3$ | 2.87–3.09 | 4.5–5 | tricl., wh., col., gray, pa. grn. | 1.616–1.640, 1.628–1.650, 1.631–1.653 |
| Wulfenite | $PbMoO_4$ | 6.5–7.0 | 2.75–3 | tetr., orange-yel. to yel., gray, grn., br., red | 2.403, 2.283 |
| Wurtzite | $ZnS$ | 3.98 | 3.5–4 | hex., brnsh. blk. | 2.356, 2.378 |
| Xenotime | $Y(PO_4)$ | 4.4–5.1 | 4–5 | tetr., yelsh. br. to redsh. br., somet. gray, wh., pa. yel., grnsh. | 1.721, 1.816 |
| Zeunerite (R) | $Cu(UO_2)_2(AsO_4)_2 \cdot 10{-}16H_2O$ | 5.64–5.68 | 4 | tetr. | 1.602–1.610 |
| Zincite | $ZnO$ | 4.6–4.7 | 7.5 | hex., orange-yel. to dk. red, somet. yel. | 2.013, 2.029 |
| Zircon | $ZrSiO_4$ | 4.6–4.7 | 7.5 | tetr., redsh. br., yel., gray, grn., col. | 1.923–1.960, 1.968–2.015 |
| Zoisite | $Ca_3Al_3Si_3O_{12}(OH)$ | 3.15–3.365 | 6 | rhomb., gray, grnsh., brnsh. | 1.685–1.705, 1.688–1.710, 1.697–1.725 |

From Kretz, R., in *Handbook of Chemistry and Physics*, 69th ed., Weast, R. C., Ed., CRC Press, Boca Raton, Fla., 1988, B-182.

### Table 3–2
### SOLUBILITY CHART

Abbreviations:

W = soluble in water

A = insoluble in water but soluble in acids

w = sparingly soluble in water but soluble in acids

a = insoluble in water and only sparingly soluble in acids

I = insoluble in both water and acids

d = decomposes in water

| No. | | Al | NH₄ | Sb | Ba | Bi | Cd | Ca | Cr | Co | Cu | Au' | Au''' | H | Fe'' | Fe''' |
|---|---|---|---|---|---|---|---|---|---|---|---|---|---|---|---|---|
| 1 | Acetates —(C₂H₃O₂) | $Al(-)_3$ W | $NH_4(-)$ W | ...... | $Ba(-)_2$ W | $Bi(-)_3$ W | $Cd(-)_2$ W | $Ca(-)_2$ W | $Cr(-)_3$ W | $Co(-)_2$ W | $Cu(-)_2$ W | ...... | ...... | $C_2H_4O_2$ W | $Fe(-)_2$ W | $Fe(-)_4$ W |
| 2 | Arsenate —(AsO₄) | $Al(-)$ a | $(NH_4)_3(-)$ W | $Sb(-)$ A | $Ba_3(-)_2$ W | $Bi(-)$ W | $Cd_3(-)_2$ W | $Ca_3(-)_2$ W | $Cr(-)$ W | $Co_3(-)_2$ A | $Cu_3(-)_2$ A | ...... | ...... | $H_3AsO_4$ W | $Fe_3(-)_2$ A | $Fe(-)$ A |
| 3 | Arsenite —(AsO₃) | ...... | $NH_4AsO_2$ W | $Sb(-)$ A | ...... | $Bi(-)$ A | $Cd(-)$ A | $Ca(-)_2$ W | ...... | $Co_3H(-)_4$ A | $CuH(-)$ A | ...... | ...... | ...... | ...... | ...... |
| 4 | Benzoate —(C₇H₅O₂) | ...... | $NH_4(-)$ W | ...... | $Ba(-)_2$ W | $Bi(-)_3$ d | $Cd(-)_2$ W | $Ca(-)_2$ W | ...... | $Co(-)_2$ W | $Cu(-)_2$ W | ...... | ...... | $C_7H_5O_2$ W | $Fe(-)_2$ W | $Fe(-)_4$ W |
| 5 | Bromide | $AlBr_3$ W | $NH_4Br$ W | ...... | $BaBr_2$ W | $BiBr_3$ d | $CdBr_2$ W | $CaBr_2$ W | $Cr_2Br_3$ W | $CoBr_2$ W | $CuBr_2$ W | $AuBr$ W | $AuBr_3$ W | $HBr$ W | $FeBr_2$ W | $FeBr_3$ W |
| 6 | Carbonate | ...... | $(NH_4)_2CO_3$ W | ...... | $BaCO_3$ W | ...... | $CdCO_3$ A | $CaCO_3$ A | $CrCO_3$ W | $CoCO_3$ A | ...... | ...... | ...... | ...... | $FeCO_3$ W | ...... |

Table 3–2 (continued)
## SOLUBILITY CHART

| No. | | Al | NH4 | Sb | Ba | Bi | Cd | Ca | Cr | Co | Cu | Au' | Au''' | H | Fe'' | Fe''' |
|---|---|---|---|---|---|---|---|---|---|---|---|---|---|---|---|---|
| 7 | Chlorate —(ClO3) | W Al(—)3 | W NH4(—) | …… | W Ba(—)2 | W Bi(—)3 | W Cd(—)2 | W Ca(—)2 | | W Co(—)2 | W Cu(—)2 | | | W HClO3 | W Fe(—)2 | W Fe(—)3 |
| 8 | Chloride | W AlCl3 | W NH4Cl | W SbCl3 | W BaCl2 | d BiCl3 | W CdCl2 | W CaCl2 | I CrCl3 | W CoCl2 | W CuCl2 | W AuCl | W AuCl3 | W HCl | W FeCl2 | W FeCl3 |
| 9 | Chromate —(CrO4) | | W (NH4)2(—) | …… | A BaCr2 | …… | W Cd(—) | W Ca(—) | …… | A Co(—) | …… | | | W H2CrO4 | …… | A Fe2(—)3 |
| 10 | Citrate —(C6H5O7) | W Al(—) | W (NH4)3(—) | …… | W Ba3(—)2 | A Bi(—) | A Cd3(—)2 | W Ca3(—)2 | …… | W Co3(—)2 | …… | | | W C6H8O7 | …… | W Fe(—) |
| 11 | Cyanide | …… | W NH4CN | …… | W Ba(CN)2 | W Bi(CN)3 | W Cd(CN)2 | W Ca(CN)2 | A Cr(CN)3 | I Co(CN)2 | A Cu(CN)2 | W AuCN | W Au(CN)3 | W HCN | a Fe(CN)2 | a …… |
| 12 | Ferricy'de —(Fe(CN)6) | a Al(—)3 | W (NH4)3(—) | …… | W Ba3(—)2 | …… | A Cd3(—)2 | W Ca3(—)2 | A Cr2(—)3 | I Co3(—)2 | I Cu(—)2 | | | W H3(—) | I Fe3(—)2 | I …… |
| 13 | Ferrocy'de —(Fe(CN)6) | W Al4(—)3 | W (NH4)4(—) | …… | W Ba(—) | …… | A Cd(—)2 | W Ca(—) | …… | W Co3(—) | I Cu2(—) | | | W H4(—) | I Fe2(—) | I …… |
| 14 | Fluoride | W AlF3 | W NH4F | W SbF3 | W BaF2 | W BiF3 | W CdF2 | A CaF2 | W(a) CrF3 | W CoF2 | W CuF2 | | | W HF | W FeF2 | W FeF3 |
| 15 | Formate —(CHO2) | W Al(—)3 | W NH4(—) | …… | W Ba(—)2 | W Bi(—)3 | W Cd(—)2 | W Ca(—)2 | …… | W Co(—)2 | W Cu(—)2 | W AuOH | A Au(OH)3 | W CH2O2 | W Fe(—)2 | W Fe(—)3 |
| 16 | Hydroxide | A Al(OH)3 | W NH4OH | …… | W Ba(OH)2 | A Bi(OH)3 | A Cd(OH)2 | W Ca(OH)2 | A Cr(OH)3 | A Co(OH)2 | A Cu(OH)2 | | | | A Fe(OH)2 | A Fe(OH)3 |
| 17 | Iodide | W AlI3 | W NH4I | d SbI3 | W BaI2 | A BiI3 | W CdI2 | W CaI2 | W CrI3 | W CoI2 | a CuI | A AuI | A AuI3 | W HI | W FeI2 | W FeI3 |
| 18 | Nitrate | W Al(NO3)3 | W NH4NO3 | …… | W Ba(NO3)2 | d Bi(NO3)3 | W Cd(NO3)2 | W Ca(NO3)2 | A Cr(NO3)3 | W Co(NO3)2 | W Cu(NO3)2 | | | W HNO3 | W Fe(NO3)2 | W Fe(NO3)3 |
| 19 | Oxalate —(C2O4) | a Al2(—)3 | W (NH4)2(—) | …… | W Ba(—) | A Bi2(—)3 | A Cd(—) | A Ca(—) | W Cr2(—)3 | A Co(—) | A Cu(—) | | | W C2H2O4 | A Fe(—) | a Fe2(—)3 |
| 20 | Oxide | a Al2O3 | …… | W Sb2O3 | a BaO | A Bi2O3 | A CdO | W CaO | A Cr2O3 | A CoO | A CuO | Au2O | A Au2O3 | W H2O | A FeO | A Fe2O3 |
| 21 | Phosphate | A AlPO4 | W NH4H2PO4 | …… | A Ba3(PO4)2 | A BiPO4 | A Cd3(PO4)2 | A Ca3(PO4)2 | A Cr3(PO4)2 | A Co3(PO4)2 | A Cu3(PO4)2 | | | W H3PO4 | A Fe3(PO4)2 | A FePO4 |
| 22 | Silicate —(SiO3) | | | | a BaSiO3 | …… | A CdSiO3 | A CaSiO3 | | A Co2SiO4 | A Cu(—) | | | I H2SiO3 | | |
| 23 | Sulfate | W Al2(SO4)3 | W (NH4)2SO4 | A Sb2(SO4)3 | d BaSO4 | d Bi2(SO4)3 | W CdSO4 | W CaSO4 | W(I) Cr2(SO4)3 | W CoSO4 | W CuSO4 | | | W H2SO4 | W FeSO4 | W Fe2(SO4)3 |
| 24 | Sulfide | d Al2S3 | W (NH4)2S | A Sb2S3 | W BaS | A Bi2S3 | A CdS | W CaS | d Cr2S3 | A CoS | A CuS | I Au2S | I Au2S3 | W H2S | W FeS | d FeS3 |
| 25 | Tartrate —(C4H4O6) | W Al(—) | W (NH4)2(—) | W Sb(—)3 | W Ba(—)2 | A Bi(—)3 | A Cd(—) | W Ca(—) | | W Co(—) | d Cu(—) | | | W C4H6O6 | W Fe(—) | W Fe(—)3 |
| 26 | Thiocy'te | …… | W NH4CNS | …… | W Ba(CNS)2 | W Bi(—)3 | W Cd(CNS)2 | W Ca(CNS)2 | | W Co(CNS)2 | d CuCNS | | | W CNSH | W Fe(CNS)2 | W Fe(CNS)3 |

| No. | | Pb | Mg | Mn | Hg' | Hg'' | Ni | K | Ag | Na | Sn'''' | Sn'' | Sr | Zn | Pt |
|---|---|---|---|---|---|---|---|---|---|---|---|---|---|---|---|
| 1 | Acetate —(C2H3O2) | W Pb(C2H3O2)2 | W Mg(—)2 | W Mn(—)2 | W Hg(—) | W Hg(—)2 | W Ni(—)2 | W K(—) | W Ag(—) | W Na(—) | W Sn(—)4 | d Sn(—)2 | W Sr(—)2 | W Zn(—)2 | …… |
| 2 | Arsenate —(AsO4) | A PbH(—) | A Mg3(—) | A MnH(—) | A Hg2(—) | A Hg3(—)2 | A Ni3(—)2 | W K3(—) | A Ag3(—) | W Na3(—) | | | A SrH(—) | | …… |
| 3 | Arsenite —(AsO3) | …… | …… | A Mn3H(—)4 | a Hg(—) | A Hg(—) | A Ni3H(—)4 | W K3AsO3 | A Ag3(—) | W Na3H(—) | | A Sn3(—)2 | …… | …… | …… |
| 4 | Benzoate —(C7H5O2) | W Pb(—)2 | W Mg(—)2 | W Mn(—)2 | W Hg(—)2 | W Hg(—)2 | W Ni(—)2 | W K(—) | W Ag(—) | W Na(—) | | | W Sr(—)2 | W Zn(—)2 | …… |
| 5 | Bromide | W PbBr2 | W MgBr2 | W MnBr2 | W HgBr | A HgBr2 | W NiBr2 | W KBr | a AgBr | W NaBr | W SnBr4 | W SnBr2 | W SrBr2 | W ZnBr2 | W PtBr4 |

Table 3–2 (continued)
SOLUBILITY CHART

| No. | | Pb | Mg | Mn | Hg' | Hg'' | Ni | K | Ag | Na | Sn''' | Sn'' | Sr | Zn | Pt |
|---|---|---|---|---|---|---|---|---|---|---|---|---|---|---|---|
| 6 | Carbonate | A $PbCO_3$ | W $MgCO_3$ | W $MnCO_3$ | A $Hg_2CO_3$ | .... | W $NiCO_3$ | W $K_2CO_3$ | A $Ag_2CO_3$ | W $Na_2CO_3$ | .... | .... | W $SrCO_3$ | W $ZnCO_3$ | .... |
| 7 | Chlorate —$(ClO_3)$ | W $Pb(—)_2$ | W $Mg(—)_2$ | W $Mn(—)_2$ | W $Hg(—)$ | W $Hg(—)_2$ | W $Ni(—)_2$ | W $K(—)$ | W $Ag(—)$ | W $Na(—)$ | .... | W $Sn(—)_2$ | W $Sr(—)_2$ | W $Zn(—)_2$ | W |
| 8 | Chloride | W $PbCl_2$ | W $MgCl_2$ | W $MnCl_2$ | a $HgCl$ | W $HgCl_2$ | W $NiCl_2$ | W $KCl$ | a $AgCl$ | W $NaCl$ | W $SnCl_4$ | W $SnCl_2$ | W $SrCl_2$ | W $ZnCl_2$ | W $PtCl_4$ |
| 9 | Chromate —$(CrO_4)$ | A $Pb(—)$ | W $Mg(—)$ | .... | W $Hg_2(—)$ | W $Hg(—)$ | W $Ni(—)$ | W $K_2(—)$ | A $Ag_2(—)$ | W $Na_2(—)$ | .... | .... | W $Sr(—)$ | W $Zn(—)$ | .... |
| 10 | Citrate —$(C_6H_5O_7)$ | W $Pb_3(—)_2$ | W $Mg_3(—)_2$ | A $MnH(—)$ | A $Hg_2(—)$ | W $Hg(—)$ | W $Ni_3(—)_2$ | W $K_3(—)$ | A $Ag_3(—)$ | W $Na_3(—)$ | .... | .... | A $Sr_3(—)_2$ | A $Zn_3(—)_2$ | .... |
| 11 | Cyanide | W $Pb(CN)_2$ | W $Mg(CN)_2$ | .... | A $HgCN$ | W $Hg(CN)_2$ | a $Ni(CN)_2$ | W $KCN$ | A $AgCN$ | W $NaCN$ | .... | .... | W $Sr(CN)_2$ | A $Zn(CN)_2$ | I $Pt(CN)_2$ |
| 12 | Ferricy'de —$Fe(CN)_6$ | W $Pb_3(—)_2$ | W $Mg_3(—)_2$ | A $Mn_3(—)_2$ | .... | I $Hg_3(—)_2$ | A $Ni_3(—)_2$ | W $K_3(—)$ | I $Ag_3(—)$ | W $Na_3(—)$ | W $Sn(—)_4$ | A $Sn_3(—)_2$ | W $Sr_3(—)_2$ | A $Zn_3(—)_2$ | I |
| 13 | Ferrocy'de —$Fe(CN)_6$ | W $Pb_2(—)$ | W $Mg_2(—)$ | A $Mn_2(—)$ | .... | I $Hg_2(—)$ | I $Ni_2(—)$ | W $K_4(—)$ | I $Ag_4(—)$ | W $Na_4(—)$ | .... | a $Sn_2(—)$ | W $Sr_2(—)$ | I $Zn_2(—)$ | .... |
| 14 | Fluoride | W $PbF_2$ | W $MgF_2$ | A $MnF_2$ | d $HgF$ | d $HgF_2$ | W $NiF_2$ | W $KF$ | W $AgF$ | W $NaF$ | W $SnF_4$ | W $SnF_2$ | W $SrF_2$ | W $ZnF_2$ | W $PtF_4$ |
| 15 | Formate —$(CHO_2)$ | W $Pb(—)_2$ | W $Mg(—)_2$ | W $Mn(—)_2$ | d $Hg(—)$ | W $Hg(—)_2$ | W $Ni(—)_2$ | W $K(—)$ | A $Ag(—)$ | W $Na(—)$ | .... | a $Sn(—)_2$ | W $Sr(—)_2$ | W $Zn(—)_2$ | .... |
| 16 | Hydroxide | W $Pb(OH)_2$ | W $Mg(OH)_2$ | A $Mn(OH)_2$ | A $Hg(—)$ | A $Hg(OH)_2$ | W $Ni(OH)_2$ | W $KOH$ | I | W $NaOH$ | W $Sn(OH)_4$ | A $Sn(OH)_2$ | W $Sr(OH)_2$ | A $Zn(OH)_2$ | A $Pt(OH)_4$ |
| 17 | Iodide | W $PbI_2$ | W $MgI_2$ | W $MnI_2$ | I $Hg_2I$ | W $HgI_2$ | W $NiI_2$ | W $KI$ | I $AgI$ | W $NaI$ | d $SnI_4$ | d $SnI_2$ | W $SrI_2$ | W $ZnI_2$ | I $PtI_4$ |
| 18 | Nitrate | W $Pb(NO_3)_2$ | W $Mg(NO_3)_2$ | W $Mn(NO_3)_2$ | W $Hg_2NO_3$ | W $Hg(NO_3)_2$ | W $Ni(NO_3)_2$ | W $KNO_3$ | W $AgNO_3$ | W $NaNO_3$ | W $Sn(NO_3)_4$ | d $Sn(NO_3)_2$ | W $Sr(NO_3)_2$ | W $Zn(NO_3)_2$ | W $Pt(NO_3)_4$ |
| 19 | Oxalate —$(C_2O_4)$ | A $Pb(—)$ | W $Mg(—)$ | W $Mn(—)$ | a $Hg_2(—)$ | A $Hg(—)$ | A $Ni(—)$ | W $K_2(—)$ | A $Ag_2(—)$ | W $Na_2(—)$ | .... | A $Sn(—)_2$ | A $Sr(—)$ | A $Zn(—)$ | A |
| 20 | Oxide | W $PbO$ | A $MgO$ | W $MnO$ | d $HgO$ | W $HgO$ | W $NiO$ | W $K_2O$ | A $Ag_2O$ | d $Na_2O$ | A $SnO_2$ | A $SnO$ | W $SrO$ | W $ZnO$ | A $PtO$ |
| 21 | Phosphate | A $Pb_3(PO_4)_2$ | W $Mg_3(PO_4)_2$ | W $Mn_3(PO_4)_2$ | W $Hg_3PO_4$ | A $Hg_3(PO_4)_2$ | W $Ni_3(PO_4)_2$ | W $K_3PO_4$ | A $Ag_3PO_4$ | W $Na_3PO_4$ | A $Sn_3(PO_4)_4$ | A $Sn_3(PO_4)_2$ | A $Sr_3(PO_4)_2$ | A $Zn_3(PO_4)_2$ | .... |
| 22 | Silicate —$(SiO_3)$ | A $Pb(—)$ | A $Mg(—)$ | W $Mn(—)$ | .... | .... | .... | W $K_2(—)$ | .... | W $Na_2(—)$ | W $Sn(—)_2$ | W $Sn(—)$ | A $Sr(—)$ | A $Zn(—)$ | W |
| 23 | Sulfate | A $PbSO_4$ | W $MgSO_4$ | W $MnSO_4$ | W $Hg_2SO_4$ | d $HgSO_4$ | W $NiSO_4$ | W $K_2SO_4$ | a $Ag_2SO_4$ | W $Na_2SO_4$ | W $Sn(SO_4)_2$ | W $SnSO_4$ | A $SrSO_4$ | W $ZnSO_4$ | W $Pt(SO_4)_2$ |
| 24 | Sulfide | A $PbS$ | d $MgS$ | A $MnS$ | I $Hg_2S$ | I $HgS$ | A $NiS$ | W $K_2S$ | A $Ag_2S$ | W $Na_2S$ | A $SnS_2$ | A $SnS$ | W $SrS$ | A $ZnS$ | I $PtS$ |
| 25 | Tartrate —$(C_4H_4O_6)$ | A $Pb(—)$ | W $Mg(—)$ | W $Mn(—)$ | I $Hg_2(—)$ | .... | W $Ni(—)$ | W $K_2(—)$ | A $Ag_2(—)$ | W $Na_2(—)$ | .... | W $Sn(—)$ | W $Sr(—)$ | A $Zn(—)$ | .... |
| 26 | Thiocy'te | W $Pb(CNS)_2$ | W $Mg(CNS)_2$ | W $Mn(CNS)_2$ | A $Hg_2CNS$ | $Hg(CNS)_2$ | .... | W $KCNS$ | I $AgCNS$ | W $NaCNS$ | .... | .... | W $Sr(CNS)_2$ | W $Zn(CNS)_2$ | I $PtS$ |

ªCertain salts occur in two modifications.

From Weast, R. C. Ed., *Handbook of Chemistry and Physics*, 69th ed., CRC Press, Boca Raton, Fla., 1988, D-142.

## Table 3—3
## THERMAL CONDUCTIVITY OF MATERIALS

D = density in pounds per cubic foot.

K = thermal conductivity in Btu per hour, square foot, and temperature gradient of 1°F per inch thickness. The lower the conductivity, the greater the insulating values.

### Soft Flexible Materials in Sheet Form

|  |  | D | K |
|---|---|---|---|
| Dry zero | Kapok between burlap or paper | 1.0 | 0.24 |
|  |  | 2.0 | 0.25 |
| Cabots quilt | Eel grass between kraft paper | 3.4 | 0.25 |
|  |  | 4.6 | 0.26 |
| Hair felt | Felted cattle hair | 11.0 | 0.26 |
|  |  | 13.0 | 0.26 |
| Balsam wool | Chemically treated wood fiber | 2.2 | 0.27 |
| Hairinsul® | 75% hair, 25% jute | 6.3 | 0.27 |
|  | 50% hair, 50% jute | 6.1 | 0.26 |
| Linofelt® | Flax fibers between paper | 4.9 | 0.28 |
| Thermofelt | Jute and asbestos fibers, felted | 10.0 | 0.37 |
|  | Hair and asbestos fibers, felted | 7.8 | 0.28 |

### Loose Materials

|  |  | D | K |
|---|---|---|---|
| Rock wool | Fibrous material made from rock, | 6.0 | 0.26 |
|  | also made in sheet form, felted | 10.0 | 0.27 |
|  | and confined with wire netting | 14.0 | 0.28 |
|  |  | 18.0 | 0.29 |
| Glass wool | Pyrex glass, curled | 4.0 | 0.29 |
|  |  | 10.0 | 0.29 |
| Sil-O-Cel® | Powdered diatomaceous earth | 10.6 | 0.31 |
| Regranulated cork | Fine particles | 9.4 | 0.30 |
|  | About 3/16-in. particles | 8.1 | 0.31 |
| Thermofill® | Gypsum in powdered form | 26 | 0.52 |
|  |  | 34 | 0.60 |
| Sawdust | Various | 12.0 | 0.41 |
|  | Redwood | 10.9 | 0.42 |
| Shavings | Various, from planer | 8.8 | 0.41 |
| Charcoal | From maple, beech and birch, coarse | 13.2 | 0.36 |
|  | 6 mesh | 15.2 | 0.37 |
|  | 20 mesh | 19.2 | 0.39 |

### Semiflexible Materials in Sheet Form

|  |  | D | K |
|---|---|---|---|
| Flaxlinum | Flax fiber | 13.0 | 0.31 |
| Fibrofelt® | Flax and rye fiber | 13.6 | 0.32 |

### Semirigid Materials in Board Form

|  |  | D | K |
|---|---|---|---|
| Corkboard | No added binder; very low density | 5.4 | 0.25 |
| Corkboard | No added binder; low density | 7.0 | 0.27 |
| Corkboard | No added binder; medium density | 10.6 | 0.30 |
| Corkboard | No added binder; high density | 14.0 | 0.34 |
| Eureka | Corkboard with asphaltic binder | 14.5 | 0.32 |
| Rock cork | Rock wool block with binder | 14.5 | 0.326 |
|  | (also called "Tucork") |  |  |
| Lith | Board containing rock wool, flax and straw pulp | 14.3 | 0.40 |

## Table 3—3 (continued)
## THERMAL CONDUCTIVITY OF MATERIALS

| Stiff Fibrous Materials in Sheet Form | | D | K |
|---|---|---|---|
| Insulite® | Wood pulp | 16.2 | 0.34 |
| | | 16.9 | 0.34 |
| Celotex® | Sugar cane fiber | 13.2 | 0.34 |
| | | 14.8 | 0.34 |

| | |
|---|---|
| Masonite® | K = 0.33 |
| Inso-board® | 0.33 |
| Maizewood | 0.33—0.39 |
| Cornstalk pith board | 0.24—0.30 |
| Maftex | 0.34 |

### Cellular Gypsum

| | D | K |
|---|---|---|
| Insulex or Pyrocell® | 8 | 0.35 |
| | 12 | 0.44 |
| | 18 | 0.59 |
| | 24 | 0.77 |
| | 30 | 1.00 |

### Woods (Across Grain)

| | D | K |
|---|---|---|
| Balsa | 7.3 | 0.33 |
| | 8.8 | 0.38 |
| | 20 | 0.58 |
| Cypress | 29 | 0.67 |
| White pine | 32 | 0.78 |
| Mahogany | 34 | 0.90 |
| Virginia pine | 34 | 0.98 |
| Oak | 38 | 1.02 |
| Maple | 44 | 1.10 |

### Miscellaneous Building Materials

| | | |
|---|---|---|
| Cinder concrete | 2—3 Limestone | 4—9 |
| Building gypsum | About 3 Concrete | 6—9 |
| Plaster | 2—5 Sandstone | 8—16 |
| Building brick | 3—6 Marble | 14—20 |
| Glass | 5—6 Granite | 13—28 |

From Bureau of Standards Letter Circular No. 227.

## Table 3—4
## THERMAL CONDUCTIVITY DATA ON CERAMIC MATERIALS

| Description[a] | Classification[b] | Water absorption,% | Bulk density, g/cm³ | Thermal conductivity[c] | | |
|---|---|---|---|---|---|---|
| | | | | 100°F | 200°F | 300°F |
| Single crystals | | | | | | |
| Silicon carbide | 5 | – | – | 52.0 | 50.2 | 49.0 |
| Periclase | 5 | – | – | 26.7 | 22.5 | 19.5 |
| Sapphire, c-axis | 5 | – | – | 20.2 | 16.0 | 14.0 |
| Sapphire, a-axis | 5 | – | – | 18.7 | 15.0 | 12.9 |
| Topaz, a-axis | 5 | – | – | 10.8 | 9.4 | 7.9 |
| Kyanite, c-axis | 5 | – | – | 10.00 | 8.6 | 7.4 |
| Kyanite, b-axis | 5 | – | – | 9.6 | 8.3 | 7.1 |
| Spinel, $MgO \cdot Al_2O_3$ | 5 | – | – | 6.80 | 6.20 | 5.50 |
| Quartz, c-axis | 4 | – | – | 6.40 | 5.40 | 5.02 |
| Quartz, a-axis | 4 | – | – | 3.40 | 3.00 | 2.60 |
| Rutile, c-axis | 5 | – | – | 5.60 | 4.80 | 4.40 |
| Rutile, a-axis | 5 | – | – | 3.20 | 3.20 | 3.20 |
| Fluorite | 5 | – | – | 5.30 | 4.37 | 3.45 |
| Beryl, aquamarine, c-axis | 4 | – | – | 3.18 | 3.15 | 3.12 |
| Beryl, aquamarine, a-axis | 4 | – | – | 2.52 | 2.52 | 2.52 |
| Zircon, a-axis | 4 | – | – | 2.45 | 2.45 | 2.45 |
| Zircon, c-axis | 4 | – | – | 2.34 | 2.34 | 2.35 |
| | | | | | | |
| Polycrystalline Oxide Ceramics | | | | | | |
| Pure beryllium oxide, hot pressed | 2 | 0.03 | 2.97 | 125.0 | 104.0 | 92.0 |
| Magnesium oxide, spec. pure | 1 | 0.83 | 3.21 | 21.2 | 18.4 | 16.0 |
| Stannic oxide, 98% | 1 | 0.03 | 6.62 | 17.5 | 15.0 | 12.7 |
| Zinc oxide, yellow | 1 | 0.00 | 5.28 | 16.8 | 14.6 | 12.5 |
| Zinc oxide, gray | 1 | 0.03 | 5.20 | 13.6 | 11.8 | 10.2 |
| Cupric oxide, 100% | 1 | 0.04 | 6.76 | 10.2 | 9.00 | 7.80 |
| Thorium dioxide, hot pressed | 2 | – | 9.58 | 8.00 | 7.02 | 6.50 |
| Ceric oxide | 1 | 0.00 | 6.20 | 6.63 | 6.29 | 5.20 |
| Manganic-manganous oxide | 1 | 0.02 | 4.21 | 4.18 | 3.80 | 3.41 |
| Lead oxide, mono-#, 100% | 1 | 0.38 | 7.98 | 1.6 | 1.25 | 0.98 |

[a]Composition: 90% MgO, 10% $Al_2O_3$ designates
weight percent. $Li_2$ O: 4 $B_2$ $O_3$ designates mole

[a]Composition: 90% MgO, 10% $Al_2O_3$ designates weight percent. $Li_2O$:4$B_2O_3$ designates mole composition, does not indicate compound formation.

[b]Classification: 1, research body; 2, industrial research body; 3, commercial body; 4, natural mineral; 5, synthetic mineral.

[c]Thermal conductivity: units given in Btu/hr $\cdot$ ft $\cdot$ °F; to convert to cal/sec $\cdot$ cm $\cdot$ °C, multiply by 0.00413.

From *Engineering Research Bulletin No. 40,* Rutgers, the State University, New Brunswick, N.J., 1958. With permission.

## Table 3–5
## DENSITY OF VARIOUS SOLIDS

Approximate Density of Various Solids at Ordinary Atmospheric Temperature

In the case of substances with voids such as paper or leather the bulk density rather than the density of the solid portion is indicated.

| Substance | Grams per cu. cm | Pounds per cu. ft. | Substance | Grams per cu. cm | Pound per cu. ft. | Substance | Grams per cu. cm | Pounds per cu. ft. |
|---|---|---|---|---|---|---|---|---|
| Agate | 2.5–2.7 | 156–168 | Cork | 0.22–0.26 | 14–16 | Mica | 2.6–3.2 | 165–200 |
| Alabaster, carbonate | 2.69–2.78 | 168–173 | Cork linoleum | 0.54 | 34 | Muscovite | 2.76–3.00 | 172–187 |
| sulfate | 2.26–2.32 | 141–145 | Corundum | 3.9–4.0 | 245–250 | Ochre | 3.5 | 218 |
| Albite | 2.62–2.65 | 163–165 | Diamond | 3.01–3.52 | 188–220 | Opal | 2.2 | 137 |
| Amber | 1.06–1.11 | 66–69 | Dolomite | 2.84 | 177 | Paper | 0.7–1.15 | 44–72 |
| Amphiboles | 2.9–3.2 | 180–200 | Ebonite | 1.15 | 72 | Paraffin | 0.87–0.91 | 54–57 |
| Anorthite | 2.74–2.76 | 171–172 | Emery | 4.0 | 250 | Peat blocks | 0.84 | 52 |
| Asbestos | 2.0–2.8 | 125–175 | Epidote | 3.25–3.50 | 203–218 | Pitch | 1.07 | 67 |
| Asbestos slate | 1.8 | 112 | Feldspar | 2.55–2.75 | 159–172 | Porcelain | 2.3–2.5 | 143–156 |
| Asphalt | 1.1–1.5 | 69–94 | Flint | 2.63 | 164 | Porphyry | 2.6–2.9 | 162–181 |
| Basalt | 2.4–3.1 | 150–190 | Fluorite | 3.18 | 198 | Pressed wood pulp board | 0.19 | 12 |
| Beeswax | 0.96–0.97 | 60–61 | Galena | 7.3–7.6 | 460–470 | Pyrite | 4.95–5.1 | 309–318 |
| Beryl | 2.69–2.7 | 168–169 | Gamboge | 1.2 | 75 | Quartz | 2.65 | 165 |
| Biotite | 2.7–3.1 | 170–190 | Garnet | 3.15–4.3 | 197–268 | Resin | 1.07 | 67 |
| Bone | 1.7–2.0 | 106–125 | Gas carbon | 1.88 | 117 | Rock salt | 2.18 | 136 |
| Brick | 1.4–2.2 | 87–137 | Gelatin | 1.27 | 79 | Rubber, hard | 1.19 | 74 |
| Butter | 0.86–0.87 | 53–54 | Glass, common | 2.4–2.8 | 150–175 | Rubber, soft commercial | 1.1 | 69 |
| Calamine | 4.1–4.5 | 255–280 | flint | 2.9–5.9 | 180–370 | pure gum | 0.91–0.93 | 57–58 |
| Calcspar | 2.6–2.8 | 162–175 | Glue | 1.27 | 79 | Sandstone | 2.14–2.36 | 134–147 |
| Camphor | 0.99 | 62 | Granite | 2.64–2.76 | 165–172 | Serpentine | 2.50–2.65 | 156–165 |
| Caoutchouc | 0.92–0.99 | 57–62 | Graphite* | 2.30–2.72 | 144–170 | Silica, fused transparent | 2.21 | 138 |
| Cardboard | 0.69 | 43 | Gum arabic | 1.3–1.4 | 81–87 | translucent | 2.07 | 129 |
| Celluloid | 1.4 | 87 | Gypsum | 2.31–2.33 | 144–145 | Slag | 2.0–3.9 | 125–240 |
| Cement, set | 2.7–3.0 | 170.190 | Hematite | 4.9–5.3 | 306–330 | Slate | 2.6–3.3 | 162–205 |
| Chalk | 1.9–2.8 | 118–175 | Hornblende | 3.0 | 187 | Soapstone | 2.6–2.8 | 162–175 |
| Charcoal, oak | 0.57 | 35 | Ice | 0.917 | 57.2 | Spermaceti | 0.95 | 59 |
| pine | 0.28–0.44 | 18–28 | Ivory | 1.83–1.92 | 114–120 | Starch | 1.53 | 95 |
| Cinnabar | 8.12 | 507 | Leather, dry | 0.86 | 54 | Sugar | 1.59 | 99 |
| Clay | 1.8–2.6 | 112–162 | Lime, slaked | 1.3–1.4 | 81–87 | Talc | 2.7–2.8 | 168–174 |
| Coal, anthracite | 1.4–1.8 | 87–112 | Limestone | 2.68–2.76 | 167–171 | Tallow, beef | 0.94 | 59 |
| bituminous | 1.2–1.5 | 75–94 | Linoleum | 1.18 | 74 | mutton | 0.94 | 59 |
| Cocoa butter | 0.89–0.91 | 56–57 | Magnetite | 4.9–5.2 | 306–324 | Tar | 1.02 | 66 |
| Coke | 1.0–1.7 | 62–105 | Malachite | 3.7–4.1 | 231–256 | | | |
| Copal | 1.04–1.14 | 65–71 | Marble | 2.6–2.84 | 160–177 | | | |
| | | | Meerschaum | 0.99–1.28 | 62–80 | | | |

## Table 3–5 (continued)
## DENSITY OF VARIOUS SOLIDS

| Substance | Grams per cu. cm | Pounds per cu. ft. |
|---|---|---|
| Topaz | 3.5–3.6 | 219–223 |
| Tourmaline | 3.0–3.2 | 190–200 |
| Wax, sealing | 1.8 | 112 |
| Wood (seasoned) | | |
| alder | 0.42–0.68 | 26–42 |
| apple | 0.66–0.84 | 41–52 |
| ash | 0.65–0.85 | 40–53 |
| balsa | 0.11–0.14 | 7–9 |
| bamboo | 0.31–0.40 | 19–25 |
| basswood | 0.32–0.59 | 20–37 |
| beech | 0.70–0.90 | 43–56 |
| birch | 0.51–0.77 | 32–48 |
| blue gum | 1.00 | 62 |
| box | 0.95–1.16 | 59–72 |
| butternut | 0.38 | 24 |

| Substance | Grams per cu. cm | Pound per cu. ft. |
|---|---|---|
| cedar | 0.49–0.57 | 30–35 |
| cherry | 0.70–0.90 | 43–56 |
| dogwood | 0.76 | 47 |
| ebony | 1.11–1.33 | 69–83 |
| elm | 0.54–0.60 | 34–37 |
| hickory | 0.60–0.93 | 37–58 |
| holly | 0.76 | 47 |
| juniper | 0.56 | 35 |
| larch | 0.50–0.56 | 31–35 |
| lignum vitae | 1.17–1.33 | 73–83 |
| locust | 0.67–0.71 | 42–44 |
| logwood | 0.91 | 57 |
| mahogany | | |
| Honduras | 0.66 | 41 |
| Spanish | 0.85 | 53 |

| Substance | Grams per cu. cm | Pounds per cu. ft. |
|---|---|---|
| maple | 0.62–0.75 | 39–47 |
| oak | 0.60–0.90 | 37–56 |
| pear | 0.61–0.73 | 38–45 |
| pine, pitch | 0.83–0.85 | 52–53 |
| white | 0.35–0.50 | 22–31 |
| yellow | 0.37–0.60 | 23–37 |
| plum | 0.66–0.78 | 41–49 |
| poplar | 0.35–0.5 | 22–31 |
| satinwood | 0.95 | 59 |
| spruce | 0.48–0.70 | 30–44 |
| sycamore | 0.40–0.60 | 24–37 |
| teak, Indian | 0.66–0.88 | 41–55 |
| African | 0.98 | 61 |
| walnut | 0.64–0.70 | 40–43 |
| water gum | 1.00 | 62 |
| willow | 0.40–0.60 | 24–37 |

[a]Some values reported as low as 1.6.

From Weast, R. C. Ed., *Handbook of Chemistry and Physics*, 69th ed., CRC Press, Boca Raton, Fla., 1988, F-1.

## Table 3—6
## ELECTRICAL RESISTIVITY OF ALLOYS

| Composition | Name | Resistivity | | | | Temp coef of expansion/ deg C | Melting point, deg C |
|---|---|---|---|---|---|---|---|
| | | Microhm-cm | Ohm/mil-ft | Relative, pure copper = 1 | Temp coef of resist/ deg C | | |
| **ALUMINUM ALLOYS** | | | | | | | |
| 97 Al, 2 Mg, .5 Cr | Alloy 5052 | 5 | 30 | 2.9 | .004 | .000 025 | 625 |
| | | | | | | | |
| **COPPER ALLOYS** | | | | | | | |
| | Soft copper wire | 1.72 | 10.3 | 1.01 | .004 | .000 017 | 1 080 |
| Hard copper wire | 6101-B317-64 | 1.8 | 10.8 | 1.06 | .004 | | 1 085 |
| Copper-clad steel | 30 HS, B227-65 | 35 | 210 | 20.6 | .005 | | |
| 98 Cu, 2 Ni | Alloy 30 | 5 | 30 | 2.9 | .001 4 | .000 017 | 1 090 |
| 94 Cu, 6 Ni | Cuprothal 60[a] | 10 | 60 | 5.9 | .001 4 | .000 018 | 1 100 |
| 91 Cu, 8 Sn, .25 P | Phosphor bronze | 13 | 65 | 7.65 | | .000 018 | 1 050 |
| 89 Cu, 11 Ni | Alloy 90 | 15 | 90 | 8.8 | .000 5 | .000 02 | 1 110 |
| 87 Cu, 13 Mn | Manganin | 48 | 290 | 28.4 | .000 01 | .000 02 | 1 020 |
| 78 Cu, 22 Ni | Midohm[a] Cuprothal 180[a] | 30 | 180 | 17.6 | .000 2 | .000 02 | 1 130 |
| 65 Cu, 35 Zn | Brass | 7 | 42 | 4.1 | .002 | .000 018 | 940 |
| 64 Cu, 18 Zn, 18 Ni | Nickel, silver | 28 | 168 | 16.5 | .000 3 | .000 018 | 1 110 |
| 57 Cu, 43 Ni | Constantan | 49 | 294 | 28.8 | .000 01 | .000 015 | 1 220 |
| | | | | | | | |
| **IRON ALLOYS** | | | | | | | |
| | Soft steel 1010 | 12 | 72 | 7.1 | .006 | .000 011 | 1 450 |
| 99 Fe, 1 C | Carbon steel | 20 | 120 | 11.8 | .005 | .000 011 | 1 430 |
| 96 Fe, 4 Si | High silicon iron | 59 | 354 | 34.7 | .002 | .000 013 | 1 410 |
| 81 Fe, 15 Cr, 4 Al | Alloy 750 | 125 | 750 | 73.5 | .000 15 | .000 015 | 1 520 |
| 74 Fe, 18 Cr, 8 Ni | 18-8 Stainless steel | 73 | 440 | 43.0 | .000 94 | .000 018 | 1 400 |
| 72 Fe, 22.5 Cr, 5.5 Al | Alloy 875 | 146 | 875 | 85.8 | .000 02 | .000 017 | 1 520 |
| 65 Fe, 35 Ni | Invar | 81 | 485 | 47.6 | .001 35 | .000 001 | 1 425 |
| 62 Fe, 21 Ni, 12 Al, 5 Co | Alnico I[a] | 75 | 450 | 44.1 | .002 | .000 015 | |
| 58 Fe, 42 Ni | Alloy 142 | 67 | 400 | 39.3 | .001 2 | .000 005 | 1 425 |
| 55 Fe, 37.5 Cr, 7.5 Al | High-resistance alloy | 166 | 1 000 | 97.6 | .001 | .000 015 | 1 500 |
| 45 Fe, 35 Ni, 20 Cr | Chromax,[a] Chromel D[a] | 100 | 600 | 58.8 | .000 36 | .000 016 | 1 380 |
| | | | | | | | |
| **NICKEL ALLOYS (> 50%)** | | | | | | | |
| 80 Ni, 20 Cr | Nichrome V[a] Nikrothal 8[a] Chromel A[a] | 108 | 650 | 63.6 | .000 1 | .000 016 | 1 375 |
| 72 Ni, 38 Fe | Hytemco[a] | 20 | 120 | 11.8 | .004 2 | .000 015 | 1 425 |
| 68 Ni, 20 Cr, 8 Fe | Chromel AA[a] | 117 | 700 | 68.7 | .000 11 | .000 014 | 1 390 |
| 60 Ni, 16 Cr, 24 Fe | Nichrome[a] Chromel C[a] Nicrothal 6[a] | 112 | 675 | 66.0 | .000 15 | .000 015 | 1 350 |
| 95 Ni, 4 Mn, 1 Si | Alloy R63 | 22 | 135 | 13.0 | .003 | .000 015 | 1 400 |
| 67 Ni, 30 Cu, 1.4 Fe, 1 Mn | Monel | 42 | 252 | 24.7 | .002 | .000 014 | 1 330 |
| | | | | | | | |
| **SILVER ALLOYS** | | | | | | | |
| 92.5 Ag, 7.5 Cu | Sterling silver | 2 | 12 | 1.18 | .004 | .000 018 | 905 |
| 85 Ag, 15 Cd | Contact alloy | 5 | 30 | 2.9 | .004 | .000 019 | 875 |

[a]Proprietary name.

From Bolz, R. E. and Tuve, G. L., Eds., *Handbook of Tables for Applied Engineering Science*, 2nd ed., CRC Press, Cleveland, 1973, 224.

## Table 3—7
## MELTING POINTS OF METALLIC COMPOUNDS

Refractory and Ceramic Materials and Salts

| Metal | Boride | | Bromide | | Carbide | | Chloride | | Fluoride | | Iodide | |
|---|---|---|---|---|---|---|---|---|---|---|---|---|
| | K | deg F | K | deg F | K | deg F | K | deg F | K | deg F | K | deg F |
| Ag | | | AgBr 703 | 806 | | | AgCl 728 | 851 | AgF 708 | 815 | AgI 831 | 1036 |
| Al | | | AlBr$_3$ 371 | 207 | Al$_4$C$_3$ 2000$^a$ | 3600$^a$ | AlCl$_3$ 465 | 377 | AlF$_3$ 1564$^a$ | 2356$^a$ | AlI 464 | 376 |
| B | | | BBr$_3$ 227 | −51 | B$_4$C 2720 | 4440 | BCl$_3$ 166 | −161 | BF$_3$ 146 | −196 | | |
| Ba | BaB$_4$ 2543 | 4118 | BaBr$_2$ 1123 | 1562 | | | BaCl$_2$ 1235 | 1764 | BaF$_2$ 1627 | 2470 | BaI$_2$ 1013 | 1364 |
| Be | BeB$_2$ >2243 | >3357 | BeBr$_2$ 793 | 968 | Be$_2$C >2375$^a$ | >3815$^a$ | BeCl$_2$ 713 | 824 | BeF$_2$ 813 | 1004 | BeI$_2$ 783 | 950 |
| Bi | | | BiBr$_3$ 491 | 424 | | | BiCl$_3$ 507 | 452 | BiF$_3$ 1000 | 1341 | BiI$_3$ 681 | 784 |
| Ca | | | CaBr$_2$ 1003$^a$ | 1346$^a$ | | | CaCl$_2$ 1055 | 1440 | CaF$_2$ 1675 | 2555 | CaI$_2$ 848 | 1067 |
| Cd | | | CdBr$_2$ 841 | 1054 | | | CdCl$_2$ 841 | 1054 | CdF$_2$ 1373 | 2012 | CdI$_2$ 423 | 302 |
| Ce | CeB$_6$ 2463 | 3975 | | | | | CeCl$_3$ 1095 | 1512 | CeF$_3$ 1710 | 2618 | CeI$_3$ 1025 | 1386 |
| Cr | CrB$_2$ 2123 | 3362 | | | Cr$_3$C$_2$ 2168 | 3440 | | | | | | |
| Cu | | | CuBr 777 | 939 | | | CuCl 695 | 792 | CuF$_2$ 1129 | 1573 | CuI 878 | 1121 |
| Fe | | | FeBr$_2$ 955 | 1754$^a$ | Fe$_3$C 2110 | 3339 | FeCl$_2$ 945 | 1242 | FeF$_3$ >1275 | >1835 | | |
| In | | | InBr$_3$ 709 | 817 | | | InCl 498 | 437 | InF$_3$ 1443 | 2138 | InI$_3$ 483 | 410 |
| K | | | KBr 1008 | 1355 | | | KCl 1043 | 1418 | KF 1131 | 1576 | KI 958 | 1265 |
| Li | | | LiBr 823 | 1022 | | | LiCl 883 | 1130 | LiF 1119 | 1554 | LiI 722 | 840 |
| Mg | | | MgBr$_2$ 984 | 1312 | | | MgCl$_2$ 987 | 1317 | MgF$_2$ 1536 | 2305 | MgI$_2$ <910 | <1078 |
| Mn | | | | | | | MnCl$_2$ 923 | 1202 | MnF$_2$ 1129 | 1573 | | |
| Mo | MoB 2625 | 4250 | | | Mo$_2$C 2963 | 4875 | | | MoF$_6$ 290 | 63 | MoI$_4$ 373$^a$ | 212 |
| Na | | | NaBr 1023 | 1382 | NaC$_2$ 973 | 1292 | NaCl 1073 | 1472 | NaF 1267 | 1821 | NaI 935 | 1224 |

## Table 3—7 (continued)
## MELTING POINTS OF METALLIC COMPOUNDS

| Metal | Nitrate K | deg F | Nitride K | deg F | Oxide K | deg F | Silicide K | deg F | Sulfate K | deg F | Sulfide K | deg F |
|---|---|---|---|---|---|---|---|---|---|---|---|---|
| Ag | $AgNO_3$ 483 | 410 | | | $Ag_2O$ 573$^a$ | 572 | | | $Ag_2SO_4$ 933 | 1220 | $Ag_2S$ 1098 | 1517 |
| Al | | | AlN >2475 | >4000 | $Al_2O_3$ 2322 | 3720 | | | $Al_2(SO_4)_3$ 1043$^a$ | 1418$^a$ | $Al_2S_3$ 1373 | 2012 |
| B | | | BN 3000 | 4945 | $B_2O_3$ 723 | 841 | | | | | $B_2S_4$ 663 | 734 |
| Ba | $Ba(NO_3)_2$ 865 | 1098 | | | BaO 2283 | 3649 | | | $BaSO_4$ 1853 | 2876 | BaS 1473 | 2192 |
| Be | | | $Be_3N_2$ 2513$^a$ | 4064$^a$ | BeO 2725 | 4445 | | | $BeSO_4$ 848$^a$ | 1067$^a$ | | |
| Bi | | | | | $Bi_2O_3$ 1098 | 1516 | | | $Bi(SO_4)_3$ 678$^a$ | 761$^a$ | $Bi_2S_3$ 1020 | 1377 |
| Ca | $Ca(NO_3)_2$ 834 | 1042 | $Ca_3N_2$ 1468 | 2183 | CaO 3183 | 5269 | | | $CaSO_4$ 1723 | 3542 | | |
| Cd | $Cd(NO_3)_2$ 623 | 662 | | | CdO 1773 | 2731 | | | $CdSO_4$ 1273$^b$ | 1832$^b$ | CdS 2023$^b$ | 3182$^b$ |
| Ce | | | | | $CeO_2$ >2873 | >4711 | | | $Ce(SO_4)_2$ 468$^a$ | 383$^a$ | CeS 2400 | 3860 |
| Cr | | | CrN 1770$^a$ | 2730$^a$ | $Cr_2O_3$ 2603 | 4225 | $CrSi_2$ 1843 | 2858 | | | | |
| Cu | | | $Cu_3N$ 573$^a$ | 572 | $Cu_2O$ 1508 | 2254 | $Cu_4Si$ 1123 | 1562 | $Cu_2S$ 1400 | 2060 | | |
| Fe | | | | | $Fe_2O_3$ 1864 | 2895 | | | $Fe_2(SO_4)_3$ 753$^a$ | 896$^a$ | FeS 1468 | 2183 |
| In | | | | | $In_2O_3$ 2183 | 3469 | | | | | $In_2S_3$ 1323 | 1922 |
| K | $KNO_3$ 610 | 639 | | | $K_2O_3$ 703 | 806 | | | $K_2SO_4$ 1342 | 1956 | $K_2S$ 1113 | 1544 |
| Li | $LiNO_3$ 527 | 489 | $Li_3N$ 1118$^a$ | 1553 | $Li_2O$ >1975 | >3095 | | | $Li_2SO_4$ 1132 | 1578 | $Li_2S$ 1198 | 1697 |
| Mg | | | | | MgO 3098 | 5116 | $Mg_2Si$ 1375 | 2016 | MgS >2275$^a$ | >3635$^a$ | $MgSO_4$ 1397$^a$ | 2055$^a$ |
| Mn | | | | | MnO 1840 | 2852 | | | | | | |
| Mo | | | | | $MoO_3$ 1068 | 1462 | $MoSi_2$ 2553 | 3595 | | | $MoS_2$ 1458 | 2165 |
| Na | $NaNO_3$ 583 | 590 | $Na_2N$ 573$^a$ | 572$^a$ | | | | | $Na_2SO_4$ 1157 | 1632 | $Na_2S$ 1453 | 2156 |

Table 3—7 (continued)
## MELTING POINTS OF METALLIC COMPOUNDS

| Metal | Compound | | | | | | | | | | |
|-------|---|---|---|---|---|---|---|---|---|---|---|
| | Boride | | Bromide | | Carbide | | Chloride | | Fluoride | | Iodide | |
| | *K* | *deg F* | *K* | *deg F* | *K* | *deg F* | *K* | *deg F* | *K* | *deg F* | *K* | *deg F* |
| **Nb** | **NbB** > 2270 | > 3630 | | | **NbC** 3770 | 6330 | | | | | | |
| **Ni** | | | **NiBr$_2$** 1236 | 1765 | | | **NiCl$_3$** 1274 | 1834 | **NiF$_2$** 1273[a] | 1832 | **NiI$_2$** 1070 | 1467 |
| **Pb** | | | **PbBr$_2$** 643 | 698 | | | **PbCl$_2$** 771 | 928 | **PbF$_2$** 1095 | 1512 | **PbI$_2$** 675 | 756 |
| **Pt** | | | **PtBr$_2$** 523[a] | 482[a] | | | **PtCl$_2$** 854[a] | 1078[a] | | | **PtI$_2$** 633[a] | 680[a] |
| **Sb** | | | **SbBr$_3$** 370 | 207 | | | **SbCl$_3$** 346 | 164 | **SbF$_3$** 565 | 558 | **SbI$_3$** 443 | 338 |
| **Si** | | | | | **SiC** 2970 | 4890 | | | **SiF$_4$** 183 | − 130 | | |
| **Sn** | | | **SnBr$_2$** 488 | 420 | | | **SnCl$_2$** 518 | 473 | **SnF$_4$** 978[a] | 1300[a] | **SnI$_2$** 593 | 608 |
| **Sr** | **SrB$_6$** 2508 | 4055 | **SrBr$_2$** 916 | 1189 | **SrC$_2$** > 1970 | > 3100 | **SrCl$_2$** 1148 | 1607 | **SrF$_2$** 1736 | 2665 | **SrI$_2$** 788 | 959 |
| **Ta** | **TaB** > 2270[a] | > 3630[a] | **TaBr$_5$** 538 | 509 | **TaC** 3813 | 6403 | **TaCl$_5$** 489 | 421 | **TaF$_5$** 370 | 206 | | |
| **Te** | | | **TeBr$_2$** 612 | 642 | | | **TeCl$_2$** 448 | 347 | | | | |
| **Th** | **ThB$_4$** > 2770 | > 4530 | **ThBr$_4$** 883 | 1130[a] | **ThC** 2898 | 5250 | **ThCl$_4$** 1043 | 1418 | **ThF$_4$** 1375 | 2015 | | |
| **Ti** | **TiB$_2$** 3253 | 5396 | **TiBr$_4$** 312 | 102 | **TiC** 3433 | 5720 | **TiCl$_4$** 250 | − 9 | **TiF$_3$** 1475 | 2195 | **TiI$_2$** 873 | 1112 |
| **U** | **UB$_2$** > 1770 | > 2730 | **UBr$_4$** 789 | 961 | **UC** 2863 | 4693 | **UCl$_4$** 843 | 1058 | **UF$_4$** 1233 | 1760 | **UI$_4$** 779 | 943 |
| **V** | **VB$_2$** 2373 | 3812 | | | **VC** 3600 | 5120 | **VCl$_4$** 245 | − 18 | **VF$_3$** > 1075 | > 1475 | **VI$_2$** 1048[a] | 1427 |
| **W** | **WB** 3133 | 5180 | **WBr$_5$** 549 | 529 | **WC** 2900[a] | 4760[a] | **WCl$_6$** 548 | 527 | | | | |
| **Zn** | | | **ZnBr$_2$** 667 | 741 | | | **ZnCl$_2$** 548 | 527 | **ZnF$_2$** 1145 | 1602 | **ZnI$_2$** 719 | 835 |
| **Zr** | **ZrB$_2$** 3313 | 5505 | **ZrBr$_2$** > 625[a] | > 660[a] | **ZrC** 3533 | 5900 | **ZrCl$_2$** 623 | 662 | **ZrF$_4$** 873[a] | 1112[a] | **ZrI$_4$** 772 | 930 |

## Table 3—7 (continued)
## MELTING POINTS OF METALLIC COMPOUNDS

| Metal | Nitrate | | Nitride | | Oxide | | Silicide | | Sulfate | | Sulfide | |
|---|---|---|---|---|---|---|---|---|---|---|---|---|
| | *K* | *deg F* | *K* | *deg F* | *K* | *deg F* | *K* | *deg F* | *K* | *deg F* | *K* | *deg F* |
| Nb | | | NbN 2323 | 3722 | $Nb_2O_5$ 1764 | 2715 | $NbSi_2$ 2203 | 3505 | | | | |
| Ni | | | | | NiO 2257 | 3603 | | | $NiSO_4$ 1121[a] | 1558[a] | NiS 1070 | 1466 |
| Pb | $Pb(NO_3)_2$ 743[a] | 878[a] | | | PbO 1159 | 1626 | | | $PbSO_4$ 1443 | 2138 | PbS 1387 | 2037 |
| Pt | | | | | | | | | | | $PtS_2$ 508[a] | 455[a] |
| Sb | | | | | $Sb_2O_3$ 928 | 1211 | | | | | $SbS_3$ 820 | 1016 |
| Si | | | $Si_3N_4$ 2175[b] | 3450 | $SiO_2$ 1978 | 3100 | | | | | | |
| Sn | | | | | SnO 1353[a] | 1973[a] | | | $SnSO_4$ >635 | >680 | SnS 1153 | 1616 |
| Sr | $Sr(NO_3)_2$ 643 | 1058 | | | SrO 2933 | 4819 | | | $SrSO_4$ 1878 | 2921 | SrS >2275 | >3635 |
| Ta | | | $Ta_2N$ 3360 | 5595 | $Ta_2O_5$ 2100 | 3325 | $TaSi_2$ 2670 | 4350 | | | $TaS_4$ >1575 | >2375 |
| Te | | | | | $TeO_2$ 1006 | 1351 | | | | | | |
| Th | | | ThN 2903 | 4765 | $ThO_2$ 3493 | 5827 | | | | | $ThS_2$ 2198 | 3497 |
| Ti | | | TiN 3200[a] | 5790[a] | $TiO_2$ 2113 | 3344 | $TiSi_2$ 1813 | 2804 | | | | |
| U | | | UN 3123 | 5161 | $UO_2$ 3151 | 5212 | $USi_2$ 1970 | 3090 | | | $US_2$ >1375 | >2015 |
| V | | | VN 2593 | 4208 | $V_2O_5$ 947 | 1245 | $VSi_2$ 2023 | 3182 | | | $V_2S_3$ >875[a] | >1115[a] |
| W | | | | | $WO_3$ 1744 | 2679 | $WSi_2$ 2320 | 3720 | | | $WS_2$ 1523[a] | 2282[a] |
| Zn | | | | | ZnO 2248 | 3586 | | | $ZnSO_4$ 873[a] | 1112[a] | | |
| Zr | | | ZrN 3250 | 5400 | $ZrO_2$ 3123 | 5161 | | | $Zr(SO_4)_2$ 683 | 770 | $ZrS_2$ 1823 | 2822 |

[a]Decomposes or sublimes.
[b]Pressure above 1 atm.

## Table 3—7 (continued)
## MELTING POINTS OF METALLIC COMPOUNDS

Other Compounds Melting Point Temperatures in K (with °F in Parentheses)

$Ag_2CO_3$—491 (424); $BaCO_3$—1653 (2516); $CaCO_3$—1613[b] (2444)[b]; $CdCO_3$—<775[a] (<930[a]); $K_2CO_3$—1170 (1647); $Li_2CO_3$—950 (1250); $Na_2CO_3$—1127 (1569); $PbCO_3$—588[a] (600[a]); $SrCO_3$—1770 (2726)

$AgNO_2$—413[a] (284[a]); $Ba(NO_2)_2$—540 (513); $KNO_2$—692 (786); $LiNO_2$—493 (428); $NaNO_2$—558 (545)

$Ag_2Te$—1228 (1750); $Bi_2Te_3$—861 (1091); $CdTe$—1314 (1906); $InTe$—965 (1277); $PtTe_2$—1523 (2282); $Sb_2Te_3$—891 (1145); $SnTe$—1053 (1436)

$BaTiO_3$—1891 (2944); $FeTiO_3$—1640 (2492)

$Ce_2(WO_4)_3$—1362 (1992); $K_2WO_4$—1203 (1706); $LiWO_4$—1015 (1368); $NaWO_4$—971 (1288)

### REFERENCES

Weast, R. C., Ed., *Handbook of Chemistry and Physics,* 55th ed., CRC Press, Cleveland, 1974.

Charlesworth, J. H., *Melting Points of Metallic Elements and Selected Compounds,* Air Force Materials Laboratory Technical Report AFML-TR-70-137, 1970.

Janz, G. J., Dampier, F. W., Lakshminarayanan, G. R., Lorenz, P. K., and Tompkins, R. P. T., *Molten Salts: Electrical Conductance, Density, and Viscosity Data,* Vol. 1, NSRDS-NBS 15, National Bureau of Standards, October 1968.

From Bolz, R. E. and Tuve, G. L., Eds., *Handbook of Tables for Applied Engineering Science,* 2nd ed., CRC Press, Cleveland, 1973, 338.

# Table 3—8
## SOLUBILITY PRODUCT

The solubility product (or ion product constant) is the product of the concentrations of the ions in the saturated solution of a difficultly soluble salt. The concentrations are expressed as moles per liter of solution. The number of cations (or anions) resulting from the dissociation of one molecule of the salt appears in the formula for calculations of the solubility product as the exponent of the concentration of the cation (or anion).

If two solutions, each containing one of the ions of a difficultly soluble salt, are mixed, no precipitation takes place unless the product of the ion concentrations in the mixture is greater than the solubility product.

In a solution containing two salts that yield a common ion, the ratio of solubilities of the two salts is the ratio of the solubility products.

| Substance | Solubility product at temperature noted, °C | Substance | Solubility product at temperature noted, °C |
|---|---|---|---|
| Aluminum hydroxide | $4 \times 10^{-13}$ (15°) | Lead iodide | $7.47 \times 10^{-9}$ (15°) |
| Aluminum hydroxide | $1.1 \times 10^{-15}$ (18°) | Lead iodide | $1.39 \times 10^{-8}$ (25°) |
| Aluminum hydroxide | $3.7 \times 10^{-15}$ (25°) | Lead oxalate | $2.74 \times 10^{-11}$ (18°) |
| Barium carbonate | $7 \times 10^{-9}$ (16°) | Lead sulfate | $1.06 \times 10^{-8}$ (18°) |
| Barium carbonate | $8.1 \times 10^{-9}$ (25°) | Lead sulfide | $3.4 \times 10^{-28}$ (18°) |
| Barium chromate | $1.6 \times 10^{-10}$ (18°) | Lithium carbonate | $1.7 \times 10^{-3}$ (25°) |
| Barium chromate | $2.4 \times 10^{-10}$ (28°) | Magnesium ammonium phosphate | $2.5 \times 10^{-13}$ (25°) |
| Barium fluoride | $1.6 \times 10^{-6}$ (9.5°) | Magnesium carbonate | $2.6 \times 10^{-5}$ (12°) |
| Barium fluoride | $1.7 \times 10^{-6}$ (18°) | Magnesium fluoride | $7.1 \times 10^{-9}$ (18°) |
| Barium fluoride | $1.73 \times 10^{-6}$ (25.8°) | Magnesium fluoride | $6.4 \times 10^{-9}$ (27°) |
| Barium iodate, $Ba(IO_3) \cdot 2H_2O$ | $8.4 \times 10^{-11}$ (10°) | Magnesium hydroxide | $1.2 \times 10^{-11}$ (18°) |
| Barium iodate, $Ba(IO_3) \cdot 2H_2O$ | $6.5 \times 10^{-10}$ (25°) | Magnesium oxalate | $8.57 \times 10^{-5}$ (18°) |
| Barium oxalate, $BaC_2O_4 \cdot 3\frac{1}{2}H_2O$ | $1.62 \times 10^{-7}$ (18°) | Manganese hydroxide | $4 \times 10^{-14}$ (18°) |
| Barium oxalate, $BaC_2O_4 \cdot 2H_2O$ | $1.2 \times 10^{-7}$ (18°) | Manganese sulfide | $1.4 \times 10^{-15}$ (18°) |
| Barium oxalate, $BaC_2O_4 \cdot \frac{1}{2}H_2O$ | $2.18 \times 10^{-7}$ (18°) | Mercuric sulfide | $4 \times 10^{-53}$ to $2 \times 10^{-49}$ (18°) |
| Barium sulfate | $0.87 \times 10^{-10}$ (18°) | Mercurous bromide | $1.3 \times 10^{-21}$ (25°) |
| Barium sulfate | $1.08 \times 10^{-10}$ (25°) | Mercurous chloride | $2 \times 10^{-18}$ (25°) |
| Barium sulfate | $1.98 \times 10^{-10}$ (50°) | Mercurous iodide | $1.2 \times 10^{-28}$ (25°) |
| Cadmium oxalate $CdC_2O_4 \cdot 3H_2O$ | $1.53 \times 10^{-8}$ (18°) | Nickel sulfide | $1.4 \times 10^{-24}$ (18°) |
| Cadmium sulfide | $3.6 \times 10^{-29}$ (18°) | Potassium acid tartrate $[K^+]$ $[HC_4H_4O_6^-]$ | $3.8 \times 10^{-4}$ (18°) |
| Calcium carbonate (calcite) | $0.99 \times 10^{-8}$ (15°) | Silver bromate | $3.97 \times 10^{-5}$ (20°) |
| Calcium carbonate (calcite) | $0.87 \times 10^{-8}$ (25°) | Silver bromate | $5.77 \times 10^{-5}$ (25°) |
| Calcium fluoride | $3.4 \times 10^{-11}$ (18°) | Silver bromide | $4.1 \times 10^{-13}$ (18°) |
| Calcium fluoride | $3.95 \times 10^{-11}$ (26°) | Silver bromide | $7.7 \times 10^{-13}$ (25°) |
| Calcium iodate, $Ca(IO_3)_2 \cdot 6H_2O$ | $22.2 \times 10^{-8}$ (10°) | Silver carbonate | $6.15 \times 10^{-12}$ (25°) |
| Calcium iodate, $Ca(IO_3)_2 \cdot 6H_2O$ | $64.4 \times 10^{-8}$ (18°) | Silver chloride | $0.21 \times 10^{-10}$ (4.7°) |
| Calcium oxalate, $CaC_2O_4 \cdot H_2O$ | $1.78 \times 10^{-9}$ (18°) | Silver chloride | $0.37 \times 10^{-10}$ (9.7°) |
| Calcium oxalate, $CaC_2O_4 \cdot H_2O$ | $2.57 \times 10^{-9}$ (25°) | Silver chloride | $1.56 \times 10^{-10}$ (25°) |
| Calcium sulfate | $1.95 \times 10^{-4}$ (10°) | Silver chloride | $13.2 \times 10^{-10}$ (50°) |
| Calcium tartrate, $CaC_4H_4O_6 \cdot 2H_2O$ | $0.77 \times 10^{-6}$ (18°) | Silver chloride | $215 \times 10^{-10}$ (100°) |
| Cobalt sulfide | $3 \times 10^{-26}$ (18°) | Silver chromate | $1.2 \times 10^{-12}$ (14.8°) |
| Cupric iodate | $1.4 \times 10^{-7}$ (25°) | Silver chromate | $9 \times 10^{-12}$ (25°) |
| Cupric oxalate | $2.87 \times 10^{-8}$ (25°) | Silver cyanide $[Ag^+][Ag(CN)_2^-]$ | $2.2 \times 10^{-12}$ (20°) |
| Cupric sulfide | $8.5 \times 10^{-45}$ (18°) | Silver dichromate | $2 \times 10^{-7}$ (25°) |
| Cuprous bromide | $4.15 \times 10^{-8}$ (18–20°) | Silver hydroxide | $1.52 \times 10^{-8}$ (20°) |
| Cuprous chloride | $1.02 \times 10^{-6}$ (18–20°) | Silver iodate | $0.92 \times 10^{-8}$ (9.4°) |
| Cuprous iodide | $5.06 \times 10^{-12}$ (18–20°) | Silver iodide | $0.32 \times 10^{-16}$ (13°) |
| Cuprous sulfide | $2 \times 10^{-47}$ (16–18°) | Silver iodide | $1.5 \times 10^{-16}$ (25°) |
| Cuprous thiocyanate | $1.6 \times 10^{-11}$ (18°) | Silver sulfide | $1.6 \times 10^{-49}$ (18°) |
| Ferric hydroxide | $1.1 \times 10^{-36}$ (18°) | Silver thiocyanate | $0.49 \times 10^{-12}$ (18°) |
| Ferrous hydroxide | $1.64 \times 10^{-14}$ (18°) | Silver thiocyanate | $1.16 \times 10^{-12}$ (25°) |
| Ferrous oxalate | $2.1 \times 10^{-7}$ (25°) | Strontium carbonate | $1.6 \times 10^{-9}$ (25°) |
| Ferrous sulfide | $3.7 \times 10^{-19}$ (18°) | Strontium fluoride | $2.8 \times 10^{-9}$ (18°) |
| Lead carbonate | $3.3 \times 10^{-14}$ (18°) | Strontium oxalate | $5.61 \times 10^{-8}$ (18°) |
| Lead chromate | $1.77 \times 10^{-14}$ (18°) | Strontium sulfate | $2.77 \times 10^{-7}$ (2.9°) |
| Lead fluoride | $2.7 \times 10^{-8}$ (9°) | Strontium sulfate | $3.81 \times 10^{-7}$ (17.4°) |
| Lead fluoride | $3.2 \times 10^{-8}$ (18°) | Zinc hydroxide | $1.8 \times 10^{-14}$ (18–20°) |
| Lead fluoride | $3.7 \times 10^{-8}$ (26.6°) | Zinc oxalate, $ZnC_2O_4 \cdot 2H_2O$ | $1.35 \times 10^{-9}$ (18°) |
| Lead iodate | $5.3 \times 10^{-14}$ (9.2°) | Zinc sulfide | $1.2 \times 10^{-23}$ (18°) |
| Lead iodate | $1.2 \times 10^{-13}$ (18°) | | |
| Lead iodate | $2.6 \times 10^{-13}$ (25.8°) | | |

From Weast, R. C., Ed., *Handbook of Chemistry and Physics*, 55th ed., CRC Press, Cleveland, 1974, B-232.

**Table 3—9**
## CONCENTRATION OF ACIDS AND BASES

### Common Commercial Strengths

| | Molecular weight | Moles per liter | Grams per liter | Percent by weight | Specific gravity [a] |
|---|---|---|---|---|---|
| **ACIDS** | | | | | |
| Acetic acid, glacial | 60.05 | 17.4 | 1,045 | 99.5 | 1.05 |
| Acetic acid | 60.05 | 6.27 | 376 | 36 | 1.045 |
| Butyric acid | 88.1 | 10.3 | 912 | 95 | 0.96 |
| Formic acid | 46.02 | 23.4 | 1,080 | 90 | 1.20 |
| | — | 5.75 | 264 | 25 | 1.06 |
| Hydriodic acid | 127.9 | 7.57 | 969 | 57 | 1.70 |
| | — | 5.51 | 705 | 47 | 1.50 |
| | — | 0.86 | 110 | 10 | 1.1 |
| Hydrobromic acid | 80.92 | 8.89 | 720 | 48 | 1.50 |
| | — | 6.82 | 552 | 40 | 1.38 |
| Hydrochloric acid | 36.5 | 11.6 | 424 | 36 | 1.18 |
| | — | 2.9 | 105 | 10 | 1.05 |
| Hydrocyanic acid | 27.03 | 25 | 676 | 97 | 0.697 |
| | — | 0.74 | 19.9 | 2 | 0.996 |
| Hydrofluoric acid | 20.01 | 32.1 | 642 | 55 | 1.167 |
| | — | 28.8 | 578 | 50 | 1.155 |
| Hydrofluosilicic acid | 144.1 | 2.65 | 382 | 30 | 1.27 |
| Hypophosphorous acid | 66.0 | 9.47 | 625 | 50 | 1.25 |
| | — | 5.14 | 339 | 30 | 1.13 |
| | — | 1.57 | 104 | 10 | 1.04 |
| Lactic acid | 90.1 | 11.3 | 1,020 | 85 | 1.2 |
| Nitric acid | 63.02 | 15.99 | 1,008 | 71 | 1.42 |
| | — | 14.9 | 938 | 67 | 1.40 |
| | — | 13.3 | 837 | 61 | 1.37 |
| Perchloric acid | 100.5 | 11.65 | 1,172 | 70 | 1.67 |
| | — | 9.2 | 923 | 60 | 1.54 |
| Phosphoric acid | 98 | 14.7 | 1,445 | 85 | 1.70 |
| Sulfuric acid | 98.1 | 18.0 | 1,766 | 96 | 1.84 |
| Sulfurous acid | 82.1 | 0.74 | 61.2 | 6 | 1.02 |
| **BASES** | | | | | |
| Ammonia water | 17.0 | 14.8 | 252 | 28 | 0.898 |
| Potassium hydroxide | 56.1 | 13.5 | 757 | 50 | 1.52 |
| | — | 1.94 | 109 | 10 | 1.09 |
| Sodium carbonate | 106.0 | 1.04 | 110 | 10 | 1.10 |
| Sodium hydroxide | 40.0 | 19.1 | 763 | 50 | 1.53 |
| | — | 2.75 | 111 | 10 | 1.11 |

[a]For density in kg/m$^3$, multiply by 1000.

### REFERENCE

For vapor-pressure curves for dilute solutions of hydrochloric, nitric, and sulfuric acids up to about 1000 psia, consult Staples, B. G., Procopio, J. M., Jr., and Su, G. J., Vapor-pressure data for common acids at high temperatures, *Chem. Eng.*, 77(25), 113, 1970. The acid concentrations represented by these curves are for hydrochloric acid, 10–35% by weight; for nitric acid, 30–70% by weight; and for sulfuric acid, 0–70% by weight.

From *The Merck Index*, 10th ed., copyright 1983, Merck & Co., Inc., Rahway, N. J., U.S.A. With permission.

## Table 3—10
## PROPERTIES OF COMMON LIQUIDS[a]

### At 1.0 atm Pressure, 77°F (25°C), Except as Noted

For viscosity in $N \cdot s/m^2$ (= kg/m·s), multiply values in centipoises by 0.001. For surface tension in N/m, multiply values in dyn/cm by 0.001.

| Common name | Chemical formula | Molecular weight | Density, $\frac{lb}{ft^3}$ | Specific gravity | Viscosity $lb_m/ft\ sec \times 10^4$ | Viscosity cp | Sound velocity, $\frac{meters}{sec}$ | Surface tension, $\frac{dynes}{cm}$ | Dielectric constant | Refractive index |
|---|---|---|---|---|---|---|---|---|---|---|
| Acetic acid | $C_2H_4O_2$ | 60.0537 | 65.493 | 1.049 | 7.76 | 1.155 | 1584[50] | 27.3 | 6.15 | 1.37 |
| Acetone | $C_3H_6O$ | 58.081 | 48.98 | .787 | 2.12 | 0.316 | 1174 | 23.1 | 20.7 | 1.36 |
| Alcohol, ethyl | $C_2H_5OH$ | 46.070 | 49.01 | .787 | 7.36 | 1.095 | 1144 | 22.33 | 24.3 | 1.36 |
| Alcohol, methyl | $CH_3OH$ | 32.043 | 49.10 | .789 | 3.76 | 0.56 | 1103 | 22.2 | 32.6 | 1.33 |
| Alcohol, propyl | $C_3H_8O$ | 60.098 | 49.94 | .802 | 12.9 | 1.92 | 1205 | 23.5 | 20.1 | 1.38 |
| Ammonia (aqua) | — | 17.698 | 51.411 | .826 | — | — | — | — | 16.9 | — |
| Benzene | $C_6H_6$ | 78.117 | 54.55 | .876 | 4.04 | 0.601 | 1298 | 28.18 | 2.2 | 1.50 |
| Bromine | $Br_2$ | 159.818 | — | — | 6.38 | 0.95 | — | 41.5 | 3.20 | — |
| Carbon disulfide | $CS_2$ | 76.140 | 78.72 | 1.265 | 2.42 | 0.36 | 1149 | 32.33 | 2.64 | 1.63 |
| Carbon tetrachloride | $CCl_4$ | 153.824 | 98.91 | 1.59 | 6.11 | 0.91 | 924 | 26.3 | 2.23 | 1.46 |
| Castor oil | — | — | 59.69 | 0.960 | — | 650 | 1474 | — | 4.7 | — |
| Chloroform | $CHCl_3$ | 119.378 | 91.44 | 1.47 | 3.56 | 0.53 | 995 | 27.14 | 4.8 | 1.44 |
| Decane | $C_{10}H_{22}$ | 142.290 | 45.34 | .728 | 5.77 | 0.859 | — | 23.43 | 2.0 | 1.41 |
| Dodecane | $C_{12}H_{26}$ | 170.345 | 47.11 | — | 9.23 | 1.374 | — | — | — | 1.41 |
| Ether | $C_4H_{10}O$ | 74.125 | 44.54 | 0.715 | 1.50 | 0.223 | 985 | 16.42 | 4.3 | 1.35 |
| Ethylene glycol | $C_2H_6O_2$ | 62.070 | 68.47 | 1.100 | 109 | 16.2 | 1644 | 48.2 | 37.7 | 1.43 |
| Fluorine refrigerant R–11 | $CCl_3F$ | 137.369 | 92.14 | 1.480 | 2.82 | 0.42 | — | 18.3 | 2.0 | 1.37 |
| Fluorine refrigerant R–12 | $CCl_2F_2$ | 120.914 | 81.84 | 1.315 | — | — | — | 8.87 | 2.0 | 1.29 |
| Fluorine refrigerant R–22 | $CHF_2Cl$ | 86.469 | 74.53 | 1.197 | — | — | — | 8.35 | 2.0 | 1.26 |
| Glycerine | $C_3H_8O_3$ | 92.096 | 78.62 | 1.263 | 6380 | 950 | 1909 | 63.0 | 40 | 1.47 |
| Heptane | $C_7H_{16}$ | 100.208 | 42.42 | .681 | 2.53 | 0.376 | 1138 | 19.9 | 1.92 | 1.38 |
| Hexane | $C_6H_4$ | 86.181 | 40.88 | .657 | 2.00 | 0.297 | 1203 | 18.0 | — | 1.37 |
| Iodine | $I_2$ | 253.809 | — | — | — | — | — | — | 11 | — |
| Kerosene | — | — | 51.2 | 0.823 | 11.0 | 1.64 | 1320 | — | — | — |
| Linseed oil | — | — | 58.0 | 0.93 | 222 | 33.1 | — | — | 3.3 | — |
| Mercury | Hg | 200.59 | — | 13.633 | 10.3 | 1.53 | 1450 | 484 | — | — |
| Octane | $C_8H_{18}$ | 114.235 | 43.61 | .701 | 3.43 | 0.51 | 1171 | 21.14 | — | 1.40 |
| Phenol | $C_6H_6O$ | 94.116 | 66.94 | 1.071 | 54 | 8.0 | 1274[100] | 40.4 | 9.8 | — |
| Propane | $C_3H_8$ | 44.098 | 30.81 | .495 | 0.74 | 0.11 | — | 6.6 | 1.27 | 1.34 |
| Propylene | $C_3H_6$ | 42.082 | 32.11 | .516 | 0.60 | 0.09 | — | 7.0 | — | 1.36 |
| Propylene glycol | $C_3H_8O_2$ | 76.097 | 60.26 | .968 | — | 42 | — | 36.3 | — | 1.43 |
| Sea water | — | 18.52 | 64.0 | 1.03 | — | — | 1535 | — | — | — |
| Toluene | $C_7H_8$ | 92.144 | 53.83 | 0.865 | 3.70 | 0.550 | 1275[30] | 27.3 | 2.4 | 1.49 |

## Table 3—10 (continued)
## PROPERTIES OF COMMON LIQUIDS

At 1.0 atm Pressure, 77°F (25°C), Except as Noted

| Common name | Chemical formula | Molecular weight | Density, $\frac{lb}{ft^3}$ | Specific gravity | Viscosity | | Sound velocity, meters/sec | Surface tension, dynes/cm | Dielectric constant | Refractive index |
|---|---|---|---|---|---|---|---|---|---|---|
| | | | | | $lb_m/ft\,sec \times 10^4$ | cp | | | | |
| Turpentine | $C_{10}H_{16}$ | 136.242 | 54.2 | 0.87 | 9.24 | 1.375 | 1240 | — | — | 1.47 |
| Water | $H_2O$ | 18.0153 | 62.247 | 1.00 | 6.0 | 0.89 | 1498 | 71.97 | 78.54[a] | 1.33 |

[a]For thermal properties see Table 3–22. For properties of liquids in SI units see Table 3–23.
[b]The dielectric constant of water near the freezing point is 87.8; it decreases with increase in temperature to about 55.6 near the boiling point.

From Bolz, R. E. and Tuve, G. L., Eds., *Handbook of Tables for Applied Engineering Science*, 2nd ed., CRC Press, Cleveland, 1973, 90.

## Table 3—11
## THERMAL PROPERTIES OF COMMON LIQUIDS[a]

At 1.0 atm Pressure, 77°F (25°C), Except as Noted

| Common name | Specific heat, Btu/lbm °F | Thermal conductivity, Btu/ft hr °F | Freezing point, °F | Latent heat of fusion, Btu/lb | Boiling point, °F | Latent heat of evaporation, Btu/lb | Coefficient of cubical expansion per °F |
|---|---|---|---|---|---|---|---|
| Acetic acid | 0.522 | 0.099 | 62 | 77.7 | 245 | 173 | 0.0006 |
| Acetone | 0.514 | 0.093 | − 137.4 | 42.3 | 133 | 223 | 0.00082 |
| Alcohol, ethyl | 0.584 | 0.099 | − 174.2 | 46.4 | 172.96 | 364 | 0.0006 |
| Alcohol, methyl | 0.606 | 0.117 | − 143.7 | 42.5 | 148.4 | 474 | 0.00075 |
| Alcohol, propyl | 0.567 | 0.093 | − 197 | 37.2 | 208 | 335 | — |
| Ammonia (aqua) | 1.047 | 0.204 | — | — | — | — | — |
| Benzene | 0.414 | 0.083 | 41.96 | 54.4 | 176.2 | 168 | 0.0007 |
| Bromine | 0.113 | — | − 17.15 | 28.7 | 137.3 | 83 | 0.00065 |
| Carbon disulfide | 0.237 | 0.093 | − 169.5 | 24.80 | 115.26 | 151.2 | 0.0007 |
| Carbon tetrachloride | 0.207 | 0.060 | − 9.04 | 74.8 | 169.7 | 83.5 | 0.0007 |
| Castor oil | 0.47 | 0.104 | 14.1 | — | — | — | —– |
| Chloroform | 0.25 | 0.068 | − 82.3 | 33.14 | 142.2 | 106.4 | 0.00073 |
| Decane | 0.528 | 0.085 | − 21.4 | 86.6 | 345.3 | 113 | — |
| Dodecane | 0.528 | 0.081 | − 14.74 | 93.0 | 421.3 | 110 | — |
| Ether | 0.529 | 0.075 | − 177 | 41.4 | 94.2 | 160 | 0.0009 |
| Ethylene glycol | 0.565 | 0.149 | 8.6 | 77.9 | 387 | 344 | — |
| Fluorine refrigerant R–11 | 0.208[b] | 0.054[b] | − 168 | — | 74.9 | 77.58[c] | — |
| Fluorine refrigerant R–12 | 0.232[b] | 0.041[b] | − 252 | 14.8 | − 21.6 | 71.04[c] | — |
| Fluorine refrigerant R–22 | 0.300[b] | 0.050[b] | − 256 | 78.7 | − 41.4 | 100.05[c] | — |
| Glycerine | 0.627 | 0.166 | 17.0 | 86 | 554.5 | 419 | 0.0003 |

## Table 3—11 (continued)
## THERMAL PROPERTIES OF COMMON LIQUIDS[a]

### At 1.0 atm Pressure, 77°F (25°C), Except as Noted

| Common name | Specific heat, Btu/lbm °F | Thermal conductivity, Btu/ft hr °F | Freezing point, °F | Latent heat of fusion, Btu/lb | Boiling point, °F | Latent heat of evaporation, Btu/lb | Coefficient of cubical expansion per °F |
|---|---|---|---|---|---|---|---|
| Heptane | 0.536 | 0.074 | −131.1 | 60.2 | 209.1 | 137 | — |
| Hexane | 0.541 | 0.072 | −139.3 | 65.3 | 155.65 | 157 | — |
| Iodine | 0.513 | — | 236.3 | 26.74 | 363.8 | 70.71 | — |
| Kerosene | 0.5 | 0.084 | — | — | — | 108 | — |
| Linseed oil | 0.44 | — | −4 | — | 549 | — | — |
| Mercury | 0.0333 | — | −38.0 | 5.0 | 674 | 126.9 | 0.0001 |
| Octane | 0.514 | 0.076 | −70.2 | 78.0 | 257 | 128 | 0.0004 |
| Phenol | 0.342 | 0.11 | 109.4 | 52.2 | 360 | — | 0.0005 |
| Propane, R–290 | 0.576† | — | −305.8 | 34.38 | −43.73 | 184[c] | — |
| Propylene | 0.682 | — | −301.5 | 30.70 | −53.86 | 147 | — |
| Propylene glycol | 0.598 | — | −76 | — | 369 | 393 | — |
| Sea water | 0.90–.98 | — | 27.5 | — | — | — | — |
| Toluene | 0.4† | 0.077 | −139 | 30.90 | 230.8 | 156 | — |
| Turpentine | 0.425 | 0.070 | −75 | — | 320 | 126 | 0.00055 |
| Water | 0.998 | 0.352 | 32 | 143.3 | 212 | 970.3 | 0.00011 |

[a]For other properties of liquids see Table 3−21. For properties of liquids in SI units see Table 3−23.
[b]At 75°F, liquid.
[c]At 14.7 psia, saturation temperature.

From Bolz, R. E. and Tuve, G. L., Eds., *Handbook of Tables for Applied Engineering Science,* 2nd ed., CRC Press, Cleveland, 1973, 91.

**Table 3—12**

# DIELECTRIC CONSTANTS OF WATER

| °C | $\epsilon^1$ | $\epsilon^2$ |
|-----|-------|-------|
| 0   | 87.74 | 87.90 |
| 5   | 85.76 | 85.90 |
| 10  | 83.83 | 83.95 |
| 15  | 81.95 | 82.04 |
| 18  | 80.84 | 80.93 |
| 20  | 80.10 | 80.18 |
| 25  | 78.30 | 78.36 |
| 30  | 76.55 | 76.58 |
| 35  | 74.83 | 74.85 |
| 38  | 73.82 | 73.83 |
| 40  | 73.15 | 73.15 |
| 45  | 71.51 | 71.50 |
|     |       |       |
| 50  | 69.91 | 69.88 |
| 55  | 68.34 | 68.30 |
| 60  | 66.81 | 66.76 |
| 65  | 65.32 | 65.25 |
| 70  | 63.86 | 63.78 |
| 75  | 62.43 | 62.34 |
| 80  | 61.03 | 60.93 |
| 85  | 59.66 | 59.55 |
| 90  | 58.32 | 58.20 |
| 95  | 57.01 | 56.88 |
| 100 | 55.72 | 55.58 |

## REFERENCES

1. **Malmberg, C. G. and Maryott, A. A.,** *J. Res. Natl. Bur. Stand.,* 56, 1, 1956.
2. **Owen, B. B., Miller, R. C., Milner, C. E., and Cogan, H. L.,** *J. Phys. Chem.,* 65, 2065, 1961.

From Hamer, W. J. and DeWane, H. J., *NSRDS-NBS,* 33, 20, 1970.

## Table 3—13
## APPROXIMATE pH VALUES OF SOME ACIDS, BASES, FOODS, AND BIOLOGIC MATERIALS[a]

| Substance | pH | Substance | pH |
|---|---|---|---|
| **Acids** | | | |
| Acetic acid, $N$ | 2.4 | Hydrochloric acid, 0.01$N$ | 2.0 |
| Acetic acid, 0.1$N$ | 2.9 | Hydrocyanic acid, 0.1$N$ | 5.1 |
| Acetic acid, 0.01$N$ | 3.4 | Hydrogen sulfide, 0.1$N$ | 4.1 |
| Alum, 0.1$N$ | 3.2 | Lactic acid, 0.1$N$ | 2.4 |
| Arsenious acid, saturated | 5.0 | Malic acid, 0.1$N$ | 2.2 |
| Benzoic acid, 0.01$N$ | 3.1 | Orthophosphoric acid, 0.1$N$ | 1.5 |
| Boric acid, 0.1$N$ | 5.2 | Oxalic acid, 0.1$N$ | 1.6 |
| Carbonic acid, saturated | 3.8 | Sulfuric acid, $N$ | 0.3 |
| Citric acid, 0.1$N$ | 2.2 | Sulfuric acid, 0.1$N$ | 1.2 |
| Formic acid, 0.1$N$ | 2.3 | Sulfuric acid, 0.01$N$ | 2.1 |
| Hydrochloric acid, $N$ | 0.1 | Sulfurous acid, 0.1$N$ | 1.5 |
| Hydrochloric acid, 0.1$N$ | 1.1 | Tartaric acid, 0.1$N$ | 2.2 |
| **Bases** | | | |
| Ammonia, $N$ | 11.6 | Potassium hydroxide, 0.1$N$ | 13.0 |
| Ammonia, 0.1$N$ | 11.1 | Potassium hydroxide, 0.01$N$ | 12.0 |
| Ammonia, 0.01$N$ | 10.6 | Sodium bicarbonate, 0.1$N$ | 8.4 |
| Borax, 0.1$N$ | 9.2 | Sodium carbonate, 0.1$N$ | 11.6 |
| Calcium carbonate, saturated | 9.4 | Sodium hydroxide, $N$ | 14.0 |
| Ferrous hydroxide, saturated | 9.5 | Sodium hydroxide, 0.1$N$ | 13.0 |
| Lime, saturated | 12.4 | Sodium hydroxide, 0.01$N$ | 12.0 |
| Magnesia, saturated | 10.5 | Sodium metasilicate, 0.1$N$ | 12.6 |
| Potassium cyanide, 0.1$N$ | 11.0 | Sodium sesquicarbonate, 0.1$M$ | 10.1 |
| Potassium hydroxide, $N$ | 14.0 | Trisodium phosphate, 0.1$N$ | 12.0 |
| **Biologic Materials** | | | |
| Bile, human | 6.8–7.0 | Gastric contents, human | 1.0–3.0 |
| Blood, plasma, human | 7.3–7.5 | Milk, human | 6.6–7.6 |
| Blood, whole, dog | 6.9–7.2 | Saliva, human | 6.5–7.5 |
| Duodenal contents, human | 4.8–8.2 | Spinal fluid, human | 7.3–7.5 |
| Feces, human | 4.6–8.4 | Urine, human | 4.8–8.4 |
| **Foods** | | | |
| Apples | 2.9–3.3 | Milk, cows | 6.3–6.6 |
| Apricots | 3.6–4.0 | Olives | 3.6–3.8 |
| Asparagus | 5.4–5.8 | Oranges | 3.0–4.0 |
| Bananas | 4.5–4.7 | Oysters | 6.1–6.6 |
| Beans | 5.0–6.0 | Peaches | 3.4–3.6 |
| Beers | 4.0–5.0 | Pears | 3.6–4.0 |
| Beets | 4.9–5.5 | Peas | 5.8–6.4 |
| Blackberries | 3.2–3.6 | Pickles, dill | 3.2–3.6 |
| Bread, white | 5.0–6.0 | Pickles, sour | 3.0–3.4 |
| Butter | 6.1–6.4 | Pimento | 4.6–5.2 |
| Cabbage | 5.2–5.4 | Plums | 2.8–3.0 |
| Carrots | 4.9–5.3 | Potatoes | 5.6–6.0 |
| Cheeses | 4.8–6.4 | Pumpkin | 4.8–5.2 |
| Cherries | 3.2–4.0 | Raspberries | 3.2–3.6 |
| Cider | 2.9–3.3 | Rhubarb | 3.1–3.2 |
| Corn | 6.0–6.5 | Salmon | 6.1–6.3 |
| Crackers | 6.5–8.5 | Sauerkraut | 3.4–3.6 |
| Dates | 6.2–6.4 | Shrimp | 6.8–7.0 |
| Eggs, fresh white | 7.6–8.0 | Soft drinks | 2.0–4.0 |

## Table 3—13 (continued)
### APPROXIMATE pH VALUES OF SOME ACIDS, BASES, FOODS, AND BIOLOGIC MATERIALS

| Substance | pH | Substance | pH |
|---|---|---|---|
| **Foods (cont.)** | | | |
| Flour, wheat | 5.5–6.5 | Spinach | 5.1–5.7 |
| Gooseberries | 2.8–3.0 | Squash | 5.0–5.4 |
| Grapefruit | 3.0–3.3 | Strawberries | 3.0–3.5 |
| Grapes | 3.5–4.5 | Sweet potatoes | 5.3–5.6 |
| Hominy (lye) | 6.8–8.0 | Tomatoes | 4.0–4.4 |
| Jams, fruit | 3.5–4.0 | Tuna | 5.9–6.1 |
| Jellies, fruit | 2.8–3.4 | Turnips | 5.2–5.6 |
| Lemons | 2.2–2.4 | Vinegar | 2.4–3.4 |
| Limes | 1.8–2.0 | Water, drinking | 6.5–8.0 |
| Maple syrup | 6.5–7.0 | Wines | 2.8–3.8 |

[a]All values are based on measurements made at 25°C and are rounded off to the nearest tenth.

From *Modern pH and Chlorine Control*, Taylor Chemicals, Baltimore. With permission.

## Table 3—14
### IONIZATION CONSTANTS FOR WATER

| °C | $-\log_{10} K_w$ | °C | $-\log_{10} K_w$ |
|---|---|---|---|
| 0 | 14.9435 | 30 | 13.8330 |
| 5 | 14.7338 | 35 | 13.6801 |
| 10 | 14.5346 | 40 | 13.5348 |
| 15 | 14.3463 | 45 | 13.3960 |
| 20 | 14.1669 | 50 | 13.2617 |
| 24 | 14.0000 | 55 | 13.1369 |
| 25 | 13.9965 | 60 | 13.0171 |

From Weast, R. C. Ed., *Handbook of Chemistry and Physics*, 69th ed., CRC Press, Boca Raton, Fla., 1988, D-164.

## Table 3—15
### DISSOCIATION CONSTANTS ($K_b$) OF AQUEOUS AMMONIA FROM 0 TO 50°C[a]

| °C | $pK_b$ | $K_b$ | °C | $pK_b$ | $K_b$ | °C | $pK_b$ | $K_b$ |
|---|---|---|---|---|---|---|---|---|
| 0 | 4.862 | $1.374 \times 10^{-5}$ | 20 | 4.767 | $1.710 \times 10^{-5}$ | 35 | 4.733 | $1.849 \times 10^{-5}$ |
| 5 | 4.830 | $1.479 \times 10^{-5}$ | 25 | 4.751 | $1.774 \times 10^{-5}$ | 40 | 4.730 | $1.862 \times 10^{-5}$ |
| 10 | 4.804 | $1.570 \times 10^{-5}$ | 30 | 4.740 | $1.820 \times 10^{-5}$ | 45 | 4.726 | $1.879 \times 10^{-5}$ |
| 15 | 4.782 | $1.652 \times 10^{-5}$ | | | | 50 | 4.723 | $1.892 \times 10^{-5}$ |

[a]Values of $K_b$ were determined by the emf method and are accurate to ±0.005.

Reprinted with permission from Bates, R. G. and Pinching, G. D., *J. Am. Chem. Soc.*, 72, 1393, 1950. Copyright by the American Chemical Society.

## Table 3—16
### MOLECULAR ELEVATION OF THE BOILING POINT[a]

| Solvent | $K_B$ | Barometric correction per millimeter | Solvent | $K_B$ | Barometric correction per millimeter |
|---|---|---|---|---|---|
| Acetic acid | 3.07 | 0.0008 | Ethyl acetate | 2.77 | 0.0007 |
| Acetone | 1.71 | 0.0004 | Ethyl ether | 2.02 | 0.0005 |
| Aniline | 3.52 | 0.0009 | *n*-Hexane | 2.75 | 0.0007 |
| Benzene | 2.53 | 0.0007 | Methanol | 0.83 | 0.0005 |
| Bromobenzene | 6.26 | 0.0016 | Methyl acetate | 2.15 | 0.0002 |
| Carbon bisulfide | 2.34 | 0.0006 | Nitrobenzene | 5.24 | 0.0013 |
| Carbon tetrachloride | 5.03 | 0.0013 | *n*-Octane | 4.02 | 0.0010 |
| Chloroform | 3.63 | 0.0009 | Phenol | 3.56 | 0.0009 |
| Cyclohexane | 2.79 | 0.0007 | Toluene | 3.33 | 0.0008 |
| Ethanol | 1.22 | 0.0003 | Water | 0.512 | 0.0001 |

[a]Elevation of the boiling point ($K_B$) due to the addition of 1 g molecular weight of solute to 1,000 g of solvent is given in °C. Barometric correction indicates the number of °C to be subtracted for each millimeter of difference between the barometric reading and 760 mm.

From Weast, R. C. Ed., *Handbook of Chemistry and Physics*, 69th ed., CRC Press, Boca Raton, Fla., 1988, D-186.

## Table 3—17
### MOLECULAR DEPRESSION OF THE FREEZING POINT[a]

| Solvent | Depression | Solvent | Depression |
|---|---|---|---|
| Acetic acid | 39.0 | Naphthalene | 68−69 |
| Benzene | 49.0 | Nitrobenzene | 70.0 |
| Benzophenone | 98.0 | Phenol | 74.0 |
| Diphenyl | 80.0 | Stearic acid | 45.0 |
| Diphenylamine | 86.0 | Triphenyl methane | 124.5 |
| Ethylene dibromide | 118.0 | Urethane | 51.4 |
| Formic acid | 27.7 | Water | 18.5−18.7 |

[a]Depression of the freezing point due to the addition of 1 g molecular weight of solute to 100 g of solvent is given in °C.

From Weast, R. C. Ed., *Handbook of Chemistry and Physics*, 69th ed., CRC Press, Boca Raton, Fla., 1988, D-186.

## Table 3—18A
## VAPOR PRESSURE OF WATER BELOW 100°C

| °C | 0.0 | 0.2 | 0.4 | 0.6 | 0.8 |
|---|---|---|---|---|---|
| −15 | 1.436 | 1.414 | 1.390 | 1.368 | 1.345 |
| −14 | 1.560 | 1.534 | 1.511 | 1.485 | 1.460 |
| −13 | 1.691 | 1.665 | 1.637 | 1.611 | 1.585 |
| −12 | 1.834 | 1.804 | 1.776 | 1.748 | 1.720 |
| −11 | 1.987 | 1.955 | 1.924 | 1.893 | 1.863 |
| −10 | 2.149 | 2.116 | 2.084 | 2.050 | 2.018 |
| − 9 | 2.326 | 2.289 | 2.254 | 2.219 | 2.184 |
| − 8 | 2.514 | 2.475 | 2.437 | 2.399 | 2.362 |
| − 7 | 2.715 | 2.674 | 2.633 | 2.593 | 2.553 |
| − 6 | 2.931 | 2.887 | 2.843 | 2.800 | 2.757 |
| − 5 | 3.163 | 3.115 | 3.069 | 3.022 | 2.976 |
| − 4 | 3.410 | 3.359 | 3.309 | 3.259 | 3.211 |
| − 3 | 3.673 | 3.620 | 3.567 | 3.514 | 3.461 |
| − 2 | 3.956 | 3.898 | 3.841 | 3.785 | 3.730 |
| − 1 | 4.258 | 4.196 | 4.135 | 4.075 | 4.016 |
| 0 − | 4.579 | 4.513 | 4.448 | 4.385 | 4.320 |
| 0 | 4.579 | 4.647 | 4.715 | 4.785 | 4.855 |
| 1 | 4.926 | 4.998 | 5.070 | 5.144 | 5.219 |
| 2 | 5.294 | 5.370 | 5.447 | 5.525 | 5.605 |
| 3 | 5.685 | 5.766 | 5.848 | 5.931 | 6.015 |
| 4 | 6.101 | 6.187 | 6.274 | 6.363 | 6.453 |
| 5 | 6.543 | 6.635 | 6.728 | 6.822 | 6.917 |
| 6 | 7.013 | 7.111 | 7.209 | 7.309 | 7.411 |
| 7 | 7.513 | 7.617 | 7.722 | 7.828 | 7.936 |
| 8 | 8.045 | 8.155 | 8.267 | 8.380 | 8.494 |
| 9 | 8.609 | 8.727 | 8.845 | 8.965 | 9.086 |
| 10 | 9.209 | 9.333 | 9.458 | 9.585 | 9.714 |
| 11 | 9.844 | 9.976 | 10.109 | 10.244 | 10.380 |
| 12 | 10.518 | 10.658 | 10.799 | 10.941 | 11.085 |
| 13 | 11.231 | 11.379 | 11.528 | 11.680 | 11.833 |
| 14 | 11.987 | 12.144 | 12.302 | 12.462 | 12.624 |
| 15 | 12.788 | 12.953 | 13.121 | 13.290 | 13.461 |
| 16 | 13.634 | 13.809 | 13.987 | 14.166 | 14.347 |
| 17 | 14.530 | 14.715 | 14.903 | 15.092 | 15.284 |
| 18 | 15.477 | 15.673 | 15.871 | 16.071 | 16.272 |
| 19 | 16.477 | 16.685 | 16.894 | 17.105 | 17.319 |
| 20 | 17.535 | 17.753 | 17.974 | 18.197 | 18.422 |
| 21 | 18.650 | 18.880 | 19.113 | 19.349 | 19.587 |
| 22 | 19.827 | 20.070 | 20.316 | 20.565 | 20.815 |
| 23 | 21.068 | 21.324 | 21.583 | 21.845 | 22.110 |
| 24 | 22.377 | 22.648 | 22.922 | 23.198 | 23.476 |
| 25 | 23.756 | 24.039 | 24.326 | 24.617 | 24.912 |
| 26 | 25.209 | 25.509 | 25.812 | 26.117 | 26.426 |
| 27 | 26.739 | 27.055 | 27.374 | 27.696 | 28.021 |
| 28 | 28.349 | 28.680 | 29.015 | 29.354 | 29.697 |
| 29 | 30.043 | 30.392 | 30.745 | 31.102 | 31.461 |
| 30 | 31.824 | 32.191 | 32.561 | 32.934 | 33.312 |
| 31 | 33.695 | 34.082 | 34.471 | 34.864 | 35.261 |
| 32 | 35.663 | 36.068 | 36.477 | 36.891 | 37.308 |
| 33 | 37.729 | 38.155 | 38.584 | 39.018 | 39.457 |
| 34 | 39.898 | 40.344 | 40.796 | 41.251 | 41.710 |
| 35 | 42.175 | 42.644 | 43.117 | 43.595 | 44.078 |
| 36 | 44.563 | 45.054 | 45.549 | 46.050 | 46.556 |
| 37 | 47.067 | 47.582 | 48.102 | 48.627 | 49.157 |
| 38 | 49.692 | 50.231 | 50.774 | 51.323 | 51.879 |
| 39 | 52.442 | 53.009 | 53.580 | 54.156 | 54.737 |
| 40 | 55.324 | 55.91 | 56.51 | 57.11 | 57.72 |
| 41 | 58.34 | 58.96 | 59.58 | 60.22 | 60.86 |

| °C | 0.0 | 0.2 | 0.4 | 0.6 | 0.8 |
|---|---|---|---|---|---|
| 42 | 61.50 | 62.14 | 62.80 | 63.46 | 64.12 |
| 43 | 64.80 | 65.48 | 66.16 | 66.86 | 67.56 |
| 44 | 68.26 | 68.97 | 69.69 | 70.41 | 71.14 |
| 45 | 71.88 | 72.62 | 73.36 | 74.12 | 74.88 |
| 46 | 75.65 | 76.43 | 77.21 | 78.00 | 78.80 |
| 47 | 79.60 | 80.41 | 81.23 | 82.05 | 82.87 |
| 48 | 83.71 | 84.56 | 85.42 | 86.28 | 87.14 |
| 49 | 88.02 | 88.90 | 89.79 | 90.69 | 91.59 |
| 50 | 92.51 | 93.5 | 94.4 | 95.3 | 96.3 |
| 51 | 97.20 | 98.2 | 99.1 | 100.1 | 101.1 |
| 52 | 102.09 | 103.1 | 104.1 | 105.1 | 106.2 |
| 53 | 107.20 | 108.2 | 109.3 | 110.4 | 111.4 |
| 54 | 112.51 | 113.6 | 114.7 | 115.8 | 116.9 |
| 55 | 118.04 | 119.1 | 120.3 | 121.5 | 122.6 |
| 56 | 123.80 | 125.0 | 126.2 | 127.4 | 128.6 |
| 57 | 129.82 | 131.0 | 132.3 | 133.5 | 134.7 |
| 58 | 136.08 | 137.3 | 138.5 | 139.9 | 141.2 |
| 59 | 142.60 | 143.9 | 145.2 | 146.6 | 148.0 |
| 60 | 149.38 | 150.7 | 152.1 | 153.5 | 155.0 |
| 61 | 156.43 | 157.8 | 159.3 | 160.8 | 162.3 |
| 62 | 163.77 | 165.2 | 166.8 | 168.3 | 169.8 |
| 63 | 171.38 | 172.9 | 174.5 | 176.1 | 177.7 |
| 64 | 179.31 | 180.9 | 182.5 | 184.2 | 185.8 |
| 65 | 187.54 | 189.2 | 190.9 | 192.6 | 194.3 |
| 66 | 196.09 | 197.8 | 199.5 | 201.3 | 203.1 |
| 67 | 204.96 | 206.8 | 208.6 | 210.5 | 212.3 |
| 68 | 214.17 | 216.0 | 218.0 | 219.9 | 221.8 |
| 69 | 223.73 | 225.7 | 227.7 | 229.7 | 231.7 |
| 70 | 233.7 | 235.7 | 237.7 | 239.7 | 241.8 |
| 71 | 243.9 | 246.0 | 248.2 | 250.3 | 252.4 |
| 72 | 254.6 | 256.8 | 259.0 | 261.2 | 263.4 |
| 73 | 265.7 | 268.0 | 270.2 | 272.6 | 274.8 |
| 74 | 277.2 | 279.4 | 281.8 | 284.2 | 286.6 |
| 75 | 289.1 | 291.5 | 294.0 | 296.4 | 298.8 |
| 76 | 301.4 | 303.8 | 306.4 | 308.9 | 311.4 |
| 77 | 314.1 | 316.6 | 319.2 | 322.0 | 324.6 |
| 78 | 327.3 | 330.0 | 332.8 | 335.6 | 338.2 |
| 79 | 341.0 | 343.8 | 346.6 | 349.4 | 352.2 |
| 80 | 355.1 | 358.0 | 361.0 | 363.8 | 366.8 |
| 81 | 369.7 | 372.6 | 375.6 | 378.8 | 381.8 |
| 82 | 384.9 | 388.0 | 391.2 | 394.4 | 397.4 |
| 83 | 400.6 | 403.8 | 407.0 | 410.2 | 413.6 |
| 84 | 416.8 | 420.2 | 423.6 | 426.8 | 430.2 |
| 85 | 433.6 | 437.0 | 440.4 | 444.0 | 447.5 |
| 86 | 450.9 | 454.4 | 458.0 | 461.6 | 465.2 |
| 87 | 468.7 | 472.4 | 476.0 | 479.8 | 483.4 |
| 88 | 487.1 | 491.0 | 494.7 | 498.5 | 502.2 |
| 89 | 506.1 | 510.0 | 513.9 | 517.8 | 521.8 |
| 90 | 525.76 | 529.77 | 533.80 | 537.86 | 541.95 |
| 91 | 546.05 | 550.18 | 554.35 | 558.53 | 562.75 |
| 92 | 566.99 | 571.26 | 575.55 | 579.87 | 584.22 |
| 93 | 588.60 | 593.00 | 597.43 | 601.89 | 606.38 |
| 94 | 610.90 | 615.44 | 620.01 | 624.61 | 629.24 |
| 95 | 633.90 | 638.59 | 643.30 | 648.05 | 652.82 |
| 96 | 657.62 | 662.45 | 667.31 | 672.20 | 677.12 |
| 97 | 682.07 | 687.04 | 692.05 | 697.10 | 702.17 |
| 98 | 707.27 | 712.40 | 717.56 | 722.75 | 727.98 |
| 99 | 733.24 | 738.53 | 743.85 | 749.20 | 754.58 |
| 100 | 760.00 | 765.45 | 770.93 | 776.44 | 782.00 |
| 101 | 787.57 | 793.18 | 798.82 | 804.50 | 810.21 |

*Note:* Values are given in mm of Hg for temperatures ranging from −15.8°C to 101.8°C. The values for fractional degrees between 50 and 89 were obtained by interpolation.

From Weast, R. C. Ed., *Handbook of Chemistry and Physics,* 69th ed., CRC Press, Boca Raton, Fla., 1988, D-189.

## Table 3—18B
## VAPOR PRESSURE OF WATER ABOVE 100°C

| Temp. °C | Pressure (mm) | Pressure (Pounds per sq. in.) | Temp. °F |
|---|---|---|---|
| 100 | 760. | 14.696 | 212.0 |
| 101 | 787.51 | 15.228 | 213.8 |
| 102 | 815.86 | 15.776 | 215.6 |
| 103 | 845.12 | 16.342 | 217.4 |
| 104 | 875.06 | 16.921 | 219.2 |
| 105 | 906.07 | 17.521 | 221.0 |
| 106 | 937.92 | 18.136 | 222.8 |
| 107 | 970.60 | 18.768 | 224.6 |
| 108 | 1004.42 | 19.422 | 226.4 |
| 109 | 1038.92 | 20.089 | 228.2 |
| 110 | 1074.56 | 20.779 | 230.0 |
| 111 | 1111.20 | 21.487 | 231.8 |
| 112 | 1148.74 | 22.213 | 233.6 |
| 113 | 1187.42 | 22.961 | 235.4 |
| 114 | 1227.25 | 23.731 | 237.2 |
| 115 | 1267.98 | 24.519 | 239.0 |
| 116 | 1309.94 | 25.330 | 240.8 |
| 117 | 1352.95 | 26.162 | 242.6 |
| 118 | 1397.18 | 27.017 | 244.4 |
| 119 | 1442.63 | 27.896 | 246.2 |
| 120 | 1489.14 | 28.795 | 248.0 |
| 121 | 1536.80 | 29.717 | 249.8 |
| 122 | 1586.04 | 30.669 | 251.6 |
| 123 | 1636.36 | 31.642 | 253.4 |
| 124 | 1687.81 | 32.637 | 255.2 |
| 125 | 1740.93 | 33.664 | 257.0 |
| 126 | 1795.12 | 34.712 | 258.8 |
| 127 | 1850.83 | 35.789 | 260.6 |
| 128 | 1907.83 | 36.891 | 262.4 |
| 129 | 1966.35 | 38.023 | 264.2 |
| 130 | 2026.16 | 39.180 | 266.0 |
| 131 | 2087.42 | 40.364 | 267.8 |
| 132 | 2150.42 | 41.582 | 269.6 |
| 133 | 2214.64 | 42.824 | 271.4 |
| 134 | 2280.76 | 44.103 | 273.2 |
| 135 | 2347.26 | 45.389 | 275.0 |
| 136 | 2416.34 | 46.724 | 276.8 |
| 137 | 2488.16 | 48.111 | 278.6 |
| 138 | 2560.67 | 49.515 | 280.4 |
| 139 | 2634.84 | 50.950 | 282.2 |
| 140 | 2710.92 | 52.421 | 284.0 |
| 141 | 2788.44 | 53.920 | 285.8 |
| 142 | 2867.48 | 55.448 | 287.6 |
| 143 | 2948.80 | 57.030 | 289.4 |
| 144 | 3031.64 | 58.622 | 291.2 |
| 145 | 3116.76 | 60.268 | 293.0 |
| 146 | 3203.40 | 61.944 | 294.8 |
| 147 | 3292.32 | 63.663 | 296.6 |
| 148 | 3382.76 | 65.412 | 298.4 |
| 149 | 3476.24 | 67.220 | 300.2 |
| 150 | 3570.48 | 69.042 | 302.0 |
| 151 | 3667.00 | 70.908 | 303.8 |
| 152 | 3766.56 | 72.833 | 305.6 |
| 153 | 3866.88 | 74.773 | 307.4 |
| 154 | 3970.24 | 76.772 | 309.2 |
| 155 | 4075.88 | 78.815 | 311.0 |
| 156 | 4183.80 | 80.901 | 312.8 |
| 157 | 4293.24 | 83.018 | 314.6 |
| 158 | 4404.96 | 85.178 | 316.4 |
| 159 | 4519.72 | 87.397 | 318.2 |
| 160 | 4636.00 | 89.646 | 320.0 |
| 161 | 4755.32 | 91.953 | 321.8 |
| 162 | 4876.92 | 94.304 | 323.6 |
| 163 | 5000.04 | 96.685 | 325.4 |
| 164 | 5126.96 | 99.139 | 327.2 |
| 165 | 5256.16 | 101.638 | 329.0 |
| 166 | 5386.88 | 104.165 | 330.8 |
| 167 | 5521.40 | 106.766 | 332.6 |
| 168 | 5658.20 | 109.412 | 334.4 |
| 169 | 5798.04 | 112.116 | 336.2 |
| 170 | 5940.92 | 114.879 | 338.0 |
| 171 | 6085.32 | 117.671 | 339.8 |
| 172 | 6233.52 | 120.537 | 341.6 |
| 173 | 6383.24 | 123.432 | 343.4 |
| 174 | 6538.28 | 126.130 | 345.2 |
| 175 | 6694.08 | 129.442 | 347.0 |
| 176 | 6852.92 | 132.514 | 348.8 |
| 177 | 7015.56 | 135.659 | 350.6 |
| 178 | 7180.48 | 138.848 | 352.4 |
| 179 | 7349.20 | 142.110 | 354.2 |
| 180 | 7520.20 | 145.417 | 356.0 |
| 181 | 7694.24 | 148.782 | 357.8 |
| 182 | 7872.08 | 152.221 | 359.6 |
| 183 | 8052.96 | 155.719 | 361.4 |
| 184 | 8236.88 | 159.275 | 363.2 |
| 185 | 8423.84 | 162.890 | 365.0 |
| 186 | 8616.12 | 166.609 | 366.8 |
| 187 | 8809.92 | 170.356 | 368.6 |
| 188 | 9007.52 | 174.177 | 370.4 |
| 189 | 9208.16 | 178.057 | 372.2 |
| 190 | 9413.36 | 182.025 | 374.0 |
| 191 | 9620.08 | 186.022 | 375.8 |
| 192 | 9831.36 | 190.107 | 377.6 |
| 193 | 10047.20 | 194.281 | 379.4 |
| 194 | 10265.32 | 198.499 | 381.2 |
| 195 | 10488.76 | 202.819 | 383.0 |
| 196 | 10715.24 | 207.199 | 384.8 |
| 197 | 10944.76 | 211.637 | 386.6 |
| 198 | 11179.00 | 216.178 | 388.4 |
| 199 | 11417.48 | 220.778 | 390.2 |
| 200 | 11659.16 | 225.451 | 392.0 |
| 201 | 11905.40 | 230.213 | 393.8 |
| 202 | 12155.44 | 235.048 | 395.6 |
| 203 | 12408.52 | 239.942 | 397.4 |
| 204 | 12666.16 | 244.924 | 399.2 |
| 205 | 12929.12 | 250.008 | 401.0 |
| 206 | 13197.40 | 255.196 | 402.8 |
| 207 | 13467.96 | 260.428 | 404.6 |
| 208 | 13742.32 | 265.733 | 406.4 |
| 209 | 14022.76 | 271.156 | 408.2 |
| 210 | 14305.48 | 276.623 | 410.0 |
| 211 | 14595.04 | 282.222 | 411.8 |
| 212 | 14888.40 | 287.895 | 413.6 |
| 213 | 15184.80 | 293.626 | 415.4 |
| 214 | 15488.04 | 299.490 | 417.2 |
| 215 | 15792.80 | 305.383 | 419.0 |
| 216 | 16104.40 | 311.408 | 420.8 |
| 217 | 16420.56 | 317.522 | 422.6 |
| 218 | 16742.04 | 323.738 | 424.4 |
| 219 | 17067.32 | 330.028 | 426.2 |
| 220 | 17395.64 | 336.377 | 428.0 |
| 221 | 17731.56 | 342.872 | 429.8 |
| 222 | 18072.80 | 349.471 | 431.6 |
| 223 | 18417.84 | 356.143 | 433.4 |
| 224 | 18766.68 | 362.888 | 435.2 |
| 225 | 19123.12 | 369.781 | 437.0 |
| 226 | 19482.60 | 376.732 | 438.8 |
| 227 | 19848.92 | 383.815 | 440.6 |
| 228 | 20219.80 | 390.987 | 442.4 |
| 229 | 20596.76 | 398.276 | 444.2 |
| 230 | 20978.28 | 405.654 | 446.0 |
| 231 | 21365.12 | 413.134 | 447.8 |
| 232 | 21757.28 | 420.717 | 449.6 |
| 233 | 22154.00 | 428.388 | 451.4 |
| 234 | 22558.32 | 436.207 | 453.2 |
| 235 | 22967.96 | 444.128 | 455.0 |
| 236 | 23382.92 | 452.152 | 456.8 |
| 237 | 23802.44 | 460.264 | 458.6 |
| 238 | 24229.56 | 468.523 | 460.4 |
| 239 | 24661.24 | 476.871 | 462.2 |
| 240 | 25100.52 | 485.365 | 464.0 |
| 241 | 25543.60 | 493.933 | 465.8 |
| 242 | 25994.28 | 502.647 | 467.6 |
| 243 | 26449.52 | 511.450 | 469.4 |
| 244 | 26912.36 | 520.400 | 471.2 |
| 245 | 27381.28 | 529.467 | 473.0 |
| 246 | 27855.52 | 538.638 | 474.8 |
| 247 | 28335.84 | 547.926 | 476.6 |
| 248 | 28823.76 | 557.360 | 478.4 |
| 249 | 29317.00 | 566.898 | 480.2 |
| 250 | 29817.84 | 576.583 | 482.0 |
| 251 | 30324.00 | 586.370 | 483.8 |
| 252 | 30837.76 | 596.305 | 485.6 |
| 253 | 31356.84 | 606.342 | 487.4 |
| 254 | 31885.04 | 616.556 | 489.2 |
| 255 | 32417.80 | 626.858 | 491.0 |
| 256 | 32957.40 | 637.292 | 492.8 |
| 257 | 33505.36 | 647.888 | 494.6 |
| 258 | 34059.40 | 658.601 | 496.4 |
| 259 | 34618.76 | 669.417 | 498.2 |
| 260 | 35188.00 | 680.425 | 500.0 |
| 261 | 35761.80 | 691.520 | 501.8 |
| 262 | 36343.20 | 702.763 | 503.6 |
| 263 | 36932.20 | 714.152 | 505.6 |
| 264 | 37529.56 | 725.703 | 507.2 |
| 265 | 38133.00 | 737.372 | 509.0 |
| 266 | 38742.52 | 749.158 | 510.8 |
| 267 | 39361.92 | 761.135 | 512.6 |
| 268 | 39986.64 | 773.215 | 514.4 |
| 269 | 40619.72 | 785.457 | 516.2 |
| 270 | 41261.16 | 797.861 | 518.0 |
| 271 | 41910.20 | 810.411 | 519.8 |
| 272 | 42566.08 | 823.094 | 521.6 |
| 273 | 43229.56 | 835.923 | 523.4 |
| 274 | 43902.16 | 848.929 | 525.2 |
| 275 | 44580.84 | 862.053 | 527.0 |
| 276 | 45269.40 | 875.367 | 528.8 |
| 277 | 45964.04 | 888.799 | 530.6 |
| 278 | 46669.32 | 902.437 | 532.4 |
| 279 | 47382.20 | 916.222 | 534.2 |
| 280 | 48104.20 | 930.183 | 536.0 |
| 281 | 48833.80 | 944.291 | 537.8 |
| 282 | 49570.24 | 958.532 | 539.6 |
| 283 | 50316.56 | 972.963 | 541.4 |
| 284 | 51072.76 | 987.586 | 543.2 |
| 285 | 51838.08 | 1002.385 | 545.0 |
| 286 | 52611.76 | 1017.345 | 546.8 |
| 287 | 53395.32 | 1032.497 | 548.6 |
| 288 | 54187.24 | 1047.810 | 550.4 |
| 289 | 54989.04 | 1063.314 | 552.2 |
| 290 | 55799.20 | 1078.980 | 554.0 |
| 291 | 56612.40 | 1094.705 | 555.8 |
| 292 | 57448.40 | 1110.871 | 557.6 |
| 293 | 58284.40 | 1127.036 | 559.4 |
| 294 | 59135.60 | 1143.496 | 561.2 |
| 295 | 59994.40 | 1160.102 | 563.0 |
| 296 | 60860.80 | 1176.856 | 564.8 |
| 297 | 61742.40 | 1193.903 | 566.6 |
| 298 | 62624.00 | 1210.950 | 568.4 |
| 299 | 63528.40 | 1228.439 | 570.2 |
| 300 | 64432.80 | 1245.927 | 572.0 |
| 301 | 65352.40 | 1263.709 | 573.8 |
| 302 | 66279.60 | 1281.638 | 575.6 |
| 303 | 67214.40 | 1299.714 | 577.4 |
| 304 | 68156.80 | 1317.937 | 579.2 |
| 305 | 69114.40 | 1336.454 | 581.0 |
| 306 | 70072.00 | 1354.971 | 582.8 |
| 307 | 71052.40 | 1373.929 | 584.6 |
| 308 | 72048.00 | 1393.181 | 586.4 |
| 309 | 73028.40 | 1412.139 | 588.2 |
| 310 | 74024.00 | 1431.390 | 590.0 |
| 311 | 75042.40 | 1451.083 | 591.8 |
| 312 | 76076.00 | 1471.070 | 593.6 |
| 313 | 77117.20 | 1491.203 | 595.4 |
| 314 | 78166.00 | 1511.484 | 597.2 |
| 315 | 79230.00 | 1532.058 | 599.0 |
| 316 | 80294.00 | 1552.632 | 600.8 |
| 317 | 81373.20 | 1573.501 | 602.6 |
| 318 | 82467.60 | 1594.663 | 604.4 |
| 319 | 83569.60 | 1615.972 | 606.2 |
| 320 | 84686.80 | 1637.575 | 608.0 |
| 321 | 85819.20 | 1659.472 | 609.8 |
| 322 | 86959.20 | 1681.516 | 611.6 |
| 323 | 88114.40 | 1703.854 | 613.4 |
| 324 | 89277.20 | 1726.339 | 615.2 |
| 325 | 90447.60 | 1748.971 | 617.0 |
| 326 | 91633.20 | 1771.897 | 618.8 |
| 327 | 92826.40 | 1794.969 | 620.6 |
| 328 | 94042.40 | 1818.483 | 622.4 |
| 329 | 95273.60 | 1842.291 | 624.2 |
| 330 | 96512.40 | 1866.245 | 626.0 |
| 331 | 97758.80 | 1890.346 | 627.8 |
| 332 | 99020.40 | 1914.742 | 629.6 |
| 333 | 100297.20 | 1939.431 | 631.4 |
| 334 | 101581.60 | 1964.267 | 633.2 |
| 335 | 102881.20 | 1989.398 | 635.0 |
| 336 | 104196.00 | 2014.822 | 636.8 |
| 337 | 105526.00 | 2040.540 | 638.6 |
| 338 | 106871.20 | 2066.552 | 640.4 |
| 339 | 108224.00 | 2092.710 | 642.2 |
| 340 | 109592.00 | 2119.163 | 644.0 |
| 341 | 110967.60 | 2145.763 | 645.8 |
| 342 | 112358.40 | 2172.657 | 647.6 |
| 343 | 113749.20 | 2199.550 | 649.4 |
| 344 | 115178.00 | 2227.179 | 651.2 |
| 345 | 116614.40 | 2254.954 | 653.0 |
| 346 | 118073.60 | 2283.171 | 654.8 |
| 347 | 119532.80 | 2311.387 | 656.6 |
| 348 | 121014.80 | 2340.044 | 658.4 |
| 349 | 122504.40 | 2368.848 | 660.2 |
| 350 | 124001.60 | 2397.799 | 662.0 |
| 351 | 125521.60 | 2427.191 | 663.8 |
| 352 | 127049.20 | 2456.730 | 665.6 |
| 353 | 128599.60 | 2486.710 | 667.4 |
| 354 | 130157.60 | 2516.837 | 669.2 |
| 355 | 131730.80 | 2547.258 | 671.0 |
| 356 | 133326.80 | 2578.119 | 672.8 |
| 357 | 134945.60 | 2609.422 | 674.6 |
| 358 | 136579.60 | 2641.018 | 676.4 |
| 359 | 138228.80 | 2672.908 | 678.2 |
| 360 | 139893.20 | 2705.093 | 680.0 |
| 361 | 141572.80 | 2737.571 | 681.8 |
| 362 | 143275.20 | 2770.490 | 683.6 |
| 363 | 144992.80 | 2803.703 | 685.4 |
| 364 | 146733.20 | 2837.357 | 687.2 |
| 365 | 148519.20 | 2871.892 | 689.0 |
| 366 | 150320.40 | 2906.722 | 690.8 |
| 367 | 152129.20 | 2941.698 | 692.6 |
| 368 | 153960.80 | 2977.116 | 694.4 |
| 369 | 155815.20 | 3012.974 | 696.2 |
| 370 | 157692.40 | 3049.273 | 698.0 |
| 371 | 159584.80 | 3085.866 | 699.8 |
| 372 | 161507.60 | 3123.047 | 701.6 |
| 373 | 163468.40 | 3160.963 | 703.4 |
| 374 | 165467.20 | 3199.613 | 705.2 |

From Weast, R. C. Ed., *Handbook of Chemistry and Physics*, 69th ed., CRC Press, Boca Raton, Fla., 1988, D-190.

## Table 3—19
### VAPOR PRESSURE OF CARBON DIOXIDE

Critical temperature of carbon dioxide: 31.0°C
Triple point of carbon dioxide: −56.602 ± 0.005°C, 3885.2 ± 0.4 mm Hg
Density of mercury column: 13.5951 g/cm$^3$, G = 980.665 dyn

#### A. Solid CO$_2$

μm Hg

| °C | 0 | 1 | 2 | 3 | 4 | 5 | 6 | 7 | 8 | 9 |
|---|---|---|---|---|---|---|---|---|---|---|
| −180 | 0.013 | 0.008 | 0.006 | 0.004 | 0.003 | 0.0017 | 0.0011 | 0.0007 | 0.0005 | 0.0003 |
| −170 | 0.37 | 0.27 | 0.20 | 0.14 | 0.10 | 0.074 | 0.052 | 0.037 | 0.026 | 0.018 |
| −160 | 5.9 | 4.6 | 3.6 | 2.7 | 2.1 | 1.58 | 1.19 | 0.90 | 0.67 | 0.50 |
| −150 | 60.5 | 48.8 | 39.2 | 31.4 | 25.1 | 19.9 | 15.8 | 12.4 | 9.8 | 7.6 |
| −140 | 431 | 359 | 298 | 247 | 204 | 168 | 138 | 113 | 92 | 75 |

mm Hg

| °C | 0 | 1 | 2 | 3 | 4 | 5 | 6 | 7 | 8 | 9 |
|---|---|---|---|---|---|---|---|---|---|---|
| −130 | 2.31 | 1.97 | 1.68 | 1.43 | 1.22 | 1.03 | 0.87 | 0.73 | 0.61 | 0.51 |
| −120 | 9.81 | 8.57 | 7.46 | 6.49 | 5.63 | 4.88 | 4.22 | 3.64 | 3.13 | 2.69 |
| −110 | 34.63 | 30.76 | 27.27 | 24.14 | 21.34 | 18.83 | 16.58 | 14.58 | 12.80 | 11.22 |
| −100 | 104.81 | 94.40 | 84.91 | 76.27 | 68.43 | 61.30 | 54.84 | 48.99 | 43.71 | 38.94 |
| −90 | 279.5 | 254.7 | 231.8 | 210.8 | 191.4 | 173.6 | 157.3 | 142.4 | 128.7 | 116.2 |
| −80 | 672.2 | 618.3 | 568.2 | 521.7 | 478.5 | 438.6 | 401.6 | 367.4 | 335.7 | 306.5 |
| −70 | 1,486.1 | 1,377.3 | 1,275.6 | 1,180.5 | 1,091.7 | 1,008.9 | 931.7 | 859.7 | 792.7 | 730.3 |
| −60 | 3,073.1 | 2,865.1 | 2,669.7 | 2,486.3 | 2,314.2 | 2,152.8 | 2,001.5 | 1,859.7 | 1,726.9 | 1,604.6 |
| −50 | | | | | | | | 3,780.9 | 3,530.2 | 3,294.6 |

#### B. Liquid CO$_2$

mm Hg

| °C | 0 | 1 | 2 | 3 | 4 | 5 | 6 | 7 | 8 | 9 |
|---|---|---|---|---|---|---|---|---|---|---|
| −50 | 5,127.8 | 4,922.7 | 4,723.9 | 4,531.1 | 4,344.3 | 4,163.2 | 3,987.9 | 3,818.2[a] | 3,653.9[a] | 3,495.0[a] |
| −40 | 7,545 | 7,271 | 7,005 | 6,746 | 6,494 | 6,250 | 6,012 | 5,781 | 5,557 | 5,339 |
| −30 | 10,718 | 10,363 | 10,017 | 9,679 | 9,350 | 9,029 | 8,716 | 8,412 | 8,115 | 7,826 |
| −20 | 14,781 | 14,331 | 13,891 | 13,461 | 13,040 | 12,630 | 12,229 | 11,838 | 11,455 | 11,082 |
| −10 | 19,872 | 19,312 | 18,764 | 18,288 | 17,703 | 17,189 | 16,686 | 16,194 | 15,712 | 15,241 |

## Table 3—19 (continued)
## VAPOR PRESSURE OF CARBON DIOXIDE

### B. Liquid $CO_2$ (cont.)

|  °C | | | | | | mm Hg | | | | |
|---|---|---|---|---|---|---|---|---|---|---|
|  | 0 | 1 | 2 | 3 | 4 | 5 | 6 | 7 | 8 | 9 |
| 0− | 26,142 | 25,457 | 24,786 | 24,127 | 23,482 | 22,849 | 22,229 | 21,622 | 21,026 | 20,443 |
| 0+ | 26,142 | 26,840 | 27,552 | 28,277 | 29,017 | 29,771 | 30,539 | 31,323 | 32,121 | 32,934 |
| 10 | 33,763 | 34,607 | 35,467 | 36,343 | 37,236 | 38,146 | 39,073 | 40,017 | 40,980 | 41,960 |
| 20 | 43,959 | 43,977 | 45,014 | 46,072 | 47,150 | 48,250 | 49,370 | 50,514 | 51,680 | 52,871 |
| 30 | 54,086 | 55,327 | | | | | | | | |

<sup></sup>aUndercooled liquid.

From Weast, R. C., Ed., *Handbook of Chemistry and Physics*, 55th ed., CRC Press, Cleveland, 1974, D-161.

## Table 3—20
## ELECTROCHEMICAL SERIES

### Petr Vanysek

Both tables list standard reduction potentials, E° values, at 298.15 K (25°C), and at a pressure of 101.325 kPa (1 atm). Table 3—20A lists only those reduction reactions which have E° values positive to the potential of the standard hydrogen electrode. In Table 3—20A, the reactions are listed in the order of increasing positive potential and range from 0.000 V to +3.053 V. Table 3—20B lists only those reduction reactions which have E° values negative to the potential of the standard hydrogen electrode. In Table 3—20B, reactions are listed in the order of increasing negative potential and range from −0.017 to −4.10 V.

### Table 3—20A
### REDUCTION REACTIONS HAVING E° VALUES MORE POSITIVE THAN THAT OF THE STANDARD HYDROGEN ELECTRODE

| Reaction | E°, V | Reaction | E°, V |
|---|---|---|---|
| $2 H^+ + 2 e \rightleftharpoons H_2$ | 0.00000 | $PbO_2 + H_2O + 2 e \rightleftharpoons PbO + 2 OH^-$ | 0.247 |
| $CuI_2^- + e \rightleftharpoons Cu + 2 I^-$ | 0.00 | $HAsO_2 + 3 H^+ + 3_e \rightleftharpoons As + 2 H_2O$ | 0.248 |
| $Ge^{4+} + 2 e \rightleftharpoons Ge^{2+}$ | 0.00 | $Ru^{3+} + e \rightleftharpoons Ru^{2+}$ | 0.2487 |
| $NO_3^- + H_2O + 2 e \rightleftharpoons NO_2^- + 2 OH^-$ | 0.01 | $ReO_2 + 4 H^+ + 4 e \rightleftharpoons Re + 2 H_2O$ | 0.2513 |
| $Tl_2O_3 + 3 H_2O + 4 e \rightleftharpoons 2 Tl^+ + 6 OH^-$ | 0.02 | $IO_3^- + 3 H_2O + 6 e \rightleftharpoons I^- + OH^-$ | 0.26 |
| $SeO_4^{2-} + H_2O + 2 e \rightleftharpoons SeO_3^{2-} + 2 OH^-$ | 0.05 | $Hg_2Cl_2 + 2 e \rightleftharpoons 2 Hg + 2 Cl^-$ | 0.26808 |
| $UO_2^{2+} + e \rightleftharpoons UO_2^+$ | 0.062 | Calomel electrode, molal KCl | 0.2800 |
| $Pd(OH)_2 + 2 e \rightleftharpoons Pd + 2 OH^-$ | 0.07 | Calomel electrode, 1 mol/1 KCl (NCE) | 0.2801 |
| $AgBr + e \rightleftharpoons Ag + Br^-$ | 0.07133 | $Re^{3+} + 3 e \rightleftharpoons Re$ | 0.300 |
| $S_4O_6^{2-} + 2 e \rightleftharpoons 2 S_2O_3^{2-}$ | 0.08 | $BiO^+ + 2 H^+ + 3 e \rightleftharpoons Bi + H_2O$ | 0.320 |
| $AgSCN + e \rightleftharpoons Ag + SCN^-$ | 0.8951 | $UO_2^{2+} + 4 H^+ + 2 e \rightleftharpoons U^{4+} + 2 H_2O$ | 0.327 |
| $N_2 + 2 H_2O + 6 H^+ + 6 e \rightleftharpoons 2 NH_4OH$ | 0.092 | $ClO_3^- + H_2O + 2 e \rightleftharpoons ClO_2^- + 2 OH^-$ | 0.33 |
| $HgO + H_2O + 2 e \rightleftharpoons Hg + 2 OH^-$ | 0.0977 | $2 HCNO + 2 H^+ + 2 e \rightleftharpoons (CN)_2 + 2 H_2O$ | 0.330 |
| $Ir_2O_3 + 3 H_2O + 6 e \rightleftharpoons 2 Ir + 6 OH^-$ | 0.098 | Calomel electrode, 0.1 mol/1 KCl | 0.3337 |
| $2 NO + 2 e \rightleftharpoons N_2O_2^{2-}$ | 0.10 | $VO^{2+} + 2 H^+ + e \rightleftharpoons V^{3+} + H_2O$ | 0.337 |
| $[Co(NH_3)_6]^{3+} + e \rightleftharpoons [Co(NH_3)_6]^{2+}$ | 0.108 | $Cu^{2+} + 2 e \rightleftharpoons Cu$ | 0.3419 |
| $Hg_2O + H_2O + 2 e \rightleftharpoons 2 Hg + 2 OH^-$ | 0.123 | $Ag_2O + H_2O + 2 e \rightleftharpoons 2 Ag + 2 OH^-$ | 0.342 |
| $Ge^{4+} + 4 e \rightleftharpoons Ge$ | 0.124 | $Cu^{2+} + 2 e \rightleftharpoons Cu(Hg)$ | 0.345 |
| $Hg_2Br_2 + 2 e \rightleftharpoons 2 Hg + 2 Br^-$ | 0.13923 | $AgIO_3 + e \rightleftharpoons Ag + IO^-3$ | 0.354 |
| $Pt(OH)_2 + 2 e \rightleftharpoons Pt + 2 OH^-$ | 0.14 | $[Fe(CN)_6]^{3-} + e \rightleftharpoons [Fe(CN)_6]^{4-}$ | 0.358 |
| $S + 2 H^+ + 2 e \rightleftharpoons H_2S(aq)$ | 0.142 | $ClO_4^- + H_2O + 2 e \rightleftharpoons ClO_3^- + 2 OH^-$ | 0.36 |
| $Np^{4+} + e \rightleftharpoons Np^{3+}$ | 0.147 | $Ag_2SeO_3 + 2 e \rightleftharpoons 2 Ag + SeO_3^{2-}$ | 0.3629 |
| $Ag_4[Fe(CN)_6] + 4 e \rightleftharpoons 4 Ag + [Fe(CN)_6]^{4-}$ | 0.1478 | $ReO_4^- + 8 H^+ + 7 e \rightleftharpoons Re + 4 H_2O$ | 0.368 |
| $IO_3^- + 2 H_2O + 4 e \rightleftharpoons IO^- + 4 OH^-$ | 0.15 | $(CN)_2 + 2 H^+ + 2 e \rightleftharpoons 2 HCN$ | 0.373 |
| $Mn(OH)_3 + e \rightleftharpoons Mn(OH)_2 + OH^-$ | 0.15 | $[Ferricinium]^+ + e \rightleftharpoons ferrocene$ | 0.400 |
| $2 NO_2^- + 3 H_2O + 4 e \rightleftharpoons N_2O + 6 OH^-$ | 0.15 | $Tc^{2+} + 2 e \rightleftharpoons Tc$ | 0.400 |
| $Sn^{4+} + 2 e \rightleftharpoons Sn^{2+}$ | 0.151 | $O_2 + 2 H_2O + 4 e \rightleftharpoons 4 OH^-$ | 0.401 |
| $Sb_2O_3 + 6 H^+ + 6 e \rightleftharpoons 2 Sb + 3 H_2O$ | 0.152 | $AgOCN + e \rightleftharpoons Ag + OCN^-$ | 0.41 |
| $Cu^{2+} + e \rightleftharpoons Cu^+$ | 0.153 | $[RhCl_6]^{3-} + 3 e \rightleftharpoons Rh + 6 Cl^-$ | 0.431 |
| $BiOCl + 2 H^+ + 3 e \rightleftharpoons Bi + Cl^- + H_2O$ | 0.1583 | $Ag_2CrO_4 + 2 e \rightleftharpoons 2 Ag + CrO_4^{2-}$ | 0.4470 |
| $Bi(Cl)_4^- + 3 e \rightleftharpoons Bi + 4 Cl^-$ | 0.16 | $H_2SO_3 + 4 H^+ + 4 e \rightleftharpoons S + 3 H_2O$ | 0.449 |
| $Co(OH)_3 + e \rightleftharpoons Co(OH)_2 + OH^-$ | 0.17 | $Ru^{2+} + 2 e \rightleftharpoons Ru$ | 0.455 |
| $SO_4^{2-} + 4 H^+ + 2 e \rightleftharpoons H_2SO_3 + H_2O$ | 0.172 | $Ag_2MoO_4 + 2 e \rightleftharpoons 2 Ag + MoO_4^{2-}$ | 0.4573 |
| $SbO^+ + 2 H^+ + 3 e \rightleftharpoons Sb + 2 H_2O$ | 0.212 | $Ag_2C_2O_4 + 2 e \rightleftharpoons 2 Ag + C_2O_4^{2-}$ | 0.4647 |
| $AgCl + e \rightleftharpoons Ag + Cl^-$ | 0.22233 | $Ag_2WO_4 + 2 e \rightleftharpoons 2 Ag + WO_4^{2-}$ | 0.4660 |
| $As_2O_3 + 6 H^+ + 6 e \rightleftharpoons 2 As + 3 H_2O$ | 0.234 | $Ag_2CO_3 + 2 e \rightleftharpoons 2 Ag + CO_3^{2-}$ | 0.47 |
| Calomel electrode, saturated NaCl (SSCE) | 0.2360 | $TeO_4^- + 8 H^+ + 7 e \rightleftharpoons Te + 4 H_2O$ | 0.472 |
| $Ge^{2+} + 2 e \rightleftharpoons Ge$ | 0.24 | $IO^- + H_2O + 2 e \rightleftharpoons I^- + 2 OH^-$ | 0.485 |
| Calomel electrode, saturated KCl | 0.2412 | | |

## Table 3—20A (continued)
## REDUCTION REACTIONS HAVING E° VALUES MORE POSITIVE THAN THAT OF THE STANDARD HYDROGEN ELECTRODE

| Reaction | E°, V | Reaction | E°, V |
|---|---|---|---|
| $NiO_2 + 2 H_2O + 2 e \rightleftharpoons Ni(OH)_2 + 2 OH^-$ | 0.490 | $2 NO_3^- + 4 H^+ + 2 e \rightleftharpoons N_2O_4 + 2 H_2O$ | 0.803 |
| $ReO_4^- + 4 H^+ + 3 e \rightleftharpoons ReO_2 + 2 H_2O$ | 0.510 | $ClO^- + H_2O + 2 e \rightleftharpoons Cl^- + 2 OH^-$ | 0.841 |
| $Hg_2(ac)_2 + 2 e \rightleftharpoons 2 Hg + 2 (ac)^-$ | 0.51163 | $OsO_4 + 8 H^+ + 8 e \rightleftharpoons Os + 4 H_2O$ | 0.85 |
| $Cu^+ + e \rightleftharpoons Cu$ | 0.521 | $Hg^{2+} + 2 e \rightleftharpoons Hg$ | 0.851 |
| $I_2 + 2 e \rightleftharpoons 2 I^-$ | 0.5355 | $AuBr_4^- + 3 e \rightleftharpoons Au + 4 Br^-$ | 0.854 |
| $I_3^- + 2 e \rightleftharpoons 3 I^-$ | 0.536 | $SiO_2(quartz) + 4 H^+ + 4 e \rightleftharpoons Si + 2 H_2O$ | 0.857 |
| $AgBrO_3 + e \rightleftharpoons Ag + BrO_3^-$ | 0.546 | $2 HNO_2 + 4 H^+ + 4 e \rightleftharpoons H_2N_2O_2 + H_2O$ | 0.86 |
| $MnO_4^- + e \rightleftharpoons MnO_4^{2-}$ | 0.558 | $[IrCl_6]^{2-} + e \rightleftharpoons [IrCl_6]^{3-}$ | 0.8665 |
| $H_3AsO_4 + 2 H^+ + 2 e^- \rightleftharpoons HAsO_2 + 2 H_2O$ | 0.560 | $N_2O_4 + 2 e \rightleftharpoons 2 NO_2^-$ | 0.867 |
| $S_2O_6^{2-} + 4 H^+ + 2 e \rightleftharpoons 2 H_2SO_3$ | 0.564 | $HO_2^- + H_2O + 2 e \rightleftharpoons 3 OH^-$ | 0.878 |
| $AgNO_2 + e \rightleftharpoons Ag + NO_2^-$ | 0.564 | $2 Hg^{2+} + 2 e \rightleftharpoons Hg_2^{2+}$ | 0.920 |
| $Te^{4+} + 4 e \rightleftharpoons Te$ | 0.568 | $NO_3^- + 3 H^+ + 2 e \rightleftharpoons HNO_2 + H_2O$ | 0.934 |
| $Sb_2O_5 + 6 H^+ + 4 e \rightleftharpoons 2 SbO^+ + 3 H_2O$ | 0.581 | $Pd^{2+} + 2 e \rightleftharpoons Pd$ | 0.951 |
| $RuO_4^- + e \rightleftharpoons RuO_4^{2+}$ | 0.59 | $ClO_2(aq) + e \rightleftharpoons ClO_2^-$ | 0.954 |
| $[PdCl_4]^{2-} + 2 e \rightleftharpoons Pd + 4 Cl^-$ | 0.591 | $NO_3^- + 4 H^+ + 3 e \rightleftharpoons NO + 2 H_2O$ | 0.957 |
| $TeO_2 + 4 H^+ + 4 e \rightleftharpoons Te + 2 H_2O$ | 0.593 | $AuBr_2^- + e \rightleftharpoons Au + 2 Br^-$ | 0.959 |
| $MnO_4^- + 2 H_2O + 3 e \rightleftharpoons MnO_2 + 4 OH^-$ | 0.595 | $HNO_2 + H^+ + e \rightleftharpoons NO + H_2O$ | 0.983 |
| $Rh^{2+} + 2 e \rightleftharpoons Rh$ | 0.600 | $HIO + H^+ + 2 e \rightleftharpoons I^- + H_2O$ | 0.987 |
| $Rh^+ + e \rightleftharpoons Rh$ | 0.600 | $VO_2^+ + 2 H^+ + e \rightleftharpoons VO^{2+} + H_2O$ | 0.991 |
| $MnO_4^{2-} + 2 H_2O + 2 e \rightleftharpoons MnO_2 + 4 OH^-$ | 0.60 | $RuO_4 + e \rightleftharpoons RuO_4^-$ | 1.00 |
| $2 AgO + H_2O + 2 e \rightleftharpoons Ag_2O + 2 OH^-$ | 0.607 | $V(OH)_4^+ + 2 H^+ + e \rightleftharpoons VO^{2+} + 3 H_2O$ | 1.00 |
| $BrO_3^- + 3 H_2O + 6 e \rightleftharpoons Br^- + 6 OH^-$ | 0.61 | $AuCl_4^- + 3 e \rightleftharpoons Au + 4 Cl^-$ | 1.002 |
| $UO_2^+ + 4 H^+ + e \rightleftharpoons U^{4+} + 2 H_2O$ | 0.612 | $Pu^{4+} + e \rightleftharpoons Pu^{3+}$ | 1.006 |
| $Hg_2SO_4 + 2 e \rightleftharpoons 2 Hg + SO_4^{2-}$ | 0.6125 | $H_6TeO_6 + 2 H^+ + 2 e \rightleftharpoons TeO_2 + 4 H_2O$ | 1.02 |
| $ClO_3^- + 3 H_2O + 6 e \rightleftharpoons Cl^- + 6 OH^-$ | 0.62 | $N_2O_4 + 4 H^+ + 4 e \rightleftharpoons 2 NO + 2 H_2O$ | 1.035 |
| $Hg_2HPO_4 + 2 e \rightleftharpoons 2 Hg + HPO_4^{2-}$ | 0.6359 | $[Fe(phen)_3]^{3+} + e \rightleftharpoons [Fe(phen)_3]^{2+}$ (1 (mol/ℓ $H_2SO_4$) | 1.06 |
| $Ag(ac) + e \rightleftharpoons Ag + (ac)^-$ | 0.643 | $PuO_2(OH)_2 + H^+ + e \rightleftharpoons PuO_2OH + H_2O$ | 1.062 |
| $Sb_2O_5(valentinite) + 4 H^+ + 4 e \rightleftharpoons Sb_2O_3 + 2 H_2O$ | 0.649 | $N_2O_4 + 2 H^+ + 2 e \rightleftharpoons 2 HNO_2$ | 1.065 |
| $Ag_2SO_4 + 2 e \rightleftharpoons 2 Ag + SO_4^{2-}$ | 0.654 | $Br_2(1) + 2 e \rightleftharpoons 2 Br^-$ | 1.066 |
| $ClO_2^- + H_2O + 2 e \rightleftharpoons ClO^- + 2 OH^-$ | 0.66 | $IO_3^- + 6 H^+ + 6 e \rightleftharpoons I^- + 3 H_2O$ | 1.085 |
| $Sb_2O_5(senarmontite) + 4 H^+ + 4 e \rightleftharpoons Sb_2O_3 + 2 H_2O$ | 0.671 | $Br_2(aq) + 2 e \rightleftharpoons 2 Br^-$ | 1.0873 |
| $[PtCl_6]^{2-} + 2 e \rightleftharpoons [PtCl_4]^{2-} + 2 Cl^-$ | 0.68 | $Pu^{5+} + e \rightleftharpoons Pu^{4+}$ | 1.099 |
| $O_2 + 2 H^+ + 2 e \rightleftharpoons H_2O_2$ | 0.695 | $Cu^{2+} + 2 CN^- + e \rightleftharpoons [Cu(CN)_2]^-$ | |
| p-benzoquinone + 2 H⁺ + 2 e ⇌ hydroquinone | 0.6992 | $Pt^{2+} + 2 e \rightleftharpoons Pt$ | 1.118 |
| $H_3IO_6 + 2 e \rightleftharpoons IO_3^- + 3 OH^-$ | 0.7 | $RuO_2 + 4 H^+ + 2 e \rightleftharpoons Ru^{2+} + 2 H_2O$ | 1.120 |
| $Ag_2O_3 + H_2O + 2 e \rightleftharpoons 2 AgO + 2 OH^-$ | 0.739 | $[Fe(phenanthroline)_3]^{3+} + e \rightleftharpoons [Fe(phen)_3]^{2+}$ | 1.147 |
| $[PtCl_4]^{2-} + 2 e \rightleftharpoons Pt + 4 Cl^-$ | 0.755 | $SeO_4^{2-} + 4 H^+ + 2 e \rightleftharpoons H_2SeO_3 + H_2O$ | 1.151 |
| $Rh^{3+} + 3 e \rightleftharpoons Rh$ | 0.758 | $ClO_3^- + 2 H^+ + e \rightleftharpoons ClO_2 + H_2O$ | 1.152 |
| $ClO_2^- + 2 H_2O + 4 e \rightleftharpoons Cl^- + 4 OH^-$ | 0.76 | $Ir^{3+} + 3 e \rightleftharpoons Ir$ | 1.156 |
| $2 NO + H_2O + 2 e \rightleftharpoons N_2O + 2 OH^-$ | 0.76 | $ClO_4^- + 2 H^+ + 2 e \rightleftharpoons ClO_3^- + H_2O$ | 1.189 |
| $BrO^- + H_2O + 2 e \rightleftharpoons Br^- + 2 OH^-$ | 0.761 | $2 IO_3^- + 12 H^+ + 10 e \rightleftharpoons I_2 6 H_2O$ | 1.195 |
| $ReO_4^- + 2 H^+ + e \rightleftharpoons ReO_3 + H_2O$ | 0.768 | $ClO_3^- + 3 H^+ + 2 e \rightleftharpoons HClO_2 + H_2O$ | 1.214 |
| $(CNS)_2 + 2 e \rightleftharpoons 2 CNS^-$ | 0.77 | $MnO_2 + 4 H^+ + 2 e \rightleftharpoons Mn^{2+} + 2 H_2O$ | 1.224 |
| $[IrCl_6]^{3-} + 3 e \rightleftharpoons Ir + 6 Cl^-$ | 0.77 | $O_2 + 4 H^+ + 4 e \rightleftharpoons 2 H_2O$ | 1.229 |
| $Fe^{3+} + e \rightleftharpoons Fe^{2+}$ | 0.771 | $Cr_2O_7^{2-} + 14 H^+ + 6 e \rightleftharpoons 2 Cr^{3+} + 7 H_2O$ | 1.232 |
| $Ag(F) + e \rightleftharpoons Ag + F^-$ | 0.779 | $O_3 + H_2O + 2 e \rightleftharpoons O_2 + 2OH^-$ | 1.24 |
| $TcO_4^- + 4 H^+ + 3 e \rightleftharpoons TcO_2 + 2 H_2O$ | 0.782 | $Tl^{3+} + 2 e \rightleftharpoons Tl^+$ | 1.252 |
| $Hg_2^{2+} + 2 e \rightleftharpoons 2 Hg$ | 0.7973 | $N_2H_5^+ + 3 H^+ + 2 e \rightleftharpoons 2 NH_4^+$ | 1.275 |
| $Ag^+ + e \rightleftharpoons Ag$ | 0.7996 | $ClO_2 + H^+ + e \rightleftharpoons HClO_2$ | 1.277 |
| | | $[PdCl_6]^{2-} + 2 e \rightleftharpoons [PdCl_4]^{2-} + 2 Cl^-$ | 1.288 |
| | | $2 HNO_2 + 4 H^+ + 4 e \rightleftharpoons N_2O + 3 H_2O$ | 1.297 |

## Table 3—20A (continued)

# REDUCTION REACTIONS HAVING E° VALUES MORE POSITIVE THAN THAT OF THE STANDARD HYDROGEN ELECTRODE

| Reaction | E°, V | Reaction | E°, V |
|---|---|---|---|
| $PuO_2(OH)_2 + 2 H^+ + 2 e \rightleftharpoons Pu(OH)_4$ | 1.325 | $H_5IO_6 + H^+ + 2 e \rightleftharpoons IO_3 + 3 H_2O$ | 1.601 |
| $HBrO + H^+ + 2 e \rightleftharpoons Br^- + H_2O$ | 1.331 | $Ce^{4+} + e \rightleftharpoons Ce^{3+}$ | 1.61 |
| $HCrO_4^- + 7 H^+ + 3 e \rightleftharpoons Cr^{3+} + 4 H_2O$ | 1.350 | $HClO + H^+ + e \rightleftharpoons 1/2 Cl_2 + H_2O$ | 1.611 |
| $Cl_2(g) + 2 e \rightleftharpoons Cl^-$ | 1.35827 | $HClO_2 + 3 H^+ + 3 e \rightleftharpoons 1/2 Cl_2 + 2 H_2O$ | 1.628 |
| $ClO_4^- + 8 H^+ + 8 e \rightleftharpoons Cl^- + 4 H_2O$ | 1.389 | $HClO_2 + 2 H^+ + 2 e \rightleftharpoons HClO + H_2O$ | 1.645 |
| $ClO_4^- + 8 H^+ + 7 e \rightleftharpoons {}^1/_2 Cl_2 + 4 H_2O$ | 1.39 | $NiO_2 + 4 H^+ + 2 e \rightleftharpoons Ni^{2+} + 2 H_2O$ | 1.678 |
| $Au^{3+} + 2 e \rightleftharpoons Au^+$ | 1.401 | $MnO_4^- + 4 H^+ + 3 e \rightleftharpoons MnO_2 + 2 H_2O$ | 1.679 |
| $2 NH_3OH^+ + H^+ + 2 e \rightleftharpoons N_2H_5^+ + 2 H_2O$ | 1.42 | $PbO_2 + SO_4^{2-} + 4 H^+ + 2 e \rightleftharpoons PbSO_4 + 2 H_2O$ | 1.6913 |
| $BrO_3^- + 6 H^+ + 6 e \rightleftharpoons Br^- + 3 H_2O$ | 1.423 | $Au^+ + e \rightleftharpoons Au$ | 1.692 |
| $2 HIO + 2 H^+ + 2 e \rightleftharpoons I_2 + 2 H_2O$ | 1.439 | $CeOH^{3+} + H^+ + e \rightleftharpoons Ce^{3+} + H_2O$ | 1.715 |
| $Au(OH)_3 + 3 H^+ + 3 e \rightleftharpoons Au^- + 3 H_2O$ | 1.45 | $N_2O + 2 H^+ + 2 e \rightleftharpoons N_2 + H_2O$ | 1.766 |
| $3 IO_3^- + 6 H^+ + 6 e \rightleftharpoons Cl^- + 3 H_2O$ | 1.451 | $H_2O_2 + 2 H^+ + 2 e \rightleftharpoons 2 H_2O$ | 1.776 |
| $PbO_2 + 4 H^+ + 2 e \rightleftharpoons Pb^{2+} + 2 H_2O$ | 1.455 | $Co^{3+} + e \rightleftharpoons Co^{2+} (2 mol/\ell H_2SO_4)$ | 1.83 |
| $ClO_3^- + 6 H^+ + 5 e \rightleftharpoons {}^1/_2 Cl_2 + 3 H_2O$ | 1.47 | $Ag^{2+} + e \rightleftharpoons Ag^-$ | 1.980 |
| $BrO_3^- + 6 H^+ + 5 e \rightleftharpoons {}^1/_2 Br_2 + 3 H_2O$ | 1.482 | $S_2O_8^{2-} + 2 e \rightleftharpoons 2 SO_4^{2-}$ | 2.010 |
| $HClO + H^+ + 2 e \rightleftharpoons Cl^- + H_2O$ | 1.482 | $OH + e \rightleftharpoons OH^-$ | 2.02 |
| $HO_2 + H^+ + e \rightleftharpoons H_2O_2$ | 1.495 | $O_3 + 2 H^+ + 2 e \rightleftharpoons O_2 + H_2O$ | 2.076 |
| $Au^{3+} + 3 e \rightleftharpoons Au$ | 1.498 | $S_2O_8^{2-} + 2 H^+ + 2 e \rightleftharpoons 2 HSO_4$ | 2.123 |
| $MnO_4^- + 8 H^+ + 5 e \rightleftharpoons Mn^{2+} + 4 H_2O$ | 1.507 | $F_2O + 2 H^+ + 4 e \rightleftharpoons H_2O + 2 F^-$ | 2.153 |
| $Mn^{3+} + e \rightleftharpoons Mn^{2+}$ | 1.5415 | $FeO_4^{2-} + 8 H^+ + 3 e \rightleftharpoons Fe^{3+} + 4 H_2O$ | 2.20 |
| $HClO_2 + 3 H^+ + 4 e \rightleftharpoons Cl^- + 2 H_2O$ | 1.570 | $O(g) + 2 H^+ + 2 e \rightleftharpoons H_2O$ | 2.421 |
| $HBrO + H^+ + e \rightleftharpoons {}^1/_2 Br_2(aq) + H_2O$ | 1.574 | $H_2N_2O_2 + 2 H^+ + 2 e \rightleftharpoons N_2 + 2 H_2O$ | 2.65 |
| $2 NO + 2 H^+ + 2 e \rightleftharpoons N_2O + H_2O$ | 1.591 | $F_2 + 2 e \rightleftharpoons 2 F^-$ | 2.866 |
| $Bi_2O_4 + 4 H^+ + 2 e \rightleftharpoons 2 BiO^+ + 2 H_2O$ | 1.593 | $F_2 + 2 H^+ + 2 e \rightleftharpoons 2 HF$ | 3.053 |
| $HBrO + H^+ + e \rightleftharpoons 1/2 Br_2(\ell) + H_2O$ | 1.596 | | |

## Table 3—20B
## REDUCTION REACTIONS HAVING E° VALUES MORE NEGATIVE THAN THAT OF THE STANDARD HYDROGEN ELECTRODE

| Reaction | $E°$, V | Reaction | $E°$, V |
|---|---|---|---|
| $2\ H^+ + 2\ e \rightleftharpoons H_2$ | 0.00000 | $In^{2+} + e \rightleftharpoons In^+$ | −0.40 |
| $AgCN + e \rightleftharpoons Ag + CN^-$ | −0.017 | $Cd^{2+} + 2\ e \rightleftharpoons Cd$ | −0.4030 |
| $2\ WO_3 + 2\ H^+ + 2\ e \rightleftharpoons W_2O_5 + H_2O$ | −0.029 | $Cr^{3+} + e \rightleftharpoons Cr^{2+}$ | −0.407 |
| $W_2O_5 + 2\ H^+ + 2\ e \rightleftharpoons 2\ WO_2 + H_2O$ | −0.031 | $2\ S + 2\ e \rightleftharpoons S_2^{2-}$ | −0.42836 |
| $D^+ + e \rightleftharpoons 1/2D_2$ | −0.0034 | $Tl_2SO_4 + 2\ e \rightleftharpoons Tl + SO_4^{2-}$ | −0.4360 |
| $Ag_2S + 2\ H^+ + 2\ e \rightleftharpoons 2\ Ag + H_2S$ | −0.0366 | $In^{3+} + 2\ e \rightleftharpoons In^+$ | −0.443 |
| $Fe^{3+} + 3\ e \rightleftharpoons Fe$ | −0.037 | $Fe^{2+} + 2\ e \rightleftharpoons Fe$ | −0.447 |
| $Hg_2I_2 + 2\ e \rightleftharpoons 2\ Hg + 2\ I^-$ | −0.0405 | $H_3PO_3 + 3\ H^+ + 3\ e \rightleftharpoons P + 3\ H_2O$ | −0.454 |
| $2\ D^+ + 2\ e \rightleftharpoons D_2$ | −0.044 | $Bi_2O_3 + 3\ H_2O + 6\ e \rightleftharpoons 2\ Bi + 6\ OH^-$ | −0.46 |
| $Tl(OH)_3 + 2\ e \rightleftharpoons TlOH + 2\ OH^-$ | −0.05 | $NO_2^- + H_2O + e \rightleftharpoons NO + 2\ OH$ | −0.46 |
| $TiOH^{3+} + H^+ + e \rightleftharpoons Ti^{3+} + H_2O$ | −0.055 | $PbHPO_4 + 2\ e \rightleftharpoons Pb + HPO_4^{2-}$ | −0.465 |
| $2\ H_2SO_3 + H^+ + 2\ e \rightleftharpoons HS_2O_4^- + 2\ H_2O$ | −0.056 | $S + 2\ e \rightleftharpoons S^{2-}$ | −0.47627 |
| $P(white) + 3\ H^+ + 3\ e \rightleftharpoons PH_3(g)$ | −0.063 | $S + H_2O + 2\ e \rightleftharpoons HS^- + OH^-$ | −0.478 |
| $O_2^- + H_2O + 2\ e \rightleftharpoons HO_2^- + OH^-$ | −0.076 | $In^{3+} + e \rightleftharpoons In^{2+}$ | −0.49 |
| $2\ Cu(OH)_2 + 2\ e \rightleftharpoons Cu_2O + 2\ OH^- + H_2O$ | −0.080 | $H_3PO_3 + 2\ H^+ + 2\ e \rightleftharpoons H_3PO_2 + H_2O$ | −0.499 |
| $WO_3 + 6\ H^+ + 6\ e \rightleftharpoons W + 3\ H_2O$ | −0.090 | $TiO_2 + 4\ H^+ + 2\ e \rightleftharpoons Ti^{2+} + 2\ H_2O$ | −0.502 |
| $P(red) + 3\ H^+ + 3\ e \rightleftharpoons PH_3(g)$ | −0.111 | $H_3PO_2 + H^+ + e \rightleftharpoons P + 2\ H_2O$ | −0.508 |
| $GeO_2 + 2\ H^+ + 2\ e \rightleftharpoons GeO + H_2O$ | −0.118 | $Sb + 3\ H^+ + 3\ e \rightleftharpoons SbH_3$ | −0.510 |
| $WO_2 + 4\ H^+ + 4\ e \rightleftharpoons W + 2\ H_2O$ | −0.119 | $HPbO_2^- + H_2O + 2\ e \rightleftharpoons Pb + 3\ OH^-$ | −0.537 |
| $Pb^{2+} + 2\ e \rightleftharpoons Pb(Hg)$ | −0.1205 | $TlCl + e \rightleftharpoons Tl + Cl^-$ | −0.5568 |
| $Pb^{2+} + 2\ e \rightleftharpoons Pb$ | −0.1262 | $Ga^{3+} + 3\ e \rightleftharpoons Ga$ | −0.560 |
| $CrO_4^{2-} + 4\ H_2O + 3\ e \rightleftharpoons Cr(OH)_3 + 5\ OH^-$ | −0.13 | $Fe(OH)_3 + e \rightleftharpoons Fe(OH)_2 + OH^-$ | −0.56 |
| $Sn^{2-} + 2\ e \rightleftharpoons Sn$ | −0.1375 | $TeO_3^{2-} + 3\ H_2O + 4\ e \rightleftharpoons Te + 6\ OH^-$ | −0.57 |
| $In^+ + e \rightleftharpoons In$ | −0.14 | $2\ SO_3^{2-} + 3\ H_2O + 4\ e \rightleftharpoons S_2O_3^{2-} + 6\ OH^-$ | −0.571 |
| $O_2 + 2\ H_2O + 2\ e \rightleftharpoons H_2O_2 + 2\ OH^-$ | −0.146 | $PbO + H_2O + 2\ e \rightleftharpoons Pb + 2\ OH^-$ | −0.580 |
| $AgI + e \rightleftharpoons Ag + I^-$ | −0.15224 | $ReO_2^- + 4\ H_2O + 7\ e \rightleftharpoons Re + 8\ OH^-$ | −0.584 |
| $2\ NO_2^- + 2\ H_2O + 4\ e \rightleftharpoons N_2O_2^{2-} + 4\ OH^-$ | −0.18 | $SbO_3^- + H_2O + 2\ e \rightleftharpoons SbO_2^- + 2\ OH^-$ | −0.59 |
| $H_2GeO_3 + 4\ H^+ + 4\ e \rightleftharpoons Ge + 3\ H_2O$ | −0.182 | $U^{4+} + e \rightleftharpoons U^{3+}$ | −0.607 |
| $CO_2 + 2\ H^+ + 2\ e \rightleftharpoons HCOOH$ | −0.199 | $As + 3\ H^+ + 3\ e \rightleftharpoons AsH_3$ | −0.608 |
| $Mo^{3+} + 3\ e \rightleftharpoons Mo$ | −0.200 | $Nb_2O_5 + 10\ H^+ + 10\ e \rightleftharpoons 2\ Nb + 5\ H_2O$ | −0.644 |
| $2\ SO_3^{2-} + 4\ H^+ + 2\ e \rightleftharpoons S_2O_6^{2-} + H_2O$ | −0.22 | $TlBr + e \rightleftharpoons Tl + Br^-$ | −0.658 |
| $Cu(OH)_2 + 2\ e \rightleftharpoons Cu + 2\ OH^-$ | −0.222 | $SbO_2^- + 2\ H_2O + 3\ e \rightleftharpoons Sb + 4\ OH^-$ | −0.66 |
| $CdSO_4 + 2\ e \rightleftharpoons Cd + SO_4^{2-}$ | −0.246 | $AsO_2^- + 2\ H_2O + 3\ e \rightleftharpoons As + 4\ OH^-$ | −0.68 |
| $V(OH)_4^+ + 4\ H^+ + 5\ e \rightleftharpoons V + 4\ H_2O$ | −0.254 | $Ag_2S + 2\ e \rightleftharpoons 2\ Ag + S^{2-}$ | −0.691 |
| $V^{3+} + e \rightleftharpoons V^{2+}$ | −0.255 | $AsO_4^{3-} + 2\ H_2O + 2\ e \rightleftharpoons AsO_2^- + 4\ OH^-$ | −0.71 |
| $Ni^{2+} + 2\ e \rightleftharpoons Ni$ | −0.257 | $Ni(OH)_2 + 2\ e \rightleftharpoons Ni + 2\ OH^-$ | −0.72 |
| $PbCl_2 + 2\ e \rightleftharpoons Pb + 2\ Cl^-$ | −0.2675 | $Co(OH)_2 + 2\ e \rightleftharpoons Co + 2\ OH^-$ | −0.73 |
| $H_3PO_4 + 2\ H^+ + 2\ e \rightleftharpoons H_3PO_3 + H_2O$ | −0.276 | $H_2SeO_3 + 4\ H^+ + 4\ e \rightleftharpoons Se + 3\ H_2O$ | −0.74 |
| $Co^{2+} + 2\ e \rightleftharpoons Co$ | −0.28 | $Cr^{3+} + 3\ e \rightleftharpoons Cr$ | −0.744 |
| $PbBr_2 + 2\ e \rightleftharpoons Pb + 2\ Br^-$ | −0.284 | $Ta_2O_5 + 10\ H^+ + 10\ e \rightleftharpoons 2\ Ta + 5\ H_2O$ | −0.750 |
| $Tl^+ + e \rightleftharpoons Tl(Hg)$ | −0.3338 | $TlI + e \rightleftharpoons Tl + I^-$ | −0.752 |
| $Tl^+ + e \rightleftharpoons Tl$ | −0.336 | $Zn^{2+} + 2\ e \rightleftharpoons Zn$ | −0.7618 |
| $In^{3+} + 3\ e \rightleftharpoons In$ | −0.3382 | $Zn^{2+} + 2\ e \rightleftharpoons Zn(Hg)$ | −0.7628 |
| $TlOH + e \rightleftharpoons Tl + OH^-$ | −0.34 | $Te + 2\ H^+ + 2\ e \rightleftharpoons H_2Te$ | −0.793 |
| $PbF_2 + 2\ e \rightleftharpoons Pb + 2\ F^-$ | −0.3444 | $ZnSO_4 \cdot 7H_2O + 2\ e \rightleftharpoons Zn(Hg) + SO_4^{2-}$ (Sat'd $ZnSO_4$) | −0.7993 |
| $PbSO_4 + 2\ e \rightleftharpoons Pb(Hg) + SO_4^{2-}$ | −0.3505 | $Cd(OH)_2 + 2\ e \rightleftharpoons Cd(Hg) + 2\ OH^-$ | −0.809 |
| $Cd^{2+} + 2\ e \rightleftharpoons Cd(Hg)$ | −0.3521 | $2\ H_2O + 2\ e \rightleftharpoons H_2 + 2\ OH^-$ | −0.8277 |
| $PbSO_4 + 2\ e \rightleftharpoons Pb + SO_4^{2-}$ | −0.3588 | $2\ NO_3^- + 2\ H_2O + 2\ e \rightleftharpoons N_2O_4 + 4\ OH^-$ | −0.85 |
| $Cu_2O + H_2O + 2\ e \rightleftharpoons 2\ Cu + 2\ OH^-$ | −0.360 | $H_3BO_3 + 3\ H^+ + 3\ e \rightleftharpoons B + 3\ H_2O$ | −0.8698 |
| $Eu^{3+} + e \rightleftharpoons Eu^{2+}$ | −0.36 | $P + 3\ H_2O + 3\ e \rightleftharpoons PH_3(g) + 3\ OH^-$ | −0.87 |
| $PbI_2 + 2\ e \rightleftharpoons Pb + 2\ I^-$ | −0.365 | $HSnO_2^- + H_2O + 2\ e \rightleftharpoons Sn + 3\ OH^-$ | −0.909 |
| $SeO_3^{2-} + 3\ H_2O + 4\ e \rightleftharpoons Se + 6\ OH^-$ | −0.366 | $Cr^{2+} + 2\ e \rightleftharpoons Cr$ | −0.913 |
| $Ti^{3+} + e \rightleftharpoons Ti^{2+}$ | −0.368 | | |
| $Se + 2\ H^+ + 2\ e \rightleftharpoons H_2Se(aq)$ | −0.399 | | |

## Table 3—20B (continued)
## REDUCTION REACTIONS HAVING E° VALUES MORE NEGATIVE THAN THAT OF THE STANDARD HYDROGEN ELECTRODE

| Reaction | E°, V | Reaction | E°, V |
|---|---|---|---|
| $Se + 2\,e \rightleftharpoons Se^{2-}$ | $-0.924$ | $Np^{3+} + 3\,e \rightleftharpoons Np$ | $-1.856$ |
| $SO_4^{2-} + H_2O + 2\,e \rightleftharpoons SO_3^{2-} + 2\,OH^-$ | $-0.93$ | $Th^{4+} + 4\,e \rightleftharpoons Th$ | $-1.899$ |
| $Sn(OH)_6^{2-} + 2\,e \rightleftharpoons HSnO_2^- + 3\,OH^- + H_2O$ | $-0.93$ | $Pu^{3+} + 3\,e \rightleftharpoons Pu$ | $-2.031$ |
| $NpO_2 + H_2O + H^+ + e \rightleftharpoons Np(OH)_3$ | $-0.962$ | $AlF_6^{3-} + 3\,e \rightleftharpoons Al + 6\,F^-$ | $-2.069$ |
| $PO_4^{3-} + 2\,H_2O + 2\,e \rightleftharpoons HPO_3^{2-} + 3\,OH^-$ | $-1.05$ | $Sc^{3+} + 3\,e \rightleftharpoons Sc$ | $-2.077$ |
| $Nb^{3+} + 3\,e \rightleftharpoons Nb$ | $-1.099$ | $H_2 + 2\,e \rightleftharpoons 2\,H^-$ | $-2.23$ |
| $2\,SO_3^{2-} + 2\,H_2O + 2\,e \rightleftharpoons S_2O_4^{2-} + 4\,OH^-$ | $-1.12$ | $H_2AlO_3^- + H_2O + 3\,e \rightleftharpoons Al + 4\,OH^-$ | $-2.33$ |
| $Te + 2\,e \rightleftharpoons Te^{2-}$ | $-1.143$ | $ZrO(OH)_2 + H_2O + 4\,e \rightleftharpoons Zr + 4\,OH^-$ | $-2.36$ |
| $V^{2+} + 2\,e \rightleftharpoons V$ | $-1.175$ | $Mg^{2+} + 2\,e \rightleftharpoons Mg$ | $-2.372$ |
| $Mn^{2+} + 2\,e \rightleftharpoons Mn$ | $-1.185$ | $Y^{3+} + 3\,e \rightleftharpoons Y$ | $-2.372$ |
| $CrO_2^- + 2\,H_2O + 3\,e \rightleftharpoons Cr + 4\,OH^-$ | $-1.2$ | $Eu^{3+} + 3\,e \rightleftharpoons Eu$ | $-2.407$ |
| $ZnO_2^- + 2\,H_2O + 2\,e \rightleftharpoons Zn + 4\,OH^-$ | $-1.215$ | $Nd^{3+} + 3\,e \rightleftharpoons Nd$ | $-2.431$ |
| $H_2GaO_3^- + H_2O + 3\,e \rightleftharpoons Ga + 4\,OH^-$ | $-1.219$ | $Th(OH)_4 + 4\,e \rightleftharpoons Th + 4\,OH^-$ | $-2.48$ |
| $H_2BO_3^- + 5\,H_2O + 8\,e \rightleftharpoons BH_4^- + 8\,OH^-$ | $-1.24$ | $Ce^{3+} + 3\,e \rightleftharpoons Ce$ | $-2.483$ |
| $SiF_6^{2-} + 4\,e \rightleftharpoons Si + 6\,F^-$ | $-1.24$ | $HfO(OH)_2 + H_2O + 4\,e \rightleftharpoons Hf + 4\,OH^-$ | $-2.50$ |
| $Ce^{3+} + 3\,e \rightleftharpoons Ce(Hg)$ | $-1.4373$ | $La^{3+} + 3\,e \rightleftharpoons La$ | $-2.522$ |
| $UO_2^+ + 4\,H^+ + 6\,e \rightleftharpoons U + 2\,H_2O$ | $-1.444$ | $Be_2O_3^{2-} + 3\,H_2O + 4\,e \rightleftharpoons 2\,Be + 6\,OH^-$ | $-2.63$ |
| $Cr(OH)_3 + 3\,e \rightleftharpoons Cr + 3\,OH^-$ | $-1.48$ | $Mg(OH)_2 + 2\,e \rightleftharpoons Mg + 2\,OH^-$ | $-2.690$ |
| $HfO_2 + 4\,H^+ + 4\,e \rightleftharpoons Hf + 2\,H_2O$ | $-1.505$ | $Mg^+ + e \rightleftharpoons Mg$ | $-2.70$ |
| $ZrO_2 + 4\,H^+ + 4\,e \rightleftharpoons Zr + 2\,H_2O$ | $-1.553$ | $Na^+ + e \rightleftharpoons Na$ | $-2.71$ |
| $Mn(OH)_2 + 2\,e \rightleftharpoons Mn + 2\,OH^-$ | $-1.56$ | $Ca^{2+} + 2\,e \rightleftharpoons Ca$ | $-2.868$ |
| $Ba^{2+} + 2\,e \rightleftharpoons Ba(Hg)$ | $-1.570$ | $Sr(OH)_2 + 2\,e \rightleftharpoons Sr + 2\,OH^-$ | $-2.88$ |
| $Ti^{2+} + 2\,e \rightleftharpoons Ti$ | $-1.630$ | $Sr^{2+} + 2\,e \rightleftharpoons Sr$ | $-2.89$ |
| $HPO_3^{2-} + 2\,H_2O + 2\,e \rightleftharpoons H_2PO_2^- + 3\,OH^-$ | $-1.65$ | $La(OH)_3 + 3\,e \rightleftharpoons La + 3\,OH^-$ | $-2.90$ |
| $Al^{3+} + 3\,e \rightleftharpoons Al$ | $-1.662$ | $Ba^{2+} + 2\,e \rightleftharpoons Ba$ | $-2.912$ |
| $SiO_3^{2-}\ H_2O + 4\,e \rightleftharpoons Si + 6\,OH^-$ | $-1.697$ | $Cs^+ + e \rightleftharpoons Cs$ | $-2.92$ |
| $HPO_3^{2-} + 2\,H_2O + 3\,e \rightleftharpoons P + 5\,OH^-$ | $-1.71$ | $K^+ + e \rightleftharpoons K$ | $-2.931$ |
| $HfO^{2+} + 2\,H^+ + 4\,e \rightleftharpoons Hf + H_2O$ | $-1.724$ | $Rb^+ + e \rightleftharpoons Rb$ | $-2.98$ |
| $ThO_2 + 4\,H^+ + 4\,e \rightleftharpoons Th + 2\,H_2O$ | $-1.789$ | $Ba(OH)_2 + 2\,e \rightleftharpoons Ba + 2\,OH^-$ | $-2.99$ |
| $H_2BO_3^- + H_2O + 3\,e \rightleftharpoons B + 4\,OH^-$ | $-1.79$ | $Ca(OH)_2 + 2\,e \rightleftharpoons Ca + 2\,OH^-$ | $-3.02$ |
| $Sr^{2+} + 2\,e \rightleftharpoons Sr(Hg)$ | $-1.793$ | $Li^+ + e \rightleftharpoons Li$ | $-3.0401$ |
| $U^{3+} + 3\,e \rightleftharpoons U$ | $-1.798$ | $3\,N_2 + 2\,H^+ + 2\,e \rightleftharpoons 2\,NH_3$ | $-3.09$ |
| $H_2PO_2^- + e \rightleftharpoons P + 2\,OH^-$ | $-1.82$ | $Eu^{2+} + 2\,e \rightleftharpoons Eu$ | $-3.395$ |
| $Be^{2+} + 2\,e \rightleftharpoons Be$ | $-1.847$ | $Ca^+ + e \rightleftharpoons Ca$ | $-3.80$ |
|  |  | $Sr^+ + e \rightleftharpoons Sr$ | $-4.10$ |

From Weast, R. C. Ed., *Handbook of Chemistry and Physics,* 69th ed., CRC Press, Boca Raton, Fla., 1988, D-151, 157.

# Section 4

## Conversion Tables

# Section 4

# CONVERSION TABLES

Table 4–1
### UNITS AND THEIR CONVERSION

## Policy

In each table in this handbook, the numerical values are preferably expressed in those units most commonly used by U.S. engineers and scientists working in the specific field, but sometimes SI metric units have also been added. In some cases two tables are given, one in English units, one in metric. In other tables parallel columns showing figures in both units are used, or the conversion factors are listed.

In a general materials handbook complete consistency in units, abbreviations, and symbols is hardly possible, or even desirable. Such consistency would quickly defeat the objective of providing quick access to numbers of maximum immediate usefulness. Within each special field of engineering and science, the technical societies and industry associations have developed certain uniform practices and standards; if tables and data are given only in units that are foreign to these prevailing standards, convenience is sacrificed.

The present edition of this handbook reflects the changes in abbreviations, symbols, and forms that are resulting from the efforts to reduce the diversity of practices from one specialty to another and from one nation to another. Recommendations of the International Organization for Standardization (ISO-R 1000) and of the "Metric Practice Guide," adopted by ASTM, NBS, APL, and others, have focused attention on the diversity of so-called standards.

Since the United States is the only major industrial nation that has not yet converted to metric units, some legal requirements in that direction are to be expected. It is now a contradiction to speak of the "English" system of units, and for. some time to come U.S. engineers and scientists must accommodate to a wide use of conversions from one set of units to another. The extensive conversion tables that follow are offered with this expectation.

In spite of major efforts to unify engineering practices, there are many good reasons for retaining several means of expressing a physical quantity. For ease of learning and communication a descriptive name is better than one arbitrarily assigned, such as Hz for cps, celsius for centigrade, and torr for mm Hg; an opposite trend is prevalent at this time. Numerical scales directly related to the physical phenomena and to the method of their measurement have an advantage in the laboratory or field and will not soon be abandoned. Examples are barometric pressure in millimeters or inches of mercury, viscosity in seconds Saybolt, the calorie or the Btu, and even the "coefficients" of expansion, friction, diffusion, attenuation, and reflection. Symbols, abbreviations, and even the units themselves are not infrequently subject to change; note, for example, the new preferred *dB* in place of the well-established *db*; elimination of widely used abbreviations, such as kwh, cps, gpm, cc, and psi; and revised values for the second, the calorie, or the atomic weights. Users of this handbook are invited to call attention to places where consistency could be improved without sacrificing the objectives.

For units that might be assigned more than one value, we prefer the thermochemical gram-calorie (4.184 J), the thermochemical Btu (1054.35 J), the avoirdupois pound and ounce, the statute mile (5,280 ft), the short ton (2,000 lb), the U.S. liquid gallon (231 in.$^3$), and the electrical horsepower (746 W).

Both a special condensed version of conversion factors for quick reference and a more extended table have been included. Certain specialized conversion factors and tables have also been included. This results in some redundancy, but the aim is to provide ease and utility in making unit conversions.

## The Metric International System (SI)

Moves toward an international system of metric units are now following each other in quick

**Table 4–1 (continued)**
**UNITS AND THEIR CONVERSION**

succession, so a table of conversion factors for the most common units is given herewith (see also Table 4–3). Perhaps the most definite are the moves toward the SI standards already initiated by the National Bureau of Standards, the various military services, the National Aeronautics and Space Administration, and other U.S. Government research groups. The American Society for Testing and Materials has declared in favor of SI units and will give other units only a secondary place in all newly issued ASTM Standards.[a] Other major engineering societies have committees to explore the adoption of SI units and are holding many meetings for discussion among members.

Whatever the decisions about converting to the metric system, the actual process will require many years, as can readily be seen from the experiences of other countries; in Great Britain, for example, even the single conversion to decimal monetary units and coinage moves very slowly. The practices and standards among the metric system countries are far from uniform; no real international system exists among them.

Mere conversion of present U.S. specifications, drawings, tools, machines, and stock sizes to equivalent metric units (so-called "soft" conversion) will not in any sense result in an "international" system. Instead, a "hard" conversion representing the abandonment of the 1/2-fractional system in favor of a 1/10-fractional system is necessary to attain the real advantages of the metric system. This means resizing of all round and sheet stock, lumber, bolts, screws, nails, wires,

gears, containers, modules, and subassemblies, plus all the tools and machines related thereto. A long period of double stocking must follow. The entire change is made the more difficult by the great penetration of U.S. products and materials into the markets of the world, e.g., airplanes and military equipment, production, and construction machinery. This is not to mention the problem of the individual engineer, technician, and user, who visualizes all his size relationships in inches and feet and his weights in pounds. Realistically, more than one generation will be required for the educational conversion alone.

In presenting data in international standard metric units throughout this edition, the practices and forms used in the "Metric Practice Guide" have been carefully followed.[a] Certain conventions used in the "Metric Practice Guide" are not consistent with those originally adopted for this handbook, nor with ANSI standards. Special attention is directed to the following conventions:

1.  For degrees Kelvin the degree symbol is omitted; for example, 50 K, not $50°K$.

2.  For multiplication a center point is used; for example, the unit of dynamic viscosity is abbreviated as $N \cdot s/m^2$, not $N \, s/m^2$ or $N \times s/m^2$.

3.  Symbols for SI units are not capitalized unless the unit is derived from a proper name, as N for Sir Isaac Newton; however, *unabbreviated* units are not capitalized, such as newton, kelvin, hertz.

### Table 4–1 (continued)
### UNITS AND THEIR CONVERSION

**Conversion Factors to SI Standard Units[b]**

| To convert | To | Multiply by | To convert | To | Multiply by |
|---|---|---|---|---|---|
| Acceleration | | | Power | | |
| feet/second$^2$ | meters/second$^2$ | 0.3048 | Btu/second | watt | 1,054.350 |
| Area | | | foot-pounds/second | watt | 1.355818 |
| square feet | square meters | 0.09290304 | horsepower | watt | 746. |
| Energy | | | Pressure | | |
| Btu (mean) | joule | 1,055.87 | atmosphere | newtons/meter$^2$ | 101,325.0 |
| calorie (mean) | joule | 4.19002 | bar | newtons/meter$^2$ | 100,000. |
| electron volt | joule | $1.60210 \times 10^{-19}$ | kilograms/centimeter$^2$ | newtons/meter$^2$ | 98,066.50 |
| foot-pound | joule | 1.355818 | pounds/inch$^2$ | newtons/meter$^2$ | 6,894.757 |
| watthour | joule | 3,600. | torr (mm Hg, 0°C) | newtons/meter$^2$ | 133.322 |
| Force | | | Viscosity | | |
| dyne | newton | 0.00001 | centipoise | newton-second/meter$^2$ | 0.001 |
| kilogram | newton | 9.80665 | pounds/foot second | newton-second/meter$^2$ | 1.488164 |
| pound | newton | 4.448222 | Volume | | |
| Length | | | cubic foot | cubic meter | 0.02831685 |
| foot | meter | 0.3048000 | gallon (U.S. liquid) | cubic meter | 0.003785412 |
| mil | meter | 0.0000254 | | | |
| mile (U.S. statute) | meter | 1,609.344 | | | |
| Mass | | | | | |
| pound | kilogram | 0.4535924 | | | |
| slug | kilogram | 14.59390 | | | |
| ton (2,000 lb) | kilogram | 907.1847 | | | |

[a]See *Metric Practice Guide*, ASTM Standard E 380-70, American Society for Testing and Materials, 1970.
[b]For more complete conversions to SI units, see Tables 4–3 and 4–4.

From Bolz, R. E. and Tuve, G. L., Eds., *Handbook of Tables for Applied Engineering Science*, 2nd ed., CRC Press, Cleveland, 1973, 803.

## Table 4–2

### ABBREVIATED COMPARISON OF INTERNATIONAL SYSTEM OF UNITS (SI) WITH PRE-SI SYSTEM OF METRIC UNITS

| Physical quantity | MKS nomenclature (SI) | | | cgs nomenclature (pre SI) | | |
|---|---|---|---|---|---|---|
| | Symbol | Equivalents in basic MKS units | Name | Symbol | Equivalents in basic cgs units | Name |
| **BASIC UNITS** | | | | | | |
| Length | m | | meter | cm | | centimeter |
| Mass | kg | | kilogram | g | | gram |
| Time | s | | second | sec | | second |
| Temperature | K | | kelvin | °F | | degrees fahrenheit |
| | | | | °C | | degrees centigrade (Celsius) |
| Angle | | | | | | |
| -plane (2D) | rad | | radian | ° ′ ″ | | degrees, minutes, seconds |
| -solid (3D) | sr | | steradian | | | |
| Luminous intensity | cd | | candela | | | cf. Table 4–2B |
| Electric current | A | | ampere | | | |
| **DERIVED UNITS** | | | | | | |
| Length | A | 0.1 nm | angstrom | A | $10^{-8}$ cm | angstrom |
| Frequency | Hz | 1/s | hertz | cps | 1/sec | cycle per second |
| Velocity | | m/s | meter per second | | cm/sec | centimeter per second |
| Acceleration | | $m/s^2$ | meter per sq second | | $cm/sec^2$ | centimeter per sq second |
| Density | | $kg/m^3$ | kilogram per cubic meter | | $g/cm^3$ | gram per cubic centimeter |
| Momentum | | kg·m/s | kilogram meter per second | | g·cm/sec | gram centimeter per second |
| Force | N | $kg·m/s^2$ | newton | dyn | $g·cm/sec^2$ | dyne |
| Pressure | $N/m^2$ | $kg/m·s^2$ | newton per sq meter | atm | $\sim10^6$ $dyn/cm^2$ | atmosphere |
| Dynamic viscosity | $N·s/m^2$ | kg/m·s | newton second per sq meter | poise | g/cm·sec | poise |
| Energy (work, heat) | J, (N·m) | $kg·m^2/s^2$ | joule or newton meter | erg | $g·cm^2/sec^2$ | erg or stat voltcoulomb |
| Power | W, (J/s) | $kg·m^2/s^3$ | watt or joule per second | W (VA) | $10^7$ erg/sec | watt or $10^7$ stat voltampere |
| Potential | V, (W/A) | $kg·m^2/A·s^3$ | volt or watt per ampere | | | |
| Charge | C, (J/V) | A·s | coulomb or joule per volt | | | |
| Resistance | Ω, (V/A) | $kg·m^2/A^2·s^3$ | ohm or volt per ampere | | | |
| Capacitance | F, (C/V) | $A^2·s^4/kg·m^2$ | farad or coulomb per volt | | | |
| Magnetic flux | Wb, (V·s) | $kg·m^2/A·s^2$ | weber or volt second | | | |
| Inductance | H, (Wb/A) | $kg·m^2/A^2·s^2$ | henry or weber per ampere | | | |
| Luminous flux | lm | cd·sr | lumen | | | |
| Luminous flux density | lx | $cd·sr/m^2$ | lux or illumination | | | |

Table compiled by A. Pigeaud,

## Table 4–3
## INTERNATIONAL SYSTEM (SI) METRIC UNITS

### Basic Units – MKS

| Property | Unit | Abbreviation |
|---|---|---|
| Length | meter | m |
| Mass | kilogram | kg |
| Time | second | s |
| Electric current | ampere | A |
| Thermodynamic temperature | kelvin | K |
| Luminous intensity | candela | cd |

### Derived Units

| Property | Units[a] | Abbreviations and dimensions | |
|---|---|---|---|
| Acceleration | Meter per second squared | $m/s^2$ | |
| Activity (of radioactive source) | 1 per second | $s^{-1}$ | |
| Angular acceleration | radian per second squared | $rad/s^{-1}$ | |
| Angular velocity | radian per second | $rad/s$ | |
| Area | square meter | $m^2$ | |
| Density | kilogram per cubic meter | $kg/m^3$ | |
| Dynamic viscosity | newton-second per square meter | $N \cdot s/m^2$ | |
| Electric capacitance | farad | F | $(A \cdot s/V)$ |
| Electric charge | coulomb | C | $(A \cdot s)$ |
| Electric field strength | volt per meter | $V/m$ | |
| Electric resistance | ohm | | $(V/A)$ |
| Entropy | joule per kelvin | $J/K$ | |
| Force | newton | N | $(kg \cdot m/s^2)$ |
| Frequency | hertz | hz | $(s^{-1})$ |
| Illumination | lux | lx | $(lm/m^2)$ |
| Inductance | henry | H | $(V \cdot s/A)$ |
| Kinematic viscosity | square meter per second | $m^2/s$ | |
| Luminance | candela per square meter | $cd/m^2$ | |
| Luminous flux | lumen | lm | |
| Magnetomotive force | ampere | A | $(cd \cdot sr)$ |
| Magnetic field strength | ampere per meter | $A/m$ | |
| Magnetic flux | weber | Wb | $(V \cdot s)$ |
| Magnetic flux density | tesla | T | $(Wb/m^2)$ |
| Power | watt | W | $(J/s)$ |
| Pressure | newton per square meter (Pascal) | $N/m^2$ (Pa) | |

## Table 4–3
## INTERNATIONAL SYSTEM (SI) METRIC UNITS (continued)

### Derived Units (continued)

| Property | Units[a] | Abbreviations and dimensions | |
|---|---|---|---|
| Radiant intensity | watt per steradian | W/sr | |
| Specific heat | joule per kilogram kelvin | J/kg K | |
| Thermal conductivity | watt per meter kelvin | W/m K | |
| Velocity | meter per second | m/s | |
| Volume | cubic meter | $m^3$ | |
| Voltage, potential difference, electromotive force | volt | V | (W/A) |
| Wave number | 1 per meter | $m^{-1}$ | |
| Work, energy, quantity of heat | joule | J | (N·m) |

### Prefix Names of Multiples and Submultiples of Units

| Decimal equivalent | Prefix | Pronunciation | Symbol | Exponential expression |
|---|---|---|---|---|
| 1,000,000,000,000 | tera | ter′a | T | $10^{12}$ |
| 1,000,000,000 | giga | ji′ga | G | $10^{9}$ |
| 1,000,000 | mega | meg′a | M | $10^{6}$ |
| 1,000 | kilo | kil′ō | k | $10^{3}$ |
| 100 | hecto | hek′tō | h | $10^{2}$ |
| 10 | deka | dek′a | da | 10 |
| 0.1 | deci | des′i | d | $10^{-1}$ |
| 0.01 | centi | sen′ti | c | $10^{-2}$ |
| 0.001 | milli | mil′i | m | $10^{-3}$ |
| 0.000001 | micro | mī′krō | μ | $10^{-6}$ |
| 0.000000001 | nano | nan′ō | n | $10^{-9}$ |
| 0.000000000001 | pico | pē′kō | p | $10^{-12}$ |
| 0.000000000000001 | femto | fem′tō | f | $10^{-15}$ |
| 0.000000000000000001 | atto | at′tō | a | $10^{-18}$ |

**Table 4—3 (continued)**
**INTERNATIONAL SYSTEM (SI) METRIC UNITS**

Definitions of the Most Important International System (SI) Units

The *ampere* (unit of electric current) is the constant current that, if maintained in two straight parallel conductors of infinite length, of negligible circular sections, and placed 1 m apart in a vacuum, will produce between these conductors a force equal to $2 \times 10^{-7}$ N per meter of length.

The *candela* is the luminous intensity, in the direction of the normal, of a black body surface $1/600,000$ m$^2$ in area, at the temperature of solidification of platinum under a pressure of 101,325 N per square meter.

The *coulomb* (unit of quantity of electricity) is the quantity of electricity transported in 1 s by a current of 1 A.

The *ephemeris second* (unit of time) is exactly $1/31,556,925.9747$ of the tropical year of 1900, January, 0 days, and 12 hr ephemeris time.

The *farad* (unit of electric capacitance) is the capacitance of a capacitor between the plates of which there appears a difference of potential of 1 V when it is charged by a quantity of electricity equal to 1 C.

The *henry* (unit of electric inductance) is the inductance of a closed circuit in which an electromotive force of 1 V is produced when the electric current in the circuit varies uniformly at a rate of 1 A per second.

The *International Practical Kelvin Temperature Scale* of 1960 and the *International Practical Celsius Temperature Scale* of 1960 are defined by a set of interpolation equations based on the following reference temperatures:

|  | K | deg C |
|---|---|---|
| Oxygen, liquid-gas equilibrium | 90.18 | −182.97 |
| Water, solid-liquid equilibrium | 273.15 | 0.00 |
| Water, solid-liquid-gas equilibrium | 273.16 | 0.01 |
| Water, liquid-gas equilibrium | 373.15 | 100.00 |
| Zinc, solid-liquid equilibrium | 692.655 | 419.505 |
| Sulfur, liquid-gas equilibrium | 717.75 | 444.6 |
| Silver, solid-liquid equilibrium | 1,233.95 | 960.8 |
| Gold, solid-liquid equilibrium | 1,336.15 | 1,063.0 |

The *joule* (unit of energy) is the work done when the point of application of 1 N is displaced a distance of 1 m in the direction of the force.

The *kelvin* (unit of thermodynamic temperature) is the fraction $1/273.16$ of the thermodynamic temperature of the triple point of water. The decision was made at the 13th General Conference on Weights and Measures on October 13, 1967, that the name of the unit of thermodynamic temperature would be changed from *degree Kelvin* (symbol: °K) to *kelvin* (symbol: K). The name (*kelvin*) and symbol (*K*) are to be used for expressing temperature intervals. The former convention that expressed a temperature interval in *degrees Kelvin* or, abbreviated, *deg K*, is dropped. However, the old designations are acceptable temporarily as alternatives to the new ones. One may also express temperature intervals in *degrees Celsius*.

The *kilogram* (unit of mass) is the mass of a particular cylinder of platinum iridium alloy, called the International Prototype Kilogram, which is preserved in a vault at Sevres, France, by the International Bureau of Weights and Measures.

*Length*: The name *micron*, for a unit of length equal to $10^{-6}$ m, and the symbol $\mu$ that has been used for it were dropped by action of the 13th General Conference on Weights and Measures on October 13, 1967. The symbol $\mu$ is to be used solely as an abbreviation for the prefix *micro-*, standing for the multiplication by $10^{-6}$. Thus, the length previously designated as $1\ \mu$ should be designated $1\ \mu$m.

The *lumin* (unit of luminous flux) is the luminous flux emitted in a solid angle of 1 sr by a uniform point source having an intensity of 1 cd.

The *newton* (unit of force) is that force that gives to a mass of 1 kg an acceleration of 1 m per second.

The *ohm* (unit of electric resistance) is the

**Table 4—3 (continued)**
**INTERNATIONAL SYSTEM (SI) METRIC UNITS**

electric resistance between two points of a conductor when a constant difference of potential of 1 V, applied between these two points, produces in this conductor a current of 1 A, this conductor not being the source of any electromotive force.

The *meter* (unit of length) is the length of exactly 1,650,763.73 wavelengths of the radiation in vacuum corresponding to the unperturbed transition between the levels $2p_{10}$ and $5d_5$ of the atom of krypton 86, the orange-red line. At the October 1984 General Conference of Weights and Measures, the meter was defined as the length of the path traveled by light in vacuum during a time interval of 1/299,792,458 of a second.

The *second* is the unit of time of the International System of Units. The definition adopted at the October 13, 1967, meeting of the 13th General Conference on Weights and Measures is "The second is the duration of 9,192,631,770 periods of the radiation corresponding to the transition between the two hyperfine levels of the fundamental state of the atom of cesium 133." The frequency (9,192,631,770 hz), which the definition assigns to the cesium radiation, was carefully chosen to make it impossible, by any existing experimental evidence, to distinguish the new second from the *ephemeris second* based on the earth's motion. Therefore, no changes need to be made in data stated in terms of the old standard in order to convert them to the new one. The atomic definition has two important advantages over the previous definition: (1) it can be realized (i.e., generated by a suitable clock) with sufficient

precision, ± 1 part per hundred billion ($10^{11}$) or better, to meet the most exacting demands of modern metrology, and (2) it is available to anyone who has access to or who can build an atomic clock controlled by the specified cesium radiation.[b] In addition, one can compare other high-precision clocks directly with such a standard in a relatively short time − an hour or so compared against years with the astronomical standard. Laboratory-type atomic clocks are complex and expensive, so that most clocks and frequency generators will continue to be calibrated against a standard such as the NBS Frequency Standard, controlled by a cesium atomic beam, at the Radio Standards Laboratory in Boulder, Colorado. In most cases the comparison will be by way of the standard-frequency and time-interval signals broadcast by NBS radio stations WWV, WWVH, WWVB, and WWVL.

The *volt* (unit of electric potential difference and electromotive force) is the difference of electric potential between two points of a conducting wire carrying a constant current of 1 A, when the power dissipated between these points is equal to 1 W.

The *watt* (unit of power) is the power that gives rise to the production of energy at the rate of 1 J per second.

The *weber* (unit of magnetic flux) is the magnetic flux that, linking a circuit of one turn, produces in it an electromotive force of 1 V as it is reduced to zero at a uniform rate in 1 s.

[a]According to SI terminology, the following should be treated as obsolete:

| | |
|---|---|
| angstrom (now 100 picometers or 0.1 nanometer) | kiloton (now gigagram) |
| | liter (now cubic decimeter) |
| bar (now 100 kilonewtons/meter²) | metric ton (now megagram) |
| kiloliter (now cubic meter) | micron (now micrometer) |

[b]A description of such clocks is given in "Atomic Frequency Standards," *NBS Tech. News Bull.*, 45, 8, January 1961.

*Note:* For more recent developments and technical details, see Beehler, R. E., Mockler, R. C., and Richardson, J. M., Cesium beam atomic time and frequency standards, *Metrologia*, 1, 114, July 1965.

Modified from Bolz, R. E. and Tuve, G. L., Eds., *Handbook of Tables for Applied Engineering Science*, 2nd ed., CRC Press, Cleveland, 1973, 805.

## Table 4—4
### SELECTED MATHEMATICAL AND PHYSICAL CONSTANTS

Rounded to 3 to 5 Significant Figures

| Name | Symbol (or equivalent) | Value | Units |
|---|---|---|---|
| Naperian base (natural) | $e$ | 2.718 | |
| Natural logarithm of 10 | $\ln_e 10$ | 2.303 | |
| Common logarithm of 2 | $\log_{10} 2$ | 0.301 | |
| Semicircle | $\pi$ | 3.1416 | radians |
| Cos 60° | ½ | 0.500 | |
| Cos 45° | $\sqrt{2}/2$ | 0.707 | |
| Cos 30° | $\sqrt{3}/2$ | 0.866 | |
| Ice point | $T_0$ | 273.15 | K |
| Avogadro's number | N | $6.022 \times 10^{23}$ | atoms (particles)/mole |
| Gas constant | R | $8.317 \times 10^7$ | erg/ K-mole |
| | | 1.988 | cal/ K-mole |
| | | 8.208 | l-atm/ K-mole |
| Boltzmann constant | k (R/N) | $8.617 \times 10^{-5}$ | ev/ K-atom (particles) |
| | | $1.380 \times 10^{-16}$ | erg/ K-atom |
| | | $0.33 \times 10^{-23}$ | cal/ K-atom |
| Acceleration due to gravity | $g_0$ | 9.806 | m/s² |
| Velocity of light (vacuum) | c | $2.998 \times 10^8$ | m/s |
| Planck constant | h | $6.626 \times 10^{-27}$ | erg-sec/atom (particles) |
| | | $1.58 \times 10^{-34}$ | cal-sec/atom |
| Faraday constant | $\mathscr{F}$ | $9.649 \times 10^4$ | coulomb/mole |
| | | 27 (approx.) | amp-hr/mole |
| Charge of one electron | $e^-$ ($\mathscr{F}$/N) | $4.803 \times 10^{-10}$ | (erg-cm)½, or esu |
| | | $1.602 \times 10^{-12}$ | erg/volt |
| | | $1.602 \times 10^{-19}$ | joule/volt (J/V) |
| | | $1.602 \times 10^{-20}$ | emu |
| Charge-to-mass ratio of electron | $e^-/m_e$ | $1.759 \times 10^8$ | coulomb/gram |
| Rest mass of one electron | $m_e \left(\dfrac{\mathscr{F}}{N} \Big/ \dfrac{e^-}{m_e}\right)$ | $9.109 \times 10^{-28}$ | gram |
| Compton wavelength of free electron | $\lambda_{ce}(h/m_e c)$ | $2.426 \times 10^{-10}$ | cm |

Table compiled by A. Pigeaud.

## Table 4—5
### ABBREVIATED TABLE OF CONVERSION FACTORS

**Length (meter)**

| | | | | | | | | |
|---|---|---|---|---|---|---|---|---|
| 1 light year | = | 9,460 | Tm ($10^{12}$ m) | 1 Tm | = | $0.1058 \times 10^{-3}$ | light years |
| 1 astron. unit | = | 149.5 | Gm ($10^9$ m) | 1 Tm | = | 6.69 | astron. units |
| 1 naut. mile | = | 1.852 | km ($10^3$ m) | 1 km | = | 0.54 | naut. miles |
| 1 mile (U.S.) | = | 1.609 | km | 1 km | = | 0.622 | mile (U.S.) |
| 1 fathom | = | 1.829 | m | 1 m | = | 0.547 | fathom |
| 1 yard | = | 0.9144 | m | 1 m | = | 1.0936 | yard |
| 1 foot | = | 0.3048 | m | 1 m | = | 3.281 | foot |
| 1 inch | = | 2.54 | cm ($10^{-2}$ m) | 1 cm | = | 0.3937 | inch |
| 1 mil | = | 25.4 | $\mu$m ($10^{-6}$ m) | 1 $\mu$m | ≅ | 0.04 | mil |
| 1 Kr$^{86}$ wavelength | = | 606 | nm ($10^{-9}$ m) | 1 $\mu$m | = | 1.65 | Kr$^{86}$ wavelength |
| 1 angstrom | = | 0.1 | nm | 1 nm | = | 10 | angstrom |

## Table 4—5 (continued)
### ABBREVIATED TABLE OF CONVERSION FACTORS

### Area (meter)$^2$

| | | | | | | |
|---|---|---|---|---|---|---|
| 1 sq mile (U.S.) | = | 2.59 | km$^2$ $(10^3$ m)$^2$ | 1 km$^2$ | = | 0.3861 | sq mile (U.S.) |
| 1 hectare | = | 1 | hm$^2$ $(10^2$ m)$^2$ | 1 km$^2$ | = | 100 | hectare |
| 1 acre | = | 4,047 | m$^2$ | 1 hm$^2$ | = | 2.471 | acre |
| 1 are | = | 100 | m$^2$ | 1 hm$^2$ | = | 100 | are |
| 1 sq yard | = | 0.8361 | m$^2$ | 1 m$^2$ | = | 1.196 | sq yard |
| 1 sq foot | = | 929 | cm$^2$ $(10^{-2}$ m)$^2$ | 1 m$^2$ | = | 10.764 | sq foot |
| 1 sq inch | = | 6.452 | cm$^2$ | 1 cm$^2$ | = | 0.155 | sq inch |
| 1 circ. mil | ≅ | 0.0005 | mm$^2$ $(10^{-3}$ m)$^2$ | 1 mm$^2$ | = | 1,973.5 | circ. mil |
| 1 sq angstrom | = | 10,000 | pm$^2$ $(10^{-12}$ m)$^2$ | 1 nm$^2$ | = | 100 | sq angstrom |
| 1 barn | = | 0.0001 | pm$^2$ | 1 pm$^2$ | = | 10,000 | barns |

### Density (kilogram/m$^3$)

| | | | | | | |
|---|---|---|---|---|---|---|
| 1 lb mass/cu inch | = | 27.68 | Mg/m$^3$ | 1 kg/m$^3$ | = | 36.13 | lb mass/cu inch |
| 1 gram/cc | = | 1,000 | kg/m$^3$ | 1 kg/m$^3$ | = | 0.001 | gram/cc |
| 1 slug/cu ft | = | 515.4 | kg/m$^3$ | 1 g/cm$^3$ | = | 1.94 | slug/cu ft |
| 1 lb mass/USG | = | 119.8 | kg/m$^3$ | 1 g/cm$^3$ | = | 8.345 | lb mass/USG |
| 1 lb mass/cu ft | = | 16.02 | kg/m$^3$ | 1 g/cm$^3$ | = | 62.43 | lb mass/cu ft |

### Volume (meter)$^3$

| | | | | | | |
|---|---|---|---|---|---|---|
| 1 acre foot | = | 1,233.5 | m$^3$ (kl) | 1 dam$^3$ | = | 0.8107 | acre foot |
| 1 cord | = | 3.625 | m$^3$ | 1 m$^3$ | = | 0.2759 | cord |
| 1 register ton | = | 2.832 | m$^3$ | 1 m$^3$ | = | 0.3531 | register ton |
| 1 cu yard | = | 764.5 | dm$^3$ $(10^{-1}$ m)$^3$ | 1 m$^3$ | = | 1.308 | cu yard |
| 1 barrel (42 USG) | = | 159 | l $(10^{-1}$ m)$^3$ | 1 kl | = | 6.3 | barrel (USG) |
| 1 bushel (U.S.) | = | 35.24 | l | 1 kl | = | 28.4 | bushel (U.S.) |
| 1 cu ft | = | 28.32 | dm$^3$ | 1 m$^3$ | = | 35.31 | cu ft |
| 1 board foot | = | 2.36 | dm$^3$ | 1 dm$^3$ | = | 0.424 | board foot |
| 1 gallon (Imp.) | = | 4.546 | l | 1 l | = | 0.220 | (Imp.) gallon |
| 1 gallon (U.S.) | = | 3.785 | l | 1 l | = | 0.2642 | (U.S.) gallon |
| 1 quart (U.S.) | = | 0.9463 | l | 1 l | = | 1.0567 | (U.S.) quart |
| 1 pint (U.S.) | = | 473.2 | ml $(10^{-2}$ m)$^3$ | 1 l | = | 2.113 | (U.S.) pint |
| 1 fl ounce (U.S.) | = | 29.57 | ml | 1 l | = | 33.81 | (U.S.) fl ounce |
| 1 cu inch | = | 16.39 | cc $(10^{-2}$ m)$^3$ | 1 cc | = | 0.061 | cu inch |
| 1 dram (U.S.) | = | 3.697 | ml | 1 ml | = | 0.27 | dram (U.S.) |
| 1 minim (U.S.) | = | 61.6 | μl $(10^{-3}$ m)$^3$ | 1 ml | = | 16.23 | minim (U.S.) |

### Mass (kilogram)

| | | | | | | |
|---|---|---|---|---|---|---|
| 1 long ton | = | 1.016 | Mg $(10^6$ g) | 1 Mg | = | 0.985 | long ton |
| 1 metric ton | = | 1,000 | kg $(10^3$ g) | 1 Mg | = | 1.0 | metric ton |
| 1 short ton (2,000 lb) | = | 907 | kg | 1 Mg | = | 1.102 | short ton (2,000 lb) |
| 1 pound (avoir) | = | 453.6 | g | 1 kg | = | 2.204 | pounds (avoir) |
| 1 ounce (avoir) | = | 28.35 | g | 1 kg | = | 35.27 | ounce (avoir) |
| 1 dram (avoir) | = | 1.77 | g | 1 g | = | 0.564 | dram (avoir) |
| 1 carat (metric) | = | 0.2 | g | 1 g | = | 5 | carat (metric) |
| 1 grain | = | 64.8 | mg $(10^{-3}$ g) | 1 g | = | 15.4 | grain |

### Time (mean solar seconds)

| | | | | | | |
|---|---|---|---|---|---|---|
| 1 calendar year | ≈ | 30 | Ms $(10^6$ s) | 1 Ms | ≈ | 0.032 | calendar year |
| 1 calendar month | ≈ | 2.5 | Ms | 1 Ms | ≈ | 0.4 | calendar month |
| 1 day | = | 86.40 | ks | 1 Ms | = | 11.574 | days |
| 1 hour | = | 3.60 | ks $(10^3$ s) | 1 Ms | = | 277.8 | hours |
| 1 minute | = | 60.0 | s | 1 ks | = | 16.67 | minutes |

## Table 4—5 (continued)
## ABBREVIATED TABLE OF CONVERSION FACTORS

### Planar Angle (radian)

| | | | | | | | |
|---|---|---|---|---|---|---|---|
| 360° ($2\pi$) | = | 6.283 | rad | 1 rad | = | 57.296° | (degrees) |
| 180° ($\pi$) | = | 3.1416 | rad | 1 mrad | = | 3.438' | (minutes) |
| 90° ($\frac{\pi}{2}$) | = | 1.5708 | rad | 1 $\mu$rad | = | 0.206" | (seconds) |
| 1° ($\frac{\pi}{180}$) | = | 0.01745 | rad | | | | |

### Solid Angle (steradian)

| | | | | | | | |
|---|---|---|---|---|---|---|---|
| 1 sphere | = | 12.566 | sr | 1 sr | = | 0.0796 | sphere |
| 1 hemisphere | = | 6.283 | sr | 1 sr | = | 0.159 | hemisphere |
| 1 octant | = | 1.5708 | sr | 1 sr | = | 0.636 | octant |
| 1 square degree | = | 0.3046 | msr | 1 sr | = | 3,282.8 | square degree |

### Velocity (meter/second)

| | | | | | | | |
|---|---|---|---|---|---|---|---|
| 1 (U.S.) mile/sec | = | 1.609 | km/s | 1 km/s | = | 0.622 | (U.S.) mile/sec |
| 1 (U.S.) mile/min | = | 26.82 | m/s | 1 km/s | = | 37.28 | (U.S.) mile/min |
| 1 (Int.) knot | = | 0.514 | m/s | 1 m/s | = | 1.94 | (Int.) knot |
| 1 (U.S.) mile/hr | = | 0.447 | m/s | 1 m/s | = | 2.237 | (U.S.) mile/hr |
| 1 foot/sec | = | 0.305 | m/s | 1 m/s | = | 3.28 | ft/sec |
| 1 km/hr | = | 0.278 | m/s | 1 m/s | = | 3.6 | km/hr |
| 1 inch/sec | = | 2.54 | cm/s | 1 m/s | = | 39.37 | in./sec |
| 1 foot/min | = | 5.08 | mm/s | 1 m/s | = | 196.85 | ft/min |
| 1 foot/hr | = | 84.67 | $\mu$m/s | 1 cm/s | = | 118.1 | ft/hr |

### Acceleration (meter/second$^2$)

| | | | | | | | |
|---|---|---|---|---|---|---|---|
| gravity (mean) | = | 9.806 | m/s$^2$ | 1 m/s$^2$ | = | 0.102 | gravity (mean) |
| 1 mile/hr-sec | = | 0.447 | m/s$^2$ | 1 m/s$^2$ | = | 2.237 | mile/hr-sec |
| 1 foot/sq sec | = | 0.305 | m/s$^2$ | 1 m/s$^2$ | = | 3.281 | foot/sq sec |
| 1 km/hr-sec | = | 0.278 | m/s$^2$ | 1 m/s$^2$ | = | 3.6 | km/hr-sec |
| 1 inch/sq sec | = | 0.0254 | m/s$^2$ | 1 m/s$^2$ | = | 39.37 | inch/sq sec |
| 1 galileo | = | 0.01 | m/s$^2$ | 1 m/s$^2$ | = | 100 | galileo |

### Force (newton)

| | | | | | | | |
|---|---|---|---|---|---|---|---|
| 1 kg force (kgf) | = | 9.806 | N | 1 N | = | 0.102 | kg force |
| 1 pound force (lbf) | = | 4.448 | N | 1 N | = | 0.225 | pound force (avoir) |
| 1 ounce force | = | 0.278 | N | 1 N | = | 3.6 | ounce force |
| 1 poundal | = | 0.138 | N | 1 N | = | 7.24 | poundal |
| 1 dyne | = | 10.0 | $\mu$N | 1 N | = | $10^5$ | dyne |

### Pressure (newton/m$^2$ or pascal)

| | | | | | | | |
|---|---|---|---|---|---|---|---|
| 1 atmosphere (14.7 psi) | = | 101.325 | kN/m$^2$ | 1 MN/m$^2$ | = | 9.87 | atmosphere |
| 1 bar | = | 100.0 | kN/m$^2$ | 1 MN/m$^2$ | = | 10.0 | bar |
| 1 kgf/sq cm | = | 98.06 | kN/m$^2$ | 1 MN/m$^2$ | = | 10.197 | kgf/sq cm |
| 1 lbf/sq inch (psi) | = | 6.895 | kN/m$^2$ | 1 MN/m$^2$ | = | 145. | lbf/sq inch |
| 1 inch Hg (60°F) | = | 3.377 | kN/m$^2$ | 1 kN/m$^2$ | = | 0.296 | inch Hg (60°F) |
| 1 cm Hg (0°F) | = | 1.333 | kN/m$^2$ | 1 kN/m$^2$ | = | 0.75 | cm Hg (0°F) |
| 1 inch $H_2O$ (60°F) | = | 248.84 | N/m$^2$ | 1 kN/m$^2$ | = | 4.02 | inch $H_2O$ (60°F) |
| 1 mm Hg (0°C) or torr | = | 133.32 | N/m$^2$ | 1 kN/m$^2$ | = | 7.57 | mm Hg (0°C) or torr |
| 1 millibar | = | 100.0 | N/m$^2$ | 1 kN/m$^2$ | = | 10.0 | millibar |
| 1 cm $H_2O$ (4°C) | = | 98.064 | N/m$^2$ | 1 kN/m$^2$ | = | 10.2 | cm $H_2O$ (4°C) |
| 1 lbf/sq f | = | 47.88 | N/m$^2$ | 1 kN/m$^2$ | = | 20.88 | lbf/sq foot |
| 1 kgf/sq meter | = | 9.806 | N/m$^2$ | 1 kN/m$^2$ | = | 101.97 | kgf/sq meter |
| 1 pascal (P) | = | 1.0 | N/m$^2$ | 1 N/m$^2$ | = | 1.0 | P (pascal) |

## Table 4—5 (continued)
## ABBREVIATED TABLE OF CONVERSION FACTORS

### Pressure (newton/m² or pascal) (continued)

| 1 dyne/sq cm or barye | = | 0.1 | N/m² | 1 N/m² | = | 10.0 | dyne/sq cm or barye |
|---|---|---|---|---|---|---|---|
| $10^{-6}$ torr | = | 133.32 | $\mu$N/m² | 1 mN/m² | = | 7.57 | $10^{-6}$ torr |

### Dynamic Viscosity (newton-seconds/meter²)

| 1 lbf-sec/sq foot | = | 47.88 | N·s/m² | 1 N·s/m² | = | 0.0209 | lbf-sec/sq foot |
|---|---|---|---|---|---|---|---|
| 1 poundal-sec/sq foot | = | 1.488 | N·s/m² | 1 N·s/m² | = | 0.672 | poundal-sec/sq foot |
| 1 lbm/foot-sec | = | 1.488 | N·s/m² | 1 N·s/m² | = | 0.672 | lbm/foot-sec |
| 1 poise | = | 0.1 | N·s/m² | 1 N·s/m² | = | 10 | poise |
| 1 centipoise | = | 0.001 | N·s/m² | 1 N·s/m² | = | 1,000 | centipoise |

### Energy (joule)

| 1 ton (TNT-equiv) | = | 4.2 | GJ | 1 TJ | = | 238 | ton (TNT-equiv) |
|---|---|---|---|---|---|---|---|
| 1 kWh | = | 3.6 | MJ | 1 GJ | = | 277.8 | kWh |
| 1 HP-hr | = | 2.68 | MJ | 1 GJ | = | 372.5 | HP-hr |
| 1 watt-hr | = | 3.6 | kJ | 1 MJ | = | 277.8 | watt-hr |
| 1 Btu (mean) | = | 1.0559 | kJ | 1 MJ | = | 947.0 | Btu (mean) |
| 1 l-atm | = | 101.33 | J | 1 kJ | = | 9.869 | liter-atmosphere |
| 1 kg-m | = | 9.807 | J | 1 kJ | = | 101.97 | kilogram-meter |
| 1 cal (mean) | = | 4.190 | J | 1 kJ | = | 238.89 | calorie (mean) |
| 1 cal (15°C) | = | 4.1858 | J | 1 kJ | = | 238.95 | calorie (15°C) |
| 1 cal (thermochem) | = | 4.1840 | J | 1 kJ | = | 239.00 | calorie (thermochem) |
| 1 cal (20°C) | = | 4.1819 | J | 1 J | = | 239.18 | calorie (20°C) |
| 1 foot-lbf | = | 1.356 | J | 1 J | = | 0.737 | foot-lbf |
| 1 watt-sec | = | 1.0 | J | 1 J | = | 1.0 | watt-sec |
| 1 foot-poundal | = | 42.14 | mJ | 1 J | = | 23.73 | foot-poundal |
| 1 erg (dyne-cm) | = | 0.10 | $\mu$J | 1 $\mu$J | = | 10.0 | erg |
| 1 eV | = | 0.1602 | aJ | 1 aJ | = | 6.242 | eV |

### Power (watt)

| 1 boiler horsepower | = | 9.809 | kW | 1 MW | = | 102 | boiler horsepower |
|---|---|---|---|---|---|---|---|
| 1 kcal/sec | = | 4.184 | kW | 1 MW | = | 239 | kcal/sec (thermochem) |
| 1 Btu/sec | = | 1.054 | kW | 1 kW | = | 0.9485 | Btu/sec (thermochem) |
| 1 HP (electric/water) | = | 746 | W | 1 kW | = | 1.3405 | HP (electric/water) |
| 1 HP (U.K.) | = | 745.7 | W | 1 kW | = | 1.341 | HP (550 ft-lbf/sec) |
| 1 HP (metric) | = | 735.5 | W | 1 kW | = | 1.3596 | HP (metric) |
| 1 kcal/min | = | 69.73 | W | 1 kW | = | 14.33 | kcal/min (thermochem) |
| 1 Btu/min | = | 17.57 | W | 1 kW | = | 56.88 | Btu/min (thermochem) |
| 1 cal/sec | = | 4.184 | W | 1 kW | = | 239 | cal/sec (thermochem) |
| 1 ft-lbf/sec | = | 1.356 | W | 1 kW | = | 737.56 | ft-lbf/sec |
| 1 cal/min | = | 69.73 | mW | 1 W | = | 14.33 | cal/min (thermochem) |
| 1 ft-lbf/min | = | 22.597 | mW | 1 W | = | 44.25 | ft-lbf/min |
| 1 ft-lbf/hr | = | .3766 | mW | 1 mW | = | 2.655 | ft-lbf/hr |
| 1 erg/sec | = | 0.10 | $\mu$W | 1 $\mu$W | = | 10 | erg/sec |

### Charge (coulomb)

| 1 faraday | = | 96,487 | C | 1 MC | = | 10.363 | faraday |
|---|---|---|---|---|---|---|---|
| 1 amp-hr | = | 3,600 | C | 1 MC | = | 277.8 | amp-hr |
| 1 amp-sec | = | 1.0 | C | 1 C | = | 1.0 | amp-sec |
| 1 faraday | = | 26.8 | A·hr | 1 A·hr | = | 0.0336 | faraday |

Table compiled by A. Pigeaud,

# Table 4—6

| Structural and material features | Log dimensions | Analytical and instrument resolutions |

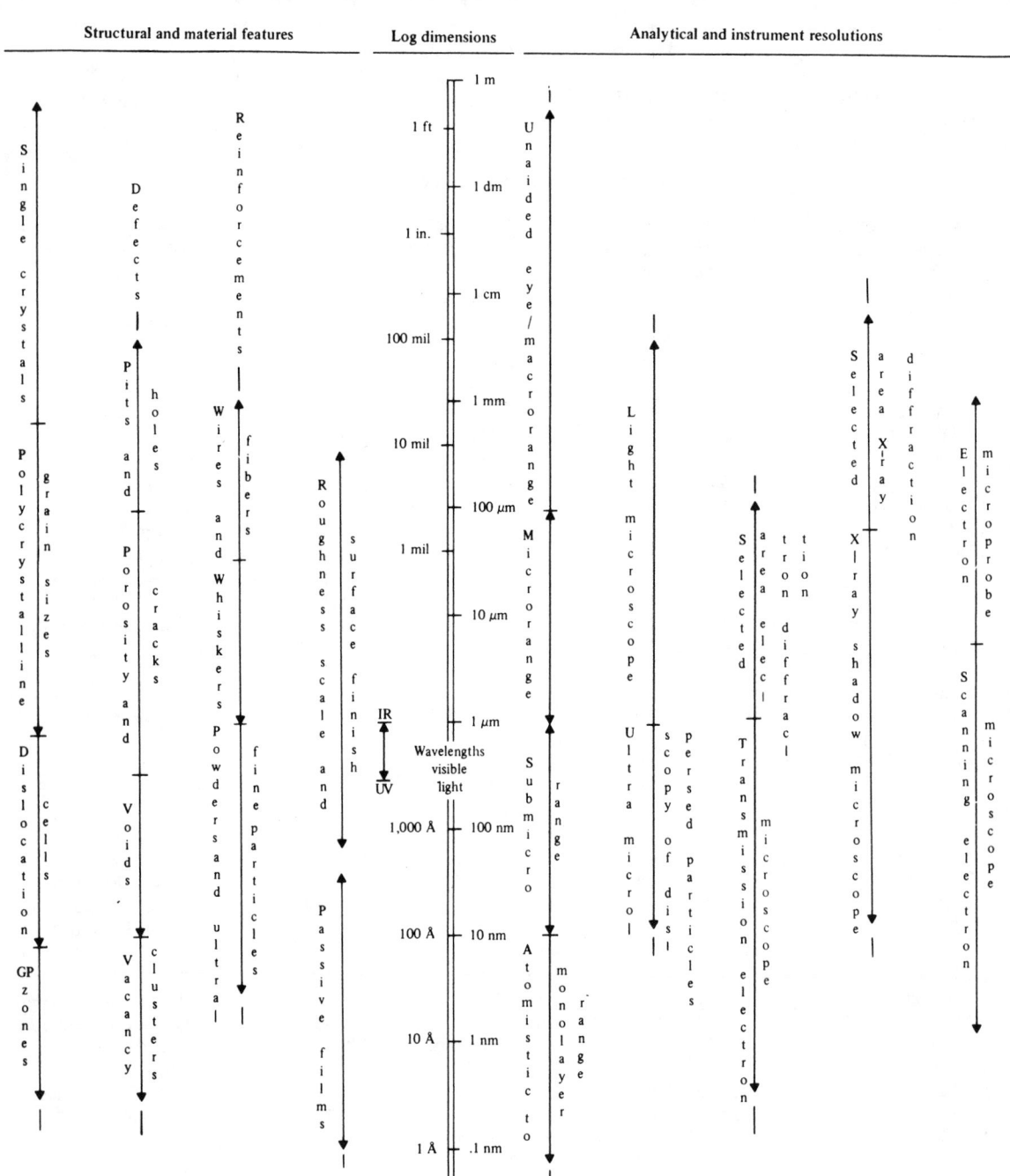

Table compiled by A. Pigeaud.

## Table 4—7

### CONVERSION FACTORS

| To convert from | To | Multiply by |
|---|---|---|
| Abamperes | Amperes | 10 |
| " | E.M. cgs. units of current | 1 |
| " | E.S. cgs. units | $2.997930 \times 10^{10}$ |
| " | Faradays (chem.)/sec | $1.036377 \times 10^{-4}$ |
| " | Faradays (phys.)/sec | $1.036086 \times 10^{-4}$ |
| " | Statamperes | $2.997930 \times 10^{10}$ |
| Abamperes/cm | E.M. cgs. units of surface charge density | 1 |
| " | E.S. cgs. units | $2.997930 \times 10^{10}$ |
| Abamperes/sq. cm | Amperes/circ. mil | $5.0670748 \times 10^{-5}$ |
| " | Amperes/sq. cm | 10 |
| " | Amperes/sq. inch | 64.516 |
| Abampere-turns | Ampere-turns | 10 |
| Abampere-turns/cm | Ampere-turns/cm | 10 |
| Abcoulombs | Ampere-hours | 0.0027777 |
| " | Coulombs | 10 |
| " | Electronic charges | $6.24196 \times 10^{19}$ |
| " | E.M. cgs. units of charge | 1 |
| " | E.S. cgs. units | $2.997930 \times 10^{10}$ |
| " | Faradays (chem.) | $1.036377 \times 10^{-4}$ |
| " | Faradays (phys.) | $1.036086 \times 10^{-4}$ |
| " | Statcoulombs | $2.997930 \times 10^{10}$ |
| Abfarads | E.M. cgs. units of capacitance | 1 |
| " | E.S. cgs. units | $8.987584 \times 10^{20}$ |
| " | Farads | $1 \times 10^{9}$ |
| " | Microfarads | $1 \times 10^{15}$ |
| " | Statfarads | $8.987584 \times 10^{20}$ |
| Abhenries | E.M. cgs. units of induction | 1 |
| " | E.S. cgs. units | $1.112646 \times 10^{-21}$ |
| " | Henries | $1 \times 10^{-9}$ |
| Abmhos | E.M. cgs. units of conductance | 1 |
| " | E.S. cgs. units | $8.987584 \times 10^{20}$ |
| " | Megamhos | 1000 |
| " | Mhos | $1 \times 10^{9}$ |
| " | Statmhos | $8.987584 \times 10^{20}$ |
| Abohms | E.M. cgs. units of resistance | 1 |
| " | Megohms | $1 \times 10^{-15}$ |
| " | Microhms | 0.001 |
| " | Ohms | $1 \times 10^{-9}$ |
| " | Statohms | $1.112646 \times 10^{-21}$ |
| Abohm-cm | Circ. mil-ohms/ft | 0.0060153049 |
| " | E.M. cgs. units of resistivity | 1 |
| " | Microhm-inches | 0.00039370079 |
| " | Ohm-cm | $1 \times 10^{-9}$ |
| Abvolts | Microvolts | 0.01 |
| " | Millivolts | $1 \times 10^{-5}$ |
| " | Volts | $1 \times 10^{-8}$ |
| " | Volts (Int.) | $9.99670 \times 10^{-9}$ |
| Abvolts/cm | E.M. cgs. units of electric field intensity | 1 |
| " | E.S. cgs. units | $3.335635 \times 10^{-11}$ |
| " | Volts/cm | $1 \times 10^{-8}$ |
| " | Volts/inch | $2.54 \times 10^{-8}$ |
| " | Volts/meter | $1 \times 10^{-6}$ |
| Acres | Sq. cm | 40468564 |
| " | Sq. ft. | 43560 |
| " | Sq. ft. (U.S. Survey) | 43559.826 |
| " | Sq. inches | 6272640 |
| " | Sq. kilometers | 0.0040468564 |
| " | Sq. links (Gunter's) | $1 \times 10^{5}$ |
| " | Sq. meters | 4046.8564 |
| " | Sq. miles (statute) | 0.0015625 |
| " | Sq. perches | 160 |
| " | Sq. rods | 160 |
| " | Sq. yards | 4840 |
| Acre-feet | Cu. feet | 43560 |
| " | Cu. meters | 1233.4818 |
| " | Cu. yards | 1613.333 |
| Acre-inches | Cu. feet | 3630 |
| " | Cu. meters | 102.79033 |

| To convert from | To | Multiply by |
|---|---|---|
| Acre-inches | Gallons (U.S.) | 27154.286 |
| Amperes | Abamperes | 0.1 |
| " | Amperes (Int.) | 1.000165 |
| " | Cgs. units of current | 1 |
| " | Mks. units of current | 1 |
| " | Coulombs/sec | 1 |
| " | Coulombs (Int.)/sec | 1.000165 |
| " | Faradays (chem.)/sec | $1.036377 \times 10^{-5}$ |
| " | Faradays (phys.)/sec | $1.036086 \times 10^{-5}$ |
| " | Statamperes | $2.997930 \times 10^{9}$ |
| Amperes (Int.) | Amperes | 0.999835 |
| " | Coulombs/sec | 0.999835 |
| " | Coulombs (Int.)/sec | 1 |
| " | Faradays (chem.)/sec* | $1.03623 \times 10^{-5}$ |
| " | Faradays (phys.)/sec* | $1.03592 \times 10^{-5}$ |
| Amperes/meter | Cgs. units of surface current density | 0.01 |
| " | E.M. cgs. units | 0.001 |
| " | E.S. cgs. units | $2.997930 \times 10^{7}$ |
| " | Mks. units | 1 |
| Amperes/sq. meter | Cgs. units of volume current density | 0.0001 |
| " | E.M. cgs. units | $1 \times 10^{-5}$ |
| " | E.S. cgs. units | 299793.0 |
| " | Mks. units | 1 |
| Amperes/sq. mil | Abamperes/sq. cm | 15500.031 |
| " | Amperes/sq. cm | $1.5500031 \times 10^{6}$ |
| Ampere-hours | Abcoulombs | 360 |
| " | Coulombs | 3600 |
| " | Faradays (chem.)* | 0.0373096 |
| " | Faradays (phys.)* | 0.0372991 |
| Ampere-turns | Cgs. units of magneto-motive force | 1.2566371 |
| " | E.M. cgs. units | 1.2566371 |
| " | E.S. cgs. units | $3.767310 \times 10^{10}$ |
| " | Gilberts | 1.2566371 |
| Ampere-turns/weber | Cgs. units of reluctance | $1.256637 \times 10^{-8}$ |
| " | E.M. cgs. units | $1.256637 \times 10^{-8}$ |
| " | E.S. cgs. units | $1.129413 \times 10^{11}$ |
| " | Gilberts/maxwell | $1.256637 \times 10^{-8}$ |
| Ångström units | Centimeters | $1 \times 10^{-8}$ |
| " | Inches | $3.9370079 \times 10^{-9}$ |
| " | Microns | 0.0001 |
| " | Millimicrons | 0.1 |
| " | Wave length of orange-red line of krypton 86 | 0.000165076373 |
| " | Wave length of red line of cadmium | 0.000155316413 |
| Ares | Acres | 0.024710538 |
| " | Sq. dekameters | 1 |
| " | Sq. feet | 1076.3910 |
| " | Sq. ft. (U.S. Survey) | 1076.3867 |
| " | Sq. meters | 100 |
| " | Sq. miles | $3.8610216 \times 10^{-5}$ |
| Atmospheres | Bars | 1.01325 |
| " | Cm. of Hg (0°C.) | 76 |
| " | Cm. of H₂O (4°C.) | 1033.26 |
| " | Dynes/sq. cm. | $1.01325 \times 10^{6}$ |
| " | Ft. of H₂O (39.2°F.) | 33.8995 |
| " | Grams/sq. cm. | 1033.23 |
| " | In. of Hg (32°F.) | 29.9213 |
| " | Kg./sq. cm. | 1.03323 |
| " | Mm. of Hg (0°C.) | 760 |
| " | Pounds/sq. inch | 14.6960 |
| " | Tons (short)/sq. ft. | 1.05811 |
| " | Torrs | 760 |
| Atomic mass units (chem.)* | Electron volts | $9.31395 \times 10^{8}$ |
| " | Grams* | $1.66024 \times 10^{-24}$ |
| Atomic mass units (phys.)* | Electron volts | $9.31141 \times 10^{8}$ |
| " | Grams* | $1.65979 \times 10^{-24}$ |

**Table 4—7 (continued)**

**CONVERSION FACTORS**

| To convert from | To | Multiply by | To convert from | To | Multiply by |
|---|---|---|---|---|---|
| Bags (Brit.)........ | Bushels (Brit.)............ | 3 | B.t.u............ | Kw.-hours (Int.)......... | 0.000292827 |
| Barns.............. | Sq. cm................. | $1 \times 10^{-24}$ | " | Liter-atm.............. | 10.4053 |
| Barrels (Brit.)*..... | Bags (Brit.)............. | 1.5 | " | Tons of refrig. (U.S. std.).. | $3.46995 \times 10^{-8}$ |
| " | Barrels (U.S., dry)........ | 1.415404 | " | Watt-seconds........... | 1054.35 |
| " | Barrels (U.S., liq.)....... | 1.372513 | " | Watt-seconds (Int.)...... | 1054.18 |
| " | Bushels (Brit.).......... | 4.5 | B.t.u. (IST.)........ | B.t.u................. | 1.00065 |
| " | Bushels (U.S.).......... | 4.644253 | B.t.u. (mean)..... | B.t.u................. | 1.00144 |
| " | Cu. feet.............. | 5.779568 | " | B.t.u. (IST.)........... | 1.00078 |
| " | Cu. meters............ | 0.1636591 | " | B.t.u. (39°F.)........... | 0.996415 |
| " | Gallons (Brit.)......... | 36 | " | B.t.u. (60°F.)........... | 1.00113 |
| " | Liters................ | 163.6546 | " | Hp.-hours............. | 0.000393317 |
| Barrels (petroleum, U.S.)............. | Cu. feet.............. | 5.614583 | " | Joules................ | 1055.87 |
| " | Gallons (U.S.)......... | 42 | " | Kg.-meters............ | 107.669 |
| " | Liters................ | 158.98284 | " | Kw.-hours............. | 0.000293297 |
| Barrels (U.S., dry).... | Barrels (U.S. liq.)....... | 0.969696 | " | Kw.-hours (Int.)......... | 0.000293248 |
| " | Bushels (U.S.).......... | 3.2812195 | " | Liter-atm.............. | 10.4203 |
| " | Cu. feet.............. | 4.083333 | " | Watt-hours............ | 0.293297 |
| " | Cu. inches............ | 7056 | " | Watt-hours (Int.)........ | 0.293248 |
| " | Cu. meters............ | 0.11562712 | B.t.u. (39°F.)........ | B.t.u................. | 1.00504 |
| " | Quarts (U.S., dry)....... | 105 | " | B.t.u. (IST.).......... | 1.00439 |
| Barrels (U.S., liq.).... | Barrels (U.S., dry)....... | 1.03125 | " | B.t.u. (mean).......... | 1.00360 |
| " | Barrels (wine).......... | 1 | " | B.t.u. (60°F).......... | 1.00473 |
| " | Cu. feet.............. | 4.2109375 | " | Joules................ | 1059.67 |
| " | Cu. inches............ | 7276.5 | B.t.u. (60°F.)........ | B.t.u................. | 1.00031 |
| " | Cu. meters............ | 0.11924047 | " | B.t.u. (IST.).......... | 0.999657 |
| " | Gallons (Brit.)......... | 26.22925 | " | B.t.u. (mean).......... | 0.998873 |
| " | Gallons (U.S., liq.)...... | 31.5 | " | B.t.u. (39°F.)......... | 0.995291 |
| " | Liters................ | 119.23713 | B.t.u./hr......... | Cal., *kg./hr*.......... | 0.251996 |
| Bars............... | Atmospheres........... | 0.986923 | " | Ergs/sec.............. | $2.928751 \times 10^{6}$ |
| " | Baryes............... | $1 \times 10^{6}$ | " | Foot-pounds/hr......... | 777.649 |
| " | Cm. of Hg (0°C.)....... | 75.0062 | " | Horsepower............ | 0.000392752 |
| " | Dynes/sq. cm.......... | $1 \times 10^{6}$ | " | Horsepower (boiler)..... | $2.98563 \times 10^{-5}$ |
| " | Ft. of H₂O (60°F.)...... | 33.4883 | " | Horsepower (electric)..... | 0.000392594 |
| " | Grams/sq. cm.......... | 1019.716 | " | Horsepower (metric)..... | 0.000398199 |
| " | In. of Hg (32°F.)....... | 29.5300 | " | Kilowatts............. | 0.000292875 |
| " | Kg./sq. cm............ | 1.019716 | " | Lb. ice melted/hr........ | 0.0069714 |
| " | Millibars.............. | 1000 | " | Tons of refrig. (U.S. comm.) | $8.32789 \times 10^{-5}$ |
| " | Pounds/sq. inch........ | 14.5038 | " | Watts................ | 0.292875 |
| Baryes............. | Atmospheres........... | $9.86923 \times 10^{-7}$ | B.t.u./min......... | Cal., *kg./min*......... | 0.251996 |
| " | Bars................. | $1 \times 10^{-6}$ | " | Ergs/sec.............. | $1.75725 \times 10^{8}$ |
| " | Dynes/sq. cm.......... | 1 | " | Foot-pounds/min....... | 777.649 |
| " | Grams/sq. cm.......... | 0.001019716 | " | Horsepower............ | 0.0235651 |
| " | Millibars.............. | 0.001 | " | Horsepower (boiler)...... | 0.00179138 |
| Bels............... | Decibels.............. | 10 | " | Horsepower (electric)..... | 0.0235556 |
| Board feet......... | Cu. cm............... | 2359.7372 | " | Horsepower (metric)..... | 0.0238920 |
| " | Cu. feet.............. | 0.083333 | " | Joules/sec............. | 17.5725 |
| " | Cu. inches............ | 144 | " | Kg.-meters/min......... | 107.514 |
| Bolts of cloth....... | Linear feet............ | 120 | " | Kilowatts............. | 0.0175725 |
| " | Meters............... | 36.576 | " | Lb. ice melted/hr........ | 0.41828 |
| Bougie decimales.... | Candles (Int.)......... | 1.00 | " | Tons of refrig. (U.S. comm.) | 0.00499673 |
| B.t.u.............. | B.t.u. (IST.)**........ | 0.999346 | " | Watts................ | 17.5725 |
| " | B.t.u. (mean).......... | 0.998563 | B.t.u. (mean)/min.. | B.t.u. (mean)/hr....... | 60 |
| " | B.t.u. (39°F.)......... | 0.994982 | " | Cal., *kg*. (mean)/hr...... | 15.1197 |
| " | B.t.u. (60°F.)......... | 0.999689 | " | Cal., *kg*. (mean)/min..... | 0.251996 |
| " | Cal. *gm*............. | 251.99576 | " | Ergs/sec.............. | $1.75978 \times 10^{8}$ |
| " | Cal., *gm*. (IST.)....... | 251.831 | " | Foot-pounds/min....... | 778.768 |
| " | Cal., *gm*. (mean)...... | 251.634 | " | Horsepower............ | 0.0235990 |
| " | Cal., *gm*. (20°C.)...... | 252.122 | " | Horsepower (boiler)...... | 0.00179396 |
| " | Cu. cm.-atm........... | 10405.6 | " | Horsepower (electric)..... | 0.0235895 |
| " | Ergs................. | $1.05435 \times 10^{10}$ | " | Horsepower (metric)..... | 0.0239264 |
| " | Foot-poundals......... | 25020.1 | " | Joules/sec............. | 17.5978 |
| " | Foot-pounds........... | 777.649 | " | Kg.-meters/min......... | 107.669 |
| " | Gram-cm.............. | $1.07514 \times 10^{7}$ | " | Kilowatts............. | 0.0175978 |
| " | Hp.-hours............. | 0.000392752 | " | Lb. ice-melted/hr....... | 0.41888 |
| " | Hp.-years............. | $4.48347 \times 10^{-8}$ | B.t.u./lb.......... | Cal., *gm*./gram........ | 0.555555 |
| " | Joules................ | 1054.35 | " | Cu. cm.-atm./gram...... | 22.9405 |
| " | Joules (Int.).......... | 1054.18 | " | Cu. ft.-atm./lb......... | 0.367471 |
| " | Kg.-meters............ | 107.514 | " | Cu. ft.-(lb./sq. in.)/lb.... | 5.40034 |
| " | Kw.-hours............. | 0.000292875 | " | Foot-pounds/lb......... | 777.649 |
| | | | " | Hp.-hr./lb............. | 0.000392752 |

\* Barrel (Brit., liq.) = Barrel (Brit., dry)
\*\* International Steam Table.

# Table 4—7 (continued)

## CONVERSION FACTORS

| To convert from | To | Multiply by | To convert from | To | Multiply by |
|---|---|---|---|---|---|
| B.t.u./lb | Joules/gram | 2.32444 | Calories, *gm.*** | B.t.u. | 0.0039683207 |
| B.t.u. (mean)/lb | Cal., *gm.* (mean)/gram | 0.555555 | " | B.t.u. (IST.) | 0.00396573 |
| " | Cu. cm.-atm./gram | 22.9735 | " | B.t.u. (mean) | 0.00396262 |
| " | Foot-pounds/lb | 778.768 | " | B.t.u. (39°F.) | 0.00394841 |
| " | Hp.-hr./lb | 0.000393317 | " | B.t.u. (60°F.) | 0.00396709 |
| " | Joules/gram | 2.32779 | " | Cal., *gm.* (IST.) | 0.999346 |
| B.t.u./sec | B.t.u./hr | 3600 | " | Cal., *gm.* (mean) | 0.998563 |
| " | B.t.u./min | 60 | " | Cal., *gm.* (15°C.) | 0.999570 |
| " | Cal., *kg.*/hr. | 907.185 | " | Cal., *gm.* (20°C.) | 1.00050 |
| " | Cal., *kg.*/min | 15.1197 | " | Cal., *kg.* | 0.001 |
| " | Cheval-vapeur | 1.43352 | " | Cal., *kg.* (IST.) | 0.000999346 |
| " | Ergs/sec | $1.05435 \times 10^{10}$ | " | Cal., *kg.* (mean) | 0.000998563 |
| " | Foot-pounds/sec | 777.649 | " | Cal., *kg.* (15°C.) | 0.000999570 |
| " | Horsepower | 1.41391 | " | Cal., *kg.* (20°C.) | 0.00100050 |
| " | Horsepower (boiler) | 0.107483 | " | Cu. cm.-atm | 41.2929 |
| " | Horsepower (electric) | 1.41334 | " | Cu. ft.-atm | 0.00145824 |
| " | Horsepower (metric) | 1.43352 | " | Ergs | $4.184 \times 10^{7}$ |
| " | Kg.-meters/sec | 107.514 | " | Foot-poundals | 99.2878 |
| " | Kilowatts | 1.05435 | " | Foot-pounds | 3.08596 |
| " | Kilowatts (Int.) | 1.05418 | " | Gram-cm | 42664.9 |
| " | Watts | 1054.35 | " | Hp.-hours | $1.55857 \times 10^{-6}$ |
| " | Watts (Int.) | 1054.18 | " | Joules | 4.184 |
| B.t.u. (mean)/sec | Ergs/sec | $1.05587 \times 10^{10}$ | " | Joules (Int.) | 4.18331 |
| " | Foot-pounds/sec | 778.768 | " | Kg.-meters | 0.426649 |
| " | Horsepower | 1.41594 | " | Kw.-hours | $1.162222 \times 10^{-6}$ |
| " | Horsepower (boiler) | 0.107637 | " | Liter-atm | 0.0412917 |
| " | Horsepower (electric) | 1.41537 | " | Watt-hours | 0.001162222 |
| " | Horsepower (metric) | 1.43558 | " | Watt-hours (Int.) | 0.00116203 |
| " | Watts | 1055.87 | " | Watt-seconds | 4.184 |
| B.t.u./sq. ft. | Cal., *gm.*/sq. cm. | 0.271246 | Calories, *gm.* (mean) | B.t.u. | 0.00397403 |
| B.t.u./sq.ft. × min.) | Hp./sq. ft. | 0.0235651 | " | Cal., *gm* | 1.00144 |
| " | Kw./sq. ft. | 0.0175725 | " | Cal., *gm.* (IST.) | 1.00078 |
| " | Watts/sq. in. | 0.122031 | " | Cal., *gm.* (20°C.) | 1.00194 |
| Buckets (Brit.) | Cu. cm. | 18184.35 | " | Cal., *kg.* (mean) | 0.001 |
| " | Gallons (Brit.) | 4 | " | Cu. cm.-atm | 41.3523 |
| Bushels (Brit.) | Bags (Brit.) | 0.333333 | " | Cu. ft.-atm | 0.00146034 |
| " | Bushels (U.S.) | 1.032056 | " | Ergs | $4.19002 \times 10^{7}$ |
| " | Cu. cm. | 36368.70 | " | Foot-poundals | 99.4308 |
| " | Cu. feet | 1.284348 | " | Foot-pounds | 3.09040 |
| " | Cu. inches | 2219.354 | " | Hp.-hours | $1.56081 \times 10^{-6}$ |
| " | Dekaliters | 3.636768 | " | Joules | 4.19002 |
| " | Gallons (Brit.) | 8 | " | Joules (Int.) | 4.18933 |
| " | Hectoliters | 0.3636768 | " | Kg.-meters | 0.427263 |
| " | Liters | 36.36768 | " | Kw.-hours | $1.16390 \times 10^{-6}$ |
| Bushels (U.S.)* | Barrels (U.S.), dry | 0.3047647 | " | Liter-atm | 0.0413511 |
| " | Bushels (Brit.) | 0.9689395 | " | Watt-seconds | 4.19002 |
| " | Cu. cm. | 35239.07 | Calories, *gm.* (15°C.) | B.t.u. | 0.00397003 |
| " | Cu. feet | 1.244456 | " | Cal., *gm* | 1.00043 |
| " | Cu. inches | 2150.42 | " | Cal., *gm.* (IST.) | 0.999776 |
| " | Cu. meters | 0.03523907 | " | Cal., *gm.* (mean) | 0.998992 |
| " | Cu. yards | 0.04609096 | " | Cal., *gm.* (20°C.) | 1.00093 |
| " | Gallons (U.S., dry) | 8 | " | Joules | 4.18580 |
| " | Gallons (U.S., liq.) | 9.309177 | " | Joules (Int.) | 4.18511 |
| " | Liters | 35.23808 | Calories, *gm.* (20°C.) | B.t.u. | 0.00396633 |
| " | Ounces (U.S., fluid) | 1191.575 | " | Cal., *gm* | 0.999498 |
| " | Pecks (U.S.) | 4 | " | Cal., *gm.* (IST.) | 0.998845 |
| " | Pints (U.S., dry) | 64 | " | Cal., *gm.* (mean) | 0.998061 |
| " | Quarts (U.S., dry) | 32 | " | Cal., *gm.* (15°C.) | 0.999068 |
| " | Quarts (U.S., liq.) | 37.23671 | " | Joules | 4.18190 |
| Butts (Brit.) | Bushels (U.S.) | 13.53503 | " | Joules (Int.) | 4.18121 |
| " | Cu. feet | 16.84375 | Calories, *kg* | B.t.u. | 3.9683207 |
| " | Cu. meters | 0.4769619 | " | B.t.u. (IST.) | 3.96573 |
| " | Gallons (U.S.) | 126 | " | B.t.u. (mean) | 3.96262 |
| | | | " | B.t.u. (60°F.) | 3.96709 |
| | | | " | Cal., *gm* | 1000 |
| Cable lengths | Fathoms | 120 | " | Cal., *kg.* (mean) | 0.998563 |
| " | Feet | 720 | " | Cal., *kg.* (15°C.) | 0.999570 |
| " | Meters | 219.456 | " | Cal., *kg.* (20°C.) | 1.00050 |
| | | | " | Cu. cm.-atm | 41292.86 |

* Stricken or struck bushel. A heaped bushel for apples of 2747.715 cu. inches was established by the U.S. Court of Customs Appeals on Feb. 15, 1912. A heaped bushel equal to 1¼ stricken bushels is also known.

** This is the calorie as defined by the U.S. National Bureau of Standards and is equal to 4.18400 joules.

## Table 4—7 (continued)

### CONVERSION FACTORS

| To convert from | To | Multiply by |
|---|---|---|
| Calories, *kg.* | Ergs | $4.184 \times 10^{10}$ |
| " | Foot-poundals | 99287.8 |
| " | Foot-pounds | 3085.96 |
| " | Gram-cm | $4.26649 \times 10^7$ |
| " | Hp.-hours | 0.00155857 |
| " | Joules | 4184 |
| " | Kw.-hours | 0.001162222 |
| " | Liter-atm | 41.2917 |
| " | Watt-hours | 1.162222 |
| Calories, *kg.* (mean) | B.t.u. | 3.97403 |
| " | B.t.u. (IST.) | 3.97144 |
| " | B.t.u. (mean) | 3.9683207 |
| " | B.t.u. (60°F.) | 3.97280 |
| " | Cal., *gm* | 1001.44 |
| " | Cal., *gm.* (IST.) | 1000.78 |
| " | Cal., *gm.* (mean) | 1000 |
| " | Cal., *gm.* (15°C.) | 1000.10 |
| " | Cal., *gm.* (20°C.) | 1001.94 |
| " | Ergs | $4.19002 \times 10^{10}$ |
| " | Foot-poundals | 99430.8 |
| " | Foot-pounds | 3090.40 |
| " | Gram-cm | $4.27263 \times 10^7$ |
| " | Hp.-hours | 0.00156081 |
| " | Joules | 4190.02 |
| " | Kg.-meters | 427.263 |
| " | Kw.-hours (Int.) | 0.00116370 |
| " | Liter-atm | 41.3511 |
| " | Watt-hours | 1.16390 |
| Cal., *gm.*/°C | B.t.u./°F | 0.00220462 |
| " | Joules/°F | 2.324444 |
| " | Joules (Int.)/°F | 2.32406 |
| Cal., *gm.*/gram | B.t.u./lb | 1.8 |
| " | Foot-pounds/lb | 1399.77 |
| " | Joules/gram | 4.184 |
| " | Watt-hours/gram | 0.001162222 |
| Cal., *gm.*/(gram × °C) | B.t.u./(lb. × °C.) | 1.8 |
| " | B.t.u./(lb, × °F.) | 1 |
| " | Cal., *kg.*/(kg. × °C.) | 1 |
| " | Joules/(gram × °C.) | 4.184 |
| " | Joules/(lb. × °F.) | 1054.35 |
| Cal., *gm.*/hr | B.t.u./hr | 0.0039683207 |
| " | Ergs/sec | 11622.222 |
| " | Watts | 0.001162222 |
| Cal., *gm.* (mean)/hr | B.t.u. (mean)/hr | 0.0039683207 |
| " | Ergs/sec | 11639.0 |
| " | Watts | 0.00116390 |
| Cal., *kg.*/hr | Watts | 1.162222 |
| Cal., *gm.*/min | B.t.u./min | 0.0039683207 |
| " | Ergs/sec | 697333.3 |
| " | Watts | 0.069733 |
| Cal., *gm.* (mean)/min | B.t.u. (mean)/min | 0.0039683207 |
| " | Ergs/sec | 698337 |
| " | Joules/sec | 0.0698337 |
| " | Watts | 0.0698337 |
| Cal., *kg.*/min | Kg. ice melted/min | 0.012548 |
| " | Lb. ice melted/min | 0.027665 |
| " | Watts | 69.7333 |
| Cal., *gm.*/sec | B.t.u./sec | 0.0039683207 |
| " | Ergs/sec | $4.184 \times 10^7$ |
| " | Foot-pounds/sec | 3.08596 |
| " | Horsepower | 0.00561084 |
| " | Watts | 4.184 |
| Cal., *gm.* (mean)/sec | Ergs/sec | $4.19002 \times 10^7$ |
| " | Watts | 4.19002 |
| Cal., *gm.*/(sec. × sq. cm.) | B.t.u./(hr. × sq. ft.) | 13272.1 |
| " | Cal., *gm.*/(hr. × sq. cm.) | 3600 |
| " | Watts/sq. cm | 4.184 |
| Cal., *gm.*/(sec. × sq. cm. × °C.) | B.t.u./(hr. × sq. ft. × °F.) | 7373.38 |
| Cal., *gm.*/sq. cm | B.t.u./sq. ft | 3.68669 |

| To convert from | To | Multiply by |
|---|---|---|
| Cal., *gm.*-cm. (hr. × sq. cm. × °C.) | B.t.u.-ft. (hr. × sq. ft. × °F.) | 0.0671969 |
| " | B.t.u.-inch (hr. × sq. ft. × °F.) | 0.806363 |
| Cal., *gm.*-cm./sq. cm | B.t.u.-inch/sq. ft. | 1.4514530 |
| Cal., *gm.*-sec. | Planck's constant | $6.31531 \times 10^{33}$ |
| Cal., *gm.*-sec./Avog. No. (chem.)* | Planck's constant | $1.04849 \times 10^{10}$ |
| Cal., *gm.*-sec./Avog. No. (phys.)* | Planck's constant | $1.04821 \times 10^{10}$ |
| Candles (English) | Candles (Int.) | 1.04 |
| " | Hefner units | 1.16 |
| Candles (German) | Candles (English) | 1.01 |
| " | Candles (Int.) | 1.05 |
| " | Hefner units | 1.17 |
| Candles (Int.) | Candles (English) | 0.96 |
| " | Candles (German) | 0.95 |
| " | Candles (pentane) | 1.00 |
| " | Hefner units | 1.11 |
| " | Lumens (Int.)/steradian | 1 |
| Candles (pentane) | Candles (Int.) | 1.00 |
| Candles/sq. cm | Candles/sq. inch | 6.4516 |
| " | Candles/sq. meter | 10000 |
| " | Foot-lamberts | 2918.6351 |
| " | Lamberts | 3.1415927 |
| Candles/sq. ft | Candles/sq. inch | 0.0069444 |
| " | Candles/sq. meter | 10.763910 |
| " | Foot-lamberts | 3.1415927 |
| " | Lamberts | 0.0033815822 |
| Candles/sq. inch | Candles/sq. cm. | 0.15500031 |
| " | Candles/sq. foot | 144 |
| " | Foot-lamberts | 452.38934 |
| " | Lamberts | 0.48694784 |
| Candle power (spher.) | Lumens | 12.566370 |
| Carats (parts of gold per 24 of mixture) | Milligrams/gram | 41.6666 |
| Carats (1877) | Grains | 3.168 |
| " | Milligrams | 205.3 |
| Carats (metric) | Grains | 3.08647 |
| " | Grams | 0.2 |
| " | Milligrams | 200 |
| Carcel units | Candles (Int.) | 9.61 |
| Centals | Kilograms | 45.359237 |
| " | Pounds | 100 |
| Centares | Ares | 0.01 |
| " | Sq. feet | 10.763910 |
| " | Sq. inches | 1550.0031 |
| " | Sq. meters | 1 |
| " | Sq. yards | 1.1959900 |
| Centigrams | Grains | 0.15432358 |
| " | Grams | 0.01 |
| Centiliters | Cu. cm | 10 |
| " | Cu. inches | 0.6102545 |
| " | Liters | 0.01 |
| " | Ounces (U.S., fluid) | 0.3381497 |
| Centimeters | Ångström units | $1 \times 10^8$ |
| " | Feet | 0.032808399 |
| " | Feet (U.S. Survey) | 0.032808333 |
| " | Hands | 0.098425197 |
| " | Inches | 0.39370079 |
| " | Links (Gunter's) | 0.049709695 |
| " | Links (Ramden's) | 0.032808399 |
| " | Meters | 0.01 |
| " | Microns | 10000 |
| " | Miles (naut., Int.) | $5.3995680 \times 10^{-6}$ |
| " | Miles (statute) | $6.2137119 \times 10^{-6}$ |
| " | Millimeters | 10 |
| " | Millimicrons | $1 \times 10^7$ |
| " | Mils | 393.70079 |
| " | Picas (printer's) | 2.3710630 |
| " | Points (printer's) | 28.452756 |

## Table 4—7 (continued)

### CONVERSION FACTORS

| To convert from | To | Multiply by | To convert from | To | Multiply by |
|---|---|---|---|---|---|
| Centimeters | Rods | 0.0019883878 | Circumferences | Minutes | 21600 |
| " | Wave length of orange-red line of krypton 86 | 16507.6373 | " | Radians | 6.2831853 |
| | | | " | Seconds | 1296000 |
| " | Wave length of red line of cadmium | 15531.6413 | Cords | Cord-feet | 8 |
| " | Yards | 0.010936133 | " | Cu. feet | 128 |
| Cm. of Hg (0°C.) | Atmospheres | 0.013157895 | " | Cu. meters | 3.6245734 |
| " | Bars | 0.0133322 | Cord-feet | Cords | 0.125 |
| " | Dynes/sq. cm. | 13332.2 | " | Cu. feet | 16 |
| " | Ft. of $H_2O$ (4°C.) | 0.446050 | Coulombs | Abcoulombs | 0.1 |
| " | Ft. of $H_2O$ (60°F.) | 0.446474 | " | Ampere-hours | 0.0002777 |
| " | In. of Hg (0°C.) | 0.39370079 | " | Ampere-seconds | 1 |
| " | Kg./sq. meter | 135.951 | " | Coulombs (Int.) | 1.000165 |
| " | Pounds/sq. ft. | 27.8450 | " | Electronic charge | $6.24196 \times 10^{18}$ |
| " | Pounds/sq. inch | 0.193368 | " | E.M. cgs. units of electric charge | 0.1 |
| " | Torrs | 10 | | | |
| Cm. of $H_2O$ (4°C.) | Atmospheres | 0.000967814 | " | E.S. cgs. units of electric charge | $2.997930 \times 10^9$ |
| " | Dynes/sq. cm. | 980.638 | " | Faradays (chem.) | $1.036377 \times 10^{-5}$ |
| " | Pounds/sq. inch | 0.0142229 | " | Faradays (phys.) | $1.036086 \times 10^{-5}$ |
| Centimeters/sec. | Feet/min | 1.9685039 | " | Mks. units of electric charge | 1 |
| " | Feet/sec. | 0.032808399 | " | Statcoulombs | $2.997930 \times 10^9$ |
| " | Kilometers/hr. | 0.036 | Coulombs/cu. meter | E.M. cgs. units of volume charge density | $1 \times 10^{-7}$ |
| " | Kilometers/min. | 0.0006 | | | |
| " | Knots (Int.) | 0.019438445 | " | E.S. cgs. units | 2997.930 |
| " | Meters/min. | 0.6 | Coulombs/sq. cm. | Abcoulombs/sq. cm. | 0.1 |
| " | Miles/hr. | 0.022369363 | " | Cgs. units of polarization, *and* surface charge density | 1 |
| " | Miles/min. | 0.00037282272 | Cubic centimeters | Board feet | 0.00042377600 |
| Cm./(sec. × sec.) | Kilometers/(hr. × sec.) | 0.036 | " | Bushels (Brit.) | $2.749617 \times 10^{-5}$ |
| " | Miles/(hr. × sec.) | 0.022369363 | " | Bushels (U.S.) | $2.837759 \times 10^{-5}$ |
| Centimeters/year | Inches/year | 0.39370079 | " | Cu. feet | $3.5314667 \times 10^{-5}$ |
| Centipoises* | Grams/(cm. × sec.) | 0.01 | " | Cu. inches | 0.061023744 |
| " | Poises | 0.01 | " | Cu. meters | $1 \times 10^{-6}$ |
| " | Pound/(ft. × hr.) | 2.4190883 | " | Cu. yards | $1.3079506 \times 10^{-6}$ |
| " | Pounds/(ft. × sec.) | 0.00067196898 | " | Drachms (Brit., fluid) | 0.28156080 |
| Centistokes* | Stokes | 0.01 | " | Drams (U.S., fluid) | 0.27051218 |
| Chains (Gunter's) | Centimeters | 2011.68 | " | Gallons (Brit.) | 0.0002199694 |
| " | Chains (Ramden's) | 0.66 | " | Gallons (U.S., dry) | 0.00022702075 |
| " | Feet | 66 | " | Gallons (U.S., liq.) | 0.00026417205 |
| " | Feet (U.S. Survey) | 65.999868 | " | Gills (Brit.) | 0.007039020 |
| " | Furlongs | 0.1 | " | Gills (U.S.) | 0.0084535058 |
| " | Inches | 792 | " | Liters | 0.001 |
| " | Links (Gunter's) | 100 | " | Ounces (Brit., fluid) | 0.03519510 |
| " | Links (Ramden's) | 66 | " | Ounces (U.S., fluid) | 0.033814023 |
| " | Meters | 20.1168 | " | Pints (U.S., dry) | 0.0018161660 |
| " | Miles (statute) | 0.0125 | " | Pints (U.S., liq.) | 0.0021133764 |
| " | Rods | 4 | " | Quarts (Brit.) | 0.0008798775 |
| " | Yards | 22 | " | Quarts (U.S., dry) | 0.00090808298 |
| Chains (Ramden's) | Centimeters | 3048 | " | Quarts (U.S., liq.) | 0.0010566882 |
| " | Chains (Gunter's) | 1.515151 | Cu. cm./gram | Cu. ft./lb. | 0.016018463 |
| " | Feet | 100 | Cu. cm./sec. | Cu. ft./min. | 0.0021188800 |
| " | Feet (U.S. Survey) | 99.999800 | " | Gal. (U.S.)/min. | 0.015850323 |
| Cheval-vapeur | Horsepower (metric) | 1 | " | Gal. (U.S.)/sec. | 0.00026417205 |
| Cheval-vapeur-heures | Joules | 2647795 | Cu. cm.-atm. | B.t.u. | $9.61019 \times 10^{-5}$ |
| Circles | Degrees | 360 | " | B.t.u. (mean) | $9.59637 \times 10^{-5}$ |
| " | Grades | 400 | " | Cal., *gm* | 0.0242173 |
| " | Minutes | 21600 | " | Cal., *gm.* (mean) | 0.0241824 |
| " | Radians | 6.2831853 | " | Cu. ft.-atm. | $3.5314667 \times 10^{-5}$ |
| " | Signs | 12 | " | Joules | 0.101325 |
| Circular inches | Circular mm | 645.16 | " | Watt-hours | $2.81458 \times 10^{-5}$ |
| " | Sq. cm. | 5.0670748 | Cu. cm.-atm./gram | B.t.u./lb. | 0.0435911 |
| " | Sq. inches | 0.78539816 | " | Cal., *gm.*/gram | 0.0242173 |
| Circular mm | Sq. cm. | 0.0078539816 | " | Cu. ft.-(lb./sq. in.)/lb. | 0.235406 |
| " | Sq. inches | 0.0012173696 | " | Ft.-lb./lb. | 33.8985 |
| " | Sq. mm. | 0.78539816 | " | Joules/gram | 0.101325 |
| Circular mils | Circular inches | $1 \times 10^{-6}$ | " | Kg.-meters/gram | 0.0103323 |
| " | Sq. cm. | $5.0670748 \times 10^{-6}$ | " | Kw.-hr./gram | $2.81458 \times 10^{-8}$ |
| " | Sq. inches | $7.8539816 \times 10^{-7}$ | Cubic decimeters | Cu. cm. | 1000 |
| " | Sq. mm. | 0.00050670748 | " | Cu. feet | 0.035316667 |
| " | Sq. mils | 0.78539816 | " | Cu. inches | 61.023744 |
| Circumferences | Degrees | 360 | " | Cu. meters | 0.001 |
| " | Grades | 400 | | | |

## Table 4—7 (continued)

### CONVERSION FACTORS

| To convert from | To | Multiply by | To convert from | To | Multiply by |
|---|---|---|---|---|---|
| Cubic decimeters | Cu. yards | 0.0013079506 | Cubic inches | Ounces (U.S., fluid) | 0.55411255 |
| " | Liters | 1 | " | Pecks (U.S.) | 0.0018601017 |
| Cubic dekameters | Cu. decimeters | $1 \times 10^6$ | " | Pints (U.S., dry) | 0.029761628 |
| " | Cu. feet | 35314.667 | " | Pints (U.S., liq.) | 0.034632035 |
| " | Cu. inches | $6.1023744 \times 10^7$ | " | Quarts (U.S., dry) | 0.014880814 |
| " | Cu. meters | 1000 | " | Quarts (U.S., liq.) | 0.017316017 |
| " | Liters | 999972 | Cu. in. of $H_2O$ (4°C.) | Pounds of $H_2O$ | 0.0361263 |
| Cubic feet | Acre-feet | $2.2956841 \times 10^{-5}$ | Cu. in. of $H_2O$ (60°F.) | Pounds of $H_2O$ | 0.0360916 |
| " | Board feet | 12 | Cubic meters | Acre-feet | 0.00081071319 |
| " | Bushels (Brit.) | 0.7786049 | " | Barrels (Brit.) | 6.110261 |
| " | Bushels (U.S.) | 0.80356395 | " | Barrels (U.S., dry) | 8.648490 |
| " | Cords (wood) | 0.0078125 | " | Barrels (U.S., liq.) | 8.3864145 |
| " | Cord-feet | 0.0625 | " | Bushels (Brit.) | 27.49617 |
| " | Cu. centimeters | 28316.847 | " | Bushels (U.S.) | 28.377593 |
| " | Cu. meters | 0.028316847 | " | Cu. cm | $1 \times 10^6$ |
| " | Gallons (U.S., dry) | 6.4285116 | " | Cu. feet | 35.314667 |
| " | Gallons (U.S., liq.) | 7.4805195 | " | Cu. inches | 61023.74 |
| " | Liters | 28.316847 | " | Cu. yards | 1.3079506 |
| " | Ounces (Brit., fluid) | 996.6143 | " | Gallons (Brit.) | 219.9694 |
| " | Ounces (U.S., fluid) | 957.50649 | " | Gallons (U.S., liq.) | 264.17205 |
| " | Pints (U.S., liq.) | 59.844156 | " | Hogshead | 4.1932072 |
| " | Quarts (U.S., dry) | 25.714047 | " | Liters | 1000 |
| " | Quarts (U.S., liq.) | 29.922078 | " | Pints (U.S., liq.) | 2113.3764 |
| Cu. ft. of $H_2O$ (39.2°F.) | Pounds of $H_2O$ | 62.4262 | " | Quarts (U.S., liq.) | 1056.6882 |
| Cu. ft. of $H_2O$ (60°F.) | Pounds of $H_2O$ | 62.3663 | " | Steres | 1 |
| Cu. ft./hr | Acre-feet/hr | $2.2956841 \times 10^{-5}$ | Cu. meters/min | Gal. (Brit.)/min | 219.9694 |
| " | Cu. cm./sec | 7.8657907 | " | Gal. (U.S.)/min | 264.1721 |
| " | Cu. ft./day | 24 | " | Liters/min | 999.972 |
| " | Gal. (U.S.)/hr | 7.4805195 | Cu. millimeters | Cu. cm | 0.001 |
| " | Liters/hr | 28.31605 | " | Cu. inches | $6.1023744 \times 10^{-5}$ |
| Cu. ft./min | Acre-feet/hr | 0.0013774105 | " | Cu. meters | $1 \times 10^{-9}$ |
| " | Acre-feet/min | $2.2956841 \times 10^{-5}$ | " | Minims (Brit.) | 0.01689365 |
| " | Cu. cm./sec | 471.94744 | " | Minims (U.S.) | 0.016230731 |
| " | Cu. ft./hr | 60 | Cu. yards | Bushels (Brit.) | 21.02233 |
| " | Gal. (U.S.)/min | 7.4805195 | " | Bushels (U.S.) | 21.696227 |
| " | Liters/sec | 0.4719342 | " | Cu. cm | 764554.86 |
| Cu. ft./lb | Cu. cm./gram | 62.427961 | " | Cu. feet | 27 |
| " | Millimeters/gram | 62.42621 | " | Cu. inches | 46656 |
| Cu. ft./sec | Acre-inches/hr | 0.99173553 | " | Cu. meters | 0.76455486 |
| " | Cu. cm./sec | 28316.847 | " | Gallons (Brit.) | 168.1787 |
| " | Cu. yards/min | 2.222222 | " | Gallons (U.S., dry) | 173.56981 |
| " | Gal. (U.S.)/min | 448.83117 | " | Gallons (U.S., liq.) | 201.97403 |
| " | Liters/min | 1698.963 | " | Liters | 764.55486 |
| " | Liters/sec | 28.31605 | " | Quarts (Brit.) | 672.7146 |
| Cu. ft. of $H_2O$ (60°F.)/sec | Lb. of $H_2O$/min | 3741.98 | " | Quarts (U.S., dry) | 694.27926 |
| Cu. ft.-atm | B.t.u. | 2.72130 | " | Quarts (U.S., liq.) | 807.89610 |
| " | Cal., *gm* | 685.756 | Cu. yd./min | Cu. ft./sec | 0.45 |
| " | Cu. cm.-atm | 28316.847 | " | Gal. (U.S.)/sec | 3.3662338 |
| " | Cu. ft.-(lb/sq. in.) | 14.6960 | " | Liters/sec | 12.74222 |
| " | Foot-pounds | 2116.22 | Cubits | Centimeters | 45.72 |
| " | Hp.-hours | 0.00106880 | " | Feet | 1.5 |
| " | Joules | 2869.20 | " | Inches | 18 |
| " | Kg.-meters | 292.577 | | | |
| " | Kw.-hours | 0.000797001 | Daltons (chem.) | Grams | $1.66024 \times 10^{-24}$ |
| Cubic inches | Barrels (Brit.) | 0.0001001292 | Daltons (phys.) | Grams | $1.65979 \times 10^{-24}$ |
| " | Barrels (U.S., dry) | 0.00014172336 | Days (mean solar) | Days (sidereal) | 1.00273791 |
| " | Board feet | 0.0069444 | " | Hours (mean solar) | 24 |
| " | Bushels (Brit.) | 0.0004505815 | " | Hours (sidereal) | 24.065710 |
| " | Bushels (U.S.) | 0.00046502544 | " | Years (calendar) | 0.0027397260 |
| " | Cu. cm | 16.387064 | " | Years (sidereal) | 0.0027378031 |
| " | Cu. feet | 0.00057870370 | " | Years (tropical) | 0.0027379093 |
| " | Cu. meters | $1.6387064 \times 10^{-5}$ | Days (sidereal) | Days (mean solar) | 0.99726957 |
| " | Cu. yards | $2.1433470 \times 10^{-5}$ | " | Hours (mean solar) | 23.934470 |
| " | Drams (U.S., fluid) | 4.4329004 | " | Hours (sidereal) | 24 |
| " | Gallons (Brit.) | 0.003604652 | " | Minutes (mean solar) | 1436.0682 |
| " | Gallons (U.S., dry) | 0.0037202035 | " | Minute (sidereal) | 1440 |
| " | Gallons (U.S., liq.) | 0.0043290043 | " | Second (sidereal) | 86400 |
| " | Liters | 0.016387064 | " | Years (calendar) | 0.0027322454 |
| " | Milliliters | 16.387064 | " | Years (sidereal) | 0.0027303277 |
| " | Ounces (Brit., fluid) | 0.5767444 | " | Years (tropical) | 0.0027304336 |
| | | | Decibels | Bels | 0.1 |
| | | | Decimeters | Centimeters | 10 |

# Table 4—7 (continued)

## CONVERSION FACTORS

| To convert from | To | Multiply by | To convert from | To | Multiply by |
|---|---|---|---|---|---|
| Decimeters | Feet | 0.32808399 | Dynes/sq. cm | Cm. of $H_2O$ (4°C.) | 0.001019745 |
| " | Feet (U.S. Survey) | 0.328083333 | " | Grams/sq. cm | 0.001019716 |
| " | Inches | 3.9370079 | " | In. of Hg (32°F.) | $2.95300 \times 10^{-5}$ |
| " | Meters | 0.1 | " | In. of $H_2O$ (4°C.) | 0.000401474 |
| Decisteres | Cu. meters | 0.1 | " | Kg./sq. meter | 0.01019716 |
| Degrees | Circles | 0.0027777 | " | Poundals/sq. in. | 0.00046664510 |
| " | Minutes | 60 | " | Pounds/sq. in. | $1.450377 \times 10^{-5}$ |
| " | Quadrants | 0.0111111 | Dyne-centimeters | Ergs | 1 |
| " | Radians | 0.017453293 | " | Foot-poundals | $2.3730360 \times 10^{-6}$ |
| " | Seconds | 3600 | " | Foot-pounds | $7.37562 \times 10^{-8}$ |
| Degrees/cm | Radians/cm | 0.017453293 | " | Gram-cm | 0.001019716 |
| Degrees/foot | Radians/cm | 0.00057261458 | " | Inch-pounds | $8.85075 \times 10^{-7}$ |
| Degrees/inch | Radian/cm | 0.0068713750 | " | Kg.-meters | $1.019716 \times 10^{-8}$ |
| Degrees/min | Degrees/sec | 0.0166666 | " | Newton-meters | $1 \times 10^{-7}$ |
| " | Radians/sec | 0.00029088821 | Electron volts | Ergs | $1.60219 \times 10^{-12}$ |
| " | Revolutions/sec | $4.629629 \times 10^{-5}$ | " | Grams | $1.78253 \times 10^{-33}$ |
| Degrees/sec | Radians/sec | 0.017453293 | Electronic charges | Abcoulombs | $1.60209 \times 10^{-20}$ |
| " | Revolutions/min | 0.166666 | " | Coulombs | $1.60209 \times 10^{-19}$ |
| " | Revolutions/sec | 0.0027777 | " | Statcoulombs | $4.80296 \times 10^{-10}$ |
| Dekaliters | Pecks (U.S.) | 1.135136 | Electronic charges/kg. | Statcoulombs/dyne | $4.89766 \times 10^{-16}$ |
| " | Pints (U.S., dry) | 18.16217 | E.S. cgs. units of induction flux | E.M. cgs. units | $2.997930 \times 10^{10}$ |
| Dekameters | Centimeters | 1000 | E.S. cgs. units of magnetic charge | E.M. cgs. units | $2.997930 \times 10^{10}$ |
| " | Feet | 32.808399 | E.S. cgs. units of magnetic field intensity | E.M. cgs. units | $3.335635 \times 10^{-11}$ |
| " | Feet (U.S. Survey) | 32.808333 | Ells | Centimeters | 114.3 |
| " | Inches | 393.70079 | " | Inches | 45 |
| " | Kilometers | 0.01 | Ergs | B.t.u. | $9.48451 \times 10^{-11}$ |
| " | Meters | 10 | " | Cal., gm. | $2.39006 \times 10^{-8}$ |
| " | Yards | 10.93613 | " | Cal., kg. | $2.39006 \times 10^{-11}$ |
| Demals | Gram-equiv./cu. decimeter | 1 | " | Cal., kg. (20°C.) | $2.39126 \times 10^{-11}$ |
| Drachms (Brit., fluid) | Cu. cm. | 3.551631 | " | Cu. cm.-atm | $9.86923 \times 10^{-7}$ |
| " | Cu. inches | 0.2167338 | " | Cu. ft.-atm | $3.48529 \times 10^{-11}$ |
| " | Drams (U.S., fluid) | 0.9607594 | " | Cu. ft.-(lb./sq. in.) | $5.12196 \times 10^{-10}$ |
| " | Milliliters | 3.551531 | " | Dyne-cm | 1 |
| Drams (apoth. or troy) | Drams (avdp.) | 2.1942857 | " | Electron volts | $6.24145 \times 10^{11}$ |
| " | Grains | 60 | " | Foot-poundals | $2.3730360 \times 10^{-6}$ |
| " | Grams | 3887.9346 | " | Foot-pounds | $7.37562 \times 10^{-8}$ |
| " | Ounces (apoth. or troy) | 0.125 | " | Gram-cm | 0.001019716 |
| " | Ounces (avdp.) | 0.13714286 | " | Joules | $1 \times 10^{-7}$ |
| " | Scruples (apoth.) | 3 | " | Joules (Int.) | $9.99835 \times 10^{-8}$ |
| Drams (avdp.) | Drams (apoth. or troy) | 0.455729166 | " | Kw.-hours | $2.777777 \times 10^{-14}$ |
| " | Grains | 27.34375 | " | Kg.-meters | $1.019716 \times 10^{-8}$ |
| " | Grams | 1.7718452 | " | Liter-atm | $9.86895 \times 10^{-10}$ |
| " | Ounces (apoth. or troy) | 0.056966146 | " | Watt-sec | $1 \times 10^{-7}$ |
| " | Ounces (avdp.) | 0.0625 | Ergs/(gram-mol. × °C.) | Foot-pounds/(lb.-mol. × °F.) | $1.85863 \times 10^{-5}$ |
| " | Pennyweights | 1.1393229 | Ergs/sec | B.t.u./min | $5.69071 \times 10^{-9}$ |
| " | Pounds (apoth. or troy) | 0.0047471788 | " | Cal., gm./min | $1.43403 \times 10^{-6}$ |
| " | Pounds (avdp.) | 0.00390625 | " | Dyne-cm./sec | 1 |
| " | Scruples (apoth.) | 1.3671875 | " | Foot-pounds/min | $4.42537 \times 10^{-6}$ |
| Drams (U.S., fluid) | Cu. cm. | 3.6967162 | " | Gram-cm./sec | 0.001019716 |
| " | Cu. inches | 0.22558594 | " | Horsepower | $1.34102 \times 10^{-10}$ |
| " | Drachms (Brit., fluid) | 1.040843 | " | Joules/sec | $1 \times 10^{-7}$ |
| " | Gills (U.S.) | 0.03125 | " | Kilowatts | $1 \times 10^{-10}$ |
| " | Milliliters | 3.696588 | " | Watts | $1 \times 10^{-7}$ |
| " | Minims (U.S.) | 60 | Ergs/sq. cm | Dynes/cm | 1 |
| " | Ounces (U.S., fluid) | 0.125 | " | Ergs/sq. mm | 0.01 |
| " | Pints (U.S., liq.) | 0.0078125 | Ergs/sq. mm | Dynes/cm | 100 |
| Dynes | Grains | 0.01573663 | " | Ergs/sq. cm | 100 |
| " | Grams | 0.001019716 | Erg-sec | Planck's constant | $1.50932 \times 10^{26}$ |
| " | Newtons | 0.00001 | Farads | Abfarads | $1 \times 10^{-9}$ |
| " | Poundals | $7.2330138 \times 10^{-5}$ | " | E.M. cgs. units | $1 \times 10^{-9}$ |
| " | Pounds | $2.248089 \times 10^{-6}$ | " | E.S. cgs. units | $8.987584 \times 10^{11}$ |
| Dynes/cm | Ergs/sq. cm | 1 | " | Farads (Int.) | 1.000495 |
| " | Ergs/sq. mm | 0.01 | " | Microfarads | $1 \times 10^{6}$ |
| " | Grams/cm | 0.001019716 | " | Statfarads | $8.98758 \times 10^{11}$ |
| " | Poundals/inch | 0.00018371855 | Farads (Int.) | Farads | 0.999505 |
| Dynes/cu. cm | Grams/cu. cm | 0.001019716 | Fathoms | Centimeters | 182.88 |
| " | Poundals/cu. inch | 0.0011852786 | | | |
| Dynes/sq. cm | Atmospheres | $9.86923 \times 10^{-7}$ | | | |
| " | Bars | $1 \times 10^{-6}$ | | | |
| " | Baryes | 1 | | | |
| " | Cm. of Hg (0°C.) | $7.50062 \times 10^{-5}$ | | | |

## Table 4—7 (continued)

### CONVERSION FACTORS

| To convert from | To | Multiply by | To convert from | To | Multiply by |
|---|---|---|---|---|---|
| Fathoms | Feet | 6 | Feet/(sec. × sec.) | Meters/(sec. × sec.) | 0.3048 |
| " | Inches | 72 | " | Miles/(hr. × sec.) | 0.68181818 |
| " | Meters | 1.8288 | Firkins (Brit.) | Bushels (Brit.) | 1.125 |
| " | Miles (naut., Int.) | 0.00098747300 | " | Cu. cm | 40914.79 |
| " | Miles (statute) | 0.001136363 | " | Cu. feet | 1.444892 |
| " | Yards | 2 | " | Firkins (U.S.) | 1.200949 |
| Feet | Centimeters | 30.48 | " | Gallons (Brit.) | 9 |
| " | Chains (Gunter's) | 0.01515151 | " | Liters | 40.91364 |
| " | Fathoms | 0.166666 | " | Pints (Brit.) | 72 |
| " | Feet (U.S. Survey) | 0.99999800 | Firkins (U.S.) | Barrels (U.S., dry) | 0.29464286 |
| " | Furlongs | 0.00151515 | " | Barrels (U.S., liq.) | 0.28571429 |
| " | Inches | 12 | " | Bushels (U.S.) | 0.96678788 |
| " | Meters | 0.3048 | " | Cu. feet | 1.203125 |
| " | Microns | 304800 | " | Firkins (Brit.) | 0.8326747 |
| " | Miles (naut., Int.) | 0.00016457883 | " | Liters | 34.06775 |
| " | Miles (statute) | 0.000189393 | " | Pints (U.S., liq.) | 72 |
| " | Rods | 0.060606 | Foot-candles | Lumens/sq. ft | 1 |
| " | Ropes (Brit.) | 0.05 | " | Lumens/sq. meter | 10.763910 |
| " | Yards | 0.333333 | " | Lux | 10.763910 |
| Feet (U.S. Survey) | Centimeters | 30.480061 | " | Milliphots | 1.0763910 |
| " | Chains (Gunter's) | 0.015151545 | Foot-lamberts | Candles/sq. cm | 0.00034262591 |
| " | Chains (Ramden's) | 0.010000020 | " | Candles/sq. ft | 0.31830989 |
| " | Feet | 1.0000020 | " | Millilamberts | 1.0763910 |
| " | Inches | 12.000024 | " | Lamberts | 0.0010763910 |
| " | Links (Gunter's) | 1.5151545 | " | Lumens/sq. ft | 1 |
| " | Links (Ramden's) | 1.0000020 | Foot-poundals | B.t.u. | $3.99678 \times 10^{-5}$ |
| " | Meters | 0.30480061 | " | B.t.u. (IST.) | $3.99417 \times 10^{-5}$ |
| " | Miles (statute) | 0.00018939432 | " | B.t.u. (mean) | $3.99104 \times 10^{-5}$ |
| " | Rods | 0.060606182 | " | Cal., gm | 0.0100717 |
| " | Yards | 0.33333400 | " | Cal., gm. (IST.) | 0.0100651 |
| Feet of air (1 atm., 60°F.) | Atmospheres | $3.6083 \times 10^{-5}$ | " | Cal., gm. (mean) | 0.0100573 |
| " | Ft. of Hg (32°F.) | 0.00089970 | " | Cu. cm-atm | 0.415890 |
| " | Ft. of H₂O (60°F.) | 0.0012244 | " | Cu. ft.-atm | $1.46870 \times 10^{-5}$ |
| " | In. of Hg (32°F.) | 0.0010796 | " | Dyne-cm | $4.2140110 \times 10^{5}$ |
| " | Pounds/sq. inch | 0.00053027 | " | Ergs | $4.2140110 \times 10^{5}$ |
| Feet of Hg (32°F.) | Cm. of Hg (0°C.) | 30.48 | " | Foot-pounds | 0.0310810 |
| " | Ft. of H₂O (60°F.) | 13.6085 | " | Hp.-hours | $1.56974 \times 10^{-8}$ |
| " | In. of H₂O (60°F.) | 163.302 | " | Joules | 0.042140110 |
| " | Ounces/sq. inch | 94.3016 | " | Joules (Int.) | 0.0421332 |
| " | Pounds/sq. inch | 5.89385 | " | Kg.-meters | 0.00429710 |
| Feet of H₂O (4°C.) | Atmospheres | 0.0294990 | " | Kw.-hours | $1.17056 \times 10^{-8}$ |
| " | Cm. of Hg (0°C.) | 2.24192 | " | Liter-atm | 0.000415879 |
| " | Dynes/sq. cm | 29889.8 | Foot-pounds | B.t.u. | 0.00128593 |
| " | Grams/sq. cm | 30.4791 | " | B.t.u. (IST.) | 0.00128509 |
| " | In. of Hg (32°F.) | 0.882646 | " | B.t.u. (mean) | 0.00128408 |
| " | Kg./sq. meter | 304.791 | " | Cal., gm | 0.324048 |
| " | Pounds/sq. inch | 0.433515 | " | Cal., gm. (IST.) | 0.323836 |
| Feet/hour | Cm./hr | 30.48 | " | Cal., gm. (mean) | 0.323582 |
| " | Cm./min | 0.508 | " | Cal., gm. (20°C.) | 0.324211 |
| " | Cm./sec | 0.0084666 | " | Cal., kg | 0.000324048 |
| " | Feet/min | 0.0166666 | " | Cal., kg. (IST.) | 0.000323836 |
| " | Inches/hr | 12 | " | Cal., kg. (mean) | 0.000323582 |
| " | Kilometers/hr | 0.0003048 | " | Cu. ft.-atm | 0.000472541 |
| " | Kilometers/min | $5.08 \times 10^{-6}$ | " | Dyne-cm | $1.35582 \times 10^{7}$ |
| " | Knots (Int.) | 0.0001645788 | " | Ergs | $1.35582 \times 10^{7}$ |
| " | Miles/hr | 0.000189393 | " | Foot-poundals | 32.1740 |
| " | Miles/min | $3.156565 \times 10^{-6}$ | " | Gram-cm | 13825.5 |
| " | Miles/sec | $5.2609428 \times 10^{-8}$ | " | Hp.-hours | $5.05050 \times 10^{-7}$ |
| Feet/minute | Cm./sec | 0.508 | " | Joules | 1.35582 |
| " | Feet/sec | 0.0166666 | " | Kg.-meters | 0.138255 |
| " | Kilometers/hr | 0.018288 | " | Kw.-hours | $3.76616 \times 10^{-7}$ |
| " | Meters/min | 0.3048 | " | Kw.-hours (Int.) | $3.76554 \times 10^{-7}$ |
| " | Meters/sec | 0.00508 | " | Liter-atm | 0.0133805 |
| " | Miles/hr | 0.01136363 | " | Newton-meters | 1.3558180 |
| Feet/second | Cm./sec | 30.48 | " | Lb. H₂O evap. from and at 212°F. | $1.3245 \times 10^{-6}$ |
| " | Kilometers/hr | 1.09728 | " | Watt-hours | 0.000376616 |
| " | Kilometers/min | 0.018288 | Foot-pounds/hr | B.t.u./min | $2.14321 \times 10^{-5}$ |
| " | Meters/min | 18.288 | " | B.t.u. (mean)/min | $2.14013 \times 10^{-5}$ |
| " | Miles/hr | 0.68181818 | " | Cal. gm./min | 0.00540080 |
| " | Miles/min | 0.01136363 | " | Cal., gm. (mean)/min | 0.00539304 |
| Feet/(sec. × sec.) | Kilometers/(hr. × sec.) | 1.09728 | " | Ergs/min | $2.25970 \times 10^{6}$ |

# Table 4—7 (continued)

## CONVERSION FACTORS

| To convert from | To | Multiply by | To convert from | To | Multiply by |
|---|---|---|---|---|---|
| Foot-pounds/hr | Foot-pounds/min | 0.0166666 | Gallons (U.S., dry) | Cu. inches | 268.8025 |
| " | Horsepower | $5.050505 \times 10^{-7}$ | " | Gallons (U.S., liq.) | 1.16364719 |
| " | Horsepower (metric) | $5.12055 \times 10^{-7}$ | " | Liters | 4.404760 |
| " | Kilowatts | $3.76616 \times 10^{-7}$ | Gallons (U.S., liq.) | Acre-feet | $3.0688833 \times 10^{-6}$ |
| " | Watts | 0.000376616 | " | Barrels (U.S., liq.) | 0.031746032 |
| " | Watts (Int.) | 0.000376554 | " | Barrels (petroleum, U.S.) | 0.023809524 |
| Foot-pounds/min | B.t.u./sec | $2.14321 \times 10^{-5}$ | " | Bushels (U.S.) | 0.10742088 |
| " | B.t.u. (mean)/sec | $2.14013 \times 10^{-5}$ | " | Cu. centimeters | 3785.4118 |
| " | Cal., gm./sec | 0.00540080 | " | Cu. feet | 0.133680555 |
| " | Cal., gm. (mean)/sec | 0.00539304 | " | Cu. inches | 231 |
| " | Ergs/sec | $2.25970 \times 10^{5}$ | " | Cu. meters | 0.0037854118 |
| " | Foot-pounds/sec | 0.0166666 | " | Cu. yards | 0.0049511317 |
| " | Horsepower | $3.030303 \times 10^{-5}$ | " | Gallons (Brit.) | 0.8326747 |
| " | Horsepower (metric) | $3.07233 \times 10^{-5}$ | " | Gallons (U.S., dry) | 0.85936701 |
| " | Joules/sec | 0.0225970 | " | Gallons (wine) | 1 |
| " | Joules (Int.)/sec | 0.0225932 | " | Gills (U.S.) | 32 |
| " | Kilowatts | $2.25970 \times 10^{-5}$ | " | Liters | 3.7854118 |
| " | Watts | 0.0225970 | " | Minims (U.S.) | 61440 |
| Foot-pounds/lb | B.t.u./lb | 0.00128593 | " | Ounces (U.S., fluid) | 128 |
| " | B.t.u. (IST.)/lb | 0.00128509 | " | Pints (U.S., liq.) | 8 |
| " | B.t.u. (mean)/lb | 0.00128408 | " | Quarts (U.S., liq.) | 4 |
| " | Cal., gm./gm | 0.000714404 | Gallons (U.S.) of $H_2O$ (4°C.) in air | Lb. of $H_2O$ | 8.33585 |
| " | Cal., gm. (IST.)/gram | 0.000713937 | Gallons (U.S.) of $H_2O$ (60°F.) in air | Lb. of $H_2O$ | 8.32823 |
| " | Cal., gm. (mean)/gram | 0.000713377 | Gallons (U.S.)/day | Cu. ft./hr | 0.0055700231 |
| " | Hp.-hr./lb | $5.05050 \times 10^{-7}$ | Gallons (Brit.)/hr | Cu. meters/min | $7.576812 \times 10^{-5}$ |
| " | Joules/gram | 0.00298907 | Gallons (U.S.)/hr | Acre-feet/hr | $3.0688833 \times 10^{-6}$ |
| " | Kg.-meters/gram | 0.000304800 | " | Cu. ft./hr | 0.1336805 |
| " | Kw.-hr./gram | $8.30296 \times 10^{-10}$ | " | Cu. meters/min | $6.3090197 \times 10^{-5}$ |
| Foot-pounds/sec | B.t.u./min | 0.0771556 | " | Cu. yd./min | $8.2518861 \times 10^{-5}$ |
| " | B.t.u. (mean)/min | 0.0770447 | " | Liters/hr | 3.7854118 |
| " | B.t.u./sec | 0.00128593 | Gal. (Brit.)/sec | Cu. cm./sec | 4546.087 |
| " | B.t.u. (mean)/sec | 0.00128408 | Gal. (U.S.)/sec | Cu. cm./sec | 3785.4118 |
| " | Cal., gm./sec | 0.324048 | " | Cu. ft./min | 8.020833 |
| " | Cal., gm. (mean)/sec | 0.323582 | " | Cu. yd./min | 0.29706790 |
| " | Ergs/sec | $1.35582 \times 10^{7}$ | " | Liters/min | 227.1183 |
| " | Gram-cm./sec | 13825.5 | Gammas | Grams | $1 \times 10^{-6}$ |
| " | Horsepower | 0.00181818 | " | Micrograms | 1 |
| " | Joules/sec | 1.35582 | Gausses | E.M. cgs. units of magnetic flux density | 1 |
| " | Kilowatts | 0.00135582 | " | E.S. cgs. units | $3.335635 \times 10^{-11}$ |
| " | Watts | 1.35582 | " | Gausses (Int.) | 0.999670 |
| " | Watts (Int.) | 1.35559 | " | Maxwells/sq. cm | 1 |
| Furlongs | Centimeters | 20116.8 | " | Lines/sq. cm | 1 |
| " | Chains (Gunter's) | 10 | " | Lines/sq. inch | 6.4516 |
| " | Chains (Ramden's) | 6.6 | Gausses (Int.) | Gausses | 1.000330 |
| " | Feet | 660 | Gausses/oersted | E.M. cgs. units of permeability | 1 |
| " | Inches | 7920 | " | E.S. cgs. units | $1.112646 \times 10^{-21}$ |
| " | Meters | 201.168 | Geepounds | Slugs | 1 |
| " | Miles (naut., Int.) | 0.10862203 | " | Kilograms | 14.5939 |
| " | Miles (statute) | 0.125 | Gigameters | Meters | $1 \times 10^{9}$ |
| " | Rods | 40 | Gilberts | Abampere-turns | 0.079577472 |
| " | Yards | 220 | " | Ampere-turns | 0.79577472 |
| | | | " | E.M. cgs. units of mmf., or magnetic potential | 1 |
| Gallons (Brit.) | Barrels (Brit.) | 0.027777 | " | E.S. cgs. units | $2.997930 \times 10^{10}$ |
| " | Bushels (Brit.) | 0.125 | " | Gilberts (Int.) | 1.000165 |
| " | Cu. centimeters | 4546.087 | Gilberts (Int.) | Gilberts | 0.999835 |
| " | Cu. feet | 0.1605436 | Gilberts/cm | Ampere-turns/cm | 0.79577472 |
| " | Cu. inches | 277.4193 | " | Ampere-turns/in | 2.0212678 |
| " | Drachms (Brit. fluid) | 1280 | " | Oersteds | 1 |
| " | Firkins (Brit.) | 0.111111 | Gilberts/maxwell | Ampere-turns/weber | $7.957747 \times 10^{7}$ |
| " | Gallons (U.S., liq.) | 1.200949 | " | E.M. cgs. units of reluctance | 1 |
| " | Gills (Brit.) | 32 | " | E.S. cgs. units | $8.987584 \times 10^{20}$ |
| " | Liters | 4.545960 | Gills (Brit.) | Cu. cm | 142.0652 |
| " | Minims (Brit.) | 76800 | " | Gallons (Brit.) | 0.03125 |
| " | Ounces (Brit., fluid) | 160 | " | Gills (U.S.) | 1.200949 |
| " | Ounces (U.S., fluid) | 153.7215 | " | Liters | 0.1420613 |
| " | Pecks (Brit.) | 0.5 | " | Ounces (Brit., fluid) | 5 |
| " | Lb. of $H_2O$ (62°F.) | 10 | | | |
| Gallons (U.S., dry) | Barrels (U.S., dry) | 0.038095592 | | | |
| " | Barrels (U.S., liq.) | 0.036941181 | | | |
| " | Bushels (U.S.) | 0.125 | | | |
| " | Cu. centimeters | 4404.8828 | | | |
| " | Cu. feet | 0.15555700 | | | |

## Table 4—7 (continued)

### CONVERSION FACTORS

| To convert from | To | Multiply by | To convert from | To | Multiply by |
|---|---|---|---|---|---|
| Gills (Brit.)........ | Ounces (U.S., fluid)...... | 4.803764 | Grams/cu. cm...... | Pounds/gal. (U.S., dry).... | 9.7111064 |
| " | Pints (Brit.)............. | 0.25 | " | Pounds/gal. (U.S., liq.).... | 8.3454044 |
| Gills (U.S.)...... | Cu. cm............. | 118.29412 | Grams/cu. meter..... | Grains/cu. ft.......... | 0.43699572 |
| " | Cu. inches........... | 7.21875 | Grams/liter..... | Parts/million*........... | 1000 |
| " | Drams (U.S., fluid)....... | 32 | " | Lb./cu. ft. | 0.06242621 |
| " | Gallons (U.S., liq.)....... | 0.03125 | " | Lb./gal. (U.S.)...... | $8.345171 \times 10^{-3}$ |
| " | Gills (Brit.)........ | 0.8326747 | Grams/milliliter..... | Grams/cu. cm......... | 1 |
| " | Liters............. | 0.1182908 | " | Pounds/cu. ft........ | 62.42621 |
| " | Minims (U.S.).......... | 1920 | " | Pounds/gallon (U.S.)..... | 8.345171 |
| " | Ounces (U.S., fluid)...... | 4 | Grams/sq. cm...... | Atmospheres........... | 0.000967841 |
| " | Pints (U.S., liq.)......... | 0.25 | " | Bars.... | 0.000980665 |
| " | Quarts (U.S., liq.)........ | 0.125 | " | Cm. of Hg. (0°C.)....... | 0.0735559 |
| Grades....... | Circles............. | 0.0025 | " | Dynes/sq. cm......... | 980.665 |
| " | Circumferences........ | 0.0025 | " | In. of Hg (32°F.)....... | 0.0289590 |
| " | Degrees............ | 0.9 | " | Kg./sq. meter........ | 10 |
| " | Minutes............ | 54 | " | Mm. of Hg (0°C.)....... | 0.735559 |
| " | Radians............ | 0.015707963 | " | Poundals/sq. inch....... | 0.457623 |
| " | Revolutions.......... | 0.0025 | " | Pounds/sq. inch........ | 0.014223343 |
| " | Seconds............ | 3240 | Grams/ton (long).... | Milligrams/kg......... | 0.98420653 |
| Grains...... | Carats (metric)........... | 0.32399455 | Grams/ton (short).... | Milligrams/kg......... | 1.1023113 |
| " | Drams (apoth. or troy).... | 0.016666 | Grams-cm...... | B.t.u............. | $9.30113 \times 10^{-8}$ |
| " | Drams (avdp.)......... | 0.036571429 | " | B.t.u. (IST.)......... | $9.29505 \times 10^{-8}$ |
| " | Dynes............. | 63.5460 | " | B.t.u. (mean)........ | $9.28776 \times 10^{-8}$ |
| " | Grams.... | 0.06479891 | " | Cal., gm. (IST.)........ | $2.34385 \times 10^{-5}$ |
| " | Milligrams........... | 64.79891 | " | Cal., gm. (IST.)........ | $2.34231 \times 10^{-5}$ |
| " | Ounces (apoth. or troy)... | 0.0020833 | " | Cal., gm. (mean)........ | $2.34048 \times 10^{-5}$ |
| " | Ounces (avdp.)......... | 0.0022857143 | " | Cal., gm. (15°C.)....... | $2.34284 \times 10^{-5}$ |
| " | Pennyweights......... | 0.041666 | " | Cal., gm. (20°C.)....... | $2.34502 \times 10^{-5}$ |
| " | Pounds (apoth. or troy)... | 0.000173611 | " | Cal., kg............ | $2.34385 \times 10^{-5}$ |
| " | Pounds (avdp.)........ | 0.00014285714 | " | Cal., kg. (IST.)........ | $2.34231 \times 10^{-5}$ |
| " | Scruples (apoth.)........ | 0.05 | " | Cal., kg. (mean)....... | $2.34048 \times 10^{-5}$ |
| " | Tons (metric)........ | $6.479891 \times 10^{-5}$ | " | Dyne-cm............ | 980.665 |
| Grains/cu. ft. | Grams/cu. meter...... | 2.2883519 | " | Ergs.... | 980.665 |
| Grains/gal. (U.S.)... | Parts/million*........ | 17.11854 | " | Foot-poundals......... | 0.00232715 |
| " | Pounds/million gal. | 142.8571 | " | Foot-pounds.......... | $7.2330138 \times 10^{-5}$ |
| Grams...... | Carats (metric)......... | 5 | " | Hp.-hours........... | $3.65303 \times 10^{-11}$ |
| " | Decigrams........... | 10 | " | Joules............. | $9.80665 \times 10^{-5}$ |
| " | Dekagrams........... | 0.1 | " | Kw.-hours........... | $2.72407 \times 10^{-11}$ |
| " | Drams (apoth. or troy)... | 0.25720597 | " | Kw.-hours (Int.)........ | $2.72362 \times 10^{-11}$ |
| " | Drams (avdp.)......... | 0.56438339 | " | Newton-meters........ | $9.80665 \times 10^{-5}$ |
| " | Dynes............. | 980.665 | " | Watt-hours.......... | $2.72407 \times 10^{-8}$ |
| " | Grains............. | 15.432358 | Gram-cm./sec. | B.t.u./sec........... | $9.30113 \times 10^{-8}$ |
| " | Kilograms............ | 0.001 | " | Cal., gm./sec......... | $2.34385 \times 10^{-5}$ |
| " | Micrograms........... | $1 \times 10^{6}$ | " | Ergs-sec............ | 980.665 |
| " | Myriagrams........... | 0.0001 | " | Foot-pounds/sec....... | $7.2330138 \times 10^{-5}$ |
| " | Ounces (apoth. or troy)... | 0.032150737 | " | Horsepower.......... | $1.31509 \times 10^{-7}$ |
| " | Ounces (avdp.)......... | 0.035273962 | " | Joules/sec........... | $9.80665 \times 10^{-5}$ |
| " | Pennyweights......... | 0.64301493 | " | Kilowatts.......... | $9.80665 \times 10^{-8}$ |
| " | Poundals........... | 0.0709316 | " | Kilowatts (Int.)........ | $9.80503 \times 10^{-8}$ |
| " | Pounds (apoth. or troy)... | 0.0026792289 | " | Watts............. | $9.80665 \times 10^{-5}$ |
| " | Pounds (avdp.)......... | 0.0022046226 | Gram/sq. cm...... | Pounds/sq. inch....... | 0.000341717 |
| " | Scruples (apoth.)........ | 0.77161792 | Gram wt.-sec./sq. cm. | Poises............. | 980.665 |
| " | Tons (metric)........ | $1 \times 10^{-6}$ | Gravitational constants........... | Cm./(sec. × sec.)....... | 980.621 |
| Grams/cm...... | Dynes/cm............ | 980.665 | " | Ft./(sec. × sec.)........ | 32.1725 |
| " | Grams/inch........... | 2.54 | | | |
| " | Kg./km............ | 100 | Hands............ | Centimeters........... | 10.16 |
| " | Kg./meter........... | 0.1 | " | Inches............. | 4 |
| " | Poundals/inch......... | 0.180166 | Hectares....... | Acres............. | 2.4710538 |
| " | Pounds/ft........... | 0.067196898 | " | Ares.............. | 100 |
| " | Pounds/inch.......... | 0.0055997415 | " | Sq. cm............ | $1 \times 10^{8}$ |
| " | Tons (metric)/km....... | 0.1 | " | Sq. feet............ | 107639.10 |
| Grams/(cm. × sec.).. | Poises............. | 1 | " | Sq. meters.......... | 10000 |
| " | Lb./(ft. × sec.)....... | 0.06719690 | " | Sq. miles........... | 0.0038610216 |
| Grams/cu. cm...... | Dynes/cu. cm...... | 980.665 | " | Sq. rods........... | 395.36861 |
| " | Grains/milliliter........ | 15.43279 | Hectograms........ | Grams............. | 100 |
| " | Grams/milliliter........ | 1 | " | Poundals........... | 7.09316 |
| " | Poundals/cu. inch...... | 1.16236 | " | Pounds (apoth or troy).... | 0.26792289 |
| " | Pounds/circ. mil-ft...... | $3.4049170 \times 10^{-7}$ | " | Pounds (avdp.)......... | 0.22046226 |
| " | Pounds/cu. ft......... | 62.427961 | Hectoliters........ | Bushels (Brit.)........ | 2.749694 |
| " | Pounds/cu. inch....... | 0.036127292 | " | Bushels (U.S.)........ | 2.837839 |
| " | Pounds/gal. (Brit.)....... | 10.02241 | | | |

* Based on density of 1 gram/ml.

## Table 4—7 (continued)

### CONVERSION FACTORS

| To convert from | To | Multiply by | To convert from | To | Multiply by |
|---|---|---|---|---|---|
| Hectoliters | Cu. cm | 1.00028 × 10⁵ | Horsepower (boiler) | Horsepower (metric) | 13.3372 |
| " | Cu. feet | 3.531566 | " | Horsepower (water) | 13.1487 |
| " | Gallons (U.S., liq.) | 26.41794 | " | Joules/sec | 9809.50 |
| " | Liters | 100 | " | Kilowatts | 9.80950 |
| " | Ounces (U.S.) fluid | 3381.497 | " | Lb. H₂O evap. per hr. from and at 212°F | 34.5 |
| " | Pecks (U.S.) | 11.35136 | Horsepower (electric) | B.t.u./hr | 2547.16 |
| Hectometers | Centimeters | 10000 | " | B.t.u. (IST.)/hr | 2545.50 |
| " | Decimeters | 1000 | " | B.t.u. (mean)/hr | 2543.50 |
| " | Dekameters | 10 | " | Cal., gm./sec | 178.298 |
| " | Feet | 328.08399 | " | Cal., kg./hr | 641.874 |
| " | Meters | 100 | " | Ergs/sec | 7.46 × 10⁹ |
| " | Rods | 19.883878 | " | Foot-pounds/min | 33013.3 |
| " | Yards | 109.3613 | " | Foot-pounds/sec | 550.221 |
| Hectowatts | Watts | 100 | " | Horsepower | 1.00040 |
| Hefner units | Candles (English) | 0.86 | " | Horsepower (boiler) | 0.0760487 |
| " | Candles (German) | 0.85 | " | Horsepower (metric) | 1.0142777 |
| " | Candles (Int.) | 0.90 | " | Horsepower (water) | 0.999942 |
| " | 10-cp. pentane candles | 0.090 | " | Joules/sec | 746 |
| Henries | Abhenries | 1 × 10⁹ | " | Kilowatts | 0.746 |
| " | E.M. cgs. units | 1 × 10⁹ | " | Watts | 746 |
| " | E.S. cgs. units | 1.112646 × 10⁻¹² | Horsepower (metric) | B.t.u./hr | 2511.31 |
| " | Henries (Int.) | 0.999505 | " | B.t.u. (IST.)/hr | 2509.66 |
| " | Millihenries | 1000 | " | B.t.u. (mean)/hr | 2507.70 |
| " | Mks. (r or nr) units | 1 | " | Cal., gm./hr | 6.32838 × 10⁵ |
| " | Stathenries | 1.112646 × 10⁻¹² | " | Cal., gm. (IST.)/hr | 6.32425 × 10⁵ |
| Henries (Int.) | Henries | 1.000495 | " | Cal., gm. (mean)/hr | 6.31929 × 10⁵ |
| Henries/meter | Cgs. units of permeability | 795774.72 | " | Ergs/sec | 7.35499 × 10⁹ |
| " | E.M. cgs. units | 795774.72 | " | Foot-pounds/min | 32548.6 |
| " | E.S. cgs. units | 8.854156 × 10⁻¹⁶ | " | Foot-pounds/sec | 542.476 |
| " | Gausses/oersted | 795774.72 | " | Horsepower | 0.986320 |
| " | Mks. (nr) units | 0.079577472 | " | Horsepower (boiler) | 0.0749782 |
| " | Mks. (r) units | 1 | " | Horsepower (electric) | 0.985923 |
| Hogsheads | Butts (Brit.) | 0.5 | " | Horsepower (water) | 0.985866 |
| " | Cu. feet | 8.421875 | " | Kg.-meters/sec | 75 |
| " | Cu. inches | 14553 | " | Kilowatts | 0.735499 |
| " | Cu. meters | 0.23848094 | " | Watts | 735.499 |
| " | Gallons (Brit.) | 52.458505 | Horsepower (water) | Foot-pounds/min | 33015.2 |
| " | Gallons (U.S.) | 63 | " | Horsepower | 1.00046 |
| " | Gallons (wine) | 63 | " | Horsepower (boiler) | 0.0760531 |
| " | Liters | 238.47427 | " | Horsepower (electric) | 1.00006 |
| Horsepower* | B.t.u. (mean)/hr | 2542.48 | " | Horsepower (metric) | 1.01434 |
| " | B.t.u./min | 42.4356 | " | Kilowatts | 0.746043 |
| " | B.t.u. (mean)/sec | 0.706243 | Horsepower-hours | B.t.u. | 2546.14 |
| " | Cal., gm./hr | 6.41616 × 10⁵ | " | B.t.u. (IST.) | 2544.47 |
| " | Cal., gm. (IST.)/hr | 6.41196 × 10⁵ | " | B.t.u. (mean) | 2542.48 |
| " | Cal., gm. (mean)/hr | 6.40693 × 10⁵ | " | Cal., gm. | 641616 |
| " | Cal., gm./min | 10693.6 | " | Cal., gm. (IST.) | 641196 |
| " | Cal., gm. (IST.)/min | 10686.6 | " | Cal., gm. (mean) | 640693 |
| " | Cal., gm. (mean)/min | 10678.2 | " | Foot-pounds | 1.98 × 10⁶ |
| " | Ergs/sec | 7.45700 × 10⁹ | " | Joules | 2.68452 × 10⁶ |
| " | Foot-pounds/hr | 1980000 | " | Kg.-meters | 273745 |
| " | Foot-pounds/min | 33000 | " | Kw.-hours | 0.745700 |
| " | Foot-pounds/sec | 550 | " | Watt-hours | 745.700 |
| " | Horsepower (boiler) | 0.0760181 | Hp.-hr./lb | B.t.u./lb | 2546.14 |
| " | Horsepower (electric) | 0.999598 | " | Cal., gm./gram | 1414.52 |
| " | Horsepower (metric) | 1.01387 | " | Cu. ft.-(lb./sq. in.)/lb | 13750 |
| " | Joules/sec | 745.700 | " | Foot-pounds/lb | 1980000 |
| " | Kilowatts | 0.745700 | " | Joules/gram | 5918.35 |
| " | Kilowatts (Int.) | 0.745577 | Hours (mean solar) | Days (mean solar) | 0.0416666 |
| " | Tons of refrig. (U.S., comm.) | 0.21204 | " | Days (sidereal) | 0.041780746 |
| " | Watts | 745.700 | " | Hours (sidereal) | 1.00273791 |
| Horsepower (boiler) | B.t.u. (mean)/hr | 33445.7 | " | Minutes (mean solar) | 60 |
| " | Cal., gm./min | 140671.6 | " | Minutes (sidereal) | 60.164275 |
| " | Cal., gm. (mean)/min | 140469.4 | " | Seconds (mean solar) | 3600 |
| " | Cal., gm. (15°C.)/min | 140611.1 | " | Seconds (sidereal) | 3609.8565 |
| " | Cal., gm. (20°C.)/min | 140742.2 | " | Weeks (mean calendar) | 0.0059523809 |
| " | Ergs/sec | 9.80950 × 10¹⁰ | Hours (sidereal) | Days (mean solar) | 0.41552899 |
| " | Foot-pounds/min | 434107 | " | Days (sidereal) | 0.0416666 |
| " | Horsepower | 13.1548 | " | Hours (mean solar) | 0.99726957 |
| " | Horsepower (electric) | 13.1495 | " | Minutes (mean solar) | 59.836174 |

* Mechanical horsepower, equal to 550 ft.-lb./sec.

## Table 4—7 (continued)

### CONVERSION FACTORS

| To convert from | To | Multiply by | To convert from | To | Multiply by |
|---|---|---|---|---|---|
| Hours (sidereal) | Minutes (sidereal) | 60 | Joules (abs.) | Cal., *kg.* (mean) | 0.000238662 |
| Hundredweights (long) | Kilograms | 50.802345 | " " | Cu. ft.-atm | 0.000348529 |
| " | Pounds | 112 | " " | Ergs | $1 \times 10^7$ |
| " | Quarters (Brit., long) | 4 | " " | Foot-poundals | 23.730360 |
| " | Quarters (U.S., long) | 0.2 | " " | Foot-pounds | 0.737562 |
| " | Tons (long) | 0.05 | " " | Gram-cm | 10197.16 |
| Hundredweights (short) | Kilograms | 45.359237 | " " | Hp.-hours | $3.72506 \times 10^{-7}$ |
| " | Pounds (advp.) | 100 | " " | Joules (Int.) | 0.999835 |
| " | Quarters (Brit., short) | 4 | " " | Kg.-meters | 0.1019716 |
| " | Quarters (U.S., short) | 0.2 | " " | Kw.-hours | $2.7777 \times 10^{-7}$ |
| " | Tons (long) | 0.044642857 | " " | Liter-atm | 0.00986895 |
| " | Tons (metric) | 0.045359237 | " " | Volt-coulombs (Int.) | 0.999835 |
| " | Tons (short) | 0.05 | " " | Watt-hours (abs.) | 0.0002777777 |
| | | | " " | Watt-hours (Int.) | 0.000277732 |
| Inches | Ångström units | $2.54 \times 10^8$ | " " | Watt-sec | 1 |
| " | Centimeters | 2.54 | " " | Watt-sec. (Int.) | 0.999835 |
| " | Chains (Gunter's) | 0.00126262 | Joules (Int.) | B.t.u | 0.000948608 |
| " | Cubits | 0.055555 | " | B.t.u. (IST.) | 0.000947988 |
| " | Fathoms | 0.013888 | " | B.t.u. (mean) | 0.000947244 |
| " | Feet | 0.083333 | " | Cal. *gm* | 0.239045 |
| " | Feet (U.S. Survey) | 0.083333167 | " | Cal., *gm.* (IST.) | 0.238888 |
| " | Links (Gunter's) | 0.126262 | " | Cal., *gm.* (mean) | 0.238702 |
| " | Links (Ramden's) | 0.083333 | " | C.h.u. | 0.000527004 |
| " | Meters | 0.0254 | " | C.h.u. (IST.) | 0.000526660 |
| " | Mils | 1000 | " | C.h.u. (mean) | 0.000526247 |
| " | Picas (printer's) | 6.0225 | " | Cu. cm.-atm | 9.87086 |
| " | Points (printer's) | 72.27000 | " | Cu. ft.-atm | 0.000348586 |
| " | Wave length of orange-red line of krypton 86 | 41929.399 | " | Dyne-cm | $1.000165 \times 10^7$ |
| | | | " | Ergs | $1.000165 \times 10^7$ |
| " | Wave length of the red line of cadmium | 39450.369 | " | Foot-poundals | 23.73428 |
| | | | " | Foot-pounds | 0.737684 |
| " | Yards | 0.027777 | " | Gram-cm | 10198.8 |
| Inches of Hg (32°F.) | Atmospheres | 0.0334211 | " | Joules (abs.) | 1.000165 |
| " | Bars | 0.0338639 | " | Kw.-hours | $2.77824 \times 10^{-7}$ |
| " | Dynes/sq. cm | 33863.9 | " | Liter-atm | 0.00987058 |
| " | Ft. of air (1 atm., 60°F.) | 926.24 | " | Volt-coulombs | 1.000165 |
| " | Ft. of H₂O (39.2°F.) | 1.132957 | " | Volt-coulombs (Int.) | 1 |
| " | Grams/sq. cm | 34.5316 | " | Watt-sec | 1.000165 |
| " | Kg./sq. meter | 345.316 | " | Watt-sec. (Int.) | 1 |
| " | Mm. of Hg (60°C.) | 25.4 | Joules/(abcoulomb × °F.) | Joules/(coulomb × °C.) | 0.18 |
| " | Ounces/sq. inch | 7.85847 | Joules/amp.-hr | Joules/abcoulomb | 0.002777 |
| Inches of Hg (32°F.) | Pounds/sq. ft | 70.7262 | " | Joules/statcoulomb | $9.265653 \times 10^{-14}$ |
| Inches of Hg (60°F.) | Atmospheres | 0.0333269 | Joules/coulomb | Joules/abcoulomb | 10 |
| " | Dynes/sq. cm | 39768.5 | " | Volts | 1 |
| " | Grams/sq. cm | 34.4343 | Joules/(coulomb × °F.) | Joules/(coulomb × °C.) | 1.8 |
| " | Mm. of Hg (60°F.) | 25.4 | Joules/°C | B.t.u./°F | 0.000526917 |
| " | Ounces/sq. inch | 7.83633 | " | Cal., *gm.*/°C | 0.239006 |
| " | Pounds/sq. ft | 70.5269 | " | Cal., *gm.* (mean)/°C | 0.238662 |
| Inches of H₂O(4°C.) | Atmospheres | 0.0024582 | Joules/electronic charge | Joules/abcoulomb | $6.24196 \times 10^{18}$ |
| " | Dynes/sq. cm | 2490.82 | Joules/(electronic charge × °C.) | Joules/(coulomb × °C.) | $6.24196 \times 10^{18}$ |
| " | In. of Hg (32°F.) | 0.0735539 | Joules/(gram × °C.) | B.t.u./(lb. × °F.) | 0.239006 |
| " | Kg./sq. meter | 25.3993 | " | Cal., *gm.*/(gram × °C.) | 0.239006 |
| " | Ounces/sq. ft | 83.2350 | Joules (Int.)/(gram °C.) | B.t.u./(lb. × °F.) | 0.239045 |
| " | Ounces/sq. inch | 0.578020 | " | Cal., *gm.* (mean)/(gram × °C.) | 0.238702 |
| " | Pounds/sq. ft | 5.20218 | | | |
| " | Pounds/sq. inch | 0.03612628 | Joules/sec. (abs.) | B.t.u./min | 0.0569071 |
| Inches/hr | Cm./hr | 2.54 | " | Cal., *gm.*/min | 14.3403 |
| " | Feet/hr | 0.0833333 | " | Cal., *kg.*/min | 0.0143403 |
| " | Miles/hr | $1.578242 \times 10^{-5}$ | " | Cal., *kg.* (mean)/min | 0.0143197 |
| Inches/min | Cm./hr | 152.4 | " | Dyne-cm./sec | $1 \times 10^7$ |
| " | Feet/hr | 5 | " | Ergs/sec | $1 \times 10^7$ |
| " | Miles/hr | 0.000946969 | " | Foot-pounds/sec | 0.737562 |
| | | | " | Gram-cm./sec | 10197.16 |
| Joules (abs.) | B.t.u. | 0.000948451 | " | Horsepower | 0.00134102 |
| " " | B.t.u. (IST.) | 0.000947831 | " | Watts | 1 |
| " " | B.t.u. (mean) | 0.000947088 | " | Watts (Int.) | 0.999835 |
| " " | Cal., *gm* | 0.239006 | Joules (Int.)/sec | B.t.u./min | 0.0569165 |
| " " | Cal., *gm.* (IST.) | 0.238849 | | | |
| " " | Cal., *gm.* (mean) | 0.238662 | | | |
| " " | Cal., *gm.* (15°C.) | 0.238903 | | | |
| " " | Cal., *gm.* (20°C.) | 0.239126 | | | |

## Table 4—7 (continued)

### CONVERSION FACTORS

| To convert from | To | Multiply by | To convert from | To | Multiply by |
|---|---|---|---|---|---|
| Joules (Int.)/sec | B.t.u. (mean)/min | 0.0568347 | Kilogram-meters | Hp.-hours | $3.65304 \times 10^{-6}$ |
| " | Cal., *gm.*/min | 14.3427 | " | Joules | 9.80665 |
| " | Cal., *kg.*/min | 0.0143427 | " | Joules (Int.) | 9.80503 |
| " | Dyne-cm./sec | $1.000165 \times 10^7$ | " | Kw.-hours | $2.72407 \times 10^{-6}$ |
| " | Ergs/sec | $1.000165 \times 10^7$ | " | Liter-atm | 0.0967814 |
| " | Foot-pounds/min | 44.2610 | " | Newton-meters | 9.80665 |
| " | Foot-pounds/sec | 0.737684 | " | Watt-hours | 0.00272407 |
| " | Gram-cm./sec | 10198.8 | " | Watt-hours (Int.) | 0.00272362 |
| " | Horsepower | 0.00134124 | Kilogram-meters/sec. | Watts | 9.80665 |
| " | Watts | 1.000165 | Kilolines | Maxwells | 1000 |
| " | Watts (Int.) | 1 | " | Webers | $1 \times 10^{-8}$ |
| Kilderkins (Brit.) | Cu. cm | 81829.57 | Kiloliters | Cu. centimeters | $1 \times 10^6$ |
| " | Cu. feet | 2.889784 | " | Cu. feet | 35.31566 |
| " | Cu. inches | 4993.55 | " | Cu. inches | 61025.45 |
| " | Cu. meters | 0.08182957 | " | Cu. meters | 1.000028 |
| " | Gallons (Brit.) | 18 | " | Cu. yards | 1.307987 |
| Kilograms | Drams (apoth. *or* troy) | 257.20597 | " | Gallons (Brit.) | 219.9755 |
| " | Drams (avdp.) | 564.38339 | " | Gallons (U.S., dry) | 227.0271 |
| " | Dynes | 980665 | " | Gallons (U.S., liq.) | 264.1794 |
| " | Grains | 15432.358 | " | Liters | 1000 |
| " | Hundredweights (long) | 0.019684131 | Kilometers | Astronomical units | $6.68878 \times 10^{-9}$ |
| " | Hundredweights (short) | 0.022046226 | " | Centimeters | 100000 |
| " | Ounces (apoth. *or* troy) | 32.150737 | " | Feet | 3280.8399 |
| " | Ounces (avdp.) | 35.273962 | " | Feet (U.S. Survey) | 3280.833 |
| " | Pennyweights | 643.01493 | " | Light years | $1.05702 \times 10^{-13}$ |
| " | Poundals | 70.931635 | " | Meters | 1000 |
| " | Pounds (apoth. *or* troy) | 2.6792289 | " | Miles (naut., Int.) | 0.53995680 |
| " | Pounds (avdp.) | 2.2046226 | " | Miles (statute) | 0.62137119 |
| " | Quarters (Brit., long) | 0.078736522 | " | Myriameters | 0.1 |
| " | Quarters (U.S. long) | 0.0039368261 | " | Rods | 198.83878 |
| " | Scruples (apoth.) | 771.61792 | " | Yards | 1093.6133 |
| " | Slugs | 0.06852177 | Kilometers/hr | Cm./sec | 27.7777 |
| " | Tons (long) | 0.00098420653 | " | Feet/hr | 3280.8399 |
| " | Tons (metric) | 0.001 | " | Feet/min | 54.680665 |
| " | Tons (short) | 0.0011023113 | " | Knots (Int.) | 0.53995680 |
| Kilograms/cu. meter | Grams/cu. cm | 0.001 | " | Meters/sec | 0.277777 |
| " | Lb./cu. ft | 0.062427961 | " | Miles (statute)/hr | 0.62137119 |
| " | Lb./cu. inch | $3.6127292 \times 10^{-5}$ | Kilometers/(hr. × sec.) | Cm./(sec. × sec.) | 27.7777 |
| Kg. of ice melted/hr | Tons of refrig. (U.S. comm.) | 0.026336 | " | Ft./(sec. × sec.) | 0.91134442 |
| Kilograms/sq. cm | Atmospheres | 0.967841 | " | Meters/(sec. × sec.) | 0.277777 |
| " | Bars | 0.980665 | Kilometers/min | Cm./sec | 1666.666 |
| " | Cm. of Hg (0°C.) | 73.5559 | " | Feet/min | 3280.8399 |
| " | Dynes/sq. cm | 980665 | " | Kilometers/hr | 60 |
| " | Ft. of H₂O (39.2°F.) | 32.8093 | " | Knots (Int.) | 32.397408 |
| " | In. of Hg (32°F.) | 28.9590 | " | Miles/hr | 37.282272 |
| " | Pounds/sq. inch | 14.223343 | " | Miles/min | 0.62137119 |
| Kilograms/sq. meter | Atmospheres | $9.67841 \times 10^{-5}$ | Kilovolts/cm | Abvolts/cm | $1 \times 10^{11}$ |
| " | Bars | $9.80665 \times 10^{-5}$ | " | Microvolts/meter | $1 \times 10^{11}$ |
| " | Dynes/sq. cm | 98.0665 | " | Millivolts/meter | $1 \times 10^8$ |
| " | Ft. of H₂O (39.2°F.) | 0.00328093 | " | Statvolts/cm | 3.335635 |
| " | Grams/sq. cm | 0.1 | " | Volts/inch | 2540 |
| " | In. of Hg. (32°F.) | 0.00289590 | Kilowatts | B.t.u./hr | 3414.43 |
| " | Mm. of Hg (0°C.) | 0.0735559 | " | B.t.u. (IST.)/hr | 3412.19 |
| " | Pounds/sq. ft | 0.20481614 | " | B.t.u. (mean)/hr | 3409.52 |
| " | Pounds/sq. in | 0.0014223343 | " | B.t.u. (mean)/min | 56.8253 |
| Kilograms/sq. mm | Pounds/sq. ft | 204816.14 | " | B.t.u. (mean)/sec | 0.947088 |
| " | Pounds/sq. in | 1422.3343 | " | Cal., *gm.* (mean)/hr | 859184 |
| " | Tons (short)/sq. in | 0.71116716 | " | Cal., *gm.* (mean)/min | 14319.7 |
| Kilogram sq. cm | Pounds sq. ft | 0.0023730360 | " | Cal., *gm.* (mean)/sec | 238.662 |
| " | Pounds sq. in | 0.34171719 | " | Cal., *kg.* (mean)/hr | 859.184 |
| Kilogram-meters | B.t.u. (mean) | 0.00928776 | " | Cal., *kg.* (mean)/min | 14.3197 |
| " | Cal., *gm.* (mean) | 2.34048 | " | Cal., *kg.* (mean)/sec | 0.238662 |
| " | Cal., *kg.* (mean) | 0.00234048 | " | Cu. ft.-atm./hr | 1254.70 |
| " | Cu. ft.-atm | 0.00341790 | " | Ergs/sec | $1 \times 10^{10}$ |
| " | Dynes-cm | $9.80665 \times 10^7$ | " | Foot-poundals/min | $1.42382 \times 10^6$ |
| " | Ergs | $9.80665 \times 10^7$ | " | Foot-pounds/hr | $2.65522 \times 10^6$ |
| " | Foot-poundals | 232.715 | " | Foot-pounds/min | 44253.7 |
| " | Foot-pounds | 7.23301 | " | Foot-pounds/sec | 737.562 |
| " | Gram-cm | 100000 | " | Gram-cm./sec | $1.019716 \times 10^7$ |
| | | | " | Horsepower | 1.34102 |

## Table 4—7 (continued)

### CONVERSION FACTORS

| To convert from | To | Multiply by | To convert from | To | Multiply by |
|---|---|---|---|---|---|
| Kilowatts | Horsepower (boiler) | 0.101942 | Lamberts | Candles/sq. cm | 0.31830989 |
| " | Horsepower (electric) | 1.34048 | " | Candles/sq. ft | 295.71956 |
| " | Horsepower (metric) | 1.35962 | " | Candles/sq. inch | 2.0536081 |
| " | Joules/hr | $3.6 \times 10^6$ | " | Foot-lamberts | 929.0304 |
| " | Joules (IST.)/hr | $3.59941 \times 10^6$ | " | Lumens/sq. cm | 1 |
| " | Joules/sec | 1000 | Lasts (Brit.) | Liters | 2909.414 |
| " | Kg.-meters/hr | $3.67098 \times 10^5$ | Leagues (naut., Brit.) | Feet | 18240 |
| " | Kilowatts (Int.) | 0.999835 | " | Kilometers | 5.559552 |
| " | Watts (Int.) | 999.835 | " | Leagues (naut., Int.) | 1.0006393 |
| Kilowatts (Int.) | B.t.u./hr | 3414.99 | " | Leagues (statute) | 1.151515 |
| " | B.t.u. (IST.)/hr | 3412.76 | " | Miles (statute) | 3.454545 |
| " | B.t.u. (mean)/hr | 3410.08 | Leagues (naut., Int.) | Fathoms | 3038.0577 |
| " | B.t.u. (mean)/min | 56.8347 | " | Feet | 18228.346 |
| " | B.t.u. (mean)/sec | 0.947244 | " | Kilometers | 5.556 |
| " | Cal., gm. (mean)/hr | 859.326 | " | Leagues (statute) | 1.1507794 |
| " | Cal., gm. (mean)/min | 14322.1 | " | Miles (statute) | 3.4523383 |
| " | Cal., kg./hr | 860.563 | Leagues (statute) | Fathoms | 2640 |
| " | Cal., kg. (IST.)/hr | 860 | " | Feet | 15840 |
| " | Cal., kg. (mean)/hr | 859.326 | " | Kilometers | 4.828032 |
| " | Cu. cm.-atm./hr | $3.55351 \times 10^7$ | " | Leagues (naut., Int.) | 0.86897625 |
| " | Cu. ft.-atm./hr | 1254.91 | " | Miles (naut., Int.) | 2.6069287 |
| " | Ergs/sec | $1.000165 \times 10^{10}$ | " | Miles (statute) | 3 |
| " | Foot-poundals/min | $1.42406 \times 10^6$ | Light years | Astronomical units | 63279.5 |
| " | Foot-pounds/min | 44261.0 | " | Kilometers | $9.46055 \times 10^{12}$ |
| " | Foot-pounds/sec | 737.684 | " | Miles (statute) | $5.87851 \times 10^{12}$ |
| " | Gram-cm./sec | $1.01988 \times 10^7$ | Lines | Maxwells | 1 |
| " | Horsepower | 1.34124 | Lines (Brit.) | Centimeters | 0.211666 |
| " | Horsepower (boiler) | 0.101959 | " | Inches | 0.083333 |
| " | Horsepower (electric) | 1.34070 | Lines/sq. cm | Gausses | 1 |
| " | Horsepower (metric) | 1.35985 | Lines/sq. inch | Gausses | 0.15500031 |
| " | Joules/hr | $3.60059 \times 10^6$ | " | Webers/sq. inch | $1 \times 10^{-8}$ |
| " | Joules (Int.)/hr | $3.6 \times 10^6$ | Links (Gunter's) | Chains (Gunter's) | 0.01 |
| " | Kg.-meters/hr | 367158 | " | Feet | 0.66 |
| " | Kilowatts | 1.000165 | " | Feet (U.S. Survey) | 0.65999868 |
| Kilowatt-hours | B.t.u. (mean) | 3409.52 | " | Inches | 7.92 |
| " | Cal., gm. (mean) | 859184 | " | Meters | 0.201168 |
| " | Foot-pounds | $2.65522 \times 10^6$ | " | Miles (statute) | 0.000125 |
| " | Hp.-hours | 1.34102 | " | Rods | 0.04 |
| " | Joules | $3.6 \times 10^6$ | Links (Ramden's) | Centimeters | 30.48 |
| " | Kg.-meters | 367098 | " | Chains (Ramdens) | 0.01 |
| " | Lb. H₂O evap. from and at 212°F | 3.5168 | " | Feet | 1 |
| " | Watt-hours | 1000 | " | Inches | 12 |
| " | Watt-hours (Int.) | 999.835 | Liters | Bushels (Brit.) | 0.02749694 |
| Kilowatt-hours (Int.) | B.t.u. (mean) | 3410.08 | " | Bushels (U.S.) | 0.02837839 |
| " | Cal., gm. (IST.) | 860000 | " | Cu. centimeters | 1000 |
| " | Cal., gm. (mean) | 859326 | " | Cu. feet | 0.03531566 |
| " | Cu. cm.-atm | $3.55351 \times 10^7$ | " | Cu. inches | 61.02545 |
| " | Cu. ft.-atm | 1254.91 | " | Cu. meters | 0.001 |
| " | Foot-pounds | $2.65566 \times 10^6$ | " | Cu. yards | 0.001307987 |
| " | Hp.-hours | 1.34124 | " | Drams (U.S., fluid) | 270.5198 |
| " | Joules | $3.60059 \times 10^6$ | " | Gallons (Brit.) | 0.2199755 |
| " | Joules (Int.) | $3.6 \times 10^6$ | " | Gallons (U.S., dry) | 0.2270271 |
| " | Kg.-meters | 367158 | " | Gallons (U.S., liq.) | 0.2641794 |
| Kw.-hr./gram | B.t.u./lb | $1.54876 \times 10^6$ | " | Gills (Brit.) | 7.039217 |
| " | B.t.u. (IST.)/lb | $1.54774 \times 10^6$ | " | Gills (U.S.) | 8.453742 |
| " | B.t.u. (mean)/lb | $1.54653 \times 10^6$ | " | Hogsheads | 0.004193325 |
| " | Cal., gm./gram | 860421 | " | Minims (U.S.) | 16231.19 |
| " | Cal., gm. (mean)/gram | 859184 | " | Ounces (Brit., fluid) | 35.19609 |
| " | Cu. cm.-atm./gram | $3.55292 \times 10^7$ | " | Ounces (U.S., fluid) | 33.81497 |
| " | Cu. ft.-atm./lb | 569124 | " | Pecks (Brit.) | 0.1099878 |
| " | Hp.-hr./lb | 608.277 | " | Pecks (U.S.) | 0.1135136 |
| " | Joules/gram | $3.6 \times 10^6$ | " | Pints (Brit.) | 1.759804 |
| Knots (Int.) | Cm./sec | 51.4444 | " | Pints (U.S., dry) | 1.816217 |
| " | Feet/hr | 6076.1155 | " | Pints (U.S., liq.) | 2.113436 |
| " | Feet/min | 101.26859 | " | Quarts. (Brit.) | 0.8799021 |
| " | Feet/sec | 1.6878099 | " | Quarts (U.S., dry) | 0.9081084 |
| " | Kilometers/hr | 1.852 | " | Quarts (U.S., liq.) | 1.056718 |
| " | Meters/min | 30.8666 | Liters/min | Cu. ft./min | 0.03531566 |
| " | Meters/sec | 0.514444 | " | Cu. ft./sec | 0.0005885943 |
| " | Miles (naut., Int.)/hr | 1 | " | Gal. (U.S., liq.)/min | 0.2641794 |
| " | Miles (statute)/hr | 1.1507794 | Liters/sec | Cu. ft./min | 2.118939 |
| | | | " | Cu. ft./sec | 0.03531566 |

## Table 4—7 (continued)

### CONVERSION FACTORS

| To convert from | To | Multiply by | To convert from | To | Multiply by |
|---|---|---|---|---|---|
| Liters/sec.......... | Cu. yards/min.......... | 0.07847923 | Meters......... | Links (Ramden's)....... | 3.2808399 |
| " .......... | Gal. (U.S., liq.)/min..... | 15.85077 | " .......... | Megameters.......... | $1 \times 10^{-6}$ |
| " .......... | Gal. (U.S., liq.)/sec..... | 0.2641794 | " .......... | Miles (naut., Brit.)..... | 0.00053961182 |
| Liter-atm.......... | B.t.u.............. | 0.0961045 | " .......... | Miles (naut., Int.)..... | 0.00053995680 |
| " | B.t.u. (IST.).......... | 0.0960417 | " .......... | Miles (statute)........ | 0.00062137119 |
| " | B.t.u. (mean).......... | 0.0959664 | " .......... | Millimeters........... | 1000 |
| " | Cal., *gm.*.......... | 24.2179 | " .......... | Millimicrons.......... | $1 \times 10^{9}$ |
| " | Cal., *gm.* (IST.)......... | 24.2021 | " .......... | Mils............. | 39370.079 |
| " | Cal., *gm.* (mean)........ | 24.1831 | " .......... | Rods............. | 0.19883878 |
| " | Cu. ft.-atm.......... | 0.0353157 | " .......... | Yards............. | 1.0936133 |
| " | Foot-poundals.......... | 2404.55 | Meters of Hg (0°C.).. | Atmospheres.......... | 1.3157895 |
| " | Foot-pounds.......... | 74.7356 | " | Ft. of H₂O (60°F.)....... | 44.6474 |
| " | Hp.-hours.......... | $3.77452 \times 10^{-5}$ | " | In. of Hg (32°F.)....... | 39.370079 |
| " | Joules.......... | 101.328 | " | Kg./sq. cm.......... | 1.35951 |
| " | Joules (Int.).......... | 101.311 | " | Pounds/sq. inch....... | 19.3368 |
| " | Kg.-meters.......... | 10.3326 | Meters/hr........ | Feet/hr.......... | 3.2808399 |
| " | Kw.-hours.......... | $2.81466 \times 10^{-5}$ | " | Feet/min.......... | 0.054680665 |
| Liter-atm. (lat. 45°)... | Joules.......... | 101.323 | " | Knots (Int.).......... | 0.00053995680 |
| Lumens........... | Candle power (spher.).... | 0.079577472 | " | Miles (statute)/hr..... | 0.00062137119 |
| Lumens (at 5550 Å).. | Watts.......... | 0.0014705882 | Meters/min....... | Cm./sec.......... | 1.666666 |
| Lumens/sq. cm....... | Lamberts.......... | 1 | " | Feet/min.......... | 3.2808399 |
| " ...... | Phots.......... | 1 | " | Feet/sec.......... | 0.054680665 |
| Lumens/(sq. cm. × steradian)....... | Lamberts.......... | 3.1415927 | " | Kilometers/hr.......... | 0.06 |
| Lumens/sq. ft....... | Foot-candles.......... | 1 | " | Knots (Int.).......... | 0.032397408 |
| " ...... | Foot-lamberts.......... | 1 | " | Miles (statute)/hr..... | 0.037282272 |
| " ...... | Lumens/sq. meter........ | 10.763910 | Meters/sec....... | Feet/min.......... | 196.85039 |
| Lumens/(sq. ft. × steradian)....... | Millilamberts.......... | 3.3815822 | " | Feet/sec.......... | 3.2808399 |
| Lumens/sq. meter.... | Foot-candles.......... | 0.09290304 | " | Kilometers/hr.......... | 3.6 |
| " ...... | Lumens/sq. ft.......... | 0.09290304 | " | Kilometers/min.......... | 0.06 |
| " ...... | Phots.......... | 0.0001 | " | Miles (statute)/hr..... | 2.2369363 |
| Lux............ | Foot-candles.......... | 0.09290304 | Meters/(sec. × sec.).. | Kilometers/(hr. × sec.).. | 3.6 |
| " .......... | Lumens/sq. meter.......... | 1 | " | Miles/(hr. × sec.)..... | 2.2369363 |
| " .......... | Phots.......... | 0.0001 | Meter-candles....... | Lumens/sq. meter........ | 1 |
| Maxwells........... | E.M. cgs. units of induction flux.......... | 1 | Mhos............ | Abmhos.......... | $1 \times 10^{-9}$ |
| " | E.S. cgs. units.......... | $3.335635 \times 10^{-11}$ | " | Cgs. units of conductance.. | 1 |
| " | Gauss-sq. cm.......... | 1 | " | E.M. cgs. units.......... | $1 \times 10^{-9}$ |
| " | Lines.......... | 1 | " | E.S. cgs. units.......... | $8.987584 \times 10^{11}$ |
| " | Maxwells (Int.).......... | 0.999670 | " | Mhos (Int.).......... | 1.000495 |
| " | Volt-seconds.......... | $1 \times 10^{-8}$ | " | Mks. (r *or* nr) units..... | 1 |
| " | Webers.......... | $1 \times 10^{-8}$ | " | Ohms⁻¹.......... | 1 |
| Maxwells (Int.)..... | Maxwells.......... | 1.000330 | " | Siemen's units.......... | 1 |
| Maxwells/sq. cm.... | Maxwells/sq. in.......... | 6.4516 | " | Statmhos.......... | $8.987584 \times 10^{11}$ |
| " | Maxwells (Int.)/sq. cm.... | 0.999670 | Mhos (Int.)...... | Abmhos.......... | $9.99505 \times 10^{-10}$ |
| Maxwells (Int.)/sq. cm.......... | Maxwells/sq. cm.......... | 1.000330 | " | Mhos.......... | 0.999505 |
| Maxwells/sq. inch.... | Maxwells/sq. cm.......... | 0.15500031 | Mhos/meter....... | Abmhos/cm.......... | $1 \times 10^{-11}$ |
| Megalines.......... | Maxwells.......... | $1 \times 10^{6}$ | " | Mhos (Int.)/meter.......... | 1.000495 |
| Megmhos/cm....... | Abmhos/cm.......... | 0.001 | Mho-ft./circ. mil.... | Mhos/cm.......... | $6.0153049 \times 10^{6}$ |
| " | Megmhos/inch cube.......... | 2.54 | Microfarads........ | Abfarads.......... | $1 \times 10^{-15}$ |
| " | (Microhm-cm.)⁻¹.......... | 1 | " | Farads.......... | $1 \times 10^{-6}$ |
| Megmhos/inch...... | Megmhos/cm.......... | 0.39370079 | " | Statfarads.......... | $8.987584 \times 10^{5}$ |
| " | (Microhm-inches)⁻¹.......... | 1 | Micrograms........ | Grams.......... | $1 \times 10^{-6}$ |
| Megohms.......... | Microhms.......... | $1 \times 10^{12}$ | " | Milligrams.......... | 0.001 |
| " | Ohms.......... | $1 \times 10^{6}$ | Microhenries........ | Henries.......... | $1 \times 10^{-6}$ |
| " | Statohms.......... | $1.112646 \times 10^{-6}$ | " | Stathenries.......... | $1.112646 \times 10^{-18}$ |
| Megohms⁻¹........ | Micromhos.......... | 1 | Microhms.......... | Abohms.......... | 1000 |
| Meters............ | Ångström units.......... | $1 \times 10^{10}$ | " | Megohms.......... | $1 \times 10^{-12}$ |
| " | Centimeters.......... | 100 | " | Ohms.......... | $1 \times 10^{-6}$ |
| " | Chains (Gunter's).......... | 0.049709695 | " | Statohms.......... | $1.112646 \times 10^{-18}$ |
| " | Chains (Ramden's)........ | 0.032808399 | Microhm-cm....... | Abohm-cm.......... | 1000 |
| " | Fathoms.......... | 0.54680665 | " | Circ. mil-ohms/ft.......... | 6.0153049 |
| " | Feet.......... | 3.2808399 | " | Microhm-inches.......... | 0.39370079 |
| " | Feet (U.S. Survey)........ | 3.280833 | " | Ohm-cm.......... | $1 \times 10^{-6}$ |
| " | Furlongs.......... | 0.0049709695 | Microhm-inches.... | Circ. mil-ohms/ft.......... | 15.278875 |
| " | Inches.......... | 39.370079 | " | Michrom-cm.......... | 2.54 |
| " | Kilometers.......... | 0.001 | Micromicrofarads.... | Farads.......... | $1 \times 10^{-12}$ |
| " | Links (Gunter's).......... | 4.9709695 | Micromicrons...... | Ångström units.......... | 0.01 |
| | | | " | Centimeters.......... | $1 \times 10^{-10}$ |
| | | | " | Inches.......... | $3.9370079 \times 10^{-11}$ |
| | | | " | Meters.......... | $1 \times 10^{-12}$ |
| | | | " | Microns.......... | $1 \times 10^{-6}$ |
| | | | Microns........... | Ångström units.......... | 10000 |

## Table 4—7 (continued)

### CONVERSION FACTORS

| To convert from | To | Multiply by | To convert from | To | Multiply by |
|---|---|---|---|---|---|
| Microns............ | Centimeters............ | 0.0001 | Milligrams.......... | Grains............... | 0.015432358 |
| " ............ | Feet............ | $3.2808399 \times 10^{-6}$ | " .......... | Grams............... | 0.001 |
| " ............ | Inches............ | $3.9370079 \times 10^{-5}$ | " .......... | Ounces (apoth. or troy).... | $3.2150737 \times 10^{-5}$ |
| " ............ | Meters............ | $1 \times 10^{-6}$ | " .......... | Ounces (avdp.)........... | $3.5273962 \times 10^{-5}$ |
| " ............ | Millimeters............ | 0.001 | " .......... | Pennyweights........ | 0.00064301493 |
| " ............ | Millimicrons............ | 1000 | " .......... | Pounds (apoth. or troy).... | $2.6792289 \times 10^{-6}$ |
| Miles (naut., Brit.)... | Cable lengths (Brit.)...... | 8.4444 | " .......... | Pounds (avdp.)........... | $2.2046226 \times 10^{-6}$ |
| " | Fathoms............ | 1013.333 | " .......... | Scruples (apoth.)........ | 0.00077161792 |
| " | Feet............ | 6080 | Milligrams/assay ton. | Milligrams/kg.......... | 34.285714 |
| " | Meters............ | 1853.184 | " | Ounces (troy)/ton (avdp.). | 1 |
| " | Miles (Adm., Brit.)........ | 1 | Milligrams/gm...... | Dynes/cm............ | 0.980665 |
| " | Miles (naut., Int.)........ | 1.0006393 | " | Pounds/inch............ | $5.5997415 \times 10^{-6}$ |
| " | Miles (statute)........... | 1.151515 | Milligrams/gram.... | Carats (parts gold per 24 of | 0.024 |
| Miles (naut., Int.)... | Cable lengths........... | 8.4390493 | | mixture) | |
| " | Fathoms............ | 1012.6859 | " | Grams/ton (short)........ | 907.18474 |
| " | Feet............ | 6076.1155 | " | Milligrams/assay ton...... | 29.166666 |
| " | Feet (U.S. Survey)....... | 6076.1033 | " | Ounces (avdp.)/ton (long).. | 35.84 |
| " | Kilometers........... | 1.852 | " | Ounces (avdp.)/ton (short). | 32 |
| " | Leagues (naut., Int.)..... | 0.333333 | " | Ounces (troy)/ton (long)... | 32.6666 |
| " | Meters............ | 1852 | " | Ounces (troy)/ton (short).. | 29.1666 |
| " | Miles (geographical)...... | 1 | Milligrams/inch.... | Dynes/inch............ | 0.386089 |
| " | Miles (naut. Brit.)....... | 0.99936110 | " | Dynes/inch............ | 0.980665 |
| " | Miles (statute)........... | 1.1507794 | " | Grams/cm............ | 0.00039370079 |
| Miles (statute)...... | Centimeters........... | 160934.4 | " | Grams/inch............ | 0.0001 |
| " | Chains (Gunter's)........ | 80 | Milligrams/kg...... | Pounds (avdp.)/ton (short) | 0.002 |
| " | Chains (Ramden's)........ | 52.8 | Milligrams/liter.... | Grains/gal. (U.S.)........ | 0.05841620 |
| " | Feet............ | 5280 | " | Grams/liter........... | 0.001 |
| " | Feet (U.S. Survey)....... | 5279.9894 | " | Parts/million*........... | 1 |
| " | Furlongs............ | 8 | " | Lb./cu. ft............ | $6.242621 \times 10^{-5}$ |
| " | Inches............ | 63360 | Milligrams/mm..... | Dynes/cm............ | 9.80665 |
| " | Kilometers........... | 1.609344 | Millihenries.......... | Abhenries........... | $1 \times 10^{6}$ |
| " | Light years............ | $1.70111 \times 10^{-13}$ | " | Henries............ | 0.001 |
| " | Links (Gunter's)........ | 8000 | " | Stathenries............ | $1.112646 \times 10^{-15}$ |
| " | Meters............ | 1609.344 | Millilamberts........ | Candles/sq. cm........ | 0.00031830989 |
| " | Miles (naut., Brit.)....... | 0.86842105 | " | Candles/sq. inch........ | 0.0020536081 |
| " | Miles (naut., Int.)....... | 0.86897624 | " | Foot-lamberts......... | 0.9290304 |
| " | Myriameters........... | 0.1609344 | " | Lamberts........... | 0.001 |
| " | Rods............ | 320 | " | Lumens/sq. cm........ | 0.001 |
| " | Yards............ | 1760 | " | Lumens/sq. ft......... | 0.9290304 |
| Miles/hr.......... | Cm./sec............ | 44.704 | Milliliters.......... | Cu. cm............ | 1 |
| " | Feet/hr............ | 5280 | " | Cu. inches............ | 0.06102545 |
| " | Feet/min............ | 88 | " | Drams (U.S., fluid)...... | 0.2705198 |
| " | Feet/sec............ | 1.466666 | " | Gills (U.S.)........... | 0.008453742 |
| " | Kilometers/hr......... | 1.609344 | " | Liters............ | 0.001 |
| " | Knots (Int.)........... | 0.86897624 | " | Minims (U.S.)......... | 16.23119 |
| " | Meters/min........... | 26.8224 | " | Ounces (Brit., fluid)...... | 0.03519609 |
| " | Miles/min............ | 0.0166666 | " | Ounces (U.S., fluid)...... | 0.03381497 |
| Miles/(hr. × min.)... | Cm./(sec. × sec.)...... | 0.7450666 | " | Pints (Brit.).......... | 0.001759804 |
| Miles/(hr. × sec.)... | Cm./(sec. × sec.)...... | 44.704 | " | Pints (U.S., liq.)........ | 0.002113436 |
| " | Ft./(sec. × sec.)....... | 1.466666 | Millimeters.......... | Ångström units........ | $1 \times 10^{7}$ |
| " | Kilometers/(hr. × sec.).... | 1.609344 | " | Centimeters............ | 0.1 |
| " | Meters/(sec. × sec.)...... | 0.44704 | " | Decimeters............ | 0.01 |
| Miles/min.......... | Cm./sec............ | 2682.24 | " | Dekameters............ | 0.0001 |
| " | Feet/hr............ | 316800 | " | Feet............ | 0.0032808399 |
| " | Feet/sec............ | 88 | " | Inches............ | 0.039370079 |
| " | Kilometers/min......... | 1.609344 | " | Meters............ | 0.001 |
| " | Knots (Int.)........... | 52.138574 | " | Microns............ | 1000 |
| " | Meters/min........... | 1609.344 | " | Mils............ | 39.370079 |
| " | Miles/hr............ | 60 | " | Wave length of orange-red | 1650.76373 |
| Millibars.......... | Atmospheres............ | 0.000986923 | | line of krypton 86 | |
| " | Bars............ | 0.001 | " | Wave length of red line of | 1553.16413 |
| " | Baryes............ | 1000 | | cadmium | |
| " | Dynes/sq. cm........ | 1000 | Millimeters of Hg | | |
| " | Grams/sq. cm........ | 1.019716 | (0°C.)............ | Atmospheres........... | 0.0013157895 |
| " | In. of Hg (32°F.)........ | 0.0295300 | " | Bars............... | 0.00133322 |
| " | Pounds/sq. ft......... | 2.088543 | " | Dynes/sq. cm........ | 1333.224 |
| " | Pounds/sq. inch....... | 0.0145038 | " | Grams/sq. cm......... | 1.35951 |
| Milligrams.......... | Carats (1877)......... | 0.004871 | " | Kg./sq. meter........ | 13,5951 |
| " | Carats (metric)......... | 0.005 | " | Pounds/sq. ft......... | 2.78450 |
| " | Drams (apoth. or troy).... | 0.00025720597 | " | Pounds/sq. inch........ | 0.0193368 |
| " | Drams (advp.)........... | 0.00056438339 | " | Torrs................ | 1 |

* Density of 1 gram per milliliter of solvent.

## Table 4—7 (continued)

### CONVERSION FACTORS

| To convert from | To | Multiply by | To convert from | To | Multiply by |
|---|---|---|---|---|---|
| Millimicrons | Ångström units | 10 | Oersteds | Gilberts/cm | 1 |
| " | Centimeters | $1 \times 10^{-7}$ | " | Oersteds (Int.) | 1.000165 |
| " | Inches | $3.9370079 \times 10^{-8}$ | Oersteds (Int.) | Oersteds | 0.999835 |
| " | Microns | 0.001 | Ohms | Abohms | $1 \times 10^{9}$ |
| " | Millimeters | $1 \times 10^{-6}$ | " | Cgs. units of resistance | 1 |
| Milliphots | Foot-candles | 0.9290304 | " | Megohms | $1 \times 10^{-6}$ |
| " | Lumens/sq. ft | 0.9290304 | " | Microhms | $1 \times 10^{6}$ |
| " | Lumens/sq. meter | 10 | " | Ohms (Int.) | 0.999505 |
| " | Lux | 10 | " | Statohms | $1.112646 \times 10^{-12}$ |
| " | Phots | 0.001 | Ohms (Int.) | Ohms | 1.000495 |
| Millivolts | Statvolts | $3.335635 \times 10^{-6}$ | Ohms (mil, foot) | Circ. mil-ohms/ft | 1 |
| " | Volts | 0.001 | " | Ohm-cm | $1.6624261 \times 10^{-7}$ |
| Minims (Brit.) | Cu. cm | 0.05919385 | Ohm-cm | Circ. mil-ohms/ft | $6.0153049 \times 10^{6}$ |
| " | Cu. inches | 0.003612230 | " | Microhm-cm | $1 \times 10^{6}$ |
| " | Milliliters | 0.05919219 | " | Ohm-inches | 0.39370079 |
| " | Ounces (Brit., fluid) | 0.0020833333 | Ohm-inches | Ohm-cm | 2.54 |
| " | Scruples (Brit., fluid) | 0.05 | Ohm-meters | Abohm-cm | $1 \times 10^{11}$ |
| Minims (U.S.) | Cu. cm | 0.061611520 | " | E.M. cgs. units | $1 \times 10^{11}$ |
| " | Cu. inches | 0.0037597656 | " | E.S. cgs. units | $1.112646 \times 10^{-10}$ |
| " | Drams (U.S., fluid) | 0.0166666 | " | Mks. units | 1 |
| " | Gallons (U.S., liq.) | $1.6276042 \times 10^{-5}$ | " | Statohm-cm | $1.112646 \times 10^{-10}$ |
| " | Gills (U.S.) | 0.0005208333 | Ounces (apoth. or troy) | Dekagrams | 1.7554286 |
| " | Liters | $6.160979 \times 10^{-5}$ | " | Drams (apoth. or troy) | 8 |
| " | Milliliters | 0.06160979 | " | Drams (avdp.) | 17.554286 |
| " | Ounces (U.S., fluid) | 0.002083333 | " | Grains | 480 |
| " | Pints (U.S., liq.) | 0.0001302083 | " | Grams | 31.103486 |
| Minutes (angular) | Degrees | 0.0166666 | " | Milligrams | 31103.486 |
| " | Quadrants | 0.000185185 | " | Ounces (avdp.) | 1.0971429 |
| " | Radians | 0.00029088821 | " | Pennyweights | 20 |
| " | Seconds (angular) | 60 | " | Pounds (apoth. or troy) | 0.0833333 |
| Minutes (mean solar) | Days (mean solar) | 0.0006944444 | " | Pounds (avdp.) | 0.068571429 |
| " | Days (sidereal) | 0.00069634577 | " | Scruples (apoth.) | 24 |
| " | Hours (mean solar) | 0.0166666 | " | Tons (short) | $3.4285714 \times 10^{-5}$ |
| " | Hours (sidereal) | 0.016712298 | Ounces (avdp.) | Drams (apoth. or troy) | 7.291666 |
| " | Minutes (sidereal) | 1.00273791 | " | Drams (avdp.) | 16 |
| Minutes (sidereal) | Days (mean solar) | 0.00069254831 | " | Grains | 437.5 |
| " | Minutes (mean solar) | 0.99726957 | " | Grams | 28.349523 |
| " | Months (mean calendar) | $2.2768712 \times 10^{-5}$ | " | Hundredweights (long) | 0.00055803571 |
| " | Seconds (sidereal) | 60 | " | Hundredweights (short) | 0.000625 |
| Minutes/cm | Radians/cm | 0.00029088821 | " | Ounces (apoth. or troy) | 0.9114583 |
| Months (lunar) | Days (mean solar) | 29.530588 | " | Pennyweights | 18.229166 |
| " | Hours (mean solar) | 708.73411 | " | Pounds (apoth. or troy) | 0.075954861 |
| " | Minutes (mean solar) | 42524.047 | " | Pounds (avdp.) | 0.0625 |
| " | Second (mean solar) | $2.5514428 \times 10^{6}$ | " | Scruples (apoth.) | 21.875 |
| " | Weeks (mean calendar) | 4.2186554 | " | Tons (long) | $2.7901786 \times 10^{-5}$ |
| Months (mean calendar) | Days (mean solar) | 30.416666 | " | Tons (metric) | $2.8349527 \times 10^{-5}$ |
| " | Hours (mean solar) | 730 | " | Tons (short) | $3.125 \times 10^{-5}$ |
| " | Months (lunar) | 1.0300055 | Ounces (Brit., fluid) | Cu. cm | 28.41305 |
| " | Weeks (mean calendar) | 4.3452381 | " | Cu. inches | 1.733870 |
| " | Years (calendar) | 0.08333333 | " | Drachms (Brit., fluid) | 8 |
| " | Years (sidereal) | 0.083274845 | " | Drams (U.S., fluid) | 7.686075 |
| " | Years (tropical) | 0.083278075 | " | Gallons (Brit.) | 0.00625 |
| Myriagrams | Grams | 10000 | " | Milliliters | 28.41225 |
| " | Kilograms | 10 | " | Minims (Brit.) | 480 |
| " | Pounds (avdp.) | 22.046226 | " | Ounces (U.S., fluid) | 0.9607594 |
| Newtons | Dynes | 100000 | Ounces (U.S., fluid) | Cu. cm | 29.573730 |
| " | Pounds | 0.22480894 | " | Cu. inches | 1.8046875 |
| Newton-meters | Dyne-cm | $1 \times 10^{7}$ | " | Cu. meters | $2.9573730 \times 10^{-5}$ |
| " | Gram-cm | 10197.162 | " | Drams (U.S., fluid) | 8 |
| " | Kg.-meters | 0.10197162 | " | Gallons (U.S., dry) | 0.0067138047 |
| " | Pound-feet | 0.73756215 | " | Gallons (U.S., liq.) | 0.0078125 |
| Noggins (Brit.) | Cu. cm | 142.0652 | " | Gills (U.S.) | 0.25 |
| " | Gallons (Brit.) | 0.03125 | " | Liters | 0.029572702 |
| " | Gills (Brit.) | 1 | " | Minims (U.S.) | 480 |
| Oersteds | Ampere-turns/inch | 2.0212678 | " | Ounces (Brit., fluid) | 1.040843 |
| " | Ampere-turns/meter | 79.577472 | " | Pints (U.S., liq.) | 0.0625 |
| " | E.M. cgs. units of magnetic field intensity | 1 | " | Quarts (U.S., liq.) | 0.03125 |
| " | E.S. cgs. units | $2.997930 \times 10^{10}$ | Ounces/sq. inch | Dynes/sq. cm | 4309.22 |
| | | | " | Grams/sq. cm | 4.3941849 |
| | | | " | In. of $H_2O$ (39.2°F.) | 1.73004 |
| | | | " | In. of $H_2O$ (60°F.) | 1.73166 |

## Table 4—7 (continued)
### CONVERSION FACTORS

| To convert from | To | Multiply by | To convert from | To | Multiply by |
|---|---|---|---|---|---|
| Ounces/sq. inch | Pounds/sq. ft. | 9 | Phots | Foot-candles | 929.0304 |
| " | Pounds/sq. inch | 0.0625 | " | Lumens/sq. cm. | 1 |
| Ounces (avdp.)/ton (long) | Milligrams/kg. | 27.901786 | " | Lumens/sq. meter | 10000 |
| " | | | " | Lux | 10000 |
| Ounces (avdp.)/ton (short) | Milligrams/kg. | 31.25 | Picas (printer's) | Centimeters | 0.42175176 |
| | | | " | Inches | 0.166044 |
| Paces | Centimeters | 76.2 | Pints (Brit.) | Cu. cm. | 568.26092 |
| " | Chains (Gunter's) | 0.0378788 | " | Gallons (Brit.) | 0.125 |
| " | Chains (Ramden's) | 0.025 | " | Gills (Brit.) | 4 |
| " | Feet | 2.5 | " | Gills (U.S.) | 4.803797 |
| " | Hands | 7.5 | " | Liters | 0.5682450 |
| " | Inches | 30 | | | |
| " | Ropes (Brit.) | 0.125 | Pints (Brit.) | Minims (Brit.) | 9600 |
| Palms | Centimeters | 7.62 | " | Ounces (Brit., fluid) | 20 |
| " | Chains (Ramden's) | 0.0025 | " | Pints (U.S., dry) | 1.032056 |
| " | Cubits | 0.1666666 | " | Pints (U.S., liq.) | 1.200949 |
| " | Feet | 0.25 | " | Quarts (Brit.) | 0.5 |
| " | Hands | 0.75 | " | Scruples (Brit., fluid) | 480 |
| " | Inches | 3 | Pints (U.S., dry) | Bushels (U.S.) | 0.015625 |
| Parsecs | Kilometers | $3.08572 \times 10^{13}$ | " | Cu. cm. | 550.61047 |
| " | Miles (statute) | $1.91738 \times 10^{13}$ | " | Cu. inches | 33.6003125 |
| Parts/million* | Grains/gal. (Brit.) | 0.07015488 | " | Gallons, (U.S., dry) | 0.125 |
| " | Grains/gal. (U.S.) | 0.05841620 | " | Gallons (U.S., liq.) | 0.14545590 |
| " | Grams/liter | 0.001 | " | Liters | 0.5505951 |
| " | Milligrams/liter | 1 | " | Pecks (U.S.) | 0.0625 |
| Pascal | Atmosphere | $9.869233 \times 20^{-6}$ | " | Quarts (U.S., dry) | 0.5 |
| " | Bar | $1 \times 10^{-5}$ | Pints (U.S., liq.) | Cu. cm. | 473.17647 |
| " | Dyne/sq. cm. | 10 | " | Cu. feet | 0.016710069 |
| " | Foot of $H_2O$ (conv.) | $3.34552 \times 10^{-4}$ | " | Cu. inches | 28.875 |
| " | Inch of Hg (conv.) | $2.95300 \times 10^{-4}$ | " | Cu. yards | 0.00061889146 |
| " | Inch of $H_2O$ (conv.) | $4.01463 \times 10^{-3}$ | " | Drams (U.S., fluid) | 128 |
| " | Kilogram-force/sq. cm. | $1.01972 \times 10^{-5}$ | " | Gallons (U.S., liq.) | 0.125 |
| " | Millibar | 0.01 | " | Gills (U.S.) | 4 |
| " | Millimeter of Hg (conv.) | $7.50062 \times 10^{-3}$ | " | Liters | 0.4731632 |
| " | Newton/sq. m. | 1 | " | Milliliters | 473.1632 |
| " | Newton/sq. mm. | $1 \times 10^{-6}$ | " | Minims (U.S.) | 7680 |
| " | Poundal/sq. ft. | 0.671969 | " | Ounces (U.S., fluid) | 16 |
| " | Pound-force/sq. ft. | 0.0208854 | " | Pints (Brit.) | 0.8326747 |
| " | Pound-force/sq. in. | $1.45038 \times 10^{-4}$ | " | Quarts (U.S., liq.) | 0.5 |
| " | Torr | $7.50062 \times 10^{-3}$ | Planck's constant | Erg-seconds | $6.6255 \times 10^{-27}$ |
| Pecks (Brit.) | Bushels (Brit.) | 0.25 | " | Joule-seconds | $6.6255 \times 10^{-34}$ |
| " | Coombs (Brit.) | 0.0625 | " | Joule-sec./Avog. No. (chem.) | $3.9905 \times 10^{-10}$ |
| " | Cu. cm. | 9092.175 | Points (printer's) | Centimeters | 0.03514598 |
| " | Cu. inches | 554.8385 | " | Inches | 0.013837 |
| " | Gallons (Brit.) | 2 | " | Picas | 0.0833333 |
| " | Gills (Brit.) | 64 | Poises | Cgs. units of absolute viscosity | 1 |
| " | Hogsheads | 0.03812537 | " | Grams/(cm. × sec.) | 1 |
| " | Kilderkins (Brit.) | 0.111111 | Poise-cu. cm./gram | Sq. cm./sec. | 1 |
| " | Liters | 9.091920 | Poise-cu. ft./lb. | Sq. cm./sec. | 62.427960 |
| " | Pints (Brit.) | 16 | Poise-cu. in./gram | Sq. cm./sec. | 16.387064 |
| " | Quarterns (Brit., dry) | 4 | Poles/sq. cm. | E.M. cgs. units of magnetization | 1 |
| " | Quarters (Brit., dry) | 0.03125 | Pottles (Brit.) | Gallons (Brit.) | 0.5 |
| " | Quarts (Brit.) | 8 | " | Liters | 2.272980 |
| " | Quarts (U.S., dry) | 8.256449 | Poundals | Dynes | 13825.50 |
| Pecks (U.S.) | Barrels (U.S., dry) | 0.076191185 | " | Grams | 14.09808 |
| " | Bushels (U.S.) | 0.25 | " | Pounds (avdp.) | 0.0310810 |
| " | Cu. cm. | 8809.7675 | Pounds (apoth. or troy) | Drams (apoth. or troy) | 96 |
| " | Cu. feet | 0.311114005 | " | Drams (avdp.) | 210.65143 |
| " | Cu. inches | 537.605 | " | Grains | 5760 |
| " | Gallons (U.S., dry) | 2 | " | Grams | 373.24172 |
| " | Gallons (U.S., liq.) | 2.3272944 | " | Kilograms | 0.37324172 |
| " | Liters | 8.809521 | " | Ounces (apoth. or troy) | 12 |
| " | Pints (U.S., dry) | 16 | " | Ounces (avdp.) | 13.165714 |
| " | Quarts (U.S., dry) | 8 | " | Pennyweights | 240 |
| Pennyweights | Drams (apoth. or troy) | 0.4 | " | Pounds (avdp.) | 0.8228571 |
| " | Drams (avdp.) | 0.87771429 | " | Scruples (apoth.) | 288 |
| " | Grains | 24 | " | Tons (long) | 0.00036734694 |
| " | Grams | 1.55517384 | " | Tons (metric) | 0.00037324172 |
| " | Ounces (apoth. or troy) | 0.05 | " | Tons (short) | 0.00041142857 |
| " | Ounces (avdp.) | 0.054857143 | Pounds (avdp.) | Drams (apoth. or troy) | 116.6666 |
| " | Pounds (apoth. or troy) | 0.0041666 | " | Drams (avdp.) | 256 |
| " | Pounds (avdp.) | 0.0034285714 | | | |
| Perches (masonry) | Cu. feet | 24.75 | | | |

* Based on density of 1 gram/ml. for the solvent.

## Table 4—7 (continued)

### CONVERSION FACTORS

| To convert from | To | Multiply by | To convert from | To | Multiply by |
|---|---|---|---|---|---|
| Pounds (avdp.) | Grains | 7000 | " | Liters | 2.272980 |
| " | Grams | 453.59237 | " | Pecks (Brit.) | 0.25 |
| " | Hundredweights (long) | 0.00892857 | Quarterns (Brit., liq.) | Cu. cm | 142.0652 |
| " | Hundredweights (short) | 0.01 | | | |
| " | Kilograms | 0.45359237 | Quarterns (Brit., liq.) | Gallons (Brit.) | 0.03125 |
| " | Ounces (apoth. *or* troy) | 14.583333 | " | Liters | 0.1420613 |
| " | Ounces (avdp.) | 16 | Quarters (U.S., long) | Kilograms | 254.0117272 |
| " | Pennyweights | 291.6666 | " | Pounds (avdp.) | 560 |
| " | Poundals | 32.1740 | Quarters (U.S., short) | Kilograms | 226.796185 |
| " | Pounds (apoth. *or* troy) | 1.215277 | " | Pounds | 500 |
| " | Scruples (apoth.) | 350 | Quarts (Brit.) | Cu. cm | 1136.522 |
| " | Slugs | 0.0310810 | " | Cu. inches | 69.35482 |
| " | Tons (long) | 0.00044642857 | " | Gallons (Brit.) | 0.25 |
| " | Tons (metric) | 0.00045359237 | " | Gallons (U.S., liq.) | 0.3002373 |
| " | Tons (short) | 0.0005 | " | Liters | 1.136490 |
| Pounds of H₂O evap. from and at 212°F | B.t.u. | 970.9 | " | Quarts (U.S., dry) | 1.032056 |
| " | B.t.u. (IST.) | 970.2 | " | Quarts (U.S., liq.) | 1.200949 |
| " | B.t.u. (mean) | 969.4 | Quarts (U.S., dry) | Bushels (U.S.) | 0.03125 |
| " | Joules | $1.0237 \times 10^6$ | " | Cu. cm | 1101.2209 |
| " | Joules (Int.) | $1.0234 \times 10^6$ | " | Cu. feet | 0.038889251 |
| Pounds/cu. ft. | Grams/cu. cm | 0.016018463 | " | Cu. inches | 67.200625 |
| " | Kg./cu. meter | 16.018463 | " | Gallons (U.S., dry) | 0.25 |
| Pounds/cu. inch | Grams/cu. cm | 27.679905 | " | Gallons (U.S., liq.) | 0.29091180 |
| " | Grams/liter | 27.68068 | " | Liters | 1.1011901 |
| " | Kg./cu. meter | 27679.905 | " | Pecks (U.S.) | 0.125 |
| Pounds/gal. (Brit.) | Pounds/cu. ft. | 6.228839 | " | Pints (U.S., dry) | 2 |
| Pounds/gal. (U.S., liq.) | Grams/cu. cm | 0.11982643 | Quarts (U.S., liq.) | Cu. cm | 946.35295 |
| " | Pounds/cu. ft. | 7.4805195 | " | Cu. feet | 0.033420136 |
| Pounds/inch | Grams/cm | 178.57967 | " | Cu. inches | 57.75 |
| " | Grams/ft. | 5443.1084 | " | Drams (U.S., fluid) | 256 |
| " | Grams/inch | 453.59237 | " | Gallons (U.S., dry) | 0.21484175 |
| " | Ounces/cm. | 6.2992 | " | Gallons (U.S., liq.) | 0.25 |
| " | Ounces/inch | 16 | " | Gills (U.S.) | 8 |
| " | Pounds/meter | 39.370079 | " | Liters | 0.9463264 |
| Pounds/minute | Kilograms/hr | 27.2155422 | " | Ounces (U.S., fluid) | 32 |
| " | Kilograms/min | 0.45359237 | " | Pints (U.S., liq.) | 2 |
| Pounds of H₂O (39.2°F.)/min. | Cu. ft./min. | 0.01601891 | " | Quarts (Brit.) | 0.8326747 |
| " | Gal. (U.S.)/min. | 0.1198298 | " | Quarts (U.S., dry) | 0.8593670 |
| " | Liters/min. | 0.45359237 | Quintals (metric) | Grams | 100000 |
| Pounds/sq. ft. | Atmospheres | 0.000472541 | " | Hundredweights (long) | 1.9684131 |
| " | Bars | 0.000478803 | " | Kilograms | 100 |
| " | Cm. of Hg (0°C.) | 0.0359131 | " | Pounds (avdp.) | 220.46226 |
| " | Dynes/sq. cm. | 478.803 | Radians | Circumferences | 0.15915494 |
| " | Ft. of air (1 atm., 60°F.) | 13.096 | " | Degrees | 57.295779 |
| " | Grams/sq. cm. | 0.48824276 | " | Minutes | 3437.7468 |
| " | In. of Hg (32°F.) | 0.0141390 | " | Quadrants | 0.63661977 |
| " | In. of H₂O (39.2°F.) | 0.192227 | " | Revolutions | 0.15915494 |
| " | Kg./sq. meter | 4.8824276 | " | Seconds | 206264.81 |
| " | Mm. of Hg (0°C.) | 0.359131 | Radians/cm | Degrees/cm | 57.295779 |
| Pounds/sq. inch | Atmospheres | 0.0680460 | " | Degrees/ft. | 1746.3754 |
| " | Bars | 0.0689476 | " | Degrees/inch | 145.53128 |
| " | Cm. of Hg (0°C.) | 5.17149 | " | Minutes/cm | 3437.7468 |
| " | Cm. of H₂O (4°C.) | 70.3089 | Radians/sec | Degrees/sec | 57.295779 |
| " | Dynes/sq. cm. | 68947.6 | " | Revolutions/min | 9.5492966 |
| " | Grams/sq. cm. | 70.306958 | " | Revolutions/sec | 0.15915494 |
| " | In. of Hg (32°F.) | 2.03602 | Radians/(sec. × sec.) | Revolutions/(min. × min.) | 572.95779 |
| " | In. of H₂O (39.2°F.) | 27.6807 | " | Revolutions/(min. × sec.) | 9.5492966 |
| " | Kg./sq. cm. | 0.070306958 | " | Revolutions/(sec. × sec.) | 0.15915494 |
| " | Mm. of Hg (0°C.) | 51.7149 | Register tons | Cu. feet | 100 |
| Pound wt.-sec./sq. ft. | Poises | 478.803 | " | Cu. meters | 2.8316847 |
| Pound wt.-sec./sq. in. | Poises | 68947.6 | Revolutions | Degrees | 360 |
| Puncheons (Brit.) | Cu. meters | 0.31797510 | " | Grades | 400 |
| " | Gallons (Brit.) | 69.94467 | " | Quadrants | 4 |
| " | Gallons (U.S.) | 84 | " | Radians | 6.2831853 |
| | | | Reyns* | Centipoises | $6.89476 \times 10^6$ |
| Quadrants | Minutes | 5400 | Rhes | Poises⁻¹ | 1 |
| " | Radians | 1.5707963 | Rods | Centimeters | 502.92 |
| Quarterns (Brit., dry) | Buckets (Brit.) | 0.125 | " | Chains (Gunter's) | 0.25 |
| " | Bushels (Brit.) | 0.0625 | " | Chains (Ramden's) | 0.165 |
| " | Cu. cm | 2273.044 | " | Feet | 16.5 |
| " | Gallons (Brit.) | 0.5 | " | Feet (U.S. Survey) | 16.499967 |
| | | | " | Furlongs | 0.025 |
| | | | " | Inches | 198 |
| | | | " | Links (Gunter's) | 25 |

## Table 4—7 (continued)

### CONVERSION FACTORS

| To convert from | To | Multiply by | To convert from | To | Multiply by |
|---|---|---|---|---|---|
| Rods | Links (Ramden's) | 16.5 | Sq. centimeters | Sq. mm | 100 |
| " | Meters | 5.0292 | " | Sq. mils | 155000.31 |
| " | Miles (statute) | 0.003125 | " | Sq. rods | $3.9536861 \times 10^{-6}$ |
| " | Perches | 1 | " | Sq. yards | 0.00011959900 |
| " | Yards | 5.5 | Sq. chains (Gunter's) | Acres | 0.1 |
| Rods (Brit., volume) | Cu. feet | 1000 | " | Sq. feet | 4356 |
| " | Cu. meters | 28.316847 | " | Sq. ft. (U.S. Survey) | 4355.9826 |
| Roods (Brit.) | Acres | 0.25 | " | Sq. inches | 627264 |
| " | Ares | 10.117141 | " | Sq. links (Gunter's) | 10000 |
| " | Sq. perches | 40 | " | Sq. meters | 404.68564 |
| " | Sq. yards | 1210 | " | Sq. miles | 0.00015625 |
| Ropes (Brit.) | Feet | 20 | " | Sq. rods | 16 |
| " | Meters | 6.096 | " | Sq. yards | 484 |
| " | Yards | 6.6666666 | Sq. chains (Ramden's) | Acres | 0.22956841 |
| Scruples (apoth.) | Drams (apoth. or troy) | 0.333333 | " | Sq. feet | 10000 |
| " | Drams (avdp.) | 0.73142857 | " | Sq. ft. (U.S. Survey) | 9999.9600 |
| " | Grains | 20 | " | Sq. inches | $1.44 \times 10^6$ |
| " | Grams | 1.2959782 | " | Sq. links (Ramden's) | 10000 |
| " | Ounces (apoth. or troy) | 0.041666 | " | Sq. meters | 929.0304 |
| " | Ounces (avdp.) | 0.045714286 | " | Sq. miles | 0.00035870064 |
| " | Pennyweights | 0.833333 | " | Sq. rods | 36.730946 |
| " | Pounds (apoth. or troy) | 0.003472222 | " | Sq. yards | 1111.111 |
| " | Pounds (avdp.) | 0.0028571429 | Sq. decimeters | Sq. cm | 100 |
| Scruples (Brit., fluid) | Minims (Brit.) | 20 | " | Sq. inches | 15.500031 |
| Seams (Brit.) | Bushels (Brit.) | 8 | Square degrees | Steradians | 0.00030461742 |
| " | Cu. feet | 10.27479 | Sq. dekameters | Acres | 0.024710538 |
| " | Liters | 290.9414 | " | Ares | 1 |
| Seconds (angular) | Degrees | 0.000277777 | " | Sq. meters | 100 |
| " | Minutes | 0.0166666 | " | Sq. yards | 119.59900 |
| " | Radians | $4.8481368 \times 10^{-6}$ | Sq. feet | Acres | $2.295684 \times 10^{-5}$ |
| Seconds (mean solar) | Days (mean solar) | $1.1574074 \times 10^{-5}$ | " | Ares | 0.0009290304 |
| " | Days (sidereal) | $1.1605763 \times 10^{-5}$ | " | Sq. cm | 929.0304 |
| " | Hours (mean solar) | 0.0002777777 | " | Sq. chains (Gunter's) | 0.00022956841 |
| " | Hours (sidereal) | 0.00027853831 | " | Sq. ft. (U.S. Survey) | 0.99999600 |
| " | Minutes (mean solar) | 0.0166666 | " | Sq. inches | 144 |
| " | Minutes (sidereal) | 0.016712298 | " | Sq. links (Gunter's) | 2.2956841 |
| " | Seconds (sidereal) | 1.00273791 | " | Sq. meters | 0.09290304 |
| Seconds (sidereal) | Days (mean solar) | $1.1542472 \times 10^{-5}$ | " | Sq. miles | $3.5870064 \times 10^{-8}$ |
| " | Days (sidereal) | $1.1574074 \times 10^{-5}$ | " | Sq. rods | 0.0036730946 |
| " | Hours (mean solar) | 0.00027701932 | " | Sq. yards | 0.111111 |
| " | Hours (sidereal) | 0.000277777 | Sq. feet (U.S. Survey) | Acres | $2.29569330 \times 10^{-5}$ |
| " | Minutes (mean solar) | 0.016621159 | " | Sq. centimeters | 929.03412 |
| " | Minutes (sidereal) | 0.0166666 | " | Sq. chains (Ramden's) | 0.00010000040 |
| " | Seconds (mean solar) | 0.99726957 | " | Sq. feet | 1.0000040 |
| Siemen's units | *Same as* Mhos | | Sq. hectometers | Sq. meters | 10000 |
| Skeins | Feet | 360 | Sq. inches | Circ. mils | 1273239.5 |
| " | Meters | 109.728 | " | Sq. cm | 6.4516 |
| Slugs | Geepounds | 1 | " | Sq. chains (Gunter's) | $1.5942251 \times 10^{-6}$ |
| " | Kilograms | 14.5939 | " | Sq. decimeters | 0.064516 |
| " | Pounds (avdp.) | 32.1740 | " | Sq. feet | 0.0069444 |
| Slugs/cu. ft. | Grams/cu. cm | 0.515379 | " | Sq. ft. (U.S. Survey) | 0.0069444167 |
| Space (entire) | Hemispheres | 2 | " | Sq. links (Gunter's) | 0.01594225 |
| " | Steradians | 12.566371 | " | Sq. meters | 0.00064516 |
| Spans | Centimeters | 22.86 | " | Sq. miles | $2.4909767 \times 10^{-10}$ |
| " | Fathoms | 0.125 | " | Sq. mm | 645.16 |
| " | Feet | 0.75 | " | Sq. mils | $1 \times 10^6$ |
| " | Inches | 9 | Sq. inches/sec | Sq. cm./hr | 23225.76 |
| " | Quarters (Brit. linear) | 1 | " | Sq. cm./sec | 6.4516 |
| Spherical right angles | Hemispheres | 0.25 | " | Sq. ft./min | 0.416666 |
| " | Spheres | 0.125 | Sq. kilometers | Acres | 247.10538 |
| " | Steradians | 1.5707963 | " | Sq. feet | $1.0763910 \times 10^7$ |
| Sq. centimeters | Ares | $1 \times 10^{-6}$ | " | Sq. ft. (U.S. Survey) | $1.0763867 \times 10^7$ |
| " | Circ. mm | 127.32395 | " | Sq. inches | $1.5500031 \times 10^9$ |
| " | Circ. mils | 197352.52 | " | Sq. meters | $1 \times 10^6$ |
| " | Sq. chains (Gunter's) | $2.4710538 \times 10^{-7}$ | " | Sq. miles | 0.38610216 |
| " | Sq. chains (Ramden's) | $1.0763910 \times 10^{-7}$ | " | Sq. yards | $1.1959900 \times 10^6$ |
| " | Sq. decimeters | 0.01 | Sq. links (Gunter's) | Acres | $1 \times 10^{-5}$ |
| " | Sq. feet | 0.0010763910 | " | Sq. cm | 404.68564 |
| " | Sq. ft. (U.S. Survey) | 0.0010763867 | " | Sq. chains (Gunter's) | 0.0001 |
| " | Sq. inches | 0.15500031 | " | Sq. feet | 0.4356 |
| " | Sq. meters | 0.0001 | " | Sq. ft. (U.S. Survey) | 0.43559826 |
| | | | " | Sq. Inches | 62.7264 |

## Table 4—7 (continued)

### CONVERSION FACTORS

| To convert from | To | Multiply by |
|---|---|---|
| Sq. links (Ramden's).. | Acres | $2.2956841 \times 10^{-5}$ |
| " | Sq. feet | 1 |
| Sq. meters | Acres | 0.00024710538 |
| " | Ares | 0.01 |
| " | Hectares | 0.0001 |
| " | Sq. cm | 10000 |
| " | Sq. feet | 10.763910 |
| " | Sq. inches | 1550.0031 |
| " | Sq. kilometers | $1 \times 10^{-6}$ |
| " | Sq. links (Gunter's) | 24.710538 |
| " | Sq. links (Ramden's) | 10.763910 |
| " | Sq. miles | $3.8610216 \times 10^{-7}$ |
| " | Sq. mm | $1 \times 10^{6}$ |
| " | Sq. rods | 0.039536861 |
| " | Sq. yards | 1.1959900 |
| Sq. miles | Acres | 640 |
| " | Hectares | 258.99881 |
| " | Sq. chains (Gunter's) | 6400 |
| " | Sq. feet | $2.7878288 \times 10^{7}$ |
| " | Sq. ft. (U.S. Survey) | $2.78288 \times 10^{7}$ |
| " | Sq. kilometers | 2.5899881 |
| " | Sq. meters | 2589988.1 |
| " | Sq. rods | 102400 |
| " | Sq. yards | $3.0976 \times 10^{6}$ |
| Sq. millimeters | Circ. mm | 1.2732395 |
| " | Circ. mils | 1973.5252 |
| " | Sq. cm | 0.01 |
| " | Sq. inches | 0.0015500031 |
| " | Sq. meters | $1 \times 10^{-6}$ |
| Sq. mils | Circ. mils | 1.2732395 |
| " | Sq. cm | $6.4516 \times 10^{-6}$ |
| " | Sq. inches | $1 \times 10^{-6}$ |
| " | Sq. mm | 0.00064516 |
| Sq. rods | Acres | 0.00625 |
| " | Ares | 0.2529285264 |
| " | Hectares | 0.002529285264 |
| " | Sq. cm | 252928.5264 |
| " | Sq. feet | 272.25 |
| " | Sq. ft. (U.S. Survey) | 272.24891 |
| " | Sq. inches | 39204 |
| " | Sq. links (Gunter's) | 625 |
| " | Sq. links (Ramden's) | 272.25 |
| " | Sq. meters | 25.29285264 |
| " | Sq. miles | $9.765625 \times 10^{-5}$ |
| " | Sq. yards | 30.25 |
| Sq. yards | Acres | 0.00020661157 |
| " | Ares | 0.0083612736 |
| " | Hectares | $8.3612736 \times 10^{-5}$ |
| " | Sq. cm | 8361.2736 |
| " | Sq. chains (Gunter's) | 0.0020661157 |
| " | Sq. chains (Ramden's) | 0.0009 |
| " | Sq. feet | 9 |
| " | Sq. ft. (U.S. Survey) | 8.9999640 |
| " | Sq. inches | 1296 |
| " | Sq. links (Gunter's) | 20.661157 |
| " | Sq. links (Ramden's) | 9 |
| " | Sq. meters | 0.83612736 |
| " | Sq. miles | $3.228305785 \times 10^{-7}$ |
| " | Sq. perches (Brit.) | 0.033057851 |
| " | Sq. rods | 0.033057851 |
| Statamperes | Abamperes | $3.335635 \times 10^{-11}$ |
| " | Amperes | $3.335635 \times 10^{-10}$ |
| " | E.M. cgs. units of current | $3.335635 \times 10^{-11}$ |
| " | E.S. cgs. units | 1 |
| Statcoulombs | Ampere-hours | $9.265653 \times 10^{-14}$ |
| " | Coulombs | $3.335635 \times 10^{-10}$ |
| " | Electronic charges | $2.082093 \times 10^{9}$ |
| " | E.M. cgs. units of electric charge | $3.335635 \times 10^{-11}$ |
| Statfarads | E.M. cgs. units of capacitance | $1.112646 \times 10^{-21}$ |
| " | E.S. cgs. units | 1 |

| To convert from | To | Multiply by |
|---|---|---|
| Statfarads | Farads | $1.112646 \times 10^{-12}$ |
| " | Microfarads | $1.112646 \times 10^{-6}$ |
| Stathenries | Abhenries | $8.987584 \times 10^{20}$ |
| " | E.M. cgs. units of inductance | $8.987584 \times 10^{20}$ |
| " | E.S. cgs. units | 1 |
| " | Henries | $8.987584 \times 10^{11}$ |
| " | Millihenries | $8.987584 \times 10^{14}$ |
| Statohms | Abohms | $8.987584 \times 10^{20}$ |
| " | E.S. cgs. units | 1 |
| " | Ohms | $8.987584 \times 10^{11}$ |
| Statvolts | Abvolts | $2.997930 \times 10^{10}$ |
| " | Volts | 299.7930 |
| Statvolts/cm | Volts/cm | 299.7930 |
| " | Volts/inch | 761.4742 |
| Statvolts/inch | Volts/cm | 118.0287 |
| Steradians | Hemispheres | 0.15915494 |
| " | Solid angles | 0.079577472 |
| " | Spheres | 0.079577472 |
| " | Spher. right angles | 0.63661977 |
| " | Square degrees | 3282.8063 |
| Steres | Cubic meters | 1 |
| " | Decisteres | 10 |
| " | Dekasteres | 0.1 |
| " | Liters | 999.972 |
| Stilbs | Candles/sq. cm | 1 |
| " | Candles/sq. inch | 6.4516 |
| " | Lamberts | 3.1415927 |
| Stokes* | Cgs. units of kinematic viscosity | 1 |
| " | Sq. cm./sec | 1 |
| " | Sq. inches/sec | 0.15500031 |
| " | Poise cu. cm./gram | 1 |
| Stones (Brit., legal) | Centals (Brit.) | 0.14 |
| Tons (long) | Dynes | $9.96402 \times 10^{8}$ |
| " | Hundredweights (long) | 20 |
| " | Hundredweights (short) | 22.4 |
| " | Kilograms | 1016.0469 |
| " | Ounces (avdp.) | 35840 |
| " | Pounds (apoth. or troy) | 2722.22 |
| " | Pounds (avdp.) | 2240 |
| " | Tons (metric) | 1.0160469 |
| " | Tons (short) | 1.12 |
| Tons (metric) | Dynes | $9.80665 \times 10^{8}$ |
| " | Grams | $1 \times 10^{6}$ |
| " | Hundredweights (short) | 22.046226 |
| " | Kilograms | 1000 |
| " | Ounces (avdp.) | 35273.962 |
| " | Pounds (apoth. or troy) | 2679.2289 |
| " | Pounds (avdp.) | 2204.6226 |
| " | Tons (long) | 0.98420653 |
| " | Tons (short) | 1.1023113 |
| Tons (short) | Dynes | $8.89644 \times 10^{8}$ |
| " | Hundredweights (short) | 20 |
| " | Kilograms | 907.18474 |
| " | Ounces (avdp.) | 32000 |
| " | Pounds (apoth. or troy) | 2430.555 |
| " | Pounds (avdp.) | 2000 |
| " | Tons (long) | 0.89285714 |
| " | Tons (metric) | 0.90718474 |
| Tons of refrig. (U.S., comm.) | B.t.u. (IST.)/hr | 12000 |
| " | B.t.u. (IST.)/min | 200 |
| " | Cal., *kg.* (IST.)/hr | 3023.949 |
| " | Horsepower | 4.71611 |
| " | Kg. of ice melted/hr | 37.971 |
| " | Lb. of ice melted/hr | 83.711 |
| Tons of refrig. (U.S., std.) | B.t.u. (IST.) | 288000 |
| " | B.t.u. (mean) | 287774 |
| " | Cal., *kg.* (IST.) | 72574.8 |

## Table 4—7 (continued)

### CONVERSION FACTORS

| To convert from | To | Multiply by |
|---|---|---|
| Tons of refrig. (U.S., std.) | Cal., *kg.* (mean) | 72517.9 |
| " | Lb. of ice melted | 2009.1 |
| Tons (long)/sq. ft | Atmospheres | 1.05849 |
| " | Dynes/sq. cm | 1.07252 × 10⁸ |
| " | Grams/sq. cm | 1093.6638 |
| " | Pounds/sq. ft | 2240 |
| Tons (short)/sq. ft | Atmospheres | 0.945082 |
| " | Dynes/sq. cm | 957.605 |
| " | Grams/sq. cm | 976.486 |
| " | Pounds/sq. inch | 13.8888 |
| Tons (long)/sq. in | Atmospheres | 152.423 |
| " | Dynes/sq. cm | 1.54443 × 10⁸ |
| " | Grams/sq. cm | 157487.59 |
| Tons (short)/sq. in | Dynes/sq. cm | 1.37895 × 10⁸ |
| " | Kg./sq. mm | 1406.139 |
| " | Pounds/sq. inch | 2000 |
| Torrs (*or* Tors) | Millimeters of Hg (0°C.) | 1 |
| Townships (U.S.) | Acres | 23040 |
| " | Sections | 36 |
| " | Sq. miles | 36 |
| Tuns | Gallons (U.S.) | 252 |
| " | Hogsheads | 4 |
| Volts | Abvolts | 1 × 10⁸ |
| " | Mks. (r *or* nr) units | 1 |
| " | Statvolts | 0.003335635 |
| " | Volts (Int.) | 0.999670 |
| Volts (Int.) | Volts | 1.000330 |
| Volts/°C | Joules/(coulomb × °C.) | 1 |
| Volt-coulombs | Joules (Int.) | 0.999835 |
| Volt-coulombs (Int.) | Joules | 1.000165 |
| Volt-electronic charge-seconds | Planck's constant* | 2.41814 × 10¹⁴ |
| Volt-faraday (chem.)-seconds | Planck's constant* | 1.45650 × 10³³ |
| Volt-faraday (phys.)-seconds | Planck's constant* | 1.45690 × 10³³ |
| Volt-seconds | Maxwells | 1 × 10⁸ |
| Watts | B.t.u./hr | 3.41443 |
| " | B.t.u. (mean)/hr | 3.40952 |
| " | B.t.u. (mean)/min | 0.0568253 |
| " | B.t.u./sec | 0.000948451 |
| " | B.t.u. (mean)/sec | 0.000947088 |
| " | Cal., *gm.*/hr | 860.421 |
| " | Cal., *gm.* (mean)/hr | 859.184 |
| " | Cal., *gm.* (20°C.)/hr | 860.853 |
| " | Cal., *gm.*/min | 14.3403 |
| " | Cal., *gm.* (IST.)/min | 14.3310 |
| " | Cal., *gm.* (mean)/min | 14.3197 |
| " | Cal., *kh.*/min | 0.0143403 |
| " | Cal., *kg.* (IST.)/min | 0.0143310 |
| " | Cal., *kg.* (mean)/min | 0.0143197 |
| " | Ergs/sec | 1 × 10⁷ |
| " | Foot-pounds/min | 44.2537 |
| " | Horsepower | 0.00134102 |
| " | Horsepower (boiler) | 0.000101942 |
| " | Horsepower (elec.) | 0.00134048 |
| " | Horsepower (metric) | 0.00135962 |
| " | Joules/sec | 1 |
| " | Kilowatts | 0.001 |
| " | Liter-atm./hr | 35.5282 |
| Watts (Int.) | B.t.u./hr | 3.41499 |
| " | B.t.u. (mean)/hr | 3.41008 |
| " | B.t.u./min | 0.569165 |
| " | B.t.u. (mean)/min | 0.0568347 |
| " | Cal., *gm.*/hr | 860.563 |
| " | Cal., *gm.* (mean)/hr | 859.326 |
| " | Cal., *kg.*/min | 0.0143427 |
| " | Cal., *kg.* (IST.)/min | 0.0143333 |
| " | Cal., *kg.* (mean)/min | 0.0143221 |

| To convert from | To | Multiply by |
|---|---|---|
| Watts (Int.) | Ergs/sec | 1.000165 × 10⁷ |
| " | Joules (Int.)/sec | 1 |
| " | Watts | 1.000165 |
| Watts/sq. cm | B.t.u./(hr. × sq. ft.) | 3172.10 |
| " | Cal., *gm.*/(hr. × sq. cm.) | 860.421 |
| " | Ft.-lb./(min. × sq. ft.) | 41113.1 |
| Watts/sq. in | B.t.u./(hr. × sq. ft.) | 491.677 |
| " | Cal., *gm.*/(hr. × sq. cm.) | 133.365 |
| " | Ft.-lb./(min. × sq. ft.) | 6372.54 |
| Watt-hours | B.t.u. | 3.41443 |
| " | B.t.u. (mean) | 3.40952 |
| " | Cal., *gm.* | 860.421 |
| Watt-hours | Cal., *kg.* (mean) | 0.859184 |
| " | Cal., *gm.* (mean) | 859.184 |
| " | Foot-pounds | 2655.22 |
| " | Hp.-hours | 0.00134102 |
| " | Joules | 3600 |
| " | Joules (Int.) | 3599.41 |
| " | Kg.-meters | 367.098 |
| " | Kw.-hours | 0.001 |
| " | Watt-hours (Int.) | 0.999835 |
| Watt-sec | Foot-pounds | 0.737562 |
| " | Gram-cm | 10197.16 |
| " | Joules | 1 |
| " | Liter-atm | 0.00986895 |
| " | Volt-coulombs | 1 |
| Wave length of orange-red line of krypton 86 | Ångström units | 6057.80211 |
| " | Millimeters | 0.000605780211 |
| Wave length of red line of cadmium | Ångström units | 6438.4696 |
| " | Millimeters | 0.00064384696 |
| Webers | Cgs. units of induction flux | 1 × 10⁸ |
| " | E.M. cgs. units of induction flux | 1 × 10⁸ |
| " | Lines | 1 × 10⁸ |
| " | Maxwells | 1 × 10⁸ |
| " | Mks. units of induction flux | 1 |
| " | Mks. nr units of magnetic charge | 0.079577472 |
| " | Mks. r units of magnetic charge | 1 |
| " | Volt-seconds | 1 |
| Webers/sq. cm | Gausses | 1 × 10⁸ |
| " | Lines/sq. cm | 1 × 10⁸ |
| " | Lines/sq. inch | 6.4516 × 10⁸ |
| Webers/sq. in | Gausses | 1.5500031 × 10⁷ |
| Weeks (mean calendar) | Days (mean solar) | 7 |
| " | Days (sidereal) | 7.0191654 |
| " | Hours (mean solar) | 168 |
| " | Hours (sidereal) | 168.45997 |
| " | Minutes (mean solar) | 10080 |
| " | Minutes (sidereal) | 10107.598 |
| " | Months (lunar) | 0.23704235 |
| " | Months (mean calendar) | 0.23013699 |
| " | Years (calendar) | 0.019178082 |
| " | Years (sidereal) | 0.019164622 |
| " | Years (tropical) | 0.019165365 |
| Weys (Brit., mass.) | Pounds (avdp.) | 252 |
| Yards | Centimeters | 91.44 |
| " | Chains (Gunter's) | 0.4545454 |
| " | Chains (Ramden's) | 0.03 |
| " | Cubits | 2 |
| " | Fathoms | 0.5 |
| " | Feet | 3 |
| " | Feet (U.S. Survey) | 2.9999940 |
| " | Furlongs | 0.00454545 |
| " | Inches | 36 |
| " | Meters | 0.9144 |
| " | Poles (Brit.) | 0.181818 |

## Table 4—7 (continued)

### CONVERSION FACTORS

| To convert from | To | Multiply by | To convert from | To | Multiply by |
|---|---|---|---|---|---|
| Yards | Quarters (Brit., linear) | 4 | Years (sidereal) | Days (sidereal) | 366.25640 |
| " | Rods | 0.181818 | " | Years (calendar) | 1.0007024 |
| " | Spans | 4 | " | Years (tropical) | 1.0000388 |
| Years (calendar) | Days (mean solar) | 365 | Years (tropical) | Days (mean solar) | 365.24219 |
| " | Hours (mean solar) | 8760 | " | Days (sidereal) | 366.24219 |
| " | Minutes (mean solar) | 525600 | " | Hours (mean solar) | 8765.8126 |
| " | Months (lunar) | 12.360065 | " | Hours (sidereal) | 8789.8126 |
| " | Months (mean calendar) | 12 | " | Months (mean calendar) | 12.007963 |
| " | Seconds (mean solar) | $3.1536 \times 10^7$ | " | Seconds (mean solar) | $3.1556926 \times 10^7$ |
| " | Weeks (mean calendar) | 52.142857 | " | Seconds (sidereal) | $3.1643326 \times 10^7$ |
| " | Years (sidereal) | 0.99929814 | " | Weeks (mean calendar) | 52.177456 |
| " | Years (tropical) | 0.99933690 | " | Years (calendar) | 1.0006635 |
| Years (leap) | Days (mean solar) | 366 | " | Years (sidereal) | 0.99996121 |
| Years (sidereal) | Days (mean solar) | 365.25636 | | | |

From various U.S. Government and IUPAC publications and from the *Handbook of Chemistry and Physics,* 69th ed., Weast, R. C., Ed., CRC Press, Boca Raton, Fla., 1988. F-324.

**Table 4—8**
**INCH–MILLIMETER CONVERSIONS–EXACT**

| 16ths | 32nds | 64ths | Decimal equivalents, inches (exact) | Millimeters (exact) | 16ths | 32nds | 64ths | Decimal equivalents, inches (exact) | Millimeters (exact) |
|---|---|---|---|---|---|---|---|---|---|
| | | 1 | 0.015 625 | 0.396 875 | | | 33 | 0.515 625 | 13.096 875 |
| | 1 | 2 | 0.031 25 | 0.793 750 | | 17 | 34 | 0.531 25 | 13.493 750 |
| | | 3 | 0.046 875 | 1.190 625 | | | 35 | 0.546 875 | 13.890 625 |
| 1 | 2 | 4 | 0.062 5 | 1.587 500 | 9 | 18 | 36 | 0.562 5 | 14.287 500 |
| | | 5 | 0.078 125 | 1.984 375 | | | 37 | 0.578 125 | 14.684 375 |
| | 3 | 6 | 0.093 75 | 2.381 250 | | 19 | 38 | 0.593 75 | 15.081 250 |
| | | 7 | 0.109 375 | 2.778 125 | | | 39 | 0.609 375 | 15.478 125 |
| 2 | 4 | 8 | 0.125 0 | 3.175 000 | 10 | 20 | 40 | 0.625 0 | 15.875 000 |
| | | 9 | 0.140 625 | 3.571 875 | | | 41 | 0.640 625 | 16.271 875 |
| | 5 | 10 | 0.156 25 | 3.968 750 | | 21 | 42 | 0.656 25 | 16.668 750 |
| | | 11 | 0.171 875 | 4.365 625 | | | 43 | 0.671 875 | 17.065 625 |
| 3 | 6 | 12 | 0.187 5 | 4.762 500 | 11 | 22 | 44 | 0.687 5 | 17.462 500 |
| | | 13 | 0.203 125 | 5.159 375 | | | 45 | 0.703 125 | 17.859 375 |
| | 7 | 14 | 0.218 75 | 5.556 250 | | 23 | 46 | 0.718 75 | 18.256 250 |
| | | 15 | 0.234 375 | 5.953 125 | | | 47 | 0.734 375 | 18.653 125 |
| 4 | 8 | 16 | 0.250 0 | 6.350 000 | 12 | 24 | 48 | 0.750 0 | 19.050 000 |
| | | 17 | 0.265 625 | 6.746 875 | | | 49 | 0.765 625 | 19.446 875 |
| | 9 | 18 | 0.281 25 | 7.143 750 | | 25 | 50 | 0.781 25 | 19.843 750 |
| | | 19 | 0.296 875 | 7.540 625 | | | 51 | 0.796 875 | 20.240 625 |
| 5 | 10 | 20 | 0.312 5 | 7.937 500 | 13 | 26 | 52 | 0.812 5 | 20.637 500 |
| | | 21 | 0.328 125 | 8.334 375 | | | 53 | 0.828 125 | 21.034 375 |
| | 11 | 22 | 0.343 75 | 8.731 250 | | 27 | 54 | 0.843 75 | 21.431 250 |
| | | 23 | 0.359 375 | 9.128 125 | | | 55 | 0.859 375 | 21.828 125 |
| 6 | 12 | 24 | 0.375 0 | 9.525 000 | 14 | 28 | 56 | 0.875 0 | 22.225 000 |
| | | 25 | 0.390 625 | 9.921 875 | | | 57 | 0.890 625 | 22.621 875 |
| | 13 | 26 | 0.406 25 | 10.318 750 | | 29 | 58 | 0.906 25 | 23.018 750 |
| | | 27 | 0.421 875 | 10.715 625 | | | 59 | 0.921 875 | 23.415 625 |
| 7 | 14 | 28 | 0.437 5 | 11.112 500 | 15 | 30 | 60 | 0.937 5 | 23.812 500 |
| | | 29 | 0.453 125 | 11.509 375 | | | 61 | 0.953 125 | 24.209 375 |
| | 15 | 30 | 0.468 75 | 11.906 250 | | 31 | 62 | 0.968 75 | 24.606 250 |
| | | 31 | 0.484 375 | 12.303 125 | | | 63 | 0.984 375 | 25.003 125 |
| 8 | 16 | 32 | 0.500 0 | 12.700 000 | 16 | 32 | 64 | 1.000 0 | 25.400 000 |

From Bolz, R. E. and Tuve, G. L., Eds., *Handbook of Tables for Applied Engineering Science,* 2nd ed., CRC Press, Cleveland, 1973, 842.

## Table 4—9
## TEMPERATURE CONVERSION

This table permits one to convert from degrees Celsius to degrees Fahrenheit or from degrees Fahrenheit to degrees Celsius. The conversion is accomplished by first locating in a column printed in boldface type the number that is to be converted. If the number to be converted is in degrees Fahrenheit, one may find its equivalent in degrees Celsius by reading to the left. If the number to be converted is in degrees Celsius, one may find its equivalent in degrees Fahrenheit by reading to the right. Degrees Celsius are identical to degrees Centigrade; however, the word Celsius is preferred for international use.

The approved international symbolic abbreviation for degrees Celsius is °C; for degrees Fahrenheit it is °F. Absolute zero on the Celsius scale is $-273.15°C$; on the Fahrenheit scale it is $-459.67°F$. The relation between degrees Fahrenheit and degrees Celsius may be expressed by

$$°C = 5/9 \ (°F - 32) \ \text{or}$$
$$°F = 9/5 \ (°C) + 32.$$

| To convert | | | To convert | | | To convert | | |
|---|---|---|---|---|---|---|---|---|
| To °C | ←°F or °C→ | To °F | To °C | ←°F or °C→ | To °F | To °C | ←°F or °C→ | To °F |
| −273.15 | −459.67 | — | −106.67 | −160 | −256 | −33.89 | −29 | −20.2 |
| −267.78 | −450 | — | −103.89 | −155 | −247 | −33.33 | −28 | −18.4 |
| −262.22 | −440 | — | −101.11 | −150 | −238 | −32.78 | −27 | −16.6 |
| −256.67 | −430 | — | −98.33 | −145 | −229 | −32.22 | −26 | −14.8 |
| −251.11 | −420 | — | −95.56 | −140 | −220 | −31.67 | −25 | −13 |
| −245.56 | −410 | — | −92.78 | −135 | −211 | −31.11 | −24 | −11.2 |
| −240 | −400 | — | −90 | −130 | −202 | −30.56 | −23 | −9.4 |
| −234.44 | −390 | — | −87.22 | −125 | −193 | −30 | −22 | −7.6 |
| −228.89 | −380 | — | −84.44 | −120 | −184 | −29.44 | −21 | −5.8 |
| −223.33 | −370 | — | −81.67 | −115 | −175 | −28.89 | −20 | −4 |
| −217.78 | −360 | — | −78.89 | −110 | −166 | −28.33 | −19 | −2.2 |
| −212.22 | −350 | — | −76.11 | −105 | −157 | −27.78 | −18 | −0.4 |
| −206.67 | −340 | — | −73.33 | −100 | −148 | −27.22 | −17 | 1.4 |
| −201.11 | −330 | — | −70.56 | −95 | −139 | −26.67 | −16 | 3.2 |
| −195.56 | −320 | — | −67.78 | −90 | −130 | −26.11 | −15 | 5 |
| −190 | −310 | — | −65 | −85 | −121 | −25.56 | −14 | 6.8 |
| −184.44 | −300 | — | −62.22 | −80 | −112 | −25 | −13 | 8.6 |
| −178.89 | −290 | — | −59.44 | −75 | −103 | −24.44 | −12 | 10.4 |
| −173.33 | −280 | — | −56.67 | −70 | −94 | −23.89 | −11 | 12.2 |
| −167.78 | −270 | −454 | −53.89 | −65 | −85 | −23.33 | −10 | 14 |
| −162.22 | −260 | −436 | −51.11 | −60 | −76 | −22.78 | −9 | 15.8 |
| −156.67 | −250 | −418 | −48.33 | −55 | −67 | −22.22 | −8 | 17.6 |
| −151.11 | −240 | −400 | −45.56 | −50 | −58 | −21.67 | −7 | 19.4 |
| −145.56 | −230 | −382 | −42.78 | −45 | −49 | −21.11 | −6 | 21.2 |
| −140 | −220 | −364 | −40 | −40 | −40 | −20.56 | −5 | 23 |
| −134.44 | −210 | −346 | −39.44 | −39 | −38.2 | −20 | −4 | 24.8 |
| −131.67 | −205 | −337 | −38.89 | −38 | −36.4 | −19.44 | −3 | 26.6 |
| −128.89 | −200 | −328 | −38.33 | −37 | −34.6 | −18.89 | −2 | 28.4 |
| −126.11 | −195 | −319 | −37.78 | −36 | −32.8 | −18.33 | −1 | 30.2 |
| −123.33 | −190 | −310 | −37.22 | −35 | −31 | −17.78 | 0 | 32 |
| −120.56 | −185 | −301 | −36.67 | −34 | −29.2 | −17.22 | 1 | 33.8 |
| −117.78 | −180 | −292 | −36.11 | −33 | −27.4 | −16.67 | 2 | 35.6 |
| −115 | −175 | −283 | −35.56 | −32 | −25.6 | −16.11 | 3 | 37.4 |
| −112.22 | −170 | −274 | −35 | −31 | −23.8 | −15.56 | 4 | 39.2 |
| −109.44 | −165 | −265 | −34.44 | −30 | −22 | −15 | 5 | 41 |

## Table 4—9 (continued)
## TEMPERATURE CONVERSION

| To °C | ←°F or °C→ | To °F | To °C | ←°F or °C→ | To °F | To °C | ←°F or °C→ | To °F |
|---|---|---|---|---|---|---|---|---|
| −14.44 | 6 | 42.8 | 12.78 | 55 | 131 | 40.56 | 105 | 221 |
| −13.89 | 7 | 44.6 | 13.33 | 56 | 132.8 | 41.11 | 106 | 222.8 |
| −13.33 | 8 | 46.4 | 13.89 | 57 | 134.6 | 41.67 | 107 | 224.6 |
| −12.78 | 9 | 48.2 | 14.44 | 58 | 136.4 | 42.22 | 108 | 226.4 |
| −12.22 | 10 | 50 | 15 | 59 | 138.2 | 42.78 | 109 | 228.2 |
| −11.67 | 11 | 51.8 | 15.56 | 60 | 140 | 43.33 | 110 | 230 |
| −11.11 | 12 | 53.6 | 16.11 | 61 | 141.8 | 43.89 | 111 | 231.8 |
| −10.56 | 13 | 55.4 | 16.67 | 62 | 143.6 | 44.44 | 112 | 233.6 |
| −10 | 14 | 57.2 | 17.22 | 63 | 145.4 | 45 | 113 | 235.4 |
| −9.44 | 15 | 59 | 17.78 | 64 | 147.2 | 45.56 | 114 | 237.2 |
| −8.89 | 16 | 60.8 | 18.33 | 65 | 149 | 46.11 | 115 | 239 |
| −8.33 | 17 | 62.6 | 18.89 | 66 | 150.8 | 46.67 | 116 | 240.8 |
| −7.78 | 18 | 64.4 | 19.44 | 67 | 152.6 | 47.22 | 117 | 242.6 |
| −7.22 | 19 | 66.2 | 20 | 68 | 154.4 | 47.78 | 118 | 244.4 |
| −6.67 | 20 | 68 | 20.56 | 69 | 156.2 | 48.33 | 119 | 246.2 |
| −6.11 | 21 | 69.8 | 21.11 | 70 | 158 | 48.89 | 120 | 248 |
| −5.56 | 22 | 71.6 | 21.67 | 71 | 159.8 | 49.44 | 121 | 249.8 |
| −5 | 23 | 73.4 | 22.22 | 72 | 161.6 | 50 | 122 | 251.6 |
| −4.44 | 24 | 75.2 | 22.78 | 73 | 163.4 | 50.56 | 123 | 253.4 |
| −3.89 | 25 | 77 | 23.33 | 74 | 165.2 | 51.11 | 124 | 255.2 |
| −3.33 | 26 | 78.8 | 23.89 | 75 | 167 | 51.67 | 125 | 257 |
| −2.78 | 27 | 80.6 | 24.44 | 76 | 168.8 | 52.22 | 126 | 258.8 |
| −2.22 | 28 | 82.4 | 25 | 77 | 170.6 | 52.78 | 127 | 260.6 |
| −1.67 | 29 | 84.2 | 25.56 | 78 | 172.4 | 53.33 | 128 | 262.4 |
| −1.11 | 30 | 86 | 26.11 | 79 | 174.2 | 53.89 | 129 | 264.2 |
| −0.56 | 31 | 87.8 | 26.67 | 80 | 176 | 54.44 | 130 | 266 |
| 0 | 32 | 89.6 | 27.22 | 81 | 177.8 | 55 | 131 | 267.8 |
| .56 | 33 | 91.4 | 27.78 | 82 | 179.6 | 55.56 | 132 | 269.6 |
| 1.11 | 34 | 93.2 | 28.33 | 83 | 181.4 | 56.11 | 133 | 271.4 |
| 1.67 | 35 | 95 | 28.89 | 84 | 183.2 | 56.67 | 134 | 273.2 |
| 2.22 | 36 | 96.8 | 29.44 | 85 | 185 | 57.22 | 135 | 275 |
| 2.78 | 37 | 98.6 | 30 | 86 | 186.8 | 57.78 | 136 | 276.8 |
| 3.33 | 38 | 100.4 | 30.56 | 87 | 188.6 | 58.33 | 137 | 278.6 |
| 3.89 | 39 | 102.2 | 31.11 | 88 | 190.4 | 58.89 | 138 | 280.4 |
| 4.44 | 40 | 104 | 31.67 | 89 | 192.2 | 59.44 | 139 | 282.2 |
| 5 | 41 | 105.8 | 32.22 | 90 | 194 | 60 | 140 | 284 |
| 5.56 | 42 | 107.6 | 32.78 | 91 | 195.8 | 60.56 | 141 | 285.8 |
| 6.11 | 43 | 109.4 | 33.33 | 92 | 197.6 | 61.11 | 142 | 287.6 |
| 6.67 | 44 | 111.2 | 33.89 | 93 | 199.4 | 61.67 | 143 | 289.4 |
| 7.22 | 45 | 113 | 34.44 | 94 | 201.2 | 62.22 | 144 | 291.2 |
| 7.78 | 46 | 114.8 | 35 | 95 | 203 | 62.78 | 145 | 293 |
| 8.33 | 47 | 116.6 | 35.56 | 96 | 204.8 | 63.33 | 146 | 294.8 |
| 8.89 | 48 | 118.4 | 36.11 | 97 | 206.6 | 63.89 | 147 | 296.6 |
| 9.44 | 49 | 120.2 | 36.67 | 98 | 208.4 | 64.44 | 148 | 298.4 |
|  |  |  | 37.22 | 99 | 210.2 | 65 | 149 | 300.2 |
| 10 | 50 | 122 | 37.78 | 100 | 212 | 65.56 | 150 | 302 |
| 10.56 | 51 | 123.8 | 38.33 | 101 | 213.8 | 66.11 | 151 | 303.8 |
| 11.11 | 52 | 125.6 | 38.89 | 102 | 215.6 | 66.67 | 152 | 305.6 |
| 11.67 | 53 | 127.4 | 39.44 | 103 | 217.4 | 67.22 | 153 | 307.4 |
| 12.22 | 54 | 129.2 | 40 | 104 | 219.2 | 67.78 | 154 | 309.2 |

## Table 4—9 (continued)
## TEMPERATURE CONVERSION

| To convert | | | To convert | | | To convert | | |
|---|---|---|---|---|---|---|---|---|
| To °C | ←°F or °C→ | To °F | To °C | ←°F or °C→ | To °F | To °C | ←°F or °C→ | To °F |
| 68.33 | 155 | 311 | 96.11 | 205 | 401 | 123.89 | 255 | 491 |
| 68.89 | 156 | 312.8 | 96.67 | 206 | 402.8 | 124.44 | 256 | 492.8 |
| 69.44 | 157 | 314.6 | 97.22 | 207 | 404.6 | 125 | 257 | 494.6 |
| 70 | 158 | 316.4 | 97.78 | 208 | 406.4 | 125.56 | 258 | 496.4 |
| 70.56 | 159 | 318.2 | 98.33 | 209 | 408.2 | 126.11 | 259 | 498.2 |
| 71.11 | 160 | 320 | 98.89 | 210 | 410 | 126.67 | 260 | 500 |
| 71.67 | 161 | 321.8 | 99.44 | 211 | 411.8 | 127.22 | 261 | 501.8 |
| 72.22 | 162 | 323.6 | 100 | 212 | 413.6 | 127.78 | 262 | 503.6 |
| 72.78 | 163 | 325.4 | 100.56 | 213 | 415.4 | 128.33 | 263 | 505.4 |
| 73.33 | 164 | 327.2 | 101.11 | 214 | 417.2 | 128.89 | 264 | 507.2 |
| 73.89 | 165 | 329 | 101.67 | 215 | 419 | 129.44 | 265 | 509 |
| 74.44 | 166 | 330.8 | 102.22 | 216 | 420.8 | 130 | 266 | 510.8 |
| 75 | 167 | 332.6 | 102.78 | 217 | 422.6 | 130.56 | 267 | 512.6 |
| 75.56 | 168 | 334.4 | 103.33 | 218 | 424.4 | 131.11 | 268 | 514.4 |
| 76.11 | 169 | 336.2 | 103.89 | 219 | 426.2 | 131.67 | 269 | 516.2 |
| 76.67 | 170 | 338 | 104.44 | 220 | 428 | 132.22 | 270 | 518 |
| 77.22 | 171 | 339.8 | 105 | 221 | 429.8 | 132.78 | 271 | 519.8 |
| 77.78 | 172 | 341.6 | 105.56 | 222 | 431.6 | 133.33 | 272 | 521.6 |
| 78.33 | 173 | 343.4 | 106.11 | 223 | 433.4 | 133.89 | 273 | 523.4 |
| 78.89 | 174 | 345.2 | 106.67 | 224 | 435.2 | 134.44 | 274 | 525.2 |
| 79.44 | 175 | 347 | 107.22 | 225 | 437 | 135 | 275 | 527 |
| 80 | 176 | 348.8 | 107.78 | 226 | 438.8 | 135.56 | 276 | 528.8 |
| 80.56 | 177 | 350.6 | 108.33 | 227 | 440.6 | 136.11 | 277 | 530.6 |
| 81.11 | 178 | 352.4 | 108.89 | 228 | 442.4 | 136.67 | 278 | 532.4 |
| 81.67 | 179 | 354.2 | 109.44 | 229 | 444.2 | 137.22 | 279 | 534.2 |
| 82.22 | 180 | 356 | 110 | 230 | 446 | 137.78 | 280 | 536 |
| 82.78 | 181 | 357.8 | 110.56 | 231 | 447.8 | 138.33 | 281 | 537.8 |
| 83.33 | 182 | 359.6 | 111.11 | 232 | 449.6 | 138.89 | 282 | 539.6 |
| 83.89 | 183 | 361.4 | 111.67 | 233 | 451.4 | 139.44 | 283 | 541.4 |
| 84.44 | 184 | 363.2 | 112.22 | 234 | 453.2 | 140 | 284 | 543.2 |
| 85 | 185 | 365 | 112.78 | 235 | 455 | 140.56 | 285 | 545 |
| 85.56 | 186 | 366.8 | 113.33 | 236 | 456.8 | 141.11 | 286 | 546.8 |
| 86.11 | 187 | 368.6 | 113.89 | 237 | 458.6 | 141.67 | 287 | 548.6 |
| 86.67 | 188 | 370.4 | 114.44 | 238 | 460.4 | 142.22 | 288 | 550.4 |
| 87.22 | 189 | 372.2 | 115 | 239 | 462.2 | 142.78 | 289 | 552.2 |
| 87.78 | 190 | 374 | 115.56 | 240 | 464 | 143.33 | 290 | 554 |
| 88.33 | 191 | 375.8 | 116.11 | 241 | 465.8 | 143.89 | 291 | 555.8 |
| 88.89 | 192 | 377.6 | 116.67 | 242 | 467.6 | 144.44 | 292 | 557.6 |
| 89.44 | 193 | 379.4 | 117.22 | 243 | 469.4 | 145 | 293 | 559.4 |
| 90 | 194 | 381.2 | 117.78 | 244 | 471.2 | 145.56 | 294 | 561.2 |
| 90.56 | 195 | 383 | 118.33 | 245 | 473 | 146.11 | 295 | 563 |
| 91.11 | 196 | 384.8 | 118.89 | 246 | 474.8 | 146.67 | 296 | 564.8 |
| 91.67 | 197 | 386.6 | 119.44 | 247 | 476.6 | 147.22 | 297 | 566.6 |
| 92.22 | 198 | 388.4 | 120 | 248 | 478.4 | 147.78 | 298 | 568.4 |
| 92.78 | 199 | 390.2 | 120.56 | 249 | 480.2 | 148.33 | 299 | 570.2 |
| 93.33 | 200 | 392 | 121.11 | 250 | 482 | 148.89 | 300 | 572 |
| 93.89 | 201 | 393.8 | 121.67 | 251 | 483.8 | 149.44 | 301 | 573.8 |
| 94.44 | 202 | 395.6 | 122.22 | 252 | 485.6 | 150 | 302 | 575.6 |
| 95 | 203 | 397.4 | 122.78 | 253 | 487.4 | 150.56 | 303 | 577.4 |
| 95.56 | 204 | 399.2 | 123.33 | 254 | 489.2 | 151.11 | 304 | 579.2 |

## Table 4—9 (continued)
## TEMPERATURE CONVERSION

| To °C | ←°F or °C→ | To °F | To °C | ←°F or °C→ | To °F | To °C | ←°F or °C→ | To °F |
|---|---|---|---|---|---|---|---|---|
| 151.67 | 305 | 581 | 179.44 | 355 | 671 | 210 | 410 | 770 |
| 152.22 | 306 | 582.8 | 180 | 356 | 672.8 | 211.11 | 412 | 773.6 |
| 152.78 | 307 | 584.6 | 180.56 | 357 | 674.6 | 212.22 | 414 | 777.2 |
| 153.33 | 308 | 586.4 | 181.11 | 358 | 676.4 | 213.33 | 416 | 780.8 |
| 153.89 | 309 | 588.2 | 181.67 | 359 | 678.2 | 214.44 | 418 | 784.4 |
| 154.44 | 310 | 590 | 182.22 | 360 | 680 | 215.56 | 420 | 788 |
| 155 | 311 | 591.8 | 182.78 | 361 | 681.8 | 216.67 | 422 | 791.6 |
| 155.56 | 312 | 593.6 | 183.33 | 362 | 683.6 | 217.78 | 424 | 795.2 |
| 156.11 | 313 | 595.4 | 183.89 | 363 | 685.4 | 218.89 | 426 | 798.8 |
| 156.67 | 314 | 597.2 | 184.44 | 364 | 687.2 | 220 | 428 | 802.4 |
| 157.22 | 315 | 599 | 185 | 365 | 689 | 221.11 | 430 | 806 |
| 157.78 | 316 | 600.8 | 185.56 | 366 | 690.8 | 222.22 | 432 | 809.6 |
| 158.33 | 317 | 602.6 | 186.11 | 367 | 692.6 | 223.33 | 434 | 813.2 |
| 158.89 | 318 | 604.4 | 186.67 | 368 | 694.4 | 224.44 | 436 | 816.8 |
| 159.44 | 319 | 606.2 | 187.22 | 369 | 696.2 | 225.56 | 438 | 820.4 |
| 160 | 320 | 608 | 187.78 | 370 | 698 | 226.67 | 440 | 824 |
| 160.56 | 321 | 609.8 | 188.33 | 371 | 699.8 | 227.78 | 442 | 827.6 |
| 161.11 | 322 | 611.6 | 188.89 | 372 | 701.6 | 228.89 | 444 | 831.2 |
| 161.67 | 323 | 613.4 | 189.44 | 373 | 703.4 | 230 | 446 | 834.8 |
| 162.22 | 324 | 615.2 | 190 | 374 | 705.2 | 231.11 | 448 | 838.4 |
| 162.78 | 325 | 617 | 190.56 | 375 | 707 | 232.22 | 450 | 842 |
| 163.33 | 326 | 618.8 | 191.11 | 376 | 708.8 | 233.33 | 452 | 845.6 |
| 163.89 | 327 | 620.6 | 191.67 | 377 | 710.6 | 234.44 | 454 | 849.2 |
| 164.44 | 328 | 622.4 | 192.22 | 378 | 712.4 | 235.56 | 456 | 852.8 |
| 165 | 329 | 624.2 | 192.78 | 379 | 714.2 | 236.67 | 458 | 856.4 |
| 165.56 | 330 | 626 | 193.33 | 380 | 716 | 237.78 | 460 | 860 |
| 166.11 | 331 | 627.8 | 193.89 | 381 | 717.8 | 238.89 | 462 | 863.6 |
| 166.67 | 332 | 629.6 | 194.44 | 382 | 719.6 | 240 | 464 | 867.2 |
| 167.22 | 333 | 631.4 | 195 | 383 | 721.4 | 241.11 | 466 | 870.8 |
| 167.78 | 334 | 633.2 | 195.56 | 384 | 723.2 | 242.22 | 468 | 874.4 |
| 168.33 | 335 | 635 | 196.11 | 385 | 725 | 243.33 | 470 | 878 |
| 168.89 | 336 | 636.8 | 196.67 | 386 | 726.8 | 244.44 | 472 | 881.6 |
| 169.44 | 337 | 638.6 | 197.22 | 387 | 728.6 | 245.56 | 474 | 885.2 |
| 170 | 338 | 640.4 | 197.78 | 388 | 730.4 | 246.67 | 476 | 888.8 |
| 170.56 | 339 | 642.2 | 198.33 | 389 | 732.2 | 247.78 | 478 | 892.4 |
| 171.11 | 340 | 644 | 198.89 | 390 | 734 | 248.89 | 480 | 896 |
| 171.67 | 341 | 645.8 | 199.44 | 391 | 735.8 | 250 | 482 | 899.6 |
| 172.22 | 342 | 647.6 | 200 | 392 | 737.6 | 251.11 | 484 | 903.2 |
| 172.78 | 343 | 649.4 | 200.56 | 393 | 739.4 | 252.22 | 486 | 906.8 |
| 173.33 | 344 | 651.2 | 201.11 | 394 | 741.2 | 253.33 | 488 | 910.4 |
| 173.89 | 345 | 653 | 201.67 | 395 | 743 | 254.44 | 490 | 914 |
| 174.44 | 346 | 654.8 | 202.22 | 396 | 744.8 | 255.56 | 492 | 917.6 |
| 175 | 347 | 656.6 | 202.78 | 397 | 746.6 | 256.67 | 494 | 921.2 |
| 175.56 | 348 | 658.4 | 203.33 | 398 | 748.4 | 257.78 | 496 | 924.8 |
| 176.11 | 349 | 660.2 | 203.89 | 399 | 750.2 | 258.89 | 498 | 928.4 |
| 176.67 | 350 | 662 | 204.44 | 400 | 752 | 260 | 500 | 932 |
| 177.22 | 351 | 663.8 | 205.56 | 402 | 755.6 | 261.11 | 502 | 935.6 |
| 177.78 | 352 | 665.6 | 206.67 | 404 | 759.2 | 262.22 | 504 | 939.2 |
| 178.33 | 353 | 667.4 | 207.78 | 406 | 762.8 | 263.33 | 506 | 942.8 |
| 178.89 | 354 | 669.2 | 208.89 | 408 | 766.4 | 264.44 | 508 | 946.4 |

## Table 4—9 (continued)
## TEMPERATURE CONVERSION

| To °C | ←°F or °C→ | To °F | To °C | ←°F or °C→ | To °F | To °C | ←°F or °C→ | To °F |
|---|---|---|---|---|---|---|---|---|
| 265.56 | 510 | 950 | 321.11 | 610 | 1130 | 376.67 | 710 | 1310 |
| 266.67 | 512 | 953.6 | 322.22 | 612 | 1133.6 | 377.78 | 712 | 1313.6 |
| 267.78 | 514 | 957.2 | 323.33 | 614 | 1137.2 | 378.89 | 714 | 1317.2 |
| 268.89 | 516 | 960.8 | 324.44 | 616 | 1140.8 | 380 | 716 | 1320.8 |
| 270 | 518 | 964.4 | 325.56 | 618 | 1144.4 | 381.11 | 718 | 1324.4 |
| 271.11 | 520 | 968 | 326.67 | 620 | 1148 | 382.22 | 720 | 1328 |
| 272.22 | 522 | 971.6 | 327.78 | 622 | 1151.6 | 383.33 | 722 | 1331.6 |
| 273.33 | 524 | 975.2 | 328.89 | 624 | 1155.2 | 384.44 | 724 | 1335.2 |
| 274.44 | 526 | 978.8 | 330 | 626 | 1158.8 | 385.56 | 726 | 1338.8 |
| 275.56 | 528 | 982.4 | 331.11 | 628 | 1162.4 | 386.67 | 728 | 1342.4 |
| 276.67 | 530 | 986 | 332.22 | 630 | 1166 | 387.78 | 730 | 1346 |
| 277.78 | 532 | 989.6 | 333.33 | 632 | 1169.6 | 388.89 | 732 | 1349.6 |
| 278.89 | 534 | 993.2 | 334.44 | 634 | 1173.2 | 390 | 734 | 1353.2 |
| 280 | 536 | 996.8 | 335.56 | 636 | 1176.8 | 391.11 | 736 | 1356.8 |
| 281.11 | 538 | 1000.4 | 336.67 | 638 | 1180.4 | 392.22 | 738 | 1360.4 |
| 282.22 | 540 | 1004 | 337.78 | 640 | 1184 | 393.33 | 740 | 1364 |
| 283.33 | 542 | 1007.6 | 338.89 | 642 | 1187.6 | 394.44 | 742 | 1367.6 |
| 284.44 | 544 | 1011.2 | 340 | 644 | 1191.2 | 395.56 | 744 | 1371.2 |
| 285.56 | 546 | 1014.8 | 341.11 | 646 | 1194.8 | 396.67 | 746 | 1374.8 |
| 286.67 | 548 | 1018.4 | 342.22 | 648 | 1198.4 | 397.78 | 748 | 1378.4 |
| 287.78 | 550 | 1022 | 343.33 | 650 | 1202 | 398.89 | 750 | 1382 |
| 288.89 | 552 | 1025.6 | 344.44 | 652 | 1205.6 | 400 | 752 | 1385.6 |
| 290 | 554 | 1029.2 | 345.56 | 654 | 1209.2 | 401.11 | 754 | 1389.2 |
| 291.11 | 556 | 1032.8 | 346.67 | 656 | 1212.8 | 402.22 | 756 | 1392.8 |
| 292.22 | 558 | 1036.4 | 347.78 | 658 | 1216.4 | 403.33 | 758 | 1396.4 |
| 293.33 | 560 | 1040 | 348.89 | 660 | 1220 | 404.44 | 760 | 1400 |
| 294.44 | 562 | 1043.6 | 350 | 662 | 1223.6 | 405.56 | 762 | 1403.6 |
| 295.56 | 564 | 1047.2 | 351.11 | 664 | 1227.2 | 406.67 | 764 | 1407.2 |
| 296.67 | 566 | 1050.8 | 352.22 | 666 | 1230.8 | 407.78 | 766 | 1410.8 |
| 297.78 | 568 | 1054.4 | 353.33 | 668 | 1234.4 | 408.89 | 768 | 1414.4 |
| 298.89 | 570 | 1058 | 354.44 | 670 | 1238 | 410 | 770 | 1418 |
| 300 | 572 | 1061.6 | 355.56 | 672 | 1241.6 | 411.11 | 772 | 1421.6 |
| 301.11 | 574 | 1065.2 | 356.67 | 674 | 1245.2 | 412.22 | 774 | 1425.2 |
| 302.22 | 576 | 1068.8 | 357.78 | 676 | 1248.8 | 413.33 | 776 | 1428.8 |
| 303.33 | 578 | 1072.4 | 358.89 | 678 | 1252.4 | 414.44 | 778 | 1432.4 |
| 304.44 | 580 | 1076 | 360 | 680 | 1256 | 415.56 | 780 | 1436 |
| 305.56 | 582 | 1079.6 | 361.11 | 682 | 1259.6 | 416.67 | 782 | 1439.6 |
| 306.67 | 584 | 1083.2 | 362.22 | 684 | 1263.2 | 417.78 | 784 | 1443.2 |
| 307.78 | 586 | 1086.8 | 363.33 | 686 | 1266.8 | 418.89 | 786 | 1446.8 |
| 308.89 | 588 | 1090.4 | 364.44 | 688 | 1270.4 | 420 | 788 | 1450.4 |
| 310 | 590 | 1094 | 365.56 | 690 | 1274 | 421.11 | 790 | 1454 |
| 311.11 | 592 | 1097.6 | 366.67 | 692 | 1277.6 | 422.22 | 792 | 1457.6 |
| 312.22 | 594 | 1101.2 | 367.78 | 694 | 1281.2 | 423.33 | 794 | 1461.2 |
| 313.33 | 596 | 1104.8 | 368.89 | 696 | 1284.8 | 424.44 | 796 | 1464.8 |
| 314.44 | 598 | 1108.4 | 370 | 698 | 1288.4 | 425.56 | 798 | 1468.4 |
| 315.56 | 600 | 1112 | 371.11 | 700 | 1292 | 426.67 | 800 | 1472 |
| 316.67 | 602 | 1115.6 | 372.22 | 702 | 1295.6 | 427.78 | 802 | 1475.6 |
| 317.78 | 604 | 1119.2 | 373.33 | 704 | 1299.2 | 428.89 | 804 | 1479.2 |
| 318.89 | 606 | 1122.8 | 374.44 | 706 | 1302.8 | 430 | 806 | 1482.8 |
| 320 | 608 | 1126.4 | 375.56 | 708 | 1306.4 | 431.11 | 808 | 1486.4 |

## Table 4—9 (continued)
## TEMPERATURE CONVERSION

| To °C | ←°F or °C→ | To °F | To °C | ←°F or °C→ | To °F | To °C | ←°F or °C→ | To °F |
|---|---|---|---|---|---|---|---|---|
| 432.22 | 810 | 1490 | 487.78 | 910 | 1670 | 543.33 | 1010 | 1850 |
| 433.33 | 812 | 1493.6 | 488.89 | 912 | 1673.6 | 544.44 | 1012 | 1853.6 |
| 434.44 | 814 | 1497.2 | 490 | 914 | 1677.2 | 545.56 | 1014 | 1857.2 |
| 435.56 | 816 | 1500.8 | 491.11 | 916 | 1680.8 | 546.67 | 1016 | 1860.8 |
| 436.67 | 818 | 1504.4 | 492.22 | 918 | 1684.4 | 547.78 | 1018 | 1864.4 |
| 437.78 | 820 | 1508 | 493.33 | 920 | 1688 | 548.89 | 1020 | 1868 |
| 438.89 | 822 | 1511.6 | 494.44 | 922 | 1691.6 | 550 | 1022 | 1871.6 |
| 440 | 824 | 1515.2 | 495.56 | 924 | 1695.2 | 551.11 | 1024 | 1875.2 |
| 441.11 | 826 | 1518.8 | 496.67 | 926 | 1698.8 | 552.22 | 1026 | 1878.8 |
| 442.22 | 828 | 1522.4 | 497.78 | 928 | 1702.4 | 553.33 | 1028 | 1882.4 |
| 443.33 | 830 | 1526 | 498.89 | 930 | 1706 | 554.44 | 1030 | 1886 |
| 444.44 | 832 | 1529.6 | 500 | 932 | 1709.6 | 555.56 | 1032 | 1889.6 |
| 445.56 | 834 | 1533.2 | 501.11 | 934 | 1713.2 | 556.67 | 1034 | 1893.2 |
| 446.67 | 836 | 1536.8 | 502.22 | 936 | 1716.8 | 557.78 | 1036 | 1896.8 |
| 447.78 | 838 | 1540.4 | 503.33 | 938 | 1720.4 | 558.89 | 1038 | 1900.4 |
| 448.89 | 840 | 1544 | 504.44 | 940 | 1724 | 560 | 1040 | 1904 |
| 450 | 842 | 1547.6 | 505.56 | 942 | 1727.6 | 561.11 | 1042 | 1907.6 |
| 451.11 | 844 | 1551.2 | 506.67 | 944 | 1731.2 | 562.22 | 1044 | 1911.2 |
| 452.22 | 846 | 1554.8 | 507.78 | 946 | 1734.8 | 563.33 | 1046 | 1914.8 |
| 453.33 | 848 | 1558.4 | 508.89 | 948 | 1738.4 | 564.44 | 1048 | 1918.4 |
| 454.44 | 850 | 1562 | 510 | 950 | 1742 | 565.56 | 1050 | 1922 |
| 455.56 | 852 | 1565.6 | 511.11 | 952 | 1745.6 | 566.67 | 1052 | 1925.6 |
| 456.67 | 854 | 1569.2 | 512.22 | 954 | 1749.2 | 567.78 | 1054 | 1929.2 |
| 457.78 | 856 | 1572.8 | 513.33 | 956 | 1752.8 | 568.89 | 1056 | 1932.8 |
| 458.89 | 858 | 1576.4 | 514.44 | 958 | 1756.4 | 570 | 1058 | 1936.4 |
| 460 | 860 | 1580 | 515.56 | 960 | 1760 | 571.11 | 1060 | 1940 |
| 461.11 | 862 | 1583.6 | 516.67 | 962 | 1763.6 | 572.22 | 1062 | 1943.6 |
| 462.22 | 864 | 1587.2 | 517.78 | 964 | 1767.2 | 573.33 | 1064 | 1947.2 |
| 463.33 | 866 | 1590.8 | 518.89 | 966 | 1770.8 | 574.44 | 1066 | 1950.8 |
| 464.44 | 868 | 1594.4 | 520 | 968 | 1774.4 | 575.56 | 1068 | 1954.4 |
| 465.56 | 870 | 1598 | 521.11 | 970 | 1778 | 576.67 | 1070 | 1958 |
| 466.67 | 872 | 1601.6 | 522.22 | 972 | 1781.6 | 577.78 | 1072 | 1961.6 |
| 467.78 | 874 | 1605.2 | 523.33 | 974 | 1785.2 | 578.89 | 1074 | 1965.2 |
| 468.89 | 876 | 1608.8 | 524.44 | 976 | 1788.8 | 580 | 1076 | 1968.8 |
| 470 | 878 | 1612.4 | 525.56 | 978 | 1792.4 | 581.11 | 1078 | 1972.4 |
| 471.11 | 880 | 1616 | 526.67 | 980 | 1796 | 582.22 | 1080 | 1976 |
| 472.22 | 882 | 1619.6 | 527.78 | 982 | 1799.6 | 583.33 | 1082 | 1979.6 |
| 473.33 | 884 | 1623.2 | 528.89 | 984 | 1803.2 | 584.44 | 1084 | 1983.2 |
| 474.44 | 886 | 1626.8 | 530 | 986 | 1806.8 | 585.56 | 1086 | 1986.8 |
| 475.56 | 888 | 1630.4 | 531.11 | 988 | 1810.4 | 586.67 | 1088 | 1990.4 |
| 476.67 | 890 | 1634 | 532.22 | 990 | 1814 | 587.78 | 1090 | 1994 |
| 477.78 | 892 | 1637.6 | 533.33 | 992 | 1817.6 | 588.89 | 1092 | 1997.6 |
| 478.89 | 894 | 1641.2 | 534.44 | 994 | 1821.2 | 590 | 1094 | 2001.2 |
| 480 | 896 | 1644.8 | 535.56 | 996 | 1824.8 | 591.11 | 1096 | 2004.8 |
| 481.11 | 898 | 1648.4 | 536.67 | 998 | 1828.4 | 592.22 | 1098 | 2008.4 |
| 482.22 | 900 | 1652 | 537.78 | 1000 | 1832 | 593.33 | 1100 | 2012 |
| 483.33 | 902 | 1655.6 | 538.89 | 1002 | 1835.6 | 594.44 | 1102 | 2015.6 |
| 484.44 | 904 | 1659.2 | 540 | 1004 | 1839.2 | 595.56 | 1104 | 2019.2 |
| 485.56 | 906 | 1662.8 | 541.11 | 1006 | 1842.8 | 596.67 | 1106 | 2022.8 |
| 486.67 | 908 | 1666.4 | 542.22 | 1008 | 1846.4 | 597.78 | 1108 | 2026.4 |

## Table 4—9 (continued)
### TEMPERATURE CONVERSION

| To convert | | | To convert | | | To convert | | |
|---|---|---|---|---|---|---|---|---|
| To °C | ←°F or °C→ | To °F | To °C | ←°F or °C→ | To °F | To °C | ←°F or °C→ | To °F |
| 598.89 | 1110 | 2030 | 654.44 | 1210 | 2210 | 710 | 1310 | 2390 |
| 600 | 1112 | 2033.6 | 655.56 | 1212 | 2213.6 | 711.11 | 1312 | 2393.6 |
| 601.11 | 1114 | 2037.2 | 656.67 | 1214 | 2217.2 | 712.22 | 1314 | 2397.2 |
| 602.22 | 1116 | 2040.8 | 657.78 | 1216 | 2220.8 | 713.33 | 1316 | 2400.8 |
| 603.33 | 1118 | 2044.4 | 658.89 | 1218 | 2224.4 | 714.44 | 1318 | 2404.4 |
| 604.44 | 1120 | 2048 | 660 | 1220 | 2228 | 715.56 | 1320 | 2408 |
| 605.56 | 1122 | 2051.6 | 661.11 | 1222 | 2231.6 | 716.67 | 1322 | 2411.6 |
| 606.67 | 1124 | 2055.2 | 662.22 | 1224 | 2235.2 | 717.78 | 1324 | 2415.2 |
| 607.78 | 1126 | 2058.8 | 663.33 | 1226 | 2238.8 | 718.89 | 1326 | 2418.8 |
| 608.89 | 1128 | 2062.4 | 664.44 | 1228 | 2242.4 | 720 | 1328 | 2422.4 |
| 610 | 1130 | 2066 | 665.56 | 1230 | 2246 | 721.11 | 1330 | 2426 |
| 611.11 | 1132 | 2069.6 | 666.67 | 1232 | 2249.6 | 722.22 | 1332 | 2429.6 |
| 612.22 | 1134 | 2073.2 | 667.78 | 1234 | 2253.2 | 723.33 | 1334 | 2433.2 |
| 613.33 | 1136 | 2076.8 | 668.89 | 1236 | 2256.8 | 724.44 | 1336 | 2436.8 |
| 614.44 | 1138 | 2080.4 | 670 | 1238 | 2260.4 | 725.56 | 1338 | 2440.4 |
| 615.56 | 1140 | 2084 | 671.11 | 1240 | 2264 | 726.67 | 1340 | 2444 |
| 616.67 | 1142 | 2087.6 | 672.22 | 1242 | 2267.6 | 727.78 | 1342 | 2447.6 |
| 617.78 | 1144 | 2091.2 | 673.33 | 1244 | 2271.2 | 728.89 | 1344 | 2451.2 |
| 618.89 | 1146 | 2094.8 | 674.44 | 1246 | 2274.8 | 730 | 1346 | 2454.8 |
| 620 | 1148 | 2098.4 | 675.56 | 1248 | 2278.4 | 731.11 | 1348 | 2458.4 |
| 621.11 | 1150 | 2102 | 676.67 | 1250 | 2282 | 732.22 | 1350 | 2462 |
| 622.22 | 1152 | 2105.6 | 677.78 | 1252 | 2285.6 | 733.33 | 1352 | 2465.6 |
| 623.33 | 1154 | 2109.2 | 678.89 | 1254 | 2289.2 | 734.44 | 1354 | 2469.2 |
| 624.44 | 1156 | 2112.8 | 680 | 1256 | 2292.8 | 735.56 | 1356 | 2472.8 |
| 625.56 | 1158 | 2116.4 | 681.11 | 1258 | 2296.4 | 736.67 | 1358 | 2476.4 |
| 626.67 | 1160 | 2120 | 682.22 | 1260 | 2300 | 737.78 | 1360 | 2480 |
| 627.78 | 1162 | 2123.6 | 683.33 | 1262 | 2303.6 | 738.89 | 1362 | 2483.6 |
| 628.89 | 1164 | 2127.2 | 684.44 | 1264 | 2307.2 | 740 | 1364 | 2487.2 |
| 630 | 1166 | 2130.8 | 685.56 | 1266 | 2310.8 | 741.11 | 1366 | 2490.8 |
| 631.11 | 1168 | 2134.4 | 686.67 | 1268 | 2314.4 | 742.22 | 1368 | 2494.4 |
| 632.22 | 1170 | 2138 | 687.78 | 1270 | 2318 | 743.33 | 1370 | 2498 |
| 633.33 | 1172 | 2141.6 | 688.89 | 1272 | 2321.6 | 744.44 | 1372 | 2501.6 |
| 634.44 | 1174 | 2145.2 | 690 | 1274 | 2325.2 | 745.56 | 1374 | 2505.2 |
| 635.56 | 1176 | 2148.8 | 691.11 | 1276 | 2328.8 | 746.67 | 1376 | 2508.8 |
| 636.67 | 1178 | 2152.4 | 692.22 | 1278 | 2332.4 | 747.78 | 1378 | 2512.4 |
| 637.78 | 1180 | 2156 | 693.33 | 1280 | 2336 | 748.89 | 1380 | 2516 |
| 638.89 | 1182 | 2159.6 | 694.44 | 1282 | 2339.6 | 750 | 1382 | 2519.6 |
| 640 | 1184 | 2163.2 | 695.56 | 1284 | 2343.2 | 751.11 | 1384 | 2523.2 |
| 641.11 | 1186 | 2166.8 | 696.67 | 1286 | 2346.8 | 752.22 | 1386 | 2526.8 |
| 642.22 | 1188 | 2170.4 | 697.78 | 1288 | 2350.4 | 753.33 | 1388 | 2530.4 |
| 643.33 | 1190 | 2174 | 698.89 | 1290 | 2354 | 754.44 | 1390 | 2534 |
| 644.44 | 1192 | 2177.6 | 700 | 1292 | 2357.6 | 755.56 | 1392 | 2537.6 |
| 645.56 | 1194 | 2181.2 | 701.11 | 1294 | 2361.2 | 756.67 | 1394 | 2541.2 |
| 646.67 | 1196 | 2184.8 | 702.22 | 1296 | 2364.8 | 757.78 | 1396 | 2544.8 |
| 647.78 | 1198 | 2188.4 | 703.33 | 1298 | 2368.4 | 758.89 | 1398 | 2548.4 |
| 648.89 | 1200 | 2192 | 704.44 | 1300 | 2372 | 760 | 1400 | 2552 |
| 650 | 1202 | 2195.6 | 705.56 | 1302 | 2375.6 | 761.11 | 1402 | 2555.6 |
| 651.11 | 1204 | 2199.2 | 706.67 | 1304 | 2379.2 | 762.22 | 1404 | 2559.2 |
| 652.22 | 1206 | 2202.8 | 707.78 | 1306 | 2382.8 | 763.33 | 1406 | 2562.8 |
| 653.33 | 1208 | 2206.4 | 708.89 | 1308 | 2386.4 | 764.44 | 1408 | 2566.4 |

## Table 4—9 (continued)
## TEMPERATURE CONVERSION

| To °C | ←°F or °C→ | To °F | To °C | ←°F or °C→ | To °F | To °C | ←°F or °C→ | To °F |
|---|---|---|---|---|---|---|---|---|
| | To convert | | | To convert | | | To convert | |
| 765.56 | 1410 | 2570 | 843.33 | 1550 | 2922 | 1121.11 | 2050 | 3722 |
| 766.67 | 1412 | 2573.6 | 848.89 | 1560 | 2840 | 1126.67 | 2060 | 3740 |
| 767.78 | 1414 | 2577.2 | 854.44 | 1570 | 2858 | 1132.22 | 2070 | 3758 |
| 768.89 | 1416 | 2580.8 | 860 | 1580 | 2876 | 1137.78 | 2080 | 3776 |
| 770 | 1418 | 2584.4 | 865.56 | 1590 | 2894 | 1143.33 | 2090 | 3794 |
| 771.11 | 1420 | 2588 | 871.11 | 1600 | 2912 | 1148.89 | 2100 | 3812 |
| 772.22 | 1422 | 2591.6 | 876.67 | 1610 | 2930 | 1154.44 | 2110 | 3830 |
| 773.33 | 1424 | 2595.2 | 882.22 | 1620 | 2948 | 1160 | 2120 | 3848 |
| 774.44 | 1426 | 2598.8 | 887.78 | 1630 | 2966 | 1165.56 | 2130 | 3866 |
| 775.56 | 1428 | 2602.4 | 893.33 | 1640 | 2984 | 1171.11 | 2140 | 3884 |
| 776.67 | 1430 | 2606 | 898.89 | 1650 | 3002 | 1176.67 | 2150 | 3902 |
| 777.78 | 1432 | 2609.6 | 904.44 | 1660 | 3020 | 1182.22 | 2160 | 3920 |
| 778.89 | 1434 | 2613.2 | 910 | 1670 | 3038 | 1187.78 | 2170 | 3938 |
| 780 | 1436 | 2616.8 | 915.56 | 1680 | 3056 | 1193.33 | 2180 | 3956 |
| 781.11 | 1438 | 2620.4 | 921.11 | 1690 | 3074 | 1198.89 | 2190 | 3974 |
| 782.22 | 1440 | 2624 | 926.67 | 1700 | 3092 | 1204.44 | 2200 | 3992 |
| 783.33 | 1442 | 2627.6 | 932.22 | 1710 | 3110 | 1210 | 2210 | 4010 |
| 784.44 | 1444 | 2631.2 | 937.78 | 1720 | 3128 | 1215.56 | 2220 | 4028 |
| 785.56 | 1446 | 2634.8 | 943.33 | 1730 | 3146 | 1221.11 | 2230 | 4046 |
| 786.67 | 1448 | 2638.4 | 948.89 | 1740 | 3164 | 1226.67 | 2240 | 4064 |
| 787.78 | 1450 | 2642 | 954.44 | 1750 | 3182 | 1232.22 | 2250 | 4082 |
| 788.89 | 1452 | 2645.6 | 960 | 1760 | 3200 | 1237.78 | 2260 | 4100 |
| 790 | 1454 | 2649.2 | 965.56 | 1770 | 3218 | 1243.33 | 2270 | 4118 |
| 791.11 | 1456 | 2652.8 | 971.11 | 1780 | 3236 | 1248.89 | 2280 | 4136 |
| 792.22 | 1458 | 2656.4 | 976.67 | 1790 | 3254 | 1254.44 | 2290 | 4154 |
| 793.33 | 1460 | 2660 | 982.22 | 1800 | 3272 | 1260 | 2300 | 4172 |
| 794.44 | 1462 | 2663.6 | 987.78 | 1810 | 3290 | 1265.56 | 2310 | 4190 |
| 795.56 | 1464 | 2667.2 | 993.33 | 1820 | 3308 | 1271.11 | 2320 | 4208 |
| 796.67 | 1466 | 2670.8 | 998.89 | 1830 | 3326 | 1276.67 | 2330 | 4226 |
| 797.78 | 1468 | 2674.4 | 1004.44 | 1840 | 3344 | 1282.22 | 2340 | 4244 |
| 798.89 | 1470 | 2678 | 1010 | 1850 | 3362 | 1287.78 | 2350 | 4262 |
| 800 | 1472 | 2681.6 | 1015.56 | 1860 | 3380 | 1293.33 | 2360 | 4280 |
| 801.11 | 1474 | 2685.2 | 1021.11 | 1870 | 3398 | 1298.89 | 2370 | 4298 |
| 802.22 | 1476 | 2688.8 | 1026.67 | 1880 | 3416 | 1304.44 | 2380 | 4316 |
| 803.33 | 1478 | 2692.4 | 1032.22 | 1890 | 3434 | 1310 | 2390 | 4334 |
| 804.44 | 1480 | 2696 | 1037.78 | 1900 | 3452 | 1315.56 | 2400 | 4352 |
| 805.56 | 1482 | 2699.6 | 1043.33 | 1910 | 3470 | 1321.11 | 2410 | 4370 |
| 806.67 | 1484 | 2703.2 | 1048.89 | 1920 | 3488 | 1326.67 | 2420 | 4388 |
| 807.78 | 1486 | 2706.8 | 1054.44 | 1930 | 3506 | 1332.22 | 2430 | 4406 |
| 808.89 | 1488 | 2710.4 | 1060 | 1940 | 3524 | 1337.78 | 2440 | 4424 |
| 810 | 1490 | 2714 | 1065.56 | 1950 | 3542 | 1343.33 | 2450 | 4442 |
| 811.11 | 1492 | 2717.6 | 1071.11 | 1960 | 3560 | 1348.89 | 2460 | 4460 |
| 812.22 | 1494 | 2721.2 | 1076.67 | 1970 | 3578 | 1354.44 | 2470 | 4478 |
| 813.33 | 1496 | 2724.8 | 1082.22 | 1980 | 3596 | 1360 | 2480 | 4496 |
| 814.44 | 1498 | 2728.4 | 1087.78 | 1990 | 3614 | 1365.56 | 2490 | 4514 |
| 815.56 | 1500 | 2732 | 1093.33 | 2000 | 3632 | 1371.11 | 2500 | 4532 |
| 821.11 | 1510 | 2750 | 1098.89 | 2010 | 3650 | 1385 | 2525 | 4577 |
| 826.67 | 1520 | 2768 | 1104.44 | 2020 | 3668 | 1398.89 | 2550 | 4622 |
| 832.22 | 1530 | 2786 | 1110 | 2030 | 3686 | 1412.78 | 2575 | 4667 |
| 837.78 | 1540 | 2804 | 1115.56 | 2040 | 3704 | 1426.67 | 2600 | 4712 |

## Table 4—9 (continued)
## TEMPERATURE CONVERSION

| To °C | To convert ←°F or °C→ | To °F | To °C | To convert ←°F or °C→ | To °F | To °C | To convert ←°F or °C→ | To °F |
|---|---|---|---|---|---|---|---|---|
| 1440.56 | 2625 | 4757 | 1996.11 | 3625 | 6557 | 2551.67 | 4625 | 8357 |
| 1454.44 | 2650 | 4802 | 2010.00 | 3650 | 6602 | 2565.56 | 4650 | 8402 |
| 1468.33 | 2675 | 4847 | 2023.89 | 3675 | 6647 | 2579.44 | 4675 | 8447 |
| 1482.22 | 2700 | 4892 | 2037.78 | 3700 | 6692 | 2593.33 | 4700 | 8492 |
| 1496.11 | 2725 | 4937 | 2051.67 | 3725 | 6737 | 2607.22 | 4725 | 8537 |
| 1510 | 2750 | 4982 | 2065.56 | 3750 | 6782 | 2621.11 | 4750 | 8582 |
| 1523.89 | 2775 | 5027 | 2079.44 | 3775 | 6827 | 2635.00 | 4775 | 8627 |
| 1537.78 | 2800 | 5072 | 2093.33 | 3800 | 6872 | 2648.89 | 4800 | 8672 |
| 1551.67 | 2825 | 5117 | 2107.22 | 3825 | 6917 | 2662.78 | 4825 | 8717 |
| 1565.56 | 2850 | 5162 | 2121.11 | 3850 | 6962 | 2676.67 | 4850 | 8762 |
| 1579.44 | 2875 | 5207 | 2135.00 | 3875 | 7007 | 2690.55 | 4875 | 8807 |
| 1593.33 | 2900 | 5252 | 2148.89 | 3900 | 7052 | 2704.44 | 4900 | 8852 |
| 1607.22 | 2925 | 5297 | 2162.78 | 3925 | 7097 | 2718.33 | 4925 | 8897 |
| 1621.11 | 2950 | 5342 | 2176.67 | 3950 | 7142 | 2732.22 | 4950 | 8942 |
| 1635 | 2975 | 5387 | 2190.56 | 3975 | 7187 | 2746.11 | 4975 | 8987 |
| 1648.89 | 3000 | 5432 | 2204.44 | 4000 | 7232 | 2760.00 | 5000 | 9032 |
| 1662.78 | 3025 | 5477 | 2218.33 | 4025 | 7277 | 2787.78 | 5050 | 9122 |
| 1676.67 | 3050 | 5522 | 2232.22 | 4050 | 7322 | 2815.56 | 5100 | 9212 |
| 1690.56 | 3075 | 5567 | 2246.11 | 4075 | 7367 | 2843.33 | 5150 | 9302 |
| 1704.44 | 3100 | 5612 | 2260.00 | 4100 | 7412 | 2871:11 | 5200 | 9392 |
| 1718.33 | 3125 | 5657 | 2273.89 | 4125 | 7457 | 2898.89 | 5250 | 9482 |
| 1732.22 | 3150 | 5702 | 2287.78 | 4150 | 7502 | 2926.67 | 5300 | 9572 |
| 1746.11 | 3175 | 5747 | 2301.67 | 4175 | 7547 | 2954.44 | 5350 | 9662 |
| 1760 | 3200 | 5792 | 2315.56 | 4200 | 7592 | 2982.22 | 5400 | 9752 |
| 1773.89 | 3225 | 5837 | 2329.44 | 4225 | 7637 | 3010.00 | 5450 | 9842 |
| 1787.78 | 3250 | 5882 | 2343.33 | 4250 | 7682 | 3037.78 | 5500 | 9932 |
| 1801.67 | 3275 | 5927 | 2357.22 | 4275 | 7727 | 3065.56 | 5550 | 10022 |
| 1815.56 | 3300 | 5972 | 2371.11 | 4300 | 7772 | 3093.33 | 5600 | 10112 |
| 1829.44 | 3325 | 6017 | 2385.00 | 4325 | 7817 | 3121.11 | 5650 | 10202 |
| 1843.33 | 3350 | 6062 | 2398.89 | 4350 | 7862 | 3148.89 | 5700 | 10292 |
| 1857.22 | 3375 | 6107 | 2412.78 | 4375 | 7907 | 3176.67 | 5750 | 10382 |
| 1871.11 | 3400 | 6152 | 2426.67 | 4400 | 7952 | 3204.44 | 5800 | 10472 |
| 1885.00 | 3425 | 6197 | 2440.56 | 4425 | 7997 | 3232.22 | 5850 | 10562 |
| 1898.89 | 3450 | 6242 | 2454.44 | 4450 | 8042 | 3260.00 | 5900 | 10652 |
| 1912.78 | 3475 | 6287 | 2468.33 | 4475 | 8087 | 3287.78 | 5950 | 10742 |
| 1926.67 | 3500 | 6332 | 2482.22 | 4500 | 8132 | 3315.56 | 6000 | 10832 |
| 1940.56 | 3525 | 6377 | 2496.11 | 4525 | 8177 | 3593.33 | 6500 | 11732 |
| 1954.44 | 3550 | 6422 | 2510.00 | 4550 | 8222 | 3871.11 | 7000 | 12632 |
| 1968.33 | 3575 | 6467 | 2523.89 | 4575 | 8267 | 4148.89 | 7500 | 13532 |
| 1982.22 | 3600 | 6512 | 2537.78 | 4600 | 8312 | 4426.67 | 8000 | 14432 |

Condensed from Weast, R.C., Ed., *Handbook of Chemistry and Physics,* 53rd ed., Chemical Rubber, Cleveland, 1972.

## Table 4—10
### TEMPERATURE CONVERSION – DEGREES FAHRENHEIT TO KELVIN

$$K = (5/9)(\deg F + 459.67)$$

| Deg F | K | Deg F | K | Deg F | K | Deg F | K | Deg F | K | Deg F | K |
|---|---|---|---|---|---|---|---|---|---|---|---|
| −459.67 | 0 | −412 | 26.48 | −220 | 133.14 | −28 | 239.82 | 20 | 266.48 | 68 | 293.15 |
| −459 | .37 | −411 | 27.04 | −215 | 135.91 | −27 | 240.37 | 21 | 267.04 | 69 | 293.71 |
| −458 | .93 | −410 | 27.59 | −210 | 138.69 | −26 | 240.93 | 22 | 267.59 | 70 | 294.26 |
| −457 | 1.48 | −409 | 28.15 | −205 | 141.47 | −25 | 241.48 | 23 | 268.15 | 71 | 294.82 |
| −456 | 2.04 | −408 | 28.70 | −200 | 144.25 | −24 | 242.04 | 24 | 268.71 | 72 | 295.37 |
| −455 | 2.59 | −407 | 29.26 | −195 | 147.02 | −23 | 242.59 | 25 | 269.26 | 73 | 295.93 |
| −454 | 3.15 | −406 | 29.81 | −190 | 149.80 | −22 | 243.15 | 26 | 269.82 | 74 | 296.48 |
| −453 | 3.70 | −405 | 30.37 | −185 | 152.58 | −21 | 243.71 | 27 | 270.37 | 75 | 297.04 |
| −452 | 4.26 | −404 | 30.92 | −180 | 155.36 | −20 | 244.26 | 28 | 270.93 | 76 | 297.59 |
| −451 | 4.82 | −403 | 31.48 | −175 | 158.13 | −19 | 244.82 | 29 | 271.48 | 77 | 298.15 |
| −450 | 5.37 | −402 | 32.04 | −170 | 160.91 | −18 | 245.37 | 30 | 272.04 | 78 | 298.71 |
| −449 | 5.93 | −401 | 32.59 | −165 | 163.69 | −17 | 245.93 | 31 | 272.59 | 79 | 299.26 |
| −448 | 6.48 | −400 | 33.15 | −160 | 166.47 | −16 | 246.48 | 32 | 273.15 | 80 | 299.82 |
| −447 | 7.04 | −395 | 35.92 | −155 | 169.24 | −15 | 247.04 | 33 | 273.71 | 81 | 300.37 |
| −446 | 7.59 | −390 | 38.70 | −150 | 172.02 | −14 | 247.59 | 34 | 274.26 | 82 | 300.93 |
| −445 | 8.15 | −385 | 41.48 | −145 | 174.80 | −13 | 248.15 | 35 | 274.82 | 83 | 301.48 |
| −444 | 8.70 | −380 | 44.26 | −140 | 177.58 | −12 | 248.71 | 36 | 275.37 | 84 | 302.04 |
| −443 | 9.26 | −375 | 47.03 | −135 | 180.35 | −11 | 249.26 | 37 | 275.93 | 85 | 302.59 |
| −442 | 9.82 | −370 | 49.81 | −130 | 183.13 | −10 | 249.82 | 38 | 276.48 | 86 | 303.15 |
| −441 | 10.37 | −365 | 52.59 | −125 | 185.91 | −9 | 250.37 | 39 | 277.04 | 87 | 303.71 |
| −440 | 10.93 | −360 | 55.37 | −120 | 188.69 | −8 | 250.93 | 40 | 277.59 | 88 | 304.26 |
| −439 | 11.48 | −355 | 58.14 | −115 | 191.46 | −7 | 251.48 | 41 | 278.15 | 89 | 304.82 |
| −438 | 12.04 | −350 | 60.92 | −110 | 194.24 | −6 | 252.04 | 42 | 278.71 | 90 | 305.37 |
| −437 | 12.59 | −345 | 63.70 | −105 | 197.02 | −5 | 252.59 | 43 | 279.26 | 91 | 305.93 |
| −436 | 13.15 | −340 | 66.48 | −100 | 199.80 | −4 | 253.15 | 44 | 279.82 | 92 | 306.48 |
| −435 | 13.70 | −335 | 69.25 | −95 | 202.57 | −3 | 253.71 | 45 | 280.37 | 93 | 307.04 |
| −434 | 14.26 | −330 | 72.03 | −90 | 205.35 | −2 | 254.26 | 46 | 280.93 | 94 | 307.59 |
| −433 | 14.82 | −325 | 74.81 | −85 | 208.13 | −1 | 254.82 | 47 | 281.48 | 95 | 308.15 |
| −432 | 15.37 | −320 | 77.59 | −80 | 210.91 | 0 | 255.37 | 48 | 282.04 | 96 | 308.71 |
| −431 | 15.93 | −315 | 80.36 | −75 | 213.68 | 1 | 255.93 | 49 | 282.59 | 97 | 309.26 |
| −430 | 16.48 | −310 | 83.14 | −70 | 216.46 | 2 | 256.48 | 50 | 283.15 | 98 | 309.82 |
| −429 | 17.04 | −305 | 85.92 | −65 | 219.24 | 3 | 257.04 | 51 | 283.71 | 99 | 310.37 |
| −428 | 17.59 | −300 | 88.70 | −60 | 222.02 | 4 | 257.59 | 52 | 284.26 | 100 | 310.93 |
| −427 | 18.15 | −295 | 91.47 | −55 | 224.79 | 5 | 258.15 | 53 | 284.82 | 101 | 311.48 |
| −426 | 18.70 | −290 | 94.25 | −50 | 227.57 | 6 | 258.71 | 54 | 285.37 | 102 | 312.04 |
| −425 | 19.26 | −285 | 97.03 | −45 | 230.35 | 7 | 259.26 | 55 | 285.93 | 103 | 312.59 |
| −424 | 19.81 | −280 | 99.81 | −40 | 233.13 | 8 | 259.82 | 56 | 286.48 | 104 | 313.15 |
| −423 | 20.37 | −275 | 102.58 | −39 | 233.71 | 9 | 260.37 | 57 | 287.04 | 105 | 313.71 |
| −422 | 20.92 | −270 | 105.36 | −38 | 234.26 | 10 | 260.93 | 58 | 287.59 | 106 | 314.26 |
| −421 | 21.48 | −265 | 108.14 | −37 | 234.82 | 11 | 261.48 | 59 | 288.15 | 107 | 314.82 |
| −420 | 22.04 | −260 | 110.92 | −36 | 235.37 | 12 | 262.04 | 60 | 288.71 | 108 | 315.37 |
| −419 | 22.59 | −255 | 113.69 | −35 | 235.93 | 13 | 262.59 | 61 | 289.26 | 109 | 315.93 |
| −418 | 23.15 | −250 | 116.47 | −34 | 236.48 | 14 | 263.15 | 62 | 289.82 | 110 | 316.48 |
| −417 | 23.70 | −245 | 119.25 | −33 | 237.04 | 15 | 263.71 | 63 | 290.37 | 111 | 317.04 |
| −416 | 24.26 | −240 | 122.03 | −32 | 237.59 | 16 | 264.26 | 64 | 290.93 | 112 | 317.59 |
| −415 | 24.81 | −235 | 124.80 | −31 | 238.15 | 17 | 264.82 | 65 | 291.48 | 113 | 318.15 |
| −414 | 25.37 | −230 | 127.58 | −30 | 238.71 | 18 | 265.37 | 66 | 292.04 | 114 | 318.71 |
| −413 | 25.92 | −225 | 130.36 | −29 | 239.26 | 19 | 265.93 | 67 | 292.59 | 115 | 319.26 |

## Table 4—10 (continued)
## TEMPERATURE CONVERSION – DEGREES FAHRENHEIT TO KELVIN

$$K = (5/9)(deg\ F + 459.67)$$

| Deg F | K | Deg F | K | Deg F | K | Deg F | K | Deg F | K | Deg F | K |
|---|---|---|---|---|---|---|---|---|---|---|---|
| 116 | 319.82 | 166 | 347.59 | 216 | 375.37 | 266 | 403.15 | 460 | 510.93 | 960 | 788.71 |
| 117 | 320.37 | 167 | 348.15 | 217 | 375.93 | 267 | 403.71 | 470 | 516.48 | 970 | 794.26 |
| 118 | 320.93 | 168 | 348.71 | 218 | 376.48 | 268 | 404.26 | 480 | 522.04 | 980 | 799.82 |
| 119 | 321.48 | 169 | 349.26 | 219 | 377.04 | 269 | 404.82 | 490 | 527.59 | 990 | 805.37 |
| 120 | 322.04 | 170 | 349.82 | 220 | 377.59 | 270 | 405.37 | 500 | 533.15 | 1 000 | 810.93 |
| 121 | 322.59 | 171 | 350.37 | 221 | 378.15 | 271 | 405.93 | 510 | 538.71 | 1 010 | 816.48 |
| 122 | 323.15 | 172 | 350.93 | 222 | 378.71 | 272 | 406.48 | 520 | 544.26 | 1 020 | 822.04 |
| 123 | 323.71 | 173 | 351.48 | 223 | 379.26 | 273 | 407.04 | 530 | 549.82 | 1 030 | 827.59 |
| 124 | 324.26 | 174 | 352.04 | 224 | 379.82 | 274 | 407.59 | 540 | 555.37 | 1 040 | 833.15 |
| 125 | 324.82 | 175 | 352.59 | 225 | 380.37 | 275 | 408.15 | 550 | 560.93 | 1 050 | 838.71 |
| 126 | 325.37 | 176 | 353.15 | 226 | 380.93 | 276 | 408.71 | 560 | 566.48 | 1 060 | 844.26 |
| 127 | 325.93 | 177 | 353.71 | 227 | 381.48 | 277 | 409.26 | 570 | 572.04 | 1 070 | 849.82 |
| 128 | 326.48 | 178 | 354.26 | 228 | 382.04 | 278 | 409.82 | 580 | 577.59 | 1 080 | 855.37 |
| 129 | 327.04 | 179 | 354.82 | 229 | 382.59 | 279 | 410.37 | 590 | 583.15 | 1 090 | 860.93 |
| 130 | 327.59 | 180 | 355.37 | 230 | 383.15 | 280 | 410.93 | 600 | 588.71 | 1 100 | 866.48 |
| 131 | 328.15 | 181 | 355.93 | 231 | 383.71 | 281 | 411.48 | 610 | 594.26 | 1 110 | 872.04 |
| 132 | 328.71 | 182 | 356.48 | 232 | 384.26 | 282 | 412.04 | 620 | 599.82 | 1 120 | 877.59 |
| 133 | 329.26 | 183 | 357.04 | 233 | 384.82 | 283 | 412.59 | 630 | 605.37 | 1 130 | 883.15 |
| 134 | 329.82 | 184 | 357.59 | 234 | 385.37 | 284 | 413.15 | 640 | 610.93 | 1 140 | 888.71 |
| 135 | 330.37 | 185 | 358.15 | 235 | 385.93 | 285 | 413.71 | 650 | 616.48 | 1 150 | 894.26 |
| 136 | 330.93 | 186 | 358.71 | 236 | 386.48 | 286 | 414.26 | 660 | 622.04 | 1 160 | 899.82 |
| 137 | 331.48 | 187 | 359.26 | 237 | 387.04 | 287 | 414.82 | 670 | 627.59 | 1 170 | 905.37 |
| 138 | 332.04 | 188 | 359.82 | 238 | 387.59 | 288 | 415.37 | 680 | 633.15 | 1 180 | 910.93 |
| 139 | 332.59 | 189 | 360.37 | 239 | 388.15 | 289 | 415.93 | 690 | 638.71 | 1 190 | 916.48 |
| 140 | 333.15 | 190 | 360.93 | 240 | 388.71 | 290 | 416.48 | 700 | 644.26 | 1 200 | 922.04 |
| 141 | 333.71 | 191 | 361.48 | 241 | 389.26 | 291 | 417.04 | 710 | 649.82 | 1 210 | 927.59 |
| 142 | 334.26 | 192 | 362.04 | 242 | 389.82 | 292 | 417.59 | 720 | 655.37 | 1 220 | 933.15 |
| 143 | 334.82 | 193 | 362.59 | 243 | 390.37 | 293 | 418.15 | 730 | 660.93 | 1 230 | 938.71 |
| 144 | 335.37 | 194 | 363.15 | 244 | 390.93 | 294 | 418.71 | 740 | 666.48 | 1 240 | 944.26 |
| 145 | 335.93 | 195 | 363.71 | 245 | 391.48 | 295 | 419.26 | 750 | 672.04 | 1 250 | 949.82 |
| 146 | 336.48 | 196 | 364.26 | 246 | 392.04 | 296 | 419.82 | 760 | 677.59 | 1 260 | 955.37 |
| 147 | 337.04 | 197 | 364.82 | 247 | 392.59 | 297 | 420.37 | 770 | 683.15 | 1 270 | 960.93 |
| 148 | 337.59 | 198 | 365.37 | 248 | 393.15 | 298 | 420.93 | 780 | 688.71 | 1 280 | 966.48 |
| 149 | 338.15 | 199 | 365.93 | 249 | 393.71 | 299 | 421.48 | 790 | 694.26 | 1 290 | 972.04 |
| 150 | 338.71 | 200 | 366.48 | 250 | 394.26 | 300 | 422.04 | 800 | 699.82 | 1 300 | 977.59 |
| 151 | 339.26 | 201 | 367.04 | 251 | 394.82 | 310 | 427.59 | 810 | 705.37 | 1 310 | 983.15 |
| 152 | 339.82 | 202 | 367.59 | 252 | 395.37 | 320 | 433.15 | 820 | 710.93 | 1 320 | 988.71 |
| 153 | 340.37 | 203 | 368.15 | 253 | 395.93 | 330 | 438.71 | 830 | 716.48 | 1 330 | 994.26 |
| 154 | 340.93 | 204 | 368.71 | 254 | 396.48 | 340 | 444.26 | 840 | 722.04 | 1 340 | 999.82 |
| 155 | 341.48 | 205 | 369.26 | 255 | 397.04 | 350 | 449.82 | 850 | 727.59 | 1 350 | 1 005.37 |
| 156 | 342.04 | 206 | 369.82 | 256 | 397.59 | 360 | 455.37 | 860 | 733.15 | 1 360 | 1 010.93 |
| 157 | 342.59 | 207 | 370.37 | 257 | 398.15 | 370 | 460.93 | 870 | 738.71 | 1 370 | 1 016.48 |
| 158 | 343.15 | 208 | 370.93 | 258 | 398.71 | 380 | 466.48 | 880 | 744.26 | 1 380 | 1 022.04 |
| 159 | 343.71 | 209 | 371.48 | 259 | 399.26 | 390 | 472.04 | 890 | 749.82 | 1 390 | 1 027.59 |
| 160 | 344.26 | 210 | 372.04 | 260 | 399.82 | 400 | 477.59 | 900 | 755.37 | 1 400 | 1 033.15 |
| 161 | 344.82 | 211 | 372.59 | 261 | 400.37 | 410 | 483.15 | 910 | 760.93 | 1 410 | 1 038.71 |
| 162 | 345.37 | 212 | 373.15 | 262 | 400.93 | 420 | 488.71 | 920 | 766.48 | 1 420 | 1 044.26 |
| 163 | 345.93 | 213 | 373.71 | 263 | 401.48 | 430 | 494.26 | 930 | 772.04 | 1 430 | 1 049.82 |
| 164 | 346.48 | 214 | 374.26 | 264 | 402.04 | 440 | 499.82 | 940 | 777.59 | 1 440 | 1 055.37 |
| 165 | 347.04 | 215 | 374.82 | 265 | 402.59 | 450 | 505.37 | 950 | 783.15 | 1 450 | 1 060.93 |

## Table 4—10 (continued)
### TEMPERATURE CONVERSION – DEGREES FAHRENHEIT TO KELVIN

$$K = (5/9)(\deg F + 459.67)$$

| Deg F | K | Deg F | K | Deg F | K | Deg F | K | Deg F | K | Deg F | K |
|---|---|---|---|---|---|---|---|---|---|---|---|
| 1 460 | 1 066.48 | 1 960 | 1 344.26 | 3 150 | 2 005.37 | 4 400 | 2 699.82 | 6 300 | 3 755.37 | 8 800 | 5 144.26 |
| 1 470 | 1 072.04 | 1 970 | 1 349.82 | 3 175 | 2 019.26 | 4 425 | 2 713.70 | 6 350 | 3 783.15 | 8 850 | 5 172.04 |
| 1 480 | 1 077.59 | 1 980 | 1 355.37 | 3 200 | 2 033.15 | 4 450 | 2 727.59 | 6 400 | 3 810.93 | 8 900 | 5 199.82 |
| 1 490 | 1 083.15 | 1 990 | 1 360.93 | 3 225 | 2 047.04 | 4 475 | 2 741.48 | 6 450 | 3 838.71 | 8 950 | 5 227.59 |
| 1 500 | 1 088.71 | 2 000 | 1 366.48 | 3 250 | 2 060.93 | 4 500 | 2 755.37 | 6 500 | 3 866.48 | 9 000 | 5 255.37 |
| 1 510 | 1 094.26 | 2 025 | 1 380.37 | 3 275 | 2 074.82 | 4 525 | 2 769.26 | 6 550 | 3 894.26 | 9 050 | 5 283.15 |
| 1 520 | 1 099.82 | 2 050 | 1 394.26 | 3 300 | 2 088.70 | 4 550 | 2 783.15 | 6 600 | 3 922.04 | 9 100 | 5 310.93 |
| 1 530 | 1 105.37 | 2 075 | 1 408.15 | 3 325 | 2 102.59 | 4 575 | 2 797.04 | 6 650 | 3 949.82 | 9 150 | 5 338.71 |
| 1 540 | 1 110.93 | 2 100 | 1 422.04 | 3 350 | 2 116.48 | 4 600 | 2 810.93 | 6 700 | 3 977.59 | 9 200 | 5 366.48 |
| 1 550 | 1 116.48 | 2 125 | 1 435.93 | 3 375 | 2 130.37 | 4 625 | 2 824.82 | 6 750 | 4 005.37 | 9 250 | 5 394.26 |
| 1 560 | 1 122.04 | 2 150 | 1 449.82 | 3 400 | 2 144.26 | 4 650 | 2 838.70 | 6 800 | 4 033.15 | 9 300 | 5 422.04 |
| 1 570 | 1 127.59 | 2 175 | 1 463.70 | 3 425 | 2 158.15 | 4 675 | 2 852.59 | 6 850 | 4 060.93 | 9 350 | 5 449.82 |
| 1 580 | 1 133.15 | 2 200 | 1 477.59 | 3 450 | 2 172.04 | 4 700 | 2 866.48 | 6 900 | 4 088.71 | 9 400 | 5 477.59 |
| 1 590 | 1 138.71 | 2 225 | 1 491.48 | 3 475 | 2 185.93 | 4 725 | 2 880.37 | 6 950 | 4 116.48 | 9 450 | 5 505.37 |
| 1 600 | 1 144.26 | 2 250 | 1 505.37 | 3 500 | 2 199.82 | 4 750 | 2 894.26 | 7 000 | 4 144.26 | 9 500 | 5 533.15 |
| 1 610 | 1 149.82 | 2 275 | 1 519.26 | 3 525 | 2 213.70 | 4 775 | 2 908.15 | 7 050 | 4 172.04 | 9 550 | 5 560.93 |
| 1 620 | 1 155.37 | 2 300 | 1 533.15 | 3 550 | 2 227.59 | 4 800 | 2 922.04 | 7 100 | 4 199.82 | 9 600 | 5 588.71 |
| 1 630 | 1 160.93 | 2 325 | 1 547.04 | 3 575 | 2 241.48 | 4 825 | 2 935.93 | 7 150 | 4 227.59 | 9 650 | 5 616.48 |
| 1 640 | 1 166.48 | 2 350 | 1 560.93 | 3 600 | 2 255.37 | 4 850 | 2 949.82 | 7 200 | 4 255.37 | 9 700 | 5 644.26 |
| 1 650 | 1 172.04 | 2 375 | 1 574.82 | 3 625 | 2 269.26 | 4 875 | 2 963.70 | 7 250 | 4 283.15 | 9 750 | 5 672.04 |
| 1 660 | 1 177.59 | 2 400 | 1 588.70 | 3 650 | 2 283.15 | 4 900 | 2 977.59 | 7 300 | 4 310.93 | 9 800 | 5 699.82 |
| 1 670 | 1 183.15 | 2 425 | 1 602.59 | 3 675 | 2 297.04 | 4 925 | 2 991.48 | 7 350 | 4 338.71 | 9 850 | 5 727.59 |
| 1 680 | 1 188.71 | 2 450 | 1 616.48 | 3 700 | 2 310.93 | 4 950 | 3 005.37 | 7 400 | 4 366.48 | 9 900 | 5 755.37 |
| 1 690 | 1 194.26 | 2 475 | 1 630.37 | 3 725 | 2 324.82 | 4 975 | 3 019.26 | 7 450 | 4 394.26 | 9 950 | 5 783.15 |
| 1 700 | 1 199.82 | 2 500 | 1 644.26 | 3 750 | 2 338.70 | 5 000 | 3 033.15 | 7 500 | 4 422.04 | 10 000 | 5 810.93 |
| 1 710 | 1 205.37 | 2 525 | 1 658.15 | 3 775 | 2 352.59 | 5 050 | 3 060.93 | 7 550 | 4 449.82 | | |
| 1 720 | 1 210.93 | 2 550 | 1 672.04 | 3 800 | 2 366.48 | 5 100 | 3 088.71 | 7 600 | 4 477.59 | | |
| 1 730 | 1 216.48 | 2 575 | 1 685.93 | 3 825 | 2 380.37 | 5 150 | 3 116.48 | 7 650 | 4 505.37 | | |
| 1 740 | 1 222.04 | 2 600 | 1 699.82 | 3 850 | 2 394.26 | 5 200 | 3 144.26 | 7 700 | 4 533.15 | | |
| 1 750 | 1 227.59 | 2 625 | 1 713.70 | 3 875 | 2 408.15 | 5 250 | 3 172.04 | 7 750 | 4 560.93 | | |
| 1 760 | 1 233.15 | 2 650 | 1 727.59 | 3 900 | 2 422.04 | 5 300 | 3 199.82 | 7 800 | 4 588.71 | | |
| 1 770 | 1 238.71 | 2 675 | 1 741.48 | 3 925 | 2 435.93 | 5 350 | 3 227.59 | 7 850 | 4 616.48 | | |
| 1 780 | 1 244.26 | 2 700 | 1 755.37 | 3 950 | 2 449.82 | 5 400 | 3 255.37 | 7 900 | 4 644.26 | | |
| 1 790 | 1 249.82 | 2 725 | 1 769.26 | 3 975 | 2 463.70 | 5 450 | 3 283.15 | 7 950 | 4 672.04 | | |
| 1 800 | 1 255.37 | 2 750 | 1 783.15 | 4 000 | 2 477.59 | 5 500 | 3 310.93 | 8 000 | 4 699.82 | | |
| 1 810 | 1 260.93 | 2 775 | 1 797.04 | 4 025 | 2 491.48 | 5 550 | 3 338.71 | 8 050 | 4 727.59 | | |
| 1 820 | 1 266.48 | 2 800 | 1 810.93 | 4 050 | 2 505.37 | 5 600 | 3 366.48 | 8 100 | 4 755.37 | | |
| 1 830 | 1 272.04 | 2 825 | 1 824.82 | 4 075 | 2 519.26 | 5 650 | 3 394.26 | 8 150 | 4 783.15 | | |
| 1 840 | 1 277.59 | 2 850 | 1 838.70 | 4 100 | 2 533.15 | 5 700 | 3 422.04 | 8 200 | 4 810.93 | | |
| 1 850 | 1 283.15 | 2 875 | 1 852.59 | 4 125 | 2 547.04 | 5 750 | 3 449.82 | 8 250 | 4 838.71 | | |
| 1 860 | 1 288.71 | 2 900 | 1 866.48 | 4 150 | 2 560.93 | 5 800 | 3 477.59 | 8 300 | 4 866.48 | | |
| 1 870 | 1 294.26 | 2 925 | 1 880.37 | 4 175 | 2 574.82 | 5 850 | 3 505.37 | 8 350 | 4 894.26 | | |
| 1 880 | 1 299.82 | 2 950 | 1 894.26 | 4 200 | 2 588.70 | 5 900 | 3 533.15 | 8 400 | 4 922.04 | | |
| 1 890 | 1 305.37 | 2 975 | 1 908.15 | 4 225 | 2 602.59 | 5 950 | 3 560.93 | 8 450 | 4 949.82 | | |
| 1 900 | 1 310.93 | 3 000 | 1 922.04 | 4 250 | 2 616.48 | 6 000 | 3 588.71 | 8 500 | 4 977.59 | | |
| 1 910 | 1 316.48 | 3 025 | 1 935.93 | 4 275 | 2 630.37 | 6 050 | 3 616.48 | 8 550 | 5 005.37 | | |
| 1 920 | 1 322.04 | 3 050 | 1 949.82 | 4 300 | 2 644.26 | 6 100 | 3 644.26 | 8 600 | 5 033.15 | | |
| 1 930 | 1 327.59 | 3 075 | 1 963.70 | 4 325 | 2 658.15 | 6 150 | 3 672.04 | 8 650 | 5 060.93 | | |
| 1 940 | 1 333.15 | 3 100 | 1 977.59 | 4 350 | 2 672.04 | 6 200 | 3 699.82 | 8 700 | 5 088.71 | | |
| 1 950 | 1 338.71 | 3 125 | 1 991.48 | 4 375 | 2 685.93 | 6 250 | 3 727.59 | 8 750 | 5 116.48 | | |

From Bolz, R. E. and Tuve, G. L., Eds., *Handbook of Tables for Applied Engineering Science,* 2nd ed., CRC Press, Cleveland, 1973, 855.

# Section 5

# Miscellaneous Materials Properties

# Section 5

# MISCELLANEOUS MATERIALS PROPERTIES

## 5.1 BUILDING MATERIALS

### 5.1.1 Concrete

**Table 5.1.1—1**
**CONCRETE FOR VARIOUS TYPES OF CONSTRUCTION**

For cement factor in kilograms per cubic meter, multiply the tabulated value in sacks (94 lb) per cubic yard by 55.767.

For aggregate size in millimeters, multiply values in inches by 25.4.

| Type of construction | Typical structures | Consistency | Cement factor sacks per cubic yard | Maximum size of aggregate, inches |
|---|---|---|---|---|
| Massive | Dams, heavy piers, large open foundations | Stiff | $2\frac{1}{2}$–5 | 3–6 |
| Semimassive | Piers, heavy walls, foundations, heavy arches, girders | Stiff; medium | 4–6 | 2–3 |
| Heavy building | Large structural members, small piers, medium footings; wide to moderately wide spacing of reinforcement | Medium wet | 5–7 | 1–2 |
| Light | Small structural members, thin slabs, small columns, heavily reinforced sections, closely spaced reinforcement | Wet | $5\frac{1}{2}$–7 | $\frac{1}{2}$–1 |

From Troxell, G. E. and Davis, H. E., *Composition and Properties of Concrete,* 2nd ed., McGraw-Hill, New York, copyright © 1968. With permission.

## Table 5.1.1—2

## TYPES OF PORTLAND CEMENT AND INCREASE IN STRENGTH OF CONCRETE WITH AGE

### Composition and Properties

Portland cement is a mixture consisting mainly of calcium and silicon oxides (as calcium silicates) in powder form with particle sizes largely in the range 10 to 50 $\mu$m. While the composition is approximately 65% CaO, 21% $SiO_2$, and 7% $Al_2O_3$ (balance from other oxides), the actual compounds are $3CaO \cdot SiO_2$, $2CaO \cdot SiO_2$, $3CaO \cdot Al_2O_3$, and $4CaO \cdot Al_2O_3 \cdot Fe_2O_3$. The proportions of these actual compounds will vary with the type of cement. When water is added and the mixture hardens, the heat of hydration is approximately 100 cal/g cement (180 Btu/lb). The heat is released gradually as hydration proceeds, with a shrinkage of about 8% compared with the total volume of cement-plus-water in the original mix. Shrinkage is minimal if original water content is low. To accelerate the setting of cement, especially in cold weather, calcium chloride (2% or less of weight of cement) often is added. In very hot weather retarders are sometimes used to delay hardening.

| Standard ASTM classification | Character and use | Compressive strength,[a] psi/1,000, at age of | | | | |
|:---:|:---|:---:|:---:|:---:|:---:|:---:|
| | | 7 days | 28 days | 3 months | 1 year | 5 years |
| I | Common; general use | 3.0 | 4.3 | 5.1 | 5.5 | 5.7 |
| II | General use; moderate resistance to sulfate attack; lower heat of hydration | 2.6 | 4.2 | 5.2 | 5.9 | 6.4 |
| III | High early strength; shorter curing time | 3.8 | 4.7 | 5.1 | 5.4 | 5.5 |
| IV | Develops less heat; slow curing; resists cracking; high resistance to sulfate attack | 1.5 | 3.5 | 5.2 | 6.0 | 6.5 |
| V | Highest resistance to sulfate attack | 2.5 | 4.1 | 5.3 | 6.1 | 6.7 |

[a] For strength in MN/m², multiply tabulated values in thousands of psi by 6.8948.

*Note:* Although strengths will vary greatly with materials, proportions, and curing conditions, these are typical.

From Bolz, R. E. and Tuve, G. L., Eds., *Handbook of Tables for Applied Engineering Science,* 2nd ed., CRC Press, Cleveland, 1973, 644.

## Table 5.1.1—3
### RELATIONSHIPS BETWEEN
### WATER-CEMENT RATIO AND
### COMPRESSIVE STRENGTH OF CONCRETE

| Compressive strength at 28 days, psi[a] | Water-cement ratio, by weight | |
|---|---|---|
| | Non-air-entrained concrete | Air-entrained concrete |
| 6000 | 0.41 | – |
| 5000 | 0.48 | 0.40 |
| 4000 | 0.57 | 0.48 |
| 3000 | 0.68 | 0.59 |
| 2000 | 0.82 | 0.74 |

[a]Values are estimated average strengths for concrete containing not more than the percentage of air shown in Table 5.3.3 of the original source. For a constant water-cement ratio, the strength of concrete is reduced as the air content is increased.

Strength is based on 6 × 12 in. cylinders moist-cured 28 days at 73.4 ± 3°F (23 ± 1.7°C) in accordance with Section 9(b) of ASTM C31 for Making and Curing Concrete Compression and Flexure Test Specimens in the Field.

Relationship assumes maximum size of aggregate about ¾ to 1 in.; for a given source, strength produced for a given water-cement ratio will increase as maximum size of aggregate decreases.

From *Recommended Practice for Selecting Proportions for Normal and Heavyweight Concrete (ACI 211.1-74)*, American Concrete Institute, Detroit. With permission.

## Table 5.1.1—4
### PROPERTIES OF SPECIAL CONCRETES

A great many varieties of aggregates have been used for concrete, dependent largely on the materials available. In general, high density results in high strength and high thermal conductivity, and vice versa, although such variables as water/cement ratio, percentage of fines, and curing conditions may result in wide differences in properties with the same materials. The following table gives typical examples.

| Description; type of aggregate | Approximate density, lb/ft³ | | Compressive strength, lb/in.² | Thermal conductivity, Btu/hr·ft·deg F |
|---|---|---|---|---|
| | Aggregate | Concrete | | |
| Frost resisting; 1% CaCl₂; normal aggregates | 110 | 140 | 4,500 | 1.0 |
| Frost-resisting porous; 6% air entrainment | 100 | 130 | 3,500 | 0.85 |
| Lightweight, with expanded shale or clay | 50 | 75 | 1,500 | 0.25 |
| Lightweight, with foamed slag | 40 | 75 | 1,000 | 0.20 |
| Cinder concrete, fine and coarse | 50 | 80 | 1,000 | 0.25 |
| Pulverized fuel ash | 60 | 85 | 1,200 | 0.25 |
| Lightweight refractory concrete with aluminous cement | 35 | 65 | 3,500 | 0.20 |
| Lightweight, insulating, with perlite | 10 | 35 | 250 | 0.15 |
| Lightweight, insulating, with expanded vermiculite | 8 | 30 | 150 | 0.10 |

From Bolz, R. E. and Tuve, G. L., Eds., *Handbook of Tables for Applied Engineering Science*, 2nd ed., CRC Press, Cleveland, 1973, 645.

**Table 5.1.1—5**

**EFFECTS OF VARIOUS SUBSTANCES ON UNPROTECTED CONCRETE**

In addition to attack by corrosion (e.g., by acids) and other chemical combination (with sulfates and chlorides), there are many types of erosion of concrete. Concrete is highly susceptible to erosion by cavitation occurring in water-flow structures.

| Substance | Effect on unprotected concrete |
|---|---|
| Petroleum oils: heavy, light, and volatile | None |
| Coal-tar distillates | None or very slight |
| Inorganic acids | Disintegration |
| Organic materials | |
|     Acetic acid | Slow disintegration |
|     Oxalic and dry carbonic acids | None |
|     Carbonic acid in water | Slow attack |
|     Lactic and tannic acids | Slow attack |
|     Vegetable oils | Slight or very slight attack |
| Inorganic salts | |
|     Sulfates of calcium, sodium, magnesium, potassium, aluminum, iron | Active attack |
|     Chlorides of sodium, potassium | None |
|     Chlorides of magnesium, calcium | Slight attack |
| Miscellaneous | |
|     Milk | Slow attack |
|     Silage juices | Slow attack |
|     Molasses, corn syrup, and glucose | Slight attack |

From *Concrete Manual,* 7th ed., U.S. Bureau of Reclamation, 1966.

## Table 5.1.1—6
### SAFE LOADS FOR PIPE COLUMNS

Following are approximate maximum axial loads for "Standard Pipe" (Schedule 40) used as columns, plain or filled with concrete. For load in kN, multiply tabulated values in thousands of pounds by 4.4482.

| Description | Maximum load in thousands of pounds for column lengths, feet | | | | | | |
|---|---|---|---|---|---|---|---|
| | 6 | 7 | 8 | 9 | 10 | 11 | 12 |
| **3-in. pipe, OD 3.5 in., wall thickness 0.216 in.** | | | | | | | |
| Plain | 25 | 24 | 23 | 21 | 20 | 18 | 16 |
| **3½-in. pipe, OD 4.0 in., wall thickness 0.226 in.** | | | | | | | |
| Plain | 31 | 30 | 29 | 28 | 26 | 24 | 23 |
| Filled | 38 | 37 | 36 | 35 | 33 | 31 | 30 |
| **4-in. pipe, OD 4.5 in., wall thickness 0.237 in.** | | | | | | | |
| Plain | 38 | 37 | 36 | 34 | 33 | 31 | 30 |
| Filled | 46 | 45 | 44 | 43 | 41 | 40 | 38 |
| **5-in. pipe, OD 5.563 in., wall thickness 0.258 in.** | | | | | | | |
| Plain | 52 | 51 | 50 | 49 | 48 | 47 | 46 |
| Filled | 66 | 65 | 64 | 63 | 62 | 60 | 59 |
| **6-in. pipe, OD 6⅝ in., wall thickness 0.280 in.** | | | | | | | |
| Plain | 69 | 68 | 67 | 66 | 65 | 64 | 63 |
| Filled | 88 | 87 | 86 | 85 | 84 | 83 | 82 |
| **8-in. pipe, OD 8⅝ in., wall thickness 0.322 in.** | | | | | | | |
| Plain | 105 | 104 | 103 | 102 | 101 | 100 | 99 |
| Filled | 126 | 125 | 124 | 123 | 122 | 121 | 120 |
| **10-in. pipe, OD 10¾ in., wall thickness 0.365 in.** | | | | | | | |
| Plain | 150 | 149 | 148 | 147 | 146 | 145 | 144 |
| Filled | 203 | 202 | 201 | 200 | 199 | 198 | 197 |

From Bolz, R. E. and Tuve, G. L., Eds., *Handbook of Tables for Applied Engineering Science,* 2nd ed., CRC Press, Cleveland, 1973, 632.

## 5.1.2    Foundations, Rocks, and Soils

### Table 5.1.2—1
### AVERAGE AMOUNTS OF THE ELEMENTS IN THE EARTH'S CRUST

In Grams Per Metric Ton or Parts Per Million

| Element | Quantity | Element | Quantity | Element | Quantity |
|---------|----------|---------|----------|---------|----------|
| O  | 466,000 | N  | 46  | Br | 1.6 |
| Si | 277,200 | Ce | 46  | Ho | 1.2 |
| Al | 81,300  | Sn | 40  | Eu | 1.1 |
| Fe | 50,000  | Y  | 28  | Sb | 1? |
| Ca | 36,300  | Nd | 24  | Tb | 0.9 |
| Na | 28,300  | Nb | 24  | Lu | 0.8 |
| K  | 25,900  | Co | 23  | Tl | 0.6 |
| Mg | 20,900  | La | 18  | Hg | 0.5 |
| Ti | 4,400   | Pb | 16  | I  | 0.3 |
| H  | 1,400   | Ga | 15  | Bi | 0.2 |
| P  | 1,180   | Mo | 15  | Tm | 0.2 |
| Mn | 1,000   | Th | 12  | Cd | 0.15 |
| S  | 520     | Cs | 7   | Ag | 0.1 |
| C  | 320     | Ge | 7   | In | 0.1 |
| Cl | 314     | Sm | 6.5 | Se | 0.09 |
| Rb | 310     | Gd | 6.4 | Ar | 0.04 |
| F  | 300     | Be | 6   | Pd | 0.01 |
| Sr | 300     | Pr | 5.5 | Pt | 0.005 |
| Ba | 250     | Sc | 5   | Au | 0.005 |
| Zr | 220     | As | 5   | He | 0.003 |
| Cr | 200     | Hf | 4.5 | Te | 0.002? |
| V  | 150     | Dy | 4.5 | Rh | 0.001 |
| Zn | 132     | U  | 4   | Re | 0.001 |
| Ni | 80      | B  | 3   | Ir | 0.001 |
| Cu | 70      | Yb | 2.7 | Os | 0.001? |
| W  | 69      | Er | 2.5 | Ru | 0.001? |
| Li | 65      | Ta | 2.1 |    |      |

## Table 5.1.2—2
## COMPOSITION OF SEA WATER

### Table A. Elements in Solution

#### Excluding Dissolved Gases

| Element | Concentration, parts per million (approximate) | Percent by weight |
|---------|-----------------------------------------------|-------------------|
| Oxygen | 857,000. | 85.7000 |
| Hydrogen | 108,000. | 10.8000 |
| Chlorine | 19,000. | 1.9000 |
| Sodium | 10,500. | 1.0500 |
| Magnesium | 1,275. | 0.1275 |
| Sulfur | 885. | 0.0885 |
| Calcium | 400. | 0.0400 |
| Potassium | 380. | 0.0380 |
| Bromine | 65. | 0.0065 |
| Carbon | 30. | 0.0030 |
| Strontium | 13. | 0.0013 |
| Boron | 4.6 | 0.00046 |
| Silicon | 2. | 0.0002 |
| Fluorine | 1.3 | 0.00013 |
| Aluminum | 1. | 0.0001 |

### Table B. Ionic Constituents

#### Anions in Sea Water of 34.4 Salinity per Mil, or 3.44% by Weight

Tabular values are given in percent by weight.

| Chloride | 1.897 | Bromide | 0.0065 |
|----------|-------|---------|--------|
| Sulfate | 0.265 | Borate | 0.0027 |
| Bicarbonate | 0.014 | | |

### Table C. Artificial Sea Water

To simulate the physical properties, a 3.4% solution of sodium chloride or natural sea salt may be used.

For a more exact chemical reproduction the following is an average.

| Salt | Grams |
|------|-------|
| NaCl | 25. |
| $MgCl_2$ | 3. |
| $MgSO_4$ (or $NaSO_4$) | 4. |
| $CaCl_2$ | 1. |
| KCl | 0.7 |
| $NaHCO_3$ | 0.2 |
| NaBr (or KBr) | 0.1 |
| TOTAL | 34.0 |

*Note:* Add water to make 1 kg.

USN Specification 44T27b—1940 is as follows:

| | |
|--|--|
| NaCl | 23 g |
| $Na_2SO_4 \cdot 10H_2O$ | 8 g |
| Stock solution | 20 ml |
| Sterile distilled water to 1 liter. | |

The stock solution is as follows:

| | |
|--|--|
| Magnesium chloride | 550 g |
| Calcium chloride | 110 g |
| Potassium bromide | 45 g |
| Potassium chloride | 10 g |
| Sterile distilled water to 1 liter. | |

From Bolz, R. E. and Tuve, G. L., Eds., *Handbook of Tables for Applied Engineering Science*, 2nd ed., CRC Press, Cleveland, 1973, 659.

## Table 5.1.2—3
## GENERAL CLASSIFICATION OF  COMMON ROCKS

### Table A. Igneous Rocks

#### Solidified from a Molten State

| Coarse-grained crystalline | Fine-grained crystalline (or crystals and glass) | Fragmental (crystalline or glassy) |
|---|---|---|
| Origin: deep intrusion, slowly cooled | Origin: quickly cooled, volcanic or shallow intrusive | Origin: explosive volcanic fragments deposited as sediments |
| Granite | Rhyolite | Ash and pumice (volcanic dust or cinders) |
| Diorite | Andesite | Tuff (consolidated volcanic ash) |
| Gabbro | Basalt | Agglomerate (coarse and fine volcanic debris) |
|  | Obsidian and pitchstone (essentially glass—suddenly chilled, few or no crystals) |  |

### Table B. Mineral Constituents of Igneous Rocks

Minerals Key:

Q = quartz ($SiO_2$): hard, shiny, no true cleavage

O = orthoclase feldspar: silicates; regular cleavage

P = plagioclase feldspar: nearly white, good cleavage

A = amphibole (magnesium, iron, calcium) and/or biotite (black mica)

B = pyroxene; nearly black

| Coarsely crystalline | Principal constituent minerals | Finely crystalline or porphyritic |
|---|---|---|
| Granite | Q+O+A(+P)[a] | Rhyolite |
| Syenite | O+A(+P)[a] | Trachyte |
| Quartz monzonite | Q+O+P+A | Dellenite |
| Monzonite | O+P+A | Latite |
| Quartz diorite | Q+P+A or B (+O)[a] | Dacite |
| Diorite | P+A or B (+O)[a] | Andesite |
| Gabbro | P+B | Basalt |

[a]Small amount.

**Table 5.1.2—3 (continued)**
**GENERAL CLASSIFICATION OF COMMON ROCKS**

**Table C. Sedimentary Rocks**

**Sediments Transported by Water, Air, Ice, Gravity**

**Mechanically deposited**

**Chemically or biochemically deposited**

A—Unconsolidated
Clay
Silt ⎫
Sand ⎬ According to particle size
Gravel
Cobbles ⎭

A—Calcareous
Limestone ($CaCO_3$)
Dolomite ($CaCO_3 \cdot MgCO_3$)
Marl (calcareous shale)
Caliche (calcareous soil)
Coquina (shell limestone)

B—Consolidated
Shale (consolidated clay)
Siltstone (consolidated silt)
Sandstone (consolidated sand)
Conglomerate (consolidated gravel or cobbles—rounded)
Breccia (angular fragments)

B—Siliceous
Chert
Flint
Agate ⎫
Opal ⎬ Spring deposit, vein or cavity filling
Chalcedony ⎭

C—Others
Coal, phosphate, salines, etc.

**Table D. Metamorphic Rocks**

**Igneous or Sedimentary Rocks**
**Changed by Heat, Pressure**

A—Foliated
Slate: dense, dark, splits into thin plates (metamorphosed shale)
Schist: predominantly micaceous, semiparallel lamellae
Gneiss: granular, banded, subordinately micaceous

B—Massive
Marble: coarsely crystalline, calcareous (metamorphosed limestone)
Quartzite: dense, very hard, quartzose (metamorphosed sandstone)

From *Concrete Manual,* 7th ed., U.S. Bureau of Reclamation, 1966, Tables 10 and 11.

## Table 5.1.2—4
## SOIL MECHANICS—CLASSES OF SOILS

Comparison of Classification Systems

| Classification System | | | | | | | | |
|---|---|---|---|---|---|---|---|---|
| American Society for Testing and Materials | Colloids[a] | Clay | Silt | Fine sand | Coarse sand | Gravel | | Boulders |
| American Association of State Highway Officials Soil Classification | Colloids[a] | Clay | Silt | Fine sand | Coarse sand | Fine gravel / Medium gravel / Coarse gravel | | Cobbles |
| U.S. Department of Agriculture Soil Classification | Clay | | Silt | Very fine sand / Fine sand / Medium sand / Coarse sand / Very coarse sand | | Fine gravel / Coarse gravel | | |
| Federal Aviation Agency Soil Classification | Clay | | Silt | Fine sand | Coarse sand | Gravel | | |
| Unified Soil Classification (Corps of Engineers, Department of the Army, and Bureau of Reclamation) | Fines (silt or clay) | | | Fine sand | Medium sand | Coarse sand | Fine gravel / Coarse gravel | Cobbles |
| Massachusetts Institute of Technology | Clay | Silt (Fine / Medium / Coarse) | | Sand (Fine / Medium / Coarse) | | Gravel | | |

[a]Colloids are included in clay fraction in test reports.

From Bolz, R. E. and Tuve, G. L., Eds., *Handbook of Tables for Applied Engineering Science*, 2nd ed., CRC Press, Cleveland, 1973, 635.

## Table 5.1.2—5
## SOIL MECHANICS – VOLUME AND WEIGHT RELATIONSHIPS

### Density, Porosity, and Moisture

#### Table A. Representative Specific Gravities of Some Dense Soils

| Soil | Specific gravity |
|------|------------------|
| Normal inorganic clay | 2.70 |
| Silt | 2.70 |
| Loess | 2.70 |
| Sand | 2.65 |
| Diatomaceous earth | 2.65 |
| Bentonite clay | 2.34 |

From Bolz, R. E. and Tuve, G. L., Eds., *Handbook of Tables for Applied Engineering Science,* 2nd ed., CRC Press, Cleveland, 1973, 636.

#### Table B. Relative Density of Sands

| | |
|---|---|
| Very dense | 85–100% |
| Dense | 65–85% |
| Medium | 35–65% |
| Loose | 15–35% |

From Bolz, R. E. and Tuve, G. L., Eds., *Handbook of Table for Applied Engineering Science,* 2nd ed., CRC Press, Cleveland, 1973, 636.

#### Table C. Porosity of Soils in Natural State

*Porosity* is the percentage ratio, volume of voids to total volume. *Void ratio* is the ratio of volume of voids to volume of moist solids. For unit weights in kilograms per cubic meter, multiply values in grams per cubic centimeters.

| Description | Porosity, % | Void ratio | Water content, % | Unit weights g/cm³ Dry | Unit weights g/cm³ Sat. | Unit weights lb/ft³ Dry | Unit weights lb/ft³ Sat. |
|-------------|-------------|------------|------------------|-----|------|-----|------|
| Uniform sand, loose | 46 | 0.85 | 32 | 1.43 | 1.89 | 90 | 118 |
| Uniform sand, dense | 34 | 0.51 | 19 | 1.75 | 2.09 | 109 | 130 |
| Mixed-grained sand, loose | 40 | 0.67 | 25 | 1.59 | 1.99 | 99 | 124 |
| Mixed-grained sand, dense | 30 | 0.43 | 16 | 1.86 | 2.16 | 116 | 135 |
| Glacial till, very mixed grained | 20 | 0.25 | 9 | 2.12 | 2.32 | 132 | 145 |
| Soft glacial clay | 55 | 1.2 | 45 | – | 1.77 | – | 110 |
| Stiff glacial clay | 37 | 0.6 | 22 | – | 2.07 | – | 129 |

## Table 5.1.2—5 (continued)
## SOIL MECHANICS – VOLUME AND WEIGHT RELATIONSHIPS

Density, Porosity, and Moisture (continued)

### Table C. Porosity of Soils in Natural State (continued)

| | | | | Unit weights | | | |
| | | | | g/cm$^3$ | | lb/ft$^3$ | |
| Description | Porosity, % | Void ratio | Water content, % | Dry | Sat. | Dry | Sat. |
|---|---|---|---|---|---|---|---|
| Soft slightly organic clay | 66 | 1.9 | 70 | – | 1.58 | – | 98 |
| Soft very organic clay | 75 | 3.0 | 110 | – | 1.43 | – | 89 |
| Soft bentonite | 84 | 5.2 | 194 | – | 1.27 | – | 80 |

From Terzaghi, K. and Peck, R. B., *Soil Mechanics in Engineering Practice,* 2nd ed., John Wiley & Sons, New York, 1967, 28 (Table 6.3). Copyright © 1967 by John Wiley & Sons. With permission.

### Table D. Degree of Saturation, Decimal

| Description | Degree of saturation |
|---|---|
| Dry soil | 0.0 |
| Humid soil | <0.25 |
| Damp soil | 0.26–0.50 |
| Moist soil | 0.51–0.75 |
| Wet soil | 0.76–0.99 |
| Saturated | 1.00 |

From Bolz, R. E. and Tuve, G. L., Eds., *Handbook of Tables for Applied Engineering Science,* 2nd ed., CRC Press, Cleveland, 1973, 636.

## Table 5.1.2—6
### SOIL MECHANICS—PLASTICITY AND WATER CONTENT OF SOILS

Moisture, Consistency, and Atterberg Limits

Definitions:

Plastic limit: the percentage of water content by weight at which soil crumbles when rolled into thin threads.

Liquid limit: the percentage of water content by weight at which soil has a very small but measurable shearing strength.

Plasticity index: the arithmetical difference between the liquid limit and the plastic limit.

**Terminology for Degree of Plasticity**

| Description | Range of plasticity index |
|---|---|
| Nonplastic | 0–5 |
| Moderately plastic | 5–15 |
| Plastic | 15–40 |
| Highly plastic | >40 |

**Atterberg Limits for Clays**

| Clay | Adsorbed ion | Plastic limit | Liquid limit |
|---|---|---|---|
| Kaolinite | Sodium | 26 | 52 |
|  | Calcium | 36 | 73 |
| Illite | Sodium | 34 | 61 |
|  | Calcium | 40 | 90 |
| Montmorillonite | Sodium | 97 | 900 |
|  | Calcium | 63 | 177 |

From Bolz, R. E. and Tuve, G. L., Eds., *Handbook of Tables for Applied Engineering Science,* 2nd ed., CRC Press, Cleveland, 1973, 637.

## Table 5.1.2—7
### SOIL MECHANICS–COMPARATIVE PROPERTIES OF CLAY SOILS

Consistency, Sensitivity, Brittleness, and Activity

Definitions:

Consistency: the relative compressive strength of the unconfined clay.

Sensitivity: the ratio of unconfined compressive strengths (undisturbed material to remolded material).

Brittleness: the percentage of strain at failure in an unconfined compression test.

Activity: the ratio of the plasticity index to the clay fraction, i.e., (liquid limit – plastic limit)/ (percentage by weight of particles smaller than 2 $\mu$m).

| Consistency | | Sensitivity | | Brittleness | | Activity | |
|---|---|---|---|---|---|---|---|
| **Description** | **Range** | **Description** | **Range** | **Description** | **Range** | **Description** | **Range** |
| Very soft | <.25 | Insensitive | <2 | Brittle | 3–8 | Inactive | <.75 |
| Soft | .25–.50 | Moderately | | Semibrittle | 8–14 | Normal | .75–1.25 |
| Medium | .50–1.0 | sensitive | 2–4 | Plastic | 14–20 | Active | >1.25 |
| Stiff | 1.0–2.0 | Sensitive | 4–8 | | | | |
| Very stiff | 2.0–4.0 | Very sensitive | 8–16 | | | | |
| Hard | >4.0 | Slightly quick | 16–32 | | | | |
| | | Medium quick | 32–64 | | | | |
| | | Quick | >64 | | | | |

From Bolz, R. E. and Tuve, G. L., Eds., *Handbook of Tables for Applied Engineering Science,* 2nd ed., CRC Press, Cleveland, 1973, 637.

## Table 5.1.2—8
### SOIL MECHANICS–DENSITY AND PENETRATION

Density and consistency of soil may be determined during boring and sampling operations by the standard penetration test. In this test the number of blows required per foot of penetration of a standard sampling spoon is observed. A 140-lb hammer is used, with a 30-in. drop; the sampling spoon is 2.0 in. OD and 1.375 in. ID. The density and consistency are characterized in the following table.

| Relative density of sand | | Relative consistency of clay | |
|---|---|---|---|
| **Number of blows per foot** | **Description** | **Number of blows per foot** | **Description** |
| 0–4 | Very loose | 0–2 | Very soft |
| 4–10 | Loose | 2–4 | Soft |
| 10–30 | Medium | 4–8 | Medium |
| 30–50 | Dense | 8–15 | Stiff |
| >50 | Very dense | 15–30 | Very stiff |
| | | >30 | Hard |

*Note:* For sands and silts a correction for depth (and overburden pressure) or a correction for location of the water table may be required.

From Bolz, R. E. and Tuve, G. L., Eds., *Handbook of Tables for Applied Engineering Science,* 2nd ed., CRC Press, Cleveland, 1973, 638.

**Table 5.1.2—9**
## BUILDING FOUNDATIONS AND PILES

Typical Building Code Requirements for Soil Bearing
Pressures and Design of Pile Foundations

**Table A. Allowable Bearing Pressures**

| | Maximum bearing pressure[b] | |
|---|---|---|
| Soil descriptions[a] | tons/ft$^2$ | MN/m$^2$ |
| Rock, bedrock, solid, massive, sound | 100 | 1,380 |
| Rock, sound foliated (limestone, slate) | 40 | 550 |
| Rock, hard sedimentary (hard shale, sandstone, conglomerate) | 25 | 245 |
| Rock, broken bedrock | 10 | 138 |
| Hardpan | 10 | 138 |
| Gravel, very compact | 10 | 138 |
| Gravel, compact | 6 | 83 |
| Gravel, loose | 4 | 55 |
| Sand or silty gravel | 1–3 | 14–40 |

[a]Descriptive terminology used in building codes is far from uniform, but the terms here given occur in several codes.
[b]Each of these values is to be found in several of the building codes of large U.S. cities or government authorities. Modifications or corrections are recognized for conditions such as depth of stratum, depth of embedment, and extent of overburden. For gravel and sand the allowable bearing pressures are sometimes stated in terms of on-site soil test results.

**Table B. Allowable Design Stresses for Loaded Piles**

Widely Used Specifications in Building Codes

Key:
   BS = basic stress for clear timber as per ASTM D25 and D2555
   CS = compressive strength of concrete
   YS = yield strength of steel

| | | Allowable unit stress, psi[a] | | |
|---|---|---|---|---|
| Type of pile | Number of components | Core | Reinforcing steel | Steel shell |
| Steel H-section | 1 | 35% YS | – | – |
| Timber | 1 | 60% BS | – | – |
| Precast concrete | 2 | 22.5% CS | 35% YS | – |
| Cast-in-place concrete | 3 | 22.5% CS | 35% YS | 35% YS |
| Steel pipe (filled) | 3 | 22.5% CS | 35% YS | 35% YS |

[a]Many city codes specifying numerical values for allowable stresses and variations among cities are occasionally 2 to 1 or even more. A few codes use percentages other than those here quoted. A limit of 1,000 psi (6,895 MN/m$^2$) is often specified for timber piling.

**Table 5.1.2—9 (continued)**
**BUILDING FOUNDATIONS AND PILES**

Typical Building Code Requirements for Soil Bearing
Pressures and Design of Pile Foundations (continued)

*Notes:* Most piles are designed as "short" columns, in compression only; for long
piles in soft soil, a "long-column" treatment may be necessary.

Estimation of axial-load distribution along the pile may be included in a
building code, but there is little agreement on the basis for estimating. The
percentage of the total load being carried at the tip of the pile is usually
assumed to be less than one third; for shorter piles in coarse material on
bedrock, it may even be 100%.

Possible pile damage, due to overdriving, obstructions, or defects, should
be considered in the design decisions.

Pile spacing is often specified in the building codes, with 24- or 30-in.
(0.61–0.76 m) as minimum for cylindrical piles.

Success of a pile foundation depends on the ability of the soil to support
the pile, which, in turn, depends on the adhesion and shear properties of
the soil. The use of tapered piles and rigid-body caps is common. In any
case the limitations on eventual settling should be examined, especially in
locations where settling has been a problem, as in Mexico City.

For soil conditions producing very heavy corrosion, it may be necessary to
avoid steel piles.

For extreme loads the caisson-type pile, terminating in a recess in bedrock,
is recommended.

**REFERENCES**

*Building Code Requirements for Reinforced Concrete,* ACI 318, American
Concrete Institute.
*Design Manual,* DM7, Bureau of Yards and Docks, U.S. Navy.
*Foundation Piling: A Survey of Practice,* Building Research Advisory Board,
Report No. 4, National Academy of Sciences-National Research Council.
*Southern Standard Building Code* and *Uniform Building Code.*
*Specification for the Design Fabrication and Erection of Structural Steel for
Buildings,* American Institute of Steel Construction.
Consult also (1) *Journal of the Soil Mechanics and Foundations Division,
Proceedings of the ASCE,* (2) *Proceedings of the International Conferences on
Soil Mechanics and Foundation Engineering,* and (3) city building codes.

From Bolz, R. E. and Tuve, G. L., Eds., *Handbook of Tables for Applied
Engineering Science,* 2nd ed., CRC Press, Cleveland, 1973, 639.

### 5.1.3    Wood

### Table 5.1.3—1
### AMERICAN WOODS—PROPERTIES AND USES

For weight-density in kg/m³, multiply value in lb/ft³ by 16.02.

| Species | Specific gravity | | Characteristics | Uses | Weight | | |
|---|---|---|---|---|---|---|---|
| | Green | Dry | | | lb/ft³, green | lb/ft³, air-dry 12% | lb/1,000 board ft, air-dry 12% |
| Alder, red | 0.37 | 0.41 | Low shrinkage; moderate in strength, shock resistance, hardness, and weight[a] | Furniture; sash; doors; millwork | 46 | 28 | 2,330 |
| Ash, black | 0.45 | 0.49 | Light in weight[a] | Cabinets; veneer; cooperage, containers | 52 | 34 | 2,830 |
| Ash, Oregon | 0.50 | 0.55 | Similar to but lighter than white ash[a] | Similar to white ash | 46 | 38 | 3,160 |
| Ash, white | 0.54 | 0.58 | Heavy; hard; stiff; strong; high shock resistance[a] | Handles; ladder rungs; baseball bats; farm implements; car parts | 48 | 41 | 3,420 |
| Bald cypress (Southern cypress) | | | Moderate in strength, weight, hardness, and shrinkage[c] | Building construction; beams; posts; ties; tanks; ships; paneling | 51 | 32 | 2,670 |
| Beech, American | 0.56 | 0.64 | Heavy; high strength, shock resistance, and shrinkage; uniform texture[a] | Flooring; furniture; handles; kitchenwear; ties (treated) | 54 | 45 | 3,750 |
| Birch | 0.57 | 0.63 | Heavy; high strength, shock resistance, and shrinkage; uniform texture[a] | Interior finish; dowels; ties (treated); veneer; musical instruments | 57 | 44 | 3,670 |
| Cottonwood | 0.37 | 0.40 | Uniform texture; does not split readily; moderate in weight, strength, hardness, and shrinkage | Crates; trunks; car parts; farm implements | 49 | 28 | 2,330 |
| Douglas fir | 0.41 | 0.44 | Moderate in strength, weight, shock resistance, and shrinkage[b] | Building and construction; poles; veneer; plywood; ships; furniture; boxes | 38 | 34 | 2,830 |
| Elm | 0.57 | 0.63 | Moderate in strength, weight, and hardness; high in shock resistance and shrinkage; good in bending[a] | Cooperage; baskets; crates; veneer; vehicle parts | 54 | 34 | 2,920 |
| Hemlock, Eastern | 0.38 | 0.40 | Moderate in weight, strength, and hardness[a] | Building and construction; boxes | 50 | 28 | 2,330 |
| Hemlock, Western | 0.38 | 0.42 | Moderate in weight, strength, and hardness[a] | Sash; doors; posts; piles; building and construction | 41 | 29 | 2,420 |
| Hickory, true | 0.65 | 0.73 | High toughness, hardness, shock resistance, strength, and shrinkage[a] | Dowels; spokes; poles; shafts; gymnasium equipment | 63 | 51 | 4,250 |
| Incense cedar | 0.35 | | Uniform texture; easy to season; low shrinkage, shock resistance, weight, and stiffness[c] | Lumber; fence posts; ties; poles; shingles | 45 | | |
| Larch, Western | 0.48 | 0.52 | Moderate in strength, weight, shock resistance, hardness, and shrinkage[b] | Doors; sash; posts; pilings; building and construction | 48 | 36 | 3,000 |

## Table 5.1.3—1 (continued)
## AMERICAN WOODS–PROPERTIES AND USES

| Species | Specific gravity | | Characteristics | Uses | Weight | | |
| | Green | Dry | | | lb/ft³, green | lb/ft³, air-dry 12% | lb/1,000 board ft, air-dry 12% |
| --- | --- | --- | --- | --- | --- | --- | --- |
| Locust, black | 0.66 | 0.69 | High in shock resistance, weight, and hardness; very high strength; moderate shrinkage[c] | Mine timbers; posts; poles; ties | 58 | 48 | 4,000 |
| Maple | 0.44 | 0.48 | High in hardness, weight, strength, shock resistance, and shrinkage; uniform texture[a] | Flooring; furniture; trim; spools; farm implements | 54 | 40 | 3,330 |
| Oak, red and white | 0.57 | 0.63 | High in hardness, weight, strength, shock resistance, and shrinkage; red,[a] white[b] | Trim; ships; flooring; ties; furniture; cooperage; piles | 64 | 44 | 3,670 |
| Pine, jack | | | Coarse texture; low strength, stiffness, shock resistance, and shrinkage | Box lumber; fuel; mine timber; ties; poles; posts | | | |
| Pine, lodgepole | 0.38 | 0.41 | Moderate in weight, hardness, strength, shock resistance, and shrinkage; easy to work[b] | Poles; mine timber; ties; construction | 39 | 29 | 2,420 |
| Pine | | | High shrinkage; moderate strength, stiffness, hardness, and shock resistance | General construction; ties; poles; posts | | | |
| Pine, Ponderosa | 0.38 | 0.40 | Moderate in weight, shock resistance, shrinkage, and hardness; easy to work[a] | Building; paneling; sash; frames | 45 | 28 | 2,330 |
| Pine, S. yellow | 0.47 | 0.51 | Moderate in shock resistance, shrinkage, and hardness; high in strength[b] | Building and construction; poles; pilings; boxes | 55 | 41 | 3,420 |
| Pine, sugar | 0.35 | 0.36 | Low shock resistance; easy to work; moderate strength[a] | Sash; counters; blinds; patterns | 52 | 25 | 2,080 |
| Pine, Western white | 0.36 | 0.38 | Moderate in strength, shock resistance, shrinkage, and hardness; easy to work[a] | Building and construction; patterns; boxes | 35 | 27 | 2,250 |
| Red cedar, Eastern and Western | 0.44 | 0.47 | High shock resistance; low stiffness and shrinkage; moderate in strength and hardness[c] | Fence posts; closet liners; chests; flooring | 37 | 37 | 2,750 |
| Redwood | 0.38 | 0.40 | Low shrinkage; medium in weight, strength, hardness, and shock resistance[c] | Posts; doors; interiors; cooling towers | 50 | 28 | 2,330 |
| Spruce, Eastern | 0.38 | 0.40 | Moderate in hardness, shock resistance, weight, shrinkage, and strength[a] | Building; millwork; boxes; ladders | 34 | 28 | 2,330 |
| Spruce, Engelmann | 0.31 | 0.33 | Generally straight grained; light in weight; low strength as a beam or post; low shock resistance; moderate shrinkage | Mine timber; ties; poles; flooring; studding; paper | 39 | 23 | 1,920 |

**Table 5.1.3—1 (continued)**
**AMERICAN WOODS–PROPERTIES AND USES**

| Species | Specific gravity | | Characteristics | Uses | Weight | | |
|---------|-------|-----|-----------------|------|------------------|------------------|------------------------------|
| | Green | Dry | | | lb/ft³, green | lb/ft³, air-dry 12% | lb/1,000 board ft, air-dry 12% |
| Spruce, Sitka | 0.37 | 0.40 | Moderate in weight, hardness, strength, shock resistance, and shrinkage[a] | Important in boat and plane construction; sash; doors; boxes; siding | 33 | 28 | 2,330 |
| Sycamore | 0.46 | 0.49 | High shrinkage; moderate in weight, strength, hardness, and shock resistance[a] | Boxes; ties; posts; veneer; flooring; butcher blocks | 52 | 34 | 2,830 |
| Tamarack | 0.49 | 0.53 | Coarse texture; moderate in strength, hardness, shrinkage, and shock resistance | Ties; mine timber; posts; poles; tanks; scaffolding | 47 | 37 | 3,080 |
| Tupelo | | | Uniform texture; moderate in strength, hardness, shock resistance; high shrinkage; interlocked grain makes splitting difficult[a] | Flooring; planking; crates; furniture | | | |
| Walnut, black | 0.51 | 0.55 | Moderate shrinkage; high weight, strength, hardness, and shock resistance; easily worked and glued[c] | Gun stocks; cabinets; plywood; furniture; veneer | 58 | 38 | 3,170 |
| White cedar | 0.31 | 0.32 | Low shrinkage, weight, shock resistance, and strength; soft; easily worked[c] | Poles; posts; ties; tanks; ships | 24 | 23 | 1,920 |
| Willow, black | | | High strength and shock resistance; low beam strength and weight; interlocked grain | Lumber; veneer; charcoal; furniture; subflooring; studding | | | |

[a]Decay resistance low.
[b]Decay resistance medium.
[c]Decay resistance high.

From Parker, E. R., *Materials Data Book*, McGraw-Hill, New York, copyright © 1967, 252. With permission.

## Table 5.1.3—2
## ALLOWABLE UNIT STRESSES FOR LUMBER

Grading and Specifications of Softwood Lumber

**American Softwood Lumber Standard** — A voluntary standard for softwood lumber has been developing since 1922. Five editions of Simplified Practice Recommendation R16 were issued from 1924—53 by the Department of Commerce; the present NBS Voluntary Product Standard PS 20-70, *American Softwood Lumber Standard,* was issued in 1970. It was supported by the American Lumber Standards Committee, which functions through a widely representative National Grading Rule Committee.

The *American Softwood Lumber Standard,* PS 20-70, gives the size and grade provisions for American Standard lumber and describes the organization and procedures for compliance enforcement and review. It lists commercial name classifications and complete definitions of terms and abbreviations.

PS 20-70 lists 11 softwood species, viz., cedar, cypress, fir, hemlock, juniper, larch, pine, redwood, spruce, tamarack, and yew. Five dimensional tables show the standard dressed (surface planed) sizes for almost all types of lumber, including matched tongue-and-grooved and shiplapped flooring, decking, siding, etc. Dry or seasoned lumber must have 19% or less moisture content, with an allowance for shrinkage of 0.7 to 1.0% for each four points of moisture content below this maximum. Green lumber has more than 19% moisture. Table A illustrates the relation between nominal size and dressed or green sizes.

## Table 5.1.3—2 (continued)
## ALLOWABLE UNIT STRESSES FOR LUMBER

### Table A. Nominal and Minimum-dressed Sizes

This table applies to boards, dimensional lumber, and timbers. The thicknesses apply to all widths and all widths to all thicknesses.

| Item | Thicknesses | | | Face widths | | |
|------|-------------|-----------------|-------------------|-------------|-----------------|-------------------|
| | | Minimum dressed | | | Minimum dressed | |
| | Nominal | Dry,[a] inches | Green, inches | Nominal | Dry,[a] inches | Green, inches |
| Boards[b] | | | | 2 | 1 1/2 | 1 9/16 |
| | | | | 3 | 2 1/2 | 2 9/16 |
| | | | | 4 | 3 1/2 | 3 9/16 |
| | | | | 5 | 4 1/2 | 4 5/8 |
| | 1 | 3/4 | 25/32 | 6 | 5 1/2 | 5 5/8 |
| | | | | 7 | 6 1/2 | 6 5/8 |
| | 1 1/4 | 1 | 1 1/32 | 8 | 7 1/4 | 7 1/2 |
| | | | | 9 | 8 1/4 | 8 1/2 |
| | 1 1/2 | 1 1/4 | 1 9/32 | 10 | 9 1/4 | 9 1/2 |
| | | | | 11 | 10 1/4 | 10 1/2 |
| | | | | 12 | 11 1/4 | 11 1/2 |
| | | | | 14 | 13 1/4 | 13 1/2 |
| | | | | 16 | 15 1/4 | 15 1/2 |
| Dimension | | | | 2 | 1 1/2 | 1 9/16 |
| | | | | 3 | 2 1/2 | 2 9/16 |
| | | | | 4 | 3 1/2 | 3 9/16 |
| | 2 | 1 1/2 | 1 9/16 | 5 | 4 1/2 | 4 5/8 |
| | 2 1/2 | 2 | 2 1/16 | 6 | 5 1/2 | 5 5/8 |
| | 3 | 2 1/2 | 2 9/16 | 8 | 7 1/4 | 7 1/2 |
| | 3 1/2 | 3 | 3 1/16 | 10 | 9 1/4 | 9 1/2 |
| | | | | 12 | 11 1/4 | 11 1/2 |
| | | | | 14 | 13 1/4 | 13 1/2 |
| | | | | 16 | 15 1/4 | 15 1/2 |
| Dimension | | | | 2 | 1 1/2 | 1 9/16 |
| | | | | 3 | 2 1/2 | 2 9/16 |
| | | | | 4 | 3 1/2 | 3 9/16 |
| | | | | 5 | 4 1/2 | 4 5/8 |
| | 4 | 3 1/2 | 3 9/16 | 6 | 5 1/2 | 5 5/8 |
| | 4 1/2 | 4 | 4 1/16 | 8 | 7 1/4 | 7 1/2 |
| | | | | 10 | 9 1/4 | 9 1/2 |
| | | | | 12 | 11 1/4 | 11 1/2 |
| | | | | 14 | | 13 1/2 |
| | | | | 16 | | 15 1/2 |
| Timbers | 5 and thicker | | 1/2 off | 5 and wider | | 1/2 off |

[a]Maximum moisture content of 19% or less.

[b]Boards less than the minimum thickness for 1-in. nominal but 5/8-in. or greater thickness dry (11/16 in. green) may be regarded as American Standard Lumber, but such boards shall be marked to show the size and condition of seasoning at the time of dressing. They shall also be distinguished from 1-in. boards on invoices and certificates.

From *American Softwood Lumber Standard,* NBS 20-70, National Bureau of Standards, Washington, D.C., 1970 (available from Superintendent of Documents).

## Table 5.1.3—2 (continued)
### ALLOWABLE UNIT STRESSES FOR LUMBER

**National Design Specification** — Table B is condensed from the 1971 edition of *National Design Specification for Stress-grade Lumber and Its Fastenings,* as recommended and published by the National Forest Products Association, Washington, D.C. This specification was first issued by the National Lumber Manufacturers Association in 1944; subsequent editions have been issued as recommended by the Technical Advisory Committee. The 1971 edition is a 65-page bulletin with a 20-page supplement giving "Allowable Unit Stresses, Structural Lumber," from which Table B has been condensed. The data on working stresses in this Supplement have been determined in accordance with the corresponding ASTM Standards, D245-70 and D2555-70.

#### Table B. Species, Sizes, Allowable Stresses, and Modulus of Elasticity

Normal loading conditions: moisture content not over 19%, No. 1 grade, visual grading. To convert psi to $N/m^2$, multiply by 6,895.

| Species[a] | Sizes, nominal | Typical grading agency, 1971[b] | Allowable unit stresses, psi[d] | | | | Modulus of elasticity, psi |
|---|---|---|---|---|---|---|---|
| | | | Extreme fiber in bending[c] | Tension parallel to grain | Compression perpendicular | Compression parallel | |
| **Cedar** | | | | | | | |
| Northern white | 2 × 4 | NL, NH | 1,100 | 600 | 205 | 675 | 800,000 |
| | 2 or 4 × 6+ | NL, NH | 1,000 | 575 | 205 | 675 | 800,000 |
| Western | 2 × 4 | NC | 1,450 | 725 | 285 | 975 | 1,100,000 |
| | 2 or 4 × 6+ | NC, WW | 1,250 | 725 | 285 | 975 | 1,100,000 |
| **Fir** | | | | | | | |
| Balsam | 2 × 4 | NL, NH | 1,300 | 675 | 170 | 825 | 1,200,000 |
| | 2 or 4 × 6+ | NL, NH | 1,150 | 650 | 170 | 825 | 1,200,000 |
| Douglas (larch) | 2 × 4 | WC, NC | 2,400 | 1,200 | 385 | 1,250 | 1,800,000 |
| | 2 or 4 × 6+ | WC, NC | 1,750 | 1,000 | 385 | 1,250 | 1,800,000 |
| **Hemlock** | | | | | | | |
| Eastern (tamarack) | 2 × 4 | NL, NH | 1,750 | 900 | 365 | 1,050 | 1,300,000 |
| | 2 or 4 × 6+ | NL, NH | 1,500 | 875 | 365 | 1,050 | 1,300,000 |
| Hem-fir | 2 × 4 | WC, NC | 1,600 | 825 | 245 | 1,000 | 1,500,000 |
| | 2 or 4 × 6+ | WC, NC | 1,400 | 800 | 245 | 1,000 | 1,500,000 |
| Mountain | 2 × 4 | WC, WW | 1,700 | 850 | 370 | 1,000 | 1,300,000 |
| | 2 or 4 × 6+ | WC, WW | 1,450 | 850 | 370 | 1,000 | 1,300,000 |
| **Pine** | | | | | | | |
| Idaho white | 2 × 4 | WW | 1,400 | 725 | 240 | 925 | 1,400,000 |
| | 2 or 4 × 6+ | WW | 1,200 | 700 | 240 | 925 | 1,400,000 |
| Lodgepole | 2 × 4 | WW | 1,500 | 750 | 250 | 900 | 1,300,000 |
| | 2 or 4 × 6+ | WW | 1,300 | 750 | 250 | 900 | 1,300,000 |
| Northern | 2 × 4 | NL, NH | 1,600 | 825 | 280 | 975 | 1,400,000 |
| | 2 or 4 × 6+ | NL, NH | 1,400 | 800 | 280 | 975 | 1,400,000 |
| Ponderosa (sugar) | 2 × 4 | WW, NC | 1,400 | 700 | 250 | 850 | 1,200,000 |
| | 2 or 4 × 6+ | WW, NC | 1,200 | 700 | 250 | 850 | 1,200,000 |
| Red | 2 × 4 | NC | 1,350 | 700 | 280 | 825 | 1,300,000 |
| | 2 or 4 × 6+ | NC | 1,150 | 675 | 280 | 825 | 1,300,000 |
| Southern | 2 × 4 | SP | 2,000 | 1,000 | 405 | 1,250 | 1,800,000 |
| | 2 or 4 × 6+ | SP | 1,750 | 1,000 | 405 | 1,250 | 1,800,000 |

### Table 5.1.3—2 (continued)
### ALLOWABLE UNIT STRESSES FOR LUMBER

Table B. Species, Sizes, Allowable Stresses, and Modulus of Elasticity (continued)

| Species[a] | Sizes, nominal | Typical grading agency, 1971[b] | Allowable unit stresses, psi[d] | | | | Modulus of elasticity, psi |
|---|---|---|---|---|---|---|---|
| | | | Extreme fiber in bending[c] | Tension parallel to grain | Compression perpendicular | Compression parallel | |
| Redwood | | | | | | | |
| California | 2 or 4 × 2 or 4 | RI | 1,950 | 1,000 | 425 | 1,250 | 1,400,000 |
| | 2 or 4 × 6 to 12 | RI | 1,700 | 1,000 | 425 | 1,250 | 1,400,000 |
| Spruce | | | | | | | |
| Eastern | 2 × 4 | NL, NH | 1,500 | 750 | 255 | 900 | 1,400,000 |
| | 2 or 4 × 6+ | NL, NH | 1,250 | 750 | 255 | 900 | 1,400,000 |
| Engelmann | 2 × 4 | WW | 1,300 | 675 | 195 | 725 | 1,200,000 |
| | 2 or 4 × 6+ | WW | 1,150 | 650 | 195 | 725 | 1,200,000 |
| Sitka | 2 × 4 | WC | 1,550 | 775 | 280 | 925 | 1,500,000 |
| | 2 or 4 × 6+ | WC | 1,300 | 775 | 280 | 925 | 1,500,000 |

[a]Grade designations are not entirely uniform. Values in the table apply approximately to "No. 1." There is seldom more than one better grade than No. 1, and this may be designated as select, select structural, dense, or heavy. In addition to lower grades 2 and 3, there may be other lower grades, designated as construction, standard, stud, and utility. In bending and tension the allowable unit stresses in the lowest recognized grade (utility) are of the order of one eighth to one sixth of the allowable stresses for grade No. 1. The tabular values for allowable bending stress are for the extreme fiber in "repetitive member uses" and edgewise use. The original tables give correction factors, which are less than unity for moist locations and for short-time loading; they are greater than unity if the moisture content of the wood in service is 15% or less. In general, all data apply to uses within covered structures. From the extensive tables, only the No. 1 grade in nominal 2 × 4 size and 2-in. or 4-in. planks, 6 in., and wider have been selected for illustration.

In a few cases the allowable stresses specified for the Canadian products will vary slightly from those given here for the same species by the U.S. agencies.

[b]Grading agencies represented by letters in this column are as follows:

    NC  =  National Lumber Grades Authority (a Canadian agency)
    NH  =  Northern Hardwood and Pine Manufacturers Association
    NL  =  Northeastern Lumber Manufacturers Association
    RI  =  Redwood Inspection Service
    SP  =  Southern Pine Inspection Bureau
    WC  =  West Coast Lumber Inspection Bureau
    WW  =  Western Wood Products Association

[c]It is assumed that all members are so framed, anchored, tied, and braced that they have the necessary rigidity.

[d]For short term loads, these values may be increased: add 15% for 2-month snow load; add 33% for wind or earthquake; add 100% for impact load.

*Note:* Allowable unit stresses in horizontal shear are in the range 60 to 100 psi for No. 1 grade.

### REFERENCES

*Wood Handbook,* Handbook No. 72, U.S. Department of Agriculture, Washington, D.C., 1955.
*Timber Construction Manual,* American Institute of Timber Construction, John Wiley & Sons, New York, 1966.
*National Design Specification for Stress-grade Lumber and Its Fastenings,* National Forest Products Association, Washington, D.C., 1971.

## Table 5.1.3—3
## STRESS-GRADE LUMBER–MAXIMUM END LOADS

Allowable Unit Stresses of Wood for End Grain in Bearing Parallel to Grain

These allowable unit stresses apply to the net area in bearing. When the stress in end-grain bearing exceeds 75% of the adjusted allowable unit stresses, bearing shall be on a metal plate, strap, or other durable, rigid, homogeneous material of adequate strength.

To convert stresses to $N/m^2$, multiply by 6,895.

| Species | Unseasoned, psi | Seasoned, psi |
|---|---|---|
| Ash, commercial white | 1,510 | 2,060 |
| Balsam fir | 980 | 1,330 |
| Beech | 1,310 | 1,790 |
| Birch, sweet and yellow | 1,260 | 1,720 |
| California redwood (close grain) | 1,720 | 2,340 |
| California redwood (open grain) | 1,270 | 1,730 |
| Cottonwood, Eastern | 840 | 1,150 |
| Douglas fir–larch (dense) | 1,730 | 2,360 |
| Douglas fir–larch | 1,480 | 2,020 |
| Douglas fir, South | 1,340 | 1,820 |
| Eastern hemlock–tamarack | 1,270 | 1,730 |
| Eastern spruce | 1,060 | 1,450 |
| Eastern white pine | 990 | 1,360 |
| Engelmann spruce | 860 | 1,170 |
| Hem-fir | 1,230 | 1,680 |
| Hickory and pecan | 1,510 | 2,050 |
| Idaho white pine | 1,080 | 1,470 |
| Lodgepole pine | 1,060 | 1,450 |
| Maple, black and sugar | 1,260 | 1,710 |
| Mountain hemlock | 1,170 | 1,600 |
| Northern pine | 1,150 | 1,570 |
| Northern white cedar | 810 | 1,110 |
| Oak, red and white | 1,160 | 1,590 |
| Ponderosa pine–sugar pine | 1,000 | 1,360 |
| Red pine and northern species | 970 | 1,320 |
| Sitka spruce | 1,090 | 1,480 |
| Southern cypress | 1,460 | 1,990 |
| Southern pine (dense) | 1,730 | 2,360 |
| Southern pine (med. grain) | 1,480 | 2,020 |
| Southern pine (open grain) | 1,260 | 1,720 |
| Spruce–pine–fir | 1,040 | 1,410 |
| Subalpine fir | 840 | 1,150 |
| Sweetgum and tupelo | 1,120 | 1,530 |
| Western cedars | 1,140 | 1,560 |
| Western white pine | 1,030 | 1,400 |
| Yellow poplar | 980 | 1,340 |

From *National Design Specification for Stress-grade Lumber and Its Fastenings*, National Forest Products Association, Washington, D.C., 1973, II-3. With permission.

## Table 5.1.3—4
### ELECTRICAL RESISTANCE OF VARIOUS SPECIES OF WOOD

This table gives the average of measurements made along the grain between two pairs of needle electrodes 1 1/4 in. apart and driven to a depth of 5/16 in., measured at 80°F.

| | Electrical resistance, megohms, for various moisture contents | | | | | | |
|---|---|---|---|---|---|---|---|
| **Species** | **7%** | **8%** | **9%** | **10%** | **12%** | **16%** | **20%** |
| **SOFTWOODS** | | | | | | | |
| Cypress, Southern | 12,600 | 3,980 | 1,410 | 630 | 120 | 11.2 | 1.78 |
| Douglas fir (coast region) | 22,400 | 4,780 | 1,660 | 630 | 120 | 11.2 | 2.14 |
| Fir, white | 57,600 | 15,850 | 3,980 | 1,120 | 180 | 16.6 | 3.02 |
| Hemlock, Western | 22,900 | 5,620 | 2,040 | 850 | 185 | 16.2 | 2.52 |
| Larch, Western | 39,800 | 11,200 | 3,980 | 1,445 | 250 | 19.9 | 3.39 |
| Pine | | | | | | | |
|   Eastern white | 20,900 | 5,620 | 2,090 | 850 | 200 | 19.9 | 3.31 |
|   Ponderosa | 39,800 | 8,910 | 3,310 | 1,410 | 300 | 25.1 | 3.55 |
|   Southern longleaf | 25,000 | 8,700 | 3,160 | 1,320 | 270 | 24.0 | 3.72 |
|   Southern shortleaf | 43,600 | 11,750 | 3,720 | 1,350 | 255 | 22.4 | 3.80 |
|   Sugar | 22,900 | 5,250 | 1,660 | 645 | 140 | 15.9 | 3.02 |
| Redwood | 22,400 | 4,680 | 1,550 | 615 | 100 | 7.2 | 1.74 |
| Spruce, Sitka | 22,400 | 5,890 | 2,140 | 830 | 165 | 15.5 | 3.02 |
| **HARDWOODS** | | | | | | | |
| Ash, commercial white | 12,000 | 2,190 | 690 | 250 | 55 | 5.0 | 0.89 |
| Birch | 87,000 | 19,950 | 4,470 | 1,290 | 200 | 18.2 | 3.55 |
| Gum, red | 38,000 | 6,460 | 2,090 | 815 | 160 | 15.1 | 2.63 |
| Hickory, true | | 31,600 | 2,190 | 340 | 50 | 3.7 | 0.71 |
| Maple, sugar | 72,400 | 13,800 | 3,160 | 690 | 105 | 10.2 | 2.24 |
| Oak | | | | | | | |
|   Commercial red | 14,400 | 4,790 | 1,590 | 630 | 125 | 11.3 | 2.09 |
|   Commercial white | 17,400 | 3,550 | 1,100 | 415 | 80 | 7.2 | 1.15 |

From *Wood Handbook,* Handbook No. 72, U.S. Department of Agriculture, Washington, D.C., 1955.

## Table 5.1.3—5
### LOADS FOR NAILS AND SCREWS IN WOOD

#### Table A. Grouping of Wood Species

The holding power of wood fastenings depends largely on the density of the wood, hard to soft. Loads increase with the specific gravity of wood.

| Group No. | Typical species | Specific gravity |
|---|---|---|
| 1 | Hickory, hard maple, oak, beech, birch | .62–.75 |
| 2 | Douglas fir (larch), Southern pine, gum | .51–.55 |
| 3 | Hemlock, spruce, Northern pine, cypress, redwood | .42–.48 |
| 4 | Balsam fir, cedar, Engelmann spruce, soft white pine | .31–.41 |

## Table 5.1.3—5 (continued)
## LOADS FOR NAILS AND SCREWS IN WOOD

### Table B. Sizes and Allowable Design Loads

The following table gives the normal range of loading in pounds for joints secured by nails, spikes, wood screws, lag screws, or bolts. For load in kilograms, multiply tabular values by 0.453 6.

**Symbols:**

L = length; D = diameter

| Fastener description (seasoned wood only) | Nominal gage | Length and diameter, inches | Wood species, group 1 | 2 | 3 | 4 |
|---|---|---|---|---|---|---|
| **Nails,**[a] allowable withdrawal load per individual nail, per in. penetration into side grain | 6 d | 2 × .113 | 47–76 | 29–34 | 18–25 | 9–17 |
| | 8 d | 2.5 × .131 | 55–87 | 34–39 | 21–29 | 10–20 |
| | 10 d | 3 × .148 | 62–78 | 38–44 | 23–33 | 12–22 |
| | 12 d | 3.25 × .148 | 62–78 | 38–44 | 23–33 | 12–22 |
| | 16 d | 3.5 × .162 | 68–85 | 42–49 | 25–36 | 13–24 |
| **Spikes,** allowable withdrawal load per individual spike per in. penetration into side grain | 12 | 3.25 × .192 | 80–127 | 49–57 | 30–42 | 15–29 |
| | 16 | 3.5 × .207 | 86–138 | 53–61 | 33–46 | 16–31 |
| | 20 | 4 × .225 | 94–150 | 58–67 | 35–50 | 18–33 |
| | 40 | 5 × .263 | 110–174 | 68–79 | 41–58 | 21–39 |
| **Nails and spikes,** allowable lateral load in total lb shear for penetration from 10 diam for Group 1 to 14 diam for Group 4, into final member | 6 d | 2 × .113 | 78 | 63 | 51 | 41 |
| | 8 d | 2.5 × .131 | 97 | 78 | 64 | 51 |
| | 10 d | 3 × .148 | 116 | 94 | 77 | 62 |
| | 12 d | 3.25 × .148 | 116 | 94 | 77 | 62 |
| | 16 | 3.5 × .207 | 191 | 155 | 126 | 101 |
| | 20 | 4 × .225 | 218 | 176 | 144 | 116 |
| | 40 | 5 × .623 | 276 | 223 | 182 | 146 |
| **Wood screws,**[b] allowable withdrawal load per in. penetration to full length of threaded section | 6 | .67 L × .138 | 151–222 | 102–118 | 69–91 | 38–66 |
| | 8 | .67 L × .164 | 180–263 | 121–141 | 82–108 | 45–79 |
| | 10 | .67 L × .190 | 208–306 | 141–164 | 95–125 | 53–91 |
| | 12 | .67 L × .216 | 237–347 | 160–186 | 109–142 | 59–103 |
| | 16 | .67 L × .268 | 294–430 | 199–231 | 135–176 | 73–128 |
| **Wood screws,**[c] allowable lateral load for penetration 7 times the shank diameter into final member | 8 | 7 D × .164 | 129 | 106 | 87 | 68 |
| | 10 | 7 D × .190 | 173 | 143 | 117 | 91 |
| | 12 | 7 D × .216 | 224 | 185 | 151 | 118 |
| | 16 | 7 D × .268 | 345 | 284 | 233 | 181 |
| **Lag screws,**[d] allowable withdrawal load lb per in. penetration into side grain | 3/8 | 10 D × .375 | 421–561 | 313–356 | 235–287 | 145–226 |
| | 1/2 | 10 D × .500 | 523–697 | 389–443 | 291–357 | 180–281 |
| | 5/8 | 10 D × .625 | 618–824 | 460–524 | 344–421 | 213–332 |
| | 3/4 | 10 D × .75 | 708–944 | 528–601 | 395–483 | 244–380 |
| **Lag bolts,** allowable lateral load using 1/2-in. metal side pieces; penetration into side grain, single shear; load parallel to grain[e] | 5/16 | 4 × .312 | 410 | 355 | 290 | 235 |
| | 3/8 | 6 × .375 | 630 | 545 | 490 | 430 |
| | 1/2 | 8 × .5 | 1 140 | 985 | 880 | 775 |
| | 3/4 | 10 × .75 | 2 540 | 2 190 | 1 970 | 1 625 |
| | 1 | 12 × 1 | 4 520 | 3 900 | 3 290 | 2 630 |
| **Bolted joints,** double shear (joint consisting of 3 members; 2 side members, each 1/2 the thickness of main member. Bolt length given for main-member thickness only.) | 1/2 | 1.5 × .5 | 820–1 120 | | | |
| | 5/8 | 3 × .625 | 1 700–2 340 | | | |
| | 3/4 | 4.5 × .75 | 2 500–3 440 | | | |
| | 7/8 | 7.5 × .875 | 3 390–4 670 | | | |
| | 1 | 11.5 × 1 | 4 460–6 140 | | | |

**Table 5.1.3—5 (continued)**
**LOADS FOR NAILS AND SCREWS IN WOOD**

[a]Loads for threaded, hardened steel nails, in 6 d to 20 d sizes, are the same as for common nails.
[b]Wood screws shall not be driven with a hammer. Soap or other lubricant may be used, and lead holes are permitted, usually 70% of root diameter of thread. Spacing of screws shall be such as to prevent splitting.
[c]Tabular values are for screws inserted into side grain. With metal side plates loads may be increased 25%. If screw is inserted in end grain, allowable loads are to be reduced 33%.
[d]Penetration of threaded portion about 10 D, but withdrawal resistance approximates tensile (root) strength for penetrations as follows: species group 1 = 7 D; group 2 = 8 D; group 3 = 10 D; group 4 = 11 D. For penetration into end grain, allowable loads are reduced 25%. Lead holes for threaded section about 75% of shank diameter. Lag screws shall not be driven with a hammer.
[e]For load perpendicular to grain, the allowable loads are much less.

*Note*: The examples in Table 4–65 indicate the ranges of the allowable loads specified in the extensive tables in National Design Specification for Stress-grade Lumber and Its Fastenings, National Forest Products Association, Washington, D.C., 1971.

From Bolz, R. E. and Tuve, G. L., Eds., *Handbook of Tables for Applied Engineering Science,* 2nd ed., CRC Press, Cleveland, 1973, 632.

## Table 5.1.3—6
## UNIT STRESSES FOR LAMINATED TIMBERS

The following table gives the range of allowable unit stresses for "structural glued-laminated timber," in which the grain of all laminations is approximately parallel, net finished widths 2 1/4 to 14 1/2 in. These values are for dry conditions of use (moisture content <16%) and normal loading duration. Lower unit stresses are specified for wet conditions of use, for curved members, and for deep and slender beams (see References). Ranges indicate various grades and also numbers of laminations.

For stress and modulus of elasticity in $MN/m^2$, multiply values in psi by 0.0068948.

| Laminations of | Modulus of elasticity, millions of psi | Allowable unit stresses, psi | | | | | |
| --- | --- | --- | --- | --- | --- | --- | --- |
| | | Bending, load parallel[a] | Bending, load perpendicular | Tension, parallel to grain | Compression, parallel to grain | Compression, perpendicular to grain | Horizontal shear |
| Douglas fir or larch | 1.82 | 900–2,300 | 1,500–2,600 | 1,800–2,400 | 1,200–1,800 | 385–450 | 195 |
| Douglas fir, Coast region | 1.80 | 1,200–2,600 | 2,200–2,600 | 1,600 | 1,500 | 450 | 165 |
| Southern pine | 1.80 | 1,000–2,200 | 1,800–2,600 | 2,200–2,600 | 1,800–2,000 | 385–450 | 200 |
| California redwood | 1.30 | 1,100–2,300 | 1,400–2,200 | 1,800–2,200 | 1,800–2,200 | 325 | 125 |
| Ash, commercial white | 1.60 | 1,200–2,450 | — | 1,400–2,300 | 1,600–2,200 | 610 | 230 |
| Birch, sweet or yellow | 1.80 | 1,200–2,450 | — | 1,500–2,450 | 1,800–2,400 | 610 | 230 |
| Cottonwood, Eastern | 1.10 | 600–1,250 | — | 750–1,250 | 900–1,200 | 180 | 110 |
| Hickory | 2.00 | 1,500–3,100 | — | 1,900–3,100 | 2,200–3,000 | 730 | 260 |
| Maple, hard | 1.80 | 1,200–2,450 | — | 1,500–2,450 | 1,800–2,400 | 610 | 230 |
| Oak, red or white | 1.60 | 1,100–2,300 | — | 1,350–2,300 | 1,500–2,000 | 610 | 230 |

[a]Parallel or perpendicular with respect to the wide face of the laminations.

## Table 5.1.3—6 (continued)
## UNIT STRESSES FOR LAMINATED TIMBERS

*Standard Specification for the Design and Fabrication of Hardwood Glued-laminated Lumber for Structural, Marine, and Vehicular Uses*, Southern Hardwood Producers, Inc., Appalachian Hardwood Manufacturers, Inc., and Northern Hardwood and Pine Manufacturers Association.

*Standard Specifications for Structural Glued-laminated Timber*, American Institute of Timber Construction, 1970.

From Bolz, R. E. and Tuve, G. L., Eds., *Handbook of Tables for Applied Engineering Science*, 2nd ed., CRC Press, Cleveland, 1973, 634.

### REFERENCES

*National Design Specification for Stress-grade Lumber and Its Fastenings*, National Forest Productions Association, Washington, D.C., 1971.

*Standards for Structural Glued-laminated Members Assembled with WWPA Grades of Douglas Fir and Larch Lumber*, Western Wood Products Association, 1966.

*Standard Specifications for Structural Glued-laminated Douglas Fir Timber*, West Coast Lumber Inspection Bureau, 1963.

*Standard Specifications, Structural Glued-laminated California Redwood Timber*, California Redwood Association, 1965.

### 5.1.4     Other (Cryogenic, Flame Retardant, Etc.)

### Table 5.1.4—1
### STRENGTHS OF METALS AT CRYOGENIC TEMPERATURES

Test Data on 25 Alloys

Key:     UTS = ultimate tensile strength, kpsi
         YS = yield strength, kpsi
       Elong = percentage elongation in 2 in.
   Notch ratio = strength ratio: notched/unnotched
     Joint eff = percentage strength ratio welded/clear specimen

Specimens were mostly $\frac{1}{16}$-in. sheet, 2-in. gage length, cut longitudinal to rolling direction. Tests were made in triplicate. The test temperatures were as follows: 297 to 300 K = room temperature; 200 K = -100 deg F; 144 K = -200 deg F; 77 K = -320 deg F; 20 K = -423 deg F; 5 K = -450 deg F.

For MN/m$^2$, multiply kpsi by 6.8948.

| Temp, K | UTS, kspi | YS, kpsi | Elong, %, 2 in. | Notch ratio | Joint eff, % | Temp, K | UTS, kpsi | YS, kpsi | Elong, %, 2 in. | Notch ratio | Joint eff, % |
|---|---|---|---|---|---|---|---|---|---|---|---|
| **ALUMINUM ALLOYS** | | | | | | | | | | | |
| **2014: Al + Cu 4.5%, Mn 1%, Si 1%, Mg 0.5%** | | | | | | **5086: Al + Mg 4.0%, Fe 0.5%, Mn 0.45%** | | | | | |
| 300 | 70 | 63 | 9.7 | 0.99 | 66 | 300 | 47 | 37 | 10.4 | 1.00 | 90 |
| 200 | 73 | 68 | 9.5 | 1.00 | 60 | 200 | 48 | 38 | 12.0 | 0.99 | 86 |
| 144 | 76 | 71 | 9.3 | 0.99 | 59 | 144 | 52 | 39 | 15.4 | 0.98 | 89 |
| 77 | 84 | 76 | 11.7 | 0.93 | 63 | 77 | 64 | 44 | 25.0 | 0.89 | 93 |
| 20 | 96 | 79 | 13.6 | 0.88 | 82 | 20 | 85 | 47 | 20.2 | 0.76 | 80 |
| 5 | 97 | 82 | 10.4 | 0.83 | 74 | 5 | 80 | 49 | 23.4 | 0.74 | 86 |
| **2020: Al + Cu 4.5%, Li 1.1%, Mn 0.5%** | | | | | | **5456: Al + Mg 5.0%, Mn 0.75%, Zr <0.25%** | | | | | |
| 300 | 79 | 75 | 8.0 | 0.67 | — | 300 | 57 | 45 | 8.7 | 0.92 | 84 |
| 200 | 82 | 76 | 6.3 | 0.65 | — | 200 | 57 | 45 | 9.3 | 0.90 | 84 |
| 144 | 86 | 83 | 3.0 | 0.60 | — | 144 | 62 | 47 | 11.7 | 0.91 | 86 |
| 77 | 95 | 88 | 4.0 | 0.52 | — | 77 | 74 | 53 | 13.0 | 0.79 | 85 |
| 20 | 101 | 93 | 2.3 | 0.50 | — | 20 | 87 | 57 | 8.7 | 0.75 | 74 |
| 5 | 104 | 95 | 3.6 | 0.50 | — | 5 | 86 | 57 | 9.5 | 0.73 | 74 |
| **2119: Al + Cu 5.9%, Fe 0.15%, Ti 0.15%** | | | | | | **7002: Al + Zn 3.35%, Mg 2.07%, Cu 0.88%** | | | | | |
| 300 | 60 | 43 | 9.0 | 0.93 | — | 300 | 67 | 57 | 16.7 | 1.05 | 74 |
| 200 | 63 | 44 | 9.5 | 0.92 | — | 200 | 71 | 61 | 18.0 | 1.07 | 73 |
| 144 | 66 | 47 | 10.2 | 0.90 | — | 144 | 74 | 64 | 18.8 | 1.08 | 73 |
| 77 | 76 | 53 | 12.2 | 0.86 | — | 77 | 83 | 70 | 19.8 | 1.03 | 68 |
| 20 | 88 | 43 | 16.5 | 0.62 | — | 20 | 104 | 77 | 18.9 | 0.86 | 56 |
| **2219: Al + Cu 6.0%, Mn 0.33%, Fe 0.16%** | | | | | | **7075: Al + Zn 5.6%, Mg 2.5%, Cu 1.6%, Cr 0.3%** | | | | | |
| 300 | 65 | 52 | 9.8 | 0.92 | 66 | 300 | 79 | 74 | 9.2 | 0.90 | — |
| 200 | 69 | 55 | 9.3 | 0.92 | 67 | 200 | 83 | 77 | 8.7 | 0.82 | — |
| 144 | 73 | 58 | 10.0 | 0.92 | 70 | 144 | 85 | 80 | 6.7 | 0.78 | — |
| 77 | 83 | 64 | 12.1 | 0.90 | 75 | 77 | 94 | 88 | 5.2 | 0.68 | — |
| 20 | 96 | 79 | 15.3 | 0.81 | 77 | 20 | 101 | 95 | 3.2 | 0.56 | — |
| 5 | 94 | 69 | 12.0 | 0.68 | 80 | | | | | | |
| **5052: Al + Mg 2.5%, Fe 0.45%, Cr 0.25%** | | | | | | **7079: Al + Zn 4.5%, Mg 3.3%, Cu 0.6%** | | | | | |
| 300 | 34 | 25 | 10.6 | 0.97 | 95 | 300 | 76 | 67 | 9.0 | 1.00 | — |
| 200 | 36 | 26 | 15.1 | 1.01 | 93 | 200 | 80 | 68 | 9.0 | 0.84 | — |
| 144 | 39 | 27 | 20.1 | 0.96 | 98 | 144 | 86 | 75 | 7.0 | 0.78 | — |
| 77 | 52 | 29 | 30.0 | 0.93 | 98 | 77 | 93 | 84 | 4.0 | 0.68 | — |
| 20 | 73 | 37 | 26.5 | 0.88 | 92 | 20 | 101 | 94 | 3.0 | 0.56 | — |
| 5 | 72 | 33 | 27.0 | 0.83 | 95 | 5 | 102 | 93 | 2.5 | 0.53 | — |

## Table 5.1.4—1 (continued)
## STRENGTHS OF METALS AT CRYOGENIC TEMPERATURES

### ALUMINUM ALLOYS (continued)

**7178: Al + Zn 7%, Mg 3%, Cu 2%, Cr 0.3%**

| Temp, K | UTS, kpsi | YS, kpsi | Elong, %, 2 in. | Notch ratio | Joint eff, % |
|---|---|---|---|---|---|
| 300 | 94 | 88 | 7.5 | 0.67 | — |
| 200 | 96 | 93 | 4.0 | 0.65 | — |
| 144 | 100 | 96 | 4.0 | 0.60 | — |
| 77 | 109 | 104 | 1.2 | 0.41 | — |
| 20 | 117 | 113 | 1.0 | 0.32 | — |

### MAGNESIUM ALLOYS

**LA-91: Mg + Li 9.0%, Al 1.0%**

| Temp, K | UTS, kpsi | YS, kpsi | Elong, %, 2 in. | Notch ratio | Joint eff, % |
|---|---|---|---|---|---|
| 297 | 23 | 20 | 36.8 | 0.95 | 100 |
| 200 | 32 | 23 | 11.2 | 0.82 | — |
| 144 | 34 | 23 | 14.3 | 0.77 | — |
| 77 | 36 | 25 | 15.5 | 0.77 | 85 |
| 20 | 46 | 36 | 20.5 | 0.75 | 88 |
| 5 | 48 | 41 | 11.5 | — | — |

**LA-141: Mg + Li 14.5%, Al 1.5%**

| Temp, K | UTS, kpsi | YS, kpsi | Elong, %, 2 in. | Notch ratio | Joint eff, % |
|---|---|---|---|---|---|
| 297 | 20 | 18 | 23.7 | 1.06 | 95 |
| 200 | 28 | 21 | 10.8 | 0.94 | 88 |
| 144 | 30 | 23 | 13.7 | 0.94 | 87 |
| 77 | 33 | 28 | 13.8 | 0.95 | 87 |
| 20 | 43 | 39 | 14.3 | 0.86 | 98 |

### ALLOY STEELS

**20-Cb, Carpenter: Fe + Ni 25%, Cr 20%, Cu 3.5%, Mo 2.5%**

| Temp, K | UTS, kpsi | YS, kpsi | Elong, %, 2 in. | Notch ratio | Joint eff, % |
|---|---|---|---|---|---|
| 300 | 95 | 55 | 33.3 | 0.92 | 101 |
| 200 | 109 | 63 | 36.2 | 0.93 | 102 |
| 144 | 120 | 71 | 35.7 | 0.92 | 105 |
| 77 | 154 | 87 | 64.0 | 0.82 | 101 |
| 20 | 163 | 104 | 30.1 | 0.89 | 117 |

**A286-N:[a] Fe + Ni 26%, Cr 16%, Ti 2%**

| Temp, K | UTS, kpsi | YS, kpsi | Elong, %, 2 in. | Notch ratio | Joint eff, % |
|---|---|---|---|---|---|
| 300 | 93 | 42 | 37.3 | 0.86 | 100 |
| 200 | 104 | 48 | 38.8 | 0.88 | 101 |
| 144 | 115 | 57 | 43.0 | 0.87 | 101 |
| 77 | 144 | 68 | 71.0 | 0.80 | 99 |
| 20 | 161 | 81 | 47.3 | 0.82 | 96 |

**A286-H:[a] (Same as above)**

| Temp, K | UTS, kpsi | YS, kpsi | Elong, %, 2 in. | Notch ratio | Joint eff, % |
|---|---|---|---|---|---|
| 300 | 140 | 94 | 22.0 | 0.94 | 71 |
| 200 | 153 | 101 | 25.7 | 0.92 | 74 |
| 144 | 162 | 110 | 28.2 | 0.90 | 79 |
| 77 | 191 | 122 | 40.7 | 0.82 | 72 |
| 20 | 218 | 137 | 28.5 | 0.83 | 71 |

**AISI 202-N:[a] Fe + Cr 18%, Mn 8%, Ni 5%**

| Temp, K | UTS, kpsi | YS, kpsi | Elong, %, 2 in. | Notch ratio | Joint eff, % |
|---|---|---|---|---|---|
| 300 | 101 | 47 | 56.8 | — | 106 |
| 200 | 156 | 70 | 40.7 | — | 101 |
| 144 | 176 | 78 | 43.7 | — | 100 |
| 77 | 231 | 88 | 51.7 | — | 100 |
| 20 | — | — | — | — | — |
| 5 | 206 | 111 | 25.0 | — | 91 |

**Maraging H:[a] Fe + Ni 18%, Co 8%, Mo 5%**

| Temp, K | UTS, kpsi | YS, kpsi | Elong, %, 2 in. | Notch ratio | Joint eff, % |
|---|---|---|---|---|---|
| 300 | 254 | 245 | 2.8 | 1.09 | 94 |
| 200 | 275 | 266 | 2.8 | 1.08 | 100 |
| 144 | 288 | 274 | 3.0 | 1.06 | 102 |
| 77 | 321 | 309 | 2.5 | 0.90 | 99 |
| 20 | 365 | 355 | 3.2 | 0.41 | 86 |

### NICKEL ALLOYS

**Inconel X-H:[a] Ni + Cr 15%, Fe 7%, Ti 2.5%, Mn 1.0%, Al 0.7%**

| Temp, K | UTS, kpsi | YS, kpsi | Elong, %, 2 in. | Notch ratio | Joint eff, % |
|---|---|---|---|---|---|
| 300 | 180 | 125 | 25.3 | 0.90 | 67 |
| 200 | 194 | 132 | 26.7 | 0.87 | 69 |
| 144 | 203 | 136 | 27.2 | 0.84 | 71 |
| 77 | 220 | 139 | 32.0 | 0.79 | 72 |
| 20 | 224 | 140 | 28.0 | 0.82 | 79 |

**Waspaloy-H:[a] Ni + Cr 19%, Co 14%, Mo 4.3%, Ti 3.0%, Al 1.3%**

| Temp, K | UTS, kpsi | YS, kpsi | Elong, %, 2 in. | Notch ratio | Joint eff, % |
|---|---|---|---|---|---|
| 300 | 178 | 116 | 26.3 | 0.81 | 79 |
| 200 | 193 | 128 | 20.5 | 0.81 | 86 |
| 144 | 203 | 139 | 18.8 | 0.79 | 83 |
| 77 | 205 | 142 | 15.0 | 0.80 | 92 |
| 20 | 197 | 154 | 10.2 | 0.85 | 97 |

**K Monel-H:[a] Ni + Cu 29%, Al 3%, Fe 1.5%, Mn 1.0%**

| Temp, K | UTS, kpsi | YS, kpsi | Elong, %, 2 in. | Notch ratio | Joint eff, % |
|---|---|---|---|---|---|
| 300 | 148 | 106 | 22.7 | 0.92 | 67 |
| 200 | 156 | 111 | 24.0 | 0.96 | 70 |
| 144 | 165 | 119 | 24.7 | 0.93 | 72 |
| 77 | 177 | 128 | 30.7 | 0.92 | 76 |
| 20 | 192 | 137 | 28.3 | 0.90 | 83 |

**René 41-H: Ni + Cr 20%, Co 10%, Mo 10%, Ti 3.0%, Fe 3.0%**

| Temp, K | UTS, kpsi | YS, kpsi | Elong, %, 2 in. | Notch ratio | Joint eff, % |
|---|---|---|---|---|---|
| 300 | 199 | 147 | 16.0 | — | 77 |
| 200 | 201 | 152 | 14.0 | — | 86 |
| 144 | 210 | 167 | 12.0 | — | 81 |
| 77 | 229 | 179 | 12.0 | — | 85 |
| 20 | 239 | 199 | 9.0 | — | 85 |

### TITANIUM ALLOYS

**TiAlV: Ti + Al 5.9%, V 4%, Fe 0.12%**

| Temp, K | UTS, kpsi | YS, kpsi | Elong, %, 2 in. | Notch ratio | Joint eff, % |
|---|---|---|---|---|---|
| 297 | 140 | 133 | 11.0 | 1.02 | 100 |
| 200 | 161 | 158 | 9.3 | 1.00 | 99 |
| 144 | 178 | 177 | 6.5 | 0.99 | 100 |
| 77 | 218 | 214 | 13.0 | 0.82 | 102 |
| 20 | 240 | 240 | 1.7 | 0.61 | 96 |
| 5 | 242 | — | 0.2 | 0.62 | 100 |

**TiAlSn: Ti + Al 5.2%, Sn 2.4%, Fe 0.32%**

| Temp, K | UTS, kpsi | YS, kpsi | Elong, %, 2 in. | Notch ratio | Joint eff, % |
|---|---|---|---|---|---|
| 297 | 134 | 128 | 12.8 | 1.20 | 102 |
| 200 | 157 | 152 | 11.8 | 1.10 | 100 |
| 144 | 172 | 169 | 9.3 | 1.10 | 100 |
| 77 | 213 | 207 | 14.0 | 0.81 | 101 |
| 20 | 234 | 234 | 5.0 | 0.66 | 104 |
| 5 | 235 | — | 1.3 | 0.62 | 100 |

**TiVCr: Ti + V 13.4%, Cr 11.3%, Al 2.8%, Fe 0.18%**

| Temp, K | UTS, kpsi | YS, kpsi | Elong, %, 2 in. | Notch ratio | Joint eff, % |
|---|---|---|---|---|---|
| 297 | 137 | 137 | 13.3 | 1.20 | 106 |
| 200 | 182 | 182 | 6.0 | 1.10 | 106 |
| 144 | 218 | 215 | 4.5 | 0.96 | 106 |
| 77 | 285 | 282 | 2.5 | 0.54 | 60 |
| 20 | 289 | — | 0.7 | 0.40 | 34 |
| 5 | 301 | — | 0.0 | — | — |

[a]N: annealed; H; thermally age hardened, welded before aging.

**Table 5.1.4—1 (continued)**
**STRENGTHS OF METALS AT CRYOGENIC TEMPERATURES**

*Notes:* Only the major alloying elements are listed herewith; for minor alloying elements, see original data.

Data on specimens cut transverse to rolling direction are given in original source, as are data on aged and cold-worked alloys.

Note that 5, 20, and 77 K the approximately the atmospheric boiling points of liquid helium, hydrogen, and nitrogen, respectively; 200 K is just above the evaporation temperature of dry ice.

Notch ratios of less than one indicate that the material is weakened by the presence of stress-raising defects.

From *Effects of Low Temperature on Structural Metals,* NASA SP-5012, National Aeronautics and Space Administration, 1964 (revised edition, 1968).

**Table 5.1.4—2**
**CRYOGENIC THERMAL INSULATION[a]**

Description and Advantages

Classes

1. **Liquid and vapor shields** – Very low-temperature, valuable, or dangerous liquids such as helium or fluorine are often shielded by an intermediate cryogenic liquid or vapor container that must in turn be insulated by one of the methods described below.

2. **Multilayer reflecting shields** – Foil or aluminized plastic alternated with paper-thin glass- or plastic-fiber sheets; lowest conductivity, low density, and heat storage; good stability; minimum support structure.

3. **Opacified evacuated powders** – Contain metallic flakes to reduce radiation; conform to irregular shapes.

4. **Evacuated dielectric powders** – Very fine powders of low-conductivity adsorbent; moderate vacuum requirement; minimum fire hazard in oxygen.

5. **Vacuum flasks (Dewar)** – Tight shield-space with highly reflecting walls and high vacuum; minimum heat capacity; rugged; small thickness.

6. **Gas-filled powders** – Same powders as Class 4 but with air or inert gas; low cost; easy application; no vacuum requirement.

7. **Expanded foams** – Very light foamed plastic; inexpensive; minimum weight but bulky; self supporting.

8. **Porous fiber blankets** – Blanket material of fine fibers, usually glass; minimum cost and easy installation but not an adequate insulation for most cryogenic applications.

Insulation Properties

| Class | Descriptive name | Approximate density lbm/ft³ | kg/m³ | Approximate specific heat Btu/lbm·deg F | kJ/kg·K | Range of mean conductivities Btu/hr·ft·deg F | mW/m·K | Interspace pressure, mm Hg[b] |
|-------|------------------|------|------|------|------|------|------|------|
| 2 | Multilayer | 5 | 80 | .22 | 0.92 | .000023–.00012 | 0.04–0.2 | $10^{-4}$ |
| 3 | Opacified powder | 7 | 110 | .23 | 0.96 | .00015–.0004 | 0.26–0.7 | $10^{-4}$ |
| 4 | Evacuated powder | 6 | 100 | .25 | 1.05 | .00057–.00115 | 1.0–2.0 | $10^{-4}$ |
| 5 | Vacuum flask | – | – | – | – | .0029 | 5.0 | $10^{-6}$ |
| 6 | Gas-filled powder | 6 | 100 | .25 | 1.05 | .001–.004 | 1.7–7.0 | 760 |
| 7 | Expanded foam | 2 | 30 | 0.4 | 1.67 | .0029–.020 | 5.0–35 | 760 |
| 8 | Fiber blanket | 8 | 130 | 0.5 | 2.09 | .02–.026 | 35–45 | 760 |

## Table 5.1.4—2 (continued)
## CRYOGENIC THERMAL INSULATION

### Structural Support

For those insulating materials and constructions requiring structural support, the relative strengths, weights, heat capacities, and conductivities of the supporting materials are important.

| Material | Tensile yield strength S, 1000's psi[c] | Density, $\rho$ lbm/ft$^3$ | Density, $\rho$ kg/m$^3$ | Specific heat, $c_p$ Btu/lbm·deg F | Specific heat, $c_p$ kJ/kg·K | Mean thermal conductivity k,[d] 20–300°K Btu/hr·ft·deg F | Mean thermal conductivity k,[d] 20–300°K W/m·K | Relative S/k | Relative $\rho$/k | Relative $\frac{c_p \rho}{S}$ |
|---|---|---|---|---|---|---|---|---|---|---|
| Aluminum alloy | 50 | 170 | 2720 | .22 | 0.92 | 50 | 86 | 1 | 3 | .75 |
| "K" Monel® | 100 | 520 | 8330 | .13 | 0.54 | 10 | 17 | 10 | 52 | .68 |
| Stainless steel | 100 | 500 | 8010 | .12 | 0.50 | 5.4 | 9.3 | 18 | 93 | .60 |
| Titanium alloy | 100 | 625 | 10010 | .06 | 0.25 | 3.5 | 6.1 | 29 | 180 | .37 |
| Nylon | 15 | 70 | 1120 | .4 | 1.67 | .17 | 0.29 | 88 | 41 | 1.9 |
| Teflon | 2 | 120 | 1920 | .25 | 1.05 | .14 | 0.24 | 14 | 86 | 15.0 |

[a]For other thermal conductivities, see Tables 4–51 and 4–71.
[b]For N/m² multiply by 133.32.
[c]For MN/m² multiply tabulated values in 1000's psi by 6.8948.
[d]For solid members; perforation and lamination used to reduce condition.

### REFERENCE

*Thermal Insulation Systems,* NASA SP-5027, National Aeronautics and Space Administration, 1967.

From Bolz, R. E. and Tuve, G. L., Eds., *Handbook of Tables for Applied Engineering Science,* 2nd ed., CRC Press, Cleveland, 1973, 529.

## Table 5.1.4—3
## LOW-TEMPERATURE COOLING BATHS

### Cryogenic and Refrigerating Fluids — Atmospheric Pressure

| Liquid | Boiling point K | Boiling point °C | Boiling point °F | Liquid | Boiling point K | Boiling point °C | Boiling point °F |
|---|---|---|---|---|---|---|---|
| Helium | 4.2 | −268.95 | −452.1 | Refrigerant 14 (CF$_4$) | 145. | −128. | −198. |
| Hydrogen | 20.4 | −252.7 | −422.8 | Refrigerant 13 (CClF$_3$) | 192. | −81. | −114. |
| Neon | 27.1 | −246. | −410.8 | Carbon dioxide (CO$_2$)[a] | 195. | −78. | −108.5 |
| Nitrogen | 77.4 | −195.8 | −320.4 | Propylene (C$_3$H$_6$) | 225. | −48. | −54. |
| Air | 79. | −194. | −317. | Refrigerant 502 (Azeotrope) | 227. | −46. | −50. |
| Argon | 87.3 | −185.9 | −302.6 | Propane (C$_3$H$_8$) | 225. | −48. | −54. |
| Oxygen | 90.2 | −183.0 | −297.5 | Refrigerant 22 (CHClF$_2$) | 232. | −41. | −41. |
| Methane (CH$_4$) | 111. | −162. | −259. | Refrigerant 12 (CCl$_2$F$_2$) | 243. | −30. | −22. |

[a]Solid; sublimes.

*Notes:*  Low-temperature baths are conveniently prepared by adding dry ice or liquid nitrogen to acetone, ether, chloroform, or one of the alcohols, stirring the mixture until a slush is formed. Bath

## Table 5.1.4—3 (continued)
## LOW-TEMPERATURE COOLING BATHS

temperatures will then be fixed by the freezing point and will range from about $-82°$ F for chloroform to $-197°$ F for propyl alcohol. The lowest bath temperature with dry ice is $-108°$ F. Minimum temperatures attainable (with difficulty) using water-ice and salt mixtures are as follows: NaCl (23.3%), $-6°$ F; $CaCl_2$ (30%), $-67°$ F.

**Warning:** Adequate safety precautions are necessary when using low-temperature baths. In addition to skin "burns," toxic and explosion hazards may be present. In the initial cooling of a bath from room temperature, using dry ice or liquid nitrogen, the evolution of gas may be so rapid as to approach an explosion. Adequate venting and room ventilation are necessary. Liquid nitrogen absorbs and condenses oxygen (shown by bluish color) and may also produce "liquid air" in a heat-transfer device such as a vacuum cold trap. Later reevaporation of the air calls for adequate venting, or explosive forces are produced. Bath liquids and vapors may be toxic or irritant to skin or eyes (e.g., acetone or ketone) and contact should be avoided.

From Bolz, R. E. and Tuve, G. L., Eds., *Handbook of Tables for Applied Engineering Science,* 2nd ed., CRC Press, Cleveland, 1973, 587.

## Table 5.1.4—4
## CRYOGENIC AND REFRIGERATING LIQUIDS–SI UNITS

| Fluid | Boiling K | Boiling °R | Boiling °C | Boiling °F | Density, kg/m³ | Volume ratio (to room temp), gas/liq | Latent heat of vaporization, kJ/kg | Specific heat, $c_p$ Liquid, kJ/kg·K | Specific heat, $c_p$ Gas, kJ/kg·K | Viscosity Liquid, mN·s/m² | Viscosity Gas, mN·s/m² | Thermal conductivity Liquid, mW/m·K | Thermal conductivity Gas, mW/m·K | Dielectric constant |
|---|---|---|---|---|---|---|---|---|---|---|---|---|---|---|
| Air | 79 | 142 | −194 | −318 | 875 | 740:1 | 205 | 1.97 | 1.02 | 0.080 6 | 0.006 61 | | 7.44 | 1.52 |
| Argon | 87 | 157 | −186 | −303 | 1 400 | 840:1 | 162 | 1.14 | 0.531 | 0.252 | 0.008 27 | 123 | 6.05 | 1.59[c] |
| Carbon dioxide[a] | 195 | 350 | −79 | −110 | 1 560 | 730:1 | 572 | 1.33 | 0.795 | | | | 14.7[c] | |
| Ethane | 185 | 334 | −88 | −126 | 548 | 420:1 | 488 | 2.51 | 1.05 | 0.245 | 0.007 36 | | 25.1[d] | 1.43 |
| Fluorine | 85 | 154 | −188 | −306 | 1 500 | 880:1 | 166 | 1.55 | 0.812 | | | 135 | 7.20 | |
| Helium 3 | 3.20 | 5.76 | −270 | −454 | 58.9 | 600:1 | 8.48 | 4.60 | 6.82 | 0.001 62 | 0.001 28 | 17.1 | | |
| Helium 4 | 4.215 | 7.59 | −269 | −452 | 125 | 600:1 | 20.7 | 4.56 | 11.9 | 0.003 57 | 0.001 05 | 27.0 | 9.69 | 1.049 2 |
| Hydrogen | 20 | 36.7 | −253 | −423 | 71.0 | 800:1 | 446 | 9.79 | 1.66 | 0.013 1 | | 118 | 15.6 | 1.226 |
| Methane | 111 | 201 | −162 | −259 | 424 | 550:1 | 509 | 3.45 | 1.17 | 0.119 | | 111 | | 1.68 |
| Neon | 27 | 48.8 | −246 | −411 | 1 200 | 1 400:1 | 86.7 | 1.84 | | 0.124 | 0.004 55 | 130 | 9.86 | |
| Nitrogen | 77.4 | 139.2 | −196 | −320 | 810 | 700:1 | 198 | 2.04 | 1.08 | 0.158 | 0.005 54 | 139 | 7.18 | 1.434 |
| Oxygen[e] | 90.2 | 162. | −183 | −297 | 1 140 | 860:1 | 213 | 1.90 | 1.40 | | | | | 1.51 |
| Propane | 231 | 416 | −42 | −44 | 581 | 310:1 | 425 | 2.20 | 1.21 | | | | | 1.27 |
| Propylene | 225 | 406 | −48 | −54 | 614 | 330:1 | 437 | 2.59 | 1.52 | 0.371 | 0.010 9 | | | |
| Refrigerant 12 | 243 | 438 | −30 | −22 | 1 490 | 280:1 | 165 | 0.891 | 0.456 | | | | | 2.13 |
| Refrigerant 13 | 192 | 346 | −81 | −114 | 1 520 | 350:1 | 149 | 0.895 | 0.413 | | | 60.5 | | |
| Refrigerant 13B1 | 215 | 388 | −58 | −72 | 1 990 | 310:1 | 119 | 1.69 | 0.849 | | | | | |
| Refrigerant 22 | 232 | 419 | −41 | −41 | 1 410 | 380:1 | 234 | 1.05 | 0.464 | 0.351 | 0.010 5 | 110[b] | 7.78 | 2.44 |

[a]Sublimes.
[b]At 253.15 K.
[c]At 273.15 K.
[d]At 324.82 K.
[e]Solidifies at −218° C, 55 K.

From Bolz, R. E. and Tuve, G. L., Eds., *Handbook of Tables for Applied Engineering Science*, 2nd ed., CRC Press, Cleveland, 1973, 590.

## Table 5.1.4—5
## CRYOGENIC AND REFRIGERATING LIQUIDS

| Fluid | Boiling point | | | | Density (lb/ft³) | Volume ratio (to room temp.), gas/liq. | Latent heat of vaporization (Btu/lbm) |
|---|---|---|---|---|---|---|---|
| | K | °R | °C | °F | | | |
| Air | 79 | 142 | −194 | −318 | 54.6 | 740:1 | 88.2 |
| Argon | 87 | 157 | −186 | −303 | 87.6 | 840:1 | 69.5 |
| Carbon dioxide[a] | 195 | 350 | −79 | −110 | 97.6 | 730:1 | 246 |
| Ethane | 185 | 334 | −88 | −126 | 34.2 | 420:1 | 210 |
| Fluorine | 85 | 154 | −188 | −306 | 94.0 | 880:1 | 71.6 |
| Helium 3 | 3.20 | 5.76 | −270 | −454 | 3.68 | 600:1 | 3.65 |
| Helium 4 | 4.215 | 7.59 | −269 | −452 | 7.80 | 600:1 | 8.92 |
| Hydrogen | 20 | 36.7 | −253 | −423 | 4.43 | 800:1 | 192 |
| Methane | 111 | 201 | −162 | −259 | 26.5 | 550:1 | 219 |
| Neon | 27 | 48.8 | −246 | −411 | 75.2 | 1400:1 | 37.3 |
| Nitrogen | 77.4 | 139.2 | −196 | −320 | 50.6 | 700:1 | 85.3 |
| Propane | 231 | 416 | −42 | −44 | 36.3 | 310:1 | 183 |
| Propylene | 225 | 406 | −48 | −54 | 38.3 | 330:1 | 188 |
| Refrigerant 12 | 243 | 438 | −30 | −22 | 92.9 | 280:1 | 71.1 |
| Refrigerant 13 | 192 | 346 | −81 | −114 | 95.0 | 350:1 | 64 |
| Refrigerant 13B1 | 215 | 388 | −58 | −72 | 124 | 310:1 | 51.1 |
| Refrigerant 22 | 232 | 419 | −41 | −41 | 88.2 | 380:1 | 100.5 |

| Fluid | Specific heat, $c_p$ | | Viscosity | | Thermal conductivity | | Dielectric constant |
|---|---|---|---|---|---|---|---|
| | liquid (Btu/lbm °R) | gas (Btu/lbm °R) | liquid (lbm/ft hr) | gas (lbm/ft hr) | liquid (Btu/hr ft °F) | gas (Btu/hr ft °F) | |
| Air | .470 | .245 | .195 | .016 | | .0043 | |
| Argon | .272 | .127 | .610 | .0200 | .0712 | .00350 | 1.52 |
| Carbon Dioxide[a] | .318 | .190 | | | | .0085[c] | 1.59[c] |
| Ethane | .600 | .250 | .592 | .0178 | | .0145[d] | |
| Fluorine | .37 | .194 | | | .078 | .00416 | 1.43 |
| Helium 3 | 1.10 | | .00392 | | .0099 | | |
| Helium 4 | 1.09 | 1.63 | .00864 | .00309 | .0156 | .00560 | 1.0492 |
| Hydrogen | 2.34 | 2.85 | .0316 | .00254 | .0683 | .0090 | 1.226 |

## Table 5.1.4—5 (continued)
## CRYOGENIC AND REFRIGERATING LIQUIDS

| Fluid | Specific heat, $c_p$ | | Viscosity | | Thermal conductivity | | Dielectric constant |
| --- | --- | --- | --- | --- | --- | --- | --- |
| | liquid (Btu/lb$_m$ °R) | gas (Btu/lb$_m$ °R) | liquid (lb$_m$/ft hr) | gas (lb$_m$/ft hr) | liquid $\dfrac{\text{Btu}}{\text{hr ft °F}}$ | gas $\dfrac{\text{Btu}}{\text{hr ft °F}}$ | |
| Methane | .825 | .398 | .287 | | .0642 | | 1.68 |
| Neon | 0.44 | .280 | .30 | .0110 | .075 | .0057 | |
| Nitrogen | .487 | .259 | .382 | .0134 | .0804 | .00415 | 1.434 |
| Propane | .526 | .290 | | | | | 1.27 |
| Propylene | .619 | .363 | | | | | |
| Refrigerant 12 | .213 | .109 | .897 | .0264 | | | 2.13 |
| Refrigerant 13 | .214 | .0987 | | | .035 | | |
| Refrigerant 13B1 | .405 | .203 | | | | | |
| Refrigerant 22 | .252 | .111 | .849 | .0254 | .0635[b] | .00450 | 2.44 |

[a]Sublimes.
[b]At −4°F.
[c]At 32°F.
[d]At 125°F.

From Bolz, R. E. and Tuve, G. L., Eds., *Handbook of Tables for Applied Engineering Science*, 2nd ed., CRC Press, Cleveland, 1973, 97.

## Table 5.1.4—6
## HEAT TRANSMISSION THROUGH BUILDING STRUCTURES[a]

Temperature control within structures requires the use of accurate and consistent data on heat transfer coefficients. Extensive tables have been compiled by the American Society of Heating, Refrigerating and Air-conditioning Engineers (*ASHRAE Handbook of Fundamentals,* chap. 26), and these are accepted as standard. The following tables give some of the basic coefficients (Table A) and examples of overall coefficients for common wall and roof constructions (Table B) from this source.

### Table A. Basic Coefficients

**Emissivities** — Aluminum foil, 1 surface, 0.05; 2 surfaces, 0.03. Aluminum paint, 0.50. Nonmetallic surface, 0.90. (Blackbody = 1.0.)

**Coefficients in Btu/hr·ft²·deg F (W/m²·K)** — Indoor nonmetallic surfaces, still air; vertical, 1.46 (8.29); horizontal ceiling, winter, 1.63 (9.26); summer, 1.08 (6.13). Vertical air spaces: non-metallic, 1.04 (5.91); foil on both surfaces, 0.34 (1.93). Horizontal air space: nonmetallic surfaces, winter, 1.18 (6.70); summer, 1.01 (5.74). Outdoor surface coefficients, winter, 6.0 (34) for 15 m/h (6.7 m/s) wind; summer, 4.0 (23) for 7.5 m/h (3.35 m/s) wind.

For density in kg/m³, multiply values in lbm/ft³ by 16.018.

Conductivity, k, Btu in./ft²·hr·°F (at 75°F)

| Material | Density, lbm ft³ | Thermal conductivity Btu·in. hr·ft²·°F | Thermal conductivity W m·K |
|---|---|---|---|
| Asbestos cement board | 120 | 4.0 | 0.58 |
| Blanket of batt insulation | 3 | 0.27 | 0.039 |
| Brick, common | 120 | 5.0 | 0.72 |
| Brick, face | 130 | 9.0 | 1.30 |
| Cement mortar | 116 | 5.0 | 0.72 |
| Concrete, gypsum fiber | 51 | 1.66 | 0.24 |
| Concrete, lightweight | 40 | 1.15 | 0.17 |
| Corkboard | 7 | 0.27 | 0.039 |
| Hardboard, wood fiber | 65 | 1.40 | 0.20 |
| Mineral fiberboard | 18 | 0.35 | 0.05 |
| Mineral wool fill | 3 | 0.28 | 0.04 |
| Plaster, cement sand | 116 | 5.0 | 0.72 |
| Plaster, gypsum perlite | 45 | 1.5 | 0.22 |
| Plaster, vermiculite | 45 | 1.7 | 0.25 |
| Plywood | 34 | 0.80 | 0.115 |
| Redwood bark fill | 4 | 0.27 | 0.039 |
| Sheathing, wood fiber | 22 | 0.41 | 0.059 |
| Shredded wood, cemented | 22 | 0.60 | 0.086 |
| Vermiculite | 8 | 0.47 | 0.068 |
| Wood, oak, maple | 45 | 1.10 | 0.159 |
| Wood or cane fiberboard | 15 | 0.35 | 0.050 |
| Wood, pine, fir | 32 | 0.80 | 0.115 |

## Table 5.1.4—6 (continued)
## HEAT TRANSMISSION THROUGH BUILDING STRUCTURES

### Table B. Overall Coefficients (Transmittance) Air to Air

| Structure | U, W/m² · K | U, Btu/hr · ft² · °F |
|---|---|---|
| Single-glass window | 6.2 | 1.1 |
| Double-insulating glass with ½-in. air space, or storm window | 3.2 | 0.56 |
| Frame wall: wood sheathing, lath, and sand plaster | 1.5 | 0.26 |
| Frame wall: insulating sheathing, lath, and lightweight plaster | 1.1 | 0.19 |
| Frame wall: full fibrous insulating between studs | 0.40 | 0.07 |
| Brick veneer, insulating sheathing, lightweight plaster | 1.2 | 0.21 |
| Brick wall, 8-in. solid, sand plaster | 2.6 | 0.45 |
| Brick wall, 8-in., gypsum lath, furred, lightweight plaster | 1.5 | 0.27 |
| Clay tile wall, 8-in. hollow, gypsum lath, furred, lightweight plaster | 1.3 | 0.23 |
| Masonry cavity wall: 4-in. face brick, air space, 4-in. cinder block, gypsum lath, furred, lightweight plaster, two aluminum foil surfaces or ¾-in. insulation | 0.74 | 0.13 |
| Roof, pitched, shingle, unventilated rafter space, lath and plaster ceiling | 1.7 | 0.30 |
| Roof, pitched, asbestos cement, slate or tile shingles on wood sheathing, plastered ceiling, 3-in. insulation between joists | 0.40 | 0.07 |
| Roof, flat, metal deck, plaster ceiling on gypsum board | 1.9 | 0.33 |
| Roof, flat, preformed 2-in. insulating slab deck, acoustical ceiling on gypsum board | 0.74 | 0.13 |

[a]For other data on thermal conductivities, see Tables 4–44 and 4–71.

From Bolz R. E and Tuve. G. L , Eds., *Handbook of Tables for Applied Engineering Science*, 2nd ed., CRC Press, Cleveland, 1973, 682.

## Table 5.1.4—7
## DUST EXPLOSION CHARACTERISTICS

The following table is based on laboratory test results by the U.S. Bureau of Mines on dried samples of fine dusts (passing 200-mesh sieve).[a]    The values below probably represent "the most hazardous conditions" for these materials.

| Type of dust | Ignition temperature of dust cloud, °C | Minimum igniting energy, J | Minimum explosive concentration, oz/ft³ | Maximum explosion pressure, psig | Maximum rate of pressure rise, psi/sec | Terminal oxygen concentration, %[b] | Relative explosion hazard |
|---|---|---|---|---|---|---|---|
| Agricultural | | | | | | | |
| Alfalfa | 530 | 0.320 | 0.105 | 92 | 2,200 | — | Moderate |
| Cereal grass | 550 | 0.800 | 0.250 | 52 | 500 | — | Weak |
| Coffee | 720 | 0.160 | 0.085 | 53 | 300 | — | Weak |
| Corn | 400 | 0.040 | 0.055 | 95 | 6,000 | — | Strong |
| Corncob | 480 | 0.080 | 0.040 | 110 | 3,100 | — | Strong |
| Cornstarch | 390 | 0.030 | 0.040 | 115 | 9,000 | — | Severe |
| Cotton linters | 520 | 1,920 | 0.500 | 48 | 150 | — | Moderate |
| Cottonseed | 530 | 0.120 | 0.055 | 96 | 3,000 | — | Moderate |
| Grain, mixed | 430 | 0.030 | 0.055 | 115 | 5,500 | — | Strong |
| Grass seed | 490 | 0.260 | 0.290 | 34 | 400 | — | Weak |
| Malt, brewers | 400 | 0.035 | 0.055 | 92 | 4,400 | — | Strong |
| Milk, skim | 490 | 0.050 | 0.050 | 83 | 2,100 | — | Strong |
| Peanut hull | 460 | 0.050 | 0.045 | 82 | 4,700 | — | Strong |
| Peat, sphagnum | 460 | 0.050 | 0.045 | 84 | 2,200 | — | Strong |
| Pecan nutshell | 440 | 0.050 | 0.030 | 106 | 4,400 | — | Strong |
| Potato starch | 440 | 0.025 | 0.045 | 97 | 8,000 | — | Severe |
| Rice | 440 | 0.050 | 0.050 | 93 | 2,600 | — | Strong |
| Soy flour | 550 | 0.100 | 0.060 | 111 | 1,600 | 15 | Moderate |
| Sugar, powdered | 370 | 0.030 | 0.045 | 91 | 1,700 | — | Strong |
| Wheat flour | 440 | 0.060 | 0.050 | 104 | 4,400 | — | Strong |
| Wheat straw | 470 | 0.050 | 0.055 | 99 | 6,000 | — | Strong |
| Carbonaceous | | | | | | | |
| Charcoal, hardwood mix, volatile content 27.1% | 530 | 0.020 | 0.140 | 100 | 1,800 | 18 | Strong |
| Coal, Ill., No. 7, volatile content 48.6% | 600 | 0.050 | 0.040 | 84 | 1,800 | 15 | Strong |

## Table 5.1.4—7 (continued)
## DUST EXPLOSION CHARACTERISTICS

| Type of dust | Ignition temperature of dust cloud, °C | Minimum igniting energy, J | Minimum explosive concentration, oz/ft³ | Maximum explosion pressure, psig | Maximum rate of pressure rise, psi/sec | Terminal oxygen concentration, %[b] | Relative explosion hazard |
|---|---|---|---|---|---|---|---|
| Coal, Pa. (Pittsburgh), volatile content 37.0% | 610 | 0.060 | 0.055 | 83 | 2,300 | 17 | Strong |
| Gilsonite, Utah, volatile content 86.5% | 580 | 0.025 | 0.020 | 78 | 3,700 | – | Severe |
| Lignite, Cal., volatile content 60.4% | 450 | 0.030 | 0.030 | 90 | 8,000 | – | Severe |
| Pitch, coal tar, volatile content 58.1% | 710 | 0.020 | 0.035 | 88 | 6,000 | – | Severe |
| **Metals** | | | | | | | |
| Aluminum | 650 | 0.015 | 0.045 | 100 | 10,000[b] | 2 | Severe |
| Copper | 900 | – | – | – | – | – | Fire |
| Iron | 420 | 0.020 | 0.100 | 46 | 6,000 | 10 | Strong |
| Magnesium | 520 | 0.020 | 0.020 | 94 | 10,000[b] | 0 | Severe |
| Tin | 630 | 0.080 | 0.190 | 37 | 1,300 | 15 | Moderate |
| Titanium | 460 | 0.010 | 0.045 | 80 | 10,000[b] | 0 | Severe |
| Uranium | 20 | 0.045 | 0.060 | 53 | 3,400 | 0 | Severe |
| Zinc | 600 | 0.640 | 0.480 | 48 | 1,800 | 9 | Weak |
| **Plastics** | | | | | | | |
| Acetal resin (polyformaldehyde) | 440 | 0.020 | 0.035 | 89 | 4,100 | 11 | Severe |
| Acrylic polymer resin / Methyl methacrylate-ethyl acrylate | 480 | 0.010 | 0.030 | 85 | 6,000 | 11 | Severe |
| Alkyd resin / Alkyd molding compound | 500 | 0.120 | 0.155 | 15 | 150 | 15 | Weak |
| Amino resin / Urea-formaldehyde molding compound | 450 | 0.080 | 0.075 | 89 | 3,600 | 17 | Strong |
| Cellulose fillers / Wood flour | 430 | 0.020 | 0.035 | 110 | 5,500 | 17 | Severe |
| Cellulose resin / Ethyl cellulose molding compound | 320 | 0.010 | 0.025 | 102 | 6,000 | 11 | Severe |

## Table 5.1.4—7 (continued)
## DUST EXPLOSION CHARACTERISTICS

| Type of dust | Ignition temperature of dust cloud, °C | Minimum igniting energy, J | Minimum explosive concentration, oz/ft³ | Maximum explosion pressure, psig | Maximum rate of pressure rise, psi/sec | Terminal oxygen concentration, %[b] | Relative explosion hazard |
|---|---|---|---|---|---|---|---|
| Epoxy resin | 530 | 0.020 | 0.020 | 86 | 6,000 | 12 | Severe |
| Phenolic resin | | | | | | | |
| Phenol-formaldehyde molding compound | 500 | 0.020 | 0.030 | 92 | 10,000[b] | 14 | Severe |
| Rayon (viscose) flock | 520 | 0.240 | 0.055 | 88 | 1,700 | — | Moderate |
| Rubber, synthetic | 320 | 0.030 | 0.030 | 93 | 3,100 | 15 | Severe |

[a]See U.S. Bureau of Mines Reports of Investigations No. 5624, 5753, 5971, 6516.

[b]The terminal oxygen concentration is the limiting oxygen concentration in air-$CO_2$ atmosphere required to prevent ignition of dust clouds by electric spark.

From Bolz, R. E. and Tuve  G. L., Eds., *Handbook of Tables for Applied Engineering Science*, 2nd ed., CRC Press, Cleveland, 1973, 784.

## Table 5.1.4—8
## TYPICAL FIRE-RESISTANCE RATINGS FOR REINFORCED CONCRETE CONSTRUCTIONS

Ratings are dependent on protective cover of concrete over steel. For length in millimeters, multiply values in inches by 25.4.

| Fire-resistance rating, hours[a] | Structural members | Cover over steel, inches |
|---|---|---|
| 1 | Beams, medium size | 3/4–7/8 |
|   | Slabs, unrestrained | 1 |
| 2 | Beams, medium size | 1–1 1/4 |
|   | Slabs, unrestrained | 1 1/2 |
| 3 | Beams, medium size | 1 1/2 |
|   | Columns, 12–14 in. | 1 1/2 |
|   | Slabs, unrestrained | 2 |
|   | (Solid concrete walls, 6 in.) | – |
| 4 | Columns, 12–14 in. | 2 |
|   | Columns, > 16 in. | 1 1/2 |
|   | Slabs, unrestrained | 2 |

[a]The standard test for fire resistance rating for a large wall specifies measurement of the time required for a temperature rise of 250°F on the unexposed face when the other face is subjected to a "standard fire."

*Note:* Concrete with aggregates containing more than 30% quartz, chert, flint, or granite has an inferior fire rating unless mesh is used.
High cement content or light aggregates improve the fire rating.
Cover is defined as the minimum distance from steel to exposed surface.
Prestressed concrete members require more cover for a given rating.

From Bolz, R. E. and Tuve, G. L., Eds., *Handbook of Tables for Applied Engineering Science,* 2nd ed., CRC Press, Cleveland, 1973, 782.

## Table 5.1.4—9
## COMBUSTION RETARDANTS FOR SOLID MATERIALS

Fire and Flame Retardants for Combustible, Structural Materials, Sheets, Fabrics, and Plastics

**Variables** — Conditions for the combustion of solids in air are so diverse that most data and tests provide only relative values. Among the variables, in addition to the chemical composition, are size, shape, and orientation of the materials and the heat source, initial and ambient temperatures, humidity and moisture content, intensity of radiation, air motion, and duration of the exposure. In addition, the material itself may agglomerate, melt, vaporize, decompose, distort, or intumesce (swell), thus greatly altering the access of oxygen to support combustion.

**Fire retardants** — In most cases a fire-retardant treatment also will raise the maximum service temperature, and both objectives deserve attention. Oxygen supply is largely determined by surface/volume ratio; three classes are quickly recognized: (1) bulk solids and shapes, (2) sheets and fabrics, and (3) particulates and dusts. The common combustible solids — wood, paper, fabrics, and plastics — each present their special problems. The methods of application also are distinctive, e.g., compounding (filling), coating, and impregnating.

The chemistry of fire retardants deals largely with six elements — phosphorus, chlorine, bromine, boron, antimony, and nitrogen. As many treatments involve two or more retardants, a great variety of compounds of these six elements have been used. Inert and refractory coatings of oxides

## Table 5.1.4—9 (continued)
## COMBUSTION RETARDANTS FOR SOLID MATERIALS

Fire and Flame Retardants for Combustible, Structural Materials, Sheets, Fabrics, and Plastics (continued)

or mineral powders comprise another class of fire retardants, merging into the insulating protective coverings as the thickness and porosity are increased. Many kinds of fillers are used in plastics and rubbers; these usually reduce the flammability, but residues of plasticizers and solvents have opposite effects.

**Standard tests** — Arbitrary methods of testing for flammability, fire resistance, or fire endurance have been prescribed by the ASTM,[a] the Underwriters' Laboratories, and similar agencies in other countries.

**Tabular summary** — The following table presents a partial list of the fire-retardant treatments for wood, paper, textiles, and plastics. Many of these same materials are used in fire-retardant treatment or compounding for rubbers, asphalts, and bitumens. Not included in this table are several of the incombustible, or reflective, coatings

that may be applied to the surfaces of combustible materials.

A single table hardly can suggest the many complexities of the methods for fire retardation. Flammability is reduced by stabilizing a material to reduce the decomposition into volatile, combustible products. Some treatments increase the residual char, which then acts as a barrier against propagation of the flame. Certain additives melt and form a hard skin; others decompose and give a protective blanket of inert gas. These are complex processes that are not easily analyzed, but there is a vast literature reporting the efforts and the specific results. It should be mentioned that when certain fire-retarding compounds are used together, their effectiveness increases. Examples are combinations of phosphorus compounds with chlorine compounds and antimony in combination with halogens.

Chemical Fire Retardants for Solids

Key for methods of treatment:

I = immersion or impregnation;
M = mixture compounded;
S = surface application

| | | Fire retardant for | | | |
|---|---|---|---|---|---|
| Class of compound | Examples | Wood[b] | Paper[c] | Cellulose fabrics | Plastics |
| Phosphorus compounds | | | | | |
| Phosphoric acid compounds | TCP,DCP,TPP,THPC | I, S | | I | M |
| Ammonium phosphate | $(NH_4)_2PO_4$ ; $NH_4H_2PO_4$ | M | I, M | I | M |
| p-Halogen compounds | $PCl_3$ | I | | | M |
| Bromine compounds | | | | | |
| Organic | CHBr aromatics | | M, S | I | M |
| Bromides | $MgBr_2$ ; $ZnBr_2$ ; $NH_4Br$ | I | | | |
| Chlorine compounds | | | | | |
| Organic | Chlorinated paraffin | M | M, S | I | M |
| Zinc chloride | $ZnCl_2$ | I | | | |
| Boron compounds | | | | | |
| Borax; boric acid | $Na_2B_2O_7$ ; $H_3BO_3$ | S | M, S | I | |
| Nitrogen compounds | | | | | |
| Ammonium compounds | $(NH_4)_2SO_4$ ; $(NH_4)_2HPO_4$ | S | M | | M |
| Miscellaneous materials | | | | | |
| Silicates | Sodium silicate | I, S | S, M | | |
| Fire-retardant paints | | S | | | |
| Titanium salts | $(TiCl_4)$ ; $Ti_2(SO_4)_5$ | | | | M |

## Table 5.1.4—9 (continued)
## COMBUSTION RETARDANTS FOR SOLID MATERIALS

Chemical Fire Retardants for Solids (continued)

| | | Fire retardant for | | | |
|---|---|---|---|---|---|
| Class of compound | Examples | Wood[b] | Paper[c] | Cellulose fabrics | Plastics |
| Antimony compounds (often with halogens) | | | | | |
| Oxide | $Sb_4O_6$ $(Sb_2O_3)$ | M | M, S | I | M |
| Chloride | $SbCl_3$ | I | | | |
| Organic | | M | M | I | |

[a]See the ASTM Index for flammability tests such as No. D568, D635, D777, D1230, D2859, E162, and E286. See also fire tests such as D1360–61, E84, etc.
[b]Including fiber wallboards.
[c]Including cover stock, cardboards, and boxboards.

### REFERENCES

Lyons, J. W., *The Chemistry and Uses of Fire Retardants,* John Wiley & Sons, New York, 1970.
Goundry, J. H., *Fireproofing,* American Elsevier, New York, 1970.

From Bolz, R. E. and Tuve, G. L., Eds., *Handbook of Tables for Applied Engineering Science,* 2nd ed., CRC Press, Cleveland, 1973, 781.

## Table 5.1.4—10
## FLAMMABILITY LIMITS FOR GASES AND VAPORS IN AIR

At Atmospheric Pressure and Approximately Room Temperature[a]

| Compound | Formula | Flammability limits in air, % of total volume | | Compound | Formula | Flammability limits in air, % of total volume | |
|---|---|---|---|---|---|---|---|
| | | Lower (lean) | Upper (rich) | | | Lower (lean) | Upper (rich) |
| Acetaldehyde | $C_2H_4O$ | 4. | 57. | Ethyl bromide | $C_2H_5Br$ | 6.7 | 11.3 |
| Acetone | $C_3H_6O$ | 2.6 | 12.8 | (Bromoethane) | | | |
| (2-Propanone) | | | | Ethyl chloride | $C_2H_5Cl$ | 3.6 | 15.4 |
| Acetonitrile | $C_2H_3N$ | 4.5 | 16. | (Chloroethane) | | | |
| Acrolein | $C_3H_4O$ | 2.7 | 40. | Ethyl ether | $C_4H_{10}O$ | 1.7 | 48. |
| Acrylonitrile | $C_3H_3N$ | 3. | 17. | (Diethyl ether) | | | |
| Allyl alcohol | $C_3H_6O$ | 2.5 | 18. | Ethyl formate | $C_3H_6O_2$ | 2.7 | 16.5 |
| Allyl bromide | $C_3H_5Br$ | 4.35 | 7.3 | Ethyl nitrite | $C_2H_5NO_2$ | 3. | 50. |
| (3-Bromopropene) | | | | Ethylene dichloride | $C_2H_4Cl_2$ | 6.2 | 16. |
| Allyl chloride | $C_3H_5Cl$ | 3.3 | 11.2 | (1,2-Dichloroethane) | | | |
| (3-Chloropropene) | | | | Ethylene oxide | $C_2H_4O$ | 3. | 90. |
| Ammonia | $NH_3$ | 15.5 | 27. | (Oxirane) | | | |
| Amyl chloride | $C_5H_{11}Cl$ | 1.5 | 8.6 | Hydrocyanic acid | HCN | 5.6 | 41. |
| (1-Chloropentane) | | | | Hydrogen | $H_2$ | 4. | 75. |
| 1,3-Butadiene | $C_4H_6$ | 2. | 12. | Hydrogen sulfide | $H_2S$ | 4.3 | 45.5 |
| 2-Butoxyethanol | $C_6H_{14}O_2$ | 1.1 | 12.7 | Isopropyl acetate | $C_5H_{10}O_2$ | 1.8 | 7.8 |
| Butyl acetate | $C_6H_{12}O_2$ | 1.7 | 10. | Methyl acetate | $C_3H_6O_2$ | 3.1 | 16. |
| Butyl alcohol | $C_4H_{10}O$ | 1.4 | 18. | Methyl alcohol | $CH_4O$ | 7.3 | 40. |
| (1-Butanol) | | | | Methylamine | $CH_5N$ | 4.9 | 20.8 |

## Table 5.1.4—10 (continued)
## FLAMMABILITY LIMITS FOR GASES AND VAPORS IN AIR

At Atmospheric Pressure and Approximately Room Temperature (continued)

| Compound | Formula | Flammability limits in air, % of total volume Lower (lean) | Upper (rich) | Compound | Formula | Flammability limits in air, % of total volume Lower (lean) | Upper (rich) |
|---|---|---|---|---|---|---|---|
| Butylamine | $C_4H_{11}N$ | 1.7 | 9.8 | Methyl bromide | $CH_3Br$ | 10. | 16. |
| Butyl chloride | $C_4H_9Cl$ | 1.8 | 10.1 | (Bromomethane) | | | |
| (1-Chlorobutane) | | | | Methyl butyl ketone | $C_6H_{12}O$ | 1.25 | 8. |
| Butyl ether | $C_8H_{18}O$ | 1.5 | 7.6 | Methyl chloride | $CH_3Cl$ | 8.1 | 19.5 |
| Carbon disulfide | $CS_2$ | 1. | 50. | Methyl ether | $C_2H_6O$ | 2. | 20. |
| Carbon monoxide | CO | 12. | 74. | Methyl ethyl ketone | $C_4H_8O$ | 1.8 | 11. |
| Chloroprene | $C_4H_5Cl$ | 4. | 20. | Methyl formate | $C_2H_4O_2$ | 5. | 22.7 |
| Crotonaldehyde | $C_4H_6O$ | 2.1 | 15.5 | n-Propyl acetate | $C_5H_{10}O_2$ | 2.0 | 8. |
| Cyanogen | $C_2N_2$ | 6. | 42. | n-Propyl alcohol | $C_3H_8O$ | 2.1 | 13.5 |
| Cyclohexane | $C_6H_{12}$ | 1.3 | 8.3 | Propylamine | $C_3H_9N$ | 2. | 10.4 |
| Cyclopropane | $C_3H_6$ | 2.4 | 10.4 | Propyl chloride | $C_3H_7Cl$ | 2.6 | 11.1 |
| n-Decane | $C_{10}H_{22}$ | .7 | 5.4 | (1-Chloropropane) | | | |
| Deuterium | $D_2$ | 5. | 75. | Propylene dichloride | $C_3H_6Cl_2$ | 3.4 | 14.5 |
| Diborane | $B_2H_6$ | .9 | 98. | (1,2-Dichloropropane) | | | |
| Dichlorobenzene | $C_6H_4Cl_2$ | 2.2 | 9.2 | Propylene oxide | $C_3H_6O$ | 2. | 22. |
| Diethylamine | $C_4H_{11}N$ | 1.8 | 10.1 | Pyridine | $C_5H_5N$ | 1.8 | 12.4 |
| Dimethylamine | $C_2H_7N$ | 2.8 | 14.4 | Triethylamine | $C_6H_{15}N$ | 1.2 | 8. |
| Dioxane | $C_4H_8O_2$ | 2. | 22.2 | Trimethylamine | $C_3H_9N$ | 2. | 11.6 |
| 2-Ethoxyethanol | $C_4H_{10}O_2$ | 1.7 | 15.6 | Vinyl acetate | $C_6H_6O_2$ | 2.6 | 21.7 |
| Ethyl acetate | $C_4H_8O_2$ | 2.5 | 11.5 | Vinyl chloride | $C_2H_3Cl$ | 4. | 22. |
| Ethyl alcohol | $C_2H_6O$ | 4.3 | 19. | (Chloroethylene) | | | |
| Ethylamine | $C_2H_7N$ | 3.5 | 14. | Xylene | $C_8H_{10}$ | 1. | 7. |

[a]Flammable or explosive limits in pure air will differ greatly at other temperatures or pressures. In general, the effect of increasing temperature or pressure is to widen the flammable range.

### REFERENCE

Williams-Steiger Occupational Safety and Health Act of 1970, Chapter XVII, Title 29, *Code of Federal Regulations,* April 13, 1971 (amended August 13, 1971), Part 1910 (Occupational Safety and Health Standards); published in *Fed. Register,* 36 (105), May 29, 1971.

From Bolz, R. E. and Tuve, G. L., Eds., *Handbook of Tables for Applied Engineering Science,* 2nd ed., CRC Press, Cleveland, 1973, 783.

## Table 5.1.4—11
## SURFACE FLAMMABILITY OF WOODS

Tests on Bare Wood, Plywood, Wallboards, and Painted Wood Surfaces
Basis of Test Data

A report titled "Surface Flammability of Various Wood-Base Building Materials" was issued by the Forest Products Laboratory in 1959.[a] A similar report on painted wood surfaces appeared in 1963 in the *Official Digest* of the FSPT,[b] and this material was reprinted in the *NFPA Quarterly* in 1964.[c] All tests were made in a tunnel furnace by a uniform method.[c] From these tests three index numbers were computed and tabulated. In each case an index number of 100 was assigned to the performance of a standard red oak specimen; the comparative index numbers were determined as follows:

1. Flame-spread index. If the flame spread was faster than for the standard red-oak specimen, the index was determined by the length of time for the flames to reach the end of the test specimen, as compared with the standard time (18.4 min) for the flame to reach the end of the red-oak specimen. For a flame spread slower than that for red oak, the index was based on the ratio of the distances reached by the flames on the two specimens in the standard 18.4 min period (87 in. for red oak).

2. Index of heat contributed. This index was based on the readings of thermocouples in the furnace stack.

3. Smoke density index. This index was obtained from the readings of photoelectric smoke meter in the furnace stack.

A zero reference for both the smoke-density and heat-contributed indices was established by using a test specimen of asbestos millboard.

It is apparent that this is an arbitrary test procedure that gives approximate comparative results only. Tests of identical specimens often varied ten points or more on the index scales (sometimes with one above 100 and the other below 100). It must be concluded that the various woods and finishes cannot be ranked in absolute order and that the tests serve rather to establish classes or categories of sufrace-flammability performance. This also is to be expected from the fact that the specimens within a class will vary, i.e., no two trees are exactly alike and no two manufacturers of wallboard of the same description will produce identical specimens.

Although specific index numbers are quoted in the following table, they must be considered as approximate test results rather than absolute ratings.

| Description of test specimens | lb/ft³ | Moisture % | Index numbers | | | Evaluation of flammability |
|---|---|---|---|---|---|---|
| | | | Flame spread | Heat contributed | Smoke density | |
| **Tests on 1-in. lumber** | | | | | | |
| Alder | 29.7 | 6.5 | 121 | 121 | 112 | 8 |
| Aspen | 27.3 | 6.1 | 121 | 124 | 72 | 7 |
| Bald cypress | 29.6 | 6.9 | 112 | 109 | 389 | 14 |
| Basswood | 28.0 | 5.4 | 128 | 128 | 79 | 3 |
| Beech | 46.8 | 5.8 | 101 | 140 | 132 | 24 |
| Birch | 39.4 | 5.6 | 96 | 94 | 86 | 26 |
| Cedar, red | 33.8 | 8.9 | 109 | 94 | 224 | 16 |
| Chestnut | 29.0 | 6.1 | 120 | 92 | 17 | 10 |
| Cottonwood | 27.3 | 5.0 | 134 | 135 | 125 | 1 |
| Elm, slippery | 38.5 | 6.1 | 89 | 108 | 151 | 29 |
| Fir, Douglas | 27.1 | 5.9 | 116 | 77 | 81 | 11 |
| Fir, white | 29.9 | 6.7 | 115 | 96 | 109 | 12 |
| Hemlock | 29.0 | 7.4 | 108 | 95 | 78 | 17 |

## Table 5.1.4—11 (continued)
## SURFACE FLAMMABILITY OF WOODS

Tests on Bare Wood, Plywood, Wallboards, and Painted Wood Surfaces (continued)

| Description of test specimens | lb/ft³ | Moisture % | Index numbers Flame spread | Index numbers Heat contributed | Index numbers Smoke density | Evaluation of flammability |
|---|---|---|---|---|---|---|
| **Tests on 1-in. lumber** (continued) | | | | | | |
| Larch | 34.9 | 6.8 | 106 | 98 | 56 | 19 |
| Mahogany | 29.7 | 6.6 | 104 | 70 | 34 | 22 |
| Mahogany, Philippine | 35.3 | 5.4 | 106 | 81 | 55 | 20 |
| Maple, sugar | 41.7 | 6.2 | 95 | 83 | 76 | 28 |
| Oak, red | 39.0 | 5.0 | 100 | 100 | 100 | 25 |
| Oak, white | 40.8 | 6.4 | 95 | 91 | 40 | 27 |
| Pine, Northern white | 23.9 | 5.7 | 132 | 104 | 193 | 2 |
| Pine, ponderosa | 27.7 | 6.5 | 114 | 102 | 230 | 13 |
| Pine, Southern yellow | 29.2 | 6.8 | 102 | 115 | 158 | 23 |
| Pine, sugar | 24.0 | 6.2 | 125 | 98 | 262 | 4 |
| Pine, western white | 25.9 | 6.0 | 123 | 113 | 274 | 6 |
| Poplar, yellow | 31.0 | 5.7 | 124 | 125 | 155 | 5 |
| Redwood | 25.6 | 6.4 | 121 | 68 | 188 | 9 |
| Spruce, sitka | 26.8 | 6.8 | 112 | 80 | 57 | 15 |
| Sweet gum | 32.3 | 6.8 | 105 | 105 | 68 | 21 |
| Walnut, black | 37.4 | 5.1 | 107 | 114 | 85 | 18 |
| **Tests on plywood** | | | | | | |
| ¼-in., 3 ply, fir, interior, protein glue | 30.7 | 6.3 | 123 | 119 | 136 | |
| ¼-in., 3 ply, fir, interior, resin glue | 34.6 | 5.0 | 121 | 122 | 81 | |
| ⅜-in., 3 ply, fir, exterior, resin glue | 33.6 | 4.8 | 114 | 112 | 96 | |
| ⅜-in., 3 ply, fir, exterior, paint A, 1 coat | 34.8 | 4.8 | 83 | 58 | 477 | |
| ⅜-in., 3 ply, fir, exterior, paint A, 2 coats | 33.9 | 5.1 | 53 | 19 | 968 | |
| ⅜-in., 3 ply, fir, exterior, paint B, 1 coat | 34.1 | 5.0 | 65 | 21 | 747 | |
| ⅜-in., 3 ply, fir, exterior, paint B, 2 coats | 34.1 | 5.0 | 34 | 10 | 1 143 | |
| ⅜-in., 3 ply, fir, exterior, paint C, 2 coats | 34.1 | 5.0 | 81 | 38 | 1 000 | |
| ⅜-in., 3 ply, fir, exterior, paper, plastic overlay | 36. | 5.0 | 105 | 112 | 141 | |
| **Tests on fiberboard** | | | | | | |
| Four different ½-in. insulating fiberboards | 19. | 5.0 | 125 | 91 | 105 | |
| Three heavier fiberboards | 32. | 4.0 | 122 | 161 | — | |
| **Tests on hardboard** | | | | | | |
| Willow, oak, wax, resin, 0.22 in. | 51.4 | 3.2 | 119 | 177 | 131 | |
| Fir, pine, wax, resin (4 makes) | 61.4 | 4.3 | 96 | 169 | 407 | |
| Fir, redwood, dense, oil-tempered | 70.8 | 3.1 | 90 | 165 | 657 | |
| **Tests on particle boards** | | | | | | |
| Nine wood-chip boards | 36–67 | 3.8–6.6 | 85–104 | 80–150 | 55–650 | |

**Table 5.1.4—11 (continued)**
## SURFACE FLAMMABILITY OF WOODS
Conclusions From Test Results

Index numbers obtained by flame-spread tests of 29 species of structural lumber by the tunnel-furnace method gave a graduated scale from the 130 range for flammable softwoods, such as white pine, cottonwood, basswood, and poplar, to the 90 range for the dense hardwoods, such as white oak, birch, elm, and sugar maple. Although the "heat-contributed" index followed the flame-spread index roughly, the smoke density was dependent on other factors. The flame-spread index of 1/4- and 3/8-in. fir plywood was somewhat higher than for a 1-in. fir board. The flame-spread index for either structural or insulating fiberboard differed little from those for the softwoods.

The flammability of pressed hardboards varied from the softwood range for the lighter boards (50 lb/ft$^3$) to the hardwood range for oil-tempered board of high density (65 lb/ft$^3$).

The particle boards tested were no more flammable than hardwood; however, the range was rather wide, as there is a great variety of such boards.

Many tests were made on painted specimens, most of them with 3/8-in. Douglas fir plywood as a base. For some coatings such as interior oil-base paints and enamels, varnish, shellac, lacquer, and asphalt paint the flammability was changed little from that of bare plywood, but the smoke and heat were increased. The effectiveness of fire-retardant paints was clearly demonstrated, especially if two coats were used or an intumescent (foaming) composition was employed. Most of the fire-retardant paints reduced the flammability of the fir-plywood specimen to that of the least-flammable hardwood. The best of the fire-retardant paints reduced the flammability index to the 30 range, on a scale of red oak = 100 and asbestos millboard = 0.

[a]Bruce, H. D. and Downs, L. E., Surface Flammability of Various Wood-base Building Materials, Forest Products Laboratory Report No. 2140, Forest Service, U.S. Department of Agriculture, 1959.
[b]*Official Digest,* Federation of Societies for Paint Technology, August 1963.
[c]Eickner, W. C. and Peters, C. C., *Natl. Fire Protection Q.,* April 1964.

From Bolz, R. E. and Tuve, G. L., Eds., *Handbook of Tables for Applied Engineering Science,* 2nd ed., CRC Press, Cleveland, 1973, 778.

# 5.2    PAINTS AND COATINGS

## Table 5.2—1
## CLASSIFICATION OF PAINTS AND COATINGS[a]

Materials and practices in the field of paints and coatings are diverse and changing rapidly with the shift from natural to synthetic materials. This table and those following quote data on well-known practices and materials only, using industrial terminology. For more comprehensive data consult the references.

Classifications of coating materials are related to their uses, as indicated in the following table. Typical organic coatings are prepared with pigments and drying oils, with resins and solvents, or with latex or resin emulsions. The pigments are mixed for color, and dyes are sometimes added. Minor components may accelerate or retard

drying. Many resin varnishes and lacquers are nearly transparent, but dyes, pigments, and metallic powders may be added for color effects. Plasticizers are used to improve the elastic qualities of the film.

The inexpensive coatings used in large quantities are simple formulations, but special uses may demand highly complex mixtures. Among the recognized and desirable properties are hiding power or opaqueness, light, heat, and chemical resistance, bleeding resistance, and ease of application. Other identified properties are penetration, chalking, gloss or texture, hardness, mold resistance, stain resistance, and package stability.

| Class | Description or source | Forms and properties | Typical uses |
|---|---|---|---|
| Oil paint | Vegetable oil and mineral pigment | Linseed (or other) drying oil with opaque but colored (or white) pigment | Exterior and general painting; interior trim; industrial products |
| Latex paint | Partially polymerized liquid resins in water | Water paint for porous and absorbent materials; alkali resistant | Wood and masonry surfaces; wall board; sheathings and sidings; interiors |
| Varnish | Drying oil, resin, solvent, and drier | Usually transparent to amber in color; glossy and impervious | Furniture, woodwork, floors, boats, trim, sealers |
| Lacquer | Nitrocellulose, acrylic, or other resins in volatile solvent | Quick-drying by solvent evaporation | Automobiles; transport equipment; metal products; oven-dried finishes |
| Whitewash | Lime, water, and additives | Age-old water paint for decorative and reflective purposes | Inexpensive outdoor or barn paint |
| Cement-water | White cement, lime, and pigment | Water paint for smooth finish on coarse masonry | Concrete or tile finish or grout-coat |
| Glue or size | Gelatin, skin, bone, starch | Water solutions | Sealer for plaster, etc. |
| Bituminous | Petroleum; coal tar | Paints; hot mastic; emulsions | Underground; waterproofing; roofs; tanks |
| Fire retardant | Brominated resins; insulation; glass | Flame resistant and/or insulating or glazing | Prevention of flame spread over combustible surfaces or textiles |
| Chlorinated rubber | 65% chlorine with poly-isoprene and plasticizer | Paints; chemical and water resistant | Traffic, swimming pool, and masonry paints; quick-drying industrial coatings |
| Strippable coating | Cellulose esters, or other resins, solvent and oil | Spray or hot-dip (oil migrates to interface) | Protection of machine parts and assemblies; "mothballing" |
| Fluorescent | Added fluorescent dyes | Absorbed violet reemitted as yellow, orange | High visibility for signs, hazard protection, advertising |
| Wax polish | Hard wax plus resin | Organic solvent, water emulsion, or both | Floors, furniture, automobiles, paneling |

**Table 5.2—1 (continued)**
## CLASSIFICATION OF PAINTS AND COATINGS

### REFERENCES

Roberts, A. G., *Organic Coatings – Properties, Selection, and Use,* National Bureau of Standards, Washington, D.C., 1968.
Morgans, W. M., *Outlines of Paint Technology,* Griffin, London, 1969.
Martens, C. R., *Technology of Paints, Varnishes, and Lacquers,* Van Nostrand Reinhold, New York, 1968.
Myers, R. R. and Long, J. S., Eds., *Treatise on Coatings,* Vol. 1, Marcel Dekker, New York, 1967.
Parker, D. H., *Principles of Surface Coating Technology,* John Wiley & Sons, New York, 1965.
Payne, H. F., *Organic Coating Technology,* John Wiley & Sons, New York, 1954.
Von Fischer, W. and Bobalek, E. G., Eds., *Organic Protective Coatings,* Reinhold, New York, 1953.
Abraham, H., *Asphalts and Allied Substances,* D. Van Nostrand, Princeton, N.J., 1945.
National Paint, Varnish, and Lacquer Association, *Raw Materials Index,* Washington, D.C.
Bennet, H., Bishop, J. L., Wulfinghoff, M. F., *Practical Emulsions,* Vol. 2, Chemical Publishing, New York, 1968.

From Bolz, R. E. and Tuve, G. L., Eds., *Handbook of Tables for Applied Engineering Science,* 2nd ed., CRC Press, Cleveland, 1973, 352.

**Table 5.2—2**
## ORGANIC COATINGS – DRYING OILS AND DRIERS

Drying oils contain glycerides or esters from a combination of glycerol with unsaturated fatty acids. The drying of a thin oil film is a complex chemical process. Oxygen plays a major part, and the process is retarded at low temperatures and in the absence of light. It is accelerated by the catalytic action compounds of certain metals (e.g., lead).

### Typical Properties of Drying Oils

| Oil | Specific gravity | Viscosity, cp | Refractive index | Saponification value[a] | Acid value[b] | Iodine value[c] |
|---|---|---|---|---|---|---|
| Linseed (flax) | | | | | | |
| Raw | 0.93 | 40 | 1.48 | 192 | 2 | 177 |
| Boiled | 0.95 | 100 | 1.48 | | 3 | 165 |
| Blown | 0.97 | 300 | 1.48 | | 5 | 120 |
| Tung | 0.94 | | 1.52 | 193 | 7 | 162 |
| Fish | 0.93 | | 1.48 | 191 | 6 | 165 |
| Tobacco seed | 0.92 | | 1.48 | 190 | 5 | 140 |
| Soybean | 0.93 | | 1.47 | 190 | 2 | 135 |
| Tall oil | | | | 169 | 160 | |

[a]Saponification value is the quantity of potassium hydroxide required to saponify 1 g of oil.
[b]Acid value is a quantitative index of the amount of free organic acid in an oil.
[c]Iodine value is a measure of the degree of unsaturation of the oil and, hence, of its drying properties.

*Notes:* Raw linseed oil dries slowly. Boiled oil, which has been heated with lead, cobalt, or manganese compounds to reduce drying time, is preferable for interior paints. Blown oil has been thickened by blowing air through hot oil.
Tung-nut oil is fast-drying and has high-water resistance.
Fish oil dries slowly and has a strong odor and darker color unless refined. It is heat resistant and elastic and is used in high-temperature and roofing paints.
Soybean and other seed oils are semidrying but are used in oil blends to give elasticity and durability to the paint film.
Tall oil is a semidrying oil, a by-product of the sulfite papermaking process. It is expensive and is used with various synthetic resins.

From Bolz, R. E. and Tuve, G. L., Eds., *Handbook of Tables for Applied Engineering Science,* 2nd ed., CRC Press, Cleveland, 1973, 353.

## Table 5.2—3
## ORGANIC COATINGS – SOLVENTS AND THINNERS

Typical Designations and Properties

| Name | Alternative name | Specific gravity | Boiling range, deg C | Flash point, deg C (closed) | ASTM No. |
|---|---|---|---|---|---|
| PETROLEUM PRODUCTS | | | | | |
| Mineral spirits | Petroleum ether; ligroin | 0.66 | 60–80 | 5 | D484 |
| Naphtha thinner | White spirits; turpentine substitute | | 150–190 | 40 | D235 |
| Aromatic naphtha | | 0.85 | | | |
| COAL-TAR PRODUCTS | | | | | |
| Toluene | Toluol | 0.87 | 110 | 5 | D362 |
| Xylene (3 isomers) | | 0.86 | 137–148 | 27 | D364 |
| Coal-tar naphtha | Heavy naphtha | | <190 | 37 | |
| CONIFEROUS TREE PRODUCTS | | | | | |
| Turpentine (gum, wood) | | 0.85 | 150–180 | 30–36 | D13 |
| Dipentene | | | 175–195 | 54 | |
| Pine oil | | | 200–230 | | D802 |
| ORGANIC COMPOUNDS | | | | | |
| Acetone | | 0.79 | 56 | −18 | D329 |
| Amyl acetate | Banana oil | 0.88 | 146 | 24 | |
| Butanol | Butyl alcohol | 0.81 | 118 | 25 | D304 |
| Denatured alcohol | Methylated ethanol | 0.79 | 78 | 14 | |
| Ethyl acetate | Acetic ester | 0.90 | 77 | 5 | |
| Isopropyl alcohol | 2-Propanol | 0.79 | 82 | 14 | D770 |
| Methanol | Wood alcohol | 0.79 | 65 | 14 | D1152 |
| Propyl alcohol | *n*-Propanol | 0.80 | 98 | 22 | |
| Trichloroethylene | | 1.46 | 87 | — | |

Compiled from several sources and presented in Bolz, R. E. and Tuve, G. L., Eds., *Handbook of Tables for Applied Engineering Science,* 2nd ed., CRC Press, Cleveland, 1973, 353.

## Table 5.2—4
## ORGANIC COATINGS – NATURAL RESINS AND MATERIALS[a]

Natural and Processed Materials Used in Paint, Varnish, and Protective Coatings and Finishes

| Common name | Alternate name | Source | Properties | Uses |
|---|---|---|---|---|
| Rosin | Colophony; gum and wood rosins; tall oil rosin; phenolic blend | U.S.A., France, Portugal, Spain; tapped from coniferous trees | Depends on processing; dispersion, adhesion, gloss | Wide variety of rosin products, e.g., ester gum, limed rosin, zinc resinate, maleic resin; for varnishes, enamels, lacquers |
| Copal resin | Class name for several resins | Africa, East Indies, Philippines | Hard, glossy film for varnish or enamel | Largely displaced by synthetic resins |
| Damar resin | Singapore; dipterocarpus | East Indies | Soft, non-yellowing, fume-resistant film | Specialty finishes; paper varnish (not widely used) |
| Lac | Shellac base; seed lac | India; insect excretion from tree sap | M.p., about 80 deg C; acid value, 70; iodine value, 20; soluble in alcohols; color, orange or darker; tough; fast-drying | Sealer and base coat; foundry patterns; white shellac on floors; largely displaced by synthetics |

**Table 5.2—4 (continued)**
## ORGANIC COATINGS – NATURAL RESINS AND MATERIALS

| Common name | Alternate name | Source | Properties | Uses |
|---|---|---|---|---|
| Rubber | Chlorinated rubber; cyclized rubber | Natural rubber latex | Tough; resistant to abrasion, moisture, chemicals, mildew, petroleum; quick-drying | In blended paints for chemical apparatus, marine uses, traffic areas, swimming pools, rust protection |
| Asphalt | Bitumen; gilsonite; manjak | Petroleum residues; some natural deposits, as in Trinidad | Softens at 46 deg C; good adhesion; waterproof and chemical-resistant | Roof and waterproof paints; corrosion protection; sound control |
| Pitch | Coal tar | Residue from coal-tar distillation | Impervious to water; high dielectric strength; heavy coating | Ship-bottom paint; moisture and corrosion protection; chemically resistant paint |
| Gum | Examples: gum arabic, gum tragacanth | Africa and Asia | Thickening agent and base for water colors | Seldom used except in artists' paints |
| Glue | Glue size | Skin, bone, animal, and fish scrap | Binder for water paint and ceiling white; adhesion; reduces "suction" | Older types of water paint and wall-size; sealers |
| Casein | Milk protein | Skimmed milk | Adhesion; thickening | Gel and thickener for water paints, including latex; wallboard base coat |
| Dextrin | Starch protein | Grain and seeds; potatoes | Adhesion; stiffener | Sizing and coating for fabric, paper, wallboard |
| Wax | Beeswax, paraffin wax, carnauba wax, etc. | Natural waxes; petroleum | Resistant to moisture, spray, marring, light abrasion; high gloss | Widely used in protective polishes; small amounts added to decorative enamels and varnishes |

See also Tables 5.2—5 and 5.2—6.

From Bolz, R. E. and Tuve, G. L., Eds., *Handbook of Tables for Applied Engineering Science*, 2nd ed., CRC Press, Cleveland, 1973, 354.

## Table 5.2—5
## ORGANIC COATINGS – SYNTHETIC RESINS

Classes, Modifications, Properties, and Applications of Synthetic Resins for Paints and Coatings

| Class | Kinds and modifications | Properties or advantages | Usage |
|---|---|---|---|
| Acrylic | Emulsions and water-soluble polymers; thermoplastic or thermosetting, soluble in organic solvents | Thermoplastic; resistant to age, light, water, chemicals, oil; hardness; flexibility | Very widely used in both solution and emulsion; clear coatings; enamels; latex paints; automobile and metal finishes |
| Alkyd | Alkyd-oil; styrenated; modified; combined with other resins | Solubility; compatibility; durability; hardness; toughness; gloss retention; adhesion | Most widely used of all resins; air-dry or baked; paints, varnishes, enamels, primers |
| Amino | Urea and melamine; blends | Hardness; color retention; fast baking; durability | Durable finishes for automobiles, appliances, metal surfaces |
| Cellulose | Cellulose nitrate and others; also combinations with alkyd and amino resins | Flammable; compatibility with oils, plasticizers, and other resins | Quick-drying lacquers, enamels, dopes |
| Epoxy | Epoxy polyamide; esterified; amine catalyzed | Adhesion; toughness; chemical resistance; quick drying | Baking enamels; marine varnish; can coatings; chemical equipment |
| Fluorocarbon | Polytetrafluoroethylene; polychlorotrifluoroethylene | Resistant to moisture, heat, chemicals, fungi, and abrasion | Chemical-resistant linings for tanks and industrial equipment; electrical insulation; weatherproofing |
| Phenolic | Bakelite; modified oil-soluble | Resistant to chemicals and solvents; water- and weather-proof | Primer, sealer, varnish; structural and marine paints |
| Polyamide | Nylon; thermoplastic polyamides | Strength, toughness, abrasion resistance; oil and solvent resistance; good adhesion | Wear-resistant and chemical-resistant coatings for metal textiles, leather, paper, industrial equipment |
|  | Glycol esters of unsaturated and saturated dibasic acids, such as with phthalic acid isomers or maleic or fumaric acids. Unsaturated acid types are usually used as copolymers with styrene or acrylic monomers | No volatile thinners need to be expelled in forming thick or thin films that are hard and adhesive | High-build coatings with or without fillers for wood, concrete, and other non-metallic substrates |
| Polyethylene | Low- and high-density; chlorinated unsaturated | Odorless, tasteless, non-toxic; waterproof; high strength; toughness; flexibility | Coatings for food cartons; wire insulation; flame-sprayed coatings on metals; textile coatings |
| Polystyrene | Used largely in copolymers and latex | Flexibility; adhesion; toughness; weather resistance | With butadiene or alkyd for latex paints; solution-type outdoor paints |
| Polyurethane | Urethane oil or alkyd; moisture-cured and two-package compositions | Flexibility; high gloss; abrasion and chemical resistance; toughness; dielectric strength | Varnishes for severe service; enamels; marine paints; textile coatings; wire coatings; concrete finish |
| Rubbers | Neoprene; butyl; nitrile; SBR, and copolymers of isoprene, butadiene, polypropylene | Variety of properties including chemical resistance, resilience, abrasion resistance, oil resistance | Coatings where resilience, elasticity, and abrasion resistance are needed |
| Silicone | Copolymer or modified with metal powder or frit | Stable in the 200–500 deg C range with heat-resistant pigments | Heat-stable and high temperature paints; corrosion-resistant paints |
| Vinyl | Acetate, chloride; copolymers; plastisols | Toughness; flexibility; chemical resistance; wear resistance; dielectric strength | Masonry finish; textured coatings; outdoor uses |

From Bolz, R. E. and Tuve, G. L., Eds., *Handbook of Tables for Applied Engineering Science,* 2nd ed., CRC Press, Cleveland, 1973, 355.

## Table 5.2—6
## PIGMENTS FOR PAINTS AND COATINGS[a]

The term *pigments* was formerly applied to natural color materials such as those used by artists. Surface coatings today are increasingly dependent on the synthetic resins. The following table covers most of the common classes of paint solids other than the synthetic resins (Table 5.2—4, the natural resins and gums (Table 5.2—5), and the water paint powders such as lime and cement.

A paint mixture to be applied as a surface coating is often described in terms of two constituents — the pigment and the vehicle. The liquid vehicle wets the surface and dries or cures into a film, while the pigment, dispersed in the vehicle, functions largely as a radiation absorber and reflector, providing opacity or color. Common insoluble pigments are mixtures of very fine powders or flakes of metal oxides or salts and sometimes the metals themselves. Other constituents are added to the pigment or to the vehicle for special purposes such as controlling the gloss, hardness, adhesion, abrasion resistance, and weather resistance of the film or providing corrosion resistance or increasing the dielectric strength.

For latex paints, and for solution paints such as shellac or asphalt paint, the dual-constituent classification in terms of a film-forming vehicle and a surface-hiding pigment is not applicable. The same is true for such water paints as whitewash, and for many of the complex formulations that often include both dispersions and solutions and even emulsions.

### Common Paint Pigments

| Name and composition | Typical particle size, $\mu$m | Specific gravity | Refractive index | Hiding power | Properties and uses | ASTM No. |
|---|---|---|---|---|---|---|
| **WHITE PIGMENTS (OPAQUE)** | | | | | | |
| Titanium dioxide (anatase or rutile) | 0.25 | 3.5 | 2.7 | Excellent | Brilliant white; anatase form is chalking; non-toxic | D476 |
| White lead (basic carbonate, about 68%), $2PbCO_3 \cdot Pb(OH)_2$; flake white | | 6.7 | 2.0 | Good | Durable film; tends to darken; weather and water resistant; toxic; primer and undercoat | D81 |
| Zinc oxide, $ZnO$; Chinese white | 0.2–0.3 | 5.65 | 2.0 | Good | Mildew resistant; highly durable; outdoor oil paints | D79 |
| Zinc sulfide | | | 2.37 | Very good | Not widely used | D477 |
| Antimony oxide | 0.5–2.0 | 5.5 | 2.1 | Good | Little used except in fire-retardant paint | |
| Lithopone (regular), approx 70% $BaSO_4$ and 30% $ZnS$ | 0.2–2.0 | 4.2 | 1.9 | Fair | Interior oil and emulsion paints | D477 |
| **WHITE EXTENDER PIGMENTS (LOW REFRACTIVE INDEX)** | | | | | | |
| Calcium carbonate, $CaCO_3$; precipitated chalk; whiting | 2–5 | 2.7 | 1.58 | Poor | Ceiling white; undercoat | D1199 |
| Calcium sulfate, $CaSO_4 \cdot 2H_2O$; gypsum | | 2.35 | 1.53 | Poor | Filler; limited use in paints | |
| Magnesium silicate, $H_2Mg_3(SiO_3)_4$; talc | | 2.7 | 1.57 | Poor | Flatting; anti-settling; chemically inert | D605 |
| Barium sulfate, $BaSO_4$; barytes; barite | | 4.4 | 1.64 | Poor | Undercoats; fillers; chemically stable | D602 |
| China clay, $Al_2O_3 \cdot 2SiO_2 \cdot 2H_2O$; kaolin | 1.0 (avg) | 2.5 | 1.56 | Poor | Undercoats; thickening agent | |
| Hydrated silica | 1–10 | 2.2 | | | Flatting; consistency control | D604 |

## Table 5.2—6 (continued)
## PIGMENTS FOR PAINTS AND COATINGS

| Name and composition | Typical particle size, $\mu$m | Specific gravity | Refractive index | Hiding power | Properties and uses | ASTM No. |
|---|---|---|---|---|---|---|
| **BLACK, GRAY, BROWN PIGMENTS** | | | | | | |
| Carbon black; furnace black; impingement black | .02–.09 | | Opaque | Good | Very widely used; jet-black gloss | |
| Lampblack, C; oil black | | 1.8 | Opaque | Good | Undercoats; primers; high adhesion; blue undertone; matte | D209 |
| Graphite (amorphous) | | 2.3 | Opaque | Good | Topcoat for steel structures; high coverage; durable | |
| Iron oxide black, $Fe_3O_4$; magnetite (synthetic) | 0.5 | | Opaque | Good | Metal paints and fillers | D769 |
| Bone black; ivory drop black | 325 mesh (solution) | 2.3 | Opaque | Good | Black undercoat and filler | D210 |
| Asphaltum; cut-back asphalt; petroleum paint (black to brown) | | | Opaque | Fair | High adhesion; automotive uses; roof paint; chemical and water resistance; also emulsions | |
| Blue lead (45% lead sulfate, 30% lead oxide); basic lead sulfate, sublimed | | | Opaque | Good | Structural steel primer (gray) | D405 |
| Sienna and umber, largely $Fe_2O_3$; range of brown and gray (to red synthetics) | | 3.3 | 1.9 | Fair | Widely used tinting colors; low cost | |
| **COLORS** | | | | | | |
| Prussian blue (ferrocyanides); iron blue | | 1.8 | 1.55 | Fair | Low cost; non-bleeding; high-temperature baking; light tints may fade | D261 |
| Cobalt blue (cobalt and aluminum oxides) | | 3.8 | 1.7 | Fair | Expensive; art finishes; highly stable and resistant | |
| Cobalt green (cobalt and zinc oxides) | | 4.0 | 1.95 | Fair | Highly stable and resistant; not widely used | |
| Chrome green (yellow $PbCrO_4$ with $PbSO_4$ blended with iron blue) | | 5.1 | 2.5 | Excellent | Widely used; wide range of greens; good color retention | D263; D212 |
| Red lead (synthetic), $Pb_3O_4$ >85% | 2.0 | 8.7 | 2.4 | Excellent | Durable; good adhesion; corrosion protection; primer for steel; fades in sun | D83; D49 |
| Iron oxide, $Fe_2O_3$; red hematite; red to maroon and brown | | 5.2 | 2.5 | Excellent | Inexpensive; widely used | |
| Cadmium red, largely CdS | | 4.5 | 2.7 | Excellent | Bright tones; non-bleeding high-temperature bake | |
| Manganese violet | | | 1.7 | Good | | |
| Chrome yellow and orange, largely $PbCrO_4$ (with $PbSO_4$ light and PbO orange) | | 5.9 | 2.3 | Excellent | Bright tints | D211 |
| Zinc-yellow, ZnO and $CrO_3$; zinc chromate | | 3.5 | 1.9 | Good | Exterior paints; resists darkening | D478 |
| Yellow ochre, $Fe_2O_3$; ferrite yellow; natural or synthetic | 0.5–1.5 | 3.5 | 2.0 | Good | Inexpensive; low-temperature bake | D85 |
| Cadmium yellow, largely CdS with CdSe; also lithopone with $BaSO_4$ | | 4.3 | 2.4 | Excellent | Interior oil and water paints | |

**Table 5.2—6 (continued)**
**PIGMENTS FOR PAINTS AND COATINGS**

| Name and composition | Typical particle size, $\mu$m | Specific gravity | Refractive index | Hiding power | Properties and uses | ASTM No. |
|---|---|---|---|---|---|---|
| ORGANIC TONERS | 0.01–0.1 | Variable | Varies with wavelength | Poor to excellent | Cover spectral range from red to violet; some with hiding power are used in mass tone or solid colors; most with low hiding power give vivid colors and are used in mixtures with titanium dioxide or aluminum flake | |
| METAL PIGMENTS | | | | | | |
| Aluminum (leafing—flakes overlap) | | | Opaque | Excellent | Brilliant metallic finish; durable | D962 |
| Aluminum (non-leafing) | 100–200 mesh | | Opaque | Good | Less brilliant; widely used; chemical and heat resistance; durable | D962 |
| Aluminum (extra fine) | 400 mesh | | Opaque | Excellent | High hiding power; durable | D962 |
| Zinc dust | 325 mesh | | Opaque | Good | Rust-inhibitive primer for steel and galvanized iron; weatherproof paints | D520 |
| Copper bronze, 2% zinc | | | Opaque | Good | Decorative copper finish; fungicidal paint | |
| Gold bronze, Cu, Zn, and Al | | | Opaque | Good | Decorative color range, pale gold to red gold; exterior and interior | D267 |
| FLUORESCENT PIGMENTS | | | | | | |
| Organic-dyed resins (ultraviolet reflectors) | | | Variable | Fair | Maximum visibility in yellow-orange range; high-visibility coatings; safety paints | |

[a]   This table has been compiled from several sources. While it includes many common pigments, it is in no sense a complete list. The synthetic organic pigments are entirely omitted, and many common mixtures and blends are not included. A more complete treatment should also give attention to the science of color, to phenomena of spectral absorption and reflectance, and to the various methods for specifying, matching, and measuring colors (see References, Table 5.2—1.

From Bolz, R. E. and Tuve, G. L., Eds., *Handbook of Tables for Applied Engineering Science,* 2nd ed., CRC Press, Cleveland, 1973, 356.

## Table 5.2—7
## PAINTS AND COATINGS FOR CORROSION RESISTANCE

*Corrosion* is a very general term applied to a variety of processes by which metal surfaces are attacked and converted to oxides, sulfides, or other compounds. As much corrosion is electrochemical, methods of protection are based on isolation of the metal from the environment that contains oxygen, moisture, sulfur, etc., and on minimizing galvanic potentials. The methods by which these steps are accomplished are so diverse that no single summary can adequately cover them.

Galvanic corrosion is prevented by elimination of contact between metals, whether direct or through an electrolyte. A most common case is

copper and iron. Surface treatment of steel with a phosphate wash or a primer such as red lead or zinc chrome may form a thin protective layer of stable compound, over which an impervious and durable coating will adhere.

The following table lists many of the common protective coatings, most of which are used on iron and steel, since they present the corrosion problem of greatest economic importance. Asphalt and red lead have long been widely used, but there is an increasing use of protective metals and of synthetic resins. For detailed discussions consult the references listed in Table 5.2—1.

| Name or protective constituent | Description or composition | Advantages | Typical application |
|---|---|---|---|
| Aluminum | Leafing or non-leafing paints, varnishes and primers (hot dipping) | High hiding; fume-resistant; low cost; high coverage; ultraviolet protection | One-coat protection for wide use; additive for asphalt and phenolic resin paints |
| Asphalt | Petroleum residue and solvents; with asbestos, vermiculite, and perlite | Low cost; moisture and chemical resistance; good adhesion | Thinned or hot; pipe lines, tanks, undercoat; buried structures |
| Barium potassium chromate | With barium and chromium oxides | High-strength film; elastic and durable | For steel or light metals; air drying or baking types |
| Calcium plumbate | Calcium and lead oxides | Quick-drying; smooth; hard; salt-resistant | Primers for structural and galvanized steel |
| Carbon black | Furnace black or amorphous graphite | High hiding; low cost; high coverage; high chemical resistance | With black iron oxide for shop and foundry paint; topcoat varnish |
| Chlorinated rubber | 10–25% in oil or alkyd paints or lacquers | Chemical resistance; low permeability; wide color range | Marine paint; machinery finish; high-build paints; chemical equipment |
| Etch primer | Phosphoric acid or phosphate; often with zinc chromate | Versatile water-solution wash as quick-dry primer | Clean-metal primers for good overcoat adhesion; various metals, including galvanized |
| Lacquer | Resin lacquers, as nitrocellulose and acrylic | Transparent or colors; quick drying; cold or hot application | For highly finished metal surfaces, including brass |
| Lead cyanamide | Yellow $Pb(CN)_2$ | Anodic protection | Primer for steel; with linseed oil and pigments |
| Pitch, coal-tar | Coal tar in aromatic solvent | Black, brown, green; glossy; hard; adhesive; chemical resistance | Emulsion topcoat; with epoxy or other resin for sea-water immersion |
| Phenolic resin | Air-dry varnish and baking lacquer | Very low permeability; good baking enamel; good adhesion | Chemically resistant paints; electric insulation varnish; tank linings |
| Red lead | $Pb_3O_4$ and $PbO$ with linseed oil | Readily available; weather-resistant; good adhesion | Most widely used where linseed oil is common; outdoor ferrous structures |
| Red iron oxide | 30–50% in oil or chlorinated rubber paints | Low cost; one or two coats without primer; high-hiding; durable | Ferrous metal primer or one-coat protection |

## Table 5.2—7 (continued)
### PAINTS AND COATINGS FOR CORROSION RESISTANCE

| Name or protective constituent | Description or composition | Advantages | Typical applications |
|---|---|---|---|
| Stainless steel flake | Barrier coat | Inherently corrosion resistant; good appearance | Metallic barrier for chemical resistance |
| Zinc chrome | 10% or more in non-ferrous primers; with red iron oxide for ferrous metals | High-hiding, sealing, and adhesion; salt-resistant | On light metals; with red iron oxide on ferrous metals; alkyd paints |

*Notes:* *Surface preparation* is very important prior to coating treatment. Solvent degreasing, alkali washing, sand or shot blasting, pickling, acid or phosphate dipping, flame treatment, wire brushing, and abrasive cleaning are among the methods that should receive consideration for surface preparation.

*Dual protective films* and even complex mixtures are often used for final-coat or one-coat protection. Examples are pigment with drying oil, pigment and alkyd or phenolic vehicle, powdered or leafing metal with polyvinyl chloride or other resin, pigment and chlorinated rubber, and pigment mixtures with both chlorinated rubber and drying oil. In each case there is at least a dual barrier coat, sometimes over a primer.

*High pigment content* (40–90%) is characteristic of most protective paints, since these solids act as barrier coats.

*Synthetic resins* are now being used in many corrosion-protective coatings. Alkyd, PVC, epoxy, vinyl, and certain copolymers are common in paints with various protective pigments.

### REFERENCE

Burns, R. M. and Bradley, W. W., *Protective Coatings for Metals,* 3rd ed., Van Nostrand Reinhold, New York, 1967.

From Bolz, R. E. and Tuve, G. L., Eds., *Handbook of Tables for Applied Engineering Science,* 2nd ed., CRC Press, Cleveland, 1973, 359.

## Table 5.2—8
### INORGANIC SURFACE COATINGS[a]

Functions, Materials, and Application Methods

**Functions** – Inorganic coatings are applied to metals and to some other materials to protect and to *modify* the surface properties, as follows:

Mechanical: for resistance to abrasion, erosion, shock, and for control of hardness, texture, smoothness, lubrication.

Chemical: for resistance to oxidation, corrosion, chemical reaction, diffusion.

Thermal: for heat insulation, ablation, control of heat transfer by conduction and by radiation.

Electrical: for control of electrical resistance, magnetic properties, thermionic performance.

Radiation: for control of light and radiative properties including reflection, absorption, diffusion or transmission of radiation (total or spectral); for optical imaging by phosphor.

**Materials** – Coating materials are used singly and in numerous combinations. Most coatings are chemical compounds of the metallic elements.

Oxides of aluminum, silicon, calcium, magnesium, zirconium, chromium, beryllium, thorium, hafnium, nickel, etc.

Other oxygen compounds: the aluminates, silicates, chromates, zirconates.

Hard metal compounds: the carbides, nitrides, borides, silicides.

Intermetallics: the aluminides, beryllides, stannides.

Combinations of materials: classified under such common class names as enamels, glasses, ceramics, cermets, refractories, composites.

**Application methods** – These may include more than one of the following steps or processes:

**Table 5.2—8 (continued)**
**INORGANIC SURFACE COATINGS**

| | | |
|---|---|---|
| Adhesion | Flame spraying | Spreading |
| Atomized spraying | Fusion or firing | Trowelling |
| Dipping (hot or cold) | Packed retorts | Vapor deposition (pyrolysis) |
| Electrophoresis | Painting; brushing | |
| Electroplating | Slip or slurry (fired) | |

**Examples** – These might include a very long list and a wide variety of uses, but those listed in the following table are illustrative.

### Common Inorganic Coatings

| Coating name or function | Base or substrate | Coating description, processes, and typical compositions | Typical uses |
|---|---|---|---|
| Enamel (porcelain) | Steel | Surface coated with slip or slurry of powdered glasses and colors. Oxides of Si, Al, Na, B, and Ca predominate | Cookware, signs, tanks, tubs, pails, housewares |
| Enamel (vitreous) | Cast iron | Similar to above | Sanitary ware, structural decoration |
| Enamel | Aluminum | Low-melting frit (below 1050 deg F) containing PbO, $Li_2O$, SrO, $TiO_2$, etc. Aluminum alloy must be enameling grade | Architectural trim, siding, wall coverings; highway and advertising signs |
| Anodized | Aluminum | Anodic oxidation and coloring. Processes use phosphoric, sulfuric, or chromic acid | Architectural and vehicle trim; housewares; aircraft materials; handrails |
| Reflector | Glass, metal, or crystal | Mirror metals or paints for control of light or electromagnetic radiation | Solar reflection, optical instruments, thermal radiation control |
| Photo converter | Metal or glass | Amplifying, photoemissive, and photovoltaic surfaces with compounds of Pb, Cd, Ge, and Cs | Solar cells; photomultipliers, amplifiers, detection, and measurement |
| Phosphor | Glass | Optical imaging, using zinc and cadmium sulfides, silicates | Cathode-ray and television screens; instrument displays, illumination tubes; particle detectors |
| Ablation coating | Metal | Wound or woven glass fibers with phenolic resin binder | Rockets, nosecones, re-entry vehicles |
| Lubricant | Metal pair | Lubricating powders and mixtures, usually oxides of Pb, Bi, Cd, or disulfides of Mo or W | High-temperature and space applications; journal and ball bearings; extrusion dies |
| Electrical insulation | Metal wire or surface | Enameled conductors baked from slurry-covered surface; thermionic layers | Large coils and magnets; structural separators |
| Diffusion coating | Metals | Reactive coating material diffused into substrate by heat treatment. Compounds of Al, Cr, Si, and Ti most common | Oxidation-resistant or hard-surface layers on metals and alloys |
| Ceramics and cermets | Metals and alloys | Sprayed ceramic mixtures and electrodeposited particles from stirred aqueous bath | High-temperature applications with short-term erosion |
| Fire retardants | Wood, fiber, or plastic | Compounds of P, Cl, Sb, Br, and B | Fireproofing of combustible surfaces |
| Thermal insulation | Metals | Thick, porous refractory and fiber coatings and bonded ceramic foams; refractory cements | High-temperature protection for structural materials |

ª   For data on organic coatings, see Tables 5.2—1 through 5.2—7.

## Table 5.2—8 (continued)
### INORGANIC SURFACE COATINGS

#### REFERENCES

**Plunkett, J. D.,** *NASA Contributions to the Technology of Inorganic Coatings,* NASA SP-5014, National Aeronautics and Space Administration, 1964.

**Huminik, J.,** *High-temperature Inorganic Coatings,* Van Nostrand Reinhold, New York, 1963.

**Campbell, I. E. and Sherwood, E. M., Eds.,** *High-temperature Materials and Technology,* John Wiley & Sons, New York, 1967.

**Van Horn, K. R., Ed.,** *Aluminum, Fabrication and Finishing,* Vol. 3, American Society for Metals, Metals Park, O., 1967.

**Andrews, A. I.,** *Porcelain Enamels,* Garrard, Champaigne, Ill., 1961.

From Bolz, R. E. and Tuve, G. L., Eds., *Handbook of Tables for Applied Engineering Science,* 2nd ed., CRC Press, Cleveland, 1973, 360.

## Table 5.2—9
### SUMMARY OF ELECTROPLATING PRACTICE

### Average Operating Conditions

| Metal | Principal uses | Type of solution | Principal ingredients | Temp, °F | CD ASF | Volts | Cathode efficiency, % | Time to deposit, 0.001 in. |
|---|---|---|---|---|---|---|---|---|
| Cadmium | Protection | Cyanide | CdO, NaCN, brighteners | 70–95 | 15–45 | 1–4 | 90 | 20 min |
| Chromium | Decorative Engineering (hard) Cylinder liners (porous) | Chromic acid | $CrO_3$, $H_2SO_4$ | 120 | 250 | 6–8 | 15 | 2 hr |
| Copper | Electroforming | Acid | $CuSO_4 \cdot 5H_2O$, $H_2SO_4$ | 75–120 | 15–40 | 1–2 | 100 | 35 min |
| | Undercoat for other metals | Cyanide | CuCN, NaCN, $Na_2CO_3$ | 75–100 | 5–15 | 1.5–3 | 50 | 90 min |
| | Stop-off in case-hardening, etc. | Rochelle | Above + rochelle salts | 140–160 | 20–60 | 2–3 | 60 | 45 min |
| | | Many other types, *e.g.*, fluoborate, pyrophosphate, amine, all-potassium cyanide | | — | — | — | — | — |
| Gold | Decorative | Cyanide | $KAu(CN)_2$, $K_2CO_3$, KCN (Solutions vary considerably, depending on color wanted) | 120–160 | 5–15 | 2–6 | 80 | — |
| Indium | Bearing surfaces | Cyanide | $InCl_3$, NaCN, addition agent | Room | 10–150 | — | 40 | — |
| | | Sulfate | $In_2(SO_4)_3$, $Na_2SO_4$ | Room | 20 | — | 75 | — |
| | | Fluoborate | $In(BF_4)_3$, $H_3BO_3$, $NH_4BF_4$ | 70–90 | 50–100 | — | 50 | — |
| Iron | Electroforming | Chloride | $FeCl_2$, $CaCl_2$ | 190 | 60 | — | 95 | 20 min |
| | Repair | Sulfate | $FeSO_4(NH_4)_2SO_4$ | Room | 20 | — | 95 | 1 hr |
| Lead | Protection | Fluoborate | $Pb(BF_4)_2$, $HBF_4$, glue | Room | 10–80 | 0.5 | 100 | 40 min |
| | Bearing surfaces | Sulfamate | Pb sulfamate, sulfamic acid, addition agents | 75–120 | 5–40 | 3–8 | 100 | 20 min |
| Nickel | Protection Decorative Electroforming Undercoat for Cr, etc. | Sulfate-chloride | $NiCl_2$, $NiSO_4$, $NH_4$ ion, $H_3BO_3$ (Formulations differ widely, depending on purpose) | 75–100 | Varies greatly | 0.5–3 | 95 | 30 min |
| Rhodium | Decorative Optical | Sulfate Phosphate | Prepared salts | 110–120 | 10–80 | 2.5–5 | 15 | -- |
| Silver | Decorative Protective Bearing surfaces | Cyanide | AgCN, KCN, $K_2CO_3$, $CS_2$ (Or Na in place of K) | 80 | 5–15 | 1 | 100 | — |
| Tin | Protection Food and dairy Bearings Electrical To enable easy soldering | Sulfate | $SnSO_4$, $H_2SO_4$, addition agents | Room | 40 | 1–3 | 90 | 15 min |
| | | Fluoborate | $Sn(BF_4)_2$, $HBF_4$, addition agents | 75–100 | 50 | — | 100 | 10 min |
| | | Other acid electrolytes | | — | — | — | — | — |
| | | Stannate | $Na_2$- or $K_2Sn(OH)_6$, Na- or KOH | 150–190 | 40 | 4–8 | 80 | 30 min |
| Zinc | Protection | Sulfate | $ZnSO_4$, $NH_4Cl$, addition agents | 75–100 | 15–400 | — | 99 | 10 min |
| | | Cyanide | $Zn(CN)_2$, NaCN, NaOH, brighteners | 100 | 10–50 | — | 85 | 40 min |

## Table 5.2—9 (continued)
## SUMMARY OF ELECTROPLATING PRACTICE

### ALLOYS

| Metal | Principal uses | Type of solution | Principal ingredients | Temp, °F | CD ASF | Volts | Cathode efficiency, % | Time to deposit, 0.001 in. |
|-------|---------------|-----------------|----------------------|----------|--------|-------|----------------------|---------------------------|
| Brass | Rubber-bonding Decorative | Cyanide | $CuCN$, $Zn(CN)_2$, $NaCN$, $Na_2CO_3$ | 75–100 | 3–10 | 2–3 | 75 | — |
| Bronze | Decorative Undercoat for chromium Stop-off for steel | Cyanide-stannate | $CuCN$, $KCN$, $KOH$, $K_2Sn(OH)_6$, rochelle salt | 155 | 20–100 | 3–6 | 70 | 30 min |
|  |  | Pyrophosphate-cyanide | $Sn_2P_2O_7$, $KCN$, $CuCN$, $K_4P_2O_7$, addition agents | 140–180 | 20–70 | 2–5 | 70 | 30 min |
| Lead-tin | Bearings Solderability Electrotyping | Fluoborate | $Sn(BF_4)_2$, $Pb(BF_4)_2$, $HBF_4$, addition agents | Room | 60 | 1–2 | 100 | — |
| Tin-zinc | Solderability | Cyanide-stannate | $Zn(CN)_2$, $KCN$, $KOH$, $K_2Sn(OH)_6$ | 150 | 10–75 | 4–5 | 80–95 | 30 min |

### REFERENCES

Gray, A., Ed., *Modern Electroplating,* John Wiley & Sons, New York, 1953.

*Plating and Finishing Guidebook-Directory,* Finishing Publications (published yearly).

Bandes, H., *Trans. Electrochem. Soc.,* 88, 263, 1945.

Technical Data Sheet No. 127, Metal and Thermit Corp., New York, 1954.

Safranek, W. H. and Faust, C. L., *Proc. Am. Electroplat. Soc.,* 41, 201, 1954.

From Knowlton, A. E., Ed., *Standard Handbook for Electrical Engineers,* 9th ed., McGraw-Hill, New York, copyright © 1957. With permission.

## 5.3     TEXTILES

### Table 5.3—1
### CLASSIFICATION OF NATURAL FIBERS

| Name | Source | Principal producers | Approximate production, $10^6$ lb/year | Typical uses |
|---|---|---|---|---|
| **ANIMAL ORIGIN** | | | | |
| Wool (pure, dry) | Sheep | Australia, Argentina | 3 000 | Warm clothing, carpet, felt |
| Silk | Silkworm cater-pillar | Japan, China | 70 | Fine fabrics |
| Cashmere | Goat | India, Tibet | 70 | Quality clothes |
| Mohair | Goat | U.S.A., Turkey | 30 | Upholstery, linings, suitings, rugs |
| Camel hair | Camel | China, Mongolia | 2 | Overcoats, knits |
| **VEGETABLE ORIGIN** | | | | |
| Cotton | Seed | U.S.A., U.S.S.R. | 25 000 | Almost all textile uses |
| Jute | Bast | India, Pakistan | 4 500 | Sacking, bale wrapping, curtains, bags, oakum |
| Sisal | Leaf | Mexico, Brazil | 1 300 | Rope, twine, rugs |
| Flax | Bast | U.S.S.R., Poland | 1 200 | Strong fabrics, paper |
| Kenaf | Bast | India, Pakistan | 1 100 | Rope, twine, carpet, canvas, bags |
| Hemp | Bast | U.S.S.R., Yugoslavia | 500 | Rope, sacking, canvas |
| Henequen | Leaf | Mexico, Cuba | 300 | Rope, twine, canvas, bags |
| Abaca (Manila) | Leaf | Phillipines, Guatemala | 200 | Rope, marine cable |
| Sunn | Bast | India, Pakistan | 150 | Rope, twine, carpet, paper |
| Ramie | Bast | China, Japan | 25 | Canvas |
| **MINERAL ORIGIN** | | | | |
| Asbestos | Ore | Canada, U.S.S.R. | 2 000 | Brakes and clutches, building materials, packings, fire-proofing |
| Glass[a] | Sand | U.S.A. | — | Composites, insulation, draperies, tire cord, filters |
| Aluminum silicate[a] | Ore | U.S.A. | — | Packings and insulation for high temperatures |

[a]Here classified as natural fibers for convenience, although they are man-made by processing.

### REFERENCE

**Mauersberger, H. R., Ed.,** *Matthews' Textile Fibers,* John Wiley & Sons, New York, 1954.

From Bolz, R. E. and Tuve, G. L., Eds., *Handbook of Tables for Applied Engineering Science,* 2nd ed., CRC Press, Cleveland, 1973, 171.

## Table 5.3—2
## PROPERTIES OF NATURAL FIBERS

Because there are great variations within a given fiber class, average properties may be misleading. The following typical values are only a rough comparative guide.

| Name | Specific gravity | Tenacity, g/denier | Tensile strength, $10^3$psi | Elongation at break (dry), % | Standard regain, % of dry[b] | Fiber diameter, microns | Fiber length, inches | Fiber shape and kind | Resistant to |
|---|---|---|---|---|---|---|---|---|---|
| **ANIMAL ORIGIN** | | | | | | | | | |
| Wool | 1.32 | 1.0–1.7 | 17–29 | 23–35 | 15–18 | 17–40 | 1.5–5 | Oval, crimped, scales | Age, weak acids, solvents |
| Silk | 1.25 | 3.5–5 | 90 | 20–25 | 10 | 10–13 | | Flexible, soft, smooth | Heat, solvents, weak acids, wear |
| Cashmere | | | | | | 15–16 | 1–4 | Round, scales, soft | |
| Mohair | 1.32 | 1.2–1.5 | | 30 | 13 | 24–50 | 6–12 | Round, silky | Wear, age, solvents, weak acids |
| Camel hair | 1.32 | 1.8 | | 40 | 13 | 10–40 | 1–6 | Oval, striated | Age, solvents |
| **VEGETABLE ORIGIN** | | | | | | | | | |
| Cotton | 1.54 | 2–5 | 30–120 | 5–11 | 7.5–8.5 | 10–20 | 0.5–2 | Flat, convoluted, ribbon | Age, heat, washing, wear, solvents, alkalies, insects |
| Jute (bast) | 1.5 | | 50 | 1–1.5 | 14 | 15–20 | | Woody, rough, polygon | |
| Sisal (leaf) | 1.49 | 2.2 | 75 | 2–2.5 | 13 | 10–30 | Strand 30–40 | Stiff, straight | |
| Flax (bast) | 1.52 | 4–7 | | 2–3 | 12 | 15–18 | Strand 40–50 | Soft, fine | Age, solvents, washing, insects, weak acids, and alkalies |
| Kenaf (bast) | | | 45 | | | 15–30 | | Polygon or oval | |
| Hemp (bast) | 1.48 | | | 2 | | 18–25 | Strand 30–70 | Polygon or oval, irregular | |
| Henequen (leaf) | | | 60 | | | | Strand 30–60 | Finer than sisal | |
| Abaca (leaf) (Manila) | 1.48 | 2.3–2.9 | 100 | 2–3 | 13 | | Strand 30–120 | | |
| **MINERAL ORIGIN** | | | | | | | | | |
| Asbestos | 2.5 | | 40–200 | | | Various | 0.5–10 | Smooth, straight | Heat to 400 deg C, acids, chemicals, organisms |
| Glass[a] | 2.5 | 7–12 | 200–500 | 3–4.5 | 0 | Various | | Circular, smooth | Chemicals, insects |
| Silicate[a] (Ca, Al, Mg) | 2.85 | | | | 0 | | | | Heat to 900 deg C, most chemicals, insects, rot |

[a]Here classified as natural fibers for convenience, although they are man-made by processing.
[b]Expected equilibrium moisture regain of dry fiber, in percent of dry weight, when exposed in air at 70 deg F, 65% relative humidity.

*Note*: Wide variations may be expected, especially for different grades of cotton. Wet strength is lower (for rayon, very much lower), but it depends on the duration of soaking. The strength of yarn is only a fraction of the cumulative strength of all individual fibers.

Most fibers exhibit relaxation of stress at constant strain and also increase in elongation at constant load (creep). The stress-strain curve is greatly affected by the rate of extension. When the stress is removed, there is a quick elastic recovery, a delayed recovery, and a permanent set. Hence the elastic behavior of any fiber depends on its

**Table 5.3—2 (continued)**
**PROPERTIES OF NATURAL FIBERS**

stress-strain history. The elastic recoveries of nylon and wool are high; those of cotton, flax, and rayon are much lower.

The heat capacity (specific heat) of most fibers is about one third that of water.

Other fibers: Fur hair is slightly coarser than silk fibers. Camel and llama hairs are almost as coarse as wool but only about one third the size of human hair. Horse hair is over 100 $\mu$m; hog bristles, over 200 $\mu$m. Jute, sisal, and hemp are intermediate between cotton and wool. These are rough average sizes, and many natural fibers range 50% above or below such averages.

From Bolz, R. E. and Tuve, G. L., Eds., *Handbook of Tables for Applied Engineering Science,* 2nd ed., CRC Press, Cleveland, 1973, 172.

## Table 5.3—3
## CLASSIFICATION OF MAN-MADE FIBERS AND FABRICS

Key to U.S. Manufacturers:

A – American Viscose Corp.
B – Beaunit Fibers, Div. of Beaunit Corp.
C – Celanese Fibers Co., Div. of Celanese Corp.
E – Tennessee Eastman Co., Div. of Eastman Kodak Co.
F – Fiber Industries, Inc.

G – W. R. Grace & Co., Dawbarn Div.
H – Hercules Powder Co.
K – American Enka Corp.
M – Monsanto Co., Textiles Div. (Chemstrand)
N – Vectra Co., Div. of National Plastics Products Co., Inc.

P – E. I. du Pont de Nemours & Co., Inc.
S – Firestone Synthetic Fibers Co.
U – Uniroyal, Inc., Textile Div.
UC – Union Carbide Corp.
V – FMC Corp, American Viscose Div.

| Chemical class; common name (sources) | Typical proprietary names (and manufacturer) | Resistant to | Damaged by[a] | Typical uses |
|---|---|---|---|---|
| **Cellulose fibers (natural)** | | | | |
| Acetate | Esteron®, etc. (E) Celacloud®, etc. (C) | Petroleum chemicals, dilute acids, weak alkalies, mildew, moths | Oxidizers, many solvents, strong alkalies, heat (above 140°C) | Clothing, satins, drapes, linings, knits |
| Triacetate | Arnel® (C) | Petroleum solvents, bleaches, insects, mildew, heat | Strong acids, strong alkalies, most solvents | Pleated garments, knits, drip-dry wear, table covers |
| Viscose rayon | Enka®, etc. (K) Avisco®, etc. (A) | Dilute alkalies, insects, solvents | Acids, strong alkalies, heat (above 150°C), moisture, mildew | Clothing, carpets, curtains, upholstery, linings |
| High-tenacity viscose | Tenasco® (V) | Moisture, solvents | Strong acids, strong alkalies, heat (above 150°C) | Tire cord, belting, hose |
| Polynosic viscose | Avril® (A) | Moisture, solvents, insects | Strong acids, strong alkalies, heat (above 150°C) | Dress fabrics, knits, drapes |
| Cuprammonium rayon (cupro) | Bemberg®, etc. (B) | Bleaches, weak alkalies | Strong oxidizers, mildews, some insects, strong acids, heat (above 230°C) | Sheer fabrics, drapes, upholstery, satins |
| **Protein fibers (natural)** | | | | |
| Animal: casein (milk) | Now little used | Solvents | Strong acids, alkalies, mildew, heat (above 100°C) | Largely for blending with wool, cotton, rayon, etc. |
| Vegetable–seed: soybeans, peanuts, corn | Now little used | Acids, solvents, moths, mildew | Alkalies, heat above 150°C | Blends with wool, cotton, etc. |
| Vegetable–latex: rubber (vulcanized) | Lastex® (U) | Moisture, insects | Heat (above 110°C), oxidation, ozone, oils, hydrocarbons, fats, solvents | Corsetry, swimwear, footwear, supports |

## Table 5.3—3 (continued)
## CLASSIFICATION OF MAN-MADE FIBERS AND FABRICS

| Chemical class; common name (sources) | Typical proprietary names (and manufacturer) | Resistant to | Damaged by[a] | Typical uses |
|---|---|---|---|---|
| **Synthetic fibers** | | | | |
| Polyacrylonitrile (acrylic) | Orlon® (P), Acrilan® (M), Creslan® (AC), Cantrece® (P), Verel®-copolymer (E), Dynel®-copolymer (P) | Dilute acids and alkalies, solvents, insects, mildew, weather | Strong alkalies and acids, heat (above 180°C), acetone, ketones | Outdoor fabrics, carpets, knits, furlike fabrics, blankets |
| Polyamide | Nylon[b] (P) 6, 6.6, etc. Chemstrand nylon® (M) | Alkalies, molds, solvents, moths | Strong acids, phenol, bleaches, heat (above 170°C) | Tire cord, carpet, upholstery, apparel, belting, hose, tents |
| Polyester | Dacron® (P), Kodel® (E), Fortrel (F) | Weak acids and alkalies, solvents, oils, mildew, moths | Phenol; heat (above 200°C) | Apparel, curtains, rope, twine, sailcloth, belting, fiberfill |
| Polyethylene (olefin, low density) | DLP® (G) | Alkalies, acids (except nitric), insects, mildew | Oil and grease; heat (above 90°C) oxidizers | Outdoor fabrics; filter fabrics; decorative coverings |
| Polyethylene (olefin, high density) | DLP (G) | Alkalies, acids (except nitric), insects, mildew | Oil and grease; heat (above 100°C) oxidizers | Rope, twine, fishnets |
| Polypropylene (olefin) | Herculon® (H), Polycrest® (U) | Alkalies, acids, solvents, insects, mildew | Heat (above 110°C) | Rope, twine, outdoor fabrics, carpets, upholstery |
| Polyurethane; spandex[b] | Lycra® (P), Spandelle® (S) | Solvents, oils, alkalies, insects, oxidation | Heat (above 140°C); strong acids | Elastic garments, swimwear, hosiery, tricot fabrics, knits |
| Polyvinyl chloride (PVC) | Vinyon[b] HH (V), Dynel®-copolymer (U) | Acids and alkalies, insects, mildew, alcohol, oils | Ethers, esters, aromatic hydrocarbons, ketones; hot acids; heat (above 70°C) | Nonwoven materials; felts; filters; blends with other fibers |
| Polyvinyl alcohol (PVA) | (Foreign manufacture) | Acids, alkalies, insects, mildew, oils | Heat (above 160°C); phenol, cresol, formic acid | Wide range of industrial and apparel uses; rope, work clothes; fish nets |
| Polyvinylidene chloride | Saran[b] by Vectra (N) | Acids, most alkalies, alcohol, bleaches, insects, mildew, weather | Heat (above 90°C); many solvents | Outdoor fabrics; insect screen; curtains, upholstery; carpet; work clothes |
| Polytetrafluoroethylene (PTFE) | Teflon® (P) | Almost all chemicals, solvents, insects, mildew | Heat (above 250°C); fluorine at high temperature | Corrosion-resistant packings, etc., tapes, filters, bearings |

## Table 5.3—3 (continued)
## CLASSIFICATION OF MAN-MADE FIBERS AND FABRICS

aFabrics are often damaged by heat in ironing. The synthetics will not withstand the usual 400°F that is used for cotton and linen. For acetate and olefins the ironing temperature should be below 250°F, as for silk and wool. For most other synthetics a 300°F ironing temperature is recommended, but triacetate will tolerate higher temperature without damage.

bThe names *nylon, vinyon, saran,* and *spandex* are recognized as generic terms rather than proprietary names.

From Bolz, R. E. and Tuve, G. L., Eds., *Handbook of Tables for Applied Engineering Science*, 2nd ed., CRC Press, Cleveland, 1973, 173.

## Table 5.3—4
## PROPERTIES OF MAN-MADE FIBERS[a]

| Chemical class; common name (sources) | Specific gravity | Tenacity, g/denier | Tensile strength, 10³ psi | Elonga-tion at break, % | Regain (standard) | Soften-ing point, °C | Melting point, °C | Flamma-bility | Brittleness temp, °C |
|---|---|---|---|---|---|---|---|---|---|
| **CELLULOSE FIBERS (NATURAL)** | | | | | | | | | |
| Acetate | 1.30 | 1.–1.3 | 18–25 | 20–30 | 6.5 | 140 | 230 | Melts and burns | |
| Triacetate | 1.32 | 1.2–1.4 | 20–28 | 25–30 | 3–4.5 | 225 | 300 | Melts and burns | |
| Viscose rayon | 1.51 | 2–2.6 | 30–46 | 17–25 | 13. | | 200[b] | Burns readily | |
| High-tenacity viscose | 1.53 | 3–5 | 60–80 | 10–12 | 10 | | 200[b] | Burns readily | <–114 |
| Polynosic viscose | 1.53 | 3–5 | 60–80 | 8–20 | 7 | | 200[b] | Burns readily | |
| Cuprammonium rayon (cupro) | 11.52 | 1.7–2.3 | 30–45 | 10–17 | 12.5 | | 250[b] | Burns readily | |
| **PROTEIN FIBERS (NATURAL)** | | | | | | | | | |
| Animal: casein (milk) | 1.3 | 1.0 | 15 | 60–70 | 14 | 100 | 150 | Slow | |
| Vegetable—seed: soybeans, peanuts, corn | 1.3 | 0.7–0.9 | 11–14 | 40–60 | 11–15 | 150 | 250 | Slow | |
| Vegetable—latex: rubber (vulcanized) | 1.0 | 0.4–0.6 | 4–7 | 700–900 | 0 | 300 | | Burns | –60 |

## Table 5.3—4 (continued)
## PROPERTIES OF MAN-MADE FIBERS

| Chemical class; common name (sources) | Specific gravity | Tenacity, g/denier | Tensile strength, 10³ psi | Elongation at break, % | Regain (standard) | Softening point, °C | Melting point, °C | Flammability | Brittleness temp, °C |
|---|---|---|---|---|---|---|---|---|---|
| **SYNTHETIC FIBERS** | | | | | | | | | |
| Polyacrylonitrile (acrylic) | 1.17 | 2–5 | 50–75 | 25–40 | 2 | 190 | 260 | Burns | |
| Polyamide (nylon) | 1.14 | 4–9 | 70–120 | 20–40 | 4 | 200 | 215–250 | Slow | < –100 |
| Polyester (PET dacron) | 1.38 | 4–8 | 70–120 | 10–50 | 0.4 | 225 | 250–290 | Low | |
| Polyethylene (olefin, low density) | 0.92 | 3–6 | 40–70 | 25–40 | 0.15 | 90–120 | 120 | Slow | –114 |
| Polyethylene (olefin, high density) | 0.95 | 5–7 | 60–80 | 10–20 | 0.01 | 120–130 | 140 | Slow | –114 |
| Polypropylene (olefin) | 0.91 | 4.5–8 | 45–80 | 15–30 | 0–0.5 | 145 | 160–170 | Self-ext. low | –70 |
| Polyurethane (spandex) | 1.1 | 0.5–1.0 | 7–16 | 500–700 | 1.0 | 190 | 250 | Burns | |
| Polyvinyl chloride (PVC) | 1.38 | 0.7–2 | 12–17 | 100–125 | 0.1 | 70 | 140[b] | No; chars | < –100 |
| Polyvinyl alcohol (PVA) | 1.3 | 3–7 | 60–90 | 15–28 | 5 | 230 | 240 | Slow | |
| Polyvinylidene chloride (saran) | 1.7 | 2 | 40 | 20–30 | 0.1 | 115–135 | 170 | No | |
| Polytetrafluoroethylene (PTFE) | 2.1 | 1.2–1.4 | 33 | 15–30 | 0 | 225 | 300[b] | No | |

[a] For additional properties of fibers, see Table 4–87.
[b] Decomposition; does not melt.

*Note:* Mechanical properties are for room temperature and humidity and based on cross section.

From Bolz, R. E. and Tuve, G. L., Eds., *Handbook of Tables for Applied Engineering Science*, 2nd ed., CRC Press, Cleveland, 1973, 175.

Table 5.3—5
FIBERS FOR SPECIAL USES

| Desired characteristics | Fibers that are superior for these requirements | Desired characteristics | Fibers that are superior for these requirements |
|---|---|---|---|
| Moisture resistant<br>Nonabsorbent, fast drying, high wet strength, nonswelling, nonshrinking | Glass, Teflon®, saran, PVC, rubber, polyethylene, polypropylene saran, spandex, acrylic polyester | Fire resistant<br>Nonflammable or very slow burning; flame resistant | Asbestos, Teflon, PVC, saran polypropylene, polyvinyl alcohol |
| Climate resistant<br>Minimum deterioration from sun, rain, sea water, insects, mildew, and other environmental factors | Glass, Teflon, saran, PVC, rubber, acrylic, polypropylene, spandex, polyester, polyvinyl alcohol | Lightweight<br>Low density and high strength-weight ratio | Polypropylene, polyethylene, polyurethane, nylon |
| | | High tenacity<br>High breaking strength for given diameter; high ultimate tensile strength | Glass, nylon, polyester, ramie, polypropylene, polyvinyl alcohol, flax, silk, high-tensile viscose |
| Chemical resistant<br>Unharmed by acids, alkalies, salts, oils, common solvents | Asbestos, Teflon, saran, polypropylene, polyvinyl alcohol, polyester | Hard wearing<br>High resistance to friction and abrasion | Flax, silk, cotton, kenef, sunn, PVC, polyvinyl alcohol, polyester, polypropylene, nylon, acrylic |
| Temperature resistant<br>Retains properties at high and low temperatures; heat resistant | Asbestos, Teflon, nylon, polyester, cupro, triacetate | Elastic and resilient<br>No permanent set after large deformation; springy | Rubber, spandex, polyester, wool, PVC, nylon, acrylic, silk |

From Bolz, R. E. and Tuve, G. L., Eds., *Handbook of Tables for Applied Engineering Science*, 2nd ed., CRC Press, Cleveland, 1973, 176.

## Table 5.3—6
## TEXTILE YARN AND FIBER SIZES

The *fineness* of silk, cotton, rayon, and other man-made yarns is usually expressed in terms of the length of yarn per unit weight, or the weight of a given length. The *denier* is the weight in grams of 9,000 m of yarn. (Sometimes the *international denier* is used; it is defined as the weight in grams of 10,000 m of yarn.) A similar unit is the *count,* which is the number of hanks or skeins per pound. The hank, for silk or cotton, is 840 yd, but for wool, worsted, linen, or man-made fiber, it may be different. In the metric system the *count* is the number of meters per gram of yarn.

### Conversion Table

| To convert from | To | Multiply by |
|---|---|---|
| Gram/meter | Denier | .0001111 |
| Milligram/kilometer | Denier | 111.11 |
| Gram/meter | International denier | .0001 |
| Denier | Ounce/1,000 yd | 0.003584 |
| Denier | Pound/10,000 yd | 0.002240 |
| Count (cotton, silk) | Yard/pound | 840 |

*Note:* There is no simple conversion from these units of fineness to the diameter in microns, nor between breaking strength per denier (tenacity) and tensile strength per unit area.

From Bolz, R. E. and Tuve, G. L., Eds., *Handbook of Tables for Applied Engineering Science,* 2nd ed., CRC Press, Cleveland, 1973, 176.

## 5.4 OTHER (ABSORBENTS, ACOUSTIC MATERIALS, ETC.)

### Table 5.4—1
### EFFICIENCY OF DRYING AGENTS

#### A. Drying Agents Depending on Chemical Action (Absorption)[a]

| Substance | Formula | Residual water, milligrams per liter of dry air[b] | Reference |
|---|---|---|---|
| Phosphorus pentoxide | $P_2O_5$ | <1 mg in 40,000 l | 1 |
| Magnesium perchlorate, anhydrous | $Mg(ClO_4)_2$ | Unweighable in 210 l | 2 |
| Barium oxide | $BaO$ | 0.00065 | 3 |
| Potassium hydroxide, fused | $KOH$ | 0.002 | 4 |
| Calcium oxide | $CaO$ | 0.003 | 3 |
| Sulfuric acid | $H_2SO_4$ | 0.003 | 4 |
| Calcium sulfate, anhydrous | $CaSO_4$ | 0.005 | 3 |
| Aluminum oxide | $Al_2O_3$ | 0.005 | 3 |
| Potassium hydroxide, sticks | $KOH$ | 0.014 | 3 |
| Sodium hydroxide, fused | $NaOH$ | 0.16 | 4 |
| Calcium bromide | $CaBr_2$ | 0.18 | 5 |
| Calcium chloride, fused | $CaCl_2$ | 0.34 | 4 |
| Sodium hydroxide, sticks | $NaOH$ | 0.80 | 3 |
| Barium perchlorate | $Ba(ClO_4)_2$ | 0.82 | 3 |
| Zinc chloride | $ZnCl_2$ | 0.85 | 5 |
| Zinc bromide | $ZnBr_2$ | 1.16 | 5 |
| Calcium chloride, granular | $CaCl_2$ | 1.5 | 3 |
| Cupric sulfate, anhydrous | $CuSO_4$ | 2.8 | 3 |

#### B. Drying Agents Depending on Physical Action (Adsorption)[a]

| | |
|---|---|
| Alumina, low-temperature-fired | Clay and porcelain, low-temperature-fired |
| Asbestos | Kieselguhr |
| Charcoal | Silica gel |
| Glass wool | Refrigeration |

[a]It should be noted that the efficiency of some drying agents depends on both absorption and adsorption, e.g., aluminum oxide and calcium chloride, and probably also barium oxide, magnesium perchlorate, barium perchlorate, and calcium sulfate.

[b]At 25°C or room temperature, except for the values determined in Reference 3, which were taken at 30°C.

From Yoe, J. H., in *Handbook of Chemistry and Physics*, 69th ed., Weast, R. C., Ed., CRC Press, Boca Raton, Fla., 1988, E-37.

### REFERENCES

1. **Morley,** *Am. J. Sci.,* 34, 199, 1887; *J. Am. Chem. Soc.,* 26, 1171, 1904.
2. **Willard and Smith,** *J. Am. Chem. Soc.,* 44, 2255, 1922.
3. **Bower,** *Bur. Stand. J. Res.,* 12, 241, 1934.
4. **Baxter and Starkweather,** *J. Am. Chem. Soc.,* 38, 2038, 1916.
5. **Baxter and Warren,** *J. Am. Chem. Soc.,* 33, 340, 1911.

## Table 5.4—2
## CONSTANT HUMIDITY

The following table shows the percent humidity and the aqueous tension at the given temperature within a closed space when an excess of the substance indicated is in contact with a saturated aqueous solution of the given solid phase.

| Solid Phase | °C | Percent Humidity | Aqueous Tension, mm Hg | Solid Phase | °C | Percent Humidity | Aqueous Tension, mm Hg |
|---|---|---|---|---|---|---|---|
| $H_3PO_4 \cdot \frac{1}{2}H_2O$ | 24 | 9 | 1.99 | $NH_4Cl$ and $KNO_3$ | 20 | 72.6 | 12.6 |
| $KC_2H_3O_2$ | 168 | 13 | 738 | $NaClO_3$ | 20 | 75 | 13.0 |
| $LiCl \cdot H_2O$ | 20 | 15 | 2.60 | $(NH_4)_2SO_4$ | 108 | 75 | 754 |
| $KC_2H_3O_2$ | 20 | 20 | 3.47 | $NaC_2H_3O_2 \cdot 3H_2O$ | 20 | 76 | 13.2 |
| $KF$ | 100 | 22.9 | 174 | $H_2C_2O_4 \cdot 2H_2O$ | 20 | 76 | 13.2 |
| $NaBr$ | 100 | 22.9 | 174 | $NH_4Cl$ | 30 | 77.5 | 24.4 |
| $NaCl$, $KNO_3$ and $NaNO_3$ | 16.39 | 30.49 | 4.23 | $Na_2S_2O_3 \cdot 5H_2O$ | 20 | 78 | 13.5 |
| | | | | $NH_4Cl$ | 25 | 79.3 | 18.6 |
| $CaCl_2 \cdot 6H_2O$ | 24.5 | 31 | 7.08 | $NH_4Cl$ | 20 | 79.5 | 13.8 |
| $CaCl_2 \cdot 6H_2O$ | 20 | 32.3 | 5.61 | $(NH_4)_2SO_4$ | 20 | 81 | 14.1 |
| $CaCl_2 \cdot 6H_2O$ | 18.5 | 35 | 5.54 | $(NH_4)_2SO_4$ | 25 | 81.1 | 19.1 |
| $CrO_3$ | 20 | 35 | 6.08 | $(NH_4)_2SO_4$ | 30 | 81.1 | 25.6 |
| $CaCl_2 \cdot 6H_2O$ | 10 | 38 | 3.47 | $KBr$ | 20 | 84 | 14.6 |
| $CaCl_2 \cdot 6H_2O$ | 5 | 39.8 | 2.59 | $Tl_2SO_4$ | 104.7 | 84.8 | 768 |
| $Zn(NO_3)_2 \cdot 6H_2O$ | 20 | 42 | 7.29 | $KHSO_2$ | 20 | 86 | 14.9 |
| $K_2CO_3 \cdot 2H_2O$ | 24.5 | 43 | 9.82 | $Na_2CO_3 \cdot 10H_2O$ | 24.5 | 87 | 20.9 |
| $K_2CO_3 \cdot 2H_2O$ | 18.5 | 44 | 6.96 | $BaCl_2 \cdot 2H_2O$ | 24.5 | 88 | 20.1 |
| $KNO_2$ | 20 | 45 | 7.81 | $K_2CrO_4$ | 20 | 88 | 15.3 |
| $KCNS$ | 20 | 47 | 8.16 | $Pb(NO_3)_2$ | 103.5 | 88.4 | 760 |
| $NaI$ | 100 | 50.4 | 383 | $ZnSO_4 \cdot 7H_2O$ | 20 | 90 | 15.6 |
| $Ca(NO_3)_2 \cdot 4H_2O$ | 24.5 | 51 | 11.6 | $Na_2CO_3 \cdot 10H_2O$ | 18.5 | 92 | 14.6 |
| $NaHSO_4 \cdot H_2O$ | 20 | 52 | 9.03 | $NaBrO_3$ | 20 | 92 | 16.0 |
| $Na_2Cr_2O_7 \cdot 2H_2O$ | 20 | 52 | 9.03 | $K_2HPO_4$ | 20 | 92 | 16.0 |
| $Mg(NO_3)_2 \cdot 6H_2O$ | 24.5 | 52 | 11.9 | $NH_4H_2PO_4$ | 30 | 92.9 | 29.3 |
| $NaClO_3$ | 100 | 54 | 410 | $NH_4H_2PO_4$ | 25 | 93 | 21.9 |
| $Ca(NO_3)_2 \cdot 4H_2O$ | 18.5 | 56 | 8.86 | $Na_2SO_4 \cdot 10H_2O$ | 20 | 93 | 16.1 |
| $Mg(NO_3)_2 \cdot 6H_2O$ | 18.5 | 56 | 8.86 | $NH_4H_2PO_4$ | 20 | 93.1 | 16.2 |
| $KI$ | 100 | 56.2 | 427 | $ZnSO_4 \cdot 7H_2O$ | 5 | 94.7 | 6.10 |
| $NaBr \cdot 2H_2O$ | 20 | 58 | 10.1 | $Na_2SO_3 \cdot 7H_2O$ | 20 | 95 | 16.5 |
| $Mg(C_2H_3O_2)_2 \cdot 4H_2O$ | 20 | 65 | 11.3 | $Na_2HPO_4 \cdot 12H_2O$ | 20 | 95 | 16.5 |
| $NaNO_2$ | 20 | 66 | 11.5 | $NaF$ | 100 | 96.6 | 734 |
| $NH_4Cl$ and $KNO_3$ | 30 | 68.6 | 21.6 | $Pb(NO_3)_2$ | 20 | 98 | 17.0 |
| $KBr$ | 100 | 69.2 | 526 | $CuSO_4 \cdot 5H_2O$ | 20 | 98 | 17.0 |
| $NH_4Cl$ and $KNO_3$ | 25 | 71.2 | 16.7 | $TlNO_3$ | 100.3 | 98.7 | 759 |
| $TlCl$ | 100.1 | 99.7 | 761 | | | | |

From Weast, R. C. Ed., *Handbook of Chemistry and Physics*, 69th ed., CRC Press, Boca Raton, Fla., 1988, E-42.

## Table 5.4—3

## PERMEANCE AND PERMEABILITY OF MATERIALS TO WATER VAPOR[a]

| Material | Thickness (in.) | Permeance (Perm) | Resistance[b] (Rep) | Permeability (Perm-in.) | Resistance/in.[b] (Rep/in.) |
|---|---|---|---|---|---|
| **Materials used in construction** | | | | | |
| Concrete (1:2:4 mix) | | | | 3.2 | 0.31 |
| Brick masonry | 4 | 0.8[c] | 1.3 | | |
| Concrete block (cored, limestone aggregate) | 8 | 2.4[c] | 0.4 | | |
| Tile masonry, glazed | 4 | 0.12[c] | 8.3 | | |
| Asbestos cement board | 0.12 | 4—8[d] | 0.1—0.2 | | |
| With oil base finishes | | 0.3—0.5[d] | 2—3 | | |
| Plaster on metal lath | 0.75 | 15[c] | 0.067 | | |
| Plaster on wood lath | | 11[c] | 0.091 | | |
| Plaster on plain gypsum lath (with studs) | | 20[c] | 0.050 | | |
| Gypsum wall board (plain) | 0.375 | 50[c] | 0.020 | | |
| Gypsum sheathing (asphalt impregnated) | 0.5 | | | 20[d] | 0.050 |
| Structural insulating board (sheathing quality) | | | | 20—50[c] | 0.050—0.020 |
| Structural insulating board (interior, uncoated) | 0.5 | 50—90[c] | 0.020—0.011 | | |
| Hardboard (standard) | 0.125 | 11[c] | 0.091 | | |
| Hardboard (tempered) | 0.125 | 5[c] | 0.2 | | |
| Built-up roofing (hot mopped) | | 0.0 | | | |
| Wood, sugar pine | | | | 0.4—5.4[c,f] | 2.5—0.19 |
| Plywood (douglas fir, exterior glue) | 0.25 | 0.7[c] | 1.4 | | |
| Plywood (douglas fir, interior glue) | 0.25 | 1.9[c] | 0.53 | | |
| Acrylic, glass fiber reinforced sheet | 0.056 | 0.12[d] | 8.3 | | |
| Polyester, glass fiber reinforced sheet | 0.048 | 0.05[d] | 20 | | |
| **Thermal insulations** | | | | | |
| Air (still) | | | | 120[c] | 0.0083 |
| Cellular glass | | | | 0.0[d] | ∞ |
| Corkboard | | | | 2.1—2.6[d] | 0.48—0.38 |
| | | | | 9.5[c] | 0.11 |
| Mineral wool (unprotected) | | | | 116[e] | 0.0086 |
| Expanded polyurethane (R-11 blown) board stock | | | | 0.4—1.6[d] | 2.5—0.62 |
| Expanded polystyrene (extruded) | | | | 1.2[d] | 0.83 |

## Table 5.4—3 (continued)
## PERMEANCE AND PERMEABILITY OF MATERIALS TO WATER VAPOR[a]

| Material | Thickness (in.) | Permeance (Perm) | Resistance[b] (Rep) | Permeability (Perm-in.) | Resistance/in.[b] (Rep/in.) |
|---|---|---|---|---|---|
| Expanded polystyrene (bead) | | | | 2.05—5.8[d] | 0.50—0.17 |
| Phenolic foam (covering removed) | | | | 26 | 0.038 |
| Unicellular synthetic flexible rubber foam | | | | 0.02—0.15[d] | 50—6.7 |
| **Plastic and metal foils and films[g]** | | | | | |
| Aluminum foil | 0.001 | 0.0[d] | | | |
| Aluminum foil | 0.00035 | 0.05[d] | 20 | | |
| Polyethylene | 0.002 | 0.16[d] | 6.3 | | 3100 |
| Polyethylene | 0.004 | 0.08[d] | 12.5 | | 3100 |
| Polyethylene | 0.006 | 0.06[d] | 17 | | 3100 |
| Polyethylene | 0.008 | 0.04[d] | 25 | | 3100 |
| Polyethylene | 0.010 | 0.03[d] | 33 | | 3100 |
| Polyvinylchloride, unplasticized | 0.002 | 0.68[d] | 1.5 | | |
| Polyvinylchloride, plasticized | 0.004 | 0.8—1.4[d] | 1.3—0.72 | | |
| Polyester | 0.001 | 0.73[d] | 1.4 | | |
| Polyester | 0.0032 | 0.23[d] | 4.3 | | |
| Polyester | 0.0076 | 0.08[d] | 12.5 | | |
| Cellulose acetate | 0.01 | 4.6[d] | 0.2 | | |
| Cellulose acetate | 0.125 | 0.32[d] | 3.1 | | |

| Material | Weight[h] | Permeance (Perms) | | | Resistance[b] (Rep) | | |
|---|---|---|---|---|---|---|---|
| | | Dry-Cup | Wet-Cup | Other | Dry-Cup | Wet-Cup | Other |
| **Building paper, felts, roofing papers[i]** | | | | | | | |
| Duplex sheet, asphalt laminated, aluminum foil one side | 8.6 | 0.002 | 0.176 | | 500 | 5.8 | |
| Saturated and coated roll roofing | 65 | 0.05 | 0.24 | | 20 | 4.2 | |
| Kraft paper and asphalt laminated, reinforced 30-120-30 | 6.8 | 0.3 | 1.8 | | 3.3 | 0.55 | |
| Blanket thermal insulation back-up paper, asphalt coated | 6.2 | 0.4 | 0.6—4.2 | | 2.5 | 1.7—0.24 | |
| Asphalt-saturated and coated vapor retarder paper | 8.6 | 0.2—0.3 | 0.6 | | 5.0—3.3 | 1.7 | |
| Asphalt-saturated but not coated sheathing paper | 4.4 | 3.3 | 20.2 | | 0.3 | 0.05 | |
| 15-lb asphalt felt | 14 | 1.0 | 5.6 | | 1.0 | 0.18 | |
| 15-lb tar felt | 14 | 4.0 | 18.2 | | 0.25 | 0.055 | |
| Single-kraft, double | 3.2 | 31 | 42 | | 0.032 | 0.024 | |

## Table 5.4—3 (continued)
## PERMEANCE AND PERMEABILITY OF MATERIALS TO WATER VAPOR[a]

| | Inches | Permeance (Perms) | | | Resistance[b] (Rep) | | |
|---|---|---|---|---|---|---|---|
| | | Dry-Cup | Wet-Cup | Other | Dry-Cup | Wet-Cup | Other |
| **Liquid-applied coating materials** | | | | | | | |
| Commercial latex paints (dry film thickness)[j] | | | | | | | |
| Vapor retarder paint | 0.0031 | | | 0.45 | | | 2.22 |
| Primer-sealer | 0.0012 | | | 6.28 | | | 0.16 |
| Vinyl acetate/acrylic primer | 0.002 | | | 7.42 | | | 0.13 |
| Vinyl-acrylic primer | 0.0016 | | | 8.62 | | | 0.12 |
| Semi-gloss vinyl-acrylic enamel | 0.0024 | | | 6.61 | | | 0.15 |
| Exterior acrylic house and trim | 0.0017 | | | 5.47 | | | 0.18 |
| Paint — 2 coats | | | | | | | |
| Asphalt paint on plywood | | 0.3—0.5 | | | 3.3—2.0 | | |
| Aluminum varnish on wood | | | 0.4 | | | 2.5 | |
| Enamels on smooth plaster | | | | 0.5—1.5 | | | 2.0—0.66 |
| Primers and sealers on interior insulation board | | | | 0.9—2.1 | | | 1.1—0.48 |
| Various primers plus 1 coat flat oil paint on plaster | | | | 1.6—3.0 | | | 0.63—0.33 |
| Flat paint on interior insulation board | | | | 4 | | | 0.25 |
| Water emulsion on interior insulation board | | | | 30—85 | | | 0.02—0.012 |
| Paint — 3 coats | **Oz/ft²** | | | | | | |
| Exterior paint, white lead and oil on wood siding | | 0.3—1.0 | | | 3.3—1.0 | | |
| Exterior paint, white lead-zinc oxide and oil on wood | | 0.9 | | | 1.1 | | |
| Styrene-butadiene latex coating | 2 | 11 | | | 0.09 | | |
| Polyvinyl acetate latex coating | 4 | 5.5 | | | 0.18 | | |
| Chloro-sulfonated polyethylene mastic | 3.5 | 1.7 | | | 0.59 | | |
| | 7.0 | 0.06 | | | 16 | | |
| Asphalt cut-back mastic | | | | | | | |
| 1/16 in., dry | | 0.14 | | 7.2 | | | |
| 3/16 in., dry | | 0.0 | | — | | | |
| Hot melt asphalt | 2 | 0.5 | | | 2 | | |
| | 3.5 | 0.1 | | | 10 | | |

## Table 5.4—3 (continued)
## PERMEANCE AND PERMEABILITY OF MATERIALS TO WATER VAPOR[a]

*Note:* In this Table, the permeance, resistance, permeability and resistance per unit thickness values are given in the following units:

| | | |
|---|---|---|
| Permeance | Perm | = gr/h·ft²·in. Hg |
| Resistance | Rep | = in. Hg·ft²·h/gr |
| Permeability | Perm-in. | = gr/h·ft²·(in. Hg/in.) |
| Resistance/unit thickness | Rep/in. | = (in. Hg·ft²·h/gr)/in. |

[a]  Table gives the water vapor transmission rates of some representative materials. The data are provided to permit comparisons of materials; but in the selection of vapor retarded materials, exact values for permeance or permeability should be obtained from the manufacturer of the materials under consideration or secured as a result of laboratory test. A range of values shown in the table indicates variations among mean values for materials that are similar but of different density, orientation, lot, or source. The values are intended for design guidance and should not be used as design or specification data. The compilation is from a number of sources; values from dry-cup and wet-cup methods were usually obtained from investigations using ASTM E96 and C355; values shown under *others* were obtained from investigations using such techniques as *two-temperature, special cell,* and *air-velocity.* Values included were obtained from the Building Research Div., National Research Council of Canada.

[b]  Resistance and resistance/in. values have been calculated as the reciprocal of the permeance and permeability values.

[c]  Other than dry- or wet-cup method.

[d]  Dry-cup method.

[e]  Wet-cup method.

[f]  Depending on construction and direction of vapor flow.

[g]  Usually installed as vapor retarders, although sometimes used as exterior finish and elsewhere near cold side where special considerations are then required for warm side barrier effectiveness.

[h]  Basic weight in lb per 100 ft² (lb per square ft).

[i]  Low permeance sheets used as vapor retarders. High permeance used elsewhere in construction.

[j]  Cast at 10 mils wet film thickness.

Reprinted by permission from *ASHRAE Handbook — 1985 Fundamentals,* American Society of Heating Refrigerating, and Air Conditioning Engineers, Atlanta, 1985.

## Table 5.4—4
## PROPERTIES AND USES OF ADSORBENTS

For density in kg/m$^3$, multiply lb/ft$^3$ by 16.02.

| Adsorbent | Shape of particles[a] | Bulk dry density, lb/ft$^3$ | Surface area sq m/g | Uses and method of regeneration |
|---|---|---|---|---|
| Active alumina | G | 50 | 250 | Drying gases and liquids; catalyst; cat- |
|  | S | 55 | 350 | alyst support; defluoridation of alky-lates; neutralization of lube oils. Can be regenerated. |
| Activated bauxite | C, G | ~53 | — | Decolorizing petroleum products and dry-ing of gases. Regeneration by heating. |
| Aluminosilicates (Molecular sieves) | C, S, P | ~44 | 770 | Selective adsorption based on molecular size and shape; drying of gases and liquids; catalyst support. Regenera-tion by heating or elution. |
| Carbon or charcoal: |  |  |  | Water treatment; air and gas purifica- |
| Bone | G | 20–30 |  | tion; gas masks and smoke filters; sol- |
| Coal | G | 20–30 | 500–1200 | vent recovery and purification; sugar |
| Petroleum | C | 28–34 | 800–1100 | refining; decolorizing of solutions; |
| Shell | G | 10–20 |  | decolorizing natural products. Ad- |
| Wood | G | 10–35 | 625–1400 | sorbed gases can be evaporated. |
| Clay | P | 30–45 | 225–300 | Refining petroleum fractions; purifying vegetable oils, juices; catalyst base. |
| Fuller's earth | G | 30–40 | 130–250 | Uses same as for clay. Regeneration by washing and burning adsorbed organic matter. |
| Silica gel | G, P | ~25 | 320 | Drying of gases; adsorption, from solu- |
|  | S | 50 | 650 | tions; hydrocarbons; catalyst base. Regeneration by evaporation of adsorbed liquid. |

[a]Shape of particles indicated as follows: C, cylindrical pellets; G, granular; P, powder; S, spherical beads.

*Notes*: Both surface areas and pore sizes are important in adsorption.

Distinction should be made between adsorption of gases by nonporous solids such as smooth metals and by porous solids such as the adsorbents listed in this table. Nonporous inorganic block-solids have surface areas in the range below 10 m$^2$/g; their adsorption is correspondingly small but definitely measurable.

Gas and vapor adsorption increases as the temperature is reduced and the pressure of the gas is increased. The usual method for quantitative expression of this relationship is in terms of the adsorption at constant temperature. The adsorption isotherms are plotted with adsorption (e.g., g of adsorbate/g of adsorbent) on the ordinate and pressure (actual pressure/ saturation pressure) on the abscissas. These isotherms have several typical shapes, but in any case the adsorption is much higher if the gas is not highly superheated, i.e., it increases as saturation is approached at the given temperature. The BET classification system (proposed by S. Brunauer, P. H. Emmet, and E. Teller in *J. Am. Chem. Soc.*, 60, 309, 1938, but still widely quoted) for adsorption isotherms recognizes five types of such curves.

### REFERENCES

**Gregg, S. J. and Sing, K. S. W.,** *Adsorption, Surface Area, and Porosity,* Academic Press, New York, 1967.
**Hassler, J. W.,** *Activated Carbon,* Chemical Publishing, New York, 1963.
*Symposium on Activated Carbon,* Atlas Chemical Industries, Wilmington, Del., 1968.

From Bolz, R. E. and Tuve, G. L., Eds., *Handbook of Tables for Applied Engineering Science,* 2nd ed., CRC Press, Cleveland, 1973, 363.

Table 5.4—5

## SOUND FREQUENCIES AND SCALES

A pure tone or musical note is produced if the vibrations consist of a single frequency within the audible range. A "harmonic" is a partial tone, the frequency of which is an integral multiple of the fundamental or lowest frequency. A harmonic series consists of integral multiples.

If the frequency is doubled, the tone rises one octave on the musical scale. The common musical scale in Western countries has 12 half tones, with an interval or frequency ratio between successive tones of $\sqrt[12]{2}$. In this equally tempered scale the intervals are given the following names.

| Name | Frequency ratio | Name | Frequency ratio |
|---|---|---|---|
| Semitone | 1.05946 | Perfect fifth | 1.49831 |
| Whole tone or major second | 1.12246 | Minor sixth | 1.58740 |
| Minor third | 1.18921 | Major sixth | 1.68179 |
| Major third | 1.25992 | Minor seventh | 1.78180 |
| Perfect fourth | 1.33484 | Major seventh | 1.88775 |
| Augmented fourth[a] | 1.41421 | Octave | 2.00000 |

In a complex tone the ear judges the pitch by the lowest frequency, or fundamental, and interprets the tone quality in terms of the accompanying higher frequencies, or overtones. The audible range, which is about 16 to 20,000 hz, varies among individuals and among animals.

Very complex sounds may be given special names such as white noise. A "noise" is defined as an unwanted sound, and it may or may not have a prevailing frequency. Electronic generators are available for generating sound combinations of almost any desired pattern.

Engineering specifications and measurements are most likely to deal with sound intensity or energy, but the direct measurement is that of sound pressure in decibels expressed on a logarithmic scale.[b] Loudness is the human subjective interpretation of sound intensity and is expressed in phons (or sones). Sounds are analyzed with respect to frequency through the use of electronic band-pass filters, e.g., octave bands or 1/3-octave bands, with special attention to five octaves in the auditory range, 125 to 4,000 hz.

The following diagram provides a comparison of ranges in the acoustic spectrum, as represented by average human voices and by the usual string and wind instruments of the orchestra. Frequencies for the C-scale on the piano are also given, based on the tuning of this instrument to American Standard Pitch, A = 440 hz (formerly C at 256 hz).

[a] Also called diminished fifth.
[b] Sound measurement techniques are covered by standard codes; see ANSI-S1.2 and entire ANSI S-series of over 30 standards.

# Table 5.4—5
# MUSICAL SCALES

## EQUAL TEMPERED CHROMATIC SCALE
### $A_4 = 440$

American Standard pitch. Adopted by the American Standards Association in 1936

| Note | Frequency | Note | Frequency | Note | Frequency | Note | Frequency |
|---|---|---|---|---|---|---|---|
| $C_0$ | 16.35 | $C_2$ | 65.41 | $C_4$ | 261.63 | $C_6$ | 1046.50 |
| $C\#_0$ | 17.32 | $C\#_2$ | 69.30 | $C\#_4$ | 277.18 | $C\#_6$ | 1108.73 |
| $D_0$ | 18.35 | $D_2$ | 73.42 | $D_4$ | 293.66 | $D_6$ | 1174.66 |
| $D\#_0$ | 19.45 | $D\#_2$ | 77.78 | $D\#_4$ | 311.13 | $D\#_6$ | 1244.51 |
| $E_0$ | 20.60 | $E_2$ | 82.41 | $E_4$ | 329.63 | $E_6$ | 1318.51 |
| $F_0$ | 21.83 | $F_2$ | 87.31 | $F_4$ | 349.23 | $F_6$ | 1396.91 |
| $F\#_0$ | 23.12 | $F\#_2$ | 92.50 | $F\#_4$ | 369.99 | $F\#_6$ | 1479.98 |
| $G_0$ | 24.50 | $G_2$ | 98.00 | $G_4$ | 392.00 | $G_6$ | 1567.98 |
| $G\#_0$ | 25.96 | $G\#_2$ | 103.83 | $G\#_4$ | 415.30 | $G\#_6$ | 1661.22 |
| $A_0$ | 27.50 | $A_2$ | 110.00 | $A_4$ | 440.00 | $A_6$ | 1760.00 |
| $A\#_0$ | 29.14 | $A\#_2$ | 116.54 | $A\#_4$ | 466.16 | $A\#_6$ | 1864.66 |
| $B_0$ | 30.87 | $B_2$ | 123.47 | $B_4$ | 493.88 | $B_6$ | 1975.00 |
| $C_1$ | 32.70 | $C_3$ | 130.81 | $C_5$ | 523.25 | $C_7$ | 2093.00 |
| $C\#_1$ | 34.65 | $C\#_3$ | 138.59 | $C\#_5$ | 554.37 | $C\#_7$ | 2217.46 |
| $D_1$ | 36.71 | $D_3$ | 146.83 | $D_5$ | 587.33 | $D_7$ | 2349.32 |
| $D\#_1$ | 38.89 | $D\#_3$ | 155.56 | $D\#_5$ | 622.25 | $D\#_7$ | 2489.02 |
| $E_1$ | 41.20 | $E_3$ | 164.81 | $E_5$ | 659.26 | $E_7$ | 2637.02 |
| $F_1$ | 43.65 | $F_3$ | 174.61 | $F_5$ | 698.46 | $F_7$ | 2793.83 |
| $F\#1$ | 46.25 | $F\#_3$ | 185.00 | $F\#_5$ | 739.99 | $F\#_7$ | 2959.96 |
| $G_1$ | 49.00 | $G_3$ | 196.00 | $G_5$ | 783.99 | $G_7$ | 3135.96 |
| $G\#_1$ | 51.91 | $G\#_3$ | 207.65 | $G\#_5$ | 830.61 | $G\#_7$ | 3322.44 |
| $A_1$ | 55.00 | $A_3$ | 220.00 | $A_5$ | 880.00 | $A_7$ | 3520.00 |
| $A\#_1$ | 58.27 | $A\#_3$ | 233.08 | $A\#_5$ | 932.33 | $A\#_7$ | 3729.31 |
| $B_1$ | 61.74 | $B_3$ | 246.94 | $B_5$ | 987.77 | $B_7$ | 3951.07 |
|  |  |  |  |  |  | $C_8$ | 4186.01 |

## EQUAL TEMPERED CHROMATIC SCALE
### $A_4 = 435$

International Pitch, adopted 1891

| Note | Frequency | Note | Frequency | Note | Frequency | Note | Frequency |
|---|---|---|---|---|---|---|---|
| $C_0$ | 16.17 | $C_2$ | 64.66 | $C_4$ | 258.65 | $C_6$ | 1034.61 |
| $C\#_0$ | 17.13 | $C\#_2$ | 68.51 | $C\#_4$ | 274.03 | $C\#_6$ | 1096.13 |
| $D_0$ | 18.15 | $D_2$ | 72.58 | $D_4$ | 290.33 | $D_6$ | 1161.31 |
| $D\#_0$ | 19.22 | $D\#_2$ | 76.90 | $D\#_4$ | 307.59 | $D\#_6$ | 1230.37 |
| $E_0$ | 20.37 | $E_2$ | 81.47 | $E_4$ | 325.88 | $E_6$ | 1303.53 |
| $F_0$ | 21.58 | $F_2$ | 86.31 | $F_4$ | 345.26 | $F_6$ | 1381.04 |
| $F\#_0$ | 22.86 | $F\#_2$ | 91.45 | $F\#_4$ | 365.79 | $F\#_6$ | 1463.16 |
| $G_0$ | 24.22 | $G_2$ | 96.89 | $G_4$ | 387.54 | $G_6$ | 1550.16 |
| $G\#_0$ | 25.66 | $G\#_2$ | 102.65 | $G\#_4$ | 410.59 | $G\#_6$ | 1642.34 |
| $A_0$ | 27.19 | $A_2$ | 108.75 | $A_4$ | 435.00 | $A_6$ | 1740.00 |
| $A\#_0$ | 28.80 | $A\#_2$ | 115.22 | $A\#_4$ | 460.87 | $A\#_6$ | 1843.47 |
| $B_0$ | 30.52 | $B_2$ | 122.07 | $B_4$ | 488.27 | $B_6$ | 1953.08 |
| $C_1$ | 32.33 | $C_3$ | 129.33 | $C_5$ | 517.31 | $C_7$ | 2069.22 |
| $C\#_1$ | 34.25 | $C\#_3$ | 137.02 | $C\#_5$ | 548.07 | $C\#_7$ | 2192.26 |
| $D_1$ | 36.29 | $D_3$ | 145.16 | $D_5$ | 580.66 | $D_7$ | 2322.62 |
| $D\#_1$ | 38.45 | $D\#_3$ | 153.80 | $D\#_5$ | 615.18 | $D\#_7$ | 2460.73 |
| $E_1$ | 40.74 | $E_3$ | 162.94 | $E_5$ | 651.76 | $E_7$ | 2607.05 |
| $F_1$ | 43.16 | $F_3$ | 172.63 | $F_5$ | 690.52 | $F_7$ | 2762.08 |
| $F\#_1$ | 45.72 | $F\#_3$ | 182.89 | $F\#_5$ | 731.58 | $F\#_7$ | 2926.32 |
| $G_1$ | 48.44 | $G_3$ | 193.77 | $G_5$ | 775.08 | $G_7$ | 3100.33 |
| $G\#_1$ | 51.32 | $G\#_3$ | 205.29 | $G\#_5$ | 821.17 | $G\#_7$ | 3284.68 |
| $A_1$ | 54.38 | $A_3$ | 217.50 | $A_5$ | 870.00 | $A_7$ | 3480.00 |
| $A\#_1$ | 57.61 | $A\#_3$ | 230.43 | $A\#_5$ | 921.73 | $A\#_7$ | 3686.93 |
| $B_1$ | 61.03 | $B_3$ | 244.14 | $B_5$ | 976.54 | $B_7$ | 3906.17 |
|  |  |  |  |  |  | $C_8$ | 4138.44 |

## SCIENTIFIC OR JUST SCALE
### $C_4 = 256$

| Note | Frequency | Note | Frequency | Note | Frequency | Note | Frequency |
|---|---|---|---|---|---|---|---|
| $C_0$ | 16 | $C_2$ | 64 | $C_4$ | 256 | $C_6$ | 1024 |
| $D_0$ | 18 | $D_2$ | 72 | $D_4$ | 288 | $D_6$ | 1152 |
| $E_0$ | 20 | $E_2$ | 80 | $E_4$ | 320 | $E_6$ | 1280 |
| $F_0$ | 21.33 | $F_2$ | 85.33 | $F_4$ | 341.33 | $F_6$ | 1365.33 |
| $G_0$ | 24 | $G_2$ | 96 | $G_4$ | 384 | $G_6$ | 1536 |
| $A_0$ | 26.67 | $A_2$ | 106.67 | $A_4$ | 426.67 | $A_6$ | 1706.67 |
| $B_0$ | 30 | $B_2$ | 120 | $B_4$ | 480 | $B_6$ | 1920 |
| $C_1$ | 32 | $C_3$ | 128 | $C_5$ | 512 | $C_7$ | 2048 |
| $D_1$ | 36 | $D_3$ | 144 | $D_5$ | 576 | $D_7$ | 2304 |
| $E_1$ | 40 | $E_3$ | 160 | $E_5$ | 640 | $E_7$ | 2560 |
| $F_1$ | 42.67 | $F_3$ | 170.67 | $F_5$ | 682.67 | $F_7$ | 2730.67 |
| $G_1$ | 48 | $G_3$ | 192 | $G_5$ | 768 | $G_7$ | 3072 |
| $A_1$ | 53.33 | $A_3$ | 213.33 | $A_5$ | 853.33 | $A_7$ | 3413.33 |
| $B_1$ | 60 | $B_3$ | 240 | $B_5$ | 960 | $B_7$ | 3840 |
|  |  |  |  |  |  | $C_8$ | 4096 |

From Weast, R. C., Ed., *Handbook of Chemistry and Physics,* 69th ed., CRC Press, Boca Raton, Fla., 1988, E-44.

## Table 5.4—6
### CHARACTERISTICS OF PARTICLES AND PARTICLE DISPERSOIDS

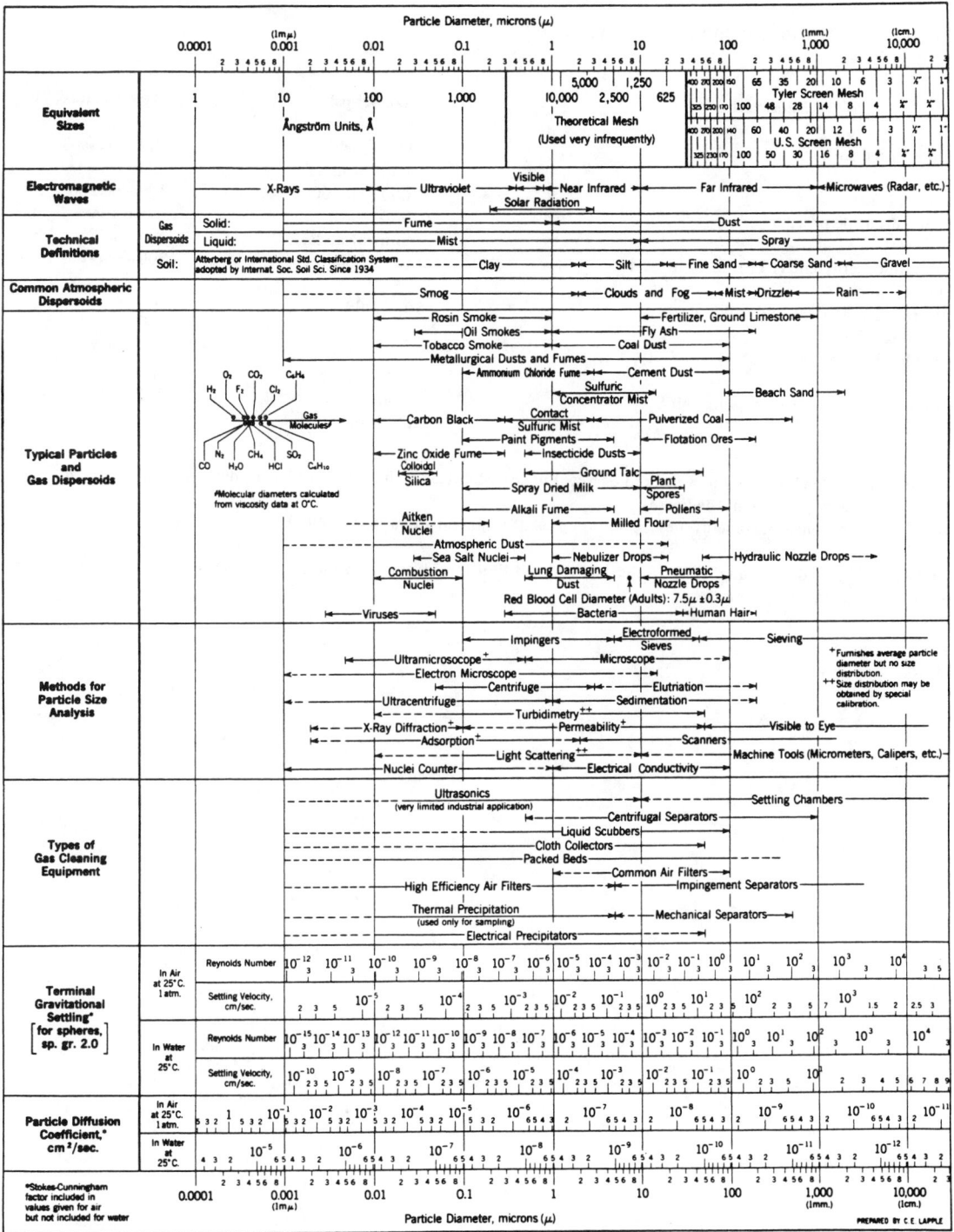

From Lapple, C. E., *SRI Journal*, 5, 94 (Third Quarter), 1961. With permission.

## Table 5.4—7
## USEFUL RANGE OF SOLIDS FOR DRY LUBRICATION

For load in $MN/m^2$, multiply values in psi by
$6.8948 \times 10^{-3}$

| Lubricant | Load,[a] psi | Temperature, °F |
|---|---|---|
| Molybdenum disulfide | 2,000–yield point of metal | –300–+750[b] |
| Graphite powder | 2,000–100,000 | –300–+1,200 |
| Tungsten disulfide powder | 2,000–yield point of metal | Low not investigated, max. 850[b] |
| Polytetrafluoroethylene (PTFE) powder and sintered shapes | 150–3,000 | –300–+500 |
| Polytetrafluoroethylene and other plastics containing fillers | 50–4,000 | Max. 500 |
| Organic binder coatings, graphite–$MoS_2$ type | 2,000–yield point of metal | –100–500 |
| Inorganic binder coatings, graphite–$MoS_2$ type | 2,000–yield point of metal | –300–1,000 |

[a] Speed affects the load capacity to a marked degree, but solid lubricants should only be considered at low speeds. As speed increases, load-carrying capacity decreases.
[b] In inert and reducing atmospheres these solids are used at temperatures above 2,000° F.

From DiSapio, A. and Gerstung, H. S., Solid and bonded-film lubricants, *Machine Design* (The Bearings Reference Issue), 38(6), 8, 1966. With permission.

## Table 5.4—8
## PROPERTIES OF COMMON SOLID MATERIALS

| Material | Specific gravity | Specific heat $\frac{Btu}{lbm \cdot deg\ R} = \frac{cal}{g \cdot K}$ | $\frac{kJ}{kg \cdot K}$ | Thermal conductivity $\frac{Btu}{hr \cdot ft \cdot deg\ F}$ | $\frac{cal}{sec \cdot cm \cdot deg\ C}$ | $\frac{W}{m \cdot K}$ |
|---|---|---|---|---|---|---|
| Asbestos cement board | 1.4 | 0.2 | .837 | 0.35 | .001 45 | 0.607 |
| Asbestos millboard | 1.0 | 0.2 | .837 | 0.08 | .000 33 | 0.14 |
| Asphalt | 1.1 | 0.4 | 1.67 | | | |
| Beeswax | 0.95 | 0.82 | 3.43 | | | |
| Brick, common | 1.75 | 0.22 | .920 | 0.42 | .001 7 | 0.71 |
| Brick, hard | 2.0 | 0.24 | 1.00 | 0.75 | .003 1 | 1.3 |
| Chalk | 2.0 | 0.215 | .900 | 0.48 | .002 0 | 0.84 |
| Charcoal, wood | 0.4 | 0.24 | 1.00 | 0.05 | .000 21 | 0.088 |
| Coal, anthracite | 1.5 | 0.3 | 1.26 | | | |
| Coal, bituminous | 1.2 | 0.33 | 1.38 | | | |
| Concrete, light | 1.4 | 0.23 | .962 | 0.25 | .001 0 | 0.42 |
| Concrete, stone | 2.2 | 0.18 | .753 | 1.0 | .004 1 | 1.7 |
| Corkboard | 0.2 | 0.45 | 1.88 | 0.025 | .000 1 | 0.04 |
| Earth, dry | 1.4 | 0.3 | 1.26 | 0.85 | .003 5 | 1.5 |
| Fiberboard, light | 0.24 | 0.6 | 2.51 | 0.035 | .000 14 | 0.058 |
| Fiber hardboard | 1.1 | 0.5 | 2.09 | 0.12 | .000 5 | 0.2 |
| Firebrick | 2.1 | 0.25 | 1.05 | 0.8 | .003 3 | 1.4 |
| Glass, window | 2.5 | 0.2 | .837 | 0.55 | .002 3 | 0.96 |
| Gypsum board | 0.8 | 0.26 | 1.09 | 0.1 | .000 41 | 0.17 |
| Hairfelt | 0.1 | 0.5 | 2.09 | 0.03 | .000 12 | 0.050 |
| Ice (32°) | 0.9 | 0.5 | 2.09 | 1.25 | .005 2 | 2.2 |
| Leather, dry | 0.9 | 0.36 | 1.51 | 0.09 | .000 4 | 0.2 |
| Limestone | 2.5 | 0.217 | .908 | 1.1 | .004 5 | 1.9 |
| Magnesia (85%) | 0.25 | 0.2 | .837 | 0.04 | .000 17 | 0.071 |
| Marble | 2.6 | 0.21 | .879 | 1.5 | .006 2 | 2.6 |
| Mica | 2.7 | 0.12 | .502 | 0.4 | .001 7 | 0.71 |
| Mineral wool blanket | 0.1 | 0.2 | .837 | 0.025 | .000 1 | 0.04 |
| Paper | 0.9 | 0.33 | 1.38 | 0.07 | .000 3 | 0.1 |
| Paraffin wax | 0.9 | 0.69 | 2.89 | 0.15 | .000 6 | 0.2 |
| Plaster, light | 0.7 | 0.24 | 1.00 | 0.15 | .000 6 | 0.2 |
| Plaster, sand | 1.8 | 0.22 | .920 | 0.42 | .001 7 | 0.71 |
| Plastics, foamed | 0.2 | 0.3 | 1.26 | 0.02 | .000 08 | 0.03 |
| Plastics, solid | 1.2 | 0.4 | 1.67 | 0.11 | .000 45 | 0.19 |
| Porcelain | 2.5 | 0.22 | .920 | 0.9 | .003 7 | 1.5 |
| Sandstone | 2.3 | 0.22 | .920 | 1.0 | .004 1 | 1.7 |
| Sawdust | 0.15 | 0.21 | .879 | 0.05 | .000 2 | 0.08 |
| Silica aerogel | 0.11 | 0.2 | .837 | 0.015 | .000 06 | 0.02 |
| Vermiculite | 0.13 | 0.2 | .837 | 0.035 | .000 14 | 0.058 |
| Wood, balsa | 0.16 | 0.7 | 2.93 | 0.03 | .000 12 | 0.050 |
| Wood, oak | 0.7 | 0.5 | 2.09 | 0.10 | .000 41 | 0.17 |
| Wood, white pine | 0.5 | 0.6 | 2.51 | 0.07 | .000 29 | 0.12 |
| Wool, felt | 0.3 | 0.33 | 1.38 | 0.04 | .000 17 | 0.071 |
| Wool, loose | 0.1 | 0.3 | 1.26 | 0.02 | .000 8 | 0.3 |

From Bolz, R. E. and Tuve, G. L., Eds., *Handbook of Tables for Applied Engineering Science*, 2nd ed., CRC Press, Cleveland, 1973, 177.

## Table 5.4—9
## PROPERTIES OF MOLTEN SALTS

For density in kg/m$^3$, multiply values in g/cm$^3$ by 1,000. For viscosity in N·s/m$^2$, multiply values in centipoise by 0.001.

| Salts | Melting point, °C | Near melting point | | | | +100° (approx.) | | | |
|---|---|---|---|---|---|---|---|---|---|
| | | Temperature, °C | Density, g/cm$^3$ | Electrical conductivity per ohm-cm | Viscosity, centipoise | Temperature, °C | Density, g/cm$^3$ | Electrical conductivity per ohm-cm | Viscosity, centipoise |
| AgBr | 430 | 447 | 5.562 | 2.896 | 3.30 | 547 | 5.458 | 3.113 | 2.53 |
| AgCl | 455 | 467 | 4.861 | 3.868 | 2.24 | 567 | 4.774 | 4.225 | 1.78 |
| AgI | 558 | 607 | 5.526 | 2.43 | 2.95 | 707 | 5.425 | 2.60 | 2.28 |
| AgNO$_3$ | 210 | 217 | 3.954 | 0.692 | — | 317 | 3.852 | 1.122 | 2.55 |
| AlCl$_3$ | 192 | 207 | 1.209 | — | 0.320 | 307 | 0.938 | — | — |
| B$_2$O$_3$ | 450 | 1137 | 1.508 | — | 5020. | 1237 | 1.499 | — | 3580. |
| BeF | 540 | 697 | — | $0.61 \times 10^{-5}$ | 2.62 | 797 | — | $13.77 \times 10^{-5}$ | 0.154 |
| Bu$_4$NBF$_4$ | 162 | 167 | 0.935 | — | 9.00 | 267 | 0.877 | — | 1.91 |
| CaBr$_2$ | 730 | 747 | 3.108 | 1.409 | — | 847 | 3.058 | 1.716 | — |
| CaCl$_2$ | 782 | 787 | 2.078 | 2.059 | 3.34 | 887 | 2.036 | 2.501 | 1.96 |
| CdBr$_2$ | 568 | 587 | 4.054 | 1.097 | 2.73 | 687 | 3.946 | 1.301 | — |
| CdCl$_2$ | 568 | 577 | 3.381 | 1.884 | — | 677 | 3.299 | 2.150 | 1.91 |
| CsCl | 645 | 667 | 2.768 | 1.167 | 1.28 | 767 | 2.662 | 1.474 | 0.94 |
| CsI | 621 | 657 | 3.140 | 0.707 | 1.72 | 757 | 3.022 | 0.916 | 1.26 |
| CuCl | 422 | 527 | 3.618 | 3.703 | 2.54 | 627 | 3.542 | 3.841 | 1.80 |
| HgCl$_2$ | 277 | 287 | 4.336 | $3.49 \times 10^{-5}$ | 1.74 | 387 | 4.050 | $7.447 \times 10^{-5}$ | — |
| HgI$_2$ | 257 | 277 | 5.164 | 0.0282 | 2.53 | 377 | 4.841 | 0.0180 | — |
| KBr | 735 | 747 | 2.116 | 1.639 | 1.18 | 847 | 2.034 | 1.886 | 0.92 |
| KCl | 770 | 787 | 1.518 | 2.203 | 1.15 | 887 | 1.460 | 2.439 | 0.86 |
| K$_2$CO$_3$ | 896 | 907 | 1.892 | 2.053 | — | 1007 | 1.848 | 2.342 | — |
| K$_2$Cr$_2$O$_7$ | 398 | 417 | 2.273 | 0.247 | 11.87 | 517 | 2.204 | 0.507 | — |
| KI | 685 | 727 | 2.404 | 1.369 | 1.45 | 827 | 2.308 | 1.569 | 1.13 |
| KNO$_3$ | 337 | 357 | 1.856 | 0.691 | 2.63 | 457 | 1.783 | 0.986 | 1.61 |
| KOH | 360 | 407 | 1.714 | 2.56 | 2.21 | 507 | 1.670 | 3.14 | 1.25 |
| LaCl$_3$ | 870 | 877 | 3.196 | 1.521 | — | 977 | 3.118 | 1.810 | 3.94 |
| LiBr | 550 | 597 | 2.499 | 4.951 | 1.52 | 697 | 2.433 | 5.470 | 1.10 |
| LiCl | 610 | 637 | 1.490 | 5.864 | 1.59 | 737 | 1.447 | 6.354 | 1.09 |
| Li$_2$CO$_3$ | 618 | 737 | 1.826 | 4.097 | — | 837 | 1.789 | 4.892 | 2 91 |
| LiI | 449 | 487 | 3.093 | 3.967 | 2.17 | 587 | 3.001 | 4.427 | 1.53 |
| LiNO$_3$ | 254 | 277 | 1.768 | 0.928 | 5.85 | 377 | 1.713 | — | 2.95 |
| NaBr | 750 | 787 | 2.309 | 3.018 | 1.28 | 887 | 2.227 | 3.319 | 1.02 |
| NaCl | 800 | 817 | 1.547 | 3.629 | 1.38 | 917 | 1.493 | 3.926 | 0.95 |
| Na$_2$CO$_3$ | 854 | 867 | 1.968 | 2.900 | — | 967 | 1.923 | 3.288 | 1.63 |
| NaF | 980 | 997 | 1.944 | 4.937 | — | 1097 | 1.888 | 5.211 | — |
| NaI | 662 | 677 | 2.726 | 2.292 | 1.45 | 777 | 2.631 | 2.603 | 1.09 |
| Na$_2$MoO$_4$ | 687 | 747 | 2.765 | 1.243 | — | 847 | 2.703 | 1.575 | — |
| NaNO$_2$ | 285 | 297 | 1.801 | 1.329 | 3.04 | 397 | 1.726 | 1.872 | — |
| NaNO$_3$ | 310 | 317 | 1.898 | 1.015 | 2.86 | 417 | 1.827 | 1.453 | 1.77 |
| NaOH | 318 | 357 | 1.767 | 2.44 | 3.79 | 457 | 1.719 | 3.34 | 2.11 |
| Na$_2$WO$_4$ | 698 | 777 | 3.792 | 1.145 | — | 877 | 3.712 | 1.35 | — |
| PbCl$_2$ | 498 | 507 | 4.942 | 1.486 | 4.41 | 607 | 4.792 | 1.957 | 2.69 |
| RbBr | 680 | 697 | 2.699 | 1.125 | 1.45 | 797 | 2.592 | 1.359 | 1.13 |
| RbCl | 715 | 737 | 2.229 | 1.549 | 1.29 | 837 | 2.141 | 1.818 | 0.98 |
| RbI | 640 | 657 | 2.886 | 0.879 | 1.39 | 757 | 2.772 | 1.056 | 1.06 |
| RbNO$_3$ | 316 | 327 | 2.466 | 0.457 | 3.68 | 427 | 2.369 | 0.673 | — |

## Table 5.4—9 (continued)
## PROPERTIES OF MOLTEN SALTS

| Salts | Melting point, °C | Near melting point | | | | +100° (approx.) | | | |
|---|---|---|---|---|---|---|---|---|---|
| | | Temper-ature, °C | Density, g/cm³ | Electrical conductivity per ohm-cm | Viscosity, centipoise | Temper-ature, °C | Density, g/cm³ | Electrical conductivity per ohm-cm | Viscosity, centipoise |
| SnCl₂ | 245 | 267 | 3.339 | 0.884 | — | 367 | 3.214 | 1.493 | — |
| SnCl₄ | −33 | 37 | 2.186 | — | 0.73 | 137 | 1.917 | — | 0.34 |
| SrCl₂ | 875 | 897 | 2.713 | 2.082 | 3.18 | 997 | 2.655 | 2.466 | — |
| TlCl | 429 | 447 | 5.597 | 1.154 | — | 547 | 5.417 | 1.515 | — |
| ZnCl₂ | 275 | 327 | 2.514 | 0.00268 | 2900. | 427 | 2.469 | 0.0323 | — |

*Note*: Tables in the original source are in 10-degree increments, and several other salts are included. "Best equations" for each property are listed, and over 250 keyed references are included.

From Janz, G. J., Dampier, F. W., Lakshiminarayanan, G. R., Lorenz, P. K., and Tomkins, R. P. T., *Molten Salts: Electrical Conductance, Density, and Viscosity Data*, Vol. 1, National Standard Reference Data Series—NBS 15, National Bureau of Standards, Washington, D. C., October 1968.

## Table 5.4—10
## PROPERTIES OF CLEAR FUSED QUARTZ

| Property | English or metric system value | International system of units (SI) value |
|---|---|---|
| Density | 2.2 g/cm$^3$ | 2.2 × 10$^3$ kg/m$^3$ |
| Hardness | 4.9 Mohs' scale | |
| Tensile strength | 7,000 psi | 4.8 × 10$^7$ N/m$^2$ |
| Compressive strength | > 160,000 psi | > 1.1 × 10$^9$ N/m$^2$ |
| Bulk modulus | 5.3 × 10$^6$ psi | 3.7 × 10$^{10}$ N/m$^2$ |
| Rigidity modulus | 4.5 × 10$^6$ psi | 3.1 × 10$^{10}$ N/m$^2$ |
| Young's modulus | 10.4 × 10$^6$ psi | 7.17 × 10$^{10}$ N/m$^2$ |
| Poisson's ratio | .16 | .16 |
| Coefficient of thermal expansion | 5.5 × 10$^{-7}$ cm/cm °C (20°C–320°C) | 5.5 × 10$^{-7}$ m/m °K (293°K–593°K) |
| Thermal conductivity | 3.3 × 10$^{-3}$ g cal cm/cm$^2$ sec °C | 1.4 W/m$^2$ °K |
| Specific heat | .18 g cal/g | 750 J/kg |
| Fusion temperature | 1800°C | 2070°K |
| Softening point | 1670°C | 1940°K |

| Property | English or metric system value | International system of units (SI) value |
|---|---|---|
| Annealing point | 1140°C | 1410°K |
| Strain point | 1070°C | 1340°K |
| Electrical resistivity | 10$^{9.5}$ ohm cm (350°C) | |
| Dielectric properties | (20°C and 1 Mc) | (293°K and 1 Mhz) |
| Constant | 3.75 | 3.75 |
| Strength | 410 volts/mil | 1.6 × 10$^7$ V/m |
| Loss factor | <4 × 10$^{-4}$ | <4 × 10$^{-4}$ |
| Dissipation factor | <1 × 10$^{-4}$ | <1 × 10$^{-4}$ |
| Index of refraction | 1.4585 | 1.4585 |
| Velocity of sound-shear wave | 3.75 × 10$^5$ cm/sec | 3.75 × 10$^3$ m/s |
| Velocity of sound-compressive wave | 5.90 × 10$^5$ cm/sec | 5.90 × 10$^3$ m/s |
| Sonic attenuation | <.033 db/ft Mc | <.11 db/m Mhz |

*Note:* These data apply to a specific grade of commercially available material. The term "fused silica" is often used to include the entire group of materials made by fusing of silica ($SiO_2$). All such products contain small amounts of impurities, but the clear varieties can have a purity of 99.98 Alumina ($Al_2O_3$) is the major impurity. Fused silica has an extremely low coefficient of expansion and does not react with most acids, metals, chlorine, or bromide at ordinary temperatures. It has good mechanical and electrical properties and is almost perfectly elastic. Its radiant transmission is high in the ultraviolet as well as in the visible region. Fused quartz or silica products are available in rod, ribbon, and other solid forms such as tubing, chemical glassware, and other fabricated quartzware.

From Bolz, R. E. and Tuve, G. L., Eds., *Handbook of Tables for Applied Engineering Science*, 2nd ed., CRC Press, Cleveland, 1973, 187. Properties data from *Fused Quartz Catalog*, courtesy of The General Electric Company.

## Table 5.4—11
## PROPERTIES OF PLASTER OF PARIS

The following table gives variations in mechanical properties of pure gypsum (hydrous calcium sulfate) plaster with proportions of water in mix.[a] For density in $kg/m^3$, multiply values in $lbm/ft^3$ by 16.018. For Young's modulus and strength in $MN/m^2$, multiply values in psi by 0.0068948. The unit grams per liter is identical to kilograms per cubic meter.

| Density, lb/ft₃ | Proportion by mass, % | | | Retarder,[b] grams per liter of water | Young's modulus, psi | Compressive strength, psi | Tensile strength, psi | Set time, minutes |
|---|---|---|---|---|---|---|---|---|
| | Water | Plaster | Diatomaceous earth | | | | | |
| 75 | 60 | 100 | 0 | 0.50 | 1,000,000 | 2000 | — | — |
| 71 | 65 | 100 | 0 | 0.25 | 870,000 | 1700 | 274 | 7 |
| 67 | 70 | 100 | 0 | 0.25 | 730,000 | 1500 | 230 | 7 |
| 62 | 80 | 100 | 0 | 0.20 | 600,000 | 1100 | — | 11 |
| 58 | 90 | 100 | 0 | 0 | 500,000 | 800 | 145 | 6.5 |
| 53 | 100 | 100 | 0 | 0 | 440,000 | 700 | — | 10 |
| 48 | 110 | 100 | 0 | 0 | 368,000 | 470 | 103 | 7 |
| 46 | 120 | 100 | 0 | 0 | 316,000 | 400 | 82 | 9 |
| 43 | 130 | 100 | 0 | 0 | 273,000 | 330 | 72 | 9.5 |
| 40 | 140 | 100 | 0 | 0 | 225,000 | 280 | 69 | 9.5 |
| 38 | 150 | 100 | 0 | 0 | 190,000 | 240 | — | 10 |
| 37 | 160 | 100 | 8 | 0 | 164,000 | 200 | 51 | 12 |
| 36 | 170 | 100 | 8 | 0 | 131,000 | 170 | — | 12 |

[a]Poisson's ratio 0.24 and independent of mix.
[b]Sodium citrate.

From Barron, K. and Larocque, G. E., The development of a model for a mine structure, in *Proc. Rock. Mech. Symp.*, Mines Branch of the Department of Energy, Mines, and Resources, Ontario, Canada. With permission.

## Table 5.4—12
## REFLECTANCE AND APPEARANCE OF COLORS

The following table gives the percentage reflectance in daylight of typical glossy- or smooth-surface finishes on wood, paper, metal, and other materials. So-called 'fluorescent" colors are not included.

| Descriptive name | Light reflected, percent | Federal color No.[a] |
|---|---|---|
| White | 85 | – |
| Light colors | | |
| Light ivory | 75 | 13711 |
| Soft yellow | 75 | 13695 |
| Cream; off-white | 75 | – |
| Light buff; light gray | 70 | – |
| Peach | 64 | 12648 |
| Suntan | 60 | 13613 |
| Light green | 55 | 14516 |
| Light blue | 50 | 15526 |
| Medium colors | | |
| Brilliant yellow | 58 | 13538 |
| Highlight buff | 55 | 13578 |
| Clear green | 50 | – |
| Pearl gray | 46 | 16492 |
| Wood finish, maple | 40 | – |
| Dark colors | | |
| Light navy gray | 28 | 16251 |
| Medium green | 25 | 14277 |
| Vivid orange | 23 | 12246 |
| Clear blue | 19 | 15177 |
| Radiation purple | 15 | 17142 |
| Wood finish, oak or walnut | 15 | – |
| Medium navy gray | 14 | 16187 |
| Light red | 13 | – |
| Medium brown | 10 | – |
| Wood finish, red mahogany | 9 | – |
| Fire red | 7 | 11105 |
| Passive green | 7 | 14077 |
| Maroon; dark green | 7 | – |
| Deep navy gray | 7 | 16081 |
| Marine Corps green | 4 | 14052 |
| Dark brown | 4 | – |

[a]Federal Color Standards No. 595. Color numbers beginning with digit 1 are gloss finish.

*Appearance of colors*

Intense or saturated hues protrude.

Incandescent lighting tends to dull the green, blue, lavender, and purple shades. Olive green appears brown.

"Cool white" fluorescent lamps make reds appear less bright, and they somewhat dull yellow and blue shades. Illuminated by "warm white" fluorescent lighting, the shades of gray and olive green are changed, and the blues and purples appear dull.

Colored light sources change the hue of most pigment colors, except those that are nearly the same color as the light.

From Bolz, R. E. and Tuve, G. L., Eds., *Handbook of Tables for Applied Engineering Science*, 2nd ed., CRC Press, Cleveland, 1973, 682.

## Table 5.4—13
## WAXES AND RELATED MATERIALS

Most waxes are commercially available in several grades, depending on source, refinement, bleaching, or processing. Thus, the properties of a single type of wax may vary over a rather wide range, and the values given in this table should be considered only as typical.

| Name, source, type, color | Specific gravity | Melting point, °C | Flash point, °C | Acid value[a] | Iodine value[b] | Saponifi- cation value[a] | Dielectric constant, at $10^6$ Hz |
|---|---|---|---|---|---|---|---|
| **VEGETABLE WAXES** | | | | | | | |
| Bayberry (myrtle) | .93 | 50 | | 3.5 | 3 | 212 | |
| Candelilla (brown) from Mexico | .98 | 68 | 241 | 16 | 24 | 57 | 2.5 |
| Carnauba (palm) from Brazil | 1.00 | 85 | 300 | 6 | 12 | 83 | 2.9 |
| Castor oil (hydrogenated) | .98 | 86 | 315 | 2 | 4 | 180 | |
| Cotton (yellow) | .96 | 80 | | 30 | 24 | 70 | |
| Esparto (grass) | .99 | 74 | 255 | 25 | 16 | 68 | |
| Japan (sumac) | .98 | 51 | 200 | 10 | 9 | 220 | 3.1 |
| Ouricuri (palm) from Brazil | 1.02 | 82 | 280 | 12 | 8 | 80 | |
| Sugar cane (brown) | .97 | 80 | | 15 | 20 | 60 | 2.9 |
| **ANIMAL WAXES** | | | | | | | |
| Beeswax, yellow | .96 | 64 | 245 | 20 | 10 | 93 | 2.8 |
| Chinese insect | .96 | 82 | | 10 | 2 | 90 | 3.7 |
| Lanolin (sheepswool) refined | .94 | 40 | | 8 | 25 | 107 | |
| Shellac (insect) | .97 | 77 | | 15 | 5 | 110 | |
| Spermaceti (whale) | .93 | 45 | 245 | 2 | 4 | 123 | 8. |
| **MINERAL, SYNTHETIC, WAXLIKE** | | | | | | | |
| Ceresin | .90 | 70 | | 0 | 8 | 0 | |
| Microcrystalline[c] | .93 | 70 | 260 | | | | 2.3 |
| Moutan (lignite) | 1.02 | 85 | | 40 | | 100 | |
| Ozocerite | .90 | 75 | | 0 | 8 | 0 | 2.4 |
| Paraffin[c] | .92 | 60 | 205 | | | | 2.2 |
| Polyethylene | .93 | 100 | 315 | 15 | | 25 | |
| Polyethylene glycol | 1.1 | 45 | 250 | | | | |

[a]*Acid value* is milligrams of KOH to neutralize the free fatty acids. *Saponification value* is milligrams of KOH for complete saponification.

[b]*Iodine value* is a measure of unsaturated linkages present and is expressed as grams iodine absorbed per 100 g of sample.

[c]There are so many grades of paraffin wax and microcrystalline petroleum wax that the properties of any specific wax should be checked.

*Note*: The *index of refraction* of most waxes is between 1.4 and 1.5

### REFERENCES

Bennett, H., *Industrial Waxes,* Vol. 1, Chemical Publishing, New York, 1963.

Weast, R. C., Ed., *Handbook of Chemistry and Physics,* 52nd ed., Chemical Rubber, Cleveland, 1971.

Lange, N. A., Ed., *Handbook of Chemistry,* 9th ed., Handbook Publishers, Sandusky, Oh., 1956.

Davidsohn, A. and Milwidsky, B. M., *Polishes,* 4th ed., CRC Press, Cleveland, 1968.

Knaggs, N. S., *Adventures in Man's First Plastic,* Reinhold, New York, 1947.

From Bolz, R. E. and Tuve, G. L., Eds., *Handbook of Tables for Applied Engineering Science,* 2nd ed., CRC Press, Cleveland, 1973, 159.

# Section 6

# Ceramic Materials

# Section 6

# CERAMIC MATERIALS

## 6.1    GLASSES AND GLASS-CERAMICS*

### D. E. Campbell and H. E. Hagy

Oxide glasses and glass-ceramics comprise a host of materials of widely diverse compositions and properties. It is important to remember that although the properties of a glass or glass-ceramic are intrinsically a function of composition, other factors can often have important, sometimes overriding, effects on observed properties. Such factors include atmospheric weathering and thermal history. Exposure to weathering conditions obviously can drastically modify the surface chemistry behavior of these materials. Thermal history, on the other hand, can alter molecular structure and phase constitution. Therefore, chemical durability, density, refractive index, electrical resistivity, etc. can be measurably affected by the time-temperature relationship a glass has experienced on cooling from high temperatures. An important instance of the effect of thermal history is found on prolonged heating of certain glasses, e.g., borosilicates, above the annealing temperature, when degradation of the durability is observed, owing to phase separation into two or more glasses, one of which is highly soluble. Glass-ceramics represent an extreme case because the material, originally a glass, is deliberately heat treated to transform it into a new material whose polycrystalline structure gives rise to a totally different set of properties. Because of these dependencies, glass properties are listed for glasses in the annealed state, and the properties of glass-ceramics are given on the basis of the manufacturer's standard production process. In addition, it should be recognized that the compositions shown are nominal, and, owing to a variety of circumstances such as limitations on raw materials, environmental restrictions, and normal manufacturing fluctuations, changes from the stated compositions are quite possible.

To assist the materials scientist and technologist, the following tables have been organized as follows. Compositions have been listed in Table 6.1—1 primarily according to application with sublisting according to manufacturer, and manufacturer's code

(if it exists). In Table 6.1—2 properties have been listed primarily according to manufacturer. The materials selected for inclusion were chosen largely on the basis of their commercial importance.

## PHYSICAL PROPERTIES

Selected physical properties of major interest are presented in Table 6.1—2. Some explanation is required for those properties unique to glass. In addition, general glass behavior is discussed with regard to the more universally applicable properties.

### 1. Viscosity Reference Tests

**Softening point** — the temperature at which a glass fiber viscously extends under its own weight at a rate of 1 mm/min in a specific test apparatus as described in ASTM Designation C-338. The approximate viscosity level for the temperature is $10^{7.6}$ P.

**Annealing point** — the temperature at which a fiber subjected to a tensile stress viscously elongates, or, alternately, a beam of glass subjected to simple three-point loading viscously bends at specified rates, according to ASTM Designations C-336 and C-598, respectively. The viscosity at the annealing point is approximately $10^{13}$ P. Stresses in glass are relieved by viscous flow at the annealing point in a matter of minutes.

**Strain point** — a temperature derived by extrapolation of the data obtained in the annealing point tests, where the elongation or bending rates are a factor of 31.6 smaller than those obtained at the annealing point. The corresponding temperature is indicative of a viscosity of approximately $10^{14.5}$ P for the test glass. Stresses are substantially relieved by viscous flow at the strain point in a matter of hours.

### 2. Thermal Expansion

The value tabulated represents the average

* Tables follow text in this subsection.

coefficient for the glass over the temperature range of 0 to 300°C. Values quoted are for glass in the well-annealed state, since thermal expansion, like many other physical properties, is thermal history sensitive.

For sealing purposes, the average coefficient from room temperature to the vicinity of the strain point is of primary interest. In general, this coefficient is about 10% higher than the 0 to 300°C value. However, for high reliability in seal design, complete expansion curves should be consulted with expert guidance.

The ASTM measurement procedure adopted for glass and glass-ceramics is designation E-228.

### 3. Young's Modulus

Room temperature values are listed. The change with temperature is small, generally being less than 10% over a very wide temperature range. Measurements are made in accordance with ASTM Designation C-623.

### 4. Electrical Properties

Glasses are used in the electrical industry for insulators, lamp and electronic tube enclosures, substrates under conductive coatings, sealing beads, and fuse bodies. The desirable properties of electrical glasses include:

> High dielectric strength
> High volume resistivity
> Low power and loss factors
> Transparency
> Ability to hold a vacuum.

For a better acquaintance with these terms, a brief discussion of each is presented.

**Dielectric strength** — Researchers in the dielectric strength field believe there are two types of dielectric failure — one thermal, the other intrinsic. In thermal breakdown, the sample temperature increases, thereby lowering the sample's electrical resistance. This lower electrical resistance causes greater sample heating until dielectric failure occurs. In disruptive dielectric failure, the sample temperature does not increase. This type of failure is usually associated with voids and defects in the material.

Measurements are made according to ASTM Designation D-149, and the values are very high compared to most other materials.

**Volume resistivity** — The dependence of volume resistivity on temperature is expressed as

$$\log \rho = A + B/T, \tag{1}$$

where A and B are constants and T is in degrees Kelvin.

Measurements are made according to ASTM Designation C-657, and as with the dielectric strength dc $\rho$ values are high.

**Dielectric properties** — Glass is a dielectric which passes a displacement current when ac fields are applied. If the dielectric is ideal, this current is 90 degrees ahead of the applied voltage. Actually, the current through the capacitor is

$$I = V \left( \frac{\omega \epsilon_0 A}{t} \right) (K'' + jK'). \tag{2}$$

In this equation, V is the voltage, $\omega$ is the angular frequency, $+ j$ designates a phase 90 degrees ahead of the applied voltage, $\epsilon_0$ is the dielectric constant of air, A is the capacitor area, t is the capacitor thickness, $K'$ is the relative dielectric constant, and $K''$ is the relative loss factor.

Thus, a good dielectric has a low $K''/K'$ ratio. This ratio is called the loss tangent (tan $\delta$). The power consumed is equal to the product of voltage and current of the real part of the equation

$$W = VI = V^2 \left( \frac{\omega \epsilon_0}{t} \right) (K'' + jK'). \tag{3}$$

Table 2–2 lists some of the electrical properties of glasses.

The dielectric properties were measured according to ASTM Designation D-150.

In determining a particular material for a dielectric application, it may be worthwhile to observe that, in addition to the obvious effect of high dielectric constant on the capacitance of the circuit element, the dielectric strength may be more important. According to Equation 3, the amount of energy a capacitor can store varies as the first power of the dielectric constant and the second power of the voltage. Thus, a material with twice the dielectric strength is as effective as a material with four times the dielectric constant.

## CHEMICAL PROPERTIES — CORROSION RESISTANCE

More commonly referred to as chemical durability, this property is defined by ASTM as "the lasting quality (both physical and chemical) of a glass surface. It is frequently evaluated, after prolonged weathering in storage, in terms of chemical and physical changes in the glass surface, or in terms of change in contents of a vessel."

Chemical durability is strongly dependent upon composition. In addition, previous thermal history and mode of forming can have pronounced effects. For glass-ceramics, the amount and composition of the crystalline and glass phases, as well as the microstructure, have effects that may well substantially alter the intrinsic durability expected on the basis of the overall bulk composition.

Generally, with but a few exceptions, silicate glasses and glass-ceramics have superior resistance to chemical attack compared to most materials. Indeed, these materials are relatively inert to the oxidizing action of all the gaseous nonmetals, except fluorine, with which they will react. Neutral or acid salts, e.g., NaCl, $K_2SO_4$, $Na_2S_2O_7$, $KNO_3$, etc., in their molten or aqueous dissolved state are usually without detrimental effects. On the other hand, under the reducing conditions of some molten metals, e.g., aluminum, alkalies, and even hydrogen at elevated temperatures, glasses and glass-ceramics may be seriously degraded. With respect to liquids, these materials are resistant to a wide variety of solutions and solvents. As a rough guide, the degree of attack of glasses and glass-ceramics by the following reagents decreases in this order: hydrofluoric acid, strong alkalies, weak alkalies, chelates, acid, water, neutral salts, organic solvents.

Attack of glass and glass-ceramic surfaces may be evidenced in a number of ways, including hazing, iridescence, etching or pitting of a surface, surface electrical leakage, change in wettability, and contamination of stored solutions. In most instances, such degradation, whether by atmospheric gases or by solution attack, is associated with the presence of water. Thus, tests for evaluating chemical durability usually involve water and include acid, base, water, and high humidity (weathering).

Acid attack by most acids except hydrofluoric involves selective extraction of the alkali (fluxing or network modifying) elements, principally by an exchange of ions in the glass for hydrogen ions in the solution. Since the protons are hydrated, some hydration of the surface layer is concomitant. Ideally, the process is diffusion-controlled as a result of buildup of the surface reaction layer so that the amount of material extracted decreases with the square root of time. In practice, the buildup of constituents in the attacking solution attenuates the square root of time dependence. The temperature effect is small; the rate of attack increases only about one and a half times for each 10°C. The effect of pH and, aside from hydrofluoric and phosphoric acids, the kind of acid is largely without effect. In the case of hydrofluoric acid[a] and, to a considerably lesser extent, phosphoric acid,[b] the attack involves complete disintegration of the glass structure. In these instances, the rate of attack increases linearly with time, and the temperature effect may increase the attack by as much as twofold per 10°C temperature rise.

Strong alkali attack for most silicate glasses and glass-ceramics is quite severe. Most glasses and glass-ceramics will lose between 75 and 330 $\times$ $10^{-4}$ mm/day when exposed to 5 wt/wt % NaOH at 95°C. Since the process involves destruction of the basic silica network much like the action of hydrofluoric acid, the attack rate is quite dependent on alkali concentration (pH) and is linear with time. The reaction rate approximately doubles for each unit pH rise during the attack of a durable chemically resistant borosilicate. The temperature effect is about a twofold increase in rate for each 10°C temperature rise. Weak alkalies such as ammonium hydroxide do not have nearly so drastic an effect, although the attack is much more severe than by acids.

Water attack is not nearly so harsh as alkaline attack, but it can approach the severity of acid degradation. The mechanism is thought to consist initially of leaching, similar to that of acids, followed by alkaline attack as the water becomes alkaline with buildup of alkaline leach products. The temperature effect on the rate of water attack will vary accordingly, corresponding to that of acid at pH levels of less than 7 and approaching that of alkaline attack, as the pH trends higher. In the vapor state, water, i.e., steam, when properly handled, is usually without effect on chemically resistant compositions. Wet steam (untrapped), on the other hand, can be severely corrosive.

Weathering refers to corrosion by atmospheric agents such as water vapor and other reactive gases

[a]Chemically resistant borosilicate glass dissolves in 50% (wt/wt) hydrofluoric acid at a rate of about 2.5 cm/day at room temperature.
[b]Hot 80% (wt/wt) phosphoric acid dissolves chemically resistant borosilicate glass at the rate of about 127 $\times$ $10^{-4}$ mm/day at 180°C.

such as sulfur dioxide. The mechanism initially involves selective removal of alkaline components, which, unless neutralized by acidic atmospheric components such as $SO_2$, accumulate on the surface and become the focus of an alkaline attack. Cleaning will often delay any appearance of visible degradation.

To be meaningful, tests for the chemical durability of glass and glass-ceramics should be performance related. If the material is to be used as a container, the release of glass components into the contents of the vessel is important; if it is to be used architecturally, appearance is important; if employed as a camera lens, the conditions for testing would be expected to be mild. Thus, evaluation of all glasses and glass-ceramics using the same test is not practical. Nevertheless, some rough universal guide is essential to the materials scientist. In the following tabulation, the durability ratings of the materials listed are defined:

**Definitions of Chemical Durability Ratings**

| Class | Acid<br>95°C-5 w/w % HCl–24 hr | Water<br>90°C–4 hr | Weathering<br>50°C–98% RH-3 Mo |
|-------|-------------------------------|--------------------|--------------------------------|
| 1 | $< 40 \times 10^{-7}$ cm | $< 5 \times 10^{-6}$ | No discernible surface change |
| 2 | 25 to $250 \times 10^{-7}$ | 5 to $30 \times 10^{-6}$ cm | Slight discernible surface change |
| 3 | 25 to $250 \times 10^{-6}$ | $> 30 \times 10^{-6}$ cm | Serious surface change |
| 4 | $> 25 \times 10^{-5}$ cm | – | – |

One must qualify the above effort to provide in one tabulation unified classifications of glasses of such diverse composition as estimates and, inasmuch as possible, upper limits. That is, most glasses will very likely perform better than shown. This is particularly true for the estimates for water attack, wherein the table shows the maximum depth of layer affected in 4 hr at 90°C. Since water attack, to a large extent, proceeds as the square root of time, most of the attack will be evident in the early stages of exposure. All tests upon which the estimates have been based are static tests. Except for the water test, which utilizes 40- to 50-mesh powdered glass, the tests presume exposure of surfaces as formed. The weathering classification relates to the appearance of the surface immediately upon removal from the test chamber.

For quantitative techniques for evaluating durability, one is advised to consult the ASTM Standards, C-225. The United States Pharmacopoeia and the National Formulary contain adapted versions of the same tests which are primarily oriented to container applications. Other than these and other like tests promulgated by corresponding agencies in other countries, there are no standardized test procedures. However, many specialized tests are described in the literature, and those interested are advised to consult the bibliography published by the International Glass Commission (ICG) in 1965 and 1972–73 (Parts 1 and 2).

**Acknowledgments**

The authors gratefully acknowledge the assistance of several of their colleagues, including T. S. Magliocca and G. B. Hares – Compositions; W. H. Barney – Electrical Properties; P. B. Adams – Corrosion Resistance; and G. W. McLellan for his suggestions relative to format and content. Mrs. E. L. Cross is also to be thanked for her arduous efforts, particularly in transcribing the data into tabular form.

## Table 6.1—1
## COMPOSITIONS OF GLASSES AND GLASS-CERAMICS

Electrical

| Type | Manufacturer | Manufacturing code | Application | Description | $SiO_2$ | $Al_2O_3$ | $B_2O_3$ |
|------|--------------|--------------------|-------------|-------------|---------|-----------|----------|
| Sealing | Corning | 0120 | Dumet; Ni-Fe | Potash-soda-lead | 56.2 | 1.6 | |
| | Corning | 1720 | Series; W; Mo | Lime-magnesia-alumino silicate | 60.7 | 17.3 | 5.0 |
| | Corning | 1990 | Fe | Lead-potash | 41.0 | | |
| | Corning | 3320 | W | Borosilicate | 76.5 | 2.6 | 14.0 |
| | Corning | 7040 | Mo | Soda-potash-boro-silicate | 66.9 | 2.6 | 22.4 |
| | Corning | 7050 | Mo; Ni-Fe-Co | Soda-potash-boro-silicate | 68.0 | 1.9 | 23.6 |
| | Corning | 7052 | Ni-Fe-Co | Alkali-barium-boro-silicate | 65.3 | 7.4 | 17.8 |
| | Corning | 7056 | Ni-Fe-Co | Alkali borosilicate | 69.2 | 3.6 | 17.6 |
| | Corning | 7070 | Series; W | Borosilicate | 71.0 | 1.2 | 25.7 |
| | Corning | 7570 | Solder glass | Lead borosilicate | 3.7 | 10.7 | 11.2 |
| | Corning | 7581 | Color TV seal | Lead-barium-boro-silicate | 2.0 | | 8.3 |
| | Corning | 7720 | W | Soda-lead-borosilicate | 73.5 | 1.6 | 14.4 |
| | Owens-Illinois | EN-1 | Ni-Fe-Co | Borosilicate | | | |
| | Owens-Illinois | KG-12 | Dumet; Ni-Fe | Lead | | | |
| | Owens-Illinois | EZ-1 | Series; W; Mo | Alkali boroalumino-silicate | | | |
| | Owens-Illinois | K-704 | Mo | Borosilicate | | | |
| | Owens-Illinois | K-705 | Mo; Ni-Fe-Co | Borosilicate | | | |
| | Owens-Illinois | ES-1 | Series; W | Borosilicate | | | |
| | Owens-Illinois | K-772 | W | Lead borosilicate | | | |
| | Owens-Illinois | KG-1 | Dumet; Ni-Fe, Pt | Potash-soda-lead | | | |
| TV | Corning | 0137 | Color, neck | Potash lead | 51.2 | 0.7 | |
| | Corning | 0138 | Color, funnel | Alkali-alkaline earth-lead | 50.3 | 4.7 | |
| | Corning | 9008 | Black and white | Alkali-lead-barium | 65.5 | 3.5 | |
| | Corning | 9068 | Color panel | Strontium | 63.2 | 2.0 | |
| | Owens-Illinois | EG-19 | Color neck | Lead potash | | | |
| | Owens-Illinois | TH-6 | Color | Lead | | | |
| | Owens-Illinois | TL-10 | Color | Strontium | | | |
| | Owens-Illinois | TM-5K | Black and white | Barium-lead | | | |
| Components | Corning | 1723 | Resistor | Alkaline earth aluminosilicate | 57.3 | 15.7 | 3.9 |
| | Corning | 7059 | Substrates | Barium borosilicate | 49.5 | 11.0 | 14.2 |
| | Corning | 8871 | Capacitor | Soda-potash-lead | 41.6 | 0.2 | |
| | Corning | 9362 | Reed switch | Alkali lead | 50.2 | 1.7 | |

## Table 6.1—1 (continued)
### COMPOSITIONS OF GLASSES AND GLASS-CERAMICS

Electrical

| Type | $Na_2O$ | $K_2O$ | $Li_2O$ | MgO/CaO | BaO | PbO | $As_2O_3/$ $Sb_2O_3$ | Other |
|------|------|------|------|---------|-----|-----|-----------|-------|
| Sealing | 4.1 | 8.8 | | | | 28.1 | 0.3/0.5 | |
| | 1.0 | | | 7.4/8.6 | | | | |
| | 4.8 | 11.5 | 2.0 | | | 40.0 | /1.0 | |
| | 4.2 | 1.5 | | | | | /0.5 | 0.7 $U_3O_8$ |
| | 4.8 | 3.0 | | | | | 0.2/ | |
| | 6.1 | 0.1 | | | | | 0.3/ | |
| | 2.2 | 3.1 | 0.6 | | 2.8 | | | 0.3 F; 0.1 Cl |
| | 0.7 | 8.0 | 0.7 | | | | 0.2/ | |
| | 0.5 | 0.8 | 0.7 | | | | | 0.1 Cl |
| | | | | | | 74.4 | | |
| | | | | | 2.0 | 75.2 | | 12.4 ZnO |
| | 3.8 | | | | | 5.7 | 0.9/ | |
| TV | 0.5 | 12.7 | | | | 28.6 | 0.4/0.4 | 5.2 SrO |
| | 6.1 | 8.4 | | 2.9/4.3 | 0.2 | 22.5 | 0.2/0.1 | 0.1 SrO; 0.1 F |
| | 6.2 | 6.6 | 0.5 | | 12.1 | 3.5 | 0.2/0.4 | 0.3 $Rb_2O$; 0.2 $TiO_2$, 0.05 $MnO_2$; 0.9 F |
| | 7.1 | 8.8 | | 0.8/1.8 | 2.4 | 2.2 | 0.2/0.4 | 10.2 SrO; 0.5 $TiO_2$, 0.5 $CeO_2$; 0.2 F |
| Components | 0.1 | | | 6.6/9.4 | 5.9 | | 0.1/ | |
| | | | | | 25.0 | | 0.4/ | |
| | 2.3 | 6.1 | 1.0 | | | 48.8 | | |
| | 4.2 | 8.6 | | | | 27.2 | 0.1/ | 6.95 $Fe_3O_4$; 1.1 FeO |

**Table 6.1—1 (continued)**
**COMPOSITIONS OF GLASSES AND GLASS-CERAMICS**

Optical

| Type | Manufacturer | Manufacturing code | Major use | Description | $SiO_2$ | $Al_2O_3$ | $B_2O_3$ | $P_2O_5$ |
|---|---|---|---|---|---|---|---|---|
| Transmitting | Corning | 2403 | Sharp cut red filter | Soda-zinc | 67.5 | | 11.5 | |
| | Corning | 4602 | Heat absorbing | Aluminophosphate | 18.5 | 14.1 | | 57.3 |
| | Corning | 5543 | Blue filter | Soda-potash-lead | 65.5 | 1.3 | | |
| | Corning | 7910 | UV transmitting | 96% silica | 96.6 | 0.3 | 3.0 | |
| | Corning | 7913 | IR transmitting | 96% silica | 96.6 | 0.3 | 3.0 | |
| | Corning | 7940 | Window-heat shock resistant | 100% silica | 99.9 | | | |
| | Corning | 8463 | Radiation absorbing | Lead borosilicate | 5.0 | 3.0 | 9.7 | |
| | Corning | 9741 | UV transmitting | Soda-alumina-borosilicate | 66.5 | 5.6 | 23.7 | |
| | Corning | 9863 | UV transmitting visible-absorb. | Alkaline earth phosphate | | 5.4 | | 66.4 |
| Refracting | Corning | 8039 | Ophthalmic flint | Alkali-lead | 34.8 | 2.0 | | |
| | Corning | 8097 | Photochromic ophthalmic | Lithium-lead-barium-borosilicate | 55.9 | 8.6 | 16.5 | |
| | Corning | 8260 | Camera lens | Alkali borosilicate | 68.8 | | 11.4 | |
| | Corning | 8316 | Ophthalmic Ba flint | Baria-lead-soda zirconia | 39.0 | | 3.5 | |
| | Corning | 8361 | Ophthalmic flint | Potash-soda-lime zinc | 68.0 | 2.0 | | |
| | Corning | 8371 | Lens element | Lead-potash-soda | 44.9 | 1.0 | | |
| | Corning | 8395 | Dense flint lens | Potash-lead | 31.1 | | | |
| Reflecting | Corning | 7940 | Teles. mirror | 100% silica | 99.9 | | | |
| | Corning | 7971 | Teles. mirror | Titanium sili. | 92.6 | | | |
| | Owens-Illinois | Cer-vit C-101 | Teles. mirror | β-Spodumene, s.s. | 66.4 | 21.4 | | |

## Table 6.1—1 (continued)
## COMPOSITIONS OF GLASSES AND GLASS-CERAMICS

| Type | | Optical | | | | | | | |
| --- | --- | --- | --- | --- | --- | --- | --- | --- | --- |
| | $Na_2O$ | $K_2O/Li_2O$ | CaO | BaO/PbO | ZnO/CdO | $TiO_2$ | $ZrO_2$ | $As_2O_3/$ $Sb_2O_3$ | Other |
| Transmitting | 5.5 | | | | 12.7/1.0 | | | /1.1 | 0.7 Se; 0.2 S |
| | 0.9 | /0.2 | | | 4.2/ | | | | 1.3 FeO; 2.7 SnO; 1.4 Cl; 0.1 $SO_3$ |
| | 6.0 | 3.5/ | 0.5 | /23.3 | | | | | 0.1 CoO |
| | | | | /82.0 | | | | /0.2 | |
| | 2.2 | 0.1/0.7 | 0.5 | | | | | | 1.1F |
| | | | 5.2 | 17.8/ | | | | | 2.5 Cl; 0.9 NiO; 1.8 CoO |
| Refracting | 1.6 | 7.6/ | | /50.8 | | 3.8 | 1.0 | 0.3/ | 0.2 Ag; 0.2 Cl; 0.2 Br; 0.2 F; 0.04 CuO |
| | | /2.6 | | 7.1/4.7 | | | 2.4 | | |
| | 9.3 | 8.4/ | 0.1 | 0.9/ | | 0.2 | | 0.9/ | |
| | 8.0 | | 5.0 | 20.0/14.2 | | 3.0 | 7.3 | 0.1/0.1 | |
| | 7.9 | 9.4/ | 8.2 | | 3.5/ | 0.4 | | 0.1/0.2 | |
| | 3.0 | 5.5/ | | /45.6 | | | | /0.1 | |
| | 1.0 | 2.0/ | | /65.9 | | | | 0.1/0.1 | |
| Reflecting | 0.5 | 0.1/3.9 | 3.6 | | | 7.4 | | | |
| | | | | | | 1.8 | 1.9 | /0.4 | |

## Table 6.1—1 (continued)
## COMPOSITIONS OF GLASSES AND GLASS-CERAMICS

Industrial — Lab/Pharmaceutical

| Manufacturer | Manufacturing code | Major use | Description | $SiO_2$ | $Al_2O_3$ | $B_2O_3$ |
|---|---|---|---|---|---|---|
| Corning | 1720 | Ignition tube | Lime-magnesia-aluminosilicate | 60.7 | 17.3 | 5.0 |
| Corning | 7331 | Gauge | Soda-alumino-borosilicate | 77.6 | 5.3 | 8.8 |
| Corning | 7740 | Lab, process | Soda-borosili-cate | 80.3 | 2.3 | 13.3 |
| Corning | 7800 | Pharmaceutical | Soda-barium borosilicate | 74.5 | 6.4 | 9.0 |
| Corning | 7913 | Lab, gauge | 96% silica | 96.5 | 0.5 | 3.0 |
| Owens-Illinois | R-6 | Lab | Soda-lime | | | |
| Owens-Illinois | KG-33 | Lab | Soda-borosilicate | | | |
| Owens-Illinois | N-51A | Pharmaceutical | Soda-barium-borosilicate | | | |

| Manufacturer | $Na_2O$ | $K_2O$ | $Li_2O$ | MgO | CaO | BaO | $As_2O_3$ | $Sb_2O_3$ | Other |
|---|---|---|---|---|---|---|---|---|---|
| Corning | 1.0 | 0.2 | | 7.4 | 8.6 | | 0.5 | | |
| Corning | 5.3 | 0.4 | 0.4 | | | | 1.0 | 0.9 | 0.1 $TiO_2$ ; 0.1 $ZrO_2$ |
| Corning | 4.0 | | | | | | | | |
| Corning | 6.2 | 0.8 | | | 0.5 | 2.2 | | | 0.1 F; 0.1 Cl |
| Corning | | | | | | | | | |
| Owens-Illinois | | | | | | | | | |
| Owens-Illinois | | | | | | | | | |
| Owens-Illinois | | | | | | | | | |

**Table 6.1—1 (continued)**
**COMPOSITIONS OF GLASSES AND GLASS-CERAMICS**

Speciality

| Type | Manufacturer | Manufacturing code | Major use | Description | $SiO_2$ | $Al_2O_3$ |
|---|---|---|---|---|---|---|
| Fiber glass | Johns Manville | E | | Lime aluminosilicate | 54.5 | 14.6 |
| | Johns Manville | 753 | | Soda-lime borosilicate | 63.5 | 5.5 |
| | Johns Manville | 475 | | Borosilicate | 58.0 | 5.8 |
| | Owens-Corning | E | Reinforcement | Lime aluminoborosilicate | 54.0 | 14.0 |
| | Owens-Corning | T | Insulation | Soda-lime | 59. | 4.5 |
| | Owens-Corning | C | Acid resistant | Soda-lime borosilicate | 65. | 4. |
| | Owens-Corning | SF | Insulation | Soda-titania-zirconia | 59.5 | 5. |
| Glass-ceramic | Corning | 0336 | Architectural | $\beta$-Spodumene, s.s. | 64.6 | 20.0 |
| | Corning | 9455 | Heat exchanger | $\beta$-Spodumene, s.s. | 71.8 | 22.9 |
| | Corning | 9606 | Radome | Cordierite-quartz-rutile | 56.0 | 19.8 |

| Type | $B_2O_3$ | $Na_2O$ | $K_2O$ | $Li_2O$ | MgO | CaO | BaO | ZnO | $TiO_2$ | $ZrO_2$ | $Fe_2O_3$ | Other |
|---|---|---|---|---|---|---|---|---|---|---|---|---|
| Fiber glass | 6.8 | 0.8 | | 0.7 | 21.2 | | | | 0.6 | | 0.4 | 0.4 F |
| | 5.5 | 14.6 | 1.0 | | 3.0 | 6.0 | | | | | 0.1 | 0.7 F; 0.2 $SO_3$ |
| | 10.6 | 10.0 | 3.0 | | 0.5 | 2.5 | 5.0 | 4.0 | | | 0.1 | 0.6 F |
| | 10.0 | | | | 4.5 | 17.5 | | | | | | |
| | 3.5 | 11. | 0.5 | | 5.5 | 16. | | | | | | |
| | 5.5 | 8. | 0.5 | | 3. | 14. | | | | | | |
| | 7 | 14.5 | | | | | | | 8. | 4. | | 1.9 F |
| Glass-ceramic | 2.0 | 0.6 | 0.2 | 3.5 | 1.8 | | | 2.2 | 4.4 | | | 0.8 $As_2O_3$ |
| | | | | 5.1 | | | | | 0.1 | | | |
| | | | | | 14.8 | | | | 9.0 | | | 0.3 $As_2O_3$ |

Housewares

| Type | Manufacturer | Manufacturing code | Major use | Description | $SiO_2$ |
|---|---|---|---|---|---|
| Cooking ware | Corning | 0281 | Lids | Soda-lime | 72.7 |
| | Corning | 1710 | Top-of-stove | Lime-magnesia-aluminosilicate | 60.2 |
| | Corning | 7740 | Ovenware | Soda borosilicate | 80.3 |
| | Corning | 9608 | Cookware | $\beta$-Spodumene, s.s. | 69.7 |
| Tableware | Corning | 6720 | Table and ovenware | Soda-zinc-lime-aluminosilicate | 59.4 |
| Appliance | Corning | 9617 | Cooktop | $\beta$-Spodumene, s.s. | 67.4 |

| Type | $Al_2O_3$ | $B_2O_3$ | $Na_2O$ | $K_2O$ | $Li_2O$ | MgO | CaO | ZnO | $TiO_2$ | $ZrO_2$ | $As_2O_3/$ $Sb_2O_3$ | Other |
|---|---|---|---|---|---|---|---|---|---|---|---|---|
| Cooking ware | 1.5 | 0.7 | 14.5 | 0.3 | | 4.1 | 5.7 | | | | /0.2 | 0.2 $SO_3$ |
| | 18.0 | 4.4 | 1.6 | | | 9.0 | 6.7 | | | | | 0.1 $ZrO_2$ |
| | 2.3 | 13.3 | 4.0 | | | | | | | | | |
| | 17.1 | | 0.4 | 0.1 | 2.5 | 2.8 | | 1.0 | 4.8 | 0.1 | 0.5/ | |
| Tableware | 10.1 | 1.4 | 8.5 | 2.2 | | | 4.7 | 9.8 | | | 0.3/ | 3.5 F |
| Appliance | 20.4 | | 0.3 | 0.1 | 3.5 | 1.6 | | 1.2 | 4.8 | | 0.4/ | 0.2 F |

# Table 6.1—1 (continued)
## COMPOSITIONS OF GLASSES AND GLASS-CERAMICS

### Flat

| Type | Manufacturer | Major use | Description | $SiO_2$ | $Al_2O_3$ | $Na_2O$ | $K_2O$ | $MgO$ | $CaO$ | $Fe_2O_3$ | $SO_3$ |
|---|---|---|---|---|---|---|---|---|---|---|---|
| Sheet | Libbey-Owens Ford | Window | Soda-lime | 72.6 | 1.1 | 13.3 | 0.1 | 3.8 | 8.6 | 0.1 | 0.3 |
|  | Pilkington | Window | Soda-lime | 72.7 | 1.4 | 12.8 | 0.7 | 3.8 | 8.2 | 0.14 | 0.2 |
| Plate | Libbey-Owens Ford | Architectural | Soda-lime | 72.2 | 0.1 | 13.9 | – | 2.1 | 11.2 | 0.1 | 0.4 |
| Float | Libbey-Owens Ford | Windshields | Soda-lime | 72.9 | 0.1 | 13.9 | – | 4.0 | 8.6 | 0.1 | 0.3 |
|  | Pilkington | Windshields | Soda-lime | 72.7 | 1.0 | 13.0 | 0.6 | 3.9 | 8.3 | 0.095 | 0.22 |
| Combination sheet-plate-float | Glaverbel | Window-windshield | Soda-lime | 72.2 | 1 | 13.6 | 0.3 | 4.2 | 8.5 | 0.1 | 0.3 |

### Container

| Type | Manufacturer | Manufacturing code | Major use | Description | $SiO_2$ | $Al_2O_3$ | $B_2O_3$ | $Na_2O$ | $K_2O$ | $MgO$ | $CaO$ | $BaO$ | $Fe_2O_3$ | $TiO_2$ | $As_2O_3/Sb_2O_3$ | $SO_3$ |
|---|---|---|---|---|---|---|---|---|---|---|---|---|---|---|---|---|
|  | Thatcher | A-3004 | Bottles | Soda-lime flint | 71.93 | 1.58 | 0.46 | 14.84 | 0.31 | 1.36 | 8.90 | 0.27 | 0.079 | 0.031 | 0.019/ | 0.22 |
|  | Owens-Illinois |  | Containers | Soda-lime |  |  |  |  |  |  |  |  |  |  |  |  |

### Lighting

| Type | Manufacturer | Manufacturing code | Major use | Description | $SiO_2$ | $Al_2O_3$ | $B_2O_3$ | $Na_2O$ | $K_2O$ | $MgO$ | $CaO$ | $BaO$ | $PbO$ | $Fe_2O_3$ | $As_2O_3/Sb_2O_3$ | $SO_3$ | $Cl/F$ |
|---|---|---|---|---|---|---|---|---|---|---|---|---|---|---|---|---|---|
| Bulb | Corning | 0081 | Lamp | Soda-lime | 73.4 | 1.4 |  | 16.2 | 0.4 | 3.4 | 5.0 |  |  |  |  | 0.1 |  |
|  | General Electric | X-4 | Lamp | Soda-lime | 72.5 | 1.3 |  | 15.9 |  | 3.0 | 6.5 |  |  | 0.05 |  | 0.3 |  |
| Tube | Corning | 0010 | Exhaust, flare | Potash-soda-lead | 61.8 | 2.2 |  | 7.0 | 7.3 |  |  |  | 21.5 |  | 0.2/ |  |  |
|  | Corning | 0080 | Fluorescent tube | Soda-lime | 73.6 | 1.4 |  | 16.2 | 0.4 | 3.4 | 4.8 |  |  |  |  | 0.2 |  |
|  | Corning | 0088 | Tubing | Soda-lime | 70.5 | 2.0 |  | 12.2 | 5.3 | 3.0 | 4.2 |  |  |  |  | 0.2 |  |
|  | Pilkington | PWM | Fluorescent tube | Soda-lime | 71.4 | 2.2 |  | 15.0 | 1.7 | 4.0 | 4.6 |  |  |  |  | 0.2 |  |
|  | Owens-Illinois | KG-1 | Exhaust, flare | Potash-soda lead |  |  | 2.6 |  |  |  |  | 0.8 |  |  |  |  |  |
| Sealed beam | Corning | 7251 | Auto head lamps | Soda-boro-silicate | 78.1 | 2.0 | 14.9 | 4.9 |  |  |  |  |  |  |  |  | 0.1/ |

Table compiled by D. E. Campbell and H. E. Hagy.

## Table 6.1—2
### PROPERTIES OF GLASSES AND GLASS-CERAMICS

| | Code | Softening point, °C | Annealing point, °C | Strain point, °C | Working point, °C | 0–300 expansion, $10^{-7}/°C$ | Young's modulus, 10 psi |
|---|---|---|---|---|---|---|---|
| Corning | 0010 | 626 | 432 | 392 | 983 | 93.5 | 9.0 |
| glasses | 0080 | 696 | 514 | 473 | 1,005 | 93.5 | 10.2 |
| | 0081 | 696 | 514 | 473 | 1,013 | 93.5 | 10.3 |
| | 0088 | 700 | 521 | 480 | 1,017 | 92. | 10.6 |
| | 0120 | 630 | 435 | 395 | 985 | 89.5 | 8.6 |
| | 0137 | 661 | 478 | 436 | 977 | 97. | – |
| | 0138 | 670 | 494 | 451 | 970 | 99. | 10.3 |
| | 0281 | 714 | 532 | 491 | 1,024 | 86. | 10.4 |
| | 1710 | 915 | 713 | 669 | 1,189 | 42. | 12.5 |
| | 1720 | 915 | 712 | 667 | 1,202 | 42. | 12.7 |
| | 1723 | 908 | 710 | 665 | 1,168 | 46. | 12.5 |
| | 1990 | 500 | 370 | 340 | 756 | 124. | 8.4 |
| | 2403 | – | – | – | – | – | – |
| | 3320 | 780 | 540 | 493 | 1,171 | 40. | 9.4 |
| | 4602 | 760 | 560 | 519 | 1,033 | 54. | 10.3 |
| | 5543 | 673 | 459 | 417 | 1,076 | 71. | – |
| | 6720 | 780 | 540 | 505 | 1,023 | 78.5 | 10.2 |
| | 7040[a] | 702 | 490 | 449 | 1,080 | 47.5 | 8.6 |
| | 7050[a] | 703 | 501 | 461 | 1,027 | 46. | 8.7 |
| | 7052[a] | 712 | 480 | 436 | 1,128 | 46. | 8.2 |
| | 7056[a] | 718 | 512 | 472 | 1,058 | 51.5 | 9.2 |
| | 7059 | 844 | 639 | 593 | 1,160 | 46. | 9.8 |
| | 7070[a] | – | 496 | 456 | 1,068 | 32. | 7.4 |
| | 7251[a] | 780 | 544 | 500 | 1,167 | 36.7 | 9.3 |
| | 7331 | 800 | 555 | 511 | 1,232 | 43. | 11.0 |
| | 7570 | 440 | 363 | 342 | – | 84[b] | 8.0 |
| | 7581 | – | – | – | – | – | – |
| | 7720[a] | 755 | 523 | 484 | 1,146 | 36. | 9.0 |
| | 7740 | 821 | 560 | 510 | 1,252 | 32.5 | 9.1 |
| | 7800 | 795 | 576 | 533 | 1,189 | 50. | 10.2 |
| | 7910 | 1,500 | 910 | 820 | – | 7.5 | 9.6 |
| | 7913 | 1,530 | 1,020 | 890 | – | 7.5 | 9.6 |
| | 7940 | ~1,550 | 1,050 | 910 | – | 5.6 | 10.5 |
| | 7971 | 1,500 | 1,000 | 890 | – | 0.3 | 9.8 |
| | 8039 | 617 | 475 | 440 | 866 | 87. | – |
| | 8097 | 675 | 511 | 473 | 947 | 51. | 10.4 |
| | 8260 | 713 | 548 | 511 | – | 82. | 12.3 |
| | 8316 | 686 | 553 | 520 | 886 | 91. | – |
| | 8361 | 726 | 543 | 500 | 1,032 | 94. | 10.5 |
| | 8371 | 598 | 429 | 391 | 905 | 88. | 8.4 |
| | 8395 | 547 | 413 | 386 | 760 | 87. | 8.0 |
| | 8463 | 377 | 316 | 300 | – | 104. | 7.5 |
| | 8871 | 527 | 384 | 350 | 783 | 102. | 8.4 |
| | 9008 | 646 | 446 | 408 | 1,002 | 89. | 10.2 |
| | 9068 | 688 | 503 | 462 | 1,005 | 99. | 10.1 |
| | 9362 | 627 | 445 | 405 | 958 | 91.5 | – |
| | 9741 | 705 | 450 | 408 | 1,161 | 39.5 | 7.2 |
| | 9863 | 596 | 482 | 453 | – | 97. | – |

**Table 6.1—2 (continued)**
PROPERTIES OF GLASSES AND GLASS-CERAMICS

| | Code | Softening point, °C | Annealing point, °C | Strain point, °C | Working point, °C | 0–300 expansion, $10^{-7}/°C$ | Young's modulus, 10 psi |
|---|---|---|---|---|---|---|---|
| Corning glass ceramics | 0336 | – | – | – | – | – | – |
| | 9455 | – | – | – | – | – | – |
| | 9606 | – | – | – | – | 57. | 17.2 |
| | 9608 | – | – | – | – | 4–20 | 12.5 |
| | 9617 | – | – | – | – | – | – |
| General Electric glasses | X4 | 710 | 520 | – | – | 93. | – |
| Glaverbel glasses | Flat glass | 724 | 545 | 517 | 1,032 | 80.1$^c$ | – |
| Johns-Manville | E | 846 | 671 | 627 | 1,071 | – | – |
| | 753 | 677 | 527 | 485 | 941 | – | – |
| | 475 | 679 | 529 | 491 | 924 | – | – |
| Owens-Corning glasses | C | 750 | – | – | – | 72. | – |
| | E | 830 | – | – | – | 60. | 10.5 |
| | T | 715 | – | – | – | 80. | – |
| | SF | 675 | – | – | – | 75. | – |
| Owens-Illinois glasses | A-3004 | 732 | 548 | 501 | 1,045 | 84. | – |
| | ES-1 | 735 | 476 | 430 | 1,095 | 33. | – |
| | EN-1 | 716 | 482 | 437 | 1,115 | 47. | – |
| | EG-19 | 665 | 484 | 440 | 990 | 95. | – |
| | EZ-1 | 915 | 715 | 670 | 1,200 | 42. | – |
| | K-704 | 713 | 485 | 443 | 1,065 | 49. | – |
| | K-705 | 715 | 503 | 464 | 1,010 | 47. | – |
| | K-772 | 755 | 520 | 478 | 1,120 | 35. | – |
| | KG-1 | 626 | 435 | 394 | 970 | 94. | – |
| | KG-12 | 632 | 438 | 400 | 980 | 90. | – |
| | KG-33 | 827 | 565 | 513 | 1,240 | 32. | – |
| | N51-A | 798 | 580 | 538 | 1,190 | 50. | – |
| | R-6 | 700 | 525 | 486 | 985 | 93. | – |
| | TH-6 | 670 | 488 | 447 | 995 | 98.5 | – |
| | TL-10 | 691 | 503 | 461 | 1,015 | 99.5 | – |
| | TM-SK | 654 | 451 | 410 | 1,025 | 90. | – |
| Owens-Illinois glass-ceramics | C-101 | – | – | – | – | – | 13.4 |
| Pilkington glasses | Float sheet | 734 | 551 | 522 | 1,047 | 80. | 10.7 |
| | PMW | 707 | 524 | 495 | 1,023 | 92. | – |
| Thatcher | Flint | 709 | 539 | 502 | 986 | – | – |

## Table 6.1—2 (continued)
## PROPERTIES OF GLASSES AND GLASS-CERAMICS

| | $\log_{10}$ dc volume resistivity (ohm-centimeter) | | Loss factor at 20°C | Dielectric constant, 20°C | Refractive index, D line | Density, grams per cubic centimeter | Corrosion resistance | | |
| --- | --- | --- | --- | --- | --- | --- | --- | --- | --- |
| | 250°C | 350°C | | | | | Acid | Water | Weather |
| Corning glasses | 8.9 | 7.0 | 0.16 | 6.7 | 1.540 | 2.86 | 2 | 2 | 2 |
| | 6.4 | 5.1 | 0.9 | 7.2 | 1.510 | 2.47 | 2 | 2 | 3 |
| | 6.4 | 5.1 | 2.1 | 5.0 | 1.510 | 2.47 | 2 | 2 | 3 |
| | 7.5 | 5.9 | 0.42 | 7.1 | 1.512 | 2.47 | 1–2 | 2 | 2 |
| | 10.1 | 8.0 | 0.12 | 6.7 | 1.560 | 3.05 | 2 | 2 | 2 |
| | 10.1 | 8.4 | 0.09 | 8.6 | 1.574 | 3.18 | 4 | 3 | 2–3 |
| | – | – | – | – | – | 3.02 | 4 | 3 | 2–3 |
| | 6.5 | 5.2 | – | – | 1.515 | 2.48 | 2 | 2 | 2–3 |
| | 10.8 | 9.0 | 0.37 | 6.3 | – | 2.52 | 3 | 1 | 1 |
| | 10.8 | 9.0 | 0.38 | 7.2 | 1.53 | 2.54 | 3 | 1 | 1 |
| | 13.5 | 11.3 | 0.16 | 6.3 | 1.545 | 2.64 | 3 | 1 | 1–2 |
| | 10.1 | – | – | – | – | 3.50 | 4 | 2–3 | 2–3 |
| | – | – | – | – | – | – | 2? | 2? | 1? |
| | 8.6 | 7.1 | 0.28 | 4.0 | 1.48 | 2.27 | 1? | 1 | 1 |
| | – | – | – | – | 1.51 | 2.52 | 4 | 3 | 3 |
| | – | – | – | – | 1.53 | 2.84 | 2? | 2? | 2? |
| | – | – | – | – | 1.51 | 2.58 | 2 | 1 | 1 |
| | 9.4 | 7.7 | 0.20 | 4.8 | 1.48 | 2.24 | 4? | 2 | 2 |
| | 8.4 | 6.8 | 0.33 | 4.9 | 1.479 | 2.24 | 4 | 2 | 2 |
| | 9.2 | 7.4 | 0.15 | 5.1 | 1.484 | 2.27 | 3 | 1 | 2 |
| | 10.3 | 8.4 | 0.27 | 5.7 | 1.487 | 2.29 | 2 | 2 | 2 |
| | 13.1 | 11.0 | 0.1 | 5.9 | 1.53 | 2.76 | 4 | 1 | 1 |
| | 11.2 | 9.1 | 0.06 | 4.1 | 1.47 | 2.13 | 2? | 2 | 2? |
| | 8.1 | 6.6 | 0.45 | 4.9 | 1.476 | 2.25 | 2 | 1 | 1 |
| | 7.1 | 5.8 | – | – | 1.486 | 2.32 | 2 | 1 | 1 |
| | 10.6 | – | 0.22 | 15. | – | 5.42 | 4 | 2 | 2 |
| | – | – | – | – | – | – | 4 | 2–3 | 2–3 |
| | 8.8 | 7.3 | 0.23 | 4.6 | 1.487 | 2.35 | 2 | 2 | 1 |
| | 8.1 | 6.6 | 0.40 | 4.6 | 1.474 | 2.23 | 1 | 1 | 1 |
| | 7.0 | 5.7 | – | – | 1.491 | 2.36 | 1 | 1 | 1 |
| | – | – | – | – | 1.458 | 2.18 | 1 | 1 | 1 |
| | 9.7 | 8.1 | 0.04 | 3.8 | 1.458 | 2.18 | 1 | 1 | 1 |
| | 12.4 | 10.7 | 0.001 | 3.8 | 1.458 | 2.20 | 1 | 1 | 1 |
| | 12.0 | 10.3 | <0.002 | 4.0 | 1.484 | 2.21 | 1 | 1 | 1 |
| | – | – | 0.16 | 10.0 | 1.700 | 4.02 | 4 | 2 | 2 |
| | – | – | – | – | 1.523 | 2.54 | 2–3 | 2 | 2 |
| | – | – | – | – | 1.517 | 2.51 | 2 | 2 | 2? |
| | – | – | – | – | 1.653 | 3.52 | 3–4 | 2 | 2 |
| | 8.7 | 6.9 | 0.24 | 7.2 | 1.523 | 2.54 | 2 | 2 | 2 |
| | – | – | – | – | 1.621 | 3.61 | 3–4 | 1–2 | 1–2 |
| | 11.1 | 9.2 | 0.10 | 11.7 | 1.751 | 4.73 | 4 | 1–2 | 1–2 |
| | – | – | – | – | 1.97 | 6.22 | 4 | 1 | 2–3 |

### Table 6.1—2 (continued)
## PROPERTIES OF GLASSES AND GLASS-CERAMICS

| | Log$_{10}$ dc volume resistivity (ohm-centimeter) | | Loss factor at 20°C | Dielectric constant, 20°C | Refractive index, D line | Density, grams per cubic centimeter | Corrosion resistance | | |
|---|---|---|---|---|---|---|---|---|---|
| | 250°C | 350°C | | | | | Acid | Water | Weather |
| Corning glasses (cont.) | 10.9 | – | 0.05 | 8.4 | 1.656 | 3.84 | 4 | 1–2 | 2 |
| | 9.1 | 7.2 | 0.18 | 6.6 | 1.513 | 2.66 | 2 | 2 | 2 |
| | – | – | – | – | – | 2.685 | 2 | 2 | 2 |
| | 9.4 | 7.7 | – | – | – | 3.12 | 2–3 | 2 | 2–3 |
| | 9.4 | 7.6 | – | – | 1.47 | 2.16 | 3 | 2 | 2 |
| | – | – | – | – | – | 2.97 | 4 | 3 | 3 |
| Corning glass-ceramics | – | – | – | – | – | – | 3 | 1 | 1 |
| | – | – | – | – | – | – | 4 | 1 | 1 |
| | 10. | 8.7 | 0.30 | 5.6 | – | 2.6 | 4 | 1–2 | 1 |
| | 8.1 | 6.8 | 0.34 | 6.9 | – | 2.5 | 2 | 1 | 1 |
| | – | – | – | – | – | – | 2 | 1 | 1 |
| General Electric glasses | – | – | – | – | – | – | – | – | – |
| Glaverbel glasses | – | – | – | – | – | – | – | – | – |
| Johns-Manville | – | – | – | – | – | 2.61 | | | |
| | – | – | – | – | – | 2.52 | – | – | – |
| | – | – | – | – | – | 2.61 | – | – | – |
| Owens-Corning glasses | – | – | – | 7.8 | 1.549 | 2.61 | – | – | – |
| | – | – | – | 6.4 | 1.548 | 2.60 | – | – | – |
| | – | – | – | 7.3 | 1.541 | 2.54 | – | – | – |
| | – | – | – | 8.3 | 1.537 | 2.57 | – | – | – |
| Owens Illinois glasses | – | – | – | – | 1.51 | 2.48 | 2 | 2 | 2 |
| | 12.6 | 10.2 | 0.4 | 4.0 | 1.47 | 2.15 | – | – | – |
| | 9.0 | 7.2 | 1.3 | 5.1 | 1.49 | 2.27 | – | – | – |
| | 10.3 | 8.3 | – | – | 1.59 | 3.23 | – | – | – |
| | 10.7 | 8.9 | 2.3 | 6.3 | 1.53 | 2.52 | – | – | – |
| | 9.2 | 7.4 | 1.5 | 5.0 | 1.48 | 2.24 | – | – | – |
| | 8.3 | 6.9 | 1.7 | 4.8 | 1.48 | 2.25 | – | – | – |
| | 9.1 | 7.5 | 0.9 | 4.6 | 1.48 | 2.35 | – | – | – |
| | 8.7 | 6.8 | – | – | 1.54 | 2.85 | – | – | – |
| | 9.9 | 7.8 | 1.0 | 6.7 | 1.56 | 3.05 | – | – | – |
| | 8.1 | 6.6 | 2.1 | 4.6 | 1.47 | 2.23 | 1 | 1 | 1 |
| | 6.8 | 5.4 | 5.7 | 5.9 | 1.49 | 2.36 | 1 | 1 | 1 |
| | 6.6 | 5.2 | 6.1 | 7.2 | 1.52 | 2.53 | 1–2 | 1–2 | 2 |
| | 9.8 | 7.8 | 0.6 | 7.8 | 1.56 | 2.96 | – | – | – |
| | 8.7 | 6.9 | 1.4 | 7.2 | 1.52 | 2.63 | – | – | – |
| | 8.9 | 7.0 | – | – | 1.51 | 2.67 | – | – | – |
| Owens-Illinois glass ceramics | – | – | – | – | – | – | – | – | – |

## Table 6.1—2 (continued)
## PROPERTIES OF GLASSES AND GLASS-CERAMICS

| | Log$_{10}$ dc volume resistivity (ohm-centimeter) | | Loss factor at 20°C | Dielectric constant, 20°C | Refractive index, D line | Density, grams per cubic centimeter | Corrosion resistance | | |
|---|---|---|---|---|---|---|---|---|---|
| | 250°C | 350°C | | | | | Acid | Water | Weather |
| Pilkington glasses | 5.4 | 4.3 | – | 7.5 | 1.518 | 2.497 | – | – | – |
| | 6.0 | 4.4 | – | – | 1.511 | 2.49 | – | – | – |
| Thatcher | – | – | – | – | – | 2.50 | 2 | 2 | 2 |

[a]Prolonged heating in the annealing temperature range may result in serious degradation of the corrosion resistance owing to appearance of a highly soluble second phase.
[b]0 to 200°C range.
[c]20 to 400°C range.

Table compiled by D. E. Campbell and H. E. Hagy.

## 6.2    ALUMINA AND OTHER REFRACTORY MATERIALS

### R. N. Kleiner

Ceramics have been an important class of materials because of their inertness and heat-resistant properties. Although these materials have been used for many years, ceramics are in a very active stage of development. Applications based on unique properties of ceramic materials range from refractory bricks and crucibles for molten metals to miniature electronic devices; from nuclear fuels and control elements to magnetic memory units. The field of application of ceramic materials is diverse; the materials are numerous.

Vast amounts of data have been generated on properties of ceramic materials, and volumes of literature have been written on the subject. In this section, data on the properties of several ceramic materials are presented in a condensed tabulation. It is beyond the scope of this work to reference all of the work done on the materials discussed. The ceramics in this section were selected to show some of the properties of commercially available ceramic materials that make them unique for many applications.

The ceramic materials described in this section are alumina, beryllia, zirconia, mullite, cordierite, silicon carbide, and silicon nitride. The properties tabulated are crystal chemical, thermodynamic, physical, mechanical, thermal, and electrical properties.

The crystal chemical and thermodynamic properties are independent of processing conditions and are listed separately from the other properties. The physical, mechanical, thermal, and electrical properties are a function of the processing conditions. These properties can be modified through process changes to meet specifications for different applications. To show these effects, data on the physical, mechanical, thermal, and electrical properties for some of the materials are given for different levels of density and purity.

### ALUMINA (Al$_2$O$_3$)

Good mechanical strength, inertness, refractoriness, and availability make alumina one of the most widely used of the ceramic materials. Some applications for alumina are electrical insulators, abrasives, cutting tools, radomes, and wear-resistant parts. Alumina is used in cements, refractories, glasses, coatings, cermets, and seals. The ability to densify this material with little or no glass phase present is advantageous in developing properties approaching the theoretical limit for this material. This has resulted in new applications for alumina such as envelopes for high-pressure sodium vapor lamps and microwave windows.

The crystal chemical and thermodynamic properties for alpha alumina are given in Table 6.2—1. Table 6.2—2 lists the physical mechanical, thermal, and electrical properties. In Table 6.2—2, properties also are included on single crystal alumina. Properties are listed for polycrystalline alumina available from some of the alumina component manufacturers. The effects that density and purity levels have on these properties can be determined from the table.

### BERYLLIA (BeO)

Beryllia is characterized by a higher thermal conductivity than any other ceramic. The thermal conductivity of beryllia is about ten times that of alumina and is approximately the same as aluminum. A good mechanical strength in combination with the high thermal conductivity gives beryllia good thermal shock-resistant characteristics.

Beryllia is used in electronic components as an electrical insulator and a heat sink. It is used as crucibles for melting uranium, thorium, and beryllium. Other applications are in the nuclear industry as moderators for fast neutrons, a matrix for fuel elements, shielding, and control rod assemblies.

Beryllia is toxic in the powder form, and special precautions must be taken in fabrication from the powder and in grinding the finished parts.

The crystal chemical and thermodynamic properties of beryllia are given in Table 6.2—3, and the physical, mechanical, thermal, and electrical properties are given in Table 6.2—4.

### ZIRCONIA (ZrO$_2$)

Zirconia has a melting point of 2,700°C in a

neutral or oxidizing atmosphere, and it is not wet by most steel alloys and noble metals. The high melting point and chemical inertness, combined with a low thermal conductivity, make zirconia a good material for refractory applications. It is also used as an opacifier for glazes and enamels and is a major constituent of lead-zirconate-titanate piezoelectrics.

Zirconia is monoclinic at room temperature and transforms to the denser tetragonal form at about 1,100 to 1,200°C, undergoing a disruptive volume change of about 9%. Zirconia is often stabilized with CaO, $Y_2O_3$ or MgO additions, which, when fired, results in a material having a cubic structure. The cubic form differs only slighly from the monoclinic and tetragonal forms and is stable above and below the transformation temperature.

The crystal chemical and thermodynamic properties of unstabilized zirconia are given in Table 6.2—5. The physical, mechanical, thermal, and electrical properties of stabilized zirconia are given in Table 6.2—6.

## MULLITE ($3Al_2O_3 \cdot 2SiO_2$)

Mullite is a refractory, mixed-oxide, ceramic material which has good thermal shock resistance properties. Mullite is used for refractory liners and laboratory ware. The silica in mullite is not present as free silica and usually does not pose a contamination problem. Mullite often appears as a separate phase in classical, triaxial porcelain whiteware ceramics. The needlelike habit of the mullite crystals adds strength to the body.

The physical, mechanical, thermal, and electrical properties of mullite are given in Table 6.2—7. The effect of different density levels on the properties of mullite can be determined from the table.

## CORDIERITE ($2MgO \cdot 2Al_2O_3 \cdot 5SiO_2$)

Cordierite is a ceramic material that has low dielectric losses and is used in high frequency insulators. Cordierite also has a low coefficient of expansion, good thermal shock resistance, and is used for heating element supports and burner tips.

Pure cordierite bodies in general have a short vitrification range so additions are often made to extend the range. Reactions to form cordierite are often sluggish, and precautions must be taken to fire the body high enough to complete the reactions.

The physical, mechanical, thermal, and electrical properties of cordierite are given in Table 6.2—8.

## SILICON CARBIDE (SiC)

Silicon carbide has a high melting point and hardness. It can be used to 1,650°C in an oxidizing atmosphere and to 2,300°C in an inert atmosphere. Its high thermal conductivity gives it high thermal shock resistance, which makes it useful in refractory furnace parts. Silicon carbide is also used as an abrasive and is currently being explored as a material for gas turbine components because of its high temperature modulus of rupture and corrosion resistance.

Silicon carbide occurs in alpha and beta forms. The alpha phase is hexagonal or rhombohedral and the beta phase is cubic. The beta phase undergoes an irreversible transformation to the alpha phase above 1,650°C. The base of the hexagonal unit cell is ∿ 3.08 Å and the c-axis is a multiple of ∿ 2.52 Å. Numerous stacking sequences can occur in the alpha form, giving rise to many polytypes. Some of these polytypes given in Ramsdell notation are 2H, 4H, 6H, 15R, 21R, and 24R (etc.). In this notation, the number refers to the number of layers in the unit cell and the letter refers to either hexagonal or rhombohedral symmetry.

Silicon carbide had not been successfully fired to near theoretical density without hot pressing in the past. The problem has been that silicon carbide is unreactive up to the temperature at which if decomposes. Silicon carbide has now been successfully densified to greater than 96% theoretical density by using boron as a sintering aid.[a] This development will promote other applications for silicon carbide which were previously limited by the restrictive geometry and cost associated with hot pressed materials. The properties of silicon carbide are given in Tables 6.2—9 and 6.2—10

[a]Prochazka, S., Investigation of Ceramics for High-temperature Turbine Vanes, General Electric Co., Quarterly Progress Report SRD-73-145, USN Contract N62269-73-C0356, June 20–September 19, 1973, General Electric Co., Schenectady, N.Y.

# SILICON NITRIDE (Si₃N₄)

Silicon nitride has high hardness, wear resistance, high corrosion resistance, high thermal shock resistance, and good high temperature strength. The material is currently used for applications such as furnace refractories and supports, bearings and seals, nozzles, thermocouple sheaths, and crucibles. Silicon nitride is currently being explored as a material for gas turbine components because of its high temperature fracture energy, corrosion resistance, and high thermal shock resistance.

Silicon nitride occurs in alpha and beta forms. These phases are hexagonal and differ only slightly in lattice parameters. The alpha phase undergoes an irreversible transformation to the beta phase above 1,500 to 1,600°C.

This material, as with silicon carbide, has not been successfully fired to near theoretical density without hot pressing. A great deal of research effort is currently being given to silicon nitride because of its potential high temperature corrosion and wear resistance properties. If developments are made with respect to densification without hot pressing, as with silicon carbide, the number of applications for this material will be greatly increased. Complex components are currently fabricated using the reaction bonded approach. This consists of fabricating parts from silicon and nitriding by firing the parts in nitrogen. The bodies formed by this process are limited in that they exhibit porosities in the order of 30%. Theoretically dense bodies have been made by hot pressing with MgO as a sintering aid.

The crystal chemical and thermodynamic properties for silicon nitride are given in Table 6.2—11. The physical, mechanical, thermal, and electrical properties are given in Table 6.2—12.

## Table 6.2—1
### CRYSTAL CHEMICAL AND THERMODYNAMIC PROPERTIES OF ALUMINA

| System | | $\alpha$-Al₂O₃ [1] Hexagonal |
|---|---|---|
| Lattice constants | | |
| a | nm | 0.4758 |
| b | nm | |
| c | nm | 1.2991 |
| $\beta$ | | |
| c/a | | 2.72 |
| Density | (kg/cm³) × 10⁻³ | 3.97 |
| Crystal lattice energy | (J/kg-mol) × 10⁻⁶ | 15,520.468 |
| Standard heat of formation, $-\Delta H°_{298}$ | (J/kg-mol) × 10⁻⁶ | 1,675.557 |
| Entropy, $S°_T$ at 298.15 K | (J/[kg-mol · deg]) × 10⁻³ | 51.020 |
| Free energy of formation of oxides, $-\Delta F$ at 298 K | (J/kg-mol) × 10⁻⁶ | 1,577.461 |
| Specific heat capacity, $C_p$ at 298.16 K | (J/kg · deg) | 774.977 |
| Heat of fusion, $\Delta H$ | (J/kg-mol) × 10⁻⁶ | 108.86 |
| Melting point | K | 2,319.7 ± 8 |
| Boiling point | K | 3,253 |

Table compiled by R. N. Kleiner.

## Table 6.2—2
### PHYSICAL, MECHANICAL, THERMAL, AND ELECTRICAL PROPERTIES OF ALUMINA

| Property | Units | Single[2] crystal | ~99%[3] | 99.9%[4] | 99.9%[5] | 99.9%[6] |
|---|---|---|---|---|---|---|
| **Physical** | | | | | | |
| Phase | | | $\alpha$-Al$_2$O$_3$ | $\alpha$-Al$_2$O$_3$ | $\alpha$-Al$_2$O$_3$ | $\alpha$-Al$_2$O$_3$ |
| Crystal structure | | | Hexagonal | | | |
| Density | g/cm³ | | 3.98 | >3.97 | 3.99 | 3.96 |
| Melting point | °C | | 2,050 | 2,040 | | |
| Water absorption | % | | | 0 | 0 | 0 |
| Gas permeability | | | | 0 | 0 | 0 |
| Grain size | μm | | | >30 | 15–45 | 1–6 |
| Surface finish | μin. (authentic average) | | | | 25 | 20 |
| Color | | | | Translucent white | Translucent white | Ivory |
| **Mechanical** | | | | | | |
| Modulus of elasticity | psi | 63 × 10⁶ | 59 × 10⁶ | 57 × 10⁶ | 57 × 10⁶ | 56 × 10⁶ |
| Modulus of rigidity | psi | | 22 × 10³ | 23 × 10⁶ | 23.5 × 10⁶ | 23 × 10⁶ |
| Bulk modulus | psi | | | | 34 × 10⁶ | 33 × 10⁶ |
| Poisson's ratio | | | 0.27 | 0.23 | 0.22 | 0.22 |
| Flexural strength | psi { 25°C | 92,000 | 67,000 | 40,000 | 41,000 | 80,000 |
| | 1,000°C | 60,000 | 50,000 | | 25,000 | 60,000 |
| Compressive strength | psi { 25°C | 60,000 | 420,000 | 325,000 | 370,000 | 550,000 |
| | 1,000°C | | 100,000 | | 70,000 | 280,000 |
| Tensile strength | psi { 25°C | | 38,000 | | 30,000 | 45,000 |
| | 1,000°C | | | | 15,000 | 32,000 |
| Transverse sonic velocity | m/sec | | 32,000 | 32,000 | 9.9 × 10³ | 9.9 × 10³ |
| Hardness | R45N | | | | 85 | 90 |
| Impact resistance | Charpy in.-lb | | | | | |
| **Thermal** | | | | | | |
| Coefficient of expansion | 10⁻⁶/°C | | | | | |
| −200– 25°C | | | | | | |
| 25– 200 | | | | | 3.4 | 3.4 |
| 25– 300 | | | | | 6.5 | 6.5 |
| 25– 400 | | | | | | |
| 25– 500 | | | | | 7.4 | 7.4 |
| 25– 600 | | | | | | |

## Table 6.2—2 (continued)
## PHYSICAL, MECHANICAL, THERMAL, AND ELECTRICAL PROPERTIES OF ALUMINA

| | 99.5%[7] | 99.5%[8] | 96%[9] | 96%[10] | 90%[11] | 85%[12] |
|---|---|---|---|---|---|---|
| **Physical** | | | | | | |
| Phase | | | | | | |
| Crystal structure | $\alpha$-Al$_2$O$_3$ | $\alpha$-Al$_2$O$_3$ | $\alpha$-Al$_2$O$_3$ | $\alpha$-Al$_2$O$_3$ | $\alpha$-Al$_2$O$_3$ | $\alpha$-Al$_2$O$_3$ |
| Density | 3.87 | 3.87 | 3.70 | 3.72 | 3.60 | 3.41 |
| Melting point | | | | | | |
| Water absorption | 0 | 0 | 0 | 0 | 0 | 0 |
| Gas permeability | 0 | 0 | 0 | 0 | 0 | 0 |
| Grain size | | 5–50 | | 2–20 | 2–10 | 2–12 |
| Surface finish | | 35 | | 65 | 65 | 65 |
| Color | White | Ivory | White | White | White | White |
| **Mechanical** | | | | | | |
| Modulus of elasticity | | $54 \times 10^6$ | $47 \times 10^6$ | $44 \times 10^6$ | $40 \times 10^6$ | $32 \times 10^6$ |
| Modulus of rigidity | | $22 \times 10^6$ | $19 \times 10^6$ | $18 \times 10^6$ | $17 \times 10^6$ | $14 \times 10^6$ |
| Bulk modulus | | $33 \times 10^6$ | | $25 \times 10^6$ | $23 \times 10^6$ | $20 \times 10^6$ |
| Poisson's ratio | | 0.22 | 0.22 | 0.21 | 0.22 | 0.22 |
| Flexural strength | 45,000 | 55,000 | 46,000 | 52,000 | 49,000 | 43,000 |
| | | | | 25,000 | 25,000 | 25,000 |
| Compressive strength | >300,000 | 380,000 | 375,000 | 300,000 | 360,000 | 280,000 |
| Tensile strength | | 38,000 | 25,000 | 28,000 | 32,000 | 22,500 |
| | | | | 14,000 | 15,000 | |
| Transverse sonic velocity | | $9.8 \times 10^3$ | | $9.1 \times 10^3$ | $8.8 \times 10^3$ | $8.2 \times 10^3$ |
| Hardness | 81 | 83 | 78 | 78 | 79 | 75 |
| Impact resistance | | | | | | |
| **Thermal** | | | | | | |
| Coefficient of expansion | | | | | | |
| −200− 25°C | | 3.4 | | 3.4 | 3.4 | 3.4 |
| 25− 200 | 6.9 | | | | | |
| 25− 300 | | | | | | |
| 25− 400 | | 7.1 | 6.4 | 6.0 | 6.1 | 5.3 |
| 25− 500 | | | | | | |
| 25− 600 | | 7.6 | 7.5 | 7.4 | 7.0 | 6.2 |

# Table 6.2—2 (continued)
## PHYSICAL, MECHANICAL, THERMAL, AND ELECTRICAL PROPERTIES OF ALUMINA

| | Single crystal[2] | ~99%[3] | 99.9%[4] | 99.9%[5] | 99.9%[6] |
|---|---|---|---|---|---|
| **Thermal** | | | | | |
| Coefficient of expansion (cont.) | | | | | |
| 25– 700 | | | | | |
| 25– 800 | | | | | |
| 25– 900 | | | | 7.8 | 7.8 |
| 25–1,000 | | | | | |
| 25–1,100 | | | | 8.0 | 8.0 |
| 25–1,200 | | | | 8.3 | 8.3 |
| 25–1,500 | | | | | |
| Conductivity | cal/(sec) (cm²) (°C/cm) | | | | |
| 20°C | .103 | .079 | | 0.095 | 0.093 |
| 100 | | | | 0.068 | 0.066 |
| 300 | .047 | .039 | | 0.032 | 0.032 |
| 400 | | | | | |
| 500 | .029 | .029 | | 0.015 | 0.015 |
| 800 | | | | | |
| 1,200 | | | | | |
| **Electrical** | | | | | |
| Dielectric constant at 25°C | | | | | |
| 1 kHz | | | | 10.1 | 9.9 |
| 1 MHz | | | | 10.1 | 9.8 |
| 10 MHz | | | | | |
| 100 MHz | | | | 10.1 | |
| 1 GHz | | | | | |
| 10 GHz | | | | 10.1 | 9.8 |
| 50 GHz | | | | | |
| Dissipation factor at 25°C | | | | | |
| 1 kHz | | | | 0.00050 | 0.0020 |
| 1 MHz | | | | 0.00004 | 0.0002 |
| 10 MHz | | | | | |
| 100 MHz | | | | 0.00006 | |
| 1 GHz | | | | | |
| 10 GHz | | | | 0.00009 | 0.0050 |
| 50 GHz | | | | | |

## Table 6.2—2 (continued)
### PHYSICAL, MECHANICAL, THERMAL, AND ELECTRICAL PROPERTIES OF ALUMINA

| | 99.5%[7] | 99.5%[8] | 96%[9] | 96%[10] | 90%[11] | 85%[12] |
|---|---|---|---|---|---|---|
| **Thermal** | | | | | | |
| Coefficient of expansion (cont.) | | | | | | |
| 25– 700 | | 8.0 | | 8.0 | 7.7 | 6.9 |
| 25– 800 | | | 7.9 | | | |
| 25– 900 | | 8.3 | | 8.2 | 8.1 | 7.2 |
| 25–1,000 | | | | | | |
| 25–1,100 | | | | | | |
| 25–1,200 | | | | 8.4 | 8.4 | 7.5 |
| 25–1,500 | | | | | | |
| Conductivity | | | | | | |
| 20°C | | 0.085 | 0.084 | 0.059 | 0.040 | 0.035 |
| 100 | | 0.062 | | 0.045 | 0.032 | 0.029 |
| 300 | | | 0.041 | | | |
| 400 | | 0.029 | | 0.024 | 0.019 | 0.016 |
| 500 | | | 0.026 | | | |
| 800 | | 0.015 | 0.020 | 0.013 | 0.012 | 0.010 |
| 1,200 | | | | | | |
| **Electrical** | | | | | | |
| Dielectric constant at 25°C | | | | | | |
| 1 kHz | | 9.8 | | 9.0 | 8.8 | 8.2 |
| 1 MHz | | 9.7 | 9.3 | 9.0 | 8.8 | 8.2 |
| 10 MHz | 9.58 | | | | | |
| 100 MHz | | | 9.3 | 9.0 | 8.7 | 8.2 |
| 1 GHz | 9.3 | 9.7 | 9.2 | 8.9 | 8.7 | 8.2 |
| 10 GHz | | | | 8.9 | | 8.2 |
| 50 GHz | | | | 8.7 | | |
| Dissipation factor at 25°C | | | | | | |
| 1 kHz | | 0.0002 | | 0.0011 | 0.0006 | 0.0014 |
| 1 MHz | 0.00003 | 0.0003 | 0.0003 | 0.0001 | 0.0004 | 0.0009 |
| 10 MHz | | | | | | |
| 100 MHz | 0.00014 | 0.0002 | 0.0003 | 0.0002 | 0.0004 | 0.0009 |
| 1 GHz | | | 0.0009 | 0.0001 | 0.0009 | 0.0014 |
| 10 GHz | | | | 0.0006 | | 0.0019 |
| 50 GHz | | | | 0.0068 | | |

## Table 6.2—2 (continued)
## PHYSICAL, MECHANICAL, THERMAL, AND ELECTRICAL PROPERTIES OF ALUMINA

| | | Single[2] crystal | ~99%[3] | 99.9%[4] | 99.9%[5] | 99.9%[6] |
|---|---|---|---|---|---|---|
| Loss factor at 25°C | | | | | | |
| 1 kHz | | | | | 0.0050 | 0.020 |
| 1 MHz | | | | | 0.0004 | 0.002 |
| 10 MHz | | | | | | |
| 100 MHz | | | | | 0.0006 | |
| 1 GHz | | | | | | |
| 10 GHz | | | | | 0.0010 | 0.005 |
| 50 GHz | | | | | | |
| | | | | | | |
| Dielectric strength | AC volts/mil | | | | | |
| 0.250 in. thick | (average RMS values | | | | 230 | 240 |
| 0.125 | at 60 Hz AC) | | | | 340 | 325 |
| 0.050 | | | | | 510 | 460 |
| 0.025 | | | | | 650 | 590 |
| 0.010 | | | | | | 800 |
| | | | | | | |
| Volume resistivity | (ohm-cm$^2$/cm) | | | | | |
| 25°C | | | | | >$10^{15}$ | |
| 100 | | | | | | |
| 300 | | | | | | $1.0 \times 10^{15}$ |
| 500 | | | | | | $3.3 \times 10^{12}$ |
| 700 | | | | | | $9.0 \times 10^{9}$ |
| 900 | | | | | | |
| 1,000 | | | | | | $1.1 \times 10^{7}$ |
| 1,500 | | | | | | |
| 2,000 | | | | | | |
| | | | | | | |
| Te value | °C | | | | | 1,170 |

**Table 6.2—2 (continued)**

**PHYSICAL, MECHANICAL, THERMAL, AND ELECTRICAL PROPERTIES OF ALUMINA**

| | 99.5%[7] | 99.5%[8] | 96%[9] | 96%[10] | 90%[11] | 85%[12] |
|---|---|---|---|---|---|---|
| **Loss factor at 25°C** | | | | | | |
| 1 kHz | | 0.002 | | 0.010 | 0.005 | 0.011 |
| 1 MHz | | 0.003 | 0.0028 | 0.001 | 0.004 | 0.007 |
| 10 MHz | 0.00029 | | | | | |
| 100 MHz | | | | 0.002 | 0.004 | 0.007 |
| 1 GHz | 0.00130 | 0.002 | 0.0028 | 0.001 | 0.008 | 0.011 |
| 10 GHz | | | 0.0082 | 0.005 | | 0.016 |
| 50 GHz | | | | 0.059 | | |
| **Dielectric strength** | | | | | | |
| 0.250 in. thick | | 220 | 210 | 210 | 235 | 240 |
| 0.125 | | 290 | | 275 | 320 | 340 |
| 0.050 | | 430 | | 370 | 450 | 440 |
| 0.025 | | 580 | | 450 | 580 | 550 |
| 0.010 | | 840 | | 580 | 760 | 720 |
| **Volume resistivity** | | | | | | |
| 25°C | $>10^{14}$ | $>10^{14}$ | $>10^{14}$ | $>10^{14}$ | $>10^{14}$ | $>10^{14}$ |
| 100 | | | $2.0 \times 10^{13}$ | | | |
| 300 | $2.0 \times 10^{11}$ | | $1.1 \times 10^{10}$ | $3.1 \times 10^{11}$ | $1.4 \times 10^{11}$ | $4.6 \times 10^{10}$ |
| 500 | | | $7.3 \times 10^{7}$ | $4.0 \times 10^{9}$ | $2.8 \times 10^{8}$ | $4.0 \times 10^{8}$ |
| 700 | | | $3.5 \times 10^{6}$ | $1.0 \times 10^{8}$ | $7.0 \times 10^{6}$ | $7.0 \times 10^{6}$ |
| 900 | $2.5 \times 10^{6}$ | | $6.8 \times 10^{5}$ | | | |
| 1,000 | | | | $1.0 \times 10^{6}$ | $8.6 \times 10^{5}$ | |
| 1,500 | | | | | | |
| 2,000 | | | | | | |
| **Te value** | $>975$ | | 840 | 1,000 | 960 | 850 |

Table compiled by R. N. Kleiner.

## Table 6.2—3
## CRYSTAL CHEMICAL AND THERMODYNAMIC PROPERTIES OF BERYLLIA

| System | | BeO[1] Hexagonal |
|---|---|---|
| Lattice constants | | |
| a | nm | 0.269 |
| b | nm | |
| c | nm | 0.437 |
| $\beta$ | | |
| c/a | | 1.621 |
| Density | $(kg/cm^3) \times 10^{-3}$ | 3.03 |
| Crystal lattice energy | $(J/kg\text{-}mol) \times 10^{-6}$ | 4,521.744 |
| Standard heat of formation, $-\Delta H^\circ_{298}$ | $(J/kg\text{-}mol) \times 10^{-6}$ | 599.131 |
| Entropy, $S^\circ_T$ at 298 K | $(J/[kg\text{-}mol \cdot deg]) \times 10^{-3}$ | 14.109 |
| Free energy of formation of oxides, $-\Delta F$ at 298 K | $(J/kg\text{-}mol) \times 10^{-6}$ | 581.965 |
| Specific heat capacity, $C_p$ at 298 K | $(J/kg \cdot deg)$ | 1,017.392 |
| Heat of fusion, $\Delta H$ | $(J/kg\text{-}mol) \times 10^{-6}$ | 71.217 |
| Melting point | K | 2,843 ± 30 |
| Boiling point | K | 4,123 |

Table compiled by R. N. Kleiner.

**Table 6.2—4**

**PHYSICAL, MECHANICAL, THERMAL, AND ELECTRICAL PROPERTIES OF BERYLLIA**

| | | ~99%[1,3] | 99.5%[1,4] | 96%[1,5] |
|---|---|---|---|---|
| **Physical** | | | | |
| Phase | | α-BeO | α-BeO | α-BeO |
| Crystal structure | | Hexagonal | | |
| Density | g/cm³ | 3.008 | 2.90 | 2.85 |
| Melting point | °C | 2,570 | | |
| Water absorption | % | | 0 | 0 |
| Gas permeability | | | 0 | 0 |
| Grain size | μm | | 10–40 | 15–140 |
| Surface finish | μin. (authentic average) | | 22 | |
| Color | | | White | Blue |
| **Mechanical** | | | | |
| Modulus of elasticity | psi | $56 \times 10^6$ | $51 \times 10^6$ | $44 \times 10^6$ |
| Modulus of rigidity | psi | $21.5 \times 10^6$ | $20 \times 10^6$ | $17 \times 10^6$ |
| Bulk modulus | psi | $54.5 \times 10^6$ | $35 \times 10^6$ | $31 \times 10^6$ |
| Poisson's ratio | | 0.34 | 0.26 | 0.30 |
| Flexural strength | psi { 25°C | 38,000 | 40,000 | 25,000 |
| | { 1,000°C | | | 9,000 |
| Compressive strength | psi { 25°C | 300,000 | 310,000 | 225,000 |
| | { 1,000°C | | 40,000 | |
| Tensile strength | psi { 25°C | 14,000 | 20,000 | |
| | { 1,000°C | | 5,000 | |
| Transverse sonic velocity | m/sec | | $11.1 \times 10^3$ | $10.7 \times 10^3$ |
| Hardness | R45N | | 67 | 64 |
| Impact resistance | Charpy D256 in.-lb | | | |
| **Thermal** | | | | |
| Coefficient of expansion | $10^{-6}$/°C | | | |
| −200– 25°C | | | 2.4 | 2.4 |
| 25– 200 | | | 6.4 | 6.3 |
| 25– 300 | | | | |
| 25– 400 | | 7.4 | | |
| 25– 500 | | | 7.7 | 7.5 |
| 25– 600 | | | | |
| 25– 700 | | | | |
| 25– 800 | | | 8.5 | 8.4 |

## Table 6.2—4 (continued)
### PHYSICAL, MECHANICAL, THERMAL, AND ELECTRICAL PROPERTIES OF BERYLLIA

| | ~99%[13] | 99.5%[14] | 96%[15] |
|---|---|---|---|
| Thermal | | | |
| Coefficient of expansion (cont.) | | | |
| 25– 900 | | | |
| 25–1,000 | | 8.9 | 8.9 |
| 25–1,100 | | | |
| 25–1,200 | | 9.4 | 9.2 |
| 25–1,500 | | | |
| | | | |
| Conductivity cal/(sec)(cm²)(°C/cm) | | | |
| 20°C | .47 | 0.67 | 0.38 |
| 100 | .34 | 0.48 | 0.32 |
| 300 | | | |
| 400 | .14 | 0.20 | 0.16 |
| 500 | | | |
| 800 | .07 | 0.07 | 0.06 |
| 1,200 | | | |
| | | | |
| Electrical | | | |
| Dielectric constant at 25°C | | | |
| 1 kHz | | 6.7 | |
| 1 MHz | | 6.7 | |
| 10 MHz | | | |
| 100 MHz | | 6.6 | |
| 1 GHz | | 6.6 | |
| 10 GHz | | 6.6 | |
| 50 GHz | | | |
| | | | |
| Dissipation factor at 25°C | | | |
| 1 kHz | | 0.0010 | 0.0005 |
| 1 MHz | | 0.0002 | 0.0001 |
| 10 MHz | | | |
| 100 MHz | | 0.0002 | 0.0005 |
| 1 GHz | | 0.0002 | 0.0004 |
| 10 GHz | | 0.0004 | |
| 50 GHz | | | |

## Table 6.2—4 (continued)
## PHYSICAL, MECHANICAL, THERMAL, AND ELECTRICAL PROPERTIES OF BERYLLIA

| | | ~99%[13] | 99.5%[14] | 96%[15] |
|---|---|---|---|---|
| Loss factor at 25°C | | | | |
| 1 kHz | | | 0.0070 | 0.0030 |
| 1 MHz | | | 0.0010 | 0.0007 |
| 10 MHz | | | | |
| 100 MHz | | | 0.0010 | 0.0030 |
| 1 GHz | | | 0.0010 | 0.0030 |
| 10 GHz | | | 0.0030 | |
| 50 GHz | | | | |
| Dielectric strength | AC volts/mil | | | |
| 0.250 in. thick | (average RMS values at 60 Hz AC) | | 260 | 240 |
| 0.125 | | | 340 | 340 |
| 0.050 | | | 490 | 400 |
| 0.025 | | | 610 | |
| 0.010 | | | 800 | |
| Volume resistivity | (ohm-cm²/cm) | | | |
| 25°C | | >10$^{17}$ | >10$^{17}$ | |
| 100 | | | | >10$^{17}$ |
| 300 | | >10$^{15}$ | >10$^{15}$ | >10$^{15}$ |
| 500 | | | 5 × 10$^{13}$ | 1 × 10$^{13}$ |
| 700 | | | 1.5 × 10$^{10}$ | 2 × 10$^{10}$ |
| 900 | | | | |
| 1,000 | | | 7 × 10$^{7}$ | 4 × 10$^{7}$ |
| 1,500 | | | | |
| 2,000 | | | | |
| Te value | °C | | 1,240 | 1,170 |

Table compiled by R. N. Kleiner.

## Table 6.2—5
## CRYSTAL CHEMICAL AND THERMODYNAMIC PROPERTIES OF ZIRCONIA

| System | | $ZrO_2$ [1] |
| --- | --- | --- |
| | | Monoclinic |
| Lattice constants | | |
| a | nm | 0.517 |
| b | nm | 0.526 |
| c | nm | 0.530 |
| $\beta$ | | $80°10^1$ |
| c/a | | |
| Density | $(kg/cm^3) \times 10^{-3}$ | 5.56 |
| Crystal lattice energy | $(J/kg\text{-}mol) \times 10^{-6}$ | 11,195.503 |
| Standard heat of formation, $-\Delta H^{\circ}_{298}$ | $(J/kg\text{-}mol) \times 10^{-6}$ | 1,094.848 |
| Entropy, $S^{\circ}_T$ at 298 K | $(J/[kg\text{-}mol \cdot deg]) \times 10^{-3}$ | 50.367 |
| Free energy of formation of oxides, $-\Delta F$ at 298 K | $(J/kg \cdot deg)\ 10^{-6}$ | 1,037.070 |
| Specific heat capacity, $C_p$ at 298 K | $(J/kg \cdot deg)$ | |
| Heat of fusion, $\Delta H$ | $(J/kg\text{-}mol) \times 10^{-6}$ | 87.085 |
| Melting point | K | 2,963 |
| Boiling point | K | 4,573 |

Table compiled by R. N. Kleiner.

**Table 6.2—6**
**PHYSICAL, MECHANICAL, THERMAL, AND**
**ELECTRICAL PROPERTIES OF ZIRCONIA**

| | | ~5–10% CaO[16] | Reference 17 |
|---|---|---|---|
| **Physical** | | | |
| Phase | | CaO stabilized | MgO stabilized |
| Crystal structure | | Cubic | Cubic |
| Density | g/cm³ | 5.5 | 5.43 |
| Melting point | °C | 2,500 | |
| Water absorption | % | | |
| Gas permeability | | | 0.5 |
| Grain size | μm | | |
| Surface finish | μin. (authentic average) | | |
| Color | | | |
| | | | |
| **Mechanical** | | | |
| Modulus of elasticity | psi | $22.6 \times 10^6$ | $15 \times 10^6$ |
| Modulus of rigidity | psi | $8.5 \times 10^6$ | |
| Bulk modulus | psi | $13.85 \times 10^6$ | |
| Poisson's ratio | | 0.324 | |
| Flexural strength | psi {25°C / 1,000°C | 20–35,000 | 30,000 |
| Compressive strength | psi {25°C / 1,000°C | 85–190,000 | |
| Tensile | psi {25°C / 1,000°C | 21,000 | |
| Transverse sonic velocity | m/sec | | |
| Hardness | R45N | | |
| Impact resistance | Charpy in.-lb | | |
| | | | |
| **Thermal** | | | |
| Coefficient of expansion | $10^{-6}/°C$ | | |
| –200–25°C | | | |
| 25–200 | | | |
| 25–300 | | | |
| 25–400 | | | |
| 25–500 | | | |
| 25–600 | | | |
| 25–700 | | | |
| 25–800 | | | |
| 25–900 | | | |
| 25–1,000 | | | |
| 25–1,100 | | | |
| 25–1,200 | | | |
| 25–1,500 | | | |
| | | | |
| Conductivity | cal/(sec) (cm²) (°C/cm) | | |
| 20°C | | | |
| 100 | | | |
| 300 | | | |
| 400 | | 0.0045 | |
| 500 | | | |
| 800 | | 0.0049 | |
| 1,200 | | 0.0057 | |
| | | | |
| **Electrical** | | | |
| Dielectric constant at 25°C | | | |
| 1 kHz | | | |
| 1 MHz | | | |
| 10 MHz | | | |

Table compiled by R. N. Kleiner.

## Table 6.2—7
## PHYSICAL, MECHANICAL, THERMAL, AND ELECTRICAL PROPERTIES OF MULLITE

| | | Reference 18 | Reference 19 | Reference 20 | Reference 21 |
|---|---|---|---|---|---|
| **Physical** | | | | | |
| Phase | | $3Al_2O_3 \cdot 2SiO_2$ | $3Al_2O_3 \cdot 2SiO_2$ | $3Al_2O_3 \cdot 2SiO_2$ | $3Al_2O_3 \cdot 2SiO_2$ |
| Crystal structure | | Orthorhombic | | | |
| Density | g/cm³ | 3.13–3.26 | 2.6 | 2.8 | 3.1 |
| Melting point | °C | 1,850 | | | |
| Water absorption | % | | | 0 | |
| Gas permeability | | | | | |
| Grain size | μm | | | | |
| Surface finish | μin. (authentic average) | | | | |
| Color | | | | Tan | |
| **Mechanical** | | | | | |
| Modulus of elasticity | psi | 21 × 10⁶ | | 23 × 10⁶ | |
| Modulus of rigidity | psi | | | | |
| Bulk modulus | psi | | | | |
| Poisson's ratio | | | | | |
| Flexural strength | psi {25°C / 1,000°C} | 25,000 | 20,000 | 27,000 | 6,000 |
| Compressive strength | psi {25°C / 1,000°C} | 100–190,000 | | 80,000 | |
| Tensile strength | psi {25°C / 1,000°C} | 16,000 | | | |
| Transverse sonic velocity | m/sec | | | | |
| Hardness | R45N | | | 71 | |
| Impact resistance | Charpy in.-lb | | | | |
| **Thermal** | | | | | |
| Coefficient of expansion | 10⁻⁶/°C | | | | |
| −200–25°C | | | | | |
| 25–200 | | | | | |
| 25–300 | | | | | |
| 25–400 | | | 4.6 | 3.7 | 2.3 |
| 25–500 | | | | | |
| 25–600 | | | | | |
| 25–700 | | | | | |
| 25–800 | | | 4.8 | 5.0 | 4.3 |
| 25–900 | | | | | |
| 25–1,000 | | | | | |

# Table 6.2—7 (continued)
## PHYSICAL, MECHANICAL, THERMAL, AND ELECTRICAL PROPERTIES OF MULLITE

| | | Reference 18 | Reference 19 | Reference 20 | Reference 21 |
|---|---|---|---|---|---|
| **Thermal** | | | | | |
| Coefficient of expansion (cont.) | | | | | |
| 25–1,100 | | | | | |
| 25–1,200 | | | | | |
| 25–1,500 | | | | | |
| **Conductivity** | cal/(sec) (cm²) (°C/cm) | | | | |
| 20°C | | | 0.005 | | 0.003 |
| 100 | | .143 | | | |
| 300 | | | | | |
| 400 | | | | | |
| 500 | | .102 | | | |
| 800 | | .090 | | | |
| 1,200 | | | | | |
| **Volume resistivity** | (ohm–cm²/cm) | | | | |
| 25°C | | | $>10^{14}$ | | |
| 100 | | | | | |
| 300 | | | | | |
| 500 | | | $10^{10}$ | | |
| 700 | | | $10^{8}$ | | |
| 900 | | | | | |
| 1,000 | | | | | |
| 1,500 | | | | | |
| 2,000 | | | | | |
| **Te value** | °C | | 670 | | |

Table compiled by R. N. Kleiner.

Table 6.2—8

PHYSICAL, MECHANICAL, THERMAL, AND ELECTRICAL PROPERTIES OF CORDIERITE

| | | Reference 22 | Reference 23 | Reference 24 | Reference 25 |
|---|---|---|---|---|---|
| **Physical** | | | | | |
| Phase | | $2MgO \cdot 2Al_2O_3 \cdot 5SiO_2$ | $2MgO \cdot 2Al_2O_3 \cdot 5SiO_2$ | $2MgO \cdot 2Al_2O_3 \cdot 5SiO_2$ | $2MgO \cdot 2Al_2O_3 \cdot 5SiO_2$ |
| Crystal structure | | Orthorhombic | | | |
| Density | g/cm³ | 2.51 | 2.3 | 2.1 | 1.8 |
| Melting point | °C | 1,471 | | | |
| Water absorption | % | | 0–1 | 10–15 | 14–17 |
| Gas permeability | | | | | |
| Grain size | μm | | | | |
| Surface finish | μin. (authentic average) | | | | |
| Color | | | | | |
| **Mechanical** | | | | | |
| Modulus of elasticity | psi | | $17 \times 10^6$ | $8.8 \times 10^6$ | |
| Modulus of rigidity | psi | | $7 \times 10^6$ | $3.8 \times 10^6$ | |
| Bulk modulus | psi | | | | |
| Poisson's ratio | | | 0.21 | 0.17 | |
| Flexural strength | psi { 25°C / 1,000°C | 16,000 | 15,000 | 8,000 | 3,400 |
| Compressive strength | psi { 25°C / 1,000°C | 50,000 | 50,000 | 30,000 | 18,500 |
| Tensile strength | psi { 25°C / 1,000°C | 7,800 | | 3,500 | 2,500 |
| Transverse sonic velocity | m/sec | | | | |
| Hardness | R45N | | | | |
| Impact resistance | Charpy in.-lb. | 4.3 | 4.0 | 2.5 | 2.5 |
| **Thermal** | | | | | |
| Coefficient of expansion | $10^{-6}/°C$ | | | | |
| −200–25°C | | | | | |
| 25–200 | | | | | |
| 25–300 | | | | | |
| 25–400 | | | 2.4 | 2.2 | 0.6 |
| 25–500 | | | | | |
| 25–600 | | | | | |
| 25–700 | | | 3.3 | 2.8 | 1.5 |
| 25–800 | | | | | |
| 25–900 | | | 3.7 | 2.8 | 1.7 |
| 25–1,000 | | | | | |

## Table 6.2—8 (continued)
## PHYSICAL, MECHANICAL, THERMAL, AND ELECTRICAL PROPERTIES OF CORDIERITE

| | Reference 22 | Reference 23 | Reference 24 | Reference 25 |
|---|---|---|---|---|
| **Thermal** | | | | |
| Coefficient of expansion (cont.) | | | | |
| 25–1,100 | 2.7 | | | |
| 25–1,200 | | | | |
| 25–1,500 | | | | |
| Conductivity cal/(sec) (cm²) (°C/cm) | | | | |
| 20°C | | 0.0077 | 0.0043 | |
| 100 | | | | |
| 300 | | 0.0062 | 0.0041 | |
| 400 | | | | |
| 500 | | 0.0055 | 0.0040 | |
| 800 | | 0.0055 | 0.0038 | |
| 1,200 | | | | |
| **Electrical** | | | | |
| Dielectric constant at 25°C | | | | |
| 1 kHz | | 5.3 | 5.0 | 4.1 |
| 1 MHz | | | 4.9 | |
| 10 MHz | | | | |
| 100 MHz | | | | |
| 1 GHz | | | | |
| 10 GHz | | | | |
| 50 GHz | | | | |
| Dissipation factor at 25°C | | | | |
| 1 kHz | | 0.0047 | 0.004 | 0.012 |
| 1 MHz | | | 0.003 | |
| 10 MHz | | | | |
| 100 MHz | | | | |
| 1 GHz | | | | |
| 10 GHz | | | | |
| 50 GHz | | | | |
| Loss factor at 25°C | | | | |
| 1 kHz | | 0.025 | 0.020 | 0.048 |
| 1 MHz | | | 0.015 | |
| 10 MHz | | | | |

## Table 6.2—8 (continued)
## PHYSICAL, MECHANICAL, THERMAL, AND ELECTRICAL PROPERTIES OF CORDIERITE

| | | Reference 22 | Reference 23 | Reference 24 | Reference 25 |
|---|---|---|---|---|---|
| Loss factor at 25°C (cont.) | | | | | |
| 100 MHz | | | | | |
| 1 GHz | | | | | |
| 10 GHz | | | | | |
| 50 GHz | | | | | |
| Dielectric strength | AC volts/mil (average RMS values at 60 Hz AC) | | | | |
| 0.250 in. thick | | | 200 | 100 | 60 |
| 0.125 | | | | | |
| 0.050 | | | | | |
| 0.025 | | | | | |
| 0.010 | | | | | |
| Volume resistivity | (ohm-cm$^2$/cm) | | | | |
| 25°C | | | $1.0 \times 10^{14}$ | $>10^{14}$ | $1.0 \times 10^{14}$ |
| 100 | | | $2.5 \times 10^{11}$ | $3.0 \times 10^{13}$ | $1.0 \times 10^{13}$ |
| 300 | | | $3.3 \times 10^{7}$ | $2.0 \times 10^{10}$ | $3.0 \times 10^{9}$ |
| 500 | | | $7.7 \times 10^{5}$ | $9.0 \times 10^{7}$ | $4.9 \times 10^{7}$ |
| 700 | | | $8.0 \times 10^{4}$ | $3.0 \times 10^{6}$ | $4.7 \times 10^{6}$ |
| 900 | | | $1.9 \times 10^{4}$ | $3.5 \times 10^{5}$ | $7.0 \times 10^{5}$ |
| 1,000 | | | | | |
| 1,500 | | | | | |
| 2,000 | | | | | |
| Te value | °C | | 485 | 780 | 850 |

Table compiled by R. N. Kleiner.

**Table 6.2—9**

## CRYSTAL CHEMICAL AND THERMODYNAMIC PROPERTIES
## OF SILICON CARBIDE

|  |  | $\alpha$-SiC | $\beta$-SiC |
|---|---|---|---|
| System |  | Hexagonal-6H | Cubic |
| Lattice constants |  |  |  |
|   a | nm | 3.073[26] | 4.358[27] |
|   b | nm |  |  |
|   c | nm | 15.08[26] |  |
|   c/a |  |  |  |
| Density | $(kg/cm^3) \times 10^{-3}$ | 3.218[26] |  |
| Crystal lattice energy | $(J/kg\text{-}mol) \times 10^{-6}$ |  |  |
| Standard heat of formation, $-\Delta H^{\circ}_{298}$ | $(J/kg\text{-}mol) \times 10^{-6}$ |  |  |
| Entropy, $S^{\circ}_{T}$ at 298 K | $(J/[kg\text{-}mol \cdot deg]) \times 10^{-3}$ |  |  |
| Free energy of formation, $-\Delta F$ at 298 K | $(J/kg\text{-}mol) \times 10^{-6}$ |  |  |
| Specific heat capacity, $C_p$ at 298 K | $(J/kg \cdot deg)$ |  |  |
| Heat of fusion, $\Delta H$ | $(J/kg\text{-}mol) \times 10^{-6}$ |  |  |
| Melting point | K |  |  |
| Boiling point | K |  |  |

Table compiled by R. N. Kleiner.

## Table 6.2—10
## PHYSICAL, MECHANICAL, THERMAL, AND ELECTRICAL PROPERTIES OF SILICON CARBIDE

| | | Reference 28 | | Reference 29 | Reference 30 | Reference 31 |
|---|---|---|---|---|---|---|
| **Physical** | | | | | | |
| Phase | | α-SiC | β-SiC | | | |
| Crystal structure | | Hexagonal | Cubic | | | |
| Density | g/cm³ | 3.208 | 3.21 | 3.1 | 2.6 | 3.10 |
| Melting point | °C | | | | | |
| Water absorption | % | | | 0 | | |
| Gas permeability | | | | 0 | | |
| Grain size | μm | | | | | |
| Surface finish | μin. (authentic average) | | | | 150 | |
| Color | | | | | | |
| **Mechanical** | | | | | | |
| Modulus of elasticity | psi | 70 × 10⁶ | | 69 × 10⁶ | 30 × 10⁶ | 60 × 10⁶ |
| Modulus of rigidity | psi | 24 × 10⁶ | | | | |
| Bulk modulus | psi | 14 × 10⁶ | | | | |
| Poisson's ratio | | 0.19 | | | | 0.24 |
| Flexural strength | psi {25°C / 1,000°C} | 25,000 | | 25,000 | 14–18,000 | 76,000 |
| Compressive strength | psi {25°C / 1,000°C} | 200,000 | | 200,000 | | |
| Tensile strength | psi {25°C / 1,000°C} | 5–20,000 | | | | |
| Transverse sonic velocity | m/sec | | | | | |
| Hardness | R45N | | | | | |
| Impact resistance | Charpy in.-lb | 0.80 | | | | |
| **Thermal** | | | | | | |
| Coefficient of expansion | 10⁻⁶/°C | | | | | |
| −200–25°C | | 4.34 | | | | |
| 25–200 | | | | | | |
| 25–300 | | | | | | |
| 25–400 | | | | 3.8 | | |
| 25–500 | | | | | | |
| 25–600 | | | | | | |
| 25–700 | | | | | | |
| 25–800 | | | | | | |
| 25–900 | | | | | | |
| 25–1,000 | | | | | | 4.3 |

**Table 6.2—10 (continued)**

**PHYSICAL, MECHANICAL, THERMAL, AND ELECTRICAL PROPERTIES OF SILICON CARBIDE**

| | | Reference 28 | Reference 29 | Reference 30 | Reference 31 |
|---|---|---|---|---|---|
| Thermal | | | | | |
| Coefficient of expansion (cont.) | | | | | |
| 25–1,100 | | | 4.0 | | |
| 25–1,200 | | | | 4.8 | |
| 25–1,500 | | | | | |
| Conductivity | cal/(sec) (cm²) (°C/cm) | | | | |
| 20°C | | | | | |
| 100 | | 3.26 | | | |
| 300 | | 2.04 | | | |
| 400 | | | | | |
| 500 | | 1.19 | | .612 | 0.2 |
| 800 | | | | .560 | |
| 1,200 | | | | .493 | 0.093 |

Table compiled by R. N. Kleiner.

**Table 6.2—11**
## CRYSTAL CHEMICAL AND THERMODYNAMIC PROPERTIES
## OF SILICON NITRIDE

| | | $\alpha$-Si$_3$N$_4$ | $\beta$-Si$_3$N$_4$ |
|---|---|---|---|
| System | | Hexagonal[32] | Hexagonal[33] |
| Lattice constants | | | |
| a | nm | 7.758[32] | 7.608[33] |
| b | nm | | |
| c | nm | 5.623[32] | 2.911[33] |
| $\beta$ | | | |
| c/a | | | |
| Density | (kg/cm$^3$) $\times$ 10$^{-3}$ | | |
| Crystal lattice energy | (J/kg-mol) $\times$ 10$^{-6}$ | | |
| Standard heat of formation, $-\Delta H^\circ_{298}$ | (J/kg-mol) $\times$ 10$^{-6}$ | | |
| Entropy, $S^\circ_T$ at 298 K | (J/[kg-mol$\cdot$deg]) $\times$ 10$^{-3}$ | | |
| Free energy of formation, $-\Delta F$ at 298 K | (J/kg-mol) $\times$ 10$^{-6}$ | | |
| Specific heat capacity, $C_p$ at 298 K | (J/kg$\cdot$deg) | | |
| Heat of fusion, $\Delta H$ | (J/kg-mol) $\times$ 10$^{-6}$ | | |
| Melting point | K | | |
| Boiling point | K | | |

Table compiled by R. N. Kleiner.

**Table 6.2—12**

## PHYSICAL, MECHANICAL, THERMAL, AND ELECTRICAL PROPERTIES OF SILICON NITRIDE

| | | Reference 34 | Reference 35 | Reference 36 | Reference 37 | Reference 38 | Reference 39 |
|---|---|---|---|---|---|---|---|
| **Physical** | | | | | | | |
| Phase | | $\alpha$-Si$_3$N$_4$ | | $\alpha$ 0–30%<br>$\beta$ 100–70% | | | |
| Crystal structure | | Hexagonal | | | | | |
| Density | g/cm³ | | 3.2 | 3.12–3.18 | 2.6 | 2.5–2.6 | 2.0–2.7 |
| Melting point | °C | 1,870 | | | | | |
| Water absorption | % | | | 0–0.1 | | | |
| Gas permeability | | | | | | | |
| Grain size | μm | | | | | | |
| Surface finish | μin. (authentic average) | | | | | | |
| Color | | | | | | | |
| **Mechanical** | | | | | | | |
| Modulus of elasticity | psi | 8–31 × 10⁶ | 35 × 10⁶ | 31.5 × 10⁶ | 32 × 10⁶ | 24 × 10⁶ | 13.9–31.6 × 10⁶ |
| Modulus of rigidity | psi | | | | | | |
| Bulk modulus | psi | | | | | | |
| Poisson's ratio | | | 0.27 | | 0.27 | | 0.257 |
| Flexural strength | psi {25°C<br>1,000°C | 10–100,000 | 100,000 | 80–100,000 | 35,000 | 35,400 | 10–30,000<br>10–30,000 |
| Compressive strength | psi {25°C<br>1,000°C | 72–90,000 | | | | | 77–110,000 |
| Tensile strength | psi {25°C<br>1,000°C | | | | | | |
| Transverse sonic velocity | m/sec | | | | | | |
| Hardness | R45N | | | | | | |
| Impact resistance | Charpy in.-lb | | | | | | |
| **Thermal** | | | | | | | |
| Coefficient of expansion | 10⁻⁶/°C | | | | | | |
| 25–1,000 | | 2.85 | 3.2 | | 3.2 | | 3.2 |
| 25–1,100 | | | | | | | |
| 25–1,200 | | | | | | | |
| 25–1,500 | | | | | | | |

## Table 6.2—12 (continued)
## PHYSICAL, MECHANICAL, THERMAL, AND ELECTRICAL PROPERTIES OF SILICON NITRIDE

| | Reference 34 | Reference 35 | Reference 36 | Reference 37 | Reference 38 | Reference 39 |
|---|---|---|---|---|---|---|
| Conductivity cal/(sec) (cm²) (°C/cm) | | | | | | |
| 20°C | | | | | | |
| 100 | | | | | | |
| 300 | | | | | | |
| 400 | | | | | | |
| 500 | | 0.042 | | 0.036 | | |
| 800 | | 0.033 | | 0.034 | | |
| 1,200 | | | | | | |

Table compiled by R. N. Kleiner.

## REFERENCES

1. **Samsonov, G. V., Ed.,** *The Oxide Handbook,* IFI/Plenum, Arlington, Va., 1972.
2. *Engineering Properties of Selected Ceramic Materials,* American Ceramic Society, Columbus, Oh., 1966, 5.4.1 – 1.
3. *Engineering Properties of Selected Ceramic Materials,* American Ceramic Society, Columbus, Oh., 1966, 5.4.1 – 1.
4. General Electric Bulletin L-4R – 9-70.
5. *Coors Ceramic Handbook,* Bulletin 952, revised January 1972, Coors Porcelain Co., Golden, Colo. (Vistol).
6. *Coors Ceramic Handbook,* Bulletin 952, revised January 1972, Coors Porcelain Co., Golden, Colo.(AD-999).
7. *Alumina Ceramics,* Wesgo Catalog No. C118, Ceramic Div., Western Gold and Platinum Co., Belmont, Cal.(AL-995).
8. *Coors Ceramic Handbook,* Bulletin 952, revised January 1972, Coors Porcelain Co., Golden, Colo.(AD-995).
9. American Lava Corp. Chart No. 711 (Alsimag 614).
10. *Coors Ceramic Handbook,* Bulletin 952, revised January 1972, Coors Porcelain Co., Golden, Colo.(AD-96).
11. *Coors Ceramic Handbook,* Bulletin 952, revised January 1972, Coors Porcelain Co., Golden, Colo.(AD-90).
12. *Coors Ceramic Handbook,* Bulletin 952, revised January 1972, Coors Porcelain Co., Golden, Colo.(AD-85).
13. *Engineering Properties of Selected Ceramic Materials,* American Ceramic Society, Columbus, Oh., 1966, 5.4.2 – 1.
14. *Coors Ceramic Handbook,* Bulletin 952, revised January 1972, Coors Porcelain Co., Golden, Colo.(BD-995-2).
15. *Coors Ceramic Handbook,* Bulletin 952, revised January 1972, Coors Porcelain Co., Golden, Colo.(BD-96).
16. *Engineering Properties of Selected Ceramic Materials,* American Ceramic Society, Columbus, Oh., 1966, 5.4.5 – 1.
17. **King, A. G.,** *Zircoa News Focus,* November 1968 (Zircoa 1706).
18. *Engineering Properties of Selected Ceramic Materials,* American Ceramic Society, Columbus, Oh., 1966, 5.5.1 – 1.
19. Kyoto Ceramic Co. Bulletin (K-635).
20. Norton Company Bulletin 1-CTM-P3 (DHP mulnorite).
21. Kyoto Ceramic Co. Bulletin(K-692).
22. *Engineering Properties of Selected Ceramic Materials,* American Ceramic Society, Columbus, Oh., 1966, 5.4.1 – 1.
23. American Lava Chart No. 711 (Alsimag 701).
24. American Lava Chart No. 711 (Alsimag 202).
25. American Lava Chart No. 711 (Alsimag 447).
26. JCPDS X-ray Powder Data File, Card No. 22-173, Joint Committee on Powder Diffraction Standards, Swarthmore, Pa.
27. **Pearson, W. B.,** *Handbook of Lattice Spacings and Structures of Metals,* Pergamon Press, Elmsford, N.Y., 1958, 956.
28. *Engineering Properties of Selected Ceramic Materials,* American Ceramic Society, Columbus, Oh., 1966, 5.2.3 – 1.
29. Carborundum Co. Bulletin ("KT" silicon carbide).
30. Norton Co. Bulletin (Crystar).
31. United Kingdom Atomic Energy Authority (UKAEA) Bulletin, September 1973 (Refel silicon carbide).
32. JCPDS X-ray Powder Data File, Card No. 4-250, Joint Committee on Powder Diffraction Standards, Swarthmore, Pa.
33. JCPDS X-ray Powder Data File, Card No. 9-259, Joint Committee on Powder Diffraction Standards, Swarthmore, Pa.
34. *Engineering Properties of Selected Ceramic Materials,* American Ceramic Society, Columbus, Oh., 1966, 5.3.3 – 1.
35. United Kingdom Atomic Energy Authority (UKAEA) Bulletin (hot pressed silicon nitride).
36. **Deeby, G. G. et al.,** Dense silicon nitride, *Powder Metallurgy,* 8, 145, 1961.
37. United Kingdom Atomic Energy Authority (UKAEA) Bulletin (reaction-bonded silicon nitride).
38. Advanced Materials Engineering Ltd. Data Sheet TD/P/74-1 (reaction bonded silicon nitride).
39. Advanced Materials Engineering Ltd. Bulletin 5/71/3 (reaction bonded silicon nitride).

# Section 7

# Composite Materials

# Section 7

# COMPOSITE MATERIALS

## 7.1 REINFORCEMENTS*

### Table 7.1—1
### PROPERTıES OF GLASS FIBERS

| Fiber | Density (g/cm³) | Tensile strength | | Tensile modulus | |
|---|---|---|---|---|---|
| | | GPa | ksi | GPa | 10⁶ psi |
| E-Glass | 2.58 | 3.45 | 500 | 72.5 | 10.5 |
| A-Glass | 2.50 | 3.04 | 440 | 69.0 | 10.0 |
| ECR-Glass | 2.62 | 3.63 | 525 | 72.5 | 10.5 |
| S-Glass | 2.48 | 4.59 | 665 | 86.0 | 12.5 |

\* Taken from a variety of commerical sources, current literature
  and handbook values, and the *Handbook of Materials Science*,
  Volume II, 389—391.

### Table 7.1—2
### PROPERTIES OF CARBON FIBERS

| Fiber | Precursor type | Density (g/cm³) | Tensile strength | | Tensile modulus | |
|---|---|---|---|---|---|---|
| | | | GPa | ksi | GPa | 10⁶ psi |
| AS-4 | PAN | 1.78 | 4.0 | 580 | 231 | 33.5 |
| AS-6 | PAN | 1.82 | 4.5 | 652 | 245 | 35.5 |
| IM-6 | PAN | 1.74 | 4.8 | 696 | 296 | 42.9 |
| T300 | PAN | 1.75 | 3.31 | 480 | 228 | 32.1 |
| T500 | PAN | 1.78 | 3.65 | 530 | 234 | 34.0 |
| T700 | PAN | 1.80 | 4.48 | 650 | 248 | 36.0 |
| T-40 | PAN | 1.74 | 4.50 | 652 | 296 | 42.9 |
| Celion | PAN | 1.77 | 3.55 | 515 | 234 | 34.0 |
| Celion ST | PAN | 1.78 | 4.34 | 630 | 234 | 34.0 |
| XAS | PAN | 1.84 | 3.45 | 500 | 234 | 34.0 |
| HMS-4 | PAN | 1.78 | 3.10 | 450 | 338 | 49.0 |
| PAN 50 | PAN | 1.81 | 2.41 | 355 | 393 | 57.0 |
| HMS | PAN | 1.91 | 1.52 | 220 | 341 | 49.4 |
| G-50 | PAN | 1.78 | 2.48 | 360 | 359 | 52.0 |
| GY-70 | PAN | 1.96 | 1.52 | 220 | 483 | 70.0 |
| P-55 | Pitch | 2.0 | 1.73 | 250 | 379 | 55.0 |
| P-75 | Pitch | 2.0 | 2.07 | 300 | 517 | 75.0 |
| P-100 | Pitch | 2.15 | 2.24 | 325 | 724 | 100 |
| HMG-50 | Rayon | 1.9 | 2.07 | 300 | 345 | 50.0 |
| Thornel 75 | Rayon | 1.9 | 2.52 | 365 | 517 | 75.0 |

## Table 7.1—3
## PROPERTIES OF OTHER ORGANIC FIBERS

| Fiber | Density (g/cm³) | Tensile strength | | Tensile modulus | |
|---|---|---|---|---|---|
| | | GPa | ksi | GPa | 10⁶ psi |
| Aramid (Kevlar 49) | 1.44 | 2.8 | 410 | 138 | 20 |
| Cotton | 1.52 | 0.2—08 | 30—120 | 27 | 4 |
| Flax | 1.52 | 0.84 | 120 | 103 | 15 |
| Jute | 1.52 | 0.86 | 120 | 62 | 9 |
| Hemp | 1.52 | 0.92 | 130 | 69 | 10 |
| Silk | 1.34 | 0.6 | 90 | 14 | 2 |

## Table 7.1—4
## PROPERTIES OF CERAMIC FIBERS

| Fiber | Composition | Form | Density q/cm³ | Tensile strength | | Tensile modulus | |
|---|---|---|---|---|---|---|---|
| | | | | GPa | ksi | GPa | 10⁶ psi |
| Fiber FP | Al₂O₃ | C | 3.95 | 1.38 | 200 | 379 | 55 |
| Sumitomo Alumina | Al₂O₃/SiO₂ | C | 3.2 | 1.45 | 210 | 193 | 28 |
| Nextel 440 | Al₂O₃/SiO₂ | C | 3.05 | 2.07 | 300 | 193 | 28 |
| Nextel Z-11 | Al₂O₃/ZrO₂ | C | 3.7 | 1.31 | 190 | 76 | 11 |
| Astroquartz | SiO₂ | C | 2.2 | 3.45 | 500 | 69 | 10 |
| Refrasil | SiO₂ | C | 2.1 | 0.21—0.41 | 30—60 | 72 | 10.5 |
| Siltemp | SiO₂ | C | 2.2 | — | — | — | — |
| Fiber frax | Al₂O₃/SiO₂ | D | 2.73 | 1.90 | 276 | 100 | 14.6 |
| Fiber max | Al₂O₃/SiO₂ | D | 3.0 | 1.03 | 150 | 150 | 22 |
| Kaowool | Al₂O₃/SiO₂ | D | 2.56 | 1.13 | 165 | 84 | 12.2 |
| Nextel 1312 Ultfbr | AlO₃/SiO₂/B₂O₃ | D | 2.75 | 1.72 | 250 | 152 | 22 |
| Nextel 1440 Ultfbr | AlO₃/SiO₂ | D | 3.1 | 1.31 | 190 | 207—241 | 30—35 |
| Nichias | Al₂O₃/SiO₂ | D | — | 1.79 | 260 | — | — |
| Saffil RF Grade | Al₂O₃/SiO₂ | D | 3.3 | 2.0 | 290 | 310 | 45 |
| Saffil RG Grade | Al₂O₃/SiO₂ | D | 3.3—3.5 | 1.0—2.0 | 145—290 | 297 | 43 |
| Zircar | ZrO₂/Y₂O₃ | D | 5.6—5.9 | — | — | — | — |
| Boron | 4 mil on W | C | 2.57 | 3.6 | 520 | 400 | 60 |
| Boron | 5.6 mil on W | C | 2.49 | 3.6 | 520 | 400 | 60 |
| Nicalon | SiC (O) | C | 2.55 | 2.5—3.2 | 360—470 | 180—200 | 26—29 |
| Tyranno | Si-Ti-C (O, Ti) | C | 2.3 | 1.99 | 286 | 117 | 17 |
| Avco CVD Sic | (C Core) | C | — | >3.4 | >500 | 428 | 62 |
| Silar SC-9 | SiC | D | 3.2 | 6.9 | 100 | 690 | 100 |
| Silar SC-10 | SiC | D | 3.2 | 6.9 | 100 | 690 | 100 |
| Takwhisker | SiC | D | 3.19 | 3—14 | 0.44—203 | 400—700 | 58—101 |
| Tateho | Si₃N₄ | D | 3.18 | — | — | — | — |
| Experimental[a] | Si₃N₄ | D | 3.2 | 4.83 | 700 | 276 | 40 |
| Experimental[a] | BeO | D | 2.9 | 13.1 | 1,900 | 345 | 50 |
| Experimental[a] | B₄C | D | 2.5 | 13.8 | 2,000 | 483 | 70 |
| Experimental[a] | BN | C and D | 1.8—2.0 | 0.3—1.4 | 45—200 | 28—69 | 4—10 |

*Note;*   Continuous, C; discontinuous, D.

[a]   Not necessarily commercially available.

## Table 7.1—5
## PROPERTIES OF METALLIC WIRES

| Metal | Density g/cm³ | Tensile strength | | Tensile modulus | |
|---|---|---|---|---|---|
| | | GPa | ksi | GPa | 10⁶ psi |
| Al | 2.71 | 0.29 | 40 | 69 | 10 |
| Be | 1.85 | 1.10 | 160 | 310 | 45 |
| Cu | 8.9 | 0.41 | 60 | 124 | 18 |
| Fe ss | 7.9 | 2.39 | 350 | 207 | 30 |
| Nb | 8.6 | 0.34 | 50 | 103 | 15 |
| Mo | 10.2 | 2.20 | 320 | 331 | 48 |
| Ta | 16.8 | 0.33 | 48 | 186 | 27 |
| Ti | 4.5 | 0.54 | 78 | 117 | 17 |
| W | 19.3 | 2.89 | 130 | 345 | 50 |
| W-2% ThO₂ | 19.3 | 2.41 | 350 | 345 | 50 |
| W-3% Re | 19.3 | 3.24 | 470 | 345 | 50 |

## 7.2   CERAMIC MATRIX COMPOSITES

## Table 7.2—1
## PROPERTIES OF SiC REINFORCED CERAMICS

| Matrix | Reinforcement/vol% | Fracture toughness (ksi $\sqrt{in.}$) | Flexural strength (ksi) |
|---|---|---|---|
| Barium Osumilite | SiC whiskers/25 | 4.1 | 50—60 |
| Corning 1723 Glass | SiC whiskers/25 | 1.9—3.1 | 30—50 |
| Cordierite | SiC whiskers/20 | 3.4 | 40 |
| MoSi₂ | SiC whiskers/20 | 7.5 | 45 |
| Mullite | SiC whiskers/20 | 4.2 | 65 |
| Si₃N₄ | SiC whiskers/10 | 5.9—8.6 | 60—75 |
| Si₃N₄ | SiC whiskers/30 | 6.8—9.1 | 50—65 |
| Spinel | SiC whiskers/30 | — | 60 |
| Toughened Al₂O₃ | SiC whiskers/20 | 7.7—12.3 | 100—130 |

## Table 7.2—2
## PROPERTIES OF SiC REINFORCED ALUMINA

| Reinforcement/vol% | Fracture strength (ksi) | Fracture toughness (ksi $\sqrt{in.}$) | Test temperature |
|---|---|---|---|
| SiC whiskers/10 | 65 | 6.5 | RT |
| SiC whiskers/10 | 45 | — | 1830°F |
| SiC whiskers/20 | 95 | 6.8—8.2 | RT |
| SiC whiskers/20 | 85 | 6.4—7.3 | 1830°F |
| SiC whiskers/40 | 120 | 5.5 | RT |
| SiC whiskers/40 | 96 | 5.6 | 1830°F |

*Note:*   Matrix is Al₂O₃.

**Table 7.2—3**
**PROPERTIES OF MULTIDIRECTIONAL CONTINUOUS FIBER**
**CERAMIC-CERAMIC COMPOSITES**

**Material/properties**

| | $SiO_2/SiO_2$3-D | $Al_2O_3/Al_2O_3$3-D | $Al_2O_3/SiO_2$3-D | BN/Bn3-D |
|---|---|---|---|---|
| Reinforcement/(vol%) ($10^3$ psi) | $SiO_2$/50 | $Al_2O_3$/30 | $Al_2O_3$/30 | BN/40 |
| Tensile strength | 3.87 | 10.3 | 10.8 | 3.6 |
| Tensile modulus ($10^6$ psi) | 2.26 | 5.26 | 4.90 | 2.23 |
| Compressive strength ($10^3$ psi) | 21.0 | 32.6 | — | 5.29 |
| Compressive modulus ($10^6$ psi) | 3.18 | 4.55 | — | 4.23 |
| Thermal conductivity (BTU/hr/ft²/°F/in) | 4.6 | 11.2 | 4.7 | 62.4 |
| Density (g/cm³) | 1.6 | 1.9 | 2.0 | 1.6 |

## 7.3   METAL MATRIX COMPOSITES

Table 7.3—1 lists data for metal matrix composites. Most of the data are for boron fiber and BORSIC® (silicon carbide coated boron fiber) reinforced composites, so they are listed first.

The first column contains the fiber used to reinforce the composite and the next column lists the diameter of the fiber. In the case of carbon fiber, the diameter is that of a single filament in a yarn or tow.

The volume percent fiber in the composite and the matrix metal alloy are then given. The fabrication process is listed in the next column. In diffusion bonding the matrix containing the fibers is hot pressed; in braze bonding the matrix is partially molten during pressing; in the diffusion bonding of plasma sprayed monolayer tapes (Diff. B. Pl. Sp. Tape) made by plasma spraying the matrix onto the fibers are hot pressed to form a composite. The eutectic (Eut.) process is a proprietary McDonnell method for forming composites.

The condition of the composite is either as fabricated (As Fab.) or heat treated (Ht. Trt.). In the next columns, the tensile strength ($\sigma$ $UTS_{11}$), tensile modulus ($E_{11}$), and elongation ($\epsilon f_{11}$) along the fiber axis are given. Then the tensile strength ($\sigma$ $UTS_{22}$), modulus ($E_{22}$), and elongation ($\epsilon f_{22}$) perpendicular to the fiber axes are listed in the next columns.

## Table 7.3—1
## DATA FOR METAL MATRIX COMPOSITES

| Fiber | Diameter (mils) | Vol % fiber | Matrix | Fabrication process | Condition of the composite | $UTS_{11}$ (10³ psi) | $E_{11}$ (10⁶ psi) | $\epsilon f_{11}$ (%) | $UTS_{22}$ (10³ psi) | $E_{22}$ (10⁶ psi) | $\epsilon f_{22}$ (%) | Source |
|---|---|---|---|---|---|---|---|---|---|---|---|---|
| Boron | 4.0 | 42 | 6061 Al | Diff. B. Pl. Sp. Tape | Ht. Trt. | | | | 34.0 | | | |
| | | 45 | 6061 Al | Diff. B. Pl. Sp. Tape | As Fab. | | | | 15.0 | 18.0 | 0.30 | 1 |
| | | 50 | 6061 Al | Diff. B. Pl. Sp. Tape | As Fab. | 200.0 | 33.0 | 0.7 | | | | 1 |
| | | 50 | 6061 Al | Diff. B. | As Fab. | 170.0 | 33.0 | 0.6 | 15.0 | 22.0 | 0.30 | 1 |
| | | 50 | 6061 Al | Diff. B. | Ht. Trt. | 160.0 | 33.0 | 0.6 | 22.0 | 20.0 | 0.30 | 1 |
| | | 50 | 6061 Al | Diff. B. | As Fab. | 170.0 | 33.0 | 0.6 | 15.0 | 22.0 | 0.30 | 2 |
| | | 50 | 6061 Al | Diff. B. | Ht. Trt. | 160.0 | 33.0 | 0.6 | 22.0 | 20.0 | 0.30 | 2 |
| | | 40 | 6061 Al | Diff. B. | As Fab. | 175.0 | 29.0 | 0.7 | 15.5 | 16.0 | 0.30 | 3 |
| | | 40 | 6061 Al | Diff. B. | Ht. Trt. | 190.0 | | | 25.0 | | | 3 |
| | | 50 | 6061 Al | Diff. B. | As Fab. | 215.0 | 34.0 | 0.7 | 15.5 | 20.0 | 0.30 | 3 |
| | | 50 | 6061 Al | Diff. B. | Ht. Trt. | 220.0 | | | 22.3 | | | 3 |
| | | 60 | 6061 Al | Diff. B. | As Fab. | 260.0 | 38.0 | 0.7 | 15.6 | 23.0 | 0.30 | 3 |
| | | 60 | 6061 Al | Diff. B. | Ht. Trt. | 250.0 | | | 21.2 | | | 3 |
| | | 44 | 2024 Al | Diff. B. Pl. Sp. Tape | As Fab. | 194.0 | | | 16.0 | 18.0 | 0.25 | 1 |
| | | 40 | 2024 Al | Diff. B. | As Fab. | 150 / 160 (260°C) | 30.0 / 28 (260°C) | 0.5 | 25.0 / 20 (260°C) | 15 (260°C) | | 4 |
| | | 50 | 7075 Al | Diff. B. | As Fab. | 198.0 | 34.0 | 0.6 | 14.3 | | | 3 |
| | | 50 | 7075 Al | Diff. B. | Ht. Trt. | 225.0 | 32.0 | 0.7 | 33.5 | | | 3 |
| | 5.6 | 50 | 6061 Al | Diff. B. | As Fab. | 200.0 | 32.0 | 0.6 | 20.0 | 20.0 | 0.30 | 1 |
| | | 50 | 6061 Al | Diff. B. | As Fab. | 200.0 | 32.0 | 0.6 | 12.0 | 20.0 | 0.30 | 2 |
| | | 40 | 6061 Al | Diff. B. | As Fab. | 187.0 | | | 22.7 | 16.0 | 0.40 | 3 |
| | | 40 | 6061 Al | Diff. B. | Ht. Trt. | 203.0 | | | 34.0 | 17.0 | 0.50 | 3 |
| | | 50 | 6061 Al | Diff. B. | As Fab. | | | | 22.0 | 20.0 | 0.60 | 3 |
| | | 50 | 6061 Al | Diff. B. | Ht. Trt. | | | | 43.5 | 20.0 | 0.60 | 3 |
| | | 60 | 6061 Al | Diff. B. | As Fab. | 261.0 | | | 27.3 | 23.0 | 0.30 | 3 |
| | | 60 | 6061 Al | Diff. B. | Ht. Trt. | 255.0 | | | 46.8 | 24.0 | 0.40 | 3 |
| | | 50 | 1100 Al | Eut. | Ht. Trt. | 209.0 | 30.8 | | | | | 5 |
| | 8.0 | 50 | 6061 Al | Diff. B. | As Fab. | 225.0 | 33.0 | 0.8 | 19.2 | 20.0 | 0.7 | 3 |
| | | 50 | 6061 Al | Diff. B. | Ht. Trt. | 238.0 | 34.0 | 0.8 | 36.8 | 21.0 | 0.2 | 3 |

## Table 7.3—1 (continued)
## DATA FOR METAL MATRIX COMPOSITES

| Fiber | Diameter (mils) | Vol % fiber | Matrix | Fabrication process | Condition of the composite | $UTS_{11}$ (10³ psi) | $E_{11}$ (10⁶ psi) | $\epsilon f_{11}$ (%) | $UTS_{22}$ (10³ psi) | $E_{22}$ (10⁶ psi) | $\epsilon f_{22}$ (%) | Source[a] |
|---|---|---|---|---|---|---|---|---|---|---|---|---|
| BORSIC® | 4.2 | 50 | 1100Al | Diff. B. Pl. Sp. Tape | As Fab. | 172.0 | 30.9 | 0.6 | 20.7 | 18.5 | 0.47 | 5 |
|  |  | 40 | 1100Al | Eut. | As Fab. | −311.0 | 29.6 | −1.4 | −30.6 | 22.4 | −0.43 | 1 |
|  |  | 50 | 2024Al | Diff. B. Pl. Sp. Tape | As Fab. |  |  |  | 13.0  12 (320°C) |  |  | 1 |
|  |  | 50 | 2024Al | Diff. B. Pl. Sp. Tape | Ht. Trt. |  |  |  | 22.0  13 (320°C) |  |  | 1 |
|  |  | 50 | 6061Al | Diff. B. Pl. Sp. Tape | As Fab. | 160.0 | 33.0 | 0.6 | 13.0  6 (320°C) | 20.0 | 0.15 | 1 |
|  |  | 50 | 6061Al | Diff. B. Pl. Sp. Tape | Ht. Trt. | 90 (500°C) |  |  | 21.0  6 (320°C) |  |  |  |
|  | 5.7 | 54 | 1100Al | Diff. B. Pl. Sp. Tape | As Fab. | 207.0 | 30.2 | 0.7 | 13.1 | 22.0 | 0.60 | 1 |
|  |  | 50 | 1100Al | Eut. | As Fab. | 200.0 | 39.0 | 0.6 |  |  |  | 5 |
|  |  | 60 | 6061Al | Diff. B. Pl. Sp. Tape | As Fab. |  |  |  | 20.0 | 24.0 | 0.30 | 1 |
|  |  | 60 | 6061Al | Diff. B. Pl. Sp. Tape | Ht. Trt. |  |  |  | 40.0 | 24.0 | 0.25 | 1 |
|  |  | 50 | 6061Al | Braze B. | Ht. Trt. | 136.0 | 29.6 | 0.5 |  |  |  | 5 |
|  | 4.2 | 32 | Beta III | Diff. B. | Ht. Trt. | 163–173 | 29.0 | 0.58−0.70 |  |  |  | 6 |
|  |  | 27 | 6/4 Ti | Diff. B. | As Fab. | 154.0 | 26.0 | 0.6 |  |  |  | 6 |
|  |  | 50 | 6/4 Ti | Diff. B. | As Fab. | 175.0 | 34.0 | 0.6 | 40.0 |  |  | 1 |
|  |  | 30 | 6/4 Ti | Diff. B. | As Fab. | 155.0 | 28.0 | 0.6 |  |  |  | 3 |
|  |  | 40 | 6/4 Ti | Diff. B. | As Fab. | 160.0 |  |  |  |  |  | 3 |
|  |  | 50 | 6/4 Ti | Diff. B. | As Fab. | 164.0 | 38.0 | 0.5 |  |  |  | 3 |
|  |  | 60 | 6/4 Ti | Diff. B. | As Fab. | 170.0 |  |  | 40.0 | 29.0 | 0.4 | 3 |
|  | 5.7 | 30 | 6/4 Ti | Diff. B. | As Fab. | 179.0 | 29.0 | 0.7 | 85.0 | 26.0 | 2.7 | 3 |
|  |  | 50 | 6/4 Ti | Diff. B. | As Fab. | 195.0 | 38.0 | 0.6 | 65.0 | 31.0 | 0.8 | 3 |
| Be | 5 | 50 | Al | Diff. B. | As Fab. | 65.0 | 25.0 |  |  |  |  | 7 |
| Be | 5 | 50 | 6/4 Ti | Diff. B. |  | 135.0 | 32.0 | 4.0 | 50 |  | 5 | 8 |
| Be | 3 | 73 | Ti | Diff. B. |  | 83.0 | 42.0 | 2.5 |  |  |  | 9 |

**Table 7.3—1 (continued)**
**DATA FOR METAL MATRIX COMPOSITES**

| Fiber | Diameter (mils) | Vol % fiber | Matrix | Fabrication process | Condition of the composite | $UTS_{11}$ ($10^3$ psi) | $E_{11}$ ($10^6$ psi) | $ef_{11}$ (%) | $UTS_{22}$ ($10^3$ psi) | $E_{22}$ ($10^6$ psi) | $ef_{22}$ (%) | Source[a] |
|---|---|---|---|---|---|---|---|---|---|---|---|---|
| Be | | 50 | Ti | Diff. B. | | 150.0 | 34.0 | 4.0 | 50 | | | 10 |
| Be | 20 | 43 | Ti | Diff. B. | As Fab. | 140.0 | 26.0 | >2.0 | 50 | 26 | >1 | 11 |
| C-Thornel 50 | 0.3 Yarn | 28 | Al-1170 Si | Liq. Infil. | As Fab. | 106.0 90 (500°C) | | 0.5 | | | | 12 |
| Morg. I | 0.3 Tow | 50 | Ni | Diff. B. | As Fab. | 105.0 | | | | | | 1 |
| Morg. I | 0.3 Tow | 45 | Ni₃Al | Diff. B. | As Fab. | 35 (1050°C) 85.0 35 (710°C) | | | | | | 1 |
| UARL monofilament | 3.4 | 45 | Al | Diff. B. | As Fab. | 98.0 | 24.0 | | | | | 1 |
| Mo TZM | 10 | 15 | Cb | Expl. | As Fab. | 127.0 | | 3.0 | | | | 13 |
| Mo TZM | 10 | 36 | Hastelloy X | Diff. B. | As Fab. | 124.0 42 (1100°C) | | 6.0 | | | | 14 |
| SiC | 4 | 40 | Ni | Electrodep. | | 130.0 | 40.0 | | | | | 15 |
| SiC | | | Ti | Diff. B. | | 140.0 | | | 80 | | | 3 |
| Steel | 3 | 30 | Ag | Diff. B. | | 39.0 | | | | | | 16 |
| Steel | | 40 | Ag | Hot extrusion | | 60.0 | | | | | | 17 |
| Steel | 2 | 20 | Al | Liq. Infil. | | 50.0 | | | | | | |
| W | 2 | 50 | Ag | Liq. Infil. | | 72 | | 1 | | | | 16 |
| W | 5 | 50 | Cb-40Ti 10Cr-5Al | Diff. B. | | 142 58 (980°C) | 32.7 | | | | | 1 |
| W-3% Re | 10 | 24 | Cb-40Ti 9Cr-4Al | Diff. B. | As Fab. | 171.0 51 (1210°C) | | 1.6 | 79 | | | 11 |
| W(NF) | 10 | 30 | Co | Diff. B. | As Fab. | 107.0 46 (1100°C) | | | 5.8 (1210°C) | | | 14 |
| W(NF) | 10 | 29 | B-1900 | | | 137.0 | | 1 | | | | 14 |
| W(NF) | 10 | 30 | Hastelloy X | | | 45 (1100°C) 99.0 46 (1100°C) | 35.1 | | | | | 14 |

# Table 7.3—1 (continued)
## DATA FOR METAL MATRIX COMPOSITES

| Fiber | Diameter (mils) | Vol % fiber | Matrix | Fabrication process | Condition of the composite | $UTS_{11}$ ($10^3$ psi) | $E_{11}$ ($10^6$ psi) | $\epsilon f_{11}$ (%) | $UTS_{22}$ ($10^3$ psi) | $E_{22}$ ($10^6$ psi) | $\epsilon f_{22}$ (%) | Source[a] |
|---|---|---|---|---|---|---|---|---|---|---|---|---|
| W(NF) | 10 | 27 | InCO713 | | | 145.0<br>51 (1100°C) | | 2.5 | | | | 14 |
| W(NF) | 10 | 23 | L-605 | | | 93.0 | | | | | | 14 |
| W | 10 | 22 | Nichrome | | | 48 (1100°C)<br>73.0 | | 3 | | | | 14 |
| W(NF) | 10 | 24 | Udimet 700 | | | 35 (1100°C)<br>51.0 | | 1 | | | | 14 |
| W(NF) | 10 | 19 | Waspalloy | | | 38 (1100°C)<br>128.0<br>27 (1100°C) | | 4 | | | | 14 |
| Al$_2$O$_3$ | 1–3 | 30 | Al-2.5Si | Liq. Phase H.P. | As Fab. | 55.0 | 20.0 | | | | | 18 |
| Al$_2$O$_3$ | 1–3 | 18 | Al-2.5Si | Liq. Phase H.P. | As Fab. | 25 (370°C)<br>45.0 | 15 (370°C)<br>16.0 | 0.4 (370°C)<br>0.9 | 21.0 | 12.0 | 1.0 | 18 |
| Al$_2$O$_3$ | 2 | 20 | Al | Liq. Infil. | As Fab. | 45.0 | | | | | | 4 |
| Al$_2$O$_3$ | 2 | 30 | Cu | Liq. Infil. | As Fab. | 115.0 | | | | | | 4 |
| Al$_2$O$_3$ | 2 | 10 | Fe | Liq. Infil. | As Fab. | 120.0 | | | | | | 19 |
| α-SiC | 1–3 | 20 | Mg Alloy | Liq. Phase H.P. | As Fab. | 68.2<br>41.2 (150°C) | | | | | | |
| β-SiC | 1–3 | 15 | 2024Al | Liq. Phase H.P. | Ht. Trt. | −270.0 | 23.0 | −2.1 | −82.0 | 14.0 | −1.0 | 18 |
| β-SiC | 1–3 | 25 | 2024Al | Liq. Phase H.P. | Ht. Trt. | 170.8<br>−200.0 | 25.2<br>28.0 | 0.9<br>−1.0 | 56.3<br>−92.0 | 17.2<br>16.0 | 0.8<br>−0.7 | 18 |

**Table 7.3—1 (continued)**

DATA FOR METAL MATRIX COMPOSITES

[a]Sources:

1. United Aircraft Research Laboratories
2. General Dynamics
3. TRW, Incorporated
4. General Electric Company
5. McDonnell Douglas
6. North American Rockwell
7. NASA Langley
8. Naval Air Systems Command
9. Whittaker Corporation
10. Allison Division of General Motors
11. Solar Division of International Harvester Corporation
12. Aerospace Corporation
13. NASA
14. Clevite Division of Gould, Incorporated
15. General Technologies Corporation
16. IITRI
17. Massachusetts Institute of Technology
18. Artech Corporation
19. Horizons

## REFERENCES

**Bilow, G.**, McDonnell Douglas, St. Louis, Mo.
**Brennan, J.**, United Aircraft Research Laboratories, East Hartford, Conn.
**Carlson, R. G.**, General Electric Company, Cincinnati, O.
**Christian, J. L.**, General Dynamics, Convair Division, San Diego, Cal.
**Hahn, H.**, Artech Corporation, Falls Church, Va.
**Hamilton, C. H.**, North American Rockwell, Los Angeles, Cal.
**Machlin, I.**, Naval Air Systems Command, Washington, D.C.
**Metcalfe, A. G.**, Solar, San Diego, Cal.
**Pepper, R. T.**, Aerospace Corporation, Los Angeles, Cal.
**Prewo, K.**, United Aircraft Research Laboratories, East Hartford, Conn.
**Schmidt, D.**, Naval Air Systems Command, Washington, D.C.
**Signorelli, R.**, NASA Lewis Research Center, Cleveland, O.
**Toth, I. J.**, TRW, Inc., Cleveland, O.

Table compiled by F. S. Galasso.

In Tables 7.3—2 through 7.3—5 are listed data for eutectics which have been studied; attempts have been made to unidirectionally solidify most of them.

The first column in each table lists the two phases, $\alpha$ and $\beta$. The next column gives the volume percent or weight percent of one of the phases.

The third column lists the eutectic temperature; the microstructure is given in the fourth column.

The fifth column lists the ultimate tensile strength of the eutectics obtained along the axis of solidification.

# Table 7.3—2
## BINARY EUTECTIC ALLOYS[a]

| System $\alpha + \beta$ | Volume %, $\alpha$ | Eutectic temperature, °C | Microstructure | Ultimate tensile strength ($10^3$ psi) |
|---|---|---|---|---|
| Ag-Bi | 3 | 262 | Broken lamellae of Ag | |
| Ag-Cu | 74 | 799 | Lamellar and rod | |
| Ag-Ge | 78 | 651 | Abnormal | |
| Ag-Pb | 15 | 304 | Rods and broken lamellae | |
| Ag-Si | 90 | 830 | Abnormal | |
| Al-Ag$_3$ Al$_2$ | 40 | 566 | Lamellar | |
| Al-AlB$_2$ | | 660 | | |
| Al-Al$_4$ Ca | 69 | 616 | Lamellar and rod | |
| Al-Al$_4$ Ce | 88 | 638 | Lamellar and rod | |
| Al-Al$_9$ Co$_2$ | 98 | 657 | Lamellar | |
| Al-CuAl$_2$ | 54 | 548 | Lamellar | 75 |
| Al-Al$_3$ Fe | 97 | 655 | Abnormal | |
| Al-Ge | 66 | 424 | Abnormal | |
| Al-Al$_3$ Ni | 89 | 640 | Distorted hexagonal rods and broken lamellae of Al$_3$Ni | 51 |
| Al-Al$_3$ Pd | 67 | 615 | Lamellar | |
| Al-Al$_3$ Pd$_2$ | 69 | | Lamellar "Chinese script" | |
| Al-AlSb | 99 | 657 | Abnormal | |
| Al-Si | 88 | 577 | Abnormal and rods | |
| Al-Sn | 1.5 | 232 | Lamellar and rod | |
| Al-Al$_3$ Th | | 634 | Spiral lamellar | |
| Al-Al$_4$ U | 12 | 640 | Chevron | |
| Al-Al$_3$ Y | 84 | 640 | Lamellar and rod | |
| Al-Zn | 26—30 | 382 | Lamellar | |
| Au-Co | | 997 | Rod | |
| Au-Ge | 69 | 356 | Abnormal interconnected flakes | |
| Bi-Au$_2$ Bi | 18 | 241 | Broken lamellar | |
| Bi-Cd | 57 | 144 | Lamellar and/or abnormal with pyramid L/S interface | |
| Bi−MnBi | 96 | 262 | Rod | |
| Bi-Pb$_2$ Bi | 27 | 125 | Abnormal, pyramid L/S interface | |
| Bi-Sn | 40 | 139 | Abnormal, pyramid L/S interface | |
| Bi-Bi$_2$ Tl | | 198 | Abnormal | |
| Bi-Zn | 96 | 254 | Broken lamellae and rod | |
| Cd-Cd$_3$ Cu | 19 | 248 | Fiber-ribbon | |
| Cd-Pb | | 248 | Lamellar | |
| Cd-CdSb | 81 | 290 | Abnormal | |

**Table 7.3—2 (continued)**
**BINARY EUTECTIC ALLOYS**

| System $\alpha + \beta$ | Volume %, $\alpha$ | Eutectic temperature, °C | Microstructure | Ultimate tensile strength ($10^3$ psi) |
|---|---|---|---|---|
| Cd-Sn | 25 | 177 | Lamellar | |
| Cd-Zn | 83 | 266 | Lamellar | |
| Co-CoAl | 65 | 1,400 | Lamellar and rod | 72–85 |
| Co-CoBe | 77 | 1,120 | Lamellar | |
| Co-CoCr | | 1,470 | Dendrites | |
| Co-CoGe$_2$ | 33 | 1,110 | Lamellar | |
| Co-Co$_2$Nb | 61 | 1,235 | Lamellar | |
| Co-CoSb | 38 | 1,095 | Lamellar | |
| Co-Co$_2$Ta | 65 | 1,276 | Lamellar and rod | |
| Co-Co$_7$W$_6$ | 77 | 1,480 | Lamellar | 109 |
| Co-Co$_{17}$Y$_2$ | 19 | 1,340 | Rod | |
| Co-Mg$_6$Co$_7$ | 67 | 1,340 | Lamellar | |
| Cr-Cr$_{23}$C$_6$ | 39 | 1,500 | Rod | |
| Cr-Cr$_2$O$_3$ | 19 | 1,660 | Rod | |
| Cu-B | 92 | | Ribbon | |
| Cu-Cr | 98 | 1,083 | Rod | |
| Cu-Cu$_2$O | 90 | 1,065 | Rod | |
| Cu-Cu$_4$Zn | 50 | | Lamellar | |
| Cu-Cu$_3$P | | 714 | | |
| Cu-Cu$_2$S | | 1,065 | | |
| Cu-Nb | | 1,550 | | |
| Fe-Fe$_2$B | 55 | 1,149 | Square rods | |
| Fe-Fe$_3$C | 41 | 1,147 | Lamellar and rod | |
| Fe-C | 92 | 1,147 | Abnormal | |
| Fe-Fe$_x$O | 10 | 1,371 | Rod | |
| Fe-FeS | 9.5 | 988 | Hexagonal rods of Fe | |
| Fe-Fe$_x$Sb | 18 | 1,002 | Hexagonal rods of Fe | |
| Fe-FeTi | | 1,340 | | |
| Fe-FeZr | | | Lamellar | |
| In-BiIn$_2$ | 30 | 72 | Lamellar | |
| $\beta$, InSn-$\gamma$, SnIn | 25 | 117 | Lamellar and rod | |
| Mg-Mg$_{17}$Al$_{12}$ | 32 | 437 | Lamellar | |
| Mg-Mg$_2$Cu | 60 | 485 | Lamellar | |
| Mg-Mg$_2$Ni | 72 | 507 | Lamellar and rod | |

**Table 7.3—2 (continued)**
**BINARY EUTECTIC ALLOYS**

| System $\alpha + \beta$ | Volume %, $\alpha$ | Eutectic temperature, °C | Microstructure | Ultimate tensile strength ($10^3$ psi) |
|---|---|---|---|---|
| Mg-Mg$_2$Si | 97 | 632 | Faceted rods | |
| Mg-Mg$_2$Sn | 76 | 561 | Lamellar "Chinese script" | |
| Mo-Mo$_2$C | 63 | 2,200 | Lamellar | |
| Nb-Nb$_2$C | 69 | 2,335 | Rectangular rods | 155 |
| Nb-Th | 10 | 1,435 | Rod | |
| Ni-C | 90 | 1,318 | Abnormal | |
| Ni-Ni$_3$B | 35 | 1,080 | Lamellar | |
| Ni-NiBe | 60—62 | 1,157 | Lamellar | 133 |
| Ni-Ni$_3$Nb | 74 | 1,270 | Lamellar | 108 |
| Ni,Al-Ni$_3$Nb | 78 | 1,270 | Lamellar | 164 |
| Ni-Cr | 77 | 1,345 | Lamellar and rod | 104 |
| Ni-Ni$_{1.5}$Gd$_2$ | 40 | 1,290 | Rod | |
| Ni-NiMo | 50 | 1,315 | Lamellar | 181 |
| Ni-Ni$_3$Sb | 40 | 1,097 | Lamellar | |
| Ni-Ni$_5$Si$_2$ | 35 | 1,125 | Lamellar | |
| Ni-Ni$_3$Sn | 38 | 1,130 | Lamellar | |
| Ni-Ni$_3$Ta | 87 | 1,360 | Lamellar | |
| Ni-Ni$_3$Ti | 61 | 942 | Lamellar | 90—100 |
| Ni-Ni$_7$Th$_2$ | 38 | 1,300 | Rod and lamellar | |
| Ni-Pb | | | | |
| Ni-TaNi$_3$ | | 1,360 | Lamellar | |
| Ni-W | 93 | 1,500 | Rod | |
| Pb-AuPb$_2$ | 52 | 215 | Lamellar | |
| Pb-Sb | 88 | 251 | Abnormal, pyramid L/S interface | |
| Pb-Sn | 37 | 183 | Lamellar | |
| Sb-Ag$_3$Sb | 28 | 484 | Abnormal, with pyramid L/S interface | |
| Sb-CdSb | 13 | 445 | Abnormal | |
| Sb-InSb | 35 | 530 | Triangular rods | |
| Sb-MnSb | 71 | 570 | Circular rods | |
| Sb-Sb$_2$Tl$_7$ | 12 | 195 | Abnormal | |
| Sn-Ag$_3$Sn | 97 | 221 | Abnormal/lamellar | |
| Sn-Cu$_6$Sn$_3$ | 98 | 227 | Rods | |
| Sn-In$_3$Sn | | 117 | Lamellar and rods | |
| Sn-Zn | 91 | 198 | Broken lamellae and rods of Zn | |

## Table 7.3—2 (continued)
### BINARY EUTECTIC ALLOYS

| System $\alpha + \beta$ | Volume %, $\alpha$ | Eutectic temperature, °C | Microstructure | Ultimate tensile strength ($10^3$ psi) |
|---|---|---|---|---|
| Ta-Ta$_2$C | 71 | 2,800 | Rectangular rods of Ta$_2$C and lamallae | 150 |
| TaFe-TaFe$_2$ | | | Lamellar/abnormal | |
| Te-Bi$_2$Te$_3$ | 73 | 413 | Lamellae of Bi$_2$Te$_3$ | |
| Th-Ti | 75 | 1,190 | Rod | |
| Ti-TiB | 90 | 1,670 | Abnormal | |
| Ti-Ti$_5$Si$_3$ | 75 | 1,330 | Rod | |
| V-V$_2$C | | 1,650 | Rod and lamellar | |
| V-V$_3$Si | 75 | 1,840 | | |
| Zn-MgZn$_2$ | | 367 | Spiral lamellar | |
| Zn-Mg$_2$Zn$_{11}$ | 50 | 367 | Lamellar and rod with pyramid L/S interface | |
| Zn-Zn$_{15}$Ti | 96 | 419 | Rod | |

$^a$ For references, see Table 7.3—5.

Table compiled by F. S. Galasso.

## Table 7.3—3
## QUASI-BINARY EUTECTICS[a]

| System $\alpha + \beta$ | Volume %, $\alpha$ | Eutectic temperature, °C | Microstructure | Ultimate tensile strength ($10^3$ psi) |
|---|---|---|---|---|
| Al-Mg$_2$Si | 88 | | Abnormal | |
| | | | | |
| Co-Cr$_7$C$_3$ | 70 | 1300 | | 185 |
| Co-Cr$_{23}$C$_6$ | 60 | 1340 | | 180 |
| Co-HfC | 85 | | Rods "arrow feather" | |
| Co-NbC | 88 | 1365 | Rods and lamellar | 150 |
| Co-TaC | 84 | | Rods "arrow feather" | 150 |
| Co-TiC | 84 | | Rods "arrow feather" | |
| Co-VC | 80 | | Rods | |
| | | | | |
| Ni-HfC | 72–85 | | Rods "arrow feather" | |
| Ni-NbC | 89 | | Rods "arrow feather" | |
| Ni-TiC | 95 | | Rods "arrow feather" | |
| NiAl-Cr | 66–67 | 1450 | Rods | 180 |
| NiAl-Mo | 89 | | Hexagonal rod | |
| Ni$_3$Al-Mo | 74 | 1306 | | 163 |
| Ni$_3$Al-Ni$_3$Cb | 56 | 1280 | Lamellar | 180 |
| Ni$_3$Al-Ni$_3$Ta | 35 | 1330 | Lamellar | 135 |
| Ni$_3$Al-Ni$_7$Zr$_2$ | 58 | 1192 | Lamellar | |

[a] For references, see Table 7.3—5.

Table compiled by F. S. Galasso.

## Table 7.3—4
## SEMICONDUCTOR QUASI-BINARY EUTECTICS[a]

| System $\alpha + \beta$ | Weight %, $\beta$ | Eutectic temperature, °C | Microstructure | Ultimate tensile strength ($10^3$ psi) |
|---|---|---|---|---|
| GaAs-CrAs | 35.4 | | Rod and lamellar | |
| GaAs-MoAs | 9.4 | | Rectangular rod and lamellar | |
| GaAs-VAs | 8.4 | | Rectangular rod and lamellar | |
| GaSb-CrSb | 13.4 | 690 | Rod | |
| GaSb-CoGa$_{1.3}$ | 7.9 | 697 | Rod | |
| GaSb-FeGa$_{1.3}$ | 7.9 | 695 | Rod | |
| GaSb-GaV$_3$Sb$_5$ | 4.9 | 710 | Square and rectangular rods | |
| GaSb-V$_2$Ga$_5$ | 4.4 | 707 | Square and rectangular rods | |
| | | | | |
| InAs-CrAs | 1.7 | 937 | Rod | |
| InAs-FeAs | 10.5 | 520 | Rod | |
| InSb-CrSb | 0.6 | | Rod | |
| InSb-FeSb | 0.7 | 520 | Rod | |
| InSb-Mg$_3$Sb$_2$ | 2.2 | 519 | Lamellae of Mg$_3$Sb$_2$ | |
| InSb-MnSb | 6.5 | 520 | Rod | |
| InSb-NiSb | 1.8 | 517 | Rod | |

[a] For references, see Table 7.3—5.

Table compiled by F. S. Galasso.

## Table 7.3—5
## IONIC SALTS AND QUASI-BINARY EUTECTICS

| System $\alpha + \beta$ | Weight %, $\alpha$ | Eutectic temperature,°C | Microstructure | Ultimate tensile strength (10³ psi) |
|---|---|---|---|---|
| LiF-NaF | 48.6–60 | 652 | Lamellar | |
| NaF-NaCl | 22.0–23.1 | 674 | Rectangular rods | |
| NaBr-NaF | 83.4 | 642 | Rectangular rods | |
| LiF-NaCl | 25 | 680 | Rod | |
| NaF-CaF$_2$ | | 810 | Lamellar | |
| NaF-MgF$_2$ | 20.0 | 820 | Rod | |
| NaF-PbF$_2$ | 20.0 | 540 | Rod | |
| MnO-MnS | 43.5 | | Lamellar | |
| FeO-FeS | 36.1 | 940 | Rectangular rods | |
| Zn$_5$B$_4$O$_{11}$-ZnB$_2$O$_{11}$ | 50.0 | 1300 | Lamellar | |
| BaFe$_{12}$O$_{19}$-BaFe$_2$O$_4$ | 28.6 | 1370 | Abnormal | |
| PbMoO$_4$-PbO | | 1310 | Rods | |
| Al$_2$O$_3$-Y$_5$Al$_3$O$_{12}$ | | | Rods and platelets of Y$_5$Al$_3$O$_{12}$ | |
| Al$_2$O$_3$-Al$_2$O$_3 \cdot$TiO$_2$ | | | Lamellar | |
| Al$_2$O$_3 \cdot$TiO$_2$-TiO$_2$ | | | Rod | |

### REFERENCES

Galasso, F. S., *High Modulus Fibers and Composites,* Gordon and Breach, New York, 1969.

Hogan, L., Kraft, W., and Lemkey, F., *Advances in Materials Research,* Vol. 5, Wiley Interscience, New York, 1971.

Lemkey, F., United Aircraft Research Laboratories, East Hartford, Conn.

Livingston, J., General Electric Research and Development Center, Schenectady, N.Y.

Sahm, P., Brown Boveri and Company, Baden, Switzerland.

Yue, A., University of California, Los Angeles, Cal.

Table compiled by F. S. Galasso.

## Table 7.3—6
## RECENT METAL MATRIX COMPOSITES

| Fiber | Vol% fiber | Matrix | Fabrication method | Tensile strength GPa | Tensile strength ksi | Tensile modulus GPa | Tensile modulus 10⁶ psi | Remarks (orientation) |
|---|---|---|---|---|---|---|---|---|
| SCS-2 | 47 | 6061Al | Diffusing bond | 1.46 | 212 | 204 | 29.6 | 0° |
| SCS-2 | 47 | 6061Al | Diffusing bond | 0.09 | 12.5 | 118 | 17.1 | 90° |
| SCS-6 | 35 | Ti-6Al-4V | Diffusing bond | 1.69 | 245 | 186 | 27 | |
| SCS-2 | 34 | 6061Al | Investment cast | 1.03 | 150 | 172 | 25 | 0° |
| SCS-2 | 37 | Mg | Cast rod | 1.38 | 200 | 181 | 26 | |
| SCS-2 | 33 | Cu | Cast | 0.96 | 140 | 200 | 29 | |
| P55 | 40 | Mg | Cast rod | 0.72 | 105 | 172 | 25 | 0° |
| P100 | 35 | Mg | Cast rod | 0.72 | 105 | 248 | 36 | 0° |
| P55 | 40 | Mg | Cast plate | 0.48 | 70 | 159 | 23 | 0° |
| Fiber FP | 50 | Al | Cast | 1.03 | 150 | 200 | 29 | |
| SiC particulate | 40 | 6061Al | Hot-press | 0.45 | 65 | 145 | 21 | |
| SiC particulate | 40 | 2124Al | Hot-press | 0.52 | 75 | 152 | 22 | |
| SiC particulate | 40 | 7090Al | Hot-press | 0.69 | 100 | 145 | 21 | |
| SiC particulate | 40 | 7091Al | Hot-press | 0.62 | 90 | 139 | 20 | |

## 7.4 ORGANIC MATRIX COMPOSITES

### Table 7.4—1
### EPOXY MOLDING COMPOUNDS

**Material/Properties**

| Matrix | Epoxy | Epoxy | Epoxy | Epoxy | Epoxy |
|---|---|---|---|---|---|
| Reinforcement/(vol%) | Glass/60 | Carbon/60 | HS carbon/60 | HM carbon/60 | Shortglass/60 |
| Density (g/cm$^3$) | 1.86—1.92 | 1.48—1.54 | 1.48—1.54 | 1.48—1.54 | 1.78—1.83 |
| Tensile strength (10$^3$ psi) | 35 | 30 | 32 | 18 | 11 |
| Tensile modulus (10$^6$ psi) | — | — | — | — | — |
| Flexural strength (10$^3$ psi) | 85 | 54 | 58 | 53 | 18 |
| Flexural modulus (10$^6$ psi) | 4.2 | 7.2 | 8.2 | 11.8 | 2.0 |
| Compressive strength (10$^3$ psi) | 42 | 36 | 44 | 31 | 28 |
| Izod impact notched (ft lb/in.) | 45 | 20 | 25 | 15 | 0.70 |
| Coeff thermal expansion (10$^{-6}$/°F) | 14 | 1.0 | 1.0 | 1.0 | 27 |
| Conductivity (BTU/hr/ft$^2$/°F/in.) | 0.02 | — | — | — | 0.02 |
| Heat deflection temp 264 psi (°F) | 250 | 250 | 250 | 250 | 154 |
| Flammability rating, UL | — | — | — | — | 94V-1 |
| Volume resistivity (ohm-cm) | 7.5 × 10$^{14}$ | — | — | — | 9 × 10$^{15}$ |
| Water absorption, 24 hr (%) | 0.10 | 0.20 | 0.20 | 0.20 | 0.10 |

### Table 7.4—2
### THERMOSET MOLDING COMPOUNDS

**Material/properties**

| Matrix | Polyimide | Silicone | Vinyl ester | Polyester | Melamine |
|---|---|---|---|---|---|
| Reinforcement/(vol%) | Glass/60 | Glass/60 | Glass/60 | Glass/60 | Glass/60 |
| Density (g/cm$^3$) | 1.95—2.00 | 2.00—2.05 | 1.84—1.90 | 1.84—1.90 | 1.79—1.84 |
| Tensile strength (10$^3$ psi) | 21 | 4.0 | 39.0 | 8.0 | 8.0 |
| Tensile modulus (10$^6$ psi) | — | — | — | — | — |
| Flexural strength (10$^3$ psi) | 37 | 10 | 70 | 20 | 14 |
| Flexural modulus (10$^6$ psi) | 3.1 | 2.0 | 2.8 | 2.2 | 2.2 |
| Compressive strength (10$^3$ psi) | 32 | 11 | 42 | 20 | 42 |
| Izod impact notched (ft lb/in.) | 22 | 5.0 | 40 | 12 | 0.50 |
| Coeff thermal expansion (10$^{-6}$/°F) | 10 | 7.0 | 10 | — | 20 |
| Conductivity (BTU/hr/ft$^2$/°F/in.) | 0.018 | 0.011 | — | — | 0.022 |
| Heat deflection temp 264 psi (°F) | 500 | 500 | 430 | 480 | 320 |
| Flammability rating, UL | — | 94V-0 | — | — | 94V-0 |
| Volume resistivity (ohm-cm) | 2.5 × 10$^{-16}$ | — | — | — | — |
| Water absorption, 24 hr (%) | 0.30 | 0.15 | 0.15 | 0.15 | 0.15 |

### Table 7.4—3
### NYLON MOLDING COMPOUNDS

**Material/properties**

| | Nylon 6 | Nylon 6 | Nylon 6/6 | Nylon 6/10 | Nylon 6/10 | Nylon 11 |
|---|---|---|---|---|---|---|
| Matrix | Nylon 6 | Nylon 6 | Nylon 6/6 | Nylon 6/10 | Nylon 6/10 | Nylon 11 |
| Reinforcement/(vol %) | Glass/20 | Glass/40 | Glass/40 | Carbon/40 | Glass/40 | Glass/20 |
| Density (g/cm$^3$) | 1.27 | 1.46 | 1.46 | 1.33 | 1.40 | 1.18 |
| Tensile strength (10$^3$ psi) | 20 | 25 | 32 | 36 | 26.5 | 14 |
| Tensile modulus (10$^6$ psi) | 0.98 | 1.4 | 1.9 | 4.2 | 1.5 | 0.75 |
| Flexural strength (10$^3$ psi) | 23 | 31 | 40 | 52 | 38 | 17 |
| Flexural modulus (10$^6$ psi) | 0.70 | 1.3 | 1.7 | 3.4 | 1.3 | 0.53 |
| Compressive strength (10$^3$ psi) | 21 | 23 | 23 | 25 | 25 | 12.5 |
| Izod impact notched (ft lb/in.) | 1.3 | 2.5 | 2.6 | 1.6 | 3.3 | 1.4 |
| Coeff thermal expansion (10$^{-6}$/°F) | 23 | 13 | 19 | 8.0 | 11 | 40 |
| Conductivity (BTU/hr/ft$^2$/°F/in.) | 3.0 | 3.6 | 3.6 | 8.0 | 3.8 | 2.6 |
| Heat deflection temp 264 psi (°F) | 390 | 400 | 480 | 500 | 420 | 340 |
| Flammability rating, UL | HB | HB | HB | HB | HB | HB |
| Volume resistivity (ohm-cm) | 10$^{14}$ | 10$^{14}$ | 10$^{14}$ | 30 | 10$^{12}$ | 10$^{13}$ |
| Water absorption, 24 hr (%) | 1.3 | 1.0 | 0.7 | 0.4 | 0.23 | 0.19 |

### Table 7.4—4
### STYRENIC MOLDING COMPOUNDS

**Material/properties**

| | ABS | ABS | ABS | PS | SAN | SMA |
|---|---|---|---|---|---|---|
| Matrix | ABS | ABS | ABS | PS | SAN | SMA |
| Reinforcement/(vol %) | Glass/20 | Glass/40 | Carbon/40 | Glass/40 | Glass/40 | Glass/40 |
| Density (g/cm$^3$) | 1.18 | 1.38 | 1.24 | 1.38 | 1.40 | 1.40 |
| Tensile strength (10$^3$ psi) | 13 | 18 | 17 | 14 | 20 | 14 |
| Tensile modulus (10$^6$ psi) | 0.88 | 1.5 | 3.1 | 2.0 | 2.0 | 1.67 |
| Flexural strength (10$^3$ psi) | 17 | 21 | 25 | 19 | 24 | 22.5 |
| Flexural modulus (10$^6$ psi) | 0.80 | 1.3 | 2.8 | 1.6 | 1.8 | 1.37 |
| Compressive strength (10$^3$ psi) | 13.5 | 19 | 19 | 17.5 | 22.0 | — |
| Izod impact notched (ft lb/in.) | 1.4 | 1.2 | 1.0 | 1.1 | 1.1 | 1.5 |
| Coeff thermal expansion (10$^{-6}$/°F) | 20 | 13 | 12 | 17 | 15.5 | — |
| Conductivity (BTU/hr/ft$^2$/°F/in.) | 1.4 | 1.6 | 3.8 | 2.2 | 2.1 | — |
| Heat deflection temp 264 psi (°F) | 220 | 240 | 240 | 210 | 217 | 250 |
| Flammability rating, UL | HB | HB | HB | HB | HB | HB |
| Volume resistivity (ohm-cm) | 10$^{15}$ | 10$^{15}$ | 30 | 10$^{16}$ | 10$^{16}$ | — |
| Water absorption 24 hr (%) | 0.18 | 0.12 | 0.14 | 0.05 | 0.1 | 0.1 |

*Note:* ABS = Acrylonitrile-butadiene-styrene; PS = polystyrene; SAN = styrene-acrylonitrile; SMA = styrene-maleic anhydride.

## Table 7.4—5
## THERMOPLASTIC MOLDING COMPOUNDS

| Material/properties | | | | |
|---|---|---|---|---|
| Matrix | AC | AC | PC | LCP |
| Reinforcement/(vol %) | Glass/20 | Glass/40 | Glass/40 | Glass/30 |
| Density (g/cm³) | 1.55 | 1.74 | 1.52 | 1.57 |
| Tensile strength ($10^3$ psi) | 12 | 13 | 21 | 16—29 |
| Tensile modulus ($10^6$ psi) | 1.2 | 1.6 | 1.7 | 2.5—2.6 |
| Flexural strength ($10^3$ psi) | 16.5 | 17.0 | 26.0 | 25—36 |
| Flexural modulus ($10^6$ psi) | 0.9 | 1.3 | 1.4 | 2.1—2.5 |
| Compressive strength ($10^3$ psi) | 12 | 11 | 22 | — |
| Izod impact notched (ft lb/in.) | 0.9 | 0.9 | 2.2 | 1.0—2.5 |
| Coef thermal expansion ($10^{-6}$/°F) | 25 | 18 | 9.5 | — |
| Conductivity (BTU/hr/ft²/°F/in.) | 2.0 | 2.3 | 2.4 | — |
| Heat deflection temp 264 psi (°F) | 325 | 328 | 300 | 445—600 |
| Flammability rating, UL | HB | HB | Vl | — |
| Volume resistivity (ohm-cm) | $10^{14}$ | $10^{14}$ | $10^{16}$ | $10^{16}$ |
| Water absorption, 24 hr (%) | 0.5 | 1.0 | 0.07 | — |

*Note:* AC = Acetal copolymer; PC = polycarbonate; LCP = liquid crystal polymer.

## Table 7.4—6
## THERMOPLASTIC MOLDING COMPOUNDS

| Material/properties | | | | | | |
|---|---|---|---|---|---|---|
| Matrix | HDPE | HDPE | PP | PP | PBT | PET |
| Reinforcement/(vol %) | Glass/20 | Glass/40 | Glass/40 | Mica/40 | Glass/40 | Glass/55 |
| Density (g/cm³) | 1.10 | 1.28 | 1.23 | 1.26 | 1.63 | 1.80 |
| Tensile strength ($10^3$ psi) | 7.0 | 10 | 16 | 5.6 | 21.5 | 28.5 |
| Tensile modulus ($10^6$ psi) | 0.6 | 1.25 | 1.3 | 1.1 | 2.0 | 3.0 |
| Flexural strength ($10^3$ psi) | 9.0 | 12 | 19 | 9 | 30 | 43 |
| Flexural modulus ($10^6$ psi) | 0.55 | 1.0 | 0.9 | 1.0 | 1.5 | 2.6 |
| Compressive strength ($10^3$ psi) | 5.0 | 7.5 | 13.0 | 7.0 | 20.0 | 28.5 |
| Izod impact notched (ft lb/in.) | 1.2 | 1.4 | 2.0 | 0.5 | 1.8 | 1.9 |
| Coeff thermal expansion ($10^{-6}$/°F) | 28 | 25 | 17.5 | 22 | 12 | 10 |
| Conductivity (BTU/hr/ft²/°F/in.) | 2.3 | 2.7 | 2.45 | 2.2 | 1.5 | 2.3 |
| Heat deflection temp 264 psi (°F) | 240 | 250 | 300 | 230 | 415 | 450 |
| Flammability rating, UL | HB | HB | HB | HB | HB | HB |
| Volume resistivity (ohm-cm) | $10^{16}$ | $10^{16}$ | $10^{15}$ | $10^{16}$ | $10^{16}$ | $10^{16}$ |
| Water absorption 24 hr (%) | 0.01 | 0.022 | 0.06 | 0.03 | 0.08 | 0.04 |

*Note:* HDPE = High density polyethylene; PP = polypropylene; PBT = polybutylene terephthalate; PET = polyethylene terephthalate.

## Table 7.4—7
## THERMOPLASTIC MOLDING COMPOUNDS

**Material/properties**

| Matrix | PPE-PPO | PPE-PPO | PPS | PPS | PPS |
|---|---|---|---|---|---|
| Reinforcement/(vol %) | Glass/20 | Graphite/20 | Glass/20 | Glass/40 | Graphite/40 |
| Density (g/cm$^3$) | 1.21 | 1.20 | 1.49 | 1.67 | 1.46 |
| Tensile strength (10$^3$ psi) | 13.5 | 15.0 | 14.5 | 20.0 | 26.0 |
| Tensile modulus (10$^6$ psi) | 1.0 | 1.0 | 1.3 | 2.0 | 4.8 |
| Flexural strength (10$^3$ psi) | 17.5 | 20.0 | 19.0 | 30.0 | 40.0 |
| Flexural modulus (10$^6$ psi) | 0.75 | 0.98 | 1.3 | 1.6 | 4.1 |
| Compressive strength (10$^3$ psi) | — | 17.0 | 22.5 | 25.0 | 27.0 |
| Izod impact notched (ft lb/in.) | 2.0 | 1.6 | 1.4 | 1.4 | 1.2 |
| Coeff thermal expansion (10$^{-6}$/°F) | 20 | 12 | 16 | 12 | 8.0 |
| Conductivity (BTU/hr/ft$^2$/°F/in.) | 1.1 | — | 2.1 | 2.2 | 3.3 |
| Heat deflection temp 264 psi (°F) | 285 | 235 | 500 | 500 | 500 |
| Flammability rating, UL | HB | — | VO | VO | VO |
| Volume resistivity (ohm-cm) | 10$^{17}$ | 13.0 | 10$^{16}$ | 10$^{16}$ | 30 |
| Water absorption 24 hr (%) | 0.06 | — | 0.02 | 0.02 | 0.02 |

*Note:* PPE = Polyphenylene ether; PPO = polyphenylene oxide; PPS = polyphenylene sulfide; PPS and PPO are blended with high impact polystyrene.

## Table 7.4—8
## THERMOPLASTIC MOLDING COMPOUNDS

**Material/properties**

| Matrix | PAS | PSF | PSF | PSF | PES | PES |
|---|---|---|---|---|---|---|
| Reinforcement/(vol%) | Glass/20 | Glass/20 | Glass/40 | Carbon/40 | Glass/40 | Carbon/40 |
| Density (g/cm$^3$) | 1.51 | 1.38 | 1.56 | 1.42 | 1.68 | 1.52 |
| Tensile strength (10$^3$ psi) | 19 | 15 | 19 | 26 | 23 | 31 |
| Tensile modulus (10$^6$ psi) | 1.0 | 0.88 | 1.7 | 3.0 | 2.0 | 3.5 |
| Flexural strength (10$^3$ psi) | 27 | 20 | 25 | 35 | 31 | 42 |
| Flexural modulus (10$^6$ psi) | 0.9 | 0.7 | 1.2 | 2.4 | 1.6 | 3.2 |
| Compressive strength (10$^3$ psi) | — | 19 | 24 | — | 22 | — |
| Izod impact notched (ft lb/in.) | 1.1 | 1.1 | 1.6 | 1.3 | 1.5 | 1.4 |
| Coeff thermal expansion (10$^{-6}$/°F) | — | 17 | 13 | — | 14 | — |
| Conductivity (BTU/hr/ft$^2$/°F/in.) | — | 2.1 | 2.6 | — | 2.6 | — |
| Heat deflection temp 264 psi (°F) | 405 | 360 | 365 | 365 | 420 | 420 |
| Flammability rating, UL | VO | V1 | VO | V1 | VO | VO |
| Volume resistivity (ohm-cm) | 10$^{16}$ | 10$^{15}$ | 10$^{15}$ | 30 | 10$^{16}$ | 30 |
| Water absorption 24 hr (%) | 0.4 | 0.24 | 0.25 | 0.25 | 0.30 | 0.30 |

*Note:* PAS = Polyaryl sulfone; PSF = polysulfone; PES = polyether sulfone.

## Table 7.4—9
### THERMOPLASTIC MOLDING COMPOUNDS

**Material/properties**

| Matrix | PEI | PEI | PEI | PEEK | PEEK |
|---|---|---|---|---|---|
| Reinforcement/(vol %) | Glass/20 | Glass/40 | Carbon/40 | Glass/20 | Carbon/40 |
| Density (g/cm³) | 1.41 | 1.59 | 1.44 | 1.46 | 1.46 |
| Tensile strength ($10^3$ psi) | 23 | 31 | 34 | 23 | 39 |
| Tensile modulus ($10^6$ psi) | 1.1 | 1.9 | 4.1 | 2.0 | 4.4 |
| Flexural strength ($10^3$ psi) | 32 | 43 | 48 | 36 | 54 |
| Flexural modulus ($10^6$ psi) | 0.95 | 1.6 | 3.2 | 1.1 | 3.2 |
| Compressive strength ($10^3$ psi) | 24 | 24.5 | — | — | — |
| Izod impact notched (ft lb/in.) | 1.6 | 2.1 | 1.2 | 1.5 | 1.7 |
| Coeff thermal expansion ($10^{-6}$/°F) | 15 | 11 | — | 14 | — |
| Conductivity (BTU/hr/ft²/°F/in.) | 1.7 | 1.8 | — | — | — |
| Heat deflection temp 264 psi (°F) | 410 | 410 | 410 | 550 | 550 |
| Flammability rating, UL | VO | VO | VO | VO | VO |
| Volume resistivity (ohm-cm) | $10^{16}$ | $10^{16}$ | $10^{12}$ | $10^{16}$ | 30 |
| Water absorption 24 hr (%) | 0.21 | 0.18 | 0.18 | 0.12 | 0.12 |

*Note:* PEI = Polyetherimide; PEEK = polyether etherketone.

## Table 7.4—10
### PROPERTIES OF 3-DIMENSIONAL ORTHOGONAL CARBON-CARBON COMPOSITES

**Material/properties**

| Direction | X-Y | Z |
|---|---|---|
| Matrix | Carbon | |
| Reinforcement/(vol %) | Carbon/~ 50% | |
| Density (g/cm³) | 1.9 | 1.9 |
| Tensile strength ($10^3$ psi) | 15 | 45 |
| Tensile modulus ($10^6$ psi) | 90 | 22 |
| Flexural strength ($10^3$ psi) | — | — |
| Flexural modulus ($10^6$ psi) | — | — |
| Compressive strength ($10^3$ psi) | 17 | 23 |
| Izod impact notched (ft lb/in.) | — | — |
| Coeff thermal expansion ($10^{-6}$/°F) | O, RT;4, 2950°F | O, RT;3, 2950°F |
| Conductivity (BTU/hr/ft²/°F/in.) | 12, RT;4, 2950°F | 142, RT;5, 2950°F |

## Table 7.4—11
### PROPERTIES OF UNIDIRECTIONAL CARBON-CARBON COMPOSITES

**Material/properties**

| Matrix | Carbon | Carbon | Carbon | Carbon |
|---|---|---|---|---|
| Reinforcement/(vol %) | HTU-C/~55% | HMS-C/~55% | HTU-C/~55% | HMS-C/~55% |
| Direction | Long. | Long. | Trans. | Trans. |
| Tensile strength ($10^3$ psi) | 10 | 85 | 0.60 | 0.75 |
| Tensile modulus ($10^6$ psi) | 20 | 30 | — | — |
| Compressive strength ($10^3$ psi) | 40 | 55 | 4 | 7.5 |
| Bend strength ($10^3$ psi) | 180—230 | 120—145 | 3 | 4.5 |
| Shear strength ($10^3$ psi) | 3 | 4 | — | — |
| Fracture toughness (ft-lbf/ft²) | 4800 | 1370 | 30 | 55 |

# Section 8

# Electronic Materials

# Section 8

# ELECTRONIC MATERIALS

### Table 8—1
### WIRE TABLE, STANDARD ANNEALED COPPER

#### American Wire Gauge (B. and S.)

| Gauge No. | Diameter in mils at 20°C | Cross-section at 20°C | | Ohms per 1,000 ft[a] | | | |
|---|---|---|---|---|---|---|---|
| | | Circular mils | Square inches | 0°C (32°F) | 20°C (68°F) | 50°C (122°F) | 75°C (167°F) |
| 0000 | 460.0 | 211,600 | 0.1662 | 0.04516 | 0.04901 | 0.05479 | 0.05961 |
| 000 | 409.6 | 167,800 | .1318 | .05695 | .06180 | .06909 | .07516 |
| 00 | 364.8 | 133,100 | .1045 | .07181 | .07793 | .08712 | .09478 |
| 0 | 324.9 | 105,500 | .08289 | .09055 | .09827 | .1099 | .1195 |
| 1 | 289.3 | 83,690 | .06573 | .1142 | .1239 | .1385 | .1507 |
| 2 | 257.6 | 66,370 | .05213 | .1440 | .1563 | .1747 | .1900 |
| 3 | 229.4 | 52,640 | .04134 | .1816 | .1970 | .2203 | .2396 |
| 4 | 204.3 | 41,740 | .03278 | .2289 | .2485 | .2778 | .3022 |
| 5 | 181.9 | 33,100 | .02600 | .2887 | .3133 | .3502 | .3810 |
| 6 | 162.0 | 26,250 | .02062 | .3640 | .3951 | .4416 | .4805 |
| 7 | 144.3 | 20,820 | .01635 | .4590 | .4982 | .5569 | .6059 |
| 8 | 128.5 | 16,510 | .01297 | .5788 | .6282 | .7023 | .7640 |
| 9 | 114.4 | 13,090 | .01028 | .7299 | .7921 | .8855 | .9633 |
| 10 | 101.9 | 10,380 | .008155 | .9203 | .9989 | 1.117 | 1.215 |
| 11 | 90.74 | 8,234 | .006467 | 1.161 | 1.260 | 1.408 | 1.532 |
| 12 | 80.81 | 6,530 | .005129 | 1.463 | 1.588 | 1.775 | 1.931 |
| 13 | 71.96 | 5,178 | .004067 | 1.845 | 2.003 | 2.239 | 2.436 |
| 14 | 64.08 | 4,107 | .003225 | 2.327 | 2.525 | 2.823 | 3.071 |
| 15 | 57.07 | 3,257 | .002558 | 2.934 | 3.184 | 3.560 | 3.873 |
| 16 | 50.82 | 2,583 | .002028 | 3.700 | 4.016 | 4.489 | 4.884 |
| 17 | 45.26 | 2,048 | .001609 | 4.666 | 5.064 | 5.660 | 6.158 |
| 18 | 40.30 | 1,624 | .001276 | 5.883 | 6.385 | 7.138 | 7.765 |
| 19 | 35.89 | 1,288 | .001012 | 7.418 | 8.051 | 9.001 | 9.792 |
| 20 | 31.96 | 1,022 | .0008023 | 9.355 | 10.15 | 11.35 | 12.35 |
| 21 | 28.45 | 810.1 | .0006363 | 11.80 | 12.80 | 14.31 | 15.57 |
| 22 | 25.35 | 642.4 | .0005046 | 14.87 | 16.14 | 18.05 | 19.63 |
| 23 | 22.57 | 509.5 | .0004002 | 18.76 | 20.36 | 22.76 | 24.76 |
| 24 | 20.10 | 404.0 | .0003173 | 23.65 | 25.67 | 28.70 | 31.22 |
| 25 | 17.90 | 320.4 | .0002517 | 29.82 | 32.37 | 36.18 | 39.36 |
| 26 | 15.94 | 254.1 | .0001996 | 37.61 | 40.81 | 45.63 | 49.64 |
| 27 | 14.20 | 201.5 | .0001583 | 47.42 | 51.47 | 57.53 | 62.59 |
| 28 | 12.64 | 159.8 | .0001255 | 59.80 | 64.90 | 72.55 | 78.93 |
| 29 | 11.26 | 126.7 | .00009953 | 75.40 | 81.83 | 91.48 | 99.52 |
| 30 | 10.03 | 100.5 | .00007894 | 95.08 | 103.2 | 115.4 | 125.5 |
| 31 | 8.928 | 79.70 | .00006260 | 119.9 | 130.1 | 145.5 | 158.2 |
| 32 | 7.950 | 63.21 | .00004964 | 151.2 | 164.1 | 183.4 | 199.5 |
| 33 | 7.080 | 50.13 | .00003937 | 190.6 | 206.9 | 231.3 | 251.6 |
| 34 | 6.305 | 39.75 | .00003122 | 240.4 | 260.9 | 291.7 | 317.3 |
| 35 | 5.015 | 31.52 | .00002476 | 303.1 | 329.0 | 367.8 | 400.1 |
| 36 | 5.000 | 25.00 | .00001964 | 382.3 | 414.8 | 463.7 | 504.5 |
| 37 | 4.453 | 19.83 | .00001557 | 482.0 | 523.1 | 584.8 | 636.2 |
| 38 | 3.965 | 15.72 | .00001235 | 607.8 | 659.6 | 737.4 | 802.2 |
| 39 | 3.531 | 12.47 | .000009793 | 766.4 | 831.8 | 929.8 | 1,012 |
| 40 | 3.145 | 9.888 | .000007766 | 966.5 | 1,049 | 1,173 | 1,276 |

## Table 8—1 (continued)
### WIRE TABLE, STANDARD ANNEALED COPPER

American Wire Gauge (B. and S.) (continued)

| Gauge No. | Pounds per 1,000 ft | Feet per pound | Feet per ohm[b] | | | |
|---|---|---|---|---|---|---|
| | | | 0°C (32°F) | 20°C (68°F) | 50°C (122°F) | 75°C (167°F) |
| 0000 | 640.5 | 1.561 | 22,140 | 20,400 | 18,250 | 16,780 |
| 000 | 507.9 | 1.968 | 17,560 | 16,180 | 14,470 | 13,300 |
| 00 | 402.8 | 2.482 | 13,930 | 12,830 | 11,480 | 10,550 |
| 0 | 319.5 | 3.130 | 11,040 | 10,180 | 9,103 | 8,367 |
| 1 | 253.3 | 3.947 | 8,758 | 8,070 | 7,219 | 6,636 |
| 2 | 200.9 | 4.977 | 6,946 | 6,400 | 5,725 | 5,262 |
| 3 | 159.3 | 6.276 | 5,508 | 5,075 | 4,540 | 4,173 |
| 4 | 126.4 | 7.914 | 4,368 | 4,025 | 3,600 | 3,309 |
| 5 | 100.2 | 9.980 | 3,464 | 3,192 | 2,855 | 2,625 |
| 6 | 79.46 | 12.58 | 2,747 | 2,531 | 2,264 | 2,081 |
| 7 | 63.02 | 15.87 | 2,179 | 2,007 | 1,796 | 1,651 |
| 8 | 49.98 | 20.01 | 1,728 | 1,592 | 1,424 | 1,309 |
| 9 | 39.63 | 25.23 | 1,370 | 1,262 | 1,129 | 1,038 |
| 10 | 31.43 | 31.82 | 1,087 | 1,001 | 895.6 | 823.2 |
| 11 | 24.92 | 40.12 | 861.7 | 794.0 | 710.2 | 652.8 |
| 12 | 19.77 | 50.59 | 683.3 | 629.6 | 563.2 | 517.7 |
| 13 | 15.68 | 63.80 | 541.9 | 499.3 | 446.7 | 410.6 |
| 14 | 12.43 | 80.44 | 429.8 | 396.0 | 354.2 | 325.6 |
| 15 | 9.858 | 101.4 | 340.8 | 314.0 | 280.9 | 258.2 |
| 16 | 7.818 | 127.9 | 270.3 | 249.0 | 222.8 | 204.8 |
| 17 | 6.200 | 161.3 | 214.3 | 197.5 | 176.7 | 162.4 |
| 18 | 4.917 | 203.4 | 170.0 | 156.6 | 140.1 | 128.8 |
| 19 | 3.899 | 256.5 | 134.8 | 124.2 | 111.1 | 102.1 |
| 20 | 3.092 | 323.4 | 106.9 | 98.50 | 88.11 | 80.99 |
| 21 | 2.452 | 407.8 | 84.78 | 78.11 | 69.87 | 64.23 |
| 22 | 1.945 | 514.2 | 67.23 | 61.95 | 55.41 | 50.94 |
| 23 | 1.542 | 648.4 | 53.32 | 49.13 | 43.94 | 40.39 |
| 24 | 1.223 | 817.7 | 42.28 | 38.96 | 34.85 | 32.03 |
| 25 | 0.9699 | 1,031 | 33.53 | 30.90 | 27.64 | 25.40 |
| 26 | .7692 | 1,300 | 26.59 | 24.50 | 21.92 | 20.15 |
| 27 | .6100 | 1,639 | 21.09 | 19.43 | 17.38 | 15.98 |
| 28 | .4837 | 2,067 | 16.72 | 15.41 | 13.78 | 12.67 |
| 29 | .3836 | 2,607 | 13.26 | 12.22 | 10.93 | 10.05 |
| 30 | .3042 | 3,287 | 10.52 | 9.691 | 8.669 | 7.968 |
| 31 | .2413 | 4,145 | 8.341 | 7.685 | 6.875 | 6.319 |
| 32 | .1913 | 5,227 | 6.614 | 6.095 | 5.452 | 5.011 |
| 33 | .1517 | 6,591 | 5.245 | 4.833 | 4.323 | 3.974 |
| 34 | .1203 | 8,310 | 4.160 | 3.833 | 3.429 | 3.152 |
| 35 | .09542 | 10,480 | 3.299 | 3.040 | 2.719 | 2.499 |
| 36 | .07568 | 13,210 | 2.616 | 2.411 | 2.156 | 1.982 |
| 37 | .06001 | 16,660 | 2.075 | 1.912 | 1.710 | 1.572 |
| 38 | .04759 | 21,010 | 1.645 | 1.516 | 1.356 | 1.247 |
| 39 | .03774 | 26,500 | 1.305 | 1.202 | 1.075 | 0.9886 |
| 40 | .02993 | 33,410 | 1.035 | 0.9534 | 0.8529 | .7840 |

## Table 8—1 (continued)
### WIRE TABLE, STANDARD ANNEALED COPPER

American Wire Gauge (B. and S.) (continued)

| Gauge No. | Diameter in mils at 20°C | Ohms per pound | | | Pounds per ohm |
|---|---|---|---|---|---|
| | | 0°C (32°F) | 20°C (68°F) | 50°C (122°F) | 20°C (68°F) |
| 0000 | 460.0 | 0.00007051 | 0.00007652 | 0.00008554 | 13,070 |
| 000 | 409.6 | .0001121 | .0001217 | .0001360 | 8,219 |
| 00 | 364.8 | .0001783 | .0001935 | .0002163 | 5,169 |
| 0 | 324.9 | .0002835 | .0003076 | .0003439 | 3,251 |
| 1 | 289.3 | .0004507 | .0004891 | .0005468 | 2,044 |
| 2 | 257.6 | .0007166 | .0007778 | .0008695 | 1,286 |
| 3 | 229.4 | .001140 | .001237 | .001383 | 808.6 |
| 4 | 204.3 | .001812 | .001966 | .002198 | 508.5 |
| 5 | 181.9 | .002881 | .003127 | .003495 | 319.8 |
| 6 | 162.0 | .004581 | .004972 | .005558 | 201.1 |
| 7 | 144.3 | .007284 | .007905 | .008838 | 126.5 |
| 8 | 128.5 | .01158 | .01257 | .01405 | 79.55 |
| 9 | 114.4 | .01842 | .01999 | .02234 | 50.03 |
| 10 | 101.9 | .02928 | .03178 | .03553 | 31.47 |
| 11 | 90.74 | .04656 | .05053 | .05649 | 19.79 |
| 12 | 80.81 | .07404 | .08035 | .08983 | 12.45 |
| 13 | 71.96 | .1177 | .1278 | .1428 | 7.827 |
| 14 | 64.08 | .1872 | .2032 | .2271 | 4.922 |
| 15 | 57.07 | .2976 | .3230 | .3611 | 3.096 |
| 16 | 50.82 | .4733 | .5136 | .5742 | 1.947 |
| 17 | 45.26 | .7525 | .8167 | .9130 | 1.224 |
| 18 | 40.30 | 1.197 | 1.299 | 1.452 | 0.7700 |
| 19 | 35.89 | 1.903 | 2.065 | 2.308 | .4843 |
| 20 | 31.96 | 3.025 | 3.283 | 3.670 | .3046 |
| 21 | 28.46 | 4.810 | 5.221 | 5.836 | .1915 |
| 22 | 25.35 | 7.649 | 8.301 | 9.280 | .1205 |
| 23 | 22.57 | 12.16 | 13.20 | 14.76 | .07576 |
| 24 | 20.10 | 19.34 | 20.99 | 23.46 | .04765 |
| 25 | 17.90 | 30.75 | 33.37 | 37.31 | .02997 |
| 26 | 15.94 | 48.89 | 53.06 | 59.32 | .01885 |
| 27 | 14.20 | 77.74 | 84.37 | 94.32 | .01185 |
| 28 | 12.64 | 123.6 | 134.2 | 150.0 | .007454 |
| 29 | 11.26 | 196.6 | 213.3 | 238.5 | .004688 |
| 30 | 10.03 | 312.5 | 339.2 | 379.2 | .002948 |
| 31 | 8.928 | 497.0 | 539.3 | 602.9 | .001854 |
| 32 | 7.950 | 790.2 | 857.6 | 958.7 | .001166 |
| 33 | 7.080 | 1,256 | 1,364 | 1,524 | .0007333 |
| 34 | 6.305 | 1,998 | 2,168 | 2,424 | .0004612 |
| 35 | 5.615 | 3,177 | 3,448 | 3,854 | .0002901 |
| 36 | 5.000 | 5,051 | 5,482 | 6,128 | .0001824 |
| 37 | 4.453 | 8,032 | 8,717 | 9,744 | .0001147 |
| 38 | 3.965 | 12,770 | 13,860 | 15,490 | .00007215 |
| 39 | 3.531 | 20,310 | 22,040 | 24,640 | .00004538 |
| 40 | 3.145 | 32,290 | 35,040 | 39,170 | .00002854 |

From Weast, R. C., Ed., *Handbook of Chemistry and Physics,* 69th ed., CRC Press, Boca Raton, Fla. 1988, F-117.

## Table 8—2
## ALUMINUM WIRE TABLE

Hard-drawn Aluminum Wire at 20°C (or 68°F) American Wire Gauge (B. and S.), English Units

| Gauge No. | Diameter in mils | Cross-section | | Ohms per 1,000 ft | Pounds per 1,000 ft | Pounds per ohm | Feet per ohm |
|---|---|---|---|---|---|---|---|
| | | Circular mils | Square inches | | | | |
| 0000 | 460 | 212,000 | 0.166 | 0.0804 | 195 | 2,420 | 12,400 |
| 000 | 410 | 168,000 | .132 | .101 | 154 | 1,520 | 9,860 |
| 00 | 365 | 133,000 | .105 | .128 | 122 | 957 | 7,820 |
| 0 | 325 | 106,000 | .0829 | .161 | 97.0 | 602 | 6,200 |
| 1 | 289 | 83,700 | .0657 | .203 | 76.9 | 379 | 4,920 |
| 2 | 258 | 66,400 | .0521 | .256 | 61.0 | 238 | 3,900 |
| 3 | 229 | 52,600 | .0413 | .323 | 48.4 | 150 | 3,090 |
| 4 | 204 | 41,700 | .0328 | .408 | 38.4 | 94.2 | 2,450 |
| 5 | 182 | 33,100 | .0260 | .514 | 30.4 | 59.2 | 1,950 |
| 6 | 162 | 26,300 | .0206 | .648 | 24.1 | 37.2 | 1,540 |
| 7 | 144 | 20,800 | .0164 | .817 | 19.1 | 23.4 | 1,220 |
| 8 | 128 | 16,500 | .0130 | 1.03 | 15.2 | 14.7 | 970 |
| 9 | 114 | 13,100 | .0103 | 1.30 | 12.0 | 9.26 | 770 |
| 10 | 102 | 10,400 | .00815 | 1.64 | 9.55 | 5.83 | 610 |
| 11 | 91 | 8,230 | .00647 | 2.07 | 7.57 | 3.66 | 484 |
| 12 | 81 | 6,530 | .00513 | 2.61 | 6.00 | 2.30 | 384 |
| 13 | 72 | 5,180 | .00407 | 3.29 | 4.76 | 1.45 | 304 |
| 14 | 64 | 4,110 | .00323 | 4.14 | 3.78 | 0.911 | 241 |
| 15 | 57 | 3,260 | .00256 | 5.22 | 2.99 | .573 | 191 |
| 16 | 51 | 2,580 | .00203 | 6.59 | 2.37 | .360 | 152 |
| 17 | 45 | 2,050 | .00161 | 8.31 | 1.88 | .227 | 120 |
| 18 | 40 | 1,620 | .00128 | 10.5 | 1.49 | .143 | 95.5 |
| 19 | 36 | 1,290 | .00101 | 13.2 | 1.18 | .0897 | 75.7 |
| 20 | 32 | 1,020 | .000802 | 16.7 | 0.939 | .0564 | 60.0 |
| 21 | 28.5 | 810 | .000636 | 21.0 | .745 | .0355 | 47.6 |
| 22 | 25.3 | 642 | .000505 | 26.5 | .591 | .0223 | 37.8 |
| 23 | 22.6 | 509 | .000400 | 33.4 | .468 | .0140 | 29.9 |
| 24 | 20.1 | 404 | .000317 | 42.1 | .371 | .00882 | 23.7 |
| 25 | 17.9 | 320 | .000252 | 53.1 | .295 | .00555 | 18.8 |
| 26 | 15.9 | 254 | .000200 | 67.0 | .234 | .00349 | 14.9 |
| 27 | 14.2 | 202 | .000158 | 84.4 | .185 | .00219 | 11.8 |
| 28 | 12.6 | 160 | .000126 | 106. | .147 | .00138 | 9.39 |
| 29 | 11.3 | 127 | .0000995 | 134. | .117 | .000868 | 7.45 |
| 30 | 10.0 | 101 | .0000789 | 169. | .0924 | .000546 | 5.91 |
| 31 | 8.9 | 79.7 | .0000626 | 213. | .0733 | .000343 | 4.68 |
| 32 | 8.0 | 63.2 | .0000496 | 269. | .0581 | .000216 | 3.72 |
| 33 | 7.1 | 50.1 | .0000394 | 339. | .0461 | .000136 | 2.95 |
| 34 | 6.3 | 39.8 | .0000312 | 428. | .0365 | .0000854 | 2.34 |
| 35 | 5.6 | 31.5 | .0000248 | 540. | .0290 | .0000537 | 1.85 |
| 36 | 5.0 | 25.0 | .0000196 | 681. | .0230 | .0000338 | 1.47 |
| 37 | 4.5 | 19.8 | .0000156 | 858. | .0182 | .0000212 | 1.17 |
| 38 | 4.0 | 15.7 | .0000123 | 1080. | .0145 | .0000134 | 0.924 |
| 39 | 3.5 | 12.5 | .00000979 | 1360. | .0115 | .00000840 | .733 |
| 40 | 3.1 | 9.9 | .00000777 | 1720. | .0091 | .00000528 | .581 |

From Weast, R. C., Ed., *Handbook of Chemistry and Physics,* 69th ed., CRC Press, Boca Raton, Fla. 1988, F-119.

**Table 8—3**

**RESISTOR COLOR CODE AND STANDARD VALUES**

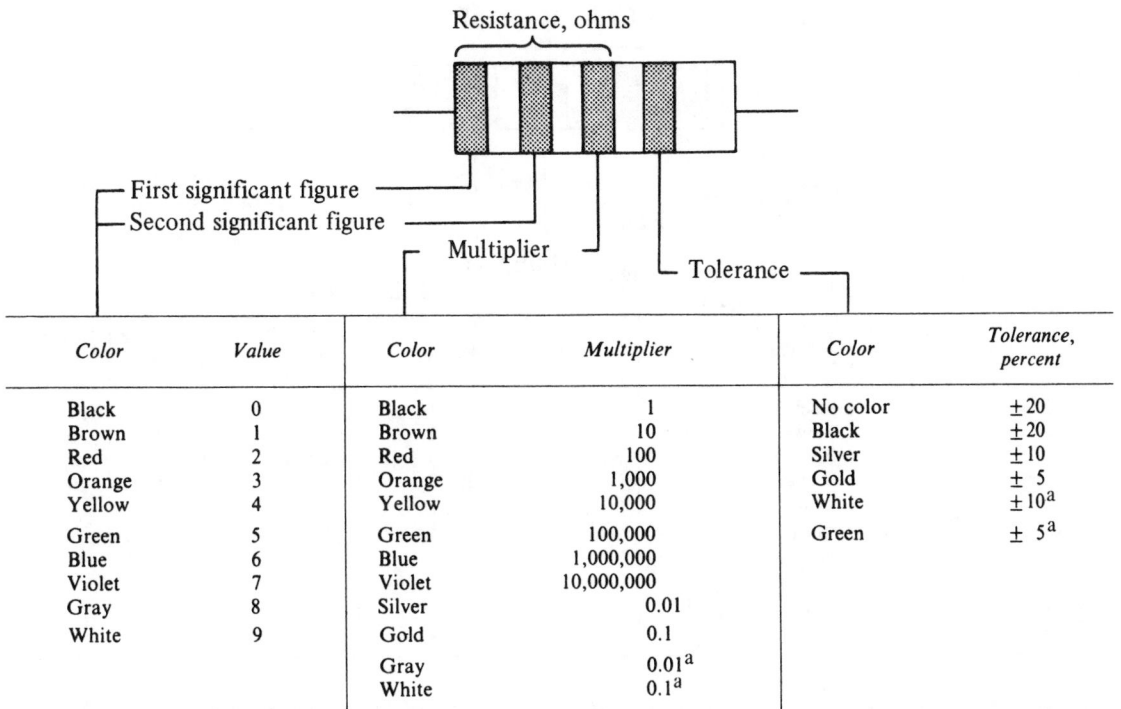

| Color | Value | Color | Multiplier | Color | Tolerance, percent |
|-------|-------|-------|------------|-------|--------------------|
| Black | 0 | Black | 1 | No color | ±20 |
| Brown | 1 | Brown | 10 | Black | ±20 |
| Red | 2 | Red | 100 | Silver | ±10 |
| Orange | 3 | Orange | 1,000 | Gold | ± 5 |
| Yellow | 4 | Yellow | 10,000 | White | ±10[a] |
| Green | 5 | Green | 100,000 | Green | ± 5[a] |
| Blue | 6 | Blue | 1,000,000 | | |
| Violet | 7 | Violet | 10,000,000 | | |
| Gray | 8 | Silver | 0.01 | | |
| White | 9 | Gold | 0.1 | | |
| | | Gray | 0.01[a] | | |
| | | White | 0.1[a] | | |

[a]Optional.

Standard Resistor Values, Significant Figures

±20%: 10, 15, 22, 33, 47, 68 (Series 6).[b]

±10%: 10, 12, 15, 18, 22, 27, 33, 39, 47, 56, 68, 82 (Series 12).[b]

±5%: 10, 11, 12, 13, 15, 16, 18, 20, 22, 24, 27, 30, 33, 36, 39, 43, 47, 51, 56, 62, 68, 75, 82, 91 (Series 24).[b]

[b]For series specifications see Bolz, R. E. and Tuve, G. L., Eds., *Handbook of Tables for Applied Engineering Science,* 2nd ed., CRC Press, Cleveland, 1973, 246.

From Bolz, R. E. and Tuve, G. L., Eds., *Handbook of Tables for Applied Engineering Science,* 2nd ed., CRC Press, Cleveland, 1973, 247.

**Table 8—4**
**CAPACITOR COLOR CODE**

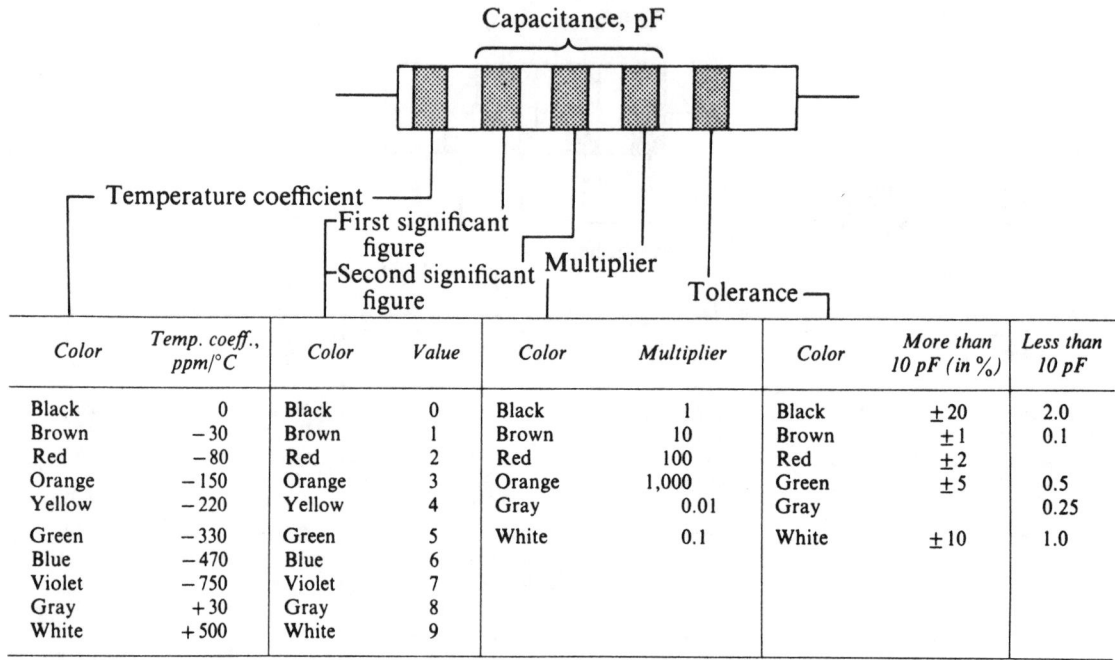

| Color | Temp. coeff., ppm/°C | Color | Value | Color | Multiplier | Color | More than 10 pF (in %) | Less than 10 pF |
|-------|------|-------|-------|-------|-----------|-------|------|------|
| Black | 0 | Black | 0 | Black | 1 | Black | ±20 | 2.0 |
| Brown | −30 | Brown | 1 | Brown | 10 | Brown | ±1 | 0.1 |
| Red | −80 | Red | 2 | Red | 100 | Red | ±2 | |
| Orange | −150 | Orange | 3 | Orange | 1,000 | Green | ±5 | 0.5 |
| Yellow | −220 | Yellow | 4 | Gray | 0.01 | Gray | | 0.25 |
| Green | −330 | Green | 5 | White | 0.1 | White | ±10 | 1.0 |
| Blue | −470 | Blue | 6 | | | | | |
| Violet | −750 | Violet | 7 | | | | | |
| Gray | +30 | Gray | 8 | | | | | |
| White | +500 | White | 9 | | | | | |

From Bolz, R. E. and Tuve, G. L., Eds., *Handbook of Tables for Applied Engineering Science,* 2nd ed., CRC Press, Cleveland, 1973, 247.

**Table 8—5**

**PROPERTIES OF SEMICONDUCTORS**

The term *semiconductor* is applied to a material in which electric current is carried by electrons or holes; its electrical conductivity when extremely pure rises exponentially with temperature and may be increased from this low "intrinsic" value by many orders of magnitude by "doping" with electrically active impurities.

Semiconductors are characterized by an energy gap in the allowed energies of electrons in the material that separates the normally filled energy levels of the *valence band* (where "missing"

electrons behave as positively charged current carriers or "holes") and the *conduction band* (where electrons behave somewhat as a gas of free negatively charged carriers with an effective mass dependent on the material and the direction of the electrons' motion). This energy gap depends on the nature of the material and varies with direction in anisotropic crystals. It is slightly dependent on temperature and pressure, and this dependence is usually almost linear at normal temperatures and pressures.

I. General Properties of Semiconductors[a]

| Substance | Lattice parameters at room temperature, Å | Density, g/cc | Melting point, K | Minimum room temperature energy gap, ev | Comparative thermal conductivity | Heat of formation, kcal/mole | Mobility (room temperature) Electrons  cm²/volt-sec | Holes | Remarks |
|---|---|---|---|---|---|---|---|---|---|
| **Part A. Tetrahedral Semiconductors** | | | | | | | | | |
| **§ A1 Diamond Structure Elements** (Strukturbericht symbol A4, Space Group Fd3m-0$_h^7$) | | | | | | | | | |
| C | 3.5597 | 3.51 | 4300 | 5.4 | 2000 | 161 | 1800 | 1400 | |
| Si | 5.43072 | 2.3283 | 1685 | 1.107 | 1240 | 77.5 | 1900 | 500 | |
| Ge | 5.65754 | 5.3234 | 1231 | 0.67 | 640 | 69.5 | 3800 | 1820 | |
| α-Sn | 6.4912 | 5.765 | 503 | 0.08 | | 64 | 2500 | 2400 | |
| **§ A2 Sphalerite (Zinc Blende) Structure Compounds** (Strukturbericht symbol B3 Space Group F $\bar{4}$ 3m-T$_d^2$) | | | | | | | | | |
| **I-VII Compounds** | | | | | | | | | |
| CuF | 4.255 | | | | | | | | |
| CuCl | 5.4057 | 3.53 | 695 | | | 115 | | | |
| CuBr | 5.6905 | 4.72 | 770 | 2.94 | | 105 | | | |
| CuI | 6.0427 | 5.63 | 878 | | | 102 | | | |
| AgBr | | | | | | 93 | 4000 | | |
| AgI | 6.473 | 5.67 | | | | | | 30 | |
| **II-VI Compounds** | | | | | | | | | |
| BeS | 4.865 | 2.36 | | | | | | | |
| BeSe | 5.139 | 4.315 | | | | | | | |
| BeTe | 5.626 | 5.090 | | | | | | | |
| BePo | 8.38 | 7.3 | | | | | | | |
| ZnO | 4.63 | | | | | | | | See § A3 |
| ZnS | 5.4093 | 4.079 | 1920 | 3.54 | 140 | 114 | 180 | 5(400°C) | See also § A3 |
| ZnSe | 5.6676 | 5.42 | 1790 | 2.58 | 140 | 101 | 540 | 28 | |
| ZnTe | 6.101 | 5.72 | 1510 | 2.26 | 140 | 90 | 340 | 100 | |
| ZnPo | | | | | | | | | See also § A3 |
| CdS | 5.5818 | | | | | | | | See § A3 |
| CdSe | 6.05 | | | | | | | | |
| CdTe | 6.477 | 5.86 | 1370 | 1.44 | 55 | 81 | 1200 | 50 | |
| CdPo | | | | | | | | | |
| HgS | 5.8517 | 7.73 | ~2020 | | | 59 | 20000 | | |
| HgSe | 6.084 | 8.25 | 1070 | 0.30 | 10 | 59 | 20000 | | |
| HgTe | 6.429 | 8.17 | 943 | 0.15 | 20 | 58 | 25000 | 350 | |
| **III-V Compounds** | | | | | | | | | |
| BN | 3.615 | 3.49 | 3000 | ~4 | 200 | 195 | | | |
| BP(L.T.) | 4.538 | 2.9 | | ~6 | | | 500 | 70 | |
| BAs | 4.777 | | | | | | | | |
| AlP | 5.451 | 2.85 | 1770 | 2.5 | | 150 | 1200 | 420 | |
| AlAs | 5.6622 | 3.81 | 1870 | 2.16 | | 140 | 200–400 | 550 | |
| AlSb | 6.1355 | 4.218 | 1330 | 1.60 | 600 | 152 | 300 | 100 | |
| GaP | 5.4505 | 4.13 | 1750 | 2.24 | 1100 | 128 | 8800 | 400 | |
| GaAs | 5.65315 | 5.316 | 1510 | 1.35 | 370 | 118 | 4000 | 1400 | |
| GaSb | 6.0954 | 5.619 | 980 | 0.67 | 270 | 134 | 4600 | 150 | |
| InP | 5.86875 | 4.787 | 1330 | 1.27 | 800 | 114 | 33000 | 460 | |
| InAs | 6.05838 | 5.66 | 1215 | 0.36 | 290 | 107 | 78000 | 750 | |
| InSb | 6.47877 | 5.775 | 798 | 0.165 | 160 | | | | |

## Table 8—5 (continued)
## PROPERTIES OF SEMICONDUCTORS

### I. General Properties of Semiconductors

| Substance | Lattice parameters at room temperature, Å | | Density, g/cc | Melting point, K | Minimum room temperature energy gap, ev | Comparative thermal conductivity | Heat of formation, kcal/mole | Mobility (room temperature) Electrons | Holes | Remarks |
|---|---|---|---|---|---|---|---|---|---|---|
| **Other Sphalerite Structure Compounds** | | | | | | | | | | |
| $\beta$-SiC | 4.348 | | 3.21 | 3070 | 2.3 | | | 4000 | | |
| $Ga_2Te_3$ | 5.899 | | 5.75 | 1063 | ~1.0 | ~14 | 65 | | | |
| $In_2Te_3$(H.T.) | 6.150 | | 5.8 | 940 | ~1.0 | ~8 | 47.4 | | ~10 | |
| $MgGeP_2$ | 5.652 | | | | | | | | | |
| $ZnSnP_2$ | 5.65 | | | | 2.1 | | | | | |
| $ZnSnAs_2$(H.T.) | 5.851 | | 5.53 | 1050 | ~0.7 | | 70 | | | |
| **§ A3 Wurtzite (Zincite) Structure Compounds (Strukturbericht symbol B4, Space Group P $6_3$ mc-$C_{6v}^4$)** | | | | | | | | | | |
| **I-VII Compounds** | | | | | | | | | | |
| CuCl | 3.91 | 6.42 | | $T_c$ 680°K | | | | | | |
| CuBr | 4.06 | 6.66 | | $T_c$ 658°K | | | | | | |
| CuI | 4.31 | 7.09 | | | | | | | | |
| AgI | 4.580 | 7.494 | | | 2.63 | | | | | |
| **II-VI Compounds** | | | | | | | | | | |
| BeO | 2.698 | 4.380 | | 2800 | | | | | | |
| MgTe | 4.54 | 7.39 | 3.85 | ~2800 | | | | | | |
| ZnO | 3.24950 | 5.2069 | 5.66 | 2250 | 3.2 | 6 | 154 | 180 | | |
| ZnS | 3.8140 | 6.2576 | 4.1 | 2100 | 3.67 | | 110 | | | |
| ZnSe | 3.996 | 6.626 | | | | | | | | |
| ZnTe | 4.27 | 6.99 | | | | | | | | |
| CdS | 4.1348 | 6.7490 | 4.82 | 2020 | 2.42 | | 96 | 400 | | |
| CdSe | 4.299 | 7.010 | 5.66 | 1530 | 1.74 | | 90 | 650 | | |
| CdTe | 4.57 | 7.47 | | | 1.50 | | | | | |
| **III-V Compounds** | | | | | | | | | | |
| BP(H.T.) | 3.562 | 5.900 | | | | | | | | |
| AlN | 3.111 | 4.978 | 3.26 | ~2500 | 6.02 | | 197 | | | |
| GaN | 3.180 | 5.166 | 6.10 | 1500 | 3.34 | | 157 | | | |
| InN | 3.533 | 5.693 | 6.88 | 1200 | 2.0 | | 133 | | | |
| **Other Wurtzite Structure Compounds** | | | | | | | | | | |
| SiC | 3.076 | 5.048 | | | | | | | | |
| MnTe | 4.078 | 6.701 | | | ~1.0 | | | | | |
| $Al_2S_3$ | 3.579 | 5.829 | 2.55 | | 4.1 | | 426 | | | |
| $Al_2Se_3$ | 3.890 | 6.30 | 3.91 | | 3.1 | | 367 | | | |
| **§ A4 Chalcopyrite Structure Compounds (Strukturbericht symbol E1$_1$, Space Group I $\bar4$ 2d $-D_{2d}^{12}$)** | | | | | | | | | | |
| **I-III-VI$_2$ Compounds** | | | | | | | | | | |
| $CuAlS_2$ | 5.323 | 10.44 | 3.47 | | 2.5 | | | | | |
| $CuAlSe_2$ | 5.617 | 10.92 | 4.70 | 1270 | 1.1 | | | | | |
| $CuAlTe_2$ | 5.976 | 11.80 | 5.50 | 1160 | 0.88 | | | | | |
| $CuGaS_2$ | 5.360 | 10.49 | 4.35 | | | | | | | |
| $CuGaSe_2$ | 5.618 | 11.01 | 5.56 | 1310 | 0.96, 1.63 | | | | | |
| $CuGaTe_2$ | 6.013 | 11.93 | 5.99 | 1150 | 0.82, 1.0 | | | | | |
| $CuInS_2$ | 5.528 | 11.08 | 4.75 | | 1.2 | | | | | |
| $CuInSe_2$ | 5.785 | 11.57 | 5.77 | 1250 | 0.86, 0.92 | 37 | | | | |
| $CuInTe_2$ | 6.179 | 12.365 | 6.10 | 970 | 0.95 | 49 | | | | |
| $CuTlS_2$ | 5.580 | 11.17 | 6.32 | | | | | | | |
| $CuTlSe_2$(L.T.) | 5.844 | 11.65 | 7.11 | 680 | 1.07 | | | | | |
| $CuFeS_2$ | 5.25 | 10.32 | | 1150 | 0.53 | | | | | |
| $CuFeSe_2$ | | | | 850 | 0.16 | | | | | |
| $CuLaS_2$ | 5.65 | 10.86 | | | | | | | | |
| $AgAlS_2$ | 5.707 | 10.28 | 3.94 | | | | | | | |
| $AgAlSe_2$ | 5.968 | 10.77 | 5.07 | 1220 | 0.7 | | | | | |
| $AgAlTe_2$ | 6.309 | 11.85 | 6.18 | 1000 | 0.56 | | | | | |
| $AgGaS_2$ | 5.755 | 10.28 | 4.72 | | | | | | | |
| $AgGaSe_2$ | 5.985 | 10.90 | 5.84 | 1120 | 1.66 | | | | | |
| $AgGaTe_2$ | 6.301 | 11.96 | 6.05 | 990 | 1.1 | 10 | | | | |
| $AgInS_2$(L.T.) | 5.828 | 11.19 | 5.00 | | 1.9 | | | | | |
| $AgInSe_2$ | 6.102 | 11.69 | 5.81 | 1053 | 1.18 | 30 | | | | |
| $AgInTe_2$ | 6.42 | 12.59 | 6.12 | 965 | 0.96, 0.52 | | | | | |
| $AgFeS_2$ | 5.66 | 10.30 | 4.53 | | | | | | | |

## Table 8—5 (continued)
## PROPERTIES OF SEMICONDUCTORS

### 1. General Properties of Semiconductors (continued)

| Substance | Lattice parameters at room temperature, Å | | Density, g/cc | Melting point, K | Minimum room temperature energy gap, ev | Comparative thermal conductivity | Heat of formation, kcal/mole | Mobility (room temperature) | | Remarks |
|---|---|---|---|---|---|---|---|---|---|---|
| | | | | | | | | Electrons | Holes | |
| | | | | | | | | cm²/volt-sec | | |
| **II-IV-V₂ Compounds** | | | | | | | | | | |
| ZnSiP₂ | 5.400 | 10.441 | 3.39 | 1640 | 2.3 | | | 1000 | | |
| ZnGeP₂ | 5.465 | 10.771 | 4.17 | 1295 | 2.2 | | | | | |
| CdSiP₂ | 5.678 | 10.431 | 4.00 | ~1470 | 2.2 | | | 1000 | | |
| CdGeP₂ | 5.741 | 10.775 | 4.48 | ~1060 | 1.8 | | | | | |
| CdSnP₂ | 5.900 | 11.518 | | | 1.5 | | | | | |
| ZnSiAs₂ | 5.61 | 10.88 | 4.70 | ~1350 | 1.7 | | | | 50 | |
| ZnGeAs₂ | 5.672 | 11.153 | 5.32 | ~1150 | 0.85 | 110 | | | | |
| ZnSnAs₂ | 5.8515 | 11.704 | 5.53 | ~ 910 | 0.65 | 150 | | | 300 | Disorders at 910°K |
| CdSiAs₂ | 5.884 | 10.882 | | | | | | | | |
| CdGeAs₂ | 5.9427 | 11.2172 | 5.60 | ~ 903 | 0.53 | 40 | | 70 | 25 | Disorders at 903°K |
| CdSnAs₂ | 6.0944 | 11.9182 | 5.72 | 880 | 0.26 | 70 | | 22000 | 250 | |

§ A5 **"Defect Chalcopyrite" Structure Compounds** (Strukturbericht symbol E3, Space Group I 4̄ — $S_4^2$)

| Substance | | | Density | Melting | Energy gap | Thermal cond | Heat | Electrons | Holes | Remarks |
|---|---|---|---|---|---|---|---|---|---|---|
| ZnAl₂Se₄ | 5.503 | 10.90 | 4.37 | | | | | | | |
| ZnAl₂Te₄(?) | 5.104 | 12.05 | 4.95 | | | | | | | |
| ZnGa₂S₄(?) | 5.274 | 10.44 | 3.80 | | | | | | | |
| ZnGa₂Se₄(?) | 5.496 | 10.99 | 5.21 | | | | | | | |
| ZnGa₂Te₄(?) | 5.937 | 11.87 | 5.67 | | 1.35 | | | | | |
| ZnIn₂Se₄ | 5.711 | 11.42 | 5.44 | 1250 | 2.6 | | | | | |
| ZnIn₂Te₄ | 6.122 | 12.24 | 5.83 | 1075 | 1.2 | | | | | |
| CdAl₂S₄ | 5.564 | 10.32 | 3.06 | | | | | | | |
| CdAl₂Se₄ | 5.747 | 10.68 | 4.54 | | | | | | | |
| CdAl₂Te₄(?) | 6.011 | 12.21 | 5.10 | | | | | | | |
| CdGa₂S₄ | 5.577 | 10.08 | 4.03 | | | | | | | |
| CdGa₂Se₄ | 5.743 | 10.73 | 5.32 | | | | | | | |
| CdGa₂Te₄ | 6.093 | 11.81 | 5.77 | | | | | | | |
| CdIn₂Te₄ | 6.205 | 12.41 | 5.9 | 1060 | (1.26 or 0.9) | 4000 | | | | |
| HgAl₂S₄ | 5.488 | 10.26 | 4.11 | | | | | | | |
| HgAl₂Se₄ | 5.708 | 10.74 | 5.05 | | | | | | | |
| HgAl₂Te₄(?) | 6.004 | 12.11 | 5.81 | | | | | | | |
| HgGa₂S₄ | 5.507 | 10.23 | 5.00 | | | | | | | |
| HgGa₂Se₄ | 5.715 | 10.78 | 6.18 | | | | | | | |
| HgIn₂Se₄ | 5.764 | 11.80 | 6.3 | 1100 | 0.6 | | | | | |
| HgIn₂Te₄(?) | 6.186 | 12.37 | 6.3 | 980 | 0.86 | | 200 | | | |

§ A6 **Other Tetrahedral Compounds**

| Substance | | | Density | Melting | Energy gap | Thermal | Heat | Electrons | Holes | Remarks |
|---|---|---|---|---|---|---|---|---|---|---|
| α-SiC | 3.0817 15.1183 | | 3.21 | 3070 | 2.86 | | | 400 | | 6H structure |
| Hg₅Ga₂Te₈ | 6.235 | | | | | | | | | B3 with super lattice |
| Hg₅In₂Te₈ | 6.328 | | | | 0.7 | | | 2000 | | B3 with super lattice |

### Part B. Octahedral Semiconductors

**Halite Structure Semiconductors** (Strukturbericht symbol B1, Space Group Fm3m — $0_h^5$)

| Substance | | | Density | Melting | Energy gap | Thermal | Heat | Electrons | Holes | Remarks |
|---|---|---|---|---|---|---|---|---|---|---|
| SnSe | 6.020 | | | 1133 | | | | | | |
| SnTe | 6.313 | | | 1080 (max) | 0.5 | 91 | | | | |
| PbS | 5.9362 | | 7.61 | 1390 | 0.37 | 23 | 104 | 600 | 600 | |
| PbSe | 6.1243 | | 8.15 | 1340 | 0.26 | 17 | 94 | 1000 | 900 | |
| PbTe | 6.454 | | 8.16 | 1180 | 0.25 | 23 | 94 | 1600 | 600 | |
| **Selected Other Binary Chalcides** | | | | | | | | | | |
| BiSe | 5.99 | | 7.98 | 880 | 0.4 | | | | | |
| BiTe | 6.47 | | | | | | | | | |
| EuSe | 6.191 | | | 2300 | 1.8 | 2.4 | | | | |
| GdSe | 5.771 | | | 2400 | | | | | | |
| NiD | 4.1684 | | 6.6 | 2260 | 2.0 or 3.7 | | | 4 | | |
| CdO | 4.6953 | | | 1700 | 2.5 | 7 | 127 | 100 | | |
| SrS | 6.0199 | | 3.643 | 3000 | 4.1 | | | | | |

**Table 8—5 (continued)**
**PROPERTIES OF SEMICONDUCTORS**

I. General Properties of Semiconductors (continued)[a]

| Substance | Lattice parameters at room temperature, Å | Density, g/cc | Melting point, K | Minimum room temperature energy gap, ev | Comparative thermal conductivity | Heat of formation, kcal/mole | Mobility (room temperature) Electrons cm³/volt-sec | Holes | Remarks |
|---|---|---|---|---|---|---|---|---|---|
| **Selected Ternary Compounds** | | | | | | | | | |
| $AgSbSe_2$ | 5.786 | 6.60 | 910 | 0.58 | 10.5 | | | | |
| $AgSbTe_2$ (or $Ag_{19}Sb_{29}Te_{52}$) | 6.078 | 7.12 | 830 | 0.7, 0.27 | 8.6, 0.3 | | | | |
| $AgBiS_2$(H.T.) | 5.648 | | | | | | | | |
| $AgBiSe_2$(H.T.) | 5.82 | | | | | | | | |
| $AgBiTe_2$(H.T.) | 6.155 | | | | | | | | |

[a]Listed by crystal structure.

II. Semiconducting Properties of Selected Materials

| Substance | Minimum energy gap, ev Room temperature | 0 K | $\dfrac{dE_g}{dT}$ $\times 10^4$ ev/°C | $\dfrac{dE_g}{dP}$ $\times 10^6$ ev cm²/kg | Density of states electron effective mass, $m_{d_n}$ $(m_o)$ | Electron mobility and temperature dependence $\mu_n$ cm²/volt-sec | $-x$ | Density of states hole effective mass, $m_{d_p}$ $(m_o)$ | Hole mobility and temperature dependence $\mu_p$ cm²/volt-sec | $-x$ | Dominant emission wavelength, 77 K, microns |
|---|---|---|---|---|---|---|---|---|---|---|---|
| Si | 1.107 | 1.153 | −2.3 | −2.0 | 0.58 | 1,900 | 2.6 | 1.06 | 500 | 2.3 | 1.274 |
| Ge | 0.67 | 0.744 | −3.7 | +7.3 | 0.35 | 3,800 | 1.66 | 0.56 | 1,820 | 2.33 | 1.770 |
| α-Sn | 0.08 | 0.094 | −0.5 | | 0.02 | 2,500 | 1.65 | 0.3 | 2,400 | 2.0 | — |
| Te | 0.33 | | | | 0.68 | 1,100 | | 0.19 | 560 | | — |
| **III-V Compounds** | | | | | | | | | | | |
| AlAs | 2.2 | 2.3 | | | | 1,200 | | | 420 | | — |
| AlSb | 1.6 | 1.7 | −3.5 | −1.6 | 0.09 | 200 | 1.5 | 0.4 | 500 | 1.8 | — |
| GaP | 2.24 | 2.40 | −5.4 | −1.7 | 0.35 | 300 | 1.5 | | 150 | 1.5 | 0.59 |
| GaAs | 1.35 | 1.53 | −5.0 | +9.4 | 0.068 | 9,000 | 1.0 | 0.5 | 500 | 2.1 | 0.84 |
| GaSb | 0.67 | 0.78 | −3.5 | +12 | 0.050 | 5,000 | 2.0 | 0.23 | 1,400 | 0.9 | 1.6 |
| InP | 1.27 | 1.41 | −4.6 | +4.6 | 0.067 | 5,000 | 2.0 | | 200 | 2.4 | 0.91 |
| InAs | 0.36 | 0.43 | −2.8 | +8 | 0.022 | 33,000 | 1.2 | 0.41 | 460 | 2.3 | 3.1 |
| InSb | 0.165 | 0.23 | −2.8 | +15 | 0.014 | 78,000 | 1.6 | 0.4 | 750 | 2.1 | 5.2 |
| **II-VI Compounds** | | | | | | | | | | | |
| ZnO | 3.2 | | −9.5 | +0.6 | 0.38 | 180 | 1.5 | | | | 0.37 |
| ZnS | 3.54 | | −5.3 | +5.7 | | 180 | | | 5 (400°C) | | 0.33 |
| ZnSe | 2.58 | 2.80 | −7.2 | +6 | | 540 | | | 28 | | — |
| ZnTe | 2.26 | | | +6 | | 340 | | | 100 | | — |
| CdO | 2.5±.1 | | −6 | | 0.1 | 120 | | | | | — |
| CdS | 2.42 | | −5 | +3.3 | 0.165 | 400 | | 0.8 | | | 0.49 |
| CdSe | 1.74 | 1.85 | −4.6 | | 0.13 | 650 | 1.0 | 0.6 | | | 0.68 |
| CdTe | 1.44 | 1.56 | −4.1 | +8 | 0.14 | 1,200 | | 0.35 | 50 | | 0.78 |
| HgSe | 0.30 | | | | 0.030 | 20,000 | 2.0 | | | | — |
| HgTe | 0.15 | | −1 | | 0.017 | 25,000 | | 0.5 | 350 | | — |
| **Halite Structure Compounds** | | | | | | | | | | | |
| PbS | 0.37 | 0.28 | +4 | | 0.16 | 800 | | 0.1 | 1,000 | 2.2 | 4.3 |
| PbSe | 0.26 | 0.16 | +4 | | 0.3 | 1,500 | | 0.34 | 1,500 | 2.2 | 8.5 |
| PbTe | 0.25 | 0.19 | +4 | −7 | 0.21 | 1,600 | | 0.14 | 750 | 2.2 | 6.5 |

## Table 8—5 (continued)
## PROPERTIES OF SEMICONDUCTORS

### II. Semiconducting Properties of Selected Materials (continued)

| Substance | Minimum energy gap, ev | | $\dfrac{dE_g}{dT}$ $\times 10^4$ ev/°C | $\dfrac{dE_g}{dP}$ $\times 10^6$ ev cm²/kg | Density of states electron effective mass, $m_{d_n}$ $(m_o)$ | Electron mobility and temperature dependence | | Density of states hole effective mass, $m_{d_p}$ $(m_o)$ | Hole mobility and temperature dependence | | Dominant emission wavelength, 77 K, microns |
|---|---|---|---|---|---|---|---|---|---|---|---|
| | Room temperature | 0 K | | | | $\mu_n$ cm²/volt-sec | $-x$ | | $\mu_p$ cm²/volt-sec | $-x$ | |
| **Others** | | | | | | | | | | | |
| ZnSb | 0.50 | 0.56 | | | 0.15 | 10 | | | | | 1.5 |
| CdSb | 0.45 | 0.57 | −5.4 | | 0.15 | 300 | | | 2,000 | | 1.5 |
| $Bi_2S_3$ | 1.3 | | | | | 200 | | | 1,100 | | |
| $Bi_2Se_3$ | 0.27 | | | | | 600 | | | 675 | | |
| $Bi_2Te_3$ | 0.13 | | −0.95 | | 0.58 | 1,200 | 1.68 | 1.07 | 510 | | 1.95 |
| $Mg_2Si$ | | 0.77 | −6.4 | | 0.46 | 400 | 2.5 | | 70 | | |
| $Mg_2Ge$ | | 0.74 | −9 | | | 280 | 2 | | 110 | | |
| $Mg_2Sn$ | 0.21 | 0.33 | −3.5 | | 0.37 | 320 | | | 260 | | |
| $Mg_3Sb_2$ | | 0.32 | | | | 20 | | | 82 | | |
| $Zn_3As_2$ | 0.93 | | | | | 10 | 1.1 | | 10 | | |
| $Cd_3As_2$ | 0.13 | | | | 0.046 | 15,000 | 0.88 | | | | |
| GaSe | 2.05 | | 3.8 | | | | | | 20 | | |
| GaTe | 1.66 | 1.80 | −3.6 | | | 14 | −5 | | | | |
| InSe | 1.8 | | | | | 900 | | | | | |
| TlSe | 0.57 | | −3.9 | | 0.3 | 30 | | 0.6 | 20 | | 1.5 |
| $CdSnAs_2$ | 0.23 | | | | 0.05 | 25,000 | 1.7 | | | | |
| $Ga_2Te_3$ | 1.1 | 1.55 | −4.8 | | | | | | | | |
| $\alpha\text{-}In_2Te_3$ | 1.1 | 1.2 | | | 0.7 | | | | 50 | | 1.1 |
| $\beta\text{-}In_2Te_3$ | 1.0 | | | | | | | | 5 | | |
| $Hg_5In_2Te_8$ | 0.5 | | | | | | | | 11,000 | | |
| $SnO_2$ | | | | | | | | | 78 | | |

### III. Valence Bands of Semiconductors[a]
#### Semiconductors with Valence Band Maximum at Center of Brillouin Zone ("Γ")

| Substance | Band curvature effective mass | | | Energy separation of "split-off" band, ev | Measured (light) hole mobility, cm²/volt-sec |
|---|---|---|---|---|---|
| | Heavy holes | Light holes | "Split-off" band holes | | |
| | (Expressed as fraction of free electron mass) | | | | |
| Si | 0.52 | 0.16 | 0.25 | 0.044 | 500 |
| Ge | 0.34 | 0.043 | 0.08 | 0.3 | 1,820 |
| Sn | 0.3 | | | | 2,400 |
| AlAs | | | | | |
| AlSb | 0.4 | | | 0.7 | 550 |
| GaP | | | | 0.13 | 100 |
| GaAs | 0.8 | 0.12 | 0.20 | 0.34 | 400 |
| GaSb | 0.23 | 0.06 | | 0.7 | 1,400 |
| InP | | | | 0.21 | 150 |
| InAs | 0.41 | 0.025 | 0.083 | 0.43 | 460 |
| InSb | 0.4 | 0.015 | | 0.85 | 750 |
| CdTe | 0.35 | | | | 50 |
| HgTe | 0.5 | | | | 350 |

#### Semiconductors with Multiple Valence Band Maxima

| Substance | Number of equivalent valleys and direction | Curvature effective masses | | Anisotropy, $\dfrac{m_L}{K = m_T}$ | Measured (light) hole mobility, cm²/volt-sec |
|---|---|---|---|---|---|
| | | Longitudinal, $m_L$ | Transverse, $m_T$ | | |
| PbSe | 4 "L" [111] | 0.095 | 0.047 | 2.0 | 1,500 |
| PbTe | 4 "L" [111] | 0.27 | 0.02 | 10 | 750 |
| $Bi_2Te_3$ | 6 | 0.207 | ~0.045 | 4.5 | 515 |

## Table 8—5 (continued)
## PROPERTIES OF SEMICONDUCTORS

### IV. Conduction Bands of Semiconductors[a]

**Single Valley Semiconductors**

| Substance | Energy gap, ev | Effective mass, $m_o$ | Mobility, $cm^2/volt\text{-}sec$ | Comments |
|---|---|---|---|---|
| GaAs | 1.35 | 0.067 | 8,500 | 3(or 6?) equivalent [100] valleys 0.36 ev above this maximum with a mobility of ~50. |
| InP | 1.27 | 0.067 | 5,000 | 3(or 6?) equivalent [100] valleys 0.4 ev above this minimum. |
| InAs | 0.36 | 0.022 | 33,000 | Equivalent valleys ~1.0 ev above this minimum. |
| InSb | 0.165 | 0.014 | 78,000 | |
| CdTe | 1.44 | 0.11 | 1,000 | 4(or 8?) equivalent [111] valleys 0.51 ev above this minimum. |

**Multivalley Semiconductors**

| Substance | Energy gap | Number of equivalent valleys and direction | Band curvature effective mass Longitudinal, $m_L$ | Band curvature effective mass Transverse, $m_T$ | Anisotropy, $K = \dfrac{m_L}{m_T}$ |
|---|---|---|---|---|---|
| Si | 1.107 | 6 in [100] "Δ" | 0.90 | 0.192 | 4.7 |
| Ge | 0.67 | 4 in [111] at "L" | 1.588 | 0.0815 | 19.5 |
| GaSb | 0.67 | as Ge (?) | ~1.0 | ~0.2 | ~5 |
| PbSe | 0.26 | 4 in [111] at "L" | 0.085 | 0.05 | 1.7 |
| PbTe | 0.25 | 4 in [111] at "L" | 0.21 | 0.029 | 5.5 |
| $Bi_2Te_3$ | 0.13 | 6 | | | ~0.05 |

[a]Room temperature data.

From Pamplin, B. R., in *Handbook of Chemistry and Physics,* 55th ed., Weast, R. C., Ed., CRC Press, Cleveland, 1974, E-99.

## Table 8—6
### LASER LINES STRONGLY ABSORBED BY THE ATMOSPHERE

| Laser | λ, microns | Absorber |
|-------|-----------|----------|
| Atomic krypton | 1.7843 | $H_2O$ |
| Atomic krypton | 1.9211 | $H_2O$ |
| $Tm^{+3}$–$CaWO_4$ | 1.911 | $H_2O$ |
| $Tm^{+3}$–$CaWO_4$ | 1.916 | $H_2O$ |
| | | |
| $U^{+3}$–$SrF_2$ | 2.472 | $H_2O$ |
| $U^{+3}$–$CaF_2$ | 2.511 | $H_2O$ |
| Atomic krypton | 2.5234 | $H_2O$ |
| $U^{+3}$–$BaF_2$ | 2.556 | $H_2O$ |
| $U^{+3}$–$CaF_2$ | 2.613 | $H_2O$ |
| | | |
| Atomic neon | 3.391317 | $CH_4$ |
| Carbon monoxide | 5.2 to 7 | $H_2O$ |
| Cesium | 7.1821 | $H_2O$ |
| Atomic neon | 18.3040 | $H_2O$ |
| Atomic neon | 20.351 | $H_2O$ |

From Eppers, W. C., Atmospheric transmission, in *Handbook of Lasers,* Pressley, R. J., Ed., The Chemical Rubber Co., Cleveland, 1971, 40.

## Table 8—7
### LASER LINES WITH WEAK TO MODERATE ABSORPTION BY THE ATMOSPHERE

| Laser | λ | Comment |
|-------|---|---------|
| Ionized argon | 4880 Å, 5145 Å | |
| Atomic neon (He–Ne) | 6328 Å | |
| GaAs | 8300 Å, 9200 Å | Close attention must be paid to temperature of operation; increased absorption occurs from approx. 8600 Å to 9250 Å |
| Ruby | 6934 → 6945 Å | Strong $H_2O$ absorptions can occur. |
| $Nd^{+3}$ | ≈1.06 μ | Very low absorption |
| Atomic neon (He–Ne) | 1.1523 μ 5 lines | Moderate $H_2O$ absorption |
| $CH_4$ Raman shift of 1.06 μ | 1.53 μ | |
| $Er^{+3}$ ($CaF_2$) (glass) | 1.55 → 1.65 μ | |
| $Ho^{+3}$–$CaWO_4$ | 2.04 μ | Mostly clear |
| $Ho^{+3}$–YAG | to | |
| $Ho^{+3}$–$CaF_2$ | 2.128 μ | |
| Atomic Xe | 3.50704 μ | |
| DF | 3.8 μ | |
| $CO_2$ | 10.6 μ | $CO_2$ Water absorption |

From Eppers, W. C., Atmospheric transmission, in *Handbook of Lasers,* Pressley, R. J., Ed., The Chemical Rubber Co., Cleveland, 1971, 40.

**Table 8—8**
**SPECTRAL RANGE COVERED BY SEMICONDUCTOR LASERS**

| | $\lambda$ ($\mu$) | $h\nu$ (eV) | Mode of excitation | | | |
|---|---|---|---|---|---|---|
| | | | Injection | Electron beam | Optical | Avalanche |
| ZnS | 0.33 | 3.8 | | + | + | |
| ZnO[a] | 0.37 | 3.4 | | + | | |
| $Zn_{1-x}Cd_xS$ | 0.49–0.32 | 2.5–3.82 | | | + | |
| ZnSe | 0.46 | 2.7 | | + | | |
| CdS[a] | 0.49 | 2.5 | | + | + | |
| ZnTe | 0.53 | 2.3 | | + | | |
| GaSe | 0.59 | 2.1 | | + | | |
| $CdSe_{1-x}S_x$ | 0.49–0.68 | 2.5–1.8 | | + | + | |
| $CdSe_{0.95}S_{0.05}$ | 0.675 | 1.8 | | | + | |
| CdSe | 0.675 | 1.8 | | + | + | |
| $Al_{1-x}Ga_xAs$ [a] | 0.63–.90 | 2.0–1.4 | + | | | |
| $GaAs_{1-x}P_x$ [a] | 0.61–0.90 | 2.0–1.4 | + | + | | |
| CdTe | 0.785 | 1.6 | | + | | |
| GaAs [a] | 0.83–.91[b] | 1.50–1.38 | + | + | + | + |
| InP | 0.91 | 1.36 | + | | | + |
| $GaAs_{1-x}Sb_x$ | 0.95–1.5 | 1.4–0.83 | c | | | |
| $CdSnP_2$ | 1.01 | 1.25 | | + | | |
| $InAs_{1-x}P_x$ | 0.9–3.2 | 1.4–0.39 | c | | | |
| $InAs_{0.94}P_{0.06}$ | 0.942 | 1.32 | + | | | |
| $InAs_{0.51}P_{0.49}$ | 1.6 | 0.78 | + | | | |
| GaSb | 1.55 | 0.80 | + | + | | |
| $In_{1-x}Ga_xAs$ | 0.58–3.1 | 2.14–0.4 | + | | | |
| $In_{0.65}Ga_{0.35}As$ | 1.77 | 0.70 | + | | | |
| $In_{0.75}Ga_{0.25}As$ | 2.07 | 0.60 | + | | | |
| $Cd_3P_2$ | 2.1 | 0.58 | | | + | |
| InAs | 3.1 | 0.39 | + | + | + | |
| $InAs_{1-x}Sb_x$ | 3.1–5.4 | 0.39–0.23 | + | | | |
| $InAs_{0.98}Sb_{0.02}$ | 3.19 | 0.39 | + | | | |
| $Cd_{1-x}Hg_xTe$ | 3–15 | 0.41–0.08 | | + | + | |
| $Cd_{0.32}Hg_{0.68}Te$ | 3.8 | 0.33 | | | + | |
| Te | 3.72 | 0.334 | | + | | |
| PbS | 4.3 | 0.29 | + | + | | |
| InSb | 5.2 | 0.236 | + | + | + | |
| PbTe | 6.5 | 0.19 | + | + | | |
| $PbS_{1-x}Se_x$ | 3.9–8.5 | 0.32–0.146 | + | + | | |
| PbSe | 8.5 | 0.146 | + | + | | |
| PbSnTe | 28 | 0.045 | + | | | |
| PbSnSe | 8–31.2 | 0.155–0.040 | + | | | |

[a] Have lased at room temperature.
[b] Depending on temperature and doping.
[c] Expected, but not observed.

From Pankove, J. I., Injection lasers, in *Handbook of Lasers,* Presley, R. J., Ed., The Chemical Rubber Co., Cleveland, 1971, 368.

## Table 8—9
## CONTINUOUS-WAVE INSULATING CRYSTAL LASERS

| Laser ion | Host | Sensitizer ion(s) | Wave-length, $\mu m$ | Tempera-ture, K | Optical pump | Power, W | Efficiency,[a] percent |
|---|---|---|---|---|---|---|---|
| **IRON GROUP IONS** | | | | | | | |
| $Cr^{3+}$ | $Al_2O_3$ | | 0.694 | 300 | Hg | 2.4 | ~0.1 |
| | | | 0.694[b] | 4.2, 77 | Ar laser | | |
| $Ni^{2+}$ | $MgF_2$ | | 1.67 | 85 | W | 1 | 0.2 |
| $Ni^{2+}$ | $MnF_2$ | | 1.93 | 85 | W | | |
| **DIVALENT RARE EARTH IONS** | | | | | | | |
| $Dy^{2+}$ | $CaF_2$ | | 2.36 | 77 | W | 1.2 | 0.06 |
| | | | 2.36 | 27 | Sunlight | | |
| $Tm^{2+}$ | $CaF_2$ | | 1.12 | 27 | Hg | | |
| **TRIVALENT RARE EARTH IONS** | | | | | | | |
| $Nd^{3+}$ | $Ca(NbO_3)_2$ | | 1.06 | 300 | Xe | 0.12 | 0.05 |
| $Nd^{3+}$ | $Ca_5(PO_4)_3F$ | | 1.06 | 300 | W | 1.3 | 0.2 |
| $Nd^{3+}$ | $CaWO_4$ | | 1.06 | 300 | Xe | <0.1 | ~0.01 |
| | | | 1.06 | 300 | Hg | ~0.01 | ~0.01 |
| | | | 1.06 | 85 | Hg | 0.5 | 0.03 |
| $Nd^{3+}$ | $LaF_3$ | | 1.04 | 300 | | | |
| $Nd^{3+}$ | $YAlO_3$ (b axis) | | 1.06 | 300 | Kr | 35 | 0.8 |
| | | | 1.08 | 300 | Kr | 100 | 1.8 |
| $Nd^{3+}$ | $YAlO_3$ (c axis) | | 1.06 | 300 | Kr | 6.5 | 0.3 |
| $Nd^{3+}$ | $Y_3Al_5O_{12}$ | | 1.06 | 300 | W | ~25 | 1.0 |
| | | | 1.06 | 300 | Kr | 250 | 2.1 |
| | | | 1.06 | 300 | Kr | 1 100[c] | 2.0 |
| | | | 1.06 | 300 | Plasma arc | 200 | 0.2 |
| | | | 1.06 | 300 | Na-doped Hg | 0.5 | 0.2 |
| | | | 1.32 | 300 | W | 0.03 | ~0.01 |
| | | $Cr^{3+}$ | 1.06 | 300 | Hg | 10 | 0.4 |
| $Ho^{3+}$ | $CaF_2$ | $Er^{3+}, Tm^{3+}, Yb^{3+}$ | 2.1 | 77 | Xe | | |
| $Ho^{3+}$ | $Er_2O_3$ | $Er^{3+}$ | 2.12 | 77 | W | | |
| $Ho^{3+}$ | $Er_{1.5}Y_{1.5}Al_5O_{12}$ | $Er^{3+}$ | 2.10 | 85 | Hg, W | | |
| $Ho^{3+}$ | $Y_3Al_5O_{12}$ | $Cr^{3+}$ | 2.10 | 85 | W | | |
| | | $Cr^{3+}$ | 2.12 | 85 | Hg, W | | |
| | | $Er^{3+}, Tm^{3+}$ | 2.12 | 85 | W | 15 | 5 |
| | | | 2.12 | 77 | W-I | 20 | 3.5 |
| $Tm^{3+}$ | $CaF_2$ | $Er^{3+}$ | 1.9 | 77 | Xe | | |
| $Tm^{3+}$ | $Er_2O_3$ | $Er^{3+}$ | 1.93 | 77 | W | | |

[a]Overall efficiency: laser output/electrical energy input to pump lamp. Because the efficiency depends on a number of factors, such as rod quality, pump cavity efficiency, and output coupling efficiency, the original references should be consulted when comparing values.

[b]Nonspiking, single-mode operation.

[c]Multiple laser rods in series inside one resonant cavity.

From Weber, M. J., Insulating Crystal Lasers, in *Handbook of Lasers,* Pressler, R. J. Ed., The Chemical Rubber Co., Cleveland, 1971, pg 393.

## Table 8—10
## PROPERTIES OF SOLID-LASER MATERIALS

The host crystal in which the best laser results were obtained is discussed in detail. Other host crystals are listed in the "Remarks" column.

*Note:* Numbers in parentheses designate temperatures in Kelvin.

| Laser material | Output wavelength, microns | Operating mode (and temperature, K) | $t_{spont}$, milliseconds | Pulse threshold in joules (and temperature, K) | Useful absorption regions, microns | Position of terminal level, $E_1$ ($cm^{-1}$) | Laser transition | Remarks |
|---|---|---|---|---|---|---|---|---|
| $Cr^{3+}:Al_2O_3$ (ruby) $10^{19}$ atoms/cc | 0.6934 ($R_1$, 77°) <br> 0.6929 ($R_2$, 290°) | cw (77), pulsed (350) <br> pulsed | 3 | ~800 (77) | 0.5–0.6 <br> 0.32–0.44 <br> 0.5–0.6 <br> 0.32–0.44 | 0 <br> 0 | $^2E(\bar{E})\rightarrow{}^4A_2$ <br> $^2E(2\bar{A})\rightarrow{}^4A_2$ | "Spiking" observed in both pulsed and cw operation |
| $Cr^{3+}:Al_2O_3$ $n\sim10^{20}$ | 0.701, 0.704 | pulsed (77) | | | | ~100 | | Due to paired chromium ions |
| $U^{3+}:CaF_2$ | 2.613 <br><br> 2.438 <br> 2.511 <br> 2.223 | pulsed (300) <br> cw (~100) <br> pulsed (77) <br> pulsed (77) <br> pulsed (77) | 0.13 (77) | 1 (77) <br><br> 6 (77) <br> 2000 (77) | ~0.9 | 609 | $^4I_{11/2}\rightarrow{}^4I_{9/2}$ | "Spiking" present in pulsed but not in cw operation. Pulsed emission also observed in $BaF_2$ at 2.556 $\mu$ and in $SrF_2$ at 2.472 $\mu$, 2.408 $\mu$ |
| $Nd^{3+}:CaWO_4$ | 1.065 <br> 1.063 <br> 1.066 <br> 1.058 <br> 1.064 | cw (300) <br> pulsed <br> pulsed <br> pulsed <br> pulsed | ~0.1 (77) | ~1 (77) <br> 14 (77) <br> 6 (77) <br> 80 (77) <br> 7 (77) | 0.57–0.6 | ~2000 | $^4F_{3/2}\rightarrow{}^4I_{11/2}$ | "Spiking" present in pulsed but not in cw operation. Laser emission from $Nd^{3+}$ also observed in $SrMO_4$, $SrWO_4$, $CaMO_4$, $PbMO_4$, $CaF_2$, $SrF_2$, $BaF_2$, and $LaF_3$ |
| $Nd^{3+}:$Glass | 1.06 | pulsed (300) | | ~50 | | ~2000 | $^4F_{3/2}\rightarrow{}^4I_{11/2}$ | |
| $Pr^{3+}:CaWO_4$ | 1.047 | pulsed (77) | 0.05 (77) | 15 (77) | 0.45–0.5 | 377 | $^1G_4\rightarrow{}^3H_4$ | Pulsed laser emission of $Pr^{3+}$ was also detected in $SrMO_4$ |

**Table 8—10 (continued)**
**PROPERTIES OF SOLID-LASER MATERIALS**

| Laser material | Output wavelength, microns | Operating mode (and temperature, K) | $t_{spont}$, milliseconds | Pulse threshold in joules (and temperature, K) | Useful absorption regions, microns | Position of terminal level, $E_1$ ($cm^{-1}$) | Laser transition | Remarks |
|---|---|---|---|---|---|---|---|---|
| $Dy^{2+}:CaF_2$ | 2.36 | cw (90) | ~10 (77) | 20 (77) | 0.8–1.0 | 35 | $^5I_7 \rightarrow {}^5I_8$ | "Spiking" in pulsed operation but not in continuous |
| $Tm^{3+}:CaWO_4$ | 1.911 1.916 | pulsed (77) pulsed (77) | | 60 (77) 73 (77) | {0.46–0.48 1.7–1.8 | ~325 | $^3H_4 \rightarrow {}^3H_6$ | 1.918 $\mu$ emission also observed. Laser emission also observed in $SrF_2$ |
| $Er^{3+}:CaWO_4$ | 1.612 | pulsed (77) | | 800 (77) | 0.38 0.52 | 375 | $^4I_{13/2} \rightarrow {}^4I_{15/2}$ | |
| $Ho^{3+}:CaWO_4$ | 2.046 2.059 | pulsed (77) pulsed (77) | | 80 (77) 250 (77) | 0.44–0.46 | ~230 | $^5I_7 \rightarrow {}^5I_8$ | |
| $Tm^{2+}:CaF_2$ | 1.116 | pulsed (~4) | 4 | 50 (4) | 0.28–0.34 0.39–0.46 0.53–0.63 | 0 | $^2F_{5/2} \rightarrow {}^2F_{7/2}$ | |
| $Sm^{2+}:CaF_2$ | 0.708 | pulsed (20) | 0.002 | 0.01 (20) | 0.425–0.5 0.59–0.65 | 263 | $^5D_0 \rightarrow {}^7F_1$ | No "spiking" in pulsed operation. Laser action at 0.6969 $\mu$ also observed in $SrF_2:Sm^{2+}$ |
| $Yb^{3+}:Glass$ | 1.015 | pulsed (77) | 1.5 | 1300 | ~0.91 ~0.95 ~0.98 | | $^2F_{5/2} \rightarrow {}^2F_{7/2}$ | |
| $Gd^{3+}:Glass$ | 0.3125 | pulsed (77) | 4 (300) | | 0.274 0.277 | | $^6P_{7/2} \rightarrow {}^8S_{7/2}$ | |
| $Ho^{3+}:Glass$ | $\lambda > 1.95\ \mu$ | pulsed (77) | ~0.7 (77) | 3600 (77) | 0.44–0.46 | | $^5I_7 \rightarrow {}^5I_8$ | |

From Yariv, A. and Gordon, J. P., *Proc. Inst. Elec. Electron. Eng.*, 51(1), 13, 1963. Copyright © 1963 IEEE. With permission.

## Table 8—11
### MAGNETIC PROPERTIES AND COMPOSITION OF PERMANENT MAGNETIC ALLOYS[a]

| Name | Composition,[b] weight percent | | | | | Remanence, $B_r$, gauss | Coercive force, $H_c$, oersteds | Maximum energy product, $(BH)_{max}$, gauss-oersteds $\times 10^{-6}$ |
|------|----|----|----|----|-------|------|------|------|
| | Al | Ni | Co | Cu | Other | | | |
| *U.S.A.* | | | | | | | | |
| Alnico I | 12 | 20–22 | 5 | | | 7,100 | 440 | 1.4 |
| Alnico II | 10 | 17 | 12.5 | 6 | | 7,200 | 540 | 1.6 |
| Alnico III | 12 | 24–26 | | 3 | | 6,900 | 470 | 1.35 |
| Alnico IV | 12 | 27–28 | 5 | | | 5,500 | 700 | 1.3 |
| Alnico V[c] | 8 | 14 | 24 | 3 | | 12,500 | 600 | 5.0 |
| Alnico V DG[c] | 8 | 14 | 24 | 3 | | 13,100 | 640 | 6.0 |
| Alnico VI[c] | 8 | 15 | 24 | 3 | 1.25 Ti | 10,500 | 750 | 3.75 |
| Alnico VII[c] | 8.5 | 18 | 24 | 3 | 5 Ti | 7,200 | 1,050 | 2.75 |
| Alnico XII | 6 | 18 | 35 | | 8 Ti | 5,800 | 950 | 1.6 |
| Carbon steel | | | | | 1 Mn 0.9 C | 10,000 | 50 | 0.2 |
| Chromium steel | | | | | 3.5 Cr 0.9 C 0.3 Mn | 9,700 | 65 | 0.3 |
| Cobalt steel | | | 17 | | 2.5 Cr 8 W 0.75 C | 9,500 | 150 | 0.65 |
| Cunico | | 21 | 29 | 50 | | 3,400 | 660 | 0.80 |
| Cunife | | 20 | | 60 | | 5,400 | 550 | 1.5 |
| Ferroxdur 1 | | | $BaFe_{12}O_{19}$ | | | 2,200 | 1,800 | 1.0 |
| Ferroxdur 2 | | | $BaF_{12}O_{19}$ (oriented) | | | 3,840 | 2,000 | 3.5 |
| Platinum-Cobalt | | | 23 | | 77 Pt | 6,000 | 4,300 | 7.5 |
| Remalloy | | | 12 | | 17 Mo | 10,500 | 250 | 1.1 |
| Silmanol | 4.4 | | | | 86.6 Ag 8.8 Mn | 550 | 6,000 | 0.075 |
| Tungsten steel | | | | | 5 W 0.3 Mn 0.7 C | 10,300 | 70 | 0.32 |
| Vicalloy I | | | 52 | | 10 V | 8,800 | 300 | 1.0 |
| Vicalloy II (wire) | | | 52 | | 14 V | 10,000 | 510 | 3.5 |
| *Germany* | | | | | | | | |
| Alni 90 | 12 | 21 | | | | 8,000 | 350 | 1.2 |
| Alni 120 | 13 | 27 | | | | 6,000 | 570 | 1.2 |
| Alnico 130 | 12 | 23 | 5 | | | 6,300 | 620 | 1.4 |
| Alnico 160 | 11 | 24 | 12 | 4 | | 6,200 | 700 | 1.6 |
| Alnico 190 | 12 | 21 | 15 | 4 | | 7,000 | 700 | 1.8 |
| Alnico 250 | 8 | 19 | 23 | 4 | 6 Ti | 6,500 | 1,000 | 2.2 |
| Alnico 400[c] | 9 | 15 | 23 | 4 | | 12,000 | 650 | 4.8 |
| Alnico 580[c] (semicolumnar) | 9 | 15 | 23 | 4 | | 13,000 | 700 | 6.0 |
| Oerstit 800 | 9 | 18 | 19 | 4 | 4 Ti | 6,600 | 750 | 1.95 |
| *Great Britain* | | | | | | | | |
| Alcomax I | 7.5 | 11 | 25 | 3 | 1.5 Ti | 12,000 | 475 | 3.5 |
| Alcomax II | 8 | 11.5 | 24 | 4.5 | | 12,400 | 575 | 4.7 |
| Alcomax IISC (semicolumnar) | 8 | 11 | 22 | 4.5 | | 12,800 | 600 | 5.15 |
| Alcomax III | 8 | 13.5 | 24 | 3 | 0.8 Nb | 12,500 | 670 | 5.10 |
| Alcomax IIISC (semicolumnar) | 8 | 13.5 | 24 | 3 | 0.8 Nb | 13,000 | 700 | 5.80 |
| Alcomax IV | 8 | 13.5 | 24 | 3 | 2.5 Nb | 11,200 | 750 | 4.30 |
| Alcomax IVSC (semicolumnar) | 8 | 13.5 | 24 | 3 | 2.5 Nb | 11,700 | 780 | 5.10 |
| Alni, high $B_r$ | 13 | 24 | | 3.5 | | 6,200 | 480 | 1.25 |
| Alni, normal | | | | | | 5,600 | 580 | 1.25 |
| Alni, high $H_c$ | 12 | 32 | | | 0–0.5 Ti | 5,000 | 680 | 1.25 |
| Alnico, high $B_r$ | 10 | 17 | 12 | 6 | | 8,000 | 500 | 1.70 |

## Table 8—11 (continued)
## MAGNETIC PROPERTIES AND COMPOSITION OF PERMANENT MAGNETIC ALLOYS[a]

| Name | Composition,[b] weight percent | | | | | Remanence, $B_r$, gauss | Coercive force, $H_c$, oersteds | Maximum energy product, $(BH)_{max}$, gauss-oersteds $\times 10^{-6}$ |
|---|---|---|---|---|---|---|---|---|
| | Al | Ni | Co | Cu | Other | | | |
| Alnico, normal | | | | | | 7,250 | 560 | 1.70 |
| Alnico, high $H_c$ | 10 | 20 | 13.5 | 6 | 0.25 Ti | 6,600 | 620 | 1.70 |
| Columax (columnar) | similar to Alcomax III or IV | | | | | 13,000–14,000 | 700–800 | 7.0–8.5 |
| Hycomax | 9 | 21 | 20 | 1.6 | | 9,500 | 830 | 3.3 |

[a]See Table 2–31 for conversion factors.
[b]Unlisted remainder of composition is either iron or iron plus trace impurities.
[c]Alloys so designated are cast anisotropic; all others are cast isotropic.

From Weast, R. C., Ed., *Handbook of Chemistry and Physics*, 69th ed., CRC Press, Boca Raton, Fla., 1988, E-122.

## Table 8—12
## CAST PERMANENT MAGNETIC ALLOYS[a]

| Name | Composition,[b] weight percent | Specific gravity, g/cc | Thermal expansion $\frac{Cm \times 10^{-6}}{cm \times °C}$ | Thermal expansion Between °C | Tensile strength $\frac{Kg[c]}{mm^2}$ | Tensile strength Form | Remarks[d] | Use |
|---|---|---|---|---|---|---|---|---|
| *U.S.A.* | | | | | | | | |
| Alnico I | Al 12; Ni 20–22; Co 5 | 6.9 | 12.6 | 20–300 | 2.9 | Cast | i | Permanent magnets |
| Alnico II | Al 10; Ni 17; Cu 6; Co 12.5 | 7.1 | 12.4 | 20–300 | 2.1 45.7 | Cast Sintered | i | Temperature controls, magnetic toys, and novelties |
| Alnico III | Al 12; Ni 24–26; Cu 3 | 6.9 | 12 | 20–300 | 8.5 | Cast | i | Tractor magnetos |
| Alnico IV | Al 12; Ni 27–28; Co 5 | 7.0 | 13.1 | 20–300 | 6.3 42.1 | Cast Sintered | i | Application requiring high coercive force |
| Alnico V | Al 8; Ni 14; Co 24; Cu 3 | 7.3 | 11.3 | | 3.8 35 | Cast Sintered | a | Application requiring high energy |
| Alnico V DG | Al 8; Ni 14; Co 24; Cu 3 | 7.3 | 11.3 | | | | a, c | |
| Alnico VI | Al 8; Ni 15; Co 24; Cu 3; Ti 1.25 | 7.3 | 11.4 | | 16.1 | Cast | a | Application requiring high energy |
| Alnico VII | Al 8.5; Ni 18; Cu 3; Co 24; Ti 5 | 7.17 | 11.4 | | | | a | |
| Alnico XII | Al 6; Ni 18; Co 35; Ti 8 | 7.2 | 11 | 20–300 | | | | Permanent magnets |
| Comol | Co 12; Mo 17 | 8.16 | 9.3 | 20–300 | 88.6 | | | Permanent magnets |
| Cunife | Cu 60; Ni 20 | 8.52 | | | 70.3 | | | Permanent magnets |
| Cunico | Cu 50; Ni 21 | 8.31 | | | 70.3 | | | Permanent magnets |
| Barium ferrite Feroxdur | Ba Fe$_{12}$O$_{19}$ | 4.7 | 10 | | 70.3 | | | Ceramics |
| *Great Britain* | | | | | | | | |
| Alcomax I | Al 7.5; Ni 11; Co 25; Cu 3; Ti 1.5 | | | | | | a | Permanent magnets |
| Alcomax II | Al 8; Ni 11.5; Co 24; Cu 4.5 | | | | | | a | Permanent magnets |

## Table 8—12 (continued)
### CAST PERMANENT MAGNETIC ALLOYS[a]

| Name | Composition,[b] weight percent | Specific gravity, g/cc | Thermal expansion | | Tensile strength | | Remarks[d] | Use |
|---|---|---|---|---|---|---|---|---|
| | | | $Cm \times 10^{-6}$ / $cm \times °C$ | Between °C | $\dfrac{Kg}{mm^2}$[c] | Form | | |
| Alcomax II SC | Al 8; Ni 11; Co 22; Cu 4.5 | 7.3 | | | | | a, sc | |
| Alcomax III | Al 8; Ni 13.5; Co 24; Nb 0.8 | 7.3 | | | | | a | Magnets for motors, loudspeakers |
| Alcomax IV | Al 8; Ni 13.5; Cu 3; Co 24; Nb 2.5 | | | | | | | Magnets for cycle-dynamos |
| Columax | Similar to Alcomax III or IV | | | | | | a, sc | Permanent magnets, heat-treatable |
| Hycomax | Al 9; Ni 21; Co 20; Cu 1.6 | | | | | | a | Permanent magnets |
| Alnico (high $H_c$) | Al 10; Ni 20; Co 13.5; Cu 6; Ti 0.25 | 7.3 | | | | | i | |
| Alnico (high $B_r$) | Al 10; Ni 17; Co 12; Cu 6 | 7.3 | | | | | i | |
| Alni (high $H_c$) | Al 12; Ni 32; Ti 0–0.5 | 6.9 | | | | | i | |
| Alni (high $B_r$) | Al 13; Ni 24; Cu 3.5 | | | | | | i | |
| *Germany* | | | | | | | | |
| Alnico 580 | Al 9; Ni 15; Co 23; Cu 4 | | | | | | i | |
| Alnico 400 | Al 9; Ni 15; Co 23; Cu 4 | | | | | | a | |
| Oerstit 800 | Al 9; Ni 18; Co 19; Cu 4; Ti 4 | | | | | | i | Permanent magnets |
| Alnico 250 | Al 8; Ni 19; Co 23; Cu 4; Ti 6 | | | | | | i | |
| Alnico 190 | Al 12; Ni 21; Cu 4; Co 15 | | | | | | i | |
| Alnico 130 | Al 12; Ni 23; Co 5 | | | | | | i | |
| Alni 120 | Al 13; Ni 27 | | | | | | i | |
| Alni 90 | Al 12; Ni 21 | | | | | | i | |
| *Austria* | | | | | | | | |
| Alnico 160 | Al 11; Ni 24; Co 12; Cu 4 | | | | | | i | Permanent magnets, sintered |

[a] For properties of permanent magnetic alloys see Table 2–34.
[b] The additional alloying metal for each of the magnets listed is iron.
[c] kg/mm² × 1422.3 = lb/in.²; kg/mm² × 9.807 = MN/m².
[d] i = isotropic; a = anisotropic; c = columnar; sc = semicolumnar.

From Weast, R. C., Ed., *Handbook of Chemistry and Physics,* 55th ed., CRC Press, Cleveland, 1974, E-117.

**Table 8—13**

**FERROMAGNETIC MATERIALS – CURIE TEMPERATURES**[a]

Upper Transition Temperature

| Material | Curie temp, K | Material | Curie temp, K | Material | Curie temp, K |
|---|---|---|---|---|---|
| AuFe | 300 | EuO | 69.5 | $GdNi_2$ | 85 |
| $Au_4Mn$ | 263 | EuS | 16 | $GdOs_2$ | 66 |
| $Au_4V$ | 43 | EuSe | 4.58 | $GdPd_2$ | 335 |
| $BaFeO_3$ | 180 | FeAl | 923 | GdZn | ~280 |
| $BiMnO_3$ | 103 | $Fe_3Al$ | 773 | $HgCr_2S_4$ | 36 |
| $CdFe_2$ | 782 | FeB | 598 | $HgCr_2Se_4$ | 120 |
| $CeAl_2$ | 8 | $Fe_3C$ | 483 | Ho | 20 |
| $CeCo_3$ | 78 | $Fe_3Cr$ | 1273 | $HoAl_2$ | 27 |
| $CeCo_5$ | 687 | Fe-Ni | | $HoCo_5$ | 1025 |
| | 737 | 4.5% Fe | 683.0 | $HoFe_2$ | 608 |
| | 464 | 19% Fe | 834.0 | $HoIr_2$ | 12 |
| $CeFe_2$ | 235 | 23% Fe | 876.0 | HoNi | 31 |
| | 878 | 50% Fe | 786.1 | $HoNi_2$ | 23 |
| $CeFe_5$ | 228 | $Fe_2O_3$ | 848 | $HoNi_3$ | 66 |
| Co | 1390 | $\varepsilon$-$Fe_2O_3$ | 483 | $HoOs_2$ | 9 |
| $Co_2B$ | 429 | $\gamma$-$Fe_2O_3$ | 743 | $LaCrO_3$ | 300 |
| | 433 | $Fe_3O_4$ | 848–858 | Mn (45% Al) | ~653 |
| $Co_3B$ | 747 | FeP | 215 | Mn-Al-Co | Co rich: 370 |
| $CoFeCoO_4$ | 450 | $Fe_2P$ (hexagonal) | 278 | | Mn rich: 466 |
| $CoFe_2O_4$ | 673–769 | FePt | 743 | $MnAu_4$ | 360 |
| CoPt | 813 | $Fe_3Pt$ | 453 | MnB | 578 |
| $CoS_2$ | 110 | FeRh | 668 | MnBi | 633 |
| $Co_2VO_4$ | 160 | Fe-Si | 1043.9–1012.6 | $Mn_3Ge$ | 28 |
| Cr | 311 | 0.9–7.4 at. % Si | | $Mn_3Ge_2$ | 300 |
| $CrO_2$ | 391 | $Fe_3Si$ | 808 | $Mn_3In$ | 583 |
| CuMnAl | 433 | $Fe_3Sn$ | 743 | $MnNi_3$ | 750 (ordered) |
| Dy | 85 | $Fe_2TiO_4$ | 142 | | 132 (dis- |
| $DyAl_2$ | 53 | Fe-53.3% V | 280 | | .ordered) |
| | 62 | $Fe_2Zr$ | 628 | $Mn_3O_4$ | 45 |
| $DyCo_5$ | 1125 | $Ga_{2-x}Fe_xO_3$ | | MnO-35.5%, | 467.5 |
| | 966 | x = 1.08 | 350 | ZnO-15%, | |
| $DyCo_3$ | 450 | x = 1.20 | 305 | $Fe_2O_3$-49.5% | |
| $DyFe_2$ | 638 | x = 0.80 | 205 | MnO-26%, | 368.5 |
| | 663 | Gd | 294 | ZnO-24%, | |
| $DyIr_2$ | 23 | $GdAl_2$ | 176 | $Fe_2O_3$-50% | |
| DyN | 22 | $GdCl_3$ | 2.20 | MnTe | 260 |
| DyNi | 48 | $GdCo_2$ | 404 | $Mn_2TiO_4$ | ~77 |
| $DyOs_2$ | 15 | $GdCo_3$ | 612 | $Mn_2VO_4$ | 62 |
| Er | 20 | Gd-Dy | | $NdAl_2$ | 65 |
| $ErAl_2$ | 21 | % Dy: 10 | 285 | $NdCo_2$ | 116 |
| $ErCo_2$ | 39, 36 | % Dy: 50 | 226 | NdGe | 28 |
| $ErCo_3$ | 401 | % Dy: 61 | 193 | $NdGe_2$ | 3.6 |
| $ErFe_2$ | 473 | % Dy: 87.5 | 120 | NdNi | 35 |
| $ErIr_2$ | 3 | $GdFe_2$ | 782 | Ni | 627 |
| ErN | 6 | | 813 | | 628.3 |
| | 5 | $Gd_3Fe_5O_{12}$ | 564 | | 633 |
| ErNi | 10 | | 574.6 | Ni-Cr | 324 |
| $ErNi_2$ | 14 | $GdIr_2$ | 88 | 5.6 at. % Cr | |
| | 21 | | 90 | $NiFe_2O_4$ | 858 |
| $ErNi_3$ | 62 | GdN | 69–72 | $NiMnO_3$ | 437 |
| $EuH_2$ | 24 | GdNi | 73 | $Pd_3Fe$ | 540 |

## Table 8—13 (continued)
### FERROMAGNETIC MATERIALS – CURIE TEMPERATURES[a]

Upper Transition Temperature (continued)

| Material | Curie temp, K | Material | Curie temp, K | Material | Curie temp, K |
|---|---|---|---|---|---|
| $PrAl_2$ | 34 | SmIG | 562 | $UFe_2$ | 172 |
| $PrCo_3$ | 349 | SmNi | 45 | USe | 160.5 |
| PrGe | 39 | Tb | 210 | $USe_2$ | 13.1 |
| $PrIr_2$ | 16 | $TbAl_2$ | 121 | UTe | 103 |
| PrNi | 20 | $TbCo_3$ | 506 | $YCo_3$ | 301 |
| $PrNi_2$ | 8 | TbGa | 155 | YIG : Nd | 548–568 |
| $SmAl_2$ | 122 | TbN | 40 | O–40% $Nd_2O_3$ | |
| $SmCo_5$ | 1020 | TbNi | 50 | $3Y_2O_3 \cdot Al_2O_3 \cdot 4Fe_2O_3$ | 415 |
| | 747 | $TbNi_2$ | 46 | $YbCo_5$ | 973 |
| | 1015 | | | $ZrFe_2$ | 633 |

[a]Curie temperature is defined as the point (temperature increasing) at which a material ceases to be ferromagnetic and becomes paramagnetic, i.e., the saturation decreases to zero. Pure iron is magnetic up about 1,455°F (790°C); above this temperature it is non-magnetic. Materials gain ferromagnetism on cooling, but not always at the same temperature.

*Note:* The values listed in this table represent a small and somewhat random sample of those reported in the quoted source (see below). Users should consult the original source, which includes a large bibliography and keyed references.

### REFERENCES

**Weast, R. C., Ed.,** Magnetic susceptibility of the elements and inorganic compounds in *Handbook of Chemistry and Physics,* 69th ed., CRC Press, Boca Raton, Fla., E-127.

**Bozorth, R. M., McGuire, T. R., and Hudson, R. P.,** Magnetic properties of materials, in *American Institute of Physics Handbook,* 2nd ed., Sect. 5g, Gray, D. E., Ed., McGraw-Hill, New York, 1963, 5-164.

From ORNL-RMIC-7 (Rev.), Research Materials Information Center, Solid State Division, Oak Ridge National Laboratory, Oak Ridge, Tenn., 1969.

**Table 8—14**
## ANTIFERROMAGNETIC MATERIALS – NEEL TEMPERATURE[a]

Upper Transition Temperature

| Material | Neel temp,[b] K | Material | Neel temp,[b] K | Material | Neel temp,[b] K | Material | Neel temp,[b] K |
|---|---|---|---|---|---|---|---|
| AuMn | 493 | CrGe | 62 | $FeWO_4$ | 66 | $\beta$-MnS | 110 |
| $Au_3Mn$ | 140 | $CrVO_4$ | 50 | GdAs | 25 | (hexagonal) | |
| $Ba_2CoWO_6$ | 17 | $Cr_2WO_6$ | 69 | GdBi | 28 | $MnSO_4$ | 11.5 |
| $BaFe_{12}O_{19}$ | 709.5 | $CuCl_2$ | 70 | GdP | 15 | MnTe | 306.7 |
| | (for $H\perp c$) | $CuF_2$ | 68.7 | $GdPO_4$ | 225 | $MnTe_2$ | 80 |
| $CdCr_2O_4$ | 9 | CuHo | 27 | GdSb | 28 | $MnUO_4$ | 12 |
| Ce | 13 | | (elec. res.) | Ho | 133 | $MnWO_4$ | 16 |
| CeS | 7 | CuO | 230 | HoGe | 18 | Nd | 12 |
| CeSb | 18 | $CuSO_4$ | 34.5 | $HoGe_2$ | 11 | $NdSn_3$ | 4.7 |
| CeSe | 12 | $CuWO_4$ | 90 | HoSb | 9 | NdTe | 13 |
| CeTe | 10 | DyCu | 64 | $KCoF_3$ | 135 | $NiCl_2$ | 50 |
| $CoBr_2$ | 19 | $DyCu_2$ | 24 | $KCrF_3$ | 40 | $NiF_2$ | ~73.2 |
| $CoCO_3$ | 18 | $DyH_2$ | 8 | $KCuF_3$ | 243 | $NiWO_4$ | 67 |
| $CoCl_2$ | 25 | Er | 84 | $KFeF_3$ | 113 | $RbMnF_3$ | 54.5 |
| $CoF_2$ | 37–45 | Eu | 103 | $KMnF_3$ | 88 | | 66 |
| $CoF_3$ | 460 | $Fe_3Al$ | 750 | $LaVO_3$ | 137 | | 82 |
| $CoMoO_3$ | 391 | $FeCl_2$ | 23.5 | $\alpha$-Mn | ~100 | | 83 |
| $CoMoO_4$ | 5 | $FeCl_3$ | 15 | $MnCl_2$ | 1.96 | | ~100 |
| CoO | 292 | $FeF_2$ | 78.11 | $MnF_2$ | 66.2 | Sm | 15 |
| $CoSO_4$ | 12 | FeO | 186 | $MnF_3$ | 47 | Tb | 230 |
| $CoUO_4$ | 12 | $Fe_2O_3$ | 259 | MnO | 118 | Tm | 51 |
| $CoWO_4$ | 55 | $\alpha$-$Fe_2O_3$ | 963 | $MnO_2$ | 92 | $UO_2$ | ~30 |
| Cr | 312 | FeS | 593 | $Mn_2O_3$ | 80 | $ZnFe_2O_4$ | 15 |
| $CrF_2$ | 53 | $FeSO_4$ | 21 | MnS | 165 | | 9 |
| $CrF_3$ | 80 | FeSn | 373 | $\beta$-MnS | 160 | | |
| CrFe | 308 | $FeTiO_3$ | 68 | (cubic) | | | |

[a]Neel temperature is defined as the point above which an antiferromagnetic material becomes paramagnetic; above the Neel point the susceptibility decreases with increasing temperature.
[b]No attempt has been made to choose a best value where more than one temperature is reported.

*Note:* The values listed in this table represent a small and somewhat random sample of those reported in the quoted source (see below). Users should consult the original source, which includes a large bibliography and keyed references.

From ORNL-RMIC-7 (Rev.), Research Materials Information Center, Solid State Division, Oak Ridge National Laboratory, Oak Ridge, Tenn., 1969.

# Table 8—15
## PROPERTIES OF PIEZOELECTRIC CERAMICS

The main properties of representative piezoelectric ceramic transducer compositions are listed. Generally the notation follows that of the IRE Standards.[1] Various electromechanical coupling factors (k), free and clamped relative permittivities ($\epsilon/\epsilon_0$), and piezoelectric d and g constants are given. The four coupling factors evaluate the ability of the material to convert energy from electrical to mechanical form (or vice versa) in the planar, transverse, parallel, and shear modes, respectively. The free (superscript T) and clamped (superscript S) permittivities govern impedance parallel and transverse to the field direction at frequencies away from the electromechanical resonances. The piezoelectric d and g constants measure charge density and field generated by an applied stress, respectively, in the parallel (33), transverse (31), and shear (15) directions of a transducer. A high d constant is valuable for generating motion and a high g constant for generating electrical signals. The table also includes tan δ, mechanical Q, and the frequency constants for a bar poled in a thin dimension ($N_1$) and a thin plate ($N_3$). The properties are intimately tied to crystal structure and chemical composition of the ceramic.

Barium titanate was the original piezoelectric ceramic. Both unmodified and modified compositions are represented. It has been largely supplanted by the lead titanate-zirconates which have higher coupling factors and can operate over a wide temperature range. The first two compositions predominate in present usage.

The last two compositions in the table have specialized uses. Lead metaniobate has strongly anisotropic piezoelectric response and very low mechanical Q, both assets in ultrasonic flaw detection. Sodium-potassium niobate, unlike the other compositions, is hot pressed rather than kiln fired. Its low relative permittivity and high value of thickness frequency constant are good for delay line transducers.

A few other modified lead titanate-zirconate compositions find substantial use. Their compositions are proprietary and they are not included here. Data on their properties may be obtained from manufacturers and elsewhere.[2] Recent sources of information[2-4] are available for a more thorough review of materials and applications.

Table 8—15 (continued)
## PROPERTIES OF PIEZOELECTRIC CERAMICS

| Quantity | Material | | | |
|---|---|---|---|---|
| | $Pb_{.94}Sr_{.06}(Ti_{.48}Sr_{.52})O_3$ | $Pb_{.988}(Ti_{.48}Zr_{.52})_{.976}Nb_{.024}O_3$ | $Pb(Ti_{.48}Zr_{.52})O_3$ | $Pb(Ti_{.46}Zr_{.54})O_3$ |
| $k_p$ | 0.58 | 0.60 | 0.53 | 0.47 |
| $k_{31}$ | 0.334 | 0.344 | 0.313 | 0.280 |
| $k_{33}$ | 0.70 | 0.705 | 0.67 | 0.626 |
| $k_{15}$ | 0.71 | 0.685 | 0.694 | 0.701 |
| $\epsilon_{33}^T/\epsilon_0$ | 1300 | 1700 | 730 | 450 |
| $\epsilon_{33}^S/\epsilon_0$ | 635 | 830 | 399 | 260 |
| $\epsilon_{11}^T/\epsilon_0$ | 1475 | 1730 | 1180 | 990 |
| $\epsilon_{11}^S/\epsilon_0$ | 730 | 916 | 612 | 504 |
| $d_{33}$, $10^{-12}$ C/N | 289 | 374 | 223 | 152 |
| $d_{31}$ | -123 | -171 | -94 | -60 |
| $d_{15}$ | 496 | 584 | 494 | 440 |
| $g_{33}$, $10^{-3}$ Vm/N | 26.1 | 24.8 | 34.5 | 38.1 |
| $g_{31}$ | -11.1 | -11.4 | -14.5 | -15.1 |
| $g_{15}$ | 39.4 | 38.2 | 47.2 | 50.3 |
| Tan δ | 0.004 | 0.02 | 0.004 | 0.003 |
| $Q_m$ | 500 | 75 | 500 | 680 |
| Density, $10^3$ kg/m³ | 7.6 | 7.75 | 7.6 | 7.6 |
| $N_1$, Hz·m | 1650 | 1400 | 1680 | 1680 |
| $N_3$ (thin plate) | 2000 | 1770 | – | 2090 |
| Curie point | 328°C | 365°C | 386°C | 370°C |
| Structure | Perovskite | Perovskite | Perovskite | Perovskite |
| Symmetry | Tetr. | Tetr. | Tetr. | Rhomb. |
| Commercial use | Ultrasonics, sonar | Microphones, hydrophones | | |

## Table 8—15 (continued)
### PROPERTIES OF PIEZOELECTRIC CERAMICS

| Quantity | Material | | | | |
|---|---|---|---|---|---|
| | $BaTiO_3$ | 95 wt % $BaTiO_3$ 5 wt % $CaTiO_3$ | 80 wt % $BaTiO_3$ 12 wt % $PbTiO_3$ 8 wt % $CaTiO_3$ | $PbNb_2O_6$ | $Na_{.5}K_{.5}NbO_3$ |
| $k_p$ | 0.36 | 0.33 | 0.19 | 0.07 | 0.45 |
| $k_{31}$ | 0.212 | 0.194 | 0.113 | 0.045 | 0.27 |
| $k_{33}$ | 0.50 | 0.48 | 0.34 | 0.38 | 0.53 |
| $k_{15}$ | 0.48 | 0.491 | 0.30 | — | — |
| $\epsilon_{33}^T/\epsilon_0$ | 1700 | 1200 | 450 | 225 | 420 |
| $\epsilon_{33}^S/\epsilon_0$ | 1260 | 910 | 395 | — | — |
| $\epsilon_{11}^T/\epsilon_0$ | 1450 | 1300 | — | — | — |
| $\epsilon_{11}^S/\epsilon_0$ | 1115 | 1000 | — | — | — |
| $d_{33}$, $10^{-12}$ C/N | 190 | 149 | 60 | 85 | 160 |
| $d_{31}$ | −78 | −58 | −20 | −9 | −49 |
| $d_{15}$ | 260 | 242 | — | — | — |
| $g_{33}$, $10^{-3}$ Vm/N | 12.6 | 14.1 | 15.1 | 42.5 | 43 |
| $g_{31}$ | −5.2 | −5.5 | −5.0 | −4.5 | −13.1 |
| $g_{15}$ | 20.2 | 21.0 | — | — | — |
| Tan δ | 0.01 | 0.006 | 0.006 | 0.01 | 0.014 |
| $Q_m$ | 300 | 400 | 1200 | 11 | 240 |
| Density, $10^3$ kg/m$^3$ | 5.7 | 5.55 | 5.4 | 6.0 | 4.46 |
| $N_1$, $H_z$·m | 2200 | 2290 | 2430 | | 2540 |
| $N_3$ (thin plate) | 2520 | 2740 | — | | 3470 |
| Curie point | 115°C | 115°C | 140°C | 570°C | 420°C |
| Structure | Perovskite | Perovskite | Perovskite | K-tungsten bronze | Perovskite |
| Symmetry | Tetr. | Tetr. | Tetr. | Orth. | Orth. |
| Commercial use | | | | Flaw detectors | Delay lines |

### REFERENCES

1. IRE standards on piezoelectric crystals: measurements of piezoelectric ceramics, 1961, *Proc. IRE,* 49, 1161, 1961.
2. **Berlincourt, D.,** Piezoelectric crystals and ceramics, in *Ultrasonic Transducer Materials,* Mattiat, O. E., Ed., Plenum Press, New York, 1971, chap. 2, p. 63.
3. **Berlincourt, D. A., Curran, D. R., and Jaffe, H.,** Piezoelectric and piezomagnetic materials, in *Physical Acoustics,* Vol. 1, Part A, Mason, W. P., Ed., Academic Press, New York, 1964, chap. 3, p. 169.
4. **Jaffe, B., Cook, W. R., Jr., and Jaffe, H.,** *Piezoelectric Ceramics,* Academic Press, New York, 1971.

Table compiled by B. Jaffe.

# Table 8—16
## COMPARISON OF BATTERY TYPES

For conversion of temperature to K, see Table 4–13 of *CRC Handbook of Materials Science*, Volume I. For output in kJ/kg, multiply the values in watt-hr/lb by 7.9367.

| Name | Type | Anode | Cathode | Electrolyte | Nominal cell voltage | Temp range, °F | Typical output, watt-hr/lb | Cycle life if recharged (50% discharge) |
|---|---|---|---|---|---|---|---|---|
| Leclanché or carbon-zinc | Primary | Zinc | $C + MnO_2$ | $NH_4Cl-ZnCl_2$ | 1.5 | 40–130 | 2–30 | – |
| Mercury | Primary | Zinc | HgO | $KOH-K_2Zn_2O_3$ | 1.35 and 1.4 | 40–130 | 50 | – |
| Silver oxide | Primary | Zinc | $Ag_2O$ or AgO | KOH or NaOH | 1.5 | 0–130 | 30–80 | 100–300 |
| Alkaline or manganese zinc | Primary and rechargeable | Zinc or zinc–Hg | $MnO_2$ | KOH or NaOH | 1.5 | 0–130 | | 50–100 |
| Lalande | Primary | Zinc | CuO | NaOH | 0.65 | | 20 | – |
| Nickel-cadmium | Rechargeable (secondary) | Cadmium | $Ni(OH)_2$ | KOH | 1.25 | 0–115 | 6–15 | 100–2,000 |
| Silver-cadmium | Rechargeable | Cadmium | $Ag_2O_2$ or AgO | KOH | 1.1 | -40–100 | 10–40 | 300–1,000 |
| Lead-acid or Planté | Rechargeable | Lead | $PbO_2$ | $H_2SO_4$ | 2.0 | -40–120 | 7–26 | 100–400 |
| Nickel-iron or Edison | Rechargeable | Iron | $NiO_2$ | KOH | 1.2 | | 10–15 | 100–3,000 |
| Cuprous chloride | Activated | Magnesium | $Cu_2Cl_2$ | Sea water | 1.2 | -80–150 | 20–40 | – |
| Silver chloride | Activated | Magnesium | AgCl | Sea water | 1.4 | -80–150 | 20–80 | – |

*Notes:* **Mercury oxide and zinc batteries** are important commercially (called RM batteries for Signal Corps walkie-talkies). They have the following characteristics: very flat voltage curve; good heavy-drain characteristics; high capacity per unit volume and weight; long dry-storage life (90% at 4 years); suited for continuous service; available in miniature (button-type); usable, at light loads, down to −40°F and up to 160°F; withstand pressure, vibration, acceleration, impact; and often used as voltage-reference sources, 1.35 V/cell.
**Silver oxide and zinc batteries** may be dry-stored and activated immediately prior to use. They have a high capacity per unit weight and are non-magnetic. Special batteries may be charged, but their high-current performance is inferior to the primary type.
**Manganese oxide and zinc batteries** sustain voltage at high current drain. They are usable to −5°F, are inexpensive, and have a long shelf-life.
**Lalande copper oxide and zinc batteries** are commercially used for crossing and semaphore signals, approach lighting, etc., in sizes of 75–1,000 Ahr (ampere hour). They have a flat voltage-time curve and allow easy field replacement. KOH electrolyte is substituted when service is much below atmospheric freezing (but above 0°F).
**Nickel-cadmium batteries** may be recharged many times and tolerate overcharge.
**Silver-cadmium batteries** have low cell voltage but high output per unit volume and weight. Their chief characteristics are long charge-cycle life, rapid charge, non-magnetic, and no residual field.

From Bolz, R. E. and Tuve, G. L., Eds., *Handbook of Tables for Applied Engineering Science*, 2nd ed., CRC Press, Cleveland, 1973, 554.

## Table 8—17
## FUEL-CELL CHARACTERISTICS

For current density in A/m², multiply values in amp/sq ft by 10.764. For output in m³/kW, multiply values in ft³/kW by 0.0283. For output in g/W, multiply values in lb/kW by 0.4536.

| Type | Fuel | Oxidant | Electrode | Electrolyte | Operating temperature | Operating pressure | Open-circuit voltage — Theoretical | Open-circuit voltage — Measured | Current density, amp/sq ft | Output, Whr/lb | Output ft³/kW | Output lb/kW | % efficiency thermal to electric |
|---|---|---|---|---|---|---|---|---|---|---|---|---|---|
| Ion-exchange membrane | Hydrogen | Oxygen or air | Activated metal | (Solid) ion-exchange membrane | 50 F above ambient −65 to +165 F | Atmospheric | 1.1 | 1 at 0.8 volt and 10 ma | 22 at 0.8 volt | 100 (measured) | 3.5–5 | 250–500 | 60 at 15–20 amp/sq ft |
| Redox | Liquefied fuel | Oxygen or air | Porous metal | Liquids | 70–85 C | Approx atmospheric | ~1 | | 200 | 1,200 with air 1,600 with pure oxygen | 5 | 50–75 | |
| Carbox | HCO-petroleum hydrocarbons (kerosene) | Oxygen or air | Porous metal | Fused carbonate | 500–800 C | Atmospheric | ~1 | 0.7 | 60 | | | | 64 |
| Hydrox | Hydrogen | Oxygen | Porous metal | | 200–250 C. 400–500 F | 10–55 atm. 400–600 psi | 1.1 | 1.0 | Up to 1,000 | | 0.25–1.2 | 40–90 | |
| Thermal regenerative | Hydrogen | Group I metals | Fuel metal and nickel | Fused group I chlorides | 608 or 1004 F | 200–500 mm. Hg abs | 0.75 | 0.72 | 245 at 0.72 volt with lithium | ~5 | | | Carnot: 40 |
| Solar regenerative | Nitric oxide | Chlorine | Carbon-disk | Liquid nitrosyl chloride | 70 F | 15 psig | 0.21 | 0.21 | 2 at 0.1 volt | | | | |
| Low temp-pressure | Hydrogen | Oxygen | Specially processed carbon | 12 molar solution of KOH | 70–150 F | 1–5 atm | 1.2 | 1.12 | 100 at 0.95 volt and 104 F and oxygen at 5 atm | 1,620 | 3.5–5 | 250–500 | 75 |

From Bolz, R. E. and Tuve, G. L., Eds., *Handbook of Tables for Applied Engineering Science*, 2nd ed., CRC Press, 1973, 559.

## Table 8—18
## ELECTROMOTIVE FORCE AND COMPOSITION OF VOLTAIC CELLS

### Standard Cells

| Name of cell | Negative pole | Positive pole | Solution | Depolarizer | Electromotive force, volts |
|---|---|---|---|---|---|
| Weston normal | Cadmium amalgam | Mercury | Saturated solution of $CdSO_4$ | Paste of $Hg_2SO_4$ and $CdSO_4$ | 1.0183 at 20°C |
| Clark standard | Zinc amalgam | Mercury | Saturated solution of $ZnSO_4$ | Paste of $Hg_2SO_4$ and $ZnSO_4$ | 1.4328 at 15°C |

Temperature equations (temperature in °C):

Clark cell: $E_t = 1.4328[1 - 0.00119(t - 15) - 0.000007(t - 15)^2]$ volt

Weston cell: $E_t = 1.0183[1 - 0.0000406(t - 20) - 0.00000095(t - 20)^2 + 0.00000001(t - 20)^3]$ volt

### Double Fluid Cells

| Name of cell | Negative pole | Positive pole | Solution | Depolarizer | Electromotive force, volts |
|---|---|---|---|---|---|
| Bunsen | Amal. zinc | Carbon | 1 part $H_2SO_4$ to 12 parts $H_2O$ | Fuming nitric acid | 1.94 |
| Bunsen | Amal. zinc | Carbon | 1 part $H_2SO_4$ to 12 parts $H_2O$ | $HNO_3$, density 1.38 | 1.86 |
| Bichromate | Amal. zinc | Carbon | 12 parts $K_2Cr_2O_7$ to 25 parts $H_2SO_4$ and 100 parts $H_2O$ | 1 part $H_2SO_4$ to 12 parts $H_2O$ | 2.00 |
| Bichromate | Amal. zinc | Carbon | 1 part $H_2SO_4$ to 12 parts $H_2O$ | 12 parts $K_2Cr_2O_7$ to 100 parts $H_2O$ | 2.03 |
| Daniell | Amal. zinc | Copper | 1 part $H_2SO_4$ to 4 parts $H_2O$ | Saturated solution of $CuSO_4 + 5H_2O$ | 1.06 |
| Daniell | Amal. zinc | Copper | 5% solution of $ZnSO_4 + 6H_2O$ | Saturated solution of $CuSO_4 + 5H_2O$ | 1.08 |
| Daniell | Amal. zinc | Copper | 1 part NaCl to 4 parts $H_2O$ | Saturated solution of $CuSO_4 + 5H_2O$ | 1.05 |
| Grove | Amal. zinc | Platinum | 1 part $H_2SO_4$ to 12 parts $H_2O$ | Fuming nitric acid | 1.93 |
| Grove | Amal. zinc | Platinum | Solution of $ZnSO_4$ | $HNO_3$, density 1.33 | 1.66 |

Reprinted by permission of the Smithsonian Institution Press from *Smithsonian Physical Tables*, 9th ed., Forsythe, W. E., Ed., Smithsonian Institution, Washington, D.C., 1956, 377.

## Table 8—19
## SUPERCONDUCTIVITY*

### B. W. Roberts**

The following tables on superconductivity include superconductive properties of chemical elements, thin films, a selected list of compounds and alloys, and high-magnetic-field superconductors.

The historically first observed and most distinctive property of a superconductive body is the near total loss of resistance at a critical temperature ($T_c$) that is characteristic of each material. Figure 1(a) below illustrates schematically two types of possible transitions. The sharp vertical discontinuity in resistance is indicative of that found for a single crystal of a very pure element or one of a few well annealed alloy compositions. The broad transition, illustrated by broken lines, suggests the transition shape seen for materials that are not homogeneous and contain unusual strain distributions. Careful testing of the resistivity limits for superconductors shows that it is less than $4 \times 10^{-23}$ ohm-cm, while the lowest resistivity observed in metals is of the order of $10^{-13}$ ohm-cm. If one compares the resistivity of a superconductive body to that of copper at room temperature, the superconductive body is at least $10^{17}$ times less resistive.

The temperature interval $\Delta T_c$, over which the transition between the normal and superconductive states takes place, may be of the order of as little as $2 \times 10^5$ K *or* several K in width, depending on the material state. The narrow transition width was attained in 99.9999% pure gallium single crystals.

A Type-I superconductor below $T_c$, as exemplified by a pure metal, exhibits perfect diamagnetism and excludes a magnetic field up to some critical field $H_c$, whereupon it reverts to the normal state as shown in the H-T diagram of Figure 1(b).

The difference in entropy near absolute zero between the superconductive and normal states relates directly to the electronic specific heat, $\gamma$:$(S_s - S_n)_{T \to 0} = -\gamma T$.

The magnetization of a typical high-field superconductor is shown in Figure 1(c). The discovery of the large current-carrying capability of $Nb_3Sn$ and other similar alloys has led to an extensive study of the physical properties of these alloys. In brief, a high-field superconductor, or Type-II superconductor, passes from the perfect diamagnetic state at low magnetic fields to a mixed state and finally to a sheathed state, before attaining the normal resistive state of the metal. The magnetic field values separating the four stages are given as $H_{c1}$, $H_{c2}$, and $H_{c3}$. The superconductive state below $H_{c1}$ is perfectly diamagnetic, identical to the state of most pure metals of the "soft" or Type-I superconductor. Between $H_{c1}$ and $H_{c2}$ a "mixed superconductive state" is found in which fluxons (a minimal unit of magnetic flux) create lines of normal superconductor in a superconductive matrix. The volume of the normal state is proportional to $-4\pi M$ in the "mixed state" region. Thus at $H_{c2}$

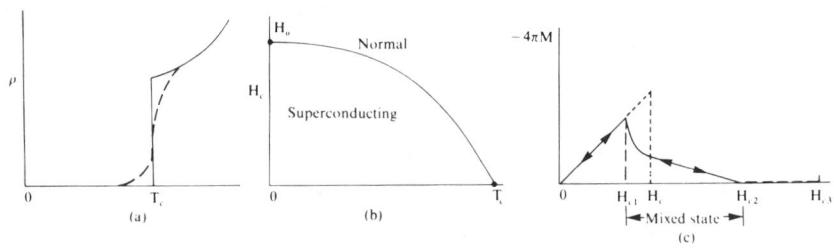

FIGURE 1. PHYSICAL PROPERTIES OF SUPERCONDUCTORS
(a) Resistivity versus temperature for a pure and perfect lattice (solid line). Impure and/or imperfect lattice (broken line).
(b) Magnetic-field temperature dependence of Type-I or "soft" superconductors.
(c) Schematic magnetization curve for "hard" or Type-II superconductors.

* Prepared for Office of Standard Reference Data, National Bureau of Standards, by Standard Reference Data Center on Superconductive Materials, Schenectady, N.Y.
** General Electric Research Laboratory, Schenectady, N.Y.

From Weast, R. C., Ed., *Handbook of Chemistry and Physics*, 69th ed., CRC Press, Boca Raton, Fla., 1988, E-93.

the fluxon density has become so great as to drive the interior volume of the superconductive body completely normal. Between $H_{c2}$ and $H_{c3}$, the superconductor has a sheath of current-carrying superconductive material at the body surface, and above $H_{c3}$ the normal state exists. With several types of careful measurement, it is possible to determine $H_{c1}$, $H_{c2}$, and $H_{c3}$.

High-field superconductive phenomena are also related to specimen dimension and configuration. For example, the Type I superconductor, Hg, has entirely different magnetization behavior in high magnetic fields when contained in the very fine sets of filamentary tunnels found in an unprocessed Vycor glass. The great majority of superconductive materials are Type II. The elements in very pure form and a very few precisely stoichiometric and well-annealed compounds are Type I, with the possible exceptions of vanadium and niobium.

**Metallurgical Aspects** — The sensitivity of superconductive properties to the material state is most pronounced and has been used in a reverse sense to study and specify the detailed state of alloys. The mechanical state, the homogeneity, and the presence of impurity atoms and other electron-scattering centers are all capable of controlling the critical temperature and the current-carrying capabilities in high-magnetic fields. Well annealed specimens tend to show sharper transitions than those that are strained or inhomogeneous. This sensitivity to mechanical state underlines a general problem in the tabulation of properties for superconductive materials. The occasional divergent values of the critical temperature and of the critical fields quoted for a Type II superconductor may lie in the variation in sample preparation. Critical temperatures of materials studied early in the history of superconductivity must be evaluated in light of the probable metallurgical state of the material, as well as the availability of less pure starting elements. It has been noted that recent work has given extended consideration to the metallurgical aspects of sample preparation.

## REFERENCES

References to the data presented in this section, to additional entries of superconductive materials, and to those materials specifically tested and found non-superconductive to some low temperature may be found in the following publications:

"Superconductive Materials and Some of Their Properties", *Progress in Cryogenics*, B. W. Roberts, Vol. IV, Heywood and Co., 1964, pp. 160—231.

"Superconductive Materials and Some of Their Properties", B. W. Roberts, National Bureau of Standards Technical Notes 408 and 482, U.S. Government Printing Office, 1966 and 1969.

**Table 8—20**

## SELECTIVE PROPERTIES OF SUPERCONDUCTIVE ELEMENTS

| Element | $T_c$(K) | $H_o$(oersted) | $\theta_D$(K) | $\gamma$(mJmol$^{-1}$K$^{-1}$) | Element | $T_c$(K) | $H_o$(oersted) | $\theta_D$(K) | $\gamma$(mJmol$^{-1}$K$^{-1}$) |
|---|---|---|---|---|---|---|---|---|---|
| Al | 1.175±0.002 | 104.9±0.3 | 420 | 1.35 | Mo | 0.915±0.005 | 96±3 | 460 | 1.83 |
| Am* ($\alpha$,?) | 0.6 | | | | Nb | 9.25±0.02 | 2060±50,HF | 276 | 7.80 |
| Am* ($\beta$,?) | 1.0 | | | | Os | 0.66±0.03 | 70 | 500 | 2.35 |
| Be | 0.026 | | | 0.21 | Pa | 1.4 | | | |
| Cd | 0.517±0.002 | 28±1 | 209 | 0.69 | Pb | 7.196±0.006 | 803±1 | 96 | 3.1 |
| Ga | 1.083±0.001 | 58.3±0.2 | 325 | 0.60 | Re | 1.697±0.006 | 200±5 | 4.5 | 2.35 |
| Ga ($\beta$) | 5.9,6.2 | 560 | | | Ru | 0.49±0.015 | 69±2 | 580 | 2.8 |
| Ga ($\gamma$) | 7 | 950,hf* | | | Sn | 3.722±0.001 | 305±2 | 195 | 1.78 |
| GA ($\Delta$) | 7.85 | 815,HF | | | Ta | 4.47±0.04 | 829±6 | 258 | 6.15 |
| Hf | 0.128 | 12.7 | | 2.21 | Tc | 7.8±0.1 | 1410,HF | 411 | 6.28 |
| Hg ($\alpha$) | 4.154±0.001 | 411±2 | 87,71.9 | 1.81 | Th | 1.38±0.02 | 160±3 | 165 | 4.32 |
| Hg ($\beta$) | 3.949 | 339 | 93 | 1.37 | Ti | 0.40±0.04 | 56 | 415 | 3.3 |
| In | 3.408±0.001 | 281.5±2 | 109 | 1.672 | Tl | 2.38±0.02 | 178±2 | 78.5 | 1.47 |
| Ir | 0.1125±0.001 | 16±0.05 | 425 | 3.19 | V | 5.40±0.05 | 1408 | 383 | 9.82 |
| La ($\alpha$) | 4.88±0.02 | 800±10 | 151 | 9.8 | W | 0.0154±0.0005 | 1.15±0.03 | 383 | 0.90 |
| La ($\beta$) | 6.00±0.1 | 1096,1600 | 139 | 11.3 | Zn | 0.85±0.01 | 54±0.3 | 310 | 0.66 |
| Lu | 0.1±0.03 | 350±50 | | | Zr | 0.61±0.15 | 47 | 290 | 2.77 |
| | | | | | Zr ($\omega$) | 0.65, 0.95 | | | |

    [a]   HF denotes high field superconductive properties.

From Weast, R. C., Ed., *Handbook of Chemistry and Physics*, 69th ed., CRC Press, Boca Raton, Fla., 1988, E-94.

**Table 8—21**

## RANGE OF CRITICAL TEMPERATURES OBSERVED FOR SUPERCONDUCTIVE ELEMENTS IN THIN FILMS CONDENSED USUALLY AT LOW TEMPERATURES

| Element | $T_c$Range (K) | $H_c$(oersted) | Element | $T_c$Range (K) | $H_c$(oersted) |
|---|---|---|---|---|---|
| | | | Mo | 3.3—8.0 | |
| Al | 1.15—~5.7 | HF* | Nb | 2.0—10.1 | |
| Be | 5—9.75 | HF | Pb | 1.8—7.5 | |
| Bi | 6.17—2.6 | | Re | 1.7—~7 | |
| Cd | | | Sn | 3.5—~6 | |
| (Disordered) | 0.79—0.91 | | Ta | <1.7—4.51 | HF |
| (Ordered) | 0.53—0.59 | | Tc | 4.6—7.7 | |
| Ga | ?.5—8.5 | HF | Ti | 1.3 Max | |
| Hg | 3.87—4.5 | | Tl | 2.33—2.96 | |
| In | 2.2—5.6 | HF | V | 1.8—6.02 | |
| La | 3.55—6.74 | | W | <1.0—4.1 | |
| | | | Zn | 0.77—1.70,~1.9 | |

    *   HF denotes high magnetic field superconductive properties.

From Weast, R. C., Ed., *Handbook of Chemistry and Physics*, 69th ed., CRC Press, Boca Raton, Fla., 1988, E-95.

## Table 8—22
## ELEMENTS EXHIBITING SUPERCONDUCTIVITY UNDER OR AFTER APPLICATION OF HIGH PRESSURE

| Element | $T_c$Range (K) | Pressure (kbar)[a] |
|---|---|---|
| Al | 1.98—0.075 | 0—62 |
| As | 0.31—0.5 | 220—140 |
|  | 0.2—0.25 | ~140—100 |
| Ba II | ~1—1.8 | ~55—85 |
| III | 1.8—5 | ~85—144 |
| IV | 4.5—5.4 | 144—190 |
| Bi II | 3.9 | 25—27 |
| III | 6.55—7.25 | 28—38 |
| IV | 7.0,8.7—6.0 | 43,43—62 |
| V | 6.7,8.3 | 48—80 |
| VI | 8.55 | 90,92—101 |
| VII(?) | 8.2 | 30 |
| Ce ($\alpha$) | 0.020—0.045 | 20—35 |
| Ce ($\alpha'$) | 1.9—1.3 | 45—125 |
| Cs V | ~1.5 | >125 |
| Ga II | 6.38 | ≥35 |
| II′ | 7.5 | ≥35 then P removed |
| Ge | 5.35 | 115 |
| La | ~5.5—12.9 | 0—210 |
| Lu | 0.022—1.0 | 45—190 |
| P | 5.8 | 170 |
| Pb II | 3.55 | 160 |
| Re II | 2.3 Max. | "Plastic" compression |
| Sb (prepared 120 kbar, held below 77K) | 2.6—2.7 | |
| Sb III | 3.55—3.40 | 85—~150 |
| Se II | 6.75,6.95 | ~130 |
| Si | 6.7—7.1 | 120—130 |
| Sn II | 5.2—4.85 | 125—160 |
| III | 5.30 | 113 |
| Te II | 2.4—5.1 | 38—55 |
| III | 4.1—4.2 | ~53—62 |
| IV | 4.72—4 | 63—80 |
| ( ) | 3.3—2.8 | 100—260 |
| Tl (cubic form) | 1.45 | 35 |
| (Hexagonal form) | 1.95 | 35 |
| U | 2.4—0.4 | 10—85 |
| Y | 2.3—1.7—2.5 | 110—125—160 |
| Zr (omega form, metastable) | 1—1.7 | 60—~130 |

[a] 1 kbar = $10^8$ newton/meter$^2$ = 0.987 katm

From Roberts, B. W., Properties of Selected Superconductive Materials, 1978 Supplement, NBS Technical Note 983, U.S. Government Printing Office, Washington, D.C.

From Weast, R. C., Ed., *Handbook of Chemistry and Physics,* 69th ed., CRC Press, Boca Raton, Fla., 1988, E-95.

## Table 8—23
## SELECTED SUPERCONDUCTIVE COMPOUNDS AND ALLOYS

All compositions are denoted on an atomic basis, i.e., AB, $AB_2$, or $AB_3$ for compounds, unless noted. Solid solutions or odd compositions may be denoted as $A_zB_{1-z}$ or $A_zB$. A series of three or more alloys is indicated as $A_xB_{1-x}$ or by actual indication of the atomic fraction range, such as $A_{0-0.6}B_{1-0.4}$. The critical temperature of such a series of alloys is denoted by a range of values or possibly the maximum value.

The selection of the critical temperature from a transition in the effective permeability, or the change in resistance, or possibly the incremental changes in frequency observed by certain techniques is not often obvious from the literature. Most authors choose the mid-point of such curves as the probable critical temperature of the idealized material, while others will choose the highest temperature at which a deviation from the normal state property is observed. In view of the previous discussion concerning the variability of the superconductive properties as a function of purity and other metallurgical aspects, it is recommended that appropriate literature be checked to determine the most probable critical temperature or critical field of a given alloy.

A very limited amount of data on critical fields, $H_o$, is available for these compounds and alloys; these values are given at the end of the table.

*SYMBOLS*:   n = number of normal carriers per cubic centimeter for semiconductor superconductors.

| Substance | $T_c$, K | Crystal structure type†† | Substance | $T_c$, K | Crystal structure type†† |
|---|---|---|---|---|---|
| $Ag_xAl_yZn_{1-x-y}$ | 0.5-0.845 | | $Al_3Th$ | 0.75 | $DO_{19}$ |
| $Ag_7BF_4O_8$ | 0.15 | Cubic | $Al_xTi_yV_{1-x-y}$ | 2.05-3.62 | Cubic |
| $AgBi_2$ | 3.0-2.78 | | $Al_{0.108}V_{0.892}$ | 1.82 | Cubic |
| $Ag_7F_{0.25}N_{0.75}O_{10.25}$ | 0.85-0.90 | | $Al_xZn_{1-x}$ | 0.5-0.845 | |
| $Ag_7FO_8$ | 0.3 | Cubic | $AlZr_3$ | 0.73 | $L1_2$ |
| $Ag_2F$ | 0.066 | | $AsBiPb$ | 9.0 | |
| $Ag_{0.8-0.3}Ga_{0.2-0.7}$ | 6.5-8 | | $AsBiPbSb$ | 9.0 | |
| $Ag_4Ge$ | 0.85 | Hex., c.p. | $As_{0.33}InTe_{0.67}$ | | |
| $Ag_{0.438}Hg_{0.562}$ | 0.64 | $D8_2$ | $(n = 1.24 \times 10^{22})$ | 0.85-1.15 | B1 |
| $AgIn_2$ | ~2.4 | C16 | $As_{0.5}InTe_{0.5}$ | | |
| $Ag_{0.1}In_{0.9}Te$ | | | $(n = 0.97 \times 10^{22})$ | 0.44-0.62 | B1 |
| $(n = 1.40 \times 10^{22})$ | 1.20-1.89 | B1 | $As_{0.50}Ni_{0.06}Pd_{0.44}$ | 1.39 | C2 |
| $Ag_{0.2}In_{0.8}Te$ | | | $AsPb$ | 8.4 | |
| $(n = 1.07 \times 10^{22})$ | 0.77-1.00 | B1 | $AsPd_2$ (low- | | |
| $AgLa$ (9.5 kbar) | 1.2 | B2 | temperature phase) | 0.60 | Hexagonal |
| $Ag_7NO_{11}$ | 1.04 | Cubic | $AsPd_2$ (high-temp. phase) | 1.70 | C22 |
| $Ag_xPb_{1-x}$ | 7.2 max. | | $AsPd_5$ | 0.46 | Complex |
| $Ag_xSn_{1-x}$ (film) | 2.0-3.8 | | $AsRh$ | 0.58 | B31 |
| $Ag_xSn_{1-x}$ | 1.5-3.7 | | $AsRh_{1.4-1.6}$ | <0.03-0.56 | Hexagonal |
| $AgTe_3$ | 2.6 | Cubic | $AsSn$ | 4.10 | |
| $AgTh_2$ | 2.26 | C16 | $AsSn$ | | |
| $Ag_{0.03}Tl_{0.97}$ | 2.67 | | $(n = 2.14 \times 10^{22})$ | 3.41-3.65 | B1 |
| $Ag_{0.94}Tl_{0.06}$ | 2.32 | | $As_{~2}Sn_{~3}$ | 3.5-3.6, | |
| $Ag_xZn_{1-x}$ | 0.5-0.845 | | | 1.21-1.17 | |
| Al (film) | 1.3-2.31 | | $As_3Sn_4$ | | |
| Al (1 to 21 katm) | 1.170-0.687 | Al | $(n = 0.56 \times 10^{22})$ | 1.16-1.19 | Rhombohedral |
| $AlAu_4$ | 0.4-0.7 | Like Al3 | $Au_5Ba$ | 0.4-0.7 | $D2_d$ |
| $Al_2CMo_3$ | 10.0 | Al3 | $AuBe$ | 2.64 | B20 |
| $Al_2CMo_3$ | 9.8-10.2 | Al3 + trace 2nd phase | $Au_2Bi$ | 1.80 | C15 |
| $Al_2CaSi$ | 5.8 | | $Au_5Ca$ | 0.34-0.38 | $C15_b$ |
| $Al_{0.131}Cr_{0.088}V_{0.781}$ | 1.46 | Cubic | $AuGa$ | 1.2 | B31 |
| $AlGe_2$ | 1.75 | | $Au_{0.40-0.92}Ge_{0.60-0.08}$ | <0.32-1.63 | Complex |
| $Al_{0.5}Ge_{0.5}Nb$ | 12.6 | A15 | $AuIn$ | 0.4-0.6 | Complex |
| $Al_{~0.8}Ge_{~0.2}Nb_3$ | 20.7 | A15 | $AuLu$ | <0.35 | B2 |
| $AlLa_3$ | 5.57 | $DO_{19}$ | $AuNb_3$ | 11.5 | A15 |
| $Al_2La$ | 3.23 | C15 | $AuNb_3$ | 1.2 | A2 |
| $Al_3Mg_2$ | 0.84 | Cubic, f.c. | $Au_{0-0.3}Nb_{1-0.7}$ | 1.1-11.0 | |
| $AlMo_3$ | 0.58 | A15 | $Au_{0.02-0.98}Nb_3Rh_{0.98-0.02}$ | 2.53-10.9 | A15 |
| $AlMo_6Pd$ | 2.1 | | $AuNb_{3(1-x)}V_{3x}$ | 1.5-11.0 | A15 |
| $AlN$ | 1.55 | B4 | $AuPb_2$ | 3.15 | |
| $Al_2NNb_3$ | 1.3 | A13 | $AuPb_2$ (film) | 4.3 | |
| $AlNb_3$ | 18.0 | A15 | $AuPb_3$ | 4.40 | |
| $Al_xNb_{1-x}$ | <4.2-13.5 | $D8_b$ | $AuPb_3$ (film) | 4.25 | |
| $Al_xNb_{1-x}$ | 12-17.5 | A15 | $Au_2Pb$ | 1.18, 6-7 | C15 |
| $Al_{0.27}Nb_{0.73-0.48}V_{0-0.25}$ | 14.5-17.5 | A15 | $AuSb_2$ | 0.58 | C2 |
| $AlNb_xV_{1-x}$ | <4.2-13.5 | | $AuSn$ | 1.25 | $B8_1$ |
| $AlOs$ | 0.39 | B2 | $Au_xSn_{1-x}$ (film) | 2.0-3.8 | |
| $Al_3Os$ | 5.90 | | $Au_5Sn$ | 0.7-1.1 | A3 |
| AlPb (films) | 1.2-7 | | $Au_3Te_5$ | 1.62 | Cubic |
| $Al_2Pt$ | 0.48-0.55 | C1 | $AuTh_2$ | 3.08 | C16 |
| $Al_5Re_{24}$ | 3.35 | A12 | $AuTl$ | 1.92 | |
| | | | $AuV_3$ | 0.74 | A15 |

††See key at end of table.

## Table 8—23 (continued)
## SELECTED SUPERCONDUCTIVE COMPOUNDS AND ALLOYS

| Substance | $T_c$, K | Crystal structure type†† | Substance | $T_c$, K | Crystal structure type†† |
|---|---|---|---|---|---|
| $Au_xZn_{1-x}$ | 0.50-0.845 | | $Bi_{0.5}Pb_{0.25}Sn_{0.25}$ | 8.5 | |
| $AuZn_3$ | 1.21 | Cubic | $BiPd_2$ | 4.0 | |
| $Au_xZr_y$ | 1.7-2.8 | A3 | $Bi_{0.4}Pd_{0.6}$ | 3.7-4 | Hexagonal, ordered |
| $AuZr_3$ | 0.92 | A15 | $BiPd$ | 3.7 | Orthorhombic |
| $BCMo_2$ | 5.4 | Orthorhombic | $Bi_2Pd$ | 1.70 | Monoclinic, α-phase |
| $B_{0.03}C_{0.51}Mo_{0.47}$ | 12.5 | | $Bi_2Pd$ | 4.25 | Tetragonal, β-phase |
| $BCMo_2$ | 5.3-7.0 | Orthorhombic | $BiPdSe$ | 1.0 | C2 |
| $BHf$ | 3.1 | Cubic | $BiPdTe$ | 1.2 | C2 |
| $B_6La$ | 5.7 | | $BiPt$ | 1.21 | $B8_1$ |
| $B_{12}Lu$ | 0.48 | | $BiPtSe$ | 1.45 | C2 |
| $BMo$ | 0.5 (extrapolated) | | $BiPtTe$ | 1.15 | C2 |
| $BMo_2$ | 4.74 | Cl6 | $Bi_2Pt$ | 0.155 | Hexagonal |
| $BNb$ | 8.25 | $B_f$ | $Bi_2Rb$ | 4.25 | Cl5 |
| $BRe_2$ | 2.80, 4.6 | | $BiRe_2$ | 1.9-2.2 | |
| $B_{0.3}Ru_{0.7}$ | 2.58 | $D10_2$ | $BiRh$ | 2.06 | $B8_1$ |
| $B_{12}Sc$ | 0.39 | | $Bi_3Rh$ | 3.2 | Orthorhombic, like $NiB_3$ |
| $BTa$ | 4.0 | $B_f$ | $Bi_4Rh$ | 2.7 | Hexagonal |
| $B_6Th$ | 0.74 | | $Bi_3Sn$ | 3.6-3.8 | |
| $BW_2$ | 3.1 | Cl6 | $BiSn$ | 3.8 | |
| $B_6Y$ | 6.5-7.1 | | $Bi_xSn_y$ | 3.85-4.18 | |
| $B_{12}Y$ | 4.7 | | $Bi_3Sr$ | 5.62 | $Ll_2$ |
| $BZr$ | 3.4 | Cubic | $Bi_3Te$ | 0.75-1.0 | |
| $B_{12}Zr$ | 5.82 | | $Bi_5Tl_3$ | 6.4 | |
| $BaBi_3$ | 5.69 | Tetragonal | $Bi_{0.26}Tl_{0.74}$ | 4.4 | Cubic, disordered |
| $Ba_xO_3Sr_{1-x}Ti$ ($n = 4.2\text{-}11 \times 10^{19}$) | <0.1-0.55 | | $Bi_{0.26}Tl_{0.74}$ | 4.15 | $Ll_2$, ordered? |
| $Ba_{0.13}O_3W$ | 1.9 | Tetragonal | $Bi_2Y_3$ | 2.25 | |
| $Ba_{0.14}O_3W$ | <1.25-2.2 | Hexagonal | $Bi_3Zn$ | 0.8-0.9 | |
| $BaRh_2$ | 6.0 | Cl5 | $Bi_{0.3}Zr_{0.7}$ | 1.51 | |
| $Be_{22}Mo$ | 2.51 | Cubic, like $Be_{22}Re$ | $BiZr_3$ | 2.4-2.8 | |
| $Be_8Nb_5Zr_2$ | 5.2 | | $CCs_x$ | 0.020-0.135 | Hexagonal |
| $Be_{0.98-0.92}Re_{0.02-0.08}$ (quenched) | 9.5-9.75 | Cubic | $C_8K$ (gold) | 0.55 | |
| $Be_{0.957}Re_{0.043}$ | 9.62 | Cubic, like $Be_{22}Re$ | $CGaMo_2$ | 3.7-4.1 | Hexagonal, H-phase |
| $BeTc$ | 5.21 | Cubic | $CHf_{0.5}Mo_{0.5}$ | 3.4 | Bl |
| $Be_{22}W$ | 4.12 | Cubic, like $Be_{22}Re$ | $CHf_{0.3}Mo_{0.7}$ | 5.5 | Bl |
| $Be_{13}W$ | 4.1 | Tetragonal | $CHf_{0.25}Mo_{0.75}$ | 6.6 | Bl |
| $Bi_3Ca$ | 2.0 | | $CHf_{0.7}Nb_{0.3}$ | 6.1 | Bl |
| $Bi_{0.5}Cd_{0.13}Pb_{0.25}Sn_{0.12}$ (weight fractions) | 8.2 | | $CHf_{0.6}Nb_{0.4}$ | 4.5 | Bl |
| $BiCo$ | 0.42-0.49 | | $CHf_{0.5}Nb_{0.5}$ | 4.8 | Bl |
| $Bi_2Cs$ | 4.75 | Cl5 | $CHf_{0.4}Nb_{0.6}$ | 5.6 | Bl |
| $Bi_xCu_{1-x}$ (electrodeposited) | 2.2 | | $CHf_{0.25}Nb_{0.75}$ | 7.0 | Bl |
| $BiCu$ | 1.33-1.40 | | $CHf_{0.2}Nb_{0.8}$ | 7.8 | Bl |
| $Bi_{0.019}In_{0.981}$ | 3.86 | | $CHf_{0.9-0.1}Ta_{0.1-0.9}$ | 5.0-9.0 | Bl |
| $Bi_{0.05}In_{0.95}$ | 4.65 | α-phase | $Ck$ (excess K) | 0.55 | Hexagonal |
| $Bi_{0.10}In_{0.90}$ | 5.05 | α-phase | $C_8K$ | 0.39 | Hexagonal |
| $Bi_{0.15-0.30}In_{0.85-0.70}$ | 5.3-5.4 | α- and β-phases | $C_{0.40-0.44}Mo_{0.60-0.56}$ | 9˙13 | |
| $Bi_{0.34-0.48}In_{0.66-0.52}$ | 4.0-4.1 | | $CMo$ | 6.5, 9.26 | |
| $Bi_3In_5$ | 4.1 | | $CMo_2$ | 12.2 | Orthorhombic |
| $BiIn_2$ | 5.65 | β-phase | $C_{0.44}Mo_{0.56}$ | 1.3 | Bl |
| $Bi_2Ir$ | 1.7-2.3 | | $C_{0.5}Mo_xNb_{1-x}$ | 10.8-12.5 | Bl |
| $Bi_2Ir$ (quenched) | 3.0-3.96 | | $C_{0.6}Mo_{4.8}Si_3$ | 7.6 | $D8_8$ |
| $BiK$ | 3.6 | | $CMo_{0.2}Ta_{0.8}$ | 7.5 | Bl |
| $Bi_2K$ | 3.58 | Cl5 | $CMo_{0.5}Ta_{0.5}$ | 7.7 | Bl |
| $BiLi$ | 2.47 | $Ll_0$, α-phase | $CMo_{0.75}Ta_{0.25}$ | 8.5 | Bl |
| $Bi_{4-9}Mg$ | 0.7-~1.0 | | $CMo_{0.8}Ta_{0.2}$ | 8.7 | Bl |
| $Bi_3Mo$ | 3-3.7 | | $CMo_{0.85}Ta_{0.15}$ | 8.9 | Bl |
| $BiNa$ | 2.25 | $Ll_0$ | $CMo_xTi_{1-x}$ | 10.2 max. | Bl |
| $BiNb_3$ (high pressure and temperature) | 3.05 | A15 | $CMo_{0.83}Ti_{0.17}$ | 10.2 | Bl |
| $BiNi$ | 4.25 | $B8_1$ | $CMo_xV_{1-x}$ | 2.9-9.3 | Bl |
| $Bi_3Ni$ | 4.06 | Orthorhombic | $CMo_xZr_{1-x}$ | 3.8-9.5 | Bl |
| $Bi_{1-0}Pb_{0-1}$ | 7.26-9.14 | | $C_{0.1-0.9}N_{0.9-0.1}Nb$ | 8.5-17.9 | |
| $Bi_{1-0}Pb_{0-1}$ (film) | 7.25-8.67 | | $C_{0-0.38}N_{1-0.62}Ta$ | 10.0-11.3 | |
| $Bi_{0.05-0.40}Pb_{0.95-0.60}$ | 7.35-8.4 | Hexagonal, c.p., to ε-phase | $CNb$ (whiskers) | 7.5-10.5 | |
| | | | $C_{0.984}Nb$ | 9.8 | Bl |
| | | | $CNb$ (extrapolated) | ~14 | |
| $BiPbSb$ | 8.9 | | $C_{0.7-1.0}Nb_{0.3-0}$ | 6-11 | Bl |
| $Bi_{0.5}Pb_{0.31}Sn_{0.19}$ (weight fractions) | 8.5 | | $CNb_2$ | 9.1 | |
| | | | $CNb_xTa_{1-x}$ | 8.2-13.9 | |
| | | | $CNb_xTi_{1-x}$ | <4.2-8.8 | Bl |
| | | | $CNb_{0.6-0.9}W_{0.4-0.1}$ | 12.5-11.6 | Bl |
| | | | $CNb_{0.1-0.9}Zr_{0.9-0.1}$ | 4.2-8.4 | Bl |

## Table 8—23 (continued)
## SELECTED SUPERCONDUCTIVE COMPOUNDS AND ALLOYS

| Substance | $T_c$, K | Crystal structure type†† | Substance | $T_c$, K | Crystal structure type†† |
|---|---|---|---|---|---|
| $CRb_x$ (gold) | 0.023–0.151 | Hexagonal | $Cr_xTi_{1-x}$ | 3.6 max. | Cr in $\alpha$-Ti |
| $CRe_{0.01-0.08}W$ | 1.3–5.0 | | $Cr_xTi_{1-x}$ | 4.2 max. | Cr in $\beta$-Ti |
| $CRe_{0.06}W$ | 5.0 | | $Cr_{0.1}Ti_{0.3}V_{0.6}$ | 5.6 | |
| CTa | ~11 (extrapolated) | | $Cr_{0.0175}U_{0.9825}$ | 0.75 | $\beta$-phase |
| | | | $Cs_{0.32}O_3W$ | 1.12 | Hexagonal |
| $C_{0.987}Ta$ | 9.7 | | $Cu_{0.15}In_{0.85}$ (film) | 3.75 | |
| $C_{0.848-0.987}Ta$ | 2.04–9.7 | | $Cu_{0.04-0.08}In_{1-x}$ | 4.4 | |
| CTa (film) | 5.09 | B1 | CuLa | 5.85 | |
| $CTa_2$ | 3.26 | $L'_3$ | $Cu_xPb_{1-x}$ | 5.7–7.7 | |
| $CTa_{0.4}Ti_{0.6}$ | 4.8 | B1 | CuS | 1.62 | B18 |
| $CTa_{1-0.4}W_{0-0.6}$ | 8.5–10.5 | B1 | $CuS_2$ | 1.48–1.53 | C18 |
| $CTa_{0.2-0.9}Zr_{0.8-0.1}$ | 4.6–8.3 | B1 | CuSSe | 1.5–2.0 | C18 |
| CTc (excess C) | 3.85 | Cubic | $CuSe_2$ | 2.3–2.43 | C18 |
| $CTi_{0.5-0.7}W_{0.5-0.3}$ | 6.7–2.1 | B1 | CuSeTe | 1.6–2.0 | C18 |
| CW | 1.0 | | $Cu_xSn_{1-x}$ | 3.2–3.7 | |
| $CW_2$ | 2.74 | $L'_3$ | $Cu_xSn_{1-x}$ (film) | | |
| $CW_2$ | 5.2 | Cubic, f.c. | (made at 10°K) | 3.6–7 | |
| $CaIr_2$ | 6.15 | C15 | $Cu_xSn_{1-x}$ (film) | | |
| $Ca_xO_3Sr_{1-x}Ti$ | | | (made at 300°K) | 2.8–3.7 | |
| ($n = 3.7–11.0 \times 10^{19}$) | <0.1–0.55 | | $CuTe_2$ | <1.25–1.3 | C18 |
| $Ca_{0.1}O_3W$ | 1.4–3.4 | Hexagonal | $CuTh_2$ | 3.49 | C16 |
| CaPb | 7.0 | | $Cu_{0-0.027}V$ | 3.9–5.3 | A2 |
| $CaRh_2$ | 6.40 | C15 | $Cu_xZn_{1-x}$ | 0.5–0.845 | |
| $Cd_{0.3-0.5}Hg_{0.7-0.5}$ | 1.70–1.92 | | $Er_xLa_{1-x}$ | 1.4–6.3 | |
| CdHg | 1.77, 2.15 | Tetragonal | $Fe_{0-0.04}Mo_{0.8}Re_{0.2}$ | 1–10 | |
| $Cd_{0.0075-0.05}In_{1-x}$ | 3.24–3.36 | Tetragonal | $Fe_{0.05}Ni_{0.05}Zr_{0.90}$ | ~3.9 | |
| $Cd_{0.97}Pb_{0.03}$ | 4.2 | | $Fe_3Th_7$ | 1.86 | D10 |
| CdSn | 3.65 | | $Fe_xTi_{1-x}$ | 3.2 max. | Fe in $\alpha$-Ti |
| $Cd_{0.17}Tl_{0.83}$ | 2.3 | | $Fe_xTi_{1-x}$ | 3.7 max. | Fe in $\beta$-Ti |
| $Cd_{0.18}Tl_{0.82}$ | 2.54 | | $Fe_xTi_{0.6}V_{1-x}$ | 6.8 max. | |
| $CeCo_2$ | 0.84 | C15 | $FeU_6$ | 3.86 | $D2_c$ |
| $CeCo_{1.67}Ni_{0.33}$ | 0.46 | C15 | $Fe_{0.1}Zr_{0.9}$ | 1.0 | A3 |
| $CeCo_{1.67}Rh_{0.33}$ | 0.47 | C15 | $Ga_{0.5}Ge_{0.5}Nb_3$ | 7.3 | A15 |
| $Ce_xGd_{1-x}Ru_2$ | 3.2–5.2 | C15 | $GaLa_3$ | 5.84 | |
| $CeIr_3$ | 3.34 | | $Ga_2Mo$ | 9.5 | |
| $CeIr_5$ | 1.82 | | $GaMo_3$ | 0.76 | A15 |
| $Ce_{0.005}La_{0.995}$ | 4.6 | | $Ga_4Mo$ | 9.8 | |
| $Ce_xLa_{1-x}$ | 1.3–6.3 | | GaN (black) | 5.85 | B4 |
| $Ce_xPr_{1-x}Ru_2$ | 1.4–5.3 | C15 | $GaNb_3$ | 14.5 | A15 |
| $Ce_xPt_{1-x}$ | 0.7–1.55 | | $Ga_xNb_3Sn_{1-x}$ | 14–18.37 | A15 |
| $CeRu_2$ | 6.0 | C15 | $Ga_{0.7}Pt_{0.3}$ | 2.9 | C1 |
| $Co_xFe_{1-x}Si_2$ | 1.4 max. | C1 | GaPt | 1.74 | B20 |
| $CoHf_2$ | 0.56 | $E9_3$ | GaSb (120 kbar, 77°K, | | |
| $CoLa_3$ | 4.28 | | annealed) | 4.24 | A5 |
| $CoLu_3$ | ~0.35 | | GaSb (unannealed) | ~5.9 | |
| $Co_{0-0.01}Mo_{0.8}Re_{0.2}$ | 2–10 | | $Ga_{0.1}Sn_{1-0}$ (quenched) | 3.47–4.18 | |
| $Co_{0.02-0.10}Nb_3Rh_{0.98-0.90}$ | 2.28–1.90 | A15 | $Ga_{0-1}Sn_{1-0}$ (annealed) | 2.6–3.85 | |
| $Co_xNi_{1-x}Si_2$ | 1.4 max. | C1 | $Ga_5V_2$ | 3.55 | Tetragonal, $Mn_2Hg_5$ type |
| $Co_{0.5}Rh_{0.5}Si_2$ | 2.5 | | $GaV_3$ | 16.8 | A15 |
| $Co_xRh_{1-x}Si_2$ | 3.65 max. | | $GaV_{2.1-3.5}$ | 6.3–14.45 | A15 |
| $Co_{-0.3}Sc_{-0.7}$ | ~0.35 | | $GaV_{4.5}$ | 9.15 | |
| $CoSi_2$ | 1.40, 1.22 | C1 | $Ga_3Zr$ | 1.38 | |
| $Co_3Th_7$ | 1.83 | $D10_2$ | $Gd_xLa_{1-x}$ | <1.0–5.5 | |
| $Co_xTi_{1-x}$ | 2.8 max. | Co in $\alpha$-Ti | $Gd_xOs_2Y_{1-x}$ | 1.4–4.7 | |
| $Co_xTi_{1-x}$ | 3.8 max. | Co in $\beta$-Ti | $Gd_xRu_2Th_{1-x}$ | 3.6 max. | C15 |
| $CoTi_2$ | 3.44 | $E9_3$ | GeIr | 4.7 | B31 |
| CoTi | 0.71 | A2 | $Ge_2La$ | 1.49, 2.2 | Orthorhombic, distorted $ThSi_2$-type |
| CoU | 1.7 | B2, distorted | | | |
| $CoU_6$ | 2.29 | $D2_c$ | | | |
| $Co_{0.28}Y_{0.72}$ | 0.34 | | $GeMo_3$ | 1.43 | A15 |
| $CoY_3$ | <0.34 | | $GeNb_2$ | 1.9 | |
| $CoZr_2$ | 6.3 | C16 | $GeNb_3$ (quenched) | 6–17 | A15 |
| $Co_{0.1}Zr_{0.9}$ | 3.9 | A3 | $Ge_{0.29}Nb_{0.71}$ | 6 | A15 |
| $Cr_{0.6}Ir_{0.4}$ | 0.4 | Hexagonal, c.p. | $Ge_xNb_3Sn_{1-x}$ | 17.6–18.0 | A15 |
| $Cr_{0.65}Ir_{0.35}$ | 0.59 | Hexagonal, c.p. | $Ge_{0.5}Nb_3Sn_{0.5}$ | 11.3 | |
| $Cr_{0.7}Ir_{0.3}$ | 0.76 | Hexagonal, c.p. | GePt | 0.40 | B31 |
| $Cr_{0.72}Ir_{0.28}$ | 0.83 | | $Ge_3Rh_5$ | 2.12 | Orthorhombic, related to $InNi_2$ |
| $Cr_3Ir$ | 0.45 | A15 | | | |
| $Cr_{0-0.1}Nb_{1-0.9}$ | 4.6–9.2 | A2 | | | |
| $Cr_{0.80}Os_{0.20}$ | 2.5 | Cubic | $Ge_2Sc$ | 1.3 | |
| $Cr_xRe_{1-x}$ | 1.2–5.2 | | $Ge_3Te_4$ | | |
| $Cr_{0.40}Re_{0.60}$ | 2.15 | $D8_b$ | ($n = 1.06 \times 10^{22}$) | 1.55–1.80 | Rhombohedral |
| $Cr_{0.8-0.6}Rh_{0.2-0.4}$ | 0.5–1.10 | A3 | $Ge_xTe_{1-x}$ | | |
| $Cr_3Ru$ (annealed) | 3.3 | A15 | ($n = 8.5–64 \times 10^{20}$) | 0.07–0.41 | B1 |
| $Cr_2Ru$ | 2.02 | $D8_b$ | $GeV_3$ | 6.01 | A15 |
| $Cr_{0.1-0.5}Ru_{0.9-0.5}$ | 0.34–1.65 | A3 | | | |

## Table 8—23 (continued)
## SELECTED SUPERCONDUCTIVE COMPOUNDS AND ALLOYS

| Substance | $T_c$, K | Crystal structure type†† | Substance | $T_c$, K | Crystal structure type†† |
|---|---|---|---|---|---|
| $Ge_2Y$ | 3.80 | $C_c$ | $Ir_{0.287}O_{0.14}Ti_{0.573}$ | 5.5 | $E9_3$ |
| $Ge_{1.62}Y$ | 2.4 | | $Ir_{0.265}O_{0.035}Ti_{0.65}$ | 2.30 | $E9_3$ |
| $H_{0.33}Nb_{0.67}$ | 7.28 | Cubic, b.c. | $Ir_xOs_{1-x}$ | 0.3–0.98 | |
| $H_{0.1}Nb_{0.9}$ | 7.38 | Cubic, b.c. | | (max.)–0.6 | |
| $H_{0.05}Nb_{0.95}$ | 7.83 | Cubic, b.c. | $IrOsY$ | 2.6 | C15 |
| $H_{0.12}Ta_{0.88}$ | 2.81 | Cubic, b.c. | $Ir_{1.5}Os_{0.5}$ | 2.4 | C14 |
| $H_{0.08}Ta_{0.92}$ | 3.26 | Cubic, b.c. | $Ir_2Sc$ | 2.07 | C15 |
| $H_{0.04}Ta_{0.96}$ | 3.62 | Cubic, b.c. | $Ir_{2.5}Sc$ | 2.46 | C15 |
| $HfN_{0.989}$ | 6.6 | B1 | $IrSn_2$ | 0.65–0.78 | C1 |
| $Hf_{0-0.5}Nb_{1-0.5}$ | 8.3–9.5 | A2 | $Ir_2Sr$ | 5.70 | C15 |
| $Hf_{0.75}Nb_{0.25}$ | >4.2 | | $Ir_{0.5}Te_{0.5}$ | ~3 | |
| $HfOs_2$ | 2.69 | C14 | $IrTe_3$ | 1.18 | C2 |
| $HfRe_2$ | 4.80 | C14 | $IrTh$ | <0.37 | $B_f$ |
| $Hf_{0.14}Re_{0.86}$ | 5.86 | A12 | $Ir_2Th$ | 6.50 | C15 |
| $Hf_{0.99-0.96}Rh_{0.01-0.04}$ | 0.85–1.51 | | $Ir_3Th$ | 4.71 | |
| $Hf_{0-0.55}Ta_{1-0.45}$ | 4.4–6.5 ~ | A2 | $Ir_3Th_7$ | 1.52 | $D10_2$ |
| $HfV_2$ | 8.9–9.6 | C15 | $Ir_5Th$ | 3.93 | $D2_d$ |
| $Hg_xIn_{1-x}$ | 3.14–4.55 | | $IrTi_3$ | 5.40 | A15 |
| $HgIn$ | 3.81 | | $IrV_2$ | 1.39 | A15 |
| $Hg_2K$ | 1.20 | Orthorhombic | $IrW_3$ | 3.82 | |
| $Hg_3K$ | 3.18 | | $Ir_{0.28}W_{0.72}$ | 4.49 | |
| $Hg_4K$ | 3.27 | | $Ir_2Y$ | 2.18, 1.38 | C15 |
| $Hg_8K$ | 3.42 | | $Ir_{0.69}Y_{0.31}$ | 1.98, 1.44 | C15 |
| $Hg_3Li$ | 1.7 | Hexagonal | $Ir_{0.70}Y_{0.30}$ | 2.16 | C15 |
| $Hg_2Na$ | 1.62 | Hexagonal | $Ir_2Y$ | 1.09 | C15 |
| $Hg_4Na$ | 3.05 | | $Ir_2Y$ | 1.61 | |
| $Hg_xPb_{1-x}$ | 4.14–7.26 | | $Ir_2Y_3$ | 1.61 | |
| $HgSn$ | 4.2 | | $Ir_xY_{1-x}$ | 0.3–3.7 | |
| $Hg_xTl_{1-x}$ | 2.30–4.109 | | $Ir_2Zr$ | 4.10 | C15 |
| $Hg_5Tl_2$ | 3.86 | | $Ir_{0.1}Zr_{0.9}$ | 5.5 | A3 |
| $Ho_xLa_{1-x}$ | 1.3–6.3 | | $K_{0.27-0.31}O_3W$ | 0.50 | Hexagonal |
| $InLa_3$ | 9.83, 10.4 | $L1_2$ | $K_{0.40-0.57}O_3W$ | 1.5 | Tetragonal |
| $InLa_3$ (0–35, kbar) | 9.75–10.55 | | $La_{0.55}Lu_{0.45}$ | 2.2 | Hexagonal, La type |
| $In_{1-0.86}Mg_{0-0.14}$ | 3.395–3.363 | | | | |
| $InNb_3$ | | | $La_{0.8}Lu_{0.2}$ | 3.4 | Hexagonal, La Type |
| (high pressure and temp.) | 4–8, 9.2 | A15 | | | |
| $In_{0-0.3}Nb_3Sn_{1-0.7}$ | 18.0–18.19 | A15 | $LaMg_2$ | 1.05 | C15 |
| $In_{0.5}Nb_3Zr_{0.5}$ | 6.4 | | $LaN$ | 1.35 | |
| $In_{0.11}O_3W$ | <1.25–2.8 | Hexagonal | $LaOs_2$ | 6.5 | C15 |
| $In_{0.95-0.85}Pb_{0.05-0.15}$ | 3.6–5.05 | | $LaPt_2$ | 0.46 | C15 |
| $In_{0.98-0.91}Pb_{0.02-0.09}$ | 3.45–4.2 | | $La_{0.28}Pt_{0.72}$ | 0.54 | C15 |
| $InPb$ | 6.65 | | $LaRh_3$ | 2.60 | |
| $InPd$ | 0.7 | B2 | $LaRh_5$ | 1.62 | |
| InSb (quenched from | | | $La_7Rh_3$ | 2.58 | $D10_2$ |
| 170 kbar into liquid $N_2$) | 4.8 | Like A5 | $LaRu_2$ | 1.63 | C15 |
| $InSb$ | 2.1 | | $La_3S_4$ | 6.5 | $D7_3$ |
| $(InSb)_{0.95-0.10}Sn_{0.05-0.90}$ | | | $La_3Se_4$ | 8.6 | $D7_3$ |
| (various heat treatments) | 3.8–5.1 | | $LaSi_2$ | 2.3 | $C_c$ |
| $(InSb)_{0-0.07}Sn_{1-0.93}$ | 3.67–3.74 | | $La_xY_{1-x}$ | 1.7–5.4 | |
| $In_3Sn$ | ~5.5 | | $LaZn$ | 1.04 | B2 |
| $In_xSn_{1-x}$ | 3.4–7.3 | | $LiPb$ | 7.2 | |
| $In_{0.82-1}Te$ | | | $LuOs_2$ | 3.49 | C14 |
| ($n = 0.83$–$1.71 \times 10^{22}$) | 1.02–3.45 | B1 | $Lu_{0.275}Rh_{0.725}$ | 1.27 | C15 |
| $In_{1.000}Te_{1.002}$ | 3.5–3.7 | B1 | $LuRh_5$ | 0.49 | |
| $In_3Te_4$ | | | $LuRu_2$ | 0.86 | C14 |
| ($n = 0.47 \times 10^{22}$) | 1.15–1.25 | Rhombohedral | $Mg_{~0.47}Tl_{~0.53}$ | 2.75 | B2 |
| $In_xTl_{1-x}$ | 2.7–3.374 | | $Mg_2Nb$ | 5.6 | |
| $In_{0.8}Tl_{0.2}$ | 3.223 | | $Mn_xTi_{1-x}$ | 2.3 max. | Mn in $\alpha$-Ti |
| $In_{0.62}Tl_{0.38}$ | 2.760 | | $Mn_xTi_{1-x}$ | 1.1–3.0 | Mn in $\beta$-Ti |
| $In_{0.78-0.69}Tl_{0.22-0.31}$ | 3.18–3.32 | Tetragonal | $MnU_6$ | 2.32 | $D2_c$ |
| $In_{0.69-0.62}Tl_{0.31-0.38}$ | 2.98–3.3 | Cubic, f.c. | $MoN$ | 12 | Hexagonal |
| $Ir_2La$ | 0.48 | C15 | $Mo_2N$ | 5.0 | Cubic, f.c. |
| $Ir_3La$ | 2.32 | $D10_2$ | $Mo_xNb_{1-x}$ | 0.016–9.2 | |
| $Ir_3La_7$ | 2.24 | $D10_2$ | $Mo_3Os$ | 7.2 | A15 |
| $Ir_5La$ | 2.13 | | $Mo_{0.62}Os_{0.38}$ | 5.65 | $D8_b$ |
| $Ir_2Lu$ | 2.47 | C15 | $Mo_3P$ | 5.31 | $DO_e$ |
| $Ir_3Lu$ | 2.89 | C15 | $Mo_{0.5}Pd_{0.5}$ | 3.52 | A3 |
| $IrMo$ | <1.0 | A3 | $Mo_3Re$ | 10.0 | |
| $IrMo_3$ | 8.8 | A15 | $Mo_xRe_{1-x}$ | 1.2–12.2 | |
| $IrMo_3$ | 6.8 | $D8_b$ | $MoRe_3$ | 9.25, 9.89 | A12 |
| $IrNb_3$ | 1.9 | A15 | $Mo_{0.42}Re_{0.58}$ | 6.35 | $D8_b$ |
| $Ir_{0.4}Nb_{0.6}$ | 9.8 | $D8_b$ | $Mo_{0.52}Re_{0.48}$ | 11.1 | |
| $Ir_{0.37}Nb_{0.63}$ | 2.32 | $D8_b$ | $Mo_{0.57}Re_{0.43}$ | 14.0 | |
| $IrNb$ | 7.9 | $D8_b$ | $Mo_{~0.60}Re_{0.395}$ | 10.6 | |
| $Ir_{0.02}Nb_3Rh_{0.98}$ | 2.43 | A15 | $MoRh$ | 1.97 | A3 |
| $Ir_{0.05}Nb_3Rh_{0.95}$ | 2.38 | A15 | $Mo_xRh_{1-x}$ | 1.5–8.2 | Cubic, b.c. |
| | | | $MoRu$ | 9.5–10.5 | A3 |

## Table 8—23 (continued)
## SELECTED SUPERCONDUCTIVE COMPOUNDS AND ALLOYS

| Substance | $T_c$, K | Crystal structure type[††] | Substance | $T_c$, K | Crystal structure type[††] |
|---|---|---|---|---|---|
| $Mo_{0.61}Ru_{0.39}$ | 7.18 | $D8_b$ | $Nb_2SnTa$ | 16.4 | A15 |
| $Mo_{0.2}Ru_{0.8}$ | 1.66 | A3 | $Nb_{2.5}SnTa_{0.5}$ | 17.6 | A15 |
| $Mo_3Sb_4$ | 2.1 | | $Nb_{2.75}SnTa_{0.25}$ | 17.8 | A15 |
| $Mo_3Si$ | 1.30 | A15 | $Nb_{3x}SnTa_{3(1-x)}$ | 6.0–18.0 | |
| $MoSi_{0.7}$ | 1.34 | | NbSnTaV | 6.2 | A15 |
| $Mo_xSiV_{3-x}$ | 4.54–16.0 | A15 | $Nb_2SnTa_{0.5}V_{0.5}$ | 12.2 | A15 |
| $Mo_xTc_{1-x}$ | 10.8–15.8 | | $NbSnV_2$ | 5.5 | A15 |
| $Mo_{0.16}Ti_{0.84}$ | 4.18, 4.25 | | $Nb_2SnV$ | 9.8 | A15 |
| $Mo_{0.913}Ti_{0.087}$ | 2.95 | | $Nb_{2.5}SnV_{0.5}$ | 14.2 | A15 |
| $Mo_{0.04}Ti_{0.96}$ | 2.0 | Cubic | $Nb_xTa_{1-x}$ | 4.4–9.2 | A2 |
| $Mo_{0.025}Ti_{0.975}$ | 1.8 | | $NbTc_3$ | 10.5 | A12 |
| $Mo_xU_{1-x}$ | 0.7–2.1 | | $Nb_xTi_{1-x}$ | 0.6–9.8 | |
| $Mo_xV_{1-x}$ | 0–~5.3 | | $Nb_{0.6}Ti_{0.4}$ | 9.8 | |
| $Mo_2Zr$ | 4.27–4.75 | C15 | $Nb_xU_{1-x}$ | 1.95 max. | |
| NNb (whiskers) | 10–14.5 | | $Nb_{0.88}V_{0.12}$ | 5.7 | A2 |
| NNb (diffusion wires) | 16.10 | | $Nb_{0.75}Zr_{0.25}$ | 10.8 | |
| NNb (film) | 6–9 | B1 | $Nb_{0.66}Zr_{0.33}$ | 10.8 | |
| $N_{0.988}Nb$ | 14.9 | B1 | $Ni_{0.3}Th_{0.7}$ | 1.98 | $D10_2$ |
| $N_{0.824-0.988}Nb$ | 14.4–15.3 | B1 | $NiZr_2$ | 1.52 | |
| $N_{0.70-0.795}Nb$ | 11.3–12.9 | Cubic and tetragonal | $Ni_{0.1}Zr_{0.9}$ | 1.5 | A3 |
| | | | $O_3Rb_{0.27-0.29}W$ | 1.98 | Hexagonal |
| $NNb_xO_y$ | 13.5–17.0 | B1 | $O_3SrTi$ | | |
| $NNb_xO_y$ | 6.0–11 | | $(n = 1.7–12.0 \times 10^{19})$ | 0.12–0.37 | |
| $N_{100-42\ w/o}Nb_{0-58\ w/o}Ti$[†] | 15–16.8 | | $O_3SrTi$ | | |
| $N_{100-75\ w/o}Nb_{0-25\ w/o}Zr$[†] | 12.5–16.35 | | $(n = 10^{18}–10^{21})$ | 0.05–0.47 | |
| $NNb_xZr_{1-x}$ | 9.8–13.8 | B1 | $O_3SrTi$ | | |
| $N_{0.93}Nb_{0.85}Zr_{0.15}$ | 13.8 | B1 | $(n = ~10^{20})$ | 0.47 | |
| $N_xO_yTi_z$ | 2.9–5.6 | Cubic | OTi | 0.58 | |
| $N_xO_yV_z$ | 5.8–8.2 | Cubic | $O_3Sr_{0.08}W$ | 2–4 | Hexagonal |
| $N_{0.34}Re$ | 4–5 | Cubic, f.c. | $O_3Tl_{0.30}W$ | 2.0–2.14 | Hexagonal |
| NTa | 12–14 | B1 | $OV_3Zr_3$ | 7.5 | $E9_3$ |
| | (extrapolated) | | $OW_3$ (film) | 3.35, 1.1 | A15 |
| NTa (film) | 4.84 | B1 | OsReY | 2.0 | C14 |
| $N_{0.6-0.987}Ti$ | <1.17–5.8 | B1 | $Os_2Sc$ | 4.6 | C14 |
| $N_{0.82-0.99}V$ | 2.9–7.9 | B1 | OsTa | 1.95 | A12 |
| NZr | 9.8 | B1 | $Os_3Th_7$ | 1.51 | $D10_2$ |
| $N_{0.906-0.984}Zr$ | 3.0–9.5 | B1 | $Os_xW_{1-x}$ | 0.9–4.1 | |
| $Na_{0.28-0.35}O_3W$ | 0.56 | Tetragonal | $OsW_3$ | ~3 | |
| $Na_{0.28}Pb_{0.72}$ | 7.2 | | $Os_2Y$ | 4.7 | C14 |
| NbO | 1.25 | | $Os_2Zr$ | 3.0 | C14 |
| $NbOs_2$ | 2.52 | A12 | $Os_xZr_{1-x}$ | 1.50–5.6 | |
| $Nb_3Os$ | 1.05 | A15 | PPb | 7.8 | |
| $Nb_{0.6}Os_{0.4}$ | 1.89, 1.78 | $D8_b$ | $PPd_{3.0-3.2}$ | <0.35–0.7 | $DO_{11}$ |
| $Nb_3Os_{0.02-0.10}Rh_{0.98-0.90}$ | 2.42–2.30 | A15 | $P_3Pd_7$ (high temperature) | 1.0 | Rhombohedral |
| $Nb_{0.6}Pd_{0.4}$ | 1.60 | $D8_f$ plus cubic | $P_3Pd_7$ (low temp.) | 0.70 | Complex |
| $Nb_3Pd_{0.02-0.10}Rh_{0.98-0.90}$ | 2.49–2.55 | A15 | PRh | 1.22 | |
| $Nb_{0.62}Pt_{0.38}$ | 4.21 | $D8_b$ | $PRh_2$ | 1.3 | C1 |
| $Nb_3Pt$ | 10.9 | A15 | $PW_3$ | 2.26 | $DO_e$ |
| $Nb_5Pt_3$ | 3.73 | $D8_b$ | $Pb_2Pd$ | 2.95 | C16 |
| $Nb_3Pt_{0.02-0.98}Rh_{0.98-0.02}$ | 2.52–9.6 | A15 | $Pb_4Pt$ | 2.80 | Related to C16 |
| $Nb_{0.38-0.18}Re_{0.62-0.82}$ | 2.43–9.70 | A12 | $Pb_2Rh$ | 2.66 | C16 |
| $Nb_3Rh$ | 2.64 | A15 | PbSb | 6.6 | |
| $Nb_{0.60}Rh_{0.40}$ | 4.21 | $D8_b$ plus other | PbTe (plus 0.1 w/o Pb)[†] | 5.19 | |
| $Nb_3Rh_{0.98-0.90}Ru_{0.02-0.10}$ | 2.42–2.44 | A15 | PbTe (plus 0.1 w/o Tl)[†] | 5.24–5.27 | |
| $Nb_xRu_{1-x}$ | 1.2–4.8 | | $PbTl_{0.27}$ | 6.43 | |
| $NbS_2$ | 6.1–6.3 | Hexagonal, $NbSe_2$ type | $PbTl_{0.17}$ | 6.73 | |
| | | | $PbTl_{0.12}$ | 6.88 | |
| $NbS_2$ | 5.0–5.5 | Hexagonal, three-layer type | $PbTl_{0.075}$ | 6.98 | |
| | | | $PbTl_{0.04}$ | 7.06 | |
| | | | $Pb_{1-0.26}Tl_{0-0.74}$ | 7.20–3.68 | |
| $Nb_3Sb_{0-0.7}Sn_{1-0.3}$ | 6.8–18 | A15 | $PbTl_2$ | 3.75–4.1 | |
| $NbSe_2$ | 5.15–5.62 | Hexagonal, $NbS_2$ type | $Pb_3Zr_5$ | 4.60 | $D8_8$ |
| | | | $PbZr_3$ | 0.76 | A15 |
| $Nb_{1-1.05}Se_2$ | 2.2–7.0 | Hexagonal, $NbS_2$ type | $Pd_{0.9}Pt_{0.1}Te_2$ | 1.65 | C6 |
| | | | $Pd_{0.05}Ru_{0.05}Zr_{0.9}$ | ~9 | |
| $Nb_3Si$ | 1.5 | $Ll_2$ | $Pd_{2.2}S$ (quenched) | 1.63 | Cubic |
| $Nb_3SiSnV_3$ | 4.0 | | $PdSb_2$ | 1.25 | C2 |
| $Nb_3Sn$ | 18.05 | A15 | PdSb | 1.50 | $B8_1$ |
| $Nb_{0.8}Sn_{0.2}$ | 18.18, 18.5 | A15 | PdSbSe | 1.0 | C2 |
| $Nb_xSn_{1-x}$ (film) | 2.6–18.5 | | PdSbTe | 1.2 | C2 |
| $NbSn_2$ | 2.60 | Orthorhombic | $Pd_4Se$ | 0.42 | Tetragonal |
| $Nb_3Sn_2$ | 16.6 | Tetragonal | $Pd_{6-7}Se$ | 0.66 | Like $Pd_4Te$ |
| $NbSnTa_2$ | 10.8 | A15 | $Pd_{2.8}Se$ | 2.3 | |
| | | | $Pd_xSe_{1-x}$ | 2.5 max. | |

[†] w/o denotes weight percent.

**Table 8—23 (continued)**
## SELECTED SUPERCONDUCTIVE COMPOUNDS AND ALLOYS

| Substance | $T_c$, K | Crystal structure type†† | Substance | $T_c$, K | Crystal structure type†† |
|---|---|---|---|---|---|
| PdSi | 0.93 | B3l | $Rh_{0-0.45}Zr_{1-0.55}$ | 2.1–10.8 | |
| PdSn | 0.41 | B3l | $Rh_{0.1}Zr_{0.9}$ | 9.0 | Hexagonal, c.p. |
| $PdSn_2$ | 3.34 | | $Ru_2Sc$ | 1.67 | C14 |
| $Pd_2Sn$ | 0.41 | C37 | $Ru_2Th$ | 3.56 | C15 |
| $Pd_3Sn_2$ | 0.47–0.64 | $B8_2$ | RuTi | 1.07 | B2 |
| PdTe | 2.3, 3.85 | $B8_1$ | $Ru_{0.05}Ti_{0.95}$ | 2.5 | |
| $PdTe_{1.02-1.08}$ | 2.56–1.88 | $B8_1$ | $Ru_{0.1}Ti_{0.9}$ | 3.5 | |
| $PdTe_2$ | 1.69 | C6 | $Ru_xTi_{0.6}V_y$ | 6.6 max. | |
| $PdTe_{2.1}$ | 1.89 | C6 | $Ru_{0.45}V_{0.55}$ | 4.0 | B2 |
| $PdTe_{2.3}$ | 1.85 | C6 | RuW | 7.5 | A3 |
| $Pd_{1.1}Te$ | 4.07 | $B8_1$ | $Ru_2Y$ | 1.52 | C14 |
| $PdTh_2$ | 0.85 | C16 | $Ru_2Zr$ | 1.84 | C14 |
| $Pd_{0.1}Zr_{0.9}$ | 7.5 | A3 | $Ru_{0.1}Zr_{0.9}$ | 5.7 | A3 |
| PtSb | 2.1 | $B8_1$ | SbSn | 1.30–1.42, | Bl or distorted |
| PtSi | 0.88 | B3l | | 1.42–2.37 | Bl |
| PtSn | 0.37 | $B8_1$ | $SbTi_3$ | 5.8 | A15 |
| PtTe | 0.59 | Orthorhombic | $Sb_2Tl_7$ | 5.2 | |
| PtTh | 0.44 | $B_f$ | $Sb_{0.01-0.03}V_{0.99-0.97}$ | 3.76–2.63 | A2 |
| $Pt_3Th_7$ | 0.98 | $D10_2$ | $SbV_3$ | 0.80 | A15 |
| $Pt_5Th$ | 3.13 | | $Si_2Th$ | 3.2 | $C_c$, α-phase |
| $PtTi_3$ | 0.58 | A15 | $Si_2Th$ | 2.4 | C32, β-phase |
| $Pt_{0.02}U_{0.98}$ | 0.87 | β-phase | $SiV_3$ | 17.1 | A15 |
| $PtV_{2.5}$ | 1.36 | A15 | $Si_{0.9}V_3Al_{0.1}$ | 14.05 | A15 |
| $PtV_3$ | 2.87–3.20 | A15 | $Si_{0.9}V_3B_{0.1}$ | 15.8 | A15 |
| $PtV_{3.5}$ | 1.26 | A15 | $Si_{0.9}V_3C_{0.1}$ | 16.4 | A15 |
| $Pt_{0.5}W_{0.5}$ | 1.45 | A1 | $SiV_{2.7}Cr_{0.3}$ | 11.3 | A15 |
| $Pt_xW_{1-x}$ | 0.4–2.7 | | $Si_{0.9}V_3Ge_{0.1}$ | 14.0 | A15 |
| $Pt_2Y_3$ | 0.90 | | $SiV_{2.7}Mo_{0.3}$ | 11.7 | A15 |
| $Pt_2Y$ | 1.57, 1.70 | C15 | $SiV_{2.7}Nb_{0.3}$ | 12.8 | A15 |
| $Pt_3Y_7$ | 0.82 | $D10_2$ | $SiV_{2.7}Ru_{0.3}$ | 2.9 | A15 |
| PtZr | 3.0 | A3 | $SiV_{2.7}Ti_{0.3}$ | 10.9 | A15 |
| $Re_{0.64}Ta_{0.36}$ | 1.46 | A12 | $SiV_{2.7}Zr_{0.3}$ | 13.2 | A15 |
| $Re_{24}Ti_5$ | 6.60 | A12 | $Si_2W_3$ | 2.8, 2.84 | |
| $Re_xTi_{1-x}$ | 6.6 max. | | $Sn_{0.174-0.104}Ta_{0.826-0.896}$ | 6.5–<4.2 | A15 |
| $Re_{0.76}V_{0.24}$ | 4.52 | $D8_b$ | $SnTa_3$ | 8.35 | A15, highly ordered |
| $Re_{0.92}V_{0.08}$ | 6.8 | A3 | | | |
| $Re_{0.6}W_{0.4}$ | 6.0 | | $SnTa_3$ | 6.2 | A15, partially ordered |
| $Re_{0.5}W_{0.5}$ | 5.12 | $D8_b$ | $SnTaV_2$ | 2.8 | A15 |
| $Re_2Y$ | 1.83 | C14 | $SnTa_2V$ | 3.7 | A15 |
| $Re_2Zr$ | 5.9 | C14 | $Sn_xTe_{1-x}$ | | |
| $Re_6Zr$ | 7.40 | A12 | $(n = 10.5-20 \times 10^{20})$ | 0.07–0.22 | Bl |
| $Rh_{17}S_{15}$ | 5.8 | Cubic | $Sn_xTl_{1-x}$ | 2.37–5.2 | |
| $Rh_{~0.24}Sc_{~0.76}$ | 0.88, 0.92 | | $SnV_3$ | 3.8 | A15 |
| $Rh_xSe_{1-x}$ | 6.0 max. | | $Sn_{0.02-0.057}V_{0.98-0.943}$ | 2.87–~1.6 | A2 |
| $Rh_2Sr$ | 6.2 | C15 | $Ta_{0.025}Ti_{0.975}$ | 1.3 | Hexagonal |
| $Rh_{0.4}Ta_{0.6}$ | 2.35 | $D8_b$ | $Ta_{0.05}Ti_{0.95}$ | 2.9 | Hexagonal |
| $RhTe_2$ | 1.51 | C2 | $Ta_{0.05-0.75}V_{0.095-0.25}$ | 4.30–2.65 | A2 |
| $Rh_{0.67}Te_{0.33}$ | 0.49 | | $Ta_{0.8-1}W_{0.2-0}$ | 1.2–4.4 | A2 |
| $Rh_xTe_{1-x}$ | 1.51 max. | | $Tc_{0.1-0.4}W_{0.9-0.6}$ | 1.25–7.18 | Cubic |
| RhTh | 0.36 | $B_f$ | $Tc_{0.50}W_{0.50}$ | 7.52 | α plus σ |
| $Rh_3Th_7$ | 2.15 | $D10_2$ | $Tc_{0.60}W_{0.40}$ | 7.88 | σ plus α |
| $Rh_5Th$ | 1.07 | | $Tc_6Zr$ | 9.7 | A12 |
| $Rh_xTi_{1-x}$ | 2.25–3.95 | | $Th_{0-0.55}Y_{1-0.45}$ | 1.2–1.8 | |
| $Rh_{0.02}U_{0.98}$ | 0.96 | | $Ti_{0.70}V_{0.30}$ | 6.14 | Cubic |
| $RhV_3$ | 0.38 | A15 | $Ti_xV_{1-x}$ | 0.2–7.5 | |
| RhW | ~3.4 | A3 | $Ti_{0.5}Zr_{0.5}$ (annealed) | 1.23 | |
| $RhY_3$ | 0.65 | | $Ti_{0.5}Zr_{0.5}$ (quenched) | 2.0 | |
| $Rh_2Y_3$ | 1.48 | | $V_2Zr$ | 8.80 | C15 |
| $Rh_3Y$ | 1.07 | C15 | $V_{0.26}Zr_{0.74}$ | ≈5.9 | |
| $Rh_5Y$ | 0.56 | C16 | $W_2Zr$ | 2.16 | C15 |
| $RhZr_2$ | 10.8 | C16 | | | |
| $Rh_{0.005}Zr$ (annealed) | 5.8 | | | | |

From Weast, R. C., Ed., *Handbook of Chemistry and Physics*, 69th ed., CRC Press, Boca Raton, Fla., 1988, E-96.

## Table 8—24
## CRITICAL FIELD DATA

| Substance | $H_o$, oersteds | Substance | $H_o$, oersteds |
|---|---|---|---|
| $Ag_2F$ | 2.5 | InSb | 1,100 |
| $Ag_7NO_{11}$ | 57 | $In_xTl_{1-x}$ | 252–284 |
| $Al_2CMo_3$ | 1,700 | $In_{0.8}Tl_{0.2}$ | 252 |
| $BaBi_3$ | 740 | $Mg_{\sim0.47}Tl_{\sim0.53}$ | 220 |
| $Bi_2Pt$ | 10 | $Mo_{0.16}Ti_{0.84}$ | <985 |
| $Bi_3Sr$ | 530 | $NbSn_2$ | 620 |
| $Bi_5Tl_3$ | >400 | $PbTl_{0.27}$ | 756 |
| CdSn | >266 | $PbTl_{0.17}$ | 796 |
| $CoSi_2$ | 105 | $PbTl_{0.12}$ | 849 |
| $Cr_{0.1}Ti_{0.3}V_{0.6}$ | 1,360 | $PbTl_{0.075}$ | 880 |
| $In_{1-0.86}Mg_{0-0.14}$ | 272.4–259.2 | $PbTl_{0.04}$ | 864 |

From Weast, R. C., Ed., *Handbook of Chemistry and Physics,* 69th ed., CRC Press, Boca Raton, Fla., 1988, E-101.

## Table 8—25
## KEY TO CRYSTAL STRUCTURE TYPES

| "Struck-turbericht" type* | Example | Class | "Struck-turbericht" type* | Example | Class |
|---|---|---|---|---|---|
| A1 | Cu | Cubic, f.c. | $C15_b$ | $AuBe_5$ | Cubic |
| A2 | W | Cubic, b.c. | C16 | $CuAl_2$ | Tetragonal, b.c. |
| A3 | Mg | Hexagonal, close packed | C18 | $FeS_2$ | Orthorhombic |
| A4 | Diamond | Cubic, f.c. | C22 | $Fe_2P$ | Trigonal |
| A5 | White Sn | Tetragonal, b.c. | C23 | $PbCl_2$ | Orthorhombic |
| A6 | In | Tetragonal, b.c. (f.c. cell usually used) | C32 | $AlB_2$ | Hexagonal |
| | | | C36 | $MgNi_2$ | Hexagonal |
| A7 | As | Rhombohedral | C37 | $Co_2Si$ | Orthorhombic |
| A8 | Se | Trigonal | C49 | $ZrSi_2$ | Orthorhombic |
| A10 | Hg | Rhombohedral | C54 | $TiSi_2$ | Orthorhombic |
| A12 | $\alpha$-Mn | Cubic, b.c. | $C_c$ | $Si_2Th$ | Tetragonal, b.c. |
| A13 | $\beta$-Mn | Cubic | $DO_3$ | $BiF_3$ | Cubic, f.c. |
| A15 | "$\beta$-W" ($W_3O$) | Cubic | $DO_{11}$ | $Fe_3C$ | Orthorhombic |
| B1 | NaCl | Cubic, f.c. | $DO_{18}$ | $Na_3As$ | Hexagonal |
| B2 | CsCl | Cubic | $DO_{19}$ | $Ni_3Sn$ | Hexagonal |
| B3 | ZnS | Cubic | $DO_{20}$ | $NiAl_3$ | Orthorhombic |
| B4 | ZnS | Hexagonal | $DO_{22}$ | $TiAl_3$ | Tetragonal |
| $B8_1$ | NiAs | Hexagonal | $DO_e$ | $Ni_3P$ | Tetragonal, b.c. |
| $B8_2$ | $Ni_2In$ | Hexagonal | $D1_3$ | $Al_4Ba$ | Tetragonal, b.c. |
| B10 | PbO | Tetragonal | $D1_c$ | $PtSn_4$ | Orthorhombic |
| B11 | $\gamma$-CuTi | Tetragonal | $D2_1$ | $CaB_6$ | Cubic |
| B17 | PtS | Tetragonal | $D2_c$ | $MnU_6$ | Tetragonal, b.c. |
| B18 | CuS | Hexagonal | $D2_d$ | $CaZn_5$ | Hexagonal |
| B20 | FeSi | Cubic | $D5_2$ | $La_2O_3$ | Trigonal |
| B27 | FeB | Orthorhombic | $D5_8$ | $Sb_2S_3$ | Orthorhombic |
| B31 | MnP | Orthorhombic | $D7_3$ | $Th_3P_4$ | Cubic, b.c. |
| B32 | NaTl | Cubic, f.c. | $D7_b$ | $Ta_3B_4$ | Orthorhombic |
| B34 | PdS | Tetragonal | $D8_1$ | $Fe_3Zn_{10}$ | Cubic, b.c. |
| $B_f$ | $\delta$-CrB | Orthorhombic | $D8_2$ | $Cu_5Zn_8$ | Cubic, b.c. |
| $B_g$ | MoB | Tetragonal, b.c. | $D8_3$ | $Cu_9Al_4$ | Cubic |
| $B_h$ | WC | Hexagonal | $D8_8$ | $Mn_5Si_3$ | Hexagonal |
| $B_i$ | $\gamma$-MoC | Hexagonal | $D8_b$ | CrFe | Tetragonal |
| C1 | $CaF_2$ | Cubic, f.c. | $D8_i$ | $Mo_2B_5$ | Rhombohedral |
| $C1_b$ | MgAgAs | Cubic, f.c. | $D10_2$ | $Fe_3Th_7$ | Hexagonal |
| C2 | $FeS_2$ | Cubic | $E2_1$ | $CaTiO_3$ | Cubic |
| C6 | $CdI_2$ | Trigonal | $E9_3$ | $Fe_3W_3C$ | Cubic, f.c. |
| C11b | $MoSi_2$ | Tetragonal, b.c. | $L1_0$ | CuAu | Tetragonal |
| C12 | $CaSi_2$ | Rhombohedral | $L1_2$ | $Cu_3Au$ | Cubic |
| C14 | $MgZn_2$ | Hexagonal | $L'_{2b}$ | $ThH_2$ | Tetragonal, b.c. |
| C15 | $Cu_2Mg$ | Cubic, f.c. | $L'_3$ | $Fe_2N$ | Hexagonal |

*See "Handbook of Lattice Spacing and Structures of Metals", W.B. Pearson, Vol. I, Pergamon Press, 1958, p. 79. and Vol. II, Pergamon Press, 1967.

From Weast, R. C., Ed., *Handbook of Chemistry and Physics,* 69th ed., CRC Press, Boca Raton, Fla., 1988, E-102.

## Table 8—26
# HIGH CRITICAL MAGNETIC-FIELD SUPERCONDUCTIVE COMPOUNDS AND ALLOYS

### With Critical Temperatures, $H_{c1}$, $H_{c2}$, $H_{c3}$, and the Temperature of Field Observations, $T_{obs}$

| Substance | $T_c$, K | $H_{c1}$, kg | $H_{c2}$, kg | $H_{c3}$, kg | $T_{obs}$, K† |
|---|---|---|---|---|---|
| $Al_2CMo_3$ | 9.8–10.2 | 0.091 | 156 | | 1.2 |
| $AlNb_3$ | | 0.375 | | | |
| $Ba_xO_3Sr_{1-x}Ti$ | <0.1–0.55 | 0.0039 max. | | | |
| $Bi_{0.5}Cd_{0.1}Pb_{0.27}Sn_{0.13}$ | | | >24 | | 3.06 |
| $Bi_xPb_{1-x}$ | 7.35–8.4 | 0.122 max. | ~30 max. | | 4.2 |
| $Bi_{0.56}Pb_{0.44}$ | 8.8 | | 15 | | 4.2 |
| $Bi_{7.5 \, w/o}Pb_{92.5 \, w/o}‡$ | | | 2.32 | | |
| $Bi_{0.099}Pb_{0.901}$ | | 0.29 | 2.8 | | |
| $Bi_{0.02}Pb_{0.98}$ | | 0.46 | 0.73 | | |
| $Bi_{0.53}Pb_{0.32}Sn_{0.16}$ | | | >25 | | 3.06 |
| $Bi_{1-0.93}Sn_{0-0.07}$ | | | 0–0.032 | | 3.7 |
| $Bi_5Tl_3$ | 6.4 | | >5.56 | | 3.35 |
| $C_8K$ (excess K) | 0.55 | | 0.160 (H⊥c) | | 0.32 |
| | | | 0.730 (H‖c) | | 0.32 |
| $C_8K$ | 0.39 | | 0.025 (H⊥c) | | 0.32 |
| | | | 0.250 (H‖c) | | 0.32 |
| $C_{0.44}Mo_{0.56}$ | 12.5–13.5 | 0.087 | 98.5 | | 1.2 |
| $CNb$ | 8–10 | 0.12 | 16.9 | | 4.2 |
| $CNb_{0.4}Ta_{0.6}$ | 10–13.6 | 0.19 | 14.1 | | 1.2 |
| $CTa$ | 9–11.4 | 0.22 | 4.6 | | 1.2 |
| $Ca_xO_3Sr_{1-x}Ti$ | <0.1–0.55 | 0.002–0.004 | | | |
| $Cd_{0.1}Hg_{0.9}$ | | 0.23 | 0.34 | | 2.04 |
| (by weight) | | | | | |
| $Cd_{0.05}Hg_{0.95}$ | | 0.28 | 0.31 | | 2.16 |
| $Cr_{0.10}Ti_{0.30}V_{0.60}$ | 5.6 | 0.071 | 84.4 | | 0 |
| $GaN$ | 5.85 | 0.725 | | | 4.2 |
| $Ga_xNb_{1-x}$ | | | >28 | | 4.2 |
| $GaSb$ (annealed) | 4.24 | | 2.64 | | 3.5 |
| $GaV_{1.95}$ | 5.3 | | 73*** | | 0 |
| $GaV_{2.1-3.5}$ | 6.3–14.45 | | 230–300** | | 0 |
| $GaV_3$ | | 0.4 | 350*** | | 0 |
| | | | 500** | | |
| $GaV_{4.5}$ | 9 15 | | 121* | | 0 |
| $Hf_xNb_y$ | | | >52–>102 | | 1.2 |
| $Hf_xTa_y$ | | | >28–>86 | | 1.2 |
| $Hg_{0.05}Pb_{0.95}$ | | 0.235 | 2.3 | | |
| $Hg_{0.101}Pb_{0.899}$ | | 0.23 | 4.3 | | 4.2 |
| $Hg_{0.15}Pb_{0.85}$ | ~6.75 | | >13 | | 2.93 |
| $In_{0.98}Pb_{0.02}$ | 3.45 | 0.1 | | 0.12 | 2.76 |
| $In_{0.96}Pb_{0.04}$ | 3.68 | 0.1 | 0.12 | 0.25 | 2.94 |
| $In_{0.94}Pb_{0.06}$ | 3.90 | 0.095 | 0.18 | 0.35 | 3.12 |
| $In_{0.913}Pb_{0.087}$ | 4.2 | ~0.17 | 0.55 | 2.65 | |
| $In_{0.316}Pb_{0.684}$ | | 0.155 | 3.7 | | 4.2 |
| $In_{0.17}Pb_{0.83}$ | | | 2.8 | 5.5 | 4.2 |
| $In_{1.000}Te_{1.002}$ | 3.5–3.7 | | 1.2* | | 0 |
| $In_{0.95}Tl_{0.05}$ | | 0.263 | 0.263 | | 3.3 |
| $In_{0.90}Tl_{0.10}$ | | 0.257 | 0.257 | | 3.25 |
| $In_{0.83}Tl_{0.17}$ | | 0.242 | 0.39 | | 3.21 |
| $In_{0.75}Tl_{0.25}$ | | 0.216 | 0.50 | | 3.16 |
| $LaN$ | 1.35 | 0.45 | | | 0.76 |
| $La_3S_4$ | 6.5 | ≈0.15 | >25 | | 1.3 |
| $La_3Se_4$ | 8.6 | ≈0.2 | >25 | | 1.25 |
| $Mo_{0.52}Re_{0.48}$ | 11.1 | | 14–21 | 22–33 | 4.2 |
| | | | 18–28 | 37–43 | 1.3 |
| $Mo_{0.6\pm0.05}Re_{0.395}$ | 10.6 | | 14–20 | 20–37 | 4.2 |
| | | | 19–26 | 26–37 | 1.3 |
| $Mo_{\sim0.5}Tc_{\sim0.5}$ | | | ~75* | | 0 |
| $Mo_{0.16}Ti_{0.84}$ | 4.18 | 0.028 | 98.7* | | 0 |
| | | | 36–38 | | 3.0 |
| $Mo_{0.913}Ti_{0.087}$ | 2.95 | 0.060 | ~15 | | 4.2 |
| $Mo_{0.1-0.3}U_{0.9-0.7}$ | 1.85–2.06 | | >25 | | |
| $Mo_{0.17}Zr_{0.83}$ | | | ~30 | | |
| $N_{(12.8 \, w/o)}Nb$ | 15.2 | | >9.5 | | 13.2 |
| $NNb$ (wires) | 16.1 | | 153* | | 0 |
| | | | 132 | | 4.2 |
| | | | 95 | | 8 |
| | | | 53 | | 12 |
| $NNb_xO_{1-x}$ | 13.5–17.0 | | ~38 | | |
| $NNb_xZr_{1-x}$ | 9.8–13.8 | | 4–>130 | | 4.2 |
| $N_{0.93}Nb_{0.85}Zr_{0.15}$ | 13.8 | | >130 | | 4.2 |
| $Na_{0.086}Pb_{0.914}$ | | 0.19 | 6.0 | | |
| $Na_{0.016}Pb_{0.984}$ | | 0.28 | 2.05 | | |
| $Nb$ | 9.15 | | 2.020 | | 1.4 |
| | | | 1.710 | | 4.2 |
| $Nb$ | | 0.4–1.1 | 3–5.5 | | 4.2 |
| $Nb$ (unstrained) | | 1.1–1.8 | 3.40 | 6–9.1 | 4.2 |
| $Nb$ (strained) | | 1.25–1.92 | 3.44 | 6.0–8.7 | 4.2 |
| $Nb$ (cold-drawn wire) | | 2.48 | 4.10 | ≈10 | 4.2 |
| $Nb$ (film) | | | >25 | | 4.2 |
| $NbSc$ | | | >30 | | |

## Table 8—26 (continued)
## HIGH  CRITICAL  MAGNETIC-FIELD  SUPERCONDUCTIVE
## COMPOUNDS  AND  ALLOYS

| Substance | $T_c$, K | $H_{c1}$, kg | $H_{c2}$, kg | $H_{c3}$, kg | $T_{obs}$, K† |
|---|---|---|---|---|---|
| $Nb_3Sn$ | | 0.170 | 221 | | 4.2 |
| | | | 70 | | 14.15 |
| | | | 54 | | 15 |
| | | | 34 | | 16 |
| | | | 17 | | 17 |
| $Nb_{0.1}Ta_{0.9}$ | | 0.084 | 0.154 | | 4.195 |
| $Nb_{0.2}Ta_{0.8}$ | | | 10 | | 4.2 |
| $Nb_{0.65-0.73}Ta_{0.02-0.10}Zr_{0.25}$ | | | >70->90 | | 4.2 |
| $Nb_xTi_{1-x}$ | | | 148 max. | | 1.2 |
| | | | 120 max. | | 4.2 |
| $Nb_{0.222}U_{0.778}$ | | 1.98 | 23 | | 1.2 |
| $Nb_xZr_{1-x}$ | | | 127 max. | | 1.2 |
| | | | 94 max. | | 4.2 |
| $O_3SrTi$ | 0.43 | .0049* | .504* | | 0 |
| $O_3SrTi$ | 0.33 | .00195* | .420* | | 0 |
| $PbSb_{1\ w/o}$ (quenched) | | | >1.5 | | 4.2 |
| $PbSb_{1\ w/o}$ (annealed) | | | >0.7 | | 4.2 |
| $PbSb_{2.8\ w/o}$ (quenched) | | | >2.3 | | 4.2 |
| $PbSb_{2.8\ w/o}$ (annealed) | | | >0.7 | | 4.2 |
| $Pb_{0.871}Sn_{0.129}$ | | 0.45 | 1.1 | | |
| $Pb_{0.965}Sn_{0.035}$ | | 0.53 | 0.56 | | |
| $Pb_{1-0.26}Tl_{0-0.74}$ | 7.20–3.68 | | 2–6.9* | | 0 |
| $PbTl_{0.17}$ | 6.73 | | 4.5* | | 0 |
| $Re_{0.26}W_{0.74}$ | | | >30 | | |
| $Sb_{0.93}Sn_{0.07}$ | | | 0.12 | | 3.7 |
| $SiV_3$ | 17.0 | 0.55 | 156*** | | |
| $Sn_xTe_{1-x}$ | | 0.00043–0.00236 | 0.005–0.0775 | | 0.012–0.079 |
| Ta (99.95%) | | 0.425 | 1.850 | | 1.3 |
| | | 0.325 | 1.425 | | 2.27 |
| | | 0.275 | 1.175 | | 2.66 |
| | | 0.090 | 0.375 | | 3.72 |
| $Ta_{0.5}Nb_{0.5}$ | | | 3.55 | | 4.2 |
| $Ta_{0.65-0}Ti_{0.35-1}$ | 4.4–7.8 | | >14–138 | | 1.2 |
| $Ta_{0.5}Ti_{0.5}$ | | | 138 | | 1.2 |
| Te | ~3.3 | 0.25* | | | 0 |
| $Tc_xW_{1-x}$ | 5.75–7.88 | | 8–44 | | 4.2 |
| Ti | | | | 2.7 | 4.2 |
| $Ti_{0.75}V_{0.25}$ | 5.3 | 0.029* | 199* | | 0 |
| $Ti_{0.775}V_{0.225}$ | 4.7 | 0.024* | 172* | | 0 |
| $Ti_{0.615}V_{0.385}$ | 7.07 | 0.050 | ~34 | | 4.2 |
| $Ti_{0.516}V_{0.484}$ | 7.20 | 0.062 | ~28 | | 4.2 |
| $Ti_{0.415}V_{0.585}$ | 7.49 | 0.078 | ~25 | | 4.2 |
| $Ti_{0.12}V_{0.88}$ | | | 17.3 | 28.1 | 4.2 |
| $Ti_{0.09}V_{0.91}$ | | | 14.3 | 16.4 | 4.2 |
| $Ti_{0.06}V_{0.94}$ | | | 8.2 | 12.7 | 4.2 |
| $Ti_{0.03}V_{0.97}$ | | | 3.8 | 6.8 | 4.2 |
| $Ti_xV_{1-x}$ | | | 108 max. | | 1.2 |
| V | 5.31 | ~0.8 | ~3.4 | | 1.79 |
| | | ~0.75 | ~3.15 | | 2 |
| | | ~0.45 | ~2.2 | | 3 |
| | | ~0.30 | ~1.2 | | 4 |
| $V_{0.26}Zr_{0.74}$ | ≈5.9 | 0 238 | | | 1.05 |
| | | 0.227 | | | 1.78 |
| | | 0.185 | | | 3.04 |
| | | 0.165 | | | 3.5 |
| W (film) | 1.7–4.1 | | >34 | | 1 |

†Temperature of critical field measurement.

‡w/o denotes weight percent.

*Extrapolated.

**Linear extrapolation.

***Parabolic extrapolation.

From Weast, R. C., Ed., *Handbook of Chemistry and Physics,* 69th ed., CRC Press, Boca Raton, Fla., 1988, E-102.

Section 9

# **Graphitic materials**

# Section 9

# GRAPHITIC MATERIALS

## PROPERTIES OF CARBON AND GRAPHITE

Manufactured carbon and graphite should not be viewed as single specific materials, but rather as families of materials. Each member of the family is essentially pure carbon, but each varies from the other in such characteristics as orientation of the crystallites, the size and number of pore spaces, grain size, degree of crystallization, and apparent density.

In the carbon industry the term *carbon* is used to refer to materials in which the small crystallites have low orientation. The term *graphite* is used to refer to materiel that has a highly ordered structure. Specific grades, having controlled characteristics, are produced by selecting raw materials and by varying processing techniques. A broad familiarity

with the properties and characteristics of the various grades is important when selecting materials for a specific application.

Tables 9—4 through 9—26 are representative of the commercially available graphites that have been produced in recent years. Approximate sizes have been given for particular grades because properties tend to show significant variance depending on this factor. Carbon-based fibers are listed in Section 7, Composite Materials.

Information regarding characteristics and recommended applications for various types of industrial graphite is available from Carbon Products Division, Union Carbide Corporation, New York.

**Table 9—1**

## TYPICAL PROPERTIES OF CARBON AND GRAPHITE AT ROOM TEMPERATURE

| Type of product | Bulk density | Porosity, percent | Strength, psi | | Elastic modulus, $10^6$ psi | Specific resistance, $10^{-5}$ ohm-in. | Thermal conductivity, Btu·ft / hr·ft²·deg F | Coefficient of thermal expansion, $10^{-7}$/deg F |
|---|---|---|---|---|---|---|---|---|
| | | | Compressive | Flexural | | | | |
| **CARBON** | | | | | | | | |
| Porous carbon, grade 45 | 1.04 | 47 | 900 | 500 | 0.4 | 700 | 1 | 16 |
| Carbon furnace lining (24″ × 30″ cross section) | 1.60 | 21 | 2 500 | 600 | 0.8 | 160 | 9 | 13 |
| Carbon refractory brick | 1.63 | 17 | 4 400 | 1 200 | 1.2 | 195 | 16 | 18 |
| Carbon chemical brick | 1.56 | 22 | 8 800 | 2 600 | 1.9 | 160 | 4 | 13 |
| Carbon pipe | 1.55 | 22 | 9 000 | 2 600 | 1.9 | 150 | 3 | 13 |
| Carbon electrodes, 8-in. diam. and equivalent rectangular | 1.57 | 21 | 2 400 | 1 100 | 1.2 | 110 | 9 | 13 |
| Carbon electrodes, 17- to 45-in. diam. and equivalent rectangular | 1.60 | 21 | 1 700 | 400 | 0.7 | 170 | 9 | 13 |
| **GRAPHITE** | | | | | | | | |
| Porous graphite, grade 45 | 1.04 | 53 | 500 | 300 | 0.3 | 130 | 45 | 11 |
| Nuclear graphite *see* Table 3.6−7 | | | | | | | | |
| Fine-grain, premium graphite | 1.73 | 23 | 8 300 | 4 000 | 1.5 | 43 | 68 | 13 |
| High-density, premium graphite | 1.84 | 19 | 8 400 | 3 700 | 1.7 | 47 | 63 | 11 |
| Recrystallized graphite | 1.95 | 13 | 7 200 | 5 400 | 2.7 | 28 | 104 | 3 |
| Graphite brick | 1.56 | 31 | 3 100 | 1 650 | 1.4 | 34 | 86 | 10 |
| Graphite pipe | 1.67 | 26 | 5 000 | 2 800 | 1.7 | 34 | 86 | 10 |
| Medium-grain, dense graphite, cylinders, and plates to 2 $\frac{3}{4}$-in. diam. and to $\frac{3}{4}$-in. thick | 1.70 | 25 | 5 600 | 4 000 | 1.8 | 27 | 100 | 7 |
| 3 to 11-in. diam. and 2 to 12-in. thick | 1.70 | 24 | 5 600 | 2 700 | 1.7 | 30 | 95 | 7 |
| 20 to 24-in. diam. and 20 × 20 cross section | 1.75 | 24 | 5 500 | 2 200 | 1.2 | 35 | 83 | 12 |

## Table 9—1 (continued)
### TYPICAL PROPERTIES OF CARBON AND GRAPHITE AT ROOM TEMPERATURE

| Type of product | Bulk density | Porosity, percent | Strength, psi | | Elastic modulus, $10^6$ psi | Specific resistance, $10^{-5}$ ohm-in. | Thermal conductivity, $\dfrac{Btu \cdot ft}{hr \cdot ft^2 \cdot deg\ F}$ | Coefficient of thermal expansion, $10^{-7}$/deg F |
|---|---|---|---|---|---|---|---|---|
| | | | Compressive | Flexural | | | | |
| Graphite electrodes, anodes, cylinders, and plates | | | | | | | | |
| to 2 $\frac{3}{4}$-in. diam. and to $\frac{3}{4}$-in. thick | 1.58 | 30 | 4 000 | 2 000 | 1.5 | 33 | 88 | 6 |
| 6 to 12-in. diam. and 6 to 12-in. thick | 1.57 | 30 | 2 900 | 1 300 | 1.5 | 33 | 88 | 7 |
| 14 to 35-in. diam. and 20 to 24-in. thick | 1.58 | 32 | 2 000 | 1 000 | 0.8 | 33 | 88 | 5 |

Table courtesy of Carbon Products Division, Union Carbide Corporation. © Union Carbide Corporation. Reproduced with permission.

## Table 9—2
### VARIATION OF PROPERTIES WITH TEMPERATURE

| Property | 500 deg C | 1 000 deg C | 1 500 deg C | 2 000 deg C | 2 500 deg C |
|---|---|---|---|---|---|
| | 932 deg F | 1 832 deg F | 2 732 deg F | 3 632 deg F | 4 532 deg F |
| Thermal expansion[a] as percent elongation from room temperature | | | | | |
| Anthracite carbon | 0.14 | 0.38 | 0.63 | — | |
| Graphite | 0.12 | 0.32 | 0.55 | — | — |
| Thermal conductivity as percent of room temperature value | | | | | |
| Fabricated anthracite carbon | 100 | 103 | 123 | — | — |
| Fabricated graphite | 60 | 40 | 30 | 25 | — |
| Instantaneous specific heat[b] as percent of room temperature value: graphite | 225 | 262 | 282 | 294 | 302 |
| Short-time breaking strength[c] as percent of room temperature value: graphite | 107 | 120 | 135 | 153 | 181 |

[a]These are longitudinal expansion; transverse expansion is 10–60% greater.

[b]The specific heat at room temperature is about 0.17.

[c]Strength increases up to 2 500 deg C, then decreases rapidly; above 2 200 deg C appreciable creep will occur at high-stress levels.

From *Carbon Products Pocket Handbook,* Carbon Products Division, Union Carbide Corporation, New York, 1964. © 1964 Union Carbide Corporation. Reproduced with permission.

**Table 9—3**

**TYPICAL ENGINEERING APPLICATIONS OF CARBON PRODUCTS**

Key to Advantageous Properties of Carbon Products:[a]

1. Stability and strength at high temperatures (to 4,500°F in nonoxidizing atmospheres)
2. High resistance to thermal shock
3. High thermal conductivity of solid; low conductivity of porous foam, cloth, and tape
4. Low coefficient of thermal expansion
5. High radiation emissivity
6. Good electrical conductivity
7. High compressive strength
8. Stiffness of solid; flexibility of filament, cloth, or tape
9. High resistance to erosion
10. Good machinability
11. Low friction; self-lubrication
12. High resistance to chemical attack and corrosion
13. High adsorption of gases and vapors
14. High moderating ratio, i.e., ratio of fast neutron slowing-down power to bulk neutron absorption coefficient
15. High ratio of thermal neutron scattering to absorption cross section

| Typical application | Type and form of carbon product | Desirable properties[b] |
|---|---|---|
| Electrodes for arc furnaces, welding, and lighting arcs | Extruded or molded carbon or graphite | 1, 2, 3, 4, 5, 6, 7, 8, 9, 10, 12 |
| Elements for electric-resistance furnaces | Resistors fabricated from round or rectangular stock, usually graphite | 1, 2, 3, 4, 5, 8, 10, 12 |
| Metallurgical crucibles, boats, trays | Castings from carbon or graphite | 1, 2, 4, 7, 8, 9, 10, 12 |
| Foundry molds, chills, cores, risers, cupola linings | Castings, rods, bricks, shapes | 1, 2, 4, 7, 8, 9, 10, 12 |
| Brazing jigs; extrusion guides; dies | Blocks; machined shapes | 1, 2, 7, 8, 9, 10, 12 |
| High-temperature refractories and insulations | Solid or foamed bricks and shapes; tapes and laminates | 1, 2, 3, 4, 7, 8, 9, 12 |
| Rocket and missile nozzles, vanes, cones, shields | Castings; built-up shapes from graphite fabric, tape, and resins | 1, 2, 3, 4, 5, 7, 8, 9, 10 |
| Chemical reactor vessels; heat exchangers; pipes and fittings | Cast, machined, or extruded graphite or carbon, pure or impregnated with resins | 1, 2, 3, 4, 7, 8, 9, 10, 12 |
| Packing and sealing rings; bushings, joint packings | Graphite blocks, sheet, cloth, and yarn | 1, 2, 3, 4, 7, 8, 9, 11, 12 |
| Electrical contacts, brushes, resistors | Blocks, rods, machined parts | 1, 3, 5, 6, 7, 10, 11 |
| Anodes, electrodes, battery components | Extruded and molded rod, granular carbon | 6, 12 |
| Air purification, gas separations, odor, and vapor removal | Activated carbon granules | 13 |
| Nuclear reactor—moderator and reflector | Purified "nuclear graphite" bars and machined shapes | 1, 2, 3, 4, 14, 15 |

[a]Protection from oxidizing atmosphere is necessary in all high-temperature uses of carbon or graphite, e.g., above 700°F.
[b]Numbers refer to items listed above.

From Bolz, R. E. and Tuve, G. L., Eds., *Handbook of Tables for Applied Engineering Science,* 2nd ed., CRC Press, Cleveland, 1973, 181.

## Table 9—4
## GRADE D-657

**Description**
Molded, fine grained; low cost

**Analysis**
Average impurity content: Ni, 400 ppm; Ca, 400 ppm; Fe, 200 ppm; Na, 200 ppm; Si, 150 ppm; Al, 150 ppm; Co, 50 ppm; Ti, 20 ppm; Mo, 20 ppm, purified grade, 50 ppm total impurities

| Properties | Units | With grain | | Against grain | | Typical high-temperature properties | |
|---|---|---|---|---|---|---|---|
| | | Average value | Standard deviation, % | Average value | Standard deviation, % | 1,300°F | 4,000°F |
| Young's modulus | $10^6$ psi | 1.5 | 15 | 1.3 | 15 | 1.6 | 2.1 |
| Tensile strength | $10^3$ psi | 3.3 | 20 | 2.8 | 20 | 3.0 | 6.3 |
| Compressive strength | $10^3$ psi | 8.6 | 20 | 10.5 | 20 | 9.0 | 13.0 |
| Flexural strength | $10^3$ psi | 4.4 | 20 | 4.0 | 20 | 4.5 | 7.5 |
| Density | $g/cm^3$ | 1.65 | 5 | | | | |
| Coefficient of thermal expansion | $10^{-6}/°C$ | 4.0 | 5 | 5.1 | 5 | 3.3 | 4.1 |
| Thermal conductivity | cal-cm/sec·cm$^2$·K | 0.350 | 15 | 0.330 | 15 | | |
| Specific resistance | $10^{-4}$ ohm-cm | 12.7 | 1 | 14.0 | 1 | | |

**Source**
Duramic Products

**Size**
24 X 20 X 9 in. maximum

**Manufacturing data**
Not available

## Table 9—5
## GRADE H205

**Description**
Molded, fine grained; high strength; high hardness

**Analysis**
Average impurity content: ash, 0.25%; Ni, 0.04%; Ca, 0.04%; Fe, 0.02%; Na, 0.02%; Si, 0.015%; Al, 0.015%

| Properties | Units | With grain | | Against grain | | Typical high-temperature properties | |
|---|---|---|---|---|---|---|---|
| | | Average value | Standard deviation, % | Average value | Standard deviation, % | 1,300°F | 4,000°F |
| Young's modulus | $10^6$ psi | 1.6 | 10 | 1.4 | 10 | | |
| Tensile strength | $10^3$ psi | 3.2 | 10 | 2.8 | 10 | | |
| Compressive strength | $10^3$ psi | 8.5 | 10 | 10.0 | 10 | | |
| Flexural strength | $10^3$ psi | 4.5 | 10 | 4.0 | 10 | | |
| Density | g/cm$^3$ | 1.75 | 2 | | | | |
| Coefficient of thermal expansion | $10^{-6}$/°C | 3.9 | 5 | 5.0 | 5 | | |
| Thermal conductivity | cal-cm/sec·cm²·K | 0.35 | 10 | 0.33 | 10 | | |
| Specific resistance | $10^{-4}$ ohm-cm | 12 | 10 | 14 | 10 | | |
| Hardness (Brinell) — 136 kg load-10 mm ball | | | | 15.0 | 10 | | |
| Permeability (D'Arcy) | | 0.2 | 10 | 0.004 | 10 | | |

**Source**
Great Lakes Carbon

**Size**
Cylinder 10—22 in.; block 9 × 20 × 24 in.

**Manufacturing data**
Graphitized over 2,500°C; Acheson electric furnace; 1—20 ton batch size

## Table 9—6
## GRADE AXF

**Description**
Molded, fine grained; high strength; high electrical resistance; high reproducibility; low porosity; chemical resistance: abrasion resistant; small sizes; isotropic

**Analysis**
Average impurity content: ash, 0.1%

| Properties | Units | With grain | | Against grain | | Typical high-temperature properties | |
|---|---|---|---|---|---|---|---|
| | | Average value | Standard deviation, % | Average value | Standard deviation, % | 1,300° F | 4,000° F |
| Young's modulus | $10^6$ psi | 1.8 | >20 | 1.6 | >20 | | 1.5 |
| Tensile strength | $10^3$ psi | 9.4 | | | | | 10.2 |
| Compressive strength | $10^3$ psi | 20.0 | 5–10 | | | | |
| Flexural strength | $10^3$ psi | 10.0 | 10–20 | 10.0 | 10–20 | | |
| Density | g/cm$^3$ | 1.80–1.88 | | | | | |
| Coefficient of thermal expansion | $10^{-6}$/°C | 9.0 | | | | | |
| Thermal conductivity | cal-cm/sec·cm²·K | 0.1–0.5 | | | | | |
| Specific resistance | $10^{-4}$ ohm-cm | 14–16 | | | | | |
| Hardness (scleroscope) | | 78 | | | | | |

**Source**
Poco Graphite, Inc.

**Size**
Rod 1/8—5/8 in.; cylinder 8 in. maximum; block 4 × 8 × 18 in. maximum

**Manufacturing data**
Not available

**Table 9—7**
**GRADE L-55**

### Description
Molded, fine grained; high strength; low coefficient thermal expansion; good electrical and thermal conductivity; high reproducibility; low friction

### Analysis
Average impurity content: ash, 0.3%

| Properties | Units | With grain | | Against grain | | Typical high-temperature properties | |
|---|---|---|---|---|---|---|---|
| | | Average value | Standard deviation, % | Average value | Standard deviation, % | 1,300° F | 4,000° F |
| Young's modulus | $10^6$ psi | <1 | | | | | |
| Tensile strength | $10^3$ psi | | | | | | |
| Compressive strength | $10^3$ psi | 3 | 15 | | | | |
| Flexural strength | $10^3$ psi | | | | | | |
| Density | g/cm³ | 1.60 | 1.5 | | | | |
| Coefficient of thermal expansion | $10^{-6}$ /°C | | | | | | |
| Thermal conductivity | cal-cm/sec·cm²·K | 75 | 7.5 | | | | |
| Specific resistance | $10^{-4}$ ohm-cm | 52 | <5 | | | | |
| Hardness (scleroscope) | | | | | | | |

### Source
Pure Carbon

### Size
Cylinder 1/8–6 in.; block 1–6 in.; rod 0.01–1/8 in.; pipe <1/2–6 in.

### Manufacturing data
Lamp black, graphite, and pitch; graphitized over 2,500°C; Acheson electric furnace; impregnated in secondary processing; finishing operations as required: 100–2,000 lb batch size

## Table 9—8
## GRADE 9RL

**Description**
Molded, fine grained; high purity; high reproducibility; high temperature oxidation resistance

**Analysis**
Average impurity content: Fe, <10 ppm; V, 1 ppm; B, <1 ppm; Si, 10 ppm; Ca, <10 ppm; Al, 5 ppm; Mg, <1.0 ppm

| Properties | Units | With grain | | Against grain | | Typical high-temperature properties | |
|---|---|---|---|---|---|---|---|
| | | Average value | Standard deviation, % | Average value | Standard deviation, % | 1,300° F | 4,000° F |
| Young's modulus | $10^6$ psi | 0.87 | | 1.18 | | 2.2 | 3.8 |
| Tensile strength | $10^3$ psi | 1.8 | | 1.6 | | 7.0 | 9.8 |
| Compressive strength | $10^3$ psi | 6.8 | | 7.2 | | 4.2 | 7.0 |
| Flexural strength | $10^3$ psi | 3.6 | | 3.1 | | | |
| Density | g/cm$^3$ | 1.68 | | | | | |
| Coefficient of thermal expansion | $10^{-6}$/°C | 3.3 | | 4.4 | | | |
| Thermal conductivity | cal-cm/sec·cm²·K | 9.6 | | | | 0.15 | |
| Specific resistance | $10^{-4}$ ohm-cm | 37 | | | | 9.1 | 10.7 |
| Hardness (scleroscope) | | | | | | | |

**Source**
Speer Carbon

**Size**
Cylinder 10 in. maximum

**Manufacturing data**
Calcined petroleum coke and coal tar pitch; graphitized over 2,500°C; machined; 1—20 ton batch size

**Table 9—9**
**GRADE 3499**

**Description**
Molded, fine grained; high strength; high reproducibility; high production

**Analysis**
Average impurity content: ash, 0.03%

| Properties | Units | With grain | | Against grain | | Typical high-temperature properties | |
|---|---|---|---|---|---|---|---|
| | | Average value | Standard deviation, % | Average value | Standard deviation, % | 1,300° F | 4,000° F |
| Young's modulus | $10^6$ psi | 1.5 | | | | | |
| Tensile strength | $10^3$ psi | 1.8 | | 1.6 | | 2.2 | 3.8 |
| Compressive strength | $10^3$ psi | 6.4 | | 6.8 | | 7.0 | 9.8 |
| Flexural strength | $10^3$ psi | 3.7 | | 3.2 | | 4.2 | 7.0 |
| Density | g/cm$^3$ | 1.65 | | | | | |
| Coefficient of thermal expansion | $10^{-6}$/°C | 3.3 | | 4.4 | | | |
| Thermal conductivity | cal-cm/sec·cm$^2$·K | 8.6 | | | | 0.12 | |
| Specific resistance | $10^{-4}$ ohm-cm | 36 | | | | 7.1 | 11.7 |
| Hardness (scleroscope) | | | | | | | |

**Source**
Speer Carbon

**Size**
Block 12 × 12 × 5 in.

**Manufacturing data**
Calcined petroleum coke and coal tar pitch; graphitized over 2,500°C; machined; 1–20 ton batch size

## Table 9—10
## GRADE 9135; 9139

**Description**
Molded, fine grained; high strength; high reproducibility; abrasion resistant

**Analysis**
Average impurity content: ash, 0.04%

| Properties | Units | With grain | | Against grain | | Typical high-temperature properties | |
|---|---|---|---|---|---|---|---|
| | | Average value | Standard deviation, % | Average value | Standard deviation, % | 1,300° F | 4,000° F |
| Young's modulus | $10^6$ psi | 1.4 | | 1.2 | | 4.0 | 5.7 |
| Tensile strength | $10^3$ psi | 2.4 | | 2.1 | | 10.5 | 14.8 |
| Compressive strength | $10^3$ psi | 9.2 | | 9.6 | | 6.0 | 11.3 |
| Flexural strength | $10^3$ psi | 4.4 | | 4.1 | | | |
| Density | g/cm$^3$ | 1.79 | | | | | |
| Coefficient of thermal expansion | $10^{-6}$/°C | 3.0 | | 4.7 | | 0.2 | |
| Thermal conductivity | cal-cm/sec·cm$^2$·K | 11.4 | | | | 8.4 | 7.6 |
| Specific resistance | $10^{-4}$ ohm-cm | 48 | | | | | |
| Hardness (scleroscope) | | | | | | | |

**Source**
Speer Carbon

**Size**
Finished shapes with <3-in. wall thickness

**Manufacturing data**
Calcined petroleum coke and coal tar pitch; graphitized over 2,500°C; 100—2,000 lb batch size

# Table 9—11
## GRADE ATJ

**Description**
Molded, fine grained; high strength; high reproducibility

**Analysis**
Average impurity content: ash, 0.15%

| Properties | Units | With grain | | Against grain | | Typical high-temperature properties | |
|---|---|---|---|---|---|---|---|
| | | Average value | Standard deviation, % | Average value | Standard deviation, % | 1,300°F | 4,000°F |
| Young's modulus | $10^6$ psi | 1.4 | 11 | 1.0 | 8 | | |
| Tensile strength | $10^3$ psi | 3.4 | 11 | 2.9 | 10 | | |
| Compressive strength | $10^3$ psi | 8.3 | 12 | 8.6 | 13 | | |
| Flexural strength | $10^3$ psi | 4.0 | 19 | 8.5 | 13 | | |
| Density | g/cm$^3$ | 1.7 | 2 | | | | |
| Coefficient of thermal expansion | $10^{-6}$/°C | 2.2 | 10 | 3.4 | 6 | | |
| Thermal conductivity | cal-cm/sec·cm$^2$·K | 0.28 | | 0.21 | | | |
| Specific resistance | $10^{-4}$ ohm-cm | 11.0 | 15 | 14.5 | 10 | | |

**Source**
Union Carbide

**Size**
Cylinder 13–17 in., block 9 × 20 × 24 in.

**Manufacturing data**
Calcined petroleum coke and coal tar pitch; graphitized over 2,500°C; Acheson electric furnace; impregnated in secondary processing; over 20-ton batch size

**Table 9—12**
**GRADE CCT**

**Description**
Molded, fine grained; high strength; high purity

**Analysis**
Average impurity content: ash, 15 ppm

| Properties | Units | With grain | | Against grain | | Typical high-temperature properties | |
| --- | --- | --- | --- | --- | --- | --- | --- |
| | | Average value | Standard deviation, % | Average value | Standard deviation, % | 1,300° F | 4,000° F |
| Young's modulus | $10^6$ psi | 1.4 | | 1.2 | | | |
| Tensile strength | $10^3$ psi | 3.5 | | 2.9 | | | |
| Compressive strength | $10^3$ psi | 8.2 | | 8.5 | | | |
| Flexural strength | $10^3$ psi | 4.0 | | 3.6 | | | |
| Density | g/cm$^3$ | 1.73 | | | | | |
| Coefficient of thermal expansion | $10^{-6}$/°C | 2.2 | | 3.4 | | | |
| Thermal conductivity | cal-cm/sec·cm$^2$·K | 0.28 | | 0.21 | | | |
| Specific resistance | $10^{-4}$ ohm-cm | 11.0 | | 1.4 | | | |

**Source**
Union Carbide

**Size**
6½ in. diameter × 24 in. long maximum

**Manufacturing data**
Calcined petroleum coke and coal tar pitch; graphitized over 2,500°C; electric resistance furnace; impregnated in secondary processing; 100—2,000 lb batch size

**Table 9—13**
**GRADE 2**

## Description

Molded, fine grained; carbon-graphite; low coefficient of friction; will stand oxidizing atmosphere to 700°F; good electrical conductor; chemical resistant

| Properties | Units | With grain | | Against grain | | Typical high-temperature properties | |
|---|---|---|---|---|---|---|---|
| | | Average value | Standard deviation, % | Average value | Standard deviation, % | 1,300° F | 4,000° F |
| Young's modulus | $10^6$ psi | 2.3 | | | | | |
| Tensile strength | $10^3$ psi | 4.5 | | | | | |
| Compressive strength | $10^3$ psi | 23.0 | | | | | |
| Flexural strength | $10^3$ psi | 5–10 | | | | | |
| Density | $g/cm^3$ | 1.8 | | | | | |
| Coefficient of thermal expansion | $10^{-6}/°C$ | | | | | | |
| Thermal conductivity | cal-cm/sec·cm$^2$·K | | | | | | |
| Specific resistance | $10^{-4}$ ohm-cm | 10–50 | | | | | |
| Hardness (scleroscope) | | 85 | | | | | |

## Source

U.S. Graphite

## Size

Cylinder 1/8–12 in., block 1–4 in., ring up to 13 3/4 × 10 × 1 3/4 in.

## Manufacturing data

Carbon and graphite powders; compacted under high pressure; furnaced at temperatures up to 4,500° F; machined or ground to tolerance

## Table 9—14
## GRADE RVA

**Description**
Molded, medium grained; high strength; high reproducibility; high density; large sizes

**Analysis**
Average impurity content: ash, 0.30%

| Properties | Units | With grain | | Against grain | | Typical high-temperature properties | |
|---|---|---|---|---|---|---|---|
| | | Average value | Standard deviation, % | Average value | Standard deviation, % | 1,300°F | 4,000°F |
| Young's modulus | $10^6$ psi | 1.7 | 9 | 1.3 | 9 | | |
| Tensile strength | $10^3$ psi | 3.0 | 15 | 2.1 | 8 | | |
| Compressive strength | $10^3$ psi | 8.4 | 13 | 8.1 | 15 | | |
| Flexural strength | $10^3$ psi | 3.7 | 8 | 3.0 | 10 | | |
| Density | g/cm$^3$ | 1.84 | 2 | | | | |
| Coefficient of thermal expansion | $10^{-6}$/°C | 1.8 | 5 | 2.7 | 3 | | |
| Thermal conductivity | cal-cm/sec·cm$^2$·K | 0.26 | | 0.21 | | | |
| Specific resistance | $10^{-4}$ ohm-cm | 12.2 | 3 | 15.7 | 6 | | |

**Source**
Union Carbide

**Size**
Cylinder 30 in

**Manufacturing data**
Calcined petroleum coke and coal tar pitch; graphitized over 2,500°C; Acheson electric furnace; impregnated in secondary processing; 1–20 ton batch size

**Table 9—15**
GRADE GSX

**Description**
Extruded, fine grained; good electrical and thermal conductor; maximum grain size 0.008 in.; high reproducibility; chemical resistant

**Analysis**
Average impurity content: ash, 0.2% maximum

| Properties | Units | With grain | | Against grain | | Typical high-temperature properties | |
| --- | --- | --- | --- | --- | --- | --- | --- |
| | | Average value | Standard deviation, % | Average value | Standard deviation, % | 1,300°F | 4,000°F |
| Young's modulus | $10^6$ psi | 1.2 | | 0.8 | | | |
| Tensile strength | $10^3$ psi | 2.1 | | 1.5 | | | |
| Compressive strength | $10^3$ psi | 6.5 | | 6.5 | | | |
| Flexural strength | $10^3$ psi | | | | | | |
| Density | g/cm$^3$ | 1.68 | | | | | |
| Coefficient of thermal expansion | $10^{-6}$/°C | | | | | | |
| Thermal conductivity | cal-cm/sec·cm$^2$·K | | | | | | |
| Specific resistance | $10^{-4}$ ohm-cm | 11.4 | | 16.0 | | | |

**Source**
Carborundum Company

**Size**
Cylinder 3/8—2 in.; pipe 1 1/4—5 1/4 in.

**Manufacturing data**
Calcined petroleum coke, coal tar pitch; graphitized over 2,500°C; electric resistance furnace; machined and ground; 100—2,000 lb batch size

**Table 9—16**
**GRADE 780GL**

**Description**
Extruded, fine grained; high strength; low coefficient of thermal expansion; high purity; good nuclear properties; high temperature oxidation resistance

**Analysis**
Average impurity content: ash, 100 ppm maximum; Si, 10 ppm; Al, <10 ppm; Fe, 10 ppm; Ca, <10 ppm; Zn, <10 ppm; Na, <10 ppm; Mg, 2ppm

| Properties | Units | With grain | | Against grain | | Typical high-temperature properties | |
|---|---|---|---|---|---|---|---|
| | | Average value | Standard deviation, % | Average value | Standard deviation, % | 1,300° F | 4,000° F |
| Young's modulus | $10^6$ psi | 2.0 | | 1.6 | | 3.5 | 5.0 |
| Tensile strength | $10^3$ psi | 2.3 | | 9.4 | | 9.3 | 14.0 |
| Compressive strength | $10^3$ psi | 9.0 | | 3.7 | | 5.8 | 8.8 |
| Flexural strength | $10^3$ psi | 4.3 | | | | | |
| Density | $g/cm^3$ | 1.79 | | | | | |
| Coefficient of thermal expansion | $10^{-6}/°C$ | 1.8 | | 2.9 | | | |
| Thermal conductivity | cal-cm/sec·cm²·K | | | | | 0.06 | |
| Specific resistance | $10^{-4}$ ohm-cm | 10.9 | | | | 9.7 | 11.7 |

**Source**
Speer Carbon

**Size**
Cylinder 2 1/2–5 in.

**Manufacturing data**
Calcined petroleum coke and coal tar pitch; graphitized over 2,500°C; machined; 100–2,000 lb batch size

## Table 9—17
## GRADE 890W

**Description**
Extruded, fine grained; high strength and purity; high-temperature oxidation resistant

**Analysis**
Average impurity content: ash, 0.03%

| Properties | Units | With grain | | Against grain | | Typical high-temperature properties | |
| --- | --- | --- | --- | --- | --- | --- | --- |
| | | Average value | Standard deviation, % | Average value | Standard deviation, % | 1,300°F | 4,000°F |
| Young's modulus | $10^6$ psi | 1.9 | | | | 2.8 | 5.0 |
| Tensile strength | $10^3$ psi | 1.6 | | 5.6 | | 6.4 | 9.6 |
| Compressive strength | $10^3$ psi | 5.4 | | 2.4 | | 4.5 | 6.5 |
| Flexural strength | $10^3$ psi | 2.7 | | | | | |
| Density | $g/cm^3$ | 1.7 | | 3.4 | | | |
| Coefficient of thermal expansion | $10^{-6}/°C$ | 1.7 | | | | | |
| Thermal conductivity | cal-cm/sec·cm²·K | | | | | 0.2 | |
| Specific resistance | $10^{-4}$ ohm-cm | 6.9 | | | | 6.6 | 10.2 |

**Source**
Speer Carbon

**Size**
Cylinder 2 1/2—9 in.; block <40 in.²

**Manufacturing data**
Calcined petroleum coke and coal tar pitch; graphitized over 2,500°C; machined; 1—20 ton batch size

## Table 9—18
## GRADE HC

**Description**
Extruded, medium grained; good electrical conductivity

**Analysis**
Average impurity content: ash, 0.30%; S, 0.10%; Si, 0.04%; Fe, 0.04%; Ca, 0.03%; Al, 0.03%; V, 70 ppm

| Properties | Units | With grain | | Against grain | | Typical high-temperature properties | |
| --- | --- | --- | --- | --- | --- | --- | --- |
| | | Average value | Standard deviation, % | Average value | Standard deviation, % | 1,300° F | 4,000° F |
| Young's modulus | $10^6$ psi | 1.2 | 10 | 1.0 | 10 | | |
| Tensile strength | $10^3$ psi | 0.8 | 10 | 0.6 | 10 | | |
| Compressive strength | $10^3$ psi | 3.5 | 10 | 3.5 | 10 | | |
| Flexural strength | $10^3$ psi | 1.2 | 10 | 1.0 | 10 | | |
| Density | g/cm$^3$ | 1.55 | 2 | | | | |
| Coefficient of thermal expansion | $10^{-6}$/°C | 1.2 | 5 | 2.4 | 5 | | |
| Thermal conductivity | cal-cm/sec·cm$^2$·K | 0.33 | 10 | 0.30 | 10 | | |
| Specific resistance | $10^{-4}$ ohm-cm | 9 | 10 | 12 | 10 | | |
| Permeability (D'Arcy) | | 0.37 | 10 | 0.34 | 10 | | |

**Source**
Great Lakes Carbon

**Size**
Cylinder 7–12 in.

**Manufacturing data**
Calcined patroleum coke and coal tar pitch; graphitized over 2,500°C; Acheson electric furnace; over 20-ton batch size

**Table 9—19**
**GRADE TL**

**Description**
Extruded, medium grained; good electrical conductivity; high purity; high reproducibility

**Analysis**
Average impurity content: ash, 0.20%; S, 0.10%; Si, 0.05%; Fe, 0.03%; Co, 0.05%; Pb, 0.04%; Ca, 0.03% Al, 0.03%; Na, 20%; Mg, 20%

| Properties | Units | With grain | | Against grain | | Typical high-temperature properties | |
|---|---|---|---|---|---|---|---|
| | | Average value | Standard deviation, % | Average value | Standard deviation, % | 1,300° F | 4,000° F |
| Young's modulus | $10^6$ psi | 1.5 | 10 | 1.2 | 10 | | |
| Tensile strength | $10^3$ psi | 0.8 | 10 | 0.6 | 10 | | |
| Compressive strength | $10^3$ psi | 3.5 | 10 | 3.5 | 10 | | |
| Flexural strength | $10^3$ psi | 2.0 | 10 | 1.8 | 10 | | |
| Density | g/cm$^3$ | 1.7 | 2 | | | | |
| Coefficient of thermal expansion | $10^{-6}$/°C | 1.8 | 5 | 2.2 | 5 | | |
| Thermal conductivity | cal-cm/sec·cm$^2$·K | 0.33 | 10 | 0.30 | 10 | | |
| Specific resistance | $10^{-4}$ ohm-cm | 9 | 10 | 12 | 10 | | |
| Porosity (apparent) | % | 16 | | | | | |

**Source**
Great Lakes Carbon

**Size**
Cylinder 3–6 in.; block 3/4–6 × 2–18 in.

**Manufacturing data**
Calcined petroleum coke and coal tar pitch; graphitized over 2,500°C; Acheson electric furnace; over 20-ton batch size

## Table 9—20
## GRADE 873 RL

**Description**

Extruded, medium grained; high strength; low coefficient of thermal expansion; high purity; good nuclear properties; high-temperature oxidation resistance

**Analysis**

Average impurity content: ash, 100 ppm maximum; Al, <10 ppm; B, <1 ppm; Ca, <1 ppm; Fe, 5 ppm; Mg, <1 ppm; Ni, <10 ppm; Si, 30 ppm; Ti, <10 ppm; V, 1 ppm

| Properties | Units | With grain | | Against grain | | Typical high-temperature properties | |
|---|---|---|---|---|---|---|---|
| | | Average value | Standard deviation, % | Average value | Standard deviation, % | 1,300°F | 4,000°F |
| Young's modulus | $10^6$ psi | 1.8 | | 1.1 | | | |
| Tensile strength | $10^3$ psi | 1.6 | | 1.4 | | 2.0 | 4.0 |
| Compressive strength | $10^3$ psi | 6.4 | | 6.8 | | 6.2 | 9.6 |
| Flexural strength | $10^3$ psi | 3.2 | | 2.6 | | 3.8 | 5.8 |
| Density | $g/cm^3$ | 1.77 | | | | | |
| Coefficient of thermal expansion | $10^{-6}/°C$ | 2.4 | | 4.2 | | | |
| Thermal conductivity | cal-cm/sec·cm$^2$·K | | | | | 0.23 | |
| Specific resistance | $10^{-4}$ ohm-cm | 6.35 | | | | 6.25 | 10.3 |

**Source**

Speer Carbon

**Size**

Cylinder 14 in. maximum diameter

**Manufacturing data**

Calcined petroleum coke and coal tar pitch; graphitized over 2,500°C; over 20-ton batch size

## Table 9—21
## GRADE AGX, AGLX

**Description**
Extruded, medium grained; long experience

**Analysis**
Average impurity content: ash, 0.42%

| Properties | Units | With grain | | Against grain | | Typical high-temperature properties | |
|---|---|---|---|---|---|---|---|
| | | Average value | Standard deviation, % | Average value | Standard deviation, % | 1,300°F | 4,000°F |
| Young's modulus | $10^6$ psi | 1.6 | 12 | 0.9 | 8 | | |
| Tensile strength | $10^3$ psi | 1.4 | 16 | 1.0 | 14 | | |
| Compressive strength | $10^3$ psi | 5.6 | 24 | 5.3 | 22 | | |
| Flexural strength | $10^3$ psi | 2.7 | 17 | 1.8 | 25 | | |
| Density | g/cm$^3$ | 1.69 | 1.5 | | | | |
| Coefficient of thermal expansion | $10^{-6}$/°C | 1.6 | 12 | | | | |
| Thermal conductivity | cal-cm/sec·cm$^2$·K | 0.38 | | 0.22 | | | |
| Specific resistance | $10^{-4}$ ohm-cm | 8.2 | 12 | 13.9 | 10 | | |

**Source**
Union Carbide

**Size**
Cylinder 3–5 3/4 in.; block 1–5 in.

**Manufacturing data**
Calcined petroleum coke and coal tar pitch; graphitized over 2,500°C; Acheson electric furnace; impregnated in secondary processing; machined; 1–20 ton batch size

## Table 9—22
## GRADE AGR

**Description**
Extruded, coarse grained; long experience; large and small sizes; thermal shock resistant

**Analysis**
Average impurity content: ash, 0.96%

| Properties | Units | With grain | | Against grain | | Typical high-temperature properties | |
|---|---|---|---|---|---|---|---|
| | | Average value | Standard deviation, % | Average value | Standard deviation, % | 1,300° F | 4,000° F |
| Young's modulus | $10^6$ psi | 0.5 | 13 | 0.5 | 21 | | |
| Tensile strength | $10^3$ psi | 0.44 | 17 | 0.42 | 11 | | |
| Compressive strength | $10^3$ psi | 1.9 | 22 | 2.0 | 18 | | |
| Flexural strength | $10^3$ psi | 0.84 | 17 | 0.84 | 17 | | |
| Density | g/cm$^3$ | 1.54 | 2.5 | | | | |
| Coefficient of thermal expansion | $10^{-6}$/°C | 1.2 | 31 | 1.9 | 16 | | |
| Thermal conductivity | cal-cm/sec·cm²·K | 0.32 | | 0.27 | | | |
| Specific resistance | $10^{-4}$ ohm-cm | 9.6 | 10 | 11.3 | | | |

**Source**
Union Carbide

**Size**
Cylinder 14—35 in. diameter; blocks 24 × 24 × 100 in. maximum

**Manufacturing data**
Calcined petroleum coke; graphitized over 2,500°C; Acheson electric furnace; machined; over 20-ton batch size

**Table 9—23**
**GRADE ZTA**

## Description

Hot worked; high density fine grained; high strength; high reproducibility; high thermal conductivity; highly oriented; low porosity; grade is certified to be free of internal cracks, voids, or other structural defects as detected by radiographic inspection

## Analysis

Average impurity content: ash, 0.1%

| Properties | Units | With grain | | Against grain | | Typical high-temperature properties | |
|---|---|---|---|---|---|---|---|
| | | Average value | Standard deviation, % | Average value | Standard deviation, % | 1,300° F | 4,000° F |
| Young's modulus | $10^6$ psi | 2.6 | 9 | 0.8 | 5 | | |
| Tensile strength | $10^3$ psi | 4.0 | 15 | 1.2 | 14 | | |
| Compressive strength | $10^3$ psi | 7.2 | 18 | 1.2 | 13 | | |
| Flexural strength | $10^3$ psi | 5.4 | 14 | 2.4 | 14 | | |
| Density | $g/cm^3$ | 1.95 | 1.5 | | | | |
| Coefficient of thermal expansion | $10^{-6}/°C$ | 0.7 | 0.35 | 8.2 | 4 | | |
| Thermal conductivity | cal-cm/sec·$cm^2$·K | 0.52 | | 0.20 | | | |
| Specific resistance | $10^{-4}$ ohm-cm | 7.1 | 7 | 19.9 | 7 | | |

## Source

Union Carbide

## Size

Cylinder 8 1/2—14 in. diameter

## Manufacturing data

Calcined petroleum coke and coal tar pitch; graphitized over 2,500°C and hot worked; 100—2,000 lb batch size

## Table 9—24
## GRADE PYROLYTIC GRAPHITE

**Description**

Pyrolytic graphite; high strength; low coefficient of thermal expansion; good electrical and thermal conductivity; high electrical resistance; good thermal insulator; high purity; good nuclear properties; low porosity; highly oriented

**Analysis**

Average impurity content: C, 99.99%

| Properties | Units | With grain Average value | With grain Standard deviation, % | Against grain Average value | Against grain Standard deviation, % | Typical high-temperature properties 1,300° F | Typical high-temperature properties 4,000° F |
|---|---|---|---|---|---|---|---|
| Young's modulus | $10^6$ psi | 4.4 | | 0.5 | | 1.89 | |
| Tensile strength | $10^3$ psi | 18.7 | | 14.5 | 7 | 22.500 | 25.000 |
| Compressive strength | $10^3$ psi | 66.1 | | | | 31.000 | |
| Flexural strength | $10^3$ psi | 23.5 | | 2.0 | 10 | | |
| Density | g/cm$^3$ | 2.20 | | 2.20 | | | |
| Coefficient of thermal expansion | $10^{-6}$/°C | 1.30 | | 23.7 | | | |
| Thermal conductivity | cal-cm/sec·cm$^2$·K | 1.24 | | 0.002 | | 0.39 | 0.0015 |
| Specific resistance | $10^{-4}$ ohm-cm | 4.79 | | 8,000 | | 2.50 | |
| | | | | | | 1,690 | |

**Source**

Raytheon Company

**Size**

Various geometric shapes

**Manufacturing data**

Gaseous hydrocarbon; processed below 2,500°C; machined; less than 100-lb batch size

# Table 9—25
## GRADE PYROLYTIC GRAPHITE

### Description

Pyrolytic graphite; high strength; low coefficient of thermal expansion; good electrical and thermal conductivity; high purity and resistance; good nuclear properties; high reproducibility; low friction; low porosity; chemical resistant; low hardness

### Analysis

Average impurity content: Al, 0.01 ppm; B, <0.01 ppm; Ca, <0.007; Co, <0.100 ppm; Cu, 0.01 ppm; Fe, 0.40 ppm; Mg, <0.001 ppm; Nb, <0.07 ppm; Ti, 0.01 ppm; Zn, <0.1 ppm; Ta, <1.00 ppm

| Properties | Units | With grain | | Against grain | | Typical high-temperature properties | |
|---|---|---|---|---|---|---|---|
| | | Average value | Standard deviation, % | Average value | Standard deviation, % | 1,300° F | 4,000° F |
| Young's modulus | $10^6$ psi | 2–5 | 10–20 | | | | |
| Tensile strength | $10^3$ psi | 10–30 | 5–10 | <1 | 5–10 | 16 | 24 |
| Compressive strength | $10^3$ psi | 10–50 | >20 | >50 | <5 | 20 | 27 |
| Flexural strength | $10^3$ psi | >20 | 10–20 | | | | |
| Density | g/cm$^3$ | 2–2.2 | <1 | | | | |
| Coefficient of thermal expansion | $10^{-6}$/°C | <2 | <2 | 10–20 | <2 | 0.7 | 1.6 |
| Thermal conductivity | cal-cm/sec·cm²·K | 0.5–1 | | <0.1 | | 0.6 | 0.2 |
| Specific resistance | $10^{-4}$ ohm-cm | <1 | <5 | >2,000 | <5 | 3.7 | 5 |
| Emissivity | | 0.8 at 2,000° F | | | | | |
| Thermal neutron abs cross section | | 3.4 mb | | | | | |

### Source

Super-temp

### Size

Plate material 1/16—1 in. thick, 16 × 65 in. maximum; cylinder 1/4—20 in. diameter, 36 in. maximum length

### Manufacturing data

Gaseous hydrocarbon; graphitized over 2,500°C; machining and grinding; 100—2,000 lb batch size

## Table 9—26
## GRADE POROUS GRAPHITE 25

**Description**
Graphite foam; good thermal insulator; high reproducibility; high porosity; chemical resistant; high permeability

**Analysis**
Average impurity content; ash, 0.1—0.5%

| Properties | Units | With grain | | Against grain | | Typical high-temperature properties | |
|---|---|---|---|---|---|---|---|
| | | Average value | Standard deviation, % | Average value | Standard deviation, % | 1,300° F | 4,000° F |
| Young's modulus | $10^6$ psi | 0.2 | | | | | |
| Tensile strength | $10^3$ psi | 0.07 | | | | | |
| Compressive strength | $10^3$ psi | 0.4 | | | | | |
| Flexural strength | $10^3$ psi | 0.2 | | | | | |
| Density | $g/cm^3$ | 1.03 | | | | | |
| Coefficient of thermal expansion | $10^{-6}/°C$ | 2.0 | | | | | |
| Thermal conductivity | $cal\text{-}cm/sec\text{-}cm^2 \cdot K$ | 16 | | | | | |
| Specific resistance | $10^{-4}$ ohm-cm | 38.0 | | | | | |
| Porosity | % | 48 | | | | | |

**Source**
Union Carbide

**Size**
Cylinder 7 1/4 in.; block 9 × 14 × 14 in., pipe 1 3/4 in. outside diameter

**Manufacturing data**
Calcined petroleum coke; graphitized over 2,500°C; Acheson electric furnace; machined; 1—20 ton batch size

*Note:* Average permeability, water (70°F, 5 psi), 1-in. thick plate: 90 gal/ft² /min.

# Section 10

# **Metallic Materials**

## Section 10

# METALLIC MATERIALS

### 10.1   ALUMINUM-BASE ALLOYS*

Aluminum base alloys are divided into two categories, wrought and cast. Wrought alloys have a systematic identification according to the alloying elements. For the Aluminum Association designation, lxxx is 99.00% aluminum or greater. The second digit indicates special purity controls and the last two digits indicate the minimum aluminum beyond 99.00%. Thus, a 1030 Al has 99.30% aluminum and no special control of individual impurities. The other series designations are

Al plus Cu as principal alloying element – 2xxx

Al plus Mn as principal alloying element – 3xxx

Al plus Si as principal alloying element – 4xxx

Al plus Mg as principal alloying element – 5xxx

Al plus Mg and Si as principal alloying elements – 6xxx

Al plus Zn as prinicipal alloying element – 7xxx

Al plus other major alloying element – 8xxx

The second digit refers to alloying modifications, and the last two digits identify the alloy, usually from its former commercial designation; thus, 75S is now 7075. The system of temper designations used for all forms of wrought and cast aluminum is briefly reviewed here. The individual company data sheets should be consulted for specific heat treatments.

F -- as fabricated

O – annealed

H – strain hardened

H1 – strain hardened only

H2 – strain hardened and partially annealed

H3 – strain hardened and stabilized by low temperature treatment. A second digit is used to indicate tempers between O (annealed) and 8 (full hard, final degree of strain hardening, although 9 is used for extra-hard tempers). A third digit indicates variations from these tempers.

W – solution heat treated, spontaneous aging at room temperature

T – thermally treated to produce stable tempers

T1 – partially solution heat treated and naturally aged to stable condition

T2 – annealed (for improved castings)

T3 – solution heat treated and cold worked

T4 – solution heat treated and naturally aged to stable condition

T5 – partially solution heat treated and artificially aged

T6 – solution heat treated and artificially aged

T7 – solution heat treated and stabilized, beyond point of maximum hardness

T8 – solution heat treated, cold worked, and artificially aged

T9 – solution heat treated, artificially aged, and cold worked

T10 – partially solution heat treated, artificially aged, and cold worked

Additional digits may be added for variation in treatments.

* Based mainly on the *Handbook of Materials Science,* Volume II, pages 177 through 217.

## 10.1.1  WROUGHT ALUMINUM ALLOYS

### Table 10.1.1—1
### WROUGHT ALUMINUM EC

Typical chemical composition, percent by weight: Al, 99.45 minimum

**Physical constants and thermal properties**
Density, lb/in.³: 0.098
Coefficient of thermal expansion, (70–200°F) in./in./°F × $10^{-6}$: 13.2
Modulus of elasticity, psi: tension, $10 × 10^6$

Poisson's ratio: 0.345
Melting range, °F: 1,195–1,215
Thermal conductivity, Btu/ft² /hr/in./°F, 70°F: 1,620
Electrical resistivity, ohms/cmil/ft, 70°F: 16.8

**Heat treatments**
Annealing temperature 650°F.

### Tensile Properties

**Annealed**

| Temperature, °F | T.S., psi | Y.S., psi | Elong., in 2 in., % | Hardness, Brinell |
|---|---|---|---|---|
| 75 | 12,000 (O) | 4,000 (O) | 23 (O) | – |
| 75 | 27,000 (H19) | 24,000 (H19) | 1.5 (H19) | – |

### Fatigue Strength

**Hard (H19)**

| Test temperature, °F | Stress, psi | Cycles to failure |
|---|---|---|
| 75 | 7,000 | $5 × 10^8$ |

### Specifications

ASTM: EC

### Table 10.1.1—2
### WROUGHT ALUMINUM 1060

Typical chemical compostion, percent by weight: Mn, 0.03; Fe, 0.35; Si, 0.25; Ti, 0.03; Al, 99.60 minimum; Cu, 0.05; Mg, 0.03; Zn, 0.05

**Physical constants and thermal properties**
Density, lb/in.³: 0.098
Coefficient of thermal expansion, (70–200°F) in./in./°F × $10^{-6}$: 13.1

Modulus of elasticity, psi: tension, $10 × 10^6$
Melting range, °F: 1,195–1,215
Thermal conductivity, Btu/ft² /hr/in./°F, 70°F: 1,536
Electrical resistivity, ohms/cmil/ft, 70°F: 16.8

**Heat treatments**
Annealing temperature 650°F.

### Tensile Properties

**Annealed**

| Temperature, °F | T.S., psi | Y.S., psi | Elong., in 2 in., % | Hardness, Brinell |
|---|---|---|---|---|
| 75 | 10,000 | 4,000 | 43 | 19 |
| 75 | 19,000 (H18) | 18,000 (H18) | 6 (H18) | 35(H18) |

### Table 10.1.1—2 (continued)
### WROUGHT ALUMINUM 1060

#### Fatigue Strength

| Test temperature, °F | Stress, psi | Cycles to failure |
|---|---|---|
| 75 | 3,000 | $5 \times 10^8$ (annealed) |
| 75 | 6,500 | $5 \times 10^8$ (H18) |

#### Specifications

ASTM: 996A; BD1S

### Table 10.1.1—3
### WROUGHT ALUMINUM 1100

**Typical chemical composition, percent by weight:** Mn, 0.05; Si + Fe, 1.0; Al, 99.0; Cu, 0.20; Zn, 0.10

**Physical constants and thermal properties**
Density, lb/in.³: 0.098
Coefficient of thermal expansion, (70–200°F) in./in./°F $\times 10^{-6}$: 13.1
Modulus of elasticity, psi: tension, $10 \times 10^6$

Melting range, °F: 1,190–1,215
Specific heat, Btu/lb/°F, 70°F: 0.22
Thermal conductivity, Btu/ft²/hr/in./°F, 70°F: 1,536
Electrical resistivity, ohms/cmil/ft, 70°F: 17.5

**Heat treatments**
Annealing temperature 650°F; hot working temperature 500–950°F.

#### Tensile Properties

| Temperature, °F | T. S., psi | Y. S., psi | Elong., in 2 in., % | Hardness, Brinell |
|---|---|---|---|---|
| **Annealed** | | | | |
| 75 | 13,000 | 5,000 | 45 | 23 |
| 212 | 11,000 | 5,000 | 45 | |
| 300 | 8,500 | 4,500 | 55 | |
| 400 | 6,000 | 3,500 | 65 | |
| 500 | 4,000 | 2,000 | 75 | |
| 600 | 2,500 | 1,500 | 80 | |
| 700 | 2,000 | 1,000 | 85 | |
| **H14 Temper** | | | | |
| 75 | 18,000 | 17,000 | 20 | 32 |
| 212 | 16,000 | 15,000 | 20 | |
| 300 | 13,000 | 12,000 | 22 | |
| 400 | 9,500 | 7,000 | 25 | |
| 500 | 4,000 | 2;500 | 75 | |
| 600 | 2,500 | 1,500 | 80 | |
| 700 | 2,000 | 1,000 | 85 | |

#### Fatigue Strength

| Test temperature, °F | Stress, psi | Cycles to failure |
|---|---|---|
| 75 | 5,000 | $5 \times 10^8$ (annealed) |
| 75 | 7,000 | $5 \times 10^8$ (H 14) |

#### Specifications

ASTM: 990A; SAE-AMS: 25; 2S

## Table 10.1.1—4
## WROUGHT ALUMINUM 2024

**Typical chemical composition, percent by weight:** Mn, 0.30–0.90; Fe, 0.5; Si, 0.5; Cr, 0.10; Al, balance; Zr, 0.10–0.25; Cu, 3.8–4.9; Cd, 0.05–0.20; Mg, 1.2–1.8; Zn, 0.25

Melting range, °F: 935–1,180
Specific heat, Btu/lb/°F, 70°F: 0.22
Thermal conductivity, Btu/ft² /hr/in./°F, 70°F: 1,310
Electrical resistivity, ohms/cmil/ft, 70°F: 20.7 (O); 34.50 (T4)

**Physical constants and thermal properties**
Density, lb/in.³ : 0.100
Coefficient of thermal expansion, (70–200°F) in./in./°F × $10^{-6}$ : 12.9
Modulus of elasticity, psi: tension, $10.6 \times 10^6$

**Heat treatments**
Annealing temperature (start) 775°F; solution temperature 930°F; aging temperature 375°F, 11–13 hr for T6.

### Tensile Properties

| Temperature, °F | T.S., psi | Y.S., psi, 0.2% offset | Elong., in 2 in., % | Hardness, Brinell |
|---|---|---|---|---|
| | | **Annealed** | | |
| 75 | 27,000 | 11,000 | 20 | 47 |
| | | **T3 Temper** | | |
| 75 | 65,000 | 45,000 | 18 | 120 |
| 300 | 55,000 | 50,000 | 11 | |
| 400 | 29,000 | 22,000 | 23 | |
| 500 | 12,000 | 9,000 | 55 | |
| 600 | 8,000 | 6,000 | 75 | |
| 700 | 5,500 | 4,000 | 100 | |

### Fatigue Strength

| Test temperature, °F | Stress, psi | Cycles to failure |
|---|---|---|
| 75 | 13,000 | $5 \times 10^8$ (annealed) |
| 75 | 20,000 | $5 \times 10^8$ (T3) |

### Specifications

ASTM: CG42A; SAE-AMS: 24; 24S

## Table 10.1.1—5
## WROUGHT ALUMINUM 3003

**Typical chemical composition, percent by weight:** Mn, 1.0–1.5; Fe, 0.7; Si, 0.6; Al, balance; Cu, 0.20; Zn, 0.10

Modulus of elasticity, psi: tension, $10 \times 10^6$
Melting range, °F: 1,190–1,210
Specific heat, Btu/lb/°F, 70°F: 0.22
Thermal conductivity, Btu/ft²/hr/in./°F, 70°F: 1,332
Electrical resistivity, ohms/cmil/ft, 70°F: 20.7

**Physical constants and thermal properties**
  Density, lb/in.³ : 0.099
  Coefficient of thermal expansion, (70–200°F) in./in./°F $\times 10^{-6}$ : 12.9

**Heat treatments**
  Annealing temperature 775°F; hot working temperature 500–950°F.

### Tensile Properties

| Temperature, °F | T.S., psi | Y.S., psi | Elong., in 2 in., % | Hardness, Brinell |
|---|---|---|---|---|
| | | Annealed | | |
| 75 | 16,000 | 6,000 | 40 | 28 |
| 212 | 13,000 | 5,500 | 43 | |
| 300 | 11,000 | 5,000 | 47 | |
| 400 | 8,500 | 4,500 | 60 | |
| 500 | 6,000 | 3,500 | 65 | |
| 600 | 4,000 | 2,500 | 70 | |
| 700 | 3,000 | 2,000 | 70 | |
| | | H14 Temper | | |
| 75 | 22,000 | 21,000 | 16 | 40 |
| 212 | 21,000 | 19,000 | 16 | |
| 300 | 18,000 | 16,000 | 16 | |
| 400 | 14,000 | 9,000 | 20 | |
| 500 | 7,500 | 4,000 | 60 | |
| 600 | 4,000 | 2,500 | 70 | |
| 700 | 3,000 | 2,000 | 70 | |

### Fatigue Strength

| Test temperature, °F | Stress, psi | Cycles to failure |
|---|---|---|
| 75 | 7,000 | $5 \times 10^8$ (annealed) |
| 75 | 9,000 | $5 \times 10^8$ (H14) |

### Specifications

ASTM: M1A; SAE-AMS: 29; 3S

## Table 10.1.1—6
## WROUGHT ALUMINUM 4032

Typical chemical composition, percent by weight: Fe, 1.0; Si, 11.0–13.5; Cr, 0.10; Ni, 0.50–1.3; Al, balance; Cu, 0.50–1.3; Mg, 0.8–1.3; Zn, 0.25

**Physical constants and thermal properties**
Density, lb/in.$^3$: 0.097
Coefficient of thermal expansion, (70–200°F) in./in./°F $\times 10^{-6}$: 10.8

Modulus of elasticity, psi: tension, $10.0 \times 10^6$
Melting range, °F: 990–1,060
Thermal conductivity, Btu/ft$^2$/hr/in./°F, 70°F: 1,070
Electrical resistivity, ohms/cmil/ft, 70°F: 26

**Heat treatments**
Anneal 775°F; solution temperature 940–960°F; age 335–345°F for 8–12 hr for T6.

### Tensile Properties

#### T6 Temper

| Temperature, °F | T.S., psi | Y.S., psi | Elong., in 2 in., % | Hardness, Brinell |
|---|---|---|---|---|
| 75 | 55,000 | 46,000 | 9 | 120 |
| 212 | 50,000 | 44,000 | 9 | |
| 300 | 37,000 | 33,000 | 9 | |
| 400 | 13,000 | 9,000 | 30 | |
| 500 | 8,000 | 5,500 | 50 | |
| 600 | 5,000 | 3,000 | 70 | |
| 700 | 3,500 | 2,000 | 90 | |

### Fatigue Strength

#### T6 Temper

| Test temperature, °F | Stress, psi | Cycles to failure |
|---|---|---|
| 75 | 16,000 | $5 \times 10^8$ |

## Table 10.1.1—7
## WROUGHT ALUMINUM 5050

**Typical chemical composition, percent by weight:** Mn, 0.10; Fe, 0.7; Si, 0.40; Cr, 0.10; Al, balance; Mg, 1.0–1.8; Cu, 0.20; Zn, 0.25

**Physical constants and thermal properties**
Density, lb/in.$^3$: 0.097
Coefficient of thermal expansion, (70–200°F) in./in./
°F × 10$^{-6}$: 13.2

Modulus of elasticity, psi: tension, $10 \times 10^6$
Melting range, °F: 1,160–1,205
Specific heat, Btu/lb/°F, 70°F: 0.22
Thermal conductivity, Btu/ft$^2$/hr/in./°F, 70°F: 1,332
Electrical resistivity, ohms/cmil/ft, 70°F: 20.4

**Heat treatments**
Annealing temperature 650°F.

### Tensile Properties

| Temperature, °F | T.S., psi | Y.S., psi | Elong., in 2 in., % | Hardness, Brinell |
|---|---|---|---|---|
| **Annealed** | | | | |
| 75 | 21,000 | 8,000 | 24 | 36 |
| 212 | 21,000 | 8,000 | 28 | |
| 300 | 19,000 | 8,000 | 38 | |
| 400 | 14,000 | 7,500 | 58 | |
| 500 | 9,000 | 6,000 | 67 | |
| 600 | 6,000 | 4,000 | 80 | |
| 700 | 4,000 | 3,000 | 95 | |
| **H34 Temper** | | | | |
| 75 | 28,000 | 24,000 | 8 | 53 |
| 212 | 32,000 | 27,000 | 18 | |
| 300 | 26,000 | 23,000 | 25 | |
| 400 | 14,000 | 7,000 | 58 | |
| 500 | 8,000 | 7,000 | 59 | |
| 600 | 6,000 | 4,000 | – | |
| 700 | 4,000 | 3,000 | – | |

### Fatigue Strength

| Test temperature, °F | Stress, psi | Cycles to failure |
|---|---|---|
| 75 | 12,000 | $5 \times 10^8$ (annealed) |
| 75 | 13,000 | $5 \times 10^8$ (H34) |

### Specifications

ASTM: 50S; R305; SAE-AMS: 207

## Table 10.1.1—8
## WROUGHT ALUMINUM 5056

**Typical chemical composition, percent by weight:** Mn, 0.05–0.20; Fe, 0.40; Si, 0.30; Cr, 0.05–0.20; Al, balance; Mg, 4.5–5.6; Cu, 0.10; Zn, 0.10

**Physical constants and thermal properties**
  Density, lb/in.$^3$: 0.095
  Coefficient of thermal expansion, (70–200°F) in./in./°F $\times$ 10$^{-6}$: 13.4

Modulus of elasticity, psi: tension, 10.3 $\times$ 10$^6$
Melting range, °F: 1,055–1,180
Specific heat, Btu/lb/°F, 70°F: 0.22
Thermal conductivity, Btu/ft/hr/in./°F, 70°F: 809
Electrical resistivity, ohms/cmil/ft, 70°F: 35.4

**Heat treatments**
  Annealing temperature 650°F.

### Tensile Properties

| Temperature, °F | T.S., psi | Y.S., psi | Elong., in 2 in., % | Hardness, Brinell |
|---|---|---|---|---|
| *Annealed* | | | | |
| 75 | 42,000 | 22,000 | 35 | 65 |
| 300 | 31,000 | 17,000 | 55 | |
| 400 | 22,000 | 13,000 | 65 | |
| 500 | 16,000 | 10,000 | 80 | |
| 600 | 11,000 | 7,000 | 100 | |
| 700 | 6,000 | 4,000 | 130 | |
| *H38 Temper* | | | | |
| 75 | 60,000 | 50,000 | 15 | 100 |
| 300 | 38,000 | 31,000 | 30 | |
| 400 | 26,000 | 18,000 | 50 | |
| 500 | 16,000 | 10,000 | 80 | |
| 600 | 11,000 | 7,000 | 100 | |
| 700 | 6,000 | 4,000 | 130 | |

### Fatigue Strength

| Test temperature, °F | Stress, psi | Cycles to failure |
|---|---|---|
| 75 | 20,000 | 5 $\times$ 10$^8$ (annealed) |
| 75 | 22,000 | 5 $\times$ 10$^8$ (H38) |

### Specifications

ASTM: GM50A; 56S

## Table 10.1.1—9
## WROUGHT ALUMINUM 6061

Typical chemical composition, percent by weight: Mn, 0.15; Fe, 0.7; Si, 0.4−0.8; Cr, 0.15−0.35; Ti, 0.15; Al, balance; Mg, 0.8−1.2; Cu, 0.15−0.40; Zn, 0.25

**Physical constants and thermal properties**

Density, lb/in.$^3$ : 0.098

Coefficient of thermal expansion, (70−200°F) in./in./°F $\times$ 10$^{-6}$ : 13.0

Modulus of elasticity, psi: tension, 10.0 $\times$ 10$^6$

Melting range, °F: 1,080−1,200

Specific heat, Btu/lb/°F, 70°F: 0.23

Thermal conductivity, Btu/ft$^2$/hr/in./°F, 70°F: 1,188

Electrical resistivity, ohms/cmil/ft, 70°F: 22.8

**Heat treatments**

Annealing temperature 775°F; solution temperature 970°F; aging temperature 320−350°F, 6−10 hr for T6.

### Tensile Properties

| Temperature, °F | T.S., psi | Y.S., psi, 0.2% offset | Elong., in 2 in., % | Hardness, Brinell |
|---|---|---|---|---|
| | | **Annealed** | | |
| 75 | 18,000 | 8,000 | 30 | 30 |
| 300 | 16,000 | 8,000 | 30 | |
| 400 | 9,000 | 6,500 | 55 | |
| 500 | 5,500 | 4,000 | 70 | |
| 600 | 4,000 | 2,500 | 85 | |
| 700 | 3,000 | 2,000 | 95 | |
| | | **T6 Temper** | | |
| 75 | 45,000 | 40,000 | 17 | 95 |
| 212 | 41,000 | 40,000 | 16 | |
| 300 | 34,000 | 31,000 | 20 | |
| 400 | 18,000 | 15,000 | 28 | |
| 500 | 10,000 | 7,000 | 50 | |
| 600 | 5,000 | 4,000 | 100 | |
| 700 | 3,000 | 2,000 | 95 | |

### Fatigue Strength

| Test temperature, °F | Stress, psi | Cycles to failure |
|---|---|---|
| 75 | 9,000 | 5 $\times$ 10$^8$ (annealed) |
| 75 | 14,000 | 5 $\times$ 10$^8$ (T6) |

### Specifications

ASTM: GS11A; SAE-AMS: 281; 61S

## Table 10.1.1—10
## WROUGHT ALUMINUM 7075

**Typical chemical composition, percent by weight:** Mn, 0.30; Fe, 0.70; Si, 0.50; Cr, 0.18–0.40; Ti, 0.20; Al, balance; Zn, 5.1–6.1; Mg, 2.1–2.9; Cu, 1.2–2.0

**Physical constants and thermal properties**
Density, lb/in.$^3$: 0.101
Coefficient of thermal expansion, (70–200°F) in./in./°F × 10$^{-6}$: 13.1

Modulus of elasticity, psi: tension, 10.4 × 10$^{-6}$
Melting range, °F: 890–1,180
Specific heat, Btu/lb/°F, 70°F: 0.23
Thermal conductivity, Btu/ft$^2$/hr/in./°F, 70°F: 840
Electrical resistivity, ohms/cmil/ft, 70°F: 34.2

**Heat treatments**
Annealing temperature 775°F; solution temperature 870°F; aging temperature 250°F, 24–28 hr for T6.

### Tensile Properties

| Temperature, °F | T.S., psi | Y.S., psi, 0.2% offset | Elong., in 2 in., % | Hardness, Brinell |
|---|---|---|---|---|
| | | **Annealed** | | |
| 75 | 33,000 | 15,000 | 16 | 60 |
| 300 | 19,000 | 13,000 | 40 | |
| 400 | 14,000 | 11,000 | 60 | |
| 500 | 11,000 | 9,000 | 65 | |
| 600 | 8,500 | 6,500 | 75 | |
| 700 | 6,500 | 5,000 | 70 | |
| | | **T6 Temper** | | |
| 75 | 83,000 | 73,000 | 11 | 150 |
| 212 | 74,000 | 68,000 | 11 | |
| 300 | 30,000 | 26,000 | 30 | |
| 400 | 18,000 | 16,000 | 45 | |
| 500 | 13,000 | 11,000 | 52 | |
| 600 | 8,500 | 6,500 | 75 | |
| 700 | 6,500 | 5,000 | 70 | |

### Fatigue Strength

**Heat Treated T6 Temper**

| Test temperature, °F | Stress, psi | Cycles to failure |
|---|---|---|
| 75 | 23,000 | 5 × 10$^8$ |

### Specifications

ASTM: 2G62A; SAE-AMS: 215; 75S

## 10.1.2 CAST ALUMINUM ALLOYS

### Table 10.1.2—1
### CAST ALUMINUM 201.0

**Typical chemical composition, percent by weight:** Cu, 4.7; Ag, 0.7; Al, balance

**Physical constants and thermal properties**
Density, lb/in.$^3$: 0.101
Coefficient of thermal expansion, (70–200°F) in./in./°F × 10$^{-6}$: 10.7
Melting range, °F: 1,200–1,060

Electrical conductivity (as cast), % IACS[a]: 27–32

[a]% IACS = percent of International Annealed Copper Standard, based on equal volume.

**Heat treatments**
Solution temperature 940–970 °F; aging temperature 305–315 °F.

### Tensile Properties

#### Solution Treated and Aged

| Temperature, °F | T.S., psi | Y.S., psi, 0.2% offset | Elong., in 2 in., % | Hardness, Brinell |
|---|---|---|---|---|
| 75 | 65,000 | 55,000 | 9 | 110 |

### Table 10.1.2—2
### CAST ALUMINUM 355.0

**Typical chemical composition, percent by weight:** Mn, 0.05 Fe, 0.16–0.30; Si, 4.5–5.5; Ti, 0.20; Al, balance; Cu, 1.0–1.5; Mg, 0.45–0.6; Zn, 0.05

**Physical constants and thermal properties**
Density, lb/in.$^3$: 0.098
Coefficient of thermal expansion, (70–200°F) in./in./°F × 10$^{-6}$: 12.4
Melting range, °F: 1,150–1,015

Thermal conductivity, Btu/ft$^2$/hr/in./°F, 70°F: 1,044
Electrical conductivity (as cast), % IACS[a]: 37

[a]% IACS = percent of International Annealed Copper Standard, based on equal volume.

**Heat treatments**
Annealing temperature 650°F; solution temperature 980°F; aging temperature 310°F.

## Table 10.1.2—2 (continued)
## CAST ALUMINUM 355.0

### Tensile Properties

| Form | Temperature, °F | T.S., psi | Y.S., psi, 0.2% offset | Elong., in 2 in., % | Hardness, Brinell |
|---|---|---|---|---|---|
| **Solution Treated and Aged** | | | | | |
| Sand casting | 75 | 35,000 | 25,000 | 3.0 | 80 |
| Permanent mold casting | 75 | 42,000 | 27,000 | 4.0 | 90 |
| **355 T6 Temper** | | | | | |
| Sand casting | 75 | 35,000 | 25,000 | 3 | |
| | 300 | 30,000 | 25,000 | 1.5 | |
| | 400 | 13,000 | 9,000 | 12 | |
| | 500 | 8,000 | 5,000 | 22 | |
| | 600 | 6,000 | 3,500 | 30 | |
| **355 T6 Temper** | | | | | |
| Permanent mold casting | 75 | 43,000 | 27,000 | 4 | |
| | 300 | 32,000 | 24,000 | 2 | |
| | 400 | 12,000 | 9,000 | 20 | |
| | 500 | 8,000 | 6,000 | 25 | |
| | 600 | 4,500 | 3,000 | 50 | |

### Fatigue Strength

#### Solution Treated and Aged

| Form | Test temperature, °F | Stress, psi | Cycles to failure |
|---|---|---|---|
| Sand casting | 75 | 9,000 | $5 \times 10^8$ |
| Permanent mold casting | 75 | 10,000 | $5 \times 10^8$ |

## Table 10.1.2—3
## CAST ALUMINUM 520.0

**Typical chemical composition, percent by weight:** Mg, 10.0; Al, balance

**Physical constants and thermal properties**
Density, lb/in.$^3$ : 0.093
Coefficient of thermal expansion, (70–200°F) in./in./°F × 10$^{-6}$ : 13.6
Melting range, °F: 1,120–840
Thermal conductivity, Btu/ft$^2$/hr/in./°F, 70°F: 612

Electrical conductivity (as cast), % IACS[a]: 21

[a]% IACS = percent of International Annealed Copper Standard, based on equal volume.

**Heat treatments**
Annealing temperature 650°F; solution temperature 810°F.

## Tensile Properties

### Solution Treated and Aged

| Temperature, °F | T.S., psi | Y.S., psi, 0.2% offset | Elong., in 2 in., % | Hardness, Brinell |
|---|---|---|---|---|
| 75 | 48,000 | 26,000 | 16 | 75 |

## 10.2 FERROUS ALLOYS*

### Table 10.2—1
### SAE ALLOY STEEL COMPOSITIONS

| SAE No. | \multicolumn{9}{c}{Composition[a], %} | | | | | | | | | Corresponding AISI No. |
|---|---|---|---|---|---|---|---|---|---|
| | C | Mn | P | S | Si | Ni | Cr | Other (Mo) | |
| 1330 | 0.28–0.33 | 1.60–1.90 | 0.035 | 0.040 | 0.20–0.35 | — | — | — | 1330 |
| 1335 | 0.33–0.38 | 1.60–1.90 | 0.035 | 0.040 | 0.20–0.35 | — | — | — | 1335 |
| 1340 | 0.38–0.43 | 1.60–1.90 | 0.035 | 0.040 | 0.20–0.35 | — | — | — | 1340 |
| 1345 | 0.43–0.48 | 1.60–1.90 | 0.035 | 0.040 | 0.20–0.35 | — | — | — | 1345 |
| 4012 | 0.09–0.14 | 0.75–1.00 | 0.035 | 0.040 | 0.20–0.35 | — | — | 0.15–0.25 | 4012 |
| 4023 | 0.20–0.25 | 0.70–0.90 | 0.035 | 0.040 | 0.20–0.35 | — | — | 0.20–0.30 | 4023 |
| 4024 | 0.20–0.25 | 0.70–0.90 | 0.035 | 0.035–0.050 | 0.20–0.35 | — | — | 0.20–0.30 | 4024 |
| 4027 | 0.25–0.30 | 0.70–0.90 | 0.035 | 0.040 | 0.20–0.35 | — | — | 0.20–0.30 | 4027 |
| 4028 | 0.25–0.30 | 0.70–0.90 | 0.035 | 0.035–0.050 | 0.20–0.35 | — | — | 0.20–0.30 | 4028 |
| 4032 | 0.30–0.35 | 0.70–0.90 | 0.035 | 0.040 | 0.20–0.35 | — | — | 0.20–0.30 | — |
| 4037 | 0.35–0.40 | 0.70–0.90 | 0.035 | 0.040 | 0.20–0.35 | — | — | 0.20–0.30 | 4037 |
| 4042 | 0.40–0.45 | 0.70–0.90 | 0.035 | 0.040 | 0.20–0.35 | — | — | 0.20–0.30 | — |
| 4047 | 0.45–0.50 | 0.70–0.90 | 0.035 | 0.040 | 0.20–0.35 | — | — | 0.20–0.30 | 4047 |
| 4118 | 0.18–0.23 | 0.70–0.90 | 0.035 | 0.040 | 0.20–0.35 | — | 0.40–0.60 | 0.08–0.15 | 4118 |
| 4130 | 0.28–0.33 | 0.40–0.60 | 0.035 | 0.040 | 0.20–0.35 | — | 0.80–1.10 | 0.15–0.25 | 4130 |
| 4135 | 0.33–0.38 | 0.70–0.90 | 0.035 | 0.040 | 0.20–0.35 | — | 0.80–1.10 | 0.15–0.25 | — |
| 4137 | 0.35–0.40 | 0.70–0.90 | 0.035 | 0.040 | 0.20–0.35 | — | 0.80–1.10 | 0.15–0.25 | 4137 |
| 4140 | 0.38–0.43 | 0.75–1.00 | 0.035 | 0.040 | 0.20–0.35 | — | 0.80–1.10 | 0.15–0.25 | 4140 |
| 4142 | 0.40–0.45 | 0.75–1.00 | 0.035 | 0.040 | 0.20–0.35 | — | 0.80–1.10 | 0.15–0.25 | 4142 |
| 4145 | 0.43–0.48 | 0.75–1.00 | 0.035 | 0.040 | 0.20–0.35 | — | 0.80–1.10 | 0.15–0.25 | 4145 |
| 4147 | 0.45–0.50 | 0.75–1.00 | 0.035 | 0.040 | 0.20–0.35 | — | 0.80–1.10 | 0.15–0.25 | 4147 |
| 4150 | 0.48–0.53 | 0.75–1.00 | 0.035 | 0.040 | 0.20–0.35 | — | 0.80–1.10 | 0.15–0.25 | 4150 |
| 4161 | 0.56–0.64 | 0.75–1.00 | 0.035 | 0.040 | 0.20–0.35 | — | 0.70–0.90 | 0.25–0.35 | 4161 |
| 4320 | 0.17–0.22 | 0.45–0.65 | 0.035 | 0.040 | 0.20–0.35 | 1.65–2.00 | 0.40–0.60 | 0.20–0.30 | 4320 |
| 4340 | 0.38–0.43 | 0.60–0.80 | 0.035 | 0.040 | 0.20–0.35 | 1.65–2.00 | 0.70–0.90 | 0.20–0.30 | 4340 |
| 4419 | 0.18–0.23 | 0.45–0.65 | 0.035 | 0.040 | 0.20–0.35 | — | — | 0.45–0.60 | 4419 |
| 4422 | 0.20–0.25 | 0.70–0.90 | 0.035 | 0.040 | 0.20–0.35 | — | — | 0.35–0.45 | — |
| 4427 | 0.24–0.29 | 0.70–0.90 | 0.035 | 0.040 | 0.20–0.35 | — | — | 0.35–0.45 | — |

* Based on section in the *Handbook of Materials Science*, Volume II, pages 4 through 118.

## Table 10.2—1 (continued)
## SAE ALLOY STEEL COMPOSITIONS

| SAE No. | Composition[a], % | | | | | | | | Corresponding AISI No. |
|---|---|---|---|---|---|---|---|---|---|
| | C | Mn | P | S | Si | Ni | Cr | Other | |
| 4615 | 0.13–0.18 | 0.45–0.65 | 0.035 | 0.040 | 0.20–0.35 | 1.65–2.00 | — | 0.20–0.30 | 4615 |
| 4617 | 0.15–0.20 | 0.45–0.65 | 0.035 | 0.040 | 0.20–0.35 | 1.65–2.00 | — | 0.20–0.30 | — |
| 4620 | 0.17–0.22 | 0.45–0.65 | 0.035 | 0.040 | 0.20–0.35 | 1.65–2.00 | — | 0.20–0.30 | 4620 |
| 4621 | 0.18–0.23 | 0.70–0.90 | 0.035 | 0.040 | 0.20–0.35 | 1.65–2.00 | — | 0.20–0.30 | 4621 |
| 4626 | 0.24–0.29 | 0.45–0.65 | 0.035 | 0.04 max | 0.20–0.35 | 0.70–1.00 | — | 0.15–0.25 | 4626 |
| 4718 | 0.16–0.21 | 0.70–0.90 | — | — | — | 0.90–1.20 | 0.35–0.55 | 0.30–0.40 | 4718 |
| 4720 | 0.17–0.22 | 0.50–0.70 | 0.035 | 0.040 | 0.20–0.35 | 0.90–1.20 | 0.35–0.55 | 0.15–0.25 | 4720 |
| 4815 | 0.13–0.18 | 0.40–0.60 | 0.035 | 0.040 | 0.20–0.35 | 3.25–3.75 | — | 0.20–0.30 | 4815 |
| 4817 | 0.15–0.20 | 0.40–0.60 | 0.035 | 0.040 | 0.20–0.35 | 3.25–3.75 | — | 0.20–0.30 | 4817 |
| 4820 | 0.18–0.23 | 0.50–0.70 | 0.035 | 0.040 | 0.20–0.35 | 3.25–3.75 | — | 0.20–0.30 | 4820 |
| 5015 | 0.12–0.17 | 0.30–0.50 | 0.035 | 0.040 | 0.20–0.35 | — | 0.30–0.50 | — | 5015 |
| 5046 | 0.43–0.48 | 0.75–1.00 | 0.035 | 0.040 | 0.20–0.35 | — | 0.20–0.35 | — | — |
| 5060 | 0.56–0.64 | 0.75–1.00 | 0.035 | 0.040 | 0.20–0.35 | — | 0.40–0.60 | — | — |
| 5115 | 0.13–0.18 | 0.70–0.90 | 0.035 | 0.040 | 0.20–0.35 | — | 0.70–0.90 | — | — |
| 5120 | 0.17–0.22 | 0.70–0.90 | 0.035 | 0.040 | 0.20–0.35 | — | 0.70–0.90 | — | 5120 |
| 5130 | 0.28–0.33 | 0.70–0.90 | 0.035 | 0.040 | 0.20–0.35 | — | 0.80–1.10 | — | 5130 |
| 5132 | 0.30–0.35 | 0.60–0.80 | 0.035 | 0.040 | 0.20–0.35 | — | 0.75–1.00 | — | 5132 |
| 5135 | 0.33–0.38 | 0.60–0.80 | 0.035 | 0.040 | 0.20–0.35 | — | 0.80–1.05 | — | 5135 |
| 5140 | 0.38–0.43 | 0.70–0.90 | 0.035 | 0.040 | 0.20–0.35 | — | 0.70–0.90 | — | 5140 |
| 5145 | 0.43–0.48 | 0.70–0.90 | 0.035 | 0.040 | 0.20–0.35 | — | 0.70–0.90 | — | 5145 |
| 5147 | 0.46–0.51 | 0.70–0.95 | 0.035 | 0.040 | 0.20–0.35 | — | 0.85–1.15 | — | 5147 |
| 5150 | 0.48–0.53 | 0.70–0.90 | 0.035 | 0.040 | 0.20–0.35 | — | 0.70–0.90 | — | 5150 |
| 5155 | 0.51–0.59 | 0.70–0.90 | 0.035 | 0.040 | 0.20–0.35 | — | 0.70–0.90 | — | 5155 |
| 5160 | 0.56–0.64 | 0.75–1.00 | 0.035 | 0.040 | 0.20–0.35 | — | 0.70–0.90 | — | 5160 |
| 6118 | 0.16–0.21 | 0.50–0.70 | 0.035 | 0.040 | 0.20–0.35 | — | 0.50–0.70 | 0.10–0.15 | 6118 |
| 6150 | 0.48–0.53 | 0.70–0.90 | 0.035 | 0.040 | 0.20–0.35 | — | 0.80–1.10 | 0.15 Mo | 6150 |
| 8115 | 0.13–0.18 | 0.70–0.90 | 0.035 | 0.040 | 0.20–0.35 | 0.20–0.40 | 0.30–0.50 | 0.08–0.15 | 8115 |
| 8615 | 0.13–0.18 | 0.70–0.90 | 0.035 | 0.040 | 0.20–0.35 | 0.40–0.70 | 0.40–0.60 | 0.15–0.25 | 8615 |
| 8617 | 0.15–0.20 | 0.70–0.90 | 0.035 | 0.040 | 0.20–0.35 | 0.40–0.70 | 0.40–0.60 | 0.15–0.25 | 8617 |
| 8620 | 0.18–0.23 | 0.70–0.90 | 0.035 | 0.040 | 0.20–0.35 | 0.40–0.70 | 0.40–0.60 | 0.15–0.25 | 8620 |

## Table 10.2—1 (continued)
## SAE ALLOY STEEL COMPOSITIONS

| SAE No. | Composition[a], % | | | | | | | | Corresponding AISI No. |
|---|---|---|---|---|---|---|---|---|---|
| | C | Mn | P | S | Si | Ni | Cr | Other | |
| 8622 | 0.20–0.25 | 0.70–0.90 | 0.035 | 0.040 | 0.20–0.35 | 0.40–0.70 | 0.40–0.60 | 0.15–0.25 | 8622 |
| 8625 | 0.23–0.28 | 0.70–0.90 | 0.035 | 0.040 | 0.20–0.35 | 0.40–0.70 | 0.40–0.60 | 0.15–0.25 | 8625 |
| 8627 | 0.25–0.30 | 0.70–0.90 | 0.035 | 0.040 | 0.20–0.35 | 0.40–0.70 | 0.40–0.60 | 0.15–0.25 | 8627 |
| 8630 | 0.28–0.33 | 0.70–0.90 | 0.035 | 0.040 | 0.20–0.35 | 0.40–0.70 | 0.40–0.60 | 0.15–0.25 | 8630 |
| 8637 | 0.35–0.40 | 0.75–1.00 | 0.035 | 0.040 | 0.20–0.35 | 0.40–0.70 | 0.40–0.60 | 0.15–0.25 | 8637 |
| 8640 | 0.38–0.43 | 0.75–1.00 | 0.035 | 0.040 | 0.20–0.35 | 0.40–0.70 | 0.40–0.60 | 0.15–0.25 | 8640 |
| 8642 | 0.40–0.45 | 0.75–1.00 | 0.035 | 0.040 | 0.20–0.35 | 0.40–0.70 | 0.40–0.60 | 0.15–0.25 | 8642 |
| 8645 | 0.43–0.48 | 0.75–1.00 | 0.035 | 0.040 | 0.20–0.35 | 0.40–0.70 | 0.40–0.60 | 0.15–0.25 | 8645 |
| 8650 | 0.48–0.53 | 0.75–1.00 | 0.035 | 0.040 | 0.20–0.35 | 0.40–0.70 | 0.40–0.60 | 0.15–0.25 | — |
| 8655 | 0.51–0.59 | 0.75–1.00 | 0.035 | 0.040 | 0.20–0.35 | 0.40–0.70 | 0.40–0.60 | 0.15–0.25 | 8655 |
| 8660 | 0.56–0.64 | 0.75–1.00 | 0.035 | 0.040 | 0.20–0.35 | 0.40–0.70 | 0.40–0.60 | 0.15–0.25 | — |
| 8720 | 0.18–0.23 | 0.70–0.90 | 0.035 | 0.040 | 0.20–0.35 | 0.40–0.70 | 0.40–0.60 | 0.20–0.30 | 8720 |
| 8740 | 0.38–0.43 | 0.75–1.00 | 0.035 | 0.040 | 0.20–0.35 | 0.40–0.70 | 0.40–0.60 | 0.20–0.30 | 8740 |
| 8822 | 0.20–0.25 | 0.75–1.00 | 0.035 | 0.040 | 0.20–0.35 | 0.40–0.70 | 0.40–0.60 | 0.30–0.40 | 8822 |
| 9254 | 0.51–0.59 | 0.50–0.80 | 0.035 | 0.040 | 1.20–1.60 | — | 0.50–0.80 | — | — |
| 9255 | 0.51–0.59 | 0.70–0.95 | 0.035 | 0.040 | 1.80–2.20 | — | — | — | 9255 |
| 9260 | 0.56–0.64 | 0.75–1.00 | 0.035 | 0.040 | 1.80–2.20 | — | — | — | 9260 |
| 9310 | 0.08–0.13 | 0.45–0.65 | 0.025 | 0.025 | 0.20–0.35 | 3.00–3.50 | 1.00–1.40 | 0.08–0.15 | — |

[a]Small quantities of other elements are present which are acceptable to the following amounts: 0.35 Cu, 0.25 Ni, 0.20 Cr, and 0.06 Mo.

Reprinted with permission of the American Iron and Steel Society.

## Table 10.2—2
### AISI-SAE STANDARD CARBON STEELS

#### A. Free-machining Grades

Composition[a], %

| AISI No. | C | Mn | P | S | SAE No. |
|---|---|---|---|---|---|
| | | | **Resulfurized** | | |
| – | 0.08–0.13 | 0.50–0.80 | 0.040 max | 0.08–0.13 | 1108 |
| 1109 | 0.08–0.13 | 0.60–0.90 | 0.040 max | 0.08–0.13 | 1109 |
| 1110 | 0.08–0.13 | 0.30–0.60 | 0.040 max | 0.08–0.13 | 1110 |
| 1116 | 0.14–0.20 | 1.10–1.40 | 0.040 max | 0.16–0.23 | 1116 |
| 1117 | 0.14–0.20 | 1.00–1.30 | 0.040 max | 0.08–0.13 | 1117 |
| 1118 | 0.14–0.20 | 1.30–1.60 | 0.040 max | 0.08–0.13 | 1118 |
| 1119 | 0.14–0.20 | 1.00–1.30 | 0.040 max | 0.24–0.33 | 1119 |
| 1132 | 0.27–0.34 | 1.35–1.65 | 0.040 max | 0.08–0.13 | 1132 |
| 1137 | 0.32–0.39 | 1.35–1.65 | 0.040 max | 0.08–0.13 | 1137 |
| 1139 | 0.35–0.43 | 1.35–1.65 | 0.040 max | 0.13–0.20 | 1139 |
| 1140 | 0.37–0.44 | 0.70–1.00 | 0.040 max | 0.08–0.13 | 1140 |
| 1141 | 0.37–0.45 | 1.35–1.65 | 0.040 max | 0.08–0.13 | 1141 |
| 1144 | 0.40–0.48 | 1.35–1.65 | 0.040 max | 0.24–0.33 | 1144 |
| 1145 | 0.42–0.49 | 0.70–1.00 | 0.040 max | 0.04–0.07 | 1145 |
| 1146 | 0.42–0.49 | 0.70–1.00 | 0.040 max | 0.08–0.13 | 1146 |
| 1151 | 0.48–0.55 | 0.70–1.00 | 0.040 max | 0.08–0.13 | 1151 |
| | | **Resulfurized and Rephosphorized** | | | |
| 1211 | 0.13 max | 0.60–0.90 | 0.07–0.12 | 0.10–0.15 | 1211 |
| 1212 | 0.13 max | 0.70–1.00 | 0.07–0.12 | 0.16–0.23 | 1212 |
| 1213 | 0.13 max | 0.70–1.00 | 0.07–0.12 | 0.24–0.33 | 1213 |
| 1215 | 0.09 max | 0.75–1.05 | 0.04–0.09 | 0.26–0.35 | 1215 |
| 12L14 | 0.15 max | 0.85–1.15 | 0.04–0.09 | 0.26–0.35 | 12L14 |

#### B. Nonresulfurized Grades

Composition[a], %

| AISI No. | C | Mn | P max | S max | SAE No. |
|---|---|---|---|---|---|
| – | 0.06 max | 0.35 max | 0.040 | 0.050 | 1005 |
| 1006 | 0.08 max | 0.25–0.40 | 0.040 | 0.050 | 1006 |
| 1008 | 0.10 max | 0.30–0.50 | 0.040 | 0.050 | 1008 |
| 1010 | 0.08–0.13 | 0.30–0.60 | 0.040 | 0.050 | 1010 |
| – | 0.08–0.13 | 0.60–0.90 | 0.040 | 0.050 | 1011 |
| 1012 | 0.10–0.15 | 0.30–0.60 | 0.040 | 0.050 | 1012 |
| – | 0.11–0.16 | 0.50–0.80 | 0.040 | 0.050 | 1013 |
| 1513 | 0.10–0.16 | 1.10–1.40 | 0.040 | 0.050 | 1513 |
| 1015 | 0.13–0.18 | 0.30–0.60 | 0.040 | 0.050 | 1015 |
| 1016 | 0.13–0.18 | 0.60–0.90 | 0.040 | 0.050 | 1016 |
| 1017 | 0.15–0.20 | 0.30–0.60 | 0.040 | 0.050 | 1017 |
| 1018 | 0.15–0.20 | 0.60–0.90 | 0.040 | 0.050 | 1018 |
| 1518 | 0.15–0.21 | 1.10–1.40 | 0.040 | 0.050 | 1518 |
| 1019 | 0.15–0.20 | 0.70–1.00 | 0.040 | 0.050 | 1019 |
| 1020 | 0.18–0.23 | 0.30–0.60 | 0.040 | 0.050 | 1020 |

**Table 10.2—2 (continued)**
**AISI-SAE STANDARD CARBON STEELS**
B. Nonresulfurized Grades (continued)
Composition[a], %

| AISI No. | C | Mn | P max | S max | SAE No. |
|---|---|---|---|---|---|
| 1022 | 0.18–0.23 | 0.70–1.00 | 0.040 | 0.050 | 1022 |
| 1522 | 0.18–0.24 | 1.10–1.40 | 0.040 | 0.050 | 1522 |
| 1023 | 0.20–0.25 | 0.30–0.60 | 0.040 | 0.050 | 1023 |
| 1524 | 0.19–0.25 | 1.35–1.65 | 0.040 | 0.050 | 1524 |
| 1025 | 0.22–0.28 | 0.30–0.60 | 0.040 | 0.050 | 1025 |
| 1525 | 0.23–0.29 | 0.80–1.10 | 0.040 | 0.050 | 1525 |
| 1026 | 0.22–0.28 | 0.60–0.90 | 0.040 | 0.050 | 1026 |
| 1526 | 0.22–0.29 | 1.10–1.40 | 0.040 | 0.050 | 1526 |
| 1527 | 0.22–0.29 | 1.20–1.50 | 0.040 | 0.050 | 1527 |
| 1030 | 0.28–0.34 | 0.60–0.90 | 0.040 | 0.050 | 1030 |
| 1035 | 0.32–0.38 | 0.60–0.90 | 0.040 | 0.050 | 1035 |
| 1536 | 0.30–0.37 | 1.20–1.50 | 0.040 | 0.050 | 1536 |
| 1037 | 0.32–0.38 | 0.70–1.00 | 0.040 | 0.050 | 1037 |
| 1038 | 0.35–0.42 | 0.60–0.90 | 0.040 | 0.050 | 1038 |
| 1039 | 0.37–0.44 | 0.70–1.00 | 0.040 | 0.050 | 1039 |
| 1040 | 0.37–0.44 | 0.60–0.90 | 0.040 | 0.050 | 1040 |
| 1541 | 0.36–0.44 | 1.35–1.65 | 0.040 | 0.050 | 1541 |
| 1042 | 0.40–0.47 | 0.60–0.90 | 0.040 | 0.050 | 1042 |
| 1044 | 0.43–0.50 | 0.30–0.60 | 0.040 | 0.050 | 1044 |
| 1045 | 0.43–0.50 | 0.60–0.90 | 0.040 | 0.050 | 1045 |
| 1046 | 0.43–0.50 | 0.70–1.00 | 0.040 | 0.050 | 1046 |
| 1547 | 0.45–0.51 | 1.35–1.65 | 0.040 | 0.050 | 1547 |
| 1548 | 0.44–0.52 | 1.10–1.40 | 0.040 | 0.050 | 1548 |
| 1049 | 0.46–0.53 | 0.60–0.90 | 0.040 | 0.050 | 1049 |
| 1050 | 0.48–0.55 | 0.60–0.90 | 0.040 | 0.050 | 1050 |
| 1551 | 0.45–0.56 | 0.85–1.15 | 0.040 | 0.050 | 1551 |
| 1552 | 0.47–0.55 | 1.20–1.50 | 0.040 | 0.050 | 1552 |
| 1053 | 0.48–0.55 | 0.70–1.00 | 0.040 | 0.050 | 1053 |
| 1055 | 0.50–0.60 | 0.60–0.90 | 0.040 | 0.050 | 1055 |
| 1060 | 0.55–0.65 | 0.60–0.90 | 0.040 | 0.050 | 1060 |
| 1561 | 0.55–0.65 | 0.75–1.05 | 0.040 | 0.050 | 1561 |
| – | 0.60–0.70 | 0.60–0.90 | 0.040 | 0.050 | 1065 |
| 1566 | 0.60–0.71 | 0.85–1.15 | 0.040 | 0.050 | 1566 |
| – | 0.65–0.75 | 0.40–0.70 | 0.040 | 0.050 | 1069 |
| 1070 | 0.65–0.75 | 0.60–0.90 | 0.040 | 0.050 | 1070 |
| 1572 | 0.65–0.76 | 1.00–1.30 | 0.040 | 0.050 | 1572 |
| – | 0.70–0.80 | 0.50–0.80 | 0.040 | 0.050 | 1074 |
| – | 0.70–0.80 | 0.40–0.70 | 0.040 | 0.050 | 1075 |
| 1078 | 0.72–0.85 | 0.30–0.60 | 0.040 | 0.050 | 1078 |
| 1080 | 0.75–0.88 | 0.60–0.90 | 0.040 | 0.050 | 1080 |

## Table 10.2—2 (continued)
### AISI-SAE STANDARD CARBON STEELS

B. Nonresulfurized Grades (continued)

Composition[a], %

| AISI No. | C | Mn | P max | S max | SAE No. |
|----------|-----------|-----------|-------|-------|---------|
| 1084 | 0.80–0.93 | 0.60–0.90 | 0.040 | 0.050 | 1084 |
| – | 0.80–0.93 | 0.70–1.00 | 0.040 | 0.050 | 1085 |
| – | 0.80–0.93 | 0.30–0.50 | 0.040 | 0.050 | 1086 |
| 1090 | 0.85–0.98 | 0.60–0.90 | 0.040 | 0.050 | 1090 |
| 1095 | 0.90–1.03 | 0.30–0.50 | 0.040 | 0.050 | 1095 |

[a]When silicon is required, it is added from 0.10 to 0.30%. In some steels, a silicon content of 0.30 to 2.20% can be specified. Silicon is not generally added to resulfurized or rephophorized steels.

Reprinted with permission of the American Iron and Steel Society.

## Table 10.2—3
## STANDARD TYPES OF STAINLESS AND HEAT-RESISTING STEELS[a]

### Chemical Ranges and Limits

Chemical composition, % (maximum unless otherwise shown)

| Type number | C | Mn, max | P, max | S, max | Si, max | Cr | Ni | Mo | Zr | Se | Ti | Nb-Ta | Ta | Al | N |
|---|---|---|---|---|---|---|---|---|---|---|---|---|---|---|---|
| 201[b] | 0.15 max | 5.50/ 7.50 | 0.060 | 0.030 | 1.00 | 16.00/ 18.00 | 3.50/ 5.50 | | | | | | | | 0.25 max |
| 202[b] | 0.15 max | 7.50/ 10.00 | 0.060 | 0.030 | 1.00 | 17.00/ 19.00 | 4.00/ 6.00 | | | | | | | | 0.25 max |
| 301[b] | 0.15 max | 2.00 | 0.045 | 0.030 | 1.00 | 16.00/ 18.00 | 6.00/ 8.00 | | | | | | | | |
| 302[b] | 0.15 max | 2.00 | 0.045 | 0.030 | 1.00 | 17.00/ 19.00 | 8.00/ 10.00 | | | | | | | | |
| 302B[b] | 0.15 max | 2.00 | 0.045 | 0.030 | 2.00/ 3.00 | 17.00/ 19.00 | 8.00/ 10.00 | | | | | | | | |
| 303[b] | 0.15 max | 2.00 | 0.20 | 0.15 min | 1.00 | 17.00/ 19.00 | 8.00/ 10.00 | 0.60 max[c] | | | | | | | |
| 303Se[b] | 0.15 max | 2.00 | 0.20 | 0.06 | 1.00 | 17.00/ 19.00 | 8.00/ 10.00 | | 0.60 max[c] | 0.15 min | | | | | |
| 304[b] | 0.08 max | 2.00 | 0.045 | 0.030 | 1.00 | 18.00/ 20.00 | 8.00/ 12.00 | | | | | | | | |
| 304L[b] | 0.03 max | 2.00 | 0.045 | 0.030 | 1.00 | 18.00/ 20.00 | 8.00/ 12.00 | | | | | | | | |
| 305[b] | 0.12 max | 2.00 | 0.045 | 0.030 | 1.00 | 17.00/ 19.00 | 10.00/ 13.00 | | | | | | | | |
| 308[b] | 0.08 max | 2.00 | 0.045 | 0.030 | 1.00 | 19.00/ 21.00 | 10.00/ 12.00 | | | | | | | | |
| 309[b] | 0.20 max | 2.00 | 0.045 | 0.030 | 1.00 | 22.00/ 24.00 | 12.00/ 15.00 | | | | | | | | |
| 309S[b] | 0.08 max | 2.00 | 0.045 | 0.030 | 1.00 | 22.00/ 24.00 | 12.00/ 15.00 | | | | | | | | |
| 310[b] | 0.25 max | 2.00 | 0.045 | 0.030 | 1.50 | 24.00/ 26.00 | 19.00/ 22.00 | | | | | | | | |
| 310S[b] | 0.08 max | 2.00 | 0.045 | 0.030 | 1.50 | 24.00/ 26.00 | 19.00/ 22.00 | | | | | | | | |
| 314[b] | 0.25 max | 2.00 | 0.045 | 0.030 | 1.50/ 3.00 | 23.00/ 26.00 | 19.00/ 22.00 | | | | | | | | |
| 316[b] | 0.08 max | 2.00 | 0.045 | 0.030 | 1.00 | 16.00/ 18.00 | 10.00/ 14.00 | 2.00/ 3.00 | | | | | | | |

# Table 10.2—3 (continued)
## STANDARD TYPES OF STAINLESS AND HEAT-RESISTING STEELS

### Chemical Ranges and Limits (continued)

Chemical composition, % (maximum unless otherwise shown)

| Type number | C | Mn, max | P, max | S, max | Si, max | Cr | Ni | Mo | Zr | Se | Ti | Nb-Ta | Ta | Al | N |
|---|---|---|---|---|---|---|---|---|---|---|---|---|---|---|---|
| 316L[b] | 0.03 max | 2.00 | 0.045 | 0.030 | 1.00 | 16.00/18.00 | 10.00/14.00 | 2.00/3.00 | | | | | | | |
| 317[b] | 0.08 max | 2.00 | 0.045 | 0.030 | 1.00 | 18.00/20.00 | 11.00/15.00 | 3.00/4.00 | | | | | | | |
| 321[b] | 0.08 max | 2.00 | 0.045 | 0.030 | 1.00 | 17.00/19.00 | 9.00/12.00 | | | | 5×C min | | | | |
| 347[b] | 0.08 max | 2.00 | 0.045 | 0.030 | 1.00 | 17.00/19.00 | 9.00/13.00 | | | | | 10×C min | | | |
| 348[b] | 0.08 max | 2.00 | 0.045 | 0.030 | 1.00 | 17.00/19.00 | 9.00/13.00 | | | | | 10×C min | 0.10 max | | |
| 403[d] | 0.15 max | 1.00 | 0.040 | 0.030 | 0.50 | 11.50/13.00 | | | | | | | | | |
| 405[e] | 0.08 max | 1.00 | 0.040 | 0.030 | 1.00 | 11.50/14.50 | | | | | | | | 0.10/0.30 | |
| 409 | 0.08 max | 1.00 | 0.045 | 0.045 | 1.00 | 10.50/11.75 | | | | | 6×C min 0.75 max | | | | |
| 410[d] | 0.15 max | 1.00 | 0.040 | 0.030 | 1.00 | 11.50/13.50 | | | | | | | | | |
| 414[d] | 0.15 max | 1.00 | 0.040 | 0.030 | 1.00 | 11.50/13.50 | 1.25/2.50 | | | | | | | | |
| 416[d] | 0.15 max | 1.25 | 0.06 | 0.15 min | 1.00 | 12.00/14.00 | | 0.60 max[c] | 0.60 max[c] | | | | | | |
| 416 Se[d] | 0.15 max | 1.25 | 0.06 | 0.06 | 1.00 | 12.00/14.00 | | | | 0.15 min | | | | | |
| 420[d] | Over 0.15 max | 1.00 | 0.040 | 0.030 | 1.00 | 12.00/14.00 | | | | | | | | | |
| 430[e] | 0.12 max | 1.00 | 0.040 | 0.030 | 1.00 | 14.00/18.00 | | | | | | | | | |
| 430F[e] | 0.12 max | 1.25 | 0.06 | 0.15 min | 1.00 | 14.00/18.00 | | 0.60 max[c] | 0.60 max[c] | | | | | | |
| 430F Se[e] | 0.12 max | 1.25 | 0.06 | 0.06 | 1.00 | 14.00/18.00 | | | | 0.15 min | | | | | |
| 431[d] | 0.20 max | 1.00 | 0.040 | 0.030 | 1.00 | 15.00/17.00 | 1.25/2.50 | | | | | | | | |

**Table 10.2—3 (continued)**
**STANDARD TYPES OF STAINLESS AND HEAT-RESISTING STEELS**

Chemical Ranges and Limits (continued)

| Type number | C | Mn, max | P, max | S, max | Si, max | Cr | Ni | Mo | Zr | Se | Ti | Nb-Ta | Ta | Al | N |
|---|---|---|---|---|---|---|---|---|---|---|---|---|---|---|---|
| | | | | | | Chemical composition, % (maximum unless otherwise shown) | | | | | | | | | |
| 440A[d] | 0.60/ 0.75 | 1.00 | 0.040 | 0.030 | 1.00 | 16.00/ 18.00 | | 0.75 max | | | | | | | |
| 440B[d] | 0.75/ 0.95 | 1.00 | 0.040 | 0.030 | 1.00 | 16.00/ 18.00 | | 0.75 max | | | | | | | |
| 440C[d] | 0.95/ 1.20 | 1.00 | 0.040 | 0.030 | 1.00 | 16.00/ 18.00 | | 0.75 max | | | | | | | |
| 446[e] | 0.20 max | 1.50 | 0.040 | 0.030 | 1.00 | 23.00/ 27.00 | | | | | | | | | 0.25 max |
| 501[d] | Over 0.10 | 1.00 | 0.040 | 0.030 | 1.00 | 4.00/ 6.00 | | 0.40/ 0.65 | | | | | | | |
| 502[d] | 0.10 max | 1.00 | 0.040 | 0.030 | 1.00 | 4.00/ 6.00 | | 0.40/ 0.65 | | | | | | | |

[a]Subject to tolerances for check, product, or verification analyses.
[b]Not heat treatable.
[c]Added at producer's option.
[d]Heat treatable.
[e]Essentially not heat treatable.

Courtesy of American Iron and Steel Institute.

## Table 10.2—4
## GRAY IRONS – CAST

### Class 20

**Typical chemical composition, percent by weight:** C, 3.5–3.8; Mn, 0.50–0.70; Fe, balance; S, 0.08–0.13; Si, 2.4–2.6; P, 0.2–0.8

**Physical constants and thermal properties**

Density, lb/in.$^3$ : 0.25

Coefficient of thermal expansion, (70–200°F) in./in./°F $\times 10^{-6}$ : 6

Modulus of elasticity, psi: tension, $12 \times 10^6$

Melting point, °F: ~2150

Thermal conductivity, Btu/ft$^2$/hr/in./°F, 70°F: 336–360

Electrical resistivity, ohms/cmil/ft, 70°F: 300–1200

### Short Time Tensile Properties

**As Cast**

| Temperature,°F | T.S., psi | Y.S., psi, 0.2% offset | Elong., in 2 in., % | Hardness, Brinell |
|---|---|---|---|---|
| 70 | 20,000–25,000 | – | – | 140–180 |

### Fatigue Strength

**As Cast**

| Test temperature, °F | Stress, psi |
|---|---|
| 70 | 10,000 endurance limit |

## Table 10.2—5
## MALLEABLE IRONS – CAST

### Ferritic 32510

**Typical chemical composition, percent by weight:** C, 2.3–2.7; Mn, 0.55 max; Fe, balance; S, < 0.15; Si, 1.5–0.8; P, < 0.18

**Physical constants and thermal properties**
Density, lb/in.$^3$: 0.25–0.263
Coefficient of thermal expansion, (70–200°F) in./in./°F × 10$^{-6}$: 5.9
Modulus of elasticity, psi: tension, 25 × 10$^{-6}$
Poisson's ratio: 0.17
Melting point, °F: ~ 2250
Thermal conductivity, Btu/ft$^2$/hr/in./°F, 70°F: 354
Electrical resistivity, ohms/cmil/ft, 70°F: 84.6
Curie temperature, °F: annealed, 2300

### Short Time Tensile Properties

**As Cast**

| Temperature, °F | T.S., psi | Y.S., psi, 0.2% offset | Elong., in 2 in., % | Hardness, Brinell |
|---|---|---|---|---|
| 70 | 50,000 | 32,500 | 10 | 110–156 |

### Fatigue Strength

**As Cast**

| Test temperature, °F | Stress, psi |
|---|---|
| 70 | 28,000 endurance limit |

### Impact Strength

**As Cast**

| Test temperature, °F | Type test | Strength, ft-lb |
|---|---|---|
| –70 | Charpy – V-notched[a] | 16.5 |
| +70 | Charpy – unnotched | 70–90 |

[a]0.394-in. square bar, 0.079-in. notch.

## Table 10.2—6
### MALLEABLE IRONS – CAST

Pearlitic 45010

**Typical chemical composition, percent by weight:** C, 2.35–2.50; Mn, 0.28–0.48; Fe, balance; S, 0.16–0.20; Si, 1.0–1.5; P, 0.06–0.10

**Physical constants and thermal properties**
Density, lb/in.$^3$: 0.265–0.268
Coefficient of thermal expansion, (70–200°F) in./in./°F $\times$ 10$^{-6}$: 7.5
Modulus of elasticity, psi: tension, 26 $\times$ 10$^{-6}$
Poisson's ratio: 0.17
Thermal conductivity, Btu/ft$^2$/hr/in./°F, 70°F: 354
Electrical resistivity, ohms/cmil/ft,70°F: 95.4
Curie temperature, °F: annealed, 430

### Short Time Tensile Properties

**As Cast**

| Temperature, °F | T.S., psi | Y.S., psi, 0.2% offset | Elong., in 2 in., % | Hardness, Brinell |
|---|---|---|---|---|
| 70 | 65,000 | 45,000 | 10 | 163–207 |

### Fatigue Strength

**As Cast**

| Test temperature, °F | Stress, psi |
|---|---|
| 70 | 28,000 endurance limit |

### Impact Strength

**As Cast**

| Test temperature, °F | Type test | Strength, ft-lb |
|---|---|---|
| –70 | Charpy – V-notched[a] | 14 |
| +70 | Charpy – unnotched | 22–35 |

[a]0.394-in. square bar, 0.079-in notch.

**Table 10.2—7**

**NODULAR OR DUCTILE IRONS**

Cast 80-55-06

**Typical chemical composition, percent by weight:** Mn, 0.2–0.5; Fe, balance; Si, 2.0–3.0; Ni, 0–1.0; P, 0.06–0.08; Mg, 0.02–0.07

**Physical constants and thermal properties**
Density, lb/in.$^3$: 0.257
Coefficient of thermal expansion, (70–200°F) in./in./°F × 10$^{-6}$: 6.6
Modulus of elasticity, psi: tension, 22–25 × 10$^6$
Melting range, °F: 2050–2150
Thermal conductivity, Btu/ft$^2$/hr/in./°F, 70°F: 216

Electrical resistivity, ohms/cmil/ft, 70°F: 408

**Heat treatments**
Ferritic – heat in 1600–1650°F range, cool to 1300°F, hold 1–3 hr, furnace cool to 1100°F, air cool.
Pearlitic-ferritic – heat to 1600–1650°F, cool rapidly, reheat at 1100–1300°F.
Pearlitic – air cool from 1600–1650°F.
Tempered – quench from 1400–1650°F, temper at 800–1300°F.

Short Time Tensile Properties

**As Cast**

| Temperature, °F | T.S., psi | Y.S., psi, 0.2% offset | Elong., in 2 in., % | Hardness, Brinell |
|---|---|---|---|---|
| 70 | 90,000–110,000 | 60,000–75,000 | 3–10 | 179–255 |

Creep Strength
(Stress, psi, to Produce 0.0001%/hr Creep)

**As Cast**

| Test temperature, °F | 10,000 hr |
|---|---|
| 1,000 | 1,750 |
| 1,200 | 480 |

Fatigue Strength

**As Cast**

Endurance ratio similar to other ferrous alloys. Notched endurance ratio from 30–40% for strong irons, up to 50–55% for soft irons.

Impact Strength

**As Cast**

| Test temperature, °F | Type test | Strength, ft-lb |
|---|---|---|
| +70 | Charpy – notched | 2–5 |
| +70 | Charpy – unnotched | 15–65 |

## Table 10.2—8
## NODULAR OR DUCTILE IRONS – CAST

### Heat Resistant

**Typical chemical composition, percent by weight:** Mn, 0.2–0.6; Fe, balance; Si, 2.5–6.0; Ni, 0–1.5; P, 0.08; Mg, 0.02–0.07

**Physical constants and thermal properties**
Density, lb/in.$^3$ : 0.25
Modulus of elasticity, psi: tension, $22-25 \times 10^6$
Melting range, °F: 2050–2150

**Heat treatments**
Ferritic – heat in 1600–1650°F range, cool to 1300°F, hold 1–3 hr, furnace cool to 1100°F, air cool.
Pearlitic-ferritic – heat to 1600–1650°F, cool rapidly, reheat at 1100–1300°F.
Pearlitic – air cool from 1600–1650°F.
Tempered – quench from 1400–1650°F, temper at 800–1300°F.

### Short Time Tensile Properties

**As Cast**

| Temperature, °F | T.S., psi | Y.S., psi, 0.2% offset | Elong., in 2 in., % | Hardness, Brinell |
|---|---|---|---|---|
| 70 | 60,000–100,000 | 45,000–75,000 | 0–20 | 140–300 |

### Fatigue Strength

Endurance ratio similar to other ferrous alloys. Notched endurance ratio from 30–40% for strong irons, up to 50–55% for soft irons.

### Impact Strength

| Test temperature, °F | Type test | Strength, ft-lb |
|---|---|---|
| +70 | Charpy – unnotched | 5–115 |

## Table 10.2—9
## WHITE AND ALLOY IRONS − CAST;
## WEAR AND ABRASION RESISTANT TYPES

White Iron

**Typical chemical composition, percent by weight:** C, 2.8−3.6; Mn, 0.4−0.9; Fe, balance; Si, 0.5−1.3

**Physical constants and thermal properties**
Density, lb/in.$^3$: 0.274−0.281
Coefficient of thermal expansion, (70−200° F) in./in./°F × 10$^{-6}$: 5.0−5.3
Electrical resistivity, ohms/cmil/ft, 70° F: 30

Short Time Tensile Properties

| Temperature, °F | T.S., psi | Y.S., psi, 0.2% offset | Elong., in 2 in., % | Hardness, Brinell |
|---|---|---|---|---|
| 70 | 20,000−50,000 | − | − | 300−575 |

Impact Strength

| Test temperature, °F | Type test | Strength, ft-lb |
|---|---|---|
| +70 | Charpy | 3.5−10.0 |

## Table 10.2—10
## WHITE AND ALLOY IRONS − CAST;
## HEAT AND CORROSION RESISTANT TYPES

High Silicon (Duriron®)

**Typical chemical composition, percent by weight:** C, 0.4−1.0; Mn, 0.4−1.0; Fe, balance; Si, 14−17; Mo, 0−3.5

**Physical constants and thermal properties**
Density, lb/in.$^3$: 0.252−0.254
Coefficient of thermal expansion, (70−200° F) in./in./°F × 10$^{-6}$: 6.70

Short Time Tensile Properties

| Temperature, °F | T.S., psi | Y.S., psi, 0.2% offset | Elong., in 2 in., % | Hardness, Brinell |
|---|---|---|---|---|
| 70 | 13,000−18,000 | − | − | 450−500 |

Impact Strength

| Test temperature, °F | Type test | Strength, ft-lb |
|---|---|---|
| +70 | Charpy | 2−4 |

## Table 10.2—11
## IRON-BASE SUPERALLOYS – CAST, WROUGHT

### Type A-286

**Typical chemical composition, percent by weight:**
C, 0.08; Mn, 1.35; Fe, balance; Si, 0.70; Cr, 15.00; Ni, 26.0; Mo, 1.25; Ti, 2.15; Al, 0.20; B. 0.003; V, 0.30

**Modulus of elasticity, psi: tension, $29.1 \times 10^6$**
Melting range, °F: 2500–2600
Specific heat, Btu/lb/°F, 70°F: 0.10–0.11
Thermal conductivity, Btu/ft$^2$/hr/in./°F, 70°F: 164.4
Curie temperature, °F: age hardened, 1325 (18 hr)

**Physical constants and thermal properties**
Density, lb/in.$^3$: 0.286
Coefficient of thermal expansion, (70–200°F) in./in./°F $\times 10^{-6}$: 10.3

**Heat treatments**
1 hr at 1800°F, oil quench, age 18 hr at 1325°F, air cool.

### Short Time Tensile Properties

| Temperature, °F | T.S., psi | Y.S., psi, 0.2% offset | Elong., in 2 in.,% | Hardness, Brinell |
|---|---|---|---|---|
| 70 | 146,000 | 100,000 | 25 | – |
| 1200 | 104,000 | 88,000 | 13 | – |
| 1400 | 64,000 | 62,000 | 19 | – |
| 1500 | 37,000 | – | 69 | – |

### Rupture Strength, 1000 hr

| Test temperature, °F | Strength, psi | Elong., in 2 in.,% | Reduction of area,% |
|---|---|---|---|
| 1350 | 21 | – | – |

### Creep Strength (Stress, psi, to Produce 0.0001%/hr Creep)

| Test temperature, °F | 10,000 hr |
|---|---|
| 1200 | 30,000 |
| 1350 | 16,000 |

### Fatigue Strength

| Test temperature, °F | Stress, psi | Cycles to failure |
|---|---|---|
| 1200 | 38,000 | $10^8$ |

### Impact Strength

| Test Temperature, °F | Type test | Strength, ft-lb |
|---|---|---|
| + 70 | Charpy | 64 |
| 1000 | Charpy | 46 |

**Table 10.2—12**
### IRON-BASE SUPERALLOYS – WROUGHT

Type 19-9DL

**Typical chemical composition, percent by weight:** C, 0.32; Mn, 1.15; Fe, balance; Si, 0.55; Cr, 18.5; Ni, 9.0; Mo, 1.40; W, 1.35; Cb + Ta, 0.40; Ti, 0.25; Cu, 0.15

**Physical constants and thermal properties**
Density, lb/in.$^3$ : 0.287

Coefficient of thermal expansion, (70–200°F) in./in./°F $\times 10^{-6}$ : 10
Modulus of elasticity, psi: tension, 29.5 $\times 10^6$
Melting range, °F: 2560–2615
Specific heat, Btu/lb/°F, 70°F: 0.10
Thermal conductivity, Btu/ft$^2$/hr/in./°F, 70°F: 146.4

### Short Time Tensile Properties

| Temperature, °F | T.S., psi | Y.S., psi, 0.2% offset | Hardness, Brinell |
|---|---|---|---|
| 70 | 114,000 | 71,000 | — |
| 1000 | 79,000 | 55,000 | — |
| 1200 | 62,000 | 52,000 | — |
| 1400 | 50,000 | 40,000 | — |

### Rupture Strength, 1000 hr

| Test temperature, °F | Strength, psi | Elong., in 2 in., % | Reduction of area, % |
|---|---|---|---|
| 1200 | 44,000 | — | — |

### Fatigue Strength

| Test temperature, °F | Stress, psi | Cycles to failure |
|---|---|---|
| 70 | 81,000 | $10^8$ |
| 1200 | 52,000 | $10^8$ |

### Impact Strength

| Test temperature, °F | Type test | Strength, ft-lb |
|---|---|---|
| +70 | Charpy | 46 |

### Specifications

SAE-AMS: 5526, 5527, 5720, 5722, 5721

## Table 10.2—13
## ALLOY STEELS – CAST

### Class 200,000

**Chemical composition:** total alloy content below 8%.

**Physical constants and thermal properties**
Density, lb/in.$^3$: 0.283
Coefficient of thermal expansion, (70–200°F) in./in./°F × 10$^{-6}$: 8.0–8.3
Modulus of elasticity, psi: tension, 29–30 × 10$^6$
Specific heat Btu/lb/°F, 70°F: 0.10–0.11

Thermal conductivity, Btu/ft$^2$/hr/in./°F, 70°F: 324
Electrical resistivity, ohms/cmil/ft, 70°F: 90–120
Curie temperature, °F: annealed – about 200°F above critical range

**Heat treatments**
Quenched and tempered; quenching temperature about 100°F above critical range.

### Short Time Tensile Properties

| Temperature, °F | T.S., psi | Y.S., psi, 0.2% offset | Elong., in 2 in., % | Hardness, Brinell |
|---|---|---|---|---|
| 70 | 205,000 | 170,000 | 8 | 401 |

### Fatigue Strength

| Test temperature, °F | Stress, psi | Cycles to failure |
|---|---|---|
| 70 | 88,000 endurance limit | – |

### Impact Strength

| Test temperature, °F | Type test | Strength, ft-lb |
|---|---|---|
| –40 | Charpy – V-notched | 8 |
| +70 | Charpy – V-notched | 14 |

## Table 10.2—14
## CARBON STEELS – CAST

### Class 100,000

**Physical constants and thermal properties**
Density, lb/in.$^3$: 0.283
Coefficient of thermal expansion, (70–200°F) in./in./°F $\times$ 10$^{-6}$: 8.3
Modulus of elasticity, psi: tension, 29.7 $\times$ 10$^6$
Specific heat, Btu/lb/°F, 70°F: 0.10–0.11
Thermal conductivity, Btu/ft$^2$/hr/in./°F, 70°F: 324

**Heat treatments**
Quenched and tempered

### Short Time Tensile Properties

| Temperature, °F | T.S., psi | Y.S., psi, 0.2% offset | Elong., in 2 in., % | Hardness, Brinell |
|---|---|---|---|---|
| 70 | 105,000 | 75,000 | 19 | 212 |

### Fatigue Strength

| Test temperature, °F | Stress, psi | Cycles to failure |
|---|---|---|
| 70 | 45,000 endurance limit | – |

### Impact Strength

| Test temperature, °F | Type test | Strength, ft-lb |
|---|---|---|
| –40 | Charpy – V-notched | 12 |
| +70 | Charpy – V-notched | 40 |

## Table 10.2—15
## CARBON STEELS – CARBURIZING GRADES

### AISI Type C1015

**Physical constants and thermal properties**
Density, lb/in.$^3$: 0.283
Coefficient of thermal expansion, (70–200°F)
   in./in./°F × 10$^{-6}$ :8.4
Modulus of elasticity, psi: tension, 29-30 × 10$^6$
Melting range, °F: 2750–2775
Specific heat, Btu/lb/°F, 70°F: 0.10–0.11

Thermal conductivity, Btu/ft$^2$/hr/in./°F, 70°F: 324

**Heat treatments**
   1-in. rounds treated as follows: carburized at 1675°F
   for 8 hr, pot cooled, reheated to 1425°F, water
   quenched, tempered at 350°F.

### Short Time Tensile Properties

| Temperature, °F | T.S., psi | Y.S., psi, 0.2% offset | Elong., in 2 in., % | Hardness, Brinell |
|---|---|---|---|---|
| 70 | 73,000 | 46,000 | 32 | 149 (core) |

### Fatigue Strength

| Test temperature, °F | Stress, psi | Cycles to failure |
|---|---|---|
| 70 | 30,000–35,000 | – |

## Table 10.2—16
## FREE-CUTTING CARBON STEELS – WROUGHT

### AISI Type B1111

**Typical chemical composition, percent by weight:** C,
0.13; Mn, 0.60–0.90; Fe, balance; S, 0.08–0.15; P,
0.07–0.12

**Physical constants and thermal properties**
Density, lb/in.$^3$: 0.283
Coefficient of thermal expansion, (70–200°F)
   in./in./°F × 10$^{-6}$: 8.4

Modulus of elasticity, psi: tension, 29 × 10$^6$
Specific heat, Btu/lb/°F, 70°F: 0.10–0.11
Thermal conductivity, Btu/ft$^2$/hr/in./°F, 70°F: 324
Electrical resistivity, ohms/cmil/ft, 70°F: 85.8

**Heat treatments**
   Tempering temperature, 300°F; case hardening
   temperature, 1450°F. Cold drawn.

### Short Time Tensile Properties

#### 1-in. Diameter

| Temperature, °F | T.S., psi | Y.S., psi, 0.2% offset | Elong., in 2 in., % | Hardness, Brinell |
|---|---|---|---|---|
| 70 | 80,000 | 70,000 | 12 | 163 |

### Fatigue Strength

Notch sensitive as cold drawn.

### Impact Strength

Relatively low impact strength at low temperatures; should not be used for shock loading applications at subzero temperatures.

### Table 10.2—17
### HIGH STRENGTH STEELS – WROUGHT;
### COLUMBIUM-BEARING CARBON STEELS

Type 50

**Typical chemical composition, percent by weight:**
C, 0.020; Mn, 0.50–1.00; Fe, balance; S, 0.05;
Si, 0.10; Cb, 0.01; P, 0.04

Short Time Tensile Properties

| Temperature, °F | T.S., psi | Y.S., psi, 0.2% offset | Elong., in 2 in., % | Hardness, Brinell |
|---|---|---|---|---|
| 70 | 65,000 | 50,000 | 22 | – |

Specifications

ASTM: A-572

### Table 10.2—18
### HIGH STRENGTH STEELS – WROUGHT;
### VANADIUM-BEARING CARBON STEELS

Type 50

**Typical chemical composition, percent by weight:**
C, 0.22; Mn, 1.25; Fe, balance; S, 0.05; P, 0.04;
N, 0.015; V, 0.02

Short Time Tensile Properties

| Temperature, °F | T.S., psi | Y.S., psi, 0.2% offset | Elong., in 2 in., % | Hardness, Brinell |
|---|---|---|---|---|
| 70 | 70,000 | 50,000 | 18 | 156 |

Specifications

ASTM: A-572

**Table 10.2—19**
## HIGH STRENGTH STEELS – WROUGHT;
## LOW ALLOY, HIGH STRENGTH STEELS

### ASTM Type A94

**Typical chemical composition, percent by weight:**
C, 0.33; Mn, 1.10–1.60; Fe, balance; S, 0.05;
Si, 0.15–0.30; P, 0.06; Cu, 0.20

### Short Time Tensile Properties

| Temperature, °F | T.S., psi | Y.S., psi, 0.2% offset | Elong., in 2 in., % | Hardness, Brinell |
|---|---|---|---|---|
| 70 | 70,000–75,000 | 45,000–50,000 | 21–20 | – |

**Table 10.2—20**
## HIGH STRENGTH STEELS – WROUGHT;
## LOW ALLOY, HIGH STRENGTH STEELS

### ASTM Type A242

**Typical chemical composition, percent by weight:**
C, 0.22; Mn, 1.25; Fe, balance; S, 0.05; Si,
0.20–0.90[a]; Cr, 0.30–1.25[a]; Zr, 0.10[a]; Cu,
0.25–0.55[a]

[a] Optional

### Short Time Tensile Properties

| Temperature, °F | T.S., psi | Y.S., psi, 0.2% offset | Elong., in 2 in., % | Hardness, Brinell |
|---|---|---|---|---|
| 70 | 63,000–70,000 | 42,000–50,000 | 24–22 | 156 |

**Table 10.2—21**
**NITRIDING STEELS – WROUGHT**

Type N

**Typical chemical composition, percent by weight:** C, 0.20–0.27; Mn, 0.40–0.70; Fe, balance; Si, 0.20–0.40; Cr, 1.00–1.50; Ni, 3.25–3.75; Mo, 0.20–0.30; Al, 0.85–1.20

Electrical resistivity, ohms/cmil/ft, 70°F: 162–174
Curie temperature, °F: annealed, 1500–1550[a]

[a]Must be cooled rapidly below 1150°F to avoid precipitation hardening.

**Physical constants and thermal properties**
Density, lb/in.$^3$: 0.283
Coefficient of thermal expansion, (70–200°F) in./in./°F × 10$^{-6}$: 6.5
Modulus of elasticity, psi: tension, 29–30 × 10$^6$
Specific heat, Btu/lb/°F, 70°F: 0.11–0.12
Thermal conductivity, Btu/ft$^2$/hr/in./°F, 70°F: 360

**Heat treatments**
Quenching temperature 1625–1675°F. Tempering temperature 1100–1300°F. Nitriding temperature 930–1050°F for periods ranging to 100 hr; 24- to 48-hr treatments are most widely used. Core properties; oil quenched from 1650°F, tempered at 1200°F before nitriding.

Short Time Tensile Properties

| Temperature, °F | T.S., psi | Y.S., psi, 0.2% offset | Elong., in 2 in., % | Hardness, Brinell |
|---|---|---|---|---|
| 70 | 132,000 | 114,000 | 22 | 277 |

**Table 10.2—22**

## AGE HARDENABLE STAINLESS STEELS – WROUGHT, CAST

### Type 17-4 PH

**Typical chemical composition, percent by weight:** C, 0.07; Mn, 1.0; Fe, balance; Si, 1.0; Cr, 16.5; Ni, 4.0; Cb + Ta, 0.30; Cu, 4.0

**Physical constants and thermal properties**
Density, lb/in.$^3$ : 0.281
Coefficient of thermal expansion, (70–200°F) in./in./°F $\times$ 10$^{-6}$ : 6.0
Modulus of elasticity, psi: tension, 28.5 $\times$ 10$^6$

Thermal conductivity, Btu/ft$^2$/hr/in./°F, 70°F: 124.8
Electrical resistivity, ohms/cmil/ft, 70°F: 462
Curie temperature, °F: annealed, 1900; age hardened, 900 (1 hr)

**Heat treatments**
1. H 900 – bar solution annealed at 1900°F, oil quench or air cool, aged 1 hr at 900°F, air cooled;
2. H 1150 – 4 hr at 1150°F, air cooled.

### Short Time Tensile Properties

| Temperature, °F | T.S., psi | Y.S., psi, 0.2% offset | Elong., in 2 in., % | Hardness, Rockwell |
|---|---|---|---|---|
| 70 | 195,000 | 180,000 | 13 | C30–45 |
| 800 | 157,000 | 138,000 | 10 | – |
| 1,000 | 100,000 | 77,000 | 15 | – |
| 1,200 | 59,000 | 42,000 | 15 | – |

### Rupture Strength, 1,000 hr

| Test temperature, °F | Strength, psi | Elong., in 2 in., % | Reduction of area, % |
|---|---|---|---|
| 800 | 128,000 | – | – |

### Creep Strength
#### (Stress, psi, to Produce 1% Creep)

| Test temperature, °F | 10,000 hr | 100,000 hr |
|---|---|---|
| 800 | 50,000 | – |

### Fatigue Strength

| Test temperature, °F | Stress, psi | Cycles to failure |
|---|---|---|
| 70 | 90,000 | 10–20 $\times$ 10$^6$ |

### Impact Strength

| Test temperature, °F | Type test | Strength, ft-lb |
|---|---|---|
| 1,000 | Charpy – unnotched | 19 |

### Specifications

ASTM: A-461, 564

## Table 10.2—23
## AGE HARDENABLE STAINLESS STEELS – WROUGHT, CAST

### Type 15-5PH

**Typical chemical composition, percent by weight:** C, 0.07; Mn, 1.00; Fe, balance; Si, 1.00; Cr, 14.00–15.00; Ni, 3.50–5.50; Cb + Ta, 0.15–0.45; Cu, 2.50–4.50

Modulus of elasticiity, psi: tension, $28.5 \times 10^6$
Electrical resistivity, ohms/cmil/ft, 70°F: 462
Curie temperature, °F: annealed, 1900; age hardened, 900 (1 hr)

**Physical constants and thermal properties**
Density, lb/in.$^3$ : 0.282
Coefficient of thermal expansion, (70–200°F) in./in./°F $\times$ $10^{-6}$: 6.2

**Heat treatments**
Annealed (1900°F, air or oil cooled), aged (900°F 1 hr) air cooled.

### Short Time Tensile Properties

| Temperature, °F | T.S., psi | Y.S., psi, 0.2% offset | Elong., in 2 in., % | Hardness, Brinell |
|---|---|---|---|---|
| 70 | 190,000 | 170,000 | 10 | – |

### Impact Strength

| Test temperature, °F | Type test | Strength, ft-lb |
|---|---|---|
| + 70 | Charpy – unnotched | 15 |

**Table 10.2—24**
## AUSTENITIC STAINLESS STEELS – WROUGHT

### AISI Type 304

**Typical chemical composition, percent by weight:**
C, 0.08; Mn, 2; Fe, balance; S, 0.030; Si, 1; Cr, 18–20; Ni, 8–12; P, 0.045

**Physical constants and thermal properties**
Density, lb/in.$^3$ : 0.29
Coefficient of thermal expansion, (70–200°F) in/in./°F × 10$^{-6}$ : 9.6
Modulus of elasticity, psi: tension, 28.0 × 10$^6$

Melting range, °F: 2550–2650
Specific heat, Btu/lb/°F, 70°F: 0.12
Thermal conductivity, Btu/ft$^2$/hr/in./°F, 70°F: 112.8
Electrical resistivity, ohms/cmil/ft, 70°F: 432
Curie temperature, °F: annealed, 1850–2050

**Heat treatments**
Annealing temperature 1850–2050°F; forging temperature (start) 2100–2300°F.

### Short Time Tensile Properties

| Temperature, °F | T.S., psi | Y.S., psi, 0.2% offset | Elong., in 2 in., % | Hardness, Brinell |
|---|---|---|---|---|
| 70 | 84,000 | 42,000 | 65 | 149 |
| +32 | 130,000 | 34,000 | 55 | – |
| –40 | 155,000 | 34,000 | 47 | – |
| –320 | 221,000 | 39,000 | 40 | – |
| –423 | 243,000 | 50,000 | 40 | – |

### Creep Strength (Stress, psi, to Produce 1% Creep)

| Test temperature, °F | 1,000 hr | 10,000 hr | 100,000 hr |
|---|---|---|---|
| 1,000 | 17,000 | 20,000 | – |
| 1,100 | – | 12,000 | – |
| 1,200 | – | 7,000 | – |
| 1,300 | 4,000 | 4,500 | – |
| 1,500 | 1,200 | 2,000 | – |

### Impact Strength

**Annealed**

| Test temperature, °F | Type test | Strength, ft-lb |
|---|---|---|
| + 32 | Izod | 110 |
| + 70 | Izod | 110 |
| – 40 | Izod | 110 |
| – 32 | Izod | 110 |

## Table 10.2—25
### AUSTENITIC STAINLESS STEELS – WROUGHT

AISI Type 304 L

**Typical chemical composition, percent by weight:** C, 0.030; Mn, 2; Fe, balance; S, 0.030; Si, 1; Cr, 18–20; Ni, 8–12; P, 0.045

**Physical constants and thermal properties**
Density, lb/in.$^3$: 0.29
Coefficient of thermal expansion, (70–200°F) in./in./°F $\times$ 10$^{-6}$: 9.6
Modulus of elasticity, psi: tension, 28.0 $\times$ 10$^6$

Melting range, °F: 2550–2650
Specific heat, Btu/lb/°F, 70°F: 0.12
Thermal conductivity, Btu/ft$^2$/hr/in./°F, 70°F: 112.8
Electrical resistivity, ohms/cmil/ft, 70°F: 432
Curie temperature, °F: annealed, 1850–2050

**Heat treatments**
Annealing temperature 1850–2050°F; forging temperature 2100–2300°F.

### Short Time Tensile Properties

| Temperature, °F | T.S., psi | Y.S., psi, 0.2% offset | Elong., in 2 in., % | Hardness, Brinell |
|---|---|---|---|---|
| 70 | 81,000 | 39,000 | 65 | 143 |
| +32 | 130,000 | 34,000 | 55 | – |
| –40 | 155,000 | 34,000 | 47 | – |
| –320 | 221,000 | 39,000 | 40 | – |
| –423 | 243,000 | 50,000 | 40 | – |

### Creep Strength
(Stress, psi, to Produce 1% Creep)

| Test temperature, °F | 10,000 hr | 100,000 hr |
|---|---|---|
| 1000 | 20,000 | – |
| 1100 | 12,000 | – |
| 1200 | 7,500 | – |
| 1300 | 4,500 | – |
| 1500 | 2,000 | – |

### Impact Strength

**Annealed**

| Test temperature, °F | Type test | Strength, ft-lb |
|---|---|---|
| +32 | Izod | 110 |
| +70 | Izod | 110 |
| –40 | Izod | 110 |
| –32 | Izod | 110 |

### Table 10.2—26
### FERRITIC STAINLESS STEELS – WROUGHT

#### AISI Type 430

**Typical chemical composition, percent by weight:** C, 0.12; Mn, 1.00; Fe, balance; S, 0.030; Si, 1.00; Cr, 14.0–18.0; P, 0.040

**Physical constants and thermal properties**
Density, lb/in.$^3$ : 0.28
Coefficient of thermal expansion, (70–200°F) in./in./ °F × 10$^{-6}$ : 5.8
Modulus of elasticity, psi: tension, 29 × 10$^6$

Melting range, °F: 2600–2750
Specific heat, Btu/lb/°F, 70°F: 0.11
Thermal conductivity, Btu/ft$^2$/hr/in./°F, 70°F: 181.2
Electrical resistivity, ohms/cmil/ft, 70°F: 360
Curie temperature, °F: annealed 1400–1500

**Heat treatments**
Forging temperature (start) 1900–2050°F; annealing temperature 1400–1500°F.

#### Short Time Tensile Properties

| Temperature, °F | T.S., psi | Y.S., psi, 0.2% offset | Elong., in 2 in., % | Hardness, Rockwell |
|---|---|---|---|---|
| 70 | 65,000–75,000 | 40,000 | 25–30 | B80 |
| +32 | 69,000 | 40,000 | 37 | – |
| –40 | 76,000 | 41,000 | 36 | – |
| –320 | 90,000 | 87,000 | 2 | – |

#### Creep Strength
#### (Stress, psi, to Produce 1% Creep)

| Test temperature, °F | 10,000 hr | 100,000 hr |
|---|---|---|
| 1000 | 8,500 | – |
| 1100 | 4,700 | – |
| 1200 | 2,600 | – |
| 1300 | 1,400 | – |

#### Impact Strength

##### Annealed

| Test temperature, °F | Type test | Strength, ft-lb |
|---|---|---|
| +32 | Izod | 20 |
| +70 | Izod | 35 |
| –40 | Izod | 10 |
| –320 | Izod | 2 |

## Table 10.2—27
## MARTENSITIC STAINLESS STEELS – WROUGHT

### AISI Type 410

**Typical chemical composition, percent by weight:** C, 0.15; Mn, 1.00; Fe, balance; S, 0.030; Si, 1; Cr, 11.5–13.5; P, 0.040

Specific heat, Btu/lb/°F, 70°F: 0.11
Thermal conductivity, Btu/ft²/hr/in./°F, 70°F: 172.8
Electrical resistivity, ohms/cmil/ft, 70°F: 420
Curie temperature, °F: annealed, 1500–1650

**Physical constants and thermal properties**
Density, lb/in.³: 0.28
Coefficient of thermal expansion, (70–200°F) in./in./°F ×10⁻⁶: 5.5
Modulus of elasticity, psi: tension, 29 × 10⁶
Melting range, °F: 2700–2790

**Heat treatments**
Annealing temperature 1500–1650°F; hardening temperature 1700–1850°F; tempering temperature 400–1400°F; forging temperature (start) 2000–2200°F.

### Short Time Tensile Properties

#### Annealed

| Temperature, °F | T.S., psi | Y.S., psi, 0.2% offset | Elong., in 2 in., % | Hardness, Brinell |
|---|---|---|---|---|
| 70 | 65,000–110,000 | 35,000–85,000 | 25–35 | 155 |
| +32 | 115,000 | 89,000 | 24 | – |
| –40 | 122,000 | 90,000 | 23 | – |
| –320 | 158,000 | 148,000 | 10 | – |

### Creep Strength
### (Stress, psi, to Produce 1% Creep)

| Test temperature, °F | 10,000 hr | 100,000 hr |
|---|---|---|
| 1,000 | 9,200 | – |
| 1,100 | 4,300 | – |
| 1,200 | 2,000 | – |
| 1,300 | 1,500 | – |

### Fatigue Strength

| Test temperature, °F | Stress, psi | Cycles to failure |
|---|---|---|
| 70 | 40,000 endurance limit | – |

### Impact Strength

#### Annealed

| Test temperature, °F | Type test | Strength, ft-lb |
|---|---|---|
| +32 | Izod | 80 |
| +70 | Izod | 85 |
| –80 | Izod | 25 |
| –320 | Izod | 5 |

## Table 10.2—28
## SPECIALTY STAINLESS STEELS – WROUGHT

### Type 18-18-2

**Typical chemical composition, percent by weight:** C, 0.06; Fe, balance; Si, 1.9; Cr, 18; Ni, 18

**Physical constants and thermal properties**
Density, lb/in.$^3$: 0.284
Coefficient of thermal expansion, $(70-200°F)$ in./in./°F $\times 10^{-6}$: 7.6
Electrical resistivity, ohms/cmil/ft, 70°F: 516

**Heat treatments**
Annealed

### Short Time Tensile Properties

| Temperature, °F | T.S., psi | Y.S., psi, 0.2% offset | Elong., in 2 in., % | Hardness, Brinell |
|---|---|---|---|---|
| 70 | 81,000 | 36,000 | 54 | – |

## Table 10.2—29
## STAINLESS STEELS – CAST

### ACI Type CH-20

**Typical chemical composition, percent by weight:** C, 0.20; Mn, 1.5; Fe, balance; S, 0.04; Si, 2.0; Cr, 22–26; Ni, 12–15; P, 0.04

Modulus of elasticity, psi: tension, $28 \times 10^6$
Melting range, °F: 2600
Specific heat, Btu/lb/°F, 70°F: 0.12
Thermal conductivity, Btu/ft$^2$/hr/in./°F, 70°F: 98.4
Electrical resistivity, ohms/cmil/ft, 70°F: 504
Permeability, 70°F, 200 Oe: annealed, 1.71

**Physical constants and thermal properties**
Density, lb/in.$^3$: 0.279
Coefficient of thermal expansion, $(70-200°F)$ in./in./°F $\times 10^{-6}$: 9.6

**Heat treatments**
Water quenched from 2000°F.

### Short Time Tensile Properties

| Temperature, °F | T.S., psi | Y.S., psi, 0.2% offset | Elong., in 2 in., % | Hardness, Brinell |
|---|---|---|---|---|
| 70 | 88,000 | 50,000 | 38 | 190 |

### Impact Strength

| Test temperature, °F | Type test | Strength, ft-lb |
|---|---|---|
| +70 | Charpy, keyhole notched | 30 |

## Table 10.2—30
## LOW TEMPERATURE STEELS – WROUGHT;
## FINE GRAIN C-Mn-Si CARBON STEELS

Type 40-50

Typical chemical composition, percent by weight: C, 0.14
or 0.2; Mn, 0.7–1.35; Fe, balance; S, 0.04 or 0.05[a];
Si, 0.15–0.5; P, 0.035 or 0.4

[a]Higher value is flange quality; lower value is firebox
quality.

**Heat treatments**
Normalized

### Short Time Tensile Properties

| Temperature, °F | T.S., psi | Y.S., psi,<br>0.2% offset | Elong., in 2 in., % | Hardness,<br>Brinell |
|---|---|---|---|---|
| 70 | 58,000–90,000 | 40,000–50,000 | 19–30 | 120–170 |

### Impact Strength

| Test temperature, °F | Type test | Strength, ft-lb |
|---|---|---|
| –100 | Charpy – V-notched | 10–16 |
| +70 | Charpy – V-notched | 40–75 |

## Table 10.2—31
## HEAT RESISTANT ALLOYS – CAST

### ACI Type HH (Ferritic)

**Typical chemical composition, percent by weight:** C, 0.20–0.50; Mn, 2.00; Fe, balance; S, 0.04; Si, 2.00; Cr, 24–28; Ni, 11–14; Mo, 0.5; P, 0.04; N, 0.2

**Physical constants and thermal properties**
Density, lb/in.$^3$: 0.279
Coefficient of thermal expansion, (70–200°F) in./in./°F × 10$^{-6}$: 10.5
Modulus of elasticity, psi: tension, 27 × 10$^6$
Melting range, °F: 2500

Specific heat, Btu/lb/°F, 70°F: 0.12
Thermal conductivity, Btu/ft$^2$/hr/in./°F, 70°F: 98.4
Electrical resistivity, ohms/cmil/ft, 70°F: 450–510
Curie temperature, °F: annealed, 1900; age hardened, 1400
Permeability, 70°F, 200 Oe: annealed, 1.0–1.9

**Heat treatments**
As cast and heat treated (aged 24 hr at 1400°F, furnace cooled); 12 hr at 1900°F may improve life.

### Short Time Tensile Properties

| Temperature, °F | T.S., psi | Y.S., psi, 0.2% offset | Elong., in 2 in., % | Hardness, Brinell |
|---|---|---|---|---|
| 70 | 80,000–86,000 | 50,000–55,000 | 25, 11 | 185, 200 |
| 1400 | 33,000 | 17,000 | 18 | – |
| 1600 | 18,500 | 13,500 | 30 | – |
| 1800 | 9,000 | 6,300 | 45 | – |

### Rupture Strength, 1000 hr

| Test temperature, °F | Strength, psi | Elong., in 2 in., % | Reduction of area, % |
|---|---|---|---|
| 1400 | 6500 | – | – |
| 1600 | 3800 | – | – |
| 2000 | – | – | – |

### Creep Strength (Stress, psi, to Produce 0.0001%/hr Creep)

| Test temperature, °F | 1000 hr | 100,000 hr |
|---|---|---|
| 1400 | 3000 | – |
| 1600 | 1700 | – |
| 2000 | 300 | – |

## Table 10.2—32
## HIGH TEMPERATURE STEELS – WROUGHT

### Type 1415 NW (Greek Ascoloy)

**Typical chemical composition, percent by weight:** C, 0.17; Mn, 0.40; Fe, balance; Si, 0.30; Cr, 12.75; Ni, 1.95; Mo, 0.15; W, 3.0; Cu, 0.13

**Physical constants and thermal properties**
Density, lb/in.$^3$: 0.284
Coefficient of thermal expansion, (70–200°F) in./in./°F × 10$^{-6}$: 6.3
Modulus of elasticity, psi: tension – 29 × 10$^6$ (70°F);
21.5 × 10$^6$ (1000°F)
Melting range, °F: 2660–2670

**Heat treatments**
Mechanical properties: austenitized ½ hr at 1800°F, oil quenched, tempered 2 hr at 1050°F, air cooled; rupture strength applies to material tempered 1½ hr at 1260°F. Hot working temperature 1700–2200°F.

### Short Time Tensile Properties

| Temperature, °F | T.S., psi | Y.S., psi, 0.2% offset | Elong., in 2 in., % | Hardness, Brinell |
|---|---|---|---|---|
| 70 | 170,000 | 150,000 | 13.3 | — |
| 800 | 135,000 | 122,000 | 13.3 | — |
| 1000 | 103,000 | 98,000 | 17.1 | — |

### Rupture Strength, 1000 hr

| Test temperature, °F | Strength, psi | Elong., in 2 in., % | Reduction of area, % |
|---|---|---|---|
| 1000 | 36,000 | — | — |

### Fatigue Strength

| Test temperature, °F | Stress, psi | Cycles to failure |
|---|---|---|
| 1000 | 53,000 | 10$^7$ |

### Impact Strength

| Test temperature, °F | Type test | Strength, ft-lb |
|---|---|---|
| +70 | Charpy – V-notched | 19.5 |

## Table 10.2—33
## ULTRA HIGH STRENGTH STEELS – WROUGHT

### Type 300-M

**Typical chemical composition, percent by weight:** C, 0.40; Mn, 0.75; Fe, balance; Si, 1.60; Cr, 0.85; Mo, 0.40; V, 0.08

**Physical constants and thermal properties**
  Coefficient of thermal expansion, (70–200°F) in./in./°F × 10⁻⁶ : 7.61 (0–600°F)

Thermal conductivity, Btu/ft²/hr/in./°F, 70°F: 260.4

**Heat treatments**
  Normalized at 1700°F, austenitized at 1600°F, oil quenched, tempered at 600°F. Hot working temperature 1700–2250°F.

### Short Time Tensile Properties

| Temperature, °F | T.S., psi | Y.S., psi, 0.2% offset | Elong., in 2 in., % | Hardness, Brinell |
|---|---|---|---|---|
| 70 | 289,000 | 242,000 | 10.0 | – |
| 500 | 270,000 | 200,000 | 13.3 | – |
| 700 | 232,000 | 178,000 | – | – |
| 800 | – | – | 15.0 | – |

### Rupture Strength, 1000 hr

| Test temperature, °F | Strength, psi | Reduction of area, % |
|---|---|---|
| – | – | 70°F: 38 |
|  |  | 500°F: 35 |
|  |  | 700°F: 52 |

### Fatigue Strength

| Test temperature, °F | Stress, psi | Cycles to failure |
|---|---|---|
| 70 | 116,000 | 10⁶ |

### Impact Strength

| Test temperature, °F | Type test | Strength, ft-lb |
|---|---|---|
| –200 | Charpy – unnotched | 11 |
| +70 | Charpy – unnotched | 22 |
| 500 | Charpy – unnotched | 23 |

**Table 10.2—34**

**ULTRA HIGH STRENGTH STEELS – WROUGHT**

Type 4340

**Typical chemical composition, percent by weight:** C,
0.40; Mn, 0.85; Fe, balance; Si, 0.20; Cr, 0.75; Ni,
1.80; Mo, 0.25

Coefficient of thermal expansion, $(70-200^\circ F)$
in./in./$^\circ F \times 10^{-6}$ : 6.3
Modulus of elasticity, psi: tension, $30 \times 10^6$

**Heat treatments**
Austenitized at $1550^\circ F$, oil quenched, tempered at
$400^\circ F$.

**Physical constants and thermal properties**
Density, lb/in.$^3$ : 0.283

Short Time Tensile Properties

| Temperature, $^\circ F$ | T.S., psi | Y.S., psi, 0.2% offset | Elong., in 2 in., % | Reduction of area, % |
|---|---|---|---|---|
| 70 | 287,000 | 270,00 | 11 | 39 |

Fatigue Strength

| Test temperature, $^\circ F$ | Stress, psi | Cycles to failure |
|---|---|---|
| 70 | 107,000 | $10^6$ |

**Table 10.2—35**

**ULTRA HIGH STRENGTH STEELS – WROUGHT**

Type 18Ni

**Typical chemical composition, percent by weight:** C,
0.026; Mn, 0.1; Fe, balance; Si, 0.11; Ni, 18.5; Co,
7.0; Mo, 4.5; Ti, 0.22; B, 0.003

Modulus of elasticity, psi: tension, $26.5 \times 10^6$
Curie temperature, $^\circ F$: annealed, $1500^\circ F$ (1 hr); age
hardened, $900^\circ F$ (3 hr)

**Physical constants and thermal properties**
Density, lb/in.$^3$ : 0.290
Coefficient of thermal expansion, $(70-200^\circ F)$
in./in./$^\circ F \times 10^{-6}$ : 5.6

**Heat treatments**
Annealed 1 hr at $1500^\circ F$, air cooled, aged 3 hr at
$900^\circ F$. Hot working temperature range
$1700-2300^\circ F$.

Short Time Tensile Properties

| Temperature, $^\circ F$ | T.S., psi | Y.S., psi, 0.2% offset | Elong., in 2 in., % | Reduction of area, % |
|---|---|---|---|---|
| 70 | 275,000 | 268,000 | 11 | 48 |
| 800 | 221,000 | 209,000 | 12 | 56 |
| 1000 | 154,000 | 138,000 | 24 | 74 |

Impact Strength

| Test temperature, $^\circ F$ | Type test | Strength, ft-lb |
|---|---|---|
| -200 | Charpy—unnotched | 20 |
| +70 | Charpy—unnotched | 23 |

## Table 10.2—36
## ULTRA HIGH STRENGTH STEELS – WROUGHT

### Type 9Ni-4Co-0.20C

**Typical chemical composition, percent by weight:** C, 0.20; Mn, 0.25; Fe, balance; Si, 0.1; Cr, 0.75; Ni, 9.0; Co, 4.5; Mo, 1.0; V, 0.08; P, 0.01

in./in./°F × $10^{-6}$: 6.4
Modulus of elasticity, psi: tension, 28 × $10^6$

**Physical constants and thermal properties**
Density, lb/in.³: 0.283
Coefficient of thermal expansion, (70–200°F)

**Heat treatments**
Austenitized at 1500–1550°F, water or oil quenched, double tempered at 1000°F. Hot working temperature range 1700–2050°F.

### Short Time Tensile Properties

| Temperature, °F | T.S., psi | Y.S., psi 0.2% offset | Elong., in 2 in., % | Reduction of area, % |
|---|---|---|---|---|
| 70 | 195,000–220,000 | 173,000–194,000 | 12–19 | 45–65 |
| 800 | 159,000 | 139,000 | 16 | 72 |
| 1000 | 139,000 | 107,000 | 18 | 80 |

### Fatigue Strength

| Test temperature, °F | Stress, psi | Cycles to failure |
|---|---|---|
| 70 | 105,000–110,000 | $10^6$ |

### Impact Strength

| Test temperature, °F | Type test | Strength, ft-lb |
|---|---|---|
| –200 | Charpy – unnotched | 40 |
| +70 | Charpy – unnotched | 40–60 |

## 10.3  LIGHT METALS*

### MAGNESIUM BASE ALLOYS

#### Nomenclature

Magnesium alloy designations are based on chemical composition. They consist of two letters representing the two alloying elements specified in the greatest amount, arranged in decreasing percentages or alphabetically if of equal percentage. The letters are followed by the respective percentages rounded off to whole numbers with a serial letter at the end. The serial letter indicates some variation in composition.

The letters used to designate various alloying elements include:

| | |
|---|---|
| A–Aluminum | M–Manganese |
| E–Rare earths | Q–Silver |
| H–Thorium | S–Silicon |
| K–Zirconium | T–Tin |
| L–Lithium | Z–Zinc |

Temper designations for magnesium alloys are separated from the alloy designations by a dash. The following designations are used to denote tempers of magnesium mill products:

| F | — as fabricated |
|---|---|
| T4 | — solution heat treated |
| T5,T51 | — artificially aged |
| T6,T61 | — solution heat treated and artificially aged |
| T7 | — solution heat treated and stabilized |
| T8 | — solution heat treated and artificially aged |
| O | — annealed |
| H24,H26 | — strain hardened, then partially annealed |

Key to mechanical property tests:

a — Permanent mold or sand castings, ½-in. diameter section
b — 0.2% offset
c — 500-kg, 10-mm ball
d — 3/16-in. diameter pin
e — Rotating beam
f — Axial load
g — Flexure

### Table 10.3—1
### AM100A

AM100A is used for pressure-tight sand and permanent mold castings. It contains aluminum and a small amount of manganese. It has a good combination of room temperature properties, similar to AZ92A, but has less tendency to crack when used for permanent mold castings.

**Typical chemical composition, percent by weight:** Al, 10; Mn, 0.10 minimum; Mg, balance

**Physical constants and thermal properties**
Density at 68°F, lb/cu in.$^{-1}$: 0.0651
Coefficient of thermal expansion, °F$^{-1}$ × $10^{-6}$, 65–212°F: 14.5
Modulus of elasticity, psi: 6.5 × $10^6$
Poisson's ratio: 0.35
Melting range, °F: 867–1,101
Specific heat, Btu/lb$^{-1}$/°F$^{-1}$, 70°F: 0.31
Electrical resistivity, microhm-cm, 70°F: 15.0

Thermal conductivity, Btu/hr$^{-1}$/ft$^{-2}$/°F$^{-1}$/ft, 70°F: 32.5; 300°F: 42.0; 500°F: 47.0

* Selected from material originally prepared by W. F. Simmons.

## Table 10.3—1 (continued)
### AM100A

### Heat Treatments

| Treatment | Temperature, °F | Time, hours | Cooling |
|---|---|---|---|
| Solution[a] | 790 | 20 | Strong air blast |
| Artificial aging (partial) | 325 | 12 | Still air |
| Artificial aging (complete) | 450 | 5 | Still air or oven |
| Stabilizing[b] | 500 | 4 | Still air |

[a]An atmosphere with at least 0.5% $SO_2$ is required.
[b]Used to minimize growth in castings to be used at elevated temperatures.

### Specifications

| | Ingot | Sand castings | Permeable mold castings |
|---|---|---|---|
| ASM | | | 4483 |
| ASTM | B93 | B80 | B199 |
| SAE | | | 502 |
| Federal | | | QQ-M-55 |

### Typical Room Temperature Mechanical Properties

| Temper | TS, ksi | YS, ksi | CYS, ksi | Elong., % | Hardness Bhn | Hardness RE | Shear strength, ksi | Charpy impact strength, ft-lb |
|---|---|---|---|---|---|---|---|---|
| F | 20/22 | 10/12 | 12 | 2 | 53 | 64 | 18 | 0.6 |
| T4 | 34/40 | 10/13 | 13 | 10 | 52 | 62 | 20 | 2.0 |
| T61 | 34/40 | 17/22 | 19 | 1 | 69 | 80 | 21 | 0.7 |
| T5 | 22 | 16 | 16 | 2 | 58 | 70 | | |
| T7 | 38 | 18 | 18 | 1 | 67 | 78 | | |
| T6 | 40 | 16 | | 4 | | | | |

| Temper | Bearing, ksi Ultimate | Bearing, ksi Yield | Fatigue strength, ksi $10^5$ cycles | Fatigue strength, ksi $10^6$ cycles | Fatigue strength, ksi $10^7$ cycles | Fatigue strength, ksi $10^8$ cycles | Fatigue strength, ksi $5 \times 10^8$ cycles |
|---|---|---|---|---|---|---|---|
| F | | | | | | | 10 |
| T4 | 69.0 | 45.0 | | | | | 11 |
| T61 | 81.0 | 68.0 | | | | | 10 |

**Table 10.3—1 (continued)**
**AM100A**

Effect of Testing Temperature on Typical Mechanical Properties

| Temper | Room temperature | | | 200°F | | | 300°F | | |
|---|---|---|---|---|---|---|---|---|---|
| | TS, ksi | TYS, ksi | Elong., % | TS, ksi | TYS, ksi | Elong., % | TS, ksi | TYS, ksi | Elong., % |
| T4 | | | | 34 | | 1.5 | 23 | | 9 |
| T6 | | | | — | | | 24 | 9 | 4 |

| Temper | 400°F | | | 500°F | | | 600°F | | |
|---|---|---|---|---|---|---|---|---|---|
| T4 | | | | 12 | | 22 | | | |
| T6 | 17 | 6.5 | 25 | 12 | 4 | 45 | 8.5 | 2.5 | 60 |

| Temper | 700°F | | | −108°F | | |
|---|---|---|---|---|---|---|
| F | | | | 22 | 18 | 1 |
| T4 | | | | 38 | 18 | 7 |
| T6 | 5.5 | 1.5 | 100 | 39 | 26 | 2 |

| | Temperature, °F | Hardness | | Charpy, ft-lb |
|---|---|---|---|---|
| | | Bhn | RE | |
| F | −108 | 63 | 75 | 0.8 |
| T4 | −108 | 60 | 73 | 2.5 |
| T6 | −108 | 85 | 90 | 0.8 |

## Table 10.3—2
## AZ91C

AZ91C is the most commonly employed alloy for sand and permanent mold castings to be used at room temperature. It combines strength and ductility with good foundry characteristics.

**Typical chemical composition, percent by weight:** Al, 8.7; Mn, 0.13 minimum; Zn, 0.7; Mg, balance

**Physical constants and thermal properties**
Density at 68°F, lb/cu in $^{-1}$ : 0.0652

Coefficient of thermal expansion, °F $^{-1}$ × 10 $^{-6}$, 65–212°F: 14.5
Modulus of elasticity, psi: 6.5 × 10$^6$
Modulus of rigidity, psi: 2.4 × 10$^6$
Poisson's ratio: 0.35
Melting range, °F: 875–1105
Specific heat, Btu/lb $^{-1}$/°F $^{-1}$, 70°F: 0.28
Electrical resistivity, microhm-cm, 70°F: 13.6; 300°F: 15.8; 500°F: 16.8
Thermal conductivity, Btu/hr $^{-1}$/ft $^{-2}$/°F $^{-1}$/ft, 70°F: 31.0; 300°F: 39.0; 500°F: 46.5

### Heat Treatments

| Treatment | Temperature, °F | Time, hours | Cooling |
|---|---|---|---|
| Solution | | | |
| T4 | 780 | 16 | Air blast |
| Aging | | | |
| T6 | 400 | 4 | |
| T7 | 450 | 5 | |

### Specifications

| | Ingot | Sand castings | Permeable mold castings |
|---|---|---|---|
| AMS | | 4437 | |
| ASTM | B93 | B80 | B199 |
| SAE | | 304 | |
| Federal | | QQ-M-56 | QQ-M-55 |

### Typical Room Temperature Mechanical Properties

| Temper | TS, ksi | YS, ksi | CYS, ksi | Elong., % | Hardness Bhn | Hardness RE | Shear strength, ksi | Charpy impact strength, ft-lb |
|---|---|---|---|---|---|---|---|---|
| F | 23/24 | 11/14 | 14 | 2 | 52 | 62 | 16 | 0.6 |
| T4 | 34/40 | 11/13 | 13 | 7/15 | 53 | 64 | 17 | 3.0 |
| T6 | 34/40 | 10/19 | 19 | 3/5 | 66 | 78 | 19 | 1.0 |

| Temper | Bearing, ksi Ultimate | Bearing, ksi Yield |
|---|---|---|
| F | 60 | 40 |
| T4 | 60 | 44 |
| T6 | 75 | 52 |

## Table 10.3—2 (continued)
### AZ91C

Effect of Testing Temperature on Typical Mechanical Properties

| Temper | Room temperature | | | 200°F | | | 300°F | | |
|---|---|---|---|---|---|---|---|---|---|
| | TS, ksi | TYS, ksi | Elong., % | TS, ksi | TYS, ksi | Elong., % | TS, ksi | TYS, ksi | Elong., % |
| T4 | 40 | 15 | 14 | 34 | 14 | 26 | 28 | 14 | 30 |
| T6 | 40 | 21 | 5 | 37 | 19 | 24 | 27 | 17 | 31 |

| Temper | 400°F | | |
|---|---|---|---|
| T4 | 20 | 13 | 30 |
| T6 | 20 | 14 | 33 |

## Table 10.3—3
## AZ31B

AZ31B is the most commonly used magnesium sheet and plate alloy. Suitable for use up to 275°F, the alloy is both formable and weldable. Weldments must be stress relieved to prevent stress-corrosion cracking. It is also used for general purpose extrusions, and sometimes for forgings.

**Typical chemical composition, percent by weight:** Al, 3.0; Mn, 0.20 minimum; Zn, 1.0; Mg, balance

**Physical constants and thermal properties**
Density at 68°F, lb/cu in.$^{-1}$: 0.0640
Coefficient of thermal expansion, °F$^{-1}$ × 10$^{-6}$, 65–212°F: 14
Modulus of elasticity, psi: 6.5 × 10$^6$
Modulus of rigidity, psi: 2.4 × 10$^6$
Poisson's ratio: 0.35
Melting range, °F: 1116–1169
Specific heat, Btu/lb$^{-1}$/°F$^{-1}$, 70°F: 0.24; 300°F: 0.26; 500°F: 0.29
Electrical resistivity, microhm-cm, 70°F: 9.2; 300°F: 11.2; 500°F: 12.9
Thermal conductivity, Btu/hr$^{-1}$/ft$^{-2}$/°F$^{-1}$/ft, 70°F: 44.5; 300°F: 53.5; 500°F: 59.0

### Heat Treatments

| | Temperature, °F | Stress relief |
|---|---|---|
| O | 650 | 500°F 15 min |
| H24 | | 300°F 60 min |
| H26 | | 300°F 60 min |
| Extrusions | | 500°F 15 min |

### Specifications

| | Sheet and plate | Extrusions Bar | Extrusions Tube | Forgings |
|---|---|---|---|---|
| AMS | 4375,4376,4377 | | | |
| ASTM | B90 | B107 | B217 | B91 |
| SAE | 510 | 510 | 510 | |
| Federal | QQ-M-44 | QQ-M-31 | WW-T-825 | |

### Typical Room Temperature Mechanical Properties

| Temper | TS, ksi | YS, ksi | CYS, ksi | Elong., % | Hardness Bhn | Hardness RE | Shear strength, ksi | Charpy impact strength, ft-lb |
|---|---|---|---|---|---|---|---|---|
| O | 32/37 | 15/22 | 16 | 9/21 | 56 | 67 | 21/29 | |
| H24 | 34/42 | 18/32 | 26 | 6/15 | 73 | 83 | 18/29 | |
| H26 | 35/40 | 21/30 | 24 | 6/16 | | | 18/28 | |
| Extruded bars | 38 | 29 | 14 | 15 | 49 | 57 | 19 | 3.2 |
| Press forged | 38 | 25 | | 15 | 50 | 59 | 19 | 3.2 |

| Temper | Bearing, ksi Ultimate | Bearing, ksi Yield | Fatigue strength, ksi 10$^5$ cycles | 10$^6$ cycles | 10$^7$ cycles | 10$^8$ cycles | 5 × 10$^8$ cycles |
|---|---|---|---|---|---|---|---|
| O | 66/70 | 37/42 | 20/24 | 19/22 | 18/21 | | |
| H24 | 77 | 47 | 22/27 | 20/24 | 19/23 | | |
| H26 | 72 | 46 | | | | | |
| Extruded bars | 56 | 33 | | | | | |
| Extruded, F | | | 23/29 | 21/26 | 19/23 | 17/21 | |
| Extruded, F | | | 23/28 | 22/25 | 20/24 | | |

**Table 10.3—3 (continued)**
**AZ31B**

Effect of Testing Temperature on Typical Mechanical Properties

| Temper | Room temperature | | | 200°F | | | 300°F | | |
|---|---|---|---|---|---|---|---|---|---|
| | TS, ksi | TYS, ksi | Elong., % | TS, ksi | TYS, ksi | Elong., % | TS, ksi | TYS, ksi | Elong., % |
| Sheet H24 | 42 | 32 | 17 | 34 | 24 | 35 | 22 | 13 | 57 |
| Extrusions, F | 38 | 28 | 15 | 35 | 21 | 22 | 24 | 14 | 28 |
| | 400°F | | | 500°F | | | 600°F | | |
| Sheet H24 | 13 | 8 | 82 | 8 | 5 | 93 | 6 | 2 | — |
| Extrusions, F | 15 | 8 | 42 | 10 | 4 | 47 | 6 | 2 | 66 |
| | −109°F | | | −320°F | | | −420°F | | |
| Sheet H24 | 46 | 33 | 6 | 57 | 37 | 3 | 66 | 40 | 2 |

Mechanical Properties of AZ31B-H24 Sheet at Elevated Temperature
After Exposure at Temperature — Limited Data

| Property | Temperature, °F | | Exposure, hours | | | | |
|---|---|---|---|---|---|---|---|
| | Exposure | Testing | 16 | 48 | 192 | 500 | 1000 |
| TS, ksi | | | 33.7 | 32.7 | 33.6 | 33.2 | 33.7 |
| TYS, ksi | 200 | 200 | 24.9 | 24.8 | 25.3 | 24.9 | 24.4 |
| Elong., % | | | 36.7 | 40.0 | 37.2 | 39.2 | 34.2 |
| CYS, ksi | | | 23.0 | 22.8 | 22.7 | 22.7 | 22.3 |
| TS, ksi | | | 28.0 | 27.7 | 28.0 | 27.5 | 27.8 |
| TYS, ksi | 250 | 250 | 20.4 | 19.9 | 20.8 | 19.6 | 21.0 |
| Elong., % | | | 48.2 | 49.5 | 49.0 | 52.5 | 50.0 |
| CYS, ksi | | | 22.3 | 21.8 | 21.4 | 21.7 | 21.7 |
| TS, ksi | | | 22.9 | 22.8 | 21.6 | 21.7 | 21.0 |
| TYS, ksi | 300 | 300 | 15.8 | 16.5 | 16.0 | 15.7 | 15.1 |
| Elong., % | | | 52.5 | 54.7 | 58.0 | 62.7 | 63.7 |
| CYS, ksi | | | 21.8 | 18.8 | 19.0 | 19.5 | 18.4 |
| TS, ksi | | | 13.3 | 13.2 | 13.0 | 13.4 | 13.3 |
| TYS, ksi | 400 | 400 | 10.4 | 10.2 | 10.1 | 10.2 | 10.8 |
| Elong., % | | | 59.5 | 83.2 | 76.2 | 73.7 | 71.5 |
| CYS, ksi | | | 12.4 | 12.1 | 12.3 | 10.7 | 12.3 |
| TS, ksi | | | 8.3 | 8.7 | 8.6 | 8.8 | 8.9 |
| TYS, ksi | 500 | 500 | 6.4 | 7.1 | 7.2 | 7.3 | 6.8 |
| Elong., % | | | 70.5 | 90.0 | 91.5 | 87.0 | 91.2 |
| CYS, ksi | | | 8.5 | 8.3 | 9.0 | 9.0 | 8.8 |
| TS, ksi | | | 6.0 | — | 5.9 | 6.3 | — |
| TYS, ksi | 600 | 600 | — | — | 4.7 | 4.5 | — |
| Elong., % | | | 115.0 | — | 112.5 | 112.5 | — |
| CYS, ksi | | | 5.4 | — | 6.1 | 5.8 | — |

## Table 10.3—3 (continued)
## AZ31B

### Typical Creep Strengths

#### Stress, ksi

| Temper | Temperature, °F | 0.1% Total extension | | | | | 0.2% Total extension | | | | |
| | | Minutes 10 | Hours 1 | 10 | 100 | 500 | Minutes 10 | Hours 1 | 10 | 100 | 500 |
|---|---|---|---|---|---|---|---|---|---|---|---|
| Sheet, O | 200 | | 6.0 | 5.0 | 4.0 | 4.0 | | 10.0 | 9.0 | 8.0 | 7.0 |
| | 250 | | 5.0 | 4.0 | 2.5 | 2.0 | | 8.0 | 7.0 | 5.0 | 3.0 |
| | 300 | | 4.0 | 2.5 | 1.0 | 0.5 | | 7.0 | 4.5 | 2.0 | 1.0 |
| | 350 | | 2.0 | 1.0 | | | | 4.0 | 2.0 | 0.5 | |
| Sheet, H24 | 200 | | 6.0 | 4.0 | 2.5 | 2.0 | | 9.0 | 7.0 | 4.0 | 3.0 |
| | 250 | | 4.0 | 2.0 | 1.0 | 1.0 | | 6.0 | 4.0 | 2.0 | 1.5 |
| | 300 | | 2.0 | 1.0 | 0.5 | | | 4.0 | 2.0 | 1.0 | 0.5 |
| | 350 | | 1.0 | 0.5 | | | | 2.0 | 1.0 | | |
| Extruded, F | 200 | | 6.0 | 5.0 | 4.0 | 3.5 | | 10.0 | 9.0 | 7.0 | 6.0 |
| | 250 | | 5.0 | 4.0 | 3.0 | 2.0 | | 9.0 | 7.0 | 5.0 | 4.0 |
| | 300 | | 4.0 | 3.0 | 1.5 | 1.0 | | 7.0 | 5.0 | 3.0 | 2.0 |
| | 350 | | 3.0 | 2.0 | | | | 5.0 | 3.0 | | |

| Temper | Temperature, °F | 0.5% Total extension | | | | | 1.0% Total extension | | | | |
| | | Minutes 10 | Hours 1 | 10 | 100 | 500 | Minutes 10 | Hours 1 | 10 | 100 | 500 |
|---|---|---|---|---|---|---|---|---|---|---|---|
| Sheet, O | 200 | | 15.0 | 14.0 | 12.0 | 10.0 | | 16.0 | 15.0 | 13.0 | 11.0 |
| | 250 | | 12.0 | 11.0 | 9.0 | 7.0 | | 14.0 | 12.0 | 10.0 | 9.0 |
| | 300 | | 10.0 | 7.0 | 4.0 | 3.0 | | 12.0 | 9.0 | 5.0 | 3.5 |
| | 350 | | 7.0 | 4.0 | | | | 8.0 | 5.0 | | |
| Sheet, H24 | 200 | | 15.0 | 11.0 | 7.0 | 5.0 | | 19.0 | 14.0 | 10.0 | 7.0 |
| | 250 | | 9.0 | 6.0 | 4.0 | 2.0 | | 13.0 | 8.0 | 5.0 | 3.0 |
| | 300 | | 6.0 | 3.0 | 1.5 | 1.0 | | 8.0 | 5.0 | 2.0 | 1.0 |
| | 350 | | 4.0 | 1.5 | | | | 6.0 | 2.0 | | |
| Extruded, F | 200 | | 16.0 | 14.0 | 12.0 | 10.0 | | 19.0 | 17.0 | 15.0 | 13.0 |
| | 250 | | 13.0 | 11.0 | 9.0 | 7.0 | | 16.0 | 14.0 | 11.0 | 9.0 |
| | 300 | | 11.0 | 8.0 | 5.0 | 4.0 | | 12.0 | 10.0 | 7.0 | 5.0 |
| | 350 | | 8.0 | 6.0 | | | | 10.0 | 8.0 | | |

## Table 10.3—4
## HM21A

HM21A is a sheet, plate, and forging alloy usable to 800°F. It is used in missile and aircraft applications.

**Typical chemical composition, percent by weight:** Mn, 0.80 minimum; Th, 2.0; Mg, balance

**Physical constants and thermal properties**
Density at 68°F, lb/cu in. $^{-1}$: 0.0643; 500°F, 0.0629

Modulus of elasticity, psi: $6.5 \times 10^6$
Modulus of rigidity, psi: $2.4 \times 10^6$
Poisson's ratio: 0.35
Melting range, °F: 1121—1202
Specific heat, Btu/lb$^{-1}$/°F$^{-1}$, 70°F: 0.24; 300°F: 0.26; 500°F: 0.27
Electrical resistivity, microhm-cm, 70°F: 5.0; 300°F: 7.2; 500°F: 9.1
Thermal conductivity, Btu/hr$^{-1}$/ft$^{-2}$/°F$^{-1}$/ft, 70°F: 79.1; 300°F: 80.3; 500°F: 81.3

### Heat Treatments

| Treatments | Temperature, °F | Time, hours |
|---|---|---|
| Solution | 850 | |
| Aging (T5, forgings) | 450 | 16 |

### Specifications

| | Sheet and plate |
|---|---|
| AMS | 4390 |
| ASTM | B90 |
| Military | MIL-M-8917 |

### Typical Room Temperature Mechanical Properties

| Temper | TS, ksi | YS, ksi | CYS, ksi | Elong., % | Hardness Bhn | Hardness RE | Shear strength, ksi | Charpy impact strength, ft-lb | Bearing, ksi Ultimate | Bearing, ksi Yield |
|---|---|---|---|---|---|---|---|---|---|---|
| T8 | 30/37 | 18/33 | 15/24 | 8/12 | | | 18/20 | | 60/67 | 36/41 |

### Effect of Testing Temperature on Typical Mechanical Properties

| Temper | Room temperature TS, ksi | Room temperature TYS, ksi | Room temperature Elong., % | 200°F TS, ksi | 200°F TYS, ksi | 200°F Elong., % | 300°F TS, ksi | 300°F TYS, ksi | 300°F Elong., % |
|---|---|---|---|---|---|---|---|---|---|
| Sheet, T8 | 35 | 27 | 12 | 27 | 24 | 15 | 22 | 21 | 20 |
| Forging, T5 | 38 | 23 | 10 | | | | 23 | 19 | 29 |

| Temper | 400°F TS, ksi | 400°F TYS, ksi | 400°F Elong., % | 500°F TS, ksi | 500°F TYS, ksi | 500°F Elong., % | 600°F TS, ksi | 600°F TYS, ksi | 600°F Elong., % |
|---|---|---|---|---|---|---|---|---|---|
| Sheet, T8 | 19 | 18 | 30 | 17 | 15 | 25 | 15 | 13 | 17 |
| Forging, T5 | 20 | 16 | 29 | 18 | 14 | 26 | 17 | 13 | 25 |

**Table 10.3—4 (continued)**
**HM21A**

Effect of Testing Temperature on Typical Mechanical Properties (continued)

| | 700° F | | | 800° F | | | 900° F | | |
|---|---|---|---|---|---|---|---|---|---|
| Sheet, T8 | 11 | 8 | 50 | 5 | 3 | 100 | 2 | 1 | — |
| Forging, T5 | 12 | 9 | 45 | | | | | | |

| | -109° F | | | -320° F | | | -420° F | | |
|---|---|---|---|---|---|---|---|---|---|
| Sheet, T8 | 43 | 26 | 6 | 50 | 26 | 5 | 52 | 27 | 5 |

Mechanical Properties of HM21A-T8 Sheet After Exposure
at Elevated Temperature – Limited Data

| Temper | Property | Temperature, ° F | | Exposure time, hours | | | | |
|---|---|---|---|---|---|---|---|---|
| | | Exposure | Testing | 0 | 1 | 100 | 1000 | 5000 |
| T8 | TS, ksi | | | 33 | 33 | 33 | 33 | 33 |
| | TYS, ksi | 600 | 70 | 25 | 25 | 25 | 25 | 25 |
| | Elong., % | | | 7 | 7 | 7 | 7 | 7 |
| | TS, ksi | | | 33 | — | 32 | — | — |
| | TYS, ksi | 700 | 70 | 25 | — | 24 | — | — |
| | Elong., % | | | 7 | — | 9 | — | — |
| | TS, ksi | | | 33 | — | 30 | — | — |
| | TYS, ksi | 800 | 70 | 25 | — | 18 | — | — |
| | Elong., % | | | 7 | — | 12 | — | — |
| | TS, ksi | 600 | 600 | 14 | No change in values | | | 14 |
| | TYS, ksi | | | 12 | | | | 12 |
| | Elong., % | | | 15 | | | | 15 |
| | CYS, ksi | | | 13 | | | | 13 |
| | TS, ksi | 700 | 700 | 11 | — | 11 | 7 | — |
| | TYS, ksi | | | 8 | — | 8 | 5 | — |
| | Elong., % | | | 50 | — | 50 | 112 | — |
| | CYS, ksi | | | 9 | — | 9 | 6 | — |

**Table 10.3—4 (continued)**
**HM21A**

Typical Creep Strengths

Stress, ksi

**0.1% Total extension**

| Temper | Temperature, °F | Minutes 10 | Hours 1 | 10 | 100 | 1000 |
|---|---|---|---|---|---|---|
| Sheet, T8 | 400 | | 5.8 | 5.8 | 5.8 | |
| | 500 | | 5.6 | 5.6 | 5.6 | |
| | 600 | | 5.2 | 4.4 | 4.3 | |
| | 700 | | 3.0 | 2.6 | 2.3 | |
| Forging, T5 | 400 | | | | 15.4ᵃ | |
| | 500 | | | | 10.8ᵃ | |
| | 600 | | | | 6.9ᵃ | |
| | 700 | | | | 3.3ᵃ | |

**0.2% Total extension**

| Temper | Temperature, °F | Minutes 10 | Hours 1 | 10 | 100 | 1000 |
|---|---|---|---|---|---|---|
| Sheet, T8 | 400 | | 11.7 | 11.6 | 11.4 | |
| | 500 | | 9.8 | 9.0 | 7.0 | |
| | 600 | | 7.2 | 5.5 | 5.0 | |
| | 700 | | 3.7 | 3.2 | 2.6 | |
| Forging, T5 | 400 | | | | 10.7 | |
| | 500 | | | | 9.0 | |
| | 600 | | | | 6.5 | |
| | 700 | | | | 3.2 | |

**0.5% Total extension**

| Temper | Temperature, °F | Minutes 10 | Hours 1 | 10 | 100 | 1000 |
|---|---|---|---|---|---|---|
| Sheet, T8 | 400 | | 16.9 | 16.6 | 13.5 | |
| | 500 | | 14.4 | 12.0 | 9.0 | |
| | 600 | | 8.7 | 6.9 | 6.0 | |
| | 700 | | 5.1 | 4.2 | 3.3 | |
| Forging, T5 | 400 | | | | 14.6 | |
| | 500 | | | | 11.0 | |
| | 600 | | | | 7.6 | |
| | 700 | | | | 4.8 | |

**1.0% Total extension**

| Temper | Temperature, °F | Minutes 10 | Hours 1 | 10 | 100 | 1000 |
|---|---|---|---|---|---|---|
| Sheet, T8 | 400 | | 17.7 | 17.5 | 13.9 | |
| | 500 | | 15.3 | 13.3 | 9.7 | |
| | 600 | | 9.6 | 8.1 | 6.8 | |
| | 700 | | 5.7 | 4.5 | 3.6 | |
| Forging, T5 | 400 | | | | | |
| | 500 | | | | | |
| | 600 | | | | | |
| | 700 | | | | | |

ᵃCreep extension

## Table 10.3—5
## LA141A

LA141A is the lightest alloy available for aerospace applications. It is 25% lighter than conventional magnesium alloys and 27% lighter than beryllium. It is easily formed and machined. It has been produced as sheet, plate, foil, extrusions, forgings, wire, and castings.

**Typical chemical composition, percent by weight:** Al, 1.2; Mn, 0.15 minimum; Li, 14.0; Mg, balance

**Physical constants and thermal properties**
Density at 68°F, lb/cu in.$^{-1}$: 0.049

Coefficient of thermal expansion,$°F^{-1} \times 10^{-6}$, 65–212°F: 21.8
Modulus of elasticity, psi: $7.2 \times 10^6$ at –105°F, $6.2 \times 10^6$ at + 70°F, $5.0 \times 10^6$ at 150°F, $3.4 \times 10^6$ at 250°F, $2.6 \times 10^6$ at 300°F
Melting range, °F (liquidus): 1075 ± 10
Specific heat Btu/lb$^{-1}$/°F$^{-1}$, 70°F: 0.346; 300°F: 0.348; 500°F: 0.350
Electrical resistivity, microhm-cm, 70°F: 15.2
Thermal conductivity, Btu/hr$^{-1}$/ft$^{-2}$/°F$^{-1}$/ft, 70°F: 301; 300°F: 287; 500°F: 280

**Heat treatments**
Stabilized at 350°F ± 25 for 3–6 hr.

### Specifications

|  | Sheet and plate |
|---|---|
| AMS | 4386 |

### Typical Room Temperature Mechanical Properties

| Temper | TS, ksi | YS, ksi | CYS, ksi | Elong., % | Hardness Bhn | Hardness RE | Shear strength, ksi | Charpy impact strength, ft-lb |
|---|---|---|---|---|---|---|---|---|
| T7 | 18/21 | 13/19 | 19 | 10/27 | 54 | 65 | 13 | 15 longitudinal |
| Extrusion | 20.2 | 15.7 |  | 22 |  |  |  | 32 transverse |
| Casting | 17.7 | 12.3 |  | 17 |  |  |  |  |

| Temper | Bearing, ksi Ultimate | Bearing, ksi Yield | Fatigue strength, ksi $10^5$ cycles | $10^6$ cycles | $10^7$ cycles | $10^8$ cycles | $5 \times 10^8$ cycles |
|---|---|---|---|---|---|---|---|
| T7 | 46.5 | 31 |  |  |  | 8.0 |  |

### Effect of Testing Temperature on Typical Mechanical Properties

| Temper | Room temperature TS, ksi | TYS, ksi | Elong., % | 200°F TS, ksi | TYS, ksi | Elong., % | 300°F TS, ksi | TYS, ksi | Elong., % |
|---|---|---|---|---|---|---|---|---|---|
| T7 | 20 | 18 | 18 | 11 |  | 33 | 5 |  | 47[a] |

| Temper | –100°F TS, ksi | TYS, ksi | Elong., % | –300°F TS, ksi | TYS, ksi | Elong., % | –420°F TS, ksi | TYS, ksi | Elong., % |
|---|---|---|---|---|---|---|---|---|---|
| T7 | 28 | 21 | 12 | 32 | 27 | 11 | 42 | 39 | 11 |

[a]At 250°F

## Table 10.3—5 (continued)
## LA141A

### LA141A Alloy Transverse and Longitudinal Bearing Strength

| Temperature, °F | Bearing strength, psi | | | |
| --- | --- | --- | --- | --- |
| | Ultimate | | Yield | |
| | Design | Typical | Design | Typical |
| Room temperature | 43,000 | – | 27,000 | – |
| Longitudinal | – | 46,500 | – | 31,000 |
| Transverse | – | 46,400 | – | 33,000 |
| 150° | 28,000 | – | 22,000 | – |
| Longitudinal | – | 30,600 | – | 26,000 |
| Transverse | – | 30,100 | – | 24,700 |

### Charpy Impact (V-notch) Toughness of LA141A Alloy

| Temperature, °F | Surface notch | | Short transverse notch | |
| --- | --- | --- | --- | --- |
| | Longitudinal, ft-lb | Transverse, ft-lb | Longitudinal, ft-lb | Transverse, ft-lb |
| –320 | 15 | 31 | 11 | 15 |
| –110 | 15 | 30 | 9 | 14 |
| –20 | 15 | 30 | 9 | 15 |
| +75 | 15 | 32 | 10 | 14 |
| +150 | 14 | 28 | 9 | 15 |

### Typical Creep Strengths

**Stress, ksi**

| Temper | Temperature, °F | 0.1% Creep extension | | | | | 0.2% Total extension | | | | |
| --- | --- | --- | --- | --- | --- | --- | --- | --- | --- | --- | --- |
| | | Minutes 10 | Hours | | | | Minutes 10 | Hours | | | |
| | | | 1 | 10 | 100 | 1000 | | 1 | 10 | 100 | 1000 |
| T7 | 70 | | | 6.8 | 6.0 | | | | | 6.6 | 6.1 |
| | 100 | | | 5.8 | 4.8 | | | | | 5.8 | 5.0 |
| | 200 | | | 2.0 | 1.5 | | | | | 2.4 | 1.4 |
| | 250 | | | 1.3 | 1.1 | | | | | 1.5 | 1.2 |

| Temper | Temperature, °F | 0.5% Total extension | | | | |
| --- | --- | --- | --- | --- | --- | --- |
| | | Minutes 10 | Hours | | | |
| | | | 1 | 10 | 100 | 1000 |
| T7 | 70 | | | 8.3 | 8.1 | |
| | 100 | | | 7.7 | 6.7 | |
| | 200 | | | 3.5 | 2.1 | |
| | 250 | | | 1.9 | 1.4 | |

## 10.4   NICKEL-BASE ALLOYS

### E. B. Fernsler

Nickel, in substantially pure form, is widely used in industry for its corrosion resistance combined with moderate mechanical strength and good thermal and electrical conductivity. The principal high nickel alloy contains approximately 99.5% nickel and small amounts of manganese, magnesium, and carbon. A wide variety of nickel alloys with widely differing properties are commercially available. These have been listed in the following six categories:

Nickel alloys
Nickel-copper alloys
Nickel-chromium alloys
Nickel-molybdenum alloys

Nickel-iron-chromium alloys
High temperature-high strength alloys

Alloys containing a high concentration of nickel in a matrix containing a larger amount of some other constituent are listed under the major element. An example of this is an austenitic stainless steel containing a higher concentration of iron than of nickel, and thus listed in Section 1.1. TD-nickel and TD-NiCr were not included because they are not currently commercially available. The new mechanically alloyed Inconel®MA753 alloy which is both dispersion and precipitation hardened has been included.

### 1.4.1 NICKEL ALLOYS

#### Table 1.4.1–1
#### NICKEL 200

**Chemical composition, percent by weight:** C, 0.08; Mn, 0.18; Fe, 0.20; S, 0.005; Si, 0.18; Ni, 99.5; Cu, 0.13

**Physical constants and thermal properties**
Density, lb/in.$^3$: 0.321
Coefficient of thermal expansion, (70–200°F) in./in./°F × 10$^{-6}$: 7.4
Modulus of elasticity, psi (dynamic): tension, 29.6 × 10$^6$; torsion, 11.7 × 10$^6$
Poisson's ratio: 0.264
Melting range, °F: 2,615–2,635

Specific heat, Btu/lb/°F, 70°F: 0.109
Thermal conductivity, Btu/ft$^2$/hr/in./°F, 70°F: 520
Electrical resistivity, ohms/cmil/ft, 70°F: 57
Curie temperature, °F: annealed, 680
Permeability (70°F, 200 Oe): annealed – ferromagnetic

**Heat treatments**
Generally used in annealed (1,300–1,700°F) condition.

### Tensile Properties

#### Annealed

| Temperature, °F | T.S., psi | Y.S., psi, 0.2% offset | Elong., in 2 in., % | Hardness, Brinell |
|---|---|---|---|---|
| 70 | 67,000 | 21,500 | 47.0 | — |
| 200 | 66,500 | 22,300 | 46.0 | — |
| 400 | 66,500 | 20,200 | 44.0 | — |
| 600 | 66,200 | 20,200 | 47.0 | — |
| 800 | 44,000 | 16,500 | 65.0 | — |
| 1,000 | 31,500 | 13,500 | 69.0 | — |
| 1,200 | 21,500 | 10,000 | 76.0 | – |
| 1,400 | 14,000 | 7,000 | 89.0 | — |
| 1,600 | 8,200 | 3,600 | 110.0 | — |
| 1,800 | 5,400 | 2,300 | 198.0 | — |
| 2,000 | 3,500 | 1,400 | 205.0 | — |

*   Based on section by E. B. Fernsler in the *Handbook of Materials Science,* Volume II.

## Table 10.4—1 (continued)
## NICKEL 200

### Creep Strength
### (Stress, psi, to Produce 1% Creep)

**Annealed**

| Test temperature, °F | 10,000 hr |
|---|---|
| 600 | 40,000 |
| 700 | 13,000 |
| 800 | 6,000 |

### Fatigue Strength

**Annealed**

| Test temperature, °F | Stress, psi | Cycles to failure |
|---|---|---|
| 70 | 33,000 | $10^8$ |

### Impact Strength

**Annealed**

| Test temperature, °F | Type test | Strength, ft-lb |
|---|---|---|
| +70 | Charpy — V-notched | 228 |

### Specifications

ASTM: B160, B161, B162, B163

## Table 10.4—2
### NICKEL 270

**Chemical composition, percent by weight:** C, 0.01; Fe, 0.003; Ni, 99.98

**Physical constants and thermal properties**
Density, lb/in.$^3$: 0.321
Coefficient of thermal expansion, (70–200°F) in./in./°F $\times$ 10$^{-6}$: 7.4
Modulus of elasticity, psi: tension, 30 $\times$ 10$^6$
Melting range, °F: 2,650

Specific heat, Btu/lb/°F, 70°F: 0.11
Thermal conductivity, Btu/ft$^2$/hr/in./°F, 70°F: 599
Electrical resistivity, ohms/cmil/ft, 70°F: 45
Curie temperature, °F: annealed, 676
Permeability (70°F, 200 Oe): annealed – ferromagnetic

**Heat treatments**
Usually used in annealed condition (800–1,000°F).

### Tensile Properties

#### Annealed

| Temperature, °F | T.S., psi | Y.S., psi 0.2% offset | Elong., in 2 in., % | Hardness, Brinell |
|---|---|---|---|---|
| 70 | 50,000 | 16,000 | 50 | 80 |

## Table 10.4—3
### MONEL® ALLOY 400

**Chemical composition, percent by weight:** C, 0.15; Mn, 1.00; Fe, 1.25; S, 0.012; Si, 0.25; Ni, 66.5; Cu, 31.5

**Physical constants and thermal properties**
Density, lb/in.$^3$:0.319
Coefficient of thermal expansion, (70–200°F) in./in./°F $\times$ 10$^{-6}$: 7.7
Modulus of elasticity, psi: tension, 26.0 $\times$ 10$^6$; torsion, 9.5 $\times$ 10$^6$

Poisson's ratio: 0.32
Melting range, °F: 2,370–2,460
Specific heat, Btu/lb/°F, 70°F: 0.105
Thermal conductivity, Btu/ft$^2$/hr/in./°F, 70°F: 151
Electrical resistivity, ohms/cmil/ft,70°F: 307
Curie temperature, °F: annealed, 20–50

**Heat treatments**
Generally used in annealed condition.

### Tensile Properties

#### Annealed

| Temperature, °F | T.S., psi | Y.S., psi, 0.2% offset | Elong., in 2 in., % | Hardness, Brinell |
|---|---|---|---|---|
| 70 | 79,000 | 30,000 | 48 | – |
| 600 | 75,000 | 21,500 | 50 | – |
| 800 | 63,500 | 21,000 | 50 | – |
| 1,000 | 45,500 | 20,000 | 26 | – |
| 1,200 | 26,500 | 14,500 | 36 | – |
| 1,400 | 17,500 | 11,000 | 44 | – |
| 1,600 | 9,000 | 6,500 | 52 | – |
| 1,800 | 5,000 | 2,500 | 60 | – |

## Table 10.4—3 (continued)
## MONEL® ALLOY 400

### Rupture Strength, 1,000 hr

**Cold Drawn, Annealed**

| Test temperature, °F | Strength, psi | Elong., in 2 in., % | Reduction of area, % |
|---|---|---|---|
| 700 | 70,000 | — | — |
| 900 | 42,000 | — | — |
| 1,100 | 17,000 | — | — |

### Creep Strength (Stress, psi, to Produce 1% Creep)

**Cold Drawn, Annealed**

| Test temperature, °F | 10,000 hr | 100,000 hr |
|---|---|---|
| 750 | 30,000 | — |
| 800 | 24,000 | — |
| 900 | 16,000 | — |
| 1,000 | 9,500 | — |

### Fatigue Strength

**Annealed**

| Test temperature, °F | Stress, psi | Cycles to failure |
|---|---|---|
| 70 | 33,500 | $10^8$ |

### Impact Strength

**Cold Drawn, Annealed**

| Test temperature, °F | Type test | Strength, ft-lb |
|---|---|---|
| −310 | Charpy−V-notched | 212 |
| −112 | Charpy−V-notched | 219 |
| 75 | Charpy−V-notched | 216 |

### Specifications

ASTM: B127, B163, B164, B165, B395; Federal QQ-N-281 Class A

## Table 10.4—4
## Ni-Cu ALLOY 410

**Chemical composition, percent by weight:** C, 0.20; Mn, 0.80; Fe, 1.00; S, 0.008; Si, 1.60; Cu, 30.5; Ni, 66.0

**Physical constants and thermal properties**
Density, lb/in.$^3$: 0.312
Coefficient of thermal expansion, (70–200°F) in./in./°F × 10$^{-6}$: 7.6
Modulus of elasticity, psi: tension, 23 × 10$^6$

Melting range, °F: 2,700–2,900
Specific heat, Btu/lb/°F, 70°F: 0.13
Thermal conductivity, Btu/ft$^2$/hr/in./°F, 70°F: 186
Electrical resistivity, ohms/cmil/ft, 70°F: 320

**Heat treatments**
Generally used as cast.

### Tensile Properties

#### As Cast

| Temperature, °F | T.S., psi | Y.S., psi, 0.2% offset | Elong., in 2 in., % | Hardness, Brinell |
|---|---|---|---|---|
| 70 | 75,000 | 35,000 | 38 | 150 |

### Fatigue Strength

#### As Cast

| Test temperature, °F | Stress, psi | Cycles to failure |
|---|---|---|
| 70 | 17,500 | 10$^8$ |

### Impact Strength

| Test temperature, °F | Type test | Strength, ft-lb |
|---|---|---|
| –210 | Charpy–V-notched | 28 |
| –110 | Charpy–V-notched | 31 |
| + 70 | Charpy–V-notched | 32 |

### Specifications

Federal: QQ-N-288 Comp. A

## Table 10.4—5
## INCONEL® 600

**Chemical composition, percent by weight:** C, 0.08; Mn, 0.5; Fe, 8.0; S, 0.008; Si, 0.25; Cr, 15.5; Ni, 76.0 Cu, 0.25; Ti, 0.35; Al, 0.25

**Physical constants and thermal properties**
Density, lb/in.$^3$: 0.304
Coefficient of thermal expansion, (70–200°F) in./in./°F $\times 10^{-6}$: 7.4
Modulus of elasticity, psi: tension, $30 \times 10^6$; torsion, $11 \times 10^6$

Poisson's ratio: 0.29
Melting range, °F: 2,470–2,575
Specific heat, Btu/lb/°F, 70°F: 0.106
Thermal conductivity, Btu/ft$^2$/hr/in./°F, 70°F: 103
Electrical resistivity, ohms/cmil/ft, 70°F: 620
Curie temperature, °F: annealed, –192
Permeability (70°F, 200 Oe): annealed, 1.010

**Heat treatments**
Used in annealed condition, 1,850°F/30 min.

### Tensile Properties

#### Hot Rolled

| Temperature, °F | T.S., psi | Y.S., psi, 0.2% offset | Elong., in 2 in., % | Hardness, Brinell |
|---|---|---|---|---|
| 70 | 90,500 | 36,500 | 47 | — |
| 600 | 90,500 | 31,000 | 46 | — |
| 800 | 88,500 | 29,500 | 49 | — |
| 1,000 | 84,000 | 28,500 | 47 | — |
| 1,200 | 65,000 | 26,500 | 39 | — |
| 1,400 | 27,500 | 17,000 | 46 | — |
| 1,600 | 15,000 | 9,000 | 80 | — |
| 1,800 | 7,500 | 4,000 | 118 | — |

### Rupture Strength, 1,000 hr

#### Solution Annealed 2,050°F/2 hr

| Test temperature, °F | Strength, psi | Elong., in 2 in., % | Reduction of area, % |
|---|---|---|---|
| 1,500 | 5,600 | — | — |
| 1,600 | 3,500 | — | — |
| 1,800 | 1,800 | — | — |
| 2,000 | 920 | — | — |

### Creep Strength (Stress, psi, to Produce 1% Creep)

#### Solution Annealed 2,050°F/2 hr

| Test temperature, °F | 10,000 hr | 100,000 hr |
|---|---|---|
| 1,300 | 5,000 | — |
| 1,500 | 3,200 | — |
| 1,600 | 2,000 | — |
| 1,700 | 1,100 | — |
| 1,800 | 560 | — |
| 2,000 | 270 | — |

**Table 10.4—5 (continued)**
**INCONEL® 600**

## Fatigue Strength

**Annealed**

| Test temperature, °F | Stress, psi | Cycles to failure |
|---|---|---|
| 70 | 39,000 | $10^8$ |

## Impact Strength

**Annealed**

| Test temperature, °F | Type test | Strength, ft-lb |
|---|---|---|
| +70 | Charpy — V-notched | 180 |
| 800 | Charpy — V-notched | 187 |
| 1,000 | Charpy — V-notched | 160 |

## Specifications

ASTM: B166, B168, B167, B163
SAE-AMS: 5665, 5540, 5580, 5687, 7232

## Table 10.4—6
## Ni-Cr 50-50

**Chemical composition, percent by weight:** C, 0.05; Cr, 48.0; Ni, balance; Ti, 0.35

**Physical constants and thermal properties**
Density, lb/in.$^3$: 0.284
Coefficient of thermal expansion, (70–200°F) in./in./°F $\times$ 10$^{-6}$ : 6.5
Melting range, °F: 2,385–2,460
Specific heat, Btu/lb/°F, 70°F: 0.109
Thermal conductivity, Btu/ft$^2$/hr/in./°F, 70°F: 109
Electrical resistivity, ohms/cmil/ft, 70°F: 523

**Heat treatments**
Anneal 2,200°F/1 hr. Air cool.

### Tensile Properties

**Annealed; 2,200°F/1 hr; Air Cool**

| Temperature, °F | T. S., psi | Y.S., psi, 0.2% offset | Elong., in 2 in., % | Hardness, Brinell |
|---|---|---|---|---|
| 400 | 103,000 | 50,500 | 19 | — |
| 600 | 102,500 | 48,300 | 19 | — |
| 800 | 101,500 | 47,500 | 22 | — |
| 1,000 | 96,700 | 44,000 | 27 | — |
| 1,200 | 80,300 | 42,500 | 37 | — |
| 1,400 | 59,000 | 43,000 | 42 | — |
| 1,600 | 27,800 | 24,000 | 67 | — |
| 1,800 | 13,100 | 10,800 | 61 | — |
| 2,000 | 6,500 | 5,400 | 42 | — |

### Rupture Strength, 1,000 hr

**Annealed 2,200°F/1 hr; Air Cool**

| Test temperature, °F | Strength, psi | Elong., in 2 in.,% | Reduction of area, % |
|---|---|---|---|
| 1,200 | 14,000 | — | — |
| 1,300 | 8,200 | — | — |
| 1,400 | 5,000 | — | — |
| 1,500 | 3,000 | — | — |
| 1,600 | 1,800 | — | — |
| 1,700 | 1,300 | — | — |
| 1,800 | 800 | — | — |

### Impact Strength

**Annealed 2,200°F/1 hr; Air Cool**

| Test temperature, °F | Type test | Strength, ft-lb |
|---|---|---|
| +70 | Charpy–V-notched | 15 |

## Table 10.4—7
## HASTELLOY® B

Chemical composition, percent by weight: C, 0.05[a]; Mn, 1.00[a]; Fe, 5.00; S, 0.030[a]; Si, 1.00[a]; Cr, 1.00[a]; Ni, balance; Co, 2.50[a]; Mo, 28.00; V, 0.30

[a]Maximum

**Physical constants and thermal properties**
Density, lb/in.$^3$ : 0.334
Coefficient of thermal expansion, (70–200°F) in./in./°F × 10$^{-6}$ : 5.6

Modulus of elasticity, psi: tension, 31.1 × 10$^6$
Melting range, °F: 2,375–2,495
Specific heat, Btu/lb/°F, 70°F: 0.091
Thermal conductivity, Btu/ft$^2$/hr/in./°F, 70°F: 85
Electrical resistivity, ohms/cmil/ft, 70°F: 811
Permeability (70°F, 200 Oe): annealed, < 1.001

**Heat treatments**
Solution treat 2,000–2,150°F, quench or rapid air cool.

### Tensile Properties

**Solution Annealed 2,000°F, Rapid Air Cool**

| Temperature, °F | T.S., psi | Y.S., psi, 0.2% offset | Elong., in 2 in., % | Hardness, Brinell |
|---|---|---|---|---|
| 70 | 134,100 | 67,000 | 51 | — |
| 1,000 | 113,500 | 48,700 | 55 | — |
| 1,200 | 106,600 | 50,400 | 50 | — |
| 1,400 | 85,300 | 47,800 | 30 | — |
| 1,600 | 71,600 | 41,100 | 22 | — |
| 1,800 | 36,200 | 14,800 | 21 | — |
| 2,000 | 25,400 | 10,100 | 20 | — |

### Rupture Strength, 1,000 hr

**Solution Annealed 2,000°F, Rapid Air Cool**

| Test temperature, °F | Strength, psi | Elong., in 2 in., % | Reduction of area, % |
|---|---|---|---|
| 1,000 | 74,000 | — | — |
| 1,200 | 36,500 | — | — |
| 1,400 | 15,500 | — | — |
| 1,500 | 9,400 | — | — |

### Fatigue Strength

**Solution Annealed, 2,000°F, Water Quench, 1,200°F/4 hr**

| Test temperature, °F | Stress, psi | Cycles to failure |
|---|---|---|
| 1,500 | 66,000 | 10$^8$ |

**Table 10.4—7 (continued)**
**HASTELLOY® B**

Impact Strength

**Solution Annealed**

| Test temperature, °F | Type test | Strength, ft-lb |
|---|---|---|
| −326 | Izod V-notch | 53 |
| −148 | Izod V-notch | 53 |
| −58 | Izod V-notch | 49 |
| +70 | Izod V-notch | 60 |

Specifications

ASTM: B333, B335, A494 (castings); SAE-AMS: 5396 (castings)

## Table 10.4—8
## HASTELLOY® ALLOY C-276

**Chemical composition, percent by weight:** C, 0.02[a]; Mn, 1.00[a]; Fe, 5.50; S, 0.03[a]; Si, 0.05[a]; Cr, 15.50; Ni, balance; Co, 2.50[a]; Mo, 16.00; W, 3.75; V, 0.35[a]; P, 0.03[a]

[a]Maximum

°F × 10$^{-6}$: 6.2
Modulus of elasticity, psi: tension, 29.8 × 10$^6$
Melting range, °F: 2,415–2,500
Specific heat, Btu/lb/°F, 70°F: 0.102
Thermal conductivity, Btu/ft$^2$/hr/in./°F, 70°F: 69
Electrical resistivity, ohms/cmil/ft, 70°F: 779

**Physical constants and thermal properties**
Density, lb/in.$^3$: 0.321
Coefficient of thermal expansion, (70–200°F) in./in./

**Heat treatments**
Solution heat treat 2,100°F, rapid quench.

### Tensile Properties

#### Solution Treated 2,100°F, Water Quench

| Temperature, °F | T.S., psi | Y.S., psi, 0.2% offset | Elong., in 2 in., % | Hardness, Brinell |
|---|---|---|---|---|
| 70 | 113,500 | 52,000 | 70 | – |
| 400 | 101,700 | 44,100 | 71 | – |
| 600 | 95,100 | 39,100 | 71 | – |
| 800 | 93,800 | 33,500 | 75 | – |
| 1,000 | 89,600 | 31,700 | 74 | – |
| 1,200 | 86,900 | 32,900 | 73 | – |
| 1,400 | 80,700 | 30,900 | 78 | – |
| 1,600 | 63,500 | 29,900 | 92 | – |
| 1,800 | 39,000 | 27,000 | 127 | – |

### Rupture Strength, 1,000 hr

#### Solution Treated 2,100°F, Water Quench

| Test temperature, °F | Strength, psi | Elong., in 2 in., % | Reduction of area, % |
|---|---|---|---|
| 1,200 | 40,000 | – | – |
| 1,400 | 18,000 | – | – |
| 1,600 | 7,000 | – | – |
| 1,800 | 3,100 | – | – |

### Impact Strength

#### Solution Treated 2,100°F, Water Quench

| Test temperature, °F | Type test | Strength, ft-lb |
|---|---|---|
| –320 | Charpy – V-notched | 181 |
| +70 | Charpy – V-notched | 238 |
| +392 | Charpy – V-notched | 239 |

## Table 10.4—9
## INCOLOY® ALLOY 800

**Chemical composition, percent by weight:** C, 0.05; Mn, 0.75; Fe, 46.00; S, 0.008; Si, 0.50; Cr, 21.00; Ni, 32.5; Ti, 0.38; Al, 0.38

**Physical constants and thermal properties**
Density, lb/in.$^3$: 0.287
Coefficient of thermal expansion, (70–200°F) in./in./°F $\times 10^{-6}$: 7.9
Modulus of elasticity, psi: tension, $28.5 \times 10^6$; torsion, $10.6 \times 10^6$
Poisson's ratio: 0.34

Melting range, °F: 2,475–2,525
Specific heat, Btu/lb/°F, 70°F: 0.12
Thermal conductivity, Btu/ft$^2$/hr/in./°F, 70°F: 80
Electrical resistivity, ohms/cmil/ft, 70°F: 595
Curie temperature, °F: annealed, –175
Permeability (70°F, 200 Oe): annealed, 1.009

**Heat treatments**
Used in annealed (1,800–1,900°) or solution treated (2,000–2,100°) condition.

### Tensile Properties

#### Annealed – 1,800°/15 min

| Temperature, °F | T. S., psi | Y. S., psi, 0.2% offset | Elong., in 2 in., % | Hardness, Brinell |
|---|---|---|---|---|
| 70 | 86,800 | 42,700 | 44 | — |
| 200 | 81,700 | 39,700 | 42.5 | — |
| 400 | 77,300 | 36,000 | 39 | — |
| 600 | 76,200 | 33,700 | 40 | — |
| 800 | 74,600 | 33,300 | 40 | — |
| 1,000 | 72,000 | 31,700 | 38.5 | — |
| 1,200 | 54,000 | 29,000 | 55.5 | — |
| 1,400 | 32,100 | 22,600 | 85 | — |

### Rupture Strength, 1,000 hr

#### Solution Treated

| Test temperature, °F | Strength, psi | Elong., in 2 in., % | Reduction of area, % |
|---|---|---|---|
| 1,000 | 50,000 | — | — |
| 1,100 | 38,000 | — | — |
| 1,200 | 25,000 | — | — |
| 1,300 | 17,000 | — | — |
| 1,400 | 9,500 | — | — |
| 1,500 | 6,000 | — | — |
| 1,600 | 3,800 | — | — |
| 1,700 | 2,500 | — | — |
| 1,800 | 1,700 | — | — |
| 2,000 | 900 | — | — |

**Table 10.4—9 (continued)**
**INCOLOY® ALLOY 800**

Creep Strength (Stress, psi, to Produce 1% Creep)

**Solution Treated**

| Test temperature, °F | 10,000 hr | 100,000 hr |
|---|---|---|
| 1,000 | 47,000 | 42,000 |
| 1,100 | 33,000 | 28,000 |
| 1,200 | 17,000 | 12,000 |
| 1,300 | 7,800 | 5,600 |
| 1,400 | 4,800 | 3,400 |
| 1,600 | 2,200 | 1,700 |
| 1,800 | 800 | — |

Fatigue Strength

**Annealed**

| Test temperature, °F | Stress, psi | Cycles to failure |
|---|---|---|
| 70 | 42,000 | $10^8$ |
| 800 | 42,000 | $10^8$ |
| 1,000 | 38,000 | $10^8$ |
| 1,400 | 22,000 | $10^8$ |

Impact Strength

| Test temperature, °F | Type test | Strength, ft-lb |
|---|---|---|
| +70 | Charpy — V-notched | 207 |
| 1,200 | Charpy — V-notched | 181 |

Specifications

ASTM: B163, B407, B408, B409

## Table 10.4—10
### ACI TYPE HX

Chemical composition, percent by weight: C, 0.55; Mn, 1.00; Fe, 14.00; S, 0.02; Si, 1.25; Cr, 17.00; Ni, 66.00

Physical constants and thermal properties
  Density, lb/in.$^3$ : 0.294
  Coefficient of thermal expansion, (70–1,000°F) in./in./°F $\times 10^{-6}$ : 7.8
  Modulus of elasticity, psi: tension, $25 \times 10^6$

Heat treatments
  Cast plus aged 1,800°F/48 hr. Air cool.

### Tensile Properties

**Cast and Aged**

| Temperature, °F | T. S., psi | Y. S., psi, 0.2% offset | Elong., in 2 in., % | Hardness, Brinell |
|---|---|---|---|---|
| 70 | 73,000 | 44,000 | 9 | 185 |

### Specifications
ASTM: A297

## Table 10.4—11
### INCONEL® 718

Chemical composition, percent by weight: C, 0.04; Mn, 0.18; Fe, 18.5; S, 0.008; Si, 0.18; Cr, 19.0; Ni, 52.5; Mo, 3.05; Cb, 5.13; Ti, 0.90; Al, 0.50

Physical constants and thermal properties
  Density, lb/in.$^3$ : 0.296
  Coefficient of thermal expansion, (70–200°F) in./in./°F $\times 10^{-6}$ : 7.2
  Modulus of elasticity, psi: tension, $29.8 \times 10^6$; torsion, $11.6 \times 10^6$
  Poisson's ratio: 0.28
  Melting range, °F: 2,300–2,437
  Specific heat, Btu/lb/°F, 70°F: 0.104

Thermal conductivity, Btu/ft$^2$/hr/in./°F, 70°F: 78
Electrical resistivity, ohms/cmil/ft, 70°F: 751
Curie temperature, °F: annealed, <–320; age hardened, <–170
Permeability (70°F, 200 Oe): annealed, 1.001; age hardened, 1.001

Heat treatments
  A. 1,800°F/1 hr/air cool plus 1,325°F/8 hr/fast cool to 1,150°F/18 hr/air cool.
  B. 1,900°F/1 hr/air cool plus 1,400°F/10 hr/fast cool to 1,200°F/20 hr/air cool.

### Tensile Properties

**A**

| Temperature, °F | T.S., psi | Y.S., psi, 0.2% offset | Elong., in 2 in., % | Hardness, Brinell |
|---|---|---|---|---|
| 70 | 208,000 | 172,000 | 21 | — |
| 1,000 | 185,000 | 154,000 | 18 | — |
| 1,200 | 178,000 | 148,000 | 19 | — |
| 1,400 | 138,000 | 107,000 | 25 | — |
| 1,600 | 49,000 | 48,000 | 88 | — |
| 1,800 | 15,000 | 15,000 | 170 | — |
| 2,000 | 8,000 | 8,000 | 125 | — |

**Table 10.4—11 (continued)**
**INCONEL® 718**

Rupture Strength, 1,000 hr

**A**

| Test temperature, °F | Strength, psi | Elong., in 2 in., % | Reduction of area, % |
|---|---|---|---|
| 1,200 | 86,000 | – | – |
| 1,300 | 53,000 | – | – |
| 1,400 | 25,000 | – | – |

Creep Strength (Stress, psi, to Produce 1% Creep)

**Aged – A**

| Test temperature, °F | 10,000 hr | 100,000 hr |
|---|---|---|
| 1,100 | 98,000 | – |
| 1,200 | 75,000 | – |
| 1,300 | 42,000 | – |

Fatigue Strength

**Aged – A**

| Test temperature, °F | Stress, psi | Cycles to failure |
|---|---|---|
| 70 | 90,000 | $10^8$ |
| 600 | 110,000 | $10^8$ |
| 1,000 | 90,000 | $10^8$ |
| 1,200 | 72,000 | $10^8$ |

Impact Strength

**Aged – B**

| Test temperature, °F | Type test | Strength, ft-lb |
|---|---|---|
| +70 | Charpy – V-notched | 35 |

Specifications

SAE-AMS: 5589, 5590, 5596, 5597, 5662, 5663, 5664, 5832

**Table 10.4—12**
**RENE 41 (AISI No. 683)**

**Chemical composition, percent by weight:** C, 0.09; Cr, 19.0; Ni, 55.0; Co, 11.0; Mo, 10.0; Ti, 3.1; Al, 1.5; B, 0.005

**Physical constants and thermal properties**
  Density, lb/in.$^3$: 0.298
  Coefficient of thermal expansion, (70–200°F) in./in./°F × 10$^{-6}$: 6.6

Modulus of elasticity, psi: tension 31.9 × 10$^6$
Melting range, °F: 2,385–2,450
Specific heat, Btu/lb/°F, 70°F: 0.108
Thermal conductivity, Btu/ft$^2$/hr/in./°F, 70°F: 62

**Heat treatments**
  Age harden: 1,950°F/4 hr/air cool plus 1,400°F/16 hr/air cool.

Tensile Properties

**Age Hardened**

| Temperature, °F | T.S., psi | Y.S., psi, 0.2% offset | Elong., in 2 in., % | Hardness, Brinell |
|---|---|---|---|---|
| 70 | 206,000 | 154,000 | 14 | — |
| 1,000 | 203,000 | 147,000 | 14 | — |
| 1,200 | 194,000 | 145,000 | 14 | — |
| 1,400 | 160,000 | 136,000 | 11 | — |
| 1,600 | 90,000 | 80,000 | 19 | — |
| 1,800 | 42,000 | 38,000 | 36 | — |

Rupture Strength, 1,000 hr

**Age Hardened**

| Test temperature, °F | Strength, psi | Elong., in 2 in., % | Reduction of area, % |
|---|---|---|---|
| 1,200 | 100,000 | — | — |
| 1,300 | 74,000 | — | — |
| 1,400 | 40,000 | — | — |
| 1,500 | 24,000 | — | — |
| 1,600 | 14,000 | — | — |

Specifications

SAE-AMS: 5399, 5545, 5712, 5713, 5800, 7469

**Table 10.4—13**
**IN 100**

**Chemical composition, percent by weight:** C, 0.18; Cr, 10.0; Ni, 60.0; Co, 15.0; Mo, 3.0; Ti, 4.7; Al, 5.5; B, 0.014; Zr, 0.06; V, 1.0

Coefficient of thermal expansion, (70–200°F) in./in./°F $\times 10^{-6}$: 7.2
Modulus of elasticity, psi: tension, $31.2 \times 10^6$
Melting range, °F: 2,305–2,435

**Physical constants and thermal properties**
Density, lb/in.$^3$: 0.280

**Heat treatments**
As cast.

## Tensile Properties

### As Cast

| Temperature, °F | T.S., psi | Y.S., psi, 0.2% offset | Elong., in 2 in., % | Hardness, Brinell |
|---|---|---|---|---|
| 70 | 147,000 | 123,000 | 9 | – |
| 1,000 | 158,000 | 128,000 | 9 | – |
| 1,200 | 161,000 | 129,000 | 6 | – |
| 1,400 | 155,000 | 125,000 | 6 | – |
| 1,600 | 128,000 | 101,000 | 6 | – |
| 1,800 | 82,000 | 54,000 | 6 | – |

## Rupture Strength, 1,000 hr

### As Cast

| Test temperature, °F | Strength, psi | Elong., in 2 in., % | Reduction of area, % |
|---|---|---|---|
| 1,400 | 75,000 | – | – |
| 1,500 | 55,000 | – | – |
| 1,600 | 37,000 | – | – |
| 1,700 | 25,000 | – | – |
| 1,800 | 15,000 | – | – |
| 1,900 | 8,500 | – | – |

## Specifications

SAE-AMS: 5397

## Table 10.4—14
## MAR-M® 200

**Chemical composition, percent by weight:** C, 0.15; Cr, 9.0; Ni, 60.0; Co, 10.0; W, 12.0; Cb, 1.0; Ti, 2.0; Al, 5.0; B, 0.015; Zr, 0.05

**Physical constants and thermal properties**
Density, lb/in.³: 0.308

Modulus of elasticity, psi: tension, $31.6 \times 10^6$
Melting range, °F: 2,400–2,500
Thermal conductivity, Btu/ft²/hr/in./°F, 70°F: 88

**Heat treatments**
Age harden: 1,600°F/50 hr/air cool.

### Tensile Properties

#### Age Hardened

| Temperature, °F | T.S., psi | Y.S., psi, 0.2% offset | Elong., in 2 in., % | Hardness, Brinell |
|---|---|---|---|---|
| 70 | 135,000 | 120,000 | 7 | – |
| 1,000 | 137,000 | 123,000 | 5 | – |
| 1,200 | 138,000 | 124,000 | 4 | – |
| 1,400 | 135,000 | 122,000 | 3 | – |
| 1,600 | 122,000 | 110,000 | 4 | – |
| 1,800 | 80,000 | 68,000 | 5 | – |

### Rupture Strength, 1,000 hr

#### Age Hardened

| Test temperature, °F | Strength, psi | Elong., in 2 in., % | Reduction of area, % |
|---|---|---|---|
| 1,400 | 84,000 | – | – |
| 1,500 | 60,000 | – | – |
| 1,600 | 43,000 | – | – |
| 1,700 | 29,000 | – | – |
| 1,800 | 18,000 | – | – |

## 10.5 OTHER ALLOYS AND MISCELLANEOUS PROPERTIES

### Table 10.5—1
### MECHANICAL PROPERTIES OF COPPER ALLOYS

Because of their corrosion resistance and the fact that copper alloys have been used for many thousands of years, the number of copper alloys available is second only to the ferrous alloys. In general, copper alloys do not have the high-strength qualities of the ferrous alloys, while their density is comparable. The cost per strength-weight ratio is high; however, they have the advantage of ease of joining by soldering, which is not shared by other metals that have reasonable corrosion resistance.

| Material | Nominal composition | | Form and condition | Typical mechanical properties | | | | Comments |
|---|---|---|---|---|---|---|---|---|
| | | | | Yield strength (0.2% offset), 1000 lb/sq in. | Tensile strength, 1000 lb/sq in. | Elongation, in 2 in., % | Hardness, Brinell | |
| Copper ASTM B152 ASTM B124, B133 ASTM B1, B2, B3 | Cu 99.9 plus | | Annealed Cold-drawn Cold-rolled | 10 40 40 | 32 45 46 | 45 15 5 | 42 90 100 | Bus-bars, switches, architectural, roofing, screens |
| Gilding metal ASTM B36 | Cu 95.0 | Zn 5.0 | Cold-rolled | 50 | 56 | 5 | 114 | Coinage, ammunition |
| Cartridge 70–30 brass ASTM B14 ASTM B19 ASTM B36 ASTM B134 ASTM B135 | Cu 70.0 | Zn 30.0 | Cold-rolled | 63 | 76 | 8 | 155 | Good cold-working properties; radiator covers, hardware, electrical |
| Phosphor bronze 10% ASTM B103 ASTM B139 ASTM B159 | Cu 90.0 P 0.25 | Sn 10.0 | Spring temper | — | 122 | 4 | 241 | Good spring qualities, high-fatigue strength |
| Yellow brass (high brass) ASTM B36 ASTM B134 ASTM B135 | Cu 65.0 | Zn 35.0 | Annealed Cold-drawn Cold-rolled (HT) | 18 55 60 | 48 70 74 | 60 15 10 | 55 115 180 | Good corrosion resistance; plumbing, architectural |
| Manganese bronze ASTM B138 | Cu 58.5 Fe 1.0 Mn 0.3 | Zn 39.2 Sn 1.0 | Annealed Cold-drawn | 30 50 | 60 80 | 30 20 | 95 180 | Forgings |

**Table 10.5—1 (continued)**
**MECHANICAL PROPERTIES OF COPPER ALLOYS**

| Material | Nominal composition | Form and condition | Yield strength (0.2% offset), 1000 lb/sq in. | Tensile strength, 1000 lb/sq in. | Elongation, in 2 in., % | Hardness, Brinell | Comments |
|---|---|---|---|---|---|---|---|
| Naval brass ASTM B21 | Cu 60.0, Zn 39.25, Sn 0.75 | Annealed | 22 | 56 | 40 | 90 | Condensor tubing; high resistance to salt-water corrosion |
|  |  | Cold-drawn | 40 | 65 | 35 | 150 |  |
| Muntz metal ASTM B111 | Cu 60.0, Zn 40.0 | Annealed | 20 | 54 | 45 | 80 | Condensor tubes; valve stress |
| Aluminum bronze ASTM B169, alloy A; B124; B150 | Cu 92.0, Al 8.0 | Annealed | 25 | 70 | 60 | 80 |  |
|  |  | Hard | 65 | 105 | 7 | 210 |  |
| Beryllium copper 25 ASTM B194; B197; B196 | Be 1.9, Co or Ni 0.25, Cu bal. | Annealed, solution-treated | 32 | 70 | 45 | B60 (Rockwell) | Bellows, fuse clips, electrical relay parts, valves, pumps |
|  |  | Cold-rolled | 104 | 110 | 5 | B81 |  |
|  |  | Cold-rolled | 70 | 190 | 3 | C40 |  |
| Free-cutting brass | Cu 62.0, Zn 35.5, Pb 2.5 | Cold-drawn | 44 | 70 | 18 | B80 (Rockwell) | Screws, nuts, gears, keys |
| Nickel silver 18% Alloy A (wrought) ASTM B122, No. 2 | Cu 65.0, Zn 17.0, Ni 18.0 | Annealed | 25 | 58 | 40 | 70 | Hardware, optical goods, camera parts |
|  |  | Cold-rolled | 70 | 85 | 4 | 170 |  |
|  |  | Cold-drawn wire | — | 105 | — | — |  |
| Nickel silver 13% (cast) 10A ASTM B149, No. 10A | Ni 12.5, Sn 2.0, Zn 20.0, Pb 9.0, Cu bal. | Cast | 18 | 35 | 15 | 55 | Ornamental castings, plumbing; good machining qualities |
| Cupronickel 55—45 (Constantan) | Cu 55.0, Ni 45.0 | Annealed | 30 | 60 | 45 | — | Electrical-resistance wire; low temperature coefficient, high resistivity |
|  |  | Cold-drawn | 50 | 65 | 30 | — |  |
|  |  | Cold-rolled | 65 | 85 | 20 | — |  |
| Cupronickel | Cu 70.0, Ni 30.0 | Wrought |  |  |  | — | Heat-exchanger process equipment, valves |
| Cupronickel 10% ASTM B111; B171 | Cu 88.35, Ni 10.0, Fe 1.25, Mn 0.4 | Annealed | 22 | 44 | 45 | — | Condensor, salt-water piping |
|  |  | Cold-drawn tube | 57 | 60 | 15 | — |  |

**Table 10.5—1 (continued)**
MECHANICAL PROPERTIES OF COPPER ALLOYS

| Material | Nominal composition | | Form and condition | Typical mechanical properties | | | | Comments |
|---|---|---|---|---|---|---|---|---|
| | | | | Yield strength (0.2% offset), 1000 lb/sq in. | Tensile strength, 1000 lb/sq in. | Elongation, in 2 in., % | Hardness, Brinell | |
| Red brass (cast) ASTM B30, No. 4A | Cu 85.0 Pb 5.0 | Zn 5.0 Sn 5.0 | As-cast | 17 | 35 | 25 | 60 | |
| Silicon bronze ASTM B30, alloy 12A | Si 4.0 Zn 4.0 Mn 1.0 | Fe 2.0 Al 1.0 | Castings | | | | | Cheaper substitute for tin bronze |
| Tin bronze ASTM B30, alloy 1B | Sn 8% | Zn 4.0 | Castings | | | | | Bearings, high-pressure bushings, pump impellers |
| Navy bronze | | | Cast | | | | | |

From Bolz, R. E. and Tuve, G. L., Eds., *Handbook of Tables for Applied Engineering Science*, 2nd ed., CRC Press, Cleveland, 1973, 109.

## Table 10.5—2
## MECHANICAL PROPERTIES OF TIN AND LEAD-BASE ALLOYS

Major uses for these alloys are as "white"-metal bearing alloys, extruded cable sheathing, and solders. Tin forms the basis of pewter used for culinary applications.

| Material | Nominal composition | Form and condition | Yield strength (0.2% offset), 1000 lb/sq in. | Tensile strength, 1000 lb/sq in. | Elongation, in 2 in., % | Hardness, Brinell | Comments |
|---|---|---|---|---|---|---|---|
| Lead-base Babbitt ASTM B23, alloy 19 | Pb 85.0　Sn 5.0<br>Sb 10.0　As 0.6<br>Cu 0.5 | Chill cast | — | 10 | 5 | 19 | Bearings, light loads and low speeds |
| Arsenical-lead Babbitt ASTM B23, alloy 15 | Pb 83.0　Sn 1.0<br>Sb 16.0　As 1.1<br>Cu 0.6 | Chill cast | — | 10.3 | 2 | 20 | Bearings, high loads and speeds, diesel engines, steel mills |
| Chemical lead | Pb 99.9　Cu 0.06<br>Bi 0.005 max | Rolled 95% | 1.9 | 2.5 | 50 | 5 | |
| Antimonial lead (hard lead) | Pb 94.0　Sb 6.0 | Chill cast<br>Rolled 95% | —<br>— | 6.8<br>4.1 | 22<br>47 | (500 kg) 9 | Good corrosion resistance and strength |
| Calcium lead | Pb 99.9　Ca 0.025<br>Cu 0.10 | Extruded and aged | — | 4.5 | 25 | — | Cable sheathing, creep-resistant pipe |
| Tin Babbitt alloy ASTM B23–61, grade 1 | Sb 4.5　Sn bal.<br>Cu 4.5 | Chill cast | — | 9.3 | 2 | 17 | General bearings and die casting |
| Tin die-casting alloy ASTM B102–52 | Sb 13.0　Sn bal.<br>Cu 5.0 | Die-cast | — | 10 | 1 | 29 | Die-casting alloy |
| Pewter | Sn 91.0　Sb 7.0<br>Cu 2.0 | Rolled sheet, annealed | — | 8.6 | 40 | 9.5 | Ornamental and household items |
| Solder 50–50 | Sn 50.0　Pb 50.0 | Cast | 4.8 | 6.1 | 60 | 14 | General-purpose solder |
| Solder | Sn 20.0　Pb 80.0 | Cast | 3.6 | 5.8 | 16 | 11 | Coating and joining, filling seams on automobile bodies |

From Bolz, R. E. and Tuve, G. L., Eds., *Handbook of Tables for Applied Engineering Science*, 2nd ed., CRC Press, Cleveland, 1973, 111.

## Table 10.5—3
## MECHANICAL PROPERTIES OF TITANIUM ALLOYS

The main application for these alloys is in the aerospace industry. Because of the low density and high strength of titanium alloys, they present excellent strength-to-weight ratios.

| Material | Nominal composition | Form and condition | Typical mechanical properties | | | | Comments |
|---|---|---|---|---|---|---|---|
| | | | Yield strength (0.2% offset), 1000 lb/sq in. | Tensile strength, 1000 lb/sq in. | Elongation, in 2 in., % | Hardness, Brinell | |
| Commercial titanium ASTM B265–58T | Ti 99.4 | Annealed at 1100 to 1350°F (593 to 732°C) | 70 | 80 | 20 | — | Moderate strength, excellent fabricability; chemical industry pipes |
| Titanium alloy ASTM B265–58T–5 Ti–6 Al–4V | | Water-quenched from 1750°F (954°C); aged at 1000°F (538°C) for 2 hr | 160 | 170 | 13 | — | High-temperature strength needed in gas-turbine compressor blades |
| Titanium alloy Ti–4 Al–4Mn | | Water-quenched from 1450°F (788°C); aged at 900°F (482°C) for 8 hr | 170 | 185 | 13 | — | Aircraft forgings and compressor parts |
| Ti-Mn alloy ASTM B265–58T–7 | Fe 0.5 Ti bal. Mn 7.0–8.0 | Sheet | 140 | 150 | 18 | — | Good formability, moderate high-temperature strength; aircraft skin |

From Bolz, R. E. and Tuve, G. L., Eds., *Handbook of Tables for Applied Engineering Science*, 2nd ed., CRC Press, Cleveland, 1973, 114.

**Table 10.5—4**
**MECHANICAL PROPERTIES OF ZINC ALLOYS**

A major use for these alloys is for low-cost die-cast products such as household fixtures, automotive parts, and trim.

| Material | Nominal composition | Form and condition | Typical mechanical properties | | | | Comments |
|---|---|---|---|---|---|---|---|
| | | | Yield strength (0.2% offset), 1000 lb/sq in. | Tensile strength, 1000 lb/sq in. | Elongation, in 2 in., % | Hardness, Brinell | |
| Zinc ASTM B69 | Cd 0.35  Zn bal. Pb 0.08 | Hot-rolled | — | 19.5 | 65 | 38 | Battery cans, grommets, lithographer's sheet |
| Zilloy-15 | Cu 1.00  Zn bal. Mg 0.010 | Hot-rolled Cold-rolled | — — | 29 36 | 20 25 | 61 80 | Corrugated roofs, articles with maximum stiffness |
| Zilloy-40 | Cu 1.00  Zn bal. | Hot-rolled Cold-rolled | — — | 24 31 | 50 40 | 52 60 | Weatherstrip, spun articles |
| Zamac-5 ASTM 25 | Zn (99.99% pure remainder)  Al 3.5–4.3  Cu 0.75–1.25  Mg 0.03–0.08 | Die-cast | — | 47.6 | 7 | 91 | Die casting for automobile parts, padlocks; used also for die material |

From Bolz, R. E. and Tuve, G. L., Eds., *Handbook of Tables for Applied Engineering Science*, 2nd ed., CRC Press, Cleveland, 1973, 114.

**Table 10.5—5**

**MECHANICAL PROPERTIES OF ZIRCONIUM ALLOYS**

These alloys have good corrosion resistance but are easily oxidized at elevated temperatures in air. The major application is for use in nuclear reactors.

| Material | Nominal composition | Form and condition | Typical mechanical properties | | | | Comments |
|---|---|---|---|---|---|---|---|
| | | | Yield strength (0.2% offset), 1000 lb/sq in. | Tensile strength, 1000 lb/sq in. | Elongation, in 2 in., % | Hardness, Brinell | |
| Zirconium, commercial | O$_2$ 0.07  C 0.15<br>Hf 1.90  Zr bal. | Annealed | 40 | 65 | 27 | B80 (Rockwell) | Nuclear power-reactor cores at elevated temperatures |
| Zircaloy 2 | Hf 0.02  Ni 0.05<br>Fe 0.15  Other 0.25<br>Sn 1.46  Zr bal. | Annealed | 50 | 75 | 22 | B90 (Rockwell) | |

From Bolz, R. E. and Tuve, G. L., Eds., *Handbook of Tables for Applied Engineering Science*, 2nd ed., CRC Press, Cleveland, 1973, 115.

**Table 10.5—6**
## SILVER AND SILVER ALLOYS

Typical Silver Alloys

| Typical composition, % | Names and uses | Important property[a] |
|---|---|---|
| Ag 99.9 + | Fine silver; contacts, chemical uses | Conductivity; ductility |
| Ag 99 +, Mg 0.25, Ni 0.2 + | High temperature contacts and spring contacts | Hardness; high conductivity |
| Ag 99 + (layer, 0.025 in.) | High-service machine bearings | Fatigue strength; thermal conductivity |
| Ag 92.5, Cu 7.5 | Sterling silver; tableware | Appearance; value |
| Ag 90, Cu 10 | Coin silver (to 1966) | Value; corrosion resistance |
| Ag 90 (outside layer) | Pressure-bonded laminates; coins | Appearance; conductivity |
| Ag 90, Pd 10 | Contacts; brazing alloys | High conductivity |
| Ag 90, CdO 10 (sintered) | Non-sticking contacts | Heat resistance; hardness |
| Ag 85, Cd 15 | Arc-quenching contacts | Wear resistance; ductility |
| Ag 80, Cu 20 | Laminated coins; tableware | Appearance; value |
| Ag 80, In 15, Cd 5 | Nuclear reactor control | Neutron absorption |
| Ag 77, Cd 22.6, Ni 0.4 | Spring contacts | Wear resistance; ductility |
| Ag 72, Cu 28 | Brazing alloy | Eutectic alloy; hardness |
| Ag 60, Ni 40 (sintered) | Circuit-breaker contacts | Hardness; heat resistance |
| Ag 49, Au 41.7, Cu 9, Zn 0.3 | 10-karat green gold; jewelry | Appearance |
| Ag 40, Mo or W 60 (sintered) | Switching contacts | Hardness; wear resistance |
| Ag 40, Au 30, Pd 30 | Corrosive-chemical apparatus (e.g., for halogens and nitric acid) | Corrosion resistance; strength |
| Ag 35, Cu 26, Zn 21, Cd 18 | Thin-joint brazing | Wetting ability; ductility |
| Ag 33, Hg 52, Sn 12.5, Cu 2, Zn 0.5 | Dental amalgam fillings | Amalgam hardening |
| Ag 30.5, Au 50, Cu 17.5, Zn 2 | Gold solder; jewelry | Appearance; strength |
| Ag 15, Au 60, Cu 10, Pd 10, Pt 4, Zn 1 | White gold denture metal | Strength; wear resistance |
| Ag 15, Au 60, Cu 15 | Precious metal solder | Appearance; strength |
| Ag 10, Au 58.3, Cu 29.7, Zn 2 | 14-karat yellow gold | Appearance; wear |
| Ag 2.5, Pb 97.5 | Soft solder | Eutectic alloy |

[a]Properties of pure silver and the high-silver alloys vary with temperature; they are subject to work hardening, and the room-temperature properties depend on the annealing temperature.

Chemical Uses of Silver

| Type of use | Form of silver | Processes and remarks |
|---|---|---|
| Batteries | Powder; oxide; chloride | Silver oxide with zinc or cadmium; silver chloride with magnesium |
| Catalysts | Shapes, screens, powder, salts | In dehydrogenation, oxidation, desulfurization |
| Coatings | Powder; vapor | Air-drying paints and fired-glass conductive coatings |
| Electroplating | Cyanide; anode shapes | Usual plate less than 0.001 5 in. thickness |
| Medicine | Nitrate; metal, colloidal | Sterilization by metal, nitrate, or colloidal suspension |
| Photography | Nitrate; halide colloidal solutions | Metallic silver from salt by radiation and "developer" |
| Reflectors (not ultraviolet) | Nitrate; metal vapor | Silver nitrate reacts with reducing solution; vacuum evaporation |

## Table 10.5—6 (continued)
## SILVER AND SILVER ALLOYS

### REFERENCES

*Metals Handbook,* Vol. I, 8th ed., American Society for Metals, Metals Park, Oh., 1961, 1174.
**Butts, A. and Coxe, C. D., Eds.,** *Silver, Economics, Metallurgy and Use,* Van Nostrand, Princeton, N.J., 1967.
**Addicks, L.,** *Silver in Industry,* Reinhold, New York, 1940.
ASTM Standards B253, B413, and E56.

From Bolz, R. E. and Tuve, G. L., Eds., *Handbook of Tables for Applied Engineering Science,* 2nd ed., CRC Press, Cleveland, 1973, 123.

## Table 10.5—7
## SPECIFIC STIFFNESS OF METALS, ALLOYS, AND CERTAIN NONMETALLICS

Specific stiffness is usually expressed as the modulus of elasticity (in tension) per unit weight-density, i.e., $E/\rho$, in units of pounds and inches. While the stiffness of similar alloys varies considerably, there are definite ranges and groups to be recognized. Since the specific stiffness of steel is about 100 million, the values in the following table are also approximately the percentage stiffness, referred to steel.

| Material | Specific stiffness, millions |
|---|---|
| Beryllium | 650 |
| Silicon carbide | 600 |
| Alumina ceramics | 400 |
| Mica | 350 |
| Titanium carbide cermet | 250 |
| Alumina cermet | 200 |
| Molybdenum and alloys; silica glass | 130 |
| Titanium and alloys; cobalt superalloys; soda-lime glass | 110 |
| Carbon and low-alloy steel; wrought iron | 105 |
| Stainless steel; nodular cast iron; magnesium and alloys; aluminum and alloys | 100 |
| Nickel and alloys; malleable iron | 95 |
| Iron silicon alloys (cast); iridium; vanadium | 90 |
| Monel alloys; tungsten | 80 |
| Gray cast iron; columbium alloys | 70 |
| Aluminum bronze; beryllium copper | 65 |
| Nickel silver; cupronickel; zirconium | 55 |
| Yellow brass; nickel cast iron; bronze; Muntz metal; antimony | 50 |
| Copper; red brass; tantalum | 45 |
| Silver and alloys; pewter; platinum and alloys; white gold | 30 |
| Tin; thorium | 25 |
| Gold | 20 |
| Tin-lead alloy | 10 |
| Lead | 5 |

From Bolz, R. E. and Tuve, G. L., Eds., *Handbook of Tables for Applied Engineering Science,* 2nd ed., CRC Press, Cleveland, 1973, 130.

# Table 10.5—8
## METAL POWDERS

Typical Metal Powders and Granules

**Varieties and uses** — The largest use of metal powders is for production of parts, shapes, and electrodes by powder metallurgy. Other major uses are in metal coatings, applied by flame spraying, or as paints, lacquers or inks, chemical catalysts, reducing agents, fuels, and explosives. Many composite materials contain metal powder fillers. Annual U.S. consumption of iron powder and alloys is well over 200 million pounds. Some 200 kinds are available. Copper and its alloys represent almost as great a variety, but the total consumption is less than one third that of iron. Aluminum, nickel, chromium, tungsten, and silver powders are used in quantity, but there are very important uses for some of the other metals also. The following table gives only a few typical examples.

**Particle size and surface** — Actual dimensions and size-distribution are not yet independent of the methods of measurement. For instance, fine particles that pass a 100-mesh sieve can be dimensionally analyzed by three or four methods with divergent results. The apparent density of a powder is not independent of the method of handling, and the constant-volume "tap density" is affected by the shape of particles. Even a single method such as a sieve analysis depends on the condition and quality of the sieves. Particle sizes above 40 $\mu$m in the table are based on standard ASTM sieve analyses and their metric equivalents in micrometers (25,400 $\mu$m = 1 in.). Particle size distribution and surface area of particles are not given in this table, but they are important for applications such as those involving either surface coverage or gas adsorption by the particles. Surface areas greater than 1 m$^2$/g are attainable for metals. Integration of the area under the size distribution curve gives a low value of surface area, since it is only practical to assume spherical particles. Other methods depend on air permeability, liquid surface spread, liquid mixture turbidity, or gas adsorption. Comparable results from the same sample are very difficult to obtain if more than one test method is used.

**Purity and analysis** — Metal purity to 99.9999% is often quoted, but the purity should be specified according to the use. Metals of purity 99% or less are much cheaper. A metal powder can be produced from an alloy of almost any composition. Although most of the common and proprietary alloys are commercially available in powder form, a special order may be required if the particle-size specification is unusual.

For suppliers of metal powders, consult the Metal Powder Industries Federation, 201 E. 42nd Street, New York, N.Y. 10017.

| Metal or alloy | Purity, % | Largest impurity | Particle size | | Specific gravity | | Designation |
|---|---|---|---|---|---|---|---|
| | | | Micrometers (av. or range) | % passing-sieve No. | Powder | Solid | |
| Aluminum | 99.9+ | Fe | 75–150 | 100–100 | 2.5 | 2.7 | Pure |
| Aluminum | 99.5 | Fe or Si | <60 | 95–325 | 2.5 | 2.7 | Atomized, fine |
| Aluminum | 99.3 | Oxygen | 30–75 | 90–325 | 1.7 | 2.7 | Atomized, spherical |
| Aluminum | | Oxygen | 6–600 | 95–40 | 1.2 | 2.7 | Granular |
| Aluminum | 95 | Oxygen | 0.03 | — | 0.2 | 2.7 | Fine, spheroidal |
| Antimony | 99.5 | | <100 | 95–325 | 2.42 | 6.7 | Fine |
| Beryllium | 99.0 | Oxygen | <80 | 98–200 | | 1.85 | Structural |
| Bismuth | 99.6 | Sb | 40–150 | 80–325 | 4.29 | 9.75 | Commercial |
| Cadmium | 99.9 | Pb | 40–150 | 95–325 | | 8.65 | Fine |
| Chromium | 99.5 | Fe | <45 | 100–325 | | 7.2 | Fine |

**Table 10.5—8 (continued)**
**METAL POWDERS**

| Metal or alloy | Purity, % | Largest impurity | Particle size Micrometers (av. or range) | % passing-sieve No. | Specific gravity Powder | Solid | Designation |
|---|---|---|---|---|---|---|---|
| Cobalt | 99.8 | Ni | 1.2 | 99.9–400 | 0.7 | 8.9 | Very fine |
| Copper | 99.9 | Sn | >1 000 | — | 5.3 | 8.95 | Granular |
| Copper | 99.8 | Sn | 15 | 95–325 | 4.55 | 8.95 | Atomized shot |
| Copper | 99.5 | Sn | 40–850 | 60–325 | 2.7 | 8.95 | Electrolytic, coarse |
| Brass | 90 Cu | Pb | 40–250 | 55–325 | 3.0 | 8.5 | 10% Zn, atomized |
| Bronze | 90 Cu | | 40–150 | 55–325 | 2.7 | 8.5 | 10% Sn |
| Gold | 99.9+ | | <35 | 100–400 | — | 19.3 | Pure, reduced |
| Gold | 99.+ | | 0.5 × 10 | 100–325 | 2.0 | 19.3 | Flake 0.5 × 10 $\mu$m |
| Iron | 99.8 | Oxygen | 7–10 | — | 3.2 | 7.9 | Very fine |
| Iron | 98.5 | SiO$_2$ | 40–150 | 30–325 | 2.9 | 7.9 | Reduced, annealed |
| Iron | 91. | C | 100–900 | 95–100 | 2.3 | 7.9 | Ground cast iron |
| Stainless steel | 71. | Si | 20–150 | 40–325 | 3.0 | 8.0 | Stainless steel 18-8 |
| Lead | 99.8 | Bi | 10–45 | 100–325 | 5.0 | 11.35 | Very fine |
| Magnesium | 99.8 | Mn | <75 | 100–200 | 0.6 | 1.74 | Commercial |
| Manganese | 99.8 | Oxygen | <100 | 80–325 | 2.8 | 7.3 | Commercial |
| Ferromanganese | 80 Mn | C | 40–300 | 30–325 | | | Medium carbon |
| Molybdenum | 99.9 | Oxygen | 3–6 | — | 1.4 | 10.2 | Hydrogen reduced |
| Nickel | 99.9+ | Co | 10–100 | 80–250 | 2.0 | 8.9 | High purity |
| Nickel | 99.5 | Co | <150 | 50–325 | 3.4 | 8.9 | Electrolytic |
| Niobium | 99.6 | Oxygen | <75 | 100–200 | | 8.57 | High purity |
| Palladium | 99.9+ | Pt | <45 | 100–325 | | 12.0 | Pure |
| Platinum | 99.0 | Ag | <420 | 100–40 | | 21.4 | Sponge |
| Silver | 99.9+ | Si | <150 | 90–325 | 1.5 | 10.5 | Precipitated, spongy |
| Silver braze | 70. | Optional | <175 | 100–80 | | | 28% Cu brazing alloy |
| Tin | 99.5 | Oxygen | <150 | 30–250 | 3.4 | 7.3 | Commercial |
| Titanium | 99. | | 8 | 98–325 | | 4.54 | Commercial |
| Tungsten | 99.9 | Mo | <175 | 100–80 | 9.8 | 19.3 | Reduced |
| Vanadium | 99.5 | Fe | <800 | 100–20 | | 6.1 | Commercial |
| Zinc | 99. | Pb | <50 | 98–325 | 7.0 | 7.0 | Reduced |
| Zirconium | 99.+ | Oxygen | <175 | 100–80 | 3.5 | 6.53 | Reactor |

REFERENCES

Barth, V. D. and McIntire, H. O., *Tungsten Powder Metallurgy*, NASA SP-5035, National Aeronautics and Space Administration, 1965.

Goetzel, C. G., *Treatise on Powder Metallurgy*, Interscience Books, New York, 1949.

Hausner, H. H., Ed., *Modern Developments in Powder Metallurgy*, Plenum Publishing, New York, 1966.

Leszynski, W., Ed., *Powder Metallurgy*, Interscience Books, New York, 1961.

Poster, A. R., *Handbook of Metal Powders*, Reinhold, New York, 1966.

Sands, R. L. and Shakespeare, C. R., *Powder Metallurgy*, Newnes, London, 1966 (distributed in the U.S.A. by The Chemical Rubber Co., Cleveland).

From Bolz, R. E. and Tuve, G. L., Eds., *Handbook of Tables for Applied Engineering Science*, 2nd ed., CRC Press, Cleveland, 1973, 131.

**Table 10.5—9**

## PRODUCTS OF POWDER METALLURGY[a]

Powder metallurgy refers to the production of parts by a process of molding metal powders and agglomerating the form by heat. The powder mixture is often hot molded under pressure (10,000 to 100,000 psi) and is sintered in an inert or a reducing atmosphere, at a temperature between 400 to 2,000°F, depending on the metal mixture. For the refractory metals higher temperatures are necessary. The methods of powder metallurgy provide a close control of the composition and allow use of mixtures that could not be fabricated by any other process. As dimensions are determined by the mold, finish machining or grinding is often eliminated, thereby reducing cost and handling, especially for large lots. Special properties of the finished product such as porosity, friction coefficient, and electrical conductivity can be varied somewhat by changing the proportions of the powder components.

| Class | Composition or constituents | Applications and uses | Desirable properties and advantages |
|---|---|---|---|
| Small, finished parts | Various ferrous, copper, and nickel alloys | Complex shapes; small parts not requiring high strength or ductility; plain bearings | Control of dimensions and finish; two-phase bearing metals; low cost in large production lots |
| Refractory metals | Pure W, Mo, Ta, Nb, Re, Ti alloys | Production of high-purity tungsten, molybdenum, tantalum, niobium, etc.; beryllium; cobalt alloys | Metals used in high-temperature service; electrical, electronic, and nuclear applications |
| Porous metals | Copper; copper-lead; bronze; stainless steel | Porous bearings, oil-impregnated, or with graphite or plastic; friction materials; metal filters; porous electrodes; catalysts; throttle plates | Interconnected pores in the size range 5–50 $\mu$m; porosity about 20–30% |
| Composite metals | Al, Cu, etc. with W, Mo, Co, or stainless steel reinforcing; reactor fuel elements | Services requiring high strength with lightness, high electrical and thermal conductivities; nuclear reactor components | High-strength materials from common metals; durability of nuclear materials |
| Metal-nonmetal composites | Filament-reinforced ceramics; dispersion strengthening by oxides | Ceramics with good structural properties; lightweight materials for high temperature (*e.g.*, SAP) | Strengthened ceramics; heat-resistant aluminum |
| Magnetic materials | Nickel-iron; cobalt mixtures; ferrites | High-permeability materials; permanent magnets; ferrite cores; magnetic storage | Very high magnetic properties and close control of magnetic properties |
| Cermets, oxide | $Al_2O_3$-Cr; $Al_2O_3$-Cr-W; $Al_2O_3$-Cr-Mo; $ThO_2$-W | Combustion and rocket nozzles; furnace muffler, tubes, seals, extrusion dies; power-tube cathodes | High-temperature strength (2 000 deg F and above); resistance to thermal shock; high thermal conductivity; corrosion resistance |

**Table 10.5—9 (continued)**
**PRODUCTS OF POWDER METALLURGY**

| Class | Composition or constituents | Applications and uses | Desirable properties and advantages |
|---|---|---|---|
| Cermets, carbide | TiC-Ni; TiC-Fe-Cr; TiC-Co-Cr-W; $Cr_3C_2$-Ni-W | High-temperature bearings, seals, and dies; gage blocks | Strength, toughness, and corrosion resistance at high temperatures (to 1 700 deg F); hardness |
| Cemented carbides | WC-Co; WC-TaC-Co; TiC-Ni; $Cr_3C_2$-WC-Ni | Tips for cutting tools, lathe centers, gages; wire-drawing dies; rock drills; crushers; blast nozzles | Very high hardness, compressive strength, and elastic modulus; wear and corrosion resistance; high conductivity; high-temperature strength |

[a]  For other data and references on powder metallurgy, see Table 10.5—8.

From Bolz, R. E. and Tuve, G. L., Eds., *Handbook of Tables for Applied Engineering Science,* 2nd ed., CRC Press, Cleveland, 1973, 133.

**Table 10.5—10**
**CORROSION OF METALS**

| Metal | Subject to corrosion by | Resistant to |
|---|---|---|
| Aluminum and alloys | Acid solutions (except concentrated nitric and acetic); caustic and mild alkalies; sea water; saturated halogen vapors; mercury and its compounds; carbon tetrachloride; cobalt; copper and nickel compounds in solution | Air; water; ammonia; combustion products; halide refrigerants; dry steam; sulfur and its compounds; concentrated ammonium hydroxide; organic acids; most organic compounds |
| Cast iron | All water solutions, moist gases; dilute acids; acid-salt solutions | Dry gases except halogens; dry air; neutral water; dry soil; concentrated acids (nitric, sulfuric, phosphoric); weak or strong alkalies; organic acids |
| Chromium and high-chrome steels | Most strong acids (limited use with acetic, nitric, sulfuric, and phosphoric); most chlorides | Air; water; steam; weak acids; most inorganic salts; most alkalies; ammonia |
| Copper, red brass, and bronze | Mercury and its salts; aqueous ammonia; saturated halogen vapors; sulfur and sulfides; oxidizing acids (nitric, concentrated sulfuric, sulfurous); oxidizing salts (Hg, Ag, Cr, Fe, Cu); cyanides | Air; water; sea water; steam; sulfate and carbonate solutions; dry halogens; moist soils; alkaline solutions; refrigerants; petrochemicals; non-oxidizing acids (acetic, hydrochloric, sulfuric) |
| Lead | Caustic solutions; halogens; acetic acid; calcium hydroxide; magnesium chloride; ferric chloride; sodium hypochlorite | Air; water; moist soil; ammonia; alcohols; sulfuric acid; ferrous chloride |
| Magnesium and alloys | Heavy metal; salts; all mineral acids (except hydrofluoric and chromic); sea water; fruit juices; milk | Most alkalies and organic compounds; air; water; soil; dry refrigerants; dry halogens |
| Monel metal and Ni-Cu alloys containing in excess of 50% nickel | Inorganic acids; sulfur; chlorine; acid solutions of ferric, stannic, or mercuric salts | Air; sea water; steam; food acids; neutral and alkaline salt solutions; dry gases; most alkalies; ammonia |
| Nickel and high-nickel steels | Inorganic acids; ammonia; mercury; oxidizing salts (Fe, Cu, Hg) | Air; water; steam; caustic and mild alkalies; organic acids; neutral and alkaline organic compounds; dry gases |
| Silver | Halogens and halogen acids; sulfur compounds; ammonia | Alkalies, including high-temperature caustic alkalies; hot concentrated organic acids; phosphoric and hydrofluoric acids |

**Table 10.5—10 (continued)**
**CORROSION OF METALS**

| Metal | Subject to corrosion by | Resistant to |
|-------|-------------------------|--------------|
| Steel, mild, low-alloy steels | Most acids; strong alkalies; salt water; sulfur and its compounds | Air; steam; ammonia; most alkalies; concentrated nitric acid; halide refrigerants |
| Tantalum | Hydrofluoric acid; concentrated sulfuric acid; strong alkalies | Nearly all salts; most acids; water; sea water; air; alcohols; hydrocarbons; sulfur |
| Tin | Inorganic acids; caustic solutions; halides | Most food acids; ammonia; neutral solutions |
| Titanium | Hydrochloric acid; sulfuric acid; hydrofluoric, oxalic, and formic acids; dangerously explosive in presence of nitric acid or liquid oxygen | Oxidizing media; air; water; sea water; aqueous chloride solutions; moist chlorine gas; sodium hypochlorite |
| Zinc | Acid or strong alkali solutions; sulfur dioxide; chlorides | Air; water; ammonia; dry common gases; refrigerants; gasoline |
| Zirconium | Concentrated sulfuric acid (hot); hydrofluoric acid; cupric and ferric chlorides | Solutions of alkalies and acids; aqua regia |

*Notes:* Polished surfaces resist corrosion.
Nonuniformity within a metal tends to increase corrosion.
Stress (especially alternating stress) tends to increase corrosion.
Dissolved gases in water (especially oxygen) accelerate corrosion.

### REFERENCES

LaQue, F. L. and Copson, H. R., Eds., *Corrosion Resistance of Metals and Alloys,* Reinhold, New York, 1963.
Romanoff, M., *Underground Corrosion,* National Bureau of Standards Circular 579, 1957.
Schweitzer, P. A., *Handbook of Corrosion Resistant Piping,* Industrial Press, New York, 1969.
Smithells, C. J., *Metals Reference Book,* Vol. 2, 4th ed., Butterworths, London, 1967.
Uhlig, H. H., *The Corrosion Handbook,* John Wiley & Sons, New York, 1948.

From Bolz, R. E. and Tuve, G. L., Eds., *Handbook of Tables for Applied Engineering Science,* 2nd ed., CRC Press, Cleveland, 1973, 350.

**Table 10.5—11**

**SOLDERING ALLOYS[a]**

Tin-lead Solders

These most common solders are available in at least 15 compositions, from 2–70% tin, balance lead. High-tin solders are expensive.

| Percent tin | Approximate melting range[b] | | Tensile strength, kpsi | Electrical resistiv-ity,[c] Cu = 1 | Corroded by | Typical uses |
|---|---|---|---|---|---|---|
| | Liquid, °F | Solid, °F | | | | |
| 2.5 | 580 | 578 | | | Sodium hypochlorite | Seams in cans; cable sheath |
| 5 | 594 | 518 | 3.4 | 11.7 | Chlorides | Filler metal; wiping solder |
| 15 | 550 | 440 | | | Potassium permanganate | Plumbing |
| 20 | 531 | 370 | 5.8 | 10.5 | | Radiators; tubing joints |
| 25[d] | 510 | 362 | | | | Torch soldering |
| 30 | 491 | 361 | | | | Machine soldering |
| 40 | 460 | 360 | | | Air, will tarnish | General-purpose joining |
| 50 | 420 | 360 | 6.1 | 9.3 | Nitric acid | Sheet metal |
| 60 | 374 | 361 | | | | Electrical |
| 63[e] | 361 | 361 | 7.5 | | | Electronic parts |
| 70 | 378 | 361 | 6.8 | 8.7 | | Coating metals |

Antimony Solders

High strength at high temperature; harder than tin-lead solders; should not be used on metals containing zinc.

| Percent tin | Percent lead | Percent antimony | Approximate melting range[b] | | Tensile strength, kpsi | Electrical resistiv-ity,[c] Cu = 1 | Typical uses |
|---|---|---|---|---|---|---|---|
| | | | Liquid, °F | Solid, °F | | | |
| 20 | 79. | 1.0 | 517 | 363 | 3.5 | | Cable sheathing; radiators |
| 30 | 68.4 | 1.6 | 482 | 364 | | 13. | Machine soldering |
| 40 | 58. | 2.0 | 448 | 365 | | | General-purpose joining |
| 95 | 0. | 5.0 | 467 | 458 | 11. | 15. | Electrical |
| 0 | 95. | 5.0 | 554 | 486 | | | Metal coating and filler; batteries |

[a]For data on hard solders, see Table 1.5–25.
[b]The number of degrees between melting and freezing is especially important for wiping and filling solders.
[c]Electrical resistivity is expressed in terms of the resistivity of copper as unity. Thermal conductivity is also roughly one tenth that of copper. Specific heat is less than that of copper.
[d]A typical low-melting solder is 25% tin, 25% lead, and 50% bismuth (liquid at 266°F).
[e]Eutectic composition, lowest melting point for tin-lead alloys.

## Table 10.5—11 (continued)
## SOLDERING ALLOYS

### REFERENCES

American Society for Testing and Materials, ASTM Special Publication 189–1956 (Symposium on Soldering), ASTM, Philadelphia.

American Society for Testing and Materials, ASTM Specification for Solder Metal, B-32-66T, ASTM, Philadelphia, 1966.

American Society for Testing and Materials, ASTM Standards, ASTM, Philadelphia.

American Welding Society, *Soldering Manual,* Miami, Fl., 1959.

Metals Handbook Committee, Metals Handbook, Vol. I, 8th ed., American Society for Metals, Metals Park, Oh., 1961.

National Bureau of Standards, Solders and Soldering, NBS Circular 492, National Bureau of Standards, U.S. Government Printing Office, Washington, D.C., 1950.

**Weast, R. C., Ed.,** *CRC Handbook of Chemistry and Physics,* 55th ed., CRC Press, Cleveland, 1974.

From Bolz, R. E. and Tuve, G. L., Eds., *Handbook of Tables for Applied Engineering Science,* 2nd ed., CRC Press, Cleveland, 1973, 1046.

# Section 11

# Nuclear Materials

# Section 11

# NUCLEAR MATERIALS*

## Table 11—1
## NUCLEAR REACTIONS

Fission Reactions

Energy Released E (MeV), Prompt Neutrons $\nu$, Ratio Values of Delayed to Prompt
Neutrons $\beta$, per Thermal Fission

| Isotope | Total energy of light fragments | Total energy of heavy fragments | Total energy of gamma rays | Total energy of fission neutron | Total energy of beta rays | Total energy | $\nu$ | $\beta$ |
|---|---|---|---|---|---|---|---|---|
| $U^{233}$ | 97 | 66 | 14 | 5 | 9 | 191 | 2.51 | 0.0026 |
| $U^{235}$ | 98 | 67 | 15 | 4.9 | 9 | 194 | 2.47 | 0.0064 |
| $Pu^{239}$ | 100 | 72 | 14 | 5.8 | 9 | 201 | 2.91 | 0.0021 |

### Breeding Processes

$U^{238} + n = U^{239} + \gamma;\ U^{239}\ \xrightarrow{23\ min.}\ Np^{239} + \beta -;\ Np^{239}\ \xrightarrow{2.3\ days}\ Pu^{239} + \beta -$

$Th^{232} + n = Th^{233} + \gamma;\ Th^{233}\ \xrightarrow{23\ min.}\ Pa^{233} + \beta -;\ Pa^{233}\ \xrightarrow{27.4\ days}\ U^{233} + \beta -$

### Thermonuclear Reactions

$D + D \Big\langle {}^{T\ +\ P\ +\ 4.0\ MeV}_{He^3\ +\ n\ +\ 3.2\ MeV}$

$D + T \rightarrow He^4 + n + 17.6\ MeV$

From Katcoff, S., *Nucleonics,* Brookhaven National Laboratory, Upton, L.I., N.Y., November 1960. With permission.

## Table 11—2
## FISSILE AND FERTILE RADIOACTIVE MATERIALS

| | Uranium 233 | Uranium 235 | Uranium 238 | Plutonium 239 | Thorium 232 | Natural uranium (>99% $U^{238}$) |
|---|---|---|---|---|---|---|
| Energy, MeV | 4.8 | 4.4 and 4.6 | | 5.1 | 4.0 | 4.2 |
| Half-life, millions of years | 0.162 | 713 | | 0.024 | 13,900 | 4,510 |
| Neutron binding energy, MeV | 6.7 | 6.4 | | 6.4 | 5.1 | 4.8 |
| Critical energy for fission, MeV[a] | 5.5 | 5.8 | | 5.5 | 5.9 | 5.9 |
| Fission cross section, barns[b] | 527. | 577. | – | 742. | | 4.2 |
| Capture cross section, barns[b] | 54. | 106. | 2.71 | 287. | | 3.5 |
| Total absorption, barns[b] | 581. | 683. | 2.71 | 1,029. | | 7.7 |
| Absorbed neutrons for fission, % | 90.5 | 84.5 | – | 72. | | |

[a]Minimum energies of photons to cause fission.
[b]Data are for 2,200 m/sec thermal neutrons.

From Bolz, R. E. and Tuve, G. L., Eds., *Handbook of Tables for Applied Engineering Science,* 2nd ed., CRC Press, Cleveland, 1973, 411.

* This section was derived from the *Handbook of Materials Science,* Volume III, Section 3, "Nuclear Materials".

## Table 11—3
## NEUTRON SOURCES

### Alpha Sources

| Alpha source | Alpha energies, MeV | Half-life |
|---|---|---|
| Radium 226 | 7.683, 5.996, 5.305 . . . ., 4.59 | 1620 years |
| Actinium 227 | 7.36, 6.617, 6.273 . . . ., 4.942 | 22 years |
| Thorium 228 | 8.780, 6.775, 6.272 . . . ., 5.338 | 1.91 years |
| Uranium 232 | 8.780, 6.775, 6.272 . . . ., 5.261 | 74 years |
| Polonium 210 | 5.305 | 138.4 days |
| Radium D (Pb$^{210}$) | 5.305 | 19.4 years |
| Plutonium 238 | 5.495, 5.452 | 86.4 years |
| Plutonium 239 | 5.147, 5.134, 5.096 | 24360 years |
| Americium 241 | 5.534, 5.50, 5.477, 5.435 | 458 years |
| Americium 242 | 6.110, 6.066 | 100 years |
| Curium 242 | 6.110, 6.066 | 162.5 days |

### Neutron Yield vs. Alpha Energy

| Alpha energy, MeV | Neutron yield from beryllium target, neutrons per million alphas |
|---|---|
| 4.0 | 24 |
| 5.0 | 54 |
| 6.0 | 105 |
| 7.0 | 185 |

### Neutron Yields of Various Targets

The following table shows yields from different target materials when bombarded with the alphas from polonium 210, which has an alpha energy of 5.30 MeV.

| Target | Yield, neutrons per million alphas |
|---|---|
| Lithium | 2.7 |
| Beryllium | 77 |
| Boron | 22 |
| Carbon | 0.1 |
| Fluorine | 12 |

### Practical Yields from Alpha-neutron Sources

| Source | Yield, n/sec/curie |
|---|---|
| Ra–Be | 1.0–1.5 × 10$^7$ |
| RaD–Be | 2.5 × 10$^6$ |
| Po–Be | 2.5 × 10$^6$ |
| Ac–Be | 2.0 × 10$^7$ |
| Pu$^{239}$–Be | 2.2 × 10$^6$ |
| RdTh–Be | 2.0 × 10$^7$ |

### Gamma Outputs of Alpha-neutron Sources

| Source | Gamma output, mrhm/curie | Gamma output, mrhm/10$^6$ n/sec |
|---|---|---|
| Ra–Be | 974 | 54 |
| RaD–Be | 33 | 13.3 |
| Po–Be | 0.1 | 0.04 |
| Ac–Be | 145 | 5.5 |
| Pu$^{239}$–Be | 11 | 4.9 |
| RdTh—Be | 944 | 34 |

### Properties of Target Materials

The following table gives the reaction energy and the approximate neutron yield for a number of target materials when the projectile particles are the 5.30 MeV alphas of polonium 210.

Also given is the Coulomb repulsion energy or electrostatic barrier: the alpha particle and the target nucleus are both positively charged; as a result, only the alpha particles that have sufficient energy to overcome the electrostatic barrier will enter the nucleus and initiate the reaction.

| Target | Isotopic abundance | Reaction energy, MeV | Coulomb repulsion, MeV | Yield, n/sec/curie |
|---|---|---|---|---|
| Lithium 6 | 7.5% | | | |
| Lithium 7 | 92.5% | −2.79 | 1.64 } | 1.0 × 10$^5$ |
| Beryllium 9 | 100 % | 5.74 | 2.10 | 2.85 × 10$^6$ |
| Boron 10 | 18.8% | 1.37 | 2.57 | |
| Boron 11 | 81.2% | 0.27 | 2.52 } | 8.1 × 10$^5$ |
| Carbon 12 | 98.9% | −8.40 | 2.97 | |
| Carbon 13 | 1.1% | 2.36 | 2.92 } | 3.7 × 10$^3$ |
| Fluorine 19 | 100 % | −1.95 | 4.05 | 4.44 × 10$^5$ |

From: Neutron Sources and Their Characteristics, Tech. Bull. NS-2, Commercial Products, Atomic Energy of Canada, Ltd., Ottawa. With permission.

## Table 11—4
## TOTAL CHAIN YIELD FROM THERMAL NEUTRON FISSIONS IN U$^{235}$ [a]

The total integrated chain yield is equivalent to the total fission yield of nuclei having a particular mass. The fission product given uniquely characterizes the total chain yield for the respective mass number.

| Mass No. | Fission product | % yield | Mass No. | Fission product | % yield |
|---|---|---|---|---|---|
| 72 | Zn$^{72}$ (49 hr) | 0.000016 | 125 | Sb$^{125}$ (2.0 yr) | 0.021 |
| 73 | Ga$^{73}$ (5.0 hr) | 0.00011 | 126 | Sb$^{126}$ (9 hr) | 0.05[b] |
| 77 | As$^{77}$ (38.7 hr) | 0.0083 | 127 | Sb$^{127}$ (91 hr) | 0.13[c] |
| 78 | As$^{78}$ (91 min) | 0.021 | 128 | Sn$^{128}$ (57 min) | 0.37 |
| 79 | As$^{79}$ (9.0 min) | 0.056 | 129 | I$^{129}$ ($1.7 \times 10^7$ yr) | 0.9 |
| 81 | Se$^{81}$ (17.6 min) | 0.14 | 130 | Sb$^{130}$ (10 min) | 2.0 |
| 83 | Kr$^{83}$ (stable) | 0.544 | 131 | Xe$^{131}$ (stable) | 2.93, 2.88[b] |
| 84 | Kr$^{84}$ (stable) | 1.00 | 132 | Xe$^{132}$ (stable) | 4.38, 4.31[b] |
| 85 | Rb$^{85}$ (stable) | 1.30 | 133 | Cs$^{133}$ (stable) | 6.59, 6.49[b] |
| 86 | Kr$^{86}$ (stable) | 2.02 | 134 | Xe$^{134}$ (stable) | 8.06, 7.9[b] |
| 87 | Rb$^{87}$ ($6 \times 10^{10}$ yr) | 2.49 | 135 | Cs$^{135}$ ($2.6 \times 10^6$ yr) | 6.41, 6.31[b] |
| 88 | Sr$^{88}$ (stable) | 3.57[c] | 136 | Xe$^{136}$ (stable) | 6.46, 6.36[b] |
| 89 | Sr$^{89}$ (51 days) | 4.79 | 137 | Cs$^{137}$ (29 yr) | 6.15, 6.05[b] |
| 90 | Sr$^{90}$ (28 yr) | 5.77[c] | 138 | Ba$^{138}$ (stable) | 5.74 |
| 91 | Zr$^{91}$ (stable) | 5.84 | 139 | Ba$^{139}$ (84 min) | 6.55[c] |
| 92 | Zr$^{92}$ (stable) | 6.03 | 140 | Ce$^{140}$ (stable) | 6.44[c,d] |
| 93 | Zr$^{93}$ ($1.1 \times 10^6$ yr) | 6.45 | 141 | Ce$^{141}$ (33 days) | ~6.0 |
| 94 | Zr$^{94}$ (stable) | 6.40 | 142 | Ce$^{142}$ (stable) | 5.95 |
| 95 | Mo$^{95}$ (stable) | 6.27 | 143 | Nd$^{143}$ (stable) | 5.98[c] |
| 96 | Zr$^{96}$ (stable) | 6.33 | 144 | Nd$^{144}$ ($5 \times 10^{15}$ yr) | 5.67[c] |
| 97 | Mo$^{97}$ (stable) | 6.09 | 145 | Nd$^{145}$ (stable) | 3.95[c] |
| 98 | Mo$^{98}$ (stable) | 5.78 | 146 | Nd$^{146}$ (stable) | 3.07[c] |
| 99 | Mo$^{99}$ (66 hr) | 6.06[c] | 147 | Sm$^{147}$ ($1.3 \times 10^{11}$ yr) | 2.38 |
| 100 | Mo$^{100}$ (stable) | 6.30 | 148 | Nd$^{148}$ (stable) | 1.70[c] |
| 101 | Ru$^{101}$ (stable) | 5.0 | 149 | Sm$^{149}$ (stable) | 1.13 |
| 102 | Ru$^{102}$ (stable) | 4.1 | 150 | Nd$^{150}$ (stable) | 0.67[c] |
| 103 | Ru$^{103}$ (39.7 days) | 3.0 | 151 | Sm$^{151}$ (80 yr) | 0.45 |
| 104 | Ru$^{104}$ (stable) | 1.8 | 152 | Sm$^{152}$ (stable) | 0.285 |
| 105 | Ru$^{105}$ (4.45 hr) | 0.90[c] | 153 | Sm$^{153}$ (47 hr) | 0.15[c] |
| 106 | Ru$^{106}$ (1.01 yr) | 0.38 | 154 | Sm$^{154}$ (stable) | 0.077 |
| 107 | Rh$^{107}$ (22 min) | 0.19 | 155 | Sm$^{155}$ (24 min) | 0.033 |
| 109 | Pd$^{109}$ (13.4 hr) | 0.030 | 156 | Eu$^{156}$ (15.4 days) | 0.014[c] |
| 111 | Ag$^{111}$ (7.6 days) | 0.019 | 157 | Eu$^{157}$ (15.4 hr) | 0.0078 |
| 112 | Pd$^{112}$ (21 hr) | 0.010[c] | 158 | Eu$^{158}$ (60 min) | 0.002 |
| 115 | Cd$^{115}$ (53 hr) + Cd$^{115}$[e] (43 days) | 0.011 | 159 | Gd$^{159}$ (18 hr) | 0.00107[b] |
| 117 | Cd$^{117}$[e] (3.0 hr) | 0.011 | 161 | Tb$^{161}$ (6.9 days) | 0.000076 |
| 121 | Sn$^{121}$ (27.5 hr) | 0.015 | | | |

[a] See original source for extensive references.
[b] Average of values in references cited in original source.
[c] Based on a yield of $(6.15 + 5.94)/2 = 6.05$ for Cs$^{137}$ (Steinberg, E. P., personal communication).
[d] Measured absolute yield is 6.32%. The number 6.44% is used to normalize other yields.
[e] Metastable.

**Table 11—4 (continued)**
**TOTAL CHAIN YIELD FROM THERMAL NEUTRON FISSIONS IN U$^{235a}$**

### REFERENCES

Petruska, J. A., Thode, H. G., and Tomlinson, R. H., The absolute fission yields of twenty-eight mass chains in the thermal neutron fission of U$^{235}$, *Can. J. Phys.*, 33(11), 693, 1955.

Steinberg, E. P. and Glendenin, L. E., Survey of radiochemical studies of the fission process, in *Proc. 1955 Geneva Conf.*, 7, Paper No. 614, 3.

From *Reactor Physics Constants,* 2nd ed., ANL-5800, Argonne National Laboratory, U.S. Atomic Energy Commission, July 1963.

**Table 11—5**
**DELAYED NEUTRON FRACTION ($\beta$) FOR FAST AND THERMAL FISSION**

Following is a summary of the more useful values of the delayed neutron fraction, i.e., the relative number of delayed neutrons per fission, for the various isotopes.

| Isotope | Fast fission | Thermal fission |
|---------|--------------|-----------------|
| U$^{233}$ | 0.0027 ± 0.0002 | 0.00264 ± 0.0002 |
| U$^{235}$ | 0.0065 ± 0.0003 | 0.0065 ± 0.0003 |
| U$^{238}$ | 0.0157 ± 0.0012 | |
| Pu$^{239}$ | 0.0021 ± 0.0002 | 0.0021 ± 0.0002 |
| Pu$^{240}$ | 0.0026 ± 0.0003 | |
| Th$^{232}$ | 0.022 | |

From *Reactor Physics Constants,* 2nd ed., ANL-5800, Argonne National Laboratory, U.S. Atomic Energy Commission, July 1963.

## Table 11—6
## RADIOACTIVE ISOTOPES

The engineering uses of radioactive isotopes are limited; for more extensive data see the Wang, Y., Ed., *CRC Handbook of Radioactive Nuclides*, Chemical Rubber Co., Cleveland, 1969.

The atomic weight of an atom depends on the number of neutrons in the nucleus as well as the number of protons (and electrons) indicated by the atomic number. For each atomic number, which defines the element, there can be several atomic weights, depending on the number of neutrons. Atoms that differ only in the number of neutrons are known as isotopes or nuclides. While there are more than 20 anisotopic elements (one atomic weight), many others have 2 to 10 isotopes. When there are too many or too few neutrons, the atom becomes unstable or radioactive, and a statistical "decay" occurs, involving the release of radiations of one kind or another. Isotopes may be produced artificially by fission or more likely by bombardment, as in a reactor or a cyclotron.

Isotope differences other than mass are very small and not easy to detect. In a few cases the physical or chemical properties of the isotopes are sufficiently different to permit separation by some common process. This is true with isotopes of boron, carbon, hydrogen, lithium, nitrogen, and oxygen. Separation methods include distillation, electrolysis, electromigration, and chemical exchange.

Mass Number and Half-life; Commercially Available Isotopes

Abbreviations:

d = days
h = hours
y = years

| Element and mass No. | | Half-life | Element and mass No. | | Half-life | Element and mass No. | | Half-life |
|---|---|---|---|---|---|---|---|---|
| **Hydrogen** | 3 | 12.3 y | **Scandium** | 43 | 3.9 h | **Copper** | 61 | 3.3 h |
| **Beryllium** | 7 | 53 d | | 44m | 2.4 d | | 64 | 12.9 h |
| | 10 | $2.7 \times 10^6$ y | | 44 | 4.0 h | | 67 | 61 h |
| **Carbon** | 14 | $5.73 \times 10^3$ y | | 46 | 84 d | **Zinc** | 65 | 245 d |
| | | | | 47 | 3.4 d | | 69m | 14 h |
| **Fluorine** | 18 | 1.8 h | **Titanium** | 44 | 48 y | **Gallium** | 66 | 9.5 h |
| **Sodium** | 22 | 2.58 y | **Vanadium** | 48 | 16.1 d | | 67 | 78 h |
| | 24 | 15.0 h | | 49 | 330 d | | 72 | 14.1 h |
| **Magnesium** | 28 | 21.3 h | **Chromium** | 51 | 27.8 d | **Germanium** | 68 | 282 d |
| **Aluminum** | 26 | $7.4 \times 10^5$ y | **Manganese** | 52 | 5.7 d | | 71 | 11 d |
| **Silicon** | 31 | 2.62 h | | 53 | $2.0 \times 10^6$ y | | 77 | 11 h |
| **Phosphorus** | 32 | 14.3 d | | 54 | 303 d | **Arsenic** | 74 | 18 d |
| | 33 | 25 d | | 56 | 2.6 h | | 76 | 26.5 h |
| **Sulfur** | 35 | 86.7 d | **Iron** | 52 | 8.3 h | | 77 | 39 h |
| **Chlorine** | 36 | $3.0 \times 10^5$ y | | 55 | 2.7 y | **Selenium** | 75 | 120 d |
| | 38 | 37.3 m | | 59 | 45 d | **Bromine** | 77 | 58 h |
| **Argon** | 37 | 35.1 d | **Cobalt** | 56 | 77.3 d | | 82 | 35.3 h |
| | | | | 57 | 267 d | **Krypton** | 79 | 34.9 h |
| **Potassium** | 42 | 12.4 h | | 58 | 71 d | | 83m | 1.86 h |
| | 43 | 22 h | | 60 | 5.26 y | | 85m | 4.4 h |
| **Calcium** | 45 | 165 d | **Nickel** | 63 | 92 y | | 85 | 10.76 y |
| | 47 | 4.7 d | | 65 | 2.56 h | | | |

## Table 11—6 (continued)
### RADIOACTIVE ISOTOPES

Mass Number and Half-life; Commercially Available Isotopes (continued)

| Element and mass No. | | Half-life | Element and mass No. | | Half-life | Element and mass No. | | Half-life |
|---|---|---|---|---|---|---|---|---|
| **Rubidium** | 83 | 83 d | **Iodine** | 123 | 13 h | **Ytterbium** | 169 | 32 d |
| | 84 | 33 d | | 124 | 4.2 d | | 175 | 4.2 d |
| | 86 | 18.7 d | | 125 | 60.2 d | **Lutetium** | 177 | 6.8 d |
| **Strontium** | 85 | 64 d | | 126 | 13.2 d | | | |
| | 87m | 2.8 h | | 129 | $1.6 \times 10^7$ y | **Hafnium** | 175 | 70 d |
| | 89 | 50.4 d | | 130 | 12.5 h | | 181 | 45 d |
| | 90 | 28 y | | 131 | 8.05 d | **Tantalum** | 182 | 115 d |
| **Yttrium** | 87 | 80 h | | 132 | 2.3 h | | | |
| | 88 | 108 d | | 133 | 21 h | **Tungsten** | 181 | 130 d |
| | 90 | 64.2 h | **Xenon** | 131m | 12 d | | 185 | 74 d |
| | 91 | 59 d | | 133 | 5.3 d | | 187 | 24 h |
| **Zirconium** | 95 | 65 d | **Cesium** | 131 | 9.7 d | **Rhenium** | 183 | 70 d |
| | 97 | 17 h | | 132 | 6.6 d | | 186 | 3.8 d |
| **Niobium** | 95 | 35 d | | 134 | 2.1 y | | 188 | 17 h |
| | | | | 137 | 30 y | **Osmium** | 185 | 94 d |
| **Molybdenum** | 99 | 66 h | **Barium** | 131 | 11.6 d | | 191m | 14 h |
| **Technetium** | 99m | 6.0 h | | 133 | 7.2 y | | 191 | 15 d |
| | 99 | $2.1 \times 10^5$ y | | 140 | 12.8 d | | 193 | 32 h |
| **Ruthenium** | 97 | 2.9 d | **Lanthanum** | 140 | 40 h | **Iridium** | 192 | 74 d |
| | 103 | 40 d | **Cerium** | 139 | 140 d | | 194 | 19 h |
| | 106 | 1.0 y | | 141 | 32.5 d | **Platinum** | 193m | 4.4 d |
| **Rhodium** | 102m | 2.9 y | | 143 | 1.4 d | | 197 | 20 h |
| | 105 | 36 h | | 144 | 285 d | **Gold** | 195 | 183 d |
| **Palladium** | 103 | 17 d | **Praseodymium** | 142 | 19.2 h | | 198 | 2.7 d |
| | 109 | 13.5 h | | 143 | 13.7 d | | 199 | 3.15 d |
| **Silver** | 105 | 40 d | **Neodymium** | 147 | 11.1 d | **Mercury** | 197m | 24 h |
| | 110m | 260 d | | 149 | 1.8 h | | 197 | 65 h |
| | 111 | 7.5 d | **Promethium** | 147 | 2.7 y | | 203 | 47 d |
| **Cadmium** | 109 | 1.3 y | | 149 | 2.2 d | **Thallium** | 202 | 12 d |
| | 115m | 43 d | | 151 | 1.2 d | | 204 | 3.8 y |
| | 115 | 2.3 d | **Samarium** | 153 | 2 d | **Lead** | 210 | 22 y |
| **Indium** | 111 | 2.8 d | **Europium** | 152m | 9.3 h | **Bismuth** | 206 | 6.24 d |
| | 114m | 50 d | | 152 | 12.4 y | | 207 | 30 y |
| **Tin** | 113 | 118 d | | 154 | 16 y | | 210m | $2.6 \times 10^6$ y |
| | 119m | 250 d | | 155 | 1.8 y | **Polonium** | 208 | 2.9 y |
| | 121 | 27 h | **Gadolinium** | 153 | 240 d | | 210 | 138 d |
| **Antimony** | 122 | 2.8 d | | 159 | 18 h | **Radon** | 222 | 3.8 d |
| | 124 | 60 d | **Terbium** | 160 | 72 d | **Radium** | 224 | 3.64 d |
| | 125 | 2.7 y | | 161 | 6.9 d | | 226 | $1.62 \times 10^3$ y |
| **Tellurium** | 125m | 58 d | **Dysprosium** | 165 | 2.3 h | | 228 | 5.7 y |
| | 127m | 105 d | **Holmium** | 166m | $1.2 \times 10^3$ y | **Actinium** | 227 | 21.8 y |
| | 127 | 9.3 h | **Erbium** | 169 | 9.4 d | **Thorium** | 228 | 1.91 y |
| | 129m | 34 d | | 171 | 7.5 h | | 230 | $7.6 \times 10^4$ y |
| | 129 | 1.1 h | **Thulium** | 170 | 130 d | **Protactinium** | 231 | $3.2 \times 10^4$ y |
| | 132 | 3.2 d | | | | | 233 | 27.4 d |
| | | | | | | | 234 | 6.7 h |

## Table 11—6 (continued)
## RADIOACTIVE ISOTOPES

Mass Number and Half-life; Commercially Available Isotopes (continued)

| Element and mass No. | | Half-life | Element and mass No. | | Half-life | Element and mass No. | | Half-life |
|---|---|---|---|---|---|---|---|---|
| **Uranium** | 232 | 72 y | **Plutonium** | 237 | 45.6 d | **Americium** | 241 | 458 y |
| | 233 | $1.6 \times 10^5$ y | | 239 | $2.4 \times 10^4$ y | **Curium** | 242 | 163 d |
| **Neptunium** | 237 | $2.1 \times 10^6$ y | | 240 | $6.7 \times 10^3$ y | | 244 | 18 y |

### REFERENCES

Wang, Y., Ed., *CRC Handbook of Radioactive Nuclides,* Chemical Rubber Co., Cleveland, 1969.

Weast, R. C., Ed., *Handbook of Chemistry and Physics,* 69th ed., CRC Press, Boca Raton, Fla., 1988, B-227.

Data from Baker, P. S., in *CRC Handbook of Radioactive Nuclides,* Wang, Y., Ed., Chemical Rubber Co., Cleveland, 1969.

## Table 11—7
## RADIOISOTOPES FOR INDUSTRIAL USE

Typical Applications and Isotopes in Common Use

| Isotopes | Applications | Isotopes | Applications |
|---|---|---|---|
| **CHEMICAL INDUSTRY** | | **ELECTRICAL INDUSTRY** | |
| $S^{35}$, $H^3$ | Efficiency of separation | $Kr^{85}$ | Leak testing |
| $Au^{198}$, $I^{131}$, $Na^{24}$, $Mn^{56}$, $Br^{82}$, $Cr^{51}$ | Thoroughness of mixing | $Hg^{197}$ | Mercury-switch studies |
| | | $H^3$, $Kr^{85}$, $Pm^{147}$ | Luminous dials |
| $I^{131}$, $Br^{82}$, $Na^{24}$, $H^3$ | Leak location | $Co^{60}$, $Ni^{63}$ | Pre-ionization of gases in electronic tubes |
| $Co^{60}$, $Cs^{137}$ | Gaging of liquid or solid levels | | |
| $Rb^{86}$ | Study of process-stream flow patterns | $Sr^{90}$ | Power for navigational lights |
| | | **METALS INDUSTRY** | |
| $Au^{198}$ | Location of pipe obstructions | $Fe^{59}$ | Tracing blast furnace operations |
| $Sb^{124}$ | Study of mass balances in refinery stream | $H^3$ | Study of hydrogen embrittlement |
| $C^{14}$, $H^3$ | Study of reaction mechanisms | | |
| $Xe^{133}$, $Kr^{85}$ | Measurement of gas-flow velocities | $Fe^{59}$, $Cu^{64}$, $Zn^{65}$, $Cr^{51}$, $Ni^{63}$ | Study of piston-ring and bearing wear |
| $ZrNb^{95}$, $Co^{60}$ | Catalyst flow studies | $Co^{60}$ | Control of discharge in coke ovens |
| $C^{14}$ | Study of carbon deposits in fuel research | | |
| $P^{32}$, $Co^{58}$, $Co^{60}$, $C^{14}$ | Drug-metabolism studies | $Sr^{90}$, $Kr^{85}$, $Tm^{170}$, $Eu^{155}$, $Ce^{144}$, $Cs^{137}$ | Thickness gaging |
| $S^{35}$ | Study of vulcanizing process and tire wear | $Co^{60}$ | Measuring wear of firebrick linings |
| $C^{14}$ | Study of frictional forces in rubber | $Co^{60}$, $Ir^{192}$, $Cs^{137}$, $Sm^{145}$, $Gd^{153}$, $Eu^{155}$, $Ce^{144}$, $Tm^{170}$, $Ta^{182}$, $Yb^{169}$ | Detecting thickness variation and defects in castings; weld inspection |
| $Zn^{65}$ | Evaluation of plastic blood bags | | |
| $Ca^{45}$, $Na^{24}$, $Cl^{38}$ | Study of diffusion in glass | | |
| $Kr^{85}$, $Xe^{133}$ | Determination of air pollution from refinery | **TRANSPORTATION INDUSTRY** | |
| $Sr^{90}$, $Kr^{85}$ | Control of rubber thickness on tire ply | $Fe^{55}$, $Zn^{65}$ | Measurement of wear in pistons and bearings |
| $Cs^{137}$ | Control of rock wool production | $Xe^{133}$ | Studies of sediment and sand movement |
| $Co^{60}$, $Cs^{137}$ | Initiation of chemical reactions; effecting of polymerization | $Au^{198}$ | Evaluation of rail life |
| $Sr^{90}$, $Kr^{85}$ | Elimination of static | $Co^{60}$, $Ir^{192}$, $Cs^{137}$ | Gaging of automobile sheet steel |
| $Co^{60}$, $Cs^{137}$ | Sterilization of medical supplies | $H^3$, $Kr^{85}$, $Pm^{147}$ | Luminous locks and dials |
| $Sr^{90}$, $Kr^{85}$ | Thickness gaging of paper and plastics | | |

From Wang, Y., Ed., *CRC Handbook of Radioactive Nuclides,* Chemical Rubber Co., Cleveland, 1969.

**Table 11—8**

## STABILITY OF RADIOACTIVE COMPOUNDS

Most users of compounds labeled with radio-isotopes recognize that such compounds decompose on storage and that the decomposition is accelerated by self irradiation. The degree of the decomposition in relation to the storage conditions of the compound and the measures that can be taken to control and minimize the rate of self radiolysis are not always so well known. Information on this subject is largely empirical. Not only are a large number of labeled compounds — particularly organic compounds — extensively used as tracers, but many applications demand a very high purity. Fractions of a percent of radio-chemical impurity can sometimes lead to incorrect deductions from a tracer investigation, and under these conditions the problem of decomposition by self irradiation becomes a very serious one.

Compounds labeled with the pure beta-emitting radioisotopes ($C^{14}$, tritium, $S^{35}$, $P^{32}$, and $Cl^{36}$) are most commonly used in tracer investigations. Compounds labeled with the gamma-emitting radioisotopes such as $I^{125}$, $I^{131}$, $Co^{57}$, $Co^{58}$, and $Se^{75}$ have special application in medicine. Some properties of these radionuclides are shown in Table A.

Table A. Physical Properties of Some Radionuclides

| Radionuclide | Half-life | Beta energy, MeV | | Specific activity, mCi/mA | | Daughter nuclide (stable) |
|---|---|---|---|---|---|---|
| | | Max. | Mean | Maximum | Common values for compounds | |
| $H^3$ | 12.26 years | 0.018 | 0.0057 | $2.9 \times 10^4$ | $10^2$–$10^4$ | $He^3$ |
| $C^{14}$ | 5700 years | 0.159 | 0.050 | 64 | 1–$10^2$ | $N^{14}$ |
| $S^{35}$ | 87.2 days | 0.167 | 0.049 | $1.5 \times 10^6$ | 1–$10^2$ | $Cl^{35}$ |
| $Cl^{36}$ | $3.03 \times 10^5$ years | 0.714 | 0.3 | 1.2 | $10^{-3}$–$10^{-1}$ | $Ar^{36}$ |
| $P^{32}$ | 14.3 days | 1.71 | 0.69 | $9.3 \times 10^6$ | 10–$10^2$ | $S^{32}$ |
| $I^{131}$ | 8.04 days | 0.81 | 0.19 | $1.7 \times 10^7$ | $10^2$–$10^4$ | $Xe^{131}$ |
| $I^{125}$ | 60 days | Electron capture | | $2.2 \times 10^6$ | $10^2$–$10^4$ | $Te^{125}$ |
| $Co^{57}$ | 270 days | Electron capture | | $4.9 \times 10^5$ | $10^3$–$10^5$ | $Fe^{57}$ |
| $Co^{58}$ | 71 days | Electron capture $+ \beta^+$ | | $1.9 \times 10^6$ | $10^3$–$10^5$ | $Fe^{58}$ |
| $Se^{75}$ | 121 days | Electron capture | | $1.1 \times 10^6$ | 10–$10^3$ | $As^{75}$ |

Decomposition depends in part on the amount of energy absorbed by the compound during its useful life, so that, for a given amount of activity, the radiation energy emitted should be a guide to the seriousness of the problem. The problem of decomposition by self irradiation might be expected to increase in magnitude as the series of pure beta emitters in Table A is descended, but, in fact, almost the reverse is true. This occurs largely for three reasons:

1. The fraction of energy absorbed is much less than unity for the more energetic beta emitters such as $P^{32}$; on the other hand, almost complete total absorption of the beta energy occurs with tritium compounds. Gamma energy is, in general, little absorbed by the compound itself or by its immediate environs.

2. The decomposition also depends on the specific activity of the compound; as can be seen from Table A, the specific activities of tritiated compounds in current use are usually much higher than those for compounds labeled with other pure beta-emitting radionuclides.

3. The absorbed energy decreases exponentially with time; this is an important factor for compounds labeled with radionuclides having a short half-life such as $I^{131}$ or $P^{32}$.

The reason why labeled compounds decompose

**Table 11—8 (continued)**
**STABILITY OF RADIOACTIVE COMPOUNDS**

is not difficult to understand: the radiation energy will be commonly absorbed by the compound itself or by its environs. If the former occurs, then the excited molecules may break up in some manner; if the latter occurs, the radiation energy can produce free radicals and other reactive species, which may then cause destruction of the molecules of the labeled compound.

The modes by which decomposition of labeled compounds can arise have been classified as shown in Table B.

Table B. Modes of Decomposition of Labeled Compounds

| Mode of decomposition | Cause | Method for control |
|---|---|---|
| Primary (internal) | Natural isotopic decay | None, for a given specific activity [a] |
| Primary (external) | Direct interaction of the radioactive emission (alpha, beta, or gamma) with molecules of the compound | Dispersal of the labeled molecules |
| Secondary | Interactions of excited products with molecules of the compound | Dispersal of active molecules; cooling to low temperatures; scavenging of free radicals |
| Chemical | Thermodynamic instability of compounds; poor choice of environment | Cooling to low temperatures; removal of harmful agents |

[a]Note that dilution with the inactive form of the compound subsequent to preparation is not beneficial in this case.

Primary decomposition is the production of an impurity due to the disintegration of the unstable nucleus. Secondary decomposition is commonly the most damaging and the most difficult to control. Chemical decomposition (by oxidation, hydrolysis, etc.) is even more likely to occur with radioactive chemicals. It may also be necessary to guard against photochemical or microbiological decomposition of the compound.

Tables reporting the various kinds of decomposition for a great variety of labeled compounds are available. Over 30 pages of such tables are given in the CRC Handbook of Radioactive Nuclides; this source also cites a large number of references.

While it is true that no measurable decomposition has occurred in many labeled compounds that have been stored for months or even years, there are other compounds showing major decomposition in a matter of days. It is thus very important to take proper account of the condition and the stability of any compounds being used.

Adapted from Wang, Y., Ed., *CRC Handbook of Radioactive Nuclides,* Chemical Rubber Co., Cleveland, 1969.

## Table 11—9
## REACTOR MATERIALS

Each engineering material used in the construction of a nuclear reactor may serve one or more functions. In most cases the physical, mechanical, and thermal properties of the material, as well as the nuclear properties, are of some importance. In the following list each function is briefly defined, and some of the important properties are noted.

**Fuel** — The fuel elements are uranium, plutonium, and thorium. The fissionable isotopes are $U^{233}$, $U^{235}$, and $Pu^{239}$. Although uranium has 14 isotopes, all radioactive, natural uranium contains 99.28% of $U^{238}$ and only 0.7% of $U^{235}$. The most common fuel is natural uranium, "enriched" with added $U^{235}$. Isotopes that can be converted into fissile fuels are called "fertile" materials (see Table 3.5—2). These include $U^{238}$, which forms $U^{239}$ by capture of a neutron and undergoes transition to $Pu^{239}$ by beta decompositions, and thorium 232, which is converted to $Th^{233}$ by neutron fluxes and forms $U^{233}$ by beta decay.

Reactor-fuel elements should be strong and corrosion resistant and have high thermal conductivity. A minimum of fission products should enter the coolant. Alloying, cladding, or plating are required to attain these properties. Stainless steel, aluminum, titanium, molybdenum, beryllium, and zirconium are used, in addition to graphite and some ceramic materials.

General physical and mechanical properties of some of these materials are given in Table A.

## Table 11—9 (continued)
## REACTOR MATERIALS

### Table A. Reactor Fuels and Associated Materials

#### Typical Properties

For specific heat in J/kg·K, multiply tabular values by 4187. For thermal conductivity in Btu/hr·ft·°F, multiply by 57.8. For N/m² multiply psi by 6895.

| Material[a] | Specific gravity | Specific heat, 25°C | Thermal conductivity, $\frac{\text{watts}}{\text{cm}°\text{C}}$ | Melting point, °C | Coefficient of linear expansion/°C ($\times 10^6$) | Tensile strength, psi | Modulus of elasticity, psi (millions) | Absorption cross section, $\sigma_a$, barns |
|---|---|---|---|---|---|---|---|---|
| Uranium | 19. | 0.03 | 0.25 | 1,132 | 46. | 60,000. | 23. | 7.6 |
| UO$_2$ | 10.96 | | 0.08 | 2,600 | 10. | — | — | 7.6 |
| Plutonium | 19.8 | 0.03 | 0.08 | 640. | 54. | — | 14. | — |
| Thorium | 11.7 | 0.03 | 0.4 | 1,750. | 11.5 | 32,000 | 9. | 7.5 |
| Beryllium | 1.85 | 0.44 | 2.18 | 1,285 | 12. | 30,000 | 42. | 0.01 |
| Zirconium | 6.5 | 0.07 | 0.21 | 1,852 | 5.8 | 30,000 | 13.7 | 0.18 |
| Cadmium | 8.65 | 0.05 | 0.93 | 321 | 32. | 10,300 | 8. | 2,450. |
| Aluminum | 2.7 | 0.21 | 2.37 | 660. | 26. | 13,000 | 10. | 0.24 |
| Stainless steel (347) | 8.0 | 0.12 | 0.2 | 1,400. | 16. | 75,000 | 29. | 3.0 |
| Titanium | 4.5 | 0.125 | 0.2 | 1,670. | 8.5 | 9,000 | 16. | 5.8 |
| Graphite | 2.2 | 0.17 | 0.24 | >3,500 | 3. | 100 | 0.7 | 0.004 |

[a]Some metals become radioactive and dangerous to handle after exposure to the neutron flux of a reactor, but these would not be used in the structure of a thermal reactor. Tungsten and cobalt are the worst in this respect, but tantalum, chromium, and manganese may also attain a high induced radioactivity, and copper a considerably lower one.

**Table 11—9 (continued)**
**REACTOR MATERIALS**

**Moderator** — For thermal reactors it is necessary to reduce the kinetic energy of the "fast" neutrons by successive scattering collisions. Materials effective for this "slowing down" of neutrons to the thermal range include graphite, water, heavy water, beryllium and beryllium oxide, and hydrocarbons (see Table 11—14 and 11—16. A moderator should be resistant to corrosion and stable at high temperatures and in high radiation flux. A moderator may serve as a fuel diluent or fuel container; in the latter case mechanical strength is required

**Reflector** — Moderating materials that also scatter neutrons back into the core to reduce "leakage" are termed reflectors. A "bare reactor" is one without such reflectors. Savings in fuel and smaller critical size are attained with reflectors.

**Control material** – Control rods or other arrangements for varying the reactor output are of three classes: (1) *shim rods* for coarse control during startup, (2) *regulating rods* for small but quick control, and (3) *safety rods* or "scram" control for emergencies. The earlier methods used materials with large capture cross sections to absorb neutrons (cadmium or boron), but it is more economical to employ uranium, cobalt, or other material that produces a secondary product such as a desired isotope. Other control methods involve the positioning of the fuel, moderator, or reflector elements. Typical properties of the more effective isotopes of the high absorption materials are given in Table B.

Table B. Properties of Neutron Absorbers (Poisons)

| Isotope | Thermal microscopic cross sections, $\sigma$, barns | Major resonance | |
|---|---|---|---|
| | | Energy, eV | $\sigma_a$, barns |
| Cadmium 113 | 20,000 | 0.18 | 7,200 |
| Boron 10 | 3,850 | None | |
| Samarium 149 | 41,000 | 0.096 | 16,000 |
| Samarium 152 | 220 | 8.2 | 15,000 |
| Gadolinium 155 | 56,000 | 2.6 | 1,400 |
| Gadolinium 157 | 240,000 | 17. | 1,000 |
| Europium 151 | 7,800 | 0.46 | 11,000 |
| Europium 153 | 450 | 2.46 | 3,000 |
| Silver 107 | 31 | 16.6 | 630 |
| Silver 109 | 87 | 5.1 | 12,500 |

**Coolant** – In some reactors it may be necessary to dissipate almost the entire heat output to a heat sink; in reactors for power generation, the primary coolant becomes the heat source for the power cycle, and two or more coolant circuits are involved. Practical coolants are water, liquid metals, and various gases. Heavy water is superior but costly. Organic liquids tend to leave surface deposits in the fuel section. High heat capacity and high thermal conductivity are desired, with low vapor pressure and stability under high temperature and radiation.

**Table 11—9 (continued)**
**REACTOR MATERIALS**

Table C. Reactor Coolants

| Liquids and liquid metals | | Gases and vapors | |
|---|---|---|---|
| Diphenyl | NaK | Air | Mercury |
| Dowtherm®a | Potassium | Carbon dioxide | Neon |
| Gallium | Rubidium | Helium | Nitrogen |
| Heavy water | Sodium | Hydrogen | Steam |
| Lithium | Tin | | |
| Mercury | Water | | |

aDowtherm Heat Transfer Agent, Dow Chemical Co., Midland, Mich.

**Shielding material** — Design of shields for personnel protection depends on objectives, space and weight requirements, and types of radiation. Table 11—16 gives further data

From Bolz, R. E. and Tuve, G. L., Eds., *Handbook of Tables for Applied Engineering Science,* 2nd ed., CRC Press, Cleveland, 1973, 432.

**Table 11—10**
**ENERGY LIBERATED BY FISSION**

The following values are approximations for the energy quantities involved in fission of any one of the three common nuclear fuels — uranium 233 or 235 or plutonium 239 — since there is a relatively small difference between these weights.

One pound of fissile material will liberate approximately 10 million kwh. The fission of 1 g/day is roughly equal to 1 mw. The "burnup" of 1,000 kg (2,200 lb) of uranium would deliver 100 mw for about 25 years.

While almost all of the fission energy ultimately appears as heat within the matter surrounding the reaction, only about 80% of the fission energy is immediately converted to heat from the kinetic energy of the fission fragments. About 95% of the energy eventually appears as heat, but the 5% represented by the neutrinos essentially escapes from the reactor.

Percentage Distribution of Fission Energy

| | |
|---|---|
| Kinetic energy of fission fragments | 82.5% |
| Instantaneous gamma-ray energy | 3.5% |
| Gamma-rays from fission products | 3.0% |
| Beta particles from fission products | 3.5% |
| Kinetic energy of fission neutrons | 2.5% |
| Neutrinos | 5.0% |

From Bolz, R. E. and Tuve, G. L., Eds., *Handbook of Tables for Applied Engineering Science,* 2nd ed., CRC Press, Cleveland, 1973, 433.

## Table 11—11
## DECAY OF REACTOR FISSION PRODUCTS

Days from Shutdown, After Infinite Operation; Various Levels of Effective Energy (EE)

Approximate decay, MeV/W–s

| Time, days | Group 1, EE = 0.4 MeV | Group 2, EE = 0.8 MeV | Group 3, EE = 1.3 MeV | Group 4, EE = 1.7 MeV | Group 5, EE = 2.2 MeV | Group 6, EE = 2.5 MeV | Group 7, EE = 2.8 MeV |
|---|---|---|---|---|---|---|---|
| 0.1 | $1.5 \times 10^9$ | $1 \times 10^{10}$ | $1.1 \times 10^9$ | $4.3 \times 10^9$ | $1.2 \times 10^9$ | $3.4 \times 10^8$ | $1.7 \times 10^9$ |
| 0.2 | $1.5 \times 10^9$ | $9.5 \times 10^9$ | $6.8 \times 10^8$ | $3.5 \times 10^9$ | $7.4 \times 10^8$ | $3 \times 10^8$ | $9 \times 10^8$ |
| 0.4 | $1.45 \times 10^9$ | $8.8 \times 10^9$ | $3.7 \times 10^8$ | $3.1 \times 10^9$ | $3 \times 10^8$ | $2.9 \times 10^8$ | $2.8 \times 10^8$ |
| 0.7 | $1.35 \times 10^9$ | $8.0 \times 10^9$ | $2 \times 10^8$ | $3 \times 10^9$ | $1.15 \times 10^8$ | $2.8 \times 10^8$ | $4.7 \times 10^7$ |
| 1.0 | $1.3 \times 10^9$ | $7.5 \times 10^9$ | $1.2 \times 10^8$ | $3 \times 10^9$ | $8 \times 10^7$ | $2.7 \times 10^8$ | |
| 2.0 | $1.1 \times 10^9$ | $6.4 \times 10^9$ | $2.3 \times 10^7$ | $3 \times 10^9$ | $5.6 \times 10^7$ | $2.6 \times 10^8$ | |
| 4.0 | $8 \times 10^8$ | $5.4 \times 10^9$ | | $2.9 \times 10^9$ | $4 \times 10^7$ | $2.4 \times 10^8$ | |
| 7.0 | $6 \times 10^8$ | $4.7 \times 10^9$ | | $2.6 \times 10^9$ | $3 \times 10^7$ | $2.1 \times 10^8$ | |
| 10. | $4.8 \times 10^8$ | $4 \times 10^9$ | | $2.2 \times 10^9$ | $2.3 \times 10^7$ | $1.7 \times 10^8$ | |
| 20. | $3 \times 10^8$ | $3.2 \times 10^9$ | | $1.3 \times 10^9$ | $1.7 \times 10^7$ | $1.0 \times 10^8$ | |
| 40. | $1.3 \times 10^8$ | $2.4 \times 10^9$ | | $4 \times 10^8$ | $1.55 \times 10^7$ | $4 \times 10^7$ | |
| 70. | $5 \times 10^7$ | $1.8 \times 10^9$ | | $1 \times 10^8$ | $1.4 \times 10^7$ | $1.3 \times 10^7$ | |
| 100 | $2.4 \times 10^7$ | $1.5 \times 10^9$ | | $3 \times 10^7$ | $1.3 \times 10^7$ | | |

*Note:* These data for infinite-power operation may be used for estimating by calculation the sources resulting from finite-time operation before shutdown.

Data from Rockwell, T., III, Ed., *Reactor Shielding Design Manual*, USAEC Report No. TID-7004, U.S. Atomic Energy Commission, March 1956.

## Table 11—12
## MODERATOR MATERIALS

Nuclear Characteristics of Potential Moderators

For density in lb/ft$^3$, multiply the value in g/cm$^3$ by 62.42.

| Element or compound | A, atomic or molecular weight | Density[a] g/cm$^3$ | N,[b] $\times 10^{-24}$ | Scattering cross section $\sigma_s$ epithermal, barns/atom | Absorption cross section $\sigma_a$, 0.025 eV, barns/atom or molecule |
|---|---|---|---|---|---|
| Hydrogen, H | 1.008 | 0.0090 | 0.0054 | 20.3 | 0.33 |
| Deuterium, D | 2.02 | 0.0180 | 0.0054 | 3.3 | 0.00046 |
| Helium, He | 4.00 | 0.0180 | 0.0027 | 0.8 | — |
| Beryllium, Be | 9.01 | 1.85 | 0.124 | 6.1 | 0.009 |
| Carbon (graphite), C | 12.0 | 1.60 | 0.080 | 4.7 | 0.0045 |
| Oxygen, O | 16.0 | 0.014 | 0.0054 | 3.8 | 0.0002 |
| Sodium, Na | 23.0 | 0.97 | 0.0254 | 3.0 | 0.49 |
| Magnesium, Mg | 24.3 | 1.74 | 0.043 | 3.4 | 0.059 |
| Aluminum, Al | 27.0 | 2.70 | 0.060 | 1.35 | 0.215 |
| Beryllia, BeO | 25.0 | 3.025 | 0.073 | 9.9 | 0.009 |
| Beryllium carbide, Be$_2$C | 30.0 | 2.4 | 0.048 | 16.9 | 0.023 |
| Beryllium fluoride, BeF$_2$ | 47.0 | 1.986 | 0.025 | 15.4 | 0.029 |
| Light water, H$_2$O | 18.0 | 1.00 | 0.033 | 44.4 | 0.66 |
| Heavy water, D$_2$O | 20.0 | 1.10 | 0.033 | 10.5 | 0.0011 |
| Sodium hydroxide, NaOH | 40.0 | 2.1 | 0.032 | 27.1 | 0.82 |
| Zirconium hydride, ZrH$_2$ | 93.2 | 5.61 | 0.036[d] | 48.6 | 0.84 |
| Biphenyl, C$_{12}$H$_{10}$ | 154.2 | 0.87 | 0.0034 | 259.4 | 2.54 |
| Polystyrene, (CH)$_n$ | 13.0 | 1.07 | 0.038[e] | 25.0[g] | 0.335[g] |
| Paraffin, (CH$_2$)$_n$ | 14.0 | 0.9 | 0.030[f] | 45.3[h] | 0.665[h] |

| Element or compound | Macroscopic absorption cross section, $\Sigma_a = N\sigma_a$ cm$^{-1}$ | Macroscopic scattering cross section, $\Sigma_s = N\sigma_a$ cm$^{-1}$ | Logarithmic mean energy loss/ collision,[c] $\xi$ | Slowing-down power, $\xi\Sigma_s$ | Moderating ratio, $\sigma_s\xi/\sigma_a$ |
|---|---|---|---|---|---|
| Hydrogen, H | 0.0018 | 0.11 | 1.000 | 0.11 | 61 |
| Deuterium, D | $2.5 \times 10^{-6}$ | 0.018 | 0.725 | 0.013 | 5,200 |
| Helium, He | — | 0.0022 | 0.425 | 0.0009 | $\infty$ |
| Beryllium, Be | 0.0011 | 0.76 | 0.206 | 0.16 | 145 |
| Carbon (graphite), C | 0.00036 | 0.38 | 0.158 | 0.060 | 165 |
| Oxygen, O | $1.1 \times 10^{-6}$ | 0.021 | 0.12 | 0.0025 | 230 |
| Sodium, Na | 0.012 | 0.076 | 0.083 | 0.0063 | 0.53 |
| Magnesium, Mg | 0.0025 | 0.15 | 0.073 | 0.011 | 4.4 |
| Aluminum, Al | 0.013 | 0.081 | 0.071 | 0.0058 | 0.45 |
| Beryllia, BeO | 0.00066 | 0.72 | 0.173 | 0.12 | 190 |
| Beryllium carbide, Be$_2$C | 0.0011 | 0.81 | 0.193 | 0.16 | 145 |
| Beryllium fluoride, BeF$_2$ | 0.00074 | 0.39 | 0.151 | 0.058 | 84 |
| Light water, H$_2$O | 0.022 | 1.47 | 0.925 | 1.36 | 62 |
| Heavy water, D$_2$O | $36 \times 10^{-6}$ | 0.35 | 0.504 | 0.18 | 5,000 |
| Sodium hydroxide, NaOH | 0.026 | 0.87 | 0.77 | 0.67 | 26 |
| Zirconium hydride, ZrH$_2$ | 0.030 | 1.75 | 0.84 | 1.47 | 49 |
| Biphenyl, C$_{12}$H$_{10}$ | 0.00862 | 0.880 | 0.812 | 0.715 | 83 |
| Polystyrene, (CH)$_n$ | 0.013 | 0.95 | 0.842 | 0.80 | 62 |
| Paraffin, (CH$_2$)$_n$ | 0.020 | 1.36 | 0.913 | 1.24 | 62 |

[a]For the gases H$_2$, D$_2$, He, O$_2$ an arbitrary density 100 times the density of NTP is assumed; i.e., a pressure on the order of 1,500 psi.

**Table 11—12 (continued)**
**MODERATOR MATERIALS**

Nuclear Characteristics of Potential Moderators (continued)

[b]Number of atoms or mole per cm³
[c]$\xi = 1 - [(A - 1)^2/2A] \ln [(A + 1)/(A - 1)]$.
[d]Below 800°C.
[e]Number of (CH) units/cm³.
[f]Number of ($CH_2$) units/cm³.
[g]Per (CH) unit.
[h]Per ($CH_2$) unit.

From Tipton, C. R., Jr., Ed., *Reactor Handbook,* 2nd ed., Vol. 1: *Materials,* Wiley Interscience, New York, 1960. Copyright © John Wiley & Sons. With permission

**Table 11—13**
**GRAPHITE FOR REACTORS**

Commercial graphite for reactors is a mixture of crystalline graphite and cross-linking inter-crystalline carbon. The physical properties that are measured are the result of contributions from both sources. Graphite for nuclear reactors is a material for neutron moderators, reflectors, thermal columns, and exponential piles. Its desirable properties for these uses include low neutron-capture cross section, high-temperature stability, and machinability.

Table A lists common properties of two typical nuclear graphite materials that differ primarily in purity. Many special grades are also available to meet such requirements as high density, low porosity, and high permeability.

Graphite is available as a matrix material containing dispersions of uranium and/or thorium for high-temperature fuel elements or boron for control-rod and neutron-shielding purposes.

The comparative position of graphite as a neutron moderating and reflecting element is indicated in Table B. *Moderator figure-of-merit* is the ratio of the life of a fast neutron before absorption to the time required to slow it down to an energy of 1 eV. *Reflector figure-of-merit* is the ratio of the probability of thermal neutron

scattering back into a reactor core to the probability of absorption.

Reactor graphite is made from petroleum coke, which is calcined, crushed, and screened, then mixed with a coal-tar pitch binder and fired at high temperature. Size of grains depends on the source of the raw material as well as on the processing. Physical properties and crystal orientation will be somewhat different for an extruded product than for a molded product. Impurities and trace elements will vary with the raw materials; the principal impurities are iron, silicon, calcium, and aluminum, each of these typically less than 500 ppm; other metals are typically less than 70 ppm.

Additional factors that affect the properties of the final product are size of the finished piece, the particle-size distribution in the original mix, the maximum processing temperature, and the temperature at which the properties are being measured. The tensile strength of graphite is about one half its flexural strength, while the crushing strength is about twice the flexural strength. Strength increases with temperature, in the usual range, as do the specific heat and the coefficient of thermal expansion. Thermal conductivity decreases with increase in temperature.

## Table 11—13 (continued)
## GRAPHITE FOR REACTORS

### Table A. Typical Properties of Nuclear Graphite

| Property | Grade A | Grade B |
|---|---|---|
| Slow neutron absorption, cross section | | |
| per carbon atom $\times$ $10^{-27}$, mb[a] | 3.95 | 4.5 |
| Total ash content, percent | 0.01 | 0.06 |
| Boron content, ppm | 0.2 | 0.3 |
| Specific resistance, microhm-meter | | |
| Longitudinal | 5.40 | 6.50 |
| Transverse | 11.75 | 11.80 |
| Thermal conductivity, Btu·ft/hr·ft²·°F | | |
| Longitudinal | 141 | 116 |
| Transverse | 62 | 69 |
| Coefficient of thermal expansion, $10^{-6}$/°F | | |
| Longitudinal | 0.2 | 0.6 |
| Transverse | 1.9 | 1.7 |
| Bulk density, g/cm³ | 1.73 | 1.71 |
| Tensile strength, psi, longitudinal | 1,400 | 1,300 |
| Flexural strength, psi, longitudinal | 3,400 | 2,700 |
| Compressive strength, psi, longitudinal | 5,000 | 5,000 |
| Elastic modulus, $10^6$ psi | | |
| Longitudinal | 2.3 | 1.8 |
| Transverse | 0.8 | 1.0 |

[a]mb = millibarns.

Courtesy of Carbon Products Division, Union Carbide Corporation, New York.

### Table B. Comparative Moderating and Reflector Characteristics

| | $H_2O$ | $D_2O$ | Be | BeO | Graphite |
|---|---|---|---|---|---|
| Moderator figure-of-merit | 209 | 17,500 | 456 | 604 | 695 |
| Reflector figure-of-merit | 172 | 15,300 | 788 | 774 | 1,430 |

Adapted from *The Industrial Graphite Engineering Handbook*, Carbon Products Division, Union Carbide Corporation, New York, 6.03.

## Table 11—14
### NUCLEAR PROPERTIES OF WATER AND HEAVY WATER

| Quantity | Light water | Heavy water |
|---|---|---|
| Abundance, % | 99.9849 to 99.9861 | 0.0139 to 0.0151 |
| Molecular weight | 18.016 | 20.028 |
| Molecular density at 20 C, cm$^{-3}$ | $0.334 \times 10^{24}$ | $0.0332 \times 10^{24}$ |
| Neutron macroscopic absorption cross section at 2,200 m/sec, cm$^{-1}$ | 0.0220 | 0.000040 |
| Thermal neutron diffusion area, cm$^2$ | 8.12 | — |
| For 0.16% $H_2O$ content | | 13,500 |
| Corrected for $H_2O$ absorption | | 25,000 |
| Thermal neutron macroscopic transport cross section, cm$^{-1}$ | 2.10 | 0.395 |
| Average cosine of neutron scattering angle | 0.34 | 0.15 |
| Epithermal neutron macroscopic scattering cross section | 1.49 | 0.349 |
| Epithermal neutron macroscopic slowing-down cross section, cm$^{-1}$ | 1.38 | 0.178 |
| Fission neutron age, cm$^2$ | | |
| To indium | 30.4 | 100 |
| To thermal | 31.4 | 125 |

From Tipton, C. R., Jr., Ed., *Reactor Handbook,* 2nd ed., Vol. 1: *Materials,* Wiley Interscience, New York, 1960. Copyright © 1960 by John Wiley & Sons. With permission

## Table 11—15
### RADIATION EXPOSURE AND SHIELDING

Data on Hazards from Alpha, Beta, and Gamma Radiation

The following miscellaneous items of information on penetration and shielding are useful in evaluating certain radiation hazards and in planning personnel protection.

**Gamma rays** — As an approximation for the dose rate at a distance of 1 ft from a point source of gamma radiation, the value in rems per hour is six times the product of total gamma energy emitted per disintegration of the parent, in MeV, times number of curies of the parent nuclide (accuracy ± 20 % from 0.07 to 4 MeV.)

The attenuation of dose rate with distance, from 100 Ci of cobalt 60, is given in the following table. Account has been taken of air absorption and build-up factor as well as inverse-square attenuation.

### Table A. Dose Rate from 100 Ci of $Co^{60}$

| Dose rate, rems/hour | Distance, feet | Dose rate, rems/hour | Distance, feet |
|---|---|---|---|
| 1,500 | 1 | 0.0075 | 400 |
| 15 | 10 | 0.0012 | 800 |
| 0.6 | 50 | 0.0006 | 1,000 |
| 0.15 | 100 | | |

*Warning note:* Hazardous beta rays may also be present. The above data provide the dose rates from gamma rays only.

Table B gives relative shield thickness for gamma radiation. The build-up factor to correct exponential to broad-beam radiation has been applied.

### Table B. Shield Thickness vs. Gamma-dose Transmission

| Broad-beam transmission | Shield thickness, inches | | |
|---|---|---|---|
| | Concrete, 147 lb/ft$^3$ | Iron | Lead |
| **Radium (11 Principal Gammas, 0.24−2.20 MeV)** | | | |
| 0.1 | 10 | 3.1 | 1.6 |
| 0.01 | 19 | 6.2 | 3.5 |
| 0.001 | 28 | 9.1 | 5.5 |
| 0.0001 | 38 | 12.0 | 7.8 |
| 0.00001 | 47 | 15.3 | 10.2 |
| **Cobalt 60 (1.33 + 1.17 MeV per Disintegration)** | | | |
| 0.1 | 11 | 3.2 | 1.7 |
| 0.01 | 19 | 6.0 | 3.3 |
| 0.001 | 27 | 8.8 | 4.8 |
| 0.0001 | 35 | 11.4 | 6.5 |
| 0.00001 | 43 | 14.6 | 8.1 |

## Table 11—15 (continued)
### RADIATION EXPOSURE AND SHIELDING

Data on Hazards from Alpha, Beta, and Gamma Radiation (continued)

Table B. Shield Thickness vs. Gamma-dose Transmission

| Broad-beam transmission | Shield thickness, inches | | | |
|---|---|---|---|---|
| | Concrete, 147 lb/ft$^3$ | Iron | | Lead |
| **Cesium 137 (0.66 MeV)** | | | | |
| 0.1 | 8.5 | 2.6 | | 0.85 |
| 0.01 | 15 | 4.7 | | 1.7 |
| 0.001 | 22 | 6.8 | | 2.5 |
| 0.0001 | 28 | 8.9 | | 3.4 |
| 0.00001 | 34 | 11.0 | | 4.2 |
| **Iridium 192 (Gammas from 0.13–0.87 MeV, Averaging 0.3 MeV)** | | | | |
| 0.1 | 7 | | | 0.48 |
| 0.01 | 13 | | | 1.1 |
| 0.001 | 18.3 | | | 1.9 |
| 0.0001 | 24 | | | 2.6 |
| 0.00001 | 30 | | | 3.5 |
| **Gold 198 (0.41-, 0.68-, and 1.1- MeV Gammas)** | | | | |
| 0.1 | 6.6 | | | 0.35 |
| 0.01 | 12.0 | | | 0.83 |
| 0.001 | 17.4 | | | 1.7 |
| 0.0001 | 22.6 | | | 2.8 |
| 0.00001 | 28.0 | | | 4.3 |
| **Iodine 131 (0.08–0.723 MeV, Predominantly 0.36 MeV)** | | | | |
| 0.1 | 6 | | | |
| 0.01 | 12 | | | |
| 0.001 | 18 | | | |
| **Barium 140 + Lanthanum 140 (0.030–2.5 MeV, Averaging about 1.6 MeV)** | | | | |
| 0.1 | 25 | 9.8 | 3.4 | 1.6 | 0.87 |
| 0.01 | 44 | 18 | 6.4 | 3.4 | 2.0 |
| 0.001 | 64 | 27 | 9.2 | 5.2 | 3.0 |
| 0.0001 | 81 | 35 | 11.8 | 7.1 | 4.1 |
| 0.00001 | 104 | 44 | 14.8 | 9.0 | 5.2 |
| 0.000001 | | 51 | | | 6.3 |

[a]After several mean free paths, every 10 in. of concrete reduces the radiation by another factor of ten.

From Brodsky, A. and Beard, G. V., *A Compendium of Information for Use in Controlling Radiation Emergencies,* TID — 8206 (rev.), U.S. Atomic Energy Commission, 1960.

**Table 11—15 (continued)**
**RADIATION EXPOSURE AND SHIELDING**

**Alpha and beta particles** — The energy required to just penetrate the 0.07-mm protective layer of skin is 7.500 keV for an alpha particle but only 70 keV for a beta particle. The alpha activity of several common sources is shown in Table C.

**Table C. Alpha Activity in Particles**
**per Minute per Microgram**

| | |
|---|---|
| Plutonium | 140,000. |
| Neptunium | 1,519. |
| Natural uranium | 1.5 |
| Uranium 238 | 0.741 |
| Thorium 232 | 0.247 |

The range of beta particles in air is about 12 ft/MeV. The air dose in rads per hour at 1 ft from a beta point source is about 200 times its value in curies (absorption neglected).

Beta-ray surface dose rates for several materials are shown in Table D.

**Table D. Beta-ray Surface Dose Rates**

| Material | mrad/hr | Material | mrad/hr |
|---|---|---|---|
| Thorium, 4—5 years after separation | 40 | Uranium slug, natural | 233 |
| Tuballoy, D-38 | 200 | $UO_2$, brown oxide | 207 |
| Oralloy | | $UF_4$, green salt | 179 |
| 40% | 180 | $UO_2(NO_3)_2 - 6H_2O$ | 111 |
| 93% | 140 | $UO_3$, orange oxide | 204 |
| Plutonium 239 | | $U_3O_8$, black oxide | 203 |
| Nickel coated | 360 | $UO_2F_2$ | 176 |
| Uncoated | 440 | $Na_2U_2O_7$ | 167 |
| Uranium 233 | | | |
| 1-month $U^{232}$ build-up | 7,000 | | |
| 1-year $U^{232}$ build-up | 58,000 | | |

From Brodsky, A. and Beard, G. V., *A Compendium of Information for Use in Controlling Radiation Emergencies,* TID-8206 (rev.), U.S. Atomic Energy Commission, 1960.

Data for Table 11—15 from Wang, Y., Ed., *CRC Handbook of Radioactive Nuclides,* Chemical Rubber Co., Cleveland, 1969.

## Table 11—16
## PROPERTIES OF SHIELDING MATERIALS

Gamma-ray Mass-absorption Coefficients for Various Materials Used in Shielding

For density in $lb/ft^3$, multiply $g/cm^3$ by 62.42

| Material | Density, $\rho$, $g/cm^3$ | Mass-absorption coefficient, $\mu$, $cm^{-1}$ | | | Material | Density, $\rho$, $g/cm^3$ | Mass-absorption coefficient, $\mu$, $cm^{-1}$ | | |
|---|---|---|---|---|---|---|---|---|---|
| | | *1 MeV* | *3 MeV* | *6 MeV* | | | *1 MeV* | *3 MeV* | *6 MeV* |
| Air | 0.001294 | 0.0000766 | 0.0000430 | 0.0000304 | Flesh[b] | 1 | 0.0699 | 0.0393 | 0.0274 |
| Aluminum | 2.7 | 0.166 | 0.0953 | 0.0718 | Fuel oil | | | | |
| Ammonia (liquid) | 0.771 | 0.0612 | 0.0322 | 0.0221 | (medium) | 0.89 | 0.0716 | 0.0350 | 0.0239 |
| Beryllium | 1.85 | 0.104 | 0.0579 | 0.0392 | Gasoline | 0.739 | 0.0537 | 0.0299 | 0.0203 |
| Beryllium carbide | 1.9 | 0.112 | 0.0627 | 0.0429 | Glass | | | | |
| Beryllium oxide | | | | | Borosilicate | 2.23 | 0.141 | 0.0805 | 0.0591 |
| (hot-pressed blocks) | 2.3 | 0.140 | 0.0789 | 0.0552 | Lead (Hi-D) | 6.4 | 0.439 | 0.257 | 0.257 |
| Bismuth | 9.80 | 0.700 | 0.409 | 0.440 | Plate (avg) | 2.4 | 0.152 | 0.0862 | 0.0629 |
| Boral | 2.53 | 0.153 | 0.0865 | 0.0678 | Iron | 7.86 | 0.470 | 0.282 | 0.240 |
| Boron (amorphous) | 2.45 | 0.144 | 0.0791 | 0.0679 | Lead | 11.34 | 0.797 | 0.468 | 0.505 |
| Boron carbide | | | | | Lithium hydride | | | | |
| (hot pressed) | 2.5 | 0.150 | 0.0825 | 0.0675 | (pressed powder) | 0.70 | 0.0444 | 0.0239 | 0.0172 |
| Bricks | | | | | Lucite®(polymethyl | | | | |
| Fire clay | 2.05 | 0.129 | 0.0738 | 0.0543 | methacrylate) | 1.19 | 0.0816 | 0.0457 | 0.0317 |
| Kaolin | 2.1 | 0.132 | 0.0750 | 0.0552 | Paraffin | 0.89 | 0.0646 | 0.0360 | 0.0246 |
| Silica | 1.78 | 0.113 | 0.0646 | 0.0473 | Rocks | | | | |
| Carbon | 2.25 | 0.143 | 0.0801 | 0.0554 | Granite | 2.45 | 0.155 | 0.0887 | 0.0654 |
| Clay | 2.2 | 0.130 | 0.0801 | 0.0590 | Limestone | 2.91 | 0.187 | 0.109 | 0.0824 |
| Cements | | | | | Sandstone | 2.40 | 0.152 | 0.0871 | 0.0641 |
| Colemanite borated | 1.95 | 0.128 | 0.0725 | 0.0528 | Rubber | | | | |
| Plain (1 Portland | | | | | Butadiene | | | | |
| cement: 3 sand | | | | | copolymer | 0.915 | 0.0662 | 0.0370 | 0.0254 |
| mixture) | 2.07 | 0.133 | 0.0760 | 0.0559 | Natural | 0.92 | 0.0652 | 0.0364 | 0.0248 |
| Concretes | | | | | Neoprene | 1.23 | 0.0813 | 0.0462 | 0.0333 |
| Barytes | 3.5 | 0.213 | 0.127 | 0.110 | Sand | 2.2 | 0.140 | 0.0825 | 0.0587 |
| Barytes-boron frits | 3.25 | 0.199 | 0.119 | 0.101 | Type 347 | | | | |
| Barytes-limonite | 3.25 | 0.200 | 0.119 | 0.0991 | stainless steel | 7.8 | 0.462 | 0.279 | 0.236 |
| Barytes-lumnite- | | | | | Steel ($1^0{}_0$C) | 7.83 | 460 | 0.276 | 0.234 |
| colemanite | 3.1 | 0.189 | 0.112 | 0.0939 | Uranium | 18.7 | 1.46 | 0.813 | 0.881 |
| Iron-Portland[a] | 6.0 | 0.364 | 0.215 | 0.181 | Uranium hydride | 11.5 | 0.903 | 0.504 | 0.542 |
| MO (ORNL mixture) | 5.8 | 0.374 | 0.222 | 0.184 | Water | 1.0 | 0.0706 | 0.0396 | 0.0277 |
| Portland (1 cement: | | | | | Wood | | | | |
| 2 sand: 4 gravel | | | | | Ash | 0.51 | 0.0345 | 0.0193 | 0.0134 |
| mixture) | 2.2 | 0.141 | 0.0805 | 0.0592 | Oak | 0.77 | 0.0521 | 0.0293 | 0.0203 |
| | 2.4 | 0.154 | 0.0878 | 0.0646 | White pine | 0.67 | 0.0452 | 0.0253 | 0.0175 |

[a]Elemental composition, wt %: hydrogen, 1.0; oxygen, 52.9; silicon, 33.7; aluminum, 3.4; iron, 1.4; calcium, 4.4; magnesium, 0.2; carbon, 0.1; sodium, 1.6; potassium, 1.3.

[b]Composition, wt%: oxygen, 65.99; carbon, 18.27; hydrogen, 10.15; nitrogen, 3.05; calcium, 1.52; phosphorus, 1.02.

From Tipton, C. R., Jr., Ed., *Reactor Handbook,* 2nd ed., Vol. 1: *Materials,* Wiley Interscience, New York, 1960. Copyright © 1960 by John Wiley & Sons. With permission

## Table 11—17
## CRITICAL MASS AND ITS PARAMETERS

Critical mass, the minimum quantity of fissile material capable of sustaining a fission chain, varies greatly with reactor design (as in the 1- to 100-kg range); thus, no simple formula involving the parameters is possible. A number of theoretical models have been set up, and the calculation procedure described, but the ultimate answers are experimental.[a]

Among the factors or conditions determining criticality are

1. The fuel and fuel enrichment
2. Absorption and leakage of neutrons
3. Size and shape of the system.

More specifically, the parameters are set up in terms of more narrowly defined factors, such as the following:

$p$ = *resonance escape probability,* the fraction of source neutrons that escape capture while being slowed down to a particular energy level (the term "resonance" refers to the resonance region of the absorber)

$f$ = *thermal utilization,* the ratio of thermal neutrons absorbed in the fuel to the total thermal neutrons absorbed

$n$ = *liberation ratio,* the number of neutrons liberated per neutron absorbed in the fuel

$e$ = *fast-fission factor,* the ratio of fast neutrons slowing down, to those produced by thermal-neutron fissions

$k\infty$ = *infinite multiplication factor,* ratio of neutrons from fission to neutrons absorbed in the preceding generation, in a system of infinite size

**Values of $n$**

|  | Natural uranium | $U^{233}$ | $U^{235}$ | $Pu^{239}$ |
|---|---|---|---|---|
| Fast neutrons | 1.09 | 2.60 | 2.18 | 2.74 |
| Thermal neutrons | 1.33 | 2.27 | 2.06 | 2.10 |

The following table illustrates the wide variation in critical mass for cylindrical and slab cores, unreflected or water reflected, contained in either stainless steel or aluminum. It will be noted that the critical mass varies in the range of 1 to 46 kg.

[a]For extensive tables of critical mass data, see Sections 3 and 4 of the original source of this table.

## Table 11—17 (continued)
## CRITICAL MASS AND ITS PARAMETERS

Cylindrical and Slab Cores Containing $UO_2F_2$-$H_2O$ Solutions, Various Concentrations; Uranium about 93% Enriched in $U^{235}$

**ALUMINUM-WALLED CYLINDERS AND SLABS**

### Water-reflected | Bare (Continued)

| Core diameter, cm | $H/U^{235}$ atom ratio | $U^{235}$ concentration, $g/cm^3$ sol. | Critical core height, cm | Critical mass, $kg\ U^{235}$ | Core diameter, cm | $H/U^{235}$ atom ratio | $U^{235}$ concentration, $g/cm^3$ sol. | Critical core height, cm | Critical mass, $kg\ U^{235}$ |
|---|---|---|---|---|---|---|---|---|---|
| 15.2 | 27.1 | 0.8288 | 89.3 | 13.47 | 30.5 | 44.3 | 0.5376 | 23.2 | 9.1 |
| | 43.2 | 0.537 | 70.1 | 6.87 | | 50.1 | 0.480 | 22.6 | 7.92 |
| | 58.8 | 0.415 | 71.8 | 5.44 | | 55.4 | 0.437 | 22.7 | 7.25 |
| 16.5 | 26.2 | 0.827 | 44.5 | 7.91 | | 60.8 | 0.402 | 22.7 | 6.67 |
| | 44.3 | 0.5376 | 38.7 | 4.45 | | 331 | 0.0779 | 32.8 | 1.86 |
| | 78.7 | 0.315 | 42.6 | 2.87 | 38.1 | 27.1 | 0.8288 | 18.5 | 17.5 |
| | 119.0 | 0.212 | 52.6 | 2.39 | | 44.3 | 0.5376 | 17.9 | 11.0 |
| 20.3 | 29.9 | 0.759 | 20.7 | 5.09 | | 50.1 | 0.480 | 17.9 | 9.79 |
| | 49.5 | 0.488 | 18.8 | 2.97 | | 74.6 | 0.3314 | 16.8 | 6.4 |
| | 78.7 | 0.315 | 19.4 | 1.98 | | 169.0 | 0.151 | 18.5 | 3.18 |
| | 192.0 | 0.134 | 28.1 | 1.22 | | 328.7 | 0.0787 | 21.7 | 1.95 |
| | 290.0 | 0.0881 | 40.1 | 1.15 | | 331 | 0.0779 | 22.9 | 2.03 |
| 25.4 | 27.1 | 0.8288 | 12.4 | 5.2 | | 499 | 0.0522 | 27.4 | 1.63 |
| | 43.2 | 0.537 | 12.5 | 3.40 | | 755 | 0.0343 | 43.6 | 1.70 |
| | 51.5 | 0.470 | 11.4 | 2.72 | 50.8 | 27.1 | 0.8288 | 15.8 | 26.5 |
| | 127 | 0.199 | 14.4 | 1.45 | | 44.3 | 0.5376 | 15.0 | 16.3 |
| | 328.7 | 0.0787 | 22.4 | 0.893 | | 50.1 | 0.480 | 15.4 | 14.9 |
| | 499 | 0.0522 | 35.2 | 0.0930 | | 60.8 | 0.402 | 15.3 | 12.5 |
| 38.1 | 27.1 | 0.8288 | 7.7 | 7.3 | | 73.4 | 0.3370 | 15.2 | 10.4 |
| | 52.9 | 0.459 | 7.90 | 4.14 | | 325 | 0.0791 | 18.7 | 2.97 |
| | 221 | 0.116 | 11.30 | 1.49 | 76.2 | 44.3 | 0.5376 | 13.7 | 33.6 |
| | 499 | 0.0522 | 16.90 | 1.01 | | 50.1 | 0.480 | 13.8 | 30.2 |
| | 755 | 0.0343 | 27.10 | 1.02 | | 72.4 | 0.3423 | 13.9±0.5 | 21.6±0.7 |
| | 999 | 0.0260 | 44.30 | 1.31 | | 331 | 0.0779 | 16.3 | 5.79 |
| 76.2 | 27.1 | 0.8288 | 5.0±1 | 18.9±3.8 | 50.8 × 50.8 sq | 27.1 | 0.8288 | 15±1 | 32±2 |
| | 44.3 | 0.5376 | 4.8±1 | 11.8±2.5 | | 72.4 | 0.3423 | 14.3 | 12.6 |
| | 72.4 | 0.3423 | 5.5±1 | 8.6±1.6 | | 331 | 0.0779 | 17.9 | 3.60 |
| 50.8 × 50.8 sq. | 27.1 | 0.8288 | 6.3±1 | 13.5±2.2 | | | | | |
| | 44.3 | 0.5376 | 4.3±1 | 6.0±1.4 | | | | | |
| | 72.4 | 0.3423 | 6.2±1 | 5.5±0.9 | | | | | |

### Bare | Partially Water-Reflected[b]

| Core diameter, cm | $H/U^{235}$ atom ratio | $U^{235}$ concentration, $g/cm^3$ sol. | Critical core height, cm | Critical mass, $kg\ U^{235}$ | Core diameter, cm | $H/U^{235}$ atom ratio | $U^{235}$ concentration, $g/cm^3$ sol. | Critical core height, cm | Critical mass, $kg\ U^{235}$ |
|---|---|---|---|---|---|---|---|---|---|
| 22.3 | 44.3 | 0.5376 | 219 | 45.8 | 15.2 | 44.3 | 0.5376 | 75.0 | 7.34 |
| | 50.1 | 0.480 | 202.2 | 37.8 | 19.1 | 44.3 | 0.5376 | 25.7 | 3.93 |
| | 55.4 | 0.437 | 171.6 | 29.2 | 20.3 | 44.3 | 0.5376 | 23.6 | 4.12 |
| | 60.8 | 0.402 | 162.5 | 25.4 | | 51.5 | 0.470 | 23.8 | 3.64 |
| | 66.1 | 0.373 | 159.8 | 23.2 | | 72.4 | 0.3423 | 23.3 | 2.59 |
| | 71.5 | 0.350 | 163.2 | 22.2 | 25.4 | 43.2 | 0.537 | 17.3 | 4.71 |
| 25.4 | 27.1 | 0.8288 | 38.9 | 16.4 | | 72.4 | 0.3423 | 16.7 | 2.90 |
| | 44.3 | 0.5376 | 35.1 | 9.6 | 38.1 | 74.6 | 0.3314 | 12.0 | 4.5 |
| | 50.1 | 0.480 | 34.8 | 8.40 | 50.8 | 72.4 | 0.3423 | 10.6 | 7.3 |
| | 52.9 | 0.459 | 34.0 | 7.90 | 76.2 | 72.4 | 0.3423 | 9.2 | 14.4 |
| | 55.4 | 0.437 | 34.3 | 7.60 | 76.2 × 152.4[c] | 57.0 | 0.4240 | 8.4 | 41.5 |
| | 60.8 | 0.402 | 34.1 | 6.96 | | | | | |
| | 66.1 | 0.373 | 34.1 | 6.45 | | | | | |
| | 73.4 | 0.3370 | 33.7 | 5.8 | | | | | |
| | 83.1 | 0.300 | 34.4 | 5.22 | | | | | |
| | 169 | 0.151 | 41.2 | 3.15 | | | | | |
| | 328 | 0.0785 | 147.8 | 5.83 | | | | | |
| | 331 | 0.0779 | 170.1 | 6.72 | | | | | |

## Table 11—17 (continued)
## CRITICAL MASS AND ITS PARAMETERS

| Core diameter, cm | $H/U^{235}$ atom ratio | $U^{235}$ concentration, g/cm³ sol. | Critical core height, cm | Critical mass, kg $U^{235}$ | Core diameter, cm | $H/U^{235}$ atom ratio | $U^{235}$ concentration, g/cm³ sol. | Critical core height, cm | Critical mass, kg $U^{235}$ |
|---|---|---|---|---|---|---|---|---|---|

**STAINLESS STEEL CYLINDERS**

| | | *Bare* | | | | | *Water-Reflected (Continued)* | | |
|---|---|---|---|---|---|---|---|---|---|
| 22.9 | 74.6 | 0.3314 | 59.0 | 8.1 | 20.3 *(Cont.)* | 226 | 0.114 | 36.3 | 1.34 |
| | | | | | | 320 | 0.0805 | 60.1 | 1.57 |
| 25.4 | 43.9 | 0.538 | 32.3 | 8.80 | | | | | |
| | 62.7 | 0.396 | 31.7 | 6.36 | 25.4 | 31.6 | 0.724 | 15.3 | 5.61 |
| | 86.4 | 0.288 | 31.9 | 4.65 | | 43.9 | 0.538 | 14.9 | 4.06 |
| | 123.2 | 0.205 | 34.3 | 3.56 | | 62.7 | 0.396 | 15.2 | 3.05 |
| | 174 | 0.148 | 38.7 | 2.90 | | 86.4 | 0.288 | 15.4 | 2.25 |
| | 226 | 0.114 | 46.6 | 2.69 | | 123.2 | 0.205 | 16.8 | 1.74 |
| | | | | | | 174 | 0.148 | 18.1 | 1.36 |
| 30.5 | 174 | 0.148 | 24.7 | 2.67 | | 226 | 0.114 | 20.0 | 1.15 |
| | 226 | 0.114 | 26.2 | 2.18 | | 320 | 0.0805 | 25.0 | 1.02 |
| | 320 | 0.0805 | 30.3 | 1.78 | | | | | |
| | 499 | 0.0522 | 48.9 | 1.86 | 30.5 | 62.6 | 0.394 | 12.3 | 3.53 |
| | | | | | | 174 | 0.148 | 14.9 | 1.61 |
| 38.1 | 56.7 | 0.424 | 17.1 | 8.26 | | 226 | 0.114 | 16.5 | 1.37 |
| | 221 | 0.116 | 19.5 | 2.58 | | 320 | 0.0805 | 18.5 | 1.09 |
| | 499 | 0.0522 | 27.0 | 1.61 | | 499 | 0.0522 | 26.3 | 1.00 |
| | 755 | 0.0343 | 41.7 | 1.63 | | 755 | 0.0343 | 48.7 | 1.22 |
| 50.8 | 119 | 0.212 | 14.3 | 6.14 | 38.1 | 56.7 | 0.424 | 10.1 | 4.88 |
| | 221 | 0.116 | 15.7 | 3.69 | | 221 | 0.116 | 13.0 | 1.72 |
| | 329 | 0.0787 | 17.4 | 2.77 | | 499 | 0.0522 | 20.0 | 1.15 |
| | 499 | 0.0522 | 20.5 | 2.17 | | 755 | 0.0343 | 28.8 | 1.13 |
| | 755 | 0.0343 | 26.7 | 1.86 | | | | | |

| | | *Water-Reflected* | | | | *Cadmium-Lined (0.44 g/cm²) Water-Reflected* | | | |
|---|---|---|---|---|---|---|---|---|---|
| 15.2 | 44.3 | 0.5376 | 118.4 | 11.59 | 20.3 | 43.9 | 0.538 | 51.5 | 8.98 |
| 16.5 | 31.6 | 0.724 | 49.0 | 7.59 | | 62.7 | 0.396 | 48.4 | 6.21 |
| | 43.9 | 0.538 | 47.1 | 5.42 | | 86.4 | 0.288 | 56.5 | 5.28 |
| | 86.4 | 0.288 | 53.8 | 3.31 | 22.9 | 31.6 | 0.724 | 29.0 | 8.62 |
| 17.8 | 31.6 | 0.724 | 34.0 | 6.11 | | 43.9 | 0.538 | 28.8 | 6.36 |
| | 43.9 | 0.538 | 32.7 | 4.37 | | 62.7 | 0.396 | 28.3 | 4.61 |
| | 86.4 | 0.288 | 36.0 | 2.57 | | 86.4 | 0.288 | 29.2 | 3.45 |
| | 174 | 0.148 | 57.3 | 2.10 | | 123.2 | 0.205 | 32.0 | 2.69 |
| | | | | | | 174 | 0.148 | 37.3 | 2.26 |
| 20.3 | 31.6 | 0.724 | 22.6 | 5.31 | 25.4 | 31.6 | 0.724 | 21.1 | 7.74 |
| | 43.9 | 0.538 | 21.9 | 3.82 | | 43.9 | 0.538 | 21.4 | 5.83 |
| | 56.7 | 0.424 | 22.2 | 3.05 | | 62.7 | 0.396 | 22.0 | 4.42 |
| | 86.4 | 0.288 | 23.8 | 2.22 | | 86.4 | 0.288 | 22.0 | 3.21 |
| | 123.2 | 0.205 | 26.0 | 1.73 | | 221 | 0.116 | 28.7 | 1.69 |
| | 174 | 0.148 | 30.1 | 1.44 | | | | | |
| | | | | | 30.5 | 56.7 | 0.424 | 15.8 | 4.89 |
| | | | | | | 174 | 0.148 | 19.5 | 2.10 |
| | | | | | | 226 | 0.114 | 20.9 | 1.74 |
| | | | | | | 499 | 0.052 | 32.8 | 1.25 |

[a] Assemblies have no top reflector.
[b] This vessel was coated with Unichrome; others were coated with Heresite.

*From: "Reactor Physics Constants", 2nd ed., Argonne National Laboratory, ANL–5800, U.S. Atomic Energy Commission, July 1963.

Section 12

# Polymeric Materials

# Section 12

# POLYMERIC MATERIALS

**Table 12—1**
**ACETALS**

| Property | Homopolymer | | Copolymer, standard |
|---|---|---|---|
| | Standard | Toughened | |
| **Physical properties** | | | |
| Specific gravity | 1.42 | 1.34 | 1.41 |
| Water absorption (24 hr) % | 0.25 | — | 0.22 |
| **Mechanical properties** | | | |
| Tensile yield strength ($10^3$ psi) | 10.0 | 6.5 | 8.8 |
| Tensile modulus ($10^5$ psi) | 5.2 | — | 4.1 |
| Flexural yield strength ($10^3$ psi) | 14.1 | — | — |
| Flexural modulus ($10^5$ psi) | 4.1—4.5 | 2.0 | 3.75 |
| Compressive strength, 2% offset ($10^3$ psi) | 5.2 | — | — |
| Izod impact strength, notched (ft-lb/in) | 1.5 | 17 | 1.0—1.5 |
| **Electrical properties** | | | |
| Dielectric strength (V/mil) | 500 | 480 | 500 |
| Dielectric constant | | | |
| @ 60 Hz | 3.7 | — | 3.7 |
| @ 1 kHz | — | — | — |
| @ 1 MHz | 3.7 | — | 3.7 |
| Arc resistance (sec) | 129 | 120 | 240 |
| **Thermal properties** | | | |
| Thermal conductivity (Btu/hr/ft$^2$/°F/in) | 1.56 | — | 1.92 |
| Coefficient of thermal expansion ($10^{-5}$/°F) | 4.5 | 6.8 | 4.7 |
| Heat deflection temperature (°F) | | | |
| 66 psi load | 340 | 293 | 316 |
| 264 psi load | 255 | 194 | 230 |
| Continuous use temperature, no load (°F) | 195 | 200 | 220 |

**Table 12—2**
**ACRYLICS**

| Property | Cast | Molding | |
|---|---|---|---|
| | | Standard | High impact |
| **Physical properties** | | | |
| Specific gravity | 1.18 | 1.19 | 1.12—1.16 |
| Water absorption (24 hr) % | 0.3 | 0.3 | 0.03 |
| **Mechanical properties** | | | |
| Tensile yield strength ($10^3$ psi) | 9 | 10.5 | 7 |
| Tensile modulus ($10^5$ psi) | 4.0 | 4.25 | 2.8 |
| Flexural yield strength ($10^3$ psi) | 14 | 16 | 10.5 |
| Flexural modulus ($10^5$ psi) | 4.0 | 4.3 | 3.2 |
| Compressive strength, 2% offset ($10^3$ psi) | 16.5 | 16.0 | 8.0—12.0 |
| Izod impact strength, notched (ft-lb/in) | 0.4 | 0.4 | 0.8—2.0 |
| **Electrical properties** | | | |
| Dielectric strength (V/mil) | 500 | 400 | 400—500 |
| Dielectric constant | | | |
| @ 60 Hz | 3.3 | 3.3 | 3.9 |
| @ 1 kHz | — | — | — |
| @ 1 MHz | 3.0 | 2.3 | 2.5—3.0 |
| Arc resistance (sec) | — | — | — |
| **Thermal properties** | | | |
| Thermal conductivity (Btu/hr/ft$^2$/°F/in) | 1.44 | 1.44 | 1.44 |
| Coefficient of thermal expansion ($10^{-5}$/°F) | 4.5 | 4.5 | 5.0 |
| Heat deflection temperature (°F) | | | |
| 66 psi load | 225 | 220 | 190 |
| 264 psi load | 200 | 200 | 185 |
| Continuous use temperature, no load (°F) | 140—160 | 155—190 | — |

## Table 12—3
## ACRYLIC-STYRENE-ACRYLONITRILE (ASA)

| | General Purpose |
|---|---|
| Physical properties | |
| Specific gravity | 1.06 |
| Water absorption (24 hr) % | 0.25 |
| Mechanical properties | |
| Tensile yield strength ($10^3$ psi) | 5.8 |
| Tensile modulus ($10^5$ psi) | — |
| Flexural yield strength ($10^3$ psi) | 8 |
| Flexural modulus ($10^5$ psi) | 2.5 |
| Compressive strength, 2% offset ($10^3$ psi) | — |
| Izod impact strength, notched (ft-lb/in) | 10.0 |
| Electrical properties | |
| Dielectric strength (V/mil) | — |
| Dielectric constant | |
| @ 60 Hz | — |
| @ 1 kHz | — |
| @ 1 MHz | — |
| Arc resistance (sec) | — |
| Thermal properties | |
| Thermal conductivity (Btu/hr/ft²/°F/in) | — |
| Coefficient of thermal expansion ($10^{-5}$/°F) | 5.9 |
| Heat deflection temperature (°F) | |
| 66 psi load | — |
| 264 psi load | 190 |
| Continuous use temperature, no load (°F) | — |

## Table 12—4
## ACRYLONITRILE BUTADIENE STYRENE (ABS)

| | Type/grade | | | |
|---|---|---|---|---|
| | Heat resistant | Flame retardant | Medium impact | High impact |
| Physical properties | | | | |
| Specific gravity | 1.04—1.08 | 1.20—1.22 | 1.05—1.06 | 1.02—1.05 |
| Water absorption (24 hr) % | 0.30 | 0.2—0.6 | 0.2—0.45 | 0.2—0.45 |
| Mechanical properties | | | | |
| Tensile yield strength ($10^3$ psi) | 6.5—7.5 | 6.0—10.0 | 6.0—8.0 | 5.0—6.0 |
| Tensile modulus ($10^5$ psi) | 3.0—4.0 | 3.2—3.7 | 3.5—4.0 | 2.0—3.3 |
| Flexural yield strength ($10^3$ psi) | 10.0—13.0 | 9.0—12.0 | 10.0—11.5 | 6.0—10.5 |
| Flexural modulus ($10^5$ psi) | 3.0—4.2 | 3.0—3.4 | 3.6—4.0 | 2.0—3.4 |
| Compressive strength, 2% offset ($10^3$ psi) | 9.3—11.0 | — | 10.5—11.0 | — |
| Izod impact strength, notched (ft-lb/in) | 4.0—8.0 | 2.5—4.0 | 3.0—5.0 | 6.0—7.0 |
| Electrical properties | | | | |
| Dielectric strength (V/mil) | 350—500 | 350—500 | 385—500 | 300—400 |
| Dielectric constant | | | | |
| @ 60 Hz | 2.7—3.5 | — | 2.8—3.2 | 2.8—3.5 |
| @ 1 kHz | — | — | — | — |
| @ 1 MHz | 2.8—3.2 | — | 2.8—3.0 | 2.4—3.0 |
| Arc resistance (sec) | 50—85 | — | 85—119 | 90—130 |
| Thermal properties | | | | |
| Thermal conductivity (Btu/hr/ft²/°F/in) | 1.4—2.4 | — | 0.96—2.1 | 0.12—1.7 |
| Coefficient of thermal expansion ($10^{-5}$/°F) | 3.7—5.1 | 3.7—4.6 | 4.0—4.6 | 5.3—6.0 |
| Heat deflection temperature (°F) | | | | |
| 66 psi load | 205—245 | 200—245 | 210—220 | 203—225 |
| 264 psi load | 220—245 | 180—220 | 185—210 | 180—210 |
| Continuous use temperature, no load (°F) | — | 150 | — | — |

## Table 12—5
## ACRYLONITRILE BUTADIENE STYRENE/POLYVINYL CHLORIDE (ABS/PVC) — RIGID

| | |
|---|---|
| Physical properties | |
| Specific gravity | 1.21 |
| Water absorption (24 hr) % | 0.22—0.33 |
| Mechanical properties | |
| Tensile yield strength ($10^3$ psi) | 5.5—7.5 |
| Tensile modulus ($10^5$ psi) | 2.9 |
| Flexural yield strength ($10^3$ psi) | 9—11 |
| Flexural modulus ($10^5$ psi) | 3.0—4.3 |
| Compressive strength, 2% offset ($10^3$ psi) | 7.4 |
| Izod impact strength, notched (ft-lb/in) | 2.0—13 |
| Electrical properties | |
| Dielectric strength (V/mil) | 600 |
| Dielectric constant | |
| @ 60 Hz | — |
| @ 1 kHz | — |
| @ 1 MHz | — |
| Arc resistance (sec) | — |
| Thermal properties | |
| Thermal conductivity (Btu/hr/ft²/°F/in) | — |
| Coefficient of thermal expansion ($10^{-5}$/°F) | 4.6 |
| Heat deflection temperature (°F) | |
| 66 psi load | 220 |
| 264 psi load | 160—210 |
| Continuous use temperature, no load (°F) | — |

## Table 12—6
## ALKYDS AND THERMOSET CARBONATE

| | Type/grade | | |
|---|---|---|---|
| | Alkyds | | Allyl diglycol carbonate thermoset carbonate |
| | Granular | Glass reinforced | |
| Physical properties | | | |
| Specific gravity | 1.60—2.30 | 2.0—2.31 | 1.32 |
| Water absorption (24 hr) % | 0.05—0.50 | 0.03—0.5 | 0.20 |
| Mechanical properties | | | |
| Tensile yield strength ($10^3$ psi) | 3.0—9.0 | 4.0—9.5 | 5—6 |
| Tensile modulus ($10^5$ psi) | 24—29 | 20—28 | 3.0 |
| Flexural yield strength ($10^3$ psi) | 6.0—17 | 8.5—26 | — |
| Flexural modulus ($10^5$ psi) | 20 | 20 | 2.8 |
| Compressive strength, 2% offset ($10^3$ psi) | 20—35 | 24—30 | 22.5 |
| Izod impact strength, notched (ft-lb/in) | 0.3—0.5 | 0.5—16 | 0.2—0.4 |
| Electrical properties | | | |
| Dielectric strength (V/mil) | 350—450 | 250—530 | 290 |
| Dielectric constant | | | |
| @ 60 Hz | 5.5—6.0 | 5.2—6.0 | 4.4 |
| @ 1 kHz | — | — | — |
| @ 1 MHz | 4.0—6.0 | 4.5—5.0 | 3.5—3.8 |
| Arc resistance (sec) | 180—240 | 180 | 185 |
| Thermal properties | | | |
| Thermal conductivity (Btu/hr/ft²/°F/in) | 4.2—7.2 | 2.4—3.6 | 1.45 |
| Coefficient of thermal expansion ($10^{-5}$/°F) | 1—3 | 1—3 | 6.0 |
| Heat deflection temperature (°F) | | | |
| 66 psi load | — | — | — |
| 264 psi load | 350—500 | 400—500 | — |
| Continuous use temperature, no load (°F) | 300 | 300 | 212 |

## Table   12—7
## AMINO TYPES

| | Type/grade and filler | | |
|---|---|---|---|
| | Urea—alpha cellulose | Melamine—alpha cellulose | Melamine—glass fiber |
| Physical properties | | | |
| Specific gravity | 1.47—1.52 | 1.47—1.52 | 1.8—2.0 |
| Water absorption (24 hr) % | 0.4—0.8 | 0.1—0.6 | 0.09—0.21 |
| Mechanical properties | | | |
| Tensile yield strength ($10^3$ psi) | 5.5—7.0 | 7.0—13.0 | 5.0—10.0 |
| Tensile modulus ($10^5$ psi) | 13—14 | 13.5 | 24 |
| Flexural yield strength ($10^3$ psi) | 11.0—18.0 | 12.0—15.0 | 13.0—24.0 |
| Flexural modulus ($10^5$ psi) | 14—15 | 11 | 24 |
| Compressive strength, 2% offset ($10^3$ psi) | — | — | — |
| Izod impact strength, notched (ft-lb/in) | 0.27—0.34 | 0.24—0.35 | 0.6—1.8 |
| Electrical properties | | | |
| Dielectric strength (V/mil) | 330—370 | 270—300 | 170—300 |
| Dielectric constant | | | |
| @ 60 Hz | — | — | — |
| @ 1 kHz | — | 7.8—9.2 | — |
| @ 1 MHz | — | — | — |
| Arc resistance (sec) | 100—135 | 125—136 | 180—186 |
| Thermal properties | | | |
| Thermal conductivity (Btu/hr/ft²/°F/in) | 10.1 | 7.0—10.1 | 11.5 |
| Coefficient of thermal expansion ($10^{-5}$/°F) | 2.2—3.6 | 2.0—5.7 | 1.5—1.7 |
| Heat deflection temperature (°F) | | | |
| 66 psi load | — | — | — |
| 264 psi load | 266 | 360 | 400 |
| Continuous use temperature, no load (°F) | — | — | — |

## Table   12—8
## CELLULASE BASE

| | Type/grade | | | |
|---|---|---|---|---|
| | Acetate | Acetate butyrate (CAB) | Acetate propionate | Ethyl cellulose |
| Physical properties | | | | |
| Specific gravity | 1.22—1.34 | 1.15—1.22 | 1.16—1.24 | 1.09—1.17 |
| Water absorption (24 hr) % | 1.7—4.5 | 0.9—2.2 | 1.2—2.8 | 0.8—1.8 |
| Mechanical properties | | | | |
| Tensile yield strength ($10^3$ psi) | 2.2—6.9 | 1.4—6.2 | 1.4—7.2 | 3.0—4.8 |
| Tensile modulus ($10^5$ psi) | 0.65—4.0 | 0.5—2.0 | 0.6—2.15 | 2.2—2.5 |
| Flexural yield strength ($10^3$ psi) | 2.5—10.4 | 1.8—9.2 | 1.7—10.6 | 4.7—6.8 |
| Flexural modulus ($10^5$ psi) | 1.2—3.6 | 0.9—3.0 | 1.15—3.7 | — |
| Compressive strength, 2% offset ($10^3$ psi) | — | — | — | — |
| Izod impact strength, notched (ft-lb/in) | 1.0—7.3 | 1.1—9.1 | 1.0—10.3 | 3.0—8.0 |
| Electrical properties | | | | |
| Dielectric strength (V/mil) | 250—600 | 250—400 | 300—500 | 350—500 |
| Dielectric constant | | | | |
| @ 60 Hz | 3.2—7.0 | 3.4—6.4 | 3.3—4.0 | 3.0—4.1 |
| @ 1 kHz | 5.1 | 4.7 | 3.55 | — |
| @ 1 MHz | — | — | — | — |
| Arc resistance (sec) | 50—310 | — | 175—190 | 60—80 |
| Thermal properties | | | | |
| Thermal conductivity (Btu/hr/ft²/°F/in) | 1.74 | 0.15 | 0.145 | — |
| Coefficient of thermal expansion ($10^{-5}$/°F) | 6.7 | 7.5 | 7.5 | — |
| Heat deflection temperature (°F) | | | | |
| 66 psi load | 120—209 | 130—227 | 147—250 | 170—180 |
| 264 psi load | 111—195 | 113—202 | 111—228 | 115—174 |
| Continuous use temperature, no load (°F) | — | — | — | — |

**Table   12—9**
## DIALLYL PHTHALATES

| | Type/grade | | | |
|---|---|---|---|---|
| | **Orlon filled** | **Dacron filled** | **Asbestos filled** | **Glass fiber filled** |
| **Physical properties** | | | | |
| Specific gravity | 1.3—1.35 | 1.4—1.6 | 1.5—1.87 | 1.6—1.9 |
| Water absorption (24 hr) % | 0.2—0.5 | 0.2—0.5 | 0.2—0.7 | 0.05—0.2 |
| **Mechanical properties** | | | | |
| Tensile yield strength ($10^3$ psi) | 4.5—6.0 | 4.6—5.5 | 4.0—10 | 5.5—9.0 |
| Tensile modulus ($10^5$ psi) | 6.0 | 6.0 | 1.2 | — |
| Flexural yield strength ($10^3$ psi) | 7.5—10.5 | 9—11.5 | 10—24 | 16—18 |
| Flexural modulus ($10^5$ psi) | 6.4 | 6.4 | 19 | 16—17 |
| Compressive strength, 2% offset ($10^3$ psi) | 20—25 | 20—30 | 18—25 | 25—31 |
| Izod impact strength, notched (ft-lb/in) | 0.5—1.2 | 1.7—4.5 | 0.5—3.6 | 0.6—6.0 |
| **Electrical properties** | | | | |
| Dielectric strength (V/mil) | 400 | 390 | 400—450 | 385—400 |
| Dielectric constant | | | | |
| @ 60 Hz | — | — | — | — |
| @ 1 kHz | 0.008 | 0.008 | 0.003—0.008 | 0.004—0.006 |
| @ 1 MHz | — | — | — | — |
| Arc resistance (sec) | 115 | 125 | 125—180 | 125—140 |
| **Thermal properties** | | | | |
| Thermal conductivity (Btu/hr/ft²/°F/in) | — | — | — | — |
| Coefficient of thermal expansion ($10^{-5}$/°F) | 5.0 | 5.2 | 4.0 | 2.2—2.6 |
| Heat deflection temperature (°F) | | | | |
| 66 psi load | — | — | — | — |
| 264 psi load | 240—266 | 270—290 | 300—500 | 325—500 |
| Continuous use temperature, no load (°F) | 300 | 300—370 | 350—450 | 400—450 |

**Table 12—10**
**EPOXIES**

| | Type/grade | | | | | |
|---|---|---|---|---|---|---|
| | Bisphenol A | | | Novolacs | | Cycloaliphatic — cast, rigid |
| | Cast, rigid | Cast, flexible | Molded, glass reinforced | Cast, rigid | Molded | |
| **Physical properties** | | | | | | |
| Specific gravity | 1.15 | 1.14—1.18 | 1.6—2.0 | 1.24 | 1.7 | 1.16—1.21 |
| Water absorption (24 hr) % | 0.1—0.2 | 0.4—1.0 | 0.05—0.20 | 0.4 | 0.11—0.2 | 0.08—0.15 |
| **Mechanical properties** | | | | | | |
| Tensile yield strength ($10^3$ psi) | 9.5—11.5 | 1.4—7.6 | 10—20 | 8—12 | 5.2—5.3 | 10—20 |
| Tensile modulus ($10^5$ psi) | 4.5 | 0.5—2.5 | 30.4 | 4—5 | — | 5—9 |
| Flexural yield strength ($10^3$ psi) | 14—18 | 1.2—12.7 | 10—60 | 11—16 | 10—12 | 15—32 |
| Flexural modulus ($10^5$ psi) | 4.5—5.4 | 0.4—3.9 | 15—25 | 4—5 | — | 4.4—4.8 |
| Compressive strength, 2% offset ($10^3$ psi) | 16—24 | — | 34—38 | 16—24 | 22—26 | 30—50 |
| Izod impact strength, notched (ft-lb/in) | 0.2—0.5 | 0.3—2.0 | 2—30 | 0.5 | 0.3—0.5 | 0.2—1.0 |
| **Electrical properties** | | | | | | |
| Dielectric strength (V/mil) | >400 | 400—410 | 300—400 | — | 280—400 | 420—440 |
| Dielectric constant | | | | | | |
| @ 60 Hz | 4.02 | 4.4—4.8 | 4.4—5.4 | 4.0 | 4.7—5.7 | 3.4 |
| @ 1 kHz | — | — | 3.5—5.0 | — | — | — |
| @ 1 MHz | 3.42 | 2.8—3.5 | 4.1—4.6 | 3.5—3.6 | 4.3—4.8 | 3.9—4.5 |
| Arc resistance (sec) | 100 | 75—98 | 120—180 | — | 180—185 | 120—180 |
| **Thermal properties** | | | | | | |
| Thermal conductivity (Btu/hr/ft²/°F/in) | 1.2—3.6 | — | 1.2—6.0 | — | — | — |
| Coefficient of thermal expansion ($10^{-5}$/°F) | 3.3 | 3.5 | 1—2 | — | 1.7—2.2 | 1.6—3.0 |
| Heat deflection temperature (°F) | | | | | | |
| 66 psi load | — | — | — | — | — | — |
| 264 psi load | 230 | 90—125 | 250—500 | 300—400 | 300—425 | 300—450 |
| Continuous use temperature, no load (°F) | 175—190 | 100—125 | 340—400 | 450 | 450—500 | 450—500 |

## Table 12—11
## FLUOROCARBONS

| | Type/grade | | | | | |
|---|---|---|---|---|---|---|
| | Polytetrafluoroethylene (PTFE or TFE) (Teflon) | Fluorinated ethylene propylene (FEP) | Perfluoroalkoxy (PFA) | Polyvinylidene-fluoride (PVDF or PVF) | Polychlorotrifluoro-ethylene (CTFE) | Ethylenetetrafluoro-ethylene (ETFE) |
| **Physical properties** | | | | | | |
| Specific gravity | 2.13—2.24 | 2.12—2.17 | 2.12—2.17 | 1.75—1.78 | 2.13 | 1.70 |
| Water absorption (24 hr) % | 0.01 | 0.01 | 0.03 | 0.04 | 0.01 | <0.03 |
| **Mechanical Properties** | | | | | | |
| Tensile yield strength ($10^3$ psi) | 3.35 | 3.0 | 4.0 | 5.2—7.4 | 5.4 | 6.5 |
| Tensile modulus ($10^5$ psi) | 0.5 | 0.5 | — | 1.7—2.0 | 1.9 | 1.2 |
| Flexural yield strength ($10^3$ psi) | — | — | — | 2 | 10.7 | No break |
| Flexural modulus ($10^5$ psi) | 0.5—0.9 | 0.95 | 1.0 | 2.0 | 2.5 | 2.0 |
| Compressive strength, 2% offset ($10^3$ psi) | 0.7—1.8 | 1.6 | — | 12.8—14.2 | — | 2.0 |
| Izod impact strength, notched (ft-lb/in) | 2.5—4.0 | No break | — | 3.8 | 3.1 | No break |
| **Electrical properties** | | | | | | |
| Dielectric strength (V/mil) | 500—600 | 500—600 | 500—600 | 260 | 490 | 400—500 |
| Dielectric constant | | | | | | |
| @ 60 Hz | 2.1 | 2.1 | 2.1 | 10.0 | — | 2.6 |
| @ 1 kHz | 2.1 | 2.1 | 2.1 | 7.5 | 2.5 | 2.6 |
| @ 1 MHz | 2.1 | 2.1 | 2.1 | 7.5 | — | 2.5 |
| Arc resistance (sec) | >200 | >180 | >180 | 50—70 | 360 | 75 |
| **Thermal Properties** | | | | | | |
| Thermal conductivity (Btu/hr/ft$^2$/°F/in) | 1.7 | 1.4 | 1.8 | 1.7 | 1.8 | 1.6 |
| Coefficient of thermal expansion ($10^{-5}$/°F) | 5.5—8.4 | 5.3—10.7 | 5.2—6.7 | 8.0—8.5 | 4.8—15 | 5—7.8 |
| Heat deflection temperature (°F) | | | | | | |
| 66 psi load | 250 | 158 | 164 | 300 | 265 | 220 |
| 264 psi load | 132 | 124 | 118 | 195 | 167 | 165 |
| Continuous use temperature, no load (°F) | 550 | 400 | 500 | 340 | 380 | 300—355 |

## Table 12—12
## NYLON (POLYAMIDES)

| | Type 6 | | | Type 6/6 | | | | | |
| | General purpose | 30% Glass reinforced | Cast | General purpose | | High impact | Type 6/9 | Type 6/12 | Type 12 |
| | | | | Molding | Extrusion | | | | |
|---|---|---|---|---|---|---|---|---|---|
| Physical properties | | | | | | | | | |
| Specific gravity | 1.13 | 1.37 | 1.15 | 1.4 | 1.13 | 1.09 | 1.07 | 1.06 | 1.01 |
| Water absorption (24 hr) % | 1.6 | 1.3 | 0.9 | 1.2 | 1.5 | — | 0.48 | 0.25 | 0.25 |
| Mechanical properties | | | | | | | | | |
| Tensile yield strength ($10^3$ psi) | 11.8 | 13—25 | 11—14 | 12 | 12.6 | — | 8.5 | 8.8 | 5.5 |
| Tensile modulus ($10^5$ psi) | 3.8 | 10—14.5 | 3.5—4.5 | 4.2 | — | — | — | 2.9 | 1.7—2.0 |
| Flexural yield strength ($10^3$ psi) | — | 26—34 | 17.5 | — | — | — | 11.8 | — | — |
| Flexural modulus ($10^5$ psi) | 4.0 | 10—15 | 5.0 | 4.1 | 1.8 | 1.3 | 3.0 | 2.9 | — |
| Compressive strength, 2% offset ($10^3$ psi) | 9.7 | 19—20 | 14 | 4.9 | 4.9 | 1.9 | — | — | — |
| Izod impact strength, notched (ft-lb/in) | 1.0 | 2.3—3.0 | 0.9 | 1.0 | 1.3 | 15 | 1.1 | 1.0 | 1.2 |
| Electrical properties | | | | | | | | | |
| Dielectric strength (V/mil) | 400 | 400—450 | 380 | 600 | — | 390 | 540 | 400 | 840 |
| Dielectric constant | | | | | | | | | |
| @ 60 Hz | 4.0—5.3 | 4.6—5.6 | 4.0 | 4.0 | — | 3.2 | 3.6 | 4.0 | — |
| @ 1 kHz | 3.7 | — | 3.7 | 3.9 | — | — | — | 4.0 | 3.6 |
| @ 1 MHz | 3.9—5.4 | 3.9—5.4 | 3.3 | 3.6 | — | 3.1 | 3.2 | 3.5 | — |
| Arc resistance (sec) | — | 93 | — | 116 | 120 | 72 | — | 121 | — |
| Thermal properties | | | | | | | | | |
| Thermal conductivity (Btu/hr/ft²/°F/in) | 1.7 | 1.2—1.7 | 1.7 | 1.7 | 1.7 | — | 1.5 | 1.5 | 1.7 |
| Coefficient of thermal expansion ($10^{-5}$/°F) | 4.5 | 1.2—3.0 | 5.0 | 4.0 | — | — | 8.3 | 5.0 | 7.2 |
| Heat deflection temperature (°F) | | | | | | | | | |
| 66 psi load | 455 | 425—430 | 425 | 455 | 470 | 420 | 330 | 356 | — |
| 264 psi load | 196 | 420 | 300—425 | 194 | 220 | 160 | 140 | 194 | — |
| Continuous use temperature, no load (°F) | — | — | — | 250—300 | 250—300 | — | 255 | 350 | — |

**Table 12—13**
**PHENOLICS**

| | Type/grade and filler | | | | | |
|---|---|---|---|---|---|---|
| | General purpose—Wood-flour | Impact—Chopped fabric/cord | Electrical—Mineral | Heat resistant—Mineral | Reinforced—Glass | Special purpose—Various |
| **Physical properties** | | | | | | |
| Specific gravity | 1.36—1.46 | 1.31—1.41 | 1.36—1.75 | 1.41—1.84 | 1.7—2.0 | 1.37—1.75 |
| Water absorption (24 hr) % | 0.6—0.7 | 0.6—0.9 | 0.05—0.20 | 0.30—0.35 | 0.05—0.2 | 0.20—0.40 |
| **Mechanical properties** | | | | | | |
| Tensile yield strength ($10^3$ psi) | 6.5—7.0 | 6.0—7.0 | 5.0—7.0 | 5.0—6.0 | 6.0—12.0 | 7.0—9.0 |
| Tensile modulus ($10^5$ psi) | 8—13 | 9—14 | 10—30 | — | 30—33 | — |
| Flexural yield strength ($10^3$ psi) | 9.0—11.0 | 10.0 | 9.0—11.0 | 10 | 12—24 | 9.5 |
| Flexural modulus ($10^5$ psi) | 11—13 | 12 | 17—25 | 14 | 20—30 | 10 |
| Compressive strength, 2% offset ($10^3$ psi) | 22—36 | 15—30 | 20 | 26.4 | 17—30 | — |
| Izod impact strength, notched (ft-lb/in) | 0.30—0.35 | 0.6—1.05 | 0.28—0.45 | 0.26 | 0.4—1.5 | 0.5 |
| **Electrical properties** | | | | | | |
| Dielectric strength (V/mil) | 350 | 350—400 | 400 | 170 | 400 | 175 |
| Dielectric constant | | | | | | |
| @ 60 Hz | 5.0—9.0 | 6.5—15.0 | 7.4 | — | 7.1—7.2 | — |
| @ 1 kHz | 5.2—5.3 | 5.2—5.4 | 4.9—6.5 | 11.7 | 4.4 | 7.8 |
| @ 1 MHz | 4.0—7.0 | 4.5—7.0 | 5.0 | 3.7 | 4.6—6.6 | — |
| Arc resistance (sec) | 100 | 50 | 184 | 181 | 181 | — |
| **Thermal properties** | | | | | | |
| Thermal conductivity (Btu/hr/ft$^2$/°F/in) | 0.10—0.3 | 0.10—0.17 | 0.24—0.34 | — | 0.20 | 0.50 |
| Coefficient of thermal expansion ($10^{-5}$/°F) | 1.7—2.5 | 1.6—2.2 | 2.60 | 2.8 | 1.8 | 3.6 |
| Heat deflection temperature (°F) | | | | | | |
| 66 psi load | — | — | — | — | — | — |
| 264 psi load | 275—360 | 270—500 | 310—400 | 530 | 600 | — |
| Continuous use temperature, no load (°F) | 300—350 | 250—300 | 400 | — | 350—450 | — |

Table 12—14
PHENYLENE OXIDE

| | Type/grade | | | |
|---|---|---|---|---|
| | Standard | Glass reinforced | Extrusion | Platable |
| Physical properties | | | | |
| Specific gravity | 1.06—1.10 | 1.21—1.36 | 1.06—1.10 | 1.05 |
| Water absorption (24 hr) % | 0.07 | 0.07 | 0.07 | 0.07 |
| Mechanical properties | | | | |
| Tensile yield strength ($10^3$ psi) | 7.0—9.0 | 14.5—17.8 | 7.8—11.0 | 7.0 |
| Tensile modulus ($10^5$ psi) | 3.6—3.8 | 9.3—12.0 | 3.6—3.8 | — |
| Flexural yield strength ($10^3$ psi) | 8.2—15.0 | 18.5—20.0 | 12.8—16.0 | 9.8 |
| Flexural modulus ($10^5$ psi) | 3.0—3.6 | 7.5—11 | 3.6 | 3.0 |
| Compressive strength, 2% offset ($10^3$ psi) | — | — | — | — |
| Izod impact strength, notched (ft-lb/in) | 5.0—10.0 | 2.3 | 3.1 | 5.0 |
| Electrical properties | | | | |
| Dielectric strength (V/mil) | 400—630 | 420—600 | 400—550 | — |
| Dielectric constant | | | | |
| @ 60 Hz | — | — | — | — |
| @ 1 kHz | — | — | — | — |
| @ 1 MHz | 2.64—2.68 | 2.85—3.11 | 2.6—2.7 | — |
| Arc resistance (sec) | 75 | 70—120 | 75 | — |
| Thermal properties | | | | |
| Thermal conductivity (Btu/hr/ft²/°F/in) | 1.50 | 1.10 | 1.5 | 3.5 |
| Coefficient of thermal expansion ($10^{-5}$/°F) | 3.3—3.8 | 1.4—2.0 | 3.3—3.8 | — |
| Heat deflection temperature (°F) | | | | |
| 66 psi load | 220—315 | 280—317 | 230—315 | — |
| 264 psi load | 190—300 | 270—300 | 185—300 | 235 |
| Continuous use temperature, no load (°F) | — | — | — | — |

Table 12—15
PHENYLENE ETHER/PHENYLENE OXIDE (PPE/PPO)

| | Type/grade | |
|---|---|---|
| | Unreinforced | Glass fiber reinforced (30% reinf.) |
| Physical properties | | |
| Specific gravity | 1.06—1.18 | 1.27 |
| Water absorption (24 hr) % | 0.06—0.07 | 0.06 |
| Mechanical properties | | |
| Tensile yield strength ($10^3$ psi) | 7.8—9.6 | 17.0 |
| Tensile modulus ($10^5$ psi) | 3.5—3.8 | 9.3—13.3 |
| Flexural yield strength ($10^3$ psi) | 7.3—13.5 | 22 |
| Flexural modulus ($10^5$ psi) | 3.6 | 10.4 |
| Compressive strength, 2% offset ($10^3$ psi) | 12—16.4 | 17.6—17.9 |
| Izod impact strength, notched (ft-lb/in) | 5.0—7.0 | 2.3 |
| Electrical properties | | |
| Dielectric strength (V/mil) | 400—500 | 1020 |
| Dielectric constant | | |
| @ 60 Hz | 2.65—2.69 | 3—11 |
| @ 1 kHz | — | — |
| @ 1 MHz | 2.64—2.68 | 3.22 |
| Arc resistance (sec) | 75 | 120 |
| Thermal properties | | |
| Thermal conductivity (Btu/hr/ft²/°F/in) | 1.1—1.5 | 1.1 |
| Coefficient of thermal expansion ($10^{-5}$/°F) | 3.3—4.4 | 1.4 |
| Heat deflection temperature (°F) | | |
| 66 psi load | 230—279 | 317 |
| 264 psi load | 200—265 | 310 |
| Continuous use temperature, no load (°F) | 212 | — |

**Table 12—16**
**POLY(AMIDE—IMIDE) (PAI)**

| | Type/grade | |
|---|---|---|
| | General purpose | Glass fiber reinforced |
| Physical properties | | |
| Specific gravity | 1.38 | 1.56 |
| Water absorption (24 hr) % | 0.33 | 0.24 |
| Mechanical properties | | |
| Tensile yield strength ($10^3$ psi) | 27.8 | 29.7 |
| Tensile modulus ($10^5$ psi) | — | — |
| Flexural yield strength ($10^3$ psi) | 34.9 | 48.3 |
| Flexural modulus ($10^5$ psi) | 7.3 | 17.0 |
| Compressive strength, 2% offset ($10^3$ psi) | 32.1 | 38.3 |
| Izod impact strength, notched (ft-lb/in) | 2.7 | 1.5 |
| Electrical properties | | |
| Dielectric strength (V/mil) | 580 | 840 |
| Dielectric constant | | |
| @ 60 Hz | — | — |
| @ 1 kHz | 3.9—4.2 | 4.4—6.5 |
| @ 1 MHz | 3.9—4.2 | 4.4—6.5 |
| Arc resistance (sec) | — | — |
| Thermal properties | | |
| Thermal conductivity (Btu/hr/ft²/°F/in) | — | — |
| Coefficient of thermal expansion ($10^{-5}$/°F) | 1.7 | 0.9 |
| Heat deflection temperature (°F) | | |
| 66 psi load | — | — |
| 264 psi load | 532 | 539 |
| Continuous use temperature, no load (°F) | — | — |

**Table 2—17**
**POLYARYLATES**

| | Type/grade | |
|---|---|---|
| | General purpose | Flame retardant |
| Physical properties | | |
| Specific gravity | 1.19—1.20 | 1.22 |
| Water absorption (24 hr) % | 0.1—0.2 | 0.1 |
| Mechanical properties | | |
| Tensile yield strength ($10^3$ psi) | 9.9—10 | 10.1 |
| Tensile modulus ($10^5$ psi) | 290 | 300 |
| Flexural yield strength ($10^3$ psi) | 11—15 | 12.3 |
| Flexural modulus ($10^5$ psi) | 3.0—3.2 | 3.1 |
| Compressive strength, 2% offset ($10^3$ psi) | 12.9 | — |
| Izod impact strength, notched (ft-lb/in) | 4.0—5.5 | 3.8 |
| Electrical properties | | |
| Dielectric strength (V/mil) | 610 | 585 |
| Dielectric constant | | |
| @ 60 Hz | 3.1 | — |
| @ 1 kHz | — | — |
| @ 1 MHz | 3.0—3.1 | 3.1 |
| Arc resistance (sec) | 78—125 | 51 |
| Thermal properties | | |
| Thermal conductivity (Btu/hr/ft²/°F/in) | 1.5 | 1.4 |
| Coefficient of thermal expansion ($10^{-5}$/°F) | 2.7—4.0 | 3.0 |
| Heat deflection temperature (°F) | | |
| 66 psi load | 340—345 | 334 |
| 264 psi load | 311—320 | 309 |
| Continuous use temperature, no load (°F) | 265 | — |

## Table 12—18
### POLYARYL SULFONE (PAS) AND POLYBUTADIENE

| | Material | |
|---|---|---|
| | Polyaryl sulfone (PAS) | Polybutadiene |
| Physical properties | | |
| Specific gravity | 1.37 | 1.6—2.0 |
| Water absorption (24 hr) % | 0.10 | 0.01 |
| Mechanical properties | | |
| Tensile yield strength ($10^3$ psi) | 12 | 5—12 |
| Tensile modulus ($10^5$ psi) | 3.85 | 4.0—12 |
| Flexural yield strength ($10^3$ psi) | 16.1 | 7—21 |
| Flexural modulus ($10^5$ psi) | 4.0 | 7.0—18 |
| Compressive strength, 2% offset ($10^3$ psi) | — | 11—20 |
| Izod impact strength, notched (ft-lb/in) | 1.6 | 1—10 |
| Electrical properties | | |
| Dielectric strength (V/mil) | 380 | 400—600 |
| Dielectric constant | | |
| @ 60 Hz | 3.5 | 3.3 |
| @ 1 kHz | — | — |
| @ 1 MHz | 3.5 | 3.3 |
| Arc resistance (sec) | 81 | — |
| Thermal properties | | |
| Thermal conductivity (Btu/hr/ft²/°F/in) | — | — |
| Coefficient of thermal expansion ($10^{-5}$/°F) | — | — |
| Heat deflection temperature (°F) | | |
| 66 psi load | — | — |
| 264 psi load | 400 | 500 |
| Continuous use temperature, no load (°F) | — | 350—500 |

## Table 12—19
### POLYCARBONATES (PC)

| | Type/grade | |
|---|---|---|
| | General purpose | High modulus |
| Physical properties | | |
| Specific gravity | 1.2 | 1.25 |
| Water absorption (24 hr) % | 0.15 | 0.12 |
| Mechanical properties | | |
| Tensile yield strength ($10^3$ psi) | 9.0—10.5 | 8.0—9.0 |
| Tensile modulus ($10^5$ psi) | 3.4 | 4.5 |
| Flexural yield strength ($10^3$ psi) | 11.0—15.0 | 15.0 |
| Flexural modulus ($10^5$ psi) | 3.0—3.4 | 5.0 |
| Compressive strength, 2% offset ($10^3$ psi) | 10—12.5 | 14 |
| Izod impact strength, notched (ft-lb/in) | 12—16 | 2 |
| Electrical properties | | |
| Dielectric strength (V/mil) | 380—400 | 450 |
| Dielectric constant | | |
| @ 60 Hz | 3.0—3.2 | 3.1 |
| @ 1 kHz | 3.0 | — |
| @ 1 MHz | 3.0—3.05 | 3.05 |
| Arc resistance (sec) | 10—120 | 5—120 |
| Thermal properties | | |
| Thermal conductivity (Btu/hr/ft²/°F/in) | 1.35 | 1.41 |
| Coefficient of thermal expansion ($10^{-5}$/°F) | 6.6—7.0 | 3.2 |
| Heat deflection temperature (°F) | | |
| 66 psi load | 280 | 295 |
| 264 psi load | 260—270 | 288 |
| Continuous use temperature, no load (°F) | — | — |

**Table   12—20**

**GLASS FIBER-REINFORCED AND POLYPHTHALATE CARBONATES**

| | Type/grade | |
|---|---|---|
| | Polycarbonate 40% glass fiber reinforced | Polyphthalate carbonate (PPC) |
| Physical properties | | |
| Specific gravity | 1.52 | 1.2 |
| Water absorption (24 hr) % | 0.12 | 0.16 |
| Mechanical properties | | |
| Tensile yield strength ($10^3$ psi) | 23 | 9.5 |
| Tensile modulus ($10^5$ psi) | 16.8 | — |
| Flexural yield strength ($10^3$ psi) | 27 | 13.8 |
| Flexural modulus ($10^5$ psi) | 14.0 | 2.94 |
| Compressive strength, 2% offset ($10^3$ psi) | 21 | — |
| Izod impact strength, notched (ft-lb/in) | 2.5 | 10 |
| Electrical properties | | |
| Dielectric strength (V/mil) | 450 | 515 |
| Dielectric constant | | |
| @ 60 Hz | 3.5 | 3.1 |
| @ 1 kHz | — | — |
| @ 1 MHz | 3.5 | 3.0 |
| Arc resistance (sec) | 120 | — |
| Thermal properties | | |
| Thermal conductivity (Btu/hr/ft²/°F/in) | 1.5 | 1.5 |
| Coefficient of thermal expansion ($10^{-5}$/°F) | 0.93 | 5.1 |
| Heat deflection temperature (°F) | | |
| 66 psi load | 310 | 320 |
| 264 psi load | 295 | 305 |
| Continuous use temperature, no load (°F) | — | — |

## Table 12—21
## POLYCARBONATE BLENDS

| | Type/grade | | | |
|---|---|---|---|---|
| | **ABS/PC** | **PBT/PC** | **PC/PET** | **PC/PU** |
| Physical properties | | | | |
| Specific gravity | 1.1—1.2 | 1.2—1.3 | 1.17—1.20 | 1.21 |
| Water absorption (24 hr) % | 0.1—0.2 | 0.1—0.14 | 0.12—0.16 | — |
| Mechanical properties | | | | |
| Tensile yield strength ($10^3$ psi) | 7.7—8.6 | 6.5—8.0 | 7.0—8.3 | 4.5—6.0 |
| Tensile modulus ($10^5$ psi) | 3.5—3.9 | — | — | — |
| Flexural yield strength ($10^3$ psi) | 12—14 | 10.5—12.5 | 10.9—11.5 | — |
| Flexural modulus ($10^5$ psi) | 3.2—3.8 | 2.75—3.25 | 3.0 | 0.3—1.5 |
| Compressive strength, 2% offset ($10^3$ psi) | — | 8.6 | — | — |
| Izod impact strength, notched (ft-lb/in) | 8.0—10.5 | 13—16 | 15—18 | No break |
| Electrical properties | | | | |
| Dielectric strength (V/mil) | 434—760 | 570—625 | — | — |
| Dielectric constant | | | | |
| @ 60 Hz | 2.9—3.0 | — | — | — |
| @ 1 kHz | — | — | — | — |
| @ 1 MHz | 2.8—2.9 | 3.3 | — | — |
| Arc resistance (sec) | 15—91 | 99—146 | — | — |
| Thermal properties | | | | |
| Thermal conductivity (Btu/hr/ft$^2$/°F/in) | — | — | — | — |
| Coefficient of thermal expansion ($10^{-5}$/°F) | 3.5—3.7 | 2.8—5.7 | 4.0 | 5.0—8.9 |
| Heat deflection temperature (°F) | | | | |
| 66 psi load | 195—250 | 223—265 | 239—282 | 131—231 |
| 264 psi load | 180—240 | 190—264 | 190—264 | 98—183 |
| Continuous use temperature, no load (°F) | — | — | — | — |

Abbreviations: ABS = Acrylonitrile-butadiene-styrene; PBT = Polybutylene terephthalate; PET = Polyethylene terephthalate; PU = Polyurethane.

**Table 12—22**
**POLYESTERS**

| | Type/grade | | | | | |
|---|---|---|---|---|---|---|
| | Polybutylene terephthalate (PBT) | | | | Polyethylene terephthalate (PET) | |
| | Unreinforced resin | 30% Glass reinforced | Mineral filled | High impact | Glass reinforced | PBT/PET Glass reinforced |
| **Physical properties** | | | | | | |
| Specific gravity | 1.31—1.43 | 1.49—1.56 | 1.36—1.47 | 1.25—1.47 | 1.56—1.69 | 1.43—1.66 |
| Water absorption (24 hr) % | 0.08—0.09 | 0.07 | 0.06—0.10 | 0.08—0.10 | 0.05—0.06 | 0.06—0.07 |
| **Mechanical properties** | | | | | | |
| Tensile yield strength ($10^3$ psi) | 8.0 | 16.5—23.0 | 7—9 | 5.8—13 | 16—28 | 12—16 |
| Tensile modulus ($10^5$ psi) | 2.8 | 12—17 | — | — | 2.8—21.1 | — |
| Flexural yield strength ($10^3$ psi) | 12—14 | 24—33 | 12.5—18 | 7—19.5 | 21.3—41.0 | 20—26 |
| Flexural modulus ($10^5$ psi) | 2.8—4.0 | 11—14 | 3.9—12.4 | 2.2—9.2 | 13—20 | 6.5—10 |
| Compressive strength, 2% offset ($10^3$ psi) | 13 | 15—18 | 9.3—13.6 | — | 13—26 | 15—22 |
| Izod impact strength, notched (ft-lb/in) | 0.5 | 1.0—2.6 | 1.0—1.9 | 3.4—18 | 1.2—2.4 | 0.7—1.5 |
| **Electrical properties** | | | | | | |
| Dielectric strength (V/mil) | 420—450 | 480—560 | 390—600 | 350—400 | 403—550 | 470—530 |
| Dielectric constant | | | | | | |
| @ 60 Hz | 3.2—3.3 | 3.6—3.8 | 3.1—4.3 | 3.2—3.6 | 4.2 | 3.6 |
| @ 1 kHz | 3.2—3.4 | 3.6—3.7 | — | — | — | — |
| @ 1 MHz | 3.1 | 3.4—3.7 | 2.9—3.9 | 3.0—3.4 | 3.5—3.9 | 3.5 |
| Arc resistance (sec) | 110—130 | 80—130 | 125—130 | 129—146 | 81—126 | 68—136 |
| **Thermal properties** | | | | | | |
| Thermal conductivity (Btu/hr/ft$^2$/°F/in) | 1.1—1.7 | 1.3—9.0 | — | — | 2.2 | — |
| Coefficient of thermal expansion ($10^{-5}$/°F) | 4.3—8.9 | 1.3—5.4 | 5.0 | — | 1.3—2.0 | 1.8—4.4 |
| Heat deflection temperature (°F) | | | | | | |
| 66 psi load | 302—354 | 435—475 | 300—420 | 250—379 | — | 400—430 |
| 264 psi load | 122—150 | 395—435 | 150—385 | 125—374 | 428—442 | 320—380 |
| Continuous use temperature, no load (°F) | — | — | — | — | — | — |

## Table 12—23
## POLYETHERKETONE (PEEK)

| | Type/grade | |
|---|---|---|
| | Unreinforced | 30% Glass reinforced |
| **Physical properties** | | |
| Specific gravity | 1.32 | 1.49 |
| Water absorption (24 hr) % | 0.5 | 0.11 |
| **Mechanical properties** | | |
| Tensile yield strength ($10^3$ psi) | 13.4 | 23.0 |
| Tensile modulus ($10^5$ psi) | 1.6 | 20 |
| Flexural yield strength ($10^3$ psi) | 24.7 | 33.8 |
| Flexural modulus ($10^5$ psi) | 5.3 | 15 |
| Compressive strength, 2% offset ($10^3$ psi) | 18.0 | — |
| Izod impact strength, notched (ft-lb/in) | 1.6 | 2.1 |
| **Electrical properties** | | |
| Dielectric strength (V/mil) | 480 | 400 |
| Dielectric constant | | |
| @ 60 Hz | — | — |
| @ 1 kHz | 3.3 | — |
| @ 1 MHz | — | — |
| Arc resistance (sec) | — | — |
| **Thermal properties** | | |
| Thermal conductivity (Btu/hr/ft²/°F/in) | 6.02 | 10.4 |
| Coefficient of thermal expansion ($10^{-5}$/°F) | 4.7 | 2.2 |
| Heat deflection temperature (°F) | | |
| 66 psi load | — | — |
| 264 psi load | 284 | 599 |
| Continuous use temperature, no load (°F) | 600 | 480 |

## Table 12—24
## POLYETHERIMIDE (PEI)

| | Type/grade | |
|---|---|---|
| | Unreinforced | 30% Glass reinforced |
| **Physical properties** | | |
| Specific gravity | 1.27 | 1.51 |
| Water absorption (24 hr) % | 0.25 | 0.18 |
| **Mechanical properties** | | |
| Tensile yield strength ($10^3$ psi) | 15.2 | 24.5 |
| Tensile modulus ($10^5$ psi) | 4.3 | 13.0 |
| Flexural yield strength ($10^3$ psi) | 21 | 33 |
| Flexural modulus ($10^5$ psi) | 4.8 | 12.0 |
| Compressive strength, 2% offset ($10^3$ psi) | 20—20.3 | 25.5 |
| Izod impact strength, notched (ft-lb/in) | 1.0 | 2.0 |
| **Electrical properties** | | |
| Dielectric strength (V/mil) | 831 | 769 |
| Dielectric constant | | |
| @ 60 Hz | — | — |
| @ 1 kHz | 3.15 | 3.7 |
| @ 1 MHz | 3.15 | 3.7 |
| Arc resistance (sec) | 128 | 85 |
| **Thermal properties** | | |
| Thermal conductivity (Btu/hr/ft²/°F/in) | 1.5 | — |
| Coefficient of thermal expansion ($10^{-5}$/°F) | 3.1 | 1.1 |
| Heat deflection temperature (°F) | | |
| 66 psi load | 410 | 414 |
| 264 psi load | 392 | 410 |
| Continuous use temperature, no load (°F) | 338 | 356 |

**Table   12—25**
**POLYIMIDE**

| | Type/grade | |
| --- | --- | --- |
| | Unreinforced | Glass fiber reinforced |
| Physical properties | | |
| Specific gravity | 1.43 | 1.9 |
| Water absorption (24 hr) % | 0.24—0.47 | 0.2 |
| Mechanical properties | | |
| Tensile yield strength ($10^3$ psi) | 7.5—18 | 27 |
| Tensile modulus ($10^5$ psi) | 4.7—7.5 | 45 |
| Flexural yield strength ($10^3$ psi) | 11—24 | 50 |
| Flexural modulus ($10^5$ psi) | 4.3—7.0 | 33 |
| Compressive strength, 2% offset ($10^3$ psi) | 13—40 | 42 |
| Izod impact strength, notched (ft-lb/in) | 0.5—1.0 | 17 |
| Electrical properties | | |
| Dielectric strength (V/mil) | 310—560 | 500 |
| Dielectric constant | | |
| @ 60 Hz | 3.6—4.1 | 4.8 |
| @ 1 kHz | — | 4.7 |
| @ 1 MHz | 3.5—3.9 | 4.7 |
| Arc resistance (sec) | 152—230 | 180 |
| Thermal properties | | |
| Thermal conductivity (Btu/hr/ft²/°F/in) | 2.5—6.8 | 3.5 |
| Coefficient of thermal expansion ($10^{-5}$/°F) | 2.5—2.8 | 1.4 |
| Heat deflection temperature (°F) | | |
| 66 psi load | — | — |
| 264 psi load | 550—680 | 660 |
| Continuous use temperature, no load (°F) | 500—800 | 500 |

## Table 12—26
## POLYOLEFINS

| | Type/grade | | | | | |
|---|---|---|---|---|---|---|
| | Polyethylene | | | | Polypropylene | |
| | Low density | Medium density | High density | Ultrahigh molecular weight | General purpose | High impact |
| **Physical properties** | | | | | | |
| Specific gravity | 0.91—0.925 | 0.926—0.940 | 0.941—0.965 | 0.928—0.941 | 0.905 | 0.89—0.91 |
| Water absorption (24 hr) % | <0.01 | <0.01 | <0.01 | <0.01 | 0.01—0.03 | 0.01—0.03 |
| **Mechanical properties** | | | | | | |
| Tensile yield strength ($10^3$ psi) | 0.6—2.3 | 1.2—3.5 | 3.1—5.5 | 4.0—6.0 | 5.0 | 2.8—4.4 |
| Tensile modulus ($10^5$ psi) | 0.14—0.38 | 0.25—0.55 | 0.6—1.8 | 0.2—1.1 | 1.6 | 1.0—1.7 |
| Flexural yield strength ($10^3$ psi) | — | — | — | — | 6—7 | 4.1 |
| Flexural modulus ($10^5$ psi) | 0.08—0.6 | 0.60—1.15 | 1.0—2.0 | 1.0—1.7 | 1.7—2.5 | 1.2—1.8 |
| Compressive strength, 2% offset ($10^3$ psi) | — | — | — | — | 5.5—6.5 | 4.4 |
| Izod impact strength, notched (ft-lb/in) | No break | 0.5—16 | 0.5—2.0 | 30 | 0.5—2.2 | 1.0—1.5 |
| **Electrical properties** | | | | | | |
| Dielectric strength (V/mil) | 460—700 | 460—650 | 450—500 | 900 | 500—660 | 500—650 |
| Dielectric constant | | | | | | |
| @ 60 Hz | 2.3 | 2.3 | 2.3 | 2.3 | 2.2—2.3 | 2.2—2.3 |
| @ 1 kHz | 2.25—2.35 | 2.25—2.35 | 2.30—2.35 | 2.3—2.35 | 2.2—2.6 | 2.3 |
| @ 1 MHz | — | — | — | — | 2.2 | 2.2—2.3 |
| Arc resistance (sec) | 135—160 | 200—235 | — | — | 160 | 100 |
| **Thermal properties** | | | | | | |
| Thermal conductivity (Btu/hr/ft$^2$/°F/in) | 8.0 | 8.0—10 | 11.0—12.4 | 11.0 | 2.8 | 3.0—4.0 |
| Coefficient of thermal expansion ($10^{-5}$/°F) | 5.6—12.2 | 7.8—8.9 | 6.1—7.2 | 7.8 | 3.2—5.7 | 3.3—4.7 |
| Heat deflection temperature (°F) | | | | | | |
| 66 psi load | 100—121 | 120—165 | 140—190 | 170 | 200—250 | 160—210 |
| 264 psi load | 90—105 | 105—120 | 110—130 | 118 | 125—140 | 120—135 |
| Continuous use temperature, no load (°F) | — | — | — | — | 230 | — |

**Table 12—27**
**POLYETHERKETONE AND POLYBUTYLENES**

| | Polyetherketone (PEK) unreinforced | Polybutylenes Copolymer | Homopolymer |
|---|---|---|---|
| | | **Type/grade** | |
| | **Polyetherketone (PEK) unreinforced** | **Polybutylenes** | |
| | | **Copolymer** | **Homopolymer** |
| Physical properties | | | |
| Specific gravity | — | 0.894—0.910 | 0.910—0.915 |
| Water absorption (24 hr) % | — | 0.01 | 0.01 |
| Mechanical properties | | | |
| Tensile yield strength ($10^3$ psi) | 16.0 | 0.9—2.2 | 2.2—2.5 |
| Tensile modulus ($10^5$ psi) | 5.8 | 0.12—0.34 | 0.34—0.36 |
| Flexural yield strength ($10^3$ psi) | — | — | — |
| Flexural modulus ($10^5$ psi) | 5.8 | — | — |
| Compressive strength, 2% offset ($10^3$ psi) | — | — | — |
| Izod impact strength, notched (ft-lb/in) | 1.5 | — | — |
| Electrical properties | | | |
| Dielectric strength (V/mil) | — | — | — |
| Dielectric constant | | | |
| @ 60 Hz | 3.5 | 2.18—2.25 | 2.25 |
| @ 1 kHz | — | — | — |
| @ 1 MHz | — | 2.18—2.25 | 2.25 |
| Arc resistance (sec) | — | — | — |
| Thermal properties | | | |
| Thermal conductivity (Btu/hr/ft²/°F/in) | — | — | 1.5 |
| Coefficient of thermal expansion ($10^{-5}$/°F) | 5.7 | — | 7.1 |
| Heat deflection temperature (°F) | | | |
| 66 psi load | — | — | 215—235 |
| 264 psi load | 330 | — | 120—140 |
| Continuous use temperature, no load (°F) | 500 | — | 225 |

Table 12—28
POLYOLEFINS

| | Type/grade | | | | | |
|---|---|---|---|---|---|---|
| | Ethylene vinyl acetate (EVA) | Ethylenebutene | Propylene-ethylene | Polymethyl pentene (PMP) | Ionomer | Polyallomer |
| **Physical properties** | | | | | | |
| Specific gravity | 0.94 | 0.95 | 0.91 | 0.83 | 0.94 | 0.898—0.904 |
| Water absorption (24 hr) % | — | — | — | 0.01 | — | — |
| **Mechanical properties** | | | | | | |
| Tensile yield strength ($10^3$ psi) | 0.36 | 0.35 | 0.40 | 4 | 0.4 | 3.0—4.3 |
| Tensile modulus ($10^5$ psi) | — | — | — | 2.1 | — | — |
| Flexural yield strength ($10^3$ psi) | — | — | — | — | — | — |
| Flexural modulus ($10^5$ psi) | — | 1.65 | 1.40 | 1.4—2.0 | — | 0.7—1.3 |
| Compressive strength, 2% offset ($10^3$ psi) | — | — | — | — | — | — |
| Izod impact strength, notched (ft-lb/in) | — | 0.4 | 0.1 | 0.8 | 9—14 | 1.5 |
| **Electrical properties** | | | | | | |
| Dielectric strength (V/mil) | 525 | — | — | 700 | 1000 | 500—650 |
| Dielectric constant | | | | | | |
| @ 60 Hz | 3.16 | — | — | 2.12 | 2.4 | 2.3 |
| @ 1 kHz | — | — | — | — | — | — |
| @ 1 MHz | — | — | — | 2.12 | — | — |
| Arc resistance (sec) | — | — | — | — | — | — |
| **Thermal properties** | | | | | | |
| Thermal conductivity (Btu/hr/ft$^2$/°F/in) | — | — | — | — | — | — |
| Coefficient of thermal expansion ($10^{-5}$/°F) | — | — | — | — | — | — |
| Heat deflection temperature (°F) | | | | | | |
| 66 psi load | — | — | 104 | — | 105 | 122—133 |
| 264 psi load | — | — | — | — | — | — |
| Continuous use temperature, no load (°F) | — | — | — | 250—320 | — | — |

Table   12—29
## POLYSTYRENES

| | Type/grade | | |
|---|---|---|---|
| | General purpose | High impact | 30% Glass fiber reinforced |
| Physical properties | | | |
| Specific gravity | 1.04 | 1.03—1.10 | 1.29 |
| Water absorption (24 hr) % | 0.03—0.10 | 0.05—0.6 | 0.07 |
| Mechanical properties | | | |
| Tensile yield strength ($10^3$ psi) | 5.0—12 | 1.5—7 | 14 |
| Tensile modulus ($10^5$ psi) | 4—6 | 1.4—5 | 12 |
| Flexural yield strength ($10^3$ psi) | 8—17 | 3—12 | 12—19 |
| Flexural modulus ($10^5$ psi) | 4.0—4.7 | 1.5—4.6 | 12 |
| Compressive strength, 2% offset ($10^3$ psi) | 11.5—16 | 4—9 | 19 |
| Izod impact strength, notched (ft-lb/in) | 0.2—0.5 | 0.5—4.0 | 1.8—2.6 |
| Electrical properties | | | |
| Dielectric strength (V/mil) | 500—700 | 300—600 | 396 |
| Dielectric constant | | | |
| @ 60 Hz | 2.5—2.7 | 2.5—4.8 | 3.1 |
| @ 1 kHz | 2.4—2.7 | 2.4—4.5 | — |
| @ 1 MHz | 2.5—2.7 | 2.5—4.0 | 3.0 |
| Arc resistance (sec) | 60—135 | 20—135 | 28 |
| Thermal properties | | | |
| Thermal conductivity (Btu/hr/ft$^2$/°F/in) | 2.4—3.3 | 1.0—3.0 | 1.4 |
| Coefficient of thermal expansion ($10^{-5}$/°F) | 3.3—4.4 | 1.9 | 1.8 |
| Heat deflection temperature (°F) | | | |
| 66 psi load | 180—230 | 180—220 | 230 |
| 264 psi load | 190—220 | 160—200 | 220 |
| Continuous use temperature, no load (°F) | 160—205 | 125—165 | 190—220 |

## Table 12—30
## STYRENE-MALEIC ANHYDRIDE AND STYRENE-ACRYLONITRILE

| | | Material | |
| --- | --- | --- | --- |
| | Styrene-maleic anhydride (SMA) | Styrene-acrylonitrile (SAN) | |
| | | Unreinforced | 30% Glass fiber reinforced |
| Physical properties | | | |
| Specific gravity | 1.07 | 1.04—1.07 | 1.35 |
| Water absorption (24 hr) % | 0.5 | 0.20—0.35 | 0.15 |
| Mechanical properties | | | |
| Tensile yield strength ($10^3$ psi) | 4—7 | 9.5—12 | 18 |
| Tensile modulus ($10^5$ psi) | 3.3—4.9 | 4—5 | 17.5 |
| Flexural yield strength ($10^3$ psi) | 9.6—14 | — | 22 |
| Flexural modulus ($10^5$ psi) | 3.2—5 | — | 14.5 |
| Compressive strength, 2% offset ($10^3$ psi) | — | — | 2.3 |
| Izod impact strength, notched (ft-lb/in) | 0.3—3.8 | 0.3—0.45 | 3.0 |
| Electrical properties | | | |
| Dielectric strength (V/mil) | — | 400—500 | 515 |
| Dielectric constant | | | |
| @ 60 Hz | — | 2.6—3.4 | 3.5 |
| @ 1 kHz | — | — | — |
| @ 1 MHz | — | 2.6—3.0 | 3.4 |
| Arc resistance (sec) | — | 100—150 | 65 |
| Thermal properties | | | |
| Thermal conductivity (Btu/hr/ft²/°F/in) | — | — | — |
| Coefficient of thermal expansion ($10^{-5}$/°F) | 3.5—4.3 | 3.6—3.7 | 1.6 |
| Heat deflection temperature (°F) | | | |
| 66 psi load | — | — | 230 |
| 264 psi load | 157—207 | 210—220 | 220 |
| Continuous use temperature, no load (°F) | — | 175—190 | — |

## Table 12—31
## POLYPHENYLENE SULFIDE (PPS)

| | Type/grade | |
| --- | --- | --- |
| | Glass and mineral filled | 40% Glass reinforced |
| Physical properties | | |
| Specific gravity | 1.8—2.1 | 1.64 |
| Water absorption (24 hr) % | 0.03 | 0.05 |
| Mechanical properties | | |
| Tensile yield strength ($10^3$ psi) | 11—13 | 19.5 |
| Tensile modulus ($10^5$ psi) | — | 11.2 |
| Flexural yield strength ($10^3$ psi) | 17—21 | 29 |
| Flexural modulus ($10^5$ psi) | 18—21 | 17 |
| Compressive strength, 2% offset ($10^3$ psi) | 16—17 | 21 |
| Izod impact strength, notched (ft-lb/in) | 0.5—1.0 | 1.4 |
| Electrical properties | | |
| Dielectric strength (V/mil) | 340—520 | 450 |
| Dielectric constant | | |
| @ 60 Hz | 3.1 | 3.8 |
| @ 1 kHz | — | — |
| @ 1 MHz | 3.2 | 3.9 |
| Arc resistance (sec) | 182—200 | 34 |
| Thermal properties | | |
| Thermal conductivity (Btu/hr/ft²/°F/in) | — | 2.0 |
| Coefficient of thermal expansion ($10^{-5}$/°F) | — | 2.2 |
| Heat deflection temperature (°F) | | |
| 66 psi load | — | — |
| 264 psi load | >500 | 485 |
| Continuous use temperature, no load (°F) | — | 500 |

# POLYSULFONES

| | | Type/grade | | |
|---|---|---|---|---|
| | Unreinforced | 30% Glass fiber reinforced | Polyether-sulfone (PES) | Polyphenyl-sulfone (PPSF) |
| **Physical properties** | | | | |
| Specific gravity | 1.24 | 1.41 | 1.37 | 1.29 |
| Water absorption (24 hr) % | 0.3 | 0.22 | 0.43 | — |
| **Mechanical properties** | | | | |
| Tensile yield strength ($10^3$ psi) | 10.2 | 17 | 12.2 | 10.4 |
| Tensile modulus ($10^5$ psi) | 3.6 | 10.9 | 3.9 | 3.1 |
| Flexural yield strength ($10^3$ psi) | 15.4 | 25 | 18.7 | 12.4 |
| Flexural modulus ($10^5$ psi) | 3.9 | 12.0 | 3.8 | 3.3 |
| Compressive strength, 2% offset ($10^3$ psi) | 13.9 | — | — | 14.4 |
| Izod impact strength, notched (ft-lb/in) | 1.3 | 1.8 | 1.6 | 1.2 |
| **Electrical properties** | | | | |
| Dielectric strength (V/mil) | 425 | 480 | 400 | 371 |
| Dielectric constant | | | | |
| @ 60 Hz | 3.03—3.07 | 3.55 | 3.5 | 3.44 |
| @ 1 kHz | — | — | — | — |
| @ 1 MHz | 3.03—3.07 | 3.41 | 3.5 | 3.44 |
| Arc resistance (sec) | 122 | 114 | 116 | 41 |
| **Thermal properties** | | | | |
| Thermal conductivity (Btu/hr/ft²/°F/in) | 6.2 | — | 3.2—4.4 | 2 |
| Coefficient of thermal expansion ($10^{-5}$/°F) | 3.1 | 1.6 | 3.1 | 3.1 |
| Heat deflection temperature (°F) | | | | |
| 66 psi load | 358 | 389 | — | — |
| 264 psi load | 345 | 365 | 398 | 400 |
| Continuous use temperature, no load (°F) | 340 | 350 | 400 | 375 |

## Table 12—33
## POLYVINYL CHLORIDE (PVC)

| | Rigid | Flexible | Rigid 30% glass coupled |
|---|---|---|---|
| | | Type/grade | |
| Physical properties | | | |
| Specific gravity | 1.30—1.58 | 1.20—1.70 | 1.53—1.57 |
| Water absorption (24 hr) % | 0.04—0.4 | 0.15—0.75 | — |
| Mechanical properties | | | |
| Tensile yield strength ($10^3$ psi) | 6.0—8.0 | 1.5—3.5 | 15 |
| Tensile modulus ($10^5$ psi) | 3.5—10 | 0.4—3.0 | 11.0 |
| Flexural yield strength ($10^3$ psi) | 11—16 | — | 21 |
| Flexural modulus ($10^5$ psi) | 3—8 | — | 1.2 |
| Compressive strength, 2% offset ($10^3$ psi) | 10—11 | — | — |
| Izod impact strength, notched (ft-lb/in) | 0.4—20 | — | 0.8 |
| Electrical properties | | | |
| Dielectric strength (V/mil) | 350—500 | 300—400 | — |
| Dielectric constant | | | |
| @ 60 Hz | 0.02—0.03 | 0.05—0.15 | — |
| @ 1 kHz | 3.0—3.8 | 4.0—8.0 | — |
| @ 1 MHz | — | — | — |
| Arc resistance (sec) | 60—80 | — | — |
| Thermal properties | | | |
| Thermal conductivity (Btu/hr/ft²/°F/in) | 3.5—5.0 | 3.0—4.0 | — |
| Coefficient of thermal expansion ($10^{-5}$/°F) | 1.2—5.6 | 3.9—13.9 | — |
| Heat deflection temperature (°F) | | | |
| 66 psi load | 135—180 | — | — |
| 264 psi load | 140—170 | — | 169 |
| Continuous use temperature, no load (°F) | 150—165 | 150—220 | — |

## Table 12—34
## VINYLIDENE CHLORIDE COPOLYMER AND CHLORINATED POLYVINYL CHLORIDE

| | Vinylidene chloride copolymer | Chlorinated polyvinyl chloride |
|---|---|---|
| | Material | |
| Physical properties | | |
| Specific gravity | 1.68—1.75 | 1.49—1.58 |
| Water absorption (24 hr) % | >0.1 | 0.02—0.15 |
| Mechanical properties | | |
| Tensile yield strength ($10^3$ psi) | 4—8 | — |
| Tensile modulus ($10^5$ psi) | 0.7—2.0 | — |
| Flexural yield strength ($10^3$ psi) | 15—17 | 14.5—17 |
| Flexural modulus ($10^5$ psi) | — | 3.8—4.5 |
| Compressive strength, 2% offset ($10^3$ psi) | 75—85 | — |
| Izod impact strength, notched (ft-lb/in) | 2—8 | 1.0—3.0 |
| Electrical properties | | |
| Dielectric strength (V/mil) | — | 1220—1500 |
| Dielectric constant | | |
| @ 60 Hz | 3.5 | 3.1 |
| @ 1 kHz | — | — |
| @ 1 MHz | — | — |
| Arc resistance (sec) | — | — |
| Thermal properties | | |
| Thermal conductivity (Btu/hr/ft²/°F/in) | 0.64 | — |
| Coefficient of thermal expansion ($10^{-5}$/°F) | 8.8 | 3.8 |
| Heat deflection temperature (°F) | | |
| 66 psi load | 190—210 | 215—247 |
| 264 psi load | 130—150 | 202—234 |
| Continuous use temperature, no load (°F) | 170—212 | 230 |

## Table 12—35
## SILICONES

| | Type/grade | | |
|---|---|---|---|
| | Molding Compound | | |
| | Fibrous filler glass | Granular filler glass/silica | Casting, no filler |
| **Physical properties** | | | |
| Specific gravity | 1.86 | 1.88 | 1.55 |
| Water absorption (24 hr) % | 0.16 | 0.09 | 0.01 |
| **Mechanical properties** | | | |
| Tensile yield strength ($10^3$ psi) | 6.5 | 6.0 | 0.4—0.5 |
| Tensile modulus ($10^5$ psi) | — | — | — |
| Flexural yield strength ($10^3$ psi) | 14.0 | 9.0 | 1.0 |
| Flexural modulus ($10^5$ psi) | 25 | 17 | — |
| Compressive strength, 2% offset ($10^3$ psi) | — | — | — |
| Izod impact strength, notched (ft-lb/in) | 10 | 0.3 | — |
| **Electrical properties** | | | |
| Dielectric strength (V/mil) | 280 | 370 | 400 |
| Dielectric constant | | | |
| @ 60 Hz | — | — | — |
| @ 1 kHz | 4.2 | 3.7 | 3.1 |
| @ 1 MHz | — | — | — |
| Arc resistance (sec) | 240 | 230 | 210 |
| **Thermal properties** | | | |
| Thermal conductivity (Btu/hr/ft²/°F/in) | 8 | 13 | 7.5 |
| Coefficient of thermal expansion ($10^{-5}$/°F) | 0.8 | 2.4 | 2.2 |
| Heat deflection temperature (°F) | | | |
| 66 psi load | — | — | — |
| 264 psi load | 900 | 520 | — |
| Continuous use temperature, no load (°F) | — | — | — |

## Table 12—36
## LIQUID CRYSTAL POLYMERS (LCP)
## (WHOLLY AROMATIC COPOLYESTERS)

| | Type/grade | |
|---|---|---|
| | Unmodified resin | 40% Glass filled |
| **Physical properties** | | |
| Specific gravity | 1.35 | 1.70 |
| Water absorption (24 hr) % | <0.1 | <0.1 |
| **Mechanical properties** | | |
| Tensile yield strength ($10^3$ psi) | 20.0 | 16.9 |
| Tensile modulus ($10^5$ psi) | 24 | 18 |
| Flexural yield strength ($10^3$ psi) | 19.0 | 19.5 |
| Flexural modulus ($10^5$ psi) | 20 | 18 |
| Compressive strength, 2% offset ($10^3$ psi) | — | 11.0 |
| Izod impact strength, notched (ft-lb/in) | 2.4 | 1.9 |
| **Electrical properties** | | |
| Dielectric strength (V/mil) | 1070 | 977 |
| Dielectric constant | | |
| @ 60 Hz | 3.6 | 4.3 |
| @ 1 kHz | — | — |
| @ 1 MHz | 3.1 | — |
| Arc resistance (sec) | 186 | 192 |
| **Thermal properties** | | |
| Thermal conductivity (Btu/hr/ft²/°F/in) | 2.0 | 1.9 |
| Coefficient of thermal expansion ($10^{-5}$/°F) | 1.8 | 0.83 |
| Heat deflection temperature (°F) | | |
| 66 psi load | — | — |
| 264 psi load | 671 | 592 |
| Continuous use temperature, no load (°F) | 464 | 428 |

## Table 12—37
## LIQUID CRYSTAL POLYMERS (LCP) (WHOLLY AROMATIC COPOLYESTERS)

|  | Type/grade | |
|---|---|---|
|  | Mineral-filled | High heat |
| **Physical properties** | | |
| Specific gravity | 1.84 | 1.57 |
| Water absorption (24 hr) % | <0.1 | — |
| **Mechanical properties** | | |
| Tensile yield strength (10³ psi) | 11.8 | 24.0 |
| Tensile modulus (10⁵ psi) | 18 | 23 |
| Flexural yield strength (10³ psi) | 16.2 | 31.8 |
| Flexural modulus (10⁵ psi) | 16 | 21 |
| Compressive strength, 2% offset (10³ psi) | 7.2 | — |
| Izod impact strength, notched (ft-lb/in) | 1.4 | 2.2 |
| **Electrical properties** | | |
| Dielectric strength (V/mil) | 1000 | 1100 |
| Dielectric constant | | |
| @ 60 Hz | 4.0 | — |
| @ 1 kHz | — | — |
| @ 1 MHz | — | — |
| Arc resistance (sec) | 241 | 137 |
| **Thermal properties** | | |
| Thermal conductivity (Btu/hr/ft²/°F/in) | 1.6 | — |
| Coefficient of thermal expansion (10⁻⁵/°F) | 1.5 | — |
| Heat deflection temperature (°F) | | |
| 66 psi load | — | — |
| 264 psi load | 601 | 465 |
| Continuous use temperature, no load (°F) | 428 | — |

## Table 12—38
## POLYURETHANE

|  | Rigid foam |
|---|---|
| **Physical properties** | |
| Specific gravity | 0.56—0.64 |
| Water absorption (24 hr) % | <1 |
| **Mechanical properties** | |
| Tensile yield strength (10³ psi) | 2.5—3.0 |
| Tensile modulus (10⁵ psi) | 1.1—1.6 |
| Flexural yield strength (10³ psi) | 6.0—7.0 |
| Flexural modulus (10⁵ psi) | 1.2—1.7 |
| Compressive strength, 2% offset (10³ psi) | — |
| Izod impact strength, notched (ft-lb/in) | 7—15 |
| **Electrical properties** | |
| Dielectric strength (V/mil) | 127—254 |
| Dielectric constant | |
| @ 60 Hz | — |
| @ 1 kHz | 1.8 |
| @ 1 MHz | — |
| Arc resistance (sec) | — |
| **Thermal properties** | |
| Thermal conductivity (Btu/hr/ft²/°F/in) | 0.4—0.8 |
| Coefficient of thermal expansion (10⁻⁵/°F) | 4.5 |
| Heat deflection temperature (°F) | |
| 66 psi load | — |
| 264 psi load | — |
| Continuous use temperature, no load (°F) | — |

## Table 12—39
## RUBBERS/ELASTOMERS

| | Type/grade | | | |
|---|---|---|---|---|
| | Natural isoprene (natural rubber) | Synthetic isoprene (synthetic NR) | Isobutylene isoprene (butyl rubber) | Chloroprene (neoprene) |
| **Physical properties** | | | | |
| Specific gravity | 0.92—1.037 | 0.93—1.037 | 0.92 | 1.23—1.25 |
| **Mechanical Properties** | | | | |
| Tensile yield strength ($10^3$ psi) | 3.5—4.6 | 2.5—4.6 | >2.0 | 0.5—3.5 |
| Modulus (100%) (psi) | 480—850 | 480—850 | 50—500 | 100—3000 |
| Elongation (%) | 500—760 | 300—750 | 300—800 | 100—800 |
| Hardness, durometer | 30A-100A | 30A-100A | 30A—100A | 30A—95A |
| Compression set, method B (%) | 10—30 | 10—30 | 25 | 20—60 |
| Resilience, Yerzley (%) | 80 | 80 | 30 | 50—80 |
| **Electrical properties** | | | | |
| Dielectric strength (V/mil) | 400—600 | — | 600—900 | 400—600 |
| Dielectric constant | | | | |
| @ 60 Hz | — | — | 2.31 | 8.0 |
| @ 1 kHz | — | — | — | — |
| @ 1 MHz | 2.9 | — | 2.25 | 6.7 |
| **Thermal properties** | | | | |
| Thermal conductivity (Btu/hr/ft²/°F/in.) | 0.082 | 0.082 | 0.053 | 0.11 |
| Coefficient of thermal expansion ($10^{-5}$/°F) | 37 | 37 | 32 | 34 |
| Continuous use temperature, °F | 250 | 180 | 300 | 225 |

# Section 13

# Materials Information

# Section 13

# MATERIALS INFORMATION

## 13.1 MAJOR COMPILATIONS

A substantial fraction of all technical literature deals with some type of information about materials. The *Directory of Information Resources in the United States* (Library of Congress) identified 2891 sources of information in 1974, other than publications, in the physical sciences and engineering. A large fraction of these also dealt with some aspect of information about materials.

The general card catalogs, lists of publications, and abstracts that cover the published literature are presumably known to readers of this *Handbook* and will not be discussed here. However, this should not be taken as a suggestion that a traditional literature search can safely be omitted when seeking information about an unfamiliar material. Some knowledge of the material is essential both to pose a question that will elicit a useful response and to judge which source is likely to contain the information needed.

Means for locating unpublished reports and other sources of information are not as widely known. Sections 13.2 and 13.3 list a number of information centers and publicly funded materials-related research laboratories. Generally, these provide a service of referring the inquirer to a source of information or specific data banks and bibliographies.

One important type of information about materials is numerical data on material properties. A major problem facing scientists and engineers today involves, first, finding, and second, evaluating the reliability of the enormous volume of property data in the literature. A number of the data centers that have been established collect and evaluate such data in defined areas.

One convenient source of bibliographies and abstracts is the Published Searches (research summaries) of the National Technical Information Service (NTIS), covering over 3000 topical subject areas. A substantial number of materials and processes are included in these searches. Over 200 new searches are conducted each year, and a master catalog is available from NTIS, 5285 Port Royal Road, Springfield, VA 22161. (See, for example,

recent catalogs PR-186, April 1987, and PR-184, November 1987 with annual updates.)

The following comprise the list of databases used in the NTIS searches and available for specific searches for materials information. The user may easily be overwhelmed by the magnitude of efforts reported since most searches are exhaustive rather than selective in nature.

**1. NTIS Bibliographic Database** — As the central information sources for federally sponsored research, NTIS receives some 250 research reports daily. Major corporations, trade associations, and university and private research facilities all contribute their results to NTIS. Some 350 Government sources include: National Air and Security Administration, National Institute of Standards and Technology (formerly the National Bureau of Standards), U.S. Department of Energy, and U.S. Department of Defense. The NTIS Bibliographic Database also contains references to Government-generated software packages, statistical data files (available on magnetic tape as well as floppy diskette), and other NTIS subscription products. The database carries over one million records. Each year 70,000 reports are added. The file is updated twice monthly and dates back to 1964.

**2. American Petroleum Institute (API)** — Covers worldwide journal and preprint literature since 1964 on topics such as petroleum processing, fuels and lubricants, petrochemicals, petroleum transportation and storage, environmental matters, and nonpetroleum energy sources.

**3. Engineering Information Inc. (EI)** — Covers international journals, articles, and reports on prominent research and applications of technical ideas relevant to the field of engineering.

**4. Food Science and Technology Abstracts (FSTA)** — FSTA is the main database of the International Food Information Service (IFIS). It covers the international literature since 1969 on all matters concerned with research and development in food science and technology.

**5. U.S. Patent Bibliographic Database** — The

---

\* Extracted in part from an article by R. S. Marvin and G. B. Sherwood in Volume III, *Handbook of Materials Science*, CRC Press, 1975.

U.S. Patent and Trademark Office (PTO) produces and leases many machine-readable patent databases that are processed and offered to the public through commercial systems. The primary patent database, the Patent Full Text Database, has several subfiles. One of the subfiles, the Patent Bibliographic Database with Exemplary Claims, is used to produce Published Searches. They are not to be construed as "legal" patent searches.

**6. International Aerospace Abstracts (IAA)** — Contains more than 850,000 citations to worldwide articles related to aerospace research and technology and allied sciences.

**7. Life Sciences Collection (LSC)** — Contains 770,000 records, updated monthly with approximately 8,000 new entries covering microbiology, biochemistry, genetics, ecology, toxicology, entomology, animal behavior, immunology, neurosciences, and chemoreception.

**8. Energy Data Base (EDB)** — Contains more than 1 million references to worldwide scientific and technical information of interest to the U.S. Department of Energy; reports, journals, patents, translations, and conference proceedings. Covers all aspects of international research on nuclear science, coal, and biomass.

**9. Computer Database** — Contains comprehensive information pertaining to the computer, telecommunication and electronic industries. The file dates back to 1983 and includes over 200,000 records. Source material drawn from journals, newsletters, and conference proceedings with 2,500 new records added monthly.

**10. Conference Papers Index (CONF)** — Contains more than 1,040,000 records, with 3,000 added monthly. Covers the literature presented annually at regional, national, and international meetings on the topics of engineering, and physical sciences.

**11. Management Contents** — Contains over 230,000 citations, with almost 2,000 new documents being added each month. Over 150 business and law journals, newsletters, tabloids, proceedings and transactions are indexed and abstracted cover-to-cover.

**12. Fluidex: (BHRA the Fluid Engineering Centre)** — Contains over 150,000 items, updated monthly, covering the international literature on fluid engineering subjects from aerodynamics, seals, and pumps to ports and off-shore technology. Source material drawn from over 1,000 scientific and technical journals scanned annually, as well as reports, conferences, books, and standards.

**13. Metals Abstracts (METADEX)** — Produced and distributed by Materials Information, a joint service of ASM International and Institute of Metals (London), this database contains approximately 650,000 citations on extractive and physical metallurgy. Major literature sources are journal articles, with 200 journals abstracted in their entirety and 1,800 journals reviewed for selected input to the database.

**14. RAPRA (the database of the Rubber and Plastics Research Association of Great Britain)** — Covers journal articles (approximately 60% of database), patents, trade literature, government reports, etc., on chemistry and the chemical engineering of polymers as well as the technical and commercial aspects of rubber and plastics. The database contains approximately 240,000 items, with 23,000 added annually.

**15. Oceanic Abstracts (OCEANICS)** — Covers the worldwide literature in the field of marine research, with topics including marine geology, biology, geophysics, pollution, and marine resources (living and nonliving). Each year approximately 10,500 items are added to the database, which at present consists of approximately 165,000 citations.

**16. Pollution Abstracts (POLLUTION)** — Contains 119,000 entries dealing with worldwide coverage of pollution-related topics such as air and water pollution, solid wastes, noise, pesticides, radiation, and general environmental quality.

**17. Research Association for the Paper & Board, Printing & Packaging Industries (PIRA)** — Database of approximately 110,000 items, covering mainly journal articles from industry-specific scientific, technical, marketing, and management literature. General coverage includes business, economics, management, education and training, forecasting, industrial relations, materials, occupational safety and health, pollution, production, and testing.

**18. Searchable Physics Information Notices (SPIN)** — Contains over 400,000 abstracts, covering approximately one third of the world physics and astronomy journal literature each year. Includes all journal articles published by the American Institute of Physics, 19 Soviet journals, selected articles from Chinese journals, as well as most of the other American physics and astronomy journals.

**19. World Textile Abstracts (WTA)** — Covers science, technology, economics, and technical management of textile and releated industries, plus all relevant U.S. and U.K. patents. Operated by the Shirley Institute, the database consists of about 150,000 entries, of which 9,000 are new annually.

**20. Information Services in Mechanical Engineering (ISMEC)** — Database is devoted to worldwide coverage of mechanical engineering, production engineering, and engineering management and contains 165,000 records. Over 500 technical journals are screened each year for citations, yielding 15,000 new additions annually.

**21. Information Services for the Physics and Engineering Communities (INSPEC)** — INSPEC dates from 1969, totals over 2.75 million items, and is being added to at the rate of 220,000 per annum. Subject breakdown is approximately 53% physics, 28% electrical and electronics, and 19% computer and control.

**22. Selected Water Resources Abstracts (SWRA)** — SWRA is the abstract journal of the Water Resources Scientific Information Center (WRSIC). The database became available in machine-readable form in 1969 and contains over 180,000 items with approximately 10,000 records being added per year. Major literature sources covered include scientific and technical journals (80%), Government reports (15%), patents (1%) and published proceedings (4%). SWRA covers all aspects of hydrology, hydraulics, water quality, and water quantity when water is used as a resource. SWRA also covers the water-related aspects of life sciences amd economic and legal aspects of the characteristics, conservation, control, use or management of water.

**23. World Surface Coatings Abstracts (WSCA)** — WSCA is a product of the Paint Research Association, Middlesex, England and contains 100,000 records dating from 1976 with 800 new citations added monthly. Subject coverage includes: pigments, synthetic resins, cellulose products, adhesives, corrosion oils, solvents, varnishes and lacquers, inks, industrial hazards, storage, and transport.

**24. Packaging Science and Technology Abstracts (PSTA)** — Contains over 18,000 abstracts gathered worldwide since 1982 on packaging economy, packaging science and institutions, packaging material, processing, equipment, packs and packages, transport and storage, testing, and stress loading.

Some of the general handbooks that are available include:

1. *International Critical Tables of Numerical Data, Physics, Chemistry, and Technology*

Published for the National Academy of Sciences-National Research Council by McGraw-Hill, 1221 Avenue of the Americas, New York. Volumes I — VII plus index, 1926 — 1933.

These Tables were initiated as a project of the International Union of Theoretical and Applied Chemistry. They were compiled by specialists throughout the world, chosen for their qualifications to select the ''best'' values in each category. Values selected are primarily from the literature published through 1923, and there is comprehensive coverage of physical and chemical properties of pure substances and of many commercial materials (alloys, clays, wood, etc.). Arrangement is not strictly logical, but the comprehensive index permits ready location of specific information. Although this source is still often quoted, subsequent improvements in purification and characterization of materials and in methods of measurements have made many of the tables obsolete. The original goal of compiling periodic revisions was never realized.

2. ''Landolt-Börnstein''

Landolt, Hans Heinrich, *Zahlenwerte und Functionen aus Physik, Chemie, Astronomie, Geophysik und Technik, Secheste Auflage*, Euken, A. and Hellwege, K. H., Eds., Springer-Verlag, Berlin (in the U.S., Springer-Verlag, New York, 175 Fifth Ave., New York 10010), 1950 —   .

This is one of the oldest and probably the most extensive of the general compilations. The first edition (one volume) was published in 1883; as of 1974 26 parts of the sixth edition had appeared, starting in 1950. It is in German with few exceptions. The original aim of ''Landolt-Börnstein'' was ''to make exact results (that can be estimated numerically) of physical, chemical, and technical research as conveniently and completely (comprising all literature data) accessible as possible...''. A consistent organization is followed with only a few exceptions.

The editors have concluded that such a systematic organization is no longer feasible. Hence, starting in 1961, the same editors and publisher initiated the "New Series": *Zahlenwerte und Functionen aus Naturwissenschaften und Technik, Neue Serie — Numerical Data and Functional Relationships in Science and Technology, New Series,* Hellwege, K. H., Ed. These tables will be published in an order dictated by the development of individual fields of specialization. In effect, it is an open-ended series with no predetermined order of publication or termination. As suggested by the title, this new series includes parallel text in German and English.

The various volumes in both series contain tables, graphs, and explanatory test (more text than the International Critical Tables). The authors of various sections are recognized authorities from throughout the world, but the sponsorship of both works is private.

3. *Table de Constantes Selectionnees — Tables of Selected Constants,* Pergamon Press, Maxwell House, Elmsford, N.Y.

Starting in 1910, a series of annual tables transcribing results from the original literature into a single series was prepared in Paris. The original coverage was broad and comprehensive rather than selective; the present series aims at selectivity, but it covers primarily nuclear, atomic, and molecular properties. Pergamon Press was the publisher of Volume 8 *et seq.* (1958 on) and also handles distribution of the earlier volumes and issues of the Annual Tables (up to 1936) still available. From 1947 through 1970, 17 new volumes were published.

4. *American Institute of Physics Handbook,* 3rd ed., Gray, D. E., Coordinating Ed., McGraw-Hill, New York, vii + 2442 pages.

Most branches of science and engineering are served by one or more single-volume handbooks, including the popular *CRC Handbook of Chemistry and Physics* and several others from the publishers of this volume. The AIP Handbook is singled out for mention here because all sections of its 3rd edition were revised or completely rewritten by recognized authorities. Therefore, both the data and the extensive bibliographies are reasonably current. The texts are brief, but generally adequate to make it a self-contined reference. The major sections, all of which (except the first) included selected property data, are Mathematics Bibliography; SI Units; Mechanics; Acoustics; Heat; Electricity and Magnetism; Optics; Atomic and Molecular Physics; Nu-

clear Physics; and Solid State Physics. Each section includes tables and extensive references, and a complete index follows the last section.

5. *Handbook of High-Temperature Materials, No. 2 — Properties Index,* Samsonov, G. V., Plenum Press, New York, 1964, 418 pages.

This is a translation from Russian of [*Refractory Compounds — Handbook of Properties and Applications*], 1963, revised to include material published in 1962 and 1963. It covers refractory compounds, mainly binary compounds between a transition metal and boron, carbon, silicon, nitrogen, sulfur, or phosphorus, and gives data on the physical, technical, mechanical, chemical, and refractory properties selected by the author as most reliable. The introduction discusses reliability of data.

6. *Metal Reference Book,* 4th ed., Smithells, C. J., Ed., Plenum Press, New York; Butterworths, London, 1967. (Three volumes, total of 1250 pages plus contents and index, repeated in each volume.)

The source contains data pertaining to metallurgy and metal physics, selected by a number of contributors as the most reliable, with bibliographies of the more important original sources. The data are presented in tables or diagrams with short monographs on particular subjects where required. Coverage includes crystal chemistry, thermochemical data, metallography, phase diagrams, physical and mechanical properties, and other properties important in production and fabrication of pure metals and alloys.

7. *Metals Handbook,* 8th ed., American Society for Metals, Metals Park, OH 44073 (12 volumes, of which Volume 1: *Properties and Selection,* 1961 [1300 pages], is most useful for general data).

A number of authorities contributed to this reference (1335 to Volume 1 alone). Most of the properties data are contained in Volume 1, whose contents are definitions and reference tables; carbon and low-alloy steels; cast irons; stainless steels and heat-resisting alloys; tool materials; magnetic, electrical, and other special purpose materials; nonferrous metals; and properties of pure metals. Tables and graphs showing the properties of major commercial alloys that are important in their principal uses are interspersed throughout sections describing their production, uses, formability, and cost; descriptions of the various standard test methods and samples employed are also included.

8. *Polymer Handbook*, Brandrup, J. and Immergut, E. H., Eds., Interscience, New York, 1966, 1249 pages.

Only fundamental constants and parameters were compiled for this handbook. This was interpreted to include data that are either physical or chemical constants of the polymer molecules within reasonable or predictable limits or constants of existing physical laws describing the properties and behavior of polymers. Constants that depend on the particular processing conditions or sample history were not compiled. No critical evaluation of published values was attempted. All data found in the literature were listed except those that were obviously erroneous. A bibliography was included at the end of each section. The sections included in this handbook are (I) Nomenclature Rules; (II) Polymerization; (III) Solid State Properties; (IV) Solution Properties; (V) Miscellaneous Properties; (VI) Physical Constants of Some Important Polymers; (VII) Physical Data of Oligomers; (VIII) Physical Properties of Monomers and Solvents; (IX) Contemporary Thermoplastic Materials, Property and Price Chart; (X) Subject Index.

9. *Handbook of Materials Science*, Lynch, C. T., Ed., CRC Press, Boca Raton, FL: Vol. I, *General Properties*, 1974, 752 pages; Vol. II, *Metals, Composites, and Refractory Materials*, 1975, 440 pages; Vol. III, *Nonmetallic Materials and Applications*, 1975, 642 pages; Vol. IV, Summit, R., and Sliker, A., Eds., *Wood*, 1980, 459 pages.

This is a comprehensive compilation of the properties of materials, both of commercial importance and new advanced high performance materials such as biomedical, composite, and laser materials. Data included were current at the time of publication.

## 13.2     MATERIALS INFORMATION CENTERS*

**1.     Aerospace Research Application Center (ARAC)**
National Aeronautics and Space
 Administration
Indianapolis Center for Advanced Research
611 N. Capitol Ave.

Indianapolis, IN 46204
Contact: (317) 262-5003

ARAC uses experienced scientists and engineers and a computerized index of worldwide scientific and technical information. Its objectives are to promote greater use of existing scientific and technical knowledge; to provide information and analysis services for users in industry, government, and education; to develop improved methods for organizing and disseminating information and for encouraging innovation; to provide special services to small and inexperienced users to help them learn to use information resources effectively; and to help NASA find new applications for research from the space program and other government projects.

**2.     Aerospace Structures Information and Analysis Center (ASIAC)**
Air Force Wright Aeronautical Laboratories
U.S. Department of the Air Force
FIBRA (ASIAC)
Wright-Patterson AFB, OH 45433
Contact: (513) 255-3688

This is a central point for the collection and dissemination of aerospace structures information and quick state-of-the-art solutions to structural problems. Its areas of interest include aerospace structural design and analysis; experimental structural data; fracture mechanics; failure analysis; structural history; fatigue; service loads; design criteria; composites.

**3.     Assessment and Information Services Center**
National Oceanic and Atmospheric
 Administration
U.S. Department of Commerce
National Environmental Satellite, Data and
Information Service
Room 511, 1825 Connecticut Ave., NW
Washington, DC 20235
Contact: (202) 673-5394

The center provides climatic and marine environmental data analysis, impact assessment, and information services to public and private sector users. Center scientists integrate and analyze multidisciplinary environmental data to provide qualitative and quantitative assessments of meterological and oceanographic impacts on human activities and important natural resources. The center's libraries,

---

*     Information extracted from Directory of Federal Laboratory and Technology Resources, 1988—1989, PB88-100011, U.S. Dept. of Commerce, NTIS, February 1988.

located in Rockville, MD; Miami, FL; and Seattle, WA maintain vast holdings of historical and recent published environmental data and reports and participate in nationwide networks for the dissemination and exchange of scientific information. The libraries manage and operate the NOAA Automated Library and Information System. The center's National Environmental Data Referral Service maintains and operates an on-line environmental data catalog which identifies and describes archived environmental data sets from around the world.

4. **Berkeley Particle Data Center**
   Lawrence Berkeley Laboratory
   U.S. Department of Energy
   Mailstop 50-308
   Berkeley, CA 94720
   Contact: (415) 486-4719

The center compiles high energy physics data and makes it available to the physics community in periodic publications and in computer searchable databases. The major publications of the Center are the Review of Particle Properties (LBL-100) and the associated data booklet and pocket diary, Current Experiments in Elementary Particle Physics (LBL-91); Major Detectors in Elementary Particle Physics (LBL-91 Supplement); A Guide to Data in Elementary Particle Physics (LBL-90); and the Compilation of High Energy Physics Reaction data (LBL-92). The computer databases corresponding to these publications are available on the Stanford Linear Accelerator Center SPIRES system and are described in A User's Guide to Particle Physics Computer-Searchable Databases on the SLAC-SPIRES System (LBL-19173). The work of the center is done in collaboration with a large number of institutions, including the European Laboratory for Particle Physics (CERN), The University of Durham, The University of Helsinki, the Institute for High Energy Physics (Serpukhov, U.S.S.R.), and the Stanford Linear Accelerator Center (SLAC).

5. **Central Industrial Applications Center (CIAC)**
   National Aeronautics and Space Administration
   Southeastern Oklahoma State University
   Station A Box 2584
   Durant, OK 74701
   Contact: (405) 924-6822

The center functions as part of the NASA network for technology transfer. It is a nonprofit organization that seeks to expand and expedite the transfer of new technology to individuals, firms, and agencies in the Texas, Oklahoma, Kansas, Nebraska, South Dakota, and North Dakota regions. Because of small business emphasis, it has been especially concerned with making state-of-the-art technology available to small and medium-sized businesses.

6. **Chemical Effects Information**
   Oak Ridge National Laboratory
   U.S. Department of Energy
   P. O. Box X, Bldg. 2001
   Oak Ridge, TN 37831
   Contact: (615) 576-7568

The Chemical Effects Information Task Group provides information support to the scientific and administrative communities concerning health and environmental effects of chemical pollutants. Projects can be grouped into three principal areas: (1) state-of-the-art knowlege reviews; (2) health and environmental assessment reports; and (3) specializing data bases and catalogs.

7. **Coastal Engineering Information Analysis Center (CEIAC)**
   Army Engineer Waterways Experiment Station
   U.S. Department of the Army
   P. O. Box 631
   Vicksburg, MS 39180
   Contact: (601) 634-2012

The center was established to collect, analyze, and disseminate information on coastal engineering research and technology. It deals primarily with the following coastal engineering subjects: wave action in coastal waters, coastal processes, tides, surges, and long period waves, inlet dynamics, coastal works evaluation, functional and structural design of coastal works, coastal construction techniques, coastal environmental data collection, and coastal harbors and channels.

8. **Cold Regions Science and Technology Information Analysis Center**
   Cold Regions Research and Engineering Laboratory
   U.S. Department of the Army
   72 Lyne Rd.
   Hanover, NH 03755

Contact: (603) 643-4221

The center provides information on snow, ice, and frozen ground; materials, facilities, systems, and operations in cold environments; methods and techniques of using various energy forms and systems to obtain information about surface and subsurface features in all environments for engineering, military, and related scientific purposes; ecology; and pollution control in arctic environments. The laboratory sponsors the Bibliography on Cold Regions Science Technology which has been prepared at the Library of Congress since 1951. This bibliography is available for interactive computerized literature searching as file COLD via the Orbit Search Service, McLean, VA.

**9.    Computer Software and Management Information Center (COSMIC)**
National Aeronautics and Space
  Administration
382 E. Broad Street, University of Georgia
Athens, GA 30602
Contact: (404) 542-3265

As NASA's software dissemination center, COSMIC makes available to business and industry over 1400 computer programs covering all areas of NASA project involvement — including structural analysis, thermal engineering, energy conservation, computer graphics, remote sensing and turbomachinery. Source code is supplied for each program along with detailed use  documentation.

**10.    Concrete Technology Information Analysis Center (CTIAC)**
Army Engineer Waterways Experiment
  Station
U.S. Department of the Army
P. O. Box 631
Vicksburg, MS 39180
Contact: (601) 634-3264

CTIAC acquires, analyzes, evaluates, and condenses world literature in the subject area of concrete technology. Its areas of interest are concrete materials, properties, construction methods, and tests; concrete composition; chemistry and physics of concrete and concrete materials; analytical procedures and test methods; portland cement grout mixtures. Services include providing information on research in progress, making referrals to other sources of information, and permitting onsite use of its collection.

**11.    Controlled Fusion Atomic Data Center**
Oak Ridge National Laboratory
U.S. Department of Energy, Bldg. 6003
P. O. Box X
Oak Ridge, TN 37831
Contact: (615) 574-4701

The Data Center maintains a detailed bibliography of atomic data measurements and calculations for collision processes of interest to the fusion community. One hundred nineteen journals are regularly searched for papers of interest with entries categorized by author, process, reactants, energy range, and theory/experiment. Because the total field of atomic and molecular collision processes is enormous, the center has restricted the coverage of molecules to $H_2$, $H_3$, HeH, $N_2$, $O_2$, CO, $CO_2$, OH, $H_2O$, CH, $CH_2$, $CH_3$, and $CH_4$ and their ions. The major categories used in classifying the Center's bibliographical input include: heavy particle — heavy particle interactions; interactions of atomic particles with electromagnetic fields; particle penetration in macroscopic matter (ions, neutrals, and electrons); particle interactions with solid surfaces; electron-particle interactions; and photon collisions with heavy particles and electrons (hv < 100 keV).

**12.    Crude Oil Property Data Base**
National Institute for Petroleum and Energy
  Research
U.S. Department of Energy
P. O. Box 2128
Bartlesville, OK 74005
Contact: (918) 336-2400

Crude oils, as produced, have widely different properties because of their variation in composition. This makes some petroleum much more adaptable to a given refining process and thus more suitable for a specific use. Consequently, there is much interest in the properties of a crude oil produced from a specific deposit. The center has the largest collection of such information on crude oils available. A data bank contains approximately 9000 analyses. An analysis contains the general properties: gravity, sulfur content, nitrogen content, viscosity, color, and pour point as well as a location description of the source of the oil. Properties of oil fractions are obtained where appropriate.

**13.    Data & Analysis Center for Software (DACS)**
Rome Air Development Center

U.S. Department of the Air Force
Attn: RADC/COED
Griffiss AFB, NY 13441-5700
Contact: (315) 330-3395

This information analysis center was established in response to a recognized need for a facility to serve as a centralized source for current, readily usable data and information concerning software technology. The major functions of the DACS are (1) to develop and maintain a computer database of empirical data collected on the development and maintenance of computer software; (2) to produce and distribute subsets of the database; (3) to maintain a software technology information base of technical documents, and evaluation data; (4) to analyze the data and information and produce technical reports; (5) to maintain a current awareness program which will include dissemination of technical information, assessments of technological developments, and publication of a quarterly newsletter; and (6) provide rapid response to inquiries for technical information and assistance.

**14. Earth Resources Observation Systems (EROS) Data Center**
U.S. Geological Survey
U.S. Department of the Interior
Sioux Falls, SD 57198
Contact: (605) 594-6511

The center is operated as part of the U.S. Department of the Interior's Geological Survey National Mapping Division Program. It is a national collection, production, and research facility for remotely sensed data and other forms of geographic information. The center was built in the early 1970s to receive, process, and distribute data from the U.S. Landsat satellite sensors and from airborne mapping cameras. Today it holds over 2,000,000 worldwide scenes of Earth acquired by Landsat satellites and over 5,000,000 aerial photographs of U.S. sites. In addition, increasing amounts of other kinds of remotely sensed Earth data are entering the Data Center archives, such as radar images acquired by aircraft, digital stream sediment data, and airborne radiometric and magnetic survey data in digital format. Nationwide, the Data Center is linked to the U.S. Geological Survey's National Cartographic Information Center Network.

**15. Earth Science Information Network**
U.S. Geological Survey

Department of the Interior
801 National Center
Reston, VA 22092
Contact: (703) 648-7112

The network is a cooperative service which provides earth science information to the private and public sectors. It is designed to provide better coordination between existing public service programs of the Geological Survey; to improve the sharing of data, information, and resources of these programs. Services include the sales and distribution of the geologic, hydrologic, geographic, and topographic maps and reports of the U.S. Geological Survey as well as the dissemination of data from various earth science data systems. A network of regional centers offers easy access by the public.

The Earth Science Data Directory (ESDD) is a nationwide system for readily determining the availability of specific earth science and natural resource data. It offers online access to a U.S. Geological Survey mainframe computer repository of information about earth science and natural-resource databases.

**16. Environmental Carcinogen Information Center**
Oak Ridge National Laboratory
U.S. Department of Energy, Bldg. 9207, MS 003
P. O. Box Y
Oak Ridge, TN 37831
Contact: (615) 574-7871

The Center is responsible for a special computerized database, the Environmental Carcinogen Information File. Formation of this file was originally constructed to support the Environmental Protection Agency Gene-Tox Program's Committee on Carcinogens. The material selected included the literature concerning carcinogenesis of short-term tested chemicals (selected by the Gene-Tox Panel). Work has continued on the acquisition and the technical indexing of all literature in this research area with an attempt being made for complete coverage from 1980. Selection of material includes any publication in which a chemical or physical agent was tested under suitable conditions of time and length of observation to have induced tumors.

**17. Environmental Technical Information System (ETIS)**
Army Engineer Construction Engineering

Research Laboratory
U.S. Department of the Army
ETIS Support Center
909 W. Nevada St.
Urbana, IL 61801
Contact: (217) 333-1369

ETIS is a computer-based information bank developed to assist either in the development of environmental impact statements (EIS) and environmental assessments (EA) or environmental management activities. It consists of five subsystems: the Environmental Impact Computer System (EICS), the Economic Impact Forecast System (EIFS), the Computer-Aided Environmental Legislative Data System (CELDS), the Soils Information Retrieval System (SIRS), and the Multiple Parameter Series Search (MPSS), EICS allows the user to identify the potential consequences of activities and guides the discussion of these consequences in an EA/EIS.

## 18. Federal Computer Products Center

National Technical Information Service
U.S. Department of Commerce
5285 Port Royal Road
Springfield, VA 22161
Contact: (703) 487-4763

The Center was established to improve access to the variety of products produced by the Government that are machine processable. This is being accomplished through extensive contact with the various Federal agencies. The center is organized into four major product areas: bibliographic data bases; computer software; numeric and textural source data files; and statistical services. A number of bibliographic files produced by other agencies have been made available to the public. A wide variety of software and data files pertinent to business and scientific interests are available for sale or lease. The present collection contains over 2,500 data files and computer programs from more than 100 Federal agencies, covering a vast array of subject fields: labor statistics, economics, education, demography and population, health statistics, building technology, energy sources, environmental pollution and control, and much more. The statistical services program provides users with two types of services: the Statistical Data Reference Service (SDRS) and the Statistical Data Tabulation Service.

## 19. Fossil Energy Information

Oak Ridge National Laboratory
U.S. Department of Energy
P.O. Box X
Oak Ridge, TN 37831-6050
Contact: (615) 574-4955

The center provides information on fossil energy. Among its products are the following: a computerized inventory of acid rain deposition research by federal and state agencies; a current awareness service on environmental regulatory and legislative issues of interest to DOE Fossil Energy; a newsletter on issues of interest to the acid rain deposition research community; reviews and other publications on the health effects of coal conversion technologies; and an on-line, interactive, computerized system of environmental, health, and safety information in support of the role of the U.S. Department of Energy as technical consultant to the Synthetic Fuels Corporation and as monitor of the Great Plains Coal Gasification Project. Products and services of the task group are only available to outside groups by prior arrangement. The center is partially funded by EPA, NOAA, and HHS.

## 20. Government-Industry Data Exchange Program (GIDEP)

U.S. Department of the Navy
Naval Fleet Analysis Center
Corona, CA 91720
Contact: (714) 736-4677

GIDEP (Government-Industry Data Exchange Program) is a cooperative activity between government and industry participants seeking to reduce or eliminate expenditures of time and money by making maximum use of existing knowledge. The program provides a means to exchange certain types of technical data essential in the research, design, development, production and operational phases of the life cycle of systems and equipment. As a result of government emphasis on commercial off-the-shelf items, any activity which uses and/or generates the types of data GIDEP exchanges may be considered for membership. The program specifically excludes classified and proprietary information. Participants in GIDEP are provided access to the four major data interchanges: the Engineering Data Interchange, the Metrology Data Interchange, the Reliability-Maintainability Data Interchange, and the Failure Experience Data Interchange. A new database on value engineering (VEDISARS) has been created.

**21. Hazardous Materials Technical Center (HMTC)**
Defense Logistics Agency
U.S. Department of Defense
P.O. Box 8168
Rockville, MD 20856-8168
Contact: (301) 468-8858

The Center was established to provide expertise in technology and regulations dealing with hazardous materials and hazardous wastes. It gathers, analyzes, stores, and disseminates information related to hazardous materials management. HMTC's staff can provide information and guidance on safety and health aspects, as well as the proper handling, storage, transportation and disposal procedures of hazardous components, substances and wastes. HMTC provides many products and services including a bi-monthly newsletter, the HMTC UP-DATE, and the quarterly HMTC Abstract Bulletin. A comprehensive bibliographic search service is also offered by the HMTC. A search on a specific chemical or topic can be done using over 250 available databases including HMTC's own hazardous materials database.

**22. High Temperature Materials Information Analysis Center (HTMIAC)**
U.S. Department of Defense
CINDAS, Purdue University
2595 Yeager Road
West Lafayette, IN 47906
Contact: (317) 494-9393 or (317) 463-1581

The mission of the Center is to provide comprehensive, authoritative, and timely scientific and technical information analysis services on thermophysical, thermomechanical, and optical properties of materials. It evaluates, analyzes, and summarizes the available scientific and technical data from worldwide sources on the various properties of materials so as to maintain a comprehensive, authoritative, and up-to-date national data base. It responds to requests for material properties data and information and publishes major reference works such as data books, handbooks and retrieval guides. Materials and properties covered are as follows:

- Materials — graphite/epoxy composites, graphite/polymimide composites, Kev-lar/epoxy composites, carbon/carbon composites, carbon/phenolic composites, fiberglass/epoxy composites, graphite/bis-maleimide composites, silica/phenolic composites, selected aluminum and titanium alloys and stainless steels, selected detector materials, selected e-m transparent materials, selected thin films, and selected switching materials.

- Thermophysical and optical properties — ablation energy, ablation temperature, absorptance, absorption coefficient, boiling point, density, electrical resistivity, emittance, heat capacity, heat of fusion, heat of vaporization, melting point, reflectance, refractive index, thermal conductivity, thermal diffusivity, thermal expansion, and transmittance.

- Thermomechanical properties — compressive modulus, compressive strength, flexural modulus, flexural strength, fracture toughness, Poison's ratio, shear modulus, shear strength, stress-strain curves, tensile (Young's) modulus, tensile strength, ultimate strain, and yield strength.

**23. Housing Technology Information Service**
HUD User
U.S. Department of Housing and Urban Development
P.O. Box 280
Germantown, MD 20874
Contact: (301) 251-5154

Information is provided on research concerning housing and urban topics. These topics include affordable housing, building technology, community development, energy conservation and utilization, services for the elderly and handicapped, and neighborhood rehabilitation and conservation.

**24. Hydraulic Engineering Information Analysis Center (HEIAC)**
Army Engineer Waterways Experiment Station
U.S. Department of the Army
P.O. Box 631
Vicksburg, MS 39180
Contact: (601) 634-3368

Information is collected, analyzed, and distributed in the field of hydraulic engineering. Topics covered include river, harbor, and tidal hydraulics; flow-through pipes, conduits, channels, and spillways, as related to flood control and navigation; hydraulic design and performance of dams, locks, channels, and other navigation structures; and water waves and underwater shock effects. Services provided by the center include literature searches, development of state-of-the-art reviews, and responses to requests for specific data.

### 25. Infrared Information and Analysis Center (IRIA)

U.S. Department of the Navy
Environmental Research Institute of Michigan
P.O. Box 8618
Ann Arbor, MI 48107
Contact: (313) 994-1200

IRIA collects, analyzes, and disseminates information on infrared and electrooptical technology, with emphasis on the military applications. The subject areas covered include: radiation sources emitting in the UV through IR regions; radiation characteristics of natural and man-made targets; optical properties of materials; detection materials, elements and arrays; lasers; image tubes and sensors; optical systems and components; detector coolers and electronics; atmospheric propagation including absorption, emission, scattering and turbulence effects; and search, homing, tracking, ranging, counter-measures, reconnaissance, and other military infrared and laser systems. Responds to technical inquiries of qualified users.

### 26. Legislative Database (Radioactive Materials)

U.S. Department of Energy
Office of Transportation and Waste Systems
505 King Street
Columbus, OH 43201
Contact: (614) 424-4370

The database is a comprehensive source of information on federal, state, and local legislation pertinent to the shipment of radioactive materials. The database currently contains citations for approximately 1900 legislative actions; it includes references to specific legislative actions that have been introduced, enacted, or denied at federal, state, and local levels of government. The database is updated twice a week.

### 27. Manufacturing Technology Information Analysis Center (MTIAC)

U.S. Department of Defense

10 West 35th Street
Chicago, IL 60616
Contact: (312) 567-4733

MTIAC is responsible for the collection, analysis, and dissemination of manufacturing technology information and data in the following areas: metals, nonmetals, inspection and testing, electronics and munitions, and computer-aided design/computer-aided manufacturing (CAD/CAM). This manufacturing technology information and data are acquired and disseminated in the following defense-related fields: machine tools and manufacturing equipment, robots and special machines, material-handling equipment, controls, software and databases, communication lines and networks, sensors and inspection or checkout procedures, signal processing, materials and materials treatment, production processes, specific defense-related products, and the management aspects of manufacturing technology.

### 28. Metal Matrix Composites Information Analysis Center (MMCIAC)

U.S. Department of Defense
Kaman-TEMPO, P.O. Drawer QQ
Santa Barbara, CA 93102
Contact: (805) 963-6482

The service provides scientific and technical information analysis in the area of metal matrix composite materials. It provides the facilities and capabilities to: (1) identify, collect, process, store, and disseminate authoritative information; (2) prepare or sponsor the preparation of the necessary products and services to communicate this information to researchers, practicing specialists, manufacturers, and other users with interests and concerns in metal matrix composites; and (3) coordinate and augment existing information activities to improve the transmittal of this information to interested organizations and individuals in the government, military and private sector. The subject matter coverage includes continuous fibers, wires, discontinuous whiskers with L/D 10, directionally solidified eutectics; boron, graphite, silicon carbide, nitride, alumina, boron carbide and titanium diboride fibers, stainless steel, tungsten, molybdenum, beryllium, and titanium, niobium alloy wires; alumina, silicon carbide, and silicon nitride whiskers. Technical areas include: manufacturing of composites; fabrication process development; defense systems applications; performance computations; cost, test and evaluation techniques and methods; materials characteristics and properties; vulnerability and survivability; operational serviceability and repair.

### 29. Metals and Ceramics Information Center (MCIC)
U.S. Department of Defense
Battelle-Columbus Division
505 King Ave.
Columbus, OH 43201-2693
Contact: (614) 424-5000

The Center provides timely, authoritative technical information on the characteristics and utilization of the advanced metals and ceramics. Materials include: titanium and titanium-base alloys, beryllium and beryllium-base alloys, high-strength aluminum alloys, magnesium, superalloys, refractory metals, coatings for these metals, other metals and alloys used in critical structural applications, single crystal and polycrystalline metals, oxides, sulfides, carbides, borides, nitrides, silicides, intermetallics, metalloid elements, glasses, carbons, graphites, and coatings for these materials. The following four basic functions are conducted: maintenance of a comprehensive, up-to-date, authoritative technical information base; response to requests for technical advice and assistance; issuance of a monthly bulletin on new technical developments; and publication of technical reports, handbooks, and related documents.

### 30. Mineral Resources Data System
U.S. Geological Survey
U.S. Department of the Interior
920 National Center
Reston, VA 22092
Contact: (703) 648-6126

The data system consists of a set of records on the mineral deposits and mineral commodities of the U.S. and, to a certain extent, the world. The file is arranged so as to accept the basic information needed to describe a mineral deposit or a group of related deposits, such as a mining district. The file is flexible so that what constitutes a record can be decided largely by the user. However, certain minimal data are required to establish a record. Retrieved information can be printed in any of three arrangements, or it can be passed to a subsequent program for further processing, such as maps and other graphics.

### 31. NASA Industrial Applications Center (NIAC)*
National Aeronautics and Space
  Administration
University of Pittsburgh
823 William Pitt Union
Pittsburgh, PA 15260
Contact: (412) 648-7010

The NASA Industrial Application Center (NIAC) has been working for over 20 years to facilitate the transfer of technology and technical knowledge from NASA to business, industrial, and government clients. As a component of NASA's technology utilization network, NIAC is a major scientific and technological resource providing information retrieval services and applied technology transfer projects. Intended to prevent wasteful duplication of research already accomplished, NIAC works to broaden and expedite the transfer of technology. NIAC provides services including market intelligence studies, engineering studies, management support, database development activities, and patent/licensing. NIAC works with all the NASA research centers, including the Lewis Research Center in Cleveland, Ohio; the Goddard Space Flight Center in Greenbelt, Maryland; and the Langley Research Center in Hampton, Virginia. Through NASA membership in the Federal Laboratory Consortium, this access has been increased to include virtually all Federal labs. As an integral part of the University of Pittsburgh, NIAC frequently utilizes the unique talents of the faculty. In addition, NIAC also works with many scientists and researchers at other academic institutions.

### 32. NASA Industrial Applications Center (NIAC)**
National Aeronautics and Space
  Administration
Denney Research Bldg.
University of Southern California
Los Angeles, CA 90007
Contact: (213) 743-6132

The center offers the locating of information through on-line searches of computerized bibliographic databases. Information is accessed for clients with problems in research and development, patent innovation, quality control, manufacturing, and other diverse technical areas. Special searches are also offered where a technical coordinator, lo-

* See also next item.
** See also preceding item.

cated at the NASA Ames Research Center, contacts experts in both government and industry for opinions on solutions to technical problems. NIAC also offers the Remote Interactive Search (RIS) where clients with the proper computer terminal can follow the informational search in real time from the convenience of their offices. The NIAC affiliates network extends the services to statewide businesses in 15 western states: businesses can contact the Alaska Economic Development Center at the University of Alaska in Juneau; NIAC at the University of Southern California; the Small Business Assistance Center at the University of Colorado; the High Technology Development Corporation in Honolulu, Hawaii; the Idaho Business Development Center in Boise State University in Boise, Idaho; CIRAS in Iowa State University, Ames, Iowa; the University Technology Assistance Center at the University of Nebraska; the Technology Applications Center, University of New Mexico; the regional Eastern Oregon State College, La Grande, Oregon; and the Small Business Development Center, Washington State University, Washington.

33. **NASA-Southern Technology Applications Center**
   National Aeronautics and Space
    Administration
   The Progress Center
   One Progress Blvd., Box 24
   Alachua, FL 32615
   Contact: (800) 225-0308. For Florida call (800) 354-4832

STAC provides assistance in transferring and commercializing technology generated by NASA, other federal laboratories, and by university researchers. STAC has developed a variety of services, including information research, database development, direct assistance to entrepreneurs, and problem-solving for virtually any type of private sector business. Headquartered at the University of Florida in Gainesville, STAC has area offices in the Colleges of Engineering or Business at each of the universities in the State University System, as well as area offices in Alabama, Georgia, Arkansas, Tennessee, and South Carolina. Affiliations with technology transfer groups also exist in Mississippi and Louisiana. STAC is a key resource for Florida's high technology economic growth initiatives and works closely with state agencies, including the High Technology and Industry Council, the Department of Education, the Department of

Commerce, and the Board of Regents to foster high technology economic development in Florida. STAC staff members are professionals with expertise including engineering (all areas), physics, medicine, biology, computer science, chemistry, marine science, and computerized information retrieval. The foundation for many of STAC's activities and services, the Information Research Center (IRC) is a world class resource with access to more than 1200 computerized databases worldwide containing over 500 million records.

34. **National Appropriate Technology Assistance Service (NATAS)**
   U.S. Department of Energy
   P.O. Box 2525
   Butte, MI 59702-2525
   Contact: (800) 428-2525

Appropriate technology is the term for those practical technologies that lend themselves to self help and provide solutions to individual energy problems. The technologies usually include use of the sun, wind and water resources. Appropriate technologies also include energy and resource conservation methods. NATAS provides three primary services: tailored information responses, engineering/scientific technical assistance, and commercialization technical assistance. NATAS is available to assist anyone in the U.S. through a toll-free number.

35. **National Center for Standards and Certification Information (NCSCI)**
   National Institute of Standards and
    Technology
   U.S. Department of Commerce
   Gaithersburg, MD 20899
   Contact: (301) 975-4030

The Center provides up-to-date information on standards, rules of certification and regulations. NCSCI maintains (1) a computerized database of U.S. voluntary engineering standards; (2) a reference collection which includes directories, technical and scientific dictionaries, encyclopedias, and handbooks; (3) microform files containing military and Federal specifications, U.S. industry and national standards, and international and foreign national standards; and (4) directories of private and Federal certification programs, standards activities of U.S. organizations and regional and international standards activities. NCSCI also provides information on proposed foreign regulations that may effect trade with signatories to the General Agreement on Tariffs and Trade (GATT) Agreement on Technical Barriers to Trade.

### 36. National Center for the Thermodynamic Data of Minerals
U.S. Geological Survey
U.S. Department of the Interior
959 National Center
Reston, VA 22092
Contact: (703) 648-6755

The data center critically evaluates all published literature on the physical and thermodynamic properties of naturally occurring phases in the ranges in temperature, pressure, and composition as found in the geological environment. It provides reference services; distributes publications; and makes referrals to other sources of information. It interacts closely with the Standard Reference Data Centers of the National Institute of Standards and Technology.

### 37. National Coal Resources Data System (NCRDS)
U.S. Geological Survey
U.S. Department of the Interior
Stop 6, 956 National Center
Reston, VA 22092
Contact: (703) 648-6435

The system is a computerized storage, retrieval, and display system developed by the Branch of Coal Resources. NCRDS correlates and standardizes coal data from federal and state agencies, universities, and the private sector. The system is constantly updated and expanded to serve current and future needs in U.S. coal resource inventory and analysis. Eight files of geologic and geographic information are currently available. The files cover the following: published coal resource estimates for coal-bearing states listed by state, county, coalfield, geologic age and formation, rank, thickness of coal, and thickness of overburden; standard USBM chemical analysis (proximate, ultimate, BTU, ash-softening temperature, free swelling index, and Hargrove grindability index) by state, county, bed code and name, and mine and nearest town codes; major-, minor-, and trace element analyses by USGS laboratories; stratigraphic data from drill holes, outcrops, and mines; petrographic analyses of selected coal samples by USGS laboratories; U.S. Bureau of Mines summary re-

serve/base analysis data; and major-, minor-, and trace-element analyses by USGS laboratories of coal from several foreign countries.

### 38. National Energy Information Center (NEIC)
U.S. Department of Energy
Energy Information Administration
Room 1F-48, Forrestal Bldg.
Washington, DC 20585
Contact: (202) 586-8800

The Center provides statistical and analytical energy data, information, and referral assistance to the Government and private sectors, as well as the general public. It also provides location assistance for sources of energy information outside EIA programs. The data are collected from 75 data-gathering surveys and are disseminated in published form, magnetic tapes, and on-line data files. Most data are available in publications — both periodicals and one-time analytic reports. Currently NEIC produces 2 weeklies, 5 monthlies, 4 quarterlies, and 50 semi-annual, annual, and biennial publications. Most current publications are available for sale by NEIC. Major data files and most of the computer models are available on magnetic tape through NTIS. One database contains statistics on petroleum, coal, and electricity and is free of charge. The types of data gathered cover production, consumption, prices, supply, disposition, imports, exports, reserves, distribution, sales, generation, inventories and storage for petroleum, natural gas, coal, electricity, nuclear power, and renewable energy.

### 39. National Energy Software Center
U.S. Department of Energy
Argonne National Laboratory
9700 South Cass Ave.
Argonne, IL 60439
Contact: (312) 972-7250

The Center serves as a computer software information and resource center for management and sharing of scientific and technical computer software developed under DOE funding. It collects, packages, maintains, and distributes computer software: checks library software for transportability and executes test problems to verify successful implementation; prepares and publishes abstracts describing the packages in the NESC collection; advises users of library packages of corrections, modifications, and replacement packages as they are processed; and assists users in implementing and using library software.

**40.  National Environmental Data Referral Service (NEDRES)**
National Oceanic and Atmospheric Administration
U.S. Department of Commerce
National Environmental Satellite, Data, and Information Service
1825 Connecticut Ave., NW
Contact: (202) 673-5404

As a computer-searchable catalog and index of environmental data, NEDRES is a ''yellow pages'' directory to private, academic, local, state, federal, and NOAA data. NEDRES does not contain the actual data, but describes them and refers the researcher to the holder of the data. It is a publicly available service which identifies the existence, location, characteristics, and availability conditions of environmental data and information. The service is accessible on a commercial computer system using a computer terminal and national and international telecommunications systems.

**41.  National Geophysical Data Center (NGDC)**
National Oceanic and Atmospheric Administration
U.S. Department of Commerce
National Environmental Satellite, Data, and Information Service
325 Broadway
Boulder, CO 80303
Contact: (303) 497-6215

The Center collects, manages, archives and distributes data in the fields of solid earth geophysics, marine geology and geophysics, and solar-terrestrial physics. NGDC data sources include NDAA's observing programs and, through cooperative arrangements, data from universities, other government agencies and foreign organizations. NGDC also serves as World Data Center-A for the above disciplines. NGDC can provide specialized data services on a reimbursable basis using geographic information systems and tabular database processing.

**42.  National Nuclear Data Center**
Brookhaven National Laboratory
U.S. Department of Energy
Upton, NY 11973
Contact: (516) 282-2901

Services are provided to the entire low energy nuclear science community. These include information on neutron physics, charged-particle reactions, nuclear structure, and decay data. The Evaluated Nuclear Structure Data file (ENSDF) and the Nuclear Data Sheets (NDS) are prepared by the center. The Nuclear Structure References Bibliographics file has been used to produce recent references issues of the Nuclear Data Sheets. The Evaluated Nuclear Data File (ENDF/B) effort is coordinated by the Center. Contributions are also made to the Computer Index of Neutron Data (CINDA) publication, which is a bibliographic index to neutron data.

**43.  National Oceanographic Data Center**
National Oceanic and Atmospheric Administration
U.S. Department of Commerce
National Environmental Satellite, Data and Information Service
1825 Connecticut Ave., NW
Washington, DC 20235
Contact: (202) 673-5594

NODC is the U.S. national facility established to acquire, process, archive, and disseminate global oceanographic data. NODC's digital databases include data collected by federal, state, and local government agencies; universities and research institutions; and private industry. It also acquires data from foreign sources and operates World Data Center A for Oceanography, a part of the World Data Center systems that facilitates international exchange of scientific data. Global, deep-ocean data bases include: (1) oceanographic station data-measurements of temperature, salinity, oxygen, phosphate, phosphorus, nitrite, nitrate, silicate, and pH at the surface and serial depths; (2) temperature-depth profiles from expendable and mechanical bathythermographs; and (3) surface current (ship drift) data; and (4) sea level, wind speed, and wave height data from the U.S. Navy Geodetic Satellite (GEOSAT). In addition, NODC holds environmental assessment data collected primarily on the U.S. outer continental shelf to support studies of the effects on marine ecosystems of offshore development. These data include: measurements of hydrocarbons, metals, and other pollutants and toxic substances. Services include: data inventory searches; selective retrieval and formatted output and other computer-generated data summaries, analyses, and graphic displays.

**44.  National Technical Information Service (NTIS)**
U.S. Department of Commerce
5285 Port Royal Road
Springfield, VA 22161

The NTIS provides access to the results of U.S. and foreign government-sponsored R&D and engineering activities through special information products. As the U.S. government's central technical and scientific information service, NTIS: annually announces more than 60,000 summaries of the results of U.S. and foreign government-sponsored R&D and engineering activities; can provide the complete technical reports for most of these announced results; and manages the Federal Computer Products Center. The Center provides access to software, data files, and databases produced by Federal agencies. NTIS manages the Center for the Utilization Federal Technology (CUFT) which prepares a variety of directories, catalogs, and other information products linking U.S. firms to key U.S. Government technologies, inventions available for licensing, and laboratory contacts. NTIS brings U.S. firms access to more than 15,000 foreign government-sponsored research and engineering results annually. NTIS is the lead U.S. government agency for cooperation in technical information exchange. Through numerous agreements new international R&D efforts are announced.

There are three major NTIS subscription services that alert businesses to newly issued technical reports or technologies. A weekly bulletin is available on a subscription basis in each of 27 different subject areas. The bulletin, called *Abstract Newsletters,* announces summaries of newly released government R&D reports and provides complete coverage of broad areas of government research. Another weekly bulletin, *Government Inventions for Licensing,* announces annually more than 1,500 U.S. government-owned inventions available for licensing, often on an exclusive basis. The third major current awareness service is a monthly subscription alerting companies to new federal technology having practical or commercial potential. This awareness service, called *Tech Notes,* offers its readers one- or two-page fact sheets, often illustrated, of new processes, equipment, software, and materials. More than 100 fact sheets are pub-

lished monthly. Special bibliographies, called Published Searches, have been prepared on more than 3500 different topics, designed to help companies inexpensively find information about past research in specific areas. For further information write to NTIS, Springfield, VA 22161 and ask for brochures PR-205, PR-365, PR-750, PR-154, and PR-701.

**45.  National Water Data Exchange (NAWDEX)**
U.S. Geological Survey
U.S. Department of the Interior
421 National Center
Reston, VA 22092
Contact: (703) 648-4000

The service is a computerized data system that identifies sources of water data and that indexes the types of water data these sources collect. The primary purpose of the system is to facilitate the exchange of data between the organizations that gather water data and the organizations that need the data. The NAWDEX office also provides direct access service to and disseminates data from the National Water Data Storage and Retrieval System (WATSTORE) of the U.S. Environmental Protection Agency.

**46.  New England Research Application Center (NERAC)**
National Aeronautics and Space
  Administration
Mansfield Professional Park
Storrs, CT 06268
Contact: (203) 429-3000

NERAC aids and promotes the transfer of technology by helping business and industry, colleges and universities, and local governments locate appropriate technical and business information. Its mission is to help industry benefit by using previously developed technology. NERAC makes use of sources from throughout the world. More than 100 databases are available on NERAC's in-house computer. One area of specialization at NERAC is the availability of current awareness profiles which offer users a means of keeping up-to-date on a particular topic.

**47.  Nitinol Technology Center**
Naval Surface Weapons Center
U.S. Department of the Navy
Code R-32, White Oak

Silver Spring, MD 20903

Contact: (301) 394-2468

Nitinol is the novel engineering alloy among metals exhibiting shape memory effects. New applications for Nitinol continue to appear where designers are able to use small amounts of heat to reestablish a prior shape, produce a mechanical force, or motivate devices. The center has been created to emphasize the manufacturing technology of Nitinol and to assist in the development of Nitinol-using devices. The center offers guidance and/or prototype development programs for the development of Nitinol-using devices.

**48.  Nondestructive Testing Information
       Analysis Center (NTIAC)**

U.S. Department of Defense

Southwest Research Institute

P.O. Drawer 28510

San Antonio, TX 78284

Contact: (512) 522-2362

NTIAC functions to collect, review, analyze, appraise, summarize, and disseminate pertinent and timely literature on the processes, techniques, and associated technologies concerning nondestructive testing. All types of techniques are covered such as radiographic, holographic, acoustic, and magnetic techniques. It is also concerned with the economic aspects of the NDE industry and industry trends in applying current NDE technologies in research and development, production, maintenance, safety monitoring, and failure prevention of in-service material. Services include technical handbooks, data books, state-of-the-art reports, technology assessments, and bibliographies.

**49.  North Carolina Science and Technology
       Research Center**

National Aeronautics and Space
  Administration

P.O. Box 12235

Research Triangle Park, NC 27709

Contact: (919) 549-0671

The Center is one of the oldest NASA information centers, serving business and industry throughout the Southeast since 1964. Using remote online terminals, its staff of information specialists have access to more than 400 million reports and documents from worldwide sources. Beacuse of industry needs, the center has specialized in textiles, food science and technology, education and

training, electronics, biomedics, and chemistry/toxicology. The recent influx of high tech industries such as semiconductors, pharmaceuticals, and telecommunications has led the Center to expand its resources and areas of expertise. Services provided by the Center include development of bibliographies on specific topics, conferences, workshops, and a monthly technical bulletin.

**50.  Nuclear Operations Analysis Center
       (NOAC)**

Oak Ridge National Laboratory

Nuclear Regulatory Commission

P.O. Box Y, Bldg. 9711-1

Oak Ridge, TN 37830

Contact: (615) 574-0394

NOAC performs analysis tasks involving many aspects of nuclear power reactor operations and safety for the Nuclear Regulatory Commission (NRC). These include analyses of nuclear power experience, including generic case studies, plant operating assessments, and risk assessments. NOAC has developed and designed a number of major databases for the NRC which collect diverse types of information on nuclear power reactors from the construction phase through routine and off-normal operation. These databases make extensive use of reactor-operator-submitted reports, such as the Licensee Event Reports (LERs). The database developed to capture operating experience information reported in LERs is the Sequence Coding and Search System Data base (SCSS). This database, developed for NRC's Office for Analysis and Evaluation of Operational Data (AEOD) and maintained by NOAC, contains all current LERs submitted by the nuclear power utilities after January 1, 1980. It is updated on a continual basis with new LERs. NOAC services include consultation with staff specialists, access to the Center for use of documents, retrospective searches of various databases, technical inquiry service, and the technical progress review, Nuclear Safety.

**51.  Nuclear Reactor Safety Data Bank**

Idaho National Engineering Laboratory

U.S. Department of Energy

P.O. Box 1625

Idaho Falls, ID 83415

Contact: (208) 526-9507

This service collects, stores, and makes available data from the many domestic and foreign water reactor safety research programs. Users can access data for safety code development and code assessment. The administrative portion of the service provides data entry documentation, training, and advisory services to users and the U.S. Nuclear Regulatory Commission.

### 52. Ocean Pollution Data and Information Network (OPDIN)

National Oceanographic Data Center
National Oceanographic and Atmospheric Administration
U.S. Department of Commerce
National Environmental Statellite, Data and Information Service
1825 Connecticut Ave.
Washington, DC 20235
Contact: (202) 673-5539

The Network was established to improve dissemination of data and information resulting from ocean pollution programs conducted or sponsored by the U.S. government. The OPDIN is developed as a coordinating mechanism to make these data and information more accessible and useful, in response to the National Ocean Pollution Planning Act of 1978 (P.L. 950-273). The Network is intended to supplement (rather than replace) existing agency data and information sources and utilizes existing facilities where possible. Tasks to improve accessibility of ocean pollution data and information completed or underway include: Regional Coordination and Referral Offices; a "Handbook of Federal Ocean Pollution Data and Information Systems"; the National Marine Pollution Information System (NMPIS); a Marine Toxic Substances and Pollutants Data System; the Coastal Ocean Pollution Assessment News (COPAS), quarterly newsletter; and the Coastal Information System, a prototype personal computer system that provides coastal marine pollution and related information to regional users.

### 53. Pavements and Soil Trafficability Information Analysis Center

Army Engineer Waterways Experiment Station
U.S. Department of the Army
P.O. Box 631
Vicksburg, MS 39180
Contact: (601) 634-2734

Subjects covered by the Center include flexible and rigid pavements, expedient surfacing, ground flotation, and research in surface vehicle mobility, trafficability, and terrain analysis. This work is directed primarily toward military needs. Services provided by the center include literature searches in specific areas of interest, state-of-the-art reviews and bibliographies, and responses to requests for specific data and technical information searches.

### 54. Plastics Technical Evaluation Center (PLASTEC)

Army Armament Research, Development and Engineering Center
U.S. Department of the Army
Army Armament R and D Company
Picatinny Arsenal, NJ 07806-5000
Contact: (201) 724-4222

The Center is responsible for the generation, evaluation and exchange of technical information related to plastics, adhesives and organic matrix composites. It covers technology from applied research through fabrication with emphasis on properties and performance. Subject areas include structural, electrical, electronic, and packaging applications. This includes molded, formed, foamed, and laminated materials. A computerized data file is maintained on compatibility of polymers with propellants and explosives. It also maintains a complete file of standards, specifications, and handbooks in subject areas. Services include: technical inquiries, state-of-the-art studies, data compilations, handbooks, bibliographic, and literature searches.

### 55. Power Information Center

Interagency Advanced Power Group (IAPG)
U.S. Department of Defense
Suite 600, 1400 I St., NW
Washington, DC 20005
Contact: (202) 842-7600

The Interagency Advanced Power Group was initiated to help government researchers in advanced power exchange information and avoid unnecessary and costly duplication of their R&D efforts. The IAPG is supported by its member agencies — the U.S. Army, U.S. Navy, U.S. Air Force, NASA, and DOE. The Power Information Center, the support group of the IAPG, collects information on this R&D and offers a series of alerting services. The R&D results of interest to the Center include:

● Chemical — all electrochemical systems, such as chemical batteries, biochemical devices, simple fuel cell systems, chemical regenerative fuel cell systems (regardless of the sources of energy for regeneration), and chemical and thermal energy storage.

- Electrical — R&D programs in power conditioning and distribution, superconductivity, generators, motors, and power transmission.
- MHD — all energy conversion systems involving the magnetohydrodynamic (MHD) interaction between magnetic fields and electrically conducting fluids. Interest areas of fuels, materials, plasma dynamics, and combustion specifically related to MHD energy conversion are also covered.
- Mechanical — conversion to mechanical, hydraulic, and pneumatic energy. This includes working fluids, materials, heat transfer processes, heat transfer and storage equipment, and other components of mechanical conversion systems regardless of energy sources.
- Nuclear, TE, and TI — energy conversion from fission, fusion, and radioisotope power sources and other thermal systems which use thermoelectric or thermionic systems. This also includes investigations of related plasma dynamics.
- Solar — collection of solar radiation, its storage, and its conversion to heat or other energy forms.
- Systems — complete electric power/energy systems, from source to user.

### 56. Radiation Shielding Information Center (RSIC)
Oak Ridge National Laboratory
U.S. Department of Energy
P.O. Box X
Oak Ridge, TN 37831
Contact: (615) 574-6176

The Center was established to serve the U.S. shielding community. The Center serves to collect, organize, evaluate, and disseminate radiation protection and shielding information related to radiation from reactors and accelerators and to radiation occurring in space. It examines and analyzes radiation protection, radiation transport, and shielding information obtained through surveys of pertinent books, journals, and reports. RSIC maintains an information retrieval system which contains abstracted and indexed shielding information, and an archival microfiche file containing copies

of the analyzed literature, and reviews specific areas of shielding to determine the state-of-the-art. The Center also answers technical inquiries, publishes a monthly newsletter, and packages and distributes computer code systems and data libraries.

### 57. Radiopharmaceutical Internal Dose Information Center
Oak Ridge Associated Universities
U.S. Department of Energy
Manpower Education, Research, and
  Training Division
P.O. Box 117
Oak Ridge, TN 37831-0117
Contact: (615) 576-3448

The work at the Center involves calculating the radiation dose from administered radiopharmaceuticals and to collecting, interpreting, and correlating information about internal dosimetry of radioactive compounds. The Center is co-funded by the Food and Drug Administration. The data bank contains more than 31,000 references pertaining to radiopharmaceuticals as well as other pertinent decay scheme data, calculation techniques, physiologic behavior, phantoms and mathematical models.

### 58. Reliability Analysis Center (RAC)
Rome Air Development Center
U.S. Department of the Air Force
Griffiss AFB, NY 13441-5700
Contact: (315) 330-4151

RAC functions as a focal point for the recovery of reliability test data and experience information on microcircuit and related component parts. It collects, analyzes, formats, and disseminates reliability information on microcircuit, discrete semiconductor, and certain electrical/electromechanical components and the equipment/systems in which these components are used. Critically analyzed and evaluated reliability experience information is disseminated through reliability data compilations, handbooks, and appropriate special publications to upgrade and support defense systems reliability.

### 59. Remedial Action Program Information Center (Nuclear Facility Decommissioning)
Oak Ridge National Laboratory
U.S. Department of Energy
P.O. Box X, Bldg. 2001

Oak Ridge, TN 37831-6050

Contact: (615) 576-0568

The Center maintains a database which serves as a comprehensive source of technical information pertinent to the U.S. Department of Energy (DOE) Remedial Action Program, which comprises the Surplus Facilities Management Program (SFMP), the Formerly Utilized Sites Remedial Action Program (FUSRAP), the Uranium Mill Tailings Remedial Action Program (UMTRAP), and the Grand Junction Remedial Action (GJRAP). The database was developed and is managed by staff of the Remedial Action Program Information Center (RAPIC). Topics covered include: characterization surveys of radioactively contaminated facilities or sites; ongoing security and surveillance programs; preventive maintenance actions to ensure containment of radioactivity while awaiting permanent disposition; and assessment of environmental and engineering aspects of proposed remedial action alternatives.

### 60. Soil Mechanics Information Analysis Center (SMIAC)

Army Engineer Waterways Experiment Station

U.S. Department of the Army

P.O. Box 631, Vicksburg, MS 39180

Contact: (601) 634-3475

Information is provided on soil mechanics, engineering geology, rock mechanics, seismology, geophysics, and earthquake engineering. Services include literature searches in specific areas of interest, development of state-of-the-art reviews and bibliographies, and responses to requests for specific data.

### 61. Standard Reference Data

National Institute of Standards and Technology

U.S. Department of Commerce

Gaithersburg, MD 20899

Contact: (301) 975-2200

Critically evaluated numerical, physical, chemical, and materials properties data are required for a wide range of applications in industry, government, and academia. To provide the scientific and engineering community with reliable and accessible sources of data, databases of evaluated physical and chemical properties of substances have been developed. Experienced researchers assess the accuracy of the data reported in the literature, prepare compilations, and recommend the best values. The data bases are then made available through publications, magnetic tapes, and on-line systems. The following is a list of active data centers: Alloy Data Center; Aqueous Electrolyte Data Center; Atomic Collision Cross Section Information Center; Atomic Energy Levels Data Center; Atomic Transition Probabilities and Atomic Line Shapes and Shifts Data Center; Chemical Kinetics Information Center; Chemical Thermodynamics Data Center; Center for Information and Numerical Data Analysis and Synthesis; Crystal Data Center; Diffusion in Metals Data Center; Fluid Mixtures Data Center; Fundamental Constants Data Center; Ion Energetics Data Center; Molecular Spectra Data Center; Molten Salts Data Center; Phase Diagrams for Ceramists Data Center; Photonuclear Data Center; Radiation Chemistry Data Center; Thermodynamics Research Center; and X-Ray and Ionizing Radiation Data Center.

### 62. Toxicology Information Program

National Library of Medicine, National Institutes of Health

U.S. Department of Health and Human Services

8600 Rockville Pike

Bethesda, MD 20894

Contact: (301) 496-3147

The program is responsible for the Library's on-line chemical and toxicological files. CHEMLINE (Chemical Dictionary Online) is an online chemical dictionary and direct file that facilitates chemical searching in other NLM databases. TOXLINE (Toxicology Information Online) is an online bibliographic retrieval service, containing approximately 2,000,000 literature citations. DIRLINE (Directory of Information Services Online) is a computerized directory that refers users to organizations, agencies, and other sources of information on specific subject areas. RTEC (Registry of Toxic Effects of Chemical Substances) is the Library's online version of the publication by the National Institute for Occupational Safety and Health; this database with information on acute and chronic toxicity will soon be transferred to the Library's new TOXNET system. At present the TOXNET system contains the Hazardous Substances Data Bank (HSDB) and the CCRIS (Chemical Carcinogenesis Research Information System) file of the National Cancer Institute, CCRIS focuses on carcinogenicity, tumor production, and mutagenicity test results for over 1000 chemicals. HSDB provides comprehensive information for over 4200 hazardous, or potentially hazardous, chemicals. The scope of HSDB includes toxicity, safety and handling, emergency measures, environmental fate and exposure, standards and regulations, and analytic laboratory methods. The TOXNET system will eventually house a cluster of toxicology-related data banks and will provide user-friendly searching.

### 63. Toxicology Information Response Center
Oak Ridge National Laboratory, National
   Institutes of Health
U.S. Department of Health and Human
   Services
P.O. Box X, Building 2001
Oak Ridge, TN 37831
Contact: (615) 576-1743

This is a national center for toxicology information on environmental pollutants, industrial chemicals, food additives, pesticides, pharmaceuticals, and other topics of toxicologic concern. The center acquires, selects, stores, analyzes, and synthesizes comprehensive literature packages according to a user's specific request. Searches are provided on a fee-for-service basis. The center is sponsored by the National Library of Medicine.

### 64. Water Data Storage and Retrieval System (WATSTORE)
U.S. Geological Survey
U.S. Department of the Interior
437 National Center
Reston, VA 22092
Contact: (703) 648-5015

All types of water data are accessed through WATSTORE. Data are grouped and stored on the basis of common characteristics and data collection frequencies. These data are organized into five files: (1) Station Header File — an index file for the 220,000 sites for which data are stored; (2) Daily Values File — contains over 200 million daily values for water parameters such as streamflow, ground-water levels, specific conductance, and water temperatures; (3) Peak Flow File — 400,000 records on annual maximum streamflow and gage height values; (4) Water Quality File — stores results of 1.4 million analyses that describe 185 different biological, chemical, physical, and

radiochemical characteristics of surface and ground waters; and (5) Ground-Water Site-Inventory File — independent but cross-referenced to Daily Values and Water Quality Files — contains data on 700,000 sites.

## 13.3 MATERIALS RESEARCH FACILITIES*

### 1. Advanced Combustion, Coal Utilization Technology, and Alternative Fuels Program
Pittsburgh Energy Technology Center
U.S. Department of Energy
P.O. Box 10940
Pittsburgh, PA 15236
Contact: (412) 892-6251

The center has taken a lead role in supplanting the use of oil with coal by testing coal-water, coal-oil, and coal-methanol mixtures in combustors originally designed to burn oil. Currently, coal utilization research focuses on combustion of coal slurry fuels, alternative fuels such as dry microfine coal, and environmental control technologies to control emissions during combustion processes. Research is performed at laboratory-, bench-, and engineering-scales along with physical and mathematical modeling.

Outside contractor work has included combustion characterization, burner testing, and ash deposition evaluations. It has also expanded to predicting performance of existing utility and industrial boilers. Test facilities at PETC available for cooperative use include the alternative fuel mixture combustion laboratory, advanced coal combustor test facility, flue gas cleanup sorbent life-cycle test facility, spray dryer and dry injection FGC test facility, fuel rheology laboratory, coal transport facility, solid fuel combustion research facility, and advanced coal utilization projects laboratory.

### 2. Albany Research Center
U.S. Bureau of Mines
U.S. Department of the Interior
P.O. Box 70
Albany, OR 97321
Contact: Director (503) 967-5893

The Center has metallurgical expertise that encompasses the entire spectrum of mineral/metal

---

* Based upon information extracted from Directory of Federal Laboratory and Technology Resources, 1988—1989, PB 88-100011, U.S. Dept. of Commerce, NTIS, February 1988.

processing form the time ore leaves a mine until a final product is produced. The Center's research program is investigating benefication and hydro-metallurgical treatment of ores; solvent extraction and chemical processing (especially chlorine processing) of concentrates; pyrometallurgical treatment of both virgin and recycle metal; wear; and alloy development.

3. **Alloy Preparation Laboratory**
   National Institute of Standards and
   Technology
   U.S. Department of Commerce
   Gaithersburg, MD 20899
   Contact: (301) 975-6157

Research-grade samples of rapidly solidifed alloys are prepared as part of cooperative research programs when such samples are not readily available commercially and difficult otherwise for users to obtain. Capabilities include equipment for high pressure inert gas atomization and electrohydro-dynamic atomization to produce rapidly solidified alloy powders, melt-spinning to produce rapidly solidified alloy ribbons, and electron beam surface melting to produce rapidly solidified surface layers.

4. **Animal Biomaterials Research Unit**
   Eastern Regional Research Center,
   Agricultural Research Service
   U.S. Department of Agiculture
   600 E. Mermaid Lane
   Philadelphia, PA 19118
   Contact: (215) 233-6585

A diversified program of basic and applied research is carried out on a variety of normally inedible materials of agricultural origin. The major objective of this program is the conversion of these materials into useful products for consumers as well as agricultural and industrial users, both domestic and foreign. The current program is largely concerned with the by-products of the meat industry: hides, tallow, and wool. Research topics include the modification of the chemistry of leather processing, enhanced quality of the U.S. hides by elimination of grain defects or modification of fiber structure, removal of contaminants in American wool, enzyme processing of tallow and tallow-related chemicals, novel uses of enzymes for processing of agricultural products, the study of lipase structure and function, and production of oleo-chemical-modifying enzymes by recombinant DNA technology.

5. **Automated Manufacturing Facility**
   National Institute of Standards and
   Technology
   U.S. Department of Commerce
   Room B112 Metrology Bldg.
   Gaithersburg, MD 20899
   Contact: (301) 975-3414

The Facility (AMRF) is a major national laboratory for technical work related to interfaces and standards for the next generation of computer-automated manufacturing. It supports research in machine tool and robot metrology, sensors and sensory processing, robot safety, robot control, software accuracy enhancement of machine tools, process planning and data preparation for machine tools and robots, parts routing and handling, realtime control of robots and aggregations of devices, workstation control, cell control, and materials handling control. The AMRF is unique in the opportunities it provides for studies of an integrated system of significant size. The AMRF consists of three machining centers, a coordinate measuring machine, and a cleaning and deburring station, each tended by an industrial robot and served by a materials handling system based on an automated wire-guided vehicle and an internal buffer storage system for tools, materials, and work in progress.

6. **Bulk Crystal Growth Facility**
   Naval Research Laboratory
   U.S. Department of the Navy
   ORTA Code 1005.4
   4555 Overlook Avenue
   Washington, DC 20375
   Contact: (202) 767-3744

This site has two high pressure crystal pulling furnaces as well as atmospheric pressure units. There are capabilities for flux growth, a computer controlled gradient freeze furnace as well as various purification and material preparation capabilities. Emphasis has been on III-V compound semiconductors. Growth techniques available include Bridgman, Czochralski, gradient freeze, and flux.

7. **Center for Advanced Materials**
   Lawrence Berkeley Laboratory
   U.S. Department of Energy
   One Cyclotron Road
   Berkeley, CA 94720
   Contact: (415) 486-4755

The mission of the center is to perform basic research in materials science that is expected to benefit American industry and strengthen its competitive position in world markets. This effort is supported through advisory boards for each program, exchange of personnel (including an industrial fellows program), and research collaborations. The six multidisciplinary programs currently pursued are

(1) Electronic materials — to solve major scientific problems impeding the development of very-large-scale digital integrated circuits and opto-electronic devices based on gallium arsenide and other compound semiconductors.

(2) Ceramic and metal interfaces — to develop the theory of adherence between dissimilar materials and to demonstrate the metallurgical principles relevant to improving resistance to decohesion.

(3) Structural materials — to create new engineering alloys for advanced technological needs with particular emphasis on the metallurgy of high strength, low density aluminum alloys.

(4) Surface science and catalysis — to apply phenomena occurring at surfaces and interfaces and on high surface-to-volume ratio materials to existing and emerging technologies.

(5) Instrumentation for surface science — to develop new techniques and instruments for the study of surfaces and interfaces.

(6) Polymers and composites — to develop a sound scientific basis for the production of high performance polymeric materials through the prediction and control of microstructure.

**8.    Center for Building Science**
Lawrence Berkeley Laboratory
U.S. Department of Energy
90-3058, One Cyclotron Road
Berkeley, CA 94720
Contact: (415) 486-4508

The Center provides an umbrella so that groups in different programs of building science but with similar interest can combine to perform joint research, develop new research areas, share resources, and produce joint publications. It is involved in planning future building-science research, obtaining funding for research, transferring building science R&D results to industry, and serves as a point of contact for building science work at the laboratory. There are about 140 people at the Center working in four different programs. The Energy Analysis Program is involved in gathering and analyzing data that relate to present and future energy policies. The Building Energy Systems Program is concerned with potential savings of various approaches toward efficient energy use. It is active in three research areas: passive approaches to heating, cooling, and lighting; computer simulation of energy performance of buildings; and active solar cooling. The Windows and Daylighting Group is involved in developing a technical basis for understanding the energy-related performance of windows. The objective of the Lighting Systems Groups is to assist the lighting community (manufacturers, designers and users) in achieving a more efficient lighting economy. It has been responsible for development of energy efficient fluorescent lamps.

**9.    Center for Building Technology (CBT)**
National Institute of Standards and
  Technology
U.S. Department of Commerce
Gaithersburg, MD 20899
Contact: (301) 975-5900

The Center is the national building research laboratory. It works cooperatively with other organizations, private and public, to improve building practices and increase the productivity and international competitiveness of the construction industry. It conducts laboratory, field, and analytical research. It develops technologies to predict, measure, and test the performance of building materials, components, systems, and practices. CBT provides technologies needed by the building community to achieve the benefits of advanced computation and automation. CBT does not promulgate building standards of regulations but its research results are widely used in the building industry and adopted by governmental and private organizations that have standards and code responsibilities.

**10.   Center for Manufacturing Engineering**
National Institute of Standards and
  Technology
U.S. Department of Commerce
Gaithersburg, MD 20899
Contact: (301) 975-3400

The Center provides and develops technical data, findings, and standards in manufacturing engineering, mechanical metrology, automation and control technology, and industrial and mechanical engineering to support the discrete parts manufacturing industries. Its activities include: maintaining competence in CAD/CAM automated process planning, and shop management systems; development of competence in engineering measurements and sensors (both static and dynamic) of dimensions, force, mass, sound, vibration, and other parameters needed for inspection, quality control, and process control and monitoring in the discrete parts industry; study of machine tool dynamics and robotics; incorporation of metrology into the precision metalworking processes, including the standards necessary for integration of equipment up to the manufacturing cell level; and development of control systems, software, interface standards, sensors, measurement techniques, and information processing related to advanced industrial robots, computer-aided-manufacturing, and automatic factories.

## 11. Center for Fire Research
National Institute of Standards and
  Technology
U.S. Department of Commerce
Room A247 Polymers Bldg.
Gaithersburg, MD 20899
Contact: (301) 975-6850

All aspects of fire are studied. This includes basic and applied research on the fundamental processes at work in fires, research into the factors affecting fire risk and human victims of fire, and operational tests, demonstration projects and fire investigations in support of such activities. The Center develops engineering data, methods, and practices, measurement and test methods, and the scientific and technical basis for fire safety design and technology practices for use by commerce and industry, state and local governments, federal agencies, and the general public. Programs are conducted to evaluate technologies for suppression and extinguishment, to mitigate the effects of smoke and toxic gases and reduce their impacts, to develop physical and scientific models to predict fire growth processes involving buildings, facilities and vehicles, and to translate these findings into design data and performance criteria and into practical fire safety systems. Means are developed to identify potentially harmful combustion products and measure their effects on living organisms.

## 12. Ceramic Powder Characterization Laboratory
National Institute of Standards and
  Technology

U.S. Department of Commerce
A247, Bldg. 223
Gaithersburg, MD 20899
Contact: (301) 921-2910

Advanced ceramics are being used in high temperature engines and turbines, electrical capacitors, semiconductors, and other aerospace and electronic technologies. To help manufacturers, the fine powder laboratory enables scientists to conduct basic research on key powder signatures that will permit them to quantitatively correlate powder characteristics with the physical and chemical properties in finished ceramics. The laboratory contains several clean room stations for use in determining the physical properties of ceramic powders as small as 0.003 micrometers. It is equipped to permit researchers to make small particle materials; to determine particle size and measure distribution of powders in unfired materials or measure particle grains in sintered (fired) materials; to classify or size fractionation of powders; and to measure physical properties of powders, green state (unfired), and finished ceramic products.

## 13. Ceramics and Coatings Development and Evaluation Laboratories
George C. Marshall Space Flight Center
National Aeronautics and Space
  Administration
MSFC, Mail Code AT01
Marshall Space Flight Center, AL 35812
Contact: (205) 544-8962

The laboratories occupy approximately 4,200 ft$^2$ of space with diverse capabilities in the fields of coating application, measurement, testing, and evaluation. The laboratories also have the capabilities to fabricate, process, and evaluate common as well as special ceramic materials. The coatings area provides equipment for preparation, application, cure, measurement, and evaluation of paints, porcelain enamels, spacecraft thermal control coatings, thermal oxidation coatings for solar energy utilization, and a variety of specialty coatings. Equipment for process control and for the measurement of coating optical properties is also included. This optical properties measurement capability includes spectral and total solar absorptivity, emissivity, chromaticity, and gloss. In the ceramic materials area, equipment is available for the batching, melting, casting, and annealing of glasses, and for the formulation and fabrication of oxide ceramic bodies. Fabrication by cold press and sinter techniques or by hot pressing can be performed. Material specimens can be prepared by any of the following: diamond sawing and grinding; airbrasive machining; or lapping and polishing. A petrographic microscope attachment and a polariscope/polarimeter is available. Equipment for fracture mechanics properties measurements of glasses and ceramic materials and a high temperature insulation test facility where ceramic, organic foam, and ablative materials can be exposed to a programmed radiant heat flux while undergoing ascent depressurization are also available.

### 14. Corrosion and Metallurgy Section
Pacific Northwest Laboratory
U.S. Department of Energy
P.O. Box 999
Richland, WA 99352
Contact: (509) 376-0989

Studies range from basic corrosion and electrochemistry, the development of electrochemical sensors, and basic fracture mechanics to full scale component testing, field testing, and the development of integrated real-time process monitoring and expert systems. A major fracture and mechanics and creep laboratory allows for strain acoustic emission and closed circuit TV data acquisition.

### 15. Corrosion Laboratory
National Fertilizer Development Center
Tennessee Valley Authority, ATTN: T107
Muscle Shoals, AL 35660
Contact: (205) 386-2551

The laboratory conducts studies and experiments to determine corrosion resistance and for failure analysis on both metals and plastics related to the production and handling of solid and fluid fertilizers. Equipment and instrumentation supporting these activities include: Versamet II metallograph with capability to produce slides and prints; ultrasonic and mechanical thickness gauges; portable and lab type hardness testers; small heat treating furnace; a tensile test machine (20,000 lb max.); five sets of thermal blocks capable of handling temperatures up to boiling point of most liquids; and equipment for mounting and polishing samples.

### 16. Drop Tower and Tube Facility
George C. Marshall Space Flight Center
National Aeronautics and Space
  Administration

MSFC, Mail Code AT01
Marshall Space Flight Center, AL 35812
Contact: (205) 544-8962

The facility is designed to provide a simulated low gravity force field for R&D experiments and studies. The drop tower is 333 ft. Deceleration of the drag shield and test package results when the shield enters an open 40-ft-long cylindrical tube. The air compresses within the tube and thereby provides the stopping force for the shield. The drop tube is used to perform low gravity tests on different materials in support of materials processing in space programs. Material samples are melted in a furnace at the top of a 10-in. diameter, 100-meter, vertical evacuated tube. A device in the furnace releases the molten material as a droplet. The droplet solidifies as it free-falls the length of the tube. The sample is then recovered and analyzed.

### 17. Dynamic Hot Corrosion Test Facility
Army Materials Technology Laboratory
U.S. Department of the Army
Arsenal Street
Watertown, MA 02172
Contact: (617) 923-5527

The facility is used to simulate complex hot corrosion environments comparable to that experienced in aircraft gas turbine engines. Installation of the AMMRC-modified Pratt and Whitney dynamic hot corrosion burner test rig with auxiliary controls represents the first such facility to be located at an Army research center. Effects of combined hot oxidation and sulfidation on experimental material systems can be realistically evaluated only after exposure to complex dynamic test conditions. This test rig consists of a source of combusted fuels into which a sea-salt corrodent is introduced, impinging on the surface of test specimens. Combustion conditions are maintained on the oxidizing side of stoichiometry and the specimen may or may not be cycled thermally out of the flowing gas stream. The velocity of gases impinging on the blade specimen ranges between 300 and 400 ft/sec. The direct flame impingement generates speciment temperatures between 1650° and 2200°F. The choice of specimen geometry is open-ended; variations can be selected that will simulate gas turbine engine components.

### 18. Engineering Support of Adhesive, Sealant, Coatings, and Lubricant Technology

Army Armament Research, Development, and Engineering Center
U.S. Department of the Army
Organic Material Branch, Armament Technology
Division SMCAR-AET-O, Bldg. 183
Picatinny Arsenal, NJ 07806-5000
Contact: (201) 724-3183

The branch is staffed with personnel who are recognized authorities on adhesive bonding, sealant, organic coating, and lubrication technologies. They recommend materials, detail process specifications, and provide assistance during item design and production start up. They determine causes of failure and recommend corrective action. Available to the engineers of the section is a fully equipped bonding laboratory where designs and processing procedures are evaluated. There are extensive environmental testing facilities and testing machines to determine the strength and endurance of different materials and bonds. Also available is a complete microscopic laboratory, including a scanning electron and an optical microscope, dedicated to the analysis of material failure.

### 19. Fire Research Facilities
National Bureau of Standards
U.S. Department of Commerce
Room A247 Polymers Bldg.
Gaithersburg, MD 20899
Contact: (301) 975-6850

Special test facilities include:

- A large building where full-scale fire tests can be carried out in mockups of rooms, vehicles, etc.
- A furniture calorimeter which measures the rate of heat release of upholstered chairs or similar sized items of furnishings
- A cone calorimeter which is a bench-scale rate of heat release calorimeter utilizing the oxygen consumption principles
- A flame spread apparatus, used to determine flame spread properties of materials based on time of ignition and flame spread rate in the vertical direction
- An airtight 1/2 scale room, $8.5 \times 7.5 \times 6.5$ ft to conduct smoke movement and coagulation experiments
- A "townhouse" burn facility which has two stories, each $22 \times 22 \times 10$ ft, connected by an open staircase within the structure —

smoke and toxic gas transport between levels and fire spread beyond the room of origin can be measured

- An industrial fire test facility building underground in a concrete bunker with a $9 \times 22$ ft ceiling vent at ground level — it is used for studies of water spray extinguishment of simulated gas well blowout fires
- A room/corridor facility is used for evaluation of analytical models developed for prediction of the transport of smoke and toxic gases
- A two-story structural steel facility is available for use as a burn compartment — it is a four-bay frame measuring $32 \times 40$ ft in plan, and has a steel frame sized equivalent to that found at midheight of a 20-story building
- An $18 \times 28 \times 11$ ft fire suppression systems test facility, where regulated water flow rates up to 150 GPM are provided from a 3000-gallon tank to test spray equipment — specialized instruments used to measure spray penetration through fires, are operated in this facility
- A computer fire simulation laboratory for hands-on usage of predictive fire growth computer programs — video color graphic capabilities are available
- A fire research library which supports the staff and the fire community — it houses a computerized fire research publications database that will soon contain the entire collection of 30,000 technical documents

### 20. Forest Products Laboratory
Forest Service
U.S. Department of Agriculture
One Gifford Lane
Madison, WI 53705-2398
Contact: (608) 264-5600

The mission of the laboratory is to improve the use of wood through research that leads to improved management and use of the timber resource, thus meeting the needs of the U.S. and contributing to the international community. Industry, university, and government cooperate in the laboratory's research and technology transfer programs, which fall into five broad areas: wood products, process and protection, chemistry and paper products, microbial and biochemical conversion, and energy from wood. Wood products investigations cover more efficient use of lumber in building construction, the properties, design, and performance of engineered structural components, and improving the structural integrity and fire safety of wood structures. Timber and wood products use trends are also assessed. Process and protection research investigates how to improve wood sawing, drying, and milling technologies and extend the useful life of wood products through effective protection. This area also explores the use and properties of composite wood products and the development of authoritative information on anatomical features of domestic and foreign species of wood, including an international computerized wood identification system. Chemistry and paper products research covers basic wood chemistry and paper physics, the expanded use of hardwood and recycled fibers, the role of chemicals in overcoming papermaking problems and enhancing properties of paper and paperboard, the production of high yield pulp at minimum energy, the improved performance of conventional adhesives, and the development of new adhesives from renewable resources. Microbial and biochemical conversion research focuses on harnessing the enzymes of fungi and microorganisms to degrade lignin, produce wood-derived chemicals, and to produce pulp by biological means. Production of alcohol from wood and how wood combustion works are the subjects of the energy from wood research. Improved accuracy of classification and identification of wood-rotting and wood-inhabiting fungi is another laboratory objective.

### 21. Glass Development Center
Pacific Northwest Laboratory
U.S. Department of Energy
P.O. Box 999
Richland, WA 99352
Contact: (509) 376-0882

The center has grown through 20 years' work developing and characterizing complex glasses for vitrification of nuclear wastes. An interdisciplinary group of scientists and engineers study chemical durability, melt viscosity and electrical conductivity, devitrification, and batch melting behavior. An efficient, statistically based method has been developed to empirically model the properties of glasses as functions of composition over wide composition regions. The resulting models are used to optimize glass compositions. The group is qualified

to apply these development and characterization methods to most oxide glass systems.

### 22. Handwear, Footwear, Rainwear, and Tentage Facilities (HMTL)
Army Combat Systems Test Activity
U.S. Department of the Army,
STECS-CC-SF
Aberdeen Proving Ground, MD 21005
Contact: (301) 278-3565

Accelerated wear test courses have been developed to test materials, designs, and construction used in military handwear and footwear. The handwear course includes controlled work conditions or tasks contributing to handwear failure. The activities are designed to create strain and abrasion in varying degrees on the seams and surface of gloves. Data can be obtained as to resistance to wear, retention of size and shape under hard use, and the degree of moisture absorption and drying of the gloves. The footwear course consists of various terrain and abrasive surfaces over which footwear is worn. The course is 1/6 mile long and includes stretches of crushed quartz, chipped and cubed granite, cinders, sand, mud, water, gravel, slag, and smooth and rough concrete. The rainwear and tentage facility is designed to simulate overall rainfall for testing the adequacy of foul weather garments, tentage, and all types of general military material and equipment. High-pressure showerheads projecting from parapets 35 ft high produce simulated rainfall of varying intensities from 1/10 in. to 3 in./h. The facility covers an area of 86 × 50 ft. It is composed of a dynamic rain course and a static rain course. On the dynamic course, personnel wearing test clothing traverse obstacles and other obstructions simulating physical situations that confront troops in combat.

### 23. High Temperature Materials Laboratory
Oak Ridge National Laboratory
U.S. Department of Energy
Bldg. 4515, P.O. Box X
Oak Ridge, TN 37831
Contact: (615) 574-5124

The laboratory is a new research facility designed to address the high-temperature materials problems that limit the efficiency and reliability of advanced energy conversion systems and thereby assist U.S. industry in meeting the current challenge of foreign competition in this area. The HTML serves as a focal point for investigating new materials, both ceramics and alloys, having potential for high-temperature applications. A major function of the HTML is its role as a user facility. There are four user centers within the HTML: (1) Materials Analysis, which encompasses ultrahigh resolution and analytical transmission electron microscopy, high resolution scanning electron microscopy and high resolution small spot ESCA and Auger spectroscopy; (2) X-Ray Diffraction, with a 2700°C-theta-theta high-temperature diffractometer; (3) Physical Properties, with TGA, DTA, and dilatometry, and (4) Mechanical Properties, with automated high temperature static fatigue testing, high temperature ceramic tensile testing, and an ultrafine probe microhardness test system. Researchers from industry and universities may use these facilities without charge when the research is relevant to the above stated mission and the results are intended for open literature publication. Proprietary investigations are also possible on a full-cost-recovery basis.

### 24. House Component and Equipment Field Test Facility
Oak Ridge National Laboratory
U.S. Department of Energy
Oak Ridge, TN 37830
Contact: (615) 574-0330

This site, operated in cooperation with the University of Tennessee, is known as the Tennessee Energy Conservation in Housing (TECH) complex. It includes a sophisticated instrumentation and data acquisition system. There are three main houses (167 $m^2$ each), originally one solar, one annual cycle energy system, and one control; and currently two other experimental buildings. These houses are being used to obtain performance data on existing and new types of HVAC equipment and appliances and their interactions with the building structure.

### 25. Individual Protection Directorate
Army Natick Research, Development and Engineering Center
U.S. Department of the Army
Natick, MA 01760-5014
Contact: (617) 651-5296

Research and development is conducted on uniforms, protective clothing, personnel armor, and life support equipment systems to protect the individual combat soldier against battlefield chemical, biological, ballistic, surveillance, directed high energy, and nuclear hazards. The Directorate's divisions conduct research and exploratory development of new and improved specialty chemicals, elastomers, plastics, fibers, dyeing, and finishing to increase survivability of the soldier in terms of protection from environmental and battlefield threats. Combat uniforms, life support clothing, protective devices and life support equipment are developed to protect the soldier from environments and threats ranging from arctic to tropic; from explosive to toxic; from flammable to chemical; from surveillance to directed high energy; and for special purpose missions. Items developed include handwear, headwear, footwear, rainwear, chemical protective clothing, combat and dress uniforms and all forms of equipage, including canteens, load carrying equipment, sleeping gear, climbing equipment, entrenching tools, body armor, helmets and microclimate cooling equipment. The Directorate conducts human factors, anthropometric and physiological studies for items of personal wear and use. The Directorate also develops and provides cleaning and decontamination procedures for all Army clothing and individual equipment.

### 26. Indoor Environment Program
Lawrence Berkeley Laboratory
U.S. Department of Energy, 90-3058
One Cyclotron Road
Berkeley, CA 94720
Contact: (415) 486-6591

The obvious way to improve energy efficiency of buildings is to lower exchange rates between inside and outside air. This may have the undesirable side effect of increasing levels of indoor pollution. This program investigates factors affecting pollutant concentrations including sources, ventilation rates, and removal processes. It consists of six groups. The Indoor Radon Group is investigating methods of predicting or identifying geographical areas that contain houses with high radon concentrations, improving understanding of mechanisms for production and transport of radon through soils into buildings, and developing means of controlling high indoor radon concentrations. The Indoor Organic Chemistry Group has constructed an environmental chamber that is used for a wide variety of indoor air studies (e.g., exposures during the use of paint removers, emissions from assembled building materials, etc.), is involved in testing samples for volatile organic compounds, and develops models that allow exposures to be calculated under various room conditions. The Indoor Exposure Assessment Group is concerned with pollutant exposure characterization and examination of health effects. A framework for evaluating health effects is being developed that considers multiple types of exposures, toxicities of pollutants, and reliability of information. The Ventilation and Indoor Air Quality Control Group uses multitracer techniques to determine ventilation rates, ages of air, and ventilation efficiencies in large buildings. It also studies residential exhaust ventilation and heat pump recovery. The Energy Performance of Buildings Group studies the flow of energy through all elements of a building. It tests air infiltration rates, studies thermal characteristics of structural elements, and develops models of the behavior of complete buildings. Research is conducted in the laboratory and in single family and multifamily buildings. New techniques for measuring air leakage using AC pressurization are under investigation.

### 27. Institute for Materials Science and Engineering
National Institute of Standards and
 Technology
U.S. Department of Commerce
Gaithersburg, MD 20899
Contact: (301) 975-5658

The Institute conducts research and provides measurements, data, standards, reference materials, quantitative understanding, and other technical information fundamental to the processing, structure, properties, and performance of materials; addresses the scientific basis for new advanced materials technologies; plans research around crosscutting scientific themes such as nondestructive evaluation (NDE), phase diagram development, automated materials processing; overseas Bureauwide technical programs in nuclear reactor radiation research and nondestructive evaluation; and broadly disseminates generic technical information resulting from its programs. The focus of the Institute is increasingly on new materials: advanced ceramics, polymer blends, metals (for extremes — viz., high- or low-temperature use), composites (polymer, metal, ceramics), optically or magnetooptically active materials. The Institute's new programs on intelligent processing will be addressing the development of process modeling (relating specific materials properties at each stage of the process to the final realized properties), real time

sensors of materials properties for in-process NDE, materials property data to be coupled with the limited real time measurements common in process control models, and the coupling of sensors and modeling with automation technology to achieve an integrated process system. These new programs will focus on generic science challenges offered by low-dimensional regimes and the complex multiparameter, nonlinear problems of solidification, phase transformations, crystallization, chain dynamics, and multiparameter process modeling. The generic technology challenge includes integrating the design, performance, and processing with materials (e.g., computer integrated manufacturing, in-process sensing, new process methodology). The institute consists of the following divisions: Nondestructive Evaluation, Ceramics, Polymers, Metallurgy, Fracture and Deformation, and Reactor Radiation.

### 28. Ion Implantation and Ion Beam Analysis Facility
Naval Research Laboratory
U.S. Department of the Navy
ORTA Code 1005.4, 4555 Overlook Avenue
Washington, DC 20375
Contact: (202) 767-3744

A 3-MV tandem Van de Graaff accelerator and a 200-KeV ion implanter are available. Experimental investigations are carried out involving the implantation of ions into the near-surface regions of solids and the concomitant changes in surface properties such as corrosion resistance, sliding wear resistance, fatigue resistance, and optical and superconductivity characteristics. Since ion implantation is not subject to the laws of thermodynamics governing equilibrium conditions of solid solubility and diffusivity, it is possible to produce by ion implantation alloys that are not producible by conventional means. The ion beams are also used as probes in making materials analysis measurements of the near-surface implanted regions.

### 29. Large-Scale Structures Testing Facility
National Institute of Standards and
 Technology
U.S. Department of Commerce
Room B168 Building Research
Gaithersburg, MD 20899
Contact: (301) 975-6048

Static and dynamic testing is accomplished by use of a heavily reinforced tie-down floor permitting mounting of complete structural members. Hydraulic actuators provide test loads in static test while closed-loop electro-hydraulic actuators provide test loads in dynamic tests. Automatic recording of up to 200 channels of sensor data is accomplished by a mini-computer-controlled data acquisition system. A 53-ft section has a 12,000 ft-kip bending moment capacity, and its 25-ft extension has a bending moment capacity of 8,000 ft-kips. The crosswise section 47 ft in length will withstand a total bending moment of 21,000 ft-kips. The floor will withstand a total horizontal shear force of 1,800 kips in either direction and a vertical shear force of 2,000 kips. Beams, slabs, frames or complete structures can be subjected to static loads as limited by test floor capacity or cyclic loads up to 50,000 pounds with programmed amplitude and frequency. Two mini-computer-controlled-data acquisition systems are available.

**30.  Laser and Laser Spectroscopy Facilities**
Los Alamos National Laboratory
U.S. Department of Energy
Los Alamos, NM 87545
Contact: (505) 667-6900

Extensive, well-staffed laser facilities are available that can provide laser output energy across the ultraviolet, visible, and infrared portions of the electromagnetic spectrum. There are tunable laser-pumped dye lasers; electrically discharged, rare-gas-excimer lasers; solid-state and semiconductor lasers; HF, $CF_4$ and $NH_3$ gas lasers, and both pulsed and continuous-wave $CO_2$ lasers. Some of the latter attain powers as high as one kilowatt at one kilohertz repetition rate. Various frequency shifting and Raman scattering devices are available. Hardware and software are available to theoretically model the laser physics of ultraviolet and infrared systems and to study geometrical and physical optics, radiation transport, and propagation effects. There is a facility to fabricate and test optical components and a facility to evaluate optical materials — including the evaluation of optical damage under conditions of high reflectivity or transmission at high average or high peak powers.

**31.  Laser-Based Optical Instrumentation**
Los Alamos National Laboratory
U.S. Department of Energy, MS-E548
Los Alamos, NM 87545
Contact: (505) 667-6900

A variety of laser-based optical instrument systems have been and are being developed for point and remote detection and identification of chemical species. These instruments rely on the selective reaction of the species to laser output at various wavelengths. In experiments, concentrations below 10 ppb have been detected. Past efforts include remote detection of chemical elements and compounds and real-time monitoring of gas streams. The techniques rely on modulated, tunable lasers covering a wide range of wavelengths.

**32.  Laser Laboratory**
Brookhaven National Laboratory
U.S. Department of Energy
Instrumentation Division, 535-B
Upton, NY 11973
Contact: (516) 282-5072

The laboratory supports research on photoemission of metals under high electric fields using picosecond laser pulses in UV, green, and infrared fast electrical switching. Photoconductive materials are irradiated with picosecond light pulses and the switched pulse is then probed to determine its shape and dispersion.

**33.  Laser Program**
Lawrence Livermore National Laboratory
U.S. Department of Energy
Technology Transfer and Exchange Office,
  P.O. Box 808
Livermore, CA 94550
Contact: (415) 422-6416

The Program is undertaking a lead in laser research as part of DOE's fusion research efforts. This research covers glass lasers, rare gas halide lasers, optical coatings, and other laser equipment and materials.

**34.  Liquid Metal Corrosion Laboratory**
Argonne National Laboratory
U.S. Department of Energy
9700 South Cass Avenue
Argonne, IL 60439
Contact: (312) 972-5191

Studies are conducted on the corrosion/mass transfer of structural materials in liquid metal (Li, LiPb, Na) environments. The laboratory includes a capability for testing the effects of liquid metal environments on the mechanical properties (fatigue, creep) of structural alloys.

### 35. Low Temperature Neutron Irradiation Facility

Oak Ridge National Laboratory
U.S. Department of Energy
P.O. Box X
Oak Ridge, TN 37831-6061
Contact: (615) 574-6270

The facility is used for studying radiation effects in materials under high radiation intensities — 2 × 10 to the 17 n/m² s (E > 0.1 MeV). It provides special environmental and testing conditions: temperatures between 3.2 and 800 K, magnetic fields up to 12 T. A closed-cycle liquid helium refrigeration system provides cooling to samples irradiated in the core of a 2-MW research reactor.

### 36. Materials Analysis Laboratory

Idaho National Engineering Laboratory
U.S. Department of Energy
P.O. Box 1625
Idaho Falls, ID 83415
Contact: (208) 526-8318

Extensive capability is available for materials analysis, including scanning Auger microprobe (with ESCA), scanning electron microscope with energy-dispersive X-ray analyzer, X-ray diffraction, scanning transmission electron microscope (STEM) with energy dispersive X-ray analyzer and energy-loss spectrometer, optical metallography, Fourier transform infrared spectrophotometer, and micro Raman/fluorescence spectrometer. Mechanical testing also is provided including tensile, compression, creep, fatigue, and impact property determinations at room and elevated temperatures.

### 37. Materials Division

Lewis Research Center
National Aeronautics and Space
 Administration
21000 Brookpark Rd. MS 49-1
Cleveland, OH 44135
Contact: (216) 433-3193

The Division addresses materials technologies critical to the performance of high temperature systems, e.g., gas turbines, rocket and solar dynamic engines. Experimental materials are tested in environments simulating those found in application. The division research interests include solidification fundamentals, microstructure vs. property models, rapid solidification technology, toughened ceramics, advanced ceramic fibers, high temperature polymers, toughened polymers, photostable/photocuring polymers, carbon/carbon precursors, composite processing, surface/interface phenomena, solid film lubricants for use from cryogenic to high temperatures, liquid lubricants, environmental attack prediction, thermal barrier, coatings, chemical vapor deposition, polymer, metallic, intermetallic, and ceramic composites, and space power materials. It has expertise in chemical and microstructural characterization and has a number of special facilities — burner rigs, high temperature creep temperature X-ray diffractometer, an extrusion press, directional solidification apparatus, thermal shock apparatus, arc spray apparatus, hot isostatic press, extensive polymer characterization facilities, a high pressure mass spectrometric sampling system, an X-ray photoelectron spectroscope and a scanning Auger microscope.

### 38. Materials Laboratory

Air Force Wright Aeronautical Laboratories
U.S. Department of the Air Force
ORTA AFWAL/XP
Wright-Patterson AFB, OH 45433
Contact: (513) 255-3570

The Laboratory has unique areas of expertise in (1) nonmetallic materials: adhesive bonding, composite structural materials, protective coatings, mechanics and surface interaction; (2) laser-resistant materials; (3) metals and ceramics: high-temperature resistant materials, nondestructive evaluation, powder metallurgy; and (4) manufacturing technology: computer-aided manufacturing, technology modernization, manufacturing methods. It is fully equipped to perform synthesis, processing, characterization and analysis, and failure analysis of all types of organic, inorganic, metallic, and refractory materials including high-performance composites.

### 39. Materials Preparation Center (MPC)

Ames Laboratory
U.S. Department of Energy
Iowa State Univeristy
Ames, IA 50011
Contact: (515) 294-5236

The Center consists of the Materials Preparation Section, the Analytical Section, and the Materials Referral System and Hotline (MRSH). Through the Center, researchers from other governmental, university, and industrial laboratories can gain the benefit of the unique capabilities of the Ames Laboratory for the preparation, purification, fabrication, and characterization of metals and the preparation of single crystals of certain materials which are not available from private, commercial suppliers. The major categories of materials with which the MPC is currently involved include: refractory metals, the alkaline earths, rare earth metals and compounds, and certain actinide metals.

**40. Materials Science Research and Analytical Services**
Army Armament Research, Development, and Engineering Center
U.S. Department of the Army
Metallic Materials Branch, Armament Technology Division
SMCAR-AET-M, Bldg. 355
Picatinny Arsenal, NJ 07806-5000
Contact: (201) 724-5746/2767

The major facilities are: Philips 420 STEM electron microscope with EELS and EDAX attachments; Forgflo EMU-3C (RCA) electron microscope: Philips electronic X-ray diffraction/fluorescence unit; Siemens Kristalloflex IV X-ray diffraction unit. The X-ray diffraction and electron microscopy laboratory is capable of conducting high resolution (2.5 Å) electron microscopy analyses, selective area electron diffraction and microdiffraction analyses of defects and second phase particles and substructures. Ribbons of amorphous materials prepared by rapid solidification technology (RST) may be studied with the scanning transmission mode of STEM. EDAX and EELS are respectively used in the analyses for high and low atomic number (Z = 11 or less) elements. The Siemens Kristalloflex IV unit is fully computer controlled. When texture analyses are made, the computerized X-ray diffraction unit processes the X-ray diffraction data and displays the results in multiple color stereographic plots for texture determination. The Philips X-ray diffraction/fluorescence unit is also computer controlled. The computer is capable of printing out wavelength dispersive data of elements with Z = 11 and greater. When studies are being made on large grain samples, the crystallographic orientation of the grains is identified by back reflection methods. The standard Debye-Scherrer cameras are also available for X-ray diffraction analyses of powdered polycrystalline samples, e.g., expended propellent residues from gun-barrels.

**41. Materials Technology Group**
Idaho National Engineering Laboratory
U.S. Department of Energy
P.O. Box 1625
Idaho Falls, ID 83415
Contact: (208) 526-9727

Materials science and engineering activities extend from basic research to field applications and include metallurgy, ceramics, materials joining, fracture mechanics, irradiation effects, powder metallurgy, rapid solidification, material joining, and plasma processing.

**42. Mechanical Fabrication — R&D Support**
Los Alamos National Laboratory
U.S. Department of Energy
Los Alamos, NM 87545
Contact: (505) 667-4849

In support of R&D activities but without regard to the sponsor's identity, the services of the facility are available. The staff of this facility have extensive experience with manufacturing prototype, close-tolerance components and apparatus, and in fabrication of radioactive and toxic materials. This facility also has the capability to machine metal laser mirrors by diamond turning.

**43. Mechanical Test Laboratory**
George C. Marshall Space Flight Center
National Aeronautics and Space Administration
MSFC, Mail Code ATO1
Marshall Space Flight Center, AL 35812
Contact: (205) 544-8962

This 2,146-ft$^2$ facility provides a full range capability for the mechanical testing and evaluation of materials over a wide temperature range. It is used in support of studies to generate design data on engineering materials, to evaluate process effects on materials, and for special component testing in which service loads and environments must be simulated. It includes modern computer-controlled equipment for axial fatigue and mechanical property testing of specimens or components and special equipment for reproducing process- or service-induced heat effects in samples and for high strain rate tensile testing. Standard equipment includes tensile test machines, fatigue test machines, creep test machines, impact testers, and hardness testers.

## 44. Mechanical Testing Facility

Army Materials Technology Laboratory
U.S. Department of the Army
Arsenal Street
Watertown, MA 02172
Contact: (617) 923-5527

The facility was established to provide a means of determining mechanical properties of materials under dynamic loading conditions. The primary objective is to study the strain rate effect on the mechanical properties of materials. The range of strain rate covers from 0.00001/s to 10/s. A laboratory computer was incorporated with the mechanical testing machine to form an automated closed-loop servo-controlled testing system. In summary, this facility consists of a biaxial medium strain rate mechanical test machine, a mechanical impactor (split Hopkinson bar apparatus), a high-temperature furnace system, and an automated recording, analyzing, and feedback system (laboratory computer). The range of high temperature is from room temperature to 4000°C.

## 45. Metallography Laboratory

Pacific Northwest Laboratory
U.S. Department of Energy
P.O. Box 999
Richland, WA 99352
Contact: (509) 375-5231

The facility offers metallographic facilities for the examination and nonradioactive materials. Materials examined in the facility included glasses, metals, minerals, and combinations of these materials. Examination techniques include a wide spectrum of optical illuminations, these being Nomarski interference contrast, polarized light, and dark field, in both reflected and transmitted light. Specialized preparation techniques have been developed to produce ultrathin (1- to 5-μm) doubly polished petrographic sections. Ultrathin sections have been applied to study of interfacial boundaries, phase constituents, and microstructural properties in lunar and terrestrial materials. These techniques are also compatible with the requirements for electron beam analytical instruments.

## 46. Metallurgical Laboratory

Tennessee Valley Authority, Central
  Laboratories
LA-PSC-1C, 1101 Market Street
Chattanooga, TN 37402-2801
Contact: (615) 751-4290

This organization provides all aspects of TVA's power program complete metallurgical support, including failure analysis, analytical and testing services, chemical analysis of metals, and applied research. Personnel have expertise in metallurgy, metallurgical engineering, metallography (*in situ* and laboratory), optical microscopy, electron microscopy, mechanical testing, chemical analyses, heat treating, corrosion, and welding. One area of special emphasis is failure analysis.

The laboratory has complete metallographic capabilities. Optical microscopy is performed using compound microscopes with bright field, dark field, phase contrast, polarized light, or interference contrast at magnifications from $12 \times$ to $2500 \times$. Scanning electron microscopy, providing resolution on the order of 50 to 100 Å, with a useful magnification range from $5 \times$ to $100,000 \times$ and a depth of focus of 10 μm, is available. This enables secondary electron imaging, back-scattered electron imaging, specimen current, X-ray analyses, and photographic recording of images. The laboratory also maintains an array of equipment to test various mechanical properties and to perform chemical analysis of metals.

## 47. Metallurgy Laboratory

Army Material Test and Evaluation
  Command
U.S. Department of the Army,
  STEWS-TE—AE
White Sands Missile Range, NM 88002-5178
Contact: (505) 678-5632

The Laboratory offers a high degree of competence in metallurgy, nondestructive inspection (NDI), and advanced engineering materials testing and acceptance. Metallurgical capabilities include failure and corrosion analysis, metallography and heat treating (ferrous and nonferrous). NDI methods include magnetic particle, liquid and fluorescent penetrant, ultrasonics, and X-ray and isotope radiography. Radiographic equipment includes 250-kV and 420-kV X-ray machines, 300-kV and 4-MeV portable X-ray machines, and a 100-Ci cobalt-60 isotope camera. Other equipment includes light and scanning electron microscopy (SEM). The SEM includes an energy dispersive X-ray (EDX) system. A 50-kip servo-hydraulic material test system (MTS) is available for materials testing and is equipped with a crack correlator unit for crack growth rate studies or stress intensity programming. Further equipment includes a portable metal analyzer, hardness testers, metallographic preparation equipment, a galvanopotentiometer, and a controlled atmosphere retort furnace.

## 48.    Micrewgravity Materials Science Laboratory

Lewis Research Center
National Aeronautics and Space
  Administration
21000 Brookpark Rd., MS 105-1
Cleveland, OH 44132
Contact: (216) 433-5013

The Laboratory (MMSL) is a nationally available laboratory open to scientists and engineers from industry, universities, and the government. It is equipped with experimental hardware which functionally duplicates that flown on the Space Shuttle or being developed for use on future Shuttle/Space Station missions. The laboratory is designed for easy, no-cost access to qualified experiments (for nonproprietary research) and aims for quick turnaround on experiments. The researcher can evaluate the results of a simple "proof of concept" experiment before a formal program at his or her home laboratory is begun. In this way, the researcher's sponsoring organization is not forced to commit itself to a major research program until the feasibility of the experiments has been demonstrated. The MMSL can also be used to perform precursor research for possible Shuttle flight experiments. The MMSL currently provides experimental capabilities to support research in crystal growth and in the solidification of metals and alloys. The experimental equipment available includes a general purpose furnace, electromagnetic levitator and instrumented drop tube, transparent and high temperature directional solidification furnaces, bulk undercooling furnace, and dendrite growth apparatus. Laboratory and experimental facilities for materials research involving ceramics and glasses and polymer processing are available.

## 49.    Navy Clothing and Textile Research Facility

U.S. Department of the Navy
21 Strathmore Road
Natick, MA 01760
Contact: (617) 651-4680

The Facility designs and develops all protective clothing, dress uniforms and utility garments worn by most Navy personnel. It conducts research on fabrics and materials which it develops into clothing items that it tests and evaluates before introduction. Fire-preventive clothing, cold-weather garments, women's wear, deep-sea swimsuit materials, boots and shoes, insignia, buoyant-ballistic vests, decompression-chamber clothing, handwear, and dress and work uniforms are developed. Research is carried out on fibers, yarns, coatings, films, laminates, dyes, and finishes used in all types of general- and special-purpose protective clothing and textiles. Colorfastness, breaking strength, tear resistance, abrasion resistance, aging, weather resistance, water repellency, air permeability, adhesive, stiffness and crease resistance are tested. The effects of laundering on fabrics and clothing are studied to determine dimensional stability, colorfastness, appearance, and durability of fabrics and fabric finishes with such properties as water repellency, soil release, flame retardance, anti-stats, and softeners.

Special facilities include the Environmental Test Chamber that reproduces temperature and relative humidity extremes ranging from $-40$ to $200°F$ at 5 to 100% RH; the Hydro-Environment Simulator that reproduces air-sea surface temperature conditions existing anywhere on earth; and the Thermal Flammability Simulator that provides controlled thermal energy in the forms of infrared and conductive heat for firefighters' clothing.

## 50.    Nondestructive Evaluation

National Institute of Standards and
  Technology
U.S. Department of Commerce
Gaithersburg, MD 20899
Contact: (301) 975-5727

This office was established to assist industry and government agencies in improving the reliability of materials and structures. It is working to help industry develop methods for accurate and reproducible NDE measurements. This includes technical investigations, development of standards (both measurement standards and procedural documents), characterization of instruments, and assessments of the meaning of NDE measurements in relation to material performance. The principal focus of the program is on the application of NDE expertise to automated materials processing; specifically, on the development of nondestructive sensors for real-time measurement of important material properties and characteristics during processing and for the elucidation of process models.

**51. Oak Ridge Bioprocessing Research Facility User Resource**
Oak Ridge National Laboratory
U.S. Department of Energy
P.O. Box X, Bldg. 4505 MS-227
Oak Ridge, TN 37831-6227
Contact: (615) 576-4853

The facility includes laboratories for investigation of advanced bioprocessing concepts, employing stirred-tank and columnar bioreactors and a fermentation pilot plant for large-scale fermentation experiments. Using the facility's advanced systems and other state-of-the-art equipment, researchers can engage in: pretreating and fractionating chemical feedstocks; selecting and improving microbial cultures; manipulating genes to produce designed proteins; designing advanced bioreactors; developing advanced analytical concepts; determining the feasibility of using a bioprocess and scaling it up; monitoring and controlling experimental bioprocesses; and conducting biochemical separations.

**52. Optical Fiber Drawing and Measurement Facility**
Naval Research Laboratory
U.S. Department of the Navy
ORTA Code 1005.4, 4555 Overlook Avenue
Washington, DC 20375
Contact: (202) 767-3744

Optical fibers for communications and signal processing are fabricated and evaluated. A facility for drawing silica and fluoride fibers and on-line coating of glass fibers with polymeric materials is available. The optical properties of these fibers, including spectral attenuation and numerical aperture, can be evaluated using facility apparatus.

**53. Plasma Diagnostics Laboratory**
Idaho National Engineering Laboratory
U.S. Department of Energy
P.O. Box 1625
Idaho Falls, ID 83415
Contact: (208) 526-9818

The Plasma Diagnostics Laboratory is devoted to temporal characterization of heat and mass transport in plasma processing. The endeavors include temperature distributions using an optical multichannel analysis system interfaced to an emission spectrograph for plasmas, material temperature measurements (greater than 2000 K) using infrared thermography; flow dynamics and gas density using high-speed holographic interferometry; and gas velocity measurements using laser Doppler velocimetry.

**54. Plate Accelerator (Gas Gun) Facility**
Army Materials Technology Laboratory
U.S. Department of the Army
Arsenal Street
Watertown, MA 02172
Contact: (617) 923-5527

The facility was established to study material response under shock wave conditions. The objective is to generate a plane wave with a very high stress level in materials by impacting one disk with another disk of either the same or different material. Impact velocity, material velocity, or stress levels are measured and the shock wave theory is employed to determine the equation of state of the material of interest. The facility consists of three major components; namely, a plate accelerator, an electronic measuring system, and a data analyzer. The plate accelerator includes a 4-in diameter and 26-ft-long gun barrel, a catcher and tank, and a breech designed for maximum pressure of 6,000 psi.

**55. Plumbing Research Laboratory**
National Institute of Standards and Technology
U.S. Department of Commerce
Gaithersburg, MD 20899
Contact: (301) 975-5851

With a high-speed, computerized data acquisition system and hot and cold water supplies with precisely controlled pressures and temperatures over a wide range of demands, as well as standard laboratory fluid services, this laboratory offers unique opportunities to advance the state of the art both in the design and in the evaluation of plumbing equipment and systems. Capabilities include tests on systems up to 45 ft in height and up to 50 ft in length; water supply of up 1000 gpm at constant head in gravity mode; up to 300 gpm in automatically controlled pressure mode up to 70 psi; and hot water at volumes up to 200 gph at 180°F. Applications include the development of improved criteria for general hydraulic design of and for prediction of loads on plumbing systems; the development of test methodology for calibration and performance evaluation of innovative plumbing equipment and systems; investigations of performance of piping materials as affected by pressure, flow, and contact with hot and cold water; methodology for cost effective approaches in needed national programs to update information on plumbing loads and to develop a modern data bank on performance characteristics of innovative equipment and systems.

### 56.  Polymer-Concrete Development Laboratory

Brookhaven National Laboratory
U.S. Department of Energy
Department of Applied Sciences,
  Building 526
Upton, NY 11973
Contact: (516) 282-3036

A well-equipped laboratory and expert staff are available to perform basic and applied research on materials consisting of a composite of polymer and aggregate. These rapid-setting, high strength and high durability materials were originated and developed here and have numerous applications. The laboratory facilities include resin storage, preparation, and polymerization equipment. Aggregate storage and preparation equipment for formulating and producing monomer/aggregate mixes, for measurement of material properties and for casting are also available. BNL staff can provide consultation services on all aspects of polymer concrete materials research, development, and implementation.

### 57.  Quantitative Image Analyzer

Naval Research Laboratory
U.S. Department of the Navy
ORTA Code 1005.4, 4555 Overlook Avenue
Washington, DC 20375
Contact: (202) 767-3744

This fully automated image analyzer efficiently quantifies microstructural features in composite materials. A television scanner and monitor detects and digitizes contrasting features observed directly in a specimen through an optical microscope or in any photograph through a macroviewer. Interfaced with a computer, the system performs complex quantitative analysis of composite material features in just a few seconds. Some of these features include fiber volume fraction, volume fraction, and average thickness of fiber/matrix interface layer, average fiber spacing, and fiber distribution.

### 58.  Radiographic Facility, Naval Surface Weapons Facility

U.S. Department of the Navy
Code D21, White Oak
Silver Spring, MD 20903
Contact: (301) 394-1505

The Facility's capabilities extend over a broad range from 10 KeV microfocus to 10 MeV. With highly specialized personnel and equipment, the facility has the capacity to evaluate very thin plastics and aluminum; and on the other end of the scale, the facility is capable of generating X-rays that can penetrate up to 12 in. of steel. Highly trained radiographic personnel can report interpretive results ranging from minute discontinuities in very thin materials to flaws in the steel welds of submarines. NSWC's Radiographic Facility is known both nationally and internationally as one of the leaders in radiographic R&D efforts. In this facility, numerous radiographic standards have been developed and are maintained for welds associated with various metals. This Navy facility is also known for worldwide R&D efforts in penetrameters of image quality indicator studies. This Radiographic Facility has provided the American Society for Testing and Materials (ASTM) experiments aimed at developing a method for evaluating and classifying X-ray films.

### 59.  Rolla Research Center

U.S. Bureau of Mines
U.S. Department of the Interior
P.O. Box 280
Rolla, MO 65401
Contact: (314) 364-3169

Sphalerite concentrate is leached with strong sulfuric acid at atmospheric pressure to produce zinc sulfate and elemental sulfur. A process research unit is being completed to validate bench-scale experiments. Environmentally acceptable techniques are being developed to recover accessory and other metal values from lead and zinc milling and smelter wastes. Hydrometallurgical techniques are applied to recover the preponderant and accessory metals, such as toxic cadmium, from milling and smelting wastes generated in primary lead and zinc production. Research on soldering systems which reduce toxic lead emissions, brazing alloy systems which reduce toxic cadmium emissions, and nonpolluting fluxing technology is conducted. An industrial-scale process research unit using diaphragm cells is operated to regenerate chromic acid etchants for recycle. Technology involving pelletizing and smelting is conducted to recover Cr, Ni, Mo, and Fe from specialty steelmaking (flue dusts, mill scale, etc.) and other Cr-bearing wastes.

### 60.  Roof Research Center, Oak Ridge National Laboratory

U.S. Department of Energy, P.O. Box X
Oak Ridge, TN 37831
Contact: (615) 574-1945

The Center is a center for industry/government collaborative research on issues of thermal efficiency and durability of roof systems. A climate simulator for controlled testing of large scale roof sections (up to 12 × 12 ft) and an outdoors test station for smaller specimens (4 × 8 ft) are available for thermal, hygric, and mechanical measurements. The emphasis of the center is on short-time testing with carefully controlled environments and accurate measurements. Mathematical modeling of performance and generalization of test results are key functions of the center. Research of low-slope roofs is featured; however, the apparatus can accommodate some types of pitched water-shedding systems.

### 61.  Rural Housing Research Unit

Agricultural Research Service
U.S. Department of Agriculture
P.O. Box 792
Clemson, SC 29633
Contact: (803) 654-3646

The unit is a national program directed toward reducing housing, farm structures, and service building construction and operating costs for low-income rural residents. It conducts research projects in collaboration with other USDA agencies, academic institutions, and the U.S. Department of Energy. Areas of research include low-cost housing; solar heating; and energy conservation.

### 62.  Semiconductor Detector Laboratory

Brookhaven National Laboratory
U.S. Department of the Navy
Instrumentation Division, 535-B
Upton, NY 11973
Contact: (516) 282-4238

This laboratory has developed capabilities to fabricate semiconductor radiation detectors (mainly silicon) of a wide range of types for various applications. Facilities include oxidation, lithography, annealing, bonding and packaging, test of

techniques, and evaluation of detector performance.

### 63.  Semiconductor Detector Technology Laboratory

Lawrence Berkeley Laboratory
U.S. Department of Energy
University of California, Bldg. 70A
Berkeley, CA 94720
Contact: (415) 486-6432

This laboratory conducts a broad program in materials and device technology for semiconductor detectors. Extensive capabilities exist for purification and doping of germanium, for growth of extremely pure crystals, and for characterizing the material as to quantity and type of impurities, traps, and defects of various types. Facilities exist for fabrication and testing of silicon and germanium detectors of all types including ion implanted, Schottky barrier, diffused and lithium drifted. Advanced design of signal processing electronics for these detectors is also carried out. Expertise in gamma ray, X-ray, and charged particle spectroscopy and also in infrared measurements exists in the group.

### 64.  Sputter-Coating Facility

Pacific Northwest Laboratory
U.S. Department of Energy
P.O. Box 999
Richland, WA 99352
Contact: (509) 375-2085

The laboratory includes facilities for glow discharge (diode), magnetically enhanced discharge (magnetron), thermionically supported discharge (triode), and enhanced thermionically supported discharge (ETSD) sputtering. Radio-frequency and dc target and substrate power supplies are used to deposit high integrity optical, metallic, ceramic, and compound semiconductor coatings. Optical coatings composed of oxides, nitrides, and hydrogenated compounds can be deposited on glass, ceramics, and temperature-sensitive plastics. Amorphous and very fine-grain metallic deposits are routinely produced in complex or free-standing configurations. Special bond layers are produced by grading or layering ceramic-to-metallic, metallic-to-ceramic combinations. Currently research and development is being carried out in the following areas: fine-grained and amorphous metal liners, radiofrequency and laser hardening of missile domes, variable density and multiple wavelength filters, fundamental studies of thin film optical materials, high temperature superconducting materials, diamond-like and diamond coatings, development of wear-resistant and hard coatings, glassy metals for corrosion protection, special high temperature protective coatings, Be and Be alloy coatings, and development of custom high-rate sputtering equipment with deposition rates up to 2 $\mu$m/min.

### 65. Standard Reference Materials

National Institute of Standards and
  Technology
U.S. Department of Commerce
Gaithersburg, MD 20899
Contact: (301) 975-2012

Standard reference materials (SRMs), play a major role in increasing and ensuring the accuracy of measurements. SRMs are well-characterized, homogeneous materials with specific properties measured and certified by NIST. SRMs are used widely throughout the U.S. and the world to help develop test methods of proven accuracy, to calibrate instruments and measurement systems used to maintain quality control of the production of materials and goods, to help assure equity in buyer-seller transactions, and to assure the long-term reliability and integrity of the measurement process. NIST issues SRMs in 70 major categories, and over 1,000 different SRMs are presently available from NIST. SRMs serve major segments of industry including those specializing in ferrous and nonferrous metals, mining, glass, rubber, plastics, primary chemicals, nuclear power, electronics, and automotive and computer instrumentation. Types of SRMs include: steels, cast irons, nonferrous alloys, gases in metals, high-purity metals, electron probe microanalytical, primary chemicals, clinicals, biologicals, botanicals, environmentals, industrial hygiene, metalloorganic compounds, fertilizers, ores, minerals, refractories, carbides, glasses, cements, trace elements, nuclear materials, radioactivity, X-ray diffraction, isotopics, ion activity, mechanical and metrology, superconducting, freezing point materials, thermodynamic properties, thermal conductivity, thermal expansion, thermal resistance, thermocouple materials, magnetic, optical, gases, reference fuels, and computer tapes.

### 66. Surface Analytical Facilities

Naval Research Laboratory
U.S. Department of the Navy

ORTA Code 1005.4
4555 Overlook Avenue
Washington, DC 20375
Contact: (202) 767-3744

Research is directed in part to improving existing or developing new techniques that will provide more powerful analytical tools for characterizing surfaces and studying surface reactions. The available instrumentation includes PHI model scanning Auger microscopy, Surface Science Laboratory small spot X-ray photoelectron spectroscopy, secondary ion mass spectrometry, infrared reflection/absorption spectroscopy and scanning tunneling microscopy. The surface analytical instrumentations constitute an outstanding surface analytical facility for the study of a wide range of material surface problems. The facility is currently being used to study corrosion, tribology (friction wear), electronic, and engineering materials.

### 67. Surface Evaluation Facility

Naval Surface Weapons Center
U.S. Department of the Navy
Code D21, White Oak
Silver Spring, MD 20903
Contact: (301) 394-1505

The center has equipped and staffed the facility to provide a unique Navy capability for determining surface properties. The facility engages in a full-spectrum approach to the Navy's materials problems through three major functions. Development of new materials and devices is supported through programmatic studies of surface-related problems in collaboration with the scientists and engineers working on the primary development project. Research on the fundamental properties of solid surfaces maintains a forefront expertise in surface science. Analytical services are provided for short-term trouble-shooting of materials problems. High-technology areas in which surface evaluation is necessary are: materials and device development; manufacturing technology; quality control; failure analysis; and environmental protection.

### 68. Surface Modification and Characterization Facility

Oak Ridge National Laboratory
U.S. Department of Energy
P.O. Box X
Oak Ridge, TN 37831-6048
Contact: (615) 574-6174

Research at the facility involves the use of particle accelerators and high-powered lasers to alter a few atomic layers at the surface of metals, ceramics, and semiconductors. Such modifications are designed to change physical characteristics in beneficial ways, including improved hardness and corrosion and wear resistance. The modifications also can enhance electrical conductivity, enabling scientists to "tailor" the properties of semiconductors, for example, to particular electronic applications or devices — or to improve the efficiency and fabrication economy of solar cells. The "nonequilibrium" fabrication processes lead to new, often unique properties. The facility now hosts some 40 academic and industrial groups and permits scientists to create unique new materials with equipment and techniques not widely available. It also allows industrial researchers to perform proof-of-principle experiments before making large capital investments in particle accelerators or lasers.

### 69.   Tension Testing Laboratory

Army Materials Technology Laboratory
U.S. Department of the Army
Arsenal Street
Watertown, MA 02172
Contact: (617) 923-5527

The laboratory consists of four universal tension-testing machines of different capacities capable of attaining loads from 0 to 120,000 lb in various increments and at speeds of 0.001 to 40 in./min. The system can test at temperatures ranging from $-425°F$ to $+500°F$, utilizing specialized auxiliary equipment. The tensile test consists of gripping a standard tensile test specimen between the movable heads of a test machine and recording the extension load, head speed, and other pertinent data related to the test. These machines are very versatile and can be adapted to test many complex sizes and shapes. Compressive loads may be applied.

### 70.   Transducer Calibration Facility

Naval Air Test Center
U.S. Department of the Navy
Patuxent River, MD 20870-5304
Contact: (301) 863-1141

The facility consists of the technical personnel, calibration standards, environmental simulators and test equipment required to calibrate, evaluate and repair special flight test instrumentation components and systems used on Navy aircraft test programs. The facility is unique in that it is located in a hangar, with the capability to interface type II and III calibration standards and environmental simulators directly to the aircraft system. The laboratory uses computerized data acquisition and analysis systems with computer controlled calibration standards and environmental simulators. Calibration capabilities include fuel flow, strain, shock, vibration, pressure, temperature and acceleration. This facility also has the capability to instrument aircraft structures and to provide structure loading as required for calibration. Printed circuit cards also are manufactured.

### 71.   Thin Film Growth Facility

Naval Research Laboratory
U.S. Department of the Navy
ORTA Code 1005.4, 4555 Overlook Avenue
Washington, DC 20375
Contact: (202) 767-3744

This site features three (3) computer controlled molecular beam epitaxial reactors, three (3) organometallic vapor phase deposition systems including "in situ" diagnostics as well as characterization capabilities. The facilities are currently engaged in the growth of Si, SiC, and III—V compound semiconductors. The facility is integrated so that all steps from initial substrate preparations to final film characterization can be performed.

### 72.   Ultra High Sputtering System

Naval Research Laboratory
U.S. Department of the Navy
ORTA Code 1005.4, 4555 Overlook Avenue
Washington, DC 20375
Contact: (202) 767-3744

The facility can be used for RF, de, and bias sputtering. The typical vacuum obtained in the system is $10^{-18}$ torr. The sputtering conditions can be varied extensively: substrate temperature, ambient to $1100°C$; pressure of reactive gases, to 300 mtorr; and power density, to 20 $W/cm^2$. The system is capable of handling toxic gases and it has a drive shaft which can be used in coating cylinders. Applications from this work include Josephson detectors, phonon bolometers, and X-ray monochrometers. The Kosterlitz-Thouless 2D phase transition has been studied with granular NbN films made in this system.

### 73.   Vacuum Deposition/Materials Processing Laboratory

Brookhaven National Laboratory

U.S. Department of Energy
Instrumentation Division, 535-B
Upton, NY 11973
Contact: (516) 282-4204

The laboratory can produce electrical, optical, and multilayer coatings for X-ray and neutron mirrors. Materials can be ultrasonically machined, plated, heat treated and/or brazed.

74. **Vacuum Electronics Processing Facility**
Naval Research Laboratory
U.S. Department of the Navy
ORTA Code 1005.4, 4555 Overlook Avenue
Washington, DC 20375
Contact: (202) 767-3744

The facility provides a complete capability for the assembly of experimental RF vacuum electronic devices. Specific capabilities available for the preparation of ultra-high-vacuum devices include computer-aided design, chemical cleaning, and vacuum by hydrogen brazing.

75. **Viscoelastic Materials Characterization Facility**

Naval Research Laboratory
U.S. Department of the Navy
ORTA Code 1005.4, 4555 Overlook Avenue
Washington, DC 20375
Contact: (202) 767-3744

A facility for measuring the mechanical properties for polymeric materials has been designed and constructed in the Chemistry Division of the Naval Research Laboratory. This equipment can measure the viscoelastic properties of polymers over a wide range of times/frequencies and temperatures, and strains in a variety of deformation modes (e.g., tension, compression, simple shear, pare shear, torsion). Materials examined with this equipment include adhesive and composite resins (glassy polymers), elastomers, and propellants. Mixing and sample preparation capabilities are suitable for a broad range of materials. On-line computer data acquisition and processing equipment with interactive graphics makes possible indepth analysis of results in real time. This provides the information necessary to help formulate improved materials and predict performance in practical applications.

# Index

# INDEX

## A

ABS, see Acrylonitrile butadiene styrene
ABS/PVC, see Acrylonitrile butadiene styrene/
    polyvinyl chloride
Acetals, 547, see also specific types
Acetic acid, 13, see also Acids; Liquids
Acetylene, 13
Acids, 14, 24, 43, 44, 52, see also specific types
  cold, 42
  concentration of, 119
  dilute, 37
  hot, 42
  mineral, 16, 21, 54
  organic, 51
  pH values of, 124
  strong, 50
  in wolframite, 52
ACI Type HX, 496
Acoustics, 567—568
Acrylics, 547, see also specific materials
Acrylic-styrene-acrylonitrile, 548
Acrylonitrile butadiene styrene, 548
Acrylonitrile butadiene styrene/polyvinyl chloride -
    rigid, 549
Actinides, 5, see also specific compounds
Actinium, see also Elements
  description of, 6
Actinium 227, 6
Actinon, 40
Adiabatic demagnetization, 23
Adsorbents, see also specific types
  properties and uses of, 266
Advanced Combustion, Coal Utilization Technology,
    and Alternative Fuels program, 597
Aerosols, 35, see also specific types
Aerospace Research Application Center, 581
Aerospace Structures Information and Analysis
    Center, 581
Agate, 44
Agriculture, U.S. Department of, 598, 602, 613
Agricultural Research Service, 613
Air, 34
Air Force, U.S. Department of, 581, 583, 595, 607
Air Force Wright Aeronautical Laboratories, 581,
    607
Albany Research Center, 597
Aldebaranium, 29, 54
Alkaline earth groups, 9, see also specific elements
Alkaline elements, 12, 41, see also specific elements
Alkaline metals, 3, 29, 37, 40, 45, 52, see also
    specific types
Alkalis, 37, 50, 52, 53, see also specific types
  caustic, 36
  dilute, 44
Alkyds and thermoset carbonate, 549
Allanite, 13, 27
Alloy Preparation Laboratory, 598

Alloys, see also specific types
  Alnico, 15
  aluminum, 7, 12, 51, 55
  aluminum-base, see Aluminum-base alloys
  antimony, 8
  beryllium, 10, 12
  binary eutectic, 336—339
  carbon, 27, 33
  cesium, 41
  chromium, 21
  cobalt, 15, 27, 36
  copper, see Copper alloys
  corrosion resistant, 32, see also specific types
  die casting, 50
  electrical resistivity of, 112
  ferromagnetic, 30
  ferrous, see Ferrous alloys
  gadolinium, 21
  gallium, 21
  gold, 30, 35, 41
  hafnium, 23
  heat resistant, see Heat resistant alloys
  indium, 25
  iridium, 26
  iron, see Ferrous alloys
  lead base, see Lead-base alloys
  lithium, 29
  low-melting, 10, 11
  magnesium, 12, 29, 55
  magnesium base, see Magnesium base alloys
  magnetic, 366—368
  manganese, 30, 51
  mercury, 30
  molybdenum, 31, 41, 42
  nickel, see Nickel-base alloys
  niobium, 23, 33
  osmium, 34
  palladium, 35, 41
  platinum, 26, 36, 41, 42
  polonium, 37
  potassium, 41, 46
  rhenium, 41
  rhodium, 41
  ruthenium, 42
  silver, see Silver alloys
  sodium, 41, 46
  soldering, 515
  specific stiffness of, 509
  superconductive, see Superconductive alloys
  tantalum, 23, 47
  tin, 30, 50
  tin base, see Tin-base alloys
  tin-niobium, 51
  titanium, see Titanium alloys
  tungsten, 52
  vanadium, 53—54
  yttrium, 55
  zinc, see Zinc alloys

# Q

# R